Handbook of Optoelectronic Device Modeling and Simulation

Series in Optics and Optoelectronics

Series Editors: **Robert G W Brown**
University of California, Irvine, USA

E Roy Pike
Kings College, London, UK

Handbook of Optoelectronic Device Modeling and Simulation
Fundamentals, Materials, Nanostructures, LEDs, and Amplifiers

VOLUME ONE

Edited by
Joachim Piprek

CRC Press
Taylor & Francis Group
Boca Raton London New York

CRC Press is an imprint of the
Taylor & Francis Group, an **informa** business

CRC Press
Taylor & Francis Group
6000 Broken Sound Parkway NW, Suite 300
Boca Raton, FL 33487-2742

First issued in paperback 2019

© 2018 by Taylor & Francis Group, LLC
CRC Press is an imprint of Taylor & Francis Group, an Informa business

No claim to original U.S. Government works

ISBN-13: 978-1-4987-4946-6 (hbk)
ISBN-13: 978-0-367-87560-2 (pbk)

Library of Congress Cataloging-in-Publication Data

Names: Piprek, Joachim, editor.
Title: Handbook of optoelectronic device modeling and simulation / edited by Joachim Piprek.
Other titles: Series in optics and optoelectronics (CRC Press) ; 27.
Description: Boca Raton, FL : CRC Press, Taylor & Francis Group, [2017] |
Series: Series in optics and optoelectronics ; 27 | Includes bibliographical references and index.
Contents: volume 1. Fundamentals, materials, nanostructures, LEDS, and amplifiers – volume 2. Lasers, modulators, photodetectors, solar cells, and numerical methods.
Identifiers: LCCN 2016058063 | ISBN 9781498749466 (v. 1; hardback ; alk. paper) |
ISBN 1498749461 (v. 1; hardback ; alk. paper) | ISBN 9781498749565 (v. 2 ; hardback ; alk. paper) |
ISBN 1498749569 (v. 2 ; hardback ; alk. paper)
Subjects: LCSH: Optoelectronic devices–Mathematical models–Handbooks, manuals, etc. |
Optoelectronic devices–Simulation methods–Handbooks, manuals, etc. |
Semiconductors–Handbooks, manuals, etc. | Nanostructures–Handbooks, manuals, etc.
Classification: LCC TK8304 .H343 2017 | DDC 621.381/045011–dc23
LC record available at https://lccn.loc.gov/2016058063

Visit the Taylor & Francis Web site at
http://www.taylorandfrancis.com

and the CRC Press Web site at
http://www.crcpress.com

Contents

PART I Fundamentals

PART II   Novel Materials

PART III   Nanostructures

PART IV   Light-Emitting Diodes (LEDs)

PART V   Semiconductor Optical Amplifiers (SOAs)

Series Preface

This international series covers all aspects of theoretical and applied optics and optoelectronics. Active since 1986, eminent authors have long been choosing to publish with this series, and it is now established as a premier forum for high-impact monographs and textbooks. The editors are proud of the breadth and depth showcased by the published works, with levels ranging from advanced undergraduate and graduate student texts to professional references. Topics addressed are both cutting edge and fundamental, basic science and applications-oriented, on subject matter that includes lasers, photonic devices, nonlinear optics, interferometry, waves, crystals, optical materials, biomedical optics, optical tweezers, optical metrology, solid-state lighting, nanophotonics, and silicon photonics. Readers of the series are students, scientists, and engineers working in optics, optoelectronics, and related fields in the industry.

Proposals for new volumes in the series may be directed to Lu Han, senior publishing editor at CRC Press/Taylor & Francis Group (lu.han@taylorandfrancis.com).

Preface

Optoelectronic devices have become ubiquitous in our daily lives. For example, light-emitting diodes (LEDs) are used in almost all household appliances, in traffic and streetlights, and in full-color displays. Laser diodes, optical modulators, and photodetectors are key components of the Internet. Solar cells are core elements of energy supply systems. Optoelectronic devices are typically based on nanoscale semiconductor structures that utilize the interaction of electrons and photons. The underlying and highly complex physical processes require mathematical models and numerical simulation for device design, analysis, and performance optimization. This handbook gives an introduction to modern optoelectronic devices, models, and simulation methods.

Driven by the expanding diversity of available and envisioned practical applications, mathematical models and numerical simulation software for optoelectronic devices have experienced a rapid development in recent years. In the past, advanced modeling and simulation was the domain of a few specialists using proprietary software in computational research groups. The increasing user-friendliness of commercial software now also opens the door for nontheoreticians and experimentalists to perform sophisticated modeling and simulation tasks. However, the ever-growing variety and complexity of devices, materials, physical mechanisms, theoretical models, and numerical techniques make it often difficult to identify the best approach to a given project or problem. This book presents an up-to-date review of optoelectronic device models and numerical techniques. The handbook format is ideal for beginners but also gives experienced researchers an opportunity to renew and broaden their knowledge in this expanding field.

Semiconductors are the key material of optoelectronic devices, as they enable propagation and interaction of electrons and photons. The handbook starts with an overview of fundamental semiconductor device models, which apply to almost all device types, followed by sections on novel materials and nanostructures. The main part of the handbook is ordered by device type (LED, amplifier, laser diode, photodetector, and solar cell). For each device type, an introductory chapter is followed by chapters on specialized device designs and applications, describing characteristic effects and models. Finally, novel device concepts and applications are reviewed. At the end of the handbook, an overview of numerical techniques is provided, both for electronic and photonic simulations.

I would like to thank the publisher for initiating this important handbook project and for giving me the opportunity to serve as editor. Many years of organizing the annual international conference on *Numerical Simulation of Optoelectronic Devices (NUSOD)* enabled me to attract a large number of experts from all over the world to write handbook chapters on their research area. I sincerely thank all authors for their valuable contributions.

Joachim Piprek
Newark, Delaware, USA

MATLAB® is a registered trademark of The MathWorks, Inc. For product information, please contact:

The MathWorks, Inc.
3 Apple Hill Drive
Natick, MA 01760-2098 USA
Tel: 508-647-7000
Fax: 508-647-7001
Email: info@mathworks.com
Web: www.mathworks.com

Editor

 **Joachim Piprek** received his diploma and doctoral degrees in physics from the Humboldt University in Berlin, Germany. For more than two decades, he worked in industry and academia on modeling, simulation, and analysis of various semiconductor devices used in optoelectronics. Currently, he serves as president of the NUSOD Institute, Newark, Delaware (see http://www.nusod.org). During his previous career in higher education, Dr. Piprek taught various graduate courses at universities in Germany, Sweden, and the United States. Since 2001, he has been organizing the annual international conference on *Numerical Simulation of Optoelectronic Devices (NUSOD)*. Thus far, Dr. Piprek has published three books, six book chapters, and about 250 research papers, which have received more than 6000 citations. He was an invited guest editor for several journal issues on optoelectronic device simulation and currently serves as an executive/associate editor of two research journals in this field.

Contributors

Doyeol Ahn
Department of Electrical and Computer
    Engineering
University of Seoul
Seoul, Republic of Korea

Denis Andrienko
Max Planck Institute for Polymer Research
Mainz, Germany

Enrico Bellotti
Photonics Center
Boston University
Boston, Massachusetts

Francesco Bertazzi
Department of Electronics and
    Telecommunications
Politecnico di Torino
Torino, Italy

and

Istituto di Elettronica e
    di Ingegneria dell'Informazione e
    delle Telecomunicazioni
    Consiglio Nazionale delle Ricerche
Torino, Italy

Peter Bobbert
Department of Applied Physics
Eindhoven University of Technology
Eindhoven, the Netherlands

Christopher A. Broderick
Department of Electrical and Electronic
    Engineering
University of Bristol
Bristol, United Kingdom

Steve Bull
Department of Electrical and Electronic
    Engineering
University of Nottingham
Nottingham, United Kingdom

Jih-Yuan Chang
Center for Teacher Education
National Changhua University of
    Education
Changhua, Taiwan

Fang-Ming Chen
Institute of Photonics
National Changhua University of
    Education
Changhua, Taiwan

Reinder Coehoorn
Department of Applied Physics
Eindhoven University of Technology
Eindhoven, the Netherlands

Michael J. Connelly
Department of Electronic and Computer
    Engineering
University of Limerick
Limerick, Ireland

Pierre Corfdir
Paul Drude Institute
Berlin, Germany

Pierluigi Debernardi
Istituto di Elettronica e
    di Ingegneria dell'Informazione e
    delle Telecomunicazioni
    Consiglio Nazionale delle Ricerche
Torino Italy

Gonzalo del Pozo
Center of Advanced Materials and Devices for
    ICT Applications
Universidad Politécnica de Madrid
Madrid, Spain

Michal Dlubek
ELMAT Group
Podkarpackie, Poland

Marcus Duelk
Exalos AG
Schlieren, Switzerland

Ignacio Esquivias
Department of Photonics Technology and
    Bioengineering
Center of Advanced Materials and Devices for
    ICT Applications
Universidad Politécnica de Madrid
Madrid, Spain

Yue Fu
Crosslight Software, Inc.
Vancouver, Canada

Giovanni Ghione
Department of Electronics and
    Telecommunication
Politecnico di Torino
Torino, Italy

Michele Goano
Department of Electronics and
    Telecommunications
Politecnico di Torino
Torino, Italy

and

Istituto di Elettronica e
    di Ingegneria dell'Informazione e
    delle Telecomunicazioni
    Consiglio Nazionale delle Ricerche
Torino, Italy

Yousong Gu
School of Materials Science and Engineering
University of Science and Technology
Beijing, China

Dejan Gvozdić
School of Electrical Engineering
University of Belgrade
Belgrade, Serbia

Weiguo Hu
Beijing Institute of Nanoenergy and
    Nanosystems
Chinese Academy of Sciences
Beijing, China

Xin Huang
Beijing Institute of Nanoenergy and Nanosystems
Chinese Academy of Sciences
Beijing, China

Vladimir M. Kaganer
Paul Drude Institute
Berlin, Germany

Sergey Yu. Karpov
Device Modelling Department
STR Group – Soft-Impact, Ltd.
Saint-Petersburg, Russia

Simeon N. Kaunga-Nyirenda
Department of Electrical and Electronic
    Engineering
University of Nottingham
Nottingham, United Kingdom

Pascal Kordt
Max Planck Institute for
    Polymer Research
Mainz, Germany

Yen-Kuang Kuo
Department of Physics
National Changhua University of
    Education
Changhua, Taiwan

Eric Larkins
Department of Electrical and Electronic
    Engineering
University of Nottingham
Nottingham, United Kingdom

Christian Lennartz
TrinamiX GmbH
Ludwigshafen, Germany

Simon Z. M. Li
Crosslight Software, Inc.
Shanghai, China

Jun-Jun Lim
OSRAM Opto Semiconductors
Penang, Malaysia

Benjamin Lingnau
Institute of Theoretical Physics
Technical University Berlin
Berlin, Germany

Kathy Lüdge
Institute of Theoretical Physics
Technical University Berlin
Berlin, Germany

Roderick C. I. MacKenzie
Faculty of Engineering
University of Nottingham
Nottingham, United Kingdom

Igor P. Marko
Advanced Technology Institute and
    Department of Physics
University of Surrey
Guildford, United Kingdom

Oliver Marquardt
Paul Drude Institute
Berlin, Germany

Giovanni Mascali
Department of Mathematics
University of Calabria and INFN-Gruppo c
Cosenza, Italy

Nicolai Matuschek
Exalos AG
Schlieren, Switzerland

Falk May
BASF SE, GME/MC
Ludwigshafen, Germany

Max A. Migliorato
School of Electrical and Electronic
    Engineering
University of Manchester
Manchester, United Kingdom

Eoin P. O'Reilly
Photonics Theory Group
Tyndall National Institute
Cork, Ireland

and

Department of Physics
University College Cork
Cork, Ireland

Joydeep Pal
School of Electrical and Electronic
    Engineering
University of Manchester
Manchester, United Kingdom

Seoung-Hwan Park
Department of Electronics Engineering
Catholic University of Daegu
Kyeongbuk, Republic of Korea

Antonio Pérez-Serrano
Center of Advanced Materials and Devices for
    ICT Applications
Universidad Politécnica de Madrid
Madrid, Spain

Andrew Phillips
Department of Electrical and Electronic
    Engineering
University of Nottingham
Nottingham, United Kingdom

Zoe V. Rizou
Department of Electrical and Computer
    Engineering
Democritus University of Thrace
Xanthi, Greece

Vittorio Romano
Department of Mathematics and Computer
    Science
University of Catania
Catania, Italy

Judy M. Rorison
Department of Electrical and
    Electronic Engineering
University of Bristol
Bristol, United Kingdom

Fabio Sacconi
Tiberlab Srl
Rome, Italy

Stefan Schulz
Photonics Theory Group
Tyndall National Institute
Cork, Ireland

Masoud Seifikar
Integrated Photonics Group
Tyndall National Institute
Cork, Ireland

Ya-Hsuan Shih
Department of Photonics
National Cheng Kung University
Tainan, Taiwan

Slawomir Sujecki
Department of Electrical and Electronic
    Engineering
University of Nottingham
Nottingham, United Kingdom

and

Department of Telecommunications and
    Teleinformatics
Faculty of Electronics
Wroclaw University of Science and Technology
Wroclaw, Poland

Stephen J. Sweeney
Advanced Technology Institute and Department
    of Physics
University of Surrey
Guildford, United Kingdom

Angela Thränhardt
Institute of Physics
Chemnitz University of Technology
Chemnitz, Germany

Alberto Tibaldi
Istituto di Elettronica e
    di Ingegneria dell'Informazione e
delle Telecomunicazioni
    Consiglio Nazionale delle Ricerche
Torino, Italy

José-Manuel G. Tijero
Center of Advanced Materials and Devices for
    ICT Applications
Universidad Politécnica de Madrid
Madrid, Spain

Stanko Tomić
Joule Physics Laboratory
University of Salford
Manchester, United Kingdom

Angelina Totović
School of Electrical Engineering
University of Belgrade
Belgrade, Serbia

Miao-Chan Tsai
Institute of Photonics
National Changhua University of
    Education
Changhua, Taiwan

Nenad Vukmirović
Scientific Computing Laboratory
Institute of Physics Belgrade
University of Belgrade
Belgrade, Serbia

Morten Willatzen
Department of Photonics Engineering
Technical University of Denmark
Lyngby, Denmark

Chen-Kuo Wu
Department of Electrical Engineering
National Taiwan University
Taipei, Taiwan

Yuh-Renn Wu
Department of Electrical Engineering
National Taiwan University
Taipei, Taiwan

Changsheng Xia
Crosslight Software, Inc.
Shanghai, China

Tsung-Jui Yang
Department of Electrical
    Engineering
National Taiwan University
Taipei, Taiwan

Kyriakos E. Zoiros
Department of Electrical and Computer
    Engineering
Democritus University of Thrace
Xanthi, Greece

I

Fundamentals

1

I

Fundamentals

1

Electronic Band Structure

Stefan Schulz

and

Eoin P. O'Reilly

1.1 Introduction and Overview of Importance of Semiconductor Materials

Semiconductors are probably the most technologically important materials in widespread use today. The "information age" in which we now live is entirely dependent on the modern computer, which has at its heart a microprocessor constituted mainly of silicon (Si). Another example is the reading and recording of data onto CDs, DVDs, and Blu-Rays that rely on using light, in the form of laser emission, produced by the semiconductor compounds based on gallium arsenide (GaAs) and gallium nitride (GaN), for example. These advances have undoubtedly had a huge impact on human society and are a testament to the power of scientific research.

These transformational developments are primarily due to both our fundamental understanding of the electronic band structure of semiconductor materials and the precision with which semiconductor structures can now be formed using epitaxial growth techniques—methods which allow a crystal lattice to "grow" layer-by-layer [1,2]. This fundamental understanding combined with advanced experimental techniques has driven worldwide research efforts to engineer the electronic band structure of semiconductor materials, including the use of alloys, heterostructures, quantum confinement, and strain [1,2].

A wide range of different semiconductor materials can now be used in heterostructures, increasing their versatility and enabling tunability of their optical and electrical properties. Overall, a very important class of semiconductor heterostructure, the "quantum well" (QW), is the mainstay of modern optoelectronics [1,2]. QWs consist of a thin layer of one semiconductor compound sandwiched between thicker layers of another. Typical QW thicknesses are very small (only a few nanometers), with their physical properties depending very sensitively on the layer thickness. By confining charge carriers on these length scales, one enters the quantum mechanical regime. This link between quantum mechanics and real-world "observable" properties, for example, emission wavelength, provides a fascinating and important topic not only for basic scientific research but also for designing novel and energy-efficient devices [1,2].

In this chapter, we discuss the fundamental properties of semiconductor materials and review different theoretical methods that provide insight into the electronic structure of these systems to understand and predict their properties. The methods addressed here range from first-principles calculations up to continuum-based models. This chapter is organized as follows. In Section 1.2, we focus on the consequences of the fact that solid-state materials are crystalline with a periodic lattice. This leads to the Bloch theorem and first insights into the basic band structure of solids. Subsequently, in Section 1.3, we turn our attention to the specifics of semiconductor band structures and discuss in detail similarities and differences in the band structures of two of the most often used materials for device applications, namely Si and GaAs. Having discussed the general details and aspects of the electronic structure, we introduce in Section 1.4 the basics of first-principles techniques for calculating the band structure of solid-state materials. This is followed, in Section 1.5, by an overview of current state-of-the-art techniques to address the electronic and optical properties of semiconductor materials. Since these techniques are computationally very expensive, they are typically restricted to structures containing at most several hundreds of atoms. Therefore, to describe the electronic and optical properties of semiconductor heterostructures, (semi)empirical methods are required. We start with $\mathbf{k} \cdot \mathbf{p}$ and envelope-function models in Section 1.6 and turn then to introduce the basic ideas of atomistic approaches such as tight-binding (TB) and empirical pseudopotential theory in Section 1.7. A summary of the chapter content is given in Section 1.8.

1.2 Bloch's Theorem and Band Structure

In a typical solid, around 10^{23} valence electrons contribute to the bonding in each cubic centimeter. Therefore, when one is interested in calculating and studying the electronic structure of a solid, in general, a complex many-body problem has to be solved, since the exact wave function and energy of each electron depends on those of all the others. However, a factor that considerably simplifies the calculation of the energy spectrum is the fact that most solid-state materials are crystalline with a periodic lattice. It can be shown that this symmetry property must also be reflected in the individual electron wave functions, satisfying therefore a condition referred to as *Bloch's theorem*.

We start by considering a solid with a periodic potential $V(\mathbf{r} + \mathbf{R}) = V(\mathbf{r})$, where $\mathbf{R}$ is a *lattice vector*. When studying the energy levels of a single electron in this periodic potential, the system can be described by the one-electron Hamiltonian:

$$H = -\frac{\hbar^2}{2m} \nabla^2 + V(\mathbf{r}), \tag{1.1}$$

with the kinetic energy part $T = -\frac{\hbar^2}{2m} \nabla^2$ and the potential energy contribution $V(\mathbf{r})$. *Bloch's theorem* states that the corresponding wave functions can be chosen to have the form of a plane wave times a function with the periodicity of the lattice:

$$\psi_{n\mathbf{k}}(\mathbf{r}) = e^{i\mathbf{k} \cdot \mathbf{r}} u_{n\mathbf{k}}(\mathbf{r}), \tag{1.2}$$

with

$$u_{n\mathbf{k}}(\mathbf{r} + \mathbf{R}) = u_{n\mathbf{k}}(\mathbf{r}). \tag{1.3}$$

The subscript n denotes the nth state with wave vector $\mathbf{k}$. Combining Equations 1.2 and 1.3, Bloch's theorem can also be written as

$$\psi_{n\mathbf{k}}(\mathbf{r} + \mathbf{R}) = e^{i\mathbf{k}\cdot\mathbf{R}}\psi_{n\mathbf{k}}(\mathbf{r}). \tag{1.4}$$

The proof of Bloch's theorem can be found in many textbooks [3,4]. Instead of proving Bloch's theorem, here we rather make its proof plausible by noting two consequences of Equations 1.2 and 1.4:

1. *Periodic electron density:* In a periodic solid, one expects that the probability density, $|\psi_{n\mathbf{k}}(\mathbf{r})|^2$, normally varies between different points within the unit cell under consideration. Equation 1.2 allows for this since

$$|\psi_{n\mathbf{k}}(\mathbf{r})|^2 = |e^{i\mathbf{k}\cdot\mathbf{r}}|^2|u_{n\mathbf{k}}(r)|^2 = |u_{n\mathbf{k}}(\mathbf{r})|^2. \tag{1.5}$$

The periodic function $u_{n\mathbf{k}}(\mathbf{r})$ does not have to be constant, consequently it can vary within a given unit cell. At a given point $\mathbf{r}$ within one unit cell, the charge density should be equivalent to the charge density at the point $\mathbf{r} + \mathbf{R}$ within another unit cell. This is consistent with Bloch's theorem, since from Equation 1.4 it follows that the probability density $\rho_{n\mathbf{k}}(\mathbf{r} + \mathbf{R})$ can be written as:

$$\rho_{n\mathbf{k}}(\mathbf{r} + \mathbf{R}) = |\psi_{n\mathbf{k}}(\mathbf{r} + \mathbf{R})|^2 = |e^{i\mathbf{k}\cdot\mathbf{R}}|^2|\psi_{n\mathbf{k}}(\mathbf{r})|^2 = \rho_{n\mathbf{k}}(\mathbf{r}). \tag{1.6}$$

This shows that one is left with an equal probability of finding an electron at $\mathbf{r}$ or at $\mathbf{r} + \mathbf{R}$. Obviously, this implies equal charge density at the two points, as one would expect.

2. *Empty lattice model:* In free space ($V(\mathbf{r}) \equiv 0$), the electron wave functions can be chosen to take the form of plane waves. The unnormalized wave function $\psi_{\mathbf{k}}(\mathbf{r}) = e^{i\mathbf{k}\cdot\mathbf{r}}$ describes a state with energy $E = \hbar^2 k^2/2m$. If we partition free space into a periodic array of identical boxes of volume L^3, which, which is referred to as the "empty lattice," the free space wave functions can be expressed as a product of a plane wave times a constant function, which is obviously periodic:

$$\psi_{\mathbf{k}}(\mathbf{r}) = e^{i\mathbf{k}\cdot\mathbf{r}} \cdot \frac{1}{\sqrt{L^3}}. \tag{1.7}$$

Consequently, Bloch's theorem describes wave functions that reduce to the correct form in the case where the periodic potential $V(\mathbf{r}) \equiv 0$.

Having discussed Bloch's theorem, we consider in the next section its consequences for the electronic structure of a crystalline solid and introduce the concept of the *electronic band structure* of the system under consideration.

1.2.1 Electronic Band Structure

From Bloch's theorem, it follows that in a periodic solid, we can associate a wave vector $\mathbf{k}$ with each energy state, $E_n(\mathbf{k})$. Thus, it is useful to display the energies $E_n(\mathbf{k})$ as a function of the wave vector $\mathbf{k}$. The resulting diagram is referred to as the *electronic band structure* of the given solid. Figure 1.1a shows the electronic band structure for an electron in an empty lattice (free space). As described above, the wave function can be expressed as a plane wave and the resulting $E(\mathbf{k})$ is described by the parabola $E = \hbar^2 k^2/2m$.

In a periodic solid, due to the presence of the potential $V(\mathbf{r})$, the free electron band structure is modified in several ways. In addition, the wave vector $\mathbf{k}$ associated with a given energy state is no longer uniquely defined. To illustrate this in more detail, we assume in the following a one-dimensional (1-D) periodic structure. A unit cell of length L shall be considered. In this case, we can write the wave function for the nth state with wave number k as follows:

$$\psi_{nk}(x) = e^{ikx}u_{nk}(x). \tag{1.8}$$

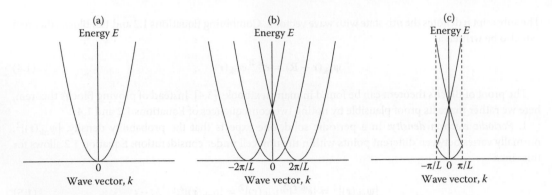

FIGURE 1.1 (a) Free electron "band structure". Here the energy E depends quadratically on the wave vector k. (b) Band structure of a one-dimensional (1-D) lattice with period L. Here wave numbers k and $k + 2\pi m/L$, where m is an integer, are equivalent. In the repeated zone scheme, all wave numbers $k + 2\pi m/L$ associated with each energy state are plotted. (c) In the reduced zone scheme, the wave number for each energy state is chosen in such a way that the magnitude of k is minimized. For the 1-D lattice this implies $-\pi/L < k < \pi/L$.

Here we have a plane wave e^{ikx} of wave number k and $u_{nk}(x)$ is a periodic function, with $u_{nk}(x) = u_{nk}(x + L)$. We wish to show that the wave number k in this case is not uniquely defined. For this we multiply Equation 1.8 by a plane wave with the periodicity of the lattice, $e^{i2\pi mx/L}$, and by its complex conjugate, $e^{-i2\pi mx/L}$, where m is an integer. This yields

$$\psi_{nk}(x) = e^{ikx} e^{\frac{i2\pi m}{L}x} e^{\frac{-i2\pi m}{L}x} u_{nk}(x)$$

$$= e^{i(k+2\pi m/L)x} \left(e^{-\frac{i2\pi m}{L}x} u_{nk}(x) \right). \tag{1.9}$$

Here $e^{i(k+2\pi m/L)x}$ is a plane wave but a different one compared to the original choice. The term $e^{-i2\pi mx/L} u_{nk}(x)$ is still a periodic function with period L. Introducing the notation $G_m = 2\pi m/L$ that we refer to as a *reciprocal lattice vector*, we observe that the wave number k is then equivalent to the wave vector $k + G_m$ in the given 1-D periodic system.

If we now consider dividing 1-D free space into unit cells of length L to create an empty lattice, we have several choices of how to plot the free electron band structure:

- In the *extended zone scheme*, one tries to associate a single, "correct" wave number k with each state, as displayed in Figure 1.1a. For the empty lattice, this is probably the best approach. However, in a periodic crystal it becomes very difficult to assign a unique, "correct" k to each state.
- In the *repeated zone scheme*, for a given energy state, one can include several (in principle, all) wave numbers k associated with this state. In doing so one can circumvent the problem of choosing the "correct" wave number k for each state. This repeated zone scheme is illustrated in Figure 1.1b. However, it is clear that this approach contains a lot of redundant information.
- In the *reduced zone scheme*, for each energy state we select the wave number k in such a way that the magnitude of the wave number k is minimized. In doing so one provides a simple rule for assigning a preferred k value to each state. This gives a simple prescription for displaying the band structure, as illustrated in Figure 1.1c. The reduced zone is also widely referred to as the *first Brillouin zone* for the given crystal structure. Thus, when plotting the electronic band structure of a periodic crystal, we can restrict our choice of wave vectors to those in the first Brillouin zone.

In summary, we see how Bloch's theorem significantly simplifies the calculation of the electronic band structure, since it allows that in a periodic solid, a wave vector $\mathbf{k}$ can be associated to each energy state, $E_n(\mathbf{k})$. Furthermore, the wave vector can be restricted to the first Brillouin zone. As an example we discussed the "band structure" of a free electron. However, the presence of the potential arising from the crystal lattice has not been addressed. In the next section, we do so by considering as an example the band structure of the two technologically very important semiconductor materials, Si and GaAs. Differences and similarities between the band structures of the two materials will be discussed before coming back in Section 1.4 to the question of how to calculate these band structures.

1.3 Semiconductor Band Structure: Direct and Indirect Gap Materials (GaAs versus Si)

As discussed in Section 1.1, the electronic band structure of semiconductor materials and their specific features are of central importance for determining their electronic and optical properties. This is not only of interest for a fundamental understanding of material properties; it also relates directly to their usefulness for devices such as transistors, lasers, or light-emitting diodes to name only a few. This has resulted in intense research efforts around the world to modify and tailor the electronic band structure of semiconductor materials, including the use of alloys, heterostructures, quantum confinement, and strain. Some of the most advanced devices now employ all of these techniques. Here, we focus our attention on some of the central features of the electronic band structure of bulk semiconductors.

When plotting the band structure of a bulk semiconductor, it is not possible to plot the energy E for each three-dimensional wave vector $\mathbf{k}$. Instead, we plot how the energy varies with wave vector $\mathbf{k}$ along different high-symmetry directions. Different high-symmetry directions and points are given specific names. In particular, the [001] direction between the Γ point ($\mathbf{k} = \mathbf{0}$) and the Brillouin zone edge at the X point is referred to as the Δ direction, while the [111] direction is denoted as the Λ-direction, joins Γ and what is referred to as the L point at the Brillouin zone edge.

Figure 1.2 displays the band structures of bulk GaAs and Si, which are two of the most technologically important semiconductors. For most applications, the bands near the energy gap, E_g, are of particular interest. The energy gap E_g is defined as the energetic separation between the highest filled valence (bonding) band state and the lowest empty conduction (antibonding) band state. From Figure 1.2 one can infer that the lowest energy state in the GaAs conduction band, E_c, is at the Γ-point ($\mathbf{k} = \mathbf{0}$). E_c is directly above the highest energy state in the valence band, E_v. For this reason, GaAs is called a *direct gap semiconductor*. This result is key to the optical properties of GaAs-based systems. If an electron relaxes to the lowest possible state in the conduction band it can recombine directly with a hole at the top of the valence band by emitting a photon—note that on the scale of Figure 1.2 a photon has negligible momentum. To realize efficient light-emitting diodes or laser structures, semiconductors with a direct band gap are utilized [1].

The band structure of GaAs can be contrasted with the band structure of Si (cf. Figure 1.2b). Si is referred to as an *indirect band gap semiconductor*, since the bottom of the conduction band is not directly above the valence band maximum at the Γ point. In Si, the conduction band minima are along the six equivalent Δ (cubic axis) directions near to the X points. For the electron in the lowest conduction band state to recombine with a hole at the top of the valence band to emit a photon, phonon absorption or emission processes are required. This stems from the fact that the momentum must be conserved in such a process. A phonon is the name given to a quantized lattice vibration [3,6]. On the scale of Figure 1.2, phonons have a negligible energy but significant momentum. This originates from the fact that the mass of the nucleus is large compared to the electron mass. The light emission from an indirect gap semiconductor material is therefore less efficient when compared to direct gap materials since it requires the simultaneous occurrence of two processes (phonon absorption/emission plus photon emission/absorption) [1].

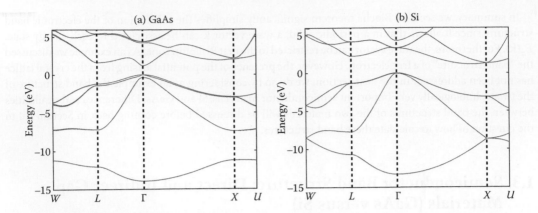

FIGURE 1.2 Electronic band structure of (a) GaAs and (b) Si calculated using the Heyd–Scuseria–Ernzerhof hybrid-functional approach [5] in the framework of density functional theory. For GaAs, the plane wave cutoff energy was set to 600 eV and the Ga semicore d-states are treated as valence electrons. A Γ-centered $6 \times 6 \times 6$ k-point mesh was chosen. For Si similar settings have been used. In this case, the cutoff energy was 245 eV and the k-point mesh is $10 \times 10 \times 10$. We have set the zero of energy in both cases to be at the valence band maximum, E_v. The calculated conduction band minimum is at energy $E_c = 1.2$ eV in GaAs and at $E_c = 1.1$ eV in Si, giving an energy gap of magnitude $E_g = 1.2$ eV and 1.1 eV, respectively.

In tetrahedrally bonded semiconductor materials, the valence band maximum is always at the Γ-point. When neglecting the electron spin, the valence band maximum consists of three degenerate p-like bonding states. However, in reality, we need to include the electron spin in the consideration. Starting from an isolated atom, the electron experiences an additional potential due to the interaction of the spin magnetic moment with its orbital angular momentum. This interaction is the so-called *spin–orbit interaction*, which splits the p-states in an isolated atom [7]. In solids, it splits the six states (including spin-up and spin-down components) of the otherwise degenerate valence band maximum, with two states shifted energetically below the other four states by an energy of Δ_{so}. The magnitude of Δ_{so} increases with atomic number [8]. It should be noted that the spin–orbit interaction not only affects the band structure of the valence band maximum, but can also lift other degeneracies in the band structure [9].

We have reviewed thus far only the general properties of bulk band structures. However, the question of how to calculate the band structure for real materials has not been addressed. To make predictions of the band structure of materials and also then to be able to tailor their properties, one needs tools that allow for the accurate calculation of the electronic structure. Different theoretical approaches that allow for a calculation of the electronic structure of semiconductor materials will be discussed and reviewed in the next sections. We introduce methods that range from first-principles calculations up to semiempirical models. We start in the next section with the "exact" solution of the Schrödinger equation to predict the band structure of a semiconductor material.

1.4 Schrödinger Equation and "Exact" Solution Using LDA–DFT

In general, for an accurate calculation of the electronic structure of materials, the Coulomb interaction between the carriers has to be considered. As already mentioned earlier, one is left with a highly complex many-body problem, even when making use of Bloch's theorem. However, the ground-breaking ideas and approaches of Hohenberg and Kohn as well as of Kohn and Sham paved the way for *density functional theory* (DFT), which has become the workhorse for first-principles electron structure calculations. Especially, the so-called *local density approximation* (LDA) has become a widely used approach to calculate the electronic structure of materials within DFT. In this section, we review the basic ideas of DFT, including how one can derive the Kohn–Sham equations and the approach taken to obtain the LDA.

1.4.1 Preliminaries

We focus in a first step on a nonrelativistic Hamiltonian for an N-electron system that accounts for the repulsive electron–electron interaction and that includes an external potential, which in solids can be assumed to be the periodic crystal lattice. The corresponding stationary many-body Hamiltonian $\hat{H}$ reads

$$\hat{H} = \underbrace{\sum_i^N \frac{-\hbar^2}{2m} \nabla_i^2}_{\hat{T}} + \underbrace{\sum_i^N v(\mathbf{r}_i)}_{\hat{V}} + \underbrace{\frac{1}{2} \sum_{i,j}^N \frac{e^2}{|\mathbf{r}_i - \mathbf{r}_j|}}_{\hat{W}}. \tag{1.10}$$

Here $\hat{T}$ is the operator for the kinetic energy, $\hat{V}$ for the potential energy, and $\hat{W}$ is the operator for the (repulsive) two-particle interaction. The kinetic energy operator $\hat{T}$ and the operator for the Coulomb interaction $\hat{W}$ are the same for all electronic systems, that is, atoms, molecules, or solids. The difference between these systems arises from differences in the form of $v(\mathbf{r})$ and also in the number of electrons in the system.

One quantity of interest for an electronic system is the ground state energy, which we denote here as E_{GS}. To calculate E_{GS} one would have to solve the full time-independent Schrödinger equation for the N particle system and the eigenvalue with the lowest energy is E_{GS}. Obviously, this is a highly nontrivial and complex problem. The basic idea of DFT is that, instead of calculating the antisymmetric ground state wave function $\psi_{GS}(\mathbf{r}_1, \ldots, \mathbf{r}_N)$, we study the corresponding electron density $\rho_{GS}(\mathbf{r})$

$$\rho_{GS}(\mathbf{r}) = \langle \psi_{GS} | \hat{\rho} | \psi_{GS} \rangle$$

$$= \int d^3 r_1 \cdots \int d^3 r_N \psi_{GS}^*(\mathbf{r}_1, \ldots, \mathbf{r}_N) \sum_i^N \delta(\mathbf{r} - \mathbf{r}_i) \psi_{GS}(\mathbf{r}_1, \ldots, \mathbf{r}_N), \tag{1.11}$$

with the density operator $\hat{\rho}$ defined as

$$\hat{\rho} = \sum_i^N \delta(\mathbf{r} - \mathbf{r}_i). \tag{1.12}$$

The benefit of studying the ground state electronic density instead of the wave function is obvious. While the ground state wave function depends on three spatial coordinates per particle, $\rho_{GS}(\mathbf{r})$ depends only on three spatial variables, since it is a collective quantity.

Hohenberg and Kohn [10] were able to prove that the ground state energy E_{GS} of a many-particle system is an unambiguous functional of the ground state electron density ρ_{GS}. This statement, however, is not obvious. We present a proof of this result, which is given in the original paper by Hohenberg and Kohn [10], in the following section.

1.4.2 Hohenberg–Kohn Theorem for Nondegenerate Ground States

We know that a wave function ψ can be connected to an energy E via [3,7]:

$$E[\psi] = \frac{\langle \psi | \hat{H} | \psi \rangle}{\langle \psi | \psi \rangle}. \tag{1.13}$$

Thus, the energy E is a functional of ψ. On the other hand, from Equation 1.11, we know also that each wave function ψ is connected to an electron density ρ. The *Hohenberg–Kohn theorem* states that for the ground state there is an unambiguous connection between the ground state energy and the density ρ_{GS} [10]. One has to prove that the connection is unique, so that two different ground states cannot have the same ρ_{GS}. Let us assume we have different ground states $|\psi_{GS}\rangle \neq |\psi'_{GS}\rangle$ but the same density $\rho_{GS} = \rho'_{GS}$. We assume here that the ground state is always nondegenerate, thus $|\psi_{GS}\rangle$ and $|\psi'_{GS}\rangle$ must belong to different Hamiltonians $\hat{H}$ and $\hat{H}'$, respectively.[†] As discussed earlier, as long as the number of particles in the systems are identical, the only difference in the Hamiltonians can arise from the potential energy operator ($\hat{V} \neq \hat{V}'$). Using Equation 1.12 we can write $\hat{V}$ in general as

$$\hat{V} = \int v(\mathbf{r})\hat{\rho}(\mathbf{r})\,\mathrm{d}^3 r. \tag{1.14}$$

With this we can write

$$
\begin{aligned}
E_{GS} = \langle \psi_{GS}|\hat{H}|\psi_{GS}\rangle &= \langle \psi_{GS}|\hat{T} + \hat{V}_1 + \hat{W}|\psi_{GS}\rangle \\
&< \langle \psi'_{GS}|\hat{H}|\psi'_{GS}\rangle = \langle \psi'_{GS}|\hat{H}' + \hat{V}_1 - \hat{V}_2|\psi'_{GS}\rangle \\
&= E'_{GS} + \int \left(v_1(\mathbf{r}) - v_2(\mathbf{r})\right)\rho'_{GS}(\mathbf{r})\mathrm{d}^3 r,
\end{aligned}
\tag{1.15}
$$

where $H' = \hat{T} + \hat{V}_2 + \hat{W}$. Along the same line we obtain the inequality

$$E'_{GS} < E_{GS} + \int \left(v_2(\mathbf{r}) - v_1(\mathbf{r})\right)\rho_{GS}(\mathbf{r})\mathrm{d}^3 r. \tag{1.16}$$

Assuming that $\rho_{GS} = \rho'_{GS}$ and adding Equations 1.15 and 1.16 results in the contradiction

$$E_{GS} + E'_{GS} < E'_{GS} + E_{GS}. \tag{1.17}$$

Consequently, using/choosing a particular ground state density ρ determines $\hat{V}$ uniquely and therefore the Hamiltonian $\hat{H}$ and the connected ground state energy E_{GS}.

According to the Rayleigh–Ritz variation principle [3], we know that

$$E_{GS} = E[\rho_{GS}(\mathbf{r})] \leq \langle \psi|H|\psi\rangle = E[\rho(\mathbf{r})]. \tag{1.18}$$

Thus, the electron density ρ for which the energy functional $E[\rho(\mathbf{r})]$ has its absolute minimum corresponds to the ground state density ρ_{GS} and the value of the energy functional is then the ground state energy E_{GS} of the Hamiltonian $\hat{H}$. In general, we are therefore left with a minimization problem. We show in the next section that by starting from the variational problem one can derive the famous *Kohn–Sham equations*.

1.4.3 Kohn–Sham Theory and LDA

As discussed in the previous section, using the Hohenberg–Kohn theorem allowed us in principle to find the ground state density $\rho_{GS}(\mathbf{r})$ and energy E_{GS} of a system of N interacting particles by minimizing the energy functional $E[\rho]$. However, even though this provides a general framework, it does not immediately

---

[†] A discussion of how to extend the Hohenberg–Kohn theorem to a degenerate ground state can be found, for example, in Ref. [11].

give the form of the different functionals, for example, how the kinetic energy T depends on ρ. Kohn and Sham [12] introduced a scheme with an effective single-particle equation in which a particular part of the effective potential, namely the *exchange-correlation part*, requires approximations. The derived effective single-particle Schrödinger equation is known as the Kohn–Sham equation. Here, we discuss the simplest and most widely used approximation for the *exchange-correlation functional*, which gives the so-called *local density approximation* (LDA) [3,13].

As discussed in the previous section, in general, for Equation 1.18, we are left with the variational problem

$$\delta E[\rho(\mathbf{r})] = 0, \tag{1.19}$$

under the constraint that we have N particles in the system:

$$\int \rho(\mathbf{r}) \, \mathrm{d}^3 r = N. \tag{1.20}$$

Including this constraint into the variational problem via a Lagrange multiplier μ, one is left with [10]

$$\delta\left(E[\rho(\mathbf{r})] - \mu\left(\int \rho(\mathbf{r}) \, \mathrm{d}^3 r - N\right)\right) = 0. \tag{1.21}$$

Based on Equation 1.10, the energy functional $E[\rho(\mathbf{r})]$ is given by

$$E[\rho(\mathbf{r})] = T[\rho(\mathbf{r})] + V[\rho(\mathbf{r})] + W[\rho(\mathbf{r})]. \tag{1.22}$$

Here one encounters the problem that the functional dependence of the kinetic energy T and the interaction part W on the density ρ is unknown. Only for the external potential we know

$$V[\rho(\mathbf{r})] = \int \mathrm{d}^3 r \, v(\mathbf{r})\rho(\mathbf{r}) . \tag{1.23}$$

A first ansatz for $W[\rho(\mathbf{r})]$ could be

$$W[\rho(\mathbf{r})] = \frac{e^2}{2} \int \mathrm{d}^3 r \int \mathrm{d}^3 r' \frac{\rho(\mathbf{r})\rho(\mathbf{r}')}{|\mathbf{r} - \mathbf{r}'|}, \tag{1.24}$$

which would give the *Hartree contribution* to the interaction functional. However, from simple *Hartree–Fock theory* we know already that corrections due to the indistinguishability of the particles are required [6]. Therefore, a more general ansatz for $W[\rho(\mathbf{r})]$ can be made [3]

$$W[\rho(\mathbf{r})] = \frac{e^2}{2} \int \mathrm{d}^3 r \int \mathrm{d}^3 r' \frac{\rho(\mathbf{r})\rho(\mathbf{r}')}{|\mathbf{r} - \mathbf{r}'|} + E_{\mathrm{xc}}[\rho(\mathbf{r})], \tag{1.25}$$

where $E_{\mathrm{xc}}[\rho(\mathbf{r})] = E_{\mathrm{x}}[\rho(\mathbf{r})] + E_{\mathrm{c}}[\rho(\mathbf{r})]$ is the *exchange-correlation functional*. The functional can be divided into an exchange part $E_{\mathrm{x}}[\rho(\mathbf{r})]$ and a correlation part $E_{\mathrm{c}}[\rho(\mathbf{r})]$. However, together with the kinetic energy functional, the dependence of $E_{\mathrm{xc}}[\rho(\mathbf{r})]$ on the density is still unknown.

Due to these obstacles Kohn and Sham introduced the idea that the density $\rho(\mathbf{r})$ of a system with interacting particles under the influence of an external potential $v(\mathbf{r})$ can be described by a non-interacting system with an effective external potential $\tilde{v}(\mathbf{r})$ that contains the contribution from the interaction part.

This concept now leads to the so-called *Kohn–Sham equations* [12], which can be solved once a choice is made for the exchange-correlation functional E_{xc}.

In the case of a noninteracting N-particle system, the corresponding Hamiltonian $\hat{H}_0$ reads

$$\hat{H}_0 = \sum_i^N \frac{-\hbar^2}{2m} \nabla_i^2 + \sum_i^N \tilde{v}(\mathbf{r}_i). \tag{1.26}$$

The ground state wave function of the non-interacting system can be written as a Slater determinant; thus it is an antisymmetrized product of single-particle orbitals $\phi_i(\mathbf{r})$. The particle density ρ_0 of the noninteracting system is given by

$$\rho_0(\mathbf{r}) = \sum_i^N |\phi_i(\mathbf{r})|^2, \tag{1.27}$$

and the corresponding energy in this state is

$$E[\rho_0(\mathbf{r})] = \sum_i^N \frac{-\hbar^2}{2m} \int d^3r\, \phi_i^*(\mathbf{r}) \nabla^2 \phi_i(\mathbf{r}) + \sum_i^N \int d^3r\, \tilde{v} |\phi_i(\mathbf{r})|^2$$

$$= \sum_i^N \frac{-\hbar^2}{2m} \int d^3r\, \phi_i^*(\mathbf{r}) \nabla^2 \phi_i(\mathbf{r}) + \int d^3r\, \tilde{v} \rho_0(\mathbf{r}). \tag{1.28}$$

Even for the much simpler case of the noninteracting system, one still does not explicitly know the dependence of the kinetic energy on the density ρ_0 explicitly. Nevertheless, in a first approximation, one can start from this assumption and use the kinetic energy functional of the noninteracting system for the interacting many-body system:

$$E[\rho(\mathbf{r})] = \sum_i^N \frac{-\hbar^2}{2m} \int d^3r\, \phi_i^*(\mathbf{r}) \nabla^2 \phi_i(\mathbf{r}) + \int d^3r\, v\rho(\mathbf{r})$$

$$+ \frac{e^2}{2} \int d^3r \int d^3r' \frac{\rho(\mathbf{r})\rho(\mathbf{r}')}{|\mathbf{r} - \mathbf{r}'|} + E_{xc}[\rho(\mathbf{r})]. \tag{1.29}$$

With this approach, we assume that the ground state density of the many-body Hamiltonian can be constructed by single-particle orbitals, which is not guaranteed. Even if that would be the case, it would not necessarily mean that the kinetic energy of the interacting system has the same form as in the noninteracting system. One can nevertheless choose this ansatz, assuming that corrections to the kinetic energy functional, introduced by the interaction between the carriers, are included in the still unknown exchange-correlation functional E_{xc}. By varying/minimizing the energy functional under the constraint that the single-particle states are normalized one can find the ground state energy [10,12]

$$\frac{\delta}{\delta \phi_i^*} \left\{ E[\rho(\mathbf{r})] - \sum_j^N \epsilon_j \left(\int d^3r |\phi_j|^2 - 1 \right) \right\} = 0. \tag{1.30}$$

This results in

$$\left\{ -\frac{\hbar^2}{2m} \nabla^2 + v(\mathbf{r}) + \int d^3r' \frac{e^2 \rho(\mathbf{r}')}{|\mathbf{r} - \mathbf{r}'|} + \frac{\delta E_{xc}[\rho(\mathbf{r})]}{\delta \rho} \right\} \phi_i(\mathbf{r}) = \left\{ -\frac{\hbar^2}{2m} \nabla^2 + v_s(\mathbf{r}) \right\} \phi_i(\mathbf{r}) = \epsilon_i \phi_i(\mathbf{r}), \tag{1.31}$$

which is obviously of the form of a single-particle Schrödinger equation with the effective potential

$$v_s\left(\mathbf{r}\right) = v(\mathbf{r}) + \int d^3 r' \frac{e^2 \rho\left(\mathbf{r}'\right)}{|\mathbf{r} - \mathbf{r}'|} + \frac{\delta E_{xc}[\rho\left(\mathbf{r}\right)]}{\delta \rho}. \tag{1.32}$$

Equations 1.31 and 1.32 together with

$$\rho(\mathbf{r}) = \sum_{i=1}^{N} |\phi_i(\mathbf{r})|^2 \tag{1.33}$$

from the so-called *Kohn–Sham equations* [12]. This set of equations has to be solved self-consistently [13].

To use the Kohn–Sham equations for practical calculations, the form of the exchange-correlation functional needs to be known. For this quantity, we have to rely on approximations. One of the first approximations for E_{xc} that has been and is still often used is the so-called local density approximation (LDA) [12,13]. In this case, one uses the homogenous electron gas as a starting point, since in this case one can calculate the dependence of exchange energy on the (**r**-independent) density ρ explicitly. Using this ansatz, the *LDA exchange-correlation functional* $E_{xc}^{LDA}[\rho(\mathbf{r})]$ is given by

$$E_{xc}^{LDA}[\rho(\mathbf{r})] = -\frac{3e^2}{4\pi} \left(3\pi^2\right)^{\frac{1}{3}} \int \left(\rho(\mathbf{r})\right)^{4/3} d^3 r. \tag{1.34}$$

The corresponding exchange-correlation potential v_{xc}^{LDA} required in Equation 1.32 is then given by

$$v_{xc}^{LDA} = \frac{\delta E_{xc}^{LDA}}{\delta \rho} = -e^2 \left(\frac{3}{\pi}\right)^{\frac{1}{3}} \left(\rho(\mathbf{r})\right)^{\frac{1}{3}}. \tag{1.35}$$

Consequently, in LDA we approximate inhomogeneous systems locally by a homogenous electron gas. Even though this is an extremely crude approximation, it turns out that this ansatz works remarkably well for many systems and properties, while it systematically fails for others [14,15]. The underestimation of the band gap of semiconductors and insulators is probably one of the best known problems of LDA [14]. To overcome problems related to LDA, different approaches have been suggested. We review and introduce some of them in the next section.

1.5 Beyond LDA: GGA, Hybrid Functional, GW, and Related Approaches

One way to go beyond LDA is the *gradient expansion approximation* (GEA) [16]. In this case, one includes also gradients of the density ρ in the exchange-correlation functional, presenting therefore a systematic nonlocal extension of the LDA. The exchange-correlation functional $E_{xc}^{GEA}[\rho(\mathbf{r})]$ in GEA can be written as [16]

$$E_{xc} = E_{xc}^{LDA}[\rho] + \int f_1\left(\rho(\mathbf{r})\right)\left(\nabla\rho(\mathbf{r})\right)^2 d^3 r + \int f_2\left(\rho(\mathbf{r})\right)\left(\nabla^2\rho(\mathbf{r})\right)^2 d^3 r \dots . \tag{1.36}$$

The functions f_i can be determined from response theory. Unfortunately, the GEA gives no real improvement compared with LDA. Furthermore, it has been shown that the GEA even violates the sum rule [17,18]

$$\int d^3 r' \left(g(\mathbf{r}, \mathbf{r}') - 1\right) \rho\left(\mathbf{r}'\right) = 1, \tag{1.37}$$

where $g(\mathbf{r}, \mathbf{r}')$ is the so-called *pair correlation function*. This quantity describes the probability of finding an electron at $\mathbf{r}'$ while there is at the same time an electron at position $\mathbf{r}$. The integrand in Equation 1.37 is usually referred to as the *exchange-correlation hole* [17,18].

To overcome these problems, exchange-correlation functionals of the following form have been suggested [17,19]:

$$E_{xc}^{GGA} = \int d^3 r f(\rho(\mathbf{r}), \nabla\rho(\mathbf{r})). \tag{1.38}$$

This approach is the so-called *generalized gradient approximation* (GGA). The function f is now chosen in a form that it satisfies the sum-rule of the exchange-correlation hole [17]. Additional constraints have also been included, leading to a variety of different GGA functionals. An overview is given in Ref. [20], for example.

However, even when going beyond LDA by using GGA, it does not solve the "band gap problem." In general, it is a misconception that the band gap cannot be calculated from DFT. For example, we can express the band gap of a system in the following way [21]:

$$E_g = \underbrace{(E_{N-1} - E_N)}_{\text{Ionization Energy}} - \underbrace{(E_N - E_{N+1})}_{\text{Electron Affinity}}. \tag{1.39}$$

Here, E_N denotes the ground state energy of the N-particle system, while E_{N+1} (E_{N-1}) denotes the energy with $N + 1$ ($N - 1$) particles. Since DFT gives the correct ground state energies, E_g should in principle be an accessible quantity from DFT calculations. For the N-particle system, the band gap E_g can also be obtained using the Kohn–Sham eigenvalues via [21],

$$E_g = E_g^{KS} + \Delta_{xc}, \tag{1.40}$$

where E_g^{KS} is the gap from Kohn–Sham theory and Δ_{xc} is given by [21]

$$\Delta_{xc} = \frac{\delta\epsilon}{\delta\rho}\Big|_{N+\delta N} - \frac{\delta\epsilon}{\delta\rho}\Big|_{N-\delta N}, \tag{1.41}$$

where the number of particles is changed by an integer number δN. However, in Kohn–Sham theory, even when going beyond LDA by, for example, using gradient corrections in the exchange-correlation functionals, $\Delta_{xc} = 0$. This originates from the fact that calculating the band gap is intrinsically a many-body effect, for which the independent electron picture underlying the Kohn–Sham approach therefore breaks down [21].

On the other hand, when dealing with isolated atoms *Hartree–Fock* (HF) *theory* allows for an accurate description of these systems. This stems from the fact that HF theory accounts for the exchange part exactly, and includes self-interaction and a derivative discontinuity [22,23]. However, correlation effects are not included in HF, which are important for solids or larger molecules. These effects are included in the local exchange-correlation functionals. Thus, combining these techniques provides a promising route to establish *hybrid functionals*. Becke introduced this idea in 1993 and considered an adiabatic connection between a noninteracting and a fully interacting system [24]

$$E_{xc}[\rho] = \int_0^1 d\lambda\, E_{xc,\lambda}[\rho]. \tag{1.42}$$

Here, $E_{xc}[\rho]$ is the exchange-correlation functional, while $E_{xc,\lambda}[\rho]$ denotes the exchange-correlation functional for a given value of λ, which determines the strength of the electron–electron interaction. In the limit of the noninteracting system, HF provides an accurate description, since the wave function is given by a single Slater determinant. In the limit of the fully interacting system, as an approximation, local exchange-correlation functionals are used. Initially, Becke [24] proposed the following expression for $E_{xc,\lambda}[\rho]$:

$$E_{xc,\lambda}[\rho] = (1 - \lambda) E_x^{HF} + \lambda E_{xc}^{LDA}. \tag{1.43}$$

The functional E_x^{HF} is the HF exchange functional and E_{xc}^{LDA} denotes the exchange-correlation functional in LDA. Evaluating Equation 1.42 with $E_{xc,\lambda}[\rho]$ from Equation 1.43, one is left with [24]

$$E_{xc,\lambda}[\rho] = \frac{1}{2}E_x^{HF} + \frac{1}{2}E_{xc}^{LDA}. \tag{1.44}$$

This results in a hybrid functional, made up of 50% HF exchange and 50% LDA exchange correlation.

Using a semiempirical approach, Becke modified this approach further by adding nonlocal GGA type contributions [25,26]. These semiempirical hybrid functionals have then been benchmarked against sets of molecules. Becke introduced different sets of these semiempirical functionals, which differ in their number of free parameters. An example is the so-called *B1X functional*, which is of the form [26]

$$E_{xc}^{B1X}[\rho] = E_{xc}^X[\rho] + a \left(E_x^{HF}[\rho] - E_x^X[\rho] \right), \tag{1.45}$$

where X denotes a particular GGA functional.

Another hybrid functional, denoted as *PBE0*, was introduced by Perdew, Burke, and Ernzerhof [27], by assuming that

$$E_{xc,\lambda}^{PBE0}[\rho] = (1 - \lambda)^{n-1} \left(E_x^{HF} - E_x^{PBE} \right) + E_{xc}^{PBE}, \tag{1.46}$$

where the GGA functional is $X = $ PBE and $n \geq 1$ is an integer, which has to be determined. Equations 1.46 and 1.42 yield

$$E_{xc}^{PBE0}[\rho] = E_{xc}^{PBE}[\rho] + \frac{1}{n} \left(E_x^{HF}[\rho] - E_x^{PBE}[\rho] \right). \tag{1.47}$$

Choosing $\frac{1}{n} = a$ gives the B1X functional (cf. Equation 1.45). The advantage of the hybrid functional $E_{xc}^{PBE0}[\rho]$ is that it can be related to the *Møller–Plesset perturbation theory* [28]. Møller–Plesset perturbation theory yields, when taken to the fourth order, very good atomization energies for many materials. Such an expansion up to fourth order results in $n = 4$ and thus [27]

$$E_{xc}^{PBE0}[\rho] = E_{xc}^{PBE}[\rho] + \frac{1}{4} \left(E_x^{HF}[\rho] - E_x^{PBE}[\rho] \right)$$

$$= E_c^{PBE}[\rho] + \frac{1}{4}E_x^{HF}[\rho] + \frac{3}{4}E_x^{PBE}[\rho]. \tag{1.48}$$

These types of hybrid functionals have been successfully applied to a variety of different materials to predict electronic and structural properties [29–32]. However, the computational effort is much higher than for conventional local functionals. This results mainly from the calculation of the long-range (LR) part of the Coulomb potential in the HF exchange [33]. Therefore, within this framework only systems with a few atoms can be treated.

However, the LR exchange part vanishes in a real system since interactions between electrons far apart are screened. In 2003, Heyd et al. introduced the idea of separating the exchange part in PBE0 functionals into LR and short-range (SR) contributions [5]:

$$E_x^{PBE0} = aE_x^{HF,SR} + aE_x^{HF,LR} + (1 - a)E_x^{PBE,SR} + (1 - a)E_x^{PBE,LR}. \tag{1.49}$$

Heyd et al. [5] proposed then to split the Coulomb interaction $v(r)$ into LR and SR contributions by applying the error function erf(x) and its complement erfc(x) = 1 − erf(x):

$$v(r) = S_\mu(r) + L_\mu(r) = \frac{erfc(\mu r)}{r} + \frac{erf(\mu r)}{r}, \tag{1.50}$$

where the screening parameter μ is adjustable and $r = |\mathbf{r} - \mathbf{r}'|$. Numerical tests revealed that for realistic screening parameter values, for example, $\mu = 0.15a_0^{-1}$, where a_0 is the Bohr radius, the contribution to the functional from the LR parts in Equation 1.49 is small and that additional HF and PBE contributions tend to cancel each other. Therefore, neglecting these terms and assuming $a = 1/4$, the resulting Heyd–Scuseria–Ernzerhof *screened hybrid functional*, referred to as HSE, is given by [5]

$$E_{xc}^{HSE} = E_c^{PBE} + aE_x^{HF,SR}(\mu) + (1 - a)E_x^{PBE,SR}(\mu) + E_x^{PBE,LR}(\mu)$$

$$= E_c^{PBE} + \frac{1}{4}E_x^{HF,SR}(\mu) + \frac{3}{4}E_x^{PBE,SR}(\mu) + E_x^{PBE,LR}(\mu). \tag{1.51}$$

As discussed by Paier et al. [34] in detail, the HSE functional can significantly reduce the computational cost. This results from the increased locality of the exchange interactions, which allows for the evaluation of the SR part on a coarser k mesh in comparison to a calculation using the full exchange operator. The chosen value for μ was refined over the years so that one finds in the literature calculations using so-called HSE03 or HSE06 functionals, where the HSE functionals differ in the value of μ [5].

Hybrid functionals are a step forward in comparison to local functionals. However, they do not present a universal improvement. This is related to the presence of the exact exchange mixing parameter a. For example, the mixing parameter a required to reproduce correct defect geometries might be very different from the value chosen to optimize the band gap. Thus, for certain systems there is not a unique value for a that describes both the bulk band edges and the ground state geometries for certain defects [35,36]. Also for band gaps of semiconductor alloys, the choice of a is problematic since the binary compounds may require different a values for an accurate band gap description [37]. Therefore, the choice for alloys is difficult.

As discussed in detail in Ref. [38], there are several ways to go beyond standard DFT that can offer better accuracy, but keeping it still systematic and general. Here, *many-body perturbation theory* (MBPT) using an efficiently achievable and trustworthy reference state has been widely used. In the solid-state community, Hedin's *GW approximation* [39], which represents such an MBPT treatment, has become very popular since it allows for calculations of quasi-particle band structures which can be compared to direct and inverse photoemission spectra [23,40]. A detailed description of MBPT approaches is beyond the scope of the present work and can be found, for example, in Ref. [38].

Overall, the basics concepts of DFT and some of the current state-of-the-art first-priciples approaches have been discussed in this section. These approaches allow in general for a description of electronic band structure over the entire first Brillouin zone. However, when looking back at the general discussion of the electronic band structure of GaAs in Section 1.3, it was highlighted that this material system is of particular interest for optoelectronic devices because of the band structure specifics around $\mathbf{k} = \mathbf{0}$. Thus, when interested in describing the optical properties of these systems this region is of central importance, while the behavior of the highest valence and lowest conduction bands at the boundaries of the first Brillouin zone is generally of secondary importance. Therefore, more empirical methods, which describe a particular part of the band structure accurately, have been developed over the years. These (semi)empirical models are computationally less demanding and, due to their semiempirical nature, they can overcome, for example, the band gap problem of standard DFT calculations. Additionally, because they are computationally relatively cheap, they become highly attractive for the calculation of electronic and optical properties of semiconductor nanostructures. In the next sections the basic concepts of these methods are reviewed, starting with the so-called *(multiband)* **k·p** *method.*

1.6 k·p and Envelope-Function Approaches

k·p theory is a perturbation method, whereby if the exact energy levels are known at a given point in the Brillouin zone, for example, say $\mathbf{k} = \mathbf{0}$, it can be used to find the band structure near the **k** in question. We outline in the following the general idea behind the **k·p** approach. More details can be found, for example, in Refs. [8,41].

We start again from the Hamiltonian, H_0, of a single electron moving in a periodic solid:

$$H_0 = -\frac{\hbar^2}{2m}\nabla^2 + V(\mathbf{r}), \tag{1.52}$$

with $V(\mathbf{r} + \mathbf{R}) = V(\mathbf{r})$. From Bloch's theorem we know that the eigenstates, $\psi_{n\mathbf{k}}(\mathbf{r})$, can be written as the product of a plane wave, $e^{i\mathbf{k}\cdot\mathbf{r}}$, times a periodic function, $u_{n\mathbf{k}}(\mathbf{r})$, with associated energy levels, $E_n(\mathbf{k})$. For a chosen value of **k**, say $\mathbf{k}_0$, Schrödinger's equation can be written as

$$H_0\psi_{n\mathbf{k}_0}(\mathbf{r}) = \left(-\frac{\hbar^2}{2m}\nabla^2 + V(\mathbf{r})\right)\left(e^{i\mathbf{k}_0\cdot\mathbf{r}}u_{n\mathbf{k}_0}(\mathbf{r})\right) = E_n(\mathbf{k}_0)\left(e^{i\mathbf{k}_0\cdot\mathbf{r}}u_{n\mathbf{k}_0}(\mathbf{r})\right). \tag{1.53}$$

In the following, we assume that the allowed energy levels $E_n(\mathbf{k}_0)$ at $\mathbf{k}_0$ are known. The aim is now to find the energy levels, $E_n(\mathbf{k})$, at a wave vector **k** close to $\mathbf{k}_0$, for which the Schrödinger equation can be written as

$$\left(-\frac{\hbar^2}{2m}\nabla^2 + V(\mathbf{r})\right)\left(e^{i\mathbf{k}\cdot\mathbf{r}}u_{n\mathbf{k}}(\mathbf{r})\right) = E_n(\mathbf{k})\left(e^{i\mathbf{k}\cdot\mathbf{r}}u_{n\mathbf{k}}(\mathbf{r})\right). \tag{1.54}$$

Since we are interested in a **k** close to $\mathbf{k}_0$, we rewrite Equation 1.54 as follows:

$$\left(-\frac{\hbar^2}{2m}\nabla^2 + V(\mathbf{r})\right)e^{i(\mathbf{k}-\mathbf{k}_0)\cdot\mathbf{r}}\left(e^{i\mathbf{k}_0\cdot\mathbf{r}}u_{n\mathbf{k}}(\mathbf{r})\right) = E_n(\mathbf{k})e^{i(\mathbf{k}-\mathbf{k}_0)\cdot\mathbf{r}}\left(e^{i\mathbf{k}_0\cdot\mathbf{r}}u_{n\mathbf{k}}(\mathbf{r})\right). \tag{1.55}$$

Multiplying both sides of Equation 1.55 from the left by $e^{-i(\mathbf{k}-\mathbf{k}_0)\cdot\mathbf{r}}$, the following modified differential equation is obtained:

$$\left[e^{-i(\mathbf{k}-\mathbf{k}_0)\cdot\mathbf{r}}H_0 e^{i(\mathbf{k}-\mathbf{k}_0)\cdot\mathbf{r}}\right]\left(e^{i\mathbf{k}_0\cdot\mathbf{r}}u_{n\mathbf{k}}(\mathbf{r})\right) = \left[e^{-i(\mathbf{k}-\mathbf{k}_0)\cdot\mathbf{r}}E_n(\mathbf{k})e^{i(\mathbf{k}-\mathbf{k}_0)\cdot\mathbf{r}}\right]e^{i\mathbf{k}_0\cdot\mathbf{r}}u_{n\mathbf{k}}(\mathbf{r})$$

$$= E_n(\mathbf{k})\left(e^{i\mathbf{k}_0\cdot\mathbf{r}}u_{n\mathbf{k}}(\mathbf{r})\right). \tag{1.56}$$

When transforming Equation 1.55 into Equation 1.56, the Hamiltonian is now a **k**-dependent Hamiltonian, which is denoted in the following by $H_{\mathbf{q}}$, where $\mathbf{q} = \mathbf{k} - \mathbf{k}_0$. Equation 1.56 can then be rewritten as

$$H_{\mathbf{q}}\phi_{n\mathbf{k}}(\mathbf{r}) = e^{-i\mathbf{q}\cdot\mathbf{r}}\left(-\frac{\hbar^2}{2m}\nabla^2 + V(\mathbf{r})\right)e^{i\mathbf{q}\cdot\mathbf{r}}\phi_{n\mathbf{k}}(\mathbf{r}), \tag{1.57}$$

where $\phi_{n\mathbf{k}}(\mathbf{r}) = e^{i\mathbf{k}_0 \cdot \mathbf{r}} u_{n\mathbf{k}}(\mathbf{r})$. With a small amount of algebra, Equation 1.57 becomes

$$H_{\mathbf{q}}\phi_{n\mathbf{k}}(\mathbf{r}) = \left(-\frac{\hbar^2}{2m}\nabla^2 + \frac{\hbar^2}{m}\mathbf{q} \cdot \frac{1}{i}\nabla + \frac{\hbar^2 q^2}{2m} + V(r) \right)\phi_{n\mathbf{k}}(\mathbf{r})$$

$$= \left[H_0 + \frac{\hbar}{m}\mathbf{q} \cdot \mathbf{p} + \frac{\hbar^2 q^2}{2m} \right]\phi_{n\mathbf{k}}(\mathbf{r}), \qquad (1.58)$$

where we have used Equation 1.52 to introduce H_0, and replaced $\frac{\hbar}{i}\nabla$ by the momentum operator, $\mathbf{p}$.

Equation 1.58 forms the foundation of the $\mathbf{k} \cdot \mathbf{p}$ method. In the case of $\mathbf{q} = 0$, Equation 1.58 reduces to the standard form of Schrödinger's time-independent equation. When dealing with a direct gap semiconductor one chooses, in general, $\mathbf{k}_0 = 0$, the Γ-point. At the Γ-point one generally knows or can estimate the values of all the relevant zone center energies, $E_n(0)$. The contribution H',

$$H' = \frac{\hbar}{m}\mathbf{q} \cdot \mathbf{p} + \frac{\hbar^2 q^2}{2m}, \qquad (1.59)$$

to the full Hamiltonian can be regarded as a perturbation to the zone center Hamiltonian, H_0. Using *second-order perturbation theory* one can evaluate the variation of the energy levels $E_n(\mathbf{k})$ with wave vector $\mathbf{k} (= \mathbf{q})$ close to the Γ-point. For a nondegenerate band, the energy of the nth band in the neighborhood of $\mathbf{k} = 0$ is given by

$$E_n(\mathbf{k}) = E_n(0) + \frac{\hbar}{m}\mathbf{k} \cdot \mathbf{p}_{nn} + \frac{\hbar^2 k^2}{2m} + \frac{\hbar^2}{m^2}\sum_{n' \neq n}\frac{|\mathbf{k} \cdot \mathbf{p}_{nn'}|^2}{E_n(0) - E_{n'}(0)}. \qquad (1.60)$$

Here $\mathbf{p}_{nn'}$ denotes the *momentum matrix element* between the nth and n'th zone center states, $u_{n0}(\mathbf{r})$ and $u_{n'0}(\mathbf{r})$,

$$\mathbf{p}_{nn'} = \int_v \mathrm{d}^3 r\, u_{n0}^*(\mathbf{r})\mathbf{p}u_{n'0}(\mathbf{r}) = \langle u_{n0}|\mathbf{p}|u_{n'0}\rangle, \qquad (1.61)$$

with the integration taken over a unit cell of the crystal structure. Equation 1.60 shows that the energy $E_n(\mathbf{k})$ at a given wave vector $\mathbf{k}$ can be expressed in terms of the known energies at $\mathbf{k} = 0$, and the interactions between the zone center states through the momentum matrix elements $\mathbf{p}_{nn'}$. From Equation 1.60 one can also infer that corrections linear in $\mathbf{k}$ ($\sim \mathbf{k} \cdot \mathbf{p}_{nn}$) and quadratic in $\mathbf{k}$ are present. It has been shown by Kane [42] that in a diamond crystal structure the linear term is by symmetry identically equal to zero. For III–V semiconductors, the effects of the linear in $\mathbf{k}$ terms are generally negligibly small and are most often ignored.

To derive Equation 1.60, we have assumed a single nondegenerate band. As discussed in Section 1.3, for a realistic description of the bulk band structure near the Γ-point of semiconductors, effects such as spin–orbit interaction and also of band degeneracies have to be included. The practical application of the $\mathbf{k} \cdot \mathbf{p}$ model, therefore, looks rather complicated. However, two main reasons help ensure its usefulness [42,43]:

- Given that in Equation 1.60 the quadratic in $\mathbf{k}$ correction to $E_n(0)$ depends inversely on $E_n(0) - E_{n'}(0)$, in most cases one has to deal with a very small number of bands, which are close to each other in energy. Higher and lower bands can be ignored, since the denominator in this case will be large, resulting in small corrections.
- When restricting the number of bands in Equation 1.60, and then starting to fit to experimental data, such as measured energy gaps $E_n(0) - E_{n'}(0)$, it is found that the matrix elements $\mathbf{p}_{nn'}$ are remarkably constant between different semiconductors, with $\mathbf{p}_{nn'}$ identically zero for many pairs of bands n and n'.

When describing the band structures of realistic semiconductor materials such as GaAs or Si, grown in the zincblende or diamond crystal structure, respectively, the valence band structure, in particular, presents a complicated situation (cf. Section 1.3). This originates from the fact that on a microscopic level the conduction band is mainly described by an *s*-like state, while the valence band states are predominantly *p*-like in character near $\mathbf{k} = \mathbf{0}$. The "Bloch" wave functions have therefore the symmetry of p_x, p_y, and p_z-like atomic orbitals and will all contribute to the valence band states. This needs to be included in Equation 1.60. Focusing on the valence band structure, from Figure 1.2 one can infer that at the Γ-point one is left with two doubly degenerate valence bands and one which is shifted downward in energy. This third band is usually referred to as the *split-off* (SO) *band*. The two doubly degenerate bands at Γ are the so-called *heavy-hole* (HH) and *light-hole* (LH) *bands*. The degeneracy between the HH and LH bands is lifted for finite values of $\mathbf{k}$. They receive their names from the fact that the energy of the HH band varies at a slower rate than the LH band when moving away from $\mathbf{k} = \mathbf{0}$.

For bulk systems, these bands are eigenstates of the system. However, in the case of a nanostructure where the potential varies spatially, this is no longer the case [44]. For example, in a QW system, the valence sub-band states can be mixtures of HH, LH, and SO bands. Thus, Hamiltonians describing this kind of mixing between different states are given by matrices or a system of coupled time-independent Schrödinger equations [44]. The solution of this set of coupled time-independent Schrödinger equations will give the energy eigenvalues and the corresponding eigenstates as a set of *envelope functions*, which are themselves given in terms of the contribution of the underlying basis states, for example, the bulk HH, LH, or SO bands. This approach is referred to as the *multiband envelope function method* or the *multiband effective mass approximation*. When the interaction between the different bands is described in terms of $\mathbf{k} \cdot \mathbf{p}$ perturbation theory, it is also called a "multiband $\mathbf{k} \cdot \mathbf{p}$ method" [44].

These multiband envelope or $\mathbf{k} \cdot \mathbf{p}$ methods can be at different levels of sophistication, basically depending on the number of bands included in the description. The number of explicitly included bulk bands in the model also describes the name of the multiband Hamiltonian [44]. For instance, four-band (HH, LH) and six-band (HH, LH, SO) Hamiltonians for valence band envelope wave functions for zincblende materials were introduced by Luttinger and Kohn [45]. Pidgeon and Brown [46] extended these models later on to an eight-band model, which includes also the conduction band. Chuang and Chang introduced in 1996 an eight-band model for wurtzite semiconductors [47]. A detailed overview of different types of multiband $\mathbf{k} \cdot \mathbf{p}$ models is given in the textbook by Lew Yan Voon and Willatzen [41].

It is important to note that all these Hamiltonians implicitly include contributions from more remote bands, which are not explicitly included. This originates from the fact that the used material parameters incorporate the effects of the omitted bands. For example, the simple conduction band effective mass model is a special case of an envelope-function approximation. Here, the existence of the other bands in the system is accounted for by using the effective electron mass instead of the free electron mass. In this single-band approach, only the conduction band envelope function is calculated.

Introductions to multiband $\mathbf{k} \cdot \mathbf{p}$ models and their applications to nanostructures can be found in several textbooks [41,44,48]. These approaches have been use extensively in the literature to study the electronic and optical properties of semiconductor heterostructures. Due to the limited number of bands normally included in such a method, these models can be applied, for example, to embedded semiconductor quantum dots of realistic size and shape, which is beyond a DFT treatment, since supercell sizes can easily reach hundreds of thousands of atoms [49]. However, one of the main drawbacks of conventional six- or eight-band models is that they do not necessarily capture the symmetry of the whole system (crystal structure plus quantum dot geometry) correctly [50–52]. Consider the example of a pyramidal, zincblende InAs/-GaAs quantum dot with a square base. Within eight-band $\mathbf{k} \cdot \mathbf{p}$ theory, neglecting piezoelectric effects, this system exhibits a C_{4v} symmetry [50,51], with 90° rotations about the growth direction, leaving the structure unchanged. However, when accounting for the underlying zincblende structure on an atomistic level (two atoms per unit cell), the symmetry is actually reduced to a C_{2v} symmetry [51,52]. This symmetry reduction leads to a lifting of degeneracies in the electronic structure and affects the optical properties of these systems [51,52]. Therefore, semiempirical models, which can treat large numbers of atoms plus provide an atomistic description of the electronic structure, are required. *Empirical tight-binding models* and *empirical pseudopotential theory* can provide this and are the topic of the next section.

1.7 Need for Semiempirical Atomistic Approaches— Tight-Binding and Pseudopotential Methods

In this section, we introduce the basic concepts of (empirical) tight-binding (TB) and empirical pseudopotential theory. Detailed descriptions of these approaches can be found in several textbooks [3,6,8,44]. Here, the basic concepts are summarized. We start with the (empirical) TB approach in the following section. In a second step, in Section 1.7.2, insight into the empirical pseudopotential theory is given. In both cases, the inclusion of the spin–orbit interaction is discussed.

1.7.1 The Empirical Tight-Binding Method

In contrast to the free-electron picture, within which the crystal potential in perturbation theory is often treated, the TB model describes the electronic band structure starting from the limit of isolated atoms. The basis states correspond to the localized orbitals of the different atoms. In this way, one obtains a description of the electronic properties, which simultaneously takes into account the microscopic structure of the solid and offers a transparent approach. The matrix elements in the TB approach have a simple physical interpretation, as they represent interactions between electrons on adjacent atoms.

The localized basis states $|\mathbf{R}, \alpha, \nu, \sigma\rangle$ are classified according to their unit cell $\mathbf{R}$, the type of atom α at which they are centered, the orbital type ν, and the spin σ. The basic assumptions of the TB model are that (1) a small number of basis states per unit cell is already sufficient to describe the bulk band structure and (2) that the overlap of the strongly localized atomic orbitals decreases rapidly with increasing distance between the atomic sites.

Since the inner electronic shells are only slightly affected by the field of all the other atoms, for the description of the bulk band structure it is sufficient to take into account the states of the outer shells. These orbitals then form the highest valence band and lowest conduction band. We start from electron wave functions of an isolated atom. The time-independent, single-electron Schrödinger equation for an atom located at the position $\mathbf{R}_l$ is given by

$$H^{\mathrm{at}}|\mathbf{R}_l, \alpha, \nu, \sigma\rangle = E^{\mathrm{at}}_{\alpha,\nu}|\mathbf{R}_l, \alpha, \nu, \sigma\rangle,$$

with

$$H^{\mathrm{at}} = \frac{p^2}{2m_0} + V^0(\mathbf{R}_l, \alpha),$$

where $V^0(\mathbf{R}_l, \alpha)$ denotes the atomic potential of the atom at the position $\mathbf{R}_l$. Due to the presence of all of the other atoms in the crystal, the wave functions are modified. The single-particle Hamiltonian of the periodic system can be written in the following way:

$$H^{\mathrm{bulk}} = H^{\mathrm{at}}(\mathbf{R}_l, \alpha) + \underbrace{\sum_{\substack{n \neq l \\ \alpha'}} V(\mathbf{R}_n, \alpha')}_{\Delta V(\mathbf{R}_l)}.$$

Here, $H^{\mathrm{at}}(\mathbf{R}_l, \alpha)$ is the Hamiltonian of the isolated atom α at the position $\mathbf{R}_l$ and $\Delta V(\mathbf{R}_l)$ is the potential generated by all the other ions in the lattice. The Schrödinger equation for the periodic solid is then

$$H^{\mathrm{bulk}}|\mathbf{k}\rangle = E(\mathbf{k})|\mathbf{k}\rangle, \tag{1.62}$$

where $\mathbf{k}$ denotes the crystal wave vector. To solve the Schrödinger equation, Equation 1.62, the electronic wave functions $|\mathbf{k}\rangle$ are approximated by linear combinations of atomic orbitals. Because of the translational symmetry of the crystal, the wave functions can be expressed in terms of Bloch functions:

$$|\mathbf{k}\rangle = \sqrt{\frac{V_0}{V}} \sum_n e^{i\mathbf{k}\cdot\mathbf{R}_n} \sum_{\alpha,\nu,\sigma} e^{i\mathbf{k}\cdot\boldsymbol{\Delta}_\alpha} u_{\alpha,\nu,\sigma}(\mathbf{k}) |\mathbf{R}_n, \alpha, \nu, \sigma\rangle \quad \text{with} \quad \mathbf{k} \in 1.\,\text{BZ}. \tag{1.63}$$

The position of the atom α in the unit cell $\mathbf{R}_n$ is denoted by $\boldsymbol{\Delta}_\alpha$. The volume of the unit cell is given by V_0 and the volume of the system is V. Due to the periodicity of the crystal, it is sufficient to restrict $\mathbf{k}$ to the first Brillouin zone (1. BZ). This ansatz fulfills the Bloch theorem. By plugging the wave function of Equation 1.63 into Schrödinger's equation

$$H^{\text{bulk}}|\mathbf{k}\rangle = \left[H^{\text{at}}(\mathbf{R}_l, \alpha) + \Delta V(\mathbf{R}_l)\right] |\mathbf{k}\rangle = E(\mathbf{k})|\mathbf{k}\rangle, \tag{1.64}$$

and applying the bra-vector $\langle \mathbf{k}, \alpha', \nu', \sigma'| = \sqrt{\frac{V_0}{V}} \sum_m e^{-i(\mathbf{k}\cdot\mathbf{R}_m + \mathbf{k}\cdot\boldsymbol{\Delta}_{\alpha'})} \langle \mathbf{R}_m, \alpha', \nu', \sigma'|$, one is left with a matrix equation

$$\sum_{\alpha,\nu,\sigma} H^{\text{bulk}}_{\alpha',\nu'\,\sigma';\alpha,\nu,\sigma}(\mathbf{k}) u_{\alpha,\nu,\sigma}(\mathbf{k}) = E(\mathbf{k}) \sum_{\alpha,\nu,\sigma} S_{\alpha',\nu'\,\sigma';\alpha,\nu,\sigma}(\mathbf{k}) u_{\alpha,\nu,\sigma}(\mathbf{k}), \tag{1.65}$$

instead of a differential equation (Equation 1.64). For each $\mathbf{k}$, this constitutes a *generalized eigenvalue problem* for the matrix $H^{\text{bulk}}(\mathbf{k})$, with a matrix S instead of the identity that occurs in an ordinary eigenvalue problem. $E(\mathbf{k})$ represents the eigenvalue, corresponding to the eigenvector $\mathbf{u}(\mathbf{k})$. The elements of the Hamiltonian matrix read:

$$H^{\text{bulk}}_{\alpha',\nu'\,\sigma';\alpha,\nu,\sigma}(\mathbf{k}) = \frac{V_0}{V} \sum_{n,m} e^{i\mathbf{k}\cdot(\mathbf{R}_n + \boldsymbol{\Delta}_\alpha - \mathbf{R}_m - \boldsymbol{\Delta}_{\alpha'})} \langle \mathbf{R}_m, \alpha', \nu', \sigma'|H^{\text{bulk}}|\mathbf{R}_n, \alpha, \nu, \sigma\rangle, \tag{1.66}$$

and the overlap-matrix elements are given by $S_{\alpha',\nu'\,\sigma';\alpha,\nu,\sigma}(\mathbf{k})$:

$$S_{\alpha',\nu'\,\sigma';\alpha,\nu,\sigma}(\mathbf{k}) = \frac{V_0}{V} \sum_{n,m} e^{i\mathbf{k}\cdot(\mathbf{R}_n + \boldsymbol{\Delta}_\alpha - \mathbf{R}_m - \boldsymbol{\Delta}_{\alpha'})} \langle \mathbf{R}_m, \alpha', \nu', \sigma'|\mathbf{R}_n, \alpha, \nu, \sigma\rangle. \tag{1.67}$$

According to the basic assumptions of a TB model, the electrons stay close to the atomic sites and the electronic wave functions centered around neighboring sites have little overlap. Consequently, there is almost no overlap between wave functions for electrons that are separated by two or more atoms (second-nearest neighbors, third-nearest neighbors, etc.). Nevertheless, the basis orbitals, and thus the Bloch sums, are in general not fully orthogonal to one another. It turns out, however, that if the localized TB basis $\mathcal{B}$ is not orthogonal, one can use a so-called *Löwdin transformation* to transform it into an orthogonal one [53]. These Löwdin orbitals are also localized and preserve the symmetry of the orbital from which they are derived. Therefore, we assume in the following an orthogonal basis set, so that the overlap-matrix elements, Equation 1.67, are given simply by

$$S_{\alpha',\alpha,\nu',\nu,\sigma',\sigma} = \delta_{\mathbf{R}_m,\mathbf{R}_n} \delta_{\alpha',\alpha} \delta_{\nu',\nu} \delta_{\sigma',\sigma}, \tag{1.68}$$

and the Schrödinger equation, Equation 1.65, is reduced to

$$\sum_{\alpha,\nu,\sigma} H^{\text{bulk}}_{\alpha',\nu'\,\sigma';\alpha,\nu,\sigma}(\mathbf{k}) u_{\alpha,\nu,\sigma}(\mathbf{k}) = E(\mathbf{k}) u_{\alpha',\nu',\sigma'}(\mathbf{k}). \tag{1.69}$$

For the sake of a simplified illustration, in the next couple of paragraphs, a system with one atom per unit cell and one orbital per site is assumed. Furthermore, a spin-independent Hamiltonian H^{bulk} is assumed. Consequently, we are left with $\sigma' = \sigma$, $\alpha' = \alpha$, and $\nu' = \nu$. In this case, the matrix elements $H^{\text{bulk}}_{\alpha',\nu',\sigma';\alpha,\nu,\sigma}(\mathbf{k}) = H^{\text{bulk}}_{\alpha,\nu,\sigma;\alpha,\nu,\sigma}(\mathbf{k})$ are given by

$$\langle \mathbf{k}, \alpha, \nu, \sigma | H^{\text{bulk}} | \mathbf{k}, \alpha, \nu, \sigma \rangle = \frac{V_0}{V} \sum_{m,n} e^{i\mathbf{k}\cdot(\mathbf{R}_n - \mathbf{R}_m)} \langle \mathbf{R}_m, \alpha, \nu, \sigma | H^{\text{bulk}} | \mathbf{R}_n, \alpha, \nu, \sigma \rangle$$

$$= \frac{V_0}{V} \sum_{m,n} e^{i\mathbf{k}\cdot(\mathbf{R}_n - \mathbf{R}_m)} I_{mn}. \tag{1.70}$$

Due to the localized structure of the atomic-like wave functions, the integrals I_{mn} become exponentially small for large $R = |(\mathbf{R}_m - \mathbf{R}_n)|$. It is therefore reasonable to ignore all integrals outside some $R_{\max}$, as they would bring only negligible corrections to the band structure $E(\mathbf{k})$. The leading contribution is $n = m$, the so-called onsite contributions, then nearest neighbor contributions are denoted by $n = m \pm 1$, and so on. Keeping only the leading order of the expression of Equation 1.70, the matrix elements I_{mn} can be written as

$$I_{mn} = \delta_{n,m} \Big[\langle \mathbf{R}_m, \alpha, \nu, \sigma | H^{at} | \mathbf{R}_n, \alpha, \nu, \sigma \rangle$$

$$+ \sum_{l \neq n} \langle \mathbf{R}_m, \alpha, \nu, \sigma | V(\mathbf{R}_l, \alpha) | \mathbf{R}_n, \alpha, \nu, \sigma \rangle \Big]$$

$$+ \delta_{n\pm 1, m} \sum_l \langle \mathbf{R}_m, \alpha, \nu, \sigma | V(\mathbf{R}_l, \alpha) | \mathbf{R}_n, \alpha, \nu, \sigma \rangle + \dots$$

$$\equiv \delta_{n,m} \tilde{E} + \delta_{n\pm 1, m} \lambda + \dots. \tag{1.71}$$

$\tilde{E}$ can be interpreted as the renormalized atomic energy level in the presence of all the other atoms in the lattice. The matrix elements containing orbitals from different atomic sites are denoted by λ. Finally, one obtains from Equation 1.70:

$$\langle \mathbf{k}, \alpha, \nu, \sigma | H^{\text{bulk}} | \mathbf{k}, \alpha, \nu, \sigma \rangle = \frac{V_0}{V} \sum_{m,n} e^{i\mathbf{k}\cdot(\mathbf{R}_n - \mathbf{R}_m)} (\delta_{n,m} \tilde{E} + \delta_{n\pm 1, m} \lambda + \dots). \tag{1.72}$$

The symmetry properties of the crystal together with the symmetries of the basis states determine which matrix elements vanish and which are equal. For the calculation of the TB matrix elements, two different approaches are possible. From the knowledge of the atomic potential V^0 and the atomic orbitals, one could calculate the matrix elements of H^{bulk} as well as the overlap matrix elements $S_{\alpha',\alpha,\nu',\nu,\sigma',\sigma}$. A different way to obtain these matrix elements is to treat them as parameters. This approach leads to the so-called *empirical* TB (ETB) model. These parameters are fitted to characteristic properties of the bulk band structure, such as band gaps and effective masses. With both methods, a microscopic description of the solid arises. The empirical approach has been widely used to model semiconductor nanostructures, which consist of different semiconductor materials [54–58]. In the following, we focus on the ETB model.

As already discussed, we assume that the basis states are Löwdin orbitals. Thus, the different TB matrix elements can be symbolized by

$$\langle \mathbf{R}', \alpha', \nu', \sigma' | H^{\text{bulk}} | \mathbf{R}, \alpha, \nu, \sigma \rangle = \underbrace{E_{\nu',\nu}(\mathbf{R} - \mathbf{R}')_{\alpha',\alpha} \delta_{\sigma,\sigma'}}_{\text{TB-parameter}}. \tag{1.73}$$

Here, the translational invariance of the crystal is already used. In addition, we have assumed that H^{bulk} is spin independent. A spin-dependent component of H^{bulk}, the spin–orbit coupling, will be introduced in the following section. In this section, we drop the spin index σ and denote the Hamiltonian by H^{bulk}_0.

The onsite matrix elements, $E_{\nu,\nu}(\mathbf{0})_{\alpha,\alpha}$, which are the expectation values of the Hamiltonian H_0^{bulk} between two identical atomic orbitals at the same site, correspond to the *orbital energies* which are renormalized and shifted due to the other atoms in the crystal. The matrix element $E_{\nu,\nu}(\mathbf{0})_{\alpha,\alpha}$ equals $\tilde{E}$ in Equation 1.71. The values taken for the onsite matrix elements tend in many cases to follow trends in the atomic orbital energies of the isolated atoms [59]. The off-diagonal matrix elements, which describe the coupling between different orbitals at different sites, are called *hopping matrix elements*, because they are related to the probability amplitude of an electron moving from one site to another [3].

To further reduce the number of relevant ETB matrix elements, one can use the so-called *two-center approximation* of Slater and Koster [60]. Slater and Koster proposed to neglect the so-called *three-center integrals*, which are considerably smaller than the so-called *two-center integrals*. The three-center integrals involve two orbitals located at different atoms, and a potential part at a third atom. This corresponds to contributions in Equation 1.71 where $l \neq n \neq m$. Thus, in the Slater and Koster approach, only the potential due to the two atoms at which the orbitals are located is taken into account. Thus, the (effective) potential is symmetric around the axis $\mathbf{d} = \mathbf{R}' + \mathbf{\Delta}_{\alpha'} - (\mathbf{R} + \mathbf{\Delta}_{\alpha})$ between the two atoms. In this approximation, all hopping matrix elements vanish, if the two involved orbitals are eigenstates of the angular momentum $L_{\mathbf{d}} = \mathbf{L} \cdot \frac{\mathbf{d}}{d}$ in $\mathbf{d}$-direction with different eigenvalues $\hbar m_{\mathbf{d}} \neq \hbar m'_{\mathbf{d}}$. Since the (effective) Hamiltonian H, which contains the axially symmetric potential of the two atoms, commutes with $L_{\mathbf{d}}$, this statement follows from

$$0 = \langle i| \left[H, L_{\mathbf{d}}\right] |i'\rangle = \langle i|HL_{\mathbf{d}}|i'\rangle - \langle i|L_{\mathbf{d}}H|i'\rangle = \hbar \left(m'_{\mathbf{d}} - m_{\mathbf{d}}\right) \langle i|H|i'\rangle. \tag{1.74}$$

For an *sp*-bonding, there are only four nonzero hopping integrals as indicated in Figure 1.3, in which σ ($m_{\mathbf{d}} = 0$) and π ($m_{\mathbf{d}} = \pm 1$) bondings are defined such that the axes of the involved p-orbitals are parallel and normal to the interatomic vector $\mathbf{d}$, respectively. So far, the electronic wave functions are expanded in terms of the p-orbitals along the Cartesian x-, y-, and z-axes, whereas the hopping integrals are parameterized for p-orbitals that are parallel or normal to the bonding directions. To construct the ETB matrix elements, in the general case it is fruitful to decompose the Cartesian p-orbitals into bond-parallel and bond-normal p-orbital components.

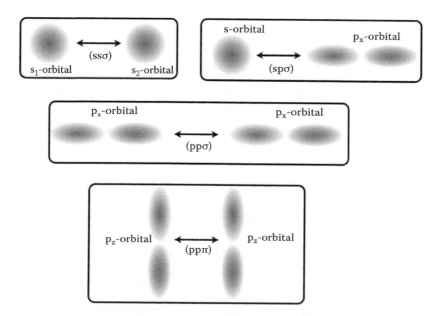

FIGURE 1.3 Schematic illustration of the nonzero hopping matrix elements $V_{ll'm}$ for *sp*-bonding. Orbital types are denoted by the indices l and l', while the z-component of the orbital angular momentum for rotation around the bonding direction is given by m.

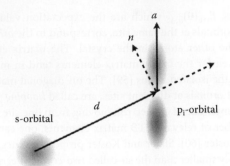

FIGURE 1.4 Schematic illustration of the overlap between *s*- and p_i-orbital along the vector **d** joining the two atoms.

As an example of this procedure, we consider the Hamiltonian matrix element $\langle s|H|p_i \rangle$, between the *s*-orbital, $|s\rangle$, and one of the *p*-orbitals, $|p_i\rangle (i = x, y, z)$, localized at different atoms. Let $\mathbf{e}_d$ be the unit vector along the bond **d** from the first atom to the second and **a** the unit vector along one of the Cartesian (*x*–, *y*–, or *z*–) axes, as shown in Figure 1.4. We first decompose the *p*-orbital along **a**, $|p_a\rangle$, into two *p*-orbitals that are parallel and normal to **d**, respectively,

$$|p_a\rangle = \mathbf{a} \cdot \mathbf{e_d}|p_\sigma\rangle + \mathbf{a} \cdot \mathbf{n}|p_\pi\rangle,$$

where **n** is the unit vector normal to **d** within the plane spanned by **d** and **a**. The Hamiltonian matrix element is then given by

$$\langle s_1|H|p_{2,a}\rangle = \mathbf{a} \cdot \mathbf{e_d}\langle s|H|p_\sigma\rangle + \mathbf{a} \cdot \mathbf{n}\langle s|H|p_\pi\rangle$$

$$= \mathbf{a} \cdot \mathbf{e_d}\langle s|H|p_\sigma\rangle$$

$$= \mathbf{a} \cdot \mathbf{e_d} V_{sp\sigma}.$$

The *s*-orbital centered around atom 1 is labeled by $|s_1\rangle$ and the *p*-orbital localized at atom 2 by $|p_{2,a}\rangle$. The matrix element $V_{sp\pi} = \langle s|H|p_\pi\rangle$ vanishes according to Equation 1.74. To obtain an explicit formula in terms of p_x, p_y, and p_z, let us introduce the directional cosines (d_x, d_y, d_z) along the *x*-, *y*-, and *z*- axes via $\mathbf{d} = |\mathbf{d}|(d_x, d_y, d_z)$. Then we obtain from the previous equation with $\mathbf{a} = \mathbf{e}_i$, where $\mathbf{e}_i$ is the unit vector along the Cartesian axes

$$\begin{pmatrix} \langle s_1|H|p_{2,x}\rangle \\ \langle s_1|H|p_{2,y}\rangle \\ \langle s_1|H|p_{2,z}\rangle \end{pmatrix} = \begin{pmatrix} d_x V_{sp\sigma}(d) \\ d_y V_{sp\sigma}(d) \\ d_z V_{sp\sigma}(d) \end{pmatrix}. \tag{1.75}$$

A similar analysis can be carried out for the matrix elements $\langle p_{1,i}|H|p_{2,j}\rangle$, which leads to [60]

$$\langle p_{1,x}|H|p_{2,x}\rangle = d_x^2 V_{pp\sigma}(d) + (1 - d_x^2)V_{pp\pi}(d),$$

$$\langle p_{1,x}|H|p_{2,y}\rangle = d_x d_y V_{pp\sigma}(d) - d_x d_y V_{pp\pi}(d). \tag{1.76}$$

The other hopping matrix elements can be obtained by cyclical permutation of the coordinates and direction cosines.

Consequently, by constructing the basis orbitals $|\mathbf{R}, \alpha, \nu\rangle$ as linear combinations of the basis states $|\mathbf{R}, \alpha, \{l, m\}\rangle$, the hopping parameters $E_{\nu',\nu}(\mathbf{R} - \mathbf{R}')_{\alpha',\alpha}$ can be described in terms of the parameters $V_{ll'm}$

and the directional cosines of **d** [60]:

$$E_{ss}(\mathbf{d})_{\alpha',\alpha} = V_{ss\sigma}(d, \alpha', \alpha),$$

$$E_{sx}(\mathbf{d})_{\alpha',\alpha} = d_x V_{sp\sigma}(d, \alpha', \alpha),$$

$$E_{xx}(\mathbf{d})_{\alpha',\alpha} = d_x^2 V_{pp\sigma}(d, \alpha', \alpha) + (1 - d_x^2)V_{pp\pi}(d, \alpha', \alpha),$$

$$E_{xy}(\mathbf{d})_{\alpha',\alpha} = d_x d_y V_{pp\sigma}(d, \alpha', \alpha) - d_x d_y V_{pp\pi}(d, \alpha', \alpha).$$

Note that interchanging the order of the indices l and l' of $V_{ll'm}$ has no effect if the sum of the parities of the two orbitals is even, but changes sign if the sum of the parities is odd [60]. Similar formulas can be derived for each combination of the localized orbitals and are listed, for example, in Ref. [60]. It is also clear from the above considerations that the hopping matrix elements in the two-center approximation depend only on the orbital type and the distance d between the two sites.

Because of the translational invariance of the bulk system, H_0^{bulk} can be divided into sub-blocks, which are diagonal in **k**:

$$H_0^{\text{bulk}}(\mathbf{k}) = \left(\langle \mathbf{k}, \alpha', \nu', \sigma' | H_0^{\text{bulk}} | \mathbf{k}, \alpha, \nu, \sigma \rangle \right). \tag{1.77}$$

The dimension of this matrix depends on the number of basis states and atoms within one unit cell. From the diagonalization of this matrix, one obtains the energy dispersion $E(\mathbf{k})$ as a function of the ETB parameters. The energy eigenstates $|n\mathbf{k}\rangle$ are related to the eigenvector $(\mathbf{u}_{n\mathbf{k}})_{\alpha,\nu,\sigma} = u_{n,\alpha,\nu,\sigma}(\mathbf{k})$ via

$$|n\mathbf{k}\rangle = \sum_{\alpha,\nu,\sigma} u_{n,\alpha,\nu,\sigma}(\mathbf{k}) |\mathbf{k}, \alpha, \nu, \sigma\rangle, \tag{1.78}$$

where n denotes the different bands.

Examples for the fitting of TB band structures to DFT band structures are given in Figure 1.5. In Figure 1.5a, a (nearest neighbor) ETB model with one s- and three p-orbitals per atom (sp^3-model) is used to describe the bulk band structure of wurtzite InN. Even with this limited basis set one obtains a very good description of the band structure around $\mathbf{k} = 0$ and also a reasonable description near the Brillouin zone boundaries. The description of the valence band structure and the first few conduction bands over the full first Brillouin zone can be further improved by including more basis states. An example is given in

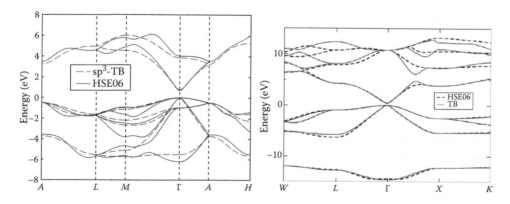

FIGURE 1.5 (a) Electronic band structure of *wurtzite* InN calculated within a nearest neighbor sp^3 TB model and HSE06-DFT. For the HSE06-DFT the cut-off energy was set to 600 eV and Γ-centered $6 \times 6 \times 4$ k-point mesh was used (figure adapted from Ref. [61]). (b) Electronic band structure of *zincblende* InN within a nearest neighbor $sp^3d^5s^*$ TB model and HSE06-DFT. The cut-off energy for the HSE06-DFT was chosen to be 600 eV and a Γ-centered $6 \times 6 \times 6$ k-point mesh was used [62].

Figure 1.5b where an $sp^3d^5s^*$ model is used to describe the band structure of *zincblende* InN. A variety of different ETB models can be found in the literature [59,63–65].

So far, we have neglected the spin of the electron in our discussion. In order to obtain an accurate description of the bulk band structure of semiconductor materials containing heavier atoms such as CdSe and InSb, relativistic effects on the electronic states in crystals have to be considered. In semiconductors containing lighter atoms such as AlP and InN, these effects are negligible for many purposes [66]. This is because the potential is very strong near the nuclei and the kinetic energy is consequently very large, so that the electron velocity is comparable to the velocity of light. Thus, relativistic corrections become more important for heavy elements. One can include these contributions for the motion of an electron in a potential $V(\mathbf{r})$ by considering the *Dirac equation* [67,68]. From a *Foldy–Wouthuysen transformation* of the Dirac equation one obtains relativistic corrections to the time-independent Schrödinger equation. These additional terms are related to (1) relativistic corrections to the kinetic energy, (2) relativistic contributions to the potential $V(\mathbf{r})$, and (3) the spin–orbit interaction [68]. The spin–orbit coupling (SOC) originates from the interaction of the electron spin magnetic moment with the magnetic field "seen" by the electron. The first two components (1) and (2) do not depend on the spin of the electron. Therefore these terms do not change the symmetry properties of the nonrelativistic Hamiltonian. However, the term corresponding to spin–orbit interaction, as we see, couples operators in spin space and ordinary spatial space, thus reducing the symmetry. The effect of the SOC in removing degeneracies of the nonrelativistic results can be deduced using group theory [9].

Therefore, for certain semiconductor materials, it is essential for an accurate bulk band structure description in the framework of an ETB model to include the contributions of the SOC. This issue will be discussed in the following subsection.

1.7.1.1 Spin–Orbit Coupling

In this section, we discuss the inclusion of the SOC in the ETB band structure formalism. Here, the approach of Chadi [69] is employed for the TB scheme. The bulk Hamiltonian of a perfect crystal, including SOC effects can be written as

$$H^{\text{bulk}} = H_0^{\text{bulk}} + H_{\text{so}}, \tag{1.79}$$

where H_0^{bulk} is the spin-independent part and H_{so} denotes the operator for the SOC. Assuming here only s- and p-orbitals, one has to investigate the action of this operator on the basis states of the previous section. The part H_0^{bulk} yields nonzero interactions only between states of the same spin. The matrix elements arising from H_{so} of the Hamiltonian H^{bulk} have the potential to connect states of different spins. To calculate these terms, we assume here that the operator H_{so} acts on the TB basis states in a manner analogous to the atomic spin–orbit operator on atomic orbitals [68]

$$H_{\text{so}}^{\text{atom}} = \frac{1}{2m^2c^2}\frac{1}{r}\frac{\partial V^0}{\partial r}\mathbf{L}\cdot\mathbf{s}. \tag{1.80}$$

Here, V^0 is the atomic potential, $\mathbf{s}$ the spin operator, and $\mathbf{L}$ denotes the angular momentum operator. All matrix elements $\langle p_i\sigma'|H_{\text{so}}|p_j\sigma\rangle$ can be determined in the following way:[†] The operators s_x and s_y can be described by the flip/flop operators $s_\pm$:

$$s_x = \frac{s_+ + s_-}{2},$$

$$s_y = -i\frac{s_+ - s_-}{2}.$$

[†] An s state with $l = 0$ is not split, since the angular part is constant.

The operators s_x, s_y, and s_z then act on the spin states $|\uparrow\rangle$ and $|\downarrow\rangle$ in the following way:

$$s_x|\uparrow\rangle = \frac{\hbar}{2}|\downarrow\rangle \quad ; \quad s_x|\downarrow\rangle = \frac{\hbar}{2}|\uparrow\rangle;$$

$$s_y|\uparrow\rangle = i\frac{\hbar}{2}|\downarrow\rangle \quad ; \quad s_y|\downarrow\rangle = -i\frac{\hbar}{2}|\uparrow\rangle; \tag{1.81}$$

$$s_z|\uparrow\rangle = \frac{\hbar}{2}|\uparrow\rangle \quad ; \quad s_z|\downarrow\rangle = -\frac{\hbar}{2}|\downarrow\rangle.$$

Furthermore, we have to define the action of the angular momentum operator **L** on the states $|p_x\rangle$, $|p_y\rangle$, and $|p_z\rangle$. The states can be written in terms of the *spherical harmonics* Y_{lm} [7]

$$|p_z\rangle = Y_{10} = \sqrt{\frac{3}{4\pi}}\cos\theta = \sqrt{\frac{3}{4\pi}}\frac{z}{r},$$

$$|p_y\rangle = \frac{i}{\sqrt{2}}(Y_{11} + Y_{1-1}) = \sqrt{\frac{3}{4\pi}}\frac{y}{r}, \tag{1.82}$$

$$|p_x\rangle = \frac{1}{\sqrt{2}}(Y_{1-1} - Y_{11}) = \sqrt{\frac{3}{4\pi}}\frac{x}{r},$$

which fulfill

$$\mathbf{L}^2 Y_{lm} = \hbar^2 l(l+1)Y_{lm},$$

$$L_z Y_{lm} = \hbar m Y_{lm}. \tag{1.83}$$

The calculation of the matrix elements will exemplarily be demonstrated for the matrix element $\langle p_x \uparrow |H_{so}|p_y \uparrow\rangle$, for which we obtain

$$\langle p_x \uparrow |H_{so}|p_y \uparrow\rangle = \langle p_x \uparrow |CL_x s_x|p_y \uparrow\rangle + \langle p_x \uparrow |CL_y s_y|p_y \uparrow\rangle + \langle p_x \uparrow |CL_z s_z|p_y \uparrow\rangle$$

$$= \frac{\hbar}{2}\langle p_x|CL_z|p_y\rangle = \frac{1}{i}\langle p_x|\frac{\hbar^2}{2}C|p_x\rangle,$$

with C given by

$$C = \frac{1}{2m^2c^2}\frac{1}{r}\frac{\partial V^0}{\partial r}.$$

Here, we have used Equation 1.81 and that the states $|\uparrow\rangle$ and $|\downarrow\rangle$ are orthogonal to each other to derive the second line from the first. The last line is obtained from Equations 1.82 and 1.83. One can show that all matrix elements can be expressed in terms of

$$\langle p_x|\frac{\hbar^2}{4m^2c^2}\frac{1}{r}\frac{\partial V_{atom}}{\partial r}|p_x\rangle = \lambda,$$

and can be deduced along the same line. To obtain a compact notation, we denote the states $|\uparrow\rangle$ and $|\downarrow\rangle$ by $|+\rangle$ and $|-\rangle$, respectively. Since the different states $|p_i\pm\rangle$ are orthogonal, many of the matrix elements will be zero. Evaluation of all possible terms gives nonzero results only for

$$\langle p_x \pm |H_{so}|p_z\mp\rangle = \pm\lambda,$$

$$\langle p_x \pm |H_{so}|p_y\pm\rangle = \mp i\lambda, \tag{1.84}$$

$$\langle p_y \pm |H_{so}|p_z\mp\rangle = -i\lambda,$$

and their complex conjugates. The magnitude of the parameter λ can then be chosen to reproduce the splitting of the valence bands in the vicinity of the Brillouin zone center.

To summarize this section, we have presented a technique that allows us to include the SOC in the ETB formalism. In semiconductor materials with a large spin–orbit splitting at the Brillouin zone center, the inclusion of the SOC is important for a more accurate calculation of the bulk band structure and therefore a more realistic description of the single-particle states in semiconductor nanostructures. However, it should be noted that when including the SOC in the calculations, the size of Hamiltonian matrix is doubled, since one has to explicitly account for the spin. This obviously increases the computational effort.

In addition to ETB models, empirical pseudopotential theory is widely used to study the electronic structure of semiconductor materials, alloys, and nanostructures [49,70]. As mentioned earlier, this approach also accounts for the underlying crystal structure on a microscopic level. The next section describes the fundamental ideas behind this method.

1.7.2 Empirical Pseudopotential Method

Another empirical approach to obtain the electronic structure of semiconductors and related nanostructures on an atomistic level is the *empirical pseudopotential method* (EPM). In a similar fashion as in the ETB approach, the EPM solves the time-independent Schrödinger equation for a bulk system without knowing the exact potential in which the electrons are moving but rather by assuming an effective potential [44,70,71].

The general idea of the EPM is to assume that core electrons are tightly bound to the nuclei and have only a small influence on conduction and valence band states and therefore the electronic properties. Thus, one is left with the problem of removing the contributions of the core electrons from the total wave function. The wave function inside the core region varies rapidly compared to the behavior of the wave function outside this region. Therefore, the idea is to introduce a pseudopotential that leads to a slowly varying wave function in the core region but reproduces the "real" wave function outside the core region. This leads to the task of finding such a pseudopotential. A schematic illustration of the situation is shown in Figure 1.6.

To achieve this, the starting point is again the time-independent Schrödinger equation:

$$\left(-\frac{\hbar^2}{2m} \nabla^2 + V(\mathbf{r}) \right) \psi_{\mathbf{k}}(\mathbf{r}) = E(\mathbf{k})\psi_{\mathbf{k}}(\mathbf{r}). \tag{1.85}$$

Due to the periodicity of the bulk crystal structure both the wave function $\psi_{\mathbf{k}}(\mathbf{r})$ and the potential $V(\mathbf{r})$ can be expanded in terms of plane waves:

$$\psi_{\mathbf{k}}(\mathbf{r}) = e^{i\mathbf{k}\cdot\mathbf{r}} \sum_m c_{\mathbf{G}_m} e^{i\mathbf{G}_m\cdot\mathbf{r}}, \tag{1.86}$$

$$V(\mathbf{r}) = \sum_l V_{\mathbf{G}_l} e^{i\mathbf{G}_l\cdot\mathbf{r}}. \tag{1.87}$$

The Fourier coefficients for a given reciprocal lattice vector $\mathbf{G}_n$ are denoted by $c_{\mathbf{G}_n}$ and $V_{\mathbf{G}_n}$, respectively. Using Equations 1.86 and 1.87 the time-independent Schrödinger equation reads

$$\frac{\hbar^2}{2m} \sum_m |\mathbf{k} + \mathbf{G}_m|^2 c_{\mathbf{G}_m} e^{i(\mathbf{k}+\mathbf{G}_m)\cdot\mathbf{r}} + \sum_m \sum_l V_{\mathbf{G}_l} c_{\mathbf{G}_m} e^{i(\mathbf{k}+\mathbf{G}_m+\mathbf{G}_l)\cdot\mathbf{r}} = E(\mathbf{k}) \sum_m c_{\mathbf{G}_m} e^{i(\mathbf{k}+\mathbf{G}_m)\cdot\mathbf{r}}. \tag{1.88}$$

Multiplying by $e^{-i(\mathbf{k}+\mathbf{G}_n)\cdot\mathbf{r}}$ and integrating over the volume of the crystal yields

$$\frac{\hbar^2}{2m} \sum_m |\mathbf{k} + \mathbf{G}_m|^2 c_{\mathbf{G}_m} \delta_{n,m} + \sum_m \sum_l V_{\mathbf{G}_l} c_{\mathbf{G}_m} \delta_{l,m-n} = E(\mathbf{k}) \sum_m c_{\mathbf{G}_m} \delta_{n,m}, \tag{1.89}$$

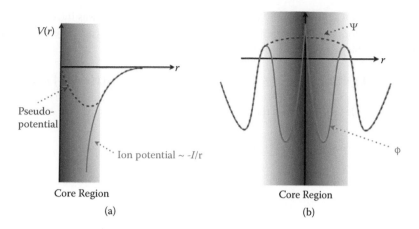

$V(r)$

Pseudo-
potential

Ion potential ~ -I/r

Core Region

(a)

Ψ

r

φ

Core Region

(b)

FIGURE 1.6 (a) Schematic illustration of the real ionic potential (solid line) and a pseudopotential (dotted line). (b) Schematic representation of a wave function φ in the real ionic potential and the corresponding pseudo-wave function ψ, when considering the pseudopotential in (a).

where $\delta_{\alpha,\beta}$ is the Kronecker delta. Thus, Equation 1.89 becomes

$$\frac{\hbar^2}{2m}|k + \mathbf{G}_n|^2 c_{\mathbf{G}_n} + \sum_m V_{\mathbf{G}_{m-n}} c_{\mathbf{G}_m} = E(\mathbf{k})c_{\mathbf{G}_n}. \tag{1.90}$$

When truncating the Fourier series $\sum_m V_{\mathbf{G}_{m-n}} c_{\mathbf{G}_m}$ one is left with a finite-size eigenvalue problem. Solving this eigenvalue problem for each **k** point gives the energy dispersion $E(\mathbf{k})$.

However, to be able to solve the eigenvalue problem by diagonalizing the Hamiltonian matrix one still has to determine the potential V. We focus our attention in the following on the zincblende or diamond structure with two atoms per unit cell. Here, following Cohen and Bergstresser [72], we assume that the origin of the coordinate system is halfway between the atoms in the cell. Additionally, the assumption is made that the potential can be split into symmetric and antisymmetric parts [72]:

$$V_{\mathbf{G}_\alpha} = V_{\mathbf{G}_\alpha}^S \cos\left(\mathbf{G}_\alpha \cdot \boldsymbol{\tau}\right) + i V_{\mathbf{G}_\alpha}^A \sin\left(\mathbf{G}_\alpha \cdot \boldsymbol{\tau}\right). \tag{1.91}$$

Here, $\boldsymbol{\tau}$ denotes the position of the atom under consideration with respect to the origin. Since in a zincblende structure we have two atoms per unit cell, we have one atom at $+\boldsymbol{\tau}$ and the second one at $-\boldsymbol{\tau}$, where $\boldsymbol{\tau}$ is given by

$$\boldsymbol{\tau} = \left(\frac{1}{8}, \frac{1}{8}, \frac{1}{8}\right) a. \tag{1.92}$$

Here a denotes the lattice constant of the material. The prefactors $V_{\mathbf{G}_\alpha}^S$ and $V_{\mathbf{G}_\alpha}^A$ in Equation 1.91 are the so-called *form factors*. Qualitatively the form factor $V_{\mathbf{G}_\alpha}^S$ can be understood as the covalent part to the bonding of the atoms, while $V_{\mathbf{G}_\alpha}^A$ describes the ionic contribution [70]. The form factors are usually determined empirically. It can be shown that the pseudopotential $V_{\mathbf{G}_\alpha}$ can be written as the product of the form factors and the *structure factor* [70]. Making use of the fact that the structure factor for a given material is only nonzero at a few of the allowed reciprocal lattice vectors, the number of unknown parameters that has to be determined empirically can be significantly reduced [70].

1.7.2.1 Spin–Orbit Coupling

As described earlier, the spin–orbit interaction plays an important role for an accurate description of the electronic structure of systems containing heavier elements (e.g., CdSe, InSb). In the following, we discuss briefly how to include SOC effects into EPM. The approach described here was introduced by Weisz [73] and later modified by Bloom and Bergstresser [74]. In reciprocal space, one can write the Hamiltonian H_{so} describing the spin–orbit interaction in the following way:

$$H_{so}^{G,G'}(\mathbf{k}) = (\mathbf{K} \times \mathbf{K}') \cdot \boldsymbol{\sigma} \left[-i\lambda^s \cos\left((\mathbf{G} - \mathbf{G}') \cdot \boldsymbol{\tau}\right) + \lambda^a \sin\left((\mathbf{G} - \mathbf{G}') \cdot \boldsymbol{\tau}\right) \right], \qquad (1.93)$$

where $\mathbf{K} = \mathbf{k} + \mathbf{G}$, $\mathbf{K}' = \mathbf{k} + \mathbf{G}'$, and $\boldsymbol{\sigma}$ are the vectors of the Pauli spin matrices. Here λ^s and λ^a are given by

$$\lambda^s = \frac{1}{2} \left(\lambda_1 + \lambda_2 \right), \qquad (1.94)$$

$$\lambda^a = \frac{1}{2} \left(\lambda_1 - \lambda_2 \right) \qquad (1.95)$$

with

$$\lambda_1 = \mu B_{1nl}(K) B_{1nl}(K'), \qquad (1.96)$$

$$\lambda_2 = \alpha\mu B_{2nl}(K) B_{2nl}(K'). \qquad (1.97)$$

Here 1 and 2 denote the two atoms in the unit cell, α is the ratio of the free atom spin–orbit splitting energies of the two atoms, $K = |\mathbf{K}|$ and $K' = |\mathbf{K}'|$ and μ is a fitting parameter. The terms B_{nl} are in general given by (the additional index $i = 1, 2$ just denotes the different atoms):

$$B_{nl} = \beta \int_0^\infty j_{nl}(Kr) R_{nl}(r) r^2 dr. \qquad (1.98)$$

Here, $j_{nl}(Kr)$ are the spherical Bessel functions and $R_{nl}(r)$ denotes the radial part of the core function. The indices n and l denote the principal quantum number and the angular moment, respectively. The prefactor β is a normalization factor and can be derived from

$$\lim_{k \to 0} k^{-l} B_{nl}(k) = 1. \qquad (1.99)$$

Similar to the ETB approach presented in the previous section, when including the SOC effects into EPM, one is left with one additional fitting parameter, namely μ (cf. Equations 1.96 and 1.97). Even though we add only one additional fitting parameter, the computational expense is increased since the size of the matrix that has to be diagonalized is doubled due to the spin degree that has to be accounted for.

1.8 Summary

In this chapter, we have presented a detailed overview of different, widely used methods to describe the electronic band structure of semiconductor (bulk and nanostructure) materials. The focus here is on crystalline solid-state materials with a periodic lattice. We have seen that this leads to the situation that individual electron wave functions can be chosen to reflect this periodicity, satisfying a condition referred to as Bloch's theorem. Its application considerably simplifies the calculation and description of the electronic structure of crystalline solids.

However, even when using Bloch's theorem, the calculation of the electronic structure of a semiconductor material is in general an extremely complicated task. There are about 10^{23} valence electrons, which contribute to the bonding in each cubic centimeter of a typical solid. This implies that the calculation of the electronic structure is in general a complex many-body problem, as the exact wave function and energy of each electron depends on those of all the others. However, when using the ideas of Hohenberg and Kohn [10], it follows that the ground state energy E_{GS} of the many-particle system is an unambiguous functional of the ground state electron density ρ_{GS}. Thus, the Hohenberg–Kohn theorem allowed us to find the ground state density $\rho_{GS}(\mathbf{r})$ and energy E_{GS} of an interacting N-particle system by minimizing the energy functional $E[\rho]$. Even though this provides a general framework, it does not give the form of the different functionals, for example, how the kinetic energy depends on ρ. We discussed in a second step the ideas of Kohn and Sham [12], which resulted in an effective single-particle equation in which a particular part of the effective potential, namely the exchange-correlation part, requires approximations. The derived effective single-particle Schrödinger equation is known as the Kohn–Sham equation, forming the foundation of density functional theory (DFT). We reviewed different approximations used in the literature for the exchange-correlation functional. This ranges from the simple so-called local density approximation (LDA) up to hybrid functionals. Ways to go beyond standard DFT that can offer better accuracy, but keeping it still systematic and general, have also been briefly discussed.

These highly accurate first-principles calculations come at the cost of high computational efforts. When interested in the electronic structure of semiconductor nanostructures with dimensions of at least several nanometers, such as quantum wells (QW) and dots, one has to deal with hundreds of thousands of atoms. These cell sizes are not manageable with standard DFT approaches. We have, therefore, discussed semiempirical models, which make use of available experimental or DFT-based data but allow for the description of realistically sized semiconductor nanostructures. These methods, namely $\mathbf{k}\cdot\mathbf{p}$ and empirical tight-binding (TB) and empirical pseudopotential theory, are commonly used to calculate and develop an understanding of the electronic structure of solids. An overview of the general concepts of these three approaches has been provided here. We also considered the main differences between standard $\mathbf{k}\cdot\mathbf{p}$ approaches and TB or empirical pseudopotential theory. While the latter two provide an atomistic treatment of the electronic structure of a semiconductor, $\mathbf{k}\cdot\mathbf{p}$ theory operates on a continuum-based level. This difference can be of critical importance when it comes to the description of the electronic and optical properties of nanostructures. The discussion of the application of the different techniques to nanostructures is beyond the scope of this chapter. Overviews can be found, for example, in Refs. [48,49,75] or in other chapters of this handbook.

References

1. J. Singh. *Electronic and Optoelectronic Properties of Semiconductor Structures*. Cambridge: Cambridge University Press, 2003.
2. D. Bimberg, M. Grundmann and N. N. Ledentsov. *Quantum Dot Heterostructures*. Chichester: Wiley, 2001.
3. G. Czycholl. *Theoretische Festkörperphysik*. Braunschweig/Wiesbaden: Vieweg, 2000.
4. N. W. Ashcroft and D. N. Mermin. *Solid State Physics*. New York: Harcourt College Publishers, 1976.
5. J. Heyd, G. E. Scuseria and M. Ernzerhof. Hybrid functionals based on a screened coulomb potential. *J. Chem. Phys.*, 118:8207, 2003.
6. O. Madelung. *Introduction to Solid-State Theory*. Solid State Sciences Series. Berlin: Springer, 1996.
7. A. I. M. Rae. *Quantum Mechanics*. Bristol: Institute of Physics, 2002.
8. E. P. O'Reilly. *Quantum Theory of Solids*. Masters Series in Physics and Astronomy. London: Taylor & Francis, 2002.
9. H. W. Streitwolf. *Group Theory in Solid-State Physics*. London: Macdonald, 1971.
10. P. Hohenberg and W. Kohn. Inhomogeneous electron gas. *Phys. Rev.*, 136:B864, 1964.
11. H. Englisch and R. Englisch. Hohenberg-Kohn theorem and non-V-representable densities. *Phy. A: Stat. Mech. Its App.*, 121:253, 1983.
12. W. Kohn and L. J. Sham. Self-consistent equations including exchange and correlation effects. *Phys. Rev.*, 140:A1133, 1965.
13. R. M. Martin. *Electronic Structure: Basic Theory and Practical Methods*. Cambridge: Cambridge University Press, 2011.

14. J. P. Perdew. Density functional theory and the band gap problem. *Int. J. Quantum Chem.*, 28:497, 1985.
15. A. J. Cohen, P. Mori-Sanchez and W. Yang. Challenges for density functional theory. *Chem. Rev.*, 112:289, 2012.
16. D. J. W. Geldart. Volume 180 of *Topics in Current Chemistry*. Berlin: Springer, 1996.
17. J. P. Perdew, K. Burke and Y. Wang. Generalized gradient approximation for the exchange-correlation hole of a many-electron system. *Phys. Rev. B*, 54:16533, 1996.
18. K. Capelle. A bird's-eye view of density-functional theory. *Braz. J. Phys.*, 36:1318, 2006.
19. J. P. Perdew, K. Burke and M. Ernzerhof. Generalized gradient approximation made simple. *Phys. Rev. Lett.*, 77:3865, 1996.
20. J. P. Perdew and S. Kurth. Density Functionals for Non-Relativistic Coulomb Systems, in *Density Functionals: Theory and Applications*, volume 500 of Lecture Notes in Physics. Berlin: Springer, 1998.
21. L. J. Sham and M. Schlüter. Density-functional theory of the energy gap. *Phys. Rev. Lett.*, 51:1888, 1983.
22. A. Seidl, A. Görling, P. Vogl, J. A. Majewski and M. Levy. Generalized Kohn–Sham schemes and the band-gap problem. *Phys. Rev. B.*, 53:3764, 1996.
23. P. Rinke, A. Qteish, J. Neugebauer, C. Freysoldt and M. Scheffler. Exciting prospects for solids: Exact-exchange based functionals meet quasiparticle energy calculation. *Phys. Stat. Sol (b)*, 245:929, 2008.
24. A. D. Becke. A new mixing of Hartree–Fock and local density-functional theories. *J. Chem. Phys.*, 98:1372, 1993.
25. A. D. Becke. Density-functional thermochemistry. III. The role of exact exchange. *J. Chem. Phys.*, 98:5648, 1993.
26. A. D. Becke. Density-functional thermochemistry. IV. A new dynamical correlation functional and implications for exact-exchange mixing. *J. Chem. Phys.*, 104:1040, 1996.
27. J. P. Perdew, M. Ernzerhof and K. Burke. Rationale for mixing exact exchange with density functional approximations. *J. Chem. Phys.*, 105:9982, 1996.
28. J. A. Pople, M. Head-Gordon, D. J. Fox, K. Raghavachari and L. A. Curtiss. Gaussian-1 theory: A general procedure for prediction of molecular energies. *J. Chem. Phys.*, 90:5622, 1989.
29. C. Adamo and V. Barone. Toward reliable density functional methods without adjustable parameters: The PBE0 model. *J. Chem. Phys.*, 110:6158, 1999.
30. T. Bredow and A. R. Gerson. Effect of exchange and correlation on bulk properties of MgO, NiO, and CoO. *Phys. Rev. B*, 61:5194, 2000.
31. J. Muscat, A. Wander and N. Harrison. On the prediction of band gaps from hybrid functional theory. *Chem. Phys. Lett.*, 342:397, 2001.
32. W. Perger. Calculation of band gaps in molecular crystals using hybrid functional theory. *Chem. Phys. Lett.*, 368:319, 2003.
33. M. Schlipf. *Heyd–Scuseria–Ernzerhof Screened-Exchange Hybrid Functional for Complex Materials: All-Electron Implementation and Application*. Schriften des Forschungszentrums Jülich Reihe Schlüssel technologien/Key Technologies. Jülich, Germany: Forschungszentrum Jülich GmbH, RWTH Aachen University, 2012.
34. J. Paier, M. Marsman, K. Hummer, I. C. Gerber G. Kresse and J. G. Angyan. Screened hybrid density functionals applied to solids. *J. Chem. Phys.*, 124:154709, 2006.
35. A. Carvalho, A. Alkauskas, Alfredo Pasquarello, A. K. Tagantsev and N. Setter. A hybrid density functional study of lithium in ZnO: Stability, ionization levels, and diffusion. *Phys. Rev. B*, 80:195205, 2009.
36. A. Carvalho, A. Alkauskas, Alfredo Pasquarello, A. K. Tagantsev and N. Setter. Li-related defects in ZnO: Hybrid functional calculations. *Physica B*, 404:4797, 2009.
37. P. G. Moses, M. Miao, Q. Yan and C. G. Van de Walle. Hybrid functional investigations of band gaps and band alignments for AlN, GaN, InN, and InGaN. *J. Chem. Phys.*, 134:084703, 2011.

38. X. Ren, P. Rinke, V. Blum, J. Wieferink, A. Tkatchenko, A. Sanfilippo, K. Reuter and M. Scheffler. Resolution-of-identity approach to Hartree-Fock, hybrid density functionals, RPA, MP2 and GW with numeric atom-centered orbital basis functions. *New J. Phys.*, 14:053020, 2012.

39. L. Hedin. New method for calculating the one-particle Green's function with application to the electron–gas problem. *Phys. Rev.*, 139:A796, 1965.

40. P. Rinke, A. Qteish, J. Neugebauer, C. Freysoldt and M. Scheffler. Combining GW calculations with exact-exchange density-functional theory: An analysis of valence-band photoemission for compound semiconductors. *New J. Phys.*, 7:126, 2005.

41. L. C. Lew Yan Voon and M. Willatzen. *The $\mathbf{k}\cdot\mathbf{p}$ Method: Electronic Properties of Semiconductors.* Heidelberg: Springer, 2009.

42. E. O. Kane. Volume 1 of *Physics of III–V Compounds*. New York: Academic Press, 1966.

43. G. Bastard. *Wave Mechanics Applied to Semiconductor Heterostructures*. Les Editions de Physique. Courtaboeuf: EDP Sciences, 1988.

44. P. Harrison. *Quantum Wells, Wires and Dots: Theoretical and Computational Physics of Semiconductor Nanostructures.* New York: John Wiley & Sons, 2005.

45. J. M. Luttinger and W. Kohn. Motion of electrons and holes in perturbed periodic fields. *Phys. Rev.*, 97:869, 1955.

46. C. R. Pidgeon and R. N. Brown. Interband magneto-absorption and Faraday rotation in InSb. *Phys. Rev.*, 146:575, 1966.

47. S. L. Chuang and C. S. Chang. $\mathbf{k}\cdot\mathbf{p}$ method for strained wurtzite semiconductors. *Phys. Rev. B*, 54:2491, 1996.

48. M. Ehrhardt and T. Koprucki, editors. *Multi-Band Effective Mass Approximations: Advanced Mathematical Models and Numerical Techniques*, volume 94 of Lecture Notes in Computational Science and Engineering. London: Springer, 2014.

49. G. Bester. Electronic excitations in nanostructures: An empirical pseudopotential based approach. *J. Phys. Condens. Matter*, 21:023202, 2009.

50. A. Zunger. On the farsightedness (hyperopia) of the standard $\mathbf{k}\cdot\mathbf{p}$ model. *Phys. Stat. Sol. (a)*, 190:467, 2002.

51. G. Bester and A. Zunger. Cylindrically shaped zinc blende semiconductor quantum dots do not have cylindrical symmetry: Atomistic symmetry, atomic relaxation, and piezoelectric effects. *Phys. Rev. B*, 71:045318, 2005.

52. N. Baer, S. Schulz, P. Gartner, S. Schumacher, G. Czycholl and F. Jahnke. Influence of symmetry and Coulomb-correlation effects on the optical properties of nitride quantum dots. *Phys. Rev. B*, 76:075310, 2007.

53. P. O. Löwdin. On the non-orthogonality problem connected with the use of atomic wave functions in the theory of molecules and crystals. *J. Chem. Phys.*, 18:365, 1950.

54. S. Lee, O. L. Lazarenkova, P. von Allmen, F. Oyafuso and G. Klimeck. Effect of wetting layers on the strain and electronic structure of In As self-assembled quantum dots. *Phys. Rev. B*, 70:125307–1, 2004.

55. W. Sheng, S. J. Cheng and P. Hawrylak. Multiband theory of multi-exciton complexes in self-assembled quantum dots. *Phys. Rev. B*, 71:035316, 2005.

56. S. Schulz, S. Schumacher and G. Czycholl. Tight-binding model for semiconductor quantum dots with a wurtzite crystal structure: From one-particle properties to Coulomb correlations and optical spectra. *Phys. Rev. B*, 73:245327, 2006.

57. M. Zielinski, M. Korkusinski and P. Hawrylak. Atomistic tight-binding theory of multi-exciton complexes in a self-assembled InAs quantum dot. *Phys. Rev. B*, 81:085301, 2010.

58. K. Schuh, S. Barthel, O. Marquardt, T. Hickel, J. Neugebauer, G. Czycholl and F. Jahnke. Strong dipole coupling in nonpolar nitride quantum dots due to Coulomb effects. *Appl. Phys. Lett.*, 100:092103, 2012.

59. P. Vogl, H. P. Hjalmarson and J. D. Dow. A semi-empirical tight-binding theory of the electronic structure of semiconductors. *J. Phys. Chem. Solids*, 44(5):365–378, 1983.

60. J. C. Slater and G. F. Koster. Simplified LCAO method for periodic potential problem. *Phys. Rev.*, 94:1498, 1954.

61. M. A. Caro, S. Schulz and E. P. O'Reilly. Theory of local electric polarization and its relation to internal strain: Impact on the polarization potential and electronic properties of group-III nitrides. *Phys. Rev. B*, 88:214103, 2013.

62. R. Benchamekh, S. Schulz and E. P. O'Reilly. Tight-binding and **k·p** for cubic nitrides. *In Preparation*, 2016.

63. A. Kobayashi, O. F. Sankey, S. M. Volz and J. D. Dow. Semiempirical tight-binding band structures of wurtzite semiconductors: AlN, CdS, CdSe, ZnS, and ZnO. *Phys. Rev. B*, 28:935, 1983.

64. J. M. Jancu, R. Scholz, F. Beltram and F. Bassani. Empirical *spds** tight-binding calculation for cubic semiconductors: General method and material parameters. *Phys. Rev. B*, 57:6493, 1998.

65. S. Schulz and G. Czycholl. Tight-binding model for semiconductor nanostructures. *Phys. Rev. B*, 72:165317, 2005.

66. P. Y. Yu and M. Cardona. *Fundamentals of Semiconductors*. Physics and Material Properties. Berlin: Springer, 1999.

67. P. A. M. Dirac. *Principles of Quantum Mechanics*. International Series of Monographs on Physics. New York: Oxford University Press, 1958.

68. F. Schwabl. *Quantenmechanik für Fortgeschrittene*. Berlin: Springer, 2000.

69. D. J. Chadi. Spin–orbit splitting in crystalline and compositionally disordered semiconductors. *Phys. Rev. B*, 16:790–796, 1977.

70. M. L. Cohen and J. R. Chelikowski. *Electronic Structure and Optical Properties of Semiconductors*, 2nd Edition. Series in Solid-State Sciences 75. Berlin: Springer, 1989.

71. J. C. Phillips and L. Kleinman. New method for calculating wave functions in crystals and molecules. *Phys. Rev.*, 116:287, 1959.

72. M. L. Cohen and T. K. Bergstresser. Band structures and pseudopotential form factors for fourteen semiconductors of the diamond and zinc-blende structures. *Phys. Rev.*, 141:789, 1966.

73. G. Weisz. Band structure and Fermi surface of white tin. *Phys. Rev.*, 149:504, 1966.

74. S. Bloom and T. K. Bergstresser. Band structure of α-Sn, InSb and CdTe including spin–orbit effects. *Solid State Commun.*, 6:465, 1968.

75. A. Di Carlo. Microscopic theory of nanostructured semiconductor devices: Beyond the envelope-function approximation. *Semicond. Sci. Technol.*, 18:R1, 2003.

2

Electron Transport

Francesco Bertazzi

Michele Goano

Giovanni Ghione

Alberto Tibaldi

Pierluigi Debernardi

and

Enrico Bellotti

2.1 Introduction

A theory of carrier transport in highly nanostructured optoelectronic devices cannot be entirely formulated without addressing the coupling of the current-carrying extended states to the localized states of the system, and the interaction of such localized carriers with coherent fields, as obtained from the classical solution of the Maxwell's equations, or with incoherent field fluctuations, if spontaneous emission is of interest. We will illustrate this complex interplay between carrier transport and optical transitions for a specific class of light-emitting devices, but similar considerations also apply to the inverse regime of light detection and photovoltaics, as a consequence of the principle of detailed balance, which links emission of light by radiative recombination to light absorption by generation of electron–hole pairs, leading e.g., to reciprocity relations between the photovoltaic and the electroluminescent properties of solar cells and light-emitting diodes (LEDs) [1]. For illustrative purposes, we will focus on spatially resolved approaches to describe the far-above threshold, possibly multimode, lasing regime of vertical cavity surface emitting lasers (VCSELs). We will not discuss lumped models based on rate equations for spatially integrated carrier and photon densities, being these approaches restricted to the near-threshold dynamic behavior of single-mode lasers.

    One viable solution to the simulation of the temporal and three-dimensional (3D) spatial dynamics of state-of-the-art VCSELs is the coupling of a full-wave optical solver to a semiclassical carrier transport model derived from the Boltzmann transport equation (BTE). In this framework, the interaction of the optical wave with the semiconductor active region is considered from two closely related perspectives, namely the electromagnetic wave standpoint, leading to wave absorption and gain, and the semiconductor standpoint, leading to the electron–hole pair generation and recombination rates [2]. In addition to the interaction with the optical field, the localized states in the active region are also coupled to the extended states of the system, which requires complementing the carrier transport model with appropriate quantum corrections. The issue of including nonclassical corrections within classical frameworks has generated quite some confusion, at least in the context of visible LEDs. The ongoing debate over the loss mechanisms in these devices emphasizes the need for a unified treatment of both optical and electronic aspects by means of genuine quantum kinetic approaches, such as the density matrix (DM) and the nonequilibrium Green's function (NEGF) theory.

Given this context, an appraisal of the theoretical underpinnings of carrier transport in nanostructured material for optoelectronic applications seemed to us far more compelling than a comprehensive review of the models, which would be unavoidably incomplete and possibly redundant, considering the vast literature dedicated to the topic. The aim of this chapter is to review the assumptions behind the BTE, to present how a hierarchical set of equations for the moments of the distribution can be derived from it (Section 2.2), and to discuss advantages and limitations of different approaches to go beyond the semiclassical limit, from quantum-corrected semiclassical models (Section 2.3) to quantum kinetic models in which dissipative carrier transport and quantum optics are described on equal footing (Section 2.4). Our hope is to be informative, while at the same time to call attention to some of the tacit assumptions of electronic transport in optoelectronic devices which are still largely untested.

2.2 From the Boltzmann Transport Equation to the Drift-Diffusion Model

The starting point of the semiclassical picture recalled in Section 2.1 is the BTE—the first equation ever written to govern the time evolution of a probability—proposed in the late nineteenth century by Ludwig Boltzmann in his kinetic theory of gases and generalized in different fields of physics [3] including, of course, carrier transport in semiconductors, where it takes the form

$$\frac{\partial f}{\partial t} + \underbrace{v \cdot \nabla_r f}_{\text{position}} + \underbrace{\frac{1}{\hbar} F \cdot \nabla_k f}_{\text{momentum}} = \frac{df}{dt}\bigg|_{\text{collision}}. \qquad (2.1)$$

The relevant probability function here is the distribution $f(r, k, t)$ that describes the evolution in the phase space (r, k) of carriers treated as point-like particles with well-defined geometrical position r and wave vector k (corresponding to a momentum $p = \hbar k$). The symbols ∇_r and ∇_k denote differentiation in the position and momentum spaces, respectively. The description of the electronic band structure $E(k)$ is included in the BTE through the velocity term $v = \nabla_k E(k)/\hbar$, where the band index has been omitted to simplify the notation. The "position" term in Equation 2.1 accounts for the motion of carriers in real space, due to their group velocity v; the "momentum" term describes the variation of momentum impressed by the classical force F. The momentum of carriers (but not their position) is also affected by scattering events with photons, phonons, impurities, and other carriers, all included in the collision term on the right-hand side (RHS) of Equation 2.1,

$$\frac{df}{dt}\bigg|_{\text{collision}} = \sum_{k'} \left\{ f(r, k', t)[1 - f(r, k, t)] W(k', k) - f(r, k, t)[1 - f(r, k', t)] W(k, k') \right\} dk', \qquad (2.2)$$

where $W(k, k')$ is the rate at which a particle scatters from a state k to k' (the first term represents the flow of particles into state k, the second is the loss by scattering into another state). Each carrier population (electrons and holes, or possibly their subsets, e.g., heavy/light/split-off band holes) is described by its own distribution function and the corresponding BTEs are coupled together (and to the crystal lattice) through the collision terms.

Heuristic derivations[†] of the BTE by means of conservation principles may be found in [4,5]. In introducing the theory of semiclassical carrier transport, it is important to recall some of the basic hypotheses that we are forced to assume to reach the BTE.

1. The carriers have a well-defined position r and momentum p as if they were classical particles. For this approximation to be acceptable, the size of the device should be much larger than the size of the electron wavepacket given by the mean free path between collisions and the energy scale of interest should be much larger than the uncertainty implied by the spread of the electron momentum.
2. At any given point in k-space, the energy bands are well separated in energy and the electric fields are sufficiently small so that we can describe the free flight of the carriers between collisions with Newton's laws.[‡]
3. The energy bands do not change too quickly in space, so that the concept of a local band structure may be accepted and level quantization can be neglected.
4. Boltzmann's Stosszahlansatz holds, i.e., all scattering processes are independent (no memory is conserved of where and when the previous collisions happened), local (they involve no change of r), instantaneous (their duration is negligible with respect to the free-flight time between successive collisions), and are perturbations weak enough to justify the use of Fermi's golden rule [5, Appendix E] to estimate their probability per unit time.

These assumptions break down in a few important cases: very fast transients that have to be studied on a time scale comparable to the scattering rate [6]; high scattering rates, so that the initial state may decay appreciably by the time the scattering is completed (collision broadening) [6]; and presence of high electric fields, which may transfer a significant amount of energy to the carriers during the finite time required for a collision, or, in other words, renormalize the states involved in the scattering process (intracollisional field effect [7]).

An additional assumption, unrelated to the BTE derivation, is that time-domain transients are slow enough to justify the use of electrostatics rather than Maxwell's equations, meaning that the electric field may be determined only by the instantaneous distributions of the carriers via Poisson's equation, therefore neglecting propagation effects. A carrier transport model coupled to a full-wave electromagnetic solver is presented in [10] in the context of semiconductor waveguides for microwave applications; the coupling of the transport equations with an optical modal solver will be discussed in Section 2.3.2 in the context of VCSELs.

2.2.1 Direct Solution of the BTE: Monte Carlo Transport Simulation

The BTE is an integro-differential equation, nonlinear because the collision term depends on the distribution function $f(r, k, t)$, itself due to the Fermi statistical factors appearing in Equation 2.2 (that also describe phase-space filling) or, in dealing with Coulomb interactions, via the Lindhart screening formula [6]. Closed-form approximate solutions of the BTE may be obtained under homogeneous conditions and small electric fields, which allow linearizing the equation with respect to the field (relaxation-time approximation). Partial differential equations (PDEs) may be obtained from the BTE by expanding the unknown $f(r, k, t)$ into suitable basis functions (e.g., spherical harmonics [11]) and then solving for the expansion coefficients (expansion methods).

A direct statistical solution of the BTE for the distribution function is provided by the Monte Carlo method [5,12–14], which consists of the simulation of the motion of an ensemble of carriers, subject to the action of electric and magnetic fields and of given scattering mechanisms. The duration of the carrier

---

[†] Note that thus far the BTE has resisted a rigorous derivation. What remains elusive is the problem of deriving macroscopic irreversibility (that defines the "arrow of time") from microscopic reversibility [8].

[‡] In wide band-gap materials, due to the small size of the Brillouin zone, multiple bands exist in relatively narrow ranges of energy, and band-to-band tunneling at band "kissing points" may be of importance [9].

free flights, i.e., the time between two successive collisions, and the states after scattering are determined stochastically according to the scattering probabilities.

As with any time-domain technique, the Monte Carlo approach is much less efficient, at least in its self-consistent version (when Poisson's equation is solved self-consistently with the charge distribution) if different time constants are involved. This is the case in the simulation of bipolar devices, where intraband and interband (recombination) processes contribute with very different scattering times to determine the profile of the minority carriers. In addition to this limitation, the treatment of minority carriers (whose diffusion determines the current across a *p-n* junction) is problematic within a Monte Carlo framework, requiring statistical enhancement techniques that are quite cumbersome to implement; in fact, most Monte Carlo studies of optoelectronic devices are restricted to non-self-consistent approximations or to unipolar devices [15–18]. This does not imply that Monte Carlo device simulation is not viable in optoelectronics. On the contrary, it is probably the most general approach to the BTE, for its ability to incorporate physical models beyond what other techniques can handle (full-Brillouin-zone descriptions of the electronic structure [19,20], first-principles phonon dispersions [9], wave vector–dependent scattering rates), limited only by the time one is willing to wait for a converged solution. As an example of how much physics can be included in a Monte Carlo code, we mention the pioneering works [21–23] on carrier–phonon and carrier–carrier scattering, with the aim of explaining high-field carrier transport starting from the sole input of the ion pseudopotentials employed for the description of the electronic structure.[†] Another example along these lines is the inclusion of interband tunneling within full-band Monte Carlo codes (by means of the time-dependent solution of the Schrödinger equation) to describe the coherent (between scattering events) drift of the carriers at band crossing points [24].

2.2.2 Hierarchical Approximations of the BTE: The Method of Moments

From an engineering perspective, in several technologically relevant cases a good description of the device behavior can be obtained at a much reduced computational cost by solving a small set of coupled nonlinear PDEs derived from the BTE by applying the *method of moments* (MOM) [25,26]. Contrary to Monte Carlo simulation, MOM does not try to obtain the full distribution function $f(\underline{r}, \underline{k}, t)$, but settles instead for knowledge of some of its ensemble averages, closely related to relevant quantities such as carrier density, velocity, and energy. The derivation of the MOM PDEs from the BTE is to some extent comparable to the definition of the moments of a probability density in statistics (i.e., average, variance, etc.), hence its name. The BTE Equation 2.1 is multiplied by a scalar function $\psi(\underline{k})$ and then integrated over the $\underline{k}$ space, which gives

$$\int_{\underline{k}} \psi \frac{\partial f}{\partial t}\, d\underline{k} + \int_{\underline{k}} \psi \underline{v} \cdot \nabla_{\underline{r}} f\, d\underline{k} + \int_{\underline{k}} \psi \frac{1}{\hbar} \underline{F} \cdot \nabla_{\underline{k}} f\, d\underline{k} = \int_{\underline{k}} \psi \left. \frac{df}{dt} \right|_{\text{collision}} d\underline{k}.$$

Since ψ and $\underline{v}$ do not depend on time and position, we have

$$\int_{\underline{k}} \psi \frac{\partial f}{\partial t}\, d\underline{k} = \frac{\partial}{\partial t} \int_{\underline{k}} \psi f\, d\underline{k}$$

$$\int_{\underline{k}} \psi \underline{v} \cdot \nabla_{\underline{r}} f\, d\underline{k} = \nabla_{\underline{r}} \cdot \int_{\underline{k}} \underline{v} \psi f\, d\underline{k}.$$

[†] Wave vector–dependent deformation potentials that rigorously satisfy selection rules along symmetry lines were derived from nonlocal empirical pseudopotential calculations of the electronic structure and density functional calculations of the lattice dynamics [9,23]. Their use shifted the empiricism of earlier Monte Carlo investigations (based on phenomenological deformation potential parameters) to a more fundamental level, with no fitting parameters involved, save for the initial choice for the pseudopotentials. Electron–phonon interactions and impact ionization were never treated in such detail within any other carrier transport code.

Moreover, $\nabla_{\underline{k}}(\psi f) = f\nabla_{\underline{k}}\psi + \psi\nabla_{\underline{k}}f$, so

$$\int_{\underline{k}} \psi \frac{1}{\hbar}\underline{F} \cdot \nabla_{\underline{k}}f \, d\underline{k} = \frac{1}{\hbar}\underline{F} \cdot \int_{\underline{k}} \psi\nabla_{\underline{k}}f \, d\underline{k} = \frac{1}{\hbar}\underline{F} \cdot \int_{\underline{k}} \nabla_{\underline{k}}(\psi f) \, d\underline{k} - \frac{1}{\hbar}\underline{F} \cdot \int_{\underline{k}} f\nabla_{\underline{k}}\psi \, d\underline{k}.$$

Applying Green's theorem we also have

$$\int_{\underline{k}} \nabla_{\underline{k}}(\psi f) \, d\underline{k} = \oint_{\Sigma} \hat{n}\psi f \, d\sigma,$$

where the surface Σ, with normal vector $\hat{n}$, has an infinite radius. Assuming that ψ is a polynomial function of $\underline{k}$ (see discussion below) and f is well behaved (i.e., decaying exponentially, Maxwellian-like, for $\underline{k} \to \infty$), the surface integral is zero, and we are left with

$$\frac{\partial}{\partial t}\int_{\underline{k}} \psi f \, d\underline{k} + \nabla_{\underline{r}} \cdot \int_{\underline{k}} \underline{v}\psi f \, d\underline{k} - \frac{1}{\hbar}\underline{F} \cdot \int_{\underline{k}} f\nabla_{\underline{k}}\psi \, d\underline{k} = \int_{\underline{k}} \frac{d(\psi f)}{dt}\bigg|_{collision} d\underline{k}. \tag{2.3}$$

By choosing the moment functions ψ_m as integer powers of k, e.g., $\psi_0 = a_0$, $\psi_1 = a_1 k$, $\psi_2 = a_2 k^2$, etc. (where $a_0, a_1, a_2 \ldots$ are proportionality coefficients), one may formally replace the BTE with a hierarchy of balance PDEs whose unknowns are the mth-order moments

$$M_m = \int_{\underline{k}} \psi_m f \, d\underline{k}.$$

In principle, the prescription above may be applied to derive an arbitrary number of moments, from which all the relevant information about the distribution function $f(\underline{r}, \underline{k}, t)$ could be recovered. However, most device simulators account for only a few moments connected to well-defined physical quantities, such as the carrier density ($m = 0$), the average momentum ($m = 1$), and the average energy ($m = 2$). Higher-order moments are usually neglected as they lack a simple physical interpretation (save for the third-order moment, which is related to the heat flow).

In truncating the infinite hierarchy of PDEs, one has to face the so-called "closure problem," since the balance equation for any given moment also includes the next higher-order moment [27]. With the help of approximate closure relations, the PDE system can be conveniently reduced to the three lower-order equations, expressing the conservation of particle density, average momentum, and average energy. Owing to the similarities between carrier transport and fluid dynamics, the resulting system is commonly referred to as the HD model [28,29]. Further simplifications lead to the so-called energy balance (EB) model [30,31], which, in turn, can be reduced to the familiar drift-diffusion (DD) picture of carrier transport, assuming local instantaneous equilibrium between the average energy and the electric field. HD models can describe nonstationary effects such as velocity overshoot, carrier heating, and negative differential mobility, but they require additional information concerning the energy and momentum relaxation times, which can be provided by Monte Carlo simulations [32,33]. On the other hand, the DD approximation, complemented by appropriate recombination and mobility models, has proved remarkably successful in explaining the electrical behavior of devices that violate most of the assumptions under which it should be applied (small electric fields, large device size with respect to the mean free path).

In the rest of this section, unless otherwise stated, only the electron population will be considered, but the same discussion can be replicated for holes with nearly identical results.

2.2.2.1 Zeroth-Order Moment: Carrier Conservation

Setting $\psi(\underline{k}) = 1$, one has

$$M_0 = \int_{\underline{k}} f(\underline{r}, \underline{k}, t) \, d\underline{k} = n(\underline{r}, t),$$

and Equation 2.3 becomes

$$\frac{\partial}{\partial t} \int_{\underline{k}} f(\underline{r}, \underline{k}, t) \, d\underline{k} + \nabla_{\underline{r}} \cdot \int_{\underline{k}} \underline{v}(\underline{k}) f(\underline{r}, \underline{k}, t) \, d\underline{k} = \int_{\underline{k}} \frac{df(\underline{r}, \underline{k}, t)}{dt} \bigg|_{\text{collision}},$$

which is a balance equation for the electron density n

$$\frac{\partial n}{\partial t} + \nabla_{\underline{r}} \cdot (n \underline{v}_n) = \frac{dn}{dt} \bigg|_{\text{collision}},$$

where the collision term includes GR mechanisms coupling the electron and hole populations, and $\underline{v}_n$ is the average electron velocity

$$\underline{v}_n = \frac{\displaystyle\int_{\underline{k}} \underline{v}(\underline{k}) f(\underline{r}, \underline{k}, t) \, d\underline{k}}{\displaystyle\int_{\underline{k}} f(\underline{r}, \underline{k}, t) \, d\underline{k}},$$

closely related to the first-order moment of the electron distribution.

2.2.2.2 First-Order Moment: Momentum Conservation

Setting $\psi(\underline{k}) = \hbar \underline{k}$, which is the carrier momentum, and taking into account that the average momentum of the electron distribution is

$$\underline{p}_n = \frac{\displaystyle\int_{\underline{k}} \hbar \underline{k} f(\underline{r}, \underline{k}, t) d\underline{k}}{\displaystyle\int_{\underline{k}} f(\underline{r}, \underline{k}, t) d\underline{k}},$$

one has in Equation 2.3

$$\frac{\partial}{\partial t} \int_{\underline{k}} \psi f d\underline{k} = \frac{\partial}{\partial t} \int_{\underline{k}} \hbar \underline{k} f d\underline{k} = \frac{\partial (n \underline{p}_n)}{\partial t}$$

and

$$\nabla_{\underline{r}} \cdot \int_{\underline{k}} \psi \underline{v} f d\underline{k} = \nabla_{\underline{r}} \cdot \int_{\underline{k}} \hbar \underline{k} \, \underline{v} f d\underline{k} = \frac{\hbar^2}{m_n^*} \nabla_{\underline{r}} \cdot \int_{\underline{k}} \underline{k} \, \underline{k} f d\underline{k},$$

where it has been assumed (*parabolic approximation*) that the electron effective mass m_n^* is independent of the electron energy. By defining the average wave vector $\underline{k}_n = \underline{p}_n / \hbar$, one has

$$\underline{k}\underline{k} = (\underline{k} - \underline{k}_n)(\underline{k} - \underline{k}_n) + 2\underline{k}\,\underline{k}_n - \underline{k}_n \underline{k}_n,$$

hence

$$\int_{\underline{k}} \underline{k} \, \underline{k} f d\underline{k} = \int_{\underline{k}} (\underline{k} - \underline{k}_n)(\underline{k} - \underline{k}_n) f d\underline{k} + 2\underline{k}_n \int_{\underline{k}} \underline{k} f d\underline{k} - \underline{k}_n \underline{k}_n \int_{\underline{k}} f d\underline{k} =$$

$$= \int_{\underline{k}} (\underline{k} - \underline{k}_n)(\underline{k} - \underline{k}_n) f d\underline{k} + n \underline{k}_n \underline{k}_n.$$

The dyad

$$\int_{\underline{k}} (\underline{k} - \underline{k}_n)(\underline{k} - \underline{k}_n) f \, d\underline{k}$$

is proportional to the *covariance matrix* $\underline{C}[\underline{k}]$ of the $\underline{k}$ components taken in pairs,

$$\underline{C}[\underline{k}] = \frac{\displaystyle\int_{\underline{k}} (\underline{k} - \underline{k}_n)(\underline{k} - \underline{k}_n) f \, d\underline{k}}{\displaystyle\int_{\underline{k}} f \, d\underline{k}}.$$

Considering, e.g., a drifted Gaussian distribution (see Figure 2.1), symmetric with respect to its central value (the average velocity), the off-diagonal terms of the covariance matrix are identically zero (integrals of odd functions over symmetric intervals), while the diagonal terms include the variances of the distribution with respect to each spatial coordinate. If the variance of the distribution is the same along all directions, $\underline{C}$ is proportional to the identity matrix $\underline{I}$ through a coefficient that can be defined in terms of the *temperature* of the distribution, usually called the *electron temperature* T_n

$$m_n^* k_{\mathrm{B}} T_n \underline{I} = \hbar^2 \underline{C}[\underline{k}] = \underline{C}[\underline{p}], \tag{2.4}$$

where $k_{\mathrm{B}} T_n$ is proportional to the average energy of the electron gas due to the *disordered* electron motion. In the general case, it is possible to define an anisotropic *temperature tensor*

$$\mathbf{T}_n = \frac{\hbar^2}{m_n^* k_{\mathrm{B}}} \underline{C}[\underline{k}],$$

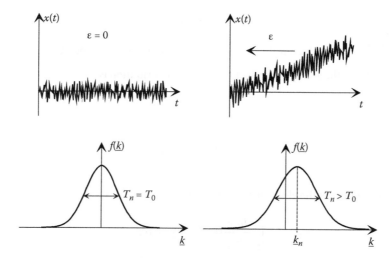

FIGURE 2.1 Drift motion and heating of an electron gas determined by an electric field $\mathcal{E}$. The carrier momentum distribution is Gaussian (under the parabolic approximation) and symmetric with respect to its central value (average momentum $\underline{k}_n$), which is zero at thermal equilibrium and shifts under the application of $\mathcal{E}$. The temperature of the electron gas T_n is proportional to the *variance* of the distribution, distribution (see Equation 2.4), which increases with $\mathcal{E}$.

but it is common to use a diagonal approximation for $\mathbf{T}_n$, so that

$$\frac{\hbar^2}{m_n^*} \int_{\underline{k}} \underline{k}\,\underline{k} f \mathrm{d}\underline{k} = \frac{\hbar^2}{m_n^*} \int_{\underline{k}} (\underline{k} - \underline{k}_n)(\underline{k} - \underline{k}_n) f \mathrm{d}\underline{k} + \frac{\hbar^2 n}{m_n^*} \underline{k}_n \underline{k}_n = n k_{\mathrm{B}} T_n \underline{I} + \frac{1}{m_n^*} n \underline{p}_n \underline{p}_n. \tag{2.5}$$

Therefore, one has

$$\nabla_{\underline{r}} \cdot \int_{\underline{k}} \hbar \underline{k}\, v f \mathrm{d}\underline{k} = \nabla_{\underline{r}} \cdot \left(n k_{\mathrm{B}} T_n \underline{I} + \frac{1}{m_n^*} n \underline{p}_n \underline{p}_n \right) = \nabla_{\underline{r}} (n k_{\mathrm{B}} T_n) + \frac{1}{m_n^*} \nabla_{\underline{r}} \cdot (n \underline{p}_n \underline{p}_n).$$

Also,

$$\int_{\underline{k}} (\nabla_{\underline{k}} \psi) f \mathrm{d}\underline{k} = \hbar \int_{\underline{k}} (\nabla_{\underline{k}} \underline{k}) f \mathrm{d}\underline{k} = \hbar \underline{I} \int_{\underline{k}} f \mathrm{d}\underline{k} = \hbar \underline{I} n.$$

Assuming that the only relevant forces are due to the electric field $\mathcal{E}$, the previous equation becomes

$$\frac{F}{\hbar} \cdot \int_{\underline{k}} \nabla_{\underline{k}} f \mathrm{d}\underline{k} = \underline{F} n = -q n \mathcal{E},$$

where q is the elementary charge. Finally, assembling all the previous partial results and introducing a phenomenological collision term, one has

$$\frac{\partial (n \underline{p}_n)}{\partial t} + \nabla_{\underline{r}} (n k_{\mathrm{B}} T_n) + \frac{1}{m_n^*} \nabla_{\underline{r}} \cdot (n \underline{p}_n \underline{p}_n) + q n \mathcal{E} = \left. \frac{\mathrm{d}(n \underline{p}_n)}{\mathrm{d}t} \right|_{\text{collision}}, \tag{2.6}$$

which is called the *momentum transport equation*, and includes a term (the electron temperature) closely related to the second-order moment (the variance) of the electron distribution.

2.2.2.3 Second-Order Moment: Energy Conservation

Setting $\psi(\underline{k}) = \dfrac{(\hbar k)^2}{2 m_n^*} = E_n$, which is the electron kinetic energy under the parabolic approximation, and taking into account that the average energy of the electron distribution is

$$\overline{E}_n = \frac{\displaystyle\int_{\underline{k}} E_n f(\underline{r}, \underline{k}, t) \mathrm{d}\underline{k}}{\displaystyle\int_{\underline{k}} f(\underline{r}, \underline{k}, t) \mathrm{d}\underline{k}},$$

the second-order moment of the electron distribution is

$$\int_{\underline{k}} \frac{(\hbar k)^2}{2 m_n^*} f(\underline{r}, \underline{k}, t) \mathrm{d}\underline{k} = n \overline{E}_n.$$

Therefore, the first term in Equation 2.3 becomes

$$\frac{\partial}{\partial t} \int_{\underline{k}} \psi f \mathrm{d}\underline{k} = \frac{\partial}{\partial t} \int_{\underline{k}} \frac{(\hbar k)^2}{2 m_n^*} f \mathrm{d}\underline{k} = \frac{\partial (n \overline{E}_n)}{\partial t}.$$

The second term in Equation 2.3,

$$\nabla_{\underline{r}} \cdot \int_{\underline{k}} \psi \underline{v} f \, d\underline{k} = \nabla_{\underline{r}} \cdot \int_{\underline{k}} \frac{(\hbar k)^2}{2m_n^*} \underline{v} f \, d\underline{k} = \nabla_{\underline{r}} \cdot \underline{S}_n,$$

now represents the divergence of the energy flow of the electron gas. In terms of $\underline{k}$, $\underline{S}_n$ can be written as

$$\underline{S}_n = \frac{\hbar^3}{2(m_n^*)^2} \int_{\underline{k}} k^2 \underline{k} f \, d\underline{k},$$

and it can be demonstrated that

$$\underline{S}_n = \frac{1}{m_n^*} n\underline{p}_{\underline{n}} \bar{E}_n + \frac{1}{m_n^*} n\underline{p}_{\underline{n}} k_B T_n + \underline{Q}, \tag{2.7}$$

where the vector $\underline{Q}$,

$$\underline{Q} = \frac{\hbar^3}{2(m_n^*)^2} \int_{\underline{k}} (\underline{k} - \underline{k}_n) \cdot (\underline{k} - \underline{k}_n)(\underline{k} - \underline{k}_n) f \, d\underline{k},$$

is the *heat flow* of the electron gas and is related to the third-order moment of the distribution. (The heat flow is identically zero when the distribution is symmetric with respect to its central value, as for example in the drifted Gaussian case.)

A closure relation for the system of three PDEs corresponding to the $m = (0, 1, 2)$ moments of the BTE is usually provided by a phenomenological approximation of the heat flow, which can be assumed proportional to the gradient of the electron temperature

$$\underline{Q} \approx -\kappa_n \nabla_{\underline{r}} T_n, \tag{2.8}$$

where the proportionality constant κ_n is the *thermal conductivity of the electron gas*. Near equilibrium, κ_n can be estimated with the *Wiedemann–Franz law*,

$$\kappa_n \approx r \left(\frac{k_B}{q} \right)^2 T_n \sigma_n,$$

where r is a unitless coefficient (close to 1), $\sigma_n = q n \mu_n$ is the electrical conductivity, and μ_n is the electron mobility (see Section 2.2.4).

2.2.2.4 The Collision Terms

Within the relaxation-time approximation, the collision term of the BTE is approximated as

$$\left. \frac{df}{dt} \right|_{\text{collision}} \approx \frac{f - f_0}{\tau},$$

where f_0 is the distribution at thermal equilibrium and τ is the *relaxation time* of the distribution, which is assumed to be either a constant or a function of some parameters defining the distribution (average, variance, etc.). A similar approach is usually adopted in all the PDEs derived from the BTE with the MOM. In the balance equation for the electron density (electron continuity equation), the collision term is the net recombination rate (see Section 2.2.5 for a more detailed discussion):

$$\left. \frac{dn}{dt} \right|_{\text{collision}} = -U_n.$$

The collision term in the momentum transport equation can be defined through a *momentum relaxation time*, possibly dependent on the average carrier energy (or temperature):

$$\left.\frac{d\underline{p}_n}{dt}\right|_{coll} \approx -\frac{\underline{p}_n}{\tau_{pn}}.$$

Finally, in the energy transport equation, the collision term is approximated by defining an *energy relaxation time*:

$$\left.\frac{d\overline{E}_n}{dt}\right|_{collision} \approx -\frac{\overline{E}_n - \overline{E}_0}{\tau_{En}}.$$

The application of the relaxation time approximation is straightforward, also in the formulation of the momentum and energy transport equations where the collision terms involve the products $n\underline{p}_n$ and $n\overline{E}_n$:

$$\left.\frac{d(n\overline{E}_n)}{dt}\right|_{collision} = n\left.\frac{d\overline{E}_n}{dt}\right|_{collision} + \overline{E}_n\left.\frac{dn}{dt}\right|_{collision} = -n\frac{\overline{E}_n - \overline{E}_0}{\tau_{En}} - \overline{E}_n U_n.$$

2.2.3 The Hydrodynamic Model

The HD model includes the transport PDEs of energy, momentum, and density for electrons and holes—or possibly for one carrier species only, the other being approximated with a simpler description. Under the relaxation time approximation, the HD transport model for electrons is

$$\frac{\partial n}{\partial t} + \nabla \cdot (n\underline{v}_n) = -U_n \tag{2.9}$$

$$\frac{\partial (n\underline{p}_n)}{\partial t} + \nabla(nk_B T_n) + \frac{1}{m_n^*}\nabla \cdot (n\underline{p}_n\underline{p}_n) + qn\underline{\mathcal{E}} = -n\frac{\underline{p}_n}{\tau_{pn}} - \underline{p}_n U_n \tag{2.10}$$

$$\frac{\partial (n\overline{E}_n)}{\partial t} + \nabla \cdot \underline{S}_n = -\frac{1}{m_n^*}qn\underline{p}_n \cdot \underline{\mathcal{E}} - n\frac{\overline{E}_n - \overline{E}_0}{\tau_{En}} - \overline{E}_n U_n, \tag{2.11}$$

where, from now on, $\nabla \equiv \nabla_r$, the energy flow $\underline{S}_n$ is reported in Equation 2.7, and the relationship between average energy and average temperature of the electron gas can be written as

$$\overline{E}_n = \frac{3}{2}k_B T_n + \frac{1}{2m_n^*}p_n^2. \tag{2.12}$$

An alternative, equivalent form of Equations 2.10 and 2.11 can be derived by using the continuity equation and the vector identity $\nabla \cdot (\underline{a}\,\underline{b}) = (\nabla \cdot \underline{a})\underline{b} + (\underline{a} \cdot \nabla)\underline{b}$:

$$\frac{\partial \underline{p}_n}{\partial t} + \frac{1}{n}\nabla(nk_B T_n) + (\underline{v}_n \cdot \nabla)\underline{p}_n + q\underline{\mathcal{E}} = -\frac{\underline{p}_n}{\tau_{pn}} \tag{2.13}$$

$$\frac{\partial \overline{E}_n}{\partial t} + \underline{v}_n \cdot \nabla\overline{E}_n + \frac{1}{n}\nabla \cdot (n\underline{v}_n k_B T_n) + \frac{1}{n}\nabla \cdot \underline{Q}_n = -q\underline{v}_n \cdot \underline{\mathcal{E}} - \frac{\overline{E}_n - \overline{E}_0}{\tau_{En}}. \tag{2.14}$$

In the HD picture, Equation 2.9 enforces carrier density conservation, whereas Equations 2.10 and 2.11 or Equations 2.13 and 2.14 impose conservation of carrier momentum and energy, respectively. The heat flow $\underline{Q}_n$, involved in the closure relation, can either be neglected or described with Equation 2.8.

2.2.4 The Energy Balance Model

The relaxation times and characteristic lengths associated with the momentum and energy transport PDEs Equations 2.10 and 2.11 or Equations 2.13 and 2.14 are often small enough to suggest the possibility of reducing the complexity of the HD model. As a first step, the momentum transport equation can be approximated to make carrier momentum a dependent variable, leading to the energy balance (or energy transport) model.

The derivation of the EB model may start by assuming that the transport phenomena under study are slow enough to allow for a quasistatic analysis of Equation 2.13, so that the time derivative can be neglected, leading to

$$\left(1 + \tau_{pn}\underline{v}_n \cdot \nabla\right)\underline{v}_n \approx -\frac{q\tau_{pn}}{m_n^*}\mathcal{E} - \frac{\tau_{pn}}{m_n^* n}\nabla(nk_B T_n), \tag{2.15}$$

where the *convection term* $\tau_{pn}\underline{v}_n \cdot \nabla\underline{v}_n$ is potentially much smaller than $\underline{v}_n$. Focusing on a spatially one-dimensional (1D) case for the sake of clarity, this condition holds if

$$\frac{dv_n}{dx} \ll \frac{1}{\tau_{pn}}.$$

Here, τ_{pn} is shorter than 500 fs [32,34,35] and the maximum Δv_n is approximately equal to the saturation velocity $v_{sat} \approx 10^7$ cm/s,[†] so that the condition becomes

$$\frac{dv_n}{dx} \approx \frac{\Delta v_n}{\Delta x} \ll \frac{1}{\tau_{pn}}.$$

Therefore, the convection term can be non-negligible only for devices with spatial extension Δx smaller than $\tau_{pn}\Delta v_n \approx 50$ nm.[‡] By neglecting this term in Equation 2.15 one obtains

$$\underline{v}_n \approx -\frac{q\tau_{pn}}{m_n^*}\mathcal{E} - \frac{\tau_{pn}}{m_n^* n}\nabla(nk_B T_n) = -\mu_n(\overline{E}_n)\mathcal{E} - \frac{\mu_n(\overline{E}_n)}{qn}\nabla(nk_B T_n), \tag{2.16}$$

where the electron mobility μ_n has been introduced:

$$\mu_n(\overline{E}_n) = \frac{q\tau_{pn}(\overline{E}_n)}{m_n^*}.$$

Since both the effective mass (if one takes into account nonparabolicity effects) and especially the momentum relaxation time depend on the average energy of the distribution, the electron mobility also is a function of $\overline{E}_n$. According to Equation 2.16, the average velocity is the superposition of two effects, like in the DD model (see Section 2.2.5): the drift-like term involves an energy-dependent (rather than

---

[†] In nonstationary conditions, the average drift velocity can be larger than v_{sat} (velocity overshoot); however, under realistic conditions, the peak velocity is of the same order of magnitude of v_{sat}.

[‡] This upper bound is material dependent, and in general the related convection term may be negligible even for smaller extensions.

field-dependent) mobility, while the diffusion-like term depends on the gradients of carrier density and temperature (rather than of carrier density only).

When substituted in Equations 2.9 and 2.14, the two surviving equations of the HD model, Equation 2.16 can be used to remove the average velocity from the independent variables, leading to the EB model,

$$\frac{\partial n}{\partial t} + \nabla \cdot \left(n \underline{v}_n \right) = -U_n \tag{2.17}$$

$$\frac{\partial \overline{E}_n}{\partial t} + \underline{v}_n \cdot \nabla \overline{E}_n + \frac{1}{n} \nabla \cdot \left(n \underline{v}_n k_B T_n \right) + \frac{1}{n} \nabla \cdot \underline{Q}_n = -q \underline{v}_n \cdot \underline{\mathcal{E}} - \frac{\overline{E}_n - \overline{E}_0}{\tau_{En}}, \tag{2.18}$$

whose closure relation is a heat flux constitutive equation (such as the Fourier law). The distinctive feature of EB is that the quantities defining the transport properties (in particular, the mobility) depend on the local average energy rather than on the local field; therefore, EB takes into account the effects of carrier heating caused by the dynamic variations of the electric field.

Another simplification can be introduced by noting that, in the relationship between average energy and temperature of the electron gas reported in Equation 2.12

$$\overline{E}_n = \frac{3}{2} k_B T_n + \frac{1}{2m_n^*} \underline{p}_n^2 = \frac{3}{2} k_B T_n + \frac{1}{2} m_n^* v_n^2,$$

a carrier population having drift velocity close (in the worst case scenario) to v_{sat} exhibits a kinetic component (≈ 10 meV) lower than its thermal one (≈ 40 meV at 300 K). For this reason, the kinetic contribution can be neglected,

$$\overline{E}_n \approx \frac{3}{2} k_B T_n,$$

and mobility is written as a function of the electron temperature.

2.2.4.1 From the Energy Balance to the Drift-Diffusion Model

Since the energy relaxation time is of the order of a picosecond, the variations of the average energy in space and time are often negligible, allowing to reduce Equation 2.18 (the energy transport equation) to its stationary form

$$\frac{\overline{E}_n - \overline{E}_0}{\tau_{En}} = -q \underline{v}_n \cdot \underline{\mathcal{E}},$$

which can be used to write the average energy as

$$\overline{E}_n = \overline{E}_0 - q \tau_{En} \underline{v}_n \cdot \underline{\mathcal{E}}. \tag{2.19}$$

From Equation 2.16, $\underline{v}_n$ includes a diffusion component depending on the carrier density gradient. Assuming that such contribution is negligible,[†] one has

$$\underline{v}_n \approx -\mu_n \underline{\mathcal{E}},$$

and Equation 2.19 becomes

$$\overline{E}_n \approx \overline{E}_0 + q \tau_{En}(\overline{E}_n) \mu_n(\overline{E}_n) \mathcal{E}^2. \tag{2.20}$$

[†] This approximation may not be appropriate whenever significant drift and diffusion components coexist, e.g., in the depletion regions of *p-n* junctions.

This local (implicit) relationship between $\overline{E}_n$ and $\mathcal{E}$ allow us to write the electron mobility as a function of the electric field; correspondingly, the current density is approximated as

$$\underline{J}_n = -qn\underline{v}_n =$$

$$= qn\mu_n(\mathcal{E})\underline{\mathcal{E}} + \mu_n(\mathcal{E})\nabla(nk_B T_n(\mathcal{E}))$$

$$= qn\mu_n(\mathcal{E})\underline{\mathcal{E}} + q\mu_n(\mathcal{E})\frac{k_B T_n}{q}\nabla n + \mu_n(\mathcal{E})n\nabla(k_B T_n(\mathcal{E})).$$

By defining the electron diffusivity through a generalized Einstein relation involving T_n,

$$D_n(\mathcal{E}) = \mu_n(\mathcal{E})\frac{k_B T_n}{q},$$

the current can be written as

$$\underline{J}_n = qn\mu_n(\mathcal{E})\left(\underline{\mathcal{E}} + \nabla\frac{k_B T_n(\mathcal{E})}{q}\right) + qD_n\nabla n.$$

In this expression, the temperature gradient contribution can be assimilated to an equivalent electric field. However, this term is to some extent inconsistent with the approximations applied so far to the energy transport equation, where temperature gradients have been neglected. Therefore, the spatial variation of T_n is neglected as well,[†] and the classical expression of the drift-diffusion (DD) current density is finally obtained:

$$\underline{J}_n \approx qn\mu_n(\mathcal{E})\underline{\mathcal{E}} + qD_n\nabla n.$$

2.2.5 The Drift-Diffusion Model

With the drastic simplifications described in Section 2.2.4 and above, we arrive at the bipolar drift-diffusion model, which includes Poisson's equation and the electron and hole continuity equations,

$$\begin{cases} -\nabla^2\phi = \frac{q}{\varepsilon}\left(p - n + N_D^+ - N_A^-\right) \\[2mm] \frac{\partial n}{\partial t} = \frac{1}{q}\nabla\cdot\underline{J}_n - U_n \\[2mm] \frac{\partial p}{\partial t} = -\frac{1}{q}\nabla\cdot\underline{J}_p - U_p \end{cases}, \tag{2.21}$$

with the following expressions for the drift and diffusion current densities:

$$\underline{J}_n = +qD_n\nabla n - q\mu_n n\nabla[\phi + \tilde{\phi}_n]$$

$$\underline{J}_p = \underbrace{-qD_p\nabla p}_{\text{diffusion}} - \underbrace{q\mu_p p\nabla[\phi + \tilde{\phi}_p]}_{\text{drift}}. \tag{2.22}$$

[†] This last approximation implies that carriers are in equilibrium with the lattice, i.e., the carrier temperature T_n is approximately equal to the lattice temperature T_0. Whenever T_0 varies in space due to heating effects, the DD model is not strictly isothermal and the temperature gradient contribution should be included, leading to the so-called *thermoelectric model* [36].

Here, ϕ is the electrostatic potential, $\varepsilon = \varepsilon_0\varepsilon_r$ is the dielectric permittivity, n and p are the electron and hole densities, $U_{n,p}$ is the net recombination rates per unit time and volume, $D_{n,p}, \mu_{n,p}$ are the field-dependent diffusivities and mobilities, and N_D^+, N_A^- are the ionized donor and acceptor densities, which differ from the total densities N_D, N_A if the ionization of the impurities is incomplete [37]. The spatially varying terms $\tilde{\phi}_{n,p}$ (in Equation 2.22)

$$\begin{aligned}\tilde{\phi}_n &= \chi/q + V_T \ln N_C + V_T \ln \gamma_n \\ \tilde{\phi}_p &= E_g/q + \chi/q - V_T \ln N_V - V_T \ln \gamma_p,\end{aligned} \qquad (2.23)$$

provide a natural treatment of heterointerfaces, i.e., of the effects of the position-dependent energy gap $E_g = E_C - E_V$, electron affinity χ, and effective density of states N_C, N_V of the conduction and valence bands.[†] Having defined the electron and hole quasi-Fermi levels $E_{F,n}, E_{F,p}$ in terms of Fermi–Dirac integrals of order 1/2 [38,39],

$$n = N_C \mathcal{F}_{\frac{1}{2}}\left(\frac{E_{F,n} - E_C}{k_B T}\right), \qquad p = N_V \mathcal{F}_{\frac{1}{2}}\left(\frac{E_V - E_{F,p}}{k_B T}\right), \qquad (2.24)$$

corrections to Boltzmann statistics are included by means of the fractions $\gamma_n = n/n_B, \gamma_p = p/p_B$ (in Equation 2.23), where n_B and p_B are the carrier densities in the nondegenerate limit ($n < N_C, p < N_V$):

$$n_B = N_C \exp\left(\frac{E_{F,n} - E_C}{k_B T}\right), \qquad p_B = N_V \exp\left(\frac{E_V - E_{F,p}}{k_B T}\right). \qquad (2.25)$$

2.2.5.1 Generation–Recombination Models

The approximations to derive the DD model discussed in Section 2.2.4 must be completed by introducing the collision term relevant in this simplified context, i.e., the expressions of the net recombination rates $U_{n,p}$ in the RHS of Equation 2.21. In a recombination process, an electron in the conduction band fills an empty state of the valence band, annihilating an electron–hole (e–h) pair and releasing energy, e.g., in the form of photons or phonons.[‡] In the reverse process (generation), an electron in the valence band is promoted to the conduction band, creating an e–h pair. As generation–recombination (GR) transitions must conserve energy and momentum, they can be phonon assisted (thermal), photon assisted (radiative), or assisted by other electrons or holes (Auger). Moreover, GR transitions can be either interband (direct mechanisms) or assisted by intermediate trap levels in the forbidden band (indirect or Shockley–Read–Hall mechanisms). In direct-gap semiconductors, direct optical GR is typically the dominant mechanism, whereas trap-assisted GR is usually prevalent in indirect-gap materials [2]. In stationary conditions, and whenever the GR process is band-to-band (i.e., involves direct and instantaneous transitions between the valence and conduction bands), $U_n = U_p = U$. In time-varying conditions, $U_n \neq U_p$ if GR takes place through intermediate traps or recombination centers acting as electron or hole reservoirs; in this case, the DD model should be complemented by additional trap equations [41].

Following the seminal work of Blakemore and Landsberg [42,43], recombination mechanisms in semiconductors have been the object of a large body of theoretical and experimental studies. Restricting our attention to steady-state carrier transport in light emitters, the net recombination rate U in the bulk

---

[†] An alternative way of treating abrupt heterointerfaces involves the use of a thermionic boundary condition (BC) that constrains the current across the interface. This approach is limited to planar structures [40].

[‡] Transitions within the same band are not normally considered recombination events.

regions of the device includes contributions from the spontaneous emission (U^{sp}), Auger (U^A), and Shockley–Read–Hall (U^{SRH}) rates,

$$U^{sp} = B(np - \gamma_n\gamma_p n_i^2)$$

$$U^{SRH} = \frac{np - \gamma_n\gamma_p n_i^2}{\tau_p(n + \gamma_n n_t) + \tau_n(p + \gamma_p p_t)}$$

$$U^A = (C_n n + C_p p)(np - \gamma_n\gamma_p n_i^2),$$

where $\gamma_{n,p}$ has been introduced to account for the effect of Fermi statistics as in Equation 2.23, B is the radiative recombination coefficient (in units of cm^3/s), n_i is the intrinsic carrier density of the material, $n_t = n_i \exp(E_{trap}/k_B T)$, $p_t = n_i \exp(-E_{trap}/k_B T)$, τ_n and τ_p are the average carrier lifetimes, E_{trap} is the difference between the defect level and the intrinsic level, C_n and C_p are the Auger coefficients (in units of cm^6/s). The electron- or hole-assisted recombination rate (Auger recombination) is proportional to $p^2 n$ or pn^2, implying proportionality not only with respect to the colliding populations (electrons and holes) but also to the population of the energy suppliers. Due to this dependence, Auger recombination is important in high-injection devices, where it can be an unwelcome competitor of radiative recombination.[†]

2.2.5.2 Boundary Conditions and Numerical Solution Schemes

All the physical models described earlier must be complemented with a suitable set of BCs. A first kind of BC concerns ideal insulating boundaries, for which the normal component of the current density is zero. This condition corresponds to zero normal derivative (also called homogeneous Neumann BC) for the potential and the carrier densities. Metallic contacts fall into two main classes: ohmic and rectifying (Schottky) contacts. In both cases, metallic contacts are assumed to be made of ideal conductors, i.e., they are equipotential. Ohmic contacts ideally behave as short circuits, since they do not support any potential drop, while rectifying Schottky contacts exhibit a diode-like behavior. For ohmic contacts, the local electron and hole densities assume the equilibrium values corresponding to the local doping, yielding a Dirichlet BC (i.e., the unknown is assigned). If the contact is controlled by an ideal voltage source, a Dirichlet BC also arises for the potential. The treatment of Schottky contacts is more involved, since several approximations can be exploited. A fairly accurate set of BCs can be derived from the Sze–Bethe thermionic-drift diffusion model [44].

The numerical solution of Equation 2.21 requires discretization in space and time. Spatial discretization reduces the physical model to a set of ordinary time-domain differential equations; the need to perform a time-domain solution depends on the regime where the device operates. In static DC conditions time derivatives can be directly set to zero, and the resulting system of nonlinear equations can be solved through the Newton method and its variants (e.g., the Newton–Richardson approach [45]). Small-signal AC excitation can be dealt with by direct linearization and transformation into the frequency domain [41]. Large-signal analysis may also be carried out in the frequency domain by means of the harmonic balance method [46]. Conventional finite-difference (FD) or finite-element (FE) techniques are ill-suited to the spatial discretization of conservation equations, including both advection (drift) and diffusion terms. Advances in the development of stable numerical schemes for semiconductor device simulation were made

---

[†] The inverse process of Auger recombination is generation by impact ionization. In high-field conditions, electrons and holes gather enough energy from the electric field between two successive scattering events (e.g., collisions with phonons or impurities) to be able to interact with another electron (hole) and promote it to the conduction (valence) band. Each electron or hole is therefore able to generate, over a certain length, a number of electron–hole pairs, that undergo in turn the same process. The resulting chain can lead to a divergent current, i.e., to avalanche breakdown in the semiconductor. Avalanche breakdown occurs for electric fields of the order of the breakdown field, and is usually an undesirable effect, with the notable exception of avalanche photodetectors [17].

in 1969, when Scharfetter and Gummel (SG) proposed a discretization scheme for the 1D current continuity equation [47], which was later extended to two-dimensional (2D) and 3D geometries by means of the generalized finite-box (FB) approach, applicable to arbitrary Delaunay grids and to axisymmetric geometries [45,48–52]. The SG–FB scheme is exploited today by most device simulators. Inspired by computational fluid dynamics [53], upwind FE schemes based on a Petrov–Galerkin method with asymmetric weight functions (equivalent to a Galerkin approach with the explicit addition of numerical diffusivity) were also proposed for the spatial discretization of the HD equations [28].

2.3 Quantum-Corrected Drift-Diffusion

For illustrative purposes, we find it convenient to focus the foregoing discussion on VCSELs. The main feature of these devices is the radiation normal to their layers, which eases fiber coupling owing to the circular emitting area. Figure 2.2a shows a typical 850-nm VCSEL with a one-λ cavity containing the quantum wells (QWs). An oxide aperture just above the QWs confines the current (see Figure 2.2b), and improves optical guiding at low bias voltages. The short gain path of VCSELs demands highly reflective (>99%) distributed Bragg reflectors (DBRs) to achieve lasing. In fact, most of the VCSEL volume is taken up by the DBRs, which include several pairs of high and low refractive index layers with $\lambda/4$ thicknesses [54]. From one perspective, VCSELs represent a paradigm of sorts in the modeling of optoelectronic devices, as their simulation requires a comprehensive multiphysics and multiscale approach to account for electrical, optical, and thermal aspects [80]. In this complex scenario, the only viable approach to carrier transport modeling seems to be a semiclassical one, complemented with appropriate models connecting the microscopic scale governed by coherent phenomena with the macroscopic scale governed by incoherent processes. These models, often referred to as "quantum corrections," account for a variety of nonclassical effects, such as quantum confinement, (quasi-)ballistic transport, tunneling, and excitonic effects in optical spectra. The most accurate approach to introducing quantum corrections in a semiclassical device simulator is to complement the carrier transport model with a Poisson–Schrödinger solver (other approaches based, e.g., on the density gradient formalism account for nonlocality effects but they are limited to dispersions with one subband only [55]).

Multiband $k \cdot p$ envelope-function models represent the traditional approach to compute subband dispersions in optoelectronics, as they represent a fair compromise between accuracy and computational cost. The unknown nanostructure wavefunction is expanded in terms of zone-center Bloch bands $u_m(\underline{\rho}, z)$

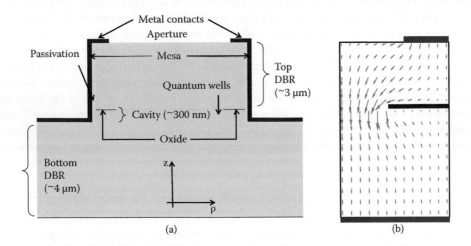

(a)   (b)

FIGURE 2.2 (a) Vertical cavity surface emitting laser (VCSEL) structure with semiconductor distributed Bragg reflectors. (b) The simulated current distribution profile.

(usually defined at the center of Brillouin zone) of the underlying bulk solids. An extremely concise representation of the nanostructure excitation spectrum is obtained by selecting a reduced set of M bulk states of interest, typically the three top valence bands and the first conduction band (inclusion of spin–orbit coupling doubles M to eight), and by removing the remaining remote bands by means of Luttinger–Kohn perturbation theory. Arresting the perturbative expansion to second-order terms, and writing the bulk Hamiltonian in a form that explicitly shows its wave vector dependence, we have

$$[H_{k \cdot p}] = \sum_{\alpha,\beta=x,y,z} [H^{(2)}_{\alpha\beta}]k_\alpha k_\beta + \sum_{\alpha=x,y,z} [H^{(1)}_\alpha]k_\alpha + [H^{(0)}], \qquad (2.26)$$

where the expressions of the matrices above depend on crystal symmetry and on the zone-center functions employed in the $k \cdot p$ representation [56–58]. Second-order terms $H^{(2)}_{ij}$ include free carrier dispersion and the perturbation treatment of remote states in perturbation theory, first-order terms $H^{(1)}_i$ stem from the direct treatment of the $k \cdot p$ interaction and spin–orbit coupling, and the zeroth–order term $H^{(0)}$ contains zone-center energies.

We assume the following *ansatz* for the wavefunction of a planar quantum well nanostructure,

$$\psi(\rho, z) = \frac{1}{\sqrt{\mathcal{A}}} e^{i k \cdot \rho} \sum_{m=1}^{M} u_{m0}(\rho, z)\zeta_{mk}(z), \qquad (2.27)$$

where the coordinates z and ρ are the symmetry broken and the translational invariant directions, respectively, $\mathcal{L}, \mathcal{A}$ their corresponding normalization volumes ($\mathcal{V} = \mathcal{A}\mathcal{L}$), k is the transversal crystal momentum, and $u_{m0}(\rho, z)$ are $k \cdot p$ orthonormal lattice-periodic functions rapidly changing over the crystal primitive cell. The slowly varying envelopes $\zeta_{mk}(z)$ describe, at every position in the symmetry broken direction z, how the lattice-periodic functions are mixed together. Upon substitution of the wave vector component k_z along the quantized direction with the corresponding operator $-i\partial_z$, one obtains a coupled equation system for the nanostructure envelopes $\{F(z)\} = \{\zeta_1(z), \zeta_2(z), ..., \zeta_M(z)\}^T$,

$$\left(-\partial_z[H^{(2)}(k, z)]\partial_z - [H^{(1)}_L(k, z)]i\partial_z - i\partial_z[H^{(1)}_R(k, z)] + [H^{(0)}(k, z)] \right) \{F(z)\} = E\{F(z)\}, \qquad (2.28)$$

parametrized for the wave vector k in the translational invariant space. Notice that $H^{(1)}$ has been split into two terms depending on the position of the operator ∂_z (terms of the type $\partial_z^2 H^{(2)}$ and $H^{(2)}\partial_z^2$ are usually neglected in $k \cdot p$ theories). Having discretized the device region into a number of elements, where the envelopes are represented with Lagrange polynomials $\zeta_m(z) = \sum_j \zeta_{mj}N_j(z)$, application of Galerkin's procedure leads to the weak form of Equation 2.28

$$\sum_j \int_e dz \left(N^z_i[H^{(2)}]N^z_j - iN_i[H^{(1)}_L]N^z_j + iN^z_i[H^{(1)}_R]N_j + N_i[H^{(0)}]N_j \right) \{F_j\}$$

$$= E \sum_j \int_e dz\, N_iN_j \{F_j\}, \qquad (2.29)$$

with $N^z_i = \partial_z N_i$. Boundary terms, arising when $[H^{(2)}]$ and $[H^{(1)}_R]$ are integrated by parts, vanish by the assumed homogeneous Dirichlet or Neumann BCs, since we are interested in the bound states of closed systems.[†] Notice that integration by parts is also an elegant way to *naturally* treat material

---

[†] The integration-by-parts boundary terms neglected in Equation 2.29 play an important role in the calculation of current-carrying scattering states of open systems, as they describe how the device interacts with the semi-infinite leads; we return to this point in Section 2.4 with the derivation of the boundary self-energy of the contacts.

discontinuities [59]. Assembling all elements, i.e., summing over e, one obtains the eigenmatrix equation

$$[H(\underline{k})]\{F\} = E[M]\{F\}, \tag{2.30}$$

where $[H]$ and $[M]$ are $MN \times MN$ sparse FE matrices, and the MN column vector $\{F\}$ is the FE representation of the nanostructure envelopes. Multiband $k \cdot p$ models are extremely concise and require only a few physically sensitive band parameters, which are usually obtained from a combination of experiments and first-principles electronic structure calculations.[†]

The Schrödinger equation should be reconciled with the solution of the Poisson equation to account for space charge effects. A simple iterative approach that computes the quantum charge from the Schrödinger equation and inserts it in the RHS of the Poisson equation, the solution of which updates in turn the Hartree potential, may be unstable due to the high sensitivity of the energy levels on the confining potential. On the other hand, an analytic expression of the Jacobian for a Newton updating scheme cannot be derived. An elegant and stable solution is provided by the predictor-corrector algorithm proposed in [65]. The idea is to define an approximate expression for the quantized charge as a function of the confining potential, so that most of the nonlinearities of the Schrödinger equation can be moved into the Poisson equation, which can be efficiently solved in an inner Newton loop. Stability is achieved by correcting the predicted results in an outer iteration loop by the exact solution of the Schrödinger's equation [65].

2.3.1 Capture and Escape Processes

Given our inability to quantize the whole VCSEL cross section, we must face the problem of letting bulk-like, 3D carriers coexist with 2D carriers populating the VCSEL active region, and connect them by means of appropriate scattering mechanisms. The choice of the energetic boundary E_u^0 between quantum and bulk regions is somewhat arbitrary [66]. A possible choice is to define the boundary by replacing the band edge of the spatial quantum region with the band edge of the barrier region in the equations for the unbound density, which leads to

$$n^{3D}(\underline{\rho}, z) = \int_{E_u^0}^{\infty} g_{3D}(E) f_{FD}(E, E_{F,n}^{3D}) \, dE \tag{2.31}$$

$$n^{2D}(\underline{\rho}, z) = \sum_i |\zeta_i(z)|^2 \int_{E_i}^{E_u^0} g_{2D}(E) f_{FD}(E, E_{F,n}^{2D}) \, dE, \tag{2.32}$$

where $g_{3D}(E)$, $g_{2D}(E)$ are the density of states of the bulk population and the ith quantized level with the envelope function $\zeta_i(z)$, respectively, and $f_{FD}(E, E_F) = 1 / \left[1 + \exp\left(\frac{E - E_F}{k_B T}\right)\right]$ is the Fermi–Dirac distribution. Having separated the two populations, the electron continuity equation in Equation 2.21 in

---

[†] The apparent simplicity of $k \cdot p$ models should not mislead the reader into thinking that first-principles/experimental information of bulk properties can be seamlessly combined to derive nanostructure properties. As it happens, band parameters are often derived from the fitting of bulk bands (sometimes with targets selected from different sources), disregarding the non-commutability of potential and position operators, and, most importantly, the ordering of the wave number operators, which ultimately means parameter splitting [58]. As terms of the type $Hk_i k_j$ with $i \neq j$ in Equation 2.26 do not have a well-defined operator ordering, they are usually split symmetrically to $-\frac{1}{2}(\partial_i H \partial_j + \partial_j H \partial_i)$ [59]. What makes parameter splitting subtle is that it is inconsequential in the calculation of bulk material dispersions, but it becomes relevant when the sub-band dispersion of a nanostructure is considered. Symmetrized operator ordering and/or an ill-advised choice of the band parameters may lead to significant deviations of subband dispersions, incorrect matrix elements, or even to the appearance of spurious solutions [60,61]. A multiband envelope-function model for wurtzite nanostructures based on a rigorous numerical procedure to determine operator ordering and band parameters from full-Brillouin-zone nonlocal empirical pseudopotential calculations is presented in [58]. For a discussion about the validity of the envelope-function approximation, see [62,63]. Atomistic models beyond the envelope-function approximation are reviewed in [64].

the active region of the VCSEL is replaced with (a similar equation holds for holes) [67]

$$\nabla \cdot \underline{J}^{n} = q \left(U^{SRH} + U^{sp} + U^{cap} + \frac{\partial n_{3D}}{\partial t} \right) \tag{2.33}$$

$$\nabla \cdot \underline{J}^{n}_{2D} = q \left(R^{nr} + R^{sp} + R^{st} - U^{cap} + \frac{\partial n_{2D}}{\partial t} \right), \tag{2.34}$$

where the stimulated and spontaneous radiative rates R^{st}, R^{sp} are defined by the optical response of the QWs (see Section 2.3.2), and R^{nr} is the nonradiative (SRH and Auger) recombination rate in the active region. The two populations are coupled by a capture rate U^{cap} that acts as a drain for the 3D charges and a source for the 2D carriers. Equation 2.34 describes the lateral currents (in the unconfined directions) of the 2D gas, whose density is determined by the localized wavefunctions given by the self-consistent solution of the Poisson–Schrödinger problem. From a physical standpoint, a capture process may be understood as a carrier propagating above a QW emitting phonons and finally relaxing into a bound state of the well.[†] Once in the bound state, the carrier may either absorb phonons and be excited back into the continuum or recombine by any of the radiative or dark recombination mechanisms. If interband scattering processes are slow enough compared to intraband processes in both the continuum and the bound states, the spectral distribution of the carriers may be approximated by Fermi distributions defined by quasi-Fermi levels.[‡] From the DD modeling standpoint, the capture of a carrier from the bulk to the quantum region may be seen as a recombination mechanism for the unbound population and as a generation mechanism for the bound population, and vice versa for an escape process. A net capture rate for electrons may be derived in a fashion similar to SRH recombination theory [68] (a similar expression holds for holes)

$$U^{cap} = \left(1 - \exp \left(\frac{E_{F,n}^{2D} - E_{F,n}^{3D}}{k_{B}T} \right) \right) \left(1 - \frac{n^{2D}}{N_{2}} \right) \frac{n^{3D}}{\tau_{scat,n}}, \tag{2.35}$$

where the parameter $N_{2} = \sum_{i} \int_{E_{i}}^{E^{0}} g_{2D}(E) \, dE$ is the sum of all QW states per unit volume, i.e., the maximum bound electron charge. The net capture rate is, therefore, proportional to the unbound carrier density, decreases with the filling of the bound states, increases with the quasi-Fermi level separation, and vanishes in the case of a completely filled bound population (phase space filling) or equal quasi-Fermi levels (thermodynamic equilibrium). In principle, the scattering rate should also depend on the distribution of the initial states of the injected carriers, but in a DD framework void of any information concerning the spectral distribution of the carriers, the complex physics of scattering can only be lumped in one average scattering time $\tau_{scat,n}$, which is usually treated as a fitting/phenomenological parameter. Relating these scattering times to structural parameters of the nanostructure is essential to understand some critical aspects related to vertical carrier transport in high-injection conditions, e.g., the possible leakage of hot carriers above the barriers, or to predict the modulation bandwidth of lasers [71,72].

2.3.2 Coupling Carrier Transport Models with Optical Solvers

From the electromagnetic standpoint, a VCSEL is an open resonator whose resonance frequency is determined by the cavity length and the mirrors design. Lasing action is achieved when the gain of the active

---

[†] NEGF simulations suggest that the process is more likely to consist of tunneling to a bound state above the barriers (i.e., with a high transverse kinetic energy) followed by its subsequent relaxation assisted by phonons [69].

[‡] This assumption breaks down at high injections, if carrier–carrier collisions are not fast enough to replenish the spectral hole burned by the state-selective stimulated emission, leading to deviations of the carriers' spectral densities from their quasiequilibrium Fermi–Dirac distributions [70].

material compensates the losses of the open structure. This condition defines the threshold gain, which, together with the operation wavelength and the light pattern emitted by the VCSEL, is the output of the optical solver. Among full-wave optical solvers (beyond scalar-based approaches, e.g., on the effective index method [73]), one may distinguish fully numerical schemes from semianalytical techniques that exploit the physics/geometry of the problem. A remarkable example of the former class is the frequency-domain FE solver LUMI (that uses a 2D body-of-revolution expansion for axisymmetric devices and perfectly matched layers BCs), also included in the Synopsys Sentaurus Device simulator [36,67]. An example belonging to the latter class is the method of lines (MoL), which is based on a spatial discretization in the radial direction and on the transmission line formalism in the longitudinal direction [74]. Other noteworthy examples of physics-based techniques rely on plane-wave expansions, where, as in MoL, the field is propagated along the longitudinal direction by admittance [75,76] or reflection matrix [77] formulations.

The authors of this chapter have proposed a 3D vectorial model of VCSELs [78–80] based on the expansion of the electromagnetic field in terms of the complete basis of the TE and TM modes of the cavity medium in cylindrical coordinates. All deviations with respect to the reference unperturbed structure, both in the transverse and longitudinal directions, are accounted by coupled-mode theory. By cascading the transmission matrices of each section of the device (evaluated by an exponential matrix when the section is longitudinally invariant), one can compute the overall transmission matrix between the two radiating sections, beyond which the device can be considered homogeneous. Resolution of the problem is obtained by introducing appropriate BCs, relating backward to forward waves at these two sections. Since the condition for an electromagnetic field to be a mode of the structure is to replicate itself after a full cavity round-trip, one finally obtains an eigenvalue problem, whose complex eigenvalues represent the modal wavelengths and the corresponding threshold gains (real and imaginary parts, respectively), and whose eigenvectors are the expansion coefficients of the field in the section where the round-trip condition is enforced.

In order to extend the semiconductor equations to model laser diodes, we still need to describe the time evolution of the electromagnetic energy in each lasing mode. The rate equation for the output stimulated power $P_{\mathrm{st},m}$ in the mth mode can be derived from quantum theory of semiconductor lasers assuming quasineutrality ($n \approx p$) and adiabatic elimination of the microscopic polarization [81]

$$\Gamma_z v_{\mathrm{ph}} \left(G_m - G_{\mathrm{th},m} \right) P_{\mathrm{st},m} + \Gamma_z \frac{\hbar\omega_m}{\tau_{\mathrm{out}}} R_{\mathrm{sp},m} = \frac{\partial P_{\mathrm{st},m}}{\partial t}. \tag{2.36}$$

The transverse modal gain G_m and the spontaneous emission rate $R_{\mathrm{sp},m}$ are

$$G_m = \int g(\underline{\rho}, \omega_m)\mathcal{E}_m^2(\underline{\rho})\,\mathrm{d}\underline{\rho} \tag{2.37}$$

$$R_{\mathrm{sp},m} = \int r_{\mathrm{sp}}(\underline{\rho}, \omega_m)\mathcal{E}_m^2(\underline{\rho})\,\mathrm{d}\underline{\rho}, \tag{2.38}$$

where $g(\underline{\rho}, \omega_m)$, $r_{\mathrm{sp}}(\underline{\rho}, \omega_m)$ are the local material gain and spontaneous emission at the lasing frequency ω_m; the longitudinal confinement factor Γ_z, the threshold gain $G_{\mathrm{th},m}$, and the photon escape time τ_{out} (s) are provided by the optical solver; $\mathcal{E}_m(\underline{\rho})$ is the mth mode field amplitude, at the QW section, whose intensity is normalized to unity. The conversion of the active region 2D carriers to coherent and incoherent photons is included in the DD transport model by the recombination terms

$$R^{\mathrm{st}}(\underline{\rho}) = \sum_m \frac{D_m}{\hbar\omega_m} g(\underline{\rho}, \omega_m)\mathcal{E}_m^2(\underline{\rho})P_{\mathrm{st},m} \tag{2.39}$$

$$R^{\mathrm{sp}}(\underline{\rho}) = \int r_{\mathrm{sp}}(\underline{\rho}, \omega)\rho^{\mathrm{opt}}(\omega)\mathrm{d}\omega, \tag{2.40}$$

where the sum is over all lasing modes, $D_m^{-1} = \int \mathcal{E}_m^2(\rho)\,d\rho\big|_{\text{output}}$ links the field intensity in the active region to the output one radiated by the VCSEL [80], and $\rho^{\text{opt}}(\omega) = (n_b^2 \omega^2)/(\pi^2 \hbar c^3)$ is the photon density of states. The requirement that the solutions of the electromagnetic problem match the optical response of the active region provides a self-consistent mechanism for the selection of the lasing modes.

2.3.2.1 The Semiconductor Bloch Equations

The starting point for the calculation of the optical response of the active region is the single-particle Hamiltonian of the nanostructure describing the free carriers interacting with a classical field $\mathcal{E}(t)$

$$H = H_0 + H_{cf} + H_{cc} \tag{2.41}$$

with the following expressions for the free-carrier H_0, the carrier-field H_{cf}, and the carrier–carrier H_{cc} contributions

$$H_0 = \sum_A E_A a_A^\dagger a_A \tag{2.42}$$

$$H_{cf} = -\sum_{A,B} \mu_{A,B} \mathcal{E}(t)\, a_A^\dagger a_B \tag{2.43}$$

$$H_{cc} = \frac{1}{2} \sum_{A,B,C,D} V_{C\,D}^{A\,B}\, a_A^\dagger a_B^\dagger a_D a_C \tag{2.44}$$

where E_A is the single-particle energy, and the a and $a^\dagger$ are the annihilation and creation operators, compound indices A, B, C, D include the transverse electron momentum $\underline{k}$ and the band index c, v. In the approximation of slowly varying envelopes, the optical dipole matrix element $\mu_{A,B}$ is related to the matrix element of the momentum operator $\hat{p}$ [59]

$$\mu_{AB} \approx -\frac{q\mathcal{A}}{i m_0 \omega} \sum_{ij} \langle u_{i0}|\hat{p}|u_{j0}\rangle \int_{\mathcal{L}} \zeta_{Ai}^*(z)\zeta_{Bj}\,dz. \tag{2.45}$$

The unscreened Coulomb matrix element in Equation 2.44 is

$$V_{C\,D}^{A\,B} = \frac{q^2}{4\pi\varepsilon_0} \iint_V \psi_A^*(\underline{\rho}, z)\psi_B^*(\underline{\rho}', z') \frac{1}{|(\underline{\rho}, z) - (\underline{\rho}', z')|} \psi_D(\underline{\rho}', z)\psi_C(\underline{\rho}, z)\, d\underline{\rho}\, d\underline{\rho}'dz\, dz'. \tag{2.46}$$

Substituting the ansatz Equation 2.27 and changing the integration over $\underline{\rho}'$ into the integration over the distance $\underline{s} = \underline{\rho} - \underline{\rho}'$, we have [59]

$$V_{C\,D}^{A\,B} = \frac{q^2}{4\pi\varepsilon_0} \iint_{\mathcal{L}} dz\,dz' \zeta_A^*(z)\zeta_B^*(z')\zeta_D(z')\zeta_C(z)$$

$$\times \iint_A \frac{1}{\sqrt{s^2 + (z - z')^2}} e^{i(\underline{k}_C + \underline{k}_D - \underline{k}_A - \underline{k}_B)\cdot\underline{\rho}}\ e^{i(\underline{k}_B - \underline{k}_D)\cdot\underline{s}}\, d\underline{\rho}\, d\underline{s} \tag{2.47}$$

The integration over $\underline{\rho}$ yields the momentum conserving function $\delta_{\underline{k}_A + \underline{k}_B, \underline{k}_C + \underline{k}_D}$, reflecting the in-plane translational invariance of the system. Replacing $1/\sqrt{s^2 + (z - z')^2}$ with its Fourier expansion $\sum_{\underline{q}} \frac{2\pi}{Aq} e^{-q|z-z'|} e^{i\underline{q}\cdot\underline{s}}$, we have

$$V_{C\,D}^{A\,B} = \frac{q^2}{4\pi\varepsilon_0} \iint_{\mathcal{L}} dz\,dz' \zeta_A^*(z)\zeta_B^*(z')\zeta_D(z')\zeta_C(z) \int_A \sum_q \frac{2\pi}{Aq} e^{-q|z-z'|} e^{i\underline{q}\cdot\underline{s}}\ e^{i(\underline{k}_2 - \underline{k}_3)\cdot\underline{s}}\, d\underline{s}. \tag{2.48}$$

The integral over $\underline{s}$ yields $\mathcal{A}\delta_{\underline{q},\underline{k}_B-\underline{k}_D}$ and we are left with

$$V_{C\;D}^{A\;B} = \frac{q^2}{2\varepsilon_s}\frac{1}{q}\iint_{\mathcal{L}}\zeta_A^*(z)\zeta_B^*(z')e^{-q|z-z'|}\zeta_D(z')\zeta_C(z)\,\mathrm{d}z\,\mathrm{d}z'. \tag{2.49}$$

Having defined the two compound indices $1 = (\underline{k}, v)$ and $2 = (\underline{k}, c)$, the dynamics of the microscopic interband polarization $p_{\underline{k}} = \langle a_1^\dagger a_2\rangle$ may be obtained from the Heisenberg equation of motion

$$i\hbar\frac{\mathrm{d}p_{\underline{k}}}{\mathrm{d}t} = [p_{\underline{k}}, H], \tag{2.50}$$

which gives for the different Hamiltonian components [82]

$$\frac{\mathrm{d}}{\mathrm{d}t}p_{\underline{k}}|_{H_0} = \frac{i}{\hbar}\left(E_1 - E_2\right)\langle a_1^\dagger a_2\rangle \tag{2.51}$$

$$\frac{\mathrm{d}}{\mathrm{d}t}p_{\underline{k}}|_{H_{cf}} = -\frac{i}{\hbar}\mu_{\underline{k}}[\langle a_2^\dagger a_2\rangle - \langle a_1^\dagger a_1\rangle]\mathcal{E}(t) \tag{2.52}$$

$$\frac{\mathrm{d}}{\mathrm{d}t}p_{\underline{k}}|_{H_{cc}} = +\frac{i}{2\hbar}\sum_{A,B,D}V_{1\;D}^{A\;B}\langle a_A^\dagger a_B^\dagger a_D a_2\rangle - \frac{i}{2\hbar}\sum_{A,B,C}V_{C\;1}^{A\;B}\langle a_A^\dagger a_B^\dagger a_C a_2\rangle$$

$$+\frac{i}{2\hbar}\sum_{A,C,D}V_{C\;D}^{A\;2}\langle a_1^\dagger a_A^\dagger a_D a_C\rangle - \frac{i}{2\hbar}\sum_{B,C,D}V_{C\;D}^{2\;B}\langle a_1^\dagger a_B^\dagger a_D a_C\rangle \tag{2.53}$$

$$\frac{\mathrm{d}}{\mathrm{d}t}p_{\underline{k}}|_{H_{cp}} = -\frac{i}{2\hbar}\sum_{3u}\left[g_{23}\langle a_1^\dagger a_3 b_u\rangle - g_{31}\langle a_3^\dagger a_2 b_u\rangle + g_{32}^*\langle a_1^\dagger a_3 b_u^\dagger\rangle - g_{13}^*\langle a_3^\dagger a_2 b_u^\dagger\rangle\right]. \tag{2.54}$$

Renaming summation indices and exploiting the Coulomb matrix element symmetries ($V_{C\;D}^{A\;B} = V_{D\;C}^{B\;A}$ and $V_{C\;D}^{A\;B} = V_{A\;B}^{C\;D*}$), we obtain

$$\frac{\mathrm{d}}{\mathrm{d}t}p_{\underline{k}}|_{H_{cc}} = \frac{i}{\hbar}\sum_{A,B,D}V_{1\;D}^{A\;B}\langle a_A^\dagger a_B^\dagger a_D a_2\rangle - \frac{i}{\hbar}\sum_{A,B,D}V_{B\;D}^{2\;A}\langle a_1^\dagger a_A^\dagger a_D a_B\rangle. \tag{2.55}$$

Because of the many-particle interaction, the equation above is not closed. The dynamics of the single-particle quantity $\sigma_{AB} = \langle a_A^\dagger a_B\rangle$ couples to two-particle contributions $C_{CD}^{AB} = \langle a_A^\dagger a_B^\dagger a_C a_D\rangle$. Applying the correlation expansion

$$C_{CD}^{AB} = \sigma_{AD}\sigma_{BC} - \sigma_{AC}\sigma_{BD} + \tilde{C}_{CD}^{AB}, \tag{2.56}$$

where $\tilde{C}_{CD}^{AB} = \langle a_A^\dagger a_B^\dagger a_C a_D\rangle^c$, the first-order Hartree–Fock (HF) contribution and the second-order correlation terms are separated:

$$\frac{\mathrm{d}}{\mathrm{d}t}p_{\underline{k}}|_{H_{cc}^{HF}} = \frac{i}{\hbar}\sum_{A,B,D}V_{1\;D}^{A\;B}[\langle a_A^\dagger a_2\rangle\langle a_B^\dagger a_D\rangle - \langle a_A^\dagger a_D\rangle\langle a_B^\dagger a_2\rangle]$$

$$-\frac{i}{\hbar}\sum_{A,B,D}V_{B\;D}^{2\;A}[\langle a_1^\dagger a_B\rangle\langle a_A^\dagger a_D\rangle - \langle a_1^\dagger a_D\rangle\langle a_A^\dagger a_B\rangle] \tag{2.57}$$

$$\frac{\mathrm{d}}{\mathrm{d}t}p_{\underline{k}}|_{H_{cc}^{corr}} = \frac{i}{\hbar}\sum_{A,B,D}V_{1\;D}^{A\;B}\tilde{C}_{D\;2}^{A\;B} - \frac{i}{\hbar}\sum_{A,B,D}V_{B\;D}^{2\;A}\tilde{C}_{D\;B}^{1\;A}. \tag{2.58}$$

In a spatially homogeneous system only DM elements that are diagonal with respect to the electronic wave vector, have a nonvanishing expectation value. Therefore, within the HF approximation, the equation of motion for the microscopic polarization $p_{\underline{k}}$ is

$$\frac{d}{dt}p_{\underline{k}}\big|_{H_{cc}^{HF}} = -\frac{i}{\hbar}\sum_{k'}[(V_{vk\ vk'}^{vk\ vk'} - V_{ck\ vk'}^{ck\ vk'})p_{\underline{k}}n_{k'}^{v} + (V_{vk\ ck'}^{vk\ ck'} - V_{ck\ ck'}^{ck\ ck'})p_{\underline{k}}n_{k'}^{c}$$

$$-\left(V_{vk'\ ck}^{ck\ vk'} - V_{vk\ vk'}^{vk'\ vk}\right)p_{\underline{k}}n_{k'}^{v} - \left(V_{ck'\ ck}^{ck\ ck'} - V_{vk\ vk'}^{ck'\ vk}\right)p_{\underline{k}}n_{k'}^{v}$$

$$-\left(V_{vk\ vk'}^{ck\ ck'} - V_{vk'\ vk}^{ck\ ck'}\right)p_{k'}^{*}(n_{k}^{c} - \rho_{k}^{v})$$

$$-\left(V_{vk\ ck'}^{ck\ vk'} - V_{ck'\ vk}^{ck\ vk'}\right)p_{k'}(n_{k}^{c} - n_{k}^{v})\,]. \tag{2.59}$$

The first line cancels for symmetry arguments. Terms that are nonlinear in the microscopic polarization have been omitted. Neglecting also terms proportional to $p_{k'}^{*} = \langle a_{2}^{\dagger}a_{1}\rangle$ involving transitions between different bands, we are left with

$$\frac{d}{dt}p_{\underline{k}}\big|_{H_{cc}^{HF}} = -\frac{i}{\hbar}\sum_{k'}\left[\left(V_{vk\ vk'}^{vk'\ vk}n_{k'}^{v} - V_{ck\ ck'}^{ck\ ck'}n_{k'}^{c}\right)p_{\underline{k}} + V_{ck'\ vk}^{ck\ vk'}p_{k'}(n_{k}^{c} - n_{k}^{v})\right]. \tag{2.60}$$

Moving to the electron–hole picture, we find the familiar semiconductor Bloch equations in the HF approximation

$$\frac{d}{dt}p_{\underline{k}}\big|_{HF} = -i\tilde{\omega}_{k}p_{k} - \gamma p_{\underline{k}} - \frac{i}{\hbar}\left\{\mathcal{E}(t)\mu_{\underline{k}} + \sum_{k'}V_{ck'\ vk}^{ck\ vk'}p_{k'}\right\}(n_{k}^{c} + n_{k}^{h} - 1), \tag{2.61}$$

where we have defined

$$\hbar\tilde{\omega}_{k} = (\varepsilon_{ck} + \varepsilon_{hk}) - \sum_{k'}\{V_{ck'\ ck}^{ck\ ck'}n_{k'}^{c} + V_{vk\ vk'}^{vk'\ vk}n_{k'}^{h}\} \tag{2.62}$$

$$\varepsilon_{hk} = -\left(\varepsilon_{vk} - \sum_{k'}V_{vk\ vk'}^{vk'\ vk}\right). \tag{2.63}$$

The second term in the RHS of Equation 2.61 is added to account for polarization decay due to carrier–carrier or carrier–phonon scattering. Formally integrating Equation 2.61 and screening the Coulomb matrix elements with the Lindhard formula, one obtains the material gain and the spontaneous emission spectra [83]

$$g(\omega) = -\frac{2\omega}{\epsilon_{0}n_{b}c\mathcal{V}\mathcal{E}}\,\text{Im}\left(\sum_{\underline{k}}(\mu_{\underline{k}})^{*}p_{\underline{k}}e^{i\omega t}\right) \tag{2.64}$$

$$r_{\text{sp}}(\omega) = \frac{1}{\hbar}\left(\frac{n_{b}\omega}{\pi c}\right)^{2}g(\omega)\left[\exp\left(\frac{\hbar\omega - (E_{F,n} - E_{F,p})}{k_{B}T}\right)\right], \tag{2.65}$$

where $\mathcal{E}$ and ω are the amplitude and frequency of the optical wave, ϵ_{0} and c are the permittivity and the speed of light in vacuum, and n_{b} is the background refractive index.

2.4　Towards a Genuine Quantum Approach

High-field nonstationary effects, phase coherence, tunneling, and other quantum features of electronic transport may have barely found mention in a handbook of optoelectronic devices, were it not for the recent developments in novel material systems, highly nanostructured to offer unprecedented performance. As characteristic length scales become smaller than the scattering mean free path, the physics of transport changes drastically from an incoherent, semiclassical picture governed by the BTE to a coherent picture described in a purely quantum mechanical framework in terms of currents associated with probability fluxes, usually injected from idealized reservoirs of carriers. Electronic transport in nanostructures is somewhere between these two limits, in a regime where quantum mechanical phase-coherent phenomena and phase-breaking and dissipative scattering processes coexist.

On the one hand, the immediate point of quantum-corrected semiclassical approaches is obvious, since, although technologically relevant optoelectronic devices do have a nanostructured active region, they really are 3D macroscopic structures that typically do not lend themselves to a fully quantum kinetic treatment. The staggering computational cost of quantum kinetic approaches becomes apparent if one considers, e.g., carrier transport across the DBR stacks of a VCSEL. More generally, a nanostructure is always a 3D system, its dimensionality referring to the free-motion reciprocal subspace only. On the other hand, no quantum correction complementing the DD equations has ever been truly validated, the underlying semiclassical framework being intrinsically local in space and void of any information about the energy distributions of the carriers.

A corollary to the above considerations is that, besides the questionable appropriateness of quantum corrections within a semiclassical framework (the complex physics described by quantum kinetic theories can hardly fit material-agnostic expressions), we are not even confident about some recombination and high-field transport properties of confined systems. For example, carrier leakage and Auger recombination imply the presence of nonequilibrium hot electrons, a concept that represents a major departure from conventional treatments based on DD approaches, which are implicitly based on the effective mass approximation. A legitimate question is therefore to what extent quantum-corrected DD models can be trusted, and what information/conclusions can be drawn from them.

Since all that may sound too abstract, let us consider a concrete example that also happens to divide the optoelectronic community, attracting journalistic attention. There is just no way that the droop debate, a lively controversy about the elusive nature of the nonthermal decline of the internal quantum efficiency in GaN-based visible light-emitting diodes (LEDs) at high current densities [84–86], could have taken place, had transport and recombination properties in nanostructured materials been precisely known.[†] Although most of the recent contributions on droop belong to the camps of proponents either of Auger recombination or of carrier leakage, alternative explanations have also been formulated, the lack of direct experimental information on critical quantities and the ambiguity of the parameters involved in ABC(D) recombination models allowing almost any interpretation. In other words, the debate clearly demonstrates that reproducing droop with a semiclassical model and an appropriate set of fitting parameters does not *eo ipso* imply any model validation.

"Fitting" is, of course, the really sensitive term here. Part of why fitting is so unwelcome in the modeling community is that we are just starting to realize how much we are still forced to rely on untested models,

---

[†] The discussion concerning the loss mechanisms in visible LEDs involves nearly every aspect of carrier transport, including carrier leakage, carrier delocalization, compositional alloy fluctuations, and Auger and defect-assisted recombination. Auger recombination was first proposed as the origin of droop on the basis of the Auger-like density-dependence of the observed losses [87], a proposal supported by first-principles theoretical calculations [88–90] and by recent experiments [91–93]. The main contra-Auger arguments are the ubiquitous use by the industry of electron-blocking layers to prevent electron escape from the active region of UV and visible LEDs, which suggests electron leakage as an important droop-inducing mechanism (see [85,94] and references therein), and the anomalous temperature dependence that the putative Auger coefficients should have to *fit* the efficiency curve [95].

which is threatening to the very concept of numerical simulation as a predictive tool for device design and optimization. Gradually emerging is the awareness that the LED droop problem is too complex and entangled for overly simplified finger-pointing (e.g., an Auger versus leakage dichotomy is simplistic and wrong, because Auger recombination would promote carriers at energies well beyond the heterobarriers and would therefore provide an important contribution to leakage) and that several axiomatic assumptions of electronic transport in optoelectronic devices could lead us to wrong conclusions. The relevant axioms here are the separation of bound and unbound carriers, usually forced by defining an arbitrary energetic boundary, the assumption that carrier confinement is strong enough so that a fully coherent quantum-mechanical description is appropriate for confined carriers, and an incoherent description may be adopted for unbound carriers.

Moving from quantum-corrected semiclassical models to genuine quantum-kinetic descriptions, one has to face the fact that two closely related aspects of the problem, namely carrier transport and optical generation, are traditionally considered from different perspectives. In the emerging field of semiconductor quantum optics, where the focus is on ultrafast transients, the DM approach is usually applied to analyze the correlations' dynamics in the coupled quantized light-semiconductor system[96], while the nonequilibrium Green's function (NEGF) approach [97] is generally considered more suited to carrier transport in the steady-state regime. Although the two formalisms may be shown to be equivalent in the equal-time limit, it is difficult to establish a connection between NEGF and DM in the general case (a discussion on this interesting topic may be found in Refs. [98,99]); and in fact it is not even possible to compare their accuracy, as the strategy to abridge the infinite hierarchy of equations arising in the many-body problem is profoundly different in the two cases (the cluster expansion method in DM theories [96], the summation over some classes of diagrams in NEGF formulations [100]). NEGF and DM are presented in the literature in different contexts probably because the one-time DM approach is admittedly more transparent for time-domain problems, while the two-times NEGF method provides a more device-oriented viewpoint of carrier transport, which seems more appropriate to the analysis of devices seen as open systems, i.e., as systems that exchange current with reservoirs.

2.4.1 Contour-Ordered NEGF

The central idea is that the analysis of the motion of strongly interacting carriers in nanostructured materials does not necessarily require detailed information about each particle in the system, i.e., the many-body wavefunction (that would be a formidable problem), but rather the *average* behavior of zero-, one- or more-particle assemblies, which may be obtained from field theoretical quantities such as no-particle (vacuum amplitude), one-particle, two-particle, and more-particles Green's functions, also known as "propagators." Examples of one-particle quantities of interest are, e.g., the local density of states (the excitation spectrum of the nanostructure), the spectral charge density, which describes how these states are occupied, and the spectral current density, which describes whether the propagating carriers emit/lose energy or tunnel through the structure. But the exact evaluation of the relevant propagators is as difficult as the solution of the many-body Schrödinger's equation. A systematic way to approximate propagators is based on the perturbative expansion of the Green's function written in the interaction picture [101] in powers of the perturbation (interacting) Hamiltonian. As propagators obey wave equations with a singular inhomogeneity, i.e., a delta function in the variables (see the following discussion), they may be considered a generalized version of classical Green's functions.

NEGF theory was developed as an extension of zero- and finite-temperature Green's function techniques, probably inspired by their common diagrammatic formulation. The zero-temperature formalism is based on the Gell–Mann and Low theorem that relates the (unknown) interacting ground state of the system to the noninteracting state defined at $t \to -\infty$, before the interactions are adiabatically switched on. The time evolution operator that generates the series evolves on the real-time axis from $t \to -\infty$ to $t \to +\infty$, when the system returns to the noninteracting ground state [101]. In the finite-temperature formalism, introduced by Matsubara (1955), the ground state average appearing in the definition of the

Green's function is replaced by an average over a grand canonical ensemble, as the system is statistically distributed over all its excited levels. Interpreting the Boltzmann weight factor that appears in the definition of the statistical average as the evolution of the system for imaginary times, the resulting Green's function can be expanded diagrammatically exactly as in the zero temperature case, but with time integrations restricted to a vertical path in the complex plane [101–103]. Although well suited for the description of reservoirs responsible for injecting carriers in a device, Matsubara Green's functions are not applicable to carrier transport, since a system driven out of equilibrium by an external (possibly time-dependent) field may not return to its initial state for asymptotically large times, i.e., the so-called "vacuum stability condition," a crucial assumption in equilibrium Green's function (EGF) theory, does not hold. A general NEGF formalism (beyond linear response theory) was established independently[†] by Kadanoff and Baym [104] (1962) and Keldysh [105] (1965) under the influence of Schwinger's school [106]. The expectation value of a given observable was reinterpreted as the system evolution on a contour along the real-time axis running from the remote past ($t = -\infty$), when the system was at equilibrium, to the highest relevant time of the evaluated Green's function, and back to the remote past. The round-trip or contour approach was the key to avoid the ill-defined state at $t = +\infty$, leading to contour-ordered Green's functions that have the usual diagrammatic expansion [107].

We summarize here only the relevant equations of the NEGF formulation, restricting our attention to steady-state conditions, and assuming that the reader is familiar with the occupation number formalism (second quantization),[‡] our objective being to give a glimpse of the general assumptions underlying NEGF, and, whenever possible, to make connections with the drift-diffusion model presented in Section 2.2.

Consider a system evolving under the Hamiltonian operator $\hat{\mathcal{H}} = \hat{H} + \hat{H}^{\text{ext}}(t)$. The time-independent Hamiltonian $\hat{H} = \hat{H}_0 + \hat{H}^i$ describes the isolated system, where $\hat{H}_0$ is the exactly solvable part that contains the lattice and the electrostatic potential, and $\hat{H}^i$ is the complicated part that contains the many-body aspects of the problem, e.g., carrier–carrier and carrier–phonon interactions. $\hat{H}^{\text{ext}}(t)$ is the explicitly time-dependent perturbation that drives the system out of equilibrium. The perturbation $\hat{H}^{\text{ext}}(t)$ is applied at times $t > t_0$, and could be, e.g., an electric field, a light excitation pulse, or an imbalance of the chemical potentials of the reservoirs coupled to the system.

The task is to calculate the expectation value of a given observable, to which one associates a quantum mechanical operator $\hat{O}$, for times $t \geq t_0$. The nonequilibrium ensemble average with respect to the grand-canonical ensemble is defined as $\langle \hat{O}(t) \rangle = \text{Tr}[\rho(\hat{H})\hat{O}_{\mathcal{H}}(t)]$. The subscript $\mathcal{H}$ indicates that the time-dependence is governed by the full Hamiltonian, i.e., $\hat{O}$ is written in the Heisenberg picture. The thermal equilibrium DM $\rho(\hat{H}) = \mathcal{Z}^{-1}e^{-\beta\hat{H}}$ describes the system before the perturbation is turned on. Here $\mathcal{Z} = \text{Tr}(e^{-\beta\hat{H}})$ is the grand partition function, $\beta = 1/k_B T$ is the inverse temperature measured in energy units, and Tr stands for trace. We can simplify the time dependence of any operator moving from the Heisenberg picture to the interaction picture with respect to the noninteracting Hamiltonian $\hat{H}_0$

$$\hat{O}_{\mathcal{H}}(t) = \hat{S}^\dagger(t, t_0)\hat{O}_{H_0}(t)\hat{S}(t, t_0). \tag{2.66}$$

[†] A brief scientific/historical review of early Russian works on the Green's functions applications to many-body theory is presented in [108].

[‡] We refer the reader to the excellent textbooks by Fetter and Walecka [102], Mahan [101], and Mattuck [103] for a detailed description of EGF theory. The NEGF formalism is described in [104,107,109–111]. Although EGF and NEGF theories are closely related, they are usually presented as separate topics, following the chronological/historical order as they originally appeared in the literature. A unified (nonchronological) description of EGF and NEGF formalisms and of their connection is presented by Stefanucci [111] and Rammer [110]. A notable attempt to make the formalism more accessible to readers not familiar with advanced many-body perturbation theory may be found in Datta [112].

The S-matrix operator evolving the interaction-picture wavefunction from t_0 to t (t_0 being the time instant at which the Heisenberg and interaction representations coincide) can be formally written as

$$\hat{S}(t, t_0) = \hat{T}\left\{ \exp\left[-\frac{\mathrm{i}}{\hbar} \int_{t_0}^{t} \mathrm{d}\tau \left(\hat{H}_{H_0}^i(\tau) + \hat{H}_{H_0}^{ext}(\tau) \right) \right] \right\}, \tag{2.67}$$

with $\hat{H}_{H_0}^{i,ext} = e^{\mathrm{i}\hat{H}_0(t-t_0)} \hat{H}^{i,ext} e^{-\mathrm{i}\hat{H}_0(t-t_0)}$. When applied to a product of operators, the chronological time-ordering operator $\hat{T}$ orders them with the earliest times appearing on the right and the latest on the left. Considering that operators commute under operator ordering, we may also write

$$\hat{S}(t, t_0) = \hat{T}\left\{ \exp\left[-\frac{\mathrm{i}}{\hbar} \int_{t_0}^{t} \mathrm{d}\tau \hat{H}_{H_0}^i(\tau) \right] \exp\left[-\frac{\mathrm{i}}{\hbar} \int_{t_0}^{t} \mathrm{d}\tau \hat{H}_{H_0}^{ext}(\tau) \right] \right\}. \tag{2.68}$$

It is understood that the exponential expressions in Equation 2.68 are a shorthand for series of time-ordered products of operators

$$\hat{T}\left\{ \exp\left[-\frac{\mathrm{i}}{\hbar} \int_{t_0}^{t} \mathrm{d}\tau \hat{H}_{H_0}^i(\tau) \right] \right\}$$
$$= \sum_{n=0}^{\infty} \left(\frac{-\mathrm{i}}{\hbar} \right)^n \frac{1}{n!} \int_{t_0}^{t} \mathrm{d}t_1 \int_{t_0}^{t} \mathrm{d}t_2 \cdots \int_{t_0}^{t} \mathrm{d}t_n \, \hat{T}\left\{ \hat{H}_{H_0}^i(t_1) \hat{H}_{H_0}^i(t_2) \cdots \hat{H}_{H_0}^i(t_n) \right\}, \tag{2.69}$$

which represents the starting point of the many-body perturbation expansion. In Equation 2.69, the term $n = 0$ is the unit operator and the $\hat{T}$ operator indicates that the products of the Hamiltonians at different times in the series expansion that defines the exponential must be kept in the correct chronological order (see [101, p. 67]). It is convenient to merge the two transformation integrals in (2.66) into the contour integral

$$\hat{O}_H(t) = \hat{T}_C\left\{ \exp\left[-\frac{\mathrm{i}}{\hbar} \int_C \mathrm{d}\tau \hat{H}_{H_0}^i(\tau) \right] \exp\left[-\frac{\mathrm{i}}{\hbar} \int_C \mathrm{d}\tau \hat{H}_{H_0}^{ext}(\tau) \right] \hat{O}_{H_0}(t) \right\}, \tag{2.70}$$

where the time-ordering operator $\hat{T}_C$ orders the operators along the contour C of Figure 2.3, running from t_0 to t (chronological branch) and then back to t_0 (antichronological branch), so that operators with time labels that occur later on the contour stand to the left of operators that occur earlier; in other words, $\hat{T}_C$ reduces to the chronological or antichronological ordering operator for the part of the contour running

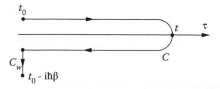

FIGURE 2.3 The contour $\tilde{C} = C \cup C_w$ for the transformation to the interaction picture where the time evolution is governed by the noninteracting Hamiltonian H_0. The contour C stretches from t_0 to t and then back to t_0. The small shift from the real axis is only meant to distinguish the two branches and to clarify the integration direction. The appendix C_w stretching into the lower complex half-plane from t_0 to $t_0 - \mathrm{i}\hbar\beta$ is required when initial correlations are of interest; in steady-state conditions, we can neglect the contribution of the imaginary appendix C_w by taking the limit $t_0 \to -\infty$. (From J. Rammer., *Quantum Field Theory of Non-equilibrium States*, Cambridge University Press, Cambridge, 2007.)

forward or backward in time, respectively. The nonequilibrium statistical average of an observable $\hat{O}_H(t)$ may now be written as

$$
\langle \hat{O}_H(t) \rangle = \mathrm{Tr}(\rho(\hat{H}) \hat{O}_H(t))
$$

$$
= \frac{\mathrm{Tr}\left(e^{-\beta\hat{H}} \hat{T}_C \left\{ \exp\left[-\frac{i}{\hbar} \int_C d\tau \hat{H}_{H_0}^i(\tau) \right] \exp\left[-\frac{i}{\hbar} \int_C d\tau \hat{H}_{H_0}^{\mathrm{ext}}(\tau) \right] \hat{O}_{H_0}(t) \right\} \right)}{\mathrm{Tr}(e^{-\beta\hat{H}})}. \tag{2.71}
$$

The Boltzmann weight factor can be expressed in terms of $\hat{H}_0$ by means of an imaginary time-evolution operator

$$
e^{-\beta\hat{H}} = e^{-\beta\hat{H}_0} \hat{T}_{C_w} \left\{ \exp\left[-\frac{i}{\hbar} \int_{t_0}^{t_0 - i\hbar\beta} d\tau \hat{H}_{H_0}^i(\tau) \right] \right\}, \tag{2.72}
$$

where $\hat{T}_{C_w}$ orders operators along the vertical track stretching into the lower complex half-plane from t_0 to $t_0 - i\hbar\beta$; see Figure 2.3. Upon substitution of Equation 2.72 in Equation 2.71, $\hat{T}_{C_w}$ and $\hat{T}_C$ can be merged into one contour operator $\hat{T}_{\tilde{C}}$ defined on $\tilde{C} = C \cup C_w$, which gives

$$
\langle \hat{O}_H(t) \rangle = \frac{\mathrm{Tr}\left(e^{-\beta\hat{H}_0} \hat{T}_{\tilde{C}} \left\{ \exp\left[-\frac{i}{\hbar} \int_{\tilde{C}} d\tau \hat{H}_{H_0}^i(\tau) \right] \exp\left[-\frac{i}{\hbar} \int_C d\tau \hat{H}_{H_0}^{\mathrm{ext}}(\tau) \right] \right\} \hat{O}_{H_0}(t) \right)}{\mathrm{Tr}\left(e^{-\beta\hat{H}_0} \hat{T}_{\tilde{C}} \left\{ \exp\left[-\frac{i}{\hbar} \int_{\tilde{C}} d\tau \hat{H}_{H_0}^i(\tau) \right] \exp\left[-\frac{i}{\hbar} \int_C d\tau \hat{H}_{H_0}^{\mathrm{ext}}(\tau) \right] \right\} \right)} \tag{2.73}
$$

$$
= \frac{\langle \hat{T}_{\tilde{C}} \left\{ \exp\left[-\frac{i}{\hbar} \int_{\tilde{C}} d\tau \hat{H}_{H_0}^i(\tau) \right] \exp\left[-\frac{i}{\hbar} \int_C d\tau \hat{H}_{H_0}^{\mathrm{ext}}(\tau) \right] \hat{O}_{H_0}(t) \right\} \rangle_0}{\langle \hat{T}_{\tilde{C}} \left\{ \exp\left[-\frac{i}{\hbar} \int_{\tilde{C}} d\tau \hat{H}_{H_0}^i(\tau) \right] \exp\left[-\frac{i}{\hbar} \int_C d\tau \hat{H}_{H_0}^{\mathrm{ext}}(\tau) \right] \right\} \rangle_0}, \tag{2.74}
$$

where all the time dependence is governed by the exactly solvable Hamiltonian $\hat{H}_0$, and $\langle ... \rangle_0$ stands for statistical average with respect to the noninteracting DM $\hat{\rho}_0$

$$
\langle ... \rangle_0 = \frac{\mathrm{Tr}\left(\exp(-\beta\hat{H}_0)... \right)}{\mathrm{Tr}\left(\exp(-\beta\hat{H}_0) \right)}. \tag{2.75}
$$

In the denominator of Equation 2.73, we have introduced the identity evolution operator defined along a closed contour that connects the initial state to itself (no operators acting at intermediate times)

$$
\hat{T}_C \left\{ \exp\left[-\frac{i}{\hbar} \int_C d\tau (\hat{H}_{H_0}^i(\tau) + \hat{H}_{H_0}^{\mathrm{ext}}(\tau)) \right] \right\} = \hat{1}, \tag{2.76}
$$

as required by combinatorial/diagrammatic arguments to cancel the disconnected diagrams of the numerator. Notice that the external perturbation is zero on the imaginary part of contour C_w. There is no restriction concerning the choice of t_0. If we are not interested in transient phenomena, we can assume that all transients have died out by letting $t_0 \to -\infty$, and the contribution from the imaginary part of the contour C_w vanishes, meaning that $\tilde{C} = C$. Moreover, we can extend the contour to $t = \infty$, since, beyond the largest time, the forward and backward evolutions simply produce the identity operator. We then obtain the Schwinger–Keldysh contour, which starts at $t = -\infty$ and proceeds to $t = \infty$ and then back again to $t = -\infty$; see Figure 2.4.

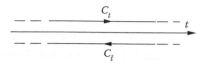

FIGURE 2.4 The Schwinger–Keldysh contour $C = C_t \cup C_{\bar{t}}$. The small shift from the real axis of the chronological and antichronological branches $C_t = (-\infty, \infty)$ and $C_{\bar{t}} = (\infty, -\infty)$ is only meant to distinguish the integration direction.

The previous considerations lead us to the analysis of the single-particle NEGF defined as the nonequilibrium ensemble average of contour-ordered field operators:

$$
\begin{aligned}
G(\underline{r}, t; \underline{r}', t') &= -\frac{i}{\hbar} \langle \hat{T}_C \{ \hat{\psi}_H(\underline{r}, t) \hat{\psi}_H^\dagger(\underline{r}', t') \} \rangle \\
&= -\frac{i}{\hbar} \left(\theta_C(t, t') \langle \hat{\psi}_H(\underline{r}, t) \hat{\psi}_H^\dagger(\underline{r}', t') \rangle \mp \theta_C(t', t) \langle \hat{\psi}_H^\dagger(\underline{r}', t') \hat{\psi}_H(\underline{r}, t) \rangle \right).
\end{aligned}
\tag{2.77}
$$

The function $\theta_C(t, t')$ is the Heaviside function defined on the contour, with $\theta_C(t, t') = 1$, if t is later on C than t', and zero otherwise. The upper (lower) sign in $\mp$ is for fermions (bosons); from now on, we restrict our attention to fermionic systems. The Heisenberg field operator $\hat{\psi}_H^\dagger(\underline{r}', t')$ ($\hat{\psi}_H(\underline{r}, t)$) creates (annihilates) a particle at position $\underline{r}'$ ($\underline{r}$) and at time t' (t). The NEGF $G(\underline{r}, t; \underline{r}', t')$ expresses a correlation between two times t and t' and two positions $\underline{r}$ and $\underline{r}'$. For t later on C than t', $G(\underline{r}, t; \underline{r}', t')$ can be interpreted as the reaction of a system to a particle created at position $\underline{r}'$ and time t', to the propagation of this perturbation to position $\underline{r}$ and time t, and finally to its annihilation there. In other words, we probe the interacting system by adding one particle, which propagates and interacts with the other particles until we remove it from the system. A similar interpretation holds if t' is later on C than t. Applying Equation 2.74 to the average in Equation 2.77 provides the starting point of the perturbation expansion of the NEGF

$$
G(\underline{r}, t; \underline{r}', t') = \frac{\langle \hat{T}_{\tilde{C}} \left\{ \exp\left[-\frac{i}{\hbar} \int_{\tilde{C}} d\tau \hat{H}_{H_0}^i(\tau)\right] \exp\left[-\frac{i}{\hbar} \int_C d\tau \hat{H}_{H_0}^{\text{ext}}(\tau)\right] \hat{\psi}_{H_0}(\underline{r}, t) \hat{\psi}_{H_0}^\dagger(\underline{r}', t') \right\} \rangle_0}{\langle \hat{T}_{\tilde{C}} \left\{ \exp\left[-\frac{i}{\hbar} \int_{\tilde{C}} d\tau \hat{H}_{H_0}^i(\tau)\right] \exp\left[-\frac{i}{\hbar} \int_C d\tau \hat{H}_{H_0}^{\text{ext}}(\tau)\right] \right\} \rangle_0}.
\tag{2.78}
$$

The following discussion is based on the assumption that initial correlations are absent, in other words, the time evolution is on the Schwinger–Keldysh contour C. The time evolution of $G(\underline{r}, t; \underline{r}', t')$ with respect to the times t and t' is obtained from the corresponding equations of motion in the Heisenberg picture. A Hamiltonian operator that includes a carrier–carrier interaction $\hat{H}^i = \hat{V}$, and a single-particle potential $\hat{H}^{\text{ext}} = \hat{U}$, reads in second quantization

$$
\begin{aligned}
\hat{H}(t) = {} & \int d\underline{r}\, \hat{\psi}_H^\dagger(\underline{r}, t)[H_0(\underline{r}) + U(\underline{r}, t)] \hat{\psi}_H(\underline{r}, t) \\
& + \frac{1}{2} \int d\underline{r} \int d\underline{r}'\, \hat{\psi}_H^\dagger(\underline{r}, t) \hat{\psi}_H^\dagger(\underline{r}', t) V(\underline{r} - \underline{r}') \hat{\psi}_H(\underline{r}', t) \hat{\psi}_H(\underline{r}, t).
\end{aligned}
\tag{2.79}
$$

The equation of motion describing the time evolution of the NEGF can be determined from the derivative of the contour-ordered pair of Heisenberg field operators

$$
\begin{aligned}
\frac{\partial}{\partial t} \langle \hat{T}_C \{ \hat{\psi}_H(\underline{r}, t) \hat{\psi}_H^\dagger(\underline{r}', t') \} \rangle &= \left\langle \hat{T}_C \left\{ \left[\frac{\partial}{\partial t} \hat{\psi}_H(\underline{r}, t) \right] \hat{\psi}_H^\dagger(\underline{r}', t') \right\} \right\rangle + \delta_C(t, t') \left[\hat{\psi}_H(\underline{r}, t) \hat{\psi}_H^\dagger(\underline{r}', t') \right]_+ \\
&= \left\langle \hat{T}_C \left\{ \left[\frac{\partial}{\partial t} \hat{\psi}_H(\underline{r}, t) \right] \hat{\psi}_H^\dagger(\underline{r}', t') \right\} \right\rangle + \delta_C(t, t') \delta(\underline{r} - \underline{r}').
\end{aligned}
\tag{2.80}
$$

In the last line of Equation 2.80 the equal-time anticommutation property $\left[\hat{\psi}_H(\underline{r}, t)\hat{\psi}_H^\dagger(\underline{r}', t)\right]_+ = \delta(\underline{r} - \underline{r}')$ was used. To evaluate the first RHS term in Equation 2.80, it is necessary to know the time evolution of the annihilation operator, whose time derivative can be calculated from the corresponding Heisenberg equation of motion

$$i\hbar\frac{\partial}{\partial t}\hat{\psi}_H(\underline{r}, t) = \left[\hat{\psi}_H(\underline{r}, t), \hat{H}(t)\right]_-$$

$$= [H_0(\underline{r}) + U(\underline{r}, t)]\hat{\psi}_H(\underline{r}, t) + \int d\underline{r}' V(\underline{r} - \underline{r}')\hat{\psi}_H^\dagger(\underline{r}', t)\hat{\psi}_H(\underline{r}', t)\hat{\psi}_H(\underline{r}, t). \tag{2.81}$$

Replacing in Equation 2.80 the time-derivative Equation 2.81 gives the equation of motion of $G(\underline{r}_1, t_1; \underline{r}_{1'}, t_{1'}) = G(11')$ relative to t_1 (a similar procedure gives the derivative relative to $t_{1'}$)

$$\left[i\hbar\frac{\partial}{\partial t_1} - H_0(\underline{r}_1) - U(1)\right] G(11') = \delta(11') - i\hbar\int_C d2 V(1 - 2)G^{(2)}(121'2^+)$$

$$\left[-i\hbar\frac{\partial}{\partial t_{1'}} - H_0(\underline{r}_{1'}) - U(1')\right] G(11') = \delta(11') - i\hbar\int_C d2 V(1' - 2)G^{(2)}(12^-1'2), \tag{2.82}$$

where the two-particle Green's function is defined as

$$G^{(2)}(121'2') = \left(\frac{-i}{\hbar}\right)^2 \left\langle \hat{T}_C\{\hat{\psi}_H(1)\hat{\psi}_H(2)\hat{\psi}_H^\dagger(2')\hat{\psi}_H^\dagger(1')\} \right\rangle. \tag{2.83}$$

The notation $2^\pm$ in Equation 2.82 is intended as a remainder that the time argument of $\hat{\psi}_H^\dagger(2)$ must be chosen to be infinitesimally larger or smaller than t_2 [104]. In writing Equation 2.82, we have used the shorthand notation $\int_C d1 = \int_C dt_1 \int d\underline{r}_1$ and the definitions

$$\delta(12) = \delta_C(t_1, t_2)\delta(\underline{r}_1 - \underline{r}_2) \tag{2.84}$$

$$V(1 - 2) = V(\underline{r}_1 - \underline{r}_2)\delta(t_1 - t_2). \tag{2.85}$$

Equation 2.82 is not closed because the interaction couples the one-particle GF to the two-particle GF, etc., leading to an infinite hierarchy of coupled equations involving GFs of ever-increasing order. Rather than adopting the hierarchical approach employed in Section 2.2 for the solution of the BTE, we can formally decouple the Martin–Schwinger hierarchy by approximating the two-particle Green's function by means of the self-energy $\Sigma(11')$ encoding interaction effects of all particles on the single-particle dynamics

$$\left[i\hbar\frac{\partial}{\partial t_1} - H_0(\underline{r}_1) - U(1)\right] G(11') = \delta(11') + \int_C d3\Sigma(13)G(31') \tag{2.86}$$

$$\left[-i\hbar\frac{\partial}{\partial t_{1'}} - H_0(\underline{r}_{1'}) - U(1')\right] G(11') = \delta(11') + \int_C d3 G(13)\Sigma(31'). \tag{2.87}$$

The self-energy may be obtained to different levels of approximations (HF, self-consistent Born approximation [SCBA], T-matrix, etc.) using the Feynman diagrammatic expansion or functional derivatives. Considering that the noninteracting GF $G_0(11')$ is the solution of Equation 2.82 without the two-particle GF

$$G_0(11') = \left[i\hbar\frac{\partial}{\partial t_1} - H_0(\underline{r}_1) - U(1)\right]^{-1} \delta(11'), \tag{2.88}$$

and integrating over one of the times to eliminate the δ-functions, one obtains the two equivalent forms of the Dyson's equations in integral form

$$G(11') = G_0(11') + \int_C d2 \int_C d3 \, G_0(12)\Sigma(23)G(31') \tag{2.89}$$

$$G(11') = G_0(11') + \int_C d2 \int_C d3 \, G(12)\Sigma(23)G_0(31'). \tag{2.90}$$

The equations of motion for the NEGF $G(11')$ include integrals over a complex time contour $\int_C$. Since it is cumbersome to keep track of the time branch in the evaluation of contour integrals, four new Green's functions with real-time arguments are introduced,

$$G(11') = \begin{cases} G^t(11') & t_1, t_{1'} \in C_t \\ G^{\bar{t}}(11') & t_1, t_{1'} \in C_{\bar{t}} \\ G^<(11') & t_1 \in C_t, t_{1'} \in C_{\bar{t}} \\ G^>(11') & t_1 \in C_{\bar{t}}, t_{1'} \in C_t, \end{cases} \tag{2.91}$$

which are named the chronological, antichronological, lesser and greater Green's functions, respectively, and are defined by

$$G^t(11') = -\frac{i}{\hbar}\langle \hat{T}_t\{\hat{\psi}_\mathcal{H}(1)\hat{\psi}_\mathcal{H}^\dagger(1')\}\rangle \tag{2.92}$$

$$G^{\bar{t}}(11') = -\frac{i}{\hbar}\langle \hat{T}_{\bar{t}}\{\hat{\psi}_\mathcal{H}(1)\hat{\psi}_\mathcal{H}^\dagger(1')\}\rangle \tag{2.93}$$

$$G^<(11') = +\frac{i}{\hbar}\langle \hat{\psi}_\mathcal{H}^\dagger(1')\hat{\psi}_\mathcal{H}(1)\rangle \tag{2.94}$$

$$G^>(11') = -\frac{i}{\hbar}\langle \hat{\psi}_\mathcal{H}(1)\hat{\psi}_\mathcal{H}^\dagger(1')\rangle. \tag{2.95}$$

The chronological and antichronological Green's functions are usually replaced by the retarded and advanced Green's functions defined by

$$G^R(11') = -\frac{i}{\hbar}\theta(t - t')\langle[\hat{\psi}_\mathcal{H}(1), \hat{\psi}_\mathcal{H}^\dagger(1')]_+\rangle = \theta(t - t')[G^>(11') - G^<(11')] \tag{2.96}$$

$$G^A(11') = \frac{i}{\hbar}\theta(t' - t)\langle[\hat{\psi}_\mathcal{H}(1), \hat{\psi}_\mathcal{H}^\dagger(1')]_+\rangle = -\theta(t' - t)[G^>(11') - G^<(11')]. \tag{2.97}$$

Subtracting the last two equations, we have

$$G^R(11') - G^A(11') = G^>(11') - G^<(11'), \tag{2.98}$$

which shows that only three of the Green's functions are linearly independent. The contour integrations in Equation 2.90 can be decomposed into real-time integrations by means of the Langreth rules [107]. Neglecting initial correlations, which are relevant only for the transient behavior, we obtain the equations of motion of the real-time Green's functions

$$G^{R(A)}(11') = G_0^{R(A)}(11') + \int d2 \int d3 G_0^{R(A)}(12)\Sigma^{R(A)}(23)G^{R(A)}(31') \tag{2.99}$$

$$G^{\lessgtr}(11') = \int d2 \int d3 G^R(12)\Sigma^{\lessgtr}(23)G^A(31'), \tag{2.100}$$

with $\int d1 = \int d\underline{r}_1 \int_{-\infty}^{+\infty} dt_1$. For steady-state calculations, only the time difference $\tau = t' - t$ is meaningful. Fourier transforming $O^\alpha = G^\alpha, \Sigma^\alpha$ ($\alpha = R, A, <, >$) to energy coordinates

$$O^\alpha(\underline{r}_1, \underline{r}_2, E) = \int_{-\infty}^{+\infty} d\tau e^{\frac{i}{\hbar}E\tau} O^\alpha(\underline{r}_1, \underline{r}_2, \tau) \tag{2.101}$$

$$O^\alpha(\underline{r}_1, \underline{r}_2, \tau) = \frac{1}{2\pi\hbar} \int_{-\infty}^{+\infty} dE e^{-\frac{i}{\hbar}E\tau} O^\alpha(\underline{r}_1, \underline{r}_2, E), \tag{2.102}$$

we may rewrite Equations 2.99 and 2.100 as

$$G^{R(A)}(\underline{r}_1, \underline{r}_{1'}, E) = G_0^{R(A)}(\underline{r}_1, \underline{r}_{1'}, E)$$

$$+ \int d\underline{r}_2 \int d\underline{r}_3 G_0^{R(A)}(\underline{r}_1, \underline{r}_2, E) \Sigma^{R(A)}(\underline{r}_2, \underline{r}_3, E) G^{R(A)}(\underline{r}_3, \underline{r}_{1'}, E) \tag{2.103}$$

$$G^{\lessgtr}(\underline{r}_1, \underline{r}_{1'}, E) = \int d\underline{r}_2 \int d\underline{r}_3 G^R(\underline{r}_1, \underline{r}_2, E) \Sigma^{\lessgtr}(\underline{r}_2, \underline{r}_3, E) G^A(\underline{r}_3, \underline{r}_{1'}, E) \tag{2.104}$$

with the noninteracting Green function given by

$$G_0^{R(A)}(\underline{r}_1, \underline{r}_{1'}, E) = [E + (-)i\eta - H_0(\underline{r}_1)]^{-1} \delta(\underline{r}_1 - \underline{r}_{1'}). \tag{2.105}$$

The small $\eta \to 0^+$ parameter provides the correct analytical properties. Time-reversal symmetry implies $G^R(\underline{r}_1, \underline{r}_{1'}, E) = [G^A(\underline{r}_1, \underline{r}_{1'}, E)]^\dagger$, and $G^{\lessgtr}(\underline{r}_1, \underline{r}_{1'}, E) = -[G^{\lessgtr}(\underline{r}_1, \underline{r}_{1'}, E)]^\dagger$, which reduces the number of independent Green's functions to two.

The Green's functions are directly related to physical observables. In particular, the expressions for the electron and hole densities in steady state follow directly from the definition of the density operator [107]

$$n(\underline{r}) = -i \int \frac{dE}{2\pi} G^<(\underline{r}, \underline{r}, E) \tag{2.106}$$

$$p(\underline{r}) = i \int \frac{dE}{2\pi} G^>(\underline{r}, \underline{r}, E). \tag{2.107}$$

Taking the difference of the equations of motion (2.86) and (2.87), in the limit $1' \to 1$, gives

$$\underbrace{\lim_{1' \to 1} \left\{ i\hbar \left(\frac{\partial}{\partial t_1} + \frac{\partial}{\partial t_{1'}} \right) G(11') \right\}}_{\frac{\partial \rho}{\partial t}(\underline{r}, t)} - \underbrace{\lim_{1' \to 1} \left\{ [H_0(\underline{r}_1) - H_0(\underline{r}_{1'})] G(11') \right\}}_{\nabla \cdot J(\underline{r}, t)} =$$

$$\underbrace{\lim_{1' \to 1} \int d2 \, [\Sigma(12)G(21') - G(12)\Sigma(21')]}_{-U_\rho(\underline{r}, t)}, \tag{2.108}$$

which may be identified with the DD continuity equation 2.21. In steady state, Equation 2.108 reads [113]

$$\nabla \cdot \underline{J}(\underline{r}) = -\int \frac{dE}{2\pi\hbar} \int d\underline{r}' [\Sigma^R(\underline{r}, \underline{r}', E)G^<(\underline{r}', \underline{r}, E) + \Sigma^<(\underline{r}, \underline{r}', E)G^A(\underline{r}', \underline{r}, E)$$

$$-G^R(\underline{r}, \underline{r}', E)\Sigma^<(\underline{r}', \underline{r}, E) - G^<(\underline{r}, \underline{r}', E)\Sigma^A(\underline{r}', \underline{r}, E)]. \tag{2.109}$$

The RHS of Equation 2.109 should vanish when it is integrated over all bands, if the self-energies are properly chosen to ensure current conservation.[†] Restricting the integration to the conduction (valence) band, Equation 2.109 can be identified as the NEGF version of the semiclassical steady-state electron (hole) continuity equation, with the RHS of Equation 2.109 corresponding to the net electron (hole) generation rate. Because the self-energies are addictive, one can separate the recombination rates due to different mechanisms.

For the numerical evaluation of the Dyson's equations, we need to express the Green's functions defined on the continuous space variables $\underline{r}$ and $\underline{r}'$ in the discrete basis adopted in Section 2.3, which consists of a product of fast-varying (over the crystal primitive cell) zone-center Bloch functions u_{m0}, plane waves in the homogenous transverse direction $\underline{\rho}$, and shape functions N_i slowly varying in the crystal growth direction z

$$\Psi_{\alpha\underline{k}}(\underline{r}) = \frac{1}{\sqrt{\mathcal{A}}}N_i(z)e^{i\underline{k}\cdot\underline{\rho}}u_{m0}(\underline{r}), \tag{2.110}$$

where the compound index $\alpha = (i, m)$ combines indices i for space and m for band, and $\mathcal{A}$ is the normalization cross-section area. Expanding the field operators on this basis

$$\hat{\Psi}_{\mathcal{H}}(\underline{r}, t) = \sum_{\alpha,\underline{k}} \Psi_{\alpha\underline{k}}(\underline{r})\hat{c}_{\alpha,\underline{k}}(t) \tag{2.111}$$

$$\hat{\Psi}_{\mathcal{H}}^{\dagger}(\underline{r}, t) = \sum_{\alpha,\underline{k}} \Psi_{\alpha\underline{k}}^{*}(\underline{r})\hat{c}_{\alpha,\underline{k}}^{\dagger}(t) \tag{2.112}$$

leads to the contravariant and covariant matrix representations $O_{\alpha\beta}(\underline{k}, E)$, $\tilde{O}_{\alpha\beta}(\underline{k}, E)$ of the steady-state fields $O = G, \Sigma^{[‡]}$

$$O(\underline{r}, \underline{r}', E) = \frac{1}{\mathcal{A}} \sum_{\underline{k}} \sum_{\alpha\beta} \Psi_{\alpha\underline{k}}(\underline{r})O_{\alpha\beta}(\underline{k}, E)\Psi_{\beta\underline{k}}^{*}(\underline{r}') \tag{2.113}$$

$$\tilde{O}_{\alpha\beta}(\underline{k}, E) = \int d\underline{r}\, d\underline{r}'\, \Psi_{\alpha\underline{k}}^{*}(\underline{r})O(\underline{r}, \underline{r}', E)\Psi_{\beta\underline{k}}(\underline{r}'). \tag{2.114}$$

Using a contravariant representation for the Green's functions and a covariant representation for the self-energies, one obtains a coupled set of linear equations that in matrix form read [115, Chapter 3]

$$[G^R(\underline{k}, E)] = (E[M] - [H(\underline{k})] - [\Sigma^R(\underline{k}, E)])^{-1} \tag{2.115}$$

$$[G^A(\underline{k}, E)] = [G^R(\underline{k}, E)]^{\dagger} \tag{2.116}$$

$$[G^{\lessgtr}(\underline{k}, E)] = [G^R(\underline{k}, E)][\Sigma^{\lessgtr}(\underline{k}, E)][G^A(\underline{k}, E)], \tag{2.117}$$

with $M_{\alpha\beta} = \int d\underline{r}\Psi_{\alpha\underline{k}}^{*}(\underline{r})\Psi_{\beta\underline{k}}(\underline{r}) \approx M_{ij}\delta_{mn}$ [69].

[†] Typical approximations of the self-energy (e.g., HF, Born, T-matrix, and random-phase-approximation) ensure the conservation for particle number, momentum, and energy. Conservation laws may be violated when an arbitrary diagram is included in the self-energy [114], or when self-energies are interpolated on an energy grid [69].

[‡] The choice of an orthonormal basis suppresses the complication of having two different representations, since in that case contravariant and covariant components coincide.

2.4.2 Boundary Self-Energies

Having partitioned the nanostructure into device and reservoir regions, boundary self-energies accounting for the openness of the system may be evaluated by augmenting Dyson's equation with open BCs relating the normal derivative of the solution at the boundary to its boundary value [116]. Consider first the simple case of the effective mass approximation, which allows us to invert the energy dispersion relation analytically. As in Section 2.3, let us assume a 1D system, which occupies the simulation region along the z-axis with left and right contacts (or reservoirs) occupying the regions $z < z_1$ and $z > z_N$, respectively, defined as fully absorbing elements attempting to maintain thermal equilibrium and charge neutrality at the z_1 and z_N boundaries. Assuming that the electrostatic potential is constant in the semi-infinite leads, the wavefunction attempting to propagate (or decay) from the device to the leads can be written in terms of plane waves. For $z < z_1$, the outgoing wave is $\zeta_{out}(z) = A^- e^{ik_z^- z}$, where A^- is the unknown outgoing coefficient, $k_z^- = -\sqrt{2m^* E}/\hbar$ is the wave number associated with the (longitudinal) energy E, and m^* is the effective mass. Integrating by parts the second-order term in the Hamiltonian, Equation 2.26 gives the boundary term $-\frac{\hbar^2}{2m^*} \partial_z \zeta(z)|_{z=z_1}$. The continuity of the wavefunction and its derivative across the boundary gives the contribution $-\frac{\hbar^2}{2m^*} ik_z^-$ to the matrix element $[H]_{11}$, which is the only term that describes the coupling of the device to the contact, and therefore defines the upper-left corner of the otherwise zero boundary self-energy matrix of the left contact: $[\Sigma_{11}^{RB}(\underline{k}, E)] = -\frac{\hbar^2}{2m^*} ik_z^-$. Similarly, at the right reservoir, we express the outgoing wave as $\zeta_{out}(z) = A^+ e^{ik_z^+ z}$, propagating or decaying to the right at energy E and wave number $k^+ = \sqrt{2m^* E}/\hbar$, which gives the lower-right corner of the otherwise zero boundary self-energy matrix of the right contact: $[\Sigma_{NN}^{RB}(\underline{k}, E)] = -\frac{\hbar^2}{2m^*} ik_z^+$.

Moving from the effective mass approximation to a multiband $k \cdot p$ description of the electronic structure, the task of describing the coupling of the device to a semi-infinite domain on either side of the region of interest becomes a more complicated affair, as the outgoing waves may occur into any of the (propagating or evanescent) states k_z and bands n such that $E_n(\underline{k}, k_z) = E$. For example, the outgoing waves at the left reservoir are found by selecting only real k_z's such that the group velocity $v_g(k_z) < 0$ (reflected waves propagating to the left) and those complex k_z's for which $\text{Im}(k_z) < 0$ (reflected waves decaying to the left). A similar argument holds for the right contact. The problem of finding the component k_z of the wave vector along the confining direction having defined the kinetic energy and the transverse wave vector is what is known as the *complex band structure problem*. In the conventional band structure problem, one fixes $(\underline{k}, k_z)$ and finds possible values of E by solving the eigenvalue problem $H(\underline{k}, k_z)\psi = E\psi$. In the complex band structure problem, one fixes the in-plane wave vector $\underline{k}$ and the energy E and finds possible values for k_z.[†] Recalling the decomposition of the bulk Hamiltonian (Equation 2.26), this procedure leads to a quadratic eigenvalue problem that can be cast into a linear one [117],

$$\begin{bmatrix} 0 & \mathbb{1} \\ H^{(0)} - E\mathbb{1} & H^{(1)} \end{bmatrix} \begin{pmatrix} \psi \\ k_z \psi \end{pmatrix} = k_z \begin{bmatrix} \mathbb{1} & 0 \\ 0 & -H^{(2)} \end{bmatrix} \begin{pmatrix} \psi \\ k_z \psi \end{pmatrix}. \tag{2.118}$$

Complex eigenvalues occur in pairs, i.e., running (evanescent) states traverse the nanostructure with the same real (imaginary) part and opposite signs. If we classify the $2M$ solutions of the complex band structure problem at each lead according to their direction of motion and organize them in the columns of matrices $[\psi^\pm]$, we obtain a boundary self-energy matrix with all zero entries, save for the upper-left and

---

[†] The following discussion is valid whenever the Hamiltonian has a polynomial dependence on k_z as in Equation 2.26; it does not apply to nonlocal plane-wave pseudopotential Hamiltonians; see [118] for a more general approach to the calculation of *ab initio* complex band structures.

lower-right $M \times M$ blocks

$$[\Sigma_{11}^{RB}(\underline{k}, E)] = -[H^{(2)}][\psi^-][\text{diag}(ik_{z1}^-, ik_{z2}^-, ..., ik_{zM}^-)][\psi^-]^{-1} + [iH_L^{(1)}] \qquad (2.119a)$$

$$[\Sigma_{NN}^{RB}(\underline{k}, E)] = -[H^{(2)}][\psi^+][\text{diag}(ik_{z1}^+, ik_{z2}^+, ..., ik_{zM}^+)][\psi^+]^{-1} + [iH_L^{(1)}]. \qquad (2.119b)$$

The expressions for $[\Sigma^{B\lessgtr}(\underline{k}, E)]$ follow from the fluctuation–dissipation theorem [101]. Assuming that the left contact is at Fermi level E_{FL}, we have

$$[\Sigma_{11}^{<B}(\underline{k}, E)] = if_{\text{FD}}(E, E_{\text{FL}})[\Gamma_1^B(\underline{k}, E)] \qquad (2.120a)$$

$$[\Sigma_{11}^{>B}(\underline{k}, E)] = -i(1 - f_{\text{FD}}(E, E_{\text{FL}}))[\Gamma_1^B(\underline{k}, E)], \qquad (2.120b)$$

where $[\Gamma_1^B(\underline{k}, E)] = i([\Sigma_{11}^{RB}(\underline{k}, E)] - [\Sigma_{11}^{RB}(\underline{k}, E)]^{\dagger})$ is the broadening function that quantifies the level broadening in the open system due to the connection to the reservoir. Similar expressions hold for the right contact.

2.4.3 Scattering Self-Energies

While boundary self-energies can be evaluated exactly within an FE scheme, scattering self-energies have to be computed perturbatively. Most of the complexity of the NEGF formalism and accordingly most of the numerical burden originate from the modeling of scattering self-energies, and it is for this reason that these quantities are usually described in NEGF studies with phenomenological models (e.g., Büttiker probes, phase- and momentum-relaxing Golizadeh self-energies) or approximated as local in space[†] and/or decoupled from Green's functions. Rather than using simplified models, we adopt the rigorous approach based on the SCBA.

The starting point is the perturbative expansion of Green's function (Equation 2.78), which includes time-ordered products of interaction-picture field operators averaged over the initial equilibrium ensemble. The equilibrium statistical average of time-ordered products can be decomposed in equilibrium statistical averages of all the possible time-ordered pairs of the field operators (Wick's theorem); the resulting terms of the series, which contain the unperturbed Green's function (i.e., Green's function related to the unperturbed, exactly solvable, Hamiltonian), can be conveniently represented in a pictorial fashion by Feynman diagrams. Introduced by Richard P. Feynman around 1948 as a graphic shorthand in his study of quantum electrodynamics, these diagrams have become an essential tool for performing calculations in modern quantum field theory.

It helps to conceive of Feynman diagrams in terms of words. Although the single diagrams may be given only a quasiphysical[‡] interpretation, the whole perturbation series forms a sentence that does have a

---

[†] Most scattering potentials do not vanish for finite distances $z - z'$. In addition, Green's functions in weakly confined systems decay relatively slowly in space, so that the effective range of the scattering self-energies may be relatively large. Although unwarranted, the local approximation (i.e., spatially diagonal self-energies) seems the only viable approach to 3D NEGF simulation [122].

[‡] Because the single terms in the expansion represent virtual rather than real scattering processes, conservative textbooks warn the reader not to give the diagrams any physical interpretation, and recommend regarding them merely as *calculational* devices, i.e., as tools designed to keep track of complicated mathematical expressions arising in a perturbative calculation. Part of the success of Feynman diagrams over alternative diagrammatic methods is related to their similarity to photographs of *real* particle tracks, although this similarity was considered ambiguous, if not misleading, since it incorrectly suggested trajectories, in contrast to the fundamental principles of quantum mechanics [119]. Made of dots, lines, and wiggles, these drawings are so charmingly expressive that it is really tempting to consider them as *representational* devices. Acknowledging the "quasi-physical" nature of the diagrams is actually the attempt by some textbooks at a compromise between calculational

real physical meaning [103]. By partially summing over selected[†] sets of diagrams, one obtains a method that goes beyond ordinary perturbation theory, which is just as well, since most of the relevant diagrams represent divergent terms, and one has to sum them up to the infinite order to obtain a convergent answer. When performing sums over diagrams, one needs to consider only topologically distinct diagrams (with an appropriate multiplication factor). Moreover, the denominator in Equation 2.78 cancels the disconnected diagrams of the numerator (i.e., those containing subunits that are not connected to the rest of the diagram by any lines). Diagrammatically speaking, Green's function is just the sum of all topologically distinct and connected diagrams. Context is essential for the interpretation of Feynman's diagrammatic language. The translation of the semantic units of the diagrams (propagation lines, interaction lines, and vertices) into mathematical expressions requires different dictionaries (also known as Feynman rules) depending on the types of quasiparticles involved, e.g., charge carriers, phonons, photons, and on the actual formalism employed, e.g., the zero- or finite-temperature formalism of systems at equilibrium (EGF) or the (finite-temperature) contour-ordered formalism of out-of-equilibrium systems (NEGF).

It is beyond the scope of this chapter to develop the full diagrammatic formulation of many-body perturbation theory. Here we just present the SCBA expressions of the carrier–phonon self-energies, see [115, Appendix E] and [69, Appendix C] for a derivation. Assuming that the phonon population N_Q is at equilibrium, and restricting our attention to bulk phonon modes (in other words, the effects of confinements enter only through the electronic part of the interaction), the carrier–phonon (Fock) self-energy in our nonorthogonal basis is

$$\Sigma_{\alpha\beta}^{\lessgtr}(\underline{k}, E) = \sum_Q |U_Q|^2 e^{iq_z(z_i - z_j)} \left[M(N_Q G^{\lessgtr}(\underline{k} - q, E \pm \hbar\omega_Q) + (N_Q + 1)G^{\lessgtr}(\underline{k} - q, E \mp \hbar\omega_Q))M \right]_{\alpha\beta}$$

(2.121)

$$\Sigma^R(\underline{k}, E) = \frac{1}{2}(\Sigma^>(\underline{k}, E) - \Sigma^<(\underline{k}, E)) - i\mathcal{P} \int \frac{dE'}{2\pi} \frac{\Sigma^>(\underline{k}, E) - \Sigma^<(\underline{k}, E)}{E - E'},$$

(2.122)

where $Q = (q, q_z)$, U_Q is the scattering strength and M is the mass matrix in Equation 2.115.[‡] The principal part is usually neglected (it just leads to a negligible energy renormalization), which gives

$$\Sigma^R(\underline{k}, E) \approx \frac{1}{2}(\Sigma^>(\underline{k}, E) - \Sigma^<(\underline{k}, E)).$$

(2.123)

A common deformation-potential model used in acoustic scattering is a linear phonon dispersion $\omega_Q = Qu_l$ and $U_Q = \sqrt{\frac{\hbar D_a^2}{2\mathcal{V}\rho u_l}} Q$, where u_l is the longitudinal sound velocity in the material, D_a is the deformation potential, ρ is the semiconductor mass density, and $\mathcal{V} = \mathcal{A}\mathcal{L}$ is the normalization volume. Since $\hbar\omega_Q$ is small compared to $k_B T$, the equipartition approximation holds, i.e., $N_Q \approx \frac{k_B T}{\hbar\omega_Q} \approx N_Q + 1$, i.e.,

---

and representational interpretations, as, e.g., "it seems a bit extreme to completely reject any sort of physical interpretation whatsoever" in [103, p. 88]. Admittedly, the intuitive appeal of Feynman diagrams seems unmatched in modern physics.

[†] Selective summation means that the perturbation series is evaluated approximately by summing to infinite order certain types of diagrams. Example: the sum over open oyster and bubble diagrams yields the HF approximation [103, p. 120].

[‡] Since photons and phonons are both bosons, Equation 2.121 also holds for carrier–photon interaction, the only difference being the expression of the photon number $N_{\lambda,q}$, which depends on the type of the radiation field considered (e.g., thermal radiation governed by the Bose–Einstein distribution, or monochromatic illumination). The carrier–photon self-energy is, of course, an important ingredient in the NEGF analysis of optoelectronic nanodevices, because it enables us to treat carrier dynamics and recombination processes on equal footing.

the acoustic self-energy is approximately elastic, which gives

$$\Sigma_{\alpha\beta}^{\gtrless R}(E) = \frac{1}{\Delta_{ij}} \frac{D_a^2 k_B T}{V \rho u_l^2} \sum_q \left[M G^{\gtrless R}(\underline{k} - \underline{q}, E) M \right]_{\alpha\beta}, \tag{2.124}$$

where α, β are compound indices that combine indices i, j for space and m, n for band, and

$$\frac{1}{\Delta_{ij}} = \sum_{q_z} e^{iq_z(z_i - z_j)} = \frac{\mathcal{L}}{2\pi} \int_{-\pi/a}^{\pi/a} dq_z \, e^{iq_z(z_i - z_j)} = \frac{\mathcal{L}}{\pi} \frac{\sin \frac{\pi}{a}(z_i - z_j)}{z_i - z_j}. \tag{2.125}$$

The Fröhlich theory of polar optical scattering assumes a dispersion-less longitudinal phonon with energy $\hbar\omega_{LO}$ and an interaction strength $U_q = \sqrt{\frac{e^2 \hbar\omega_{LO}}{2\epsilon_p V} \frac{Q}{Q^2 + q_0^2}}$, where $\epsilon_p^{-1} = \epsilon_\infty^{-1} - \epsilon_s^{-1}$, ϵ_s and ϵ_∞ are the static and optical dielectric constants of the material, and $q_0 = \sqrt{\frac{e^2 n}{\epsilon_s k_B T}}$ is the inverse Debye–Hückel screening length in the nondegenerate limit. In the axial approximation, the self-energy is

$$\Sigma_{\alpha\beta}^{\gtrless}(k, E) = \frac{e^2 \hbar\omega_{LO}}{2V\epsilon_p} \frac{V}{(2\pi)^3} \int dq \, q \int dq_z \int_0^{2\pi} d\phi \, e^{iq_z(z_i - z_j)} \frac{q_z^2 + q^2 + k^2 + 2kq\cos\phi}{(q_z^2 + q^2 + k^2 + q_0^2 + 2kq\cos\phi)^2}$$

$$\times \left[M(N_{LO} G^{\gtrless}(\underline{q}, E \pm \hbar\omega_{LO}) + (N_{LO} + 1) G^{\gtrless}(\underline{q}, E \mp \hbar\omega_{LO})) M \right]_{\alpha\beta}, \tag{2.126}$$

with the phonon occupation number $N_{LO} = \left(e^{\frac{\hbar\omega_{LO}}{kT}} - 1 \right)^{-1}$. Lengthy mathematical manipulations [120] lead to the final form of the lesser and greater self-energies

$$\Sigma_{\alpha\beta}^{\gtrless}(k, E) = \frac{e^2 \hbar\omega_{LO}}{4\pi^2 \epsilon_p} \int dq \, q F(q, s_{ij}, k, q_0)$$

$$\times \left[M(N_{LO} G^{\gtrless}(\underline{q}, E \pm \hbar\omega_{LO}) + (N_{LO} + 1) G^{\gtrless}(\underline{q}, E \mp \hbar\omega_{LO})) M \right]_{\alpha\beta} \tag{2.127}$$

with $s_{ij} = z_i - z_j$ and

$$F(q, s_{ij}, k, q_0) = \int_0^{\pi/a} dq_z \cos(q_z s_{ij})$$

$$\times \left[\frac{1}{\sqrt{(q_z^2 + q^2 + q_0^2 + k^2)^2 - 4k^2 q^2}} - q_0^2 \frac{q_z^2 + q^2 + q_0^2 + k^2}{((q_z^2 + q^2 + q_0^2 + k^2)^2 - 4k^2 q^2)^{3/2}} \right]. \tag{2.128}$$

Equations 2.115 through 2.117, together with the expressions for the boundary self-energies accounting for the coupling to the reservoir regions, and the scattering self-energies Equations 2.124 and 2.127 describing the interactions with acoustic and optical phonons, form a closed set of equations that have to be solved self-consistently with the macroscopic Poisson's equation. The self-consistent solution of Poisson's and Dyson's equations is known as the *outer loop* in NEGF theory, the *inner loop* being the one that connects Green's functions with the scattering self-energies.[†]

Summarizing, the NEGF algorithm for the calculation of physical quantities consists of the following steps (see Figure 2.5):

1. The retarded boundary self-energies $\Sigma_{L,R}^{RB}$ due to the coupling to the contacts are evaluated as in Equation 2.119; the corresponding lesser and greater quantities $\Sigma_{L,R}^{\lessgtr B}$ follow from the equilibrium relations Equation 2.120. Dyson's Equation 2.115 is solved for the retarded Green's function G^R. Conjugate transpose of G^R gives the advanced Green's function G^A.

2. The Keldysh Equation 2.117 gives the correlation functions $G^{\lessgtr}$ from the retarded Green's function G^R and the self-energies $\Sigma^{\lessgtr} = \Sigma^{\lessgtr B} + \Sigma^{\lessgtr S}$.

3. The scattering self-energies $\Sigma^{\lessgtr S}$ are updated using the previously calculated Green's functions. If the change is smaller than a given threshold, the inner loop stops; otherwise the previous steps (Dyson, Keldysh, and self-energy equations) are repeated.

4. Upon convergence of the inner loop, physical quantities (e.g., carrier densities and currents) can be computed from $G^{\lessgtr}$.

5. Given the NEGF carrier densities, space-charge effects are accounted for by an additional self-consistency loop (outer loop), in which Poisson's equation is solved for the Hartree potential U.

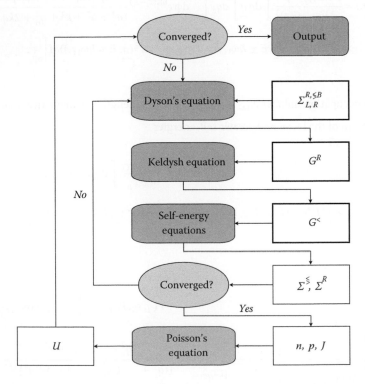

FIGURE 2.5 Flow chart of simulation. The inner self-consistency loop connects Green's functions with self-energies, while the outer loop provides the update of the Hartree potential from the solution of Poisson's equation.

[†] The outer loop can be traced back to the Hartree part of carrier–carrier interactions.

2.4.4 Conclusions

Our choice to discuss NEGF in more detail with respect to other genuine quantum approaches (e.g., the DM method) has less to do with general considerations on the merits of different formulations than with the authors' perspective on the simulation of a specific class of optoelectronic devices. In principle, a self-consistent description treating both carrier transport and optical generation on equal footing may be accomplished within an NEGF or a DM formalism. Approaching the problem from a DM perspective means solving the dynamics of diagonal (carrier populations) as well as off-diagonal (interband polarizations) elements of the reduced single-particle DM; current models usually formulated in terms of the bound states of the system have to be extended to describe their connection to the scattering states of the system. Approaching the problem from an NEGF perspective means addressing the staggering computational cost required by the calculation of Green's functions, as conventional recursive techniques are not viable when the scattering self-energies are fully nonlocal. Theoretical and numerical considerations may guide us in choosing which is the most appropriate method. As for carrier transport in GaN-based LEDs, we speculate that NEGF is probably the best choice if the objective is to clear away the uncertainty still surrounding some aspects of the droop issue, since the method gives access to energy-resolved quantities that enable us to separate different transport mechanisms. NEGF calculations provide a vivid pictorial representation of tunneling through the polarization-induced barrier, ballistic transport over the electron-blocking layer, and capture processes beyond what one-time techniques can do. As an example, Figure 2.6 shows the spectral current density in a simple LED structure that includes a single QW. Carrier–photon and carrier–photon self-energies were computed within the SCBA (carrier–carrier interactions are included at the Hartree level through coupling to Poisson's equation). Tunneling clearly plays an important role in delivering carriers to the QW.

The prevalence of FD techniques in the discretization of NEGF models of carrier transport seems, at first, simply a matter of preference/background of the NEGF demographic (a notable exception is [69]). A closer look at the problem reveals that there are actually numerical reasons motivating this seemingly arbitrary choice. For example, the use of an FE basis leads to nondiagonal terms in the self-energies, even if the

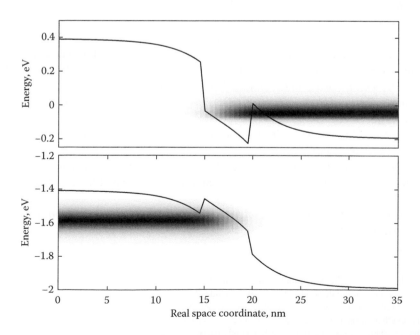

FIGURE 2.6 Spatially and energetically resolved current in a forward-biased $Al_{0.3}Ga_{0.7}As/GaAs$ *p-i-n* junction.

corresponding interactions are approximated as local in space.[†] Factor in that the use of nonorthogonal basis functions is more involved since it leads to different (contravariant and covariant) representations of Green's functions and self-energies. Factor in, too, that multiband envelope-function models widely employed in the design of optoelectronic devices may lead to spurious, unphysical solutions playing havoc in subband dispersions, especially in FE implementations. Notwithstanding the difficulties above, there are very good reasons to start using FE in the NEGF simulation of optoelectronic devices. Besides the geometrical flexibility afforded by the FE method, what makes the extra complication worthwhile is the rigorous treatment of material discontinuities in Schrödinger(-like) problems, where other approaches have to rely on special BCs to ensure current continuity. Higher-order polynomials (*p*-refinement) may be used to reduce the number of basis functions. More importantly, the FE approach allows us to unambiguously determine operator ordering and band parameters from nonlocal full-Brillouin-zone calculations, leading to well-posed, numerically stable envelope equations that provide realistic subband dispersions of quantum systems [58], while remedies proposed to eliminate spurious solutions in FD approaches have not yet found wide acceptance; the issue is also related to operator ordering and the choice of the band parameters*.

The staggering computational cost remains the main limitation of NEGF techniques. Reduced-order techniques [121], mode-space representations [122], and algorithms exploiting advanced matrix partitioning strategies (e.g., FIND [123], based on nested dissection, and SelInv [124] based on selected inversion) for the efficient calculation of Green's functions could make NEGF simulations maybe possible for realistic optoelectronic devices. Notable DM and NEGF contributions along these lines in the context of silicon nanowire field-effect transistors[125], solar cells [126] and LEDs [127,128] must be mentioned alongside promising projects such as ANGEL [69,129] and NEMO5 [130,131].

References

1. U. Rau. Reciprocity relation between photovoltaic quantum efficiency and electroluminescent emission of solar cells. *Phys. Rev. B.*, 76:085303, August 2007.

2. G. Ghione. *Semiconductor Devices for High-Speed Optoelectronics*. Cambridge: Cambridge University Press, 2009.

3. C. Cercignani and E. Gabetta, editors. *Transport Phenomena and Kinetic Theory: Applications to Gases, Semiconductors, Photons, and Biological Systems*. Boston, MA: Birkhäuser, 2007.

4. D. Vasileska, S. M. Goodnick, and G. Klimeck. *Computational Electronics. Semiclassical and Quantum Device Modeling and Simulation*. Boca Raton, FL: CRC Press, 2010.

5. C. Jacoboni. *Theory of Electron Transport in Semiconductors. A Pathway from Elementary Physics to Nonequilibrium Green Functions*. Berlin: Springer-Verlag, 2010.

6. F. Rossi and T. Kuhn. Theory of ultrafast phenomena in photoexcited semi-conductors. *Rev. Modern Phys.*, 74:895–950, August 2002.

7. T. Kuhn. Density matrix theory of coherent ultrafast dynamics. In E. Schöll, ed., *Theory of Transport Properties of Semiconductor Nanostructures*, chapter 6, pp. 173–214. Dordrecht: Springer Science+Business Media, 1998.

8. M. Fischetti and W. G. Vandenberghe. *Advanced Physics of Electron Transport in Semiconductors and Nanostructures*. Berlin: Springer-Verlag, 2016.

[†] We should explain here that the determination of the full GF would provide information on propagation and correlations between any two points of the device and thus would include all nonlocal phenomena, but in order to reduce the computational burden, a limited number of nonzero off-diagonal GF elements is usually considered. This approximation of which allows the use of efficient recursive techniques to compute selected elements of the inverse matrix (Green's function) needed for the evaluation of energy and spectrally resolved single-particle quantities.

9. F. Bertazzi, M. Moresco, and E. Bellotti. Theory of high field carrier transport and impact ionization in wurtzite GaN. Part I: A full band Monte Carlo model. *J. Appl. Phys.*, 106(6):063718, September 2009.

10. F. Bertazzi, F. Cappelluti, S. Donati Guerrieri, F. Bonani, and G. Ghione. Self-consistent coupled carrier transport full-wave EM analysis of semiconductor traveling-wave devices. *IEEE Trans. Microwave Theory Tech.*, 54(4):1611–1618, April 2006.

11. C. Jungemann and B. Meinerzhagen. *Hierarchical Device Simulation. The Monte-Carlo Perspective. Computational Microelectronics.* Vienna: Springer-Verlag, 2003.

12. C. Jacoboni and P. Lugli. *The Monte Carlo Method for Semiconductor Device Simulation. Computational Microelectronics.* Vienna: Springer-Verlag, 1989.

13. C. Moglestue. *Monte Carlo Simulation of Semiconductor Devices.* London: Chapman & Hall, 1993.

14. F. Rossi. *Theory of Semiconductor Quantum Devices. Microscopic Modeling and Simulation Strategies.* Berlin: Springer-Verlag, 2011.

15. I. Ezhov and C. Jirauschek. Influence of screening on longitudinal-optical phonon scattering in quantum cascade lasers. *J. Appl. Phys.*, 119(3):033102, 2016.

16. R. Claudia Iotti, F. Rossi, M. Serena Vitiello, G. Scamarcio, L. Mahler, and A. Tredicucci. Impact of nonequilibrium phonons on the electron dynamics in terahertz quantum cascade lasers. *Appl. Phys. Lett.*, 97(3):033110, 2010.

17. M. Moresco, F. Bertazzi, and E. Bellotti. Theory of high field carrier transport and impact ionization in wurtzite GaN. Part II: Application to avalanche photodetectors. *J. Appl. Phys.*, 106(6):063719, September 2009.

18. M. Moresco, F. Bertazzi, and E. Bellotti. GaN avalanche photodectors: A full band Monte Carlo study of gain, noise and bandwidth. *IEEE J. Quant. Electron.*, 47(4):447–454, April 2011.

19. E. Bellotti, F. Bertazzi, and M. Goano. Alloy scattering in AlGaN and InGaN: A numerical study. *J. Appl. Phys.*, 101(12):123706, 2007.

20. M. Goano, E. Bellotti, E. Ghillino, G. Ghione, and K. F. Brennan. Band structure nonlocal pseudopotential calculation of the III-nitride wurtzite phase materials system. Part I. Binary compounds GaN, AlN, and InN. *J. Appl. Phys.*, 88(11):6467–6475, December 2000.

21. M. V. Fischetti. Monte Carlo simulation of transport in technologically significant semi-conductors of the diamond and zinc-blende structures—Part I: Homogeneous trans-port. *IEEE Trans. Electron Dev.*, 38(3):634–649, March 1991.

22. M. V. Fischetti and S. E. Laux. Monte Carlo analysis of electron transport in small semiconductor devices including band-structure and space-charge effects. *Phys. Rev. B*, 38(14):9721–9745, November 1988.

23. T. Kunikiyo, M. Takenaka, Y. Kamakura, M. Yamaji, H. Mizuno, M. Morifuji, K. Taniguchi, and C. Hamaguchi. A Monte Carlo simulation of anisotropic electron transport in silicon including full band structure and anisotropic impact-ionization model. *J. Appl. Phys.*, 75(1):297–312, January 1994.

24. H.-E. Nilsson, A. Martinez, E. Ghillino, U. Sannemo, E. Bellotti, and M. Goano. Numerical modeling of hole interband tunneling in wurtzite GaN and SiC. *J. Appl. Phys.*, 90(6):2847–2852, September 2001.

25. K. Hess. *Advanced Theory of Semiconductor Devices.* Englewood Cliffs, NJ: Prentice-Hall, 1988.

26. R. Stratton. Diffusion of hot and cold electrons in semiconductor barriers. *Phys. Rev.*, 126(6):2002–2014, June 1962.

27. T. Grasser, T.-W. Tang, H. Kosina, and S. Selberherr. A review of hydrodynamic and energy-transport models for semiconductor device simulation. *Proc. IEEE*, 91(2):251–274, February 2003.

28. G. Ghione and A. Benvenuti. Discretization schemes for high-frequency semiconductor device models. *IEEE Trans. Antennas Propag.*, 45(3):443–456, March 1997.

29. M. Rudan, M. Lorenzini, and R. Brunetti. Hydrodynamic simulation of semiconductor devices. In E. Schöll, ed., *Theory of Transport Properties of Semiconductor Nanostructures*, chapter 2, pp. 27–57. Dordrecht: Springer Science+Business Media, 1998.

30. E. M. Azoff. Closed-form method for solving the steady-state generalised energy-momentum conservation equations. *COMPEL*, 6(1):25–30, 1987.

31. E. M. Azoff. Energy transport numerical simulation of graded AlGaAs/GaAs hetero-junction bipolar transistors. *IEEE Trans. Electron. Dev.*, 36(4):609–616, 1989.

32. E. Bellotti and F. Bertazzi. Transport parameters for electrons and holes. In J. Piprek, ed., *Nitride Semiconductor Devices: Principles and Simulation*, chapter 4, pp. 69–93. Weinheim: Wiley-VCH Verlag, 2007.

33. E. Furno, F. Bertazzi, M. Goano, G. Ghione, and E. Bellotti. Hydrodynamic transport parameters of wurtzite ZnO from analytic- and full-band Monte Carlo simulation. *Solid-State Electron.*, 52(11):1796–1801, 2008.

34. V. Camarchia, E. Bellotti, M. Goano, and G. Ghione. Physics based modeling of submicron GaN permeable base transistors. *IEEE Electron. Device Lett.*, EDL-23(6):303–305, June 2002.

35. B. Carnez, A. Cappy, A. Kaszynski, E. Constant, and G. Salmar. Modeling of a sub-micrometer gate field-effect transistor including effects of nonstationary electron dynamics. *J. Appl. Phys.*, 51(1):784–790, January 1980.

36. M. Streiff, A. Witzig, M. Pfeiffer, P. Royo, and W. Fichtner. A comprehensive VCSEL device simulator. *IEEE J. Sel. Top. Quantum Electron.*, 9(3):879–891, 2003.

37. S. Selberherr. MOS device modeling at 77 K. *IEEE Trans. Electron. Dev.*, 36(8):1464–1474, August 1989.

38. J. S. Blakemore. Approximation for Fermi-Dirac integrals, especially the function $F_{1/2}(\eta)$ used to describe electron density in a semiconductor. *Solid-State Electron.*, 25(11):1067–1076, 1982.

39. M. Goano. Series expansion of the Fermi-Dirac integral $F_j(x)$ over the entire domain of real j and x. *Solid-State Electron.*, 36(2):217–221, 1993.

40. D. Schroeder. *Modelling of Interface Carrier Transport for Device Simulation. Computational Microelectronics*. Vienna: Springer-Verlag, 1994.

41. F. Bonani and G. Ghione. *Noise in Semiconductor Devices. Modeling and Simulation*. Berlin: Springer-Verlag, 2001.

42. J. S. Blakemore. *Semiconductor Statistics*, Volume 3 of International Series of Monographs on Semiconductors. New York, NY: Pergamon Press, 1962.

43. P. T. Landsberg. *Recombination in Semiconductors*. Cambridge: Cambridge University Press, 1991.

44. S. M. Sze. *Physics of Semiconductor Devices*, 2nd edition. New York, NY: John Wiley & Sons, 1981.

45. S. Selberherr. *Analysis and Simulation of Semiconductor Devices*. Vienna: Springer-Verlag, 1984.

46. F. Bonani, S. Donati Guerrieri, G. Ghione, and M. Pirola. A TCAD approach to the physics-based modeling of frequency conversion and noise in semi-conductor devices under large-signal forced operation. *IEEE Trans. Electron. Dev.*, 48(5):966–977, May 2001.

47. D. L. Scharfetter and H. K. Gummel. Large-signal analysis of a silicon read diode transistor. *IEEE Trans. Electron. Dev.*, ED-16(1):64–77, January 1969.

48. M. S. Adler, V. A. K. Temple, and R. C. Rustay. Theoretical basis for field calculations on multidimensional reverse biased semiconductor devices. *Solid-State Electron.*, 25(12):1179–1186, 1982.

49. A. Deinega and S. John. Finite difference discretization of semiconductor drift-diffusion equations for nanowire solar cells. *Comp. Phys. Comm.*, 183(10):2128–2135, 2012.

50. G. A. Franz, A. F. Franz, S. Selberherr, and P. Markowich. A quasi three dimensional semiconductor device simulation using cylindrical coordinates. In *Proceedings of the NASECODE III Conference*, pp. 122–127, Dublin: Boole Press, 1983.

51. K. Matsumoto, I. Takayanagi, T. Nakamura, and R. Ohta. The operation mechanism of a charge modulation device (CMD) image sensor. *IEEE Trans. Electron. Dev.*, 38(5):989–998, 1991.

52. M. Spevak and T. Grasser. Discretization of macroscopic transport equations on non-Cartesian coordinate systems. *IEEE Trans. Comput. Aided Des. Integr. Circuits Syst.,*, 26(8):1408–1416, 2007.

53. M. Sharma and G. F. Carey. Semiconductor device modeling using flux upwind finite elements. *COMPEL*, 8(4):219–224, 1989.

54. M. Calciati, A. Tibaldi, F. Bertazzi, M. Goano, and P. Debernardi. Many-valley electron transport in AlGaAs VCSELs. *Semicond. Sci. Tech.*, 32(5):055007, 2017. doi:10.1088/1361-6641/aa66bb

55. M. G. Ancona and G. J. Iafrate. Quantum correction to the equation of state of an electron gas in a semiconductor. *Phys. Rev. B*, 39:9536–9540, May 1989.

56. S. L. Chuang and C. S. Chang. $k \cdot p$ method for strained wurtzite semiconductors. *Phys. Rev. B*, 54(4):2491–2504, July 1996.

57. R. G. Veprek, S. Steiger, and B. Witzigmann. Ellipticity and the spurious solution problem of $k \cdot p$ envelope equations. *Phys. Rev. B*, 76(16):165320, 2007.

58. X. Zhou, F. Bertazzi, M. Goano, G. Ghione, and E. Bellotti. Deriving $k \cdot p$ parameters from full-Brillouin-zone descriptions: A finite-element envelope function model for quantum-confined wurtzite nanostructures. *J. Appl. Phys.*, 116(3):033709, July 2014.

59. R. G. Veprek. *Computational Modeling of Semiconductor Nanostructures for Optoelectronics*. PhD thesis, Swiss Federal Institute of Technology, Zürich, 2009.

60. F. Mireles and S. E. Ulloa. Ordered Hamiltonian and matching conditions for heterojunctions with wurtzite symmetry: GaN/AlxGa1$-x$N quantum wells. *Phys. Rev. B*, 60(19):13659–13667, November 1999.

61. R. G. Veprek, S. Steiger, and B. Witzigmann. Operator ordering, ellipticity and spurious solutions in $k \cdot p$ calculations of III-nitride nanostructures. *Opt. Quant. Electron.*, 40(14–15):1169–1174, November 2008.

62. L.-W. Wang and A. Zunger. Linear combination of bulk bands method for largescale electronic structure calculations on strained nanostructures. *Phys. Rev. B*, 59(24):15806–15818, June 1999.

63. B. A. Foreman. Quadratic response theory for spin–orbit coupling in semiconductor heterostructures. *Phys. Rev. B.*, 72:165344, October 2005.

64. A. Di Carlo. Microscopic theory of nanostructured semiconductor devices: Beyond the envelope-function approximation. *Semiconductor Sci. Tech.*, 18(1):R1–R31, 2003.

65. A. Trellakis, A. T. Galick, A. Pacelli, and U. Ravaioli. Iteration scheme for the solution of the two-dimensional Schrödinger–Poisson equations in quantum structures. *J. Appl. Phys.*, 81(12):7880–7884, June 1997.

66. M. V. Fischetti and S. E. Laux. Monte Carlo study of electron transport in silicon inversion layers. *Phys. Rev. B*, 48(4):2244–2274, July 1993.

67. B. Witzigmann, A. Witzig, and W. Fichtner. A multidimensional laser simulator for edge-emitters including quantum carrier capture. *IEEE Trans. Electron. Dev.*, 47(10):1926–1934, October 2000.

68. M. Grupen, G. Kosinovsky, and K. Hess. The effect of carrier capture on the modulation bandwidth of quantum well lasers. In *Proceedings of the IEDM*, pp. 23.6–23.6.4, Washington, DC, 1993.

69. S. Steiger. *Modelling Nano-LEDs*. PhD thesis, Swiss Federal Institute of Technology, Zürich, 2009.

70. W. W. Chow, H. C. Schneider, S. W. Koch, C.-H. Chang, L. Chrostowski, and C. J. Chang-Hasnain. Nonequilibrium model for semiconductor laser modulation response. *IEEE J. Quant. Electron.*, 38(4):402–409, April 2002.

71. E. F. Schubert. *Physical Foundations of Solid-State Devices*. Rensselaer Polytechnic Institute, Troy, NY, 2015.

72. M. Vallone. Quantum well electron scattering rates through longitudinal optic-phonon dynamical screened interaction: An analytic approach. *J. Appl. Phys.*, 114(5):053704, August 2013.

73. G. R. Hadley, K. L. Lear, M. E. Warren, K. D. Choquette, J. W. Scott, and S. W. Corzine. Comprehensive numerical modeling of vertical-cavity surface-emitting lasers. *IEEE J. Quant. Electron.*, 32(4):607–616, 1996.

74. O. Conradi, S. Helfert, and R. Pregla. Comprehensive modeling of vertical-cavity laser diodes by the method of lines. *IEEE J. Quantum Electron.*, 37(7):928–935, 2001.

75. M. Dems, T. Czyszanowski, and K. Panajotov. Plane-wave and cylindrical-wave admittance method for simulation of classical and photonic-crystal-based VCSELs. In *Proceedings of the SPIE, Vol. 6182, Photonic Crystal Materials and Devices III*, pp. 618219-1–618219-8, Bellingham, WA: SPIE, 2006.

76. M. Dems, R. Kotynski, and K. Panajotov. Planewave admittance method—A novel approach for determining the electromagnetic modes in photonic structures. *Opt. Express.*, 13(9):3196–3207, 2005.

77. M. Dems. Modelling of high-contrast grating mirrors. The impact of imperfections on their performance in VCSELs. *Opto-Electron. Rev.*, 19(3):340–345, 2011.

78. G. P. Bava, P. Debernardi, and L. Fratta. Three-dimensional model for vectorial fields in vertical-cavity surface-emitting lasers. *Phys. Rev. A*, 63(2):23816-1–23816-13, 2001.

79. P. Debernardi and G. P. Bava. Coupled mode theory: A powerful tool for analyzing complex VCSELs and designing advanced devices features. *IEEE J. Sel. Top. Quantum Electron.*, 9(3):905–917, 2003.

80. R. Michalzik. *VCSELs: Fundamentals, Technology and Applications of Vertical-Cavity Surface-Emitting Lasers*. Berlin: Springer-Verlag, 2013.

81. G. P. Bava, P. Debernardi, and L. Fratta. Quantum noise in vertical-cavity surface-emitting lasers with polarization competition. *J. Opt. Soc. Am. B*, 16(11):2147–2157, 1999.

82. E. Malic and A. Knorr. *Graphene and Carbon Nanotubes. Ultrafast Relaxation Dynamics and Optics*. Weinheim: VCH, 2013.

83. W. W. Chow and S. W. Koch. *Semiconductor-Laser Fundamentals. Physics of the Gain Materials*. Berlin: Springer-Verlag, 1999.

84. J. Piprek. Efficiency droop in nitride-based light-emitting diodes. *Phys. Status Solidi A*, 207(10):2217–2225, October 2010.

85. G. Verzellesi, D. Saguatti, M. Meneghini, F. Bertazzi, M. Goano, G. Meneghesso, and E. Zanoni. Efficiency droop in In-GaN/GaN blue light-emitting diodes: Physical mechanisms and remedies. *J. Appl. Phys.*, 114(7):071101, August 2013.

86. C. Weisbuch, M. Piccardo, L. Martinelli, J. Iveland, J. Peretti, and J. S. Speck. The efficiency challenge of nitride light-emitting diodes for lighting. *Phys. Status Solidi A*, 212(5):899–913, 2015.

87. Y. C. Shen, G. O. Mueller, S. Watanabe, N. F. Gardner, A. Munkholm, and M. R. Krames. Auger recombination in InGaN measured by photoluminescence. *Appl. Phys. Lett.*, 91(14):141101, 2007.

88. K. T. Delaney, P. Rinke, and C. G. Van de Walle. Auger recombination rates in nitrides from first principles. *Appl. Phys. Lett.*, 94:191109, 2009.

89. E. Kioupakis, D. Steiauf, P. Rinke, K. T. Delaney, and C. G. Van de Walle. First-principles calculations of indirect Auger recombination in nitride semiconductors. 2014. http://arxiv.org/abs/1412.7555.

90. E. Kioupakis, Q. Yan, and C. G. Van de Walle. Interplay of polarization fields and Auger recombination in the efficiency droop of nitride light-emitting diodes. *Appl. Phys. Lett.*, 101(23):231107, 2012.

91. M. Binder, A. Nirschl, R. Zeisel, T. Hager, H.-J. Lugauer, M. Sabathil, D. Bougeard, J. Wagner, and B. Galler. Identification of *nnp* and *npp* Auger recombination as significant contributor to the efficiency droop in (GaIn) N quantum wells by visualization of hot carriers in photoluminescence. *Appl. Phys. Lett.*, 103(7):071108, August 2013.

92. J. Iveland, L. Martinelli, J. Peretti, J. S. Speck, and C. Weisbuch. Direct measurement of Auger electrons emitted from a semiconductor light-emitting diode under electrical injection: Identification of the dominant mechanism for efficiency droop. *Phys. Rev. Lett.*, 110:177406, April 2013.

93. A. Nirschl, M. Binder, M. Schmid, I. Pietzonka, H.-J. Lugauer, R. Zeisel, M. Sabathil, D. Bougeard, and B. Galler. Towards quantification of the crucial impact of Auger recombination for the efficiency droop in (AlInGa) N quantum well structures. *Opt. Express.*, 24(3):2971–2980, February 2016.

94. V. Avrutin, S. A. Hafiz, F. Zhang, M. Zgr, E. Bellotti, F. Bertazzi, M. Goano, A. Matulionis, A. T. Roberts, H. O. Everitt, and H. Morko. Saga of efficiency degradation at high injection in InGaN light emitting diodes. *Turk. J. Phys.*, 38(3):269–313, November 2014.

95. J. Hader, J. V. Moloney, and S. W. Koch. Beyond the *ABC*: Carrier recombination in semiconductor lasers. In *SPIE Photonics West, Physics and Simulation of Optoelectronic Devices XIV, Vol. 6115, Proceedings of the SPIE*, pp. 61151T. Bellingham, WA: SPIE, January 2006.

96. M. Kira and S. W. Koch. Many-body correlations and excitonic effects in semiconductor spectroscopy. *Prog. Quant. Electron.*, 30(5):155–296, 2006.

97. P. Vogl and T. Kubis. The non-equilibrium Green's function method: An introduction. *J. Comp. Electron.*, 9(3):237–242, 2010.

98. W. Pötz. Microscopic theory of coherent carrier dynamics and phase breaking in semi-conductors. *Phys. Rev. B*, 54(8):5647–5664, August 1996.

99. A. Wacker. Coherence and spatial resolution of transport in quantum cascade lasers. *Phys. Status Solidi C*, 5(1):215–220, 2008.

100. U. Aeberhard. Theory and simulation of quantum photovoltaic devices based on the non-equilibrium Green's function formalism. *J. Comp. Electron.*, 10:394–413, October 2011.

101. G. D. Mahan. *Many-Particle Physics*, 3rd edition. New York, NY: Kluwer Academic Publishers, 2000.

102. A. L. Fetter and J. D. Walecka. *Quantum Theory of Many-Particle Physics*, 3rd edition. New York, NY: Dover Publications, 2003.

103. R. D. Mattuck. *A Guide to Feynman Diagrams in the Many-Body Problem*, 2nd edition. New York, NY: McGraw-Hill, 1976.

104. L. P. Kadanoff and G. Baym. *Quantum Statistical Mechanics. Green's Function Methods in Equilibrium and Nonequilibrium Problems*. New York, NY: W. A. Benjamin, 1962.

105. L. V. Keldysh. Diagram technique for nonequilibrium processes. *Sov. Phys. JETP*, 20(4):1018–1026, April 1965.

106. P. C. Martin and J. Schwinger. Theory of many-particle systems. I. *Phys. Rev.*, 115(6):1342–1026, September 1959.

107. H. Haug and A.-P. Jauho. *Quantum Kinetics in Transport and Optics of Semiconductors*, 2nd edition. Berlin: Springer, 2008.

108. L. V. Keldysh. Real-time nonequilibrium Green's functions. In M. Bonitz and D. Semkat, editors, *Progress in Nonequilibrium Green's Functions*, pp. 4–17. Singapore: World Scientific, 2003.

109. M. Bonitz. *Quantum Kinetic Theory*, 2nd edition. Cham, Switzerland: Springer, 2016.

110. J. Rammer. *Quantum Field Theory of Non-Equilibrium States*. Cambridge: Cambridge University Press, 2007.

111. G. Stefanucci and R. van Leeuwen. *Nonequilibrium Many-Body Theory of Quantum Systems. A Modern Introduction*. Cambridge: Cambridge University Press, 2013.

112. S. Datta. *Quantum Transport: Atom to Transistor*. Cambridge: Cambridge University Press, 2005.

113. U. Aeberhard. Quantum-kinetic theory of photocurrent generation via direct and phonon-mediated optical transitions. *Phys. Rev. B.*, 84:035454, July 2011.

114. P. Danielewicz. Quantum theory of nonequilibrium processes. I. *Ann. Phys.*, 152:239–304, 1984.

115. U. Aeberhard. *A Microscopic Theory of Quantum Well Photovoltaics*. PhD thesis, Swiss Federal Institute of Technology, Zürich, 2008.

116. E. Polizzi and S. Datta. Multidimensional nanoscale device modeling: The finite element method applied to the non-equilibrium Green's function formalism. In *3rd IEEE Conference on Nanotechnology (IEEE-NANO 2003)*, Vol. 2, Piscataway, NJ: IEEE pp. 40–43, August 2003.

117. D. L. Smith and C. Mailhiot. Theory of semiconductor superlattice electronic structure. *Rev. Modern Phys.*, 62(1):173–234, January 1990.

118. M. G. Vergniory, C. Yang, A. Garcia-Lekue, and L.-W. Wang. Calculation of complex band structure for plane-wave nonlocal pseudopotential Hamiltonian. *Comp. Mater. Sci.*, 48(3):544–550, 2010.

119. A. Wüthrich. *The Genesis of Feynman Diagrams*. Dordrecht: Springer, 2010.

120. R. Lake, G. Klimeck, R. C. Bowen, and D. Jovanovic. Single and multiband modeling of quantum electron transport through layered semiconductor devices. *J. Appl. Phys.*, 81(12):7845–7869, February 1997.

121. J. Z. Huang, L. Zhang, P. Long, M. Povolotskyi, and G. Klimeck. Quantum transport simulation of III-V TFETs with reduced-order $k \cdot p$ method. 2015. http://arxiv.org/abs/1511.02516v1.
122. M. Luisier, A. Schenk, and W. Fichtner. Quantum transport in two-and three-dimensional nanoscale transistors: Coupled mode effects in the nonequi-librium Green's function formalism. *J. Appl. Phys.*, 100(4):043713, 2006.
123. S. Li, S. Ahmed, G. Klimeck, and E. Darve. Computing entries of the inverse of a sparse matrix using the FIND algorithm. *J. Comp. Phys.*, 227(22):9408–9427, November 2008.
124. L. Lin, J. Lu, L. Ying, R. Car, and E. Weinan. Fast algorithm for extracting the diagonal of the inverse matrix with application to the electronic structure analysis of metallic systems. *Commun. Math. Sci.*, 7(3):755–777, 2009.
125. J. Fang, W. G. Vandenberghe, B. Fu, and M. V. Fischetti. Pseudopotential-based electron quantum transport: Theoretical formulation and application to nanometer-scale silicon nanowire transistors. *J. Appl. Phys.*, 119(3):035701, 2016.
126. U. Aeberhard. Simulation of nanostructure-based and ultra-thin film solar cell devices beyond the classical picture. *J. Photon. Energy*, 4(1):042099, 2014.
127. A. Shedbalkar, Z. Andreev, and B. Witzigmann. Simulation of an indium gallium nitride quantum well light-emitting diode with the non-equilibrium Green's function method. *Phys. Status Solidi B*, 253(1):158–163, 2016.
128. R. Wang, Y. Zhang, F. Bi, G. Chen, and C. Yam. Quantum mechanical modeling of nanoscale light emitting diodes. 2015. http://arxiv.org/abs/1512.08498v1.
129. S. Steiger. ANGEL–A nonequilibrium Green function solver for LEDs. 2015. https://nanohub.org/resources/8136.
130. NEMO5. https://engineering.purdue.edu/gekcogrp/software-projects/nemo5/, 2010.
131. S. Steiger, M. Povolotskyi, H.-H. Park, T. Kubis, and G. Klimeck. NEMO5: A parallel multiscale nanoelectronics modeling tool. *IEEE Trans. Nanotech.*, 10(6):1464–1474, November 2011.

3

Electron–Photon Interaction

Angela Thränhardt

3.1 Introduction

The devices introduced in this book involve light–matter interaction. Light appears to be a trivial phenomenon known to everyone; however, its mathematical description is quite intricate and has been a major point of discussion in the physics community. Historically, this was manifested in a longstanding dispute on the nature of light, featuring Isaac Newton[†] as the advocate of a corpuscular model and Christiaan Huygens[‡] as the proponent of its wave nature.

Today, more than 300 years later, both perceptions have been reconciled and the description of light in terms of a wave–particle dualism is well established in the physics community. A full quantum-electrodynamic description, relieving the apparent contradiction between wave and particle nature, is often not required, i.e., in many cases, light can be well described as either a wave or a particle. This will be mirrored in this chapter, where both the classical and the quantum perspectives are introduced.

As often stated, the general guideline is that light may be treated as a wave when propagating in vacuum, yet it has to be quantized in the case of the interaction with matter. However, this is not always true, as, e.g., the behavior of light in a waveguide can well be understood in terms of the wave picture. In fact, quantum-optical effects only started to be discussed intensively in the 1950s and 1960s. Their investigation is closely related to the development of the laser and the requirement of solid theoretical foundations in this field. Simply speaking, the corpuscular theory of light states that it is distributed in quantized packages of energy $W = hf = \hbar\omega$, where h is Planck's constant, f is the light frequency, $\hbar = h/(2\pi)$, and ω is the angular frequency $\omega = 2\pi f$. A particle nature is thus attributed to light, where the light particle is known as a photon. The photon may be described quantum mechanically in the framework of second quantization, a formalism in which the light field and other physical field quantities are written as operators. This method was first developed by Paul Dirac[§] in 1927 in analogy to the first quantization where physical quantities

---

[†] Sir Isaac Newton, English natural philosopher (physicist and mathematician), 1643–1727.
[‡] Christiaan Huygens, Dutch physicist, astronomer, and mathematician, 1629–1695.
[§] Paul Dirac, British physicist, 1902–1984.

such as position and momentum become operators; see Ref. [1]. Further information may be found in textbooks on quantum mechanics.

In this chapter, we consider electron–photon interaction, i.e., light–matter coupling. There are different electron–photon interaction mechanisms, e.g., the photoelectric effect, Compton scattering, and pair production. In the energy range of interest in optoelectronics, the photoelectric effect is the only relevant effect and will thus be treated exclusively.

The chapter briefly introduces the theoretical treatment of electron–photon interaction for different and mostly strongly simplified cases. A large part is devoted to the two-level system in different formalisms. This allows us to concentrate on the differences and commonalities of the various approaches. We look at the two-level system in the Schrödinger and the Heisenberg representation (Section 3.2) and then move on to a solid-state system (Section 3.3). Section 3.4 deals with the quantum-optical treatment of the two-level system, while Section 3.5 is devoted to the treatment of photonic structures.

Before we enter the detailed mathematics of a quantum mechanical system, some basics will be discussed. In any ensemble of atoms or molecules in interaction with an electromagnetic field, there are three processes occurring: spontaneous emission, absorption, and stimulated emission. Although spontaneous emission is quite common in everyday life—e.g., most light sources like light bulbs or light-emitting diodes (LEDs) operate based on spontaneous emission—it is difficult to treat theoretically. The existence of spontaneous emission, however, was already postulated by Albert Einstein in 1916/1917 with his argument based on thermodynamic equilibrium. For a handwaving explanation, consider an ensemble of atoms placed inside a closed cavity whose walls are held at a constant temperature. The walls, furthermore, are assumed to be perfect absorbers and emitters of radiation. After the system has reached thermal equilibrium, the cavity will be filled with the so-called "blackbody radiation" according to the temperature T of the system. The condition of thermal equilibrium then requires that for each transition, the spontaneous emission matches the absorption of blackbody radiation, necessitating the existence of spontaneous emission.

3.2 Two-Level System

3.2.1 Optical Processes in the Two-Level System—A Discussion of the Einstein Coefficients

For a more detailed treatment, we now move to the simplest existing system, the two-level system. It includes one electron that can occupy either one of two nondegenerate states. The two levels of the model may represent, e.g., the highest occupied and lowest unoccupied level in an atom, molecule, or quantum structure. We devote this entire section to the two-level system, including a discussion of the Einstein coefficients as well as calculations in the Schrödinger (Section 3.2.2) and the Heisenberg formalism (Section 3.2.3). In Section 3.4, we again return to the two-level system for a quantum-optical discussion. For an even more detailed treatment of two-level systems, the reader is referred to the book by Allen and Eberly [2].

A schematic of the two-level system is shown in Figure 3.1. The system consists of the ground state, level 1 and one excited state, level 2, with the respective energies ϵ_1 and ϵ_2. One of the levels is occupied by an electron, described by the eigenstates $|1\rangle$ or $|2\rangle$. We consider the interaction with an incident light field with photon density Φ_{ω_0}, where ω_0 refers to the angular light frequency corresponding to the energy difference of the two-level system, $\hbar\omega_0 = \epsilon_2 - \epsilon_1$. Let us now think of an ensemble of identical two-level systems, where the number of electrons in state 1 is N_1 and the number of electrons in state 2 is N_2. The following processes indicated in the schematic may occur in any of them:

1. *Spontaneous emission*, as described earlier, lowering an electron from the excited into the ground state and emitting a photon. The rate of spontaneous emission ρ_{spont} must therefore be proportional to the number of electrons in the excited state

$$\rho_{spont} = AN_2. \tag{3.1}$$

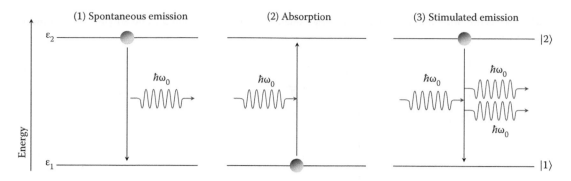

FIGURE 3.1 Relevant electron–photon interaction processes in a two-level system.

2. *Absorption*, raising an electron from the ground state into the excited state by absorbing an existing photon. The rate of absorption ρ_{abs} must be proportional to the number of electrons in the ground state and the photon density at the resonance frequency:

$$\rho_{abs} = B_{12} N_1 \Phi_{\omega_0}. \tag{3.2}$$

3. *Stimulated emission*, where an electron is stimulated by a photon to emit a second photon and drop to the ground state in consequence. Both photons agree in frequency and phase and travel in the same direction. The rate of stimulated emission ρ_{stim} is proportional to the number of electrons in the excited state and the photon density at resonance

$$\rho_{stim} = B_{21} N_2 \Phi_{\omega_0}. \tag{3.3}$$

A, B_{12}, and B_{21} are the *Einstein coefficients*, named after Albert Einstein (1879–1955), who first introduced them in 1916; see Ref. [3]. Considering all three processes, we get an expression for the overall rate of change of the occupancies of the two levels:

$$\frac{dN_1}{dt} = -\frac{dN_2}{dt} = -N_1 B_{12} \Phi_{\omega_0} + N_2 B_{21} \Phi_{\omega_0} + N_2 A. \tag{3.4}$$

Thinking of a thermal equilibrium situation, we demand

$$\frac{dN_1}{dt} = \frac{dN_2}{dt} = 0, \tag{3.5}$$

which, after substituting Equation 3.4 into 3.5 and making some algebraic conversions, renders

$$\frac{N_2}{N_1} = \frac{B_{12}}{B_{21}\Phi_{\omega_0} + A}. \tag{3.6}$$

On the other hand, thermodynamics tells us that the equilibrium distribution between ground and excited state is governed by the Boltzmann distribution,

$$\frac{N_2}{N_1} = \frac{e^{-\epsilon_2/(k_B T)}}{e^{-\epsilon_1/(k_B T)}} = e^{-\hbar\omega_0/(k_B T)}, \tag{3.7}$$

where k_B is the Boltzmann constant. Equating Equation 3.6 with Equation 3.7 and solving for the photon density Φ_{ω_0} yields

$$\Phi_{\omega_0} = \frac{A}{B_{21}} \frac{1}{\frac{B_{12}}{B_{21}} e^{\hbar\omega_0/(k_B T)} - 1}. \tag{3.8}$$

Now consider Planck's law,

$$\Phi_{\omega_0} = \frac{2\hbar\omega_0^3}{\pi c^3} \frac{1}{e^{\hbar\omega_0/(k_B T)} - 1}, \tag{3.9}$$

where c is the velocity of light. Equations 3.8 and 3.9 are very similar in form. A comparison of the two yields

$$B_{12} = B_{21} \equiv B, \tag{3.10}$$

$$A = B \frac{2\hbar\omega_0^3}{\pi c^3}. \tag{3.11}$$

Thus, the coefficients of absorption and stimulated emission are equal. This also implies that the relative strength of absorption and stimulated emission is determined exclusively by the occupation of ground and excited state. The coefficient of spontaneous emission, A, is also directly and independently of temperature related to the other two Einstein coefficients.

3.2.2 The Two-Level System in the Schrödinger Picture

We now proceed to a quantum-mechanical treatment of the electron in the two-level system. At this point, we stick to a semiclassical treatment, i.e., a classical treatment of light. As we see, this takes us a long way toward understanding absorption and stimulated emission as well as deriving the optical Bloch equations. The reader is referred to the quantum-optical treatment presented in Section 3.4 to deal with spontaneous emission as it cannot be understood semiclassically.

The two-level system is described by the Hamiltonian of the undisturbed system, H_0, and the Hamiltonian of an optical field coupling to the electronic transition dipole moment, $\hat{H}_L$:

$$\hat{H} = \hat{H}_0 + \hat{H}_L \tag{3.12}$$

with

$$\begin{aligned}
\hat{H}_0 &= \epsilon_1 |1\rangle\langle 1| + \epsilon_2 |2\rangle\langle 2|, \\
\hat{H}_L &= -\mathbf{E}(t) \cdot \left\{ \mathbf{d}_{12} |1\rangle\langle 2| + \mathbf{d}_{21} |2\rangle\langle 1| \right\},
\end{aligned} \tag{3.13}$$

using the Schrödinger wave functions of state 1 and state 2, $|1\rangle$ and $|2\rangle$, respectively. The light field couples to the transition dipole moment of the system, see also Ref. [4], using the transition dipole matrix elements $\mathbf{d}_{12} = -e\langle 1|\hat{\mathbf{r}}|2\rangle$ and $\mathbf{d}_{21} = \mathbf{d}_{12}^* = -e\langle 2|\hat{\mathbf{r}}|1\rangle$. The Schrödinger equation for the system wave function $|\psi\rangle$ reads

$$i\hbar \frac{d}{dt} |\psi\rangle = \left[\hat{H}_0 + \hat{H}_L \right] |\psi\rangle. \tag{3.14}$$

The wave function $|\psi\rangle$ may be expanded in the eigenfunctions of the system,

$$|\psi\rangle = a_1(t) e^{-i\omega_1 t} |1\rangle + a_2(t) e^{-i\omega_2 t} |2\rangle, \tag{3.15}$$

using $\omega_1 = \epsilon_1/\hbar$ and $\omega_2 = \epsilon_2/\hbar$. Inserting this expansion into Equation 3.14 and multiplying with $\langle\psi|$ from the left-hand side yields equations for the coefficients $a_1(t)$ and $a_2(t)$,

$$\dot{a}_1(t) = a_2(t)e^{-i(\omega_2-\omega_1)t}\langle 1|\hat{H}_L|2\rangle,$$

$$\dot{a}_2(t) = a_1(t)e^{-i(\omega_1-\omega_2)t}\langle 2|\hat{H}_L|1\rangle. \tag{3.16}$$

Applying a monochromatic incident field, $\mathbf{E}(t) = \frac{1}{2}\mathbf{E}_0(e^{-i\omega t} + e^{i\omega t})$, we derive

$$\dot{a}_1(t) = \frac{a_2(t)}{2} e^{-i(\omega_2-\omega_1)t}\mathbf{E}_0 \cdot \mathbf{d}_{21}(e^{-i\omega t} + e^{i\omega t}),$$

$$\dot{a}_2(t) = \frac{a_1(t)}{2} e^{-i(\omega_1-\omega_2)t}\mathbf{E}_0 \cdot \mathbf{d}_{12}(e^{-i\omega t} + e^{i\omega t}). \tag{3.17}$$

Considering an optical field near the resonance frequency of the system ω_0, one of the addends in either equation is almost time independent and the other is oscillating quickly. In the rotating wave approximation (RWA), the latter term is ignored due to averaging out and we are left with

$$\dot{a}_1(t) = \frac{a_2(t)}{2} e^{-i(\omega_0-\omega)t}\mathbf{E}_0 \cdot \mathbf{d}_{12},$$

$$\dot{a}_2(t) = \frac{a_1(t)}{2} e^{i(\omega_0-\omega)t}\mathbf{E}_0 \cdot \mathbf{d}_{21}. \tag{3.18}$$

For resonance conditions $\omega = \omega_0$, this system of coupled equations may easily be solved by time-differentiating the first equation and inserting the second, yielding

$$\frac{\mathrm{d}^2}{\mathrm{d}t^2} a_1(t) = -\frac{|\mathbf{E}_0 \cdot \mathbf{d}_{12}|^2}{4}a_1(t). \tag{3.19}$$

Defining the Rabi frequency[†]

$$\Omega = \frac{|\mathbf{E}_0 \cdot \mathbf{d}_{12}|}{\hbar}, \tag{3.20}$$

yields $a_1(t) = a_1(0)e^{\pm i\Omega t}$ as a solution to Equation 3.19, an equivalent expression for $a_2(t)$, and after resubstitution into Equation 3.15,

$$|\psi\rangle = a_1(0)e^{\pm i\Omega t}e^{-i\omega_1 t}|1\rangle + a_2(0)e^{\pm i\Omega t}e^{-i\omega_2 t}|2\rangle. \tag{3.21}$$

Starting with $a_1 = 0, a_2 = 1$, i.e., an electron in the excited state, we observe that the electron oscillates between ground and excited states with a cosine dependence and frequency Ω, describing the interplay between absorption and stimulated emission. These so-called *Rabi flops* are typical for any two-level system driven by an oscillatory field, including atoms in a light field as discussed here, but also particles in an oscillating magnetic field as investigated in nuclear magnetic resonance.

Looking at the optical processes possible in a two-level system with or without the incidence of light, we note the following: In the absence of the external perturbation represented by the laser pulse, the system is described by the Hamiltonian $\hat{H}_0$. If, at an initial time $t = 0$, it is in an eigenstate, it will remain in this state for any length of time. This applies to the ground state $|1\rangle$ as well as to the excited state $|2\rangle$.

[†] Named after American physicist Isidor Isaac Rabi (1898–1988).

It is, however, obviously not a realistic description, the reason for this being the absence of spontaneous emission in the semiclassical picture we used here. We see later that spontaneous emission enters the model with a quantization of light.

Now let us briefly consider the example of a laser system. Laser operation is basically as follows: An initial photon, usually produced by spontaneous emission, is confined to an active medium using a resonator and amplified by stimulated emission. Obviously, this will only work if stimulated emission exceeds absorption. Going back to Equation 3.10, we know that the Einstein coefficients for absorption and stimulated emission are equal and thus the occupancies of the ground and excited states govern the relative strength. Therefore, inversion has to be achieved for laser operation, i.e., the number of carriers in the excited state has to exceed the number of carriers in the ground state. Under certain conditions, quantum theory also allows lasing without inversion; since this requires intensive tailoring of the system, it will not be discussed here, but the reader is referred to the original literature, see, e.g., Refs. [5,6].

Unfortunately, inversion can, in general, not be achieved in a two-level system as there is no efficient pumping mechanism. Thus, a laser requires at least a three-level system for operation. The relevant equations for the three-level system may be derived in analogy to the case of the two-level system discussed here.

3.2.3 The Two-Level System in the Heisenberg Picture

We now introduce yet another description of the two-level system, changing from the Schrödinger picture to the Heisenberg picture. The passage from the Schrödinger picture to the Heisenberg picture is made by transferring the time dependence from the wave functions to the operator; for details, see any see any textbook on quantum mechanics. The formalism in this subsection is similar to the one in Section 3.3, which treats the many-body case. In this subsection, all calculations are presented in detail so the reader can reproduce them, while in Section 3.3, several steps are omitted for the sake of clarity and brevity.

The time evolution of the system in the Heisenberg picture is not governed by the Schrödinger equation any more, but by the Heisenberg equation of motion

$$\dot{\hat{\mathcal{O}}} = \frac{i}{\hbar}\left[\hat{H}, \hat{\mathcal{O}}\right], \tag{3.22}$$

where [.] denotes the commutator,

$$\left[\hat{H}, \hat{\mathcal{O}}\right] = \hat{H}\hat{\mathcal{O}} - \hat{\mathcal{O}}\hat{H}, \tag{3.23}$$

and $\hat{\mathcal{O}}$ is any operator.

The Heisenberg equation allows us to calculate the equation of motion for arbitrary operators. Here, we obviously only consider processes that conserve the number of carriers in the system. We revert to the formalism of second quantization mentioned in the introduction and introduce operators describing the electron. In the following, $\hat{c}_1$ ($\hat{c}_1^\dagger$) and $\hat{c}_2$ ($\hat{c}_2^\dagger$) describe annihilation (creation) operators for electrons in levels 1 and 2, respectively. The relevant particle-conserving two-operator quantities are thus $\hat{c}_1^\dagger\hat{c}_2$, $\hat{c}_2^\dagger\hat{c}_1$, $\hat{c}_1^\dagger\hat{c}_1$, and $\hat{c}_2^\dagger\hat{c}_2$. The first two quantities are the microscopic polarizations; it will be illustrated in Section 3.3 that the polarization is the relevant quantity when considering light–matter interaction. The latter two quantities $\hat{c}_1^\dagger\hat{c}_1$ and $\hat{c}_2^\dagger\hat{c}_2$ describe the carrier densities in levels 1 and 2, respectively. We observe that the equations of motion for these four quantities are coupled to each other, but form a closed system of equations.

The Hamiltonian $\hat{H} = \hat{H}_0 + \hat{H}_L$ of the two-level system may be written as

$$\hat{H}_0 = \epsilon_1\hat{c}_1^\dagger\hat{c}_1 + \epsilon_2\hat{c}_2^\dagger\hat{c}_2,$$

$$\hat{H}_L = -\mathbf{E}(t) \cdot (\mathbf{d}_{12}\hat{c}_1^\dagger\hat{c}_2 + \mathbf{d}_{21}\hat{c}_2^\dagger\hat{c}_1).$$

Inserting this Hamiltonian into the Heisenberg equation for the polarization $\hat{p} = \hat{c}_1^\dagger \hat{c}_2$, we derive the equation of motion for this quantity:

$$\dot{\hat{p}} = \frac{i}{\hbar}\left[\hat{H},\hat{p}\right] = \frac{i}{\hbar}\left(\hat{H}\hat{p} - \hat{p}\hat{H}\right)$$

$$= \frac{i}{\hbar}\left\{\epsilon_1 \hat{c}_1^\dagger \hat{c}_1 + \epsilon_2 \hat{c}_2^\dagger \hat{c}_2 - \mathbf{E}(t)\cdot(\mathbf{d}_{12}\hat{c}_1^\dagger \hat{c}_2 + \mathbf{d}_{21}\hat{c}_2^\dagger \hat{c}_1)\right\}\hat{c}_1^\dagger \hat{c}_2$$

$$- \hat{c}_1^\dagger \hat{c}_2 \left\{\epsilon_1 \hat{c}_1^\dagger \hat{c}_1 + \epsilon_2 \hat{c}_2^\dagger \hat{c}_2 - \mathbf{E}(t)\cdot(\mathbf{d}_{12}\hat{c}_1^\dagger \hat{c}_2 + \mathbf{d}_{21}\hat{c}_2^\dagger \hat{c}_1)\right\}. \tag{3.24}$$

The goal now is to relate the time derivative of the polarization to known or accessible quantities. This may also be the polarization itself, enabling an analytical or a numerical solution of the resulting differential equation. To achieve this, we need to interchange the operators, e.g., bring them into normal ordering in every term.

First, we remind ourselves of the Fermionic anticommutators to be used here

$$\left\{\hat{c}_i^\dagger, \hat{c}_j\right\}_+ = \hat{c}_i^\dagger \hat{c}_j + \hat{c}_j \hat{c}_i^\dagger = \delta_{ij}, \tag{3.25}$$

$$\left\{\hat{c}_i^\dagger, \hat{c}_j^\dagger\right\}_+ = 0, \quad \left\{\hat{c}_i, \hat{c}_j\right\}_+ = 0, \tag{3.26}$$

the rules for interchanging two operators thus being

$$\hat{c}_i \hat{c}_j = -\hat{c}_j \hat{c}_i,$$

$$\hat{c}_i^\dagger \hat{c}_j^\dagger = -\hat{c}_j^\dagger \hat{c}_i^\dagger,$$

$$i \neq j \Rightarrow \hat{c}_i^\dagger \hat{c}_j = -\hat{c}_j \hat{c}_i^\dagger,$$

$$i = j \Rightarrow \hat{c}_i^\dagger \hat{c}_i = \hat{1} - \hat{c}_i \hat{c}_i^\dagger$$

using the identity operator $\hat{1}$. The first two relations also show that the square of any Fermionic operator equals zero,

$$\hat{c}_i^\dagger \hat{c}_i^\dagger = \hat{c}_i \hat{c}_i = 0, \tag{3.27}$$

resulting from the Pauli principle[†] stating that no two fermions may occupy the same state.

From Equation 3.24, we calculate

$$-i\hbar\dot{\hat{p}} = \left(\epsilon_1 \hat{c}_1^\dagger \hat{c}_1 \hat{c}_1^\dagger \hat{c}_2 + \epsilon_2 \hat{c}_2^\dagger \hat{c}_2 \hat{c}_1^\dagger \hat{c}_2 - \mathbf{E}(t)\cdot(\mathbf{d}_{12}\hat{c}_1^\dagger \hat{c}_2 \hat{c}_1^\dagger \hat{c}_2 + \mathbf{d}_{21}\hat{c}_2^\dagger \hat{c}_1 \hat{c}_1^\dagger \hat{c}_2)\right)$$

$$- \left(\epsilon_1 \hat{c}_1^\dagger \hat{c}_2 \hat{c}_1^\dagger \hat{c}_1 + \epsilon_2 \hat{c}_1^\dagger \hat{c}_2 \hat{c}_2^\dagger \hat{c}_2 - \mathbf{E}(t)\cdot(\mathbf{d}_{12}\hat{c}_1^\dagger \hat{c}_2 \hat{c}_1^\dagger \hat{c}_2 + \mathbf{d}_{21}\hat{c}_1^\dagger \hat{c}_2 \hat{c}_2^\dagger \hat{c}_1)\right)$$

$$= \left(\epsilon_1 \hat{c}_1^\dagger \hat{c}_1 \hat{c}_1^\dagger \hat{c}_2 - \epsilon_1 \hat{c}_1^\dagger \hat{c}_2 \hat{c}_1^\dagger \hat{c}_1 + \epsilon_2 \hat{c}_2^\dagger \hat{c}_2 \hat{c}_1^\dagger \hat{c}_2 - \epsilon_2 \hat{c}_1^\dagger \hat{c}_2 \hat{c}_2^\dagger \hat{c}_2\right)$$

$$- \mathbf{E}(t)\cdot\left(\mathbf{d}_{21}\hat{c}_2^\dagger \hat{c}_1 \hat{c}_1^\dagger \hat{c}_2 - \mathbf{d}_{21}\hat{c}_1^\dagger \hat{c}_2 \hat{c}_2^\dagger \hat{c}_1\right),$$

[†] Named after Austrian-born Swiss and American physicist Wolfgang Pauli (1900–1958).

and, commuting the terms into normal order by using the Fermionic anticommutators,

$$-i\hbar\dot{\hat{p}} = \left(\epsilon_1 \hat{c}_1^\dagger (\hat{1} - \hat{c}_1^\dagger \hat{c}_1)\hat{c}_2 + \epsilon_1 \hat{c}_1^\dagger \hat{c}_1^\dagger \hat{c}_2 \hat{c}_1 - \epsilon_2 \hat{c}_2^\dagger \hat{c}_1^\dagger \hat{c}_2 \hat{c}_2 - \epsilon_2 \hat{c}_1^\dagger (\hat{1} - \hat{c}_2^\dagger \hat{c}_2)\hat{c}_2 \right)$$
$$- \mathbf{E}(t) \cdot \mathbf{d}_{21} \left(\hat{c}_2^\dagger (\hat{1} - \hat{c}_1^\dagger \hat{c}_1)\hat{c}_2 - \hat{c}_1^\dagger (\hat{1} - \hat{c}_2^\dagger \hat{c}_2)\hat{c}_1 \right). \tag{3.28}$$

The four-operator terms either cancel or disappear due to Equation 3.27 and we are left with

$$-i\hbar\dot{\hat{p}} = \left(\epsilon_1 \hat{c}_1^\dagger \hat{c}_2 - \epsilon_2 \hat{c}_1^\dagger \hat{c}_2 \right) - \mathbf{E}(t) \cdot \mathbf{d}_{21} \left(\hat{c}_2^\dagger \hat{c}_2 - \hat{c}_1^\dagger \hat{c}_1 \right)$$
$$= \left(\epsilon_1 - \epsilon_2 \right) \hat{p} + \mathbf{E}(t) \cdot \mathbf{d}_{21} \left(\hat{n}_1 - \hat{n}_2 \right). \tag{3.29}$$

Obviously, this equation couples to the equations for the occupation of the two levels, $\hat{n}_1 = \hat{c}_1^\dagger \hat{c}_1, \hat{n}_2 = \hat{c}_2^\dagger \hat{c}_2$. Deriving the equations of motion for these quantities in the same way, defining the energy difference in the system as $\hbar\omega_0 = \epsilon_2 - \epsilon_1$ as before and taking the expectation values lead to the optical Bloch equations (also called Maxwell–Bloch equations)

$$\begin{aligned}
-i\hbar\dot{p} + \hbar\omega_0 p &= \mathbf{E}(t) \cdot \mathbf{d}_{21}(n_1 - n_2), \\
-i\hbar\dot{n}_1 &= \mathbf{E}(t) \cdot (\mathbf{d}_{12}p - \mathbf{d}_{21}p^*), \\
-i\hbar\dot{n}_2 &= \mathbf{E}(t) \cdot (\mathbf{d}_{21}p^* - \mathbf{d}_{12}p).
\end{aligned}$$

These equations are completely analogous to the Bloch equations in spin-$\frac{1}{2}$-systems (where a two-level system in this chapter is not to be understood as a two-level system containing electrons *with* spin but rather as a model system whose Hilbert space is isomorphic to a spin-$\frac{1}{2}$-system).

Let us now briefly discuss the electron–hole picture at this level. Starting from the ground state where the electrons are in level 1, often only a small fraction of the electrons is lifted to the upper level. It is then frequently advantageous to count the empty spaces in the ground state rather than the electrons. These "holes" carry positive charge because the missing electron implies lack of compensation of nuclear charge. We define

$$\begin{aligned}
n_2 &= n_e &\quad &\text{upper level} \\
n_1 &= 1 - n_h &\quad &\text{lower level} \\
n_1 + n_2 &= 1 &\quad &\Rightarrow n_e = n_h \equiv n
\end{aligned}$$

and rewrite the optical Bloch equations

$$\begin{aligned}
\dot{p} &= -i\omega_0 p &&+ \frac{i}{\hbar}\mathbf{E}(t) \cdot \mathbf{d}_{21}(1 - 2n), \\
\dot{n} &= &&\frac{i}{\hbar}\mathbf{E}(t) \cdot (\mathbf{d}_{21}p - \mathbf{d}_{12}p^*).
\end{aligned}$$

For low excitation powers, where the predominant part of the carriers stays in the lower level, n may be neglected (linear case, $n \approx 0$) and a single equation of motion for the polarization remains

$$\dot{p} = -i\omega_0 p + \frac{i}{\hbar}\mathbf{E}(t) \cdot \mathbf{d}_{21}. \tag{3.30}$$

Solving this for a specific light field, one discovers that even after the end of excitation, the polarization remains forever, the reason being that relaxation processes have been neglected thus far. This may be remedied by introducing a phenomenological damping into the equations of motion

$$\begin{aligned}
\dot{p} &= -i\omega_0 p &&+ \frac{i}{\hbar}\mathbf{E}(t) \cdot \mathbf{d}_{21}(1 - 2n) - \frac{p}{T_2}, \\
\dot{n} &= &&\frac{i}{\hbar}\mathbf{E}(t) \cdot (\mathbf{d}_{21}p - \mathbf{d}_{12}p^*) - \frac{n}{T_1},
\end{aligned}$$

where T_1 and T_2 are relaxation times, which are usually derived by fits to experiment. The inclusion of relaxation times results in a broadening of the spectrum, which is called homogeneous because all emitters experience the same type of broadening mechanism; see Ref. [2]. The counterpart to this mechanism is inhomogeneous broadening, which is caused by disorder; see also Section 3.3; here, different emitters experience different environmental influences.

We now solve the polarization equation of motion in the relaxation time approach to illustrate that it leads to a Lorentzian lineshape of the polarization and absorption spectrum. Defining $\Gamma = 1/T_2$, we calculate

$$\frac{d}{dt}\left\{pe^{(i\omega_0+\Gamma)t}\right\} = \frac{i}{\hbar}\mathbf{E}(t)\cdot\mathbf{d}_{21}(1-2n(t))e^{(i\omega_0+\Gamma)t} \tag{3.31}$$

and formally integrate this equation, deriving

$$p(t) = \frac{i}{\hbar}\int_{-\infty}^{t}dt'\,e^{(i\omega_0+\Gamma)(t'-t)}\mathbf{E}(t')\cdot\mathbf{d}_{21}(1-2n(t')). \tag{3.32}$$

We now insert a monochromatic incident field

$$\mathbf{E}(t) = \frac{1}{2}\mathbf{E}_0\left\{e^{-i\omega t}+e^{i\omega t}\right\} \tag{3.33}$$

and assume a quasistationary or stationary state of the system, i.e., the carrier density changes little on the timescale of T_2. We may then replace $n(t')$ by $n(t)$ and get

$$p(t) = \frac{i}{2\hbar}\mathbf{E}_0\cdot\mathbf{d}_{21}(1-2n(t))\int_{-\infty}^{t}dt'\,e^{(i\omega_0+\Gamma)(t'-t)}\left\{e^{-i\omega t'}+e^{i\omega t'}\right\}. \tag{3.34}$$

This replacement actually means that the prediction of the system behavior is made based only on the present state and not on the history of the system. This is a so-called "memoryless" process obtained using the Markov approximation.

The integral in Equation 3.34 may now be solved, yielding

$$
\begin{aligned}
p(t) &= \frac{i}{2\hbar}\mathbf{E}_0\cdot\mathbf{d}_{21}(1-2n(t))e^{-(i\omega_0+\Gamma)t}\int_{-\infty}^{t}dt'\left\{e^{(i(\omega_0-\omega)+\Gamma)t'}+e^{(i(\omega_0+\omega)+\Gamma)t'}\right\}\\
&= \frac{i}{2\hbar}\mathbf{E}_0\cdot\mathbf{d}_{21}(1-2n(t))e^{-(i\omega_0+\Gamma)t}\left[\frac{e^{(i(\omega_0-\omega)+\Gamma)t'}}{i(\omega_0-\omega)+\Gamma}+\frac{e^{(i(\omega_0+\omega)+\Gamma)t'}}{i(\omega_0+\omega)+\Gamma}\right]_{-\infty}^{t}\\
&= \frac{i}{2\hbar}\mathbf{E}_0\cdot\mathbf{d}_{21}(1-2n(t))e^{-(i\omega_0+\Gamma)t}\left\{\frac{e^{(i(\omega_0-\omega)+\Gamma)t}}{i(\omega_0-\omega)+\Gamma}+\frac{e^{(i(\omega_0+\omega)+\Gamma)t}}{i(\omega_0+\omega)+\Gamma}\right\}.
\end{aligned}
\tag{3.35}
$$

Considering resonant or near-resonant excitation, the denominator $i(\omega_0+\omega)+\Gamma$ can be considered as large and the second addend may be neglected. This corresponds to the RWA, which was already used in getting from Equation 3.17 to 3.18 in Section 3.2.2. We are left with

$$p(t) = \frac{i}{2\hbar}\mathbf{E}_0 e^{-i\omega t}\cdot\mathbf{d}_{21}(1-2n(t))\frac{1}{i(\omega_0-\omega)+\Gamma}. \tag{3.36}$$

In the last fraction of Equation 3.36, $(i(\omega_0-\omega)+\Gamma)^{-1}$, the Lorentzian lineshape of the absorption spectrum becomes apparent, the absorption being approximately proportional to the imaginary part of the expression (see Ref. [7,8], Equation 3.53 in this chapter) and

$$\frac{1}{i(\omega_0 - \omega) + \Gamma} = \frac{\Gamma}{(\omega_0 - \omega)^2 + \Gamma^2} + i\frac{\omega_0 - \omega}{(\omega_0 - \omega)^2 + \Gamma^2}. \qquad (3.37)$$

In Section 3.3, we discuss the drawbacks of the relaxation time approach. Generally, it constitutes a simple and easily comprehensible procedure, where, however, the relaxation times must be inserted by hand and are not calculated by the theory.

3.3 The Solid-State System

We now move on to a much more complex system, the semiconductor. Not only do we encounter a system of many atoms forming an energy landscape known as band structure here (see also Chapter 1), but, in addition, we also need to take into account a large number of electrons. However, a semiconductor is what the diodes, amplifiers, laser diodes, optical modulators, and solar cells discussed in this book are made out of. In order to get a flavor of a many-body treatment of electron–photon interaction, the derivation of the semiconductor Bloch equations will be discussed briefly here. For an in-depth treatment, the reader is referred to Ref. [7], the references therein, and a wealth of original literature. Since a many-body problem can only be solved approximately, for every case the equations have to be examined separately and the proper approximations made. This is far beyond the scope of this chapter.

In optoelectronics, we customarily deal with direct-bandgap semiconductors. Here, we consider the simplest possible case, a direct-gap two-band semiconductor and derive the relevant equations of motion. It comprises a valence and a conduction band, where in the ground state, all the electrons are in the valence band and fill up its states completely. Similar to the two-level system, they may be raised to the conduction band by optical excitation. Figure 3.2 shows a simplified schematic of the direct-gap two-band semiconductor. Equations of motion for more complicated systems may be obtained by expanding the indices in the derivation below to all the existing bands; a derivation for a multiband system is thus straightforward but more cumbersome.

Looking at the semiconductor–light interaction, we need to keep in mind that the size of an atom is in the range of a single Ångström, while the wavelength of visible light is approximately 400–700 nm, i.e., 4000–7000 Ångström. The electric field is thus approximately constant over the extension of a single atom, and all quantities in Maxwell's equations[†] are averaged over a large number of atoms.

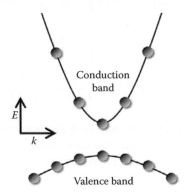

FIGURE 3.2   Schematic of a two-band direct-gap semiconductor.

[†] Named after Scottish physicist James Clerk Maxwell (1831–1879).

The Hamiltonian of the solid-state system is made up of three contributions,

$$\hat{H} = \hat{H}_{el} + \hat{H}_L = (\hat{H}_0 + \hat{H}_C) + \hat{H}_L, \qquad (3.38)$$

where $\hat{H}_{el}$ is the Hamiltonian of the electronic system, consisting of the sum of the individual single particle contributions, $\hat{H}_0$, and the Coulomb interaction between the electrons, $\hat{H}_C$. As before, $\hat{H}_L$ denotes the interaction with the light field.

The structure of an inorganic semiconductor is characterized by its crystallinity, i.e., periodicity. In other words, the environment of an electron is the same in position $\mathbf{r}$ and position $\mathbf{r} + \mathbf{R}_n$, where $\mathbf{R}_n$ is any lattice vector, i.e., a vector linking two identical sites in an infinite lattice. The periodicity is taken into account in the Bloch theorem[†] for the wave functions,

$$\psi_\lambda(\mathbf{k}, \mathbf{r} + \mathbf{R}_n) = e^{i\mathbf{k}\cdot\mathbf{R}_n}\psi_\lambda(\mathbf{k}, \mathbf{r}). \qquad (3.39)$$

Consequently, the Hamiltonian is expanded in electronic Bloch wave functions,

$$\psi_\lambda(\mathbf{k}, \mathbf{r}) \approx e^{i\mathbf{k}\cdot\mathbf{r}} \frac{u_\lambda(\mathbf{k}, \mathbf{r})}{L^{\frac{3}{2}}}. \qquad (3.40)$$

L^3 is the volume of the crystal and λ denotes the band index, comprising the conduction band $\lambda = c$ and the valence band $\lambda = v$. Bloch functions take into account the lattice periodicity of the system by including a lattice periodic function $u_\lambda(\mathbf{k}, \mathbf{r})$,

$$u_\lambda(\mathbf{k}, \mathbf{r} + \mathbf{R}_n) = u_\lambda(\mathbf{k}, \mathbf{r}). \qquad (3.41)$$

The introduction of the crystal volume L^3 necessitates further explanation, as Equation 3.39 is only valid in an infinite system. Customary crystals are so large that the assumption of infinity is quite fitting. Since for a proper normalization of the wave functions, one needs a finite crystal volume, we consider the (approximately infinite) crystal to be made up of mesoscopic cubes of size L^3 and demand periodic boundary conditions of the wave function, i.e., the wave function has to have the same value on either side of the cube.

For the single particle contribution, we derive

$$\hat{H}_0 = \sum_{\mathbf{k},\lambda=v,c} \epsilon_\lambda(\mathbf{k})\, \hat{c}^\dagger_{\lambda,\mathbf{k}} \hat{c}_{\lambda,\mathbf{k}}$$

$$= \sum_\mathbf{k} \epsilon_v(\mathbf{k})\, \hat{c}^\dagger_{v,\mathbf{k}} \hat{c}_{v,\mathbf{k}} + \sum_\mathbf{k} \epsilon_c(\mathbf{k})\, \hat{c}^\dagger_{c,\mathbf{k}} \hat{c}_{c,\mathbf{k}}. \qquad (3.42)$$

$\hat{c}_{\lambda,\mathbf{k}}$ ($\hat{c}^\dagger_{\lambda,\mathbf{k}}$) is the annihilation (creation) operator of an electron in state $\lambda, \mathbf{k}$.

Since the energies $\epsilon_\lambda(\mathbf{k})$ are rather difficult to assess, parabolic bands are often assumed, corresponding to the effective mass approximation. At the extrema of the band structure, the linear term in the Taylor expansion becomes zero, such that a parabolic approximation may be made. Formally, a constant called the "effective mass," $m_{\lambda,\mathrm{eff}} = \dfrac{\hbar^2}{\left.\frac{\partial^2\epsilon_\lambda(\mathbf{k})}{\partial k^2}\right|_{k=0}}$, may then be introduced,

$$\epsilon_\lambda(\mathbf{k}) = \epsilon_\lambda(\mathbf{k} = 0) + \frac{1}{2}\frac{\partial^2\epsilon_\lambda(\mathbf{k})}{\partial k^2} + O(k^3)$$

$$\approx \epsilon_\lambda(\mathbf{k} = 0) + \frac{\hbar^2 k^2}{2m_{\lambda,\mathrm{eff}}}. \qquad (3.43)$$

[†] Named after Swiss physicist Felix Bloch (1905–1983).

Here, $O(k^3)$ denotes higher-order terms. For the Bloch functions, this implies that the lattice-periodic part becomes momentum independent, i.e., $u_\lambda(\mathbf{k}, \mathbf{r}) \approx u_\lambda(\mathbf{k} = 0, \mathbf{r})$. This approximation is always valid in a certain range around an extremum of the band structure but is used extensively in calculations even far away from these points. In the effective-mass approximation, the electrons behave like free particles with the effective mass $m_{\lambda,\text{eff}}$ rather than the free-particle mass. $m_{\lambda,\text{eff}}$ may be positive or negative, examples being the conduction and valence band in the case of the two-band semiconductor we are looking at. When the sign is negative, this is often included into the energy by transferring to the electron–hole picture. Furthermore, the effective mass, in general, is not independent of direction and must be written as a tensor,

$$\left(\frac{1}{m_{\lambda,\text{eff}}}\right)_{\alpha\beta} = \frac{1}{\hbar^2} \frac{\partial^2 \epsilon_\lambda(\mathbf{k})}{\partial k_\alpha k_\beta}\bigg|_{k=0}. \tag{3.44}$$

The direction dependence of the effective mass is discussed, e.g., in Refs. [9–11].

Returning to light–matter coupling in semiconductors, we have seen that light couples to electrical dipoles,

$$\hat{H}_L = -\mathbf{E}(t) \cdot (e\hat{\mathbf{r}}) = -\mathbf{E}(t) \cdot \hat{\mathbf{d}}. \tag{3.45}$$

Expanding the interaction Hamiltonian in these wave functions yields

$$\begin{aligned}
\hat{H}_L &= -e\,\mathbf{E}(t) \cdot \sum_{\mathbf{k},\lambda,\lambda'} \langle \lambda', \mathbf{k} | \, \mathbf{r} \, | \lambda, \mathbf{k} \rangle \, \hat{c}^\dagger_{\lambda',\mathbf{k}} \hat{c}_{\lambda,\mathbf{k}} \\
&= -\mathbf{E}(t) \cdot \left[\sum_{\mathbf{k},\lambda,\lambda',\lambda \neq \lambda'} \mathbf{d}_{\lambda',\lambda}(\mathbf{k}) \, \hat{c}^\dagger_{\lambda',\mathbf{k}} \hat{c}_{\lambda,\mathbf{k}} \right] \\
&= -\mathbf{E}(t) \cdot \sum_{\mathbf{k}} \left[\mathbf{d}_{cv}(\mathbf{k}) \, \hat{c}^\dagger_{c,\mathbf{k}} \hat{c}_{v,\mathbf{k}} + \mathbf{d}_{vc}(\mathbf{k}) \, \hat{c}^\dagger_{v,\mathbf{k}} \hat{c}_{c,\mathbf{k}} \right] \\
&= -\mathbf{E}(t) \cdot \hat{\mathbf{P}}, \tag{3.46}
\end{aligned}$$

where $\hat{\mathbf{P}}$ is the optical interband polarization. $\mathbf{d}_{\lambda',\lambda}(\mathbf{k})$ are the band and momentum-dependent optical transition dipole matrix elements with in-plane momentum $\mathbf{k}$ and subband labels λ, λ', corresponding to the transition dipole matrix elements $\mathbf{d}_{12}, \mathbf{d}_{21}$ in the two-level system.

To obtain Equation 3.46, the following approximations are made: (1) the photon momentum is neglected and (2) intraband transitions are not considered. Rather than retracing the entire derivation,[†] let us discuss these approximations:

1. Neglect of the photon momentum: In this book, we concentrate on light in and around the visible range, i.e., with wavelengths roughly between 400 and 700 nm. For a central wavelength of 550 nm, the wave number is $k = 2\pi/\lambda = 11 \times 10^6$ m^{-1}, which is very small compared to typical wave vectors in a semiconductor band structure. Photonic transitions are consequently vertical in a band diagram.

2. Neglect of intraband transitions: Here, only interband transitions will be considered, assuming an excitation at an energy near the band gap. Suppression of intraband transitions in a regular structure becomes obvious from the vanishing photon momentum as discussed earlier.

Looking at the system Hamiltonian, Equation 3.38, we still need to calculate the term $\hat{H}_C$ describing Coulomb interaction between the carriers. $\hat{H}_C$ is specified using the so-called jellium model, which

---

† See Ref. [7] for details.

substitutes the ionic background of the lattice with a smooth background. The derivation is not included here, but may be found in Ref. [7]. The final expression of the Hamiltonian is

$$
\begin{aligned}
\hat{H}_C &= \sum_{\mathbf{k},\mathbf{k}',\mathbf{q}\neq 0,\lambda,\lambda'=v,c} V_q \, \hat{c}^\dagger_{\lambda,\mathbf{k}+\mathbf{q}} \hat{c}^\dagger_{\lambda',\mathbf{k}'-\mathbf{q}} \hat{c}_{\lambda',\mathbf{k}'} \hat{c}_{\lambda,\mathbf{k}} \\
&= \sum_{\mathbf{k},\mathbf{k}',\mathbf{q}\neq 0} V_q \left[\hat{c}^\dagger_{v,\mathbf{k}+\mathbf{q}} \hat{c}^\dagger_{v,\mathbf{k}'-\mathbf{q}} \hat{c}_{v,\mathbf{k}'} \hat{c}_{v,\mathbf{k}} + \hat{c}^\dagger_{c,\mathbf{k}+\mathbf{q}} \hat{c}^\dagger_{c,\mathbf{k}'-\mathbf{q}} \hat{c}_{c,\mathbf{k}'} \hat{c}_{c,\mathbf{k}} + 2\hat{c}^\dagger_{c,\mathbf{k}+\mathbf{q}} \hat{c}^\dagger_{v,\mathbf{k}'-\mathbf{q}} \hat{c}_{v,\mathbf{k}'} \hat{c}_{c,\mathbf{k}} \right],
\end{aligned}
\tag{3.47}
$$

where the second line spells out the interaction for our two-band model. $\hat{H}_C$ thus describes the repulsion between electrons in the valence band, electrons in the conduction band and between valence and conduction band. V_q is the Fourier transform of the Coulomb potential,

$$
V(\mathbf{r}) = \frac{e^2}{4\pi\mathcal{E}_0\mathcal{E}_r r},
\tag{3.48}
$$

$$
V_q = \int \frac{\mathrm{d}^3 r}{L^3} V(\mathbf{r}) e^{i\mathbf{q}\cdot\mathbf{r}} = \frac{e^2}{\mathcal{E}_0\mathcal{E}_r L^3 q^2},
\tag{3.49}
$$

where for the calculation a convergence generating factor needs to be used. $\mathcal{E}_0$ and $\mathcal{E}_r$ are the vacuum and relative material permittivity, respectively. The sum in Equation 3.47 omits the case $\mathbf{q} = 0$ where obviously the Coulomb matrix element has a singularity; this term is canceled by the ionic background charge in the jellium model.

The complete Hamiltonian is thus

$$
\begin{aligned}
\hat{H} &= \hat{H}_0 + \hat{H}_C + \hat{H}_L \\
&= \sum_{\mathbf{k},\lambda=v,c} \epsilon_\lambda(\mathbf{k}) \, \hat{c}^\dagger_{\lambda,\mathbf{k}} \hat{c}_{\lambda,\mathbf{k}} + \sum_{\mathbf{k},\mathbf{k}',\mathbf{q}\neq 0,\lambda,\lambda'=v,c} V_q \, \hat{c}^\dagger_{\lambda,\mathbf{k}+\mathbf{q}} \hat{c}^\dagger_{\lambda',\mathbf{k}'-\mathbf{q}} \hat{c}_{\lambda',\mathbf{k}'} \hat{c}_{\lambda,\mathbf{k}} \\
&\quad - \mathbf{E}(t) \cdot \left[\sum_{\mathbf{k},\lambda,\lambda',\lambda\neq\lambda'} \mathbf{d}_{\lambda',\lambda}(\mathbf{k},\mathbf{k}) \, \hat{c}^\dagger_{\lambda',\mathbf{k}} \hat{c}_{\lambda,\mathbf{k}} \right].
\end{aligned}
\tag{3.50}
$$

This Hamiltonian is now going to be used for the calculation of the optical polarization $\hat{\mathbf{P}}$,

$$
\hat{\mathbf{P}} = \sum_{\mathbf{k}} \left(\mathbf{d}_{vc} \, \hat{c}^\dagger_{v,\mathbf{k}} \hat{c}_{c,\mathbf{k}} + \mathbf{d}_{cv} \, \hat{c}^\dagger_{c,\mathbf{k}} \hat{c}_{v,\mathbf{k}} \right)
\tag{3.51}
$$

$$
= \sum_{\mathbf{k}} \left(\mathbf{d}_{vc} \, \hat{c}^\dagger_{v,\mathbf{k}} \hat{c}_{c,\mathbf{k}} + \text{h.c.} \right) = \sum_{\mathbf{k}} \left(\mathbf{d}_{vc} \, \hat{p}_{vc,\mathbf{k}} + \text{h.c.} \right),
$$

where $p_{vc,\mathbf{k}} = \langle \hat{p}_{vc,\mathbf{k}} \rangle$ are the microscopic interband coherences or polarizations and h.c. denotes the Hermitian conjugate. The optical polarization is a quantity that describes the reaction of matter to a light field, coupling light (electric field $\mathbf{E}(\mathbf{r}, t)$) to matter in the wave equation

$$
\Delta\mathbf{E}(\mathbf{r}, t) - \left(\frac{n}{c} \right)^2 \frac{\partial^2}{\partial t^2} \mathbf{E}(\mathbf{r}, t) = \mu_0 \frac{\partial^2}{\partial t^2} \mathbf{P}(\mathbf{r}, t)
\tag{3.52}
$$

with the index of refraction n and the vacuum permeability μ_0. The optical polarization is a fundamental parameter from which numerous optical variables can be calculated, e.g., the absorption α or laser gain g being given as

$$
\alpha(\omega) = -g(\omega) = \frac{\omega}{nc} \text{Im} \left\{ \frac{P(\omega)}{L^3 E(\omega)} \right\},
\tag{3.53}
$$

see Ref. [8]. In Equation 3.53, the index of refraction n was assumed to be constant in the region of interest. Calculation of the optical polarization thus gives us deep insight into the optical properties of matter and allows us to simulate optical devices such as the ones discussed in this book.

Returning to the actual calculation of the optical polarization, the next step is the insertion of the many-body Hamiltonian into the Heisenberg equation of motion for the microscopic polarizations

$$\frac{d}{dt}\hat{c}^\dagger_{v,\mathbf{k}}\hat{c}_{c,\mathbf{k}} = \frac{i}{\hbar}\left[\hat{H}, \hat{c}^\dagger_{v,\mathbf{k}}\hat{c}_{c,\mathbf{k}}\right]. \tag{3.54}$$

Calculating the appropriate commutators and taking the expectation values on both sides renders

$$\left[i\hbar\frac{d}{dt} - (\epsilon_c(\mathbf{k}) - \epsilon_v(\mathbf{k}))\right]p_{vc,\mathbf{k}} = (n_{c,\mathbf{k}} - n_{v,\mathbf{k}})\mathbf{d}_{cv}\cdot\mathbf{E}(t)$$

$$+ \sum_{\mathbf{k}',\mathbf{q}\neq 0} V_q\left[\langle\hat{c}^\dagger_{c,\mathbf{k}'+\mathbf{q}}\hat{c}^\dagger_{v,\mathbf{k}-\mathbf{q}}\hat{c}_{c,\mathbf{k}'}\hat{c}_{c,\mathbf{k}}\rangle + \langle\hat{c}^\dagger_{v,\mathbf{k}'+\mathbf{q}}\hat{c}^\dagger_{v,\mathbf{k}-\mathbf{q}}\hat{c}_{v,\mathbf{k}'}\hat{c}_{c,\mathbf{k}}\rangle\right.$$

$$\left. +\langle\hat{c}^\dagger_{v,\mathbf{k}}\hat{c}^\dagger_{c,\mathbf{k}'-\mathbf{q}}\hat{c}_{c,\mathbf{k}'}\hat{c}_{c,\mathbf{k}-\mathbf{q}}\rangle + \langle\hat{c}^\dagger_{v,\mathbf{k}}\hat{c}^\dagger_{v,\mathbf{k}'-\mathbf{q}}\hat{c}_{v,\mathbf{k}'}\hat{c}_{c,\mathbf{k}-\mathbf{q}}\rangle\right], \tag{3.55}$$

where $n_{c,\mathbf{k}} = \langle\hat{c}^\dagger_{c,\mathbf{k}}\hat{c}_{c,\mathbf{k}}\rangle$ and $n_{v,\mathbf{k}} = \langle\hat{c}^\dagger_{v,\mathbf{k}}\hat{c}_{v,\mathbf{k}}\rangle$ are the carrier densities in the conduction and valence band, respectively. The transition dipole matrix element was assumed to be real, $\mathbf{d}_{cv} = \mathbf{d}_{vc}$.

Obviously, the polarization as a two-operator quantity on the left-hand side of Equation 3.55 couples to four-operator quantities on the right side via the Coulomb interaction. Computing the equation of motion for these four-operator terms will in turn lead to a coupling to six-operator quantities, which will then couple to eight-operator quantities and so on, with ever-rising computational cost. This so-called *infinite hierarchy problem*, which is an expression of the many-body character of the electron system, necessitates a truncation at some level. A relatively simple approximation at two-operator level is the dynamic Hartree–Fock approximation.[†] Roughly speaking, it assumes that any particle experiences the impact of all other particles as a mean field created by them. Technically, in Hartree–Fock approximation, we split four-operator quantities into all possible combinations of two-operator quantities. For example, the first Coulomb term of Equation 3.55 yields

$$\langle\hat{c}^\dagger_{c,\mathbf{k}'+\mathbf{q}}\hat{c}^\dagger_{v,\mathbf{k}-\mathbf{q}}\hat{c}_{c,\mathbf{k}'}\hat{c}_{c,\mathbf{k}}\rangle \approx \langle\hat{c}^\dagger_{c,\mathbf{k}'+\mathbf{q}}\hat{c}_{c,\mathbf{k}}\rangle\langle\hat{c}^\dagger_{v,\mathbf{k}-\mathbf{q}}\hat{c}_{c,\mathbf{k}'}\rangle - \langle\hat{c}^\dagger_{c,\mathbf{k}'+\mathbf{q}}\hat{c}_{c,\mathbf{k}'}\rangle\langle\hat{c}^\dagger_{v,\mathbf{k}-\mathbf{q}}\hat{c}_{c,\mathbf{k}}\rangle$$

$$\approx \delta_{\mathbf{k},\mathbf{k}'+\mathbf{q}}\langle\hat{c}^\dagger_{c,\mathbf{k}}\hat{c}_{c,\mathbf{k}}\rangle\langle\hat{c}^\dagger_{v,\mathbf{k}-\mathbf{q}}\hat{c}_{c,\mathbf{k}-\mathbf{q}}\rangle - \delta_{\mathbf{k}',\mathbf{k}-\mathbf{q}}\langle\hat{c}^\dagger_{c,\mathbf{k}'}\hat{c}_{c,\mathbf{k}'}\rangle\langle\hat{c}^\dagger_{v,\mathbf{k}}\hat{c}_{c,\mathbf{k}}\rangle$$

$$= \delta_{\mathbf{k},\mathbf{k}'+\mathbf{q}}n_{c,\mathbf{k}}p_{vc,\mathbf{k}-\mathbf{q}},$$

using $\mathbf{q} \neq 0$. In the second and third lines, we have taken into account that only terms that are diagonal in momentum are excited in a spatially homogeneous system. Approximating the other terms likewise, the resulting equation of motion for $p_{vc,\mathbf{k}}$ is

$$\left[i\hbar\frac{d}{dt} - (e_c(\mathbf{k}) - e_v(\mathbf{k}))\right]p_{vc,\mathbf{k}} = [n_{c,\mathbf{k}} - n_{v,\mathbf{k}}]\left(\mathbf{d}_{cv}\cdot\mathbf{E}(t) + \sum_{\mathbf{q}\neq\mathbf{k}}V_{|\mathbf{k}-\mathbf{q}|}p_{vc,\mathbf{q}}\right), \tag{3.56}$$

[†] Named after English mathematician and physicist Douglas Rayner Hartree (1897–1958) and Soviet physicist Vladimir Aleksandrovich Fock (1898–1974).

with renormalized energies

$$e_\lambda(\mathbf{k}) = \epsilon_\lambda(\mathbf{k}) - \sum_{\mathbf{q} \neq \mathbf{k}} V_{|\mathbf{k}-\mathbf{q}|} n_{\lambda,\mathbf{q}} \equiv \epsilon_\lambda(\mathbf{k}) + \Sigma_{exc,\lambda}(\mathbf{k}), \tag{3.57}$$

where $\Sigma_{exc}(\mathbf{k})$ is the exchange self-energy.

Equation 3.56 is not yet closed, but couples to the carrier densities $n_{c,\mathbf{k}}$, $n_{v,\mathbf{k}}$. To close the system of equations, the equations of motion for these quantities are derived in analogy to the prior derivation of Equation 3.56, and the following equations are obtained:

$$\frac{d}{dt} n_{c,\mathbf{k}} = -\frac{d}{dt} n_{v,\mathbf{k}} = -\frac{2}{\hbar} \operatorname{Im} \left[\left(\mathbf{d}_{cv} \cdot \mathbf{E}(t) + \sum_{\mathbf{q} \neq \mathbf{k}} V_{|\mathbf{k}-\mathbf{q}|} p_{vc,\mathbf{q}} \right) p_{vc,\mathbf{k}}^* \right]. \tag{3.58}$$

We now have a closed system of equations for $p_{vc,\mathbf{k}}$, $n_{c,\mathbf{k}}$, and $n_{v,\mathbf{k}}$, which constitute the *semiconductor Bloch equations* in Hartree–Fock approximation. For a given electromagnetic light field, these equations may be solved numerically (starting with the unpolarized system before excitation and calculating the temporal evolution).

Let us now look at the interaction between the carriers, i.e., the Coulomb effects. These are marked by the Coulomb potential V_q and occur in the renormalized energies $e_\lambda(\mathbf{k})$ and in a renormalization to the Rabi frequency,

$$\Omega_{\text{Coulomb},\mathbf{k}} = \frac{\left| \mathbf{d}_{cv} \cdot \mathbf{E}(t) + \sum_{\mathbf{q} \neq \mathbf{k}} V_{|\mathbf{k}-\mathbf{q}|} p_{vc,\mathbf{q}} \right|}{\hbar}. \tag{3.59}$$

A characteristic of the Hartree–Fock approximation is thus that Coulomb effects may be represented as renormalization terms.

Neglecting the Coulomb interaction, we regain the *optical Bloch equations*:

$$\left[i\hbar \frac{d}{dt} - (\epsilon_c(\mathbf{k}) - \epsilon_v(\mathbf{k})) \right] p_{vc,\mathbf{k}} = [n_{c,\mathbf{k}} - n_{v,\mathbf{k}}] \mathbf{d}_{cv} \cdot \mathbf{E}(t) p_{vc,\mathbf{q}}, \tag{3.60}$$

$$\frac{d}{dt} n_{c,\mathbf{k}} = -\frac{d}{dt} n_{v,\mathbf{k}} = -\frac{2}{\hbar} \operatorname{Im} \left[\mathbf{d}_{cv} \cdot \mathbf{E}(t) p_{vc,\mathbf{k}}^* \right]. \tag{3.61}$$

Here, electrons at different $\mathbf{k}$-values do not couple to each other, i.e., the semiconductor is described as an ensemble of uncoupled two-level systems. Only the Coulomb interaction couples the different wave vectors and makes the semiconductor a true many-body system.

Using the optical Bloch equations and an approach similar to the derivation of Equation 3.36 in Section 3.2 and Equation 3.53, one obtains an often-used formula for the semiconductor absorption or gain [12],

$$\alpha(\hbar\omega) = -g(\hbar\omega) = \frac{\pi e^2}{nc\mathcal{E}_0 m_e^2 \omega} |\hat{\mathbf{e}} \cdot \mathbf{d}_{cv}|^2 \int_0^\infty dE \, \rho_r(E) \frac{\Gamma/(2\pi)}{(E_g + E - \hbar\omega)^2 + (\Gamma/2)^2} \{f_v(E) - f_c(E)\}, \tag{3.62}$$

$\rho_r(E)$ is the density of states in a three-dimensional parabolic band structure,

$$\rho_r(E) = \frac{1}{2\pi^2} \left(\frac{2m_r}{\hbar^2} \right)^{3/2} E^{1/2}, \tag{3.63}$$

which occurs when transferring from an integral over equally spaced $\mathbf{k}$-values to an energy integral [8,12]. The effective-mass approximation has been used, and m_r is the reduced mass originating from the effective masses of hole and electron. $\hat{e}$ is the unit vector pointing in the direction of the light field, and m_e is the free electron mass. The carrier densities n_v, n_c have been replaced by Fermi–Dirac distributions $f_v(E)$, $f_c(E)$, assuming that scattering is fast enough to keep the carrier distributions in thermal equilibrium. This is in general a good approximation especially in a high-density (laser) system; although hole burning or other deviations from an equilibrium distribution may occur [13,14], kinetic holes in the carrier distribution are often smoothed out by the energy integral in Equation 3.62.

Equation 3.62 is thus obtained making, apart from common assumptions such as an infinite crystal lattice, the following approximations:

- Two-band approximation
- Effective-mass approximation, including the assumption of direction-independence of the effective mass
- Neglect of Coulomb interaction between the carriers; many body effects are only included as a phenomenological broadening
- Assumption of a carrier distribution in thermal equilibrium
- Dephasing time $T_2 = 1/\Gamma$ for the decline of the polarization

Looking at laser gain, the spectrum is rather unstructured and deviations between theory and experiment are often compensated by fitting the free parameters of the theory, such as carrier density and broadening; usually, the bandgap is also obtained by fits to experiment. Many gain spectra may thus be fitted with Equation 3.62. A predictive theory, however, needs to take into account many-body effects and closely look at the impact of the other simplifications mentioned above for the structure and experimental conditions it simulates and has to go far beyond the approximations made in Equation 3.62. An excellent overview of the effects of different approximations is given in Ref. [8].

The Hartree–Fock approximation that was introduced above is a widespread method to include many-body effects while keeping computation time reasonable, yet there are many other possible approaches. These include, e.g., (1) the linear case, where the carrier densities are assumed to be zero. The polarization may then be calculated in Hartree–Fock approximation. (2) Calculations beyond Hartree–Fock, where the Coulomb terms are treated in a more elaborate way. In particular, the Coulomb terms also include carrier scattering, which is not mirrored in the Hartree–Fock approximation. Therefore, Hartree–Fock equations require an artificial relaxation to be included. Instead, scattering can be included in the second Born approximation[†] by deriving the equations of motion for four-operator terms and factorizing at this level. This avoids having to include an ad hoc-scattering time, which can only be determined experimentally, and additional changes in temperature, carrier density and other experimental parameters.

Another fact that has not been discussed so far is that realistic semiconductor structures are never perfect. There has been incredible progress in the growth of semiconductor structures since Wolfgang Pauli termed solid-state physics as "dirt physics" in 1931 [15], but there still remain some inhomogeneities and disturbances of the crystal lattice. As was mentioned previously (see Section 3.2.3), disorder introduces a different kind of broadening termed inhomogeneous, where the lineshape is often assumed to be Gaussian.

Thus, different effects are manifest in the lineshape and linewidth of resonances; in other words, the lineshape says a lot about the processes determining it. To illustrate this, Figure 3.3 [16] shows a calculation of a single absorption resonance including homogeneous and inhomogeneous broadening in differing amounts. Both lineshapes include a homogeneous phonon broadening corresponding to a temperature of 30 K; for the calculation of the solid line, an inhomogeneous broadening of 0.287 meV has been added, while for the dotted line only half the disorder strength was included and an additional phenomenological homogeneous broadening of 0.066 meV was introduced via a T_2-time approach. It becomes obvious that homogeneous broadening shifts oscillator strength to the tails of the resonance. In fact, the tails of a Lorentzian lineshape become so strong that unphysical effects may appear, e.g., absorption below the gain resonance of a laser spectrum.

[†] Named after German physicist and mathematician Max Born (1882–1970).

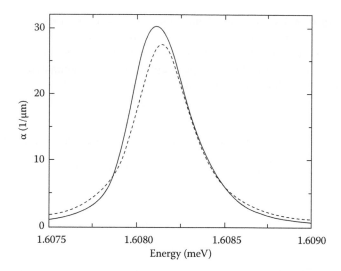

FIGURE 3.3 Absorption of a single resonance for two cases with equal total width, but differing amounts of homogeneous and inhomogeneous broadening. The solid line corresponds to phonon broadening for a temperature of 30 K and disorder broadening of 0.287 meV. For the dashed line, the disorder strength was halved and an additional phenomenological homogeneous broadening of 0.066 meV was introduced. As expected, greater homogeneous broadening results in a more symmetric line shape and stronger tails. (From A. Thränhardt et al., *Phys. Rev. B.*, 68, 035316, 2003.)

3.4 Quantization of the Light Field

Thus far, we have worked with semiclassical descriptions, i.e., including a classical light field. Yet it was shown in Section 3.2.1 that any system requires a spontaneous emission term even in a simple model, whereas in Section 3.2.2 it became apparent that the semiclassical model does not offer a term to that extent. Looking at the device aspect, we remark that the simplest optoelectronic device, the LED, emits light entirely due to spontaneous emission, and the laser diode depends on it for start-up. Quantum effects also play a role, e.g., in nanolasers and single quantum dots in resonators [17]. We thus briefly discuss quantization of the electromagnetic field. Due to the rich range of phenomena involved we leave the major discussion to the wealth of textbooks on the market, e.g., Refs. [18,19] among many others.

Consider again a two-level atom. The Hamiltonian of the fully quantized system consists of three parts,

$$\hat{H} = \hat{H}_{\text{atom}} + \hat{H}_{\text{light}} + \hat{H}_{\text{int}}, \tag{3.64}$$

where $\hat{H}_{\text{atom}}$ corresponds to the Hamiltonian of the undisturbed electron system (formerly $\hat{H}_0$), $\hat{H}_{\text{light}}$ is the Hamiltonian of the undisturbed light system, and $\hat{H}_{\text{int}}$ describes the interaction between the two.

Looking at a system in free space, we need to take into account a variety of photon modes characterized by their wave vectors $\mathbf{q}$ pointing in all directions. Spontaneous emission of a two-level atom can then be calculated. Starting from the excited two-level atom, the system wave function can be written as

$$|\Psi(t)\rangle = \hat{a}(t)e^{-i\omega_0 t}|e;0\rangle + \sum_{\mathbf{q},s} \hat{b}_{\mathbf{q},s}(t)e^{-i\omega_q t}|g;1_{\mathbf{q},s}\rangle, \tag{3.65}$$

i.e., either the system is in its initial state $|e;0\rangle$ where the atom is excited and there is no photon, or a photon with wave vector $\mathbf{q}$ and polarization s has been emitted and the atom thus returned to its ground state, yielding a wave vector $|g;1_{\mathbf{q},s}\rangle$. Here, the energy of the photon will correspond to the energy difference of the two-level system $\hbar\omega_0$, thus the absolute value of the photon wave vector $q = \omega_0/c$ is well defined, but not its direction. In 1930, Weisskopf[†] and Wigner[‡] found an irreversible exponential decay of the atom with a decay rate equal to Einstein's A coefficient (see Equation 3.11) [20].

We briefly discuss another case here, where there is only one photon mode at the energy $\hbar\omega_C$. Consider, e.g., a two-level atom in a cavity only allowing a specific photon mode, leading us to the Jaynes–Cummings model [21]. The undisturbed photon Hamiltonian is

$$\hat{H}_{light} = \hbar\omega_C\,\hat{a}^\dagger\hat{a}. \tag{3.66}$$

Here, $\hat{a}^\dagger$ ($\hat{a}$) is a photon creation (annihilation) operator. The atom Hamiltonian remains unchanged,

$$\hat{H}_{atom} = \epsilon_1|1\rangle\langle1| + \epsilon_2|2\rangle\langle2| = \hbar\omega_0|2\rangle\langle2|, \tag{3.67}$$

by setting $\epsilon_1 = 0$ and $\epsilon_2 = \hbar\omega_0$. We introduce the ladder operators

$$\hat{\sigma}^+ = |2\rangle\langle1|,$$
$$\hat{\sigma}^- = |1\rangle\langle2|, \tag{3.68}$$

expressing the transfer of the electron from level 1 to level 2 and vice versa, respectively. Next, we rewrite $\hat{H}_{atom}$ as

$$\hat{H}_{atom} = \hbar\omega_0\hat{\sigma}^+\hat{\sigma}^- \tag{3.69}$$

and express the interaction Hamiltonian in RWA as

$$\hat{H}_{int} = \hbar g(\hat{a}^\dagger\hat{\sigma}^- + \hat{a}\hat{\sigma}^+), \tag{3.70}$$

where g is the coupling strength. The RWA expresses that, whenever a photon is created, the electron has to drop from the excited state to the ground state ($\hat{a}^\dagger\hat{\sigma}^-$) and when a photon is absorbed, the electron has to jump from the ground state into the excited state ($\hat{a}\hat{\sigma}^+$).

The Jaynes–Cummings Hamiltonian may then be solved by writing it in a matrix representation and diagonalizing this matrix. Looking at Equation 3.65, the wave functions must be linear combinations of $|2, n\rangle$ (the atom is in the excited state, state 2, and there are n photons in the system) and state $|1, n + 1\rangle$ (the atom is in the ground state, state 1, and there are $n + 1$ photons in the system). However, $|2, n\rangle$ and $|1, n+1\rangle$ are not the eigenstates of the system any more. They are often called the bare states while the true eigenstates, which result from diagonalization of Hamiltonian (3.64), are referred to as dressed states. On resonance, they are found to be

$$|S\rangle = \frac{|2, n\rangle + |1, n + 1\rangle}{\sqrt{2}},$$

$$|A\rangle = \frac{|2, n\rangle - |1, n + 1\rangle}{\sqrt{2}}. \tag{3.71}$$

Similar to the semiclassical case, for a system initially in the upper state this results in an oscillation:

$$|c_{en}(t)|^2 = \cos^2\left(g\sqrt{n + 1}t\right),$$

$$|c_{gn+1}(t)|^2 = \sin^2\left(g\sqrt{n + 1}t\right). \tag{3.72}$$

[†] Victor Frederick Weisskopf, Austrian-born American physicist (1908–2002).
[‡] Eugene Paul Wigner, Hungarian-American physicist (1902–1995).

There is, however, an important difference for the case $n = 0$, i.e., when no light field is present: While from Equation (3.20), we obtain $\Omega = 0$ for the semiclassical case (i.e., no Rabi oscillations occur), the quantum optical calculation yields the so-called vacuum Rabi oscillations with nonzero vacuum Rabi frequency. The excited atom periodically emits and reabsorbs a photon.

An interesting case also occurs if the two-level system interacts not with a single photon state $|n\rangle$, but with a coherent superposition of photon states. In this case, the probability of finding the system in the excited state will dephase due to the different Rabi frequencies of the various photon states; however, at certain times there will be a revival of the excited state because of constructive interference. This "quantum revival" was predicted theoretically in 1980 [22] and experimentally observed in 1987 [23].

3.5 Role of the Electromagnetic Environment

Thus far, we have not considered the structure of the photon wave function at all. However, for the Jaynes–Cummings model (see Section 3.4), we consider the interaction of a two-level system with a single photon mode. For the preparation of this system, a high-quality microcavity is required. As we have seen, the strength of the light field at the location of the electron determines the interaction strength between light and matter. Therefore, resonators are widely used to engineer electron–photon interaction processes, and the last section deals with the theoretical handling of these structures.

In resonator structures, calculation of the appropriate photon modes is required to allow access to the strength of electron–photon coupling. This will be clarified using the example of a microcavity. Modes can be established by a transfer-matrix method, which will be briefly discussed. For simplicity, we only look at a planar structure and perpendicular light incidence in this book chapter, while for in-depth information, the reader is referred to Refs. [24,25]. Information on the transfer matrix method may also be found in Ref. [26] and the references therein.

In free space or homogeneous matter, plane waves $\mathbf{E}_0 e^{i(\mathbf{q}\cdot\mathbf{r}-\omega t)}$ are solutions to Maxwell's equations [4]. Any light field may be expanded in plane waves. The transfer matrix method uses this to express the light field as a plane wave in every layer of a microstructure and appropriately match the conditions at the interfaces.

We consider a planar structure where the growth direction is taken as the z-axis. The index of refraction then only depends on z. The homogeneous wave equation reads

$$\left\{ \frac{1}{c^2} \frac{\partial}{\partial t^2} - \Delta \right\} \mathbf{u_q}(z) = 0 \,, \tag{3.73}$$

where $\mathbf{u_q}(z)$ is the light mode we are looking for, indexed with its wave vector $\mathbf{q}$. Since we are looking at perpendicular incidence in a planar structure, the wave vector points in the z-direction. Due to the geometry of the problem, the two independent polarizations are equivalent to each other. The index of polarization is thus omitted here, and bearing in mind that we need to find two independent modes for each wave vector q. The polarization becomes important for incidence at an angle.

Consider first a single boundary between two layers L (left) and R (right). Generally, each mode $\mathbf{u_q}(z)$ is a combination of plane waves traveling in forward and backward directions in each layer,

$$\mathbf{u_q}(z) = A_j e^{iq_{zj}(z-z_j)} \mathbf{e}_j + B_j e^{-iq_{zj}(z-z_j)} \mathbf{e}'_j$$

$$\text{where} \quad q_{zj} = n(z_j)\, q \,. \tag{3.74}$$

Maxwell's equations demand continuity for the tangential components of the electric and magnetic field and thus (in this geometry) for the entire mode $\mathbf{u_q}(z)$ and its derivation $\mathrm{d}/\mathrm{d}z\mathbf{u_q}(z)$. From this condition, transmission and reflection may be derived as

$$T_j^- = \frac{2q_{z,j}}{q_{z,j} + q_{z,j+1}},$$ (3.75)

$$R_j^- = \frac{q_{z,j} - q_{z,j+1}}{q_{z,j} + q_{z,j+1}},$$ (3.76)

$$T_j^+ = \frac{2q_{z,j+1}}{q_{z,j} + q_{z,j+1}},$$ (3.77)

$$R_j^+ = \frac{q_{z,j+1} - q_{z,j}}{q_{z,j} + q_{z,j+1}};$$ (3.78)

see Ref. [4] for more details. T_j^- and R_j^- (T_j^+ and R_j^+) are transmission and reflection coefficients for light incident from the left (right) side, respectively. This may be generalized to a matrix representation, relating neighboring heterostructure layers by

$$\begin{pmatrix} A_{j+1} \\ B_{j+1} \end{pmatrix} = \frac{1}{T_j^+} \begin{pmatrix} e^{iq_{z,j}d_j} & R_j^+ e^{-iq_{z,j}d_j} \\ -R_j^- e^{iq_{z,j}d_j} & e^{-iq_{z,j}d_j} \end{pmatrix} \begin{pmatrix} A_j \\ B_j \end{pmatrix} \equiv \mathbf{M}_j \begin{pmatrix} A_j \\ B_j \end{pmatrix},$$ (3.79)

where d_j is the width of layer j, $d_j = z_j - z_{j+1}$, and $\mathbf{M}_j$ is the transfer matrix. Recursive application yields

$$\begin{pmatrix} A_{N+1} \\ B_{N+1} \end{pmatrix} = \mathbf{M}_N \mathbf{M}_{N-1} \dots \mathbf{M}_2 \mathbf{M}_1 \begin{pmatrix} A_1 \\ B_1 \end{pmatrix} \equiv \mathbf{M}_{tot} \begin{pmatrix} A_1 \\ B_1 \end{pmatrix}$$ (3.80)

and allows the calculation of modes $\mathbf{u}_\mathbf{q}^+(z)$ and $\mathbf{u}_\mathbf{q}^-(z)$ incident from the left and right of the microcavity. Figure 3.4 [27]. shows a resonant mode in a typical Galliumarsenide (GaAs)/Aluminiumarsenide (AlAs) microcavity structure with 10 layer pairs on top and 20 layer pairs on the substrate. It becomes obvious that

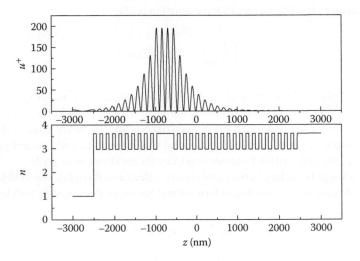

FIGURE 3.4   Top: Central mode of a GaAs/AlAs microcavity with 10 (on air) and 20 (on substrate) λ/4 layer pairs of GaAs and AlAs with background refractive indices of $n_{GaAs} = 3.63$ and $n_{AlAs} = 2.97$, respectively. Bottom: Refractive index of the microcavity along the growth direction z. (From A. Thränhardt, Equilibrium and Nonequilibrium Dynamics in Semiconductor Lasers, Habilitation Thesis, Marburg, 2006.)

the electric light field is enhanced by a factor of about 200 even in this simple structure in the peaks in the cavity. In this way, light–matter interaction may be reinforced by several orders of magnitude. Of course, photonic engineering also works for two- or three-dimensional photonic crystal structures where more elaborate theoretical approaches have to be used for the calculation of the modes, e.g., the finite difference time domain (FDTD) method [28,29], where electric and magnetic field components are calculated alternately.

References

1. P. A. M. Dirac, The quantum theory of the emission and absorption of radiation, *Proc. R. Soc. A* **114**, 243 (1927).
2. L. Allen, J. H. Eberly, *Optical Resonance and Two-Level Atoms, Dover Books on Physics*, New York, (1988) (revised edition).
3. A. Einstein, Zur Quantentheorie der Strahlung, *Phys. Zeitschrift* **18**, 121 (1917), first published in *Mitteilungen der Phys. Ges. Zürich* 18 (1916).
4. J. D. Jackson, *Classical Electrodynamics*, 3rd ed., New York: Wiley (1999).
5. J. Mompart, R. Corbalán, Lasing without inversion, *J. Opt. B: Quantum Semiclass. Opt.* **2**, R7 (2000).
6. M. D. Frogley, J. F. Dynes, M. Beck, J. Faist, C. C. Phillips, Gain without inversion in semiconductor nanostructures, *Nat. Mater.* **5**, 175 (2006).
7. H. Haug, S. W. Koch, *Quantum Theory of the Optical and Electronic Properties of Semiconductors*, 5th ed., Singapore: World Scientific (2009).
8. W. W. Chow, S. W. Koch, *Semiconductor–Laser Fundamentals*, Berlin/Heidelberg: Springer (1999).
9. S. Adachi, *GaAs and Related Materials: Bulk Semiconducting and Superlattice Properties*, Singapore: World Scientific (1994).
10. G. Franco Bassani, G. Pastori Parravicini, *Electronic States and Optical Transitions in Solids*, New York, NY: Pergamon Press (1975).
11. D. Dragoman, M. Dragoman, *Optical Characterization of Solids*, Berlin/Heidelberg: Springer (2010).
12. Shun Lien Chuang, *Physics of Optoelectronic Devices*, New York/Brisbane/Singapore: Wiley-InterScience (1995).
13. A. Thränhardt, S. Becker, C. Schlichenmaier, I. Kuznetsova, T. Meier, S. W. Koch, J. Hader, J. V. Moloney, W. W. Chow, Nonequilibrium gain in optically pumped GaInNAs laser structures, *Appl. Phys. Lett.* **85**, 5526 (2004).
14. E. Kühn, S. W. Koch, A. Thränhardt, J. Hader, J. V. Moloney, Microscopic simulation of nonequilibrium features in quantum-well pumped semiconductor disk lasers, *Appl. Phys. Lett.* **96**, 051116 (2010).
15. K. V. Meyenn (Ed.), *Wolfgang Pauli: Wissenschaftlicher Briefwechsel mit Bohr, Einstein, Heisenberg u.a. Band II: 1930–1939/Scientific Correspondence with Bohr, Einstein, Heisenberg a.o.*, Berlin/Heidelberg/New York, NY/Tokyo: Springer, Reprint (2014).
16. A. Thränhardt, C. Ell, S. Mosor, G. Rupper, G. Khitrova, H. M. Gibbs, S. W. Koch, Interplay of phonon and disorder scattering in semiconductor quantum wells, *Phys. Rev. B* **68**, 035316, 2003.
17. F. Jahnke (ed.), *Quantum Optics with Semiconductor Nanostructures*, Oxford/Cambridge/Philadelphia/New Delhi: Woodhead Publishing Series in Electronic and Optical Materials (2012).
18. P. Meystre, M. Sargent III, *Elements of Quantum Optics*, 4th ed., Berlin/Heidelberg: Springer (2010).
19. M. Kira, S. W. Koch, *Semiconductor Quantum Optics*, Cambridge: Cambridge University Press (2011).
20. V. Weisskopf, E. Wigner, Berechnung der natürlichen Linienbreite auf Grund der Diracschen Lichttheorie, *Zeitschrift für Physik* **63**, 54 (1930).
21. E. T. Jaynes, F. W. Cummings, Comparison of quantum and semiclassical radiation theories with application to the beam maser, *Proc. IEEE* **51**, 89 (1963).

22. J. H. Eberly, N. B. Narozhny, J. J. Sanchez-Mondragon, Periodic spontaneous collapse and revival in a simple quantum model, *Phys. Rev. Lett.* **44**, 1323 (1980).

23. G. Rempe, H. Walther, N. Klein, Observation of quantum collapse and revival in a one-atom MASER, *Phys. Rev. Lett.* **58**, 353 (1987).

24. M. Born, E. Wolf, *Principles of Optics: Electromagnetic Theory of Propagation, Interference and Diffraction of Light*, 7th ed., Cambridge: Cambridge University Press (2002).

25. B. E. Sernelius, *Surface Modes in Physics*, Berlin: Wiley-VCH (2001).

26. G. Khitrova, H. M. Gibbs, F. Jahnke, M. Kira, S. W. Koch, Nonlinear optics of normal-mode-coupling semiconductor microcavities, *Rev. Mod. Phys.* **71** (5), 1591 (1999).

27. A. Thränhardt, Equilibrium and Nonequilibrium Dynamics in Semiconductor Lasers, Habilitation Thesis, Marburg (2006).

28. K. Yee, Numerical solution of initial boundary value problems involving Maxwell's equations in isotropic media, *IEEE Trans. Antennas. Propag.* **14**, 302 (1966).

29. D. M. Sullivan, *Electromagnetic Simulation Using the FDTD Method*, Hoboken/New Jersey: Wiley, 2013.

4

Optical Waveguiding

Slawomir Sujecki

Optical waveguides are essential elements of many optoelectronic devices. Their primary role is shaping, confining, and guiding of an optical beam along a selected direction. The design and modeling of optical waveguides is based on the application of the macroscopic Maxwell's equations. This approach consists of approximating the microscopic electromagnetic fields that vary on the atomic scale with fields that are averaged over several atomic layers. Considering that an atom's diameter is up to several tenths of a nanometer while the optical waveguides contain structural details with dimensions smaller than 10 nm (e.g., quantum wells of a laser diode), the application of the macroscopic Maxwell equations to the design and modeling of optical waveguides is at its limit. This limitation should be borne in mind throughout this chapter.

Here, we provide an introduction to the theory of optical waveguides. In Section 4.1, we derive the wave equations and introduce a hierarchy of approximations relevant to optical waveguide modeling and design. In Section 4.2, we discuss the modal characteristics of infinitely long shift invariant optical waveguides. Since optical waveguides used in optoelectronic devices are not infinitely long and may not be straight, in Section 4.3, we outline the theory of beam propagation method, which is the standard numerical approach used for the analysis of wave propagation in optical waveguiding structures. Finally, we discuss the optical properties of bent and tapered waveguides.

4.1 Introduction to Optical Waveguide Theory

For isotropic, linear, dielectric media, the relations between the vectors of the electric and magnetic field and the electric and magnetic flux are given by Maxwell's equations

$$\nabla \times \vec{E}\left(\vec{r}, t\right) = -\frac{\partial \vec{B}\left(\vec{r}, t\right)}{\partial t} \tag{4.1}$$

$$\nabla \times \vec{H}\left(\vec{r}, t\right) = \frac{\partial \vec{D}\left(\vec{r}, t\right)}{\partial t}. \tag{4.2}$$

In the frequency domain, Equation 4.1 has the following form:

$$\nabla \times \vec{E}\left(\vec{r}\right) = -j\omega\vec{B}\left(\vec{r}\right) \tag{4.3}$$

$$\nabla \times \vec{H}\left(\vec{r}\right) = j\omega\vec{D}\left(\vec{r}\right), \tag{4.4}$$

where the angular frequency $\omega = 2\pi$ ft. Equations 4.3 and 4.4 imply that that the unknown fields depend on time according to $e^{j\omega t}$. The real field distributions of the electric and magnetic field and electric and magnetic flux $\vec{E}\left(\vec{r},t\right), \vec{H}\left(\vec{r},t\right), \vec{D}\left(\vec{r},t\right), \vec{B}\left(\vec{r},t\right)$ are obtained from the complex distributions $\vec{E}\left(\vec{r}\right), \vec{H}\left(\vec{r}\right), \vec{D}\left(\vec{r}\right), \vec{B}\left(\vec{r}\right)$ by multiplying a particular complex field distribution with $e^{j\omega t}$ term and extracting the real part of the product.

Equations 4.3 and 4.4 are complemented by the (Gauss's law) conditions

$$\nabla \cdot \vec{D}\left(\vec{r}\right) = 0 \tag{4.5}$$

$$\nabla \cdot \vec{B}\left(\vec{r}\right) = 0. \tag{4.6}$$

In addition, the dependence between the electric field and the electric flux vectors is given by

$$\vec{D}\left(\vec{r}\right) = \varepsilon\vec{E}\left(\vec{r}\right) = \varepsilon_0\varepsilon_r\vec{E}\left(\vec{r}\right), \tag{4.7}$$

where ε_0 is the electric permittivity of the free space and ε_r is the relative permittivity of the medium while the magnetic field and the magnetic flux vectors are given by the following equation:

$$\vec{B}\left(\vec{r}\right) = \mu_0\vec{H}\left(\vec{r}\right), \tag{4.8}$$

where μ_0 is the magnetic permeability of the free space. From the theoretical point of view, it is convenient to assume that optical waveguides are infinitely long and perfectly straight. Such fictitious optical waveguide structure is longitudinally shift invariant and hence the transverse and longitudinal space variables can be separated. In order to derive the wave equations, we introduce a rectangular coordinate system x, y, z and align the longitudinal waveguide dimension with the z-axis (Figure 4.1). Under such assumptions, ε does not depend on z and the spatial distribution of the unknown fields can be expressed as a product of an unknown function depending on the transverse coordinates only and a phase factor $\exp(-j\beta z)$,

$$\vec{\Phi}\left(\vec{r}\right) = \Phi\left(x,y\right)\exp(-j\beta z), \tag{4.9}$$

where $\vec{\Phi}\left(\vec{r}\right)$ stands for $\vec{E}\left(\vec{r}\right), \vec{H}\left(\vec{r}\right), \vec{D}\left(\vec{r}\right)$, or $\vec{B}\left(\vec{r}\right)$. The constant β is called the modal propagation constant while the functions $E(x,y)$, $H(x,y)$, $D(x,y)$, and $B(x,y)$ provide the field distributions for a given waveguide mode.

Substituting Equation 4.9 into Equations 4.3 through 4.8 while observing that ε does not depend on z, we obtain the following equations for the transverse components of the electric field vector

FIGURE 4.1   Schematic diagram of an optical waveguide in a rectangular coordinate system x, y, z.

$$E(x,y) = \vec{i}_x E_x(x,y) + \vec{i}_y E_y(x,y) + \vec{i}_z E_z(x,y),$$

$$\frac{\partial^2 E_x}{\partial y^2} + \frac{\partial}{\partial x}\left(\frac{1}{\varepsilon}\frac{\partial(\varepsilon E_x)}{\partial x}\right) + \left(\omega^2\mu\varepsilon - \beta^2\right)E_x + \frac{\partial}{\partial x}\left(\frac{1}{\varepsilon}\frac{\partial(\varepsilon E_y)}{\partial y}\right) - \frac{\partial^2 E_y}{\partial x\partial y} = 0 \qquad (4.10)$$

$$\frac{\partial^2 E_y}{\partial x^2} + \frac{\partial}{\partial y}\left(\frac{1}{\varepsilon}\frac{\partial(\varepsilon E_y)}{\partial y}\right) + \left(\omega^2\mu\varepsilon - \beta^2\right)E_y + \frac{\partial}{\partial y}\left(\frac{1}{\varepsilon}\frac{\partial(\varepsilon E_x)}{\partial x}\right) - \frac{\partial^2 E_x}{\partial y\partial x} = 0, \qquad (4.11)$$

where $\vec{E}(\vec{r}) = E(x,y)\exp(-j\beta z)$. Similarly, for the magnetic field vector $H(x,y) = \vec{i}_x H_x(x,y) + \vec{i}_y H_y(x,y) + \vec{i}_z H_z(x,y)$ we obtain

$$\frac{\partial^2 H_x}{\partial x^2} + \varepsilon\frac{\partial}{\partial y}\left(\frac{1}{\varepsilon}\frac{\partial H_x}{\partial y}\right) + \left(\omega^2\mu\varepsilon - \beta^2\right)H_x - \varepsilon\frac{\partial}{\partial y}\left(\frac{1}{\varepsilon}\frac{\partial H_y}{\partial x}\right) + \frac{\partial^2 H_y}{\partial y\partial x} = 0 \qquad (4.12)$$

$$\frac{\partial^2 H_y}{\partial y^2} + \varepsilon\frac{\partial}{\partial x}\left(\frac{1}{\varepsilon}\frac{\partial H_y}{\partial x}\right) + \left(\omega^2\mu\varepsilon - \beta^2\right)H_y - \varepsilon\frac{\partial}{\partial x}\left(\frac{1}{\varepsilon}\frac{\partial H_x}{\partial y}\right) + \frac{\partial^2 H_x}{\partial x\partial y} = 0. \qquad (4.13)$$

Considering that the refractive index n is defined as the ratio of the speed of light in the medium to the speed of light in the free space, we have the following relationship between the refractive index and the dielectric permittivity: $n = \sqrt{\varepsilon/\varepsilon_0}$.

We note that the equations for the transverse vector components of the electric field, Equations 4.10 and 4.11, do not contain any coupling terms with the longitudinal ones. Furthermore, the equations for both transverse components are coupled but there is no coupling between the electric and magnetic fields. Hence, Equations 4.10 and 4.11 can be solved autonomously for E_x and E_y while the remaining electromagnetic field components can be calculated from

$$E_z = -\frac{1}{j\beta\varepsilon}\left[\frac{\partial(\varepsilon E_x)}{\partial x} + \frac{\partial(\varepsilon E_y)}{\partial y}\right] \qquad (4.14)$$

$$H_x = -\frac{1}{j\omega\mu}\left(\frac{\partial E_z}{\partial y} + j\beta E_y\right) \qquad (4.15)$$

$$H_y = \frac{1}{j\omega\mu}\left(\frac{\partial E_z}{\partial x} + j\beta E_x\right) \qquad (4.16)$$

$$H_z = -\frac{1}{j\omega\mu}\left(\frac{\partial E_y}{\partial x} - \frac{\partial E_x}{\partial y}\right). \qquad (4.17)$$

Alternatively, Equations 4.12 and 4.13 can be solved to obtain the distributions of H_x and H_y while the other field components distributions can be obtained from

$$H_z = -\frac{1}{j\beta}\left(\frac{\partial H_x}{\partial x} + \frac{\partial H_y}{\partial y}\right) \qquad (4.18)$$

$$E_x = \frac{1}{j\omega\varepsilon}\left(\frac{\partial H_z}{\partial y} + j\beta H_y\right) \qquad (4.19)$$

$$E_y = -\frac{1}{j\omega\varepsilon}\left(\frac{\partial H_z}{\partial x} + j\beta H_x\right) \tag{4.20}$$

$$E_z = \frac{1}{j\omega\varepsilon}\left(\frac{\partial H_y}{\partial x} - \frac{\partial H_x}{\partial y}\right). \tag{4.21}$$

Equations 4.10 and 4.11 form an eigenvalue problem, whereby β^2 is the eigenvalue and E_x and E_y form the corresponding eigenvector. Analogous observation applies to Equations 4.12 and 4.13. In the context of the optical waveguide analysis, β is referred to as a modal propagation constant while the corresponding eigenvector provides the modal field distribution.

The solution of the eigenvalue problem formed by either Equations 4.10 and 4.11 or Equations 4.12 and 4.13 allows obtaining the exact solution of the Maxwell equations under the assumption of the longitudinal shift invariance of the optical waveguide structure. In the literature relating to the optical waveguide modeling, such a solution of the Maxwell equations is referred to as a vectorial solution. Similarly, computer software–based tools for the design of optical waveguides that solve either Equations 4.10 and 4.11 or Equations 4.12 and 4.13 are referred to as vectorial mode solvers.

For completeness, we note that there are other ways of formulating the solution of Maxwell equations for an optical waveguide (Sujecki, 2014). However, the relevance of Equations 4.10 through 4.13 stems from the fact that most computer software tools for optical waveguide design rely on the formulation of the problem in terms of either the transverse magnetic or electric field components. A particular advantage of such approach is the elimination of spurious solutions. Magnetic field distribution is smoother than that of the electric field. Hence, software–based on the magnetic field formulation tends to have better numerical convergence properties, whereas the electrical field formulation is preferred for nonlinear optics applications.

Equations 4.10 through 4.13 are relatively difficult to solve numerically. However, they can be significantly simplified for waveguides, which support modes that are nearly linearly polarized along either the x- or y-direction. For such waveguides two types of modes are guided: modes that are polarized either along the x-direction (x-polarized or quasi-transverse electric [TE] modes) and modes that are polarized along the y-direction (y-polarized or quasi-transverse magnetic [TM] modes). The main transverse components of the electromagnetic field for the x-polarized modes are E_x and H_y, while for the y-polarized ones they are E_y and H_x. The polarized approximation has proven particularly successful in the modeling of optical waveguides realized using planar technology and is therefore of great practical importance for semiconductor optoelectronic devices (Sujecki et al., 1998). The modal propagation constants and the corresponding field distributions for the x-polarized modes are obtained by solving Equation 4.10 or 4.13 while neglecting the minor field component:

$$\frac{\partial^2 E_x}{\partial y^2} + \frac{\partial}{\partial x}\left(\frac{1}{\varepsilon}\frac{\partial(\varepsilon E_x)}{\partial x}\right) + \left(\omega^2\mu\varepsilon - \beta^2\right)E_x = 0 \tag{4.22}$$

$$\frac{\partial^2 H_y}{\partial y^2} + \varepsilon\frac{\partial}{\partial x}\left(\frac{1}{\varepsilon}\frac{\partial H_y}{\partial x}\right) + \left(\omega^2\mu\varepsilon - \beta^2\right)H_y = 0. \tag{4.23}$$

Similarly, the propagation constants and field distributions of y-polarized modes are obtained from Equations 4.11 and 4.13 by leaving only the terms containing the major field component:

$$\frac{\partial^2 E_y}{\partial x^2} + \frac{\partial}{\partial y}\left(\frac{1}{\varepsilon}\frac{\partial(\varepsilon E_y)}{\partial y}\right) + \left(\omega^2\mu\varepsilon - \beta^2\right)E_y = 0 \tag{4.24}$$

$$\frac{\partial^2 H_x}{\partial x^2} + \varepsilon\frac{\partial}{\partial y}\left(\frac{1}{\varepsilon}\frac{\partial H_x}{\partial y}\right) + \left(\omega^2\mu\varepsilon - \beta^2\right)H_x = 0. \tag{4.25}$$

The solution of the partial differential equations for the major transverse field component is only much easier to implement and reduces the required computational effort. Further, improvement of computational efficiency can be achieved if ε varies slowly in the transverse plane. In such cases, the following scalar approximation can be used:

$$\left(\frac{\partial^2}{\partial x^2} + \frac{\partial^2}{\partial y^2} + \omega^2 \mu \varepsilon \right) \varphi = \beta^2 \varphi. \tag{4.26}$$

In principle, the scalar approximation is applicable only to waveguides with a small refractive index contrast. However, there are known instances of waveguides with a large refractive index contrasts for which plausible accuracy may be achieved under the scalar approximation (Riishede et al., 2003).

In summary, by deriving Equations 4.10 through 4.13, Equations 4.22 through 4.25 and 4.26, a hierarchy of approximations is established and this is applicable to optical waveguide analysis. Thus, depending upon which equation is solved by a particular solution method, the methods of calculating the modal propagation constants and the corresponding field distributions are divided into three categories. The vectorial methods attempt to solve the vectorial wave Equations 4.10 through 4.13. Methods that attempt to solve the Equations 4.22 through 4.25 are referred to as either semivectorial or using the polarized approximation. The scalar methods solve the scalar wave Equation 4.26. Compared to vectorial methods, scalar methods are more efficient but less accurate. The semivectorial methods (polarized approximation) provide a compromise between efficiency and accuracy when compared with the other two alternatives (Figure 4.2).

Finally, in order to illustrate the differences between the vectorial, semivectorial, and scalar methods, Table 4.1 presents normalized propagation constant values, which have been obtained for a rib waveguide structure with $H = 1$ μm, $W = 1.5$ μm, $n_f = 3.44$, $n_s = 3.4$, $n_c = 1.0$ (Figure 4.6). The outer slab thickness H is varied, between 0.1 and 0.9 μm. Operating wavelength is 1.15μm. The scalar and polarized results have been obtained using the finite difference method (FDM) (Sujecki et al., 1998), while the vectorial ones were taken from Hadley (2002) and Vassallo (1997b) for x- and y-polarized modes, respectively. All the results are believed to be calculated with an absolute error of 5×10^{-5}. The difference between the

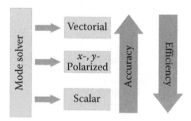

FIGURE 4.2 Diagram showing the hierarchy of approximations used in optical waveguide modeling.

TABLE 4.1 Normalized Propagation Constants B of Rib Dielectric Waveguide

h (μm)	Scalar Finite Difference	Polarized	Vectorial (Hadley, 2002)	Polarized	Vectorial (Vassallo, 1997b)
	Scalar	x-Polarized		y-Polarized	
0.1	0.3095	0.3018	0.3019	0.2674	0.2674
0.3	0.3176	0.3108	0.3110	0.2750	0.2751
0.5	0.3320	0.3267	0.3270	0.2888	0.2890
0.7	0.3542	0.3509	0.3512	0.3104	0.3107
0.9	0.3893	0.3884	0.3886	0.3453	0.3455

$H + h = 1$ μm; $W = 1.5$ μm; $n_f = 3.44$; $n_s = 3.4$; $n_c = 1.0$; $\lambda = 1.15$ μm (cf. Figure 4.6b).

polarized and vectorial analysis results does not exceed 3 on the fourth decimal place for either polarization. This illustrates the advantages of the polarized approximation in the case of rib waveguides. The results obtained by the scalar method, however, differ up to eight on the third decimal place for x-polarized modes and even in the first decimal place for the y-polarized modes. It should be noted that for $h = 0.9$ μm, the scalar approximation is accurate nearly on three decimal places for the x-polarized mode, despite a large refractive index contrast.

A survey of numerical methods used in calculations of propagation constants and field distributions of optical waveguides can be found in Sujecki (2014).

4.2 Modes of Optical Waveguides

In this section, we discuss modal characteristics of the selected longitudinally shift-invariant optical waveguides that are relevant to optoelectronic devices. We start with slab and circular waveguides, for which analytical solutions of Maxwell equations are available. Then, we consider selected index guiding optical waveguides for which Maxwell equations must be solved numerically. We then discuss the effective index method and finally provide a brief description of the concepts of radiation, leaky, and gain guided modes and also of waveguide array supermodes.

4.2.1 Slab Optical Waveguide

The refractive index distribution of a slab waveguide varies only along one selected axis of a rectangular coordinate system (cf. Figure 4.3). In any plane perpendicular to that axis, the refractive index distribution is homogenous, while the refractive index of the core n_f is larger than that of the substrate n_s and the cladding n_c and $n_s > n_c$. When a slab waveguide has a piecewise homogenous refractive index distribution (Figure 4.3), useful insight into the nature of the modes can be gained using the geometric optics. Figure 4.4 shows a cross section of a slab waveguide with an optical ray pattern, whereby γ is larger than the critical value: $\gamma = \sin^{-1}(n_s/n_f)$. The optical ray is thus totally reflected from the interfaces between the core and cladding and core and substrate. Since both interfaces are parallel, the subsequent incidence angles γ do

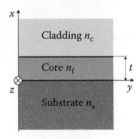

FIGURE 4.3   Three-layer slab waveguide in a rectangular coordinate system x, y, z.

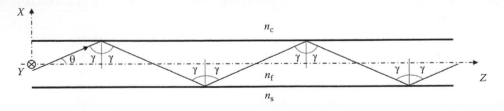

FIGURE 4.4   An optical ray pattern of a mode guided by a three-layer slab waveguide.

not change under subsequent reflections. Further, it can be observed that if the value of the incidence angle is larger than the critical angle, the optical ray is trapped within the core; and hence there is no power leakage. Furthermore, associating plane waves with optical rays leads to a realization that the waves generated through subsequent reflections interfere within the core. Thus, in order to avoid destructive interference the phase change for an optical wave upon the complete roundtrip between the core–cladding and core–substrate interfaces has to be equal to a multiple of 2π. This leads to a conclusion that the spectrum of the guided modes is discrete. The last intuitive observation that can be made with the help of ray optics is that if the plane wave associated with the ray is polarized along the y-axis, then it will not change the state of polarization upon subsequent ray reflections. This leads to a conclusion that the waveguide preserves linear polarization. All these observations are confirmed by the rigorous solution of the Maxwell equations (Sujecki, 2014).

In order to obtain the field distributions and propagation constants of the slab waveguide in the TE mode case, one needs to consider Equation 4.11. Since the derivatives along the y-direction are equal to zero, all coupling terms with E_x component can be neglected. For the three-layer waveguide (Figure 4.3), the field distribution of E_y component of the electromagnetic field therefore has the form

$$
E_y = \begin{cases} C \exp\left(-p\left(x - t\right)\right) & \text{for} \quad x > t \\ A \cos\left(ht\right) + B \sin\left(ht\right) & \text{for} \quad 0 < x < t \\ D \exp\left(qx\right) & \text{for} \quad x < 0 \end{cases} , \tag{4.27}
$$

where $p^2 = \beta^2 - n_c^2 k_0^2$, $q^2 = \beta^2 - n_s^2 k_0^2$, and $h^2 = n_f^2 k_0^2 - \beta^2$.

Substituting Equation 4.27 into Equation 4.10 and imposing the field continuity for E_y and H_z yields the dispersion equation for TE modes

$$
ht = \text{arctg}\frac{q}{h} + \text{arctg}\frac{p}{h} + m\pi. \tag{4.28}
$$

Following a similar procedure for TM modes we obtain

$$
ht = \text{arctg}\left(\frac{q}{h}\frac{n_f^2}{n_s^2}\right) + \text{arctg}\left(\frac{p}{h}\frac{n_f^2}{n_c^2}\right) + m\pi, \tag{4.29}
$$

whereby m is a parameter that sets the mode order.

Introducing the normalized frequency V, relative propagation constant B, and an asymmetry parameter A_E,

$$
V = k_0 t \sqrt{n_f^2 - n_s^2}; \quad B = \frac{\beta^2 - n_s^2 k_0^2}{n_f^2 k_0^2 - n_s^2 k_0^2}; \quad A_E = \frac{n_s^2 - n_c^2}{n_f^2 - n_s^2}
$$

allows recasting Equation 4.27 further into a more convenient normalized form

$$
V\sqrt{1 - B} - \text{arctg}\sqrt{\frac{B + A_E}{1 - B}} - \text{arctg}\sqrt{\frac{B}{1 - B}} - m\pi = 0 \tag{4.30}
$$

for the TE modes and

$$
V\sqrt{1 - B} - \text{arctg}\left(\frac{n_f^2}{n_c^2}\sqrt{\frac{B + A_E}{1 - B}}\right) - \text{arctg}\left(\frac{n_f^2}{n_s^2}\sqrt{\frac{B}{1 - B}}\right) - m\pi = 0 \tag{4.31}
$$

for the TM modes.

From Equations 4.28 through 4.31, one can obtain the values of the propagation constant for a guide mode of a slab waveguide. Once the propagation constant is calculated in the TE case, the field distribution can be obtained from Equation 4.27 while the constants A, B, C, D are calculated by solving

$$
\begin{bmatrix}
1 & 0 & 0 & -1 \\
\cos(ht) & \sin(ht) & -1 & 0 \\
0 & h & 0 & -q \\
-h\sin(ht) & h\cos(ht) & p & 0
\end{bmatrix}
\begin{bmatrix}
A \\
B \\
C \\
D
\end{bmatrix} = 0.
\tag{4.32}
$$

For the TM case, the H_y component of the electromagnetic field is obtained using the formulae given by Equation 4.27, whereby the constants A, B, C, D are calculated by solving

$$
\begin{bmatrix}
1 & 0 & 0 & -1 \\
\cos(ht) & \sin(ht) & -1 & 0 \\
0 & \frac{h}{n_f^2} & 0 & -\frac{q}{n_s^2} \\
-\frac{h}{n_f^2}\sin(ht) & \frac{h}{n_f^2}\cos(ht) & \frac{p}{n_c^2} & 0
\end{bmatrix}
\begin{bmatrix}
A \\
B \\
C \\
D
\end{bmatrix} = 0.
\tag{4.33}
$$

In summary, the rigorous solution of the Maxwell equations shows that a slab waveguide supports two types of linearly polarized modes: TE and TM. The cutoff frequency of the first-order mode is obtained by setting $B = 0$ in Equation 4.30 for the TE case: $V = arctg\sqrt{A_E} + m\pi$ and in Equation 4.31 for the TM case: $V = \arctg\left(\frac{n_f^2}{n_c^2}\sqrt{A_E}\right) + \pi$.

An analytical solution can also be also obtained in the case of a multilayered slab waveguide, if the refractive index distribution is homogenous within each layer of an isotropic (Anemogiannis and Glytsis, 1992) or a uniaxial anisotropic medium (Majewski, 1984) and also for several cases of optical waveguides with a graded index profile (Tamir, 1975). Alternatively, for multilayer slab waveguides, numerical methods can be used (Anemogiannis et al., 1994; Ding and Chan, 1997; Anemogiannis et al., 1999; Sujecki, 2010).

4.2.2 Circular Optical Waveguide

Figure 4.5 shows the cross section of a step-index circular waveguide. For this waveguide structure, the core of the refractive index n_c and radius a is surrounded by an infinitely extending cladding with refractive index n_a, whereby $n_c > n_a$. For a step-index circular waveguide, Maxwell's equations can be solved analytically (Hondros and Debye, 1910).

In order to calculate the propagation constants of a circular optical waveguide, the Hondros–Debye equation has to be solved numerically

$$
\left(X_m + Y_m\right)\left(n_c^2 Y_m + n_a^2 X_m\right) - \frac{m^2 N^2}{\left(u^2 B\right)^2} = 0,
\tag{4.34}
$$

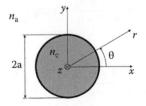

FIGURE 4.5   Step-index circular optical fiber.

where $X_m = \frac{K'_m(w)}{wK_m(w)}$, $Y_m = \frac{J'_m(u)}{uJ_m(u)}$, $B = \frac{N^2 - n_a^2}{n_c^2 - n_a^2}$ is the relative propagation constant, $N = \beta/k$ is the effective refractive index, k is the wave number, $u = \kappa a$ and $w = \gamma a$, $\kappa^2 = \omega^2 n_c^2 \varepsilon_0 \mu_0 - \beta^2$, $\gamma^2 = \beta^2 - \omega^2 n_a^2 \varepsilon_0 \mu_0$, and n_c and n_a are the refractive indices of the core and cladding, respectively. J_m and K_m denote Bessel functions of the first kind and modified Bessel functions of the second kind, while m is the function order. For $m = 0$, Equation 4.34 splits into two equations, the solutions of which give the propagation constants of the TE modes

$$\frac{J_1(u)}{uJ_0(u)} + \frac{K_1(w)}{wK_0(w)} = 0 \tag{4.35}$$

and TM modes

$$n_c^2 \frac{J_1(u)}{uJ_0(u)} + n_a^2 \frac{K_1(w)}{wK_0(w)} = 0. \tag{4.36}$$

Once the propagation constant of a mode is calculated, the longitudinal field component distribution can be obtained from

$$E_z(r, \theta, z) = A_E J_m(\kappa r) e^{j(-\beta z + m\theta)} \tag{4.37}$$

$$H_z(r, \theta, z) = A_H J_m(\kappa r) e^{j(-\beta z + m\theta)} \tag{4.38}$$

in the core and the cladding

$$E_z(r, \theta, z) = B_E K_m(\gamma r) e^{j(-\beta z + m\theta)} \tag{4.39}$$

$$H_z(r, \theta, z) = B_H K_m(\gamma r) e^{j(-\beta z + m\theta)}. \tag{4.40}$$

The transverse components are obtained from

$$E_r = -\frac{j}{\kappa^2} \left[\beta \kappa A_E J'_m(\kappa r) + j\omega\mu_0 \frac{m}{r} A_H J_m(\kappa r) \right] e^{j(-\beta z + m\theta)} \tag{4.41}$$

$$E_\theta = -\frac{j}{\kappa^2} \left[j\beta \frac{m}{r} A_E J_m(\kappa r) - \omega\mu_0 \kappa A_H J'_m(\kappa r) \right] e^{j(-\beta z + m\theta)} \tag{4.42}$$

$$H_r = -\frac{j}{\kappa^2} \left[\beta \kappa A_H J'_m(\kappa r) - j\omega\varepsilon_0 n_c^2 \frac{m}{r} A_E J_m(\kappa r) \right] e^{j(-\beta z + m\theta)} \tag{4.43}$$

$$H_\theta = -\frac{j}{\kappa^2} \left[j\beta \frac{m}{r} A_H J_m(\kappa r) + \omega\varepsilon_0 n_c^2 \kappa A_E J'_m(\kappa r) \right] e^{j(-\beta z + m\theta)} \tag{4.44}$$

in the core and

$$E_r = \frac{j}{\gamma^2} \left[\beta \gamma B_E K'_m(\gamma r) + j\omega\mu_0 \frac{m}{r} B_H K_m(\gamma r) \right] e^{j(-\beta z + m\theta)} \tag{4.45}$$

$$E_\theta = \frac{j}{\gamma^2} \left[j\beta \frac{m}{r} B_E K_m(\gamma r) - \omega\mu_0 \gamma B_H K'_m(\gamma r) \right] e^{j(-\beta z + m\theta)} \tag{4.46}$$

$$H_r = \frac{j}{\gamma^2} \left[\beta \gamma B_H K'_m(\gamma r) - j\omega\varepsilon_0 n_a^2 \frac{m}{r} B_E K_m(\gamma r) \right] e^{j(-\beta z + m\theta)} \tag{4.47}$$

$$H_\theta = \frac{j}{\gamma^2} \left[j\beta \frac{m}{r} B_H K_m(\gamma r) + \omega\varepsilon_0 n_a^2 \gamma B_E K'_m(\gamma r) \right] e^{j(-\beta z + m\theta)} \tag{4.48}$$

in the cladding, where the prime stands for the differentiation with respect to the function argument. The unknown constants A_E, A_H, B_E, and B_H are calculated by solving the homogenous set of algebraic equations:

$$\begin{bmatrix} J_m(u) & 0 & -K_m(w) & 0 \\ 0 & J_m(u) & 0 & -K_m(w) \\ j\frac{\beta m}{a\kappa^2}J_m(u) & +\frac{\omega\mu_0}{\kappa}J'_m(u) & j\beta\frac{m}{a\gamma^2}K_m(w) & +\frac{\omega\mu_0}{\gamma}K'_m(w) \\ \frac{\omega\varepsilon_0}{\kappa}n_c^2 J'_m(u) & -j\frac{\beta m}{a\kappa^2}J_m(u) & +\frac{\omega\varepsilon_0}{\gamma}n_a^2 K'_m(w) & -j\beta\frac{m}{a\gamma^2}K_m(w) \end{bmatrix} \begin{bmatrix} A_E \\ A_H \\ B_E \\ B_H \end{bmatrix} = 0. \quad (4.49)$$

Thus, a step-index circular waveguide guides three types of modes: hybrid modes, TE modes, and TM modes. The propagation constants of hybrid modes are obtained by solving the Hondros–Debye Equation 4.34, while propagation constants of TE and TM modes can be calculated by solving, respectively, Equations 4.35 and 4.36. The fundamental mode of the step-index circular waveguide is the HE_{11} mode. The single-mode region of the step-index circular optical waveguide is set by the cut-off normalized frequency of TE_{01} mode: $V = 2.40482555769577$, where $V = \frac{2\pi}{\lambda}a\sqrt{n_c^2 - n_a^2}$ and λ is the operating wavelength. The detailed description of the optical properties of step-index circular waveguides can be found in a number of textbooks (Okoshi, 1982; Snyder and Love, 1983; Majewski, 1991; Okamoto, 2006; Black and Gagnon, 2010). The analytical theory of a circular waveguide can be extended for multilayered structures (Yeh et al., 1978; Majewski, 1984, 1991).

If the refractive index contrast between the core and cladding is small, all the modes of the circular waveguide become linearly polarized. When $n_c \approx n_a$ Equation 4.34 reduces to

$$\frac{uJ_{l+1}(u)}{J_l(u)} = \frac{wK_{l+1}(w)}{K_l(w)}. \quad (4.50)$$

Equation 4.50 can also be obtained by solving the scalar wave Equation 4.26 for a step-index circular waveguide in the cylindrical coordinate system. The propagation constant of the fundamental mode in the scalar approximation is obtained from Equation 4.50 with $l = 0$. Table 4.2 shows a comparison for the normalized propagation constant $B = \frac{\beta^2 - n_a^2 k^2}{n_c^2 k^2 - n_a^2 k^2}$ of the fundamental mode obtained solving the vectorial dispersion Equation 4.34 and the scalar dispersion Equation 4.50. These results are believed to be calculated with an absolute error of 5×10^{-7}. Table 3.13 shows the relative propagation constant values that have been obtained for the circular waveguide with $n_c = 1.47$, $\lambda = 1.55$ μm, $V = ak\sqrt{n_c^2 - n_a^2}$, whereas varying the cladding refractive index. These results show that the difference between the vectorial and scalar results grows with the increase in the refractive index contrast between the core and the cladding.

TABLE 4.2 Normalized Propagation Constant of the Fundamental Mode of a Single Step-Index Circular Optical Waveguide with $V = 2.4$, $n_c = 1.47$, $\lambda = 1.55$ μm

na	B: Vectorial (4.2.4)	B: Scalar (4.2.8)	Relative Difference
1.45	0.52691	0.53003	5.9e−003
1.4	0.51882	0.53003	2.2e−002
1.3	0.50126	0.53003	5.7e−002
1.2	0.48171	0.53003	1.0e−001
1.1	0.45996	0.53003	1.5e−001
1.0	0.43582	0.53003	2.3e−001

4.2.3 Index Guiding

Slab and circular waveguides are examples of structures that guide light using the phenomenon of total internal reflection. For such structures, the light is guided by the waveguide core, which has a larger refractive index than that of the surrounding medium. Hence, such waveguides are also referred to as index guiding waveguides. There are other types of index guiding waveguides, which are relevant to semiconductor devices; for instance: rectangular, rib and in-diffused waveguides, which are discussed in this section (cf. also Tamir, 1975; Ebeling, 1993; Maerz, 1995; Hunsperger, 2002). The theory of slab and circular waveguides is well established, and the guided mode designation is consistent in the literature. This is also the case for an elliptical waveguide (Dyott, 1996). Unfortunately, in the case of optical waveguides discussed in this section, a consistent mode designation does not exist (cf. for instance Goell, 1969; Gallick et al., 1992) this generates some confusion. Figure 4.6a shows the rectangular dielectric waveguide: the light beam is guided by the core, which has the refractive index n_f that is larger than that of the cladding $-n_c$. For a square waveguide with side width w, the single mode condition is given approximately by $w \times (4\pi/\lambda) \times \sqrt{n_f^2 - n_c^2} < 2$ (Majewski and Sujecki, 1996b), where λ is the operating wavelength. For a rectangular waveguide, the single mode condition changes with the aspect ratio w/h. For large aspect ratios w/h, the single mode condition for rectangular waveguide can be approximated by the single mode condition of a slab waveguide (Marcatili, 1969). The detailed study of the modal characteristics of the rectangular waveguide modes can be found in Goell (1969) and Majewski and Sujecki (1996b,a).

Figure 4.6b shows the cross section of an optical rib waveguide. H denotes the outer slab thickness, h is the rib height, and w is the rib width. The areas between the vertical dashed lines and outside of the vertical dashed lines are referred to as the inner and outer slab, respectively. The refractive index of the core is larger than that of the substrate and cladding, i.e., $n_f > \max(n_s, n_c)$, which provides vertical index guiding. The lateral index guiding can be best explained with the help of the effective index method (Section 4.2.4). Since the thickness of the inner slab is larger than that of the outer one, the effective index of this layer is also larger than that of the outer one. Hence, it can be concluded that this structure also provides lateral confinement. An alternative way of explaining the waveguiding phenomenon in the case of rib waveguides is available in Marcatili (1974). The detailed studies of general properties of optical rib waveguides, including modal polarization, can be found in Soref et al. (1991), Chiang and Wong (1996), Majewski and Sujecki (1997), and Majewski and Sujecki (1998).

Another structure often used in integrated optics is an in-diffused waveguide (Figure 4.6c). It is fabricated by diffusion of dopants, which increase locally the refractive index in the substrate of refractive index n_s. The area with the locally increased value of the refractive index forms the core of the waveguide. The single mode properties of this waveguide strongly depend on the spatial distribution of the refractive index, which is a function of the distribution of the dopants. An experimental study of the properties of in-diffused optical waveguides can be found in Ctyroky et al. (1984).

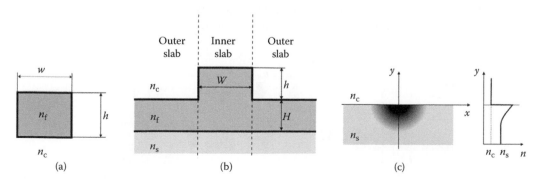

FIGURE 4.6 Rectangular waveguide (a), rib waveguide (b), and in-diffused waveguide (c).

4.2.4 Effective Index Method

It is possible to calculate a one-dimensional (1D) effective refractive index map for structures discussed in Section 4.2.4 so that the analytical techniques presented in Sections 4.2.1 and 4.2.2 can be applied to the calculation of the propagation constants of waveguides from Figure 4.6. Several techniques have been proposed in the literature for this purpose. The most popular effective index method (EIM) consists of approximating a more complex planar waveguide by a slab waveguide (Ramaswamy, 1974). This can be done for both semivectorial (4.1.7) and scalar (4.1.8) cases. EIM is based on the separation of variables, i.e., it relies on the assumption that a two-dimensional (2D) field distribution can be represented by a product of two 1D field distributions. As an illustrating example, we derive an approximate EIM solution in the case of an x-polarized fundamental mode of a rib waveguide. Setting the solution in the form $E_x(x, y, \omega) = X(x, \omega) \times Y(y, \omega)$ into Equation 4.22 gives

$$X \frac{\partial^2 Y}{\partial y^2} + Y \frac{\partial}{\partial x} \left(\frac{1}{\varepsilon} \frac{\partial (\varepsilon X)}{\partial x} \right) + \omega^2 \mu \varepsilon XY = \beta^2 XY. \tag{4.51}$$

Dividing both sides of Equation 4.51 by XY and moving the constants to the right-hand side (RHS) results in

$$\frac{1}{Y} \frac{\partial^2 Y}{\partial y^2} + \frac{1}{X} \frac{\partial}{\partial x} \left(\frac{1}{\varepsilon} \frac{\partial (\varepsilon X)}{\partial x} \right) = - \left(\omega^2 \mu \varepsilon - \beta^2 \right), \tag{4.52}$$

which implies that a separation constant can be introduced to obtain two coupled 1D equations

$$\frac{\partial^2 Y}{\partial y^2} + \omega^2 \mu \varepsilon Y = \beta_{\text{eff}}^2 Y \tag{4.53}$$

$$\frac{\partial}{\partial x} \left(\frac{1}{\varepsilon} \frac{\partial (\varepsilon X)}{\partial x} \right) + \omega^2 \mu \varepsilon_0 n_{\text{eff}}^2 X = \beta^2 X, \tag{4.54}$$

where the effective index of the slab waveguide fundamental mode $n_{\text{eff}} = \beta_{\text{eff}}/k_0$ while $k_0 = 2\pi/\lambda$. Once the dependence of n_{eff} on x is calculated using Equation 4.53, Equation 4.54 is used to obtain the propagation constant for the fundamental mode of the "effective" slab waveguide, which gives the approximate value of the propagation constant of the fundamental mode of the original 2D waveguide. The entire process is summarized in Figure 4.7.

Alternatively, one may consider a polar coordinate system and calculate the effective index along the azimuthal direction (Marcatili and Hardy, 1988). This requires a calculation of the propagation constant of a periodic waveguide (Figure 4.8). Once the effective index map is calculated, the approximate equivalent circular waveguide structure is obtained. The propagation constants of this equivalent circular waveguide can be calculated applying the analytical techniques discussed in Section 4.2.2.

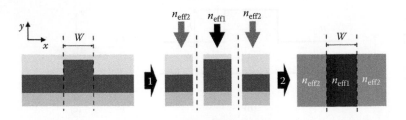

FIGURE 4.7 Schematic illustration of the effective index method in the case of a rib-loaded optical waveguide.

4.2.5 Radiation Modes, Leaky Waves, Gain Guiding, and Supermodes

If the angle of incidence for a ray shown in Figure 4.4 is smaller than the critical angle, the ray will undergo refraction at the boundary between the core and the cladding. In Figure 4.9, we show three possible scenarios for a slab waveguide with $n_f > n_s > n_c$. The case shown in Figure 4.9a corresponds to a guided mode that was already discussed in Section 4.2.1. The cases shown in Figures 4.9b and c correspond to radiation modes, i.e., modes that are not guided by the waveguide core. In fact, once the incidence angle is less than the critical angle, the ray may undergo refraction only at the interface between the core and substrate, which corresponds to a substrate radiation mode. If the ray undergoes refraction also at the cladding-core interface, the mode is referred to as free space radiation mode. Assuming that $\partial/\partial y = 0$ $\partial/\partial y = 0$, the solutions of the Maxwell equation for both modes can be obtained by solving Equations 4.24 and 4.22 for TE and TM modes, respectively (Marcuse, 1972, 1974; Ebeling, 1993). Figure 4.10 shows calculated distributions of the E_y electromagnetic field component

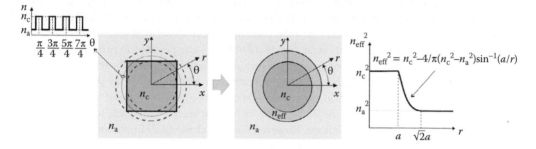

FIGURE 4.8 Schematic illustration of the azimuthal effective index method in the case of a square optical waveguide. (From Marcatili, E.A.J. and Hardy, A.A., *IEEE Journal of Quantum Electronics* 24, 766–774, 1988.)

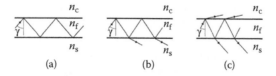

FIGURE 4.9 Optic ray paths in a slab waveguide for a guided mode (a), substrate radiation mode (b), and free space radiation mode (c).

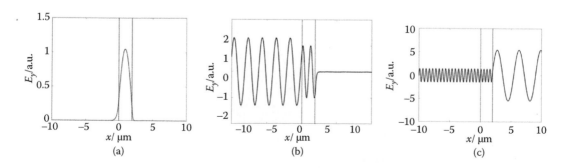

FIGURE 4.10 Field distributions of the E_y component of an electric field in a slab waveguide for a TE guided mode (a), substrate radiation mode (b), and free space radiation mode (c). The waveguide with is 2 μm, wavelength 1.55 μm, and $n_c = 1$, $n_f = 3.3$, and $n_s = 3.1$.

obtained by directly solving Maxwell equations for a guided mode and substrate and free space radiation modes.

Even if the refractive index distribution is purely real, Equations 4.28 through 4.31 admit solutions for the complex values of the propagation constant β. These modes are referred to as "leaky modes." A characteristic feature of leaky modes is an exponential growth of the field away from the core. A simple explanation of this field behavior is provided in Rozzi and Mongiardo (1997). Leaky modes are very important when studying the optical properties of waveguide modes below cut-off, bent waveguides and antireflection resonant optical waveguides (ARROWs) (Marcuse, 1974; Snyder and Love, 1983; Jablonski, 1994). Figure 4.11a shows an example cross section of an ARROW (Kubica, 2002). The top layer with refractive index n_1 guides the light and hence acts as the waveguide core. The layer with refractive index n_2 and the layer below it are designed to operate as a set of two antiresonant Fabry–Pérot resonators, which reduce the optical leakage from the core layer. ARROWs have found many applications, e.g., in laser diodes (Bhattacharya et al., 1996). Another waveguide that guides leaky waves is based on the concept of a photonic crystal (PC) and is relevant to vertical cavity surface emitting lasers (VCSELs) (Dems et al,, 2010). One example of such structure is a solid core photonic crystal waveguide (PCW), which is shown in Figure 4.11b. This waveguide consists of a set of air holes that surround the solid core. In this particular example, the air holes make a triangular lattice pattern. However, other arrangements of holes are also possible. In principle, the holes are arranged to prevent an optical wave from propagation within the air hole region at the waveguide operating wavelength. This is done by selecting the waveguide operating wavelength within the stop band of the PC (Joannopoulos et al., 1995). This results in an optical mode that is confined to the solid core (Bjarklev et al., 2003; Zolla, 2005; Poli et al., 2007). If the PC extended to infinity, such a waveguide would guide optical waves in the core region without any leakage. However, due to technological limitations, in practice only several PC layers surround the core, which results in the leakage of light.

Another wave-guiding mechanism that is relevant to optoelectronic devices is gain-guiding. Gain-guiding structures are exclusively used in semiconductor lasers. The current flowing through the active region of a laser diode creates an area of positive gain, which guides the light laterally (Petermann, 1991). The modes of gain-guided waveguides are known to have curved phase fronts (Petermann, 1991).

The final structure we discuss in this section is an optical waveguide array (Figure 4.12). Optical waveguide arrays are used in semiconductor lasers to scale up the output power of single emitter laser diodes. The modes of such structures are called supermodes to differentiate them from the modes of waveguides, which form the array. Figure 4.12 shows as an example of an array consisting of three rib waveguides. If each rib waveguide separately guides only one mode, then the entire array guides three supermodes. To illustrate the difference between the individual waveguide modes and the waveguide array supermodes, we used the EIM to calculate the field distributions $X(x)$, cf. Section 4.2.4, of all three supermodes (Figure 4.13) for an array consisting of three rib waveguides.

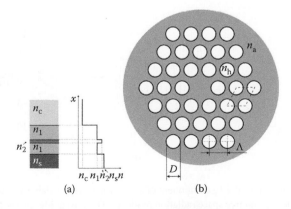

(a)   (b)

FIGURE 4.11   Antireflection resonant optical waveguides (ARROW) (a) and photonic crystal waveguide (b).

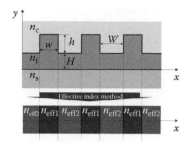

FIGURE 4.12 Optical waveguide array consisting of three rib waveguides and its effective index map.

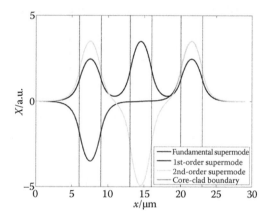

FIGURE 4.13 Lateral mode distributions $X(x)$ for an optical waveguide array consisting of three waveguides (Figure 4.12) calculated using the effective index method (EIM): $W = 4$ µm, $w = 3$ µm, $H = 4$ µm, $h = 2$ µm, $n_c = 1$, $n_f = 3.5$, $n_s = 1.45$, and $l = 1.55$ µm.

4.3 Beam Propagation Method

The operation of many optoelectronic devices relies on the phenomenon of light guiding along optical waveguides. For instance, optical waveguides are used in laser diodes, semiconductor laser amplifiers, electro-optic modulators, etc. The numerical techniques that were discussed in Section 4.2 are often insufficient for the study of light propagation in optoelectronic devices. This is because the waveguiding structures used in optoelectronic devices are not infinitely long and may not be straight. In principle, the electromagnetic field distribution within an optoelectronic device can be calculated using a general purpose Maxwell equations solver, e.g., finite difference time domain method (FDTDM). Such an approach, however, tends not to be practical because general purpose techniques calculate the field distribution in the entire domain. Hence, the resulting matrices storing all field sampling points within the computational domain are very large. For instance, 20 samples per wavelength for an optical waveguide that is 10,000 wavelengths long (a typical length of a high-power edge emitting laser diode) would result in 200,000 samples just for a 1D analysis. The beam propagation method (BPM) was developed to address these difficulties. Unlike the direct Maxwell equations solution methods, BPM allows calculating the spatial distribution of the output beam subject to the input beam shape (Figure 4.14) without the necessity of storing the field samples in the entire computational domain. In this section, we provide a short description of BPM.

In order to derive the basic equations relevant to BPM, we assume that the light propagation is taking place along the z-axis (Figure 4.1). In the rectangular coordinate system (Figure 4.3), the following set of

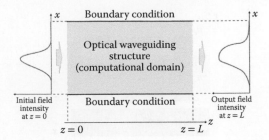

FIGURE 4.14 Schematic representation of the beam propagation method (BPM) application to an optical waveguiding structure in rectangular coordinate system.

three equations for the electric field vector components are obtained from Equation 4.3:

$$\frac{\partial^2 E_x}{\partial y^2} + \frac{\partial}{\partial x}\left(\frac{1}{\varepsilon}\frac{\partial\left(\varepsilon E_x\right)}{\partial x}\right) + \omega^2\mu\varepsilon E_x + \frac{\partial^2 E_x}{\partial z^2} + \frac{\partial}{\partial x}\left(\frac{1}{\varepsilon}\frac{\partial\left(\varepsilon E_y\right)}{\partial y}\right) - \frac{\partial^2 E_y}{\partial x\partial y} + \frac{\partial}{\partial x}\left(\frac{1}{\varepsilon}E_z\frac{\partial\varepsilon}{\partial z}\right) = 0 \quad (4.55)$$

$$\frac{\partial^2 E_y}{\partial x^2} + \frac{\partial}{\partial y}\left(\frac{1}{\varepsilon}\frac{\partial\left(\varepsilon E_y\right)}{\partial y}\right) + \omega^2\mu\varepsilon E_y + \frac{\partial^2 E_y}{\partial z^2} + \frac{\partial}{\partial y}\left(\frac{1}{\varepsilon}\frac{\partial\left(\varepsilon E_x\right)}{\partial x}\right) - \frac{\partial^2 E_x}{\partial y\partial x} + \frac{\partial}{\partial y}\left(\frac{1}{\varepsilon}E_z\frac{\partial\varepsilon}{\partial z}\right) = 0 \quad (4.56)$$

$$\frac{\partial^2 E_z}{\partial x^2} + \frac{\partial^2 E_z}{\partial y^2} + \omega^2\mu\varepsilon E_z + \frac{\partial}{\partial z}\left(\frac{1}{\varepsilon}\frac{\partial\left(\varepsilon E_z\right)}{\partial z}\right) + \frac{\partial}{\partial z}\left(\frac{1}{\varepsilon}E_x\frac{\partial\varepsilon}{\partial z}\right) + \frac{\partial}{\partial z}\left(\frac{1}{\varepsilon}E_y\frac{\partial\varepsilon}{\partial z}\right) = 0. \quad (4.57)$$

Similarly for the magnetic field vector components:

$$\frac{\partial^2 H_x}{\partial x^2} + \varepsilon\frac{\partial}{\partial y}\left(\frac{1}{\varepsilon}\frac{\partial H_x}{\partial y}\right) + \omega^2\mu\varepsilon H_x + \varepsilon\frac{\partial}{\partial z}\left(\frac{1}{\varepsilon}\frac{\partial H_x}{\partial z}\right) + \frac{1}{\varepsilon}\left(\frac{\partial\varepsilon}{\partial y}\frac{\partial H_y}{\partial x} + \frac{\partial\varepsilon}{\partial z}\frac{\partial H_z}{\partial x}\right) = 0 \quad (4.58)$$

$$\varepsilon\frac{\partial}{\partial x}\left(\frac{1}{\varepsilon}\frac{\partial H_y}{\partial x}\right) + \frac{\partial^2 H_y}{\partial y^2} + \omega^2\mu\varepsilon H_y + \varepsilon\frac{\partial}{\partial z}\left(\frac{1}{\varepsilon}\frac{\partial H_y}{\partial z}\right) + \frac{1}{\varepsilon}\left(\frac{\partial\varepsilon}{\partial x}\frac{\partial H_x}{\partial y} + \frac{\partial\varepsilon}{\partial z}\frac{\partial H_z}{\partial y}\right) = 0 \quad (4.59)$$

$$\varepsilon\frac{\partial}{\partial x}\left(\frac{1}{\varepsilon}\frac{\partial H_z}{\partial x}\right) + \varepsilon\frac{\partial}{\partial y}\left(\frac{1}{\varepsilon}\frac{\partial H_z}{\partial y}\right) + \omega^2\mu\varepsilon H_z + \frac{\partial^2 H_z}{\partial z^2} + \frac{1}{\varepsilon}\left(\frac{\partial\varepsilon}{\partial x}\frac{\partial H_x}{\partial z} + \frac{\partial\varepsilon}{\partial y}\frac{\partial H_y}{\partial z}\right) = 0. \quad (4.60)$$

We assume that the analyzed optical waveguiding structure varies slowly along the z-direction and hence $\partial\varepsilon/\partial z \approx 0$ (Vassallo, 1997a). As a consequence, the transverse components of the electromagnetic field decouple from the longitudinal ones and hence for the transverse field components, the set of Equations 4.55 through 4.57 can be approximated by the following equation:

$$\frac{\partial\vec{E}_t}{\partial z^2} = -\left(A + \omega^2\mu\varepsilon\right)\vec{E}_t, \quad (4.61)$$

and Equations 4.58 through 4.60 can be approximated by

$$\frac{\partial\vec{H}_t}{\partial z^2} = -\left(B + \omega^2\mu\varepsilon\right)\vec{H}_t, \quad (4.62)$$

where $\vec{E}_t = \vec{i}_x E_x + \vec{i}_y E_y$ and $\vec{H}_t = \vec{i}_x H_x + \vec{i}_y H_y$. The elements of matrices A and B are provided in Tables 4.3 and 4.4.

TABLE 4.3 Elements a_{ji} of Matrix A

$a_{j,i}$	$i = 1$	$i = 2$
$j = 1$	$\frac{\partial^2}{\partial y^2} + \frac{\partial}{\partial x}\left(\frac{1}{\varepsilon}\frac{\partial(\varepsilon)}{\partial x}\right)$	$\frac{\partial}{\partial x}\left(\frac{1}{\varepsilon}\frac{\partial(\varepsilon)}{\partial y}\right) - \frac{\partial^2}{\partial x\partial y}$
$j = 2$	$\frac{\partial}{\partial y}\left(\frac{1}{\varepsilon}\frac{\partial(\varepsilon)}{\partial x}\right) - \frac{\partial^2}{\partial y\partial x}$	$\frac{\partial^2}{\partial x^2} + \frac{\partial}{\partial y}\left(\frac{1}{\varepsilon}\frac{\partial(\varepsilon)}{\partial y}\right)$

TABLE 4.4 Elements b_{ji} of Matrix B

$b_{j,i}$	$i = 1$	$i = 2$
$j = 1$	$\frac{\partial^2}{\partial x^2} + \varepsilon\frac{\partial}{\partial y}\left(\frac{1}{\varepsilon}\frac{\partial}{\partial y}\right)$	$-\varepsilon\frac{\partial}{\partial y}\left(\frac{1}{\varepsilon}\frac{\partial}{\partial x}\right) + \frac{\partial^2}{\partial y\partial x}$
$j = 2$	$-\varepsilon\frac{\partial}{\partial x}\left(\frac{1}{\varepsilon}\frac{\partial}{\partial y}\right) + \frac{\partial^2}{\partial x\partial y}$	$\frac{\partial^2}{\partial y^2} + \varepsilon\frac{\partial}{\partial x}\left(\frac{1}{\varepsilon}\frac{\partial}{\partial x}\right)$

Equations 4.61 and 4.62 form the basis of most BPM algorithms. Similarly, as in Section 4.1, Equations 4.61 and 4.62 can be further simplified for waveguiding structures, which support polarized modes. For the x-polarized modes, Equations 4.61 and 4.62 reduce to

$$\frac{\partial^2 E_x}{\partial z^2} = -\left(a_{11} + \omega^2\mu\varepsilon\right)E_x \tag{4.63}$$

and

$$\frac{\partial^2 H_y}{\partial z^2} = -\left(b_{22} + \omega^2\mu\varepsilon\right)H_y, \tag{4.64}$$

while for y-polarized modes we obtain

$$\frac{\partial^2 E_y}{\partial z^2} = -\left(a_{22} + \omega^2\mu\varepsilon\right)E_y \tag{4.65}$$

and

$$\frac{\partial^2 H_x}{\partial z^2} = -\left(b_{11} + \omega^2\mu\varepsilon\right)H_x. \tag{4.66}$$

The operators a_{ji} and b_{ji} are given in Tables 4.3 and 4.4.

Furthermore, if ε varies slowly with the transverse coordinates one can use the scalar approximation

$$\frac{\partial^2 \Phi}{\partial z^2} = -\left(s + \omega^2\mu\varepsilon\right)\Phi, \tag{4.67}$$

where Φ is the scalar potential and $s = \frac{\partial^2}{\partial y^2} + \frac{\partial^2}{\partial x^2}$.

Thus, we obtain the same hierarchy of approximations as in Section 4.1. Further improvement in efficiency can be achieved using the effective index approximation to calculate the effective index map and thus reduce the number of spatial dimensions considered. It can be observed that Equations 4.61 and 4.62 through Equation 4.67 have the same generic form, namely

$$\frac{\partial^2 F}{\partial z^2} = -\left(L + \omega^2\mu\varepsilon\right)F. \tag{4.68}$$

The first step in deriving a BPM algorithm for the solution of Equation 4.68 is the derivation of the one-way wave equation. This step is usually accomplished by formally factoring Equation 4.68

$$\left(\frac{\partial}{\partial z} - j\sqrt{L + \omega^2 \mu\varepsilon} \right) \left(\frac{\partial}{\partial z} + j\sqrt{L + \omega^2 \mu\varepsilon} \right) F = 0.$$

If only one of the product components is considered, the so-called one-way wave equation is obtained

$$\frac{\partial F}{\partial z} = -j\sqrt{L + \omega^2 \mu\varepsilon} F. \tag{4.69}$$

After the introduction of the slowly varying (with z) envelope function Φ, defined by $F = \Phi(x, y, z) * \exp(-j\beta_r z)$ where β_r is a suitably selected reference propagation constant, Equation 4.69 transforms into

$$\frac{\partial \Phi}{\partial z} = j\beta_r \left(-\sqrt{1 + \frac{L + \omega^2 \mu\varepsilon - \beta_r^2}{\beta_r^2}} + 1 \right) \Phi. \tag{4.70}$$

The formal integration of Equation 4.70 yields

$$\Phi\left(z_0 + \Delta z\right) = \exp \int_{z=z_0}^{z=z_0+\Delta z} \left[j\beta_r \left(-\sqrt{1 + \frac{L + \omega^2 \mu\varepsilon - \beta_r^2}{\beta_r^2}} + 1 \right) \right] \Phi\left(z_0\right) dz. \tag{4.71}$$

If the computational domain (Figure 4.14) is divided into a concatenation of sections with a z-independent refractive index distribution, and within each section the rectangle numerical integration rule is applied, then Equation 4.71 reduces to

$$\Phi\left(z_0 + \Delta z\right) = \exp \left[j\beta_r \left(-\sqrt{1 + \frac{L + \omega^2 \mu\varepsilon - \beta_r^2}{\beta_r^2}} + 1 \right) \right] \Delta z \Phi\left(z_0\right). \tag{4.72}$$

Equation 4.72 implies an algorithm, which calculates the value of the field distribution at the position $z_0 + \Delta z$ subject to the known field distribution at the previous position z_0. Thus, the entire calculation can be started from the initial field distribution at $z = 0$ (Figure 4.14) and continued until the output field distribution at $z = L$ is obtained. At each step, only the field distribution from the previous step needs to be stored. This makes a BPM algorithm much more manageable in terms of the required computer memory than general purpose Maxwell equations solvers.

The last step left is the conversion of Equation 4.72 into an efficient computer code. For this purpose, suitable approximations for the L, exponential, and square root operators need to be found. A number of algorithms were developed for this purpose. The most often used approximations for the L operator apply the FDM, the finite element method, or local waveguide mode expansion. In the case of the exponential and square root operators, the typical approach consists of expanding the operator using either Padé or Taylor series. For instance, an approximation of the square root operator using the Taylor series truncated after the first term results in a paraxial approximation, while applying Padé (1,1) expansion to the exponential operator, results in the Crank–Nicholson scheme. A comprehensive survey and a fairly detailed description of BPM operator approximation algorithms can be found in Yamauchi (2003) and Sujecki (2014). A popular choice is the selection of paraxial approximation for the square root operator, Padé (1,1) approximation for the exponential operator, and finite difference approximation for the L operator. Such selection yields the finite difference, Crank–Nicholson paraxial BPM algorithm (FDCNP-BPM). FDCNP-BPM is very effective if the wave propagation takes place within a narrow cone centered along the z-axis. If this condition is not fulfilled, then Padé approximations are used for the square root operator, resulting in wide angle (WA) BPM algorithms. In Section 4.4, we provide examples of BPM application.

4.4 Tapered and Bent Waveguides

The waveguides used in semiconductor lasers or semiconductor optical amplifiers may be intentionally tapered or bent. For instance, tapered lasers can offer high output power without compromising on the beam quality, while bent semiconductor optical amplifiers allow reducing the coupling of the back reflected waves into the amplifier. In this section, we discuss the optical properties of tapered and bent waveguides and their design methods.

4.4.1 Tapered Waveguides

A very useful intuitive insight into the operation of tapered optical waveguides can be gained through applying ray optics. Figure 4.15 shows an optical ray entering an optical taper with an initial incidence angle θ. The taper is symmetrical with respect to the z-axis and its angle is α. By means of geometric optics, it can readily be demonstrated that after the reflection from the boundary between the core and cladding the ray will travel downward and will impinge upon the lower boundary at an incidence angle $\theta - \alpha$ (Cada et al., 1988). The same applies to the subsequent reflections: at each subsequent reflection, the angle of incidence decreases by the value of the taper angle α. The obvious consequence of this fact is that even if the ray is traveling initially at an angle significantly exceeding the critical angle, then after several reflections it might start traveling at an angle that is less than the critical angle and thus start leaking power through refraction at the boundary between the core and cladding (Zheng, 1989). If this process is continued, the ray may eventually start traveling backward, thus leading to the conclusion that a part of the incident optical power is reflected back. All these predictions, arrived at intuitively using ray optics, have been confirmed with advanced numerical models and experiments. Further, when a taper is used to connect two waveguides, then it is obvious that the ray incidence angle is different at the output when compared with the input waveguide (Figure 4.15). This leads to a conclusion that a mode conversion may occur if both waveguides are multimode (cf. Li and Lit, 1986; Ravets et al., 2013).

Further useful insight might be obtained using the wave optics and the electromagnetic field theory. Such an analysis is particularly useful when attempting to establish guidelines for the design of tapers with negligibly small losses. The formulation of the wave equation in the cylindrical coordinate system leads to a conclusion that a lossless taper can be designed if the V parameter of the tapered waveguide is kept constant (Marcatili, 1985). This rule, however, applies without ambiguity only when tapering step index slab and circular waveguides. Also such prediction is subject to an assumption that an optical taper can be approximated by an infinitely extending wedge or cone. If a taper connects two waveguides (Figure 4.16), then the simulations show that a significant portion of the loss results from the mode mismatch at the input and output of the taper (Zheng, 1989). The conformal mapping shows that such a loss can be reduced by

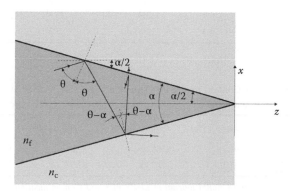

FIGURE 4.15   Optical ray path in an optical taper.

inserting a lens both at the taper input and output (Lee et al., 1997). Other advanced numerical simulations show that the loss can also be reduced by shaping the taper (Luyssaert et al., 2005).

An accurate analysis of tapered structures is complicated by the nonseparability of the wave equation in standard coordinate systems. Therefore, an application of numerical methods is necessary. BPM turned out to be particularly useful if the effect of back reflection is negligible.

Table 4.5 contains the parameters of four optical taper structures, which were studied extensively in the literature (Haes et al., 1996). Table 4.6 provides the calculated power content in the fundamental x-polarized mode at the end of the taper

$$P_c = \frac{\int E_y^*(x, z = 0) E_y(x, z = L)\, dx}{\int E_y^*(x, z = 0) E_y(x, z = 0)\, dx}. \tag{4.73}$$

Table 4.6 provides a comparison between the results calculated by FDCNP-BPM and finite difference Crank–Nicolson wide angle beam propagation method (FDCNWA-BPM), which uses Padé (1,1) approximation. The largest difference between both sets of results is one on the first decimal place. These results demonstrate the usefulness of the paraxial approximation in modeling and design of the slowly varying optical waveguiding structures.

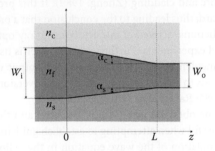

FIGURE 4.16   Optical taper connecting two optical slab waveguides.

TABLE 4.5   Slab Waveguide Taper Test Structures (Figure 4.16)

	$W_i/\mu m$	$W_o/\mu m$	$L/\mu m$	n_c	n_f	n_s	α_s
Taper 1	0.2	0.1	5.73	3.17	3.3	3.17	0
Taper 2	0.2	0.1	57.3	3.17	3.3	3.17	0
Taper 3	0.8	0.4	22.9	1.0	3.3	3.17	0
Taper 4	0.8	0.4	229	1.0	3.3	3.17	0

TABLE 4.6   Power Content in the Fundamental x-Polarized Mode at $z = L$ P_c Calculated from Equation 4.73 for Test Structures from Table 4.5 (Figure 4.16)

	Taper 1 (%)	Taper 2 (%)	Taper 3 (%)	Taper 4 (%)
FDCNP-BPM	91.18	97.29	82.24	96.63
FDCNWA-BPM	91.19	97.30	82.34	96.63

Source: Sujecki S., *International Journal of Electronics and Communications (AEU)*, 55, 185–190, 2001.

4.4.2 Bent Waveguides

Figure 4.17a shows a section of a circularly bent optical slab waveguide, which connects two straight waveguides. The bent radius R is measured with respect to the center of the waveguide. The angle α gives a measure of the azimuthal length of the bent. An optical beam experiences an optical loss, off axis offset and a polarization rotation while propagating through a bent waveguide section.

Some intuitive insight supporting the statement that an optical beam experiences power loss while propagating through a bend can be gained by observing that an optical mode of a straight waveguide after entering a bent waveguide section would have to preserve flat phase fronts in order to avoid losses. This, however, is not possible since the phase velocity within the bent waveguide section would have to grow with radius r if flat phase fronts were to be preserved. Thus, for a certain value of radius r the phase velocity would have to become larger than the light velocity of the medium thus implicating the necessity of phase front bending and the power loss through radiation from the bent (Marcatili, 1985). Alternatively, one can arrive at the same conclusion by tracking the paths of optical rays entering a bent waveguide (Figure 4.17b). An optical ray entering a bend near the outer boundary between the core and the cladding impinges upon the bend boundary at a large incidence angle and thus undergoes a total internal reflection. However, an optical ray, which enters the bend near the inner boundary between the core and cladding, has a small incidence angle and hence may also undergo refraction and hence lose power. These intuitive predictions have been confirmed by numerical simulations (cf. Hiremath et al., 2005).

The radiation into the outer space surrounding the bend is, however, not the only source of optical loss. An optical bend usually connects two straight waveguide sections. Hence, loss can be incurred by the mismatch between the field distribution of a straight waveguide mode and of the bent waveguide mode. This effect is related to the fact that the modes of an optical bend have different field distributions than those of the straight waveguide even if both straight and bent waveguides have the same transverse distribution of the refractive index (Hiremath et al., 2005). When compared with the modes of the straight waveguide, the modes of the bent waveguide have an off-axis offset. This effect can be most convincingly demonstrated using the conformal mapping to derive a refractive index distribution of an equivalent straight waveguide (Ladouceur and Labeye, 1995). In fact, conformal mapping provides another argument supporting the claim that an optical bend incurs a loss upon an optical beam. A logarithmic transformation, for instance, allows mapping of a section of a circular ring onto a straight stripe and can be used to derive the refractive index distribution of an equivalent straight waveguide (Heiblum and Harris, 1975). The resulting equivalent refractive index distribution grows away from the bent waveguide center. When the bent radius is large, the refractive index growth can be locally approximated by a linear function (Figure 4.18; Ladouceur and Labeye, 1995). Thus, at a certain value of the transverse coordinate, the value of the refractive index in the cladding exceeds the largest value of the refractive index in the core. This leads to a conclusion that all modes of such waveguide are leaky and hence inherently lost. Further, since the refractive index distribution is modified the fundamental mode intensity distribution of the bent waveguide is different from that

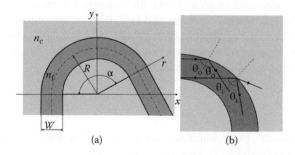

(a)   (b)

FIGURE 4.17 Bent optical slab waveguide connecting two straight waveguides (a) and the optical ray paths in a bent waveguide (b).

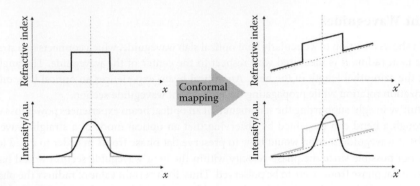

FIGURE 4.18 Schematic illustration of the mode offset in a bent waveguide.

of the straight one. The field is more concentrated in the outer part of the core and has an off-axis offset. Consequently, there is a mode mismatch between the straight and curved waveguide sections, which incurs a coupling loss. The effect of this phenomenon on the overall bent waveguide loss can be reduced by off-setting the straight waveguide (Kitoh et al., 1995). Another consequence of the fact that the modal field distributions in the straight and bent waveguides are different is the excitation of the first-order mode in the bent waveguide even if only the fundamental mode is incident from the straight waveguide. If this is the case, both the fundamental mode and the first-order mode propagate with different propagation constants along the bent waveguide and hence combine with different phases at the output. Thus, the resulting field pattern may not match the fundamental mode of the straight waveguide at the output, which results in a coupling loss. The loss resulting from the excitation of the first-order mode can be minimized either by appropriate selection of the bent radius and angle (Melloni et al., 2004) or by offsetting the straight waveguide (Hirono et al., 1998).

Finally, an optical beam, although propagating through a bent waveguide section, may undergo a change of its polarization state. This phenomenon is used to design polarization rotators (Deng et al., 2005). The calculation of the polarization rotation is not straightforward and requires either an application of the coupled mode theory or a vectorial BPM (Lui et al., 1998).

Finally, we note that a number of simple approximate techniques are available for the calculation of the bending radiation loss in various waveguide structures (Vassallo, 1991). However, a calculation of the full electromagnetic field distribution within an arbitrary bent waveguide requires an application of numerical methods. There are two ways in which the problem of calculating the electromagnetic field distribution within a bent waveguide can be formulated. One formulation of the problem consists of considering a complete ring and calculation of the complex resonant frequencies of the modes. In the other, the frequency is assumed to be a known parameter and the field distribution is calculated within the bent waveguide using for instance a BPM algorithm (Sujecki, 2014). Alternatively, for circularly bent slab waveguides an analytical solution for the propagation constants and field distributions of the guided modes is possible (Hiremath et al., 2005). The advantage of BPM is the possibility of inclusion in the model of all effects that accompany optical beam propagation through a bent waveguide. However, an optical mode solver provides a very useful insight into the bent waveguide characteristics.

References

Anemogiannis E, Glytsis EN (1992) Multilayer wave-guides—Efficient numerical-analysis of general structures. *Journal of Lightwave Technology* 10:1344–1351.

Anemogiannis E, Glytsis EN, Gaylord TK (1994) Efficient solution of eigenvalue equations of optical wave-guiding structures. *Journal of Lightwave Technology* 12:2080–2084.

Anemogiannis E, Glytsis EN, Gaylord TK (1999) Determination of guided and leaky modes in lossless and lossy planar multilayer optical waveguides: Reflection pole method and wavevector density method. *Journal of Lightwave Technology* 17:929–941.

Bhattacharya A, Mawst LJ, Nesnidal MP, Lopez J, Botez D (1996) 0.4W CW diffraction limited beam Al free 0.98 μ/m wavelength three core ARROW-type diode lasers. *Electronics Letters* 32:657–658.

Bjarklev A, Bjarklev A, Broeng J (2003) *Photonic Crystal Fibres.* London: Kluwer Academic Publishers.

Black RJ, Gagnon L (2010) *Optical Waveguide Modes: Polarization, Coupling and Symmetry.* London: McGraw-Hill.

Cada M, Feng XA, Felsen LB (1988) Intrinsic modes in tapered optical wave-guides. *IEEE Journal of Quantum Electronics* 24:758–765.

Chiang KS, Wong WP (1996) Rib waveguides with degenerate polarised modes. *Electronics Letters* 32:1098–1099.

Ctyroky J, Hofman M, Janta J, Schrofel J (1984) 3-D analysis of LiNbO$_3$-Ti channel waveguides and directional-couplers. *IEEE Journal of Quantum Electronics* 20:400–409.

Dems M, Chung I-S, Peter N, Bischoff S, Panajotov K (2010) Numerical methods for modeling photonic-crystal VCSELs. *Optics Express* 18:16042–16054.

Deng HH, Yevick DO, Chaudhuri SK (2005) Bending characteristics of asymmetric SOI polarization rotators. *IEEE Photonics Technology Letters* 17:2113–2115.

Ding H, Chan KT (1997) Solving planar dielectric waveguide equations by counting the number of guided modes. *IEEE Photonics Technology Letters* 9:215–217.

Dyott RB (1996) *Elliptical Fiber Waveguides.* Boston: Artech House.

Ebeling KJ (1993) *Integrated Optoelectronics: Waveguide Optics, Photonics, Semiconductors.* Berlin: Springer-Verlag.

Gallick AT, Kerkhoven T, Ravaioli U (1992) Iterative solution of the eigenvalue problem for a dielectric waveguide. *IEEE Transactions on Microwave Theory and Techniques* 40:699–705.

Goell JE (1969) A circular harmonic computer analysis of rectangular dielectric waveguides. *Bell System Technical Journal* 48:2133–2160.

Hadley GR (2002) High accuracy finite difference equations for dielectric waveguide analysis II: Dielectric corners. *Journal of Lightwave Technology* 20:1219–1231.

Haes J et al. (1996) A comparison between different propagative schemes for simulation of tapered step index slab waveguides. *Journal of Lightwave Technology* 14:1557–1567.

Heiblum M, Harris JH (1975) Analysis of curved optical-waveguides by conformal transformation. *IEEE Journal of Quantum Electronics QE* 11:75–83.

Hiremath K, Hammer M, Stoffer R, Prkna L, Ctyroky J (2005) Analytic approach to dielectric optical bent slab waveguides. *Optical and Quantum Electronics* 37:37–61.

Hirono T, Kohtoku M, Yoshikuni Y, Lui WW, Yokoyama K (1998) Optimized offset to eliminate first-order mode excitation at the junction of straight and curved multimode waveguides. *IEEE Photonics Technology Letters* 10:982–984.

Hondros D, Debye P (1910) Electromagnetic waves in dielectrical wires. *Annalen der Physik* 32:465–476.

Hunsperger RG (2002) *Integrated Optics.* Berlin: Springer-Verlag.

Jablonski TF (1994) Complex modes in open lossless dielectric waveguides. *Journal of the Optical Society of America A: Optics, Image Science, and Vision* 11:1272–1282.

Joannopoulos JD, Meade RD, Winn JN (1995) *Photonic Crystals: Molding the Flow of Light.* Princeton, NJ: Princeton University Press.

Kitoh T, Takato N, Yasu M, Kawachi M (1995) Bending loss reduction in silica-based wave-guides by using lateral offsets. *Journal of Lightwave Technology* 13:555–562.

Kubica JM (2002) Analysis of ARROWs with thin cores. *Optical and Quantum Electronics* 34:737–745.

Ladouceur F, Labeye P (1995) A new general-approach to optical wave-guide path design. *Journal of Lightwave Technology* 13:481–492.

Lee CT, Wu ML, Hsu JM (1997) Beam propagation analysis for tapered waveguides: Taking account of the curved phase-front effect in paraxial approximation. *Journal of Lightwave Technology* 15:2183–2189.

Li YF, Lit JWY (1986) Mode-changes in step-index multimode fiber tapers. *Journal of the Optical Society of America A: Optics, Image Science, and Vision* 3:161–164.

Lui WW, Hirono T, Yokoyama K, Huang WP (1998) Polarization rotation in semiconductor bending waveguides: A coupled-mode theory formulation. *Journal of Lightwave Technology* 16:929–936.

Luyssaert B, Bienstman P, Vandersteegen P, Dumon P, Baets R (2005) Efficient nonadiabatic planar waveguide tapers. *Journal of Lightwave Technology* 23:2462–2468.

Maerz R (1995) *Integrated Optics: Design and Modelling.* London: Artech House.

Majewski A (1984) Numerical analysis of the multistep-index fiber having the uniaxial anisotropy. *Bulletin of the Polish Academy of Sciences, Technical Sciences* 32:709–714.

Majewski A (1991) *Theory and Design of Optical Fibres* (in Polish). Warsaw: Wydawnictwo Naukowo-Techniczne.

Majewski A, Sujecki S (1996a) Analiza wlasciwosci propagacyjnych swiatlowodów prostokatnych (in Polish). *Prace Naukowe Politechniki Warszawskiej* 110:5–35.

Majewski A, Sujecki S (1996b) Modes in rectangular fibres. *Optoelectronics Review* 4:45–50.

Majewski A, Sujecki S (1997) Rib lightguides and couplers. *Bull Polish Academy of Sciences: Technical Sciences* 45:145–150.

Majewski A, Sujecki S (1998) Polarisation characteristics of optical rib waveguides. *Optica Applicata* 28:76–82.

Marcatili EAJ (1969) Dielectric rectangular waveguide and directional coupler for integrated optics. *Bell System Technical Journal* 48:2071–2102.

Marcatili EAJ (1974) Slab coupled waveguides. *Bell System Technical Journal* 53:645–674.

Marcatili EAJ (1985) Dielectric tapers with curved axes and no loss. *IEEE Journal of Quantum Electronics* 21:307–314.

Marcatili EAJ, Hardy AA (1988) The azimuthal effective-index method. *IEEE Journal of Quantum Electronics* 24:766–774.

Marcuse D (1972) *Light Transmission Optics.* New York, NY: Van Nostrand Reinhold.

Marcuse D (1974) *Theory of Dielectric Optical Waveguides.* San Diego, CA: Academic Press.

Melloni A, Cusmai G, Martinelli M (2004) Experimental confirmation of matched bends. *Optics Letters* 29:465–467.

Okamoto K (2006) *Fundamentals of Optical Waveguides.* London: Elsevier.

Okoshi T (1982) *Optical Fibres.* London: Academic Press.

Petermann K (1991) *Laser Diode Modulation and Noise.* London: Kluwer Academic Publishers.

Poli F, Cucinotta A, Selleri, S (2007) *Photonic Crystal Fibres.* Dordrecht: Springer-Verlag.

Ramaswamy W (1974) Strip-loaded film waveguides. *Bell System Technical Journal* 53:697–704.

Ravets S, Hoffman JE, Kordell PR, Wong-Campos JD, Rolston SL, Orozco LA (2013) Intermodal energy transfer in a tapered optical fiber: Optimizing transmission. *Journal of the Optical Society of America A: Optics, Image Science, and Vision* 30:2361–2371.

Riishede J, Mortensen NA, Laegsgaard J (2003) A "poor man's approach" to modelling micro-structured optical fibres. *Journal of Optics A: Pure and Applied Optics* 5:534–538.

Rozzi T, Mongiardo M (1997) *Open Electromagnetic Waveguides.* London: IEE.

Snyder AW, Love, JD (1983) *Optical Waveguide Theory.* London: Chapman & Hall.

Soref RA, Schmidtchen J, Petermann K (1991) Large single mode rib waveguides in GeSi-Si and Si-on-SiO$_2$. *IEEE Journal of Quantum Electronics* 27:1971–1974.

Sujecki S (2001) New beam propagation algorithm for optical tapers. *International Journal of Electronics and Communications (AEU)* 55:185–190.

Sujecki S (2010) Arbitrary truncation order three-point finite difference method for optical waveguides with stepwise refractive index discontinuities. *Optics Letters* 35:4115–4117.

Sujecki S (2014) *Photonics Modelling and Design.* Boca Raton, FL: CRC Press.

Sujecki S, Benson TM, Sewell P, Kendall PC (1998) Novel vectorial analysis of optical waveguides. *Journal of Lightwave Technology* 16:1329–1335.

Tamir T (1975) *Integrated Optics.* Berlin: Springer-Verlag.

Vassallo C (1991) *Optical Waveguide Concepts.* Amsterdam: Elsevier.

Vassallo C (1997a) Difficulty with vectorial BPM. *Electronics Letters* 33:61–62.

Vassallo C (1997b) Mode solvers 1993–1995 optical mode solvers. *Optical and Quantum Electronics* 29:95–114.

Yamauchi J (2003) *Propagating Beam Analysis of Optical Waveguides.* Baldock, UK: Research Studies Press.

Yeh P, Yariv A, Marom E (1978) Theory of Bragg fiber. *Journal of the Optical Society of America* 68:1196–1201.

Zheng XH (1989) Understanding radiation from dielectric tapers. *Journal of the Optical Society of America A: Optics, Image Science, and Vision* 6:190–201.

Zolla F, Reneversez G, Nicolet A, Kuhlmev B, Guenneau S, Felbacq D. (2005) *Foundations of Photonic Crystal Fibres.* London: Imperial College Press.

5

Heat Generation and Dissipation

Giovanni Mascali

and

Vittorio Romano

Thermal effects in solids are a manifestation of the vibrational motion of the molecules or atoms that make up the material. These phenomena can be described at several levels of accuracy. From a macroscopic point of view, a thermodynamical approach can be adopted, e.g., in the framework of linear irreversible thermodynamics (see e.g., [1]) or extended thermodynamics [2,3] (see e.g., [4–10,35]). More detailed models within solid-state physics introduce quasi-particles, called *phonons*, to take into account the energy transport in the crystal [11]. Phonons obey Bose–Einstein statistics and their evolution can be obtained, in a semiclassical context, by solving a system of Boltzmann equations.

In turn, thermal effects influence the electrical and optical properties of semiconductors. In particular, in micro- and nanoelectronics, the formation of hot spots and the high thermal power generation per unit area are major issues for an efficient design of semiconductor devices.

In this chapter, the basic concepts of phonon transport are introduced along with the derivation, from them, of the thermodynamical properties relevant to the electrical and optical features of a crystal lattice. Specific applications to graphene are presented.

5.1 Crystal Lattices

The structure of a solid is given by a periodic repetition of a set of atoms, called the *basis*, in the three-dimensional (3D) space [12]. In this way, a 3D crystal lattice is formed (see Figure 5.1) with a translational periodicity defined by three noncoplanar vectors a_1, a_2, a_3, such that the crystal remains identical to itself

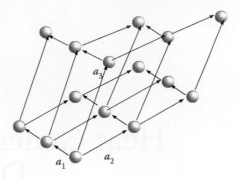

FIGURE 5.1 Example of crystal lattice.

if translated by a generic vector

$$\mathbf{T} = n_1 \boldsymbol{a}_1 + n_2 \boldsymbol{a}_2 + n_3 \boldsymbol{a}_3, \quad n_1, n_2, n_3 \in \mathbb{Z}.$$

The vectors $\boldsymbol{a}_1, \boldsymbol{a}_2, \boldsymbol{a}_2$ and the set

$$L = \{\mathbf{T} = n_1 \boldsymbol{a}_1 + n_2 \boldsymbol{a}_2 + n_3 \boldsymbol{a}_3 \,:\, n_1, n_2, n_3 \in \mathbb{Z}\} \subset \mathbb{R}^3$$

are called *primitive vectors* and *Bravais lattice*, respectively. The *primitive cell* of L is a connected set $D \subset \mathbb{R}^3$ whose volume equals that of the parallelepiped spanned by the basis vectors, i.e., $vol(D) = \mathbf{a}_1 \cdot (\mathbf{a}_2 \times \mathbf{a}_3)^\dagger$, and fills the whole space if translated by the vectors belonging to L. In Figure 5.2, a primitive cell in a bidimensional case is depicted. In particular, the special primitive cell

$$D = \left\{ \mathbf{r} \in \mathbb{R}^3 \,:\, \mathbf{r} = \sum_{n=1}^{3} \alpha_n \boldsymbol{a}_n, \ \alpha_n \in \left[-\frac{1}{2}, \frac{1}{2}\right] \right\}$$

is called the *Wigner–Seitz cell*. The lengths of the primitive vectors are the *lattice constants*. For materials having the diamond structure, like those made of carbon atoms, the primitive cell is cubic and the three lattice constants are equal.

The *reciprocal lattice* L^* of L is defined by

$$L^* = \left\{ \mathbf{G} = n_1 \boldsymbol{a}_1^* + n_2 \boldsymbol{a}_2^* + n_3 \boldsymbol{a}_3^* \,:\, n_1, n_2, n_3 \in \mathbb{Z} \right\},$$

where the *primitive vectors* $\boldsymbol{a}_1^*, \boldsymbol{a}_2^*, \boldsymbol{a}_3^* \in \mathbb{R}^3$ constitute the dual basis given by

$$\boldsymbol{a}_1^* = 2\pi \frac{\boldsymbol{a}_2 \times \boldsymbol{a}_3}{\boldsymbol{a}_1 \cdot \boldsymbol{a}_2 \times \boldsymbol{a}_3}; \quad \boldsymbol{a}_2^* = 2\pi \frac{\boldsymbol{a}_3 \times \boldsymbol{a}_1}{\boldsymbol{a}_1 \cdot \boldsymbol{a}_2 \times \boldsymbol{a}_3}; \quad \boldsymbol{a}_3^* = 2\pi \frac{\boldsymbol{a}_1 \times \boldsymbol{a}_2}{\boldsymbol{a}_1 \cdot \boldsymbol{a}_2 \times \boldsymbol{a}_3}.$$

They satisfy the relations $\boldsymbol{a}_m \cdot \boldsymbol{a}_n^* = 2\pi \delta_{mn}$ with $m, n = 1, 2, 3$ and, therefore, $e^{i\boldsymbol{a}_m \cdot \boldsymbol{a}_n^*} = 1$.

---

† The symbol × denotes the vector product.

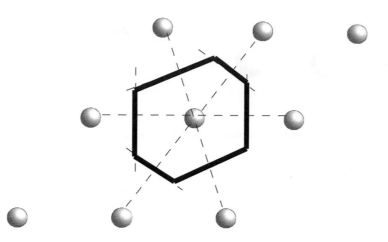

FIGURE 5.2 Primitive cell for a two-dimensional lattice.

According to the previous definitions, the volume of the reciprocal primitive cell is $a_1^* \cdot (a_2^* \times a_3^*)$, while the Wigner–Seitz cell of the reciprocal lattice is

$$B = \left\{ \mathbf{k} \in \mathbb{R}^3 : \mathbf{k} = \sum_{n=1}^{3} \beta_n a_n^*, \ \beta_n \in \left[-\frac{1}{2}, \frac{1}{2} \right] \right\} \tag{5.1}$$

and is called the first *Brillouin zone*.

Strictly speaking, the previous picture is valid only for an infinite medium. If the sample has a finite size, boundary effects should be also considered. However, for the scope of this chapter, it is reasonable to assume as valid the bulk (that is of the infinite case) properties of the crystal.

5.2 Crystal Vibrations and Phonons

The most basic manifestation of the crystal temperature is the vibration of the atoms around their equilibrium positions represented by the points of the ideal Bravais lattice. In solids, under the adiabatic approximation, thermal motion of atoms around their equilibrium positions can be described as an ensemble of normal modes of vibration of the crystal lattice.

The interactions among the atoms depend on the specific chemical bond, e.g., ionic or covalent. As the first approximation, they can be considered elastic-type interactions. In order to grasp the main features without entering into the mathematical details, let us consider the simplified case of a chain of N equal particles of mass m connected by equal massless springs with elastic constant k (see Figure 5.3, left). The equilibrium separation is the lattice constant a. The particles can vibrate along the direction of the chain producing *longitudinal waves* or along the perpendicular directions producing *transversal waves*. For simplicity, we assume that the particles can vibrate only along the longitudinal direction. Let x_j be the displacement of the jth particle from its equilibrium position.

In order to deal with an infinite chain, cyclic boundary conditions are imposed

$$x_{j+N}(t) = x_j(t), \quad t > 0,$$

with t time and $N \in \mathbb{N}$ the period. Moreover, only the interactions of each atom with the nearest neighbors are retained. Under such hypotheses, one gets the following equations of motion:

$$m\ddot{x}_j = k(x_{j+1} + x_{j-1} - 2x_j), \quad j = 1, 2, \cdots, N. \tag{5.2}$$

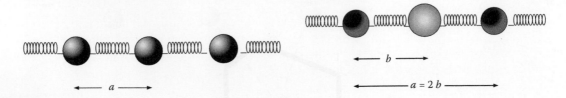

FIGURE 5.3   Schematic representation of a monoatomic (left) and a biatomic (right) linear chain.

We look for solutions of the type $x_j = \xi e^{i(qja - \omega(q)t)}$, where ξ is the amplitude, i is the imaginary unit, and q is the one-dimensional [1D] wave vector. After substituting in Equation 5.2, we have nontrivial solutions if and only if the angular frequency $\omega(q)$ satisfies the *dispersion relation*

$$\omega(q) = 2\sqrt{k/m}\left|\sin\left(\frac{qa}{2}\right)\right|.$$

By imposing the boundary conditions, we have

$$e^{iqNa} = 1 \Rightarrow q = \frac{2\pi}{Na}n = \frac{2\pi}{L}n, \quad n \in \mathbb{Z},$$

where $L = Na$ is the length of the portion of chain made of the N atoms.

The distance between two consecutive wave vectors is $\frac{2\pi}{L}$ and therefore the density of states, that is, the number of wave vectors per unit length, is $d(q) = \frac{L}{2\pi}$. Since under the transformation $q \mapsto q + l\frac{2\pi}{a}, l \in \mathbb{Z}$, the solution does not change, it is enough to limit q to the interval $[-\pi/a, \pi/a]$, which is the first Brillouin zone for the case considered. There are exactly N distinct modes. Of course, the solutions are periodic with the first Brillouin zone the interval of periodicity.

When the wavelength is much longer than the interatomic distance, $|qa| \ll 1$, the frequency is approximately given by the linear relation

$$\omega(q) = \sqrt{\frac{k}{m}}\, a|q|.$$

Note that in this limit, the waves are not dispersive and the group velocity (equal to the phase velocity) is $\sqrt{\frac{k}{m}}\, a$, which represents the sound velocity of the crystal in the continuum limit.

Let us now consider a linear chain composed of two types of different particles, with alternate masses m_1 and m_2 (see Figure 5.3, right). It is a simple model of a 1D crystal with two atoms in each unit cell. Let us assume that the lattice distance is $a = 2b$, b being the equilibrium distance between two adjacent atoms. The equations of motions are now

$$\begin{cases} m_1\ddot{x}_{2j+1} = k(x_{2j+2} + x_{2j} - 2x_{2j+1}), \\ m_2\ddot{x}_{2j} = k(x_{2j+1} + x_{2j-1} - 2x_{2j}). \end{cases}$$

We look for solutions of the form

$$x_{2j+1} = \xi\, e^{i((2j+1)bq - \omega(q)t)},$$

$$x_{2j} = \eta\, e^{i(2jbq - \omega(q)t)}.$$

The even and odd indices represent particles with mass m_1 and m_2, respectively. Proceeding as in the previous case, we obtain the dispersion relation

$$\omega^2(q) = k\left(\frac{1}{m_1} + \frac{1}{m_2}\right)\left[1 \mp \sqrt{1 - \frac{4m_1 m_2}{(m_1 + m_2)^2}\sin^2(bq)}\right].$$

In the limit $|qb| \ll 1$, it becomes

$$\omega^2(q) = \begin{cases} \dfrac{2k}{m_1 + m_2}b^2 q^2, & \text{acoustic branch,} \\[2ex] 2k\left(\dfrac{1}{m_1} + \dfrac{1}{m_2}\right), & \text{optical branch.} \end{cases}$$

Note that near $q = 0$, the acoustic branch is almost linear (*Debye approximation*), while the optical branch is almost flat (*Einstein approximation*). Therefore, near $q = 0$, the optical branch has a negligible group velocity and does not transport energy.

It is also instructive to analyze the ratio of the displacements

$$\frac{\xi}{\eta} = \frac{2k\cos(qb)}{2k - m_1\omega^2} \sim \begin{cases} 1 & \text{for acoustic modes} \\[1.5ex] -\dfrac{m_2}{m_1} & \text{for optical modes} \end{cases} \quad \text{as } q \mapsto 0.$$

Therefore, the acoustic modes produce oscillations along the same direction, while the optical modes produce oscillations in opposite directions.

In the 3D situation we have different directions of polarization with longitudinal and traversal oscillations, for both optical and acoustic modes. Each unit (primitive) cell generally consists of more than one atom, which is, therefore, characterized by four numbers, the first three, $\mathbf{n} = (n_1, n_2, n_3)$, with $n_i \in \mathbb{Z}$, $i = 1, 2, 3$, identify the cell to which it belongs, and the fourth, $s \in \mathbb{N}$, represents its number in the cell. Let $\mathbf{u}_s(\mathbf{n})$ be the displacements of the atoms in their vibrations. The index s runs from 1 to the number of atoms in each primitive cell. The Lagrangian of a crystal lattice, considered as a mechanical system of particles vibrating around their equilibrium positions, by retaining in the potential only up to the quadratic terms in the displacements (*harmonic approximation*), is given by

$$\mathcal{L} = \frac{1}{2}\sum_{\mathbf{n},s} m_s |\dot{\mathbf{u}}_s(\mathbf{n})|^2 - \frac{1}{2}\sum_{\mathbf{n},\mathbf{n}',s,s'} < \Lambda^{ss'}(\mathbf{n} - \mathbf{n}')\mathbf{u}_s(\mathbf{n}), \mathbf{u}_{s'}(\mathbf{n}') >, \tag{5.3}$$

where the dot means time derivative, $<,>$ denotes the standard scalar product, m_s is the mass of the atoms, and the matrices $\Lambda^{ss'}$ depend only on the differences $\mathbf{n} - \mathbf{n}'$, since interaction forces depend only on the relative position of the atoms. The $\Lambda^{ss'}$ satisfies the symmetry relation

$$\Lambda^{ss'}(\mathbf{n}) = \left(\Lambda^{s's}(-\mathbf{n})\right)^T, \tag{5.4}$$

along with some further relations, which derive from the fact that a parallel displacement or a rotation of the lattice as a whole gives rise to no forces (see [13]). In Equation 5.4, the superscript T indicates transposition.

The Lagrange equations of motion read

$$m_s\ddot{\mathbf{u}}_s = -\sum_{\mathbf{n}',s'} \Lambda^{ss'}(\mathbf{n} - \mathbf{n}')\mathbf{u}_{s'}(\mathbf{n}'). \tag{5.5}$$

Let us look for solutions in the form of plane waves

$$\mathbf{u}_s(\mathbf{n}) = \mathbf{e}_s(\mathbf{q})\exp[i(\mathbf{q} \cdot \mathbf{r_n} - \omega t)], \tag{5.6}$$

with $\mathbf{r_n}$ being the radius vector of any particular vertex of the cell $\mathbf{n}$, which can be used to define its position, ω being the angular frequency, $\mathbf{q}$ being the wave vector, and $\mathbf{e}_s$ being the polarization vector, which is the same for equivalent atoms in different cells. Substituting Equation 5.6 into Equation 5.5, after some algebra one gets

$$\sum_{s'} \widetilde{\Lambda}^{ss'}(\mathbf{q})\, \mathbf{e}_{s'} - \omega^2 m_s \mathbf{e}_s = 0, \tag{5.7}$$

where

$$\widetilde{\Lambda}^{ss'}(\mathbf{q}) := \sum_{\mathbf{n}} \Lambda^{ss'}(\mathbf{n}) \exp(-i\mathbf{q} \cdot \mathbf{r_n}). \tag{5.8}$$

Equation 5.7 has nontrivial solutions if and only if the determinant of the $\nu \times \nu$ block matrix, whose generic block of indices s and s' is given by

$$\widetilde{\Lambda}^{ss'}(\mathbf{q}) - \omega^2 m_{s'} \delta_{ss'} I, \tag{5.9}$$

vanishes, I being the 3×3 identity matrix. If ν is the number of atoms per cell, the order of the determinant is 3ν, therefore Equation 5.9 is an algebraic equation of degree 3ν in the unknown ω^2. Each solution determines ω as a function of the wave vector $\mathbf{q}$, which means that there are 3ν branches of such function

$$\omega = \omega_p(\mathbf{q}), \quad p = 1, \dots, 3\nu,$$

which, as said, is called dispersion relation, p labeling the various branches. From Equations 5.8 and 5.4, it follows that

$$\widetilde{\Lambda}^{ss'}(\mathbf{q}) = \widetilde{\Lambda}^{s's}(-\mathbf{q}) = [\widetilde{\Lambda}^{s's}(\mathbf{q})]^*, \tag{5.10}$$

where the star indicates complex conjugation. As a consequence, the $\widetilde{\Lambda}^{ss'}(\mathbf{q})$ forms a Hermitian matrix, where the eigenvectors corresponding to different eigenvalues are orthogonal, i.e.,

$$\sum_{s=1}^{\nu} m_s \mathbf{u}_s^{(p)} \cdot \mathbf{u}_s^{(p')*} = 0, \quad \text{for} \quad p \neq p'. \tag{5.11}$$

Due to the symmetry of the mechanical equations of motion under time reversal, the dispersion relation has to be even

$$\omega(\mathbf{q}) = \omega(-\mathbf{q}).$$

Moreover, the wave vector $\mathbf{q}$ appears in Equation 5.6 only in the exponential factor $\exp(i\mathbf{q} \cdot \mathbf{r_n})$, which does not change with the substitution

$$\mathbf{q} \to \mathbf{q} + \mathbf{G}, \qquad \mathbf{G} = n_1 a_1^* + n_2 a_2^* + n_3 a_3^*, \ n_1, n_2, n_3 \in \mathbb{Z}, \tag{5.12}$$

where $\mathbf{G}$ is an arbitrary vector of the reciprocal lattice. Therefore, the wave vector is physically indeterminate, meaning that values of $\mathbf{q}$ differing by $\mathbf{G}$ are physically equivalent. This implies that the function $\omega(\mathbf{q})$ is periodic in the reciprocal lattice

$$\omega(\mathbf{q} + \mathbf{G}) = \omega(\mathbf{q}), \tag{5.13}$$

and can be restricted to a single cell. From a geometrical point of view, the dispersion relation $\omega(\mathbf{q})$ is represented by a finite four-dimensional hypersurface, with 3ν sheets, corresponding to the various branches, which may intersect.

As seen for the simple 1D chain, it is possible to ensure also in the 3D case that some of the branches of the vibrational spectrum, for wavelengths large with respect to the physical dimension of a unit cell, correspond to sound waves in the crystal and for them the dispersion relation is a first-order homogeneous function of the components of the wave vector, for **q** small enough. These waves are called acoustic waves.

In lattices with more than one atom per cell, there are $3(\nu - 1)$ further types of waves, whose frequency does not vanish at **q** = 0, but tends to a finite constant. Such vibrations are called optical vibrations, and for them the atoms in each cell are in relative motion [13].

For example, in silicon and gallium arsenide there are two atoms per cell and, therefore, six types of elastic waves: three acoustic modes and three optical modes. Of the three acoustic modes, one is longitudinal (LA), that is the atoms are displaced in the direction of the wave propagation, and the other two are transverse (TA), in which the atoms are displaced in a transverse direction. The same is true for the optical phonons (LO and TO). The transversal modes are therefore doubly degenerate. In Figure 5.4, we represent the dispersion relations in Si along the main crystallographic directions.

What the text mentioned earlier is strictly valid only in the harmonic approximation, in which the various monochromatic waves freely propagate through the crystal without interacting. Higher-order terms are taken into account as various processes of decay and scattering among these waves. Their description is rather complex and a simplified approach based on a relaxation time approximation is usually adopted in the device simulations.

The picture we have presented thus far is based on classical mechanics. However, the applications in micro- and nanoelectronics need a quantum approach to the lattice vibrations. Since the $\widetilde{\Lambda}^{ss'}(\mathbf{q})$ is a Hermitian matrix, it is possible to write the classical equations of motion in canonical coordinates, decoupling them in independent harmonic oscillators [12,14]. As a consequence, the analogous quantum Hamiltonian is the sum of the Hamiltonians of $3\nu N$ independent quantum oscillators [12,14], with N being the total number of cells, each of them having quantized energy levels given by

$$\epsilon_n = \hbar\omega\left(n + \frac{1}{2}\right), \quad n = 0, 1, 2, \ldots,$$

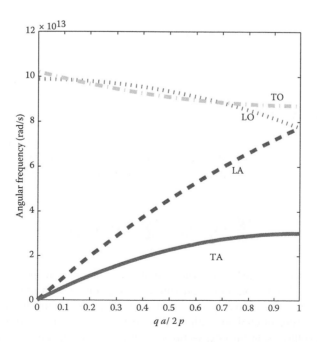

FIGURE 5.4 Phonon dispersion relations in silicon along the <100> crystallographic directions.

From the particle point of view, each quantum state can be considered as formed by n fictitious particles called *phonons* having energy $\hbar\omega$ in analogy with photons. Here $\hbar$ is the reduced Planck constant. The processes, which change the state from the energy level ϵ_n to ϵ_{n+1} (respectively, ϵ_{n-1}) can be interpreted as the *creation* (respectively, *destruction*) of a phonon.

The quasimomentum is associated with phonons:

$$\mathbf{p} = \hbar\mathbf{q}.$$

It is almost analogous to the ordinary momentum, with the important difference that it is defined up to an arbitrary constant vector $\hbar\mathbf{G}$, with $\mathbf{G}$ belonging to the reciprocal lattice. The phonon velocity is given by the group velocity of the corresponding classical waves, $\mathbf{v} = \frac{\partial\omega}{\partial\mathbf{q}}$, which can be rewritten in the form

$$\mathbf{v} = \frac{\partial\epsilon(\mathbf{p})}{\partial\mathbf{p}}.$$

In the approximation successive to the harmonic one, various elastic and inelastic phonon collision processes occur, which tends to drive the system toward equilibrium (see below for more details). In these processes, the energy and momentum conservation laws must be satisfied, the latter up to an additive vector of the form $\hbar\mathbf{G}$

$$\sum \mathbf{p} = \sum \mathbf{p}' + \hbar\mathbf{G},$$

where $\mathbf{p}$ and $\mathbf{p}'$ are the momenta of the phonons before and after the collision.

An arbitrary number of identical phonons can be created simultaneously in the lattice, which means that the phonon gas obeys the Bose statistics. Since the total number of phonons is determined by the equilibrium conditions, the phonon chemical potential is zero [13]. Therefore, the equilibrium occupation number $\bar{g}(\mathbf{p})$ of a quantum state with momentum $\mathbf{p}$ and energy $\epsilon(\mathbf{p})$ is determined by the Planck distribution function

$$\bar{g}(\mathbf{p}) = 1/(\exp(\epsilon(\mathbf{p})/T_{\mathrm{L}}) - 1),$$

where T_{L} is the lattice temperature.

5.3 Phonon Transport Equation

If the typical phonon wavelength is much shorter than the phonon mean free path, phonons can be treated as semiclassical particles that obey, as said, the Bose statistics. In this case, the state of the phonon gas can be described through the distribution function $g^p(\mathbf{q})$, in which time evolution is governed by the Boltzmann–Peierls (BP) equation

$$\frac{\partial g^p}{\partial t} + \mathbf{v}_p \cdot \nabla_{\mathbf{x}} g^p = \sum_\eta C_\eta^p(g^p) + C_{(\mathrm{oth})}^p(g^p), \tag{5.14}$$

the index p labeling the several branches. The right-hand side is the collision operator. The term $\sum_\eta C_\eta^p(g^p)$ represents the phonon–phonon collisions, and collisions of phonons with impurities, boundaries, and defects, the index η labeling the various types of scatterings. The other contribution in the collision operator is due to the interaction of phonons with other particles, e.g., electrons and photons. In addition to the phonon–phonon collisions, in the next sections we describe only the details of the phonon–electron collisions because they are of major interest in this chapter.

5.3.1 Interactions of Phonons among Themselves with Impurities and Boundaries

More specifically, phonon interaction processes, which do not involve charge carriers or photons, can be divided into intrinsic ones, which arise from the anharmonicity of the interatomic forces, and extrinsic ones, due to phonon scatterings at the boundaries of the crystal and at various types of crystal defects and imperfections [11]. The anharmonic scatterings can be *normal* processes (*N*-processes), in which the phonon total momentum after a collision is conserved, and *umklapp* processes (*U*-processes) in which the total momentum changes by a reciprocal lattice vector after a collision. On the other hand, all extrinsic processes do not conserve the total momentum after a collision and together with the umklapp processes are usually called resistive processes [15].

The anharmonic interactions involve an optical phonon, which decays into a transversal acoustic phonon and a longitudinal phonon belonging either to the acoustic or to the optical branch [16]. The expression of the corresponding collision operator in the BP equation for the optical phonons, first given by Klemens [17], reads

$$
C^{p_O}_{P_O \leftrightarrow TA + p'} = \frac{\gamma^2 y^{p_O}}{3\pi^2 \rho v_s^2} \int_B \epsilon_{p_O} \omega'_{p'} \omega_{TA}(\mathbf{q}'') \delta(\omega_{p_O} - \omega'_{p'} - \omega_{TA}(\mathbf{q}''))
$$

$$
\times \left[\left(g^{p_O}(\mathbf{q}) + 1 \right) g^{p'}(\mathbf{q}') g^{TA}(\mathbf{q}'') - g^{p_O}(\mathbf{q}) \left(g^{p'}(\mathbf{q}') + 1 \right) \left(g^{TA}(\mathbf{q}'') + 1 \right) \right] d\mathbf{q}', \quad (5.15)
$$

where $y^p = \frac{1}{(2\pi)^3}$ is the phonon density of states, $p_O = LO, TO, p' = LO, LA, \mathbf{q}'' = \mathbf{q} - \mathbf{q}' + \mathbf{G}$ (stemming from the generalized momentum conservation), γ is the Grüneisen parameter, v_s is the sound velocity, ρ is the material density, and δ is the Dirac delta function [16]. Analogous are the expressions for the corresponding collision operators appearing in the BP equation relative to the acoustic phonons. Relaxation time approximations for the operator Equation 5.15 can be found in [18].

For the interactions of the acoustic phonons with boundaries and impurities, the relaxation time approximation is commonly used for the operators, which reads [18]

$$
C^{p_A}_\eta = -\frac{g^{p_A} - g^{p_A}(0)}{\tau^\eta_p},
$$

where $\eta = B, I$, with B and I, respectively, the scattering with boundaries and impurities. The relaxation times are given by

$$
[\tau^B_{p_A}]^{-1} = \frac{v^{p_A}}{d}, \quad p_A = LA, TA,
$$

with d being the effective diameter of the sample

$$
[\tau^I_{p_A}]^{-1} = \frac{V \Gamma}{4\pi\hbar^4 (v_s^{p_A})^3} \epsilon^4_{p_A}, \quad p_A = LA, TA,
$$

where, as impurities, only the isolated defects of mass different from those of the host have been considered. In this case, V is the volume per atom and Γ is the mass-fluctuation phonon scattering parameter for a single element made up of several naturally occurring isotopes and is given by

$$
\Gamma = \sum_i c_i \left[\frac{m_i - \bar{m}}{\bar{m}} \right]^2, \quad \bar{m} = \sum_i c_i m_i,
$$

with m_i the atomic mass of the ith isotope and c_i the fractional atomic natural abundance.

5.3.2 Phonon–Carrier Interaction

Within the semiclassical approximation for electrons, we indicate by f^A the distribution function of the charge carriers occupying the valley labeled by A. The valleys are the neighborhoods of states centered in the lowest minima of the lowest conduction bands and in the highest maxima of the highest valence bands [12] in the carrier wave vector space. In fact, the electrons are occupying those states which mainly contribute to the charge transport in semiconductors. The contribution of the electrons in the valence bands can be conveniently treated in terms of the presence of particles called holes (missing electrons), which have electric charge, crystal momentum, and energy opposite to those of the missing electrons in the valence band considered [7,8,19]. The electron/hole state is characterized by the index A of the band they occupy, by their crystal momentum, $\hbar\mathbf{k}$, with $\mathbf{k}$ wave vector, and spin state, which can be up or down. An electron/hole population for each neighborhood of the lowest/highest minima/maxima of the lowest/highest conduction/valence bands has to be considered.

In order to describe phonon interactions with carriers, we consider the particular case of silicon and we limit ourselves to the electrons in the conduction valleys. More general cases involving compound semiconductors and interactions with holes can be treated similarly [8,20,24].

As regards the phonon dispersion relations in silicon, the following isotropic quadratic approximations can be used (see Figure 5.4):

$$\epsilon_p = \epsilon_0^p + \hbar v_s^p \, |\mathbf{q}| + \hbar c^p \, |\mathbf{q}|^2 \,, \ |\mathbf{q}| \in \left[0, \frac{2\pi}{a}\right], \ p = LA, \ TA, \ LO, \ TO, \tag{5.16}$$

where the first Brillouin zone is approximated by a sphere of radius $\frac{2\pi}{a}$, with a being the silicon lattice constant; A, O, L, and T stand for acoustic, optical, longitudinal, and transversal, respectively; and the coefficients in Equation 5.16 can be found in [21].

For silicon, electrons in conduction bands, which mainly contribute to the charge transport, are those in the valleys centered in the six-band minima, near the equivalent X symmetry points, of the lowest conduction band; see Figure 5.5 for an illustration of the first Brillouin zone. In the Kane ellipsoidal approximation [19], in each valley the dependence of the electron energy $\mathcal{E}$ (here and whenever possible we omit the valley index, for simplicity) on the wave vector $\mathbf{k}$ is given by

$$\mathcal{E} = \frac{\hbar^2}{2} \gamma(\mathcal{E}) \sum_{i=1}^{3} \frac{(k_i - \kappa_i)^2}{m_i}, \qquad \gamma(\mathcal{E}) := \left(1 + \alpha\mathcal{E}\right)^{-1},$$

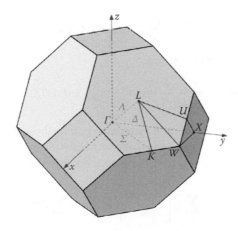

FIGURE 5.5   First Brillouin zone of silicon.

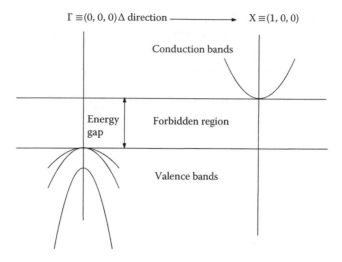

FIGURE 5.6 Schematic representation of the energy bands in silicon.

where the energy is referred to the minimum of the valley, κ_i, $i = 1, 2, 3$ are the coordinates of the minimum, α is the nonparabolicity parameter, and m_i^{-1}, $i = 1, 2, 3$ are the eigenvalues of the inverse effective valley mass tensor. Both the nonparabolicity parameter and the effective masses are temperature dependent [21]. The first Brillouin zone is usually extended to all $\mathbb{R}^3$. Figure 5.6 shows a schematic representation of the energy bands in silicon.

Since the electrons in the valleys, which correspond to minima lying on the same principal axis, have the same mass tensor [19], they can be considered a unique population. Therefore, one can consider three electron populations, depending on which axis the minimum of the valley, where the electrons live, belongs to. We label the populations by $e = x, y, z$.

The general form of the collision operators relative to the interactions of phonons with electrons, contributing to the term $C^p_{(oth)}$ in Equation 5.14, is [4,22]

$$C^p_{i, ee'}(g^p, f^e, f^{e'}) = \int_{\mathbb{R}^3} \int_{\mathbb{R}^3} \left[w^{ee'+}_{i, p}(\mathbf{k}, \mathbf{k}', \mathbf{q}) \kappa_1(g^p, f^e, f^{e'}) \right] d\mathbf{k}' d\mathbf{k},$$

with

$$\kappa_1(g^p, f^e, f^{e'}) := \left(g^p + 1 \right) f^{e'}(\mathbf{k}') - g^p f^e(\mathbf{k}),$$

$$w^{ee'\pm}_{i, p}(\mathbf{k}, \mathbf{k}', \mathbf{q}) := s^{ee'}_{i, p}(\mathbf{q}) \delta\left(\mathcal{E}_{e'}(\mathbf{k}') - \mathcal{E}_e(\mathbf{k}) \mp \epsilon_p \right) \delta\left(\mathbf{k}' - \mathbf{k} \mp \mathbf{q} + \mathbf{G} \right).$$

The phonon–electron scatterings can be intravalley ($\iota = sv$, $e = e'$, $p = LA, TA$),[†] which involve only acoustic phonons and conserve the total energy and momentum, or intervalley [19]. The latter, which pertain both to the acoustic and the optical phonons, in their turn can be distinguished in g-type ones ($\iota = dv$, $e = e'$, $p = LA, TA, LO$), which take electrons between equivalent valleys or valleys that have the transversal mass in the same direction, and f-type ones ($\iota = dv$, $e \neq e'$, $p = TA, LA/LO, TO$), taking electrons to nonequivalent valleys. Both of them are umklapp processes, which do not conserve the total momentum

---

[†] *sv* and *dv*, respectively, stand for same valley and different valleys.

and involve a reciprocal lattice vector $\mathbf{G}$ [21]. The phonon wave vectors interested in the intervalley transitions remain very close to the vector joining the minima of the initial and final valleys and therefore they are usually taken as constant and, after reduction to the first Brillouin zone, equal to $\frac{2\pi}{a}(1, 0.15, 0.15)$ for f-type scatterings and to $\frac{2\pi}{a}(0.3, 0, 0)$ for g-type scatterings. The scattering functions $s^{ee'}_{\iota,p}$ are given by [19]

$$s^{ee}_{sv,LO} = s^{ee}_{sv,TO} = 0, \quad s^{ee'}_{sv,p} = 0, \quad e = x, y, z, \ e \neq e', \ p = LA, TA, LO, TO,$$

$$s^{ee}_{sv,p} = \frac{\hbar |\mathbf{q}|^2 D^2_{a,p}}{8\pi^2 \rho \, \epsilon_p} \, I^2(|\mathbf{q}|), \quad p = LA, TA, \ e = x, y, z,$$

$$s^{ee}_{dv,p} = \frac{\hbar (\Delta^{ee}_p)^2}{8\pi^2 \rho \, \bar{\epsilon}^{ee}_p}, \quad p = LA, TA, LO, \ e = x, y, z,$$

$$s^{ee'}_{dv,p} = \frac{\hbar (\Delta^{ee'}_p)^2}{4\pi^2 \rho \, \bar{\epsilon}^{ee'}_p}, \quad p = TA, LA/LO, TO, \qquad e \neq e', \ e, e' = x, y, z,$$

where $D_{a,p}, p = LA, TA$ are the acoustic deformation potentials, $I(|\mathbf{q}|)$ is the overlap integral, ρ is the material density, $\Delta^{ee'}_p, e, e' = x, y, z, p = LA, TA, LO, TO$ are the intervalley deformation potentials, and $\bar{\epsilon}^{ee'}_p, e, e' = x, y, z, p = LA, TA, LO, TO$ are the phonon energies involved in the f-type and g-type transitions. The values of the various parameters and the expression of the overlap integral appearing in the above-mentioned scattering functions can be found in [21].

5.4 Boltzmann Equations for Carrier Transport

Before macroscopic balance equations for heat transport in semiconductors can be written, it is necessary to write down the evolution equations for the carrier distribution functions, which read

$$\frac{\partial f^A}{\partial t} + \mathbf{v}_A \cdot \nabla_{\mathbf{x}} f^A + \frac{q_A}{\hbar} \mathbf{E} \cdot \nabla_{\mathbf{k}} f^A = C^A_{im}(f^A) + \sum_{\iota, A', p} C^{AA'}_{\iota, p}(f^A, f^{A'}, g^p) + \sum_{A'} \mathcal{I}[f^A, f^{A'}],$$

$$\Delta(\varepsilon_d \, \phi) = q\big(n(\mathbf{x}, t) - p(\mathbf{x}, t) + N_a(\mathbf{x}) - N_d(\mathbf{x})\big), \tag{5.17}$$

where q_A is the charge of the carriers in the Ath valley and q its absolute value, $\mathbf{v}_A = \frac{1}{\hbar} \nabla_{\mathbf{k}} \mathcal{E}_A$ is the carrier group velocity, ι labels the type of scattering between carriers and phonons, ε_d is the material permittivity, ϕ and $\mathbf{E}$ are the electric potential and field, N_a and N_d are the acceptor and donor concentrations, n is the total electron density and p is the total hole density. The Boltzmann Equations 5.17 are coupled among them through the Poisson Equation 5.18 and some of the collision operators that appear at the right-hand side of Equation 5.17.

The operator C^A_{im} [19] takes into account scatterings between carriers and impurities, which are elastic and intravalley. Its form is

$$C^A_{im}(f^A) = \int_{\mathbb{R}^3} \big[w^A_{im}(\mathbf{k}', \mathbf{k}) f^A(\mathbf{k}') - w^A_{im}(\mathbf{k}, \mathbf{k}') f^A(\mathbf{k}) \big] d\mathbf{k}',$$

the impurity scattering transition rate being given by

$$w^A_{im}(\mathbf{k}, \mathbf{k}') = \mathcal{K}^{(im)}_A \frac{1}{\left[|\mathbf{k} - \mathbf{k}'|^2 + \lambda^2_A \right]^2} \, \delta(\mathcal{E}_A(\mathbf{k}') - \mathcal{E}_A(\mathbf{k})),$$

with $\lambda_A = \sqrt{\frac{(N_{a/d})q^2}{\varepsilon_d k_B T_L}}$ is the inverse Debye length and $\mathcal{K}_A^{(im)} = \frac{Z_{a/d}^2 N_{a/d} q^4}{4\pi\hbar\varepsilon_d^2}$, where Z is the impurity charge number and k_B is the Boltzmann constant. In the calculation of λ_A and $\mathcal{K}_A^{(im)}$, N_a, and Z_d, respectively, N_d and Z_d have to be taken according to whether interactions with donors or impurities are considered.

The collision operators $C_{1,p}^{AA'}(f^A, f^{A'}, g^p)$ describe interactions between carriers and phonons and [22]

$$C_{1p}^{AA'}(f^A, f^{A'}, g^p) = \int_B \int_{\mathbb{R}^3} \left[w_{1p}^{AA'+}(\mathbf{k}, \mathbf{k'}, \mathbf{q}) \kappa_{AA'}^1(g_p, f^A, f^{A'}) \right.$$
$$\left. + w_{1p}^{AA'-}(\mathbf{k}, \mathbf{k'}, \mathbf{q}) \kappa_{AA'}^2(g_p, f^A, f^{A'}) \right] d\mathbf{k'} d\mathbf{q},$$

where

$$\kappa_{AA'}^2(g^p, f^A, f^{A'}) := g^p f^{A'}(\mathbf{k'}) - \left(g^p + 1 \right) f^A(\mathbf{k}).$$

Earlier we reported the expressions of the κ that are valid in the case when both the initial and the final valleys belong to conduction bands and the expressions in the interband case are analogous [23]. The scattering functions $s_{1,p}^{AA'}$ are characteristic of the type of interaction of carriers, for example with acoustic and nonpolar optical phonons for elemental semiconductors like Si (see the previous section) and Ge, and also with polar optical phonons for compound semiconductors, like GaAs and SiC. For their expressions, we refer the interested reader to [19,20,24].

In the generation–recombination collision operator $\sum_{A'} \mathcal{I}[f^A, f^{A'}]$, if A is a conduction valley the sum is extended to the valence valleys and vice versa. It describes the processes in which electron–hole pairs are created by exciting an electron from the valence to the conduction band and vice versa, electrons and holes from the conduction and the valence band recombine and are annihilated. These processes are of several types; here we consider only the most important ones.

In the direct band-to-band Auger mechanism, three carriers are involved. In generation, an electron–hole pair is created at the expense of the energy of a highly energetic electron or hole. The opposite process transfers the excess energy to an electron or hole.

The Shockley–Read–Hall (SRH) generation/recombination occurs by phonon absorption/emission. This process is trap-assisted utilizing defects with energy levels within the semiconductor band-gap.

The Auger and SRH operators, in the relaxation time approximation and in the simpler case of a single conduction (n) and valence (h) valley, read [25]

$$\mathcal{I}[f^A, f^{\bar{A}}] = -\left(\Gamma_n n + \Gamma_h p\right)\left(\tilde{n} f^A - n_i^2 \mathcal{M}_A\right) - \frac{\tilde{n} f^A - n_i^2 \mathcal{M}_A}{\tau_h(n + n_i) + \tau_n(p + n_i)}, \tag{5.18}$$

where $\bar{A} = h$ if $A = e$ and vice versa, $\tilde{n} = p$ or n according to whether the Ath valley is populated by electrons or holes, the $\mathcal{M}_A$'s are the Maxwellians normalized to unit, Γ_n and Γ_h are the Auger electron and hole constant coefficients, τ_n and τ_h are the carrier lifetimes, and n_i is the intrinsic concentration [19].

In direct band-gap semiconductors, the generation/recombination processes can also involve photons. In fact, in recombinations there can be spontaneous emissions of photons, which are of fundamental importance in light-emitting optoelectronic devices [26]. Similarly, in generations, the transition energy can be provided by photons, which is the key mechanism in photodetectors and other electroabsorption devices. The operator corresponding to photon-assisted generation/recombination processes reads [26,27]

$$\int w_\uparrow(\mathbf{k'}, \mathbf{k}) f_h(\mathbf{k'})(1 - f_n(\mathbf{k})) - w_\downarrow(\mathbf{k}, \mathbf{k'}) f_n(\mathbf{k})(1 - f_h(\mathbf{k'})) d\mathbf{k'}$$

for the electrons in the conduction band, f_n being their distribution function, while f_h is the distribution function for the electrons in the valence band. In the above formula

$$w_\uparrow(\mathbf{k}', \mathbf{k}) = \frac{\hbar\pi q^2}{6m_e\omega_f\sqrt{\varepsilon_d}\left(\sqrt{\varepsilon_d} + \omega_f\frac{d\sqrt{\varepsilon_d}}{d\omega_f}\right)}\delta(\mathbf{k}' - \mathbf{k})\delta(\mathcal{E}_n(\mathbf{k}) - \mathcal{E}_h(\mathbf{k}) - \hbar\omega_f)\frac{g_f}{V_f},$$

$$w_\downarrow(\mathbf{k}', \mathbf{k}) = \frac{1 + g_f}{g_f}w_\uparrow(\mathbf{k}', \mathbf{k}),$$

where $\tilde{E}$ is a material-dependent parameter by having the dimensions of an energy [27], m_e is the free electron mass, ω_f is the photon frequency, g_f is the photon distribution function, and V_f is the photon mode volume. An analogous expression holds for holes.

5.5 Macroscopic Balance Equations for Heat Transport in Semiconductors

From a macroscopic point of view, thermal effects are described, within the standard linear irreversible thermodynamics, by the heat conduction equation with a Fourier-like law for the heat flux

$$\rho c_V \frac{\partial}{\partial t}T - \text{div}[K(T)\nabla T] = H, \tag{5.19}$$

where T is the common local equilibrium temperature of electrons, holes, and lattice, ρ is the density of the material, c_V is the specific heat at constant volume, $K(T)$ is the thermal conductivity, and H is the thermal production term. The validity of the Fourier law is rather questionable in the case of solids at low temperature in the ballistic case [2,3] but is valid if the resistive processes dominate. The dependence of K on T is usually deduced from the experimental data and can depend on the characteristic scales involved in the problem. For bulk materials, accurate analytical formulas are known [28,29]. For nanoscale devices, the surface scattering plays a major role and significant departures from the bulk case are observed [28].

The same considerations hold for the production H, which contains several effects, e.g., the Joule and Thomson ones in addition to radiative recombination and optical generation. The correct modeling of H involves several phenomenological parameters. A review of the heuristic approaches can be found in [1] where an in-depth discussion of the drift-diffusion models with varying (but equal for electrons, holes, and lattice) temperature is performed. A systematic approach to get constitutive relations directly from the phonon transport equations will be adopted in the next section of this chapter. However, in order to provide the reader with a physical intuition about what will be deduced, we first review the standard energy model based on the drift-diffusion approximation.

The standard drift-diffusion model augmented with the heat conduction equation for the common temperature T, the so-called *energy model*, is given by the following set of evolution equations:

$$\frac{\partial n}{\partial t} - \frac{1}{q}\nabla \cdot \mathbf{J}_n = -R + G, \quad \mathbf{J}_n = \mu_n k_B T\nabla n - q\, n\mu_n\nabla\phi + D_n\nabla T, \tag{5.20}$$

$$\frac{\partial p}{\partial t} + \frac{1}{q}\nabla \cdot \mathbf{J}_p = -R + G, \quad \mathbf{J}_p = -\mu_p k_B T\nabla p - q\, p\mu_p\nabla\phi - D_p\nabla T, \tag{5.21}$$

$$\rho c_V \frac{\partial T}{\partial t} - \nabla \cdot (K(T)\nabla T) = H \tag{5.22}$$

$$\nabla \cdot \left(\varepsilon_d\nabla\phi\right) = q(n - p - N_D + N_A). \tag{5.23}$$

In the expressions of the electron and hole currents, J_n and J_p, μ_n and μ_p are the mobilities that depend on the electric field and other parameters [29], while the last terms express the *Seebeck effect* (a temperature gradient induces an electric current), D_n and D_p being thermal coefficients. They are also written as $D_n = n\mu_n P_n$ and $D_p = p\mu_p P_p$ with P_n and P_p playing the role of thermoelectric powers.

The recombination and generation terms, R and G (see [29] for a complete review), can be directly deduced from Equations 5.18 and 5.19 by integration over the **k** space.

Regarding the production H, in [1], the following expression has been proposed, in the steady-state case:

$$
H = \underbrace{\frac{|J_n|^2}{q\mu_n n} + \frac{|J_p|^2}{q\mu_p p}}_{\text{Joule effect}} + T\left(\frac{k_B}{q}\log\frac{n}{n_i} + P_n\right)\nabla\cdot J_n - T\left(\frac{k_B}{q}\log\frac{p}{n_i} + P_p\right)\nabla\cdot J_p
$$

$$
+ \underbrace{T\left(J_n\cdot\nabla P_n + J_p\cdot\nabla P_p\right)}_{\text{Thomson effect}} + H_{\text{rad}}. \tag{5.24}
$$

The two intermediate terms can be interpreted by observing that in the steady state $\nabla\cdot J_n = -\nabla\cdot J_p = q(R-G)$. Thereby, the intermediate terms are the recombination-generation heat generation. The last term H_{rad} is the radiative contribution.

Equations 5.20 through 5.22 and in particular expression Equation 5.24 contain several coefficients to be prescribed, e.g., the thermoelectric powers. In the rest of the section, we proceed with a systematic way to tackle the question of determining macroscopic balance equations.

These macroscopic models can be constructed starting from the Boltzmann–Peierls system for phonons and carriers by taking suitable moments of the distribution functions. In particular, the following functions of the phonon and carrier wave vectors $\{\psi_p(\mathbf{q})\} := \{\epsilon_p, \epsilon_p \mathbf{v}^p\}$ and $\{\psi_A(\mathbf{k})\} := \{1, \mathbf{v}^A, \mathcal{E}_A, \mathcal{E}_A \mathbf{v}^A\}$, can be considered, to which the macroscopic state variables below correspond:

$$
\begin{pmatrix} W^p \\ \mathbf{Q}^p \end{pmatrix} = y^p \int_B \begin{pmatrix} \epsilon_p \\ \epsilon_p \mathbf{v}^p \end{pmatrix} g^p d\mathbf{q}, \tag{5.25}
$$

$$
\begin{pmatrix} n_A \\ W^A \end{pmatrix} = y^A \int_{\mathbb{R}^3} \begin{pmatrix} 1 \\ \mathcal{E}_A \\ n_A \end{pmatrix} f^A d\mathbf{k}, \qquad \begin{pmatrix} \mathbf{V}^A \\ \mathbf{S}^A \end{pmatrix} = \frac{y^A}{n_A}\int_{\mathbb{R}^3} \begin{pmatrix} \mathbf{v}^A \\ \mathcal{E}_A \mathbf{v}^A \end{pmatrix} f^A d\mathbf{k}, \tag{5.26}
$$

which, respectively, are the phonon energy and energy flux densities, and the carrier density, average energy, velocity, and energy flux per carrier in the valley A, $y^A = \frac{2}{(2\pi)^3}$ being the carrier density of states. The index p runs over all the phonon branches and A over the carrier valleys. The phonon momentum densities are given by

$$
\mathbf{P}^p = y^p \int_B \hbar\mathbf{q}\, g^p d\mathbf{q},
$$

where p runs over the phonon branches. Here, the minimal number of moments necessary for describing the thermal energy transport have been chosen, but this number, if required by the physical problem under study, can be easily extended to cover, e.g., an arbitrary number of scalar and vector moments both for phonons and carriers, by taking into account higher energy powers [5,15,30]. These further moments are needed to deal with situations very far from equilibrium.

The evolution equations for the state variables 5.25 and 5.26 can be obtained directly from the Boltzmann equations by integration:

$$\frac{\partial}{\partial t}\begin{pmatrix} W^p \\ Q_i^p \end{pmatrix} + \sum_j \frac{\partial}{\partial x_j}\begin{pmatrix} Q_j^p \\ T_{ij}^p \end{pmatrix} = \begin{pmatrix} C_{W^p} \\ C_{Q_i^p} \end{pmatrix}, \qquad i = 1, 2, 3. \tag{5.27}$$

$$\frac{\partial}{\partial t}\begin{pmatrix} n_A \\ n_A W^A \end{pmatrix} + \sum_j \frac{\partial}{\partial x_j}\left[n_A \begin{pmatrix} V_j^A \\ S_j^A \end{pmatrix}\right] - n_A\, q_A \begin{pmatrix} 0 \\ \sum_j E_j\, V_j^A \end{pmatrix} = n_A \begin{pmatrix} C_{n_A} \\ C_{W_A} \end{pmatrix}, \tag{5.28}$$

$$\frac{\partial}{\partial t}\left[n_A \begin{pmatrix} V_i^A \\ S_i^A \end{pmatrix}\right] + \sum_j \frac{\partial}{\partial x_j}\left[n_A \begin{pmatrix} F_{ij}^{A(0)} \\ F_{ij}^{A(1)} \end{pmatrix}\right]$$

$$- n_A\, q_A \sum_j E_j \begin{pmatrix} G_{ij}^{A(0)} \\ G_{ij}^{A(1)} \end{pmatrix} = n_A \begin{pmatrix} C_{V_i^A} \\ C_{S_i^A} \end{pmatrix}, \qquad i = 1, 2, 3. \tag{5.29}$$

The equations for the phonons are macroscopic equations for heat transport and together with those for the carriers, they constitute a generalization of the system Equations 5.20 through 5.22, where other moments of the distribution functions are considered besides the densities and the energies.

In the Equations 5.27 through 5.29, there appear extra fluxes and production terms. Those relative to phonons are defined by

$$T_{ij}^p = y^p \int \epsilon_p v_i^p\, v_j^p\, g^p(\mathbf{q})d\mathbf{q}, \qquad \text{flux of the energy flux,}$$

$$\begin{pmatrix} C_{W^p} \\ C_{Q_i^p} \end{pmatrix} = y^p \int \begin{pmatrix} \epsilon_p \\ \epsilon_p v_i^p \end{pmatrix}\left[C_{(\text{oth})}^p + \sum_\eta C_\eta^p\right]d\mathbf{q}, \qquad \begin{pmatrix} \text{energy production} \\ \text{energy flux production} \end{pmatrix}.$$

While for the carriers, we have

$$\begin{pmatrix} F_{ij}^{A(0)} \\ F_{ij}^{A(1)} \end{pmatrix} = \frac{y^A}{n_A} \int \begin{pmatrix} 1 \\ \mathcal{E}_A \end{pmatrix} v_i^A v_j^A f^A d\mathbf{k}, \qquad \begin{pmatrix} \text{velocity flux,} \\ \text{flux of the energy flux} \end{pmatrix}$$

$$\begin{pmatrix} G_{ij}^{A(0)} \\ G_{ij}^{A(1)} \end{pmatrix} = \frac{y^A}{\hbar\, n_A} \int \begin{Bmatrix} \dfrac{\partial v_i^A}{\partial k_j} \\ \dfrac{\partial \mathcal{E}_A v_i^A}{\partial k_j} \end{Bmatrix} f^A d\mathbf{k},$$

$$n_A\, C_{M_{\psi_A}} = y^A \int \psi_A(\mathbf{k})\Big\{ C_{im}(f^A) + \sum_{i, A', p} C_{i,p}^{AA'}(f^A, f^{A'}, g^p) + \sum_{A'} \mathcal{I}[f^A, f^{A'}]\Big\}d\mathbf{k}, \qquad M_{\psi_A}\text{-production,}$$

with $\{M_{\psi_A}\} := \{n_A, \mathbf{V}_A, W_A, \mathbf{S}_A\}$. Therefore, in the evolution equations, the number of the unknowns is greater than that of the equations, which means that constitutive equations are needed for the extra variables $T_{ij}^p, F_{ij}^{A(0)}, G_{ij}^{A(0)}, F_{ij}^{A(1)}, G_{ij}^{A(1)}, C_{W^p}, C_{Q^p}, C_{n_A}, C_{W^A}, C_{V^A}, C_{S^A}$.

5.5.1 Closure of the System of Macroscopic Evolution Equations

A systematic way to find the closure relations is founded on a universal physical principle: the *maximum entropy principle* (MEP) [3,31,32]. MEP states that, if a certain number of moments is known, then the least-biased distribution functions, which can be used for evaluating the unknown moments, are those maximizing the total entropy functional under the constraint that they reproduce the known moments. In the case under consideration, neglecting the mutual interactions among the subsystems, the total entropy is the sum of entropy of the several particle populations

$$S = -k_B \left\{ \sum_A y^A \int_{\mathbb{R}^3} \left(f^A \ln f^A - f^A \right) d\mathbf{k} + \sum_p y^p \int_B \left[g^p \ln g^p \right. \right.$$
$$\left. \left. - \left(1 + g^p \right) \ln \left(1 + g^p \right) \right] d\mathbf{q} \right\}.$$

The first terms are the contribution of electrons and holes according to the Fermi statistics in the non degenerate case, the latter ones are the contribution of the phonons according to Bose statistics [33].

The solution of this maximization problem, linearized with respect to the vector variables [20,24], is given by

$$g^p_{ME} = \frac{1}{\exp \left(\epsilon_p \Lambda_{W^p} \right) - 1} - \frac{\epsilon_p \exp \left(\epsilon_p \Lambda_{W^p} \right)}{\left(\exp \left(\epsilon_p \Lambda_{W^p} \right) - 1 \right)^2} \, \mathbf{v}^p \cdot \Lambda_{\mathbf{Q}^p},$$

$$f^A_{ME} = \exp \left(-\Lambda^A - \Lambda_{W^A} \mathcal{E}_A \right) \left(1 - \mathbf{v}^A \cdot \left(\Lambda_{\mathbf{V}^A} + \mathcal{E}_A \Lambda_{\mathbf{S}^A} \right) \right),$$

where the Λ's are Lagrange multipliers related to the field variables by means of the constraint relations Equations 5.25 and 5.26.

For example, in the previously considered case of silicon, the scalar constraints can be rewritten as follows:

$$W^p = 4\pi y^p d^p_1 (\Lambda_{W^p}), \qquad \begin{pmatrix} n_e \\ n_e W^e \end{pmatrix} = \frac{\sqrt{2m_e}\, m_e\, y^e\, J_0}{\hbar^3} e^{-\Lambda^e} \begin{pmatrix} d_0 (\Lambda_{W^e}) \\ d_1 (\Lambda_{W^e}) \end{pmatrix},$$

where

$$d^p_1(x) := \int_0^{\frac{2\pi}{a}} t^2 \frac{\epsilon_p(t)}{\exp \left(\epsilon_p(t) x \right) - 1} \, dt, \quad d_n(x) := \int_0^\infty t^n \exp(-xt) \sqrt{\frac{t}{\gamma^5(t)}} \left(\gamma(t) - t \frac{d\gamma(t)}{dt} \right) dt,$$

$$J_0 := \int_{S^2} \psi^{\frac{3}{2}} d\Omega, \quad \psi^{-1} := \sum_{i=1}^3 \frac{m_e k_i^2}{m_i |\mathbf{k}|^2},$$

S^2 being the unit sphere surface. Inverting, one has

$$\Lambda_{W^p} = h_p^{-1}(W^p) \quad \Lambda^e = -\log \left(\frac{\hbar^3 n_e}{\sqrt{2 m_e}\, m_e\, y^e\, J_0\, d_0} \right), \quad \Lambda_{W^e} = h_e^{-1}(W^e), \qquad (5.30)$$

h_p^{-1} and h_e^{-1} being the inverse functions of $h_p(x) := 4\pi y^p d^p_1(x)$ and $h_e(x) := \frac{d_1(x)}{d_0(x)}$, respectively.

Remark. It is important to notice that according to extended thermodynamics [3], the following relation between the energy Lagrange multiplier of a population and the corresponding local temperature holds:

$$k_B T = \frac{1}{\Lambda_W} = \frac{1}{h^{-1}(W)}.$$

Similarly, inverting the vector constraints, one obtains

$$\Lambda_{Q_i^p} = \frac{3}{4\pi y^p} \left(\frac{\partial}{\partial \Lambda_{W^p}} p_p^1(\Lambda_{W^p}) \right)^{-1} Q_i^p, \quad p_p^1(x) := \int_0^{\frac{2\pi}{a}} \frac{\epsilon_p(t) \, t^2 \, |v^p(t)|^2}{\exp(\epsilon_p(t)x) - 1} \, dt, \tag{5.31}$$

$$\Lambda_{V_i^e} = \frac{b_{11}(W^e)}{J_{1,i}^e} V_i^e + \frac{b_{12}(W^e)}{J_{1,i}^e} S_i^e, \quad \Lambda_{S_i^e} = \frac{b_{12}(W^e)}{J_{1,i}^e} V_i^e + \frac{b_{22}(W^e)}{J_{1,i}^e} S_i^e. \tag{5.32}$$

Here $J_{1,i}^e := \int_{S^2} \frac{m_e^2}{(m_i^e)^2} \psi^{\frac{5}{2}} n_i^2 \, d\Omega$, and the b_{kl} are the elements of the matrix $\mathbf{B}$, which is the inverse of the symmetric matrix $\mathbf{A}$ of elements

$$a_{kl} = -\frac{p^{k+l-2}}{m_e J_0 d_0}, \quad k, l = 1, 2, \quad \text{with } p^n = p^n(x) := \int_0^\infty \frac{2t^{n+\frac{3}{2}} \gamma^{\frac{1}{2}}(t)}{\gamma(t) - t \frac{\gamma(t)}{dt}} e^{-xt} dt.$$

Substituting Equations 5.30 through 5.32 into the MEP distributions, these latter will depend on (x, t) only through the state variables. In turn, inserting the MEP distributions into the integrals defining the extra variables, the needed closure relations can be eventually obtained.

The procedure requires the combersome computation of integrals over the wave vector space of phonons and carriers; however, it is similar for any semiconductor material and it has been done for Si in [4,7–9, 34–39], for compound semiconductors in [20,24], and for graphene in [23].

5.6 Energy-Transport Model and Its Numerical Discretization

By using MEP, the resulting equations form a hyperbolic system of balance laws [6]. In order to have a formulation which has a more evident physical interpretation and is directly comparable with system Equations 5.20 through 5.22, from the hyperbolic model, by performing a suitable scaling [10], it is possible to get a system of coupled nonlinear parabolic equations, known in the literature as the *energy-transport model* [40]. The energy-transport model can be cast in a drift-diffusion form and, therefore, after suitable modifications, the well-known Scharfetter–Gummel scheme can be employed for the numerical discretization. For a discretization of the original hyperbolic equations see [41].

Let us start by noting that, in the simpler case with one conduction and one valence band, the productions C_{Q^p} C_{V^A}, C_{S^A}, with p running over the phonon branches and $A = n, h$, appearing in Equations 5.27 through 5.29, have the general form

$$n_A \, C_{V^A} = \sum_B \left(c_{VV}^B n_B V^B + c_{VS}^B n_B S^B \right) + \sum_p c_{VQ}^p Q^p,$$

$$n_A \, C_{S^A} = \sum_B \left(c_{SV}^B n_B V^B + c_{SS}^B n_B S^B \right) + \sum_p c_{SQ}^p Q^p,$$

$$C_{Q^p} = \sum_B \left(c_{QV}^B n_B V^B + c_{QS}^B n_B S^B \right) + \sum_{p'} c_{QQ}^{pp'} Q^{p'},$$

where the c's depend on electron, hole, and phonon temperatures, through their energies. For simplicity, we assume that in the previous expressions the electron–hole mixed terms are negligible as well as the carrier–phonon terms, so that they become

$$n_A C_{\mathbf{V}^A} = c_{VV}^A n_A \mathbf{V}^A + c_{VS}^A n_A \mathbf{S}^A,$$

$$n_A C_{\mathbf{S}^A} = c_{SV}^A n_A \mathbf{V}^A + c_{SS}^A n_A \mathbf{S}^A,$$

$$C_{\mathbf{Q}^p} = \sum_{p'} c_{QQ}^{pp'} \mathbf{Q}^{p'}.$$

Moreover, we assume that all the phonons have the same local temperature T_L. Therefore, summing Equations 5.27_1 over p, we have

$$\frac{\partial W}{\partial t} + \nabla_{\mathbf{x}} \cdot \mathbf{Q} = \sum_p C_{W^p} =: H, \tag{5.33}$$

with $W := \sum_p W^p$ and $\mathbf{Q} := \sum_p \mathbf{Q}^p$. Noting that

$$\frac{\partial W}{\partial t} = \frac{dW}{dT_L} \frac{\partial T_L}{\partial t} = \rho c_V \frac{\partial T_L}{\partial t},$$

from Equations 5.28 and 5.33 one can write the following energy-transport system:

$$\frac{\partial n}{\partial t} + \operatorname{div}\left(n \, \mathbf{V}_n\right) = -R + G, \tag{5.34}$$

$$\frac{\partial p}{\partial t} + \operatorname{div}\left(p \, \mathbf{V}_h\right) = -R + G, \tag{5.35}$$

$$\frac{\partial \left(n W_n\right)}{\partial t} + \operatorname{div}\left(n \, \mathbf{S}_n\right) + nq\mathbf{V}_n \cdot \nabla\phi = nC_{W_n}, \tag{5.36}$$

$$\frac{\partial \left(p W_h\right)}{\partial t} + \operatorname{div}\left(p \, \mathbf{S}_h\right) - pq\mathbf{V}_h \cdot \nabla\phi = pC_{W_h}, \tag{5.37}$$

$$\rho c_V \frac{\partial T_L}{\partial t} + \operatorname{div}(\mathbf{Q}) = H, \tag{5.38}$$

$$\mathbf{E} = -\nabla\phi, \quad \Delta(\varepsilon_d\phi) = -q(N_d - N_a - n + p), \tag{5.39}$$

where W_n, W_h, $\mathbf{V}_n$, $\mathbf{V}_h$, $\mathbf{S}_n$, $\mathbf{S}_h$ are the electron and hole average energies, velocities, and energy fluxes, c_V is the lattice specific heat, and C_{W_n}, C_{W_h} are the electron and hole energy production terms, which in the relaxation time approximation are given by

$$C_{W_A} = -\frac{W_A - W_A^0}{\tau_{W_A}}, \quad A = n, h, \tag{5.40}$$

with $W_A^0 = W_A(1/k_B T_L) \approx 3/2 k_B T_L$ and τ_{W_A}, $A = n, h$, the energy relaxation times.

In Equations 5.36 and 5.37, the terms $-nq\mathbf{V}_n \cdot \nabla\phi$, $pq\mathbf{V}_h \cdot \nabla\phi$ represent the energy gain due to the Joule heating, while the terms nC_{W_n}, pC_{W_h} represent the energy loss and gain due to the recombination and generation of electrons and holes, and the energy exchange with the lattice.

It remains to be seen how closure relations for $\mathbf{V}_n$, $\mathbf{V}_h$, $\mathbf{S}_n$, $\mathbf{S}_h$, and $\mathbf{Q}$ can be obtained. For this reason, one can exploit the above-mentioned scaling (the interested reader is referred to [42,43] for more details), which is substantially equivalent to applying the Maxwell procedure to Equations 5.27_2 and 5.29. At the first order, this procedure consists of introducing the equilibrium values, $\mathbf{V}_n = \mathbf{V}_h = \mathbf{S}_n = \mathbf{S}_h = \mathbf{Q} = 0$, in the left-hand sides of Equations 5.27 and 5.29. So doing, one obtains a linear system, which, once solved, gives

$$\mathbf{V}_n = D_{11}^n(W_n, T_L) \nabla \log n + D_{12}^n(W_n, T_L) \nabla W_n + D_{13}^n(W_n, T_L) \nabla \phi, \tag{5.41}$$

$$\mathbf{S}_n = D_{21}^n(W_n, T_L) \nabla \log n + D_{22}^n(W_n, T_L) \nabla W_n + D_{23}^n(W_n, T_L) \nabla \phi, \tag{5.42}$$

$$\mathbf{Q} = -K(T_L) \nabla T_L, \tag{5.43}$$

with

$$D_{11}^n(W_n, T_L) = D_n^{-1} \left[c_{VS}^n F_n^{(1)} - \frac{c_{SS}^n}{\lambda_{W_n}} \right], \quad D_{12}^n(W_n, T_L) = D_n^{-1} \left[c_{VS}^n F_n^{(1)'} - \left(\frac{c_{SS}^n}{\lambda'_{W_n}} \right) \right], \tag{5.44}$$

$$D_{13}^n(W_n, T_L) = D_n^{-1} \left[c_{SS}^n q - c_{VS}^n q G_n^{(1)} \right], \tag{5.45}$$

$$D_{21}^n(W_n, T_L) = -D_n^{-1} \left[c_{VV}^n F_n^{(1)} - \frac{c_{SV}^n}{\lambda_{W_n}} \right], \quad D_{22}^n(W_n, T_L) = -D_n^{-1} \left[c_{VV}^n F_n^{(1)'} - \left(\frac{c_{SV}^n}{\lambda'_{W_n}} \right) \right], \tag{5.46}$$

$$D_{23}^n(W_n, T_L) = -D_n^{-1} \left[c_{SV}^n q - c_{VV}^n q G_n^{(1)} \right], \quad D_n(W, T_L) = c_{VS}^n c_{SV}^n - c_{SS}^n c_{VV}^n. \tag{5.47}$$

Analogous expressions are valid for $\mathbf{V}_h$ and $\mathbf{S}_h$. The coefficients c_{ij}^A depend on W_A and T_L, while the prime indicates a derivative with respect to W. The fluxes $F_A^{(1)}$ and $G_A^{(1)}$ depend on W_A. The complete expressions of c_{ij}^A, $F_A^{(1)}$, and $G_A^{(1)}$ are reported in [43,44], and an explicit expression of the lattice conductivity can be found in [5]. In this way, a generalization of system Equations 5.20 through 5.22 has been obtained where no free parameter is present.

At the source and drain contacts of devices, the Robin boundary condition

$$-K(T_L) \frac{\partial T_L}{\partial n} = R_{th}^{-1}(T_L - T_{env}) \tag{5.48}$$

is assumed, where R_{th} is the thermal resistivity of the contact, T_{env} is the environment temperature, and $\partial/\partial n$ denotes the derivative along the outer normal at the contacts. R_{th} depends on the material of the contact.

On the lateral boundaries and oxide/silicon interfaces of devices no-flux condition for the temperature is used and Dirichlet condition at the bulk contacts is assumed. The electron energy at the source, drain, and bulk contacts are taken as equal to that corresponding to the lattice temperature (Ohmic contacts). The other boundary conditions are described in [44].

The crystal lattice temperature T_L changes much more slowly than other variables. For instance, the typical time of the lattice temperature for reaching the steady state in the simulations is of the order of thousands of picoseconds, while for the other fields it is of the order of picoseconds. This double-scale behavior can be exploited by adopting the following multirate integration scheme:

- Step 1. First the balance equations for electrons and holes can be integrated, with the crystal lattice energy and the electric field frozen at the time step $k - 1$. This gives the electron and hole densities and energies at the time step k, which can be written schematically as

$$\frac{\partial \mathcal{U}^k}{\partial t} + F(\mathcal{U}^k, \phi^{k-1}, T_L^{k-1}) = 0, \tag{5.49}$$

with $\mathcal{U} = (n, p, W_n, W_h)$. Here, $k = 1, \dots, N$ is the index of the integration interval $\Delta t_k = [t_{k-1}, t_k]$, with $t_k = t_{k-1} + \Delta t$, with Δt being the time size of the synchronization window.

- Step 2. The lattice energy balance equation is integrated, with n, p, W_n, and W_h given by step 1

$$\rho c_V \frac{\partial T_L^k}{\partial t} - \operatorname{div}\left[K(T_L^k)\nabla T_L^k\right] = H(\mathcal{U}^k, T_L^k), \tag{5.50}$$

along with the Poisson equation with $n = n^k$ and $p = p^k$.

For steps 1 and 2, different time steps for the numerical integration over the interval Δt_k are used. Typically, the time step for integration of Equation 5.50 can be taken as 100 times larger than the time step for Equation 5.49.

5.6.1 Step 1

The numerical scheme is based on an exponential fitting like that employed in the Scharfetter–Gummel scheme for the drift-diffusion model of semiconductors [29]. The basic idea is to split the particle and energy density currents as the difference of two terms. Each of them is written by introducing suitable mean mobilities in order to get expressions of the currents similar to those arising in other energy-transport models known in literature [45]. The equations are spatially discretized on a regular grid. The details of the numerical scheme can be found in [44]. Here, a brief account is given.

For the sake of simplicity, the numerical method is presented only for the electron part, setting the generation–recombination term equal to zero and omitting the index n in the electron variables. The inclusion of holes and the coupling with electrons can be performed straightforwardly.

First, as said, the current density $\mathbf{J} = n\mathbf{V}$ and the energy-flux density $\mathbf{Z} = n\mathbf{S}$ are rewritten as

$$\mathbf{J} = \mathbf{J}^{(1)} - \mathbf{J}^{(2)}, \quad \mathbf{Z} = \mathbf{Z}^{(1)} - \mathbf{Z}^{(2)}, \tag{5.51}$$

and then each term is put into a drift-diffusion form

$$\mathbf{J}^{(1)} = \frac{c_{SS}}{D}\left[\nabla(n\lambda_W^{-1}) - qn\nabla\phi\right], \quad \mathbf{J}^{(2)} = \frac{c_{VS}}{D}\left[\nabla\left(nF^{(1)}\right) - qnF^{(1)}\lambda_W\nabla\phi\right], \tag{5.52}$$

$$\mathbf{Z}^{(1)} = \frac{c_{VV}}{D}\left[\nabla\left(nF^{(1)}\right) - qnF^{(1)}\lambda_W\nabla\phi\right], \quad \mathbf{Z}^{(2)} = \frac{c_{VS}}{D}\left[\nabla(n\lambda_W^{-1}) - qn\nabla\phi\right]. \tag{5.53}$$

In the two-dimensional (2D) case, the grid points (x_i, y_j) are introduced, with $x_{i+1} - x_i = h = \text{constant}$ and $y_{j+1} - y_j = k = \text{constant}$, and the middle points $(x_i, y_{j\pm1/2}) = (x_i, y_j \pm k/2)$ and $(x_{i\pm1/2}, y_j) = (x_i \pm h/2, y_j)$. A uniform time step Δt is used and $u_{i,j}^l = u(x_i, y_j, l\,\Delta t)$ is set.

By indicating with J_x and J_y the x and y components of the current density $\mathbf{J}$ and by Z_x and Z_y the x and y components of $\mathbf{Z}$, the balance Equations 5.34 and 5.36 are discretized as

$$\frac{n_i^{l+1} - n_i^l}{\Delta t} + \frac{(J_x)_{i+1/2,j} - (J_x)_{i-1/2,j}}{h} + \frac{(J_y)_{i,j+1/2} - (J_y)_{i,j-1/2}}{k} + O(h^2, k^2, \Delta t) = 0, \tag{5.54}$$

$$\frac{(n\,W)_i^{l+1} - (n\,W)_i^l}{\Delta t} + \frac{(Z_x)_{i+1/2,j} - (Z_x)_{i-1/2,j}}{h} + \frac{(Z_y)_{i,j+1/2} - (Z_y)_{i,j-1/2}}{k} +$$

$$-q\frac{(J_x)_{i+1/2,j} + (J_x)_{i-1/2,j}}{2}\frac{\phi_{i+1,j} - \phi_{i-1,j}}{2h} - q\frac{(J_y)_{i,j+1/2} + (J_y)_{i,j-1/2}}{2}\frac{\phi_{i,j+1} - \phi_{i,j-1}}{2k}$$

$$+ n_{i,j}\frac{W_{i,j} - W_0}{(\tau_W)_{i,j}} + O(h^2, k^2, \Delta t) = 0. \tag{5.55}$$

The variables without temporal index must be evaluated at time level l.

In order to calculate the components of the currents in the middle points, let us consider the sets

$$I_{i+1/2,j} = [x_i, x_{i+1}] \times [y_{j-1/2}, y_{j+1/2}], \quad I_{i,j+1/2} = [x_{i-1/2}, x_{i+1/2}] \times [y_j, y_{j+1}]$$

and expand $J_x^{(r)}$, $r = 1, 2$ in Taylor's series in $I_{i+1/2,j}$

$$J_x^{(r)}(x, y) = (J_x^{(r)})_{i+1/2,j} + (x - x_{i+1/2}) \left(\frac{\partial J_x^{(r)}}{\partial x} \right)_{i+1/2,j} + (y - y_j) \left(\frac{\partial J_x^{(r)}}{\partial y} \right)_{i+1/2,j} + o(\Delta x, \Delta y)$$

and $J_y^{(r)}$, $r = 1, 2$ in Taylor's series in $I_{i,j+1/2}$

$$J_y^{(r)}(x, y) = (J_y^{(r)})_{i,j+1/2} + (x - x_i) \left(\frac{\partial J_y^{(r)}}{\partial x} \right)_{i,j+1/2} + (y - y_{j+1/2}) \left(\frac{\partial J_y^{(r)}}{\partial y} \right)_{i,j+1/2} + o(\Delta x, \Delta y).$$

First, $U_T = \frac{1}{q\lambda_W}$ is introduced, which plays the role of a thermal potential [44] and indicates by $\overline{U}_T$ its piecewise constant approximation, which is given by $\overline{U}_T = \dfrac{\lambda_W^{-1}(W_{i,j}) + \lambda_W^{-1}(W_{i+1,j})}{2q}$ in the cell $I_{i+1/2,j}$ and by $\overline{U}_T = \dfrac{\lambda_W^{-1}(W_{i,j+1}) + \lambda_W^{-1}(W_{i,j})}{2q}$ in the cell $I_{i,j+1/2}$. Then the *local* mobilities

$$g_{11} = -\frac{\overline{c}_{SS}}{\overline{D}} n\lambda_W^{-1}, \quad g_{12} = -\frac{\overline{c}_{VS}}{\overline{D}} nF^{(1)}, \tag{5.56}$$

$$g_{21} = -\frac{\overline{c}_{VV}}{\overline{D}} nF^{(1)}, \quad g_{22} = -\frac{\overline{c}_{VS}}{\overline{D}} n\lambda_W^{-1} \tag{5.57}$$

are introduced, where $\overline{c}_{pq}$ is a piecewise constant approximation of c_{pq} $p, q = $ V, S, given by $\overline{c}_{pq} = c_{pq} \left(\frac{W_{i,j} + W_{i+1,j}}{2} \right)$ in the cell $I_{i+1/2,j}$ and by $\overline{c}_{pq} = c_{pq} \left(\frac{W_{i,j} + W_{i,j+1}}{2} \right)$ in the cell $I_{i,j+1/2}$, and, similarly to [46], the *local* Slotboom variables

$$s_{kr} = \exp\left(-\phi/\overline{U}_T\right) g_{kr} \quad k, r = 1, 2$$

that satisfy

$$\nabla s_{1r} \simeq -\exp\left(-\phi/\overline{U}_T\right) \mathbf{J}^{(r)}, \quad \nabla s_{2r} \simeq -\exp\left(-\phi/\overline{U}_T\right) \mathbf{Z}^{(r)} \quad r = 1, 2. \tag{5.58}$$

From the x component of Equation 5.58$_1$, we have

$$\frac{\partial s_{1r}(x, y_j)}{\partial x} \simeq -\exp\left(-\phi/\overline{U}_T\right) J_x^{(r)}(x, y_j)$$

$$= -\exp\left(-\phi/\overline{U}_T\right) \left\{ (J_x^{(r)})_{i+1/2,j} + (x - x_{i+1/2}) \left(\frac{\partial J_x^{(r)}}{\partial x} \right)_{i+1/2,j} + o(\Delta x, \Delta y) \right\}, \tag{5.59}$$

which, after integration over $[x_i, x_{i+1}]$, where a piecewise linear approximation is used for ϕ and some algebra, gives

$$(J_x^{(r)})_{i+1/2,j} = -z_{i+1/2,j} \coth z_{i+1/2,j} \frac{(g_{1r})_{i+1,j} - (g_{1r})_{i,j}}{h} + z_{i+1/2,j} \frac{(g_{1r})_{i+1,j} + (g_{1r})_{i,j}}{h}, \quad r = 1, 2, \quad (5.60)$$

where $z_{i+1/2,j} = \frac{\phi_{i+1,j} - \phi_{i,j}}{2\bar{U}_T}$.

Likewise, by evaluating the y component of Equation 5.58$_2$ and integrating over $[y_j, y_{j+1}]$, we find

$$(J_y^{(r)})_{i,j+1/2} = -z_{i,j+1/2} \coth z_{i,j+1/2} \frac{(g_{1r})_{i,j+1} - (g_{1r})_{i,j}}{k} + z_{i,j+1/2} \frac{(g_{1r})_{i,j+1} + (g_{1r})_{i,j}}{k}, \quad r = 1, 2, \quad (5.61)$$

where $z_{i,j+1/2} = \frac{\phi_{i,j+1} - \phi_{i,j}}{2\bar{U}_T}$. Using the same procedure, the following discrete expression for the components of the energy flux is obtained:

$$(Z_x^{(r)})_{i+1/2,j} = -z_{i+1/2,j} \coth z_{i+1/2,j} \frac{(g_{2r})_{i+1,j} - (g_{2r})_{i,j}}{h} + z_{i+1/2,j} \frac{(g_{2r})_{i+1,j} + (g_{2r})_{i,j}}{h}, \quad (5.62)$$

$$(Z_y^{(r)})_{i,j+1/2} = -z_{i,j+1/2} \coth z_{i,j+1/2} \frac{(g_{2r})_{i,j+1} - (g_{2r})_{i,j}}{k} + z_{i,j+1/2} \frac{(g_{2r})_{i,j+1} + (g_{2r})_{i,j}}{k}, \quad r = 1, 2. \quad (5.63)$$

The error in formulas Equations 5.60 through 5.63 is $o(h, k)$.

The Poisson equation is solved by replacing it with

$$\phi_t - \text{div} \left(\varepsilon_d \nabla \phi \right) = q(N_D - N_A - n). \quad (5.64)$$

The solution of Equation 5.64 as $t \mapsto +\infty$ is the same as that of the original Poisson equation, at least in the smooth case.

If a time step $\Delta \hat{t}$ is introduced and $\phi_{ij}^r = \phi(x_i, y_j, r\Delta\hat{t})$ is set, Equation 5.64 can be discretized in an explicit way as

$$\phi_{ij}^{r+1} = \phi_{ij}^r + \varepsilon_d \Delta \hat{t} \left[\frac{1}{h^2} \left(\phi_{i+1,j} - 2\phi_{i,j} + \phi_{i-1,j} \right) + \frac{1}{k^2} \left(\phi_{i,j+1} - 2\phi_{i,j} + \phi_{i,j-1} \right) + q(C_{ij} - n_{ij}) \right], \quad (5.65)$$

where $C_{ij} = N_d(x_i, y_j) - N_a(x_i, y_j)$, with the notable advantage of taking easily into account the different types of boundary conditions. The price to pay is that at each time step, one needs to reach the stationary state of Equation 5.64 by using a time step satisfying the Courant–Friedrichs–Lewy (CFL) condition, common for parabolic equations,

$$\Delta \hat{t} \leq \frac{1}{2} \frac{1}{\frac{1}{h^2} + \frac{1}{k^2}}.$$

However, the computational effort is comparable with that required by direct methods.

5.6.2 Step 2

A coordinate splitting technique [47] is used for the solution of the lattice energy equation for the variable $u = k_B T_L$, with time step Δt_T. The splitting technique allows an efficient employment of stable implicit time schemes. The procedure consists of two steps with two suboperators (with corresponding boundary conditions)

$$\rho c_V \frac{u^{n+1/2} - u^n}{\Delta t_T} = \frac{\partial}{\partial x}\left[K(T_L^n)\frac{\partial u^{n+1/2}}{\partial x}\right] + \frac{k_B}{2}H^{n+1/2} \tag{5.66}$$

and

$$\rho c_V \frac{u^{n+1} - u^{n+1/2}}{\Delta t_T} = \frac{\partial}{\partial y}\left[K(T_L^n)\frac{\partial u^{n+1}}{\partial y}\right] + \frac{k_B}{2}H^{n+1}. \tag{5.67}$$

Here, $u^n = u(x, y, n\Delta t_T)$. The scheme is absolutely stable and approximates the equation of the lattice energy with first-order accuracy in time. For the approximation of the spatial derivatives, the standard stencil with three points can been chosen. For instance, the approximation of Equation 5.67 is

$$\rho c_V u_{i,j}^{n+1} = \rho c_V u_{i,j}^{n+1/2} + \frac{\Delta t_T}{k^2}\left[\frac{\tilde{K}_{i,j} + \tilde{K}_{i,j+1}}{2}(u_{i,j+1}^{n+1} - u_{i,j}^{n+1}) - \frac{\tilde{K}_{i,j} + \tilde{K}_{i,j-1}}{2}(u_{i,j}^{n+1} - u_{i,j-1}^{n+1})\right]$$

$$+ \frac{k_B}{2}H^{n+1},$$

where $\tilde{K}_{i,j} = K(T_{L i,j})$. Of course, such a discretization is valid in the interior points of the mesh.

The Robin boundary condition Equation 5.48 is approximated as

$$-K\left(\frac{u_{i,1}^n + u_{i,0}^n}{2k_B}\right)\frac{u_{i,1}^{n+1} - u_{i,0}^{n+1}}{k} = R_{th}^{-1}(u_{i,0}^{n+1} - k_B T_{env}). \tag{5.68}$$

Here, it has been assumed that at the portion of boundary where the Robin condition holds, one has $j = 0$ and the closest interior points have $j = 1$. The resulting linear system can be solved efficiently with the tridiagonal matrix factorization procedure.

5.7 Thermal Properties of Graphene

Recent years have witnessed a great interest in 2D materials for their promising applications. The most widely investigated one is graphene, but lately the single-layer transition metal dichalcogenides (TMDCs), such as molybdenum disulfite and tungsten diselenite, and black phosphorus have received certain attention [48]. In this section, we analyze the thermal and electrical properties of suspended graphene monolayers. What follows is based on Ref. [49].

5.7.1 Semiclassical Description

Graphene is considered a potential new semiconductor material for future applications in nanoelectronic [50–55] and optoelectronic devices [48] because it has very good mechanical properties, it is an excellent heat and electricity conductor, and it also has noteworthy optical properties. It is a 2D allotrope of carbon that consists of carbon atoms tightly packed into a honeycomb hexagonal lattice (see Figure 5.7), due to their sp^2 hybridization. The atoms in graphene form two interpenetrating triangular Bravais sublattices and each unit cell contain two atoms belonging to different sublattices [51]. The first Brillouin zone is hexagonal like the Bravais lattice. There are six σ bonds per unit cell whose orbitals yield six bands, three of which (called σ) are below the Fermi energy and three of which are above (the σ^* bands).

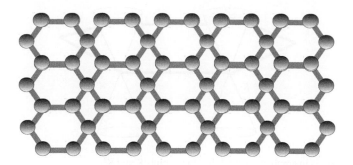

FIGURE 5.7 Honeycomb crystal lattice of graphene.

Moreover, each primitive cell contributes two $2p_z$-orbitals that participate in π bonding. The p_z electrons can be treated independently from the other valence electrons, since the overlap between the corresponding orbitals and the s or p_x and p_y orbitals is strictly zero by symmetry. Therefore, within the tight binding approximation, the energy dispersion relations of the π bands can be calculated by solving the eigenvalue problem for the approximate[†] effective tight-binding Hamiltonian

$$\mathcal{H}(\mathbf{k}) = t_O \begin{pmatrix} 0 & \gamma_{\mathbf{k}}^* \\ \gamma_{\mathbf{k}} & 0 \end{pmatrix},$$

with t_O the nearest neighbor (nn) C–C tight binding overlap energy and

$$\gamma_{\mathbf{k}} := 1 + e^{i\mathbf{k}\cdot\mathbf{a}_1} + e^{i\mathbf{k}\cdot\mathbf{a}_2}$$

the sum of the nn phase factors, $\mathbf{a}_1$ and $\mathbf{a}_2$ being the primitive vectors.

One finds

$$\mathcal{E}^\pm = \pm t_O|\gamma_{\mathbf{k}}|,$$

where the $+$ refers to conduction (π^*) band and the $-$ refers to the valence (π) band. The points where the bands touch each other are called Dirac points, at these points $\mathcal{E}^\pm = 0$. They coincide with the vertices of the first Brillouin zone and are divided into two nonequivalent sets of three points. These two sets are named K$-$ and K$'-$points.

In the proximity of the Dirac points, for $|\mathbf{k}|a << 1$ (for simplicity, here and in the following expression wave vectors are measured with respect to the K(K$'$) point), a being the distance between nn carbon atoms, the energy dispersion relation can be approximated by (see Figure 5.8)

$$\mathcal{E}^\pm = \pm\hbar v_F|\mathbf{k}|, \tag{5.69}$$

with $v_F = \dfrac{3|t_O|a}{2\hbar}$ the Fermi velocity. Electrons that mostly contribute to the charge transport in intrinsic graphene are indeed those in the two valleys around the Dirac points. Therefore, four populations of electrons will be taken into account, which will be labeled ℓ =K, K$'$ and $s = -1, 1$ (or equivalently $s = \pi, \pi^*$), respectively. On account of Equation 5.69, their energy is given by

$$\mathcal{E}_{\ell,s} = s\hbar v_F|\mathbf{k} - \mathbf{k}_\ell|,$$

[†] It does not take into account the next nearest neighbor hopping correction.

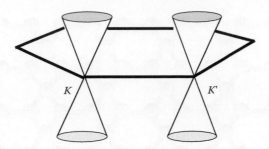

FIGURE 5.8 Energy bands in graphene around the Dirac points K and K'.

with $\mathbf{k}_\ell$ the position of the Dirac point ℓ. Because of the linearity of these relations, electrons behave as massless Dirac fermions. This linear dispersion, with zero band-gap, gives rise to the appeal of graphene for nanophotonic applications. Graphene has a high sensitivity to optical signals through various mechanisms of light–matter interaction. For this reason, it is applied in photodetectors and modulators at near-infrared, visible, and ultraviolet spectrum range, due to its broadband absorption of light through interband transitions [33]. Moreover, since graphene supports plasmons, it is also employed in plasmonics [33].

In a semiclassical kinetic setting, the charge transport in graphene is described by four Boltzmann equations, one for electrons in the valence band (π) and one for electrons in the conduction band (π^*), that in turn can belong to the K or K' valley,

$$\frac{\partial f_{\ell,s}(t,\mathbf{x},\mathbf{k})}{\partial t} + \mathbf{v}_{\ell,s} \cdot \nabla_{\mathbf{x}} f_{\ell,s}(t,\mathbf{x},\mathbf{k}) - \frac{q}{\hbar}\mathbf{E}\cdot\nabla_{\mathbf{k}}f_{\ell,s}(t,\mathbf{x},\mathbf{k}) = \left.\frac{df_{\ell,s}}{dt}(t,\mathbf{x},\mathbf{k})\right|_{e-ph}, \qquad (5.70)$$

where $f_{\ell,s}(t,\mathbf{x},\mathbf{k})$ represents the distribution function of charge carriers, in the valence or conduction band ($s = -1$ or $s = 1$) and valley ℓ (K or K'), at position $\mathbf{x}$, time t, and with wave vector $\mathbf{k}$. We denote by $\nabla_{\mathbf{x}}$ and $\nabla_{\mathbf{k}}$ the gradients with respect to the position and the wave vector, respectively. The relationship between the group velocity $\mathbf{v}_{\ell,s}$ and the band energy $\mathcal{E}_{\ell,s}$ can be expressed as

$$\mathbf{v}_{\ell,s} = \frac{1}{\hbar}\nabla_{\mathbf{k}}\mathcal{E}_{\ell,s}.$$

The electric field $\mathbf{E}$ is assumed as external here. On the right-hand sides of Equation 5.70 are the collision terms, representing the interactions of electrons with acoustic, Γ-*LO* and Γ-*TO*, respectively, and K-phonons.

Scattering with acoustic phonons is intravalley and intraband and a linear dispersion relation

$$\omega_{ac} = v_s|\mathbf{q}|$$

is assumed for them, with ω_{ac} the acoustic phonon frequency and v_s the sound speed in graphene.

The Γ-*LO* and Γ-*TO* phonons are in-plane (with respect to the plane of the lattice) modes, which give rise to electron intravalley transitions, which mostly involve phonons with wave vectors near to the center Γ of the Brillouin zone. The wave vectors $\mathbf{q}$ of these phonons are near to zero and for their dispersion relations the Einstein approximation can be used, according to which $\hbar\omega \approx$ const, with ω the phonon frequency. Such transitions can be both intraband and interband.

The K-phonons are not a real phonon branch. Their name is due to the fact that their wave vectors are close to the K- or K' point [53]. They belong to an optical branch and induce intervalley scatterings. An

Einstein approximation with a mean phonon energy is used for them. Interaction of electrons with out-of-plane phonon modes, the so-called Z phonons, or with in-plane phonons having wave vectors far from the Γ or K point is negligible [56], even though they can play some role in the thermal effects. For an accurate description of phonon dispersion relations and thermal conductivity in graphene, we refer the interested reader to [57,58].

The general form of the collision term can be written as [51,53]

$$\frac{df_{\ell,s}}{dt}(t,\mathbf{x},\mathbf{k})\bigg|_{e-ph} = \sum_{\ell',s'} \left[\int_B S_{\ell',s',\ell,s}(\mathbf{k}',\mathbf{k}) f_{\ell',s'}(t,\mathbf{x},\mathbf{k}') \left(1 - f_{\ell,s}(t,\mathbf{x},\mathbf{k})\right) d\mathbf{k}' \right.$$
$$\left. - \int_B S_{\ell,s,\ell',s'}(\mathbf{k},\mathbf{k}') f_{\ell,s}(t,\mathbf{x},\mathbf{k}) \left(1 - f_{\ell',s'}(t,\mathbf{x},\mathbf{k}')\right) d\mathbf{k}' \right],$$

where the total transition rate $S_{\ell',s',\ell,s}(\mathbf{k}',\mathbf{k})$ is given by the sum of the contributions of the several types of scatterings described earlier:

$$S_{\ell',s',\ell,s}(\mathbf{k}',\mathbf{k}) = \sum_p \left| G^{(p)}_{\ell',s',\ell,s}(\mathbf{k}',\mathbf{k}) \right|^2 \left[\left(g_p^- + 1 \right) \delta\left(\mathcal{E}_{\ell,s}(\mathbf{k}) - \mathcal{E}_{\ell',s'}(\mathbf{k}') + \hbar\omega_p \right) \right.$$
$$\left. + g_p^+ \, \delta\left(\mathcal{E}_{\ell,s}(\mathbf{k}) - \mathcal{E}_{\ell',s'}(\mathbf{k}') - \hbar\omega_p \right) \right]. \tag{5.71}$$

The index p labels the pth phonon population. The $\left| G^{(p)}_{\ell',s',\ell,s}(\mathbf{k}',\mathbf{k}) \right|^2$'s are the electron–phonon coupling matrix elements, which describe the interaction mechanism of an electron with a p-phonon, from the state of wave vector $\mathbf{k}'$ belonging to the valley ℓ' and band s' to the state $\mathbf{k}$ belonging to the valley ℓ and band s. The symbol δ denotes the Dirac distribution, ω_p is the pth phonon frequency, $g_p(\mathbf{q})$ is the phonon distribution for the p-type phonons. In Equation 5.71, $g_p^\pm = g_p\left(\mathbf{q}^\pm\right)$, where $\mathbf{q}^\pm = \pm\left(\mathbf{k}' - \mathbf{k}\right)$, stemming from the momentum conservation. For acoustic phonons, we consider the elastic approximation according to which the transition rate reads

$$\frac{1}{(2\pi)^2} \frac{\pi D_{ac}^2 \, k_B \, T}{2\hbar\,\sigma_m\,v_s^2} \left(1 + \cos\vartheta_{\mathbf{k},\mathbf{k}'}\right) \delta\left(\mathcal{E}(\mathbf{k}') - \mathcal{E}(\mathbf{k})\right), \tag{5.72}$$

where D_{ac} is the acoustic phonon-coupling constant, σ_m is the graphene areal density, and $\vartheta_{\mathbf{k},\mathbf{k}'}$ is the convex angle between $\mathbf{k}$ and $\mathbf{k}'$.

The electron–phonon coupling matrix elements for the intraband scatterings of the electrons in the conduction band with the LO, the TO, and the K-phonons are [53]

$$\left| G^{(LO)}(\mathbf{k}',\mathbf{k}) \right|^2 = \frac{1}{(2\pi)^2} \frac{\pi D_O^2}{\sigma_m\,\omega_O} \left(1 - \cos(\vartheta_{\mathbf{k},\mathbf{k}'-\mathbf{k}} + \vartheta_{\mathbf{k}',\mathbf{k}'-\mathbf{k}})\right) \tag{5.73}$$

$$\left| G^{(TO)}(\mathbf{k}',\mathbf{k}) \right|^2 = \frac{1}{(2\pi)^2} \frac{\pi D_O^2}{\sigma_m\,\omega_O} \left(1 + \cos(\vartheta_{\mathbf{k},\mathbf{k}'-\mathbf{k}} + \vartheta_{\mathbf{k}',\mathbf{k}'-\mathbf{k}})\right) \tag{5.74}$$

$$\left| G^{(K)}(\mathbf{k}',\mathbf{k}) \right|^2 = \frac{1}{(2\pi)^2} \frac{2\pi D_K^2}{\sigma_m\,\omega_K} \left(1 - \cos\vartheta_{\mathbf{k},\mathbf{k}'}\right), \tag{5.75}$$

where D_O is the optical phonon coupling constant, ω_O is the optical phonon frequency, D_K is the K-phonon coupling constant and ω_K is the K-phonon frequency. The angles $\vartheta_{\mathbf{k},\mathbf{k}'-\mathbf{k}}$ and $\vartheta_{\mathbf{k}',\mathbf{k}'-\mathbf{k}}$ denote the convex angles between $\mathbf{k}$ and $\mathbf{k}' - \mathbf{k}$ and between $\mathbf{k}'$ and $\mathbf{k}' - \mathbf{k}$, respectively.

TABLE 5.1 Physical Parameters
for the Scattering Rates

σ_m	7.6×10^{-8} g/cm^2
v_F	10^6 m/s
v_s	2×10^4 m/s
D_{ac}	6.8 eV
$\hbar \omega_O$	164.6 meV
D_O	10^9 eV/cm
$\hbar \omega_K$	124 meV
D_K	3.5×10^8 eV/cm

In the literature, there are several values for the coupling constants entering the collision terms. For example, for the acoustic deformation potential one can find values ranging from 2.6 to 29 eV. A similar degree of uncertainty is found for the optical and K-phonon coupling constants as well. In the following numerical simulations of monolayer graphene, the parameters proposed in [56,59] have been adopted. They are reported in Table 5.1. By applying a gate voltage, the Fermi energy can be varied. In the sequel, we consider only the case of high values of the Fermi energy, which is equivalent for conventional semi-conductors to an n-type doping. Under such a condition, the dynamics of the electrons belonging to the valence band can be neglected, and this is the reason why only the intraband coupling matrix elements for the electrons of the conduction band were reported earlier. In order to simplify the notation the indices s and ℓ will be omitted in the following.

From the semiclassical transport equations, using the procedure shown in the previous sections (see e.g., [24,50,60–63]), one can also formulate macroscopic models that are more suited for computer-aided design (CAD) purposes because they avoid the numerical solutions of the Boltzmann equations, even if they introduce some approximation for the closure relations that are needed.

The evolution of the phonon distributions is governed by the following Boltzmann equations:

$$\frac{\partial g_\eta}{\partial t} = C_\eta, \quad \eta = LO, TO, K, \tag{5.76}$$

$$\frac{\partial g_{ac}}{\partial t} + \mathbf{c}_{ac} \cdot \nabla_{\mathbf{x}} g_{ac} = C_{ac}, \tag{5.77}$$

with $\mathbf{c}_{ac} = \nabla_{\mathbf{q}} \omega_{ac} = v_s \frac{\mathbf{q}}{|\mathbf{q}|}$ the acoustic group velocity (on account of the Einstein approximation, the group velocity of the other phonons is negligible). The collision terms C_η and C_{ac} describe the interaction of the phonons with the electrons and the other phonons.

The phonon–phonon collision term is very complicated and represents a formidable task from a numerical point of view. For this reason, in the applications, the Bhatnagar–Gross–Krook (BGK) approximation is usually employed. It gives a simple way to describe the tendency of each species of phonons to an equilibrium distribution. We adopt the same approach as in [4] modeling the phonon–phonon interaction by requiring that each type of phonons relaxes to a local equilibrium given by a Bose–Einstein distribution evaluated, for all the phonon species, at a common local equilibrium temperature T_{LE}, determined by the procedure we specify below. Therefore, we split the phonon collision terms as follows:

$$C_\eta = C_{\eta-e} - \frac{g_\eta - g_\eta^{LE}}{\tau_\eta}, \tag{5.78}$$

$$C_{ac} = C_{ac-e} - \frac{g_{ac} - g_{ac}^{LE}}{\tau_{ac}}, \tag{5.79}$$

where $C_{\eta-e}$ and $C_{\text{ac}-e}$ describe the phonon–electron collisions, τ_{η} and τ_{ac} are the phonon relaxation times, and g_{η}^{LE} and $g_{\text{ac}}^{\text{LE}}$ are the local equilibrium phonon distributions given by

$$g_{\eta}^{\text{LE}} = \left[e^{\hbar\omega_{\eta}/k_B T_{\text{LE}}} - 1 \right]^{-1}, \tag{5.80}$$

$$g_{\text{ac}}^{\text{LE}} = \left[e^{\hbar v_s q/k_B T_{\text{LE}}} - 1 \right]^{-1}. \tag{5.81}$$

Let us introduce the average phonon energies

$$W_{LO+TO} = \frac{2}{(2\pi)^2} \int_B \hbar\omega_{LO/TO}\, g_{LO/TO}\, d\mathbf{q}, \quad W_K = \frac{1}{(2\pi)^2} \int_B \hbar\omega_K\, g_K\, d\mathbf{q}, \tag{5.82}$$

$$W_{\text{ac}} = \frac{3}{(2\pi)^2} \int_{\mathbb{R}^2} \hbar\omega_{\text{ac}}\, g_{\text{ac}}\, d\mathbf{q}, \tag{5.83}$$

where the *LO* and the *TO* phonons, having the same energy, have been considered, in the first approximation, as a unique population ($g_{LO} \approx g_{TO}$) with a density of states equal to $\frac{2}{(2\pi)^2}$, while the density of states of the acoustic phonons is equal to $\frac{3}{(2\pi)^2}$ on account of their three polarization states.

By multiplying, for each species, the phonon Boltzmann equation with the phonon energy and by integrating with respect to the wave vector $\mathbf{q}$, one gets the following macroscopic balance equations for the average phonon energies

$$\frac{\partial W_{LO+TO}}{\partial t} = C_{W_{LO+TO}}, \quad \frac{\partial W_K}{\partial t} = C_{W_K}, \tag{5.84}$$

$$\frac{\partial W_{\text{ac}}}{\partial t} = -\nabla_{\mathbf{x}} \cdot \mathbf{Q}_{\text{ac}} + C_{W_{\text{ac}}}, \tag{5.85}$$

where $\mathbf{Q}_{\text{ac}}$ is the phonon energy-flux. While

$$C_{W_{LO+TO}} = C_{W_{LO+TO}}^{p-e} - \frac{W_{LO+TO} - W_{LO+TO}^{\text{LE}}}{\tau_{LO+TO}}, \quad C_{W_K} = C_{W_K}^{p-e} - \frac{W_K - W_K^{\text{LE}}}{\tau_K} \tag{5.86}$$

$$C_{W_{\text{ac}}} = C_{W_{\text{ac}}}^{p-e} - \frac{W_{\text{ac}} - W_{\text{ac}}^{\text{LE}}}{\tau_{\text{ac}}}, \tag{5.87}$$

are the energy-production terms with

$$W_{LO+TO}^{\text{LE}} = \frac{2}{(2\pi)^2} \int_B \hbar\omega_{LO/TO} \left[e^{\hbar\omega_{\eta}/k_B T_{\text{LE}}} - 1 \right]^{-1} d\mathbf{q}, \tag{5.88}$$

$$W_K^{\text{LE}} = \frac{1}{(2\pi)^2} \int_B \hbar\omega_K \left[e^{\hbar\omega_{\eta}/k_B T_{\text{LE}}} - 1 \right]^{-1} d\mathbf{q}, \tag{5.89}$$

$$W_{\text{ac}}^{\text{LE}} = \frac{3}{(2\pi)^2} \int_B \hbar\omega_{\text{ac}} \left[e^{\hbar v_s q/k_B T_{\text{LE}}} - 1 \right]^{-1} d\mathbf{q}, \tag{5.90}$$

the average energies in local equilibrium, while $C_{W_{LO+TO}}^{p-e}$, $C_{W_K}^{p-e}$, and $C_{W_{ac}}^{p-e}$ are the contributions arising from the phonon–electron interactions. Due to energy conservation, these satisfy the relationships

$$C_{W_{LO+TO}}^{e-p} + C_{W_{LO+TO}}^{p-e} = 0, \quad C_{W_K}^{e-p} + C_{W_K}^{p-e} = 0, \tag{5.91}$$

$$C_{W_{ac}}^{e-p} + C_{W_{ac}}^{p-e} = 0, \tag{5.92}$$

where $C_{W_{LO+TO}}^{e-p}$ and $C_{W_{ac}}^{e-p}$ are the energy production terms for electrons.

Since the electron flow does not explicitly enter the phonon–phonon scatterings, the conservation of the total phonon energy in absence of an external source implies

$$\frac{W_{LO+TO} - W_{LO+TO}^{LE}}{\tau_{LO+TO}} + \frac{W_K - W_K^{LE}}{\tau_K} + \frac{W_{ac} - W_{ac}^{LE}}{\tau_{ac}} = 0. \tag{5.93}$$

Now we can define the local equilibrium temperature as follows (see [4]).

Definition 5.7.1 The temperature T_{LE} is the common temperature that we must assign to each species in order to have

$$W_{LO+TO} + W_K + W_{ac} = W_{LO+TO}^{LE} + W_K^{LE} + W_{ac}^{LE}. \tag{5.94}$$

In other words, T_{LE} is the common temperature each phonon species should to be in thermodynamic equilibrium in order to preserve the total energy.

By taking into account the expressions 5.88 through 5.90, we find that the nonlinear equation T_{LE} must satisfy

$$h(T_{LE}) := \frac{2\hbar\omega_{LO/TO} A}{(2\pi)^2} \left[e^{\hbar\omega_{LO/TO}/k_B T_{LE}} - 1 \right]^{-1} + \frac{\hbar\omega_K A}{(2\pi)^2} \left[e^{\hbar\omega_K/k_B T_{LE}} - 1 \right]^{-1}$$
$$+ \left(k_B T_{LE} \right)^3 \frac{3\,\zeta(3)}{\pi\,\hbar^2 v_s^2} = W_{LO+TO} + W_K + W_{ac}, \tag{5.95}$$

where $\zeta(\cdot)$ is the zeta function and $A = \frac{8\sqrt{3}\pi^2}{9a^2}$ is the area of the first Brillouin zone, with $a = 0.142$ nm. In Equation 5.95, W_{LO+TO}, W_K, and W_{ac} are the current values of the average phonon energies.

We observe that

$$h'(T_{LE}) = \frac{2(\hbar\omega_{LO/TO})^2 A}{(2\pi)^2 (k_B T_{LE})^2} e^{\hbar\omega_{LO/TO}/k_B T_{LE}} \left[e^{\hbar\omega_{LO/TO}/k_B T_{LE}} - 1 \right]^{-2}$$
$$+ \frac{(\hbar\omega_K)^2 A}{(2\pi)^2 (k_B T_{LE})^2} e^{\hbar\omega_K/k_B T_{LE}} \left[e^{\hbar\omega_K/k_B T_{LE}} - 1 \right]^{-2} + \left(k_B T_{LE} \right)^2 \frac{9\,\zeta(3)}{\pi\,\hbar^2 v_s^2} > 0, \quad \forall\, T_{LE} > 0,$$
$$\lim_{T_{LE} \mapsto 0^+} h(T_{LE}) = 0, \qquad \lim_{T_{LE} \mapsto +\infty} h(T_{LE}) = +\infty,$$

and therefore Equation 5.95 admits a unique solution for any assigned positive W_{LO+TO}, W_K, and W_{ac}.

By taking into account the definition of T_{LE}, a simple way to satisfy Equation 5.93 is to take

$$\tau_{LO+TO} = \tau_K = \tau_{ac} = \tau. \tag{5.96}$$

In the sequel, the phonon relaxation times will be taken as equal. Moreover, we assume what follows.

Assumption 5.7.1 The phonon distributions are considered as sums of even and odd parts, and the former are supposed to be locally of Bose–Einstein type.

As a consequence, from Equations 5.88 to 5.90 with T_{LE} substituted by the respective phonon temperatures, one finds that the temperatures of the phonon species are related to the respective average energies according to

$$k_B T_{LO/TO} = \frac{\hbar\omega_{LO/TO}}{\ln\left(1 + \dfrac{2A\hbar\omega_{LO/TO}}{(2\pi)^2 W_{LO+TO}}\right)}, \quad k_B T_K = \frac{\hbar\omega_K}{\ln\left(1 + \dfrac{A\hbar\omega_K}{(2\pi)^2 W_K}\right)}, \tag{5.97}$$

$$k_B T_{ac} = W_{ac}^{1/3}\left(\frac{3\,\zeta(3)}{\pi\,\hbar^2 v_s^2}\right)^{-1/3}. \tag{5.98}$$

5.7.2 The Simulation Scheme

For a homogeneous monolayer graphene sheet under a constant electric field **E**, the only significant components of the evolution equations are those parallel to the field and there is no dependence on the spatial variables. By choosing a reference frame in the plane of the graphene sheet with the x-axis parallel to **E**, the complete model consists of the following equations:

$$\frac{\partial f(t,\mathbf{k})}{\partial t} - \frac{q}{\hbar} E\,\frac{\partial f(t,\mathbf{k})}{\partial k_x} = \int S(\mathbf{k}',\mathbf{k})f(t,\mathbf{k}')\left(1 - f(t,\mathbf{k})\right)d\mathbf{k}',$$

$$- \int S(\mathbf{k},\mathbf{k}')f(t,\mathbf{k})\left(1 - f(t,\mathbf{k}')\right)d\mathbf{k}', \tag{5.99}$$

$$\frac{dW_{LO+TO}(t)}{dt} = C_{W_{LO+TO}}^{p-e} - \frac{W_{LO+TO} - W_{LO+TO}^{LE}}{\tau}, \quad \frac{dW_K(t)}{dt} = C_{W_K}^{p-e} - \frac{W_K - W_K^{LE}}{\tau}, \tag{5.100}$$

$$\frac{dW_{ac}(t)}{dt} = C_{W_{ac}}^{p-e} - \frac{W_{ac} - W_{ac}^{LE}}{\tau}. \tag{5.101}$$

An equation similar to Equation 5.99 holds for the K' valley. As the initial condition for the electrons we take the Fermi–Dirac distribution

$$f(0,\mathbf{k}) = \frac{1}{1 + \exp\left(\dfrac{\mathcal{E}(\mathbf{k}) - \mathcal{E}_F}{k_B T_0}\right)},$$

where T_0 is the room temperature (300 K) and $\mathcal{E}_F$ is the Fermi energy, which is related to the initial charge density by

$$\rho(0) = \frac{4}{(2\pi)^2}\int f(0,\mathbf{k})\,d\mathbf{k}. \tag{5.102}$$

In Equation 5.102, the factor 4 arises because we are considering both the two states of spin and the degeneracy (equal to 2) of the valley. Note that in the unipolar case ρ remains constant, $\rho(t) = \rho(0)$, as a consequence of the charge conservation.

Regarding the phonons, we assume that initially all the phonons are at the room temperature T_0. Therefore, the initial conditions for the phonon average energies are given by Equations 5.88 through 5.90 with $T_{LE} = T_0$.

Let us introduce a uniform time-step Δt and denote by t_n the nth time level. For each interval $[t_n, t_n + \Delta t]$ we solve Equation 5.99 by a Direct Simulation Monte Carlo scheme (DSMC) by freezing the phonon temperatures at the values they have at $t = t_n$. A crucial question is to properly take into account the degeneracy effects in the scattering terms. Usually, the approach proposed in [64] is adopted. However, by so doing, occupation numbers greater than one can be obtained [65] with the manifest violation of the Pauli exclusion principle. In order to overcome such a difficulty we employ the new DSMC scheme presented in [65], which guarantees physically correct occupation numbers. For the sake of shortness we skip the details and refer the interested reader to the paper quoted earlier.

The remaining Equations 5.100 and 5.101 are discretized by an explicit Euler method with T_{LE} evaluated at the previous time step by solving the nonlinear relation Equation 5.95.

In order to complete the numerical scheme, we have to evaluate the production terms $C^{p-e}_{W_{LO+TO}}$, $C^{p-e}_{W_K}$, and $C^{p-e}_{W_{ac}}$. They represent the rate of variation of phonon energy per unit time and are proportional to the difference between the number of emission and absorption processes per unit time due to the electron–phonon scatterings. Taking advantage of the intermediate results coming from the DSMC part, in each time window $[t_{n-1}, t_n]$ we count, for each phonon species, the number of emission scatterings C^+_η and absorption scatterings C^-_η, $\eta = LO + TO, K$. If N_e is the number of particles used in the MC method, each simulation particle has a statistical density weight given by ρ/N_e. Therefore, we can estimate the phonon energy production term as

$$C^{p-e}_{W_\eta} = \frac{\rho}{N_e \, \Delta t} \hbar \omega_\eta \left(C^+_\eta - C^-_\eta \right), \tag{5.103}$$

similar to the procedure adopted for the simulation of other semiconductors [28,66].

Since for the acoustic transition rate we have assumed the elastic approximation, it follows that $C^{p-e}_{W_{ac}} = 0$.

5.7.3 Numerical Results

Crystal heating due to an electron flow is considered rather important in graphene. It can influence the properties related to the electrical characteristics but can also create hot spots with rather high temperatures and the possibility of damaging the material. The main aim of the realized simulations is to evaluate the heating rate of a monolayer graphene under a constant electric field.

For the DSMC part, $N_e = 10^4$ particles have been used. The time step is set as $\Delta t = 2.5$ fs and a constant phonon relaxation time has been adopted, $\tau = 5$ ps, which is a value already used in the literature, also in consideration of the fact that only the phonons that more strongly interact with electrons, having as said an almost constant frequency, are significantly brought out of local equilibrium. Here we are neglecting the presence of defects, for a study of their influence on the phonon mean free path, see [67]. In [53], a temperature-dependent phonon relaxation time has been adopted but it seems to be valid only at low electric fields. More accurate expressions are still lacking. Several Fermi energies have been considered in order to investigate the dependence of the temperature increasing not only on the applied field but also on the electron density. We have also performed a comparison with the case when all phonons are kept in equilibrium at the room temperature in order to analyze the influence of the crystal heating on the characteristic curves.

First, we analyze the case with $\mathcal{E}_F = 0.3$ eV by considering several applied fields. In Figure 5.9, the evolution of the temperatures of each type of phonons along with T_{LE} is plotted. The latter can be identified as the temperature of the crystal lattice and can therefore be directly related to the measurements. The most energetic phonons are the optical ones, while the least energetic phonons are the acoustic ones. One observes that in the first 5 ps the rise in temperature increases, as expected, with the electric field. Roughly, it seems that the maximum T_{LE} after 5 ps depends on the electric field in a linear way. The temperature variation when $E = 20$ kV/cm is about 100 K, a relevant effect that should require a particular attention

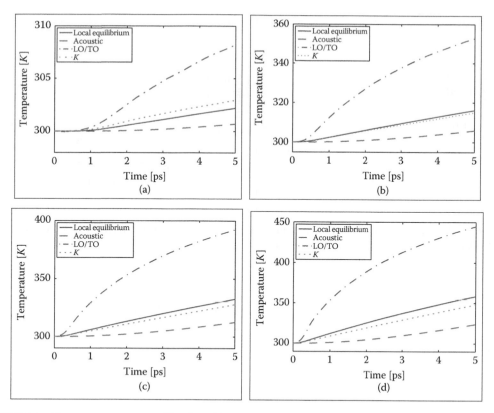

FIGURE 5.9 Phonon temperatures and local equilibrium temperature T_{LE} versus time in the case $\varepsilon_F = 0.3$ eV, when $E = 0.5$ kV/cm (a), $E = 1$ kV/cm (b), $E = 10$ kV/cm (c), and $E = 20$ kV/cm (d).

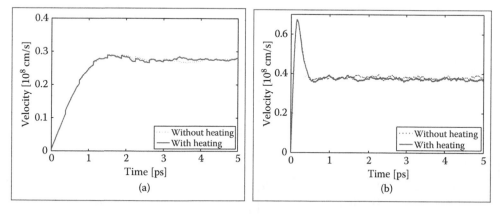

FIGURE 5.10 Average electron velocity versus time in the case $\varepsilon_F = 0.3$ eV, when $E = 1$ kV/cm (a) and $E = 20$ kV/cm (b).

in the design of future electron devices with graphene. In order to study also the influence of the heating effects on the electron transport, Figures 5.10 and 5.11 show a comparison of the average electron velocity and energy in the case when the crystal heating is considered and in the case when the phonons are at room temperature. The differences in the average electron energy and velocity are small up to fields of 20 kV/cm.

The previous cases have also been simulated with a higher Fermi energy $\mathcal{E}_F = 0.6$ eV. Now the heating effects are more evident due to the higher electron current and a consequent greater number of

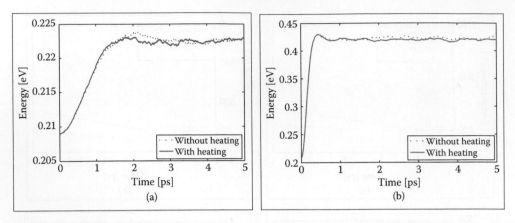

FIGURE 5.11 Average electron energy versus time in the case $\varepsilon_F = 0.3$ eV, when $E = 1$ kV/cm (a) and $E = 20$ kV/cm (b).

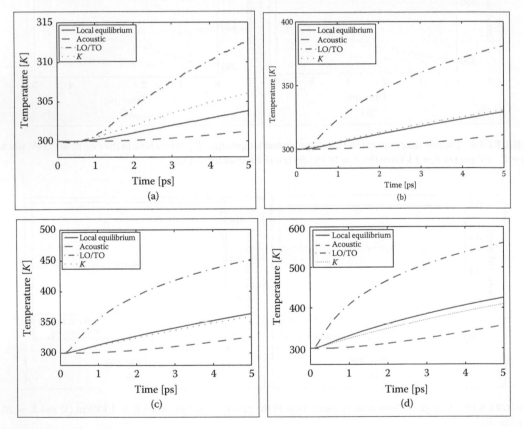

FIGURE 5.12 Phonon temperatures and local equilibrium temperature T_{LE} versus time in the case $\varepsilon_F = 0.6$ eV, when $E = 0.5$ kV/cm (a), $E = 1$ kV/cm (b), $E = 10$ kV/cm (c), and $E = 20$ kV/cm (d).

electron–phonon scatterings that transfer more energy to the lattice because the emission processes are dominant with respect to the absorption ones. The results are plotted in Figures 5.12 through 5.14 and show a qualitative trend similar to those with $\mathcal{E}_F = 0.3$. Eventually, in Figure 5.15 it is possible to notice that the effects of crystal heating on electron average energy and velocity become considerable in the presence of high electric fields.

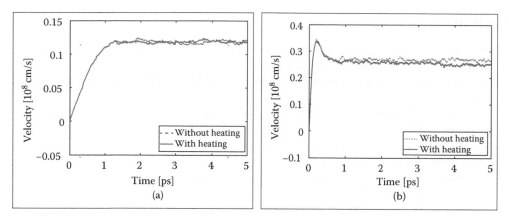

FIGURE 5.13 Average electron velocity versus time in the case $\varepsilon_F = 0.6$ eV, when $E = 1$ kV/cm (a) and $E = 20$ kV/cm (b).

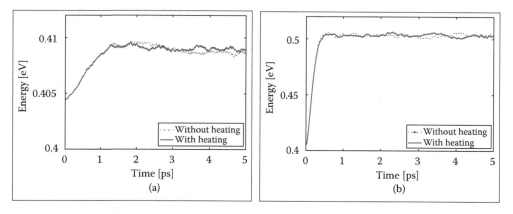

FIGURE 5.14 Average electron energy versus time in the case $\varepsilon_F = 0.6$ eV, when $E = 1$ kV/cm (a) and $E = 20$ kV/cm (b).

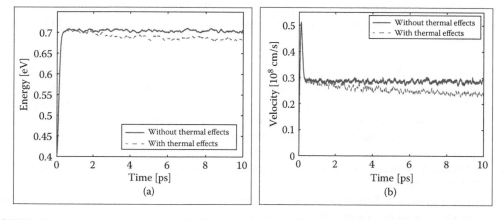

FIGURE 5.15 Average electron energy and velocity versus time in the case $\varepsilon_F = 0.6$ eV, when $E = 50$ kV/cm.

From the analysis of the simulations, it appears that the thermal effects significantly affect the electrical characteristics at high electric fields and cause a considerable rise of the lattice temperature even at medium electric fields and therefore must be taken into account if graphene is employed as semiconductor material in electron devices.

References

1. G.K. Wachutka. Rigorous thermodynamic treatment of heat generation and conduction in semi-conductor device modeling, *IEEE Trans. Comput. Aided Design*, **9**, 1141–1149, (1990).

2. D. Jou, J. Casas-Vazquez, G. Lebon. *Extended Irreversible Thermodynamics*. Berlin: Springer-Verlag, (1993).

3. I. Müller, T. Ruggeri. *Rational Extended Thermodynamics*. Berlin/Heidelberg/New York, NY: Springer, (1998).

4. G. Mascali. A hydrodynamical model for silicon semiconductors including crystal heating, *Eur. J. Appl. Math.*, **26**, 477–496, (2015).

5. G. Mascali. A new formula for thermal conductivity based on a hierarchy of hydrodynamical models, *J. Stat. Phys.*, **163**, 1268–1284, (2016).

6. G. Mascali, V. Romano. Simulation of Gunn oscillations with a non-parabolic hydrodynamical model based on the maximum entropy principle, *COMPEL*, **24**(1), 35–54, (2005).

7. G. Mascali, V. Romano. A hydrodynamical model for hole transport in silicon semiconductors: The case of non-parabolic warped bands, *Math. Comp. Mod.*, **53**(1–2), 213–229, (2011).

8. G. Mascali, V. Romano. A hydrodynamical model for hole transport in silicon semiconductors: The case of parabolic warped bands, *COMPEL*, **31**(2), 552–582, (2012).

9. V. Romano. Non parabolic band transport in semiconductors: Closure of the production terms in the moment equations, *Cont. Mech. Thermodyn.*, **12**, 31, (2000).

10. V. Romano. Non parabolic band hydrodynamical model of silicon semiconductors and simulation of electron devices, *Math. Methods Appl. Sci.*, **24**, 439–471, (2001).

11. J.M. Ziman. *Electrons and Phonons: The Theory of Transport Phenomena in Solids*. Oxford: Clarendon Press, (1960).

12. N.W. Ashcroft, N.D. Mermin. *Solid State Physics*. Philadelphia, PA: Sounders College Publishing International Edition, (1976).

13. L.D. Landau, E.M. Lifschitz. *Statistical Physics,* part 1, Vol. 5, Oxford: Pergamon Press, (1980).

14. C. Kittel. *Introduction to Solid State Physics*. New York, NY: John Wiley & Sons, (1996).

15. W.J. Dreyer, H. Struchtrup. Heat pulse experiment revisited, *Continuum Mech. Therm.*, **5**, 3, (1993).

16. Z. Aksamija, U. Ravaioli. Anharmonic decay of g-process longitudinal optical phonons in silicon, *Appl. Phys. Lett.*, **96**, 091911, (2010).

17. P.G. Klemens. Anharmonic decay of optical phonons, *Phys. Rev.*, **148**, 845, (1966).

18. J. Callaway. Model for lattice thermal conductivity at low temperatures, *Phys. Rev.*, **113** (4), 1046, (1959).

19. C. Jacoboni. *Theory of Electron Transport in Semiconductors*. Heidelberg: Springer, (2010).

20. G. Alí, G. Mascali, V. Romano, C.R. Torcasio. A hydrodynamic model for covalent semiconductors with a generalized energy dispersion relation, *Eur. J. Appl. Math.*, **25** (2), 255–276, (2014).

21. E. Pop, R.W. Dutton, K.E. Goodson. Analytic band Monte Carlo model for electron transport in Si including acoustic and optical phonon dispersion, *J. App. Phys.*, **96**, 4998–5005, (2004).

22. C. Auer, F. Schürrer, W. Koller. A semicontinuous formulation of the Bloch–Boltzmann–Peierls equations, *SIAM J. Appl. Math.*, **64**, 1457, (2004).

23. G. Mascali, V. Romano. Charge transport in graphene including thermal effects, *SIAM J. Appl. Math.*, **77**(2), 593, (2017).

24. G. Alí, G. Mascali, V. Romano, C.R. Torcasio. A hydrodynamical model for covalent semiconductors, with applications to GaN and SiC, *Acta Appl. Math.*, **122** (1), 335–348, (2012).

25. P. González, J.A. Carrillo, F. Gámiz, F., Deterministic Numerical Simulation of 1d kinetic descriptions of Bipolar Electron Devices. In A.M. Anile, G. Ali, G. Mascali (Eds.), *Scientific Computing in Electrical Engineering*, Series: Mathematics in Industry, Subseries: The European Consortium for Mathematics in Industry, Vol. 9, pp. 339–344, Berlin: Springer, (2006).

26. J. Piprek. *Semiconductor Optoelectronic Devices: Introduction to Physics and Simulation*. Cambridge, MA: Academic Press, (2013).

27. S.L. Chuang. *Physics of Optoelectronics Devices*. New York, NY: John Wiley & Sons, (1995).

28. E. Pop, S. Sinha, K.E. Goodson. Heat generation and transport in nanometer-scale transistors, *Proc. IEEE*, **94**, 1587–160, (2006).

29. S. Selberherr. *Analysis and Simulation of Semiconductor Devices*. Vienna: Springer-Verlag, (1984).

30. G. Mascali. Maximum entropy principle in relativistic radiation hydrodynamics II: Compton and double Compton scattering, *Continuum Mech. Thermodyn.*, **14**(6), 549, (2002).

31. E.T. Jaynes. Information theory and statistical mechanics, *Phys. Rev.*, **106**, 620, (1957).

32. C.D. Levermore. Moment closure hierarchies for kinetic theories, *J. Stat. Phys.*, **83**, 1021–1167, (1996).

33. K. Wang. *Statistical Mechanics*. New York, NY: John Wiley & Sons, (1987).

34. A.M. Anile, G. Mascali. Theoretical foundations for tail electron hydrodynamical models in semiconductors, *Appl. Math. Lett.*, **14** (2), 245–252, (2001).

35. A.M. Anile, V. Romano. Non parabolic band transport in semiconductors: Closure of the moment equations, *Continuum Mech. Thermodyn.*, **11**, 307, (1999).

36. S. La Rosa, G. Mascali, V. Romano. Exact maximum entropy closure of the hydrodynamical model for Si semiconductors: The 8-moment case, *SIAM J. Appl. Math.*, **70**, 710, (2009).

37. O. Muscato, V. Di Stefano. Hydrodynamic modeling of the electro-thermal transport in silicon semiconductors, *J. Phys. A: Math. Theor.*, **44**, 105501, (2011).

38. O. Muscato, V. Di Stefano. Local equilibrium and off-equilibrium thermoelectric effects in silicon semiconductors, *J. Appl. Phys.*, **110**(9), 093706, (2011).

39. V. Romano. Quantum corrections to the semiclassical hydrodynamical model of semiconductors based on the maximum entropy principle, *J. Math. Phys.*, **48**, 123504, (2007).

40. M. Brunk, A. Jungel, Numerical coupling of electric circuit equations and energy-transport models for semiconductors, *SIAM J. Sci. Comput.*, **30**, 873–894, (2008).

41. V. Romano. 2D simulation of a silicon MESFET with a nonparabolic hydrodynamical model based on the maximum entropy principle, *J. Comput. Phys.*, **176**, 7092, (2002).

42. V. Romano, A. Rusakov. 2d numerical simulations of an electron–phonon hydrodynamical model based on the maximum entropy principle, *Comput. Methods Appl. Mech. Eng.*, **199**, 2741–2751, (2010).

43. V. Romano, M. Zwierz. Electron–phonon hydrodynamical model for semiconductors, *Z. Angew. Math. Phys.*, **61**, 1111–1131, (2010).

44. V. Romano. 2D numerical simulation of the MEP energy-transport model with a finite difference scheme, *J. Comp. Phys.*, **221**, 439–468, (2007).

45. N. Ben Abdallah, P. Degond, S. Genieys. An energy-transport model for semiconductors derived from the Boltzmann equation, *J. Stat. Phys.*, **84**, 205–231 (1996).

46. P. Degond, A. Juengel, P. Pietra. Numerical discretization of energy-transport models for semiconductors with nonparabolic band structure, *SIAM J. Sci. Comput.*, **22**, 986–1007, (2000).

47. G.I. Marchuk, Splitting and alternating direction method. In *Handbook of Numerical Analysis*, Vol. 1, P. G. Ciarlet, J. L. Lions (eds.), North-Holland, Amsterdam: Wiley, pp. 197–462, (1990).

48. H. Zhao, Q. Guo, F. Xia, H. Wang. Two-dimensional materials for nanophotonics application, *Nanophotonics.*, **4**, 28–142, (2015).

49. M. Coco, G. Mascali, V. Romano. Monte Carlo analysis of thermal effects in monolayer graphene, *J. Comput. Theor. Transport*, 45 (7), 540—553, (2016).

50. L. Barletti. Hydrodynamic equations for electrons in graphene obtained from the maximum entropy principle, *J. Math. Phys.*, **55** (8), 083303, (2014).

51. A.H. Castro Neto, F. Guinea, N.M.R. Peres, K.S. Novoselov, A.K. Geim. The electronic properties of graphene, *Rev. Modern Phys.*, **81**, 109, (2009).

52. M. Coco, A. Majorana, G. Mascali, V. Romano. Comparing kinetic and hydrodynamical models for electron transport in monolayer graphene, *Proceedings of the 6th International Conference on Coupled Problems in Science and Engineering*, B. Schrefler, E. Oñate, M. Papadrakis (eds.), Venice: International Center for Numerical Methods in Engineering (CIMNE), 1003–1014, (2015).

53. P. Lichtenberger, O. Morandi, F. Schürrer. High-field transport and optical phonon scattering in graphene, *Phys. Rev. B*, **84**, 045406, (2011).

54. O. Morandi, L. Barletti. Particle dynamics in graphene: Collimated beam limit, *J. Comput. Theor. Transport.*, **43**(1–7), 418–432, (2015).

55. R. Rengel, C. Couso, M.J. Martin. A Monte Carlo study of electron transport in suspended monolayer graphene, *Spanish Conference on Electron Devices (CDE)*, Valladolid, Spain, IEEEXplore, (2013).

56. K.M. Borysenko, J.T. Mullen, E.A. Barry, S. Paul, Y.G. Semenov, J.M. Zavada, M. Buongiorno Nardelli, K.W. Kim. First-principles analysis of electron–phonon interactions in graphene, *Phys. Rev. B*, **11**, 121412(R), (2010).

57. E.N. Koukaras, G. Kalosakas, C. Galiotis, P. Konstantinos. Phonon properties of graphene derived from molecular dynamics simulations, *Sci. Rep.*, **5**, 12923, (2015).

58. D.L. Nika, A.A. Balandin. Two-dimensional phonon transport in graphene, *J. Phys. Condens. Matter*, **24**, 233203, (2012).

59. X. Li, E.A. Barry, J.M. Zavada, M. Buongiorno Nardelli, K.W. Kim. Surface polar phonon dominated electron transport in graphene, *Appl. Phys. Lett.*, **97**, 232105, (2010).

60. V.D. Camiola, G. Mascali, V. Romano. Numerical simulation of a double-gate MOSFET with a subband model for semiconductors based on the maximum entropy principle, *Continuum Mech. Therm.*, **24** (4–6), 417, (2012).

61. V.D. Camiola, V. Romano. Hydrodynamical model for charge transport in graphene, *J. Stat. Phys.*, **157**, 1114–1137, (2014).

62. G. Mascali, V. Romano. A comprehensive hydrodynamical model for charge transport in graphene, *IWCE-2014*, IEEE, Paris, June 3–6, (2016).

63. O. Morandi, F. Schürrer. Wigner model for quantum transport in graphene, *J. Phys. A: Math. Theor.*, **44**, 265301, (2011).

64. P. Lugli, D.K. Ferry. Degeneracy in the ensemble Monte Carlo method for high-field transport in semiconductors, *IEEE Trans. Elect. Dev.*, **ED-32** (11), 2431–2437, (1985).

65. V. Romano, A. Majorana, M. Coco. DSMC method consistent with the Pauli exclusion principle and comparison with deterministic solutions for charge transport in graphene, *J. Comp. Phys.*, **302**, 267–284, (2015).

66. O. Muscato, V. Di Stefano, W. Wagner. A variance-reduced electrothermal Monte Carlo method for semiconductor device simulation, *Comput. Math. Appl.*, **65**: 520–527, (2013).

67. T. Feng, X. Ruan, Z. Ye, B. Cao. Spectral phonon mean free path and thermal conductivity accumulation in defected graphene: The effects of defect type and concentration, *Phys. Rev. B*, 91, 224301-01-12, (2015).

6

Process Simulation

Simon Z. M. Li

Changsheng Xia

and

Yue Fu

6.1 Introduction

With further device simulation in mind, device process simulation aims to predict the fabrication process outcome and create the simulation mesh necessary for two-dimensional (2D) or three-dimensional (3D) simulation. The combination of process and device simulation for semiconductor technology modeling is often termed technology computer-aided design (TCAD). The pioneering work in semiconductor process simulation was the SUPREM code created by Stanford University in the 1980s and early 1990s (Hansen and Deal, 1994), mainly concerning silicon process. Commercial versions of SUPREM up to this date are still used. Compound semiconductors are the main materials used in semiconductor optoelectronics applications, and this is the focus of this chapter. Since silicon photonics (in combination with compounds) is an important research subject, we shall describe the basic theories and principles related to silicon as well. It is hoped that the elaborate theories and methods built for silicon will find use in compound semiconductors.

This chapter is organized as follows. Modeling of compound semiconductor epitaxial growth by metalorganic chemical vapor deposition (MOCVD) is described in detail. This is followed by a brief summary of implantation, diffusion, and oxidation as commonly used for silicon processes.

6.2 MOCVD Reactor Simulation

MOCVD is widely used in the epitaxial growth of compound semiconductors such as Gallium nitride (GaN) and related ternaries and quaternaries. Simulation of the MOCVD reactor not only allows understanding of the chemical reactions involved but also offers a convenient computer-aided design (CAD) tool for process optimization. The models and theories in this section have been implemented in the MOCVD reactor simulation program PROCOM from Crosslight Software Inc. (Crosslight Software Inc., 2015) and the results are generated from PROCOM simulation.

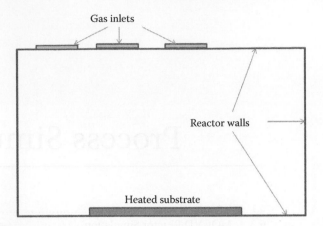

FIGURE 6.1 Schematic of a metalorganic chemical vapor deposition (MOVCD) reactor being simulated. MOVCD, metalorganic chemical vapor deposition.

We consider a simulation domain consisting of 2D or 3D space with gas, surrounded by metal walls. Gas inlets and outlets, and heated substrates are special boundary conditions. The general MOCVD reactor simulation domain is indicated in Figure 6.1. The reactor model aims to model the gas chemical reactions that result in deposition of semiconductor thin films.

6.2.1 Basic Transport Equations

For a reactor simulation program, detailed chemical kinetics and mass/heat transport models can be used to predict the growth rate, composition distribution, thickness uniformity, dopant incorporation, and defect distribution of the semiconductor films if the reactor geometry, chemical species, and growth condition parameters are given.

The processes in an MOCVD reactor belong to a special type of reactive flow: chemical species react with one another in different phases during mass transport, which has been described in the book by Oran and Boris (2001). The basic partial differential equations for the reactive flow processes in the MOCVD reactor are (Oran and Boris, 2001)

$$\frac{\partial \rho}{\partial t} = -\nabla \cdot (\rho v) \tag{6.1}$$

$$\frac{\partial n_i}{\partial t} = -\nabla \cdot \left(n_i v\right) - \nabla \cdot \left(n_i v_{di}\right) + Q_i, \quad i = 1, ..., N_s \tag{6.2}$$

$$\frac{\partial \rho v}{\partial t} = -\nabla \cdot (\rho v v) - \nabla \cdot \mathrm{P} \tag{6.3}$$

$$\frac{\partial E}{\partial t} = -\nabla \cdot (Ev) - \nabla \cdot (v \cdot \mathrm{P}) - \nabla \cdot \left(q + q_r\right), \tag{6.4}$$

which represent conservation laws of the total mass, individual species, momentum, and energy, respectively. The basic physical variables are explained in Table 6.1.

In the reactive flow processes, the motion of a species i is usually divided into convective (v) and diffusive (v_{di}) terms:

$$v_i = v + v_{di}. \tag{6.5}$$

TABLE 6.1 Definitions of Basic Variables

Symbol	Definition
E	Total energy density (erg/cm^{-3})
n_i	Number density of species i (cm^{-3})
v	Fluid velocity (cm/s)
N_s	Number of chemical species present
P	Pressure tensor (dynes/cm^2)
Q_i	Chemical production rate of species i (cm^{-3}/s)
q	Heat flux (erg/cm^2/s)
q_r	Radiative heat flux (erg/cm^2/s)
v_{di}	Diffusion velocity of species i (cm/s)
ρ	Fluid mass density (g/cm^3)

The convection velocity is the mass averaged velocity (or the fluid velocity):

$$v = \frac{\sum_i \rho_i v_i}{\rho},$$ (6.6)

where ρ_i is the mass density of individual species i. It should be noted that Equation 6.6 leads to a zero average diffusion velocity:

$$\frac{\sum_i \rho_i v_{di}}{\rho} = 0.$$ (6.7)

So, diffusion is used to describe relative motion of two or more species.

The diffusion velocity between species i and k is related to the change in species number density, pressure gradient, and temperature gradient:

$$\sum_k \frac{n_i n_k}{N D_{ik}} \left(v_{dk} - v_{ik} \right) = \nabla \left(\frac{n_i}{N} \right) - \left(\frac{\rho_i}{\rho} - \frac{n_i}{N} \right) \frac{\nabla P}{P} - K_i^T \frac{\nabla T}{T},$$ (6.8)

where N is the total species number density, P is the scalar pressure, and K_i^T is the thermal diffusion coefficient of species i. The diffusion coefficient D_{ik} of a pair of species (i, k) is estimated by the following formula (Hirschfelder et al., 1954):

$$D_{ik} = \frac{2.628 \times 10^{-3}}{P \sigma_{ik}^2 \Omega_{ik}(T_{ik}^*)} \left[\frac{T^3 \left(m_i + m_k \right)}{2 m_i m_k} \right]^{1/2},$$ (6.9)

where m is the species mass and the $\Omega_{ik}(T_{ik}^*)$ is a collision integral normalized to its rigid sphere value, which in turn depends on the reduced temperature:

$$T_{ik}^* = \frac{k_B T}{\varepsilon_{ik}}.$$ (6.10)

The intermolecular parameters σ_{ik} and ε_{ik} can be approximated with collision diameter (σ) and potential parameter (ε) of individual species within the Lennard-Jones potential model:

$$\sigma_{ik} = \frac{1}{2} \left(\sigma_i + \sigma_k \right), \quad \varepsilon_{ik} = \sqrt{\varepsilon_i \varepsilon_k}.$$ (6.11)

TABLE 6.2   Some Commonly Used Species

Gas Phase	Surface Phase	Bulk Phase
H_2	$H<s>$	$InP<d>$
N_2	$In<s>$	$InN<d>$
NH_3	$As<s>$	$AlN<d>$
PH_2	$SV<s>$	$GaP<d>$
PH_3	$AlN<s>$	$GaN<d>$
AsH	$PH_2<s>$	$InAs<d>$
AsH_2	$AsH<s>$	$GaAs<d>$
$InCH_3$	$GaN<s>$	
$GaCH_3$	$CH_3<s>$	
$Ga(CH_3)_3$	$InCH_3<s>$	
$In(CH_3)_3$	$AlCH_3<s>$	
$TMG*NH_3$	$GaCH_3<s>$	
$DMG*NH_2$		
$(DMG*NH_2)_3$		

The pressure tensor ($\mathbf{P}$) is expressed in terms of scalar pressure, shear viscosity, and bulk viscosity (Oran and Boris, 2001):

$$\mathbf{P} = P(N, T)\,\mathbf{I} - \mu_m \left[\nabla v + (\nabla v)^T\right] + \left(\frac{2}{3}\mu_m - \eta\right)(\nabla \cdot v)\,\mathbf{I}, \tag{6.12}$$

where I is the unit tensor, μ_m is the coefficient of shear viscosity for a mixture of species, and η is the bulk viscosity coefficient. The last two terms are the stress tensors which are used to include the diffusive effects of viscosity.

The heat flux can be written as (Oran and Boris, 2001)

$$q = -\lambda_m \nabla T + \sum_i \rho_i h_i v_{\mathrm{di}} + P \sum_i K_i^T v_{di}, \tag{6.13}$$

where λ_m is the thermal conductivity of the species mixture and h_i is the specific enthalpy of species i.

The total energy E consists of the kinetic energy and the internal energy (Oran and Boris, 2001):

$$E = \frac{1}{2}\rho v^2 + E_{\mathrm{int}}, \tag{6.14}$$

where E_{int} is the internal energy, which can be achieved using the temperature dependent enthalpy:

$$E_{\mathrm{int}} = \sum_i \rho_i h_i(T) - P. \tag{6.15}$$

The enthalpy as a function of temperature can be found in many thermal dynamic tables. The reactor simulator, PROCOM, tabulates the enthalpy function for many species in its species macro library so that the simulation program can use it directly. Some commonly used species are listed in Table 6.2.

6.2.2 Chemical Reactions

A full simulation of the MOCVD reactor requires an accurate description of the detailed chemical reactions which are characterized by the chemical production rate Q_i for species i. Consider N_s species involved in

N_r reactions as follows:

$$\sum_{i=1}^{N_s} v'_{ji} S_i \Leftrightarrow \sum_{i=1}^{N_s} v''_{ji} S_i \; (j = 1, ..., N_r), \tag{6.16}$$

where the left-hand side represents reagents and the right-hand side products for either reversible or irreversible reactions. S_is are the reactant and resultant in every reaction process, and v_{ji} parameters are the mole numbers before them. The production rate Q_i can be expressed in terms of the rate of progress r_j for the jth reaction:

$$Q_i = \sum_{j=1}^{N_r} \left(v''_{ji} - v'_{ji} \right) r_j, \tag{6.17}$$

where

$$r_j = k_j^f \prod_{k=1}^{N_s} [S_k]^{v'_{jk}} - k_j^r \prod_{k=1}^{N_s} [S_k]^{v''_{jk}}, \tag{6.18}$$

which is signed such that an excess amount of reagents results in a positive production rate. k_j^f and k_j^r are the forward and reverse reaction rates for the jth reaction, respectively.

The forward reaction rate is usually expressed in the form of a modified Arrhenius function:

$$k_j^f = A_j T^{B_j} \exp \left[\left(-E_j * 4.184 * 1000 \big/ R \right) \big/ T \right] (j = 1, ..., N_r), \tag{6.19}$$

where the parameters A, B, and E should be provided from experiments or theoretical calculations. The unit for E is kcal/mol. R is the mole gas constant. *In reactor simulation involving gas reactions, the three numbers A, B, and E are appended to the end of the chemical reaction formulas.*

For example, the gas phase reactions for GaN growth are written as:

$Ga(CH_3)_3 = Ga(CH_3)_2 + CH_3$	3.5e+15	0	5.95e+4
$Ga(CH_3)_3 + NH_3 => TMG*NH_3$	1.e+12	0	0
$TMG*NH_3 => Ga(CH_3)_3 + NH_3$	9.5e+9	0.	19
$TMG*NH_3 => DMG*NH_2 + CH_4$	1.e+13	0	32
$3 DMG*NH_2 => (DMG*NH_2)_3$	1.e+21	0	0
$(DMG*NH_2)_3 => 3GaN<g> + 6CH_4$	4.e+15	0	60

For pyrolysis of $TMG*NH_3$ reaction process, all v_{ji} parameters are 1, and S_is are $Ga(CH_3)_3$ and CH_3. The last three numbers are the parameters A, B, and E used in Equation 6.19 to determine the gas phase reaction rate. Figure 6.2 shows the reaction rate of every gas phase reaction for GaN growth as a function of temperature, determined by Equation 6.19. The results show that $TMG*NH_3$ tends to decompose into $Ga(CH_3)_3$ and $DMG*NH_2$ at high temperature.

For reversible reactions, the reverse reaction rate is related to the forward rate by the equilibrium constant K_j^e:

$$k_j^r = \frac{k_j^f}{K_j^e}, \tag{6.20}$$

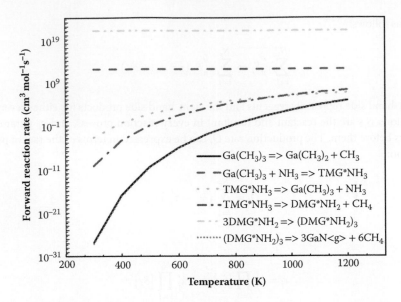

FIGURE 6.2   Reaction rates of every gas phase reaction for GaN growth as a function of temperature.

where the equilibrium constant may be obtained by setting the progress rate in Equation 6.18 to zero under the equilibrium condition (labeled with superscript 0):

$$K_j^e = \prod_{k=1}^{N_s} \left[S_k^0 \right]^{\nu_{jk}'' - \nu_{jk}'} \tag{6.21}$$

By elementary thermodynamics, the equilibrium constant can be calculated by the Gibbs energies:

$$\ln \left(K_j^e \right) = -\frac{1}{RT} \sum_{k=1}^{N_s} \left(\nu_{jk}'' - \nu_{jk}' \right) G_k^0 (T), \tag{6.22}$$

where $G_k(T)$ is the Gibbs energy of pure state species k at standard pressure P_0.

6.2.3 Surface–Gas Reactions and Film Deposition

MOCVD works by letting the chemicals in gas phase be introduced into the reactor chamber to react with one another. The purpose is to allow the gas phase to interact with the substrate to produce solid or bulk deposits of the desirable compound semiconductors. In addition to gas reactions, surface reactions between the gas species and surface vacancy are indispensable to generate the bulk material. Moreover, these reactions are usually irreversible reactions, which can be divided into four different types:

1. Adsorption reaction: A gas species collides with one or two free surface sites forming the adsorbed surface species, and, in the case of dissociative adsorption, releasing gas-phase species as well. Adsorption reaction has the form
   $$CH_3 \; + \; SV<s> \;\; => \;\; CH_3<s>,$$
   where SV denotes a surface vacancy and <s> labels a surface species.
2. Abstraction reaction: A gas species reacts with a surface species leading to the release of gas-phase species. Abstract reaction has the form
   $$CH_3 \; + \; H<s> \;\; => \;\; CH_4 \; + \; SV<s>.$$

3. Recombination reaction: Two surface species react, leading to the formation of gas-phase species. The recombination reaction has the form

$$H<s> \quad + \quad CH_3<s> \quad => \quad CH_4 \quad + \quad 2SV<s>.$$

4. Desorption reaction: A surface or bulk species becomes a gas species. It has the form

$$CH_3<s> \quad => \quad CH_3 \quad + \quad SV<s>.$$

Viscosity and thermal conductivity are very important parameters for the gases and their mixtures in MOCVD. For the most common gases such as H_2, He, and N_2, for which the experimental data on transport properties as a function of temperature are available, the viscosity μ_i and thermal conductivity λ_i usually are expressed in the following form:

$$\lambda_i = A_i T^{b_i}, \quad \mu_i = A_i T^{b_i}, \tag{6.23}$$

where A_i and b_i are parameters determined from fitting the experimental results. Table 6.3 lists the parameters we obtained for H_2, He, N_2, and Ar gases.

However, for the other gas species for which experimental data are not available, we can evaluate the viscosity of the *i*th species by the kinetic theory (Hirschfelder et al., 1954):

$$\mu_i = 8.441 \times 10^{-5} \frac{(m_i T)^{1/2}}{\sigma_i^2 \Omega (T_i^*)}, \tag{6.24}$$

and its thermal conductivity is determined by

$$\lambda_i = \left[\frac{15}{4} + 1.32 \left(\frac{c_p m_i}{R} - \frac{5}{2} \right) \right] \frac{R}{m_i} \mu_i, \tag{6.25}$$

where c_p is the heat capacity.

For gas mixtures, the following relations are usually used (Wilke, 1950):

$$\mu_m = \sum_i^N \left(\frac{f_i \mu_i}{\sum_j f_j \Phi_{ij}} \right), \tag{6.26}$$

where f is the mole fraction of species, and

$$\Phi_{ij} = \frac{1}{\sqrt{8}} \left(1 + \frac{m_i}{m_j} \right)^{-1/2} \left[1 + \left(\frac{\mu_i}{\mu_j} \right)^{1/2} \left(\frac{m_j}{m_i} \right)^{1/4} \right]^2. \tag{6.27}$$

A similar relation for the thermal conductivity of gas mixture can be used.

TABLE 6.3   Parameters for the Conductivity and Viscosity of H_2, He, N_2, and Ar

| | Conductivity (W/m · k) | Viscosity (Kg/m · s) |
	A_i b_i	A_i b_i
H_2	1.256e−3 0.8701	1.88e−7 0.676
He	3.96e−3 0.64	5.09e−7 0.64
N_2	2.11e−4 0.836	1.99e−7 0.78
Ar	1.65e−4 0.814	2.14e−7 0.809

6.2.4 Example of InGaN Multiple Quantum Well Growth Stimulation

InGaN multiple quantum well (MQW) light-emitting diodes (LEDs) have attracted considerable attention for general lighting applications and hold promise for replacing the conventional incandescent and fluorescent lamps. However, there are still some problems, which have a significant influence on further improvement of the optical and electrical performances of these LEDs such as film thickness uniformity, epitaxy quality, composition, and impurity distribution. That is to say, growth process decides the performances of the LEDs to some extent. In this section, InGaN quantum well MOCVD growth in a vertical reactor is simulated. The full chemistry model is used to describe the chemical reaction inside the reactor. The temperature, velocity, composition, and mass transport properties are obtained by solving the coupled equations described above.

In the simulation, the vertical reactor used is cylindrical. To take advantage of this symmetry, only half of the reactor is simulated, and the other half is symmetric to the simulated one with the center line being the symmetric axis. A snapshot of the simulated vertical reactor with grid mesh is given in Figure 6.3. In this reactor model, the gas inlet is at the top, the bottom right is the outlet, and the substrate is at the bottom left. The left edge is the symmetric axis and the right edge is set to be the hot wall. The widths for the inlet, substrate, and outlet are 4.98, 3.15, and 1.63 cm, respectively.

In order to grow InGaN quantum wells, the species and their flux which enter the reactor from the inlet at different times should be controlled. This can be done by defining the inlet boundary conditions of the reactor. In this example, the species N_2, NH_3, $Ga(CH_3)_3$, and $In(CH_3)_3$ with the flux of 1359.483 sccm, 453.161 sccm, 1.631 sccm, and 0.05 sccm, respectively, are used to grow InGaN barriers. When growing InGaN wells, the flux of $In(CH_3)_3$ is increased to 0.725 sccm. The growth time for each barrier is set to 20 seconds and the growth time for each well is set to 10 seconds. The simulation starts with a well-mixed gas solution as an initial solution followed by a time-dependent solution of all the equations described previously. By changing the inlet conditions with increased growth time, after 170 seconds, the InGaN MQW structure with five wells and six barriers can be deposited. The detailed input files for the reactor structure, the boundary conditions, chemical reactions, and numerical calculation parameters of this example can be found on the Crosslight website (Crosslight Software Inc., 2015).

Indium composition distribution of the InGaN MQW on the substrate along the X-direction as a function of growth time is shown in Figure 6.4. It can be seen clearly that the indium composition in every barrier and well is very uniform which is mainly relative to the flux of $In(CH_3)_3$ injected into the reactor at different growth times. We also find that the indium composition near the center of the substrate is nearly uniform along the X-direction, but at the edge of the substrate, it is increased. This can be confirmed from

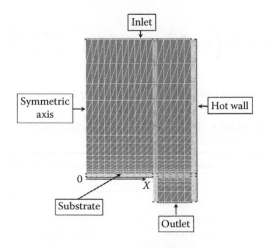

FIGURE 6.3 Snapshot of the simulated vertical reactor with grid mesh.

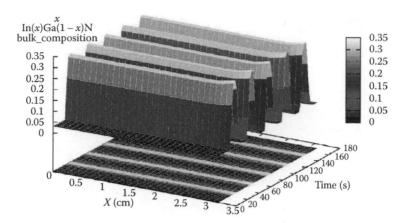

FIGURE 6.4 Indium composition distribution of InGaN multiple quantum wells on the substrate along the *X*-direction as a function of growth time.

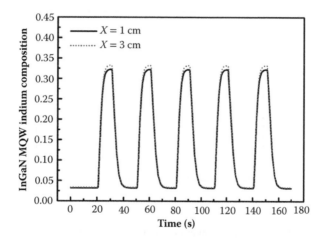

FIGURE 6.5 5 Indium composition distribution of InGaN multiple quantum wells as a function of growth time at *X* = 1 cm and *X* = 3 cm.

Figure 6.5, which shows the indium composition distributions of InGaN MQW as a function of growth time at different positions of the substrate. The indium composition in the barrier is 0.031 and that in the well is 0.32 at *X* = 1 cm. However, at *X* = 3 cm, the indium composition in the barrier is increased to 0.033 and that in the well is increased to 0.33, which indicates that the composition in the wells is more easily affected by the size of the substrate. It should be noted that the indium composition profile at the barrier/well interface is graded and that in the well is curved, which sufficiently reflects the diffusion effects of indium atoms between the barriers and wells. This result implies that graded indium composition should be used for the barrier/well interfaces and the wells in order to get good simulated results in TCAD research of GaN-based LEDs.

Figure 6.6 illustrates the thickness change of InGaN MQW on the substrate along *X*-direction as a function of growth time. It is obvious that the thickness of InGaN MQW is very nonuniform on the substrate and increase markedly from the center of the substrate to its edge. Figure 6.7 shows the thickness of InGaN MQW at *X* = 1 cm and *X* = 3 cm. As shown in this figure, at *x* = 1 cm, the barrier thickness is 11 nm and the well thickness is 8 nm. However, the barrier thickness is 38 nm and the well thickness is 28 nm at *x* = 2 cm. The total thickness of the MQW is increased about 3.4 times from the center of the substrate to its edge.

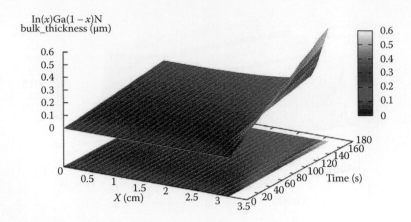

FIGURE 6.6 Thickness of InGaN multiple quantum wells on the substrate along the X-direction as a function of growth time.

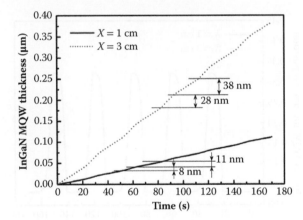

FIGURE 6.7 Thickness of InGaN multiple quantum wells as a function of growth time at $X = 1$ cm and $X = 3$ cm.

The above inhomogeneous distributions of indium composition and thickness on the substrate indicate that most species tend to aggregate at the substrate edge near the outlet. This may be caused by the nonuniform mass transport of the gas species in the reactor. So, in order to grow high-quality InGaN MQW structures on the substrate, it is significantly important to get uniform mass transport of the gas species on the substrate by controlling the flux of injected species, adjusting the pressure and temperature in the reactor, or using a rotating substrate (Cho et al., 1999; Kadinski et al., 2004).

Figures 6.8 through 6.10 illustrate the mass fraction distribution of $Ga(CH_3)_3$, TMG^*NH_3, and DMG^*NH_2 in the reactor, respectively. We can see that the mass fraction of TMG^*NH_3 focused mainly at the gas inlet and those of $Ga(CH_3)_3$ and DMG^*NH_2 are located at the region of the substrate. The results indicate the following gas reaction mechanism: Because the reaction rates are dependent on the temperature, the input gas TMG and NH_3 combine quickly to produce TMG^*NH_3 as soon as they meet at the entrance gas inlet and remain combined. They reach the substrate and then decompose according to the reactions

$$TMG^*NH_3 \implies Ga(CH_3)_3 + NH_3$$
$$TMG^*NH_3 \implies DMG^*NH_2 + CH_4$$

due to the high reaction rates at the high temperature, which indicates the important role of the TMG^*NH_3 adduct in the growth of GaN material.

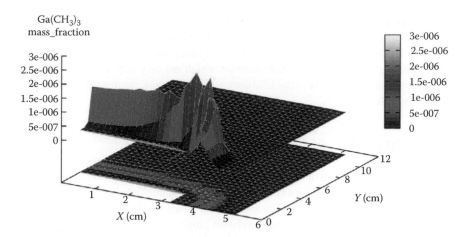

FIGURE 6.8   Mass fraction distribution of $Ga(CH_3)_3$ in the reactor.

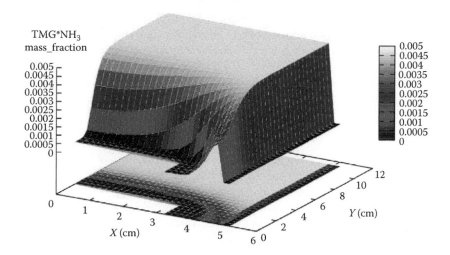

FIGURE 6.9   Mass fraction distribution of TMG^*NH_3 in the reactor.

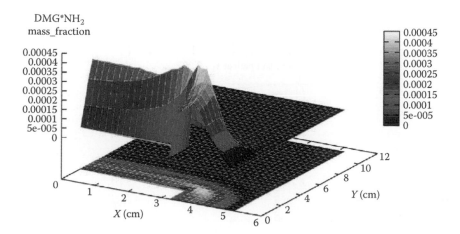

FIGURE 6.10   Mass fraction distribution of DMG^*NH_2 in the reactor.

In conclusion, the InGaN MQW growth process was simulated using PROCM. The thickness and composition distributions of InGaN MQW were analyzed. The process simulation can provide detailed information about the growth process of multiple layers directly such as layer thickness, layer composition, and their uniformity distribution on the substrate, which are very important for the researchers or the producers to optimize the structure of the LED devices and, therefore, improve the optical and electrical performances of these devices.

6.3 Implant Models

Section 6.2 dealt with epitaxial growth based on physical and chemical models. Ion implantation is, however, the most important method of introducing dopants into semiconductors, especially silicon. This section describes the basics of the ion implantation model based on the SUPREM4 code (Hansen and Deal, 1994) as implemented in the CSUPREM simulation program.

There are two approaches to ion implantation modeling. The most accurate approach is based on atomic scale interaction using Monte Carlo simulation techniques. This is usually well documented for amorphous materials with simple layer structure (Srim, 2015). The second approach is based on analytical fit to experimental secondary ion mass spectrometry (SIMS) data (or Monte Carlo data) to build up a large database. Interpolation is used for ion energies not specifically included in the SIMS database. We shall detail the second approach since this is more efficient and more commonly used in practical TCAD projects.

The theoretical basis for the SIMS-interpolation approach is to decompose the implant scattering source S as a product of two functions:

$$S(x_1) = P(d) R(r), \tag{6.28}$$

where x_1 is the entering position (see Figure 6.11 schematic). d is the depth of implant path, and r is the lateral scattering distance. Here we only consider vertical implant entry because any angle in implant may be eliminated by performing a rotation of the coordinate system.

The doping profile at any observation point (x, y) may be expressed as a contribution from lateral scattering of all implant paths coming at the same y-value. That is,

$$D(x, y) = \int P\left(d\left(x_1, y\right)\right) R\left(x_1 - x\right) dx_1. \tag{6.29}$$

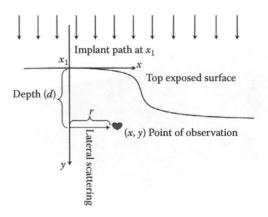

FIGURE 6.11   Schematic for ion implant model.

Please note that the lateral function $R(r)$ is normalized so that in the case of flat exposed surface and uniform material, the doping is simply given by $P(d)$, as expected. The lateral function $R(r)$ has always been assumed to be a simple Gaussian, although some sophisticated process simulator assumes it varies with depth.

Much work in the early days of TCAD focused on how best to use analytical function to construct the depth function $P(d)$ to represent the real SIMS data. Gaussian functions, Pearson functions, and dual-Pearson functions with various parameter settings and sophisticated SIMS fitting strategies have been used.

6.3.1 Gauss Model

Gaussian distribution is the most popular model based on the central moments (Selberherr, 1984). The probability density (or frequency function) $f(x)$ is given by

$$f(x) = \frac{1}{\sqrt{2\pi}\sigma_p} \exp\left[-\frac{\left(x - R_p\right)^2}{2\sigma_p^2}\right],$$ (6.30)

where R_p and σ_p are the central moment characteristic parameters, projected range and standard deviation, respectively:

$$R_p = \int_{-\infty}^{+\infty} x f(x)\, dx$$ (6.31)

$$\sigma_p = \sqrt{\int_{-\infty}^{+\infty} \left(x - R_p\right)^2 f(x)\, dx}.$$ (6.32)

6.3.2 Pearson Model

Another approach followed by Hofker (1975) for fitting a frequency function to experimental data is to use a Pearson type IV distribution function. The whole family of Pearson distributions (Gibbons and Mylroie, 1973) is based on the solutions of the Pearson differential equation:

$$\frac{df}{dy} = \frac{y - a}{b_0 + b_1 y + b_2 y^2}, \quad y = x - R_p.$$ (6.33)

The Pearson coefficients a, b_0, b_1, and b_2 can be determined by the first four characteristic parameters as follows:

$$a = b_1$$ (6.34)

$$b_0 = -\frac{\sigma_p^2 \left(4\beta - 3\gamma^2\right)}{10\beta - 12\gamma^2 - 18}$$ (6.35)

$$b_1 = -\frac{\gamma\sigma_p \left(\beta + 3\right)}{10\beta - 12\gamma^2 - 18}$$ (6.36)

$$b_2 = -\frac{\left(2\beta - 3\gamma^2 - 6\right)}{10\beta - 12\gamma^2 - 18},$$ (6.37)

where γ and β are the other two central moment characteristic parameters, skewness and the kurtosis or excess, respectively:

$$\gamma = \frac{\int_{-\infty}^{+\infty} \left(x - R_p\right)^3 f(x)\, dx}{\sigma_p^3} \tag{6.38}$$

$$\beta = \frac{\int_{-\infty}^{+\infty} \left(x - R_p\right)^4 f(x)\, dx}{\sigma_p^4}. \tag{6.39}$$

The solutions of Equation 6.33 depend on the roots of the equation

$$b_0 + b_1 y + b_2 y^2 = 0, \tag{6.40}$$

which ultimately depend on the values of γ and β. Eight different solutions—Gaussian, Pearson type I, type II, ... and type VII—can be obtained.

If $\gamma = 1$ and $\beta = 3$, we obtain the Gaussian distribution, which is the limited case for all types of Pearson distributions.

If $\gamma \neq 1$ and $\beta < 3 + 1.5\gamma^2$, we get the Pearson type II.

The case when the Equation 6.40 does not have real roots corresponds to Pearson type IV. This case arises when

$$0 < \gamma < 32 \tag{6.41}$$

$$\beta < \frac{39\gamma^2 + 48 + 6\left(\gamma^2 + 4\right)^{3/2}}{32 - \gamma^2}. \tag{6.42}$$

For the description of implantation, only Pearson type IV and type VII distribution generally can be applied. These frequency functions have a single maximum at $y = a$ and decay monotonously to zero on both sides. The type VII distribution is not skewed, which limits its application. The general solution of the Pearson differential Equation 6.33 under the condition 6.41 which characterizes the Pearson IV is given by

$$f(x) = K\left[-\left(b_0 + b_1\left(x - R_p\right) + b_2\left(x - R_p\right)^2\right)\right]^{\frac{1}{2b_2}} \exp\left[-\frac{\frac{b_1}{b_2} + 2a}{\sqrt{4b_2 b_0 - b_1^2}} \operatorname{atan}\left(\frac{2b_2\left(x - R_p\right) + b_1}{\sqrt{4b_2 b_0 - b_1^2}}\right)\right], \tag{6.43}$$

where the normalization constant K is given by

$$\int_{-\infty}^{+\infty} f(x)\, dx = 1. \tag{6.44}$$

If no values for the kurtosis are available, the following universal expression is used:

$$\beta = 2.8 + 2.4\gamma^2. \tag{6.45}$$

6.3.3 Dual Pearson Model

The Pearson type IV involves the first four moments and is thus more accurate for predicting a doping profile in many applications. However, it is not really suitable to predict the typical channeling tail observed when channeling phenomena are present. In this case, the Dual Pearson IV is used,

$$f(x) = N_d \left[(1 - \zeta) f_1 \left(x, R_p, \sigma_p, \gamma_p, \beta_2 \right) + \zeta\, f_2 \left(x, R_p', \sigma_p', \gamma_p', \beta_2' \right) \right], \tag{6.46}$$

where f_1 is the Pearson IV for the surface region and f_2 is the Pearson IV for the bulk region. R_p, σ_p, γ_p, and β_2 are the projected range, standard deviation, skewness, and the kurtosis, respectively, for the first Pearson distribution. Similarly, those with ' are for the second Pearson function distribution, describing the channeling tail. More details can be found in the works by Park et al. (1990) and Yang et al. (1995).

The use of various analytical functions originated from limitations in speed and memory of early micro-computers and these analytical functions were the only way to quickly evaluate the implant dose years ago. Moreover, the sophisticated fitting means very little since there is no atomic physics contained in those Pearson functions. However, with today's faster computers, there is no longer such a need. We can simply smooth out the SIMS data or Monte-Carlo data and import them into process simulation as depth profile. Today's fast computers will handle the interpolation easily.

Many process simulators such as Crosslight's CSUPREM have an option to use an implant table that is built using SIMS data from different energies and for different materials. No fitting of dual-Pearson function is needed and one need not worry whether there is any error from fitting analytical functions.

The above discussion is based on implant through a single material. In processing of semiconductors, it is common to implant ions through different materials. For example, a pad oxide layer is commonly formed on top of silicon before implantation. Since the implant depth profiles of the same ions in oxide and silicon are different, it is necessary to consider the situation of implant through multiple layers of materials with different implant properties. The problem is that, given implant profiles of single materials, how to combine the different profiles to form a composite doping profile for a multiple layer material system. Two different methods commonly used in the TCAD industry for such combination are the dose matching and range matching methods. Interested readers are referred to Li and Fu (2011).

6.4 Dopant Diffusion

A key simulation task in conventional TCAD process simulation is to solve the diffusion equations over the simulation mesh to predict the impurity doping profile after thermal processing. This is an essential task for silicon-based materials; the theoretical background has been detailed in Li and Fu (2011). For compound semiconductor material systems, diffusion is not often used and we shall only give a brief overview of the relevant models here with tips and hints when using commercial process simulators.

Impurity diffusion models are often based on the concept of pair-diffusion. These models not only account for the effect impurity interactions via space charge, but also accurately describe the interaction between impurities and lattice point-defects such as interstitials (I) and vacancies (V). The impurity atoms cannot diffuse by themselves and require neighboring point-defects as a diffusing vehicle. It is common to refer to impurities involved in a point-defect–related diffusion mechanism as dopant-defect pairs. We label a dopant A paired with a vacancy V as an AV pair, and a dopant A paired with interstitial I as an AI pair. So, the diffusivities of dopants are represented by the diffusivities of dopant-defect pairs. Impurity diffusion in process simulation includes several types, which we briefly describe.

6.4.1 Vacancies

Vacancies (V) can diffuse with their own diffusion constants and obey the following diffusion equation (Xia et al., 2007):

$$\frac{\partial C_V}{\partial t} = \vec{\nabla} \cdot \left(-\vec{J}_V - \sum_{imp} \vec{J}_{AV} \right) - R, \tag{6.47}$$

where C_V is the vacancy concentration, and $\vec{J}_{AV}$ is the flux of the dopant-defect pair AV described in a subsequent section. $\vec{J}_V$ is the unpaired vacancy flux, which accounts correctly for the effect of an electric field on the charged portion of the defect concentration and can be written as (Xia et al., 2007):

$$-\vec{J}_V = D_V C_V^* \vec{\nabla} \frac{C_V}{C_V^*}. \tag{6.48}$$

C_V^* is the equilibrium vacancy concentration. R is the recombination of vacancies representing simple interaction between interstitials and vacancies, which can be expressed as

$$R = K_R \left(C_I C_V - C_I^* C_V^* \right), \tag{6.49}$$

where K_R is the bulk recombination coefficient, and C_I and C_I^* are the interstitial and interstitial equilibrium concentration, respectively.

The flux balance boundary condition, which the defects obey, can be described by Hu (1985) as

$$\vec{J}_V \cdot \vec{n} + K_V \left(C_V - C_V^* \right) = g, \tag{6.50}$$

where $\vec{n}$ is the surface normal, K_V is the surface recombination constant, and g is the generation, if any, at the surface.

6.4.2 Interstitials

Interstitials can also diffuse by themselves and obey the diffusion equation expressed by (Xia et al., 2007)

$$\frac{\partial \left(C_I - C_{ET} \right)}{\partial t} = \vec{\nabla} \cdot \left(-\vec{J}_I - \sum_{imp} \vec{J}_{AI} \right) - R, \tag{6.51}$$

where C_I is the interstitial concentration, C_{ET} is the number of empty interstitial traps, and J_{AI} is the flux of paired interstitials diffusing with impurity A. J_I is the flux of the unpaired interstitials, which accounts correctly for the effect of an electric field on the charged portion of the defect concentration and can be written as (Sze, 1981):

$$-\vec{J}_I = D_I C_I^* \vec{\nabla} \frac{C_I}{C_I^*}. \tag{6.52}$$

C_I^* is the equilibrium interstitial concentration. R is all sources of bulk recombination of interstitials which is the same as described for vacancies in Section 6.4.1. Like vacancies, interstitial defects obey the following flux balance boundary condition (Hu, 1985):

$$\vec{J}_I \cdot \vec{n} + K_I \left(C_I - C_I^* \right) = g, \tag{6.53}$$

where $\vec{n}$ is the surface normal, K_I is the surface recombination constant, and g is the generation, if any, at the surface.

The trap reaction explains some of the wide variety of diffusion coefficients extracted from several different experimental conditions and is described by Griffin and Plummer (1986) as

$$\frac{\partial C_{ET}}{\partial t} = -K_T \left[C_{ET} C_I - \frac{e^*}{1 - e^*} C_I^* \left(C_T - C_{ET} \right) \right], \tag{6.54}$$

where C_T is the total trap concentration, K_T is the trap reaction coefficient, and e^* is the equilibrium trap occupancy ratio.

6.4.3 Active Impurities

Some dopants diffuse only when they are activated and the diffusion equation depends on the activated concentration of the impurities. Here, activation is used to describe how well the impurity atoms are incorporated into the host lattice. In process simulation, dopant activation is achieved via thermal annealing and this depends on solid solubility and other factors. The total dopant (before or after activation) concentration is labeled C_T while the part that has been activated is denoted C_A.

For n-type dopants, the following model applies to arsenic and selenium. The diffusion flux densities paired with V and I are given, respectively, by

$$\vec{J_{AV}} = -D_V C_A \frac{C_V}{C_V^*} \vec{\nabla} \ln \left(C_A \frac{C_V}{C_V^*} \frac{n}{n_i} \right) \tag{6.55}$$

$$\vec{J_{AI}} = -D_I C_A \frac{C_I}{C_I^*} \vec{\nabla} \ln (C_A \frac{C_I}{C_I^*} \frac{n}{n_i}). \tag{6.56}$$

The p-type dopants boron, beryllium, and magnesium follow similar equations:

$$\vec{J_{AV}} = -D_V C_A \frac{C_V}{C_V^*} \vec{\nabla} \ln \left(C_A \frac{C_V}{C_V^*} \frac{p}{n_i} \right) \tag{6.57}$$

$$\vec{J_{AI}} = -D_I C_A \frac{C_I}{C_I^*} \vec{\nabla} \ln (C_A \frac{C_I}{C_I^*} \frac{p}{n_i}). \tag{6.58}$$

In the above, n and p refer to the electron and hole concentrations, respectively, and n_i refers to the intrinsic carrier density.

6.4.4 Inactive Impurities

In some cases, inactive impurities can diffuse when they are paired with point-defects. For example, n-type dopants such as phosphorus and antimony would show this behavior. Their flux density can be defined by

$$\vec{J_{AV}} = -D_V C_T \frac{C_V}{C_V^*} \vec{\nabla} \ln (C_T \frac{C_V}{C_V^*} \frac{n}{n_i}) \tag{6.59}$$

$$\vec{J_{AI}} = D_I C_T \frac{C_I}{C_I^*} \vec{\nabla} \ln (C_T \frac{C_I}{C_I^*} \frac{n}{n_i}). \tag{6.60}$$

6.4.5 Neutral Impurities

In process simulation, many charge neutral impurities such as O_2, H_2O, gold, cesium, and germanium usually take part in the diffusion process. Some of them may result in significant influence on the performances of the devices. For example, interdiffusion of germanium in silicon can lead to mobility enhancement in nanoscale CMOS because germanium tends to produce the mechanical stress. Diffusion of charge-neutral impurities can be defined simply by Fick's law:

$$\vec{J_T} = -D \cdot \vec{\nabla} C_T \tag{6.61}$$

$$\frac{\partial C_T}{\partial t} = -\vec{\nabla} \cdot \vec{J_T}. \tag{6.62}$$

6.5 Oxidation

The oxidation model is unique for silicon, and when treated fully, it is a complicated model in process simulation because it involves consideration and coupling of oxidant diffusion, viscous fluid flow, injection of V/I defects, diffusion of existing dopants, and moving oxide/silicon interface.

We shall describe a simple 1D oxidation model proposed by Deal and Grove in the 1960s (Deal and Grove, 1965) in some detail and explain qualitatively how such a simple model can be upgraded to fully coupled 2D/3D treatments.

The idea behind the theory of the Deal–Grove model is that the oxidant is transported from the gas phase into the oxide, then diffuses through the oxide until it reach the oxide/silicon interface where chemical reaction converts the oxidant into oxide while consuming solid bulk silicon. The basic assumption of the Deal–Grove model is that chemical reaction occurs at the oxide/silicon interface instead of at the gas/oxide interface.

Referring to Figure 6.12 and assuming the concentration of oxidant in the gas phase C_g is a known quantity (proportional to gas pressure), the model assumes three density fluxes in the three regions, which can be written as

$$J_g = h(C_g - C_s) \tag{6.63}$$

$$J_o = \frac{D(C_s - C_i)}{X_o} \tag{6.64}$$

$$J_i = kC_i, \tag{6.65}$$

where C_s and C_i are oxidant concentrations at the oxide surface and oxide/silicon interface, respectively. X_o is the oxide thickness. J is the flux density in units of 1/area/s. The first equation is the gas law (Henry's law), and the equation in region two is the Fick's diffusion law, while the third equation is the chemical reaction rate equation.

For the last equation, the reaction must result in oxide deposit, and thus

$$J_i = N_i \frac{dX_o}{dt}, \tag{6.66}$$

where N_i is the number of oxidant molecules per volume.

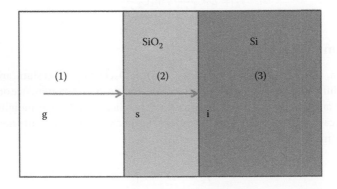

FIGURE 6.12　Schematic showing the three regions in the Deal–Grove theory of oxidation.

Usually the gas concentration is given and if steady-state condition is assumed, $J_g = J_o = J_i$ and the above three equations reduce to

$$h(C_g - C_s) = \frac{D(C_s - C_i)}{X_o} \tag{6.67}$$

$$\frac{D(C_s - C_i)}{X_o} = kC_i. \tag{6.68}$$

This enables us to solve for C_i and C_s after some lengthy algebra. Using Equation 6.66, we obtain

$$\frac{dX_o}{dt} = \frac{B}{A + 2 * X_o}, \tag{6.69}$$

where

$$B = \frac{2DC_g}{N_i} \tag{6.70}$$

$$A = 2D\left(\frac{1}{h} + \frac{1}{k}\right). \tag{6.71}$$

There is an analytical solution to Equation 6.69 for thickness versus oxidation time. It is sufficient to examine the growth behavior here. At small oxide thickness, the growth rate is a constant. As X_o becomes thicker, it is harder for the oxidant to reach the interface and growth slows down as it should.

The 1D oxidation model works very well if the device geometry is simple. The only complication would be oxidation enhanced diffusion (OED) in region 3 of Figure 6.12. It is well known that interface chemical reaction is accompanied by injection of V/I defects, which enhance the diffusivity as indicated by flux density formulas we discussed previously.

The real challenge is when the surface is not flat. Oxide behaves like a viscous fluid at high temperature and when the surface is not flat, mechanical stress equations must be solved to predict the shape of the boundary surrounding region 2 in Figure 6.12. Since diffusion and defect injection is dependent on stress, a fully coupled solution of (I, V, oxidant, impurity, and stress) must be solved under the condition of moving boundary. Due to such complexity, most process simulators can handle 2D viscous oxidation well but 3D oxidation is still considered a subject of research. The simulator CSUPREM uses quasi-3D approach with mesh plane stacking in arbitrary orientation to mimic full 3D oxidation for complicated 3D geometries such as deep 3D trench.

As a demonstration of the basic models in silicon process simulation, there is a simplified light-doped drain (LDD) NMOS example. The detailed input files can be found on the Crosslight website (Crosslight Software Inc., 2015). The net doping (defined as the log10 of n-doping minus p-doping) is displayed in Figure 6.13. Please note that the final structure contains three different impurities, As, P, and B, vacancies (V), and interstitials (I). The system contains both the total and the activated components, the latter with a letter a suffix, such as B and Ba.

6.6 Summary

We have given an overview of the theories and principles of semiconductor growth process modeling mainly for compound semiconductors. With the growing importance of compound semiconductor in both optoelectronic and microelectronic applications, we expect MOCVD reactor modeling will gain increasing importance in the near future.

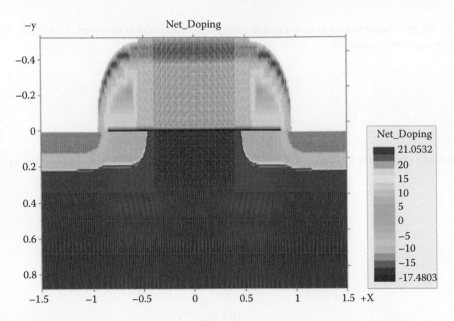

FIGURE 6.13 Results of final net-doping from CSUPREM process simulation for a lightly doped NMOS.

Silicon-related optoelectronic devices, such as CMOS image sensors and hybrid lasers, are important optoelectronic devices. We have presented a brief overview of theories behind commercial process simulators of silicon regarding ion implantation, dopant diffusion, and oxidation. In addition, other processes such as layer deposition, etching, and material-changing are typically implemented in our process simulators as simple geometric manipulations. We hope this chapter about process simulation will help users or potential users of such simulators better understand the physics behind the simulator commands.

References

Cho WK, Choi DH, Kim MU (1999) Optimization of the inlet velocity profile for uniform epitaxial growth in a vertical metalorganic chemical vapor deposition reactor. *International Journal of Heat and Mass Transfer* 42: 4143–4152.

Crosslight Software Inc. (2015) Process Simulation of Compound Semiconductors. http://www.crosslight.com.

Deal BE, Grove AS (1965) General relationship for the thermal oxidation of silicon. *Journal of Applied Physics* 36: 3770–3778.

Gibbons JF, Mylroie S (1973) Estimation of impurity profiles in ion implanted amorphous targets using joined half-Gaussian distribution. *Applied Physics Letters* 22: 568–569.

Griffin PB, Plummer JD (1986) Process physics determining 2-D impurity profiles in VLSI devices. *IEEE International Electron Devices Meeting* 32: 522–525.

Hansen SE, Deal MD (1994) SupremIV.GS, *Two Dimensional Process Simulation for Silicon and Gallium Arsenide*. http://www-tcad.stanford.edu/tcad/programs/suprem-IV.GS/Book.html.

Hirschfelder JO, Curtis CF, Bird RB (1954) *Molecular Theory of Gases and Liquids*. New York, NY: John Wiley & Sons.

Hofker WK (1975) Concentration profiles of boron implanted in amorphous and polycrystalline silicon. *Philips Research Reports Supplements* 8: 41–57.

Hu SM (1985) On interstitial and vacancy concentrations in presence of injection. *Journal of Applied Physics* 57: 1069–1075.

Kadinski L, Merai V, Parekh A, Ramer J, Armour EA, Stall R, Gurary A, Galyukov A, Makarov Y (2004) Computational analysis of GaN/InGaN deposition in MOCVD vertical rotating disk reactors. *Journal of Crystal Growth* 261: 175–181.

Li S, Fu Y (2011) *3D TCAD Simulation for Semiconductor Processes, Devices and Optoelectronics*. New York, NY: Springer.

Oran ES, Boris JP (2001) *Numerical Simulation of Reactive Flow*, 2nd Ed. Cambridge: Cambridge University Press.

Park C, Klein KM, Tasch AF (1990) Efficient modeling parameter extraction for dual Pearson approach to simulation of implanted impurity profiles in silicon. *Solid-State Electronics* 33: 645–650.

Selberherr S (1984) *Analysis and simulation of semiconductor devices*. Vienna: Springer-Verlag

Srim (2015) *The Stopping and Range of Ions in Matter*. http://www.srim.org.

Sze SM (1981) *Physics of Semiconductor Devices*, 2nd Ed. New York, NY: John Wiley & Sons.

Wilke CR (1950) A viscosity equation for gas mixtures. *Journal of Chemical Physics* 18: 517–519.

Xia G, Hoyt JL, Canonico M (2007) Si–Ge interdiffusion in strained Si/strained SiGe heterostructures and implications for enhanced mobility metal-oxide-semiconductor field-effect transistors. *Journal of Applied Physics* 101: 044901(1–11).

Yang SH, Morris S, Parab K, Tasch AF (1995) *An Accurate Monte Carlo Model for the Simulation of Arsenic and Boron Implants into Silicon. UT-MARLOWE Version 2.0*. Austin, TX: University of Texas.

II

Novel Materials

II

Novel Materials

7

Organic Semiconductors

Roderick C. I.
MacKenzie

7.1 Introduction to Organic Semiconductors

The majority of this book focuses on classical crystalline semiconductors such as silicon and gallium arsenide. Indeed, almost all electronic devices we use today from transistors to quantum well lasers (Lim 2007) are based on crystalline semiconductors, and it was the development of these materials and devices made from them that has laid the foundation of the Internet and information age. When one thinks of the atomic structure of silicon, one thinks of a highly ordered, regular arrangement of atoms spanning the physical dimensions of the sample. One could visualize this arrangement of atoms on a more human scale, as marbles of the same size neatly packed in a tin. Bloch's theorem (Bloch 1928) tells us that this regular arrangement of atoms will lead to a well-defined conduction and valance band, with a forbidden region neatly positioned between them called the band gap. Another important characteristic of crystalline semi-conductors used in semiconductor devices is that they are highly pure with typical impurity concentrations of one part in ten million (Delannoy 2012). This high purity means that conducting electrons and holes are unlikely to scatter off defects such as oxygen molecules embedded in the lattice. This results in very high-charge carrier mobility values (typically $\sim$1000 cm^2 V^{-1} s^{-1} at room temperature [Canali et al. 1975]). Practically, this high mobility means that very little waste heat is generated as the device is used; this in turn enables millions of transistors to be integrated into tiny chips and the development of highly complex information processing units. From the theoretical perspective, this lack of defect states means that we can fairly successfully imagine conducting electrons and holes in these materials as quasi-free particles that obey classical Newtonian mechanics.

While these crystalline semiconductors have excellent electrical properties, this comes at a cost (New Scientist 1979). To form pure silicon from naturally occurring quartz (SiO$_2$), one must break the highly stable Si–O bonds and remove all impurities; this requires a sophisticated and highly energetic multistage purification process. The process starts with the liquefaction of silicon to trichlorosilane (HSiCl$_3$) by application of HCl and heating to 300–400°C (Delannoy 2012), followed by the application of hydrogen at 1100°C to further purify the liquid. Numerous distillation steps follow, which end in the conversion of HSiCl$_3$ back into pure silicon (Delannoy 2012); the result is a 99.9999999% pure material. However, due to the high-energy input of this process, the resulting material is an expensive commodity. Indeed, it was recently calculated that if one added up all the energy used to make all silicon solar panels since 1970, then calculated how much energy they all produced, solar cells have only started generating net energy since around 2010 (Dale and Benson 2013). More worrying, almost all of this energy came from burning fossil fuels (Dale and Benson 2013). Therefore just by considering these simple energy arguments, we can think of typical crystalline semiconductors as excellent materials for small high-value electronic items such as CPUs and laser diodes, but not such good materials for large-area devices such as displays, solar cells, and large lighting panels. It is therefore clear that for large-area devices another approach is needed using a different, lower-energy, material system.

Over the last 20 years, the search for a low-cost alternative to Silicon has been intense with a great deal of both academic and industrial attention focused on using carbon-based conducting polymers and small organic molecules as semiconducting materials. Conducting carbon molecules are, however, not new; the synthesis of the first conducting polymer occurred by accident in 1862 while H. Letheby at the College of London Hospital was investigating two cases of fatal poisoning (Letheby 1862). He produced via electrolysis a "brilliant blue" colored pigment, which we know today to be polyaniline (or PANI for short), a well-known conducting polymer (see Figure 7.1).

The modern era of organic electronics began in 1977 when Chiang et al. (1977) reported that when the polymer polyacetylene was doped with iodine, its conductivity increased by a factor of 10 million. This resulted in a Nobel Prize in Chemistry, and the modern age of organic electronics was born. Since 1977, the field of organic electronics has exploded, resulting in a plethora of new devices and materials being developed from carbon-based molecules; indeed, today there is no inorganic device without an organic analog. Examples of such devices are the organic light-emitting diodes (OLEDs) (Yersin 2008), organic field effect transistors (OFETs) (Zhao et al. 2016), biosensors, and organic photovoltaic devices (OPVs) (Tress et al. 2011). In fact, if you have a modern smart phone in your pocket, the screen is almost certainly made from light-emitting diodes fabricated from small organic molecules (such as the one in Figure 7.2); you can tell if the screen is an OLED screen by trying to view the screen from an angle close to 90°: if at angles >85°, the image remains visible and the colors remain true, it's an OLED screen; if not, then it's an older liquid crystal display.

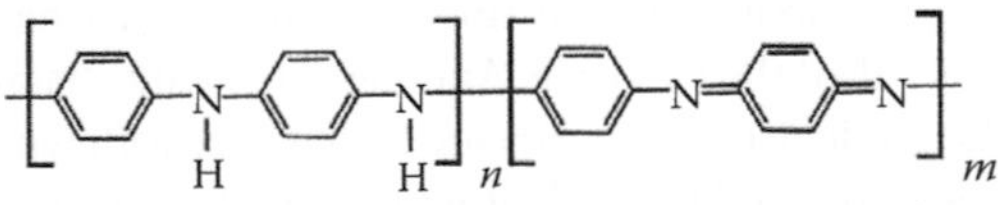

FIGURE 7.1 Polyaniline or PANI for short. The electrical properties of the material can be tuned by varying the ratio of n to m units. The material H. Letheby found probably had a high number of m units, making it blue.

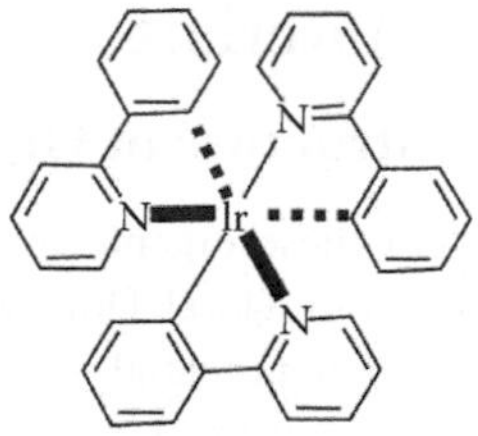

FIGURE 7.2 Tris[2-2,N]iridium(III) or IR(ppy)$_3$ for short, a green-emitting molecule used in organic light-emitting diodes (OLEDs).

The big advantage of using organic semiconductors over inorganic semiconductors is that they can be made using low-energy (read low-cost) wet chemistry and the deposition of these materials to form devices can be done at low temperature (<200°C), saving energy (read money) (Voigt et al. 2012). Another very important property of organic semiconductors is that they are flexible, and if devices are fabricated on flexible substrates, such as plastic, one can produce fully flexible devices which can easily be integrated into clothing and buildings. Although having flexible devices could be attractive from a product integration standpoint (think bendable cell phones), the real advantage of having a flexible device is that when one comes to fabricate it, one can use traditional low-cost, high-speed roll-to-roll printing techniques such as ink-jet printing. Imagine, for example, replacing the newspaper in a newspaper printing press with a flexible plastic substrate (such as indium tin oxide [ITO]-coated polyethylene terephthalate [PET]) and the newspaper ink with an organic polymer dissolved in a solvent to form an ink. One would then be able to fabricate electronic devices as cheaply and as easily as one can produce newspapers. Indeed, such processes have already been used to produce organic solar cells (Voigt et al. 2011, 2012; Adams et al. 2015). In summary, organic semiconductors offer a potential route to low-cost, low-energy, and large-area devices.

7.2 Electrical and Structural Differences between Organic and Inorganic Semiconductors

Although organic semiconductors offer many advantages over their traditional inorganic counterparts, they are intrinsically less ordered than crystalline semiconductors. If one can think of the atomic structure of a crystalline semiconductor, such as silicon, as marbles stacked neatly in a tin, the equivalent picture for an organic material would be a plate of cooked spaghetti, where each piece of pasta represented a polymer chain which can conduct charge along its backbone (see Figure 7.1 for an example of a conducting polymer). In fact, generally speaking, in organic semiconductors, there is no long-range order, with molecules very often being orientated more or less randomly. If we recall, Bloch said that the periodicity of atoms in a semiconductor is the reason for the well-defined conduction and valance band edges (Bloch 1928); we can anticipate that this lack of long-range order has important consequences for how we think of charge transport in organic materials. Indeed, in organic materials, rather than thinking of charge being transported in a band, we think of charge being localized on individual molecules; and charge is only transported from one molecule to another when a charge carrier has gained enough energy to do so, usually through thermal excitation. This process of thermally excited movement is usually referred to as hopping. Another factor limiting conduction of charge carriers in organic materials is that charge can only be transported away from a molecule, if there is another molecule (state) in range on which to hop. If two molecules are packed closely together, it will be relatively easy for charge to transfer from one molecule to the other; however, if the molecules are far apart, it will be less probable that charge will transfer. Thus, in these materials, microscopic packing of the molecules will dictate the macroscopic material mobility. In the next few pages, we explore these concepts in some depth.

7.3 Modeling of Electronic Materials

7.3.1 Calculating the Electronic Structure of Organic Semiconductors

To properly understand charge transport in these organic materials, one must first understand how molecules and charge carriers interact on a quantum level. During our undergraduate studies, we all solved Schrödinger's equation ($E\Psi = H\Psi$) in an infinite potential well to obtain the first three or so wave functions of an electron. The results of such a calculation are presented in pictorial form in Figure 7.3. If instead of solving the equation in a 1D potential well, we change the potential to represent a point charge in space,

$$V = \frac{Q}{4\pi_0 r},\tag{7.1}$$

such as a single proton, and solved Schrödinger's equation in the 3D space surrounding this point charge, we will again be able to calculate where electrons could exist. The resulting wave functions are in fact the atomic orbitals of atomic hydrogen, which electrons occupy while orbiting the atom. The results from such calculations are depicted in pictorial form in Figure 7.4. By convention, each solution to Schrödinger's equation is labeled 1s, 2s, $2p_x$, $2p_y$, and $2p_z$. Notice, just as in Figure 7.3, the wave functions depicted in Figure 7.4 have positive and negative regions.

The hydrogen atom is the simplest example of applying Schrödinger's equation to calculate the electronic orbitals of an atom. By simply rewriting the Hamiltonian to include the necessary potentials, the same basic approach can be used to calculate the electronic orbitals of other atoms, such as carbon or even whole systems of atoms such as molecules, by including multiple point charges. One could for example use this approach to calculate the delocalized electronic orbitals on polyaniline shown in Figure 7.1 or the small

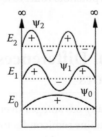

FIGURE 7.3 A diagram of the first three solutions for Schrödinger's equation in an infinite potential well.

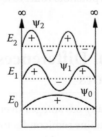

FIGURE 7.4 The 1s, 2s, $2p_x$, $2p_y$, and $2p_z$ orbitals of a hydrogen atom calculated by solving Schrödinger's equation.

molecule shown in Figure 7.2. By taking the square of the calculated wave function $|\Psi|^2$, we can then calculate the probability density function of a charge carrier, which will tell us where we are most likely to find an electron on the polymer/molecule. Clearly, once one moves away from atomic hydrogen with one proton and one electron, one not only has to include the electrostatic potentials of more protons, but also the electrostatic potentials of the other electrons in the system and the influence they will have on the electronic orbital of interest. For a molecular system containing N electrons and M nuclei we can write the general Hamiltonian as

$$H = -\sum_{i=1}^{N} \frac{1}{2}\nabla_i^2 - \sum_{A=1}^{M} \frac{1}{2M_A}\nabla_A^2 - \sum_{i=1}^{N}\sum_{A=1}^{M} \frac{Z_A}{r_{iA}} + \sum_{i=1}^{N}\sum_{j>i}^{N} \frac{1}{r_{ij}} + \sum_{A=1}^{M}\sum_{B>A}^{M} \frac{Z_A Z_B}{R_{AB}}, \tag{7.2}$$

where M_A is the ratio of the mass of the nucleus A to the mass of an electron and Z_A is the atomic number of the nucleus. The first term in Equation 7.2 represents the kinetic energy of the electrons, the second term represents the kinetic energy of the nuclei, the third term represents the Coulomb attraction between nuclei and electrons, the fourth term represents electron–electron repulsion, and the fifth term represents nuclei–nuclei repulsion (Szabo 1996). In order to make the problem more tractable, it is common to make the Born–Oppenheimer approximation (Szabo 1996), whereby one assumes that the electrons move much faster than the nuclei so one may consider the nuclei stationary. Finding a solution to the above equation is a major challenge for complex molecular systems with many atomic nuclei because each electron and proton applies a force to all other electrons and protons in the system. One must therefore not only calculate the wave functions of the electrons around the position of known nuclear potentials, but one must also first find the minimum energy positions of the nuclei of the molecule in a 3D space containing the charge clouds, and clearly every update to the positions of the nuclei will affect the distribution of the charge clouds. It is therefore a complex iterative problem requiring significant computing power. The search for efficient methods to do this has led to an entire field called computational quantum chemistry. There are also many complementary methods available today, which are all, in essence, approximations to the solution of Equation 7.2 in one form or another. These include one of the very earliest quantum chemical methods, the Hartree–Fock method (Fock 1930), dating back to 1927, and the more modern and efficient derivatives of such methods such as density functional theory (DFT) (Hohenberg and Kohn 1964; Kohn and Sham 1965), which tries to make the problem more tractable by assuming that electrons do not interact with one another. Writing such codes is not a trivial task, and fortunately today, there are many packages available that have been written to perform these calculations efficiently, such as Gaussian, General Atomic and Molecular Electronic Structure System (GAMESS) and Quantum ESPRESSO. Section 7.3.2 discusses the general results of such calculations and their importance for the field of organic electronics.

7.3.2 Electronic Structure of Organic Semiconductors

We now look at some wave functions (orbitals) of molecules commonly used in organic electronics. In Figure 7.5, a diagram of the ethene molecule is presented where only the wave functions for the 2p-orbitals have been drawn. If one compares these orbitals to those of atomic hydrogen in Figure 7.4, it can be seen that the shape of the orbitals directly above and below the carbon atoms have more or less the same form. It can also be seen from Figure 7.5 that rather than the orbitals being confined to one atom as in the case of atomic hydrogen, they stretch across the carbon bond, forming what is called an MO. A molecular orbital (MO) is the same as an atomic orbital except that wave function spans over a region of molecule rather than just one atom. The MO depicted in Figure 7.6 is called a p-orbital as it is formed from two atomic p-orbitals.

P-orbitals are of significant technological importance for organic electronics because they enable the electron wave function to spread (delocalize) over multiple carbon atoms. This means that if an electron transfers into a very delocalized p-orbital of a molecule, it will spread out in space, which in turn will

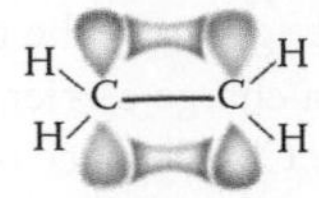

FIGURE 7.5 The wave functions of ethene.

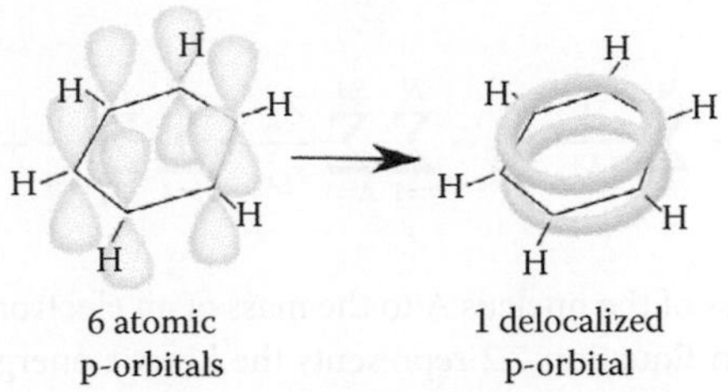

6 atomic
p-orbitals

1 delocalized
p-orbital

FIGURE 7.6 (a) The six individual atomic p-orbitals of benzene. (b) The six orbitals formed into one large delocalized orbital.

allow it to interact with many of its neighboring molecules. Thus, if an empty state exists on a neighboring molecule, a delocalized electron is likely to find it and therefore hop into it. This means that delocalized MOs are associated with high-charge carrier mobilities and good charge carrier transport.

Another example of a delocalized orbital is that found above and below the carbon ring in benzene (see Figure 7.6), and an even more extreme example can be found in Buckminsterfullerene (often called C_{60} or fullerene. This is depicted in Figure 7.7, where the MO is spread out over the entire ball-shaped carbon cage.

Fullerene is a very important organic electronic molecule that has found uses in solar cells, OLEDs, and sensors. The reason for fullerene's usefulness and popularity can be attributed to its very high electron mobility of 1 $cm^2V^{-1}s^{-1}$ (this mobility is considered high for organic electronic molecules), which can in turn be traced back to its very delocalized p-orbitals. Once an electron hops onto fullerene, it is very quickly delocalized all over the cage. Thus, if any neighboring molecule has an empty electronic state, the electron on the fullerene is highly likely to have some electronic coupling to this state and hop to it. The ball shape of fullerene also helps it pack closely with other fullerene molecules, increasing the probability of charge transfer and thus high mobility.

7.3.3 Microscopic Charge Transport in Organic Semiconductors

Many readers of this chapter will have spent time, while studying quantum mechanics, calculating the probability of an electron transferring from one quantum state to another using Fermi's golden rule. A good example of this problem is calculating the probability of stimulated emission from a quantum well laser diode. In such a case, one first calculates the wave function of an electron in the conduction band of the quantum well (Ψ_i), then calculates the wave function of a hole in the valance band of the quantum well (Ψ_f). One then uses Fermi's golden rule to give the probability of electron transitioning from its initial state to its final state

$$\Gamma = \frac{2\pi}{\hbar} |\langle \Psi_f|H|\rangle|^2 \rho_f, \tag{7.3}$$

where H is the Hamiltonian, $\langle \Psi_f |H| \Psi_i \rangle$ represents the overlap of the initial and final state, ρ_f is the density of final states, and other symbols take their usual meaning. Much the same approach is used to calculate the hopping rate of an electron between two molecules. The framework which is often used is the

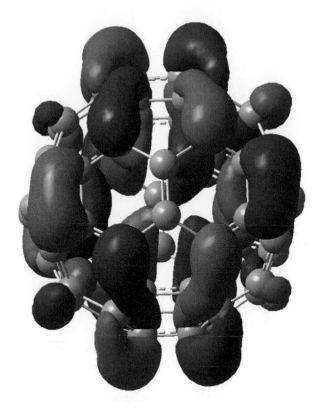

FIGURE 7.7 The delocalized electron orbital of a fullerene cage, Calculated by Dr. Anne A.Y. Guilbert of Imperial College London, using the Gaussian and the hf/6-311g(d,p) method.

semiclassical Marcus theory,

$$\Gamma = \frac{2\pi}{\hbar} |\langle \Psi_f |H| \Psi \rangle|^2 \exp\left(\frac{-\Delta E + \lambda}{4\lambda\kappa_B T}\right), \tag{7.4}$$

for which Rudolph A. Marcus won the Nobel Prize in Chemistry in 1992. If one compares Equations 7.3 and 7.4, one can see that they look more or less identical, except in the case of Marcus theory, $\Psi_i o/\Psi_f$ represent the wave function of the initial and final MO rather than a wave function in a quantum well. On the right-hand side of Equation 7.4, one can see an extra exponential term that does not appear in Fermi's golden rule. This term is in effect an activation energy, which describes the energy difference between the initial and final states, where λ is the reorganization energy, T is the temperature of the molecular system, κ_B is Boltzmann's constant, and ΔE is the energy corresponding to the driving force. These terms are discussed in detail later.

To understand how Marcus equation applies to organic molecules, for the moment let's neglect the exponential term on the right-hand side and only consider the overlap of the initial and final states. If we reexamine Figure 7.7 where the highest unoccupied orbital of C_{60} has been depicted, one can see that the MOs will in general hug the bonds between the carbon atoms. If we now consider the pair of fullerene cages in Figure 7.8, we can imagine that in some orientations the p-orbitals on fullerene-a will be well aligned to the p-orbitals on fullerene-b, and in some other orientations the p-orbitals on fullerene-a will be poorly aligned to the p-orbitals on fullerene-b. The former case represents $\langle \Psi_f |H| \Psi_i \approx 1 \rangle$, a high probability of charge transfer, and the latter case represents $\langle \Psi_f |H| \Psi_i \approx 0 \rangle$, a low probability of charge transfer.

This can be seen in Figure 7.9, where fullerene-a has been rotated around the y-axis and fullerene-b held still, and the overlap between the MOs of fullerene-a and -b, $\langle \Psi_f |H| \Psi_i \rangle$, calculated every few degrees. It

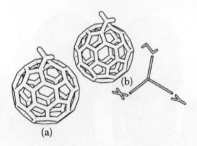

FIGURE 7.8 Two fullerene cages plotted next to each other. They are denoted fullerene-a and fullerene-b in the text. (From MacKenzie et al., *J. Chem. Phys.*, 132, 064904, 2010.)

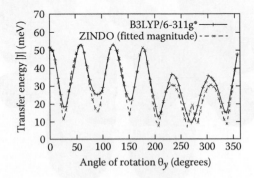

FIGURE 7.9 J as a function of rotation around the y-axis (see Figure 7.8). (From MacKenzie, R. C. I. et al., *J. Chem. Phys.*, 132, 064904, 2010.)

can be seen that at some angles of rotation, J is high and at some other angles J is almost zero. This demonstrates how very position-dependent charge transfer is between fullerenes (and all organic molecules in general) and how important it is to have well-aligned p-orbitals to obtain efficient charge transfer. One could imagine a theoretical thin film where for some reason all fullerenes were well aligned, which would have a very high-charge carrier mobility and another theoretical thin film where the fullerenes were poorly aligned resulting in a very low charge carrier mobility. Indeed, when chemists and engineers design new molecules for organic electronic applications, one key consideration influencing design is how likely the molecules are to stack in a way such that the p-orbitals line up to enable high-charge carrier mobilities

Figure 7.10 depicts J as a function of distance from the center of fullerene-a to fullerene-b. When the fullerenes are more or less touching (at a center–center distance of 10 Å) a maximum J of around 100 meV is calculated. By moving the molecules only 1 Å further apart, charge transfer probability drops by around an order of magnitude. This again highlights the need for close packing of organic molecules and good alignment to obtain efficient charge transfer.

The final thing to note about Figures 7.9 and 7.10, from a computational point of view, is that there are two lines drawn on each figure; one labeled Zerner's intermediate neglect of differential overlap (ZINDO), and one labeled B3LYP/6-311g*. ZINDO is a relatively fast quantum chemical method used to calculate MOs, which uses semiempirical parameters to speed up the calculation (Ridley and Zerner 1973). Each data point on the ZINDO lines took around 10 minutes to compute on an 8-core compute node. In comparison, the B3LYP method is a DFT method and much slower with each data point, taking around 2–3 hours to calculate on an 8-core compute node. One can see from these computational times that simulating any more than a few tens of molecules is computationally too expensive to be done with quantum chemical methods on a regular basis.

Finally, before we leave this topic let's turn our attention to the exponential term in Equation 7.4. The key terms remaining to be explained are λ and ΔE. As stated above, ΔE is the energy difference between the final

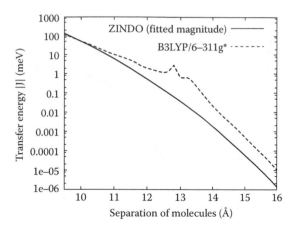

FIGURE 7.10   Transfer energy as a function of center–center fullerene separation.

and initial state. This can be caused by the molecular pair sitting in a potential field or by a dipole being induced between the two molecules. The reorganizational energy, λ, is a little more complex to explain and a full mathematical explanation can be found in Kwiatkowsk (2009). Briefly summarized, the Franck–Condon principle states that charge transfer between two molecules is much more likely if no change in momentum is required for the transition. This means for efficient transfer to occur from one molecule to another both molecules have to be in an energetic state with the same momentum. Usually both states don't have the same momentum, and λ energy is required to move one molecule into a state where no momentum transfer is required for the transition. All the parameters needed for the evaluation of Equation 7.4 can be obtained directly from quantum mechanical calculations (Kwiatkowsk 2009).

7.3.4 Molecular Dynamics

Figure 7.9 demonstrated that charge transport is highly dependent upon molecular orientation and packing. Therefore, when trying to understand microscopic organic materials, it is important to understand how molecules pack, and it is also important to be able to simulate film formation. The computational science that attempts to model how molecules pack together is called molecular dynamics (MD) (Guilbert et al. 2012, 2014, 2015; Dattani 2014). Outside of organic electronics, it is a heavily used tool to understand how biological problems, such as how proteins fold and interact with other molecules such as drugs. The general principle is to try to model the atomic bonds between atoms, and the electrostatic interactions between each molecule. One approach to doing this is by simply solving Schrödinger's equations in time domain and allowing the positions of the nuclei and charge clouds to evolve over time (Jiao et al. 2013); this is called first-principles molecular dynamics (FPMD). This approach, although attractive due to its conceptual simplicity and good approximation to reality, is extremely computationally intensive, requiring a significant cluster of computers (and electricity!) to tackle even the simplest of problems, such as one molecule's interaction with a metallic surface.

A more tractable approach is to first perform accurate quantum chemical calculations on a single test molecule (or pair of test molecules) to understand how a molecule will behave when it is stretched, put under torsion, or encounters the van der Waals attraction forces of another identical molecule, and then to use this understanding based on quantum chemistry to build simple Newtonian models of the molecules of interest. In these models, atoms are represented by Newtonian masses and molecular bonds are represented by Hook-type springs. In doing this, a complex quantum mechanical problem requiring significant CPU time, such as modeling how 200–300 molecules interact to form a film, can be sped up by many orders of magnitude.

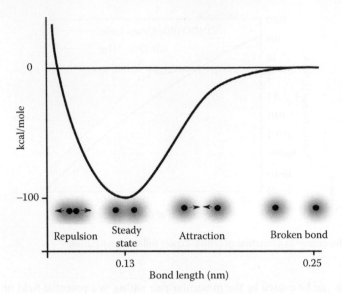

FIGURE 7.11 Potential energy surface for a carbon–carbon bond being compressed, in a relaxed state, and being broken.

Figure 7.11 depicts a typical energy surface that could have been produced by a quantum chemical calculation for the stretching and compression of a C=C double bond in a molecule of interest. One could parameterize this energy surface by fitting it to a standard Lennard-Jones potential to the curve, which is commonly used to parameterize bond energies,

$$V_{LJ} = 4\varepsilon \left[\left(\frac{\sigma}{r} \right)^{12} - \left(\frac{\sigma}{r} \right)^{6} \right], \tag{7.5}$$

where ε is the depth of the potential well and σ is the distance at which the finite potential is zero. If one now assumes that all molecular bonds in the molecule of interest can be represented by mechanical springs of form,

$$F = kx, \tag{7.6}$$

where the Hook's constant k can be calculated by taking the second derivative of Equation 7.5, one can start to construct the Newtonian molecular models.

As well as the change in potential energy as a function of bond length, one must also consider angular bond stretching (see Figure 7.12a) and the rotational potential of the bonds (see Figure 7.12b). The potential profile of angular bond stretching in Figure 7.12a is fairly self-explanatory; however, the reason for the hill-shaped potential in Figure 7.12b is less self-evident. It is in fact due to the need to break pi-bonds (see Figure 7.6) as one rotates a C=C bond.

The sets of spring constants, which describe bond stretching and rotation, along with the values used in equations such as Equation 7.5, are called force fields. Very often, one does not need to compute the various spring constants from quantum chemistry calculations directly, as one can use various libraries of precalculated force fields already available in the literature. Examples of these force fields are Assisted Model Building with Energy Refinement (AMBER), Chemistry at HARvard Macromolecular Mechanics (CHARMM), Groningen Molecular Simulation (GROMOS), and Optimized Potentials for Liquid Simulations (OPLS), but there are many more. One should, however, not use these force fields blindly, as the addition of very electronegative ions to a molecule can draw the electron clouds to one end of the molecule and alter how all bonds behave.

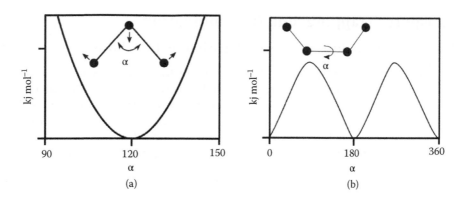

FIGURE 7.12 (a) Angular bond stretching. (b) Rotational potential.

Using MD, one can with relative ease model ensembles of a few hundred molecules without requiring a cluster of computers. To further speed up MD simulations, it is common to use a united atom approach where smaller atoms such as hydrogen are included in the weight of the carbon atoms to which they are attached. It is also possible to approximate entire units of molecules with representative weights; that is, it is common to represent a single fullerene as a single weight of 60 carbon atoms (Chen et al. 2003) rather than simulate each carbon atom in its cage. This is called coarse graining. Another example of coarse graining would be to represent a single monomer unit on a polymer as a single mass with springs connecting it to the other monomers in the polymer (To and Adams 2014). As with quantum chemistry, MD is a large established field with many of the best tools being open source packages, such as GROningen MAchine for Chemical Simulations (GROMACS) (www.gromacs.org n.d.) and Large-scale Atomic/Molecular Massively Parallel Simulator (LAMPS) (LAMPS n.d.). Indeed, the GROMACS manual is a very good starting point for learning about the practicalities of MD simulation.

7.3.5 Purity and Doping of Organic Semiconductors: Where Physicists and Chemists Disagree

Before moving away from detailed materials modeling, it is worth considering what organic semiconducting materials actually consist of. In Section 7.1, I stated that one of the advantages of organic semiconductors was that they can be made by wet chemistry without the need for expensive and energy-consuming purification steps. If one asked a device engineer or physicist what the definition of a pure semiconductor is, he would say a material that has a purity of 99.9999999% (or nine "nines" purity). The atomic density of silicon is around 5×10^{28} atoms m^{-3}; this means in nominally undoped silicon there are 5×10^{19} dopant atoms m^{-3}. In contrast, a highly doped inorganic semiconductor has around 1×10^{25} dopant atoms m^{-3}. If one then went and asked a chemist if the material he had just made is pure, he would say "Yes, it's 99.9% pure." Indeed, if one tries to buy any organic semiconductor, one will very often only be able to obtain 99.9% pure material. Thus, in general, there is a six order of magnitude disagreement between what is considered "pure" in the field of organic and inorganic electronics. It is useful to consider what this means in terms of doping and trap states. If we consider fullerene and assume it is a square box with the volume of 1 nm × 1 nm × 1 nm, it will have a density of 1×10^{27} molecules m^{-3}; if we then assume it is 99.9% pure, we can then calculate that it will have dopant density of 1×10^{24} atoms m^{-3}. Thus, from these simple calculations it can be seen that even a "pure" organic semiconductor is doped almost as much as a highly doped inorganic semiconductor. This simply highlights how contaminated organic semiconductors are when viewed from the inorganic semiconductor standpoint, which is in turn a result of the low-energy production method.

7.4 Transport, Trapping, and Recombination

7.4.1 Transport, Trapping, and Recombination in Thin Films

In Section 7.3.1, we discussed charge transport between two molecules in terms of Markov theory. Now, rather than just considering a pair of molecules, let's consider how charge moves throughout a thin film of organic material. Figure 7.13 depicts in cartoon form a film of organic semiconductor. The round balls represent fullerenes, while the filled-in squares connected by lines represent polymers, with each monomer unit being represented by a single square. Fullerene is a good electron conductor, while many polymers such as poly(3-hexylthiophene-2,5-diyl) (P3HT) (see Figure 7.14) are good hole conductors. It is therefore common to mix fullerenes with polymers to obtain a material which can transport both electrons and holes. An example of a device where such a film is used is an organic solar cell (Deschler et al. 2013)

Let's now consider the transport of an electron from one side of the material to the other. In our cartoon, an electron starts off on the left-hand side of the diagram. It is injected by an electron-injecting contact and hops between fullerenes until it reaches the far side of the film (path-a, indicated by dashed arrows). At this point, the electron will probably recombine with a hole, on the hole-injecting contact. As in an inorganic semiconductor, transport can be driven by drift or diffusion.

Path-a is the optimum route that an electron could take through the medium. Another possible path would be path-b, where after a few hops, instead of the charge proceeding across the material, it starts to travel down a dead end. This is known as a configurational dead end. If a potential were being used to drive charges across the material (see the arrow at bottom of Figure 7.13), there would be a significant

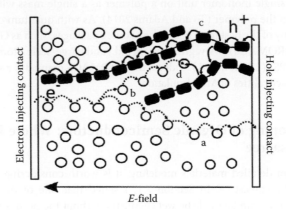

FIGURE 7.13 A cartoon of electron-conducting fullerene (balls) mixed with a hole conducting polymer (square segments). Movement of electrons is represented with dashed arrows, while movement of holes is represented with solid arrows.

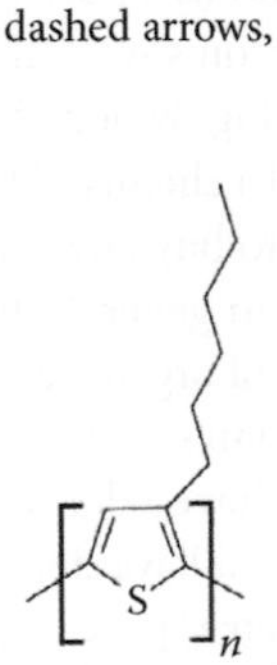

FIGURE 7.14 The hole-conducting polymer poly(3-hexylthiophene-2,5-diyl) (P3HT) is a well-understood polymer commonly used in organic solar cells.

potential gradient between the end of the dead end and the beginning of the dead end. Thus, once a charge had become trapped in the dead end, it would be very improbable that it would be able to escape. This is called an energetic dead end, and the charge carrier would be considered trapped. Now, let's consider the movement of a hole across our device, this time traveling from the right- to the left-hand side of the diagram. The charge carrier is first injected by the hole-injecting contact and hops along a polymer backbone (path-c), traveling from monomer to monomer until it reaches the left-hand side of the diagram where the hole would recombine on the electron-injecting contact. (*Note*: Sometimes when monomer units are well aligned, p-orbitals can delocalize across multiple monomer units. If this happens very fast, transport over multiple monomers can be observed.) Path-c represents an optimum hole transport path across the medium. Now, let's consider the case (path-d), where the hole nears the trapped electron that took path-b. As the hole nears the trapped electron, a finite probability exists for recombination to occur between a quasi-free hole (still moving) and the trapped electron. This free-carrier to trapped-carrier recombination process is a very important way for charges to recombine in organic semiconductors. We will discuss this process in detail later in this chapter.

As the cartoon in Figure 7.13 suggests, understanding trap states in organic semiconductors is very important for understanding both transport and recombination. In fact, in a working organic solar cell well, over 90% of the charge in the device is trapped in trap states with only around 10% being mobile at any one time (MacKenzie et al. 2011). The more disordered a material is (spaghetti like, rather than like marbles in a tin), the more trap states there will be, and the more likely it will be that a charge carrier will become trapped in configurational and energetic dead ends. Charge carrier traps also have a significant influence on net mobility because the more charge-carrier traps that exist in a material, the more probable it will be that a carrier will become trapped at some point while moving across the device, and its movement will be retarded. If the carrier does not detrap after becoming trapped, it will ultimately recombine as seen in Figure 7.13, path-d. Therefore, as a general rule one can say that the more trap states a device has, the higher the recombination rate and the lower the mobility. Very often, organic devices are thermally annealed after fabrication as this helps improve the crystallinity of the films and reduce the number of trap states, increasing mobility and decreasing the recombination rate.

7.4.2 Band Structure of Organic Materials

Figure 7.15a shows the band structure for a typical inorganic semiconductor such as silicon. It has a conduction band (E_c) and a valance band (E_v). A very small number of trap states caused by impurities is shown in the diagram as short black lines in the band gap. Figure 7.15b shows a typical band structure of an organic semiconductor. First, let's deal with the nomenclature, rather than the electron carrying states being referred to as the conduction band as for inorganic semiconductors. In organic semiconductors, they are referred to as the lowest unoccupied molecular orbital (LUMO). Rather than having a valance band, organic semiconductors have a highest occupied molecular orbital (HOMO). In Figure 7.15b, each horizontal line represents a single trap state. It can be seen that both the HOMO and LUMO consist of many individual trap states (typically about 1×10^{26}–1×10^{27} m^{-3}). Each of these trap states originates from a single MO of a molecule. The energetic distribution of these states can be attributed to a variety of factors. To understand this let's consider a film comprised entirely of fullerene molecules. Imagine that, rather than the fullerenes being packed in a film, each molecule is separated from the other by a distance large enough that the molecules cannot electrostatically interact. In this situation, all the LUMO/HOMO levels of fullerene would be identical at 4.5 and 6.1 eV. Now imagine that we slowly bring these fullerene molecules closer together to form our film. As we do this each fullerene cage starts to feel the electrostatic force of its neighboring cages. This electrostatic repulsion/attraction induces dipoles on the molecules which distort the shape of the fullerene cages. Thus, one can imagine that the solutions to Schrödinger's equation will change and so will their eigenvalues, thus the energy levels of the MOs on each molecule will also change. This is one reason for the distribution of trap states as seen in Figure 7.15b. The second reason for this distribution of trap states is the configurational/energetic disorder shown in Figure 7.13.

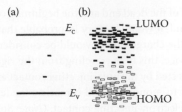

FIGURE 7.15   (a) The band structure of an inorganic semiconductor such as silicon. (b) The "band structure" of an organic semiconductor. Note the density of trap states in an organic semiconductor is typically around 1×10^{26} m^{-3}.

If one considers the distribution of trap states in Figure 7.15b, it can be seen that my cartoon follows no well-described mathematical distribution. This is deliberate as there is some debate in the literature about the exact nature of this energetic distribution. If one opens an undergraduate chemistry book describing the band structure of organic solids, one will often find a Gaussian distribution describing the distribution of trap states around the HOMO/LUMO of the form

$$\rho_{\text{LUMO}}(E) = \frac{N}{\sqrt{2\pi\sigma^2}}\exp\left[-\left(\frac{E - E_{\text{LUMO}}}{2\sigma^2}\right)^2\right], \tag{7.7}$$

where $\rho(E)$ is the density of states (DoS) at any given energy, σ is the spread of the distribution, E_{LUMO} is the center of the distribution (typically taken as the LUMO of an isolated molecule in its ground state), and N is the total number of traps in the distribution. A similar expression can be written for the HOMO. A Gaussian distribution of states is depicted in Figure 7.16a. However, many have found (e.g., To and Adams 2014; Nelson 2003; Kirchartz et al. 2012; Street 2011; MacKenzie et al. 2011) that a Gaussian distribution of traps often cannot account for all of the electrical properties of organic devices and does not always represent the experimentally measured DoS very well (Street 2011). Therefore, when modeling devices, a significant number of authors (Nelson 2003; Kirchartz et al. 2012; MacKenzie et al. 2014, n.d.) have suggested using an exponential distribution of trap states of the form

$$\rho_{\text{LUMO}}(E) = N_{\text{LUMO}}\left(\frac{E}{E_{\text{LUMO}}^{\text{u}}}\right), \tag{7.8}$$

where N_{LUMO} is the density of trap states at the LUMO edge, and $E_{\text{LUMO}}^{\text{u}}$ stands for the Urbach energy and dictates the slope of the energetic distribution of trap states (see Figure 7.16b). If E_{u} is large, the exponential distribution of trap states will slowly decay into the band gap, resulting in a large number of states deep in the band gap. If E_{u} is small, there will be few deep states in the band gap. A slope of around 30 meV is considered a small value of E_{u} and 150 meV is considered a very large value.

In reality, the exact distribution of states is probably a more complex function than a pure Gaussian or a pure exponential; there is evidence for defect states (possibly caused by impurities) deep in the band gap (MacKenzie et al. 2012; Schafferhans et al. 2010; Carr and Chaudhary 2013), meaning the DoS is probably best described as a mixture of an exponential tail and Gaussian deep states (see Figure 7.16c). However, by introducing a complex DoS distribution into any model, one introduces multiple free parameters defining the shape, which may or may not be known. Therefore, it is to some extent impractical to use such complex DoS distributions for day-to-day device simulation. It should also be noted that the exact DoS distribution will be affected by how the molecules pack and the density/type of impurities. Molecular packing will be a function of deposition technique, solvent, annealing steps, and the organic molecule under study, while impurities will be strongly dependent upon which manufacturer produced the semiconductor and from which batch it came. It is therefore hard to recommend a universal DoS function for all organic materials (Gaussian, exponential, or a more complex function), and the modeler should probably use the

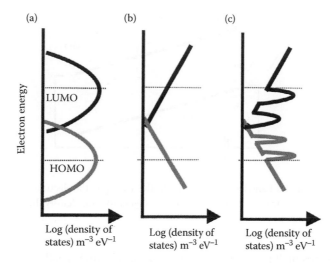

FIGURE 7.16 (a) A Gaussian distribution of trap states. (b) An exponential distribution of trap states. (c) A combination of both Gaussians and exponentials.

DoS function that fits his experimental data best (Hodgkiss et al. 2014). In my own experience, however, I have never failed to reproduce experimental results by approximating the DoS with a single exponential tail for the HOMO and another for the LUMO (see Figure 7.16b) Finally, it should be noted that here we are only discussing the states in the DoS that are deep enough to influence the electrical properties of the device, such as transport, recombination, and carrier trapping. The higher states may well have more of a Gaussian form (Boix et al. 2009); if one examines Figure 7.3 in this paper, the high-energy states look more Gaussian-like, while the lower-energy states look as if they have a more exponential form.

Before leaving this section, it is worth discussing briefly the emission and absorption models for organic materials. In inorganic devices, once one has the DoS (conduction/valance bands), it is quite common to attempt to calculate absorption and emission for the material using Fermi's golden rule. An example of this is using **k·p** theory to calculate the band structure, and thus emissions from a laser diode. As should be clear from the preceding text, the exact DoS for organic semiconductors is often not well known. Therefore, when modeling devices such as solar cells and photo detectors, this author always tries to get the experimentalists to measure the actual absorption spectra from the exact batch of organic material from which the devices were made.

7.4.3 Concept of Mobility in Organic Semiconductors

If you asked a device engineer to find out what the value of electron mobility was in a silicon wafer, without too much difficulty he/she could use a hall-effect measurement to give you *a single* value of mobility for your material. At the heart of the hall-effect measurement is the definition of mobility, which states electron velocity is equal to mobility times the electric field or

$$v = \mu E. \tag{7.9}$$

However, in organic materials, the situation is more complex because the total time it takes for a charge carrier to travel across a medium is dependent upon the density of trap states into which the charge carrier can become trapped. This is depicted in Figure 7.17 where an electron–hole pair is photogenerated on the left of the device while a potential field is applied across it. After photogeneration, the hole is quickly extracted from the left contact, while the electron travels across the medium. As it travels, it becomes

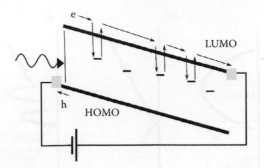

FIGURE 7.17 Photogeneration of an electron hole pair. In this example, the electron undergoes multiple capture escape events into trap states as it travels across the device. The higher the trap density the slower its net speed will be across the device.

trapped and then escapes multiple times. The more time the electron spends being trapped rather than traveling, the longer it takes to cross the device, and the lower the net mobility.

Furthermore, the velocity is dependent not only upon the density of the trap states, but also upon how filled the trap states are. If a trap state is filled, this will reduce the overall probability of a charge carrier becoming trapped (Pauli exclusion principle). Thus, generally speaking, if one deposits an organic semiconductor on a glass slide and keeps it out of the light, the trap states will be mostly empty (broadly speaking and neglecting doping), and thus its mobility will be low. If one now places this same sample in sunlight so charges can be photogenerated, the traps will start to fill and the mobility of the sample will increase. If one now deposits the same semiconductor not on glass but on a conducting transparent metal oxide contact, such as ITO, charge will be injected into the semiconductor from the contact and the semi-conductor's mobility will again be higher than when it is deposited on nonconducting glass. If one adds a second contact to the device, say an aluminum top contact to form a diode, more charge carriers will be injected into the device and the mobility will again be increased. Then finally, if one injects or extracts charge by applying a bias voltage, the carrier density within the device will again change, and so will the mobility. Thus when one sees a value of mobility in the literature for an organic semiconductor, one must always consider under what conditions it was measured, and to some extent with which method. One way of defining mobility in organic semiconductors is as follows:

$$\mu_n \left(T_e, E_{fn} \right) = \frac{\int_{E_e^{edge}}^{\infty} \mu_e^0 \rho_e \left(E \right) f \left(T_e, E, E_{fn} \right) dE}{\int_{-\infty}^{\infty} \rho_e \left(E \right) f \left(T_e, E, E_{fn} \right) dE} = \mu_e \frac{n_{free}}{n_{trap} + n_{free}}, \qquad (7.10)$$

where f is the occupation function of a state, ρ is the DoS (Equation 7.8 or 7.7), and μ_e^0 is the mobility of free (not trapped) charge carriers. Equation 7.10 assumes the existence of an imaginary energetic line, or mobility edge E_{edge}, above which all charge carriers are assumed to be free and mobile and below which charge carriers are assumed to be trapped and immobile. Thus by calculating the ratio of movable carriers to nonmovable carriers, one can calculate mobility as a function of charge carrier density. One can understand this equation by considering extreme cases: when all the carriers are highly excited and above energy E_{edge}, the net mobility will be μ_e^0 and when all the carriers are in deep trap states below E_{edge}, the net mobility will be zero. Although, the existence of a sharp mobility edge E_{edge} is somewhat of a theoretical concept, it theoretically accounts for the fact deep carriers will be immobile while highly energetic carriers will be very free. In reality, there will be a gradient of "freeness" between these two extremes. As such the concept of a sharp mobility edge has proved very useful in many models and allows mobility to vary as a function of charge-carrier density as observed experimentally.

7.4.4 The Limitations of Microscopic Charge Transport Simulations

Much of the above discussion has focused on microscopic materials and modeling of organic semiconductors. All these methods focus on how molecules interact on the molecular level and attempt to understand how molecular form affects electronic molecular function. Performing quantum chemical calculations on individual molecules can take between a few seconds to a few hours. Generally speaking, if one wants to efficiently study the motion of molecules and how they pack together one needs to parameterize the results of quantum chemical calculations to form Newtonian MD force fields. Even with simplifying the quantum mechanical problem to a Newtonian problem and the use of coarse graining the computational problem is still significant. To give a general idea of the computational task, to simulate an ensemble of 200 fullerenes for 5 ns, it would take a 16-core compute node approximately a week of compute time. This huge amount of compute time needed to simulate even simple systems means that one cannot model even simple events such as the formation of a film through evaporation because evaporation takes place on the timescale of many seconds and thus occurs on an almost unreachable timescale for these methods. Thus, even with future advances in computing power, it is difficult to see how MD/quantum chemistry can be easily used to simulate whole devices or experimentally significant timescales in the foreseeable future. It should also be noted that setting up efficient MD force fields for a given material system requires significant human input and is therefore time-consuming. If one now considers how the inorganic community models devices, they very rarely attempt to model the individual atoms to understand a device; instead they try to write general equations which approximate the movement of charges and the potential within the device as a whole. Section 7.5 deals with the adaptation of classical device models which have proved so successful for inorganic semiconductors to organic semiconductors.

7.5 Full Device Models

7.5.1 Early Device Models for Organic Semiconductors

The first attempts at device modeling of organic semiconductors borrowed significantly from inorganic device modeling (Koster et al. 2005). Rather than trying to describe the complex molecular structure of the device where electrons were transported in the LUMO of one material (usually phenyl-C61-butyric acid methyl ester [PCBM]) and holes in the HOMO of another material (often a polymer) (see Figure 7.13), the authors of the first device models used an effective medium approximation. In this approximation, one assumed that there is a conduction "band" that spans the entire device. Transport in this band represents charge carriers hopping from one fullerene molecule to another in the LUMO. They also assumed there is a single valance band that spans the entire device and transport in this band accounts for the movement of holes between the polymer chains. Note I put the word band in quotes to highlight that in organic semiconductors band-like transport does not really happen. Transport in these bands was modeled using the standard drift diffusion equations for electrons,

$$J_n = q\mu_e n \frac{\partial E_{\text{LUMO}}}{\partial x} + qD_n \frac{\partial n}{\partial x}, \tag{7.11}$$

and for holes,

$$J_p = q\mu_h p \frac{\partial E_{\text{HOMO}}}{\partial x} - qD_p \frac{\partial p}{\partial x}. \tag{7.12}$$

Poisson's equation describes how space charge alters the potential profile within the device

$$\frac{d}{dx} \cdot \varepsilon_0 \varepsilon_r \frac{d}{dx} \varphi = q\left(n - p\right), \tag{7.13}$$

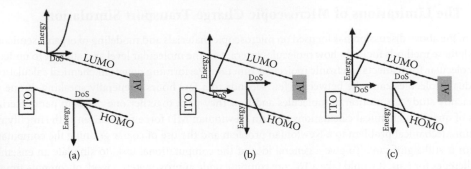

FIGURE 7.18 (a) A "band" diagram of the first organic solar cell device models using a parabolic band approximation with no traps, (b) the introduction of an exponential tail of states into device models. (c) device models based on Shockley–Read–Hall (SRH)—trapping and recombination—these have proved successful.

where ε_0 and ε_r represent the permittivity of free space and the relative permittivity of the medium, respectively, and φ represents the electrostatic potential. Such a model is depicted in Figure 7.18a. Recombination in these early device models was modeled assuming that the recombination rate is proportional to the product of the electron and hole density

$$R = knp \tag{7.14}$$

and that the rate constant could be described by a rate constant k defined as

$$k = q\left(\frac{\mu_n + \mu_p}{2}\right). \tag{7.15}$$

This description of recombination is called Langevin recombination and assumes that electrons and holes both move by Brownian motion (hence the dependence of the equation on mobility) and are annihilated when they meet.

If one contrasts this description of an organic electronic device to the general description of organic semiconductors in the first half of the chapter, one can identify a few major flaws in the model. The first flaw is that band-like transport is assumed and hopping transport via localized states is completely neglected. Furthermore, bearing in mind around 90% of the charge in organic semiconductor resides in very immobile trap states, the result is that the model will never reproduce the correct charge density, and thus the recombination rate will always be wrong (see Equation 7.14). Furthermore, the model can also never account for the change of mobility as a function of carrier density (see Equation 7.10) because mobility is a constant in this model. And more fundamentally, Langevin recombination itself assumes that electrons and holes are free and can participate in a Brownian type motion, with recombination only occurring when the charge carriers get close enough that they are sucked into each other's potential. Brownian type motion neglects carrier trapping (see Figure 7.13). It is therefore clear that to have a better device model carrier trapping must be considered.

7.5.2 Including Carrier Trapping in Organic Device Models

In an attempt to rectify the problems presented by the effective medium model, some authors (MacKenzie et al. 2011) attempted to replace the parabolic band structure with an exponential band structure starting deep in the band gap and continuing past the mobility edge (Roichman and Tessler 2002), as shown in Figure 7.18b. For simplicity, the occupation probability of this DoS was modeled with a single Fermi–Dirac distribution (Roichman and Tessler 2002). This assumed that all carriers in the DoS were able to exchange energy to relax into a distribution with a single quasi-Fermi level. If one considers the arguments above about carriers becoming trapped in deep states, it is clear that this is an over-simplification as there is no obvious reason why two deeply trapped carriers should easily be able to exchange energy However, by assigning carriers above the mobility edge a finite mobility and those below the mobility edge a mobility of zero as done in Equation 7.10, the model did allow the mobility to vary as a function of carrier density. Recombination in this model was accounted for using the Langevin expression given in Equations 7.14 and 7.15; however because now the mobility is allowed to vary as a function of carrier density (Equation 7.10), so does the Langevin prefactor (k), as was observed experimentally (see Figure 7.19),

$$k\left(n,p\right) = \frac{q\left(\mu_e\left(n\right) + \mu_h\left(p\right)\right)}{2_{0r}}. \tag{7.16}$$

Thus this model represented an improvement over the standard Langevin effective medium models which assumed parabolic bands and a constant recombination prefactor (Braun 1984; Koster et al. 2005). It also proved conceptually easy to understand and quite successful in reproducing steady-state current–voltage curves. Incidentally, it is possible to show that Equation 7.16 is comparable to Shockley–Read–Hall (SRH)-type recombination, where free carriers are annihilate trapped carriers (MacKenzie et al. 2011)

Because this book is about numerical modeling, I will now make a few notes about some practical aspects of implementing this model. With this model, it is important to note that because mobile charge carriers no longer have parabolic bands (see Figure 7.18b) the standard Einstein relation,

$$D = \mu_n \frac{k_B T}{q}, \tag{7.17}$$

no longer holds. This is because in the derivation of this relation, the assumption of parabolic bands is made (Azoff 1986; Roichman and Tessler 2002). Instead, the full form of the relation must be used namely

$$D = \frac{\mu_q n}{q(dn/dF_n)} \tag{7.18}$$

Evaluating both n and dn/dF_n requires numerical integration over energy space, which can lead to errors in the evaluation of D, if one is not careful. If numerical errors enter into D, the drift and diffusion currents will not match each other at equilibrium and dark current–voltage curves will not go through zero. The other thing to note is that quite a number of integrals over energy space are needed to make this model work. For

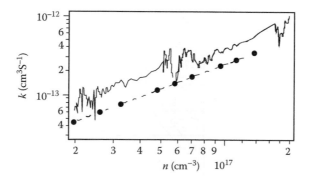

FIGURE 7.19 Variation of Langevin prefactor (k) as a function of carrier density. This shows that the recombination prefactor must vary as a function of carrier density. Solid line indicates data from Transient Absorption Spectroscopy (TAS), and dashed line indicates data from Transient Photovoltage (TPV). (From Shuttle, C. G. et al., *Phys. Rev. B.*, 78, 113201, 2008.)

example, Equation 7.10 where mobility is calculated also simply calculating the carrier density n requires another integral over energy space. Numerical integration is the bane of any engineer who wants to write efficient models because it involves loops and is therefore slow. Therefore, the most flexible and efficient way of dealing with these integrals is to tabulate them before the simulation starts. On the other hand most derivatives of the Scharfetter–Gummel current equations should be done analytically for numerical stability.

7.5.3 SRH Recombination—Splitting Up Energy Space

Even though the model described in Section 7.5.2 assumed that all carriers in the DoS at any given point are in quasi-thermal equilibrium (i.e., have a single quasi-Fermi level), the model performed well in the steady state where one could accept (if one is forgiving) that the carriers would have enough time to reach some sort of thermal equilibrium. However, this assumption of a single quasi-Fermi level describing both the carriers in the very deep traps and the very energetic carriers becomes harder to justify once one starts to perform time-domain simulations where it is clear that for a material with many trap states—not all carriers will be able to react instantaneously upon the application of a light/voltage pulse—especially if it was applied on the fs timescale. Rather than inventing new models for organic materials, the organic community borrowed the SRH model (Kirchartz et al. 2012; Shockley and Read 1952; Foertig et al. 2012), which has been used for years in the inorganic community to describe charge transport/trapping and recombination in disordered inorganic materials such as amorphous silicon.

This model is depicted in Figure 7.18c and expanded in Figure 7.20. From the band structure point of view the model assumes that free carriers move in a parabolic band (represented by parabolas in Figure 7.20, one for the free electrons one for the free holes), while trapped carriers reside in an exponential distribution of trap states (represented by two exponentials in Figure 7.20, one for the LUMO trap states one for the HOMO trap states).

From Figure 7.20 it can clearly be seen that the exponential trap distributions are broken up into a series of trap states depicted by squares. For each of these trap states the following rate equation is solved

$$\frac{dn_t}{dt} = r_{ec} - r_{ee} - r_{hc} + r_{he}, \tag{7.19}$$

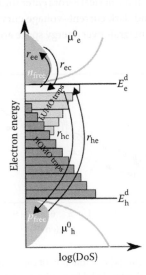

FIGURE 7.20 The SRH model depicted with a parabolic band describing free carriers and an exponential band describing trapped carriers.

where r_{ec} is the rate of electron capture into a trap, r_{ee} is the rate of electron escape from a trap, r_{hc} is the rate of hole capture into an electron trap, and r_{hc} is the rate of hole escape from an electron trap. The equations describing these are given in Table 7.1.

Where v_{th} is the thermal velocity of the carriers, n is the free electron density, σ_n is the capture cross section of free electrons into the electron trap, σ_p is the capture cross section of free holes into the electron trap, N_t is the density of trap states, and f is the occupation function of the trap state. If the capture rates for the electrons (r_{ec}) and holes (r_{hc}) into a single electron trap are examined, it can be seen that they consist of the product of the density of free charge carriers, the occupation probability of the trap state, the capture cross section, and the carrier thermal velocity. The capture rate does not, however, depend upon the depth of the energetic trap; thus, capture of charge carriers is not energetically assisted. If one examines the escape rates of carriers from the traps, r_{ee} and r_{he}, it can be seen that they depend upon the terms e_n and e_p which are defined as

$$e_n = v_{th}\sigma_n N_c \exp\left(\frac{E_t - E_c}{kT}\right) \tag{7.20}$$

and

$$e_p = v_{th}\sigma_p N_v \exp\left(\frac{E_v - E_t}{kT}\right). \tag{7.21}$$

It can be seen that these terms depend exponentially on the depth of a trap. Thus, the deeper the trap the harder it is for charge carriers to escape. In this model, there is no energetic barrier/assistance to relaxation but there is an energetic barrier to escape.

One can think of electron capture into an electron trap as simply trapping, and as free hole capture into an electron trap as charge carrier recombination. Equations 7.19 through 7.21 only describe capture/escape events for an electron trap (in this case the very highest LUMO trap level in Figure 7.20) an analogous set of equations can be written for capture and escape from hole traps. To couple this model to a drift-diffusion transport model, one must be able to calculate the total recombination rate from all trap levels. If one were solving the set of capture/escape equations in steady state, one could simply analytically integrate the equations over an exponential distribution of trap states and obtain the well-known standard steady-state SRH recombination rate equation,

$$R_{SRH} = \frac{pn - n_i^2}{\tau_n(p + n_i) + \tau_p(n + n_i)}. \tag{7.22}$$

However, because organic devices are often studied using transient experiments to learn about charge-carrier dynamics, this approach would be limiting. We, therefore, divide up the distribution of trap states into evenly spaced traps (see Figure 7.20) and solve a rate equation describing trapping/detrapping for each energetic level at each spatial mesh point in the device. Usually it is sufficient to have around 8–15 traps

TABLE 7.1 The SRH Capture and Escape Equations

Mechanism	Rate	Description
Electron capture rate	r_{ec}	$n v_{th}\sigma_n N_t(1-f)$
Electron escape rate	r_{ee}	$e_n N_t f$
Hole capture rate	r_{hc}	$p v_{th}\sigma_p N_t f$
Hole escape rate	r_{he}	$e_p N_t(1-f)$

in each band, although sometimes when investigating the structure of the DoS, up to 80 can be required. It is assumed that within each trap we can use a single quasi-Fermi level to describe the occupation of the trapped carriers. One final advantage of the SRH-based mode over the model described in Section 7.5.2 is that the SRH model uses parabolic bands to describe the movement of free carriers. Although an over-simplification, it enables the use of the usual Einstein relations which helps the stability of the model.

The SRH model, when applied to organic semiconductors, has been very successful at being able to describe a range of experimental data measured using a variety of different electrical/optical measurement techniques over a wide range of timescales. These include light/dark steady-state JV measurements (Sims et al. 2004) (see Figure 7.21), the slow transient current response of a device to a rising voltage waveform (see Figure 7.22), the microsecond transient current responses of a device due to optical excitation (see Figure 7.23), picosecond timescale pump probe responses of devices due to a femtosecond excitation (see Figure 7.24), and even frequency domain measurements (MacKenzie et al. 2016). The SRH-based drift diffusion model used to produce Figures 7.22 through 7.24 is available free to download at www.gpvdm.com.

Before we leave recombination, it is worth coming back to the Langevin recombination expression in Equations 7.14 and 7.15 once more. Equation 7.14 describes a process which is dependent upon the density of electrons and on the density of holes; this is almost always understood as *free* electrons and *holes* colliding and annihilating. This process is referred to in the literature as bimolecular recombination. Some authors understand this to be recombination between an electron on one molecule and a hole on another molecule, while others understand it to mean the process is of order two (a process depending upon the density of both electrons and holes). The latter definition seems to make more sense when compared to the inorganic literature. Although the Langevin prefactor to Equation 7.14 (7.15) comes from a derivation which is probably not valid for organic semiconductors, this does not mean free electron to free hole recombination does not happen in organic materials; it just means the prefactor is probably not correct. For the special case of highly ordered organic crystals with very few defect states, there is no reason why bimolecular recombination should not dominate SRH recombination. When modeling devices where one suspects bimolecular recombination is important, it is therefore fairly safe to assume k is simply an arbitrary model constant and neglect Equation 7.16.

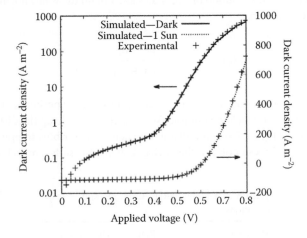

FIGURE 7.21 Steady-state JV curve measurements for a P3HT: phenyl-C61-butyric acid methyl ester (PCBM) organic solar cell reproduced by an SRH model. Note for the light JV curve, the open circuit voltage (intersect of the x-axis), short circuit voltage (intersect of the y-axis), and fill factor (maximum power point) are reproduced exactly by the model. The slope of the dark curve (ideality factor) is also reproduced by the model, suggesting recombination is being modeled correctly. (From MacKenzie, Roderick C. I. et al., *Adv. Energy Mater.*, 2012.)

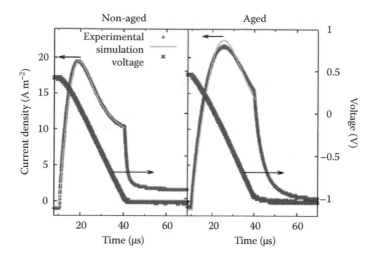

FIGURE 7.22 An example of a slow electrical transient current responses of a P3HT:PCBM diode to a 40-s-long linearly increasing voltage transient. The transient starts at 0.5 V and ends at −1 V. This measurement is called the Charge Extraction by Linearly Increasing Voltage (CELIV) measurement. By applying a voltage ramp, one can extract a significant amount of charge from a device. The time it takes the charge to leave the device can be used to calculate a mobility figure for the material. It can be seen that for this relatively slowly varying electrical transient, the SRH-based model can reproduce the results. (From Hanfland, R. et al., *Appl. Phys. Lett.*, 103, 063904, 2013.)

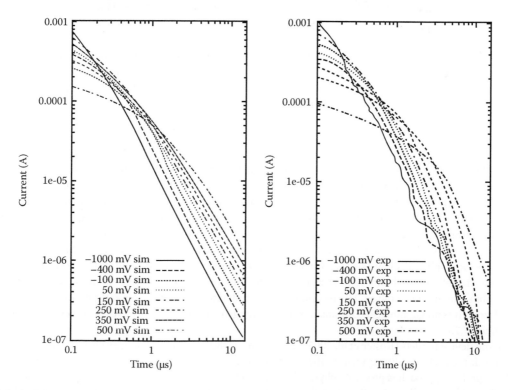

FIGURE 7.23 Right, experimental current transient responses from an organic diode after being photoexcited by a 5ns laser pulse as a function of applied bias. Left, the simulated predictions of SRH-based model of the same experiment. The model was only calibrated to the experimental data at 0 mV, but it was able to predict the transient responses up to 1V away from the calibration point. (From MacKenzie, Roderick C. I. et al., *Adv. Energy Mater.*, 2012.)

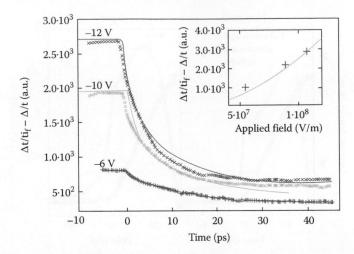

FIGURE 7.24 The simulated and experimental ultrafast (picoseconds) response of an organic diode to a 250-fs laser. The measured signal is the result of optically probing an organic diode up to 40 ps after photoexcitation. This type of pump–probe experiment can provide information on how charge carriers separate on very early timescales in these materials. It can be seen that the experiment and the result of the SRH-based model agree well. (From MacKenzie, Roderick C. I. et al., *Phys. Rev. B.*, 89, 195307, 2014.)

7.6 Geminate and Nongeminate Recombination—Role of Excitons in Organic Devices

One key difference between organic and inorganic semiconductors is that the relative permittivity of organic semiconductors is around three while the relative permittivity of inorganic semiconductors such as silicon are around 12. This becomes important when one considers photogeneration in these materials. If one considers the electrostatic force between an electron and hole,

$$F = \frac{q_1 q_2}{4\pi\varepsilon_0\varepsilon_r r^2}. \tag{7.23}$$

It can be seen that the force between the charge pairs is inversely proportional to the permittivity of the medium. Thus, the force between an electron hole pair in silicon will be three to four times weaker than in an organic material. Practically, this means after photogeneration in an inorganic semiconductor, the electron and hole can escape each other and drift off into the sea of electrons/holes. In organic semiconductors, the attractive force between newly generated electrons and holes is so strong that each can't escape the other's field. Thus, they remain bound. The bound electron and hole pair is called an exciton. Unless some external force splits this pair (such as when the pair encounters a heterojunction), the electron and hole will steadily fall into each other's field and annihilate. This annihilation process is referred to as geminate recombination in the literature. The recombination of unbound electrons/holes as described in Section 7.5 is referred to as nongeminate recombination. The charge generation process in an organic device is depicted in Figure 7.25.

Some device models (Koster et al. 2005; Fluxim n.d.) include an exciton generation/diffusion equation, then link the exciton splitting term to the standard drift diffusion equations are a charge generation rate. Although this approach is probably physically correct, exciton diffusion distances are typically quite short (5–20 nm) but usually not well known for any given material system. Therefore, when modeling photoactive diodes such as solar cells, it is usually sufficient to multiply the calculated charge generation rate (modal profile) from the optical model by a reduction parameter (0–1.0) to account for any geminate recombination which may occur.

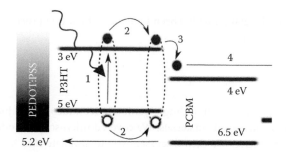

FIGURE 7.25   A diagram of the charge generation process in an organic semiconductor. A photon is absorbed in the P3HT and photoexcites an exciton (1). This bound electron hole pair diffuses to a heterojunction between the P3HT and PCBM (2) where the potential gradient splits the exciton into free charge carriers (3). Once the electron and hole are free of each other's electrostatic potential they can drift/diffuse to the contacts (4) and to the external circuit. This device is a diode with indium tin oxide (ITO) and Al contacts. The poly(3,4-ethylenedioxythiophene)-poly(styrenesulfonate) (PEDOT:PSS) layer prevents electrons from the P3HT recombining on the ITO.

7.7   Numerical Aspects of Organic Device Simulations

Many of the early papers used Gummel's method to solve the drift-diffusion equations (i.e., where one solves Poisson's equations, then solves the drift-diffusion equations, then iterates until convergence). There are quite a few papers claiming this to be a good, stable method. However, in my experience when one separates out the solution of these equations convergence is in general less stable than when one solves all sets of equations in one matrix, using Newton's method. In general, one can get away with Gummel's method when one is only solving the drift-diffusion equations, but when one starts to introduce trap states, and other nonlinearities, the problem becomes more and more nonlinear and one is forced to use Newton's method applied in a single matrix equation, of the form

$$[J] = -[\Delta]\,[f], \tag{7.24}$$

where J is the Jacobin containing the derivatives of the error function (f) and Δ is the update to the solution variables. Writing the equations in this form produces an overhead for the scientist in setting up the problem, but the time saved during computation pays back for the time invested. The only exception to this rule is when modeling thermal effects; because all parameters of the equations are highly dependent upon temperature it can often be very time-consuming to set up the Jacobin correctly. Finally, backtracking with these models, if one reads the art of scientific computing, one comes away with the impression that backtracking is useful to help convergence of these models. I have, however, never found that the computational overhead in setting up the backtracking speeds up the convergence enough to compensate for the computational overhead. In my experience, if you need to implement backtracking, your derivatives in the Newton solver are probably wrong. The matrix Equation 7.23 represents a very sparse system, and the best sparse matrix solver I have found to date is UMFPACK, published by Timothy A. Davis at the Texas A&M University.

Very often, one needs to fit these models to multiple sets of experimental data in order to extract device material. This involves a search of a large parameter space. I have found the most effective method for doing this to be the downhill simplex algorithm from the GNU scientific library; it avoids the need to evaluate derivatives of the error function and is robust. The most efficient strategy to fitting data sets is to start 100 copies of the model off on a cluster with randomly selected material parameters, then to let the downhill simplex algorithm improve the fits. After a while the fits will converge. If the result is acceptable, stop the fitting process. If not, then a local minima has been found and the fitting process should be restarted with another random set of parameters.

7.8 Conclusions

This chapter has given an overview of modeling organic semiconductors and highlighted the difference between classical materials and organic materials. We covered microscopic modeling of organic materials from the quantum perspective. We then discussed MD to simulate thin film formation. Finally, we discussed some aspects of device models used in the simulation of organic electronic devices. Organic materials are far more complex than traditional inorganic semiconductors due to the need to understand the form–function relationships, which govern their performance; this chapter represents only the briefest of introductions to the subject. Further reading on organic electronics can be found in Brutting and Adachi (2012).

References

Adams, J. et al. Air-processed organic tandem solar cells on glass: Toward competitive operating lifetimes. *Energy Environ. Sci.* 8 (2015): 169–176.

Azoff, E. M. Computer Modeling of Heterojunction Bipolar Transistors. PhD thesis. Sheffield: University of Sheffield (1986).

Berendsen, et al. (1995) *Comp. Phys. Comm.* 91: 43–56.

Bloch, F. Über die Quantenmechanik der Elektronen in Kristallgittern. *Z. Phys.* 52 (1928): 555–600.

Boix, P. P., G. Garcia-Belmonte, U. Munecas, M. Neophytou, C. Waldauf, and R. Pacios. Determination of gap defect states in organic bulk heterojunction solar cells from capacitance measurements. *Appl. Phys. Lett.* 95 (2009): 233302–233303.

Braun, C. L. Electric field assisted dissociation of charge transfer states as a mechanism of photocarrier production. *J. Chem. Phys.* 80 (1984): 4157.

Brutting, W. and C. Adachi. *Physics of Organic Semiconductors*, 2nd Ed. New York, NY: Wiley (2012).

Canali, C., C. Jacoboni, F. Nava, G. Ottaviani, and A. Alberigi-Quaranta. Electron drift velocity in silicon. *Phys. Rev. B* 12 (1975): 2265.

Carr, J. A. and S. Chaudhary. The identification, characterization and mitigation of defect states in organic photovoltaic devices: A review and outlook. *Energy Environ. Sci.* 6 (2013): 3414–3438.

Chen, B., I. Siepmann, S. Karaborni, and M. L. Klein. Vapor-liquid and vapor-solid phase equilibria of fullerenes: The role of the potential shape on the triple point. *J. Phys. Chem. B* 107 (2003): 12320–12323.

Chiang, C. K. et al. Electrical conductivity in doped polyacetylene. *Phys. Rev. Lett* 39 (1977): 1098.

Dale, M. and S. M. Benson. Energy balance of the global photovoltaic (PV) industry—Is the PV industry a net electricity producer? *Environ. Sci. Technol.* 47 (2013): 3482–3489.

Dattani, R. A general mechanism for controlling thin film structures in all-conjugated block copolymer: Fullerene blends. *J. Mater. Chem. A* 2 (2014): 14711–14719.

Delannoy, Y. Purification of silicon for photovoltaic applications. *J. Cryst. Growth* 360 (2012): 61–67.

Deschler, F, D. Riedel, B. Ecker, E. von Hauff, E. Da Como, and R. C. I. MacKenzie. Increasing organic solar cell efficiency with polymer interlayers. *Phys. Chem. Chem. Phys.* 15 (2013): 764–769.

Fluxim, Setfos "semiconducting emissive thin film optics simulator", http://www.fluxim.com/drift-diffusion-module/. (n.d).

Fock, V. Näherungsmethode zur Lösung des quantenmechanischen Mehrkörperproblems. *Z. Phys.* 61, no. 1(1930): 126–148.

Foertig, A., J. Rauh, V. Dyakonov, and C. Deibel. Shockley equation parameters of P3HT:PCBM solar cells determined by transient techniques. *Phys. Rev. B* 86 (2012): 115302.

Guilbert, A. A. Y. et al. Effect of multiple adduct fullerenes on microstructure and phase behavior of P3HT: Fullerene blend films for organic solar cells. *ACS Nano* 6 (2012): 3868–3875.

Guilbert, A. A. Y. et al. Spectroscopic evaluation of mixing and crystallinity of fullerenes in bulk heterojunctions. *Adv. Funct. Mater.* 24 (2014): 6972–6980.

Guilbert, A. A. Y. et al. Influence of bridging atom and side chains on the structure and crystallinity of cyclopentadithiophene–benzothiadiazole polymers. *Chem. Mater.* 26, no. 2 (2014): 1226–1233.

Guilbert, A. A. Y. et al. Temperature-dependent dynamics of polyalkylthiophene conjugated polymers: A combined neutron scattering and simulation study. *Chem. Mater.* 27, no. 22 (2015): 7652–7661.

Hanfland, R., M. A. Fischer, W. Brütting, U. Würfel, and R. C. I. MacKenzie. The physical meaning of charge extraction by linearly increasing voltage transients from organic solar cells. *Appl. Phys. Lett.* 103 (2013): 063904.

Hodgkiss, J., G. Conboy, G. Hutchings, S. Higgins, and I. Galbraith. Organic photovoltaics and energy: General discussion. *Faraday Discuss.* 174 (2014): 341–355.

Hohenberg, P. and W. Kohn. Inhomogeneous electron gas. *Phys. Rev.* 136 (1964): B864.

Jiao, Y., F. Zhangl, M. Grätzel, and S. Meng. Structure–property relations in all-organic dye-sensitized solar cells. *Adv. Funct. Mater.* 23 (2013): 424–429.

Kirchartz, T. et al. Sensitivity of the Mott–Schottky analysis in organic solar cells. *J. Phys. Chem. C* 116, no. 14 (2012): 7672–7680.

Kirchartz, T., B. E. Pieters, J. Kirkpatrick, U. Rau, and J. Nelson. Recombination via tail states in polythiophene: Fullerene solar cells. *Phys. Rev. B* 83 (2011): 115209

Kohn, W. and L. J. Sham. Self-consistent equations including exchange and correlation effects. *Phys. Rev.* 140 (1965): A1133.

Koster, L. J. A., E. C. P. Smits, V. D. Mihailetchi, and P. W. M. Blom. Device model for the operation of polymer/fullerene bulk heterojunction solar cells. *Phys. Rev. B* 72 (2005): 085205.

Kwiatkowsk, J. *From Molecules to Mobilities: Modelling Charge Transport in Organic Semiconductors.* Thesis London: Imperial College London (2009).

LAMPS. http://lammps.sandia.gov/. (n.d.).

Letheby, H. On the production of a blue substance by the electrolysis of sulphate of aniline. *J. Chem. Soc.* 15 (1862): 161–163.

Lim, J. J. Simulation of double quantum well GalnNAs laser diodes. *IET Optoelectron.* 1, no. 6 (2007): 259–265.

MacKenzie, R. C. I. et al. Interpreting the density of states extracted from organic solar cells using transient photocurrent measurements. *J. Phys. Chem. C* 117, no. 24 (n.d.): 12407–12414.

MacKenzie, R. C. I. et al. Loss mechanisms in high efficiency polymer solar cells. *Adv. Energy Mater.* 6, no. 4 (2016): 1501742.

MacKenzie, R. C. I., J. M. Frost, and J. Nelson. A numerical study of mobility in thin films of fullerene derivatives. *J. Chem. Phys.* 132 (2010): 064904.

MacKenzie, R. C. I., A. Göritz, S. Greedy, E. von Hauff, and J. Nelson. Theory of Stark spectroscopy transients from thin film organic semiconducting devices. *Phys. Rev. B* 89 (2014): 195307.

MacKenzie, R. C. I., T. Kirchartz, G. F. A. Dibb, and J. Nelson. Modeling nongeminate recombination in P3HT: PCBM solar cells. *J. Phys. Chem. C* 115, no. 19 (2011): 9806–9813.

MacKenzie, R. C. I., C. G. Shuttle, M. L. Chabinyc, and J. Nelson. Extracting microscopic device parameters from transient photocurrent measurements of P3HT: PCBM solar cells. *Adv. Energy Mater.* 2, no. 6 (2012): 662–669.

Nelson, J. Diffusion-limited recombination in polymer-fullerene blends and its influence on photocurrent collection. *Phys. Rev. B* 67 (2003): 155209.

New Scientist. Brighter outlook for cheaper silicon solar cells. *New Scientist* (August 1979): 522.

Ridley, J. and M. Zerner. An intermediate neglect of differential overlap technique for spectroscopy: Pyrrole and the azines. *Theor. Chim. Acta* 32 (1973): 111–134.

Roichman, Y. and N. Tessler. Generalized Einstein relation for disordered semiconductors—Implications for device performance. *Appl. Phys. Lett.* 80 (2002): 1948.

Schafferhans, J., A. Baumann, A. Wagenpfahl, C. Deibel, and V. Dyakonov. Oxygen doping of P3HT: PCBM blends: Influence on trap states, charge carrier mobility and solar cell performance. *Org. Electron.* 11, no. 10 (2010): 1693–1700.

Shockley, W. and W. T. J. Read. Statistics of the recombinations of holes and electrons. *Phys. Rev.* 87 (1952): 835.

Shuttle, C. G., B. O'Regan, A. M. Ballantyne, J. Nelson, D. D. C. Bradley, and J. R. Durrant. Bimolecular recombination losses in polythiophene: Fullerene solar cells. *Phys. Rev. B* 78 (2008): 113201.

Sims, L. et al. Investigation of the s-shape caused by the hole selective layer in bulk heterojunction solar cells. *Org. Electron.* 15 (2014): 2862–2867.

Street, R. A. Localized state distribution and its effect on recombination in organic solar cells. *Phys. Rev. B* 84 (2011): 075208.

Szabo, A. *Modern Quantum Chemistry*. Mineola, NY: Dover Books, 1996.

To, T. T. and S. Adams. Modelling of P3HT: PCBM interface using coarse-grained forcefield derived from accurate atomistic forcefield. *Phys. Chem. Chem. Phys.* 16 (2014): 4653–4663.

Tress, W, K. Leo, and M. Riede. Influence of hole-transport layers and donor materials on open-circuit voltage and shape of I–V curves of organic solar cells. *Adv. Funct. Mater.* 21, no. 11 (2011): 2140–2149.

Voigt, M. M. et al. Gravure printing for three subsequent solar cell layers of inverted structures on flexible substrates. *Sol. Energ. Mat. Sol. Cells* 95 (2011): 731–734.

Voigt, M. M. et al. Gravure printing inverted organic solar cells: The influence of ink properties on film quality and device performance. *Sol. Energ. Mat. Sol. Cells* 105 (2012): 77–85.

Yersin, H. *Highly Efficient OLEDs with Phosphorescent Materials*. New York, NY: John Wiley & Sons (2008).

Zhao, K. et al. Vertical phase separation in small molecule: Polymer blend organic thin film transistors can be dynamically controlled. *Adv. Funct. Mater.* 26 (2016): 1–10.

8

Polarization in III-N Semiconductors

Max A. Migliorato

Joydeep Pal

Xin Huang

Weiguo Hu

Morten Willatzen

and

Yousong Gu

8.1 Introduction

In recent years, because of the extraordinary technological developments in solid-state lighting, there has been increased interest in polarization effects under external strain, the so-called piezoelectric polarization. GaN and its related semiconductor materials (AlN and InN) are of crucial importance for the realization of the fundamental component of solid-state lighting, the blue light-emitting diode (LED): the Nobel Prize in Physics was jointly awarded to Isamu Akasaki, Hiroshi Amano, and Shuji Nakamura (Nobel, 2014) for this work. With LEDs in mind, here we cover all aspects of modeling the effects of piezoelectric polarization in III-N alloys. A prediction was reported in the *Proceedings of the State-of-the-Art Program on Compound Semiconductors XXXIX and Nitride and Wide Bandgap Semiconductors for Sensors, Photonics and Electronics* that the output efficiency of III-N semiconductor-based LEDs was always going to be limited by the magnitude of the internal polarization fields (Kopf, 2003). In the years since, it has become universally acknowledged that such a prediction was correct. Detailed knowledge of the piezoelectric properties of III-N semiconductors is therefore an essential element in designing more efficient LEDs.

In this chapter, we review the different models that have been proposed over the years and how they compare to experimental measurements. We outline both linear and nonlinear models and discuss the recent experimental determination of values of spontaneous polarization in polymorphic crystals.

8.2 Piezoelectric Effect

The direct piezoelectric effect is the generation of an electric dipole moment in certain crystals if a stress is applied and the electric dipole moment is proportional to the applied stress. Such an effect was first demonstrated by the Curie brothers, Jacques and Pierre, in 1880 (Curie and Curie, 1880). The word "piezo" is from the Greek "piezen (πιέζειν)" which means "to press or squeeze." The effect has been understood as the charge formation in some crystals due to the ionic displacements under some mechanical pressure. The converse piezoelectric effect, where the crystals deforms under the application of external electric field in the polarization direction, was outlined mathematically in 1881 by Lippmann (1881) from fundamental thermodynamic principles. Later, in 1910, the publication *Lehrbuch der Kristallphysik* (textbook on crystal physics) by Woldemar Voigt distinguished 20 natural piezoelectric crystal classes and for the first time treated the piezoelectric coefficients using tensor analysis. The use in this work of a contracted notation for the indexes gave rise to what is conventionally known as the Voigt notation (Voigt, 1910).

The reasons the property of piezoelectricity is closely linked to the crystal symmetry of materials is that mechanical stress must break the charge symmetry for a change in polarization to be observed. This asymmetry in the charge density leads to an electric field in the crystal which is known as the piezoelectric field.

Though it is often challenging to perform direct experimental measurements of the resulting electric field, an impact can be more easily observed on both the electrical and optical properties of the crystal.

Of interest for this chapter is that piezoelectricity is commonly found in semiconductor materials typically used in optoelectronic applications, where the crystal symmetry is either cubic (e.g., GaAs and InP) or hexagonal (GaN and ZnO), as shown in Figure 8.1. All III-nitride (III-N) semiconductors typically found in the wurtzite crystal structure (Figure 8.1d), not only GaN, strong polarization along the polar axis. This phenomenon has long been known. The most important quantitative theoretical analysis was published by Bernardini et al. (1997). At the time of this writing, the parameters proposed in their work have been cited 1500 times according to the American Physical Society (APS) database.

8.3 Piezoelectricity in Electronic Materials

Though bulk materials possess the property of piezoelectricity, the changes with pressure in the total polarization, which is the sum of spontaneous (polarization that exists even in the absence of external pressure) and strain-induced piezoelectric polarization, lead to an increase in the polarization-induced electrostatic charge density. In a structurally pure bulk semiconductor, the variation in the polarization fields will be observed at the surfaces as the electrical dipoles cancel one another throughout the structure. Nonetheless, the existence of the surface states or mobile carriers can nullify the resultant surface charges. Therefore, in conventional mainstream electronics, there is little scope for forcing stress on bulk materials for any practical purpose. This is not to say that there are no applications at all, and in fact, there is an entire field (piezotronics, from piezoelectric and electronics) which is based on using nanostructured materials (typically nanowires) and their strain–voltage relationship (Wang, 2007). Furthermore, there are some notable applications in micro- and nanoelectro mechanical systems (MEMS and NEMS). However, in mainstream optoelectronics, strain is often built into the crystal layers rather than supplied externally. This is the case of heterostructures, i.e., layers of different semiconductor materials carefully deposited through an epitaxial process (typically for III-N using metal organic chemical vapor deposition [MOCVD]), such that

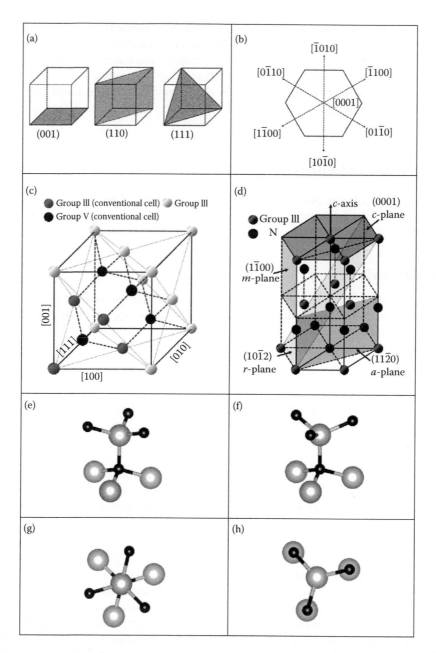

FIGURE 8.1 Zincblende (left) and wurtzite (right) crystals: (a)–(b) high symmetry planes, (c)–(d) crystal structures and atomic positions—for zincblende (c), atoms belonging to the conventional cell are given alongside the crystal directions, while for wurtzite (d), also given are the naming conventions of the high symmetry planes—(e)–(f) atomic structure side view (along the [111] and [0001] directions for the zincblende and wurtzite crystals, respectively) and (g)–(h) atomic structure top view (from the [111] and [0001] planes for the zincblende and wurtzite crystals, respectively).

the crystal maintains a high-crystal order. When two materials in a heterostructure have different lattice parameters, then the atomic crystal cells of one tend to stretch to adapt to those of the material they are being deposited over. This is called lattice-mismatched heteroepitaxy, which can produce very high-quality crystals with a very low number of crystal defects.

In a polar heterostructure, the strained layers develop a small dipole factor in each unit cell, which cancels throughout except at the interfaces. This produces a uniform distribution of polarization, associated to a net surface charge density at the interfaces of the heterostructure, leading to a net electric field which has a significant influence on the electro-optical characteristics of the material.

In order to understand this in more detail, we first need to outline the conventional formalism used to represent such strain-induced polarization effects.

8.4 Mathematical Formalism

As implied in Section 8.3, piezoelectricity can be described in terms of microscopic dipoles, and it can therefore have an atomistic description. However, because it is also a macroscopic property, a scalable mathematical description is also practical (Singh, 1992, 2003).

8.4.1 Piezoelectric Tensor

In the case of piezoelectricity, the requirement is fulfilled by describing the resulting polarization field $\mathbf{P}_{pz}$ in terms of the piezoelectric moduli $\mathbf{d}_{ijk}$ or piezoelectric coefficients $\mathbf{e}_{ijk}$ which are tensors coupling to either the stress tensor $\boldsymbol{\sigma}_{jk}$ or the strain tensor $\boldsymbol{\varepsilon}_{jk}$

$$\mathbf{P}_{pz,\,i} = \mathbf{d}_{ijk}\boldsymbol{\sigma}_{jk} = \mathbf{e}_{ijk}\boldsymbol{\varepsilon}_{jk}, \tag{8.1}$$

where the stress/strain relationship is dependent on the elastic tensor $\mathbf{c}_{ijkl}$:

$$\boldsymbol{\sigma}_{ij} = \mathbf{c}_{ijkl}\boldsymbol{\varepsilon}_{kl} \tag{8.2}$$

Understanding the various subscript indexes is not as complicated as one would imagine. The four indexes i, j, k, and l all range from 1 to 3, to represent the three directions x, y, and z of a three-dimensional (3D) space. Using Equation 8.3 we illustrate how to interpret the quantities with, e.g., three subscript indexes:

$$\mathbf{P}_{pz,\,1} = \mathbf{e}_{123}\boldsymbol{\varepsilon}_{23}. \tag{8.3}$$

This case indicates that the polarization in the x (1) direction is generated by a strain in the yz (23) direction, multiplied by the related piezoelectric coefficient, where the first index is the direction of the polarization and the other two are the direction of the strain. In this case, the strain is in a diagonal direction compared to the x-, y-, or z-axes. Assuming that the material can be taken as cubic, with sides parallel to the axes, $\boldsymbol{\varepsilon}_{23}$ would turn the square face into a diamond. This is commonly referred to as a shear deformation. If instead, the two subscript indexes of the strain component were the same, then the direction of the strain would be parallel to one of the sides of the cube and the strain would simply modify one of its dimensions. As the jk indexes are interchangeable between themselves, the nine combinations of two indexes contain three redundancies. It is then common to use a reduced notation instead. The simplification (Voigt, 1910) works by describing pairs of indexes with the following convention:

$$\mathbf{xx} = 1, \mathbf{yy} = 2, \mathbf{zz} = 3, \mathbf{yz} = 4, \mathbf{zx} = 5, \mathbf{xy} = 6. \tag{8.4}$$

This is also justified by noting that the symmetry of the tensors $\mathbf{d}_{ijk}$, $\boldsymbol{\sigma}_{jk}$, $\mathbf{e}_{ijk}$, and $\boldsymbol{\varepsilon}_{jk}$ belongs to the indexes j and k rather than i. In the reduced Voigt notation which compresses the jk indexes into a single index α, the matrix representations for the ε_α and $e_{i\alpha}$ tensors are therefore

$$\varepsilon_\alpha = \begin{pmatrix} \varepsilon_{xx} & \varepsilon_{yx} & \varepsilon_{zx} \\ \varepsilon_{xy} & \varepsilon_{yy} & \varepsilon_{zy} \\ \varepsilon_{xz} & \varepsilon_{yz} & \varepsilon_{zz} \end{pmatrix} = \begin{pmatrix} \varepsilon_1 \\ \varepsilon_2 \\ \varepsilon_3 \\ \varepsilon_4 \\ \varepsilon_5 \\ \varepsilon_6 \end{pmatrix} \tag{8.5}$$

$$e_{i\alpha} = \begin{pmatrix} e_{11} & e_{12} & e_{13} & e_{14} & e_{15} & e_{16} \\ e_{21} & e_{22} & e_{23} & e_{24} & e_{25} & e_{26} \\ e_{31} & e_{32} & e_{33} & e_{34} & e_{35} & e_{36} \end{pmatrix}. \tag{8.6}$$

Note that when using Voigt notation, the shear components are given by $\varepsilon_\alpha = \varepsilon_{jk} + \varepsilon_{kj} = 2\varepsilon_{jk}$, where in the last passage the factor of 2 arises if we assume that the indexes jk are interchangeable. Omitting the factor of 2 has been a recurring problem in the literature (Romanov et al., 2006).

Although there are many different coefficients, symmetry arguments reveal that many of the coefficients of $e_{i\alpha}$ are either vanishing or equal to others.

8.4.1.1 Piezoelectric Tensor in Zincblende Crystals

In a zincblende crystal, there are only three independent elastic constants ($c_{11} = c_{22} = c_{33}, c_{12} = c_{13} = c_{23}, c_{44} = c_{55} = c_{66}$) that reflect the isotropic cubic symmetry. Furthermore, it was shown by Nye (1957) that there are only three nonzero independent coefficients:

$$e_{i\alpha} = \begin{pmatrix} 0 & 0 & 0 & e_{14} & 0 & 0 \\ 0 & 0 & 0 & 0 & e_{25} & 0 \\ 0 & 0 & 0 & 0 & 0 & e_{36} \end{pmatrix}, \tag{8.7}$$

which are all of equal value. In such a case, it is conventional to refer to all three of them by the coefficient with the lowest index, i.e., e_{14}.

8.4.1.2 Piezoelectric Tensor in Wurtzite Crystals

In wurtzite crystals, there are five independent elastic constants $c_{11} = c_{22}, c_{33}, c_{12}, c_{13} = c_{23}, c_{44}$ which reflect the isotropic hexagonal symmetry in two directions and a different symmetry along the third direction (c-axis).

The wurtzite phase, typical of III-N materials, is a bit more complicated than zincblende in terms of piezoelectric tensor, and instead of three there are five independent coefficients:

$$e_{i\alpha} = \begin{pmatrix} 0 & 0 & 0 & 0 & e_{15} & 0 \\ 0 & 0 & 0 & e_{24} & 0 & 0 \\ e_{31} & e_{32} & e_{33} & 0 & 0 & 0 \end{pmatrix}, \tag{8.8}$$

which can be reduced to three by noting that $e_{31} = e_{32}$ and $e_{15} = e_{24}$. Again, the convention is to refer to them as e_{31} and e_{15}.

8.4.2 Crystal Vectors

Thus far, we have not specified what the x-, y-, and z-directions are for the two crystals we have considered, i.e., the zincblende and the wurtzite structures. The choices are not arbitrary.

8.4.2.1 Crystal Directions in Zincblende Crystals

For zincblende crystals, one conventionally takes x, y, and z aligned with the [100], [010], and [001] crystal directions (Figure 8.1c). This choice uniquely links the tensor formalism just described with the atomic structure, given that the crystal directions, which appear to simply describe a cubic cell, are also such that all atomic bonds are in one of four directions [111], $[\bar{1}\bar{1}1]$, $[1\bar{1}\bar{1}]$, or $[\bar{1}1\bar{1}]$. The [111] direction is known as the polar axis of the zincblende crystal, i.e., only polarization at angles less than $\pi/2$ from the polar axis is nonvanishing.

8.4.2.2 Crystal Directions in Wurtzite Crystals

In the wurtzite crystal, the crystal directions are not described as simply as in zincblende, as it is a very common convention to use a four-index representation instead of the more intuitive three-index representation. The fourth index is actually placed in third position, and instead of being independent of the other three indexes, it is the negative of the sum of the first two:

$$\vec{v} = [h, k, i = -(h + k), l]. \tag{8.9}$$

The fourth index i clearly would be redundant for describing a 3D space, but there is some advantage in using such a representation. Consider, e.g., the crystal planes (110) and $(1\bar{2}0)$. These two planes are actually equivalent, but this is not obvious in their three-index representation. If we instead use the four-index representation, then the same planes are written as $(11\bar{2}0)$ and $(1\bar{2}10)$, which are now easily identified as equivalent upon cyclic permutation of the indexes.

With the four-index representation (Figure 8.1b), the convention is that [0001], the c-axis, is the one perpendicular to the plane with hexagonal symmetry, while the vector $[10\bar{1}0]$ (and the other five vectors obtained by cyclic permutation of the first three indexes) is orthogonal to one side of the hexagon. Hence, it represents the vector normal to one of the reticular planes.

In terms of atomic bonds, one is now oriented in the [0001] direction and this bond is the one which is mostly connected to the piezoelectric polarization of wurtzite crystals. The [0001] direction or c-axis is not the only polar direction in a wurtzite crystal (polarization in the (0001) plane can also be nonzero), but it is the only polar direction for quantum wells that have the c-axis as the growth direction.

8.4.3 Classic Description of Strain-Induced Polarization

In the previous section, we have just outlined how a piezoelectric polarization field with nonzero values exists along the [111] polar direction, which is normally at an angle from the conventional growth axis (001) in zincblende structures. The polarization is present instead along the c-axis direction in an [0001] epitaxially grown coherently strained wurtzite III-N quantum well. For both of these cases, the linear strain can be easily predicted, provided one has knowledge of the elastic constants of the material.

In the remainder of this section, we will make use of two very widely diffused definitions, particularly among experimentalists. The first is the use of $\varepsilon_\perp$ to identify the strain component along the growth direction of the crystal. The second is to use $\varepsilon_\parallel$ to describe strain in the growth plane, assuming that in such a plane (orthogonal to the growth direction) the strain is perfectly isotropic. The advantage of this notation is that from the point of view of experimental measurements, it is independent of the crystal orientation or even crystal symmetry. We see that this is not the case when one has to determine the strain theoretically, from the elastic properties of the material.

8.4.3.1 Strain-Induced Polarization in Zincblende Crystals

In the case of zincblende, the strain along the growth direction is given by

$$\varepsilon_\perp = -2\frac{c_{12}}{c_{11}}\varepsilon_\parallel. \tag{8.10}$$

This expression is only valid for material with the direction $\perp$ aligned with the [001] crystal direction (and $\parallel$ aligned with either of the [100] or [010] crystal directions, assuming that the applied strain is the same for both directions). In the other common case of material with the direction $\perp$ aligned with the [111] crystal direction, which is the only one where piezoelectric polarization is present, then the correct expression is

$$\varepsilon_\perp = -2\frac{c_{11} + c_{12} + 4c_{44}}{c_{11} + 2c_{12} + c_{44}}\varepsilon_\parallel. \tag{8.11}$$

In this case, it is easy to see how the strain produced by $\varepsilon_\parallel$ is now equivalent to that produced by the off-diagonal components of the strain tensor, which are the only ones that couple with the piezoelectric coefficients of Equation 8.7:

$$P^{zb}_{pz,1} = e_{14}\varepsilon_4 \tag{8.12}$$

$$P^{zb}_{pz,2} = e_{25}\varepsilon_5 \tag{8.13}$$

$$P^{zb}_{pz,3} = e_{36}\varepsilon_6. \tag{8.14}$$

The quantity that links the parallel and perpendicular strain is usually referred to as the Poisson ratio (μ):

$$\varepsilon_\perp = -\frac{\varepsilon_\parallel}{\mu}. \tag{8.15}$$

While the expressions given indicate that μ is a constant that only depends on the crystal direction, in reality it is a nonlinear quantity in the strain. We show in later sections that most of the quantities discussed in this chapter are nonlinear in the strain.

8.4.3.2 Strain-Induced Polarization in Wurtzite Crystals

In wurtzite crystals, a free-surface boundary condition for the surface charge ($\sigma_{zz} \equiv \sigma_{33} = 0$) is expected to be present as the polarization is directed in the [0001] or z-direction. In this case, we can still use the formalism of Equations 8.10 and 8.15, but the expression for the Poisson ratio (μ) is different compared to the zincblende crystal phase:

$$\varepsilon_\perp = -2\frac{c_{13}}{c_{33}}\varepsilon_\parallel. \tag{8.16}$$

This expression is only valid for materials with the direction $\perp$ aligned with the [0001] crystal direction (and $\parallel$ aligned with either of the [10$\bar{1}$0] direction and the other five vectors obtained by cyclic permutation of the first three indexes, assuming that the applied strain is the same for all directions). This is the case for the majority of III-N semiconductor optoelectronic devices grown on either sapphire or silicon substrates. Given these definitions, the polarization along, e.g., the axis, originating from a simple tetragonal distortion of the wurtzite crystal, can be expressed as

$$P_{pz,3} = 2\left(e_{31} - \frac{c_{13}}{c_{33}}e_{33}\right)\varepsilon_1 = 2d_{31}\left(c_{11} + c_{12} - 2\frac{c_{13}^2}{c_{33}}e_{33}\right)\varepsilon_1, \tag{8.17}$$

where for the last passage we also used the piezoelectric moduli instead of the piezoelectric coefficient.

For completeness, it is also worth mentioning that an additional term can be added to Equation 8.17, which is linked to the piezoelectric coefficient $e_{15} = e_{24}$. This is linked to shear strain, whereas the normally rectangular facets orthogonal to either the [10$\bar{1}$0] direction or all the other five vectors obtained by cyclic permutation of the first three indexes become more diamond like. This can be important in 3D nanostructures such as quantum dots (Tomić et al., 2015) or flexible nanowires, but not in conventional two-dimensional (2D) structures like the quantum wells typically used for realizing LEDs.

For recommended values of the lattice and elastic parameters to be used for III-N semiconductors in Equations 8.10 through 8.17, we direct the reader to the widely used compendiums provided by Vurgaftman and Meyer (2003) and Romanov et al. (2006), although amendments to the recommended sets are routinely proposed (see, e.g., Pedesseau et al., 2012). A comprehensive discussion of the recommended values of the piezoelectric coefficients will be given in Section 8.5.1.

8.4.4 Spontaneous Polarization in Wurtzite Crystals

Though epitaxially grown wurtzite semiconductors tend to have a sizeable strain-induced piezoelectric polarization, an additional feature not present in zincblende semiconductors is spontaneous polarization, i.e., the polarization that exists even in the absence of any strain, in both III-N and II-VI semiconductors. Spontaneous polarization (like the strain-induced one) cannot be measured directly. It can only be inferred from experimental data of electro-optical properties and their change as a result of strain. This meant that for a long time the only available values were calculated ones. This has changed recently, and later we discuss the recent values of spontaneous polarization proposed in experimental measurements.

Spontaneous polarization develops from the atomic layout in the bulk materials. In the wurtzite structure, the *c*-axis marks the overlap of the neighboring dual tetrahedrons in contrast to the zincblende crystals (Figure 8.1e through h). This arrangement gives rise to closer second-nearest neighbors in the wurtzite compared to the zincblende phase. The modified interatomic forces lead also to a marginal change of the interatomic separation (embodied in the so-called *u* parameter) between the first-nearest neighbors (ionic contribution) together with a rearrangement of the electronic clouds (electronic contribution). The combination of these contributions is that all bonds that are lined up along the [0001] direction become stretched, which results in a nonzero dipole moment along the [0001] crystal direction being present in the absence of any strain. The result of this atomic scale phenomenon is the presence of a macroscopic spontaneous polarization ($P_{3,\text{sp}}$ or simply P_{sp}).

8.5 Piezoelectric Coefficients of Wurtzite III-N Semiconductors

The wurtzite III-N structure demonstrates a peculiar property in having piezoelectric coefficients similar to those of group II–VI materials and quite strikingly opposite to the other group III–V materials. The III-N materials also differ from the other III–V compounds with a larger ionic charge and the internal-strain ionic contribution becomes larger than the clamped-ion term. The fundamental contribution to the understanding of piezoelectric fields in III-N is the list of all piezoelectric coefficients and spontaneous polarization terms provided by Bernardini et al. (1997). Already cited 1500 times according to the APS database, it is impossible to reference each and every paper that has made use of the coefficients. We wish, however, to discuss in some detail the validity of the linear model in view of outlining the now generally agreed need for including nonlinear effects.

8.5.1 Proper versus Improper Coefficients

Before discussing the calculated and measured piezoelectric coefficients it is worth making a distinction between proper and improper piezoelectric coefficients. A very clear explanation is given in Caro et al. (2015b) and a more detailed discussion is available from Vanderbilt (2000) and references therein. The "improper" coefficients link the electric polarization vector and strain through Equation 8.1. An alternative yet equivalent expression is

$$e_{i\alpha} = \frac{\partial P_i}{\partial \varepsilon_\alpha}. \tag{8.18}$$

Such an expression is obviously intuitive as the coefficients express the change in the polarization vector components as a function of the change in one of the strain components.

The "proper" coefficients instead are conceptually more complex as they link the adiabatic (no change in heat) change in the components of the current density vector J in response to a "slow" deformation:

$$e_{i\alpha}^{p} = \frac{\partial J_i}{\partial \dot{\varepsilon}_{\alpha}} \tag{8.19}$$

$$\dot{\varepsilon}_{\alpha} = \frac{d\varepsilon_{\alpha}}{dt}. \tag{8.20}$$

"Proper" and "improper" coefficients are generally different and there is a set of transformations which allows one to obtain one set from another. For wurtzite this set is

$$e_{15}^{p} = e_{15} - \frac{1}{2}P_{3,sp} \tag{8.21}$$

$$e_{31}^{p} = e_{31} + P_{3,sp} \tag{8.22}$$

$$e_{33}^{p} = e_{33}. \tag{8.23}$$

The "improper" coefficient is the correct one to use when calculating polarization charges, e.g., interfacial charge accumulation in quantum wells (Bernardini et al., 2001). The "proper" coefficient instead is the quantity that can be readily compared to experiments where flowing currents are measured. For systems with small spontaneous polarization the corrections can be negligible, but that is often not the case. In the following sections, for piezoelectric coefficients we always intend improper rather than proper coefficients.

8.5.2 Linear Piezoelectric Coefficients of III-N Semiconductors

The first report on the magnitude of the piezoelectric coefficients of III-N was by Bernardini et al. (1997), followed a few years later by a second report (Bernardini et al., 2001) with updated values. The coefficients (Table 8.1) were calculated using *ab initio* density functional theory and the Berry-phase approach (Vanderbilt, 2000).

TABLE 8.1 Calculated Piezoelectric Coefficients (~ in Units of C/m^2) for Wurtzite III-N

Coefficient	GaN	AlN	InN
P_{sp} (C/m^2)	−0.034 (−0.029)	−0.090 (−0.081)	−0.042 (−0.032)
e_{31}^{p} (C/m^2)	−0.37	−0.62	−0.45
e_{31} (C/m^2)	−0.34 (−0.49)	−0.53 (−0.60)	−0.41 (−0.57)
e_{33} (C/m^2)	0.67 (0.73)	1.50 (1.46)	0.81 (0.97)
e_{15} (C/m^2)	−0.57 Pal et al. (2011a), −0.38 Shimada et al. (1998), −0.29 Pedesseau et al. (2012)	−0.6 Pal et al. (2011a), −0.48 Tsubouchi and Mikoshiba (1985), −0.35 Pedesseau et al. (2012)	−0.65 Pal et al. (2011a), −0.44 Shimada et al. (1998)

Source: Reproduced with permission from Bernardini F et al., *Physical Review B*, 56, R10024–R10027, 1997. Copyright 1997 by the **American Physical Society;** Bernardini F et al., *Physical Review B*, 63, 193201, 2001. Copyright 2001 by the **American Physical Society**.

The superscript p indicates the proper rather than improper coefficient, which for e_{33} has always the same value (Equation 8.23). To complete the table we have also added the values of e_{15} from Pal et al. (2011a) who pointed out the erroneous sign given in Vurgaftman and Meyer (2003) together with proposed values from Shimada et al. (1998), Tsubouchi and Mikoshiba (1985), and Pedesseau et al. (2012). Where available, values in brackets are from Bernardini et al. (1997).

Only the coefficients coupling with strains along the c-plane or along the c-direction were calculated, i.e., no value for e_{15} was calculated. Values of the spontaneous polarization were given instead.

To complete Table 8.1 we have also added the values of e_{15} from Pal et al. (2011a) who pointed out the erroneous sign given in Vurgaftman and Meyer (2003) together with proposed values from Shimada et al. (1998), Tsubouchi and Mikoshiba (1985), and Pedesseau et al. (2012). It is evident that there is a large spread of proposed values for e_{15} and at the time this writing, it is still unsure which should be the recommended one.

As many have pointed out, the piezoelectric coefficients are one order of magnitude larger than the typical III–V semiconductor. The largest P_{sp}, that of AlN, is only about three to five times smaller than in typical ferroelectric perovskites. Where experimental data were available, the calculated values were in agreement within 5%.

It is worth pointing out that the results were obtained for the binary III–N and device modeling required a further step. In fact, most experimental data are typically obtained for heterostructures and quantum wells or superlattices. In device modeling, the general wisdom was for many years that the properties of the alloys could be obtained by linear interpolation of the binary values, what is commonly known as Vegard's law (Vegard, 1921). Over the years, anecdotal evidence suggested that the use in itself of the linear piezoelectric coefficients of Bernardini et al. (1997), within any linear or parabolic interpolation scheme for alloys, can lead to piezoelectric fields which are substantially larger than those needed to fit experimental data. A rule of thumb employed by many in the device modeling community has, hence, been to scale down the calculated fields by multiplying by an empirical factor found to be roughly 0.7. However, it could well be that the coefficients or model are not to blame at all and that the correction applied to the piezoelectric field simply corrects other factors incorrectly or is not explicitly taken into account in the models. We discuss some of these concepts in Section 8.8.

8.5.3 Nonlinearities in III-N Semiconductors

In this section, we outline the various nonlinear models of piezoelectricity in III-N semiconductors. Such nonlinearities in the polarization were first reported for III-N by Shimada et al. (1998), expressing theoretically and calculating the dependence of the polarization on deviatoric strain, but without expressly providing a set of nonlinear coefficients. These have ever since been the subject of intense debate also in the context of zincblende semiconductors. It is still not clear whether a conclusive and universally accepted model has been achieved yet. The only thing that can be agreed upon is that nonlinearities are large and in many situations can dominate the magnitude and sign of the polarization.

8.5.3.1 Nonlinearities under Biaxial Strain

Ambacher et al. (2002) used *ab initio* calculations and calculated many polarization properties, while also reporting nonlinearities of the bulk polarization of the binary nitrides subject to biaxial strain ε, which is the difference between $\varepsilon_\perp$ and $\varepsilon_\parallel$. The quadratic dependence for [0001] crystals is given as

$$P_{pz,AlN}(\varepsilon < 0) = (-1.808\varepsilon + 5.624\varepsilon^2) \text{ C/m}^2$$

$$P_{pz,AlN}(\varepsilon > 0) = (-1.808\varepsilon - 7.888\varepsilon^2) \text{ C/m}^2 \qquad (8.24)$$

$$P_{pz,GaN} = (-0.918\varepsilon + 9.541\varepsilon^2) \text{ C/m}^2$$

$$P_{pz,InN} = (-1.373\varepsilon + 7.559\varepsilon^2) \text{ C/m}^2.$$

In 2002, Equation 8.24 constituted the first report in a usable form of nonlinear piezoelectric polarization in III–V semiconductors. However, the reported nonlinear equations are limited to biaxial strain combinations and are not a full second-order piezoelectric tensor treatment. Furthermore, e_{15} is not included as the equations are explicitly designed for c-axis-grown quantum wells where typically e_{15} would not contribute to the total polarization.

8.5.4 Full Second-Order Piezoelectric Tensor Formalism

Before discussing the reported second-order parameters found in the literature, we need to describe here the second-order piezoelectric tensor in some detail. While the 18 components of the first-order strain tensor were given in Equation 8.5, the $18 \times 6 = 108$ components of the second-order tensor couple to two strain components to generate a second-order contribution to the total polarization

$$P_i'' = B_{ijklm} \varepsilon_{jk} \varepsilon_{lm}. \tag{8.25}$$

The tensor B_{ijklm} given in Voigt notation which compresses the indexes jk and lm into the reduced ones α and β, is given by

$$B_{i\alpha\beta} = \begin{pmatrix} B_{11\beta} & B_{12\beta} & B_{13\beta} & B_{14\beta} & B_{15\beta} & B_{16\beta} \\ B_{21\beta} & B_{22\beta} & B_{23\beta} & B_{24\beta} & B_{25\beta} & B_{26\beta} \\ B_{31\beta} & B_{32\beta} & B_{33\beta} & B_{34\beta} & B_{35\beta} & B_{36\beta} \end{pmatrix}, \tag{8.26}$$

where $\beta = 1 \ldots 6$. Just like in the case of first order, symmetry arguments can be used to reduce the number of nonunique components. It is worth going into some detail to explain how this is done for a wurtzite crystal. We want to stress that in Equation 8.26 the second-order tensor components are defined using Voigt notation, using three subscripts, whereas the unreduced representation would require five subscripts, e.g., B_{ijklm}. Equation 8.25 too should therefore be expressed in Voigt notation:

$$P_i'' = B_{i\alpha\beta} \varepsilon_\alpha \varepsilon_\beta. \tag{8.27}$$

The only second-order piezoelectric coefficients that are nonzero are those fulfilling the crystal symmetry which for hexagonal crystals such as wurtzite are those invariant under a rotation of $\pi/3$ about the z-direction (c-axis). Reverting to Cartesian coordinates and initially avoiding Voigt notation, we introduce the complex variables

$$\xi = x + \Im y; \quad \eta = x - \Im y. \tag{8.28}$$

These make rotational invariance more strikingly obvious, as under a rotation of $\pi/3$ the following transformations hold:

$$\xi \to \xi e^{\Im \pi/3}; \quad \eta \to \eta e^{-\Im \pi/3}. \tag{8.29}$$

And if, e.g., in the sequence of subscripts ijklm any number of subscripts is transformed according to Equation 8.29, then the associated coefficient would be invariant and nonzero if the transformation yields overall an invariant contribution to the free energy. It is easily verified that this is fulfilled for the following 11 terms: B_{zzzzz}, $B_{z\xi\eta zz}$, $B_{zz\xi zn}$, $B_{z\xi\eta\xi\eta}$, $B_{z\xi\xi\eta\eta}$, $B_{\eta zzzz\xi}$, $B_{\eta zn\xi\xi}$, $B_{\eta z\xi\eta\xi}$, $B_{\xi zzzn}$, $B_{\xi z\xi\eta\eta}$, $B_{\xi zn\xi\eta}$. We now show that not all of these 11 terms need to be independently determined. Using the relations

$$\xi^2 = \left(x + \Im y\right)^2 = x^2 - y^2 + 2\Im xy \tag{8.30}$$

$$\eta^2 = \left(x - \Im y\right)^2 = x^2 - y^2 - 2\Im xy$$

$$\xi\eta = \left(x + \Im y\right)\left(x - \Im y\right) = x^2 + y^2$$

the associated strain components can then be assumed to transform as

$$\varepsilon_{\xi\xi} = \varepsilon_{xx} - \varepsilon_{yy} + 2\Im\varepsilon_{xy} \quad \text{Using Method} \tag{8.31}$$

$$\varepsilon_{\eta\eta} = \varepsilon_{xx} - \varepsilon_{yy} - 2\Im\varepsilon_{xy}$$

$$\varepsilon_{\xi\eta} = \varepsilon_{xx} + \varepsilon_{yy},$$

while

$$\varepsilon_{z\xi} = \varepsilon_{zx} + \Im\varepsilon_{zy} \tag{8.32}$$

$$\varepsilon_{z\eta} = \varepsilon_{zx} - \Im\varepsilon_{zy}.$$

This now allows us to write the second-order polarization in the z-direction (c-axis), which is invariant for a rotation of $\pi/3$ around the z-direction, as

$$P_z'' = B_{zzzzz}\varepsilon_{zz}^2 + B_{z\xi\eta zz}\left(\varepsilon_{xx} + \varepsilon_{yy}\right)\varepsilon_{zz} + B_{zz\xi z\eta}\left(\varepsilon_{zx} + \Im\varepsilon_{zy}\right)\left(\varepsilon_{zx} - \Im\varepsilon_{zy}\right) + B_{z\xi\eta\xi\eta}\left(\varepsilon_{xx} + \varepsilon_{yy}\right)^2$$

$$+ B_{z\xi\xi\eta\eta}\left(\varepsilon_{xx} - \varepsilon_{yy} + 2\Im\varepsilon_{xy}\right)\left(\varepsilon_{xx} - \varepsilon_{yy} - 2\Im\varepsilon_{xy}\right). \tag{8.33}$$

In order to obtain the c-plane components, we make use of Equations 8.31 and 8.32 and start from

$$P_\eta'' = B_{\eta zzz\xi}\varepsilon_{zz}\left(\varepsilon_{zx} + \Im\varepsilon_{zy}\right) + B_{\eta z\eta\xi\xi}\left(\varepsilon_{zx} - \Im\varepsilon_{zy}\right)\left(\varepsilon_{xx} - \varepsilon_{yy} + 2\Im\varepsilon_{xy}\right) + B_{\eta z\xi\eta\xi}\left(\varepsilon_{zx} + \Im\varepsilon_{zy}\right)\left(\varepsilon_{xx} + \varepsilon_{yy}\right) \tag{8.34}$$

and

$$P_\xi'' = B_{\xi zzz\eta}\varepsilon_{zz}\left(\varepsilon_{zx} - \Im\varepsilon_{zy}\right) + B_{\xi z\xi\eta\eta}\left(\varepsilon_{zx} + \Im\varepsilon_{zy}\right)\left(\varepsilon_{xx} - \varepsilon_{yy} - 2\Im\varepsilon_{xy}\right) + B_{\xi z\eta\xi\eta}\left(\varepsilon_{zx} - \Im\varepsilon_{zy}\right)\left(\varepsilon_{xx} + \varepsilon_{yy}\right). \tag{8.35}$$

The c-plane components of the polarization P_ξ'' and P_η'' obey the same transformation rules as ξ and η, respectively, for a rotation of $\pi/3$ around the z-direction and using the complex Cartesian coordinates notation of Equation 8.28 we can write

$$P_\xi'' = P_x'' + \Im P_y'' \tag{8.36}$$

$$P_\eta'' = P_x'' - \Im P_y''.$$

We can now rewrite Equations 8.33 through 8.35 as

$$2P_x'' = \left(B_{\eta z\xi\eta\xi} + B_{\xi z\eta\xi\eta} + B_{\eta z\eta\xi\xi} + B_{\xi z\xi\eta\eta}\right)\varepsilon_{zx}\varepsilon_{xx}$$

$$+ \left(B_{\eta z\xi\eta\xi} + B_{\xi z\eta\xi\eta} - B_{\eta z\eta\xi\xi} - B_{\xi z\xi\eta\eta}\right)\varepsilon_{zx}\varepsilon_{yy} \tag{8.37}$$

$$+ \left(B_{\eta zzz\xi} + B_{\xi zzz\eta}\right)\varepsilon_{zz}\varepsilon_{zx} + 2\left(B_{\eta z\eta\xi\xi} + B_{\xi z\xi\eta\eta}\right)\varepsilon_{zy}\varepsilon_{xy}$$

$$2P_y'' = -\left(B_{\eta z\xi\eta\xi} + B_{\xi z\eta\xi\eta} + B_{\eta z\eta\xi\xi} + B_{\xi z\xi\eta\eta}\right)\varepsilon_{zy}\varepsilon_{yy}$$

$$+ \left(B_{\eta z\xi\eta\xi} + B_{\xi z\eta\xi\eta} - B_{\eta z\eta\xi\xi} - B_{\xi z\xi\eta\eta}\right)\varepsilon_{zy}\varepsilon_{xx} \tag{8.38}$$

$$- \left(B_{\eta zzz\xi} + B_{\xi zzz\eta}\right)\varepsilon_{zz}\varepsilon_{zy} - 2\left(B_{\eta z\eta\xi\xi} + B_{\xi z\xi\eta\eta}\right)\varepsilon_{xy}\varepsilon_{zx}$$

$$P_z'' = \left(B_{z\xi\eta\xi\eta} + B_{z\xi\xi\eta\eta}\right)\left(\varepsilon_{xx}^2 + \varepsilon_{yy}^2\right) + B_{zzzzz}\varepsilon_{zz}^2 + B_{z\xi\eta zz}\left(\varepsilon_{xx} + \varepsilon_{yy}\right)\varepsilon_{zz}$$

$$+ 2\left(B_{z\xi\eta\xi\eta} - B_{z\xi\xi\eta\eta}\right)\varepsilon_{xx}\varepsilon_{yy} + B_{zz\xi z\eta}\left(\varepsilon_{zx}^2 + \varepsilon_{zy}^2\right) + B_{z\xi\xi\eta\eta}\varepsilon_{xy}^2. \tag{8.39}$$

Note that the polarization terms on the c-plane (*x*- and *y*-directions) can be described by only three independent parameters. Furthermore, five further independent parameters are needed for the *c*-axis polarization, making a total of eight independent second-order piezoelectric coefficients.

It is worth stressing that after we have transformed the second-order piezoelectric coefficients using subscripts ξ and η, there is no longer a direct correspondence with the coefficients B_{ijklm}. Rather, it is combinations of the transformed coefficients which equate directly to the B_{ijklm}, as it is obvious from Equations 8.37 through 8.39, with the correct choice of indexes being determined by which strain components the B_{ijklm} coefficients couple to. As an example, in Equation 8.39, the combination of transformed coefficients $\left(B_{z\xi\eta\xi\eta} + B_{z\xi\xi\eta\eta}\right)$ couple to the strain combination $\left(\varepsilon_{xx}^2 + \varepsilon_{yy}^2\right)$. Therefore, the correct choice would be $B_{311} = B_{322} = \left(B_{z\xi\eta\xi\eta} + B_{z\xi\xi\eta\eta}\right)$.

Once this is understood, a very useful strategy for visualizing the entire second-order tensor is first, to compact all coefficients B_{ijklm} using Voigt notation ($B_{i\alpha\beta}$), then order them by *i* along the rows and by $\alpha\beta$ (using the invariance between exchanging α and β to avoid repetitions) along the columns of a 3×21 matrix, given for both zincblende and wurtzite in Table 8.2, following the formalism of Grimmer (2007).

The resulting matrix contains some very interesting results. Intuitively, one could think that to have a nonzero second-order coefficient, the relevant first order should also be nonzero. That turns out to be incorrect. One example is wurtzite B_{366}. The fact that e_{36} is zero does not make B_{366} automatically zero. In fact, e_{36} is zero not because the dipole symmetry at the atomic level is not broken under strain in the *xy*-plane ε_{xy} (it actually is!) but because symmetry dictates that the polarization along the *c*-axis must be invariant for positive and negative (shear) strain in the *xy*-plane. Such condition can only be achieved by the even orders in the polarization, hence it is second (and, e.g., fourth) order that is nonzero while first (and, e.g., third) order are vanishing.

In wurtzite, all the second-order coefficients associated with shear strains obey the invariance rule consistently and the reverse is also true. Take, e.g., the first-order coefficient e_{15}. This coefficient is nonzero because the polarization in the *x*-direction is not invariant for strain in the *yz*-plane. In fact, if strain is inverted, the polarization direction is also inverted. On the other hand, second-order coefficients like, e.g., B_{155} are zero because they are instead invariant coefficients that must induce exactly the same polarization for positive and negative values of the strain, a condition that is only fulfilled if the coefficient itself is zero. This is also the case for the zincblende crystal coefficients B_{144}, B_{255}, and B_{366}, which are all vanishing despite e_{14}, e_{25}, and e_{36} being nonzero.

TABLE 8.2 Second-Order Piezoelectric Coefficients $B_{i\alpha\beta}$ for the Cubic Zincblende (Point Group $F\bar{4}3m$) and Wurtzite Crystal (Point Group [6 mm])

	11	22	33	44	55	66	12	13	23	45	36	26	16	25	15	35	46	14	24	34	56
										Zincblende Crystals											
1	–	–	–	–	–	–	–	–	–	–	–	–	–	–	–	–	–	λ	μ	μ	ν
2	–	–	–	–	–	–	–	–	–	–	–	–	–	λ	μ	μ	ν	–	–	–	–
3	–	–	–	–	–	–	–	–	–	ν	λ	μ	μ	–	–	–	–	–	–	–	–
										Wurtzite Crystals											
1	–	–	–	–	–	–	–	–	–	–	–	–	–	2b	2a	c	a–b	–	–	–	–
2	–	–	–	–	–	–	–	–	–	–	–	–	–	–	–	–	–	2b	2a	c	a–b
3	2d	2d	g	h	h	d–e	2e	f	f	–	–	–	–	–	–	–	–	–	–	–	–

Source: Reproduced with permission from Prodhomme P.-Y. et al., *Physical Review B*, 88, 121304, 2013. Copyright [2013] by the **American Physical Society**.

There are three (λ, μ, and ν) and eight (*a–h*) unique coefficients for the zincblende and wurtzite phases, respectively. A dash indicates vanishing coefficients. Rows and columns correspond to the subscript *iαβ* in $B_{i\alpha\beta}$.

It has to be mentioned that the majority of second-order coefficients typically couple with vanishing strains in the case of wurtzite c-axis grown quantum wells. Therefore, for wurtzite crystals the only coefficients of interest are often $B_{311} = B_{322}$, B_{333}, B_{312} and $B_{313} = B_{323}$. In both the zincblende and wurtzite crystals, first- or second-order coefficients, which couple to at least one shear strain component, only become relevant in nanostructures such as nanowires or quantum dots (Bimberg et al., 1998; Migliorato et al., 2002, 2004, 2005; Bester et al., 2006b; Tartakovskii et al., 2006; Ediger et al., 2007; Ding et al., 2010; Tomić et al., 2015).

8.5.5 Calculated Second-Order Piezoelectric Coefficients

After the initial report of nonlinearities in III-N semiconductors under biaxial strain Ambacher et al. (2002), many more reports appeared in the literature. We cover here a review of calculation approaches for both zincblende and wurtzite crystals but provide a compilation of values only for the III-N semiconductors.

8.5.5.1 Numerical Approaches Used for III–V and II–VI Semiconductors

A theoretical and numerical framework for evaluating the full second-order piezoelectric coefficients in III–V zincblende semiconductor materials was first proposed by Bester et al. (2006a). The method used was the linear response technique within density functional perturbation theory (DFPT). The motivation for this theoretical development was mostly that piezoelectricity in the InGaAs system appeared to be of some importance in determining the electronic properties (particularly in relation to symmetry) of self-assembled quantum dots (Bimberg et al., 1998; Cullis et al., 2002, 2005). Such nanostructures were predicted to exhibit sizeable shear strains (and hence piezoelectricity) even if the material was grown along the [001] crystal direction, where instead, for the same direction of growth, quantum wells exhibited no polarization at all (Migliorato et al., 2002, 2004, 2005; Tartakovskii et al., 2006). The nonlinear model of piezoelectricity appeared to be much more reliable than the linear model in predicting the correct magnitude and size of the resulting piezoelectric fields (Bester et al., 2006b; Ediger et al., 2007; Ding et al., 2010).

An alternative approach to estimating piezoelectric coefficients, based on a method initially proposed by Harrison (1989), was used for uniaxially strained crystals, tetragonal distortion (the strain present during pseudomorphic growth on (001) substrates), and also to determine the magnitude of nonlinearities up to third order in the diagonal terms of the strain tensor (Migliorato et al., 2006, 2014; Garg et al., 2009; Tse et al., 2013). While the only complete set of second-order coefficients is provided in Beya-Wakata et al. (2011) the validity of using the local density approximation within the Berry phase method for calculating polarization properties of semiconductors has recently been called into question. The use of a hybrid functional for the exchange correlation interaction in DFPT appears to yield results closer to experimental values (Caro et al., 2015a). Harrison's model was also recently employed to predict the nonlinear piezoelectric properties of ZnO (Al-Zahrani et al., 2013) and some wurtzite III-As and III-P semiconductors (Al-Zahrani et al., 2015).

8.5.5.2 Numerical Approaches Used for Wurtzite III-N Semiconductors

For III-N semiconductors, reports in 2011 from Pal et al. (2011a, b, c) provided a more general set of second-order coefficients (Table 8.3) compared to the case of biaxial strain of Equation 8.24, based on Harrison's method. The coefficients, numerical values relate to the following second-order equation:

$$P_{Tot, 3} = P_{3,sp} + e_{33}\varepsilon_\perp + 2e_{31}\varepsilon_\parallel + \tilde{B}_{311}\varepsilon_\parallel^2 + \tilde{B}_{333}\varepsilon_\perp^2 + \tilde{B}_{313}\varepsilon_\parallel\varepsilon_\perp \quad (8.40)$$

Equation 8.40, together with the coefficients of Table 8.3, predicts large deviations from the linear model (Figure 8.2) and appears to have had some success at explaining trends in the internal quantum efficiency (IQE) in blue and green LEDs under external pressure. In particular, it was claimed that the linear model would incorrectly predict an increase in IQE under compressive pressure (associated with a decrease of

TABLE 8.3 Linear and Partial Nonlinear Piezoelectric Coefficients for III-N Semiconductors Calculated Using Harrison's Method

Coefficient (C/m²)	GaN	AlN	InN
e_{31}	−0.55	−0.60	−0.55
e_{33}	1.05	1.47	1.07
e_{15}	−0.57	−0.60	−0.65
$\tilde{B}_{311}$	6.185	5.850	5.151
$\tilde{B}_{333}$	−8.090	−10.75	−6.680
$\tilde{B}_{313}$	1.543	4.533	1.280

Source: Adapted with permission from Pal J et al., *Physical Review B*, 84, 2011a. Copyright [2011] by the **American Physical Society**.
Please note in the original paper the $\tilde{B}_{313}$ was erroneously labeled as $\tilde{B}_{133}$.

$\mathbf{P_{Tot,3}}$), when both experiment and the second-order model instead agreed in registering a decrease in IQE, due to an increase of $\mathbf{P_{Tot,3}}$ (Crutchley et al., 2013; Pal et al., 2013).

A word of warning is necessary here. The coefficients of Table 8.3 do not directly equate to the $B_{i\alpha\beta}$ coefficients of Table 8.2 (and were obviously not intended to!), and we have used the symbol (∼) as a cap for $\tilde{B}_{i\alpha\beta}$ to emphasize this (in the original paper the notation $e_{i\alpha\beta}$ was used for the same purpose). This is because Equation 8.40 uses a simplified expression valid for the case of $\varepsilon_\parallel = \varepsilon_1 = \varepsilon_2$. The $B_{i\alpha\beta}$ and $\tilde{B}_{i\alpha\beta}$ coefficients are, however, easily linked by

$$\tilde{B}_{311} = B_{311} + 2B_{312} + B_{322} = 4B_{311} \tag{8.41}$$

$$\tilde{B}_{313} = 2B_{313} + 2B_{323} = 4B_{313} \tag{8.42}$$

$$\tilde{B}_{333} = B_{333}. \tag{8.43}$$

A complete set of second-order coefficients for AlN and GaN (Table 8.4) was provided by Pedesseau et al. (2012) who used perturbation density functional theory within the linear response method. The $B_{i\alpha\beta}$ coefficients are the same as those of Table 8.2. The authors reported that compared to the linear model, second-order piezoelectricity leads to a substantially increased prediction of the electric field in pseudomorphic AlGaN/GaN quantum wells, by up to 35% (Figure 8.3).

The full second-order tensor for III-N and ZnO was later (2013) revisited by Prodhomme et al. (2013), using a very similar method to the one used by Pedesseau et al. (2012). Coefficients are given (Table 8.5) in a slightly different style than Table 8.4 but consistent with Table 8.2. Despite the supposed equivalency of the methods used in the calculations, there are many similarities between the values of Tables 8.4 and 8.5, but also many differences in the proposed values. It is outside the scope of this chapter to discuss the merits or demerits of the different proposed coefficients, and we limit our analysis to acknowledging that such differences exist. Furthermore, on close scrutiny, the differences are not significantly large and are unlikely to result in radically different total polarizations.

8.5.6 Calculated versus Experimental Values of the Spontaneous Polarization

Different calculated values for the P_{sp} terms for III-N semiconductors are listed in Table 8.6. It is self-evident that there is a divergence between calculated values, and also between calculated and measured values. This is because they are both difficult calculations to undertake and troublesome experimental determinations to perform.

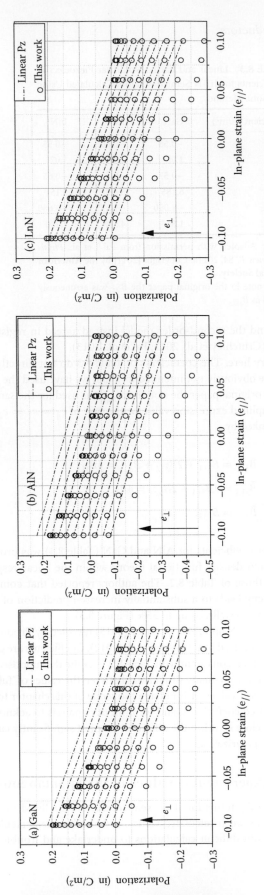

FIGURE 8.2 Comparison of the total polarization as a function of perpendicular and parallel strain calculated in this work (circles) and using the linear model with parameters from Bernardini et al. (2001) (dashed lines). The perpendicular strain varies from −0.1 to 0.1 in steps of 0.02. (Adapted with permission from Pal J et al., *Physical Review B*, 84, 2011a. Copyright [2011] by the **American Physical Society**.)

TABLE 8.4 Linear and Complete Nonlinear Piezoelectric Coefficients for III-N Semiconductors

Coefficient	GaN	AlN
e_{31} (C/m^2)	−0.41	−0.67
e_{33} (C/m^2)	0.76	1.67
e_{15} (C/m^2)	−0.29	−0.35
B_{311} (C/m^2)	6.26	3.77
B_{312} (C/m^2)	2.37	4.12
B_{313} (C/m^2)	−1.47	−8.64
B_{333} (C/m^2)	−23.43	−25.60
B_{115} (C/m^2)	6.48	8.69
B_{125} (C/m^2)	5.09	5.19
B_{135} (C/m^2)	7.02	3.20
B_{344} (C/m^2)	1.41	2.88

Source: Reproduced with permission from Pedesseau L et al., *Applied Physics Letters* 100:031903. Copyright [2012], American Institute of Physics Publishing LLC.

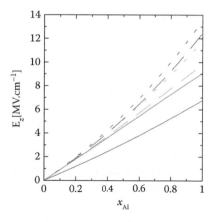

FIGURE 8.3 Variation of the electric field E_z as a function of Al mole fraction for various models. The results of the uncoupled and coupled linear models are represented by straight (first [bottom] curve) and dashed (fifth curve) lines, the coupling strongly enhancing E_z. The linear semi-coupled "standard" model (From Jogai B et al., *Journal of Applied Physics* 94, 3984, 2003) is represented by a dotted line (sixth curve). The straight line (second curve) includes all the nonlinear contributions. The electrostrictive and the nonlinear piezoelectric ones are represented by dashed (third curve) and dotted (fourth curve) lines. (Reproduced with permission from Pedesseau L et al., *Applied Physics Letters*, 100,031903, 2012. Copyright [2012], American Institute of Physics Publishing LLC.)

The values from Pal et al. (2011a) are calculated from the ionic contribution only and neglect entirely the electronic contribution, which leads to a large underestimation compared to that from, e.g., Bernardini et al. (1997, 2001), between −0.020 and −0.040 C/m^2 for the various III-N materials. Despite this, the predicted spontaneous polarization of AlN of −0.051 C/m^2 is the theoretical value that is the closest to the only reported measured one of −0.040 C/m^2 (Park and Chuang, 2000). For GaN there is a spread of calculated values but there are at least two reported experimental values (Yan et al., 2009; Lähnemann et al., 2012) which remarkably agree with each other and are also not too different from the values of both

TABLE 8.5 Piezoelectric Coefficients in C/m^2

	2*a*	**2*b***	*C*	2*d*	2*e*	f	*g*	*h*	e_{15}	e_{31}	e_{33}	P_{sp}
AlN	4.4	2.4	−0.1	3.0	3.0	3.8	−26.0	3.2	−0.35	−0.67	1.67	−0.095
GaN	3.8	2.3	2.7	6.2	3.3	0.4	−21.4	0.4	−0.31	−0.44	0.75	−0.027
InN	4.5[±0.4]	2.8[±0.2]	1.6[±0.4]	4.8	3.7	0.5[±0.1]	−18.6	0.5	−0.43	−0.59	1.14	−0.042

Source: Reproduced with permission from Prodhomme P-Y et al., *Physical Review B,* 88, 121304, 2013. Copyright [2013] by the **American Physical Society**.
Error bars for the values obtained for InN are due to the poor description of this material within DFT-LDA.

TABLE 8.6 Calculated Values of the Spontaneous Polarization (P_{sp}) in C/m^2 According to Different Reports

	Bernardini et al. (2001)	Pal et al. (2011a)	Pedesseau et al. (2012)	Prodhomme et al. (2013)	Experimental
GaN	−0.034	−0.007	0.014	−0.027	−70142 ± 0.007 Lähnemann et al. (2012) −0.022 Yan et al. (2009)
AlN	−0.090	−0.051	−0.075	−0.095	−0.040 Park and Chuang (2000)
InN	−0.042	−0.012	N/A	−0.035	N/A

The last column presents some recently reported experimental values.

Bernardini et al. (1997, 2001) and Prodhomme et al. (2013). To date, no experimental values are available for InN.

The experimental technique used by Lähnemann et al. (2012) is rather elegant. They performed optical measurements of polymorphic structures, i.e., mixed zincblende and wurtzite phases within the same material. The formation of such polymorphs arises simply from occasional stacking faults in otherwise perfect wurtzite crystals, which create zincblende inclusions typically of 1–3 atomic layers. Such naturally forming inclusions, due to a difference in the band offsets of the zincblende and wurtzite phases, can confine electrons within the zincblende regions, allowing them to be optically probed. The general strategy of experimentally measuring the electrooptical behavior of polymorphic structures is a very notable and welcome step forward, one likely in the near future to provide more quantitative data to further our understanding of piezoelectricity in semiconductors.

8.5.7 Continuity Equation in Superlattices and Quantum Well Structures

For completeness, it is worth mentioning how the coefficients outlined in the previous sections can be used to evaluate the electric field in quantum well structures or superlattices.

The conventional equations are obtained from the condition of continuity of the electrostatic potential at the interfaces, often referred to as the superlattice equations, where the subscripts *w* and *b* refer to the quantum well and barrier region, respectively, where the barrier region is the area between the wells in a periodic superlattice:

$$E_{3,w} = \frac{L_b \left(P_{sp,3,b} + P_{pz,3,b} - P_{sp,3,w} - P_{pz,3,w} \right)}{L_b \lambda_w + L_w \lambda_b}$$

(8.44)

$$E_{3,b} = -E_{3,w} \frac{L_w}{L_b}.$$

(8.45)

The subscript $3 = z$ is the *c*-axis direction, *L* indicates the width of the well or barrier region and, to avoid confusion with the symbol used for the strain components, λ indicates the relative dielectric constant $\varepsilon_r \varepsilon_0$

of the well or barrier region. Note that if the barrier or well material is lattice matched to the substrate then $P_{pz,3,b} = 0$ or $P_{pz,3,w} = 0$ (because the strain is vanishing). On the other hand, even in such a case, Equations 8.45 and 8.46 correctly indicate that both $E_{3,w}$ and $E_{3,b}$ are simultaneously nonzero. This is because a nonzero field in one region needs to be compensated by one in the other region, to maintain continuity of the displacement vector D.

Equations 8.44 and 8.45 describe superlattices but can easily be turned into the expression for a single isolated quantum well, by imposing the condition $L_b \gg L_w$. In this case, in fact, Equation 8.45 simply gives $E_{3,b} = 0$, while Equation 8.41 becomes

$$E_{3,w} = \frac{\left(P_{sp,3,b} + P_{pz,3,b} - P_{sp,3,w} - P_{pz,3,w}\right)}{\lambda_w + \left[(L_w/L_b)\lambda_b\right]} \approx \frac{\left(P_{sp,3,b} + P_{pz,3,b} - P_{sp,3,w} - P_{pz,3,w}\right)}{\lambda_w}, \tag{8.46}$$

which can assume an even simpler form in the case where the barrier material is lattice matched to the substrate and hence $P_{pz,3,b} = 0$.

8.5.8 Schemes for Dealing with Alloys

It should be noted that barrier and well regions in Equations 8.44 through 8.46, or any nanostructure for the matter, are not necessarily formed by binary III-N, and, in fact, in the majority of cases of interest at least one will be constituted by an alloy. In the case of alloys a suitable interpolation scheme needs to be employed for all the different quantities involved. Vegard's law (linear interpolation) is naturally the first scheme to consider. The validity of Vegard's law was tested for III-N semiconductor quantum wells by Ambacher et al. (2002) and found to be insufficient in many situations, which led to proposing a quadratic dependence on composition. For the sake of completeness we reproduce the proposed equations as they form the basis for device modeling of structures such as those used for fabrication of LEDs. We start with those related to the structural properties.

The lattice constants $a(x)$ and $c(x)$ of wurtzite group III nitride alloys $A_xB_{1-x}N$ ($Al_xGa_{1-x}N$, $In_xGa_{1-x}N$, and $Al_xIn_{1-x}N$), can be assumed to follow Vegard's law:

$$a_{AlGaN}(x) = (3.1986 - 0.0891x)\text{Å} \tag{8.47}$$

$$a_{InGaN}(x) = (3.1986 + 0.3862x)\text{Å}$$

$$a_{AlInN}(x) = (3.5848 - 0.4753x)\text{Å} \quad \text{Lasers}$$

$$c_{AlGaN}(x) = (5.2262 - 0.2323x)\text{Å} \tag{8.48}$$

$$c_{InGaN}(x) = (5.2262 + 0.5740x)\text{Å}$$

$$c_{AlInN}(x) = (5.8002 - 0.8063x)\text{Å}.$$

These results were found to be within 2% of experimental data obtained through a combination of high-resolution x-ray diffraction, Rutherford, and elastic recoil detection analysis. In the same work authors also recorded various types of nonlinear behavior. First, the u parameter (which expresses the deviation of the unit cell from the ideal structure) appears to have a quadratic dependence and hence does not follow Vegard's law:

$$u_{AlGaN}(x) = (0.3819x + 0.3772(1-x) - 0.0032x(1-x)) \tag{8.49}$$

$$u_{InGaN}(x) = (0.3793x + 0.3772(1-x) - 0.0057x(1-x))$$

$$u_{AlInN}(x) = (0.3819x + 0.3793(1-x) - 0.0086x(1-x)).$$

Such quadratic and hence nonlinear behavior is unsurprisingly found also in the polarization properties. The spontaneous polarization terms for random, ternary alloys with wurtzite crystal structure are fitted by high accuracy through quadratic equations (Figure 8.4), also following earlier work by Bernardini and Fiorentini (2002), and are thus given by

$$P_{sp, AlGaN}(x) = (-0.090x - 0.034(1-x) + 0.021x(1-x))\,C/m^2 \tag{8.50}$$

$$P_{sp,InGaN}(x) = (-0.042x - 0.034(1-x) + 0.037x(1-x))\,C/m^2$$

$$P_{sp,AlInN}(x) = (-0.090x - 0.042(1-x) + 0.070x(1-x))\,C/m^2,$$

while the strain-induced polarization, which obviously depends on the alloy composition but also on the substrate, is given by

$$P_{pz, AlGaN/InN}(x) = [-0.28x - 0.113(1-x) + 0.042x(1-x)]C/m^2 \tag{8.51}$$

$$P_{pz,AlGaN/GaN}(x) = [-0.0525x + 0.0282x(1-x)]C/m^2$$

$$P_{pz,AlGaN/AlN}(x) = [0.026\,(1-x) + 0.0248x\,(1-x)]C/m^2$$

$$P_{pz, InGaN/InN}(x) = [-0.113(1-x) - 0.0276x(1-x)]C/m^2 \tag{8.52}$$

$$P_{pz,InGaN/GaN}(x) = [0.148x - 0.0424x(1-x)]C/m^2$$

$$P_{pz,InGaN/AlN}(x) = [0.182x + 0.026(1-x) - 0.0456x(1-x)]C/m^2$$

$$P_{pz,AlInN/InN}(x) = [-0.28x + 0.104x(1-x)]C/m^2 \tag{8.53}$$

$$P_{pz,AlInN/GaN}(x) = [-0.0525x + 0.148(1-x) + 0.0938x(1-x)]C/m^2$$

$$P_{pz,AlInN/AlN}(x) = [0.182(1-x) + 0.092x(1-x)]C/m^2.$$

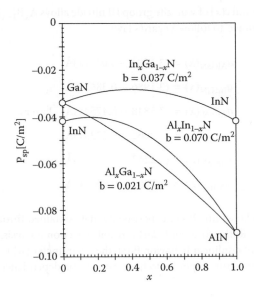

FIGURE 8.4 Predicted spontaneous polarization (P_{sp}) for random, ternary alloys with wurtzite crystal structure. The dependence of P_{sp} on x can be approximated to a high degree of accuracy by quadratic equations (Equation 8.47). The nonlinearity can be described by positive bowing parameters increasing from AlGaN to InGaN and AlInN. (Reproduced with permission Ambacher O et al., *Journal of Physics: Condensed Matter*, 2002 © IOP Publishing. All rights reserved.)

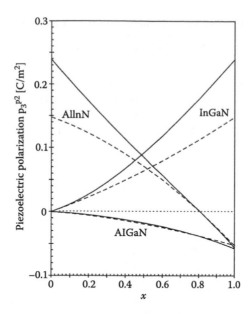

FIGURE 8.5 The piezoelectric polarization of random ternary alloys pseudomorphically grown on relaxed GaN buffer layers calculated by (1) Equation 8.17 and using linear interpolations of the physical properties for the relevant binary compounds (dashed curves); (2) taking into account the nonlinearity in the piezoelectric polarization in terms of strain (Equation 8.24) (solid curves). For alloys under high biaxial strain (layers with high In concentration), the piezoelectric polarization is underestimated by approach (1). (Reproduced with permission Ambacher O et al., *Journal of Physics: Condensed Matter*, 2002 © IOP Publishing. All rights reserved.)

The piezoelectric polarization of a random ternary III-N alloy, pseudomorphically grown on relaxed GaN buffer layers is given in Figure 8.5, where it is possible to compare the polarization calculated by (1) Equation 8.17 and that obtained using Vegard's law (dashed curves) or (2) taking into account the non-linearity in the piezoelectric polarization in terms of strain (Equation 8.24) (solid curves). In general, the polarization using Vegard's law is always underestimated.

Pal et al. (2011a, c) also proposed an interpolation scheme (Table 8.7) for the piezoelectric properties of the ternary nitride alloys ($Al_xGa_{1-x}N$, $In_xGa_{1-x}N$), which can also be applied to any nanostructure, consistent with their proposed first- and second-order coefficients and values of the spontaneous polarization. A parabolic equation was proposed.

Finally, Prodhomme et al. (2013) also provided a further mixing scheme for alloys, only for the linear coefficients (second-order coefficients are assumed to follow Vegard's law instead). The corresponding bowing parameters are listed in Table 8.8, and the effect of using this mixing scheme can be observed in Figure 8.6. The mixing scheme is quantitatively within 10% of that of Ambacher et al. (2002).

8.6 Electromechanical Coupling in Wurtzite Semiconductors

It has recently been reported by Suski et al. (2012) that when attempting to model variations of optical output in III-N quantum wells, the combination of linear elasticity with linear piezoelectricity appears to be lacking the correct framework to provide predictions that compare well with available experimental data (Figure 8.7). Linear elasticity should be replaced by the more comprehensive nonlinear elasticity (Łepkowski and Majewski, 2006), where electromechanical coupling is included.

However, this is still not sufficient. It was noted that the slope of dE_{PL}/dP, i.e., the change in peak energy (as measured from photoluminescence or calculated from **k·p** theory) for a change in applied

TABLE 8.7 Quadratic Dependence on Alloy Composition for the Ternary Nitride Alloys ($Al_x Ga_{1-x} N$, $In_x Ga_{1-x} N$)

	Y	A	B	C
$Al_x Ga_{1-x}N$	P_{sp}	−0.025	−0.019	−0.007
	e_{31}	0.064	−0.114	−0.550
	e_{33}	0.141	0.279	1.050
	$\tilde{B}_{311}$	0.674	−1.000	6.185
	$\tilde{B}_{333}$	1.055	−3.715	−8.090
	$\tilde{B}_{313}$	0.340	2.650	1.543
$In_x Ga_{1-x}N$	P_{sp}	−0.001	−0.005	−0.007
	e_{31}	−0.368	0.368	−0.550
	e_{33}	0.119	−0.099	1.050
	$\tilde{B}_{311}$	0.635	−1.669	6.185
	$\tilde{B}_{333}$	1.182	0.228	−8.090
	$\tilde{B}_{313}$	0.226	−0.489	1.543

Source: Adapted with permission from Pal J et al., *Physical Review B*, 84, 2011a; Pal J et al., *Physical Review B*, 84, 159902, 085211, 2011c. Copyright [2011] by the **American Physical Society**.
The parameters are for the equation $Y = Ax^2 + Bx + C$.

TABLE 8.8 Bowing Coefficients (MV/cm^2) for the Nonlinear Dependence on Ternary Composition for the Linear Piezoelectric Coefficients and Lattice Mismatch (Fifth Column) and Piezoelectric Coefficients Difference between Binaries (Sixth Column)

	b_{15}	b_{15}	b_{15}	$\Delta a / \langle a \rangle$	$\Delta e_{33} / \langle e_{33} \rangle$
GaN/AlN	−0.032	0.106	−0.671	2.23%	75%
GaN/InN	−0.275	0.222	−0.550	10.4%	41%
InN/AlN	−0.305	0.540	−2.178	12.65%	38%

Source: Reproduced with permission from Prodhomme P-Y et al., *Physical Review B*, 88, 121304, 2013. Copyright [2013] by the **American Physical Society**.

pressure, as a function of the unstrained peak energy $E_{PL,P=0}$, could only match experimental data by combining both nonlinear elasticity and nonlinear piezoelectricity simultaneously in the model. The nonlinear piezoelectric parameters of Pal et al. (2011a, c) were used for the demonstration (Figure 8.7). Given the importance of this development we will discuss the nonlinear elasticity model in some detail, first with linear piezoelectric coefficients, then with nonlinear ones.

8.6.1 Electromechanical Coupling with Linear Piezoelectricity

The starting point is the definition of the density of enthalpy in a material with strain-induced polarization and under external hydrostatic pressure, given by

$$H = U - E \cdot D + PdV, \tag{8.54}$$

where P is the external (hydrostatic) pressure, $dV = \text{Tr}(\varepsilon_{ij})$ is the normalized infinitesimal volume change, E is the electric field, and D is the displacement vector such that

$$E_i \cdot D_i = E_i \cdot \left(e_{ijk}\varepsilon_{jk} + \lambda_{ij}E_j + P_{sp,i} \right) = e_{ijk}\varepsilon_{jk}E_i + \lambda_{ij}E_iE_j + E_i \cdot P_{sp,i}, \tag{8.55}$$

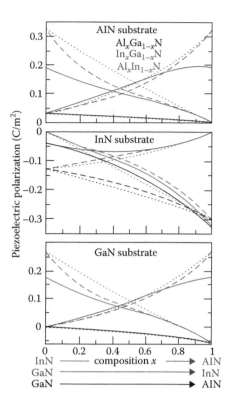

FIGURE 8.6 Piezoelectric polarization with respect to the ternary compositions for different compounds on different (0001) substrates. The curves are Ppz (dotted line), Ppz with bowing (dashed line), and Ppz with bowing + second-order Ppz (full line). (Reproduced with permission from Prodhomme P-Y et al., *Physical Review B*, 88, 121304, 2013. Copyright [2013] by the **American Physical Society**.)

where λ_{ij} is the electric permittivity tensor and the usual symbols are used for the piezoelectric coefficients and spontaneous polarization. The tensor U in Equation 8.54 is instead the elastic energy (per unit volume) defined as

$$U = \frac{1}{2} \sum_{ijkl} \left(c_{ijkl}\varepsilon_{ij}\varepsilon_{kl} + \lambda_{ij}E_iE_j \right), \tag{8.56}$$

where c_{ijkl} are the components of the elastic tensor. The stress tensor components instead are obtained as derivatives of the enthalpy with respect to the strain components, which yields

$$\sigma_{ij} = \frac{dH}{d\varepsilon_{ij}} =$$

$$= \frac{d}{d\varepsilon_{ij}} \left[\sum_{ijkl} \left(\frac{1}{2}c_{ijkl}\varepsilon_{ij}\varepsilon_{kl} - \frac{1}{2}\lambda_{ij}E_iE_j - e_{ijk}\varepsilon_{jk}E_i - E_i \cdot P_{sp,i} + PTr(\varepsilon_{ij}) \right) \right] \tag{8.57}$$

$$= \sum_{kl} \left(c_{ijkl}\varepsilon_{kl} - e_{kij}E_k \right) + P\delta_{ij},$$

where δ_{ij} is the conventional Kronecker delta. We now need to consider the symmetry of the wurtzite crystal and explicitly take into account that there are only five independent elastic constants and three

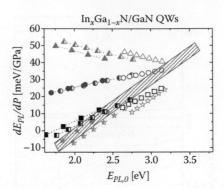

FIGURE 8.7 Comparison between the experimentally determined dependence of $\frac{dE_{PL}}{dP}$ $vs E_{PL,P=0}$ and the results of k·p calculations performed for In$_x$Ga$_{1-x}$N/GaN quantum wells with x = 0.1 (empty symbols), 0.2 (half-filled symbols), and 0.3 (solid symbols). Triangles, circles, squares, and stars correspond to theoretical results obtained using (1) linear piezoelectricity following (From Bernardini F et al., *Physical Review B*, 56, R10024–R10027, 1997; Bernardini F et al., *Physical Review B*, 63, 193201, 2001) and linear elasticity, (2) second-order piezoelectricity following (From Pal J et al., *Physical Review B*, 84, 2011a; Pal J et al., *Physical Review B*, 84, 159902, 085211, 2011c) and linear elasticity, (3) second-order piezoelectricity (Pal et al., 2011a, c) and nonlinear elasticity (From Łepkowski SP, Majewski JA, *Physical Review B*, 74, 8, 2006), (4) nonlinear piezoelectricity following (From Shimada K et al., *Japanese Journal of Applied Physics*, 37, L1421, 1998) and nonlinear elasticity (Łepkowski and Majewski, 2006), respectively. The hatched rectangles represent available experimental results. (Reproduced with permission from Suski T et al., *Journal of Applied Physics*, 112, 053509, 2012. Copyright 2012 by the **American Physical Society**.)

independent piezoelectric coefficients, as discussed in relation to Equation 8.2. Using Voigt notation the result is

$$
\begin{bmatrix} \sigma_1 \\ \sigma_2 \\ \sigma_3 \\ \sigma_4 \\ \sigma_5 \\ \sigma_6 \end{bmatrix} = \Lambda_i - \Pi'_i + P_i = \begin{bmatrix} c_{11}\varepsilon_1 + c_{12}\varepsilon_2 + c_{13}\varepsilon_3 \\ c_{12}\varepsilon_1 + c_{11}\varepsilon_2 + c_{13}\varepsilon_3 \\ c_{13}\left(\varepsilon_1 + \varepsilon_2\right) + c_{33}\varepsilon_3 \\ 2c_{44}\varepsilon_4 \\ 2c_{44}\varepsilon_5 \\ \left(c_{11} - c_{12}\right)\varepsilon_6 \end{bmatrix} - \begin{bmatrix} e_{31}E_3 \\ e_{31}E_3 \\ e_{33}E_3 \\ e_{15}E_2 \\ e_{15}E_1 \\ 0 \end{bmatrix} + \begin{bmatrix} P \\ P \\ P \\ 0 \\ 0 \\ 0 \end{bmatrix}, \tag{8.58}
$$

having indicated with Λ_i, Π'_i, and P_i the elastic, piezoelectric, and external pressure contribution, respectively.

If we assume that any stress component along the *c*-axis (*z*-direction) is equal to zero, which is reasonable, since, typically in that direction, the material is free to relax to minimize its enthalpy, then from the third row of Equation 8.55 we can write

$$
\varepsilon_3 = \frac{-c_{13}\left(\varepsilon_1 + \varepsilon_2\right)}{c_{33}} - \frac{P}{c_{33}} + \frac{e_{33}E_3}{c_{33}}. \tag{8.59}
$$

The first term of Equation 8.59 is the expected linear expression for linking parallel and perpendicular strain of Equation 8.16, the second term expresses the external pressure, while the third term, given by

$$
\Omega = \frac{e_{33}E_3}{c_{33}}, \tag{8.60}
$$

expresses the electromechanical coupling. If we now incorporate Equation 8.52 into the continuity equations (Equation 8.59), where the assumption of isotropic relative dielectric constant is made, and solve for E_z in the well region, we obtain that

$$
E_{3,w} = \frac{L_b \left[P_{sp,3,b} + \left(e_{31,b} - \frac{c_{13,b}}{c_{33,b}} e_{33,b} \right) \left(\varepsilon_{1,b} + \varepsilon_{2,b} \right) - P_{sp,3,w} - \left(e_{31,w} - \frac{c_{13,w}}{c_{33,w}} e_{33,w} \right) \left(\varepsilon_{1,w} + \varepsilon_{2,w} \right) \right]}{L_b \left(\lambda_w + \frac{e_{33,w}^2}{c_{33,w}} \right) + L_w \left(\lambda_b + \frac{e_{33,b}^2}{c_{33,b}} \right)}
$$

$$
+ P \frac{L_b \left(\frac{e_{33,w}}{c_{33,w}} - \frac{e_{33,b}}{c_{33,b}} \right)}{L_b \left(\lambda_w + \frac{e_{33,w}^2}{c_{33,w}} \right) + L_w \left(\lambda_b + \frac{e_{33,b}^2}{c_{33,b}} \right)} = E_{3,w}^* + P E_{3,w}^{**}, \tag{8.61}
$$

The field in the barrier is easily obtained from Equation 8.45. It is possible to extract the strain ε_3 from Equation 8.61, which shows how the strain is changed compared to Equation 8.59

$$
\varepsilon_{3,w} = \frac{-c_{13,w} \left(\varepsilon_{1,w} + \varepsilon_{2,w} \right)}{c_{33,w}} + \frac{e_{33,w} E_{3,w}^*}{c_{33,w}} - \frac{P}{c_{33,w}} \left(e_{33,w} E_{3,w}^{**} - 1 \right). \tag{8.62}
$$

While Equations 8.61 and 8.62 entirely define the problem, it is worth noting that such formalism simplifies considerably if we assume that L_b is much larger than L_w. Assuming also $\varepsilon_1 = \varepsilon_2$, we obtain the isolated quantum well configuration, which can be written as

$$
E_{3,w}^* = \frac{1}{\kappa_w} \left[P_{sp,3,b} + 2 \left(e_{31,b} - \frac{c_{13,b}}{c_{33,b}} e_{33,b} \right) \varepsilon_{1,b} - P_{sp,3,w} - 2 \left(e_{31,w} - \frac{c_{13,w}}{c_{33,w}} e_{33,w} \right) \varepsilon_{1,w} \right] \tag{8.63}
$$

$$
E_{3,w}^{**} = \frac{1}{\kappa_w} \left(\frac{e_{33,w}}{c_{33,w}} - \frac{e_{33,b}}{c_{33,b}} \right), \tag{8.64}
$$

having defined

$$
\kappa_w = \lambda_w + \frac{e_{33,w}^2}{c_{33,w}}. \tag{8.65}
$$

8.6.2 Electromechanical Coupling with Nonlinear Piezoelectricity

Including nonlinear piezoelectricity (to second order) in the electromechanical coupling is fairly straightforward. This is not given explicitly elsewhere and is original to this chapter.

This first step is to add to Equation 8.55 the second-order piezoelectric polarization, given by

$$
E_i \cdot D_i'' = B_{i\alpha\beta} \varepsilon_\alpha \varepsilon_\beta E_i, \tag{8.66}
$$

which adds a further tensor to the expression of the stress tensor of Equation 8.58

$$
\sigma_i = \Lambda_i - \Pi_i' + P_i - \Pi_i'', \tag{8.67}
$$

with Π_i'' given by

$$\Pi_i'' = \begin{bmatrix} B_{115}\varepsilon_5 E_1 + B_{125}\varepsilon_4 E_2 + (2B_{311}\varepsilon_1 + B_{312}\varepsilon_2 + B_{313}\varepsilon_3) E_3 \\ B_{125}\varepsilon_5 E_1 + B_{115}\varepsilon_4 E_2 + (B_{312}\varepsilon_1 + 2B_{311}\varepsilon_2 + B_{313}\varepsilon_3) E_3 \\ B_{135}(\varepsilon_5 E_1 + \varepsilon_4 E_2) + (B_{313}\varepsilon_1 + B_{313}\varepsilon_2 + 2B_{333}\varepsilon_3) E_3 \\ B_{146}\varepsilon_6 E_1 + (B_{125}\varepsilon_1 + B_{115}\varepsilon_2 + B_{135}\varepsilon_3) E_2 + 2B_{344}\varepsilon_4 E_3 \\ (B_{115}\varepsilon_1 + B_{125}\varepsilon_2 + B_{135}\varepsilon_3) E_1 + B_{146}\varepsilon_6 E_2 + 2B_{344}\varepsilon_5 E_3 \\ B_{146}(\varepsilon_4 E_1 + \varepsilon_5 E_2) + 2B_{366}\varepsilon_6 E_3 \end{bmatrix}. \tag{8.68}$$

In the absence of shear strain (or if these are sufficiently small to be neglected), and following the same steps of the previous section, the addition of second-order piezoelectric terms adds a further electromechanical coupling to Equations 8.59 and 8.60, leading to

$$\Omega = \frac{e_{33}E_3}{c_{33}} + \left(\frac{B_{313}(\varepsilon_1 + \varepsilon_2)}{c_{33}} + \frac{2B_{333}\varepsilon_3}{c_{33}} \right) E_3. \tag{8.69}$$

And Equations 8.58 through 8.62 are easily modified if you replace each first-order coefficient e_{33} with

$$\tilde{e}_{33} = e_{33} + B_{313}(\varepsilon_1 + \varepsilon_2) + 2B_{333}\varepsilon_3 \tag{8.70}$$

leading to

$$E_{3,w} = \frac{L_b \left[P_{sp,3,b} + \left(e_{31,b} - \frac{c_{13,b}}{c_{33,b}} \tilde{e}_{33,b} \right)(\varepsilon_{1,b} + \varepsilon_{2,b}) - P_{sp,3,w} - \left(e_{31,w} - \frac{c_{13,w}}{c_{33,w}} \tilde{e}_{33,w} \right)(\varepsilon_{1,w} + \varepsilon_{2,w}) \right]}{L_b \left(\lambda_w + \frac{\tilde{e}_{33,w}^2}{c_{33,w}} \right) + L_w \left(\lambda_b + \frac{\tilde{e}_{33,b}^2}{c_{33,b}} \right)}$$

$$+ P \frac{L_b \left(\frac{\tilde{e}_{33,w}}{c_{33,w}} - \frac{\tilde{e}_{33,b}}{c_{33,b}} \right)}{L_b \left(\lambda_w + \frac{\tilde{e}_{33,w}^2}{c_{33,w}} \right) + L_w \left(\lambda_b + \frac{\tilde{e}_{33,b}^2}{c_{33,b}} \right)} = E_{3,w}^* + P E_{3,w}^{**}$$

$$\tag{8.71}$$

$$\varepsilon_{3,w} = \frac{-c_{13,w}(\varepsilon_{1,w} + \varepsilon_{2,w})}{c_{33,w}} + \frac{(e_{33,w} + B_{313,w}(\varepsilon_{1,w} + \varepsilon_{2,w}) + 2B_{333,w}\varepsilon_{3,w}) E_{3,w}^*}{c_{33,w}}$$

$$- \frac{P}{c_{33,w}} \left((e_{33,w} + B_{313,w}(\varepsilon_{1,w} + \varepsilon_{2,w}) + 2B_{333,w}\varepsilon_{3,w}) E_{3,w}^{**} - 1 \right). \tag{8.72}$$

For the isolated quantum well configuration, where $\varepsilon_1 = \varepsilon_2$, we finally obtain

$$\tilde{\kappa}_w E_{3,w}^* = P_{sp,3,b} + 2 \left(e_{31,b} - \frac{c_{13,b}}{c_{33,b}} (e_{33,b} + 2B_{313,b}\varepsilon_{1,w} + 2B_{333,b}\varepsilon_{3,b}) \right) \varepsilon_{1,b}$$

$$- P_{sp,3,w} - 2 \left(e_{31,w} - \frac{c_{13,w}}{c_{33,w}} (e_{33,w} + 2B_{313,w}\varepsilon_{1,w} + 2B_{333,w}\varepsilon_{3,w}) \right) \varepsilon_{1,w} \tag{8.73}$$

$$\tilde{\kappa}_w E_{3,w}^{**} = \frac{\left(e_{33,w} + 2B_{313,w}\varepsilon_{1,w} + 2B_{333,w}\varepsilon_{3,w}\right)}{c_{33,w}}$$

$$- \frac{\left(e_{33,b} + 2B_{313,b}\varepsilon_{1,w} + 2B_{333,b}\varepsilon_{3,b}\right)}{c_{33,b}}, \tag{8.74}$$

where $\tilde{\kappa}_w$ is now given by

$$\tilde{\kappa}_w = \lambda_w + \frac{\left(e_{33,w} + 2B_{313,w}\varepsilon_{1,w} + 2B_{333,w}\varepsilon_{3,w}\right)^2}{c_{33,w}}. \tag{8.75}$$

Attempting the substitutions makes it rather complicated to obtain an analytic solution, as both $E_{3,w}^*$ and $E_{3,w}^{**}$ now depend on ε_3. Therefore, a numerical solution is the most efficient if not the only realistically possible approach.

8.7 Nonpolar Orientations in Wurtzite Semiconductors

III-N–based heterostructures grown along the polar c-axis exhibit large piezoelectric field that can localize electrons at interfaces and create reduced efficiency through Coulomb repulsion. There has thus been substantial interest in growing structures on semi- and nonpolar substrates. If the growth direction of the heterostructure forms a nonzero angle with the polar c-axis, even a complete elimination of the electrostatic built-in field can occur (Romanov et al., 2006).

With this motivation, Schulz and Marquardt (2015) have calculated a very useful set of equations to evaluate the polarization of III-N for any substrate orientation at an angle θ from the c-axis.

$$P'_1 = -P_{sp}\sin\theta - 2e_{15}\cos\theta \cdot \left[\left(\varepsilon'_1 - \varepsilon'_3\right)\cos\theta\sin\theta - \varepsilon'_5\cos 2\theta\right]$$

$$- e_{31}\sin\theta \cdot \left[\varepsilon'_2 + \varepsilon'_1\cos^2\theta + \varepsilon'_5\sin 2\theta + \varepsilon'_3\sin^2\theta\right] \tag{8.76}$$

$$- e_{33}\sin\theta \cdot \left[\varepsilon'_3\cos^2\theta - \varepsilon'_5\sin 2\theta + \varepsilon'_3\sin^2\theta\right]$$

$$P'_2 = 2e_{15} \cdot \left[\varepsilon'_4\cos\theta - \varepsilon'_6\sin\theta\right] \tag{8.77}$$

$$P'_3 = P_{sp}\cos\theta - 2e_{15}\sin\theta\left[\left(\varepsilon'_2 - \varepsilon'_3\right)\cos\theta\sin\theta - \varepsilon'_5\cos 2\theta\right]$$

$$+ e_{31}\cos\theta\left[\varepsilon'_2 + \varepsilon'_1\cos^2\theta + \varepsilon'_5\sin 2\theta + \varepsilon'_3\sin^2\theta\right] \tag{8.78}$$

$$+ e_{33}\cos\theta\left[\varepsilon'_3\cos^2\theta - \varepsilon'_3\sin 2\theta + \varepsilon'_1\sin^2\theta\right].$$

The strain components ε'_i are the ones after the θ rotation, with the i Cartesian coordinates defined in terms of the new system of reference. Equations 8.76 through 8.78 yield linear piezoelectric polarization only, but the extension to second order is straightforward.

8.8 Summary and Future Directions

Piezoelectricity in III-N semiconductors is an effect of very strong magnitude, with fields measured in the MV/cm a commonly reported feature. It affects device performance, and modeling must always take these fields into account. From the early linear models, nonlinear models have been developed, and even electromechanical coupling can now be included. DFPT and the Berry phase approach have been invaluable in obtaining reliable estimates of first- and second-order piezoelectric coefficients. At the time of this writing, a large variation in proposed values and discrepancies with experimental data still exists with regard to the values of e_{15} and spontaneous polarization, which must be addressed in the future.

The recent development in the experimental realization and characterization of polymorphic structures has already provided accurate and direct determination of the values of the spontaneous polarization of GaN (Lähnemann et al., 2012). In the near future, polymorphism might spear even more reliable experimental data to inform our theoretical understanding of piezoelectric polarization in semiconductor materials.

Though there is no reason to doubt the current models, it is still very difficult to reliably simulate full device behavior. Too often one still must rely on some level of fitting parameters to match experimental data. The reason for this is not clear. We mentioned earlier that an often empirical solution to the fitting of experimental optical data is to scale piezoelectric fields by multiplying times 0.7. Since there is hardly any scope in the models or calculated parameters for such a large correction, it is unlikely that the coefficients or models alone, whether linear or nonlinear, are responsible for less-than-accurate predictions of the optical properties. It is much more likely that the scaling of the piezoelectric field simply corrects other factors in what are otherwise complex calculations involving nonhomogeneous ternary alloys and estimates of the electronic bandstructure, quantum confined wavefunctions, and Coulomb repulsion. There might well be several factors that are incorrectly or not explicitly taken into account in the models. A nonexhaustive list includes:

1. Real quantum wells don't have sharp interfaces as typically assumed in the models.
2. Alloy disorder can have a more complex influence on electro-optical properties than the models can at present incorporate.
3. LEDs suffer from efficiency droop, which is most often simulated based on the assumption of electron leakage, which may not be correct.
4. The relationship between piezoelectric fields and Auger recombination rates is unknown.

In relation to alloy disorder, the second item on the list, some interesting discussion can be found in Yang et al. (2014). An idea that could help understand the underlying mechanisms of real device behavior is that any conduction of electrical current in alloys takes place through percolation, i.e., carriers always moving from one end to the other of the material following the path of least resistance. This means that electrons do not see the alloy as a homogeneous medium but feel predominantly the presence of the clusters of atoms that most help them continue on their path. The pockets or clusters of atoms with the largest piezoelectric fields are the most helpful in this process. However, percolation also means that the effective path from one electrode to another is no longer the distance between them. The path becomes much longer! Device behavior depends on the balance between these effects.

Though this concept of percolation-driven carrier diffusion would be a major step forward in treating alloys, one of the aspects that is still missing in evaluating polarization properties is a full atomistic treatment over realistic size models of at least tens of nanometers. Since, almost invariably, the experimental data available involves alloys and disorders which can play a significant and underestimated role in affecting real-life measurements, future work in this field must take alloy disorder fully into account.

In relation to electron leakage models and Auger recombination, the fourth item on the list, this is a very important and current topic, but outside the scope of this chapter. Interested readers should consult, e.g., Piprek (2015) and Piprek et al. (2015).

References

Al-Zahrani HYS, Pal J, Migliorato MA (2013) Non-linear piezoelectricity in wurtzite ZnO semiconductors. *Nano Energy* 2:1214–1217.

Al-Zahrani HYS, Pal J, Migliorato MA, Tse G, Yu D (2015) Piezoelectric field enhancement in III–V core–shell nanowires. *Nano Energy* 14:382–391.

Ambacher O, Majewski J, Miskys C, Link A, Hermann M, Eickhoff M, Stutzmann M, Bernardini F, Fiorentini V, Tilak V, Schaff B, Eastman LF (2002) Pyroelectric properties of Al(In)GaN/GaN hetero—And quantum well structures. *Journal of Physics: Condensed Matter* 14:3399.

Bernardini F, Fiorentini V (2002) Nonlinear behavior of spontaneous and piezoelectric polarization in III–V nitride alloys. *Physica Status Solidi (A)* 190:65–73.

Bernardini F, Fiorentini V, Vanderbilt D (1997) Spontaneous polarization and piezoelectric constants of III–V nitrides. *Physical Review B* 56:R10024–R10027.

Bernardini F, Fiorentini V, Vanderbilt D (2001) Accurate calculation of polarization-related quantities in semiconductors. *Physical Review B* 63:193201.

Bester G, Wu X, Vanderbilt D, Zunger A (2006a) Importance of second-order piezoelectric effects in zinc-blende semiconductors. *Physical Review Letters* 96:187602.

Bester G, Zunger A, Wu X, Vanderbilt D (2006b) Effects of linear and nonlinear piezoelectricity on the electronic properties of InAs/GaAs quantum dots. *Physical Review B* 74:081305.

Beya-Wakata A, Prodhomme P-Y, Bester G (2011) First- and second-order piezoelectricity in III–V semiconductors. *Physical Review B* 84:195207.

Bimberg D, Grundmann M, Ledentsov NN (1998) *Quantum Dot Heterostructures*. Chichester: Wiley.

Caro MA, Schulz S, O'Reilly EP (2015a) Origin of nonlinear piezoelectricity in III–V semiconductors: Internal strain and bond ionicity from hybrid-functional density functional theory. *Physical Review B* 91:075203.

Caro MA, Zhang S, Riekkinen T, Ylilammi M, Moram MA, Lopez-Acevedo O, Molarius J, Laurila T (2015b) Piezoelectric coefficients and spontaneous polarization of ScAlN. *Journal of Physics: Condensed Matter* 27:245901.

Crutchley BG, Marko IP, Pal J, Migliorato MA, Sweeney SJ (2013) Optical properties of InGaN-based LEDs investigated using high hydrostatic pressure dependent techniques. *Physica Status Solidi (B)* 250:698–702.

Cullis AG, Norris DJ, Migliorato MA, Hopkinson M (2005) Surface elemental segregation and the Stranski–Krastanow epitaxial islanding transition. *Applied Surface Science* 244:65–70.

Cullis AG, Norris DJ, Walther T, Migliorato MA, Hopkinson M (2002) Stranski–Krastanow transition and epitaxial island growth. *Physical Review B* 66:081305.

Curie J, Curie P (1880) Contractions et dilatations produites par des tensions dans les cristaux hémièdres à faces inclinées *Comptes Rendus de l'Académie du Sciences* 93:40.

Ding F, Singh R, Plumhof JD, Zander T, Křápek V, Chen YH, Benyoucef M, Zwiller V, Dörr K, Bester G, Rastelli A, Schmidt OG (2010) Tuning the exciton binding energies in single self-assembled InGaAs/GaAs quantum dots by piezoelectric-induced biaxial stress. *Physical Review Letters* 104:067405.

Ediger M, Bester G, Gerardot BD, Badolato A, Petroff PM, Karrai K, Zunger A, Warburton RJ (2007) Fine structure of negatively and positively charged excitons in semiconductor quantum dots: Electron-hole asymmetry. *Physical Review Letters* 98:036808.

Garg R, Hüe A, Haxha V, Migliorato MA, Hammerschmidt T, Srivastava GP (2009) Tunability of the piezoelectric fields in strained III–V semiconductors. *Applied Physics Letters* 95:041912.

Grimmer H (2007) The piezoelectric effect of second order in stress or strain: Its form for crystals and quasicrystals of any symmetry. *Acta Crystallographica Section A* 63:441–446.

Harrison W (1989) *Electronic Structure and the Properties of Solids: The Physics of the Chemical Bond.* New York, NY: Dover Publications.

Jogai B, Albrecht JD, Pan E (2003) Effect of electromechanical coupling on the strain in AlGaN/GaN heterojunction field effect transistors. *Journal of Applied Physics* 94:3984.

Kopf RF (2003) State-of-the-Art Program on Compound Semiconductors XXXIX and Nitride and Wide Bandgap Semiconductors for Sensors, Photonics and Electronics IV. *Proceedings of the International Symposia*. Pennington, NJ: Electrochemical Society.

Lähnemann J, Brandt O, Jahn U, Pfüller C, Roder C, Dogan P, Grosse F, Belabbes A, Bechstedt F, Trampert A, Geelhaar L (2012) Direct experimental determination of the spontaneous polarization of GaN. *Physical Review B* 86:081302.

Łepkowski SP, Majewski JA (2006) Effect of electromechanical coupling on the pressure coefficient of light emission in group-III nitride quantum wells and superlattices. *Physical Review B* 74:035336.

Lippmann G (1881) Principe de la conservation de l'électricité, ou second principe de la théorie des phénomènes électriques. *Journal de Physique Théorique et Appliquée* 10:14.

Migliorato MA, Cullis AG, Fearn M, Jefferson JH (2002) Atomistic simulation of $In_xGa_{1-x}As/GaAs$ quantum dots with nonuniform composition. *Physica E: Low-dimensional Systems and Nanostructures* 13:1147–1150.

Migliorato MA, Pal J, Garg R, Tse G, Al-Zahrani HYS, Monteverde U, Tomić S, Li CK, Wu YR, Crutchley BG, Marko IP, Sweeney SJ (2014) A review of non linear piezoelectricity in semiconductors. *AIP Conference Proceedings* 1590:32–41.

Migliorato MA, Powell D, Cullis AG, Hammerschmidt T, Srivastava GP (2006) Composition and strain dependence of the piezoelectric coefficients in $In_xGa_{1-x}As$ alloys. *Physical Review B* 74 Article ID: 245332.

Migliorato MA, Powell D, Liew SL, Cullis AG, Navaretti P, Steer MJ, Hopkinson M, Fearn M, Jefferson JH (2004) Influence of composition on the piezoelectric effect and on the conduction band energy levels of $In_xGa_{1-x}As/GaAs$ quantum dots. *Journal of Applied Physics* 96:5169.

Migliorato MA, Powell D, Zibik EA, Wilson LR, Fearn M, Jefferson JH, Steer MJ, Hopkinson M, Cullis AG (2005) Anisotropy of the electron energy levels in $In_xGa_{1-x}As/GaAs$ quantum dots with non uniform composition. *Physica E: Low-Dimensional Systems and Nanostructures* 26:436–440.

Nobel (2014) *Laureates in Physics*. http://www.nobelprize.org/nobel_prizes/physics/laureates/2014/

Nye JF (1957) *Physical Properties of Crystals*. Oxford: Clarendon.

Pal J, Migliorato MA, Li CK, Wu YR, Crutchley BG, Marko IP, Sweeney SJ (2013) Enhancement of efficiency of InGaN-based light emitting diodes through strain and piezoelectric field management. *Journal of Applied Physics* 114:073104.

Pal J, Tse G, Haxha V, Migliorato MA, Tomić S (2011a) Second-order piezoelectricity in wurtzite III-N semiconductors. *Physical Review B* 84 Article ID: 085211.

Pal J, Tse G, Haxha V, Migliorato MA, Tomić S (2011b) Importance of non linear piezoelectric effect in Wurtzite III-N semiconductors. *Optical and Quantum Electronics* 44:9.

Pal J, Tse G, Haxha V, Migliorato MA, Tomić S (2011c) Erratum: Second-order piezoelectricity in wurtzite III-N semiconductors. *Physical Review B* 84:159902, 085211.

Park S-H, Chuang S-L (2000) Spontaneous polarization effects in wurtzite GaN/AlGaN quantum wells and comparison with experiment. *Applied Physics Letters* 76:1981.

Pedesseau L, Katan C, Even J (2012) On the entanglement of electrostriction and non-linear piezoelectricity in non-centrosymmetric materials. *Applied Physics Letters* 100:031903.

Piprek J (2015) How to decide between competing efficiency droop models for GaN-based light-emitting diodes. *Applied Physics Letters* 107:031101.

Piprek J, Römer F, Witzigmann B (2015) On the uncertainty of the Auger recombination coefficient extracted from InGaN/GaN light-emitting diode efficiency droop measurements. *Applied Physics Letters* 106:101101.

Prodhomme P-Y, Beya-Wakata A, Bester G (2013) Nonlinear piezoelectricity in wurtzite semiconductors. *Physical Review B* 88:121304.

Romanov AE, Baker TJ, Nakamura S, Speck JS, Group EJU (2006) Strain-induced polarization in wurtzite III-nitride semipolar layers. *Journal of Applied Physics* 100:023522.

Schulz S, Marquardt O (2015) Electronic structure of polar and semipolar (1122)-oriented nitride dot-in-a-well systems. *Physical Review Applied* 3:064020.

Shimada K, Sota T, Suzuki K, Okumura H (1998) First-principles study on piezoelectric constants in strained BN, AlN, and GaN. *Japanese Journal of Applied Physics* 37:L1421.

Singh J (1992) *Physics of Semiconductors and Their Heterostructures* (McGraw Hill Series in Electrical and Computer Engineering). Blacklick, OH: McGraw-Hill College.

Singh J (2003) *Electronic and Optoelectronic Properties of Semiconductor Structures*. Cambridge: Cambridge University Press.

Suski T, Łepkowski SP, Staszczak G, Czernecki R, Perlin P, Bardyszewski W (2012) Universal behavior of photoluminescence in GaN-based quantum wells under hydrostatic pressure governed by built-in electric field. *Journal of Applied Physics* 112:053509.

Tartakovskii AI, Kolodka RS, Liu HY, Migliorato MA, Hopkinson M, Makhonin MN, Mowbray DJ, Skolnick MS (2006) Exciton fine structure splitting in dot-in-a-well structures. *Applied Physics Letters* 88:131115.

Tomić S, Pal J, Migliorato MA, Young RJ, Vukmirović N (2015) Visible spectrum quantum light sources based on $In_xGa_{1-x}N/GaN$ quantum dots. *ACS Photonics* 2:958–963.

Tse G, Pal J, Monteverde U, Garg R, Haxha V, Migliorato MA, Tomicì S (2013) Non-linear piezoelectricity in zinc blende GaAs and InAs semiconductors. *Journal of Applied Physics* 114:073515.

Tsubouchi K, Mikoshiba N (1985) Zero-temperature-coefficient SAW devices on AlN epitaxial films. *IEEE Transactions on Sonics and Ultrasonics* 32:634–644.

Vanderbilt D (2000) Berry-phase theory of proper piezoelectric response. *Journal of Physics and Chemistry of Solids* 61:147–151.

Vegard L (1921) Die Konstitution der Mischkristalle und die Raumfüllung der Atome. *Zeitschrift für Physik* 5:10.

Voigt W (1910) *Lehrbuch der Kristallphysik*. Leipzig: B. G. Teubner.

Vurgaftman I, Meyer JR (2003) Band parameters for nitrogen-containing semiconductors. *Journal of Applied Physics* 94:3675–3696.

Wang ZL (2007) Nanopiezotronics. *Advanced Materials* 19:4.

Yan WS, Zhang R, Xie ZL, Xiu XQ, Han P, Lu H, Chen P, Gu SL, Shi Y, Zheng YD, Liu ZG (2009) A thermodynamic model and estimation of the experimental value of spontaneous polarization in a wurtzite GaN. *Applied Physics Letters* 94:042106.

Yang T-J, Shivaraman R, Speck JS, Wu Y-R (2014) The influence of random indium alloy fluctuations in indium gallium nitride quantum wells on the device behavior. *Journal of Applied Physics* 116:113104.

9

Dilute Nitride Alloys

Christopher A.
Broderick

Masoud Seifikar

Eoin P. O'Reilly

and

Judy M. Rorison

9.1 Introduction

The exponential growth of optical telecommunications and the internet has been underpinned by the development of semiconductor lasers emitting at 1.3 and 1.55 µm, the respective wavelengths at which dispersion is zero and losses minimized, in standard silica optical fibers [1,2]. Lasers designed to emit at these wavelengths are based primarily on the growth of quaternary InGaAsP or AlInGaAs quantum well (QW) structures on InP substrates. However, despite their widespread applications there are a number of significant limitations associated with incumbent 1.3- and 1.55-µm technologies, several of which are associated with the constraints of growing on InP substrates [3]. This has stimulated significant interest in the development of telecom-wavelength lasers grown on GaAs substrates, which has the potential to deliver a series of key advantages. First, because GaAs is a more robust material, growth can be carried out on larger substrates. Second, growth of reliable semiconductor lasers and related devices on GaAs opens up the possibility of monolithic integration of high-performance photonic components with GaAs-based high-speed electronics. Third, from a fundamental perspective, better optical confinement can be achieved in GaAs-based structures compared to those based on InP, because of the larger refractive index contrast between GaAs and AlGaAs compared to that between InP and the InGaAsP or AlInGaAs quaternary alloys. This means that, for example, vertical-cavity surface-emitting lasers (VCSELs) can be grown monolithically on GaAs, but only with great difficulty on InP. Additionally, because AlGaAs has a considerably larger band gap than InP and can be grown lattice-matched on GaAs, it is possible to achieve significantly better carrier confinement in a GaAs-based laser structure, thereby making it possible to overcome the losses associated with carrier leakage from which InP-based devices are known to suffer under high injection currents and at high temperatures.

Considerable advantages could therefore be gained if high-quality and reliable 1.3- and 1.55-μm lasers can be developed on GaAs substrates. However, this is difficult to achieve using conventional QW structures. Highly efficient lasers based on compressively strained InGaAs QWs are well established for emission at wavelengths close to 1 μm [4], but to extend the emission to longer wavelengths requires further incorporation of indium (In) leading to a large lattice mismatch relative to GaAs. As such, critical thickness issues become important in determining the material quality, leading to a degradation in performance and a general inability to achieve reliable InGaAs/GaAs QW lasers emitting at 1.3 μm and beyond [1]. A number of approaches have been proposed and investigated to overcome these issues, in order to extend the wavelength range accessible using the GaAs platform. These include the development of InGaAs quantum dot (QD) lasers [5,6], and metamorphic QW structures incorporating relaxed InGaAs buffer layers to facilitate the growth of heterostructures having lattice constants intermediate between those of GaAs and InP [7–9].

In this chapter, we focus on an alternative approach: the use of dilute nitride alloys—III–V semiconductor alloys containing small fractions of substitutional nitrogen (N) atoms—to achieve long-wavelength emission in materials having lattice constants compatible with growth on GaAs. Interest in dilute nitride alloys for applications in photonic devices originally grew from the observation that replacing a small fraction of the arsenic (As) atoms with N in (In)GaAs leads to an extremely rapid reduction of band gap, by up to 150 meV per % N in GaN_xAs_{1-x} [10–12]. Strong interest in the fundamental aspects of the physics of dilute nitride alloys, combined with their potential for practical applications, has stimulated a significant body of research on the theory and simulation of N-containing semiconductors over the past 15 or so years. In this chapter—the first of two on highly mismatched III–V semiconductor alloys—we trace the development of the theory of dilute nitride alloys, ranging from fundamental investigations of the electronic structure of ordered and disordered alloys at an atomistic level to continuum models suited to modeling the properties of dilute nitride–based photonic devices.

Since N is much smaller and more electronegative than arsenic (As), it acts as an isovalent impurity when substituted in dilute concentrations in (In)GaAs to form the $(In)GaN_xAs_{1-x}$ alloy. Furthermore, due to its smaller size relative to As, incorporation of N results in a simultaneous reduction of the band gap and lattice constant with increasing x in $(In)GaN_xAs_{1-x}$. This trend is uncommon in III–V semiconductor alloys, in which a reduction in band gap is typically associated with an increase in lattice constant. Dilute nitride alloys therefore offer the possibility not only to extend the emission wavelength beyond that accessible using InGaAs alloys grown on GaAs substrates, but also to allow the compressive strain due to In incorporation to be compensated. This opens the possibility to achieve long-wavelength emission on GaAs at acceptable levels of epitaxial strain, and has led to the demonstration of 1.3-μm edge-emitting lasers and VCSELs [13–15], as well as 1.55-μm edge-emitting lasers [16,17].

From a fundamental perspective, the impurity-like nature of substitutional N atoms leads to a strong perturbation of the material band structure, primarily in the conduction band (CB). As a result, commonly employed approaches to modeling alloy electronic properties, such as the virtual crystal approximation, break down and a more detailed theoretical treatment is required to understand the strong impact of N incorporation on the electronic structure of GaN_xAs_{1-x} and related alloys. As such, there has been considerable theoretical interest in the causes and consequences of the unusual electronic properties of dilute nitride alloys, detailed understanding of which has largely been developed through atomistic calculations. We therefore begin this chapter by presenting, in Section 9.2, a review of atomistic theoretical calculations of the electronic structure, focusing primarily on the GaN_xAs_{1-x} and GaN_xP_{1-x} alloys. These calculations quantify the impact of N incorporation on the electronic structure, highlight the strong role played by

N-related alloy disorder, and describe the consequences of these effects for technologically relevant material properties in real N-containing alloys.

While atomistic approaches have provided significant insight into the electronic structure of N-containing alloys, they are too computationally expensive to apply directly on the length scales required to model the properties of optoelectronic devices. The significant interest in developing dilute nitride alloys for applications in semiconductor lasers and related devices has motivated the development of continuum models which are suited to calculating the properties of realistically sized N-containing device heterostructures. It has been demonstrated that the main features of the band structure of GaN_xAs_{1-x} and related alloys can be understood phenomenologically in terms of a band-anticrossing (BAC) interaction, between the unperturbed extended (Bloch) states of the (In)GaAs host matrix CB edge and localized states associated with substitutional N impurities. In Section 9.2 we review the quantitative understanding of the BAC model of the dilute nitride CB structure developed through detailed atomistic supercell calculations and outline the derivation of a 10-band $\mathbf{k} \cdot \mathbf{p}$ Hamiltonian for $(In)GaN_xAs_{1-x}$, which is suited to calculate the electronic properties of device heterostructures.

Next, in Section 9.3, we review the theory of dilute nitride semiconductor lasers. The majority of the research effort dedicated to developing GaAs-based dilute nitride semiconductor lasers has focused on using quaternary InGaNAs QW structures, to target 1.3-µm emission. As such, it is to these structures that we primarily devote our attention. We describe how the 10-band $\mathbf{k} \cdot \mathbf{p}$ Hamiltonian presented in Section 9.2 can be used to quantify the impact of N incorporation on the electronic and optical properties, and review important results outlining key trends in the predicted properties and performance of InGaNAs QW lasers. This discussion is supplemented by an outline of further investigations that have quantified the importance of various recombination pathways—including defect-related, radiative, and nonradiative (Auger) recombination—in prototypical InGaNAs devices, with comparisons between theory and experiment highlighting the progress that has been made in developing the understanding of this novel class of semiconductor lasers.

Having outlined in detail the theory and modeling of 1.3-µm InGaNAs QW lasers grown on GaAs substrates, in Section 9.4 we provide an overview of a range of alternative applications of dilute nitride alloys. These further attempts to exploit the impact of N incorporation on the band structure and facilitate the development of devices with enhanced capabilities and characteristics include the development and use of (1) InGaNAsSb structures on GaAs for 1.55-µm lasers and semiconductor optical amplifiers (SOAs), (2) InGaNAs structures grown on GaAs to develop SOAs, electro-absorption modulators (EAMs), and avalanche photodiodes (APDs) operating at 1.3 µm, and (3) GaNAsP alloys and QW structures for monolithic integration of III–V lasers on silicon (Si).

Finally, in Section 9.5 we turn our attention to the topic of carrier transport in dilute nitride alloys. Incorporation of N, which strongly perturbs the CB structure via the formation of highly localized impurity states, can be expected to have important implications for electron transport. This has been reflected in experimental measurements, which have revealed significant reductions in electron mobility even at ultra-dilute N compositions. By considering the BAC model of Section 9.2 in detail, we highlight the shortcomings of this simple picture of the CB structure from the perspective of calculating the transport properties of real materials. We provide an overview of descriptions of the band structure, based on the Anderson impurity model (AIM), which have been developed to overcome the shortcomings of the BAC model and facilitate practical calculations of density of states (DOS) and carrier transport. We then review theoretical models and calculations of electron transport in GaN_xAs_{1-x} and highlight the importance of the fine details of the N-perturbed CB structure in determining the properties of real materials. We focus in this review on calculations of the electron mobility and demonstrate the significant level of detail that must be incorporated in transport calculations in order to reliably reproduce the results of experimental measurements, and in doing so demonstrate rigorous validation of theoretical models of the CB structure in GaN_xAs_{1-x} alloys. We further highlight that N incorporation places intrinsic limits on the electron mobility in dilute nitride alloys, emphasize the dominant role in determining the transport properties played by strong resonant scattering of electrons by N-localized impurity states (associated not only with isolated

N impurities but also with N–N pairs and larger clusters of N atoms), and discuss the implications of N incorporation for the existence of a negative differential velocity regime in GaN_xAs_{1-x}.

The overall purpose of this chapter is to present a comprehensive overview of the theory and modeling of dilute nitride alloys, with a focus on applications in photonic devices. Our aim is that this chapter should provide the reader with a comprehensive introduction to the theory and simulation of dilute nitride alloys, in a manner that should enable them to engage meaningfully with the existing and emerging literature on this interesting and unusual class of semiconductor alloys.

9.2 Theory of the Electronic Structure of Dilute Nitride Alloys

In this section, we present a review of the theory and modeling of the electronic structure of dilute nitride alloys. Due to the fact that N acts as an isovalent impurity when incorporated into (In)GaAs(P) in dilute concentrations, a significant body of analysis has been undertaken to quantify the resulting strong perturbations to the band structure. A detailed exposition of the results of all of these studies is beyond the scope of this chapter. Here, we instead highlight key results from the literature, following a path that enables us to build toward a quantitative description of the properties of real, disordered dilute nitride alloys. The results and models reviewed in this section will underpin our discussions of the theory and simulation of dilute nitride materials and devices throughout the remainder of this chapter.

We begin in Section 9.2.1 with an overview of initial *ab initio* analysis of the generic properties of GaN_xAs_{1-x} alloys based on density functional theory (DFT), which served as motivation for the development of large supercell analyses of the alloy electronic structure undertaken using the empirical pseudopotential (EP) method. As described above, while atomistic analyses have provided detailed insight into the causes and consequences of the unusual properties of dilute nitride alloys, they are too computationally expensive to be applied on the length scales required to model realistic semiconductor devices. This has prompted the development of phenomenological descriptions of the dilute nitride CB structure based on the BAC model. In Section 9.2.2 we review the BAC model, beginning with the original empirical motivation and parametrization in response to experimental data before reviewing the results of atomistic tight-binding (TB) calculations, which allow the BAC and related models to be placed on a quantitative footing. We then describe the extension of the BAC approach, which has been accomplished by using TB supercell calculations to derive and parametrize an extended basis set 10-band $\mathbf{k} \cdot \mathbf{p}$ Hamiltonian for dilute nitride alloys. This 10-band model will form the basis of our discussion of the theory of $(In)GaN_xAs_{1-x}$ semiconductor lasers in Section 9.3 and has been widely employed to calculate and analyze the properties of dilute nitride heterostructures and devices.

In Section 9.2.3 we outline the linear combination of the isolated nitrogen states (LCINS) approach to describing the band structure of N-containing alloys, and review that explicit consideration and construction of the N-related localized states allows for rigorous description of the electronic structure of real, disordered dilute nitride alloys. We demonstrate that the calculations undertaken using the LCINS method are in excellent quantitative agreement with the results of experimental measurements of a range of key material properties, in particular capturing the strong impact of N-related alloy disorder and non-monotonic variation with composition of important band properties.

9.2.1 First Principles and Empirical Pseudopotential Calculations

Before discussing the electronic structure of dilute nitride alloys in detail, it is useful first to state explicitly what is meant by describing a semiconductor alloy as "highly mismatched." In conventional semiconductor alloys, such as $Al_xGa_{1-x}As$ or $In_xGa_{1-x}As_yP_{1-y}$, the material properties vary smoothly with composition. For example, the lattice and elastic constants, and deformation potentials, typically follow Végard's law. That is, they can be determined straightforwardly by interpolating linearly between those of the binary compounds constituting the alloy. For example, the lattice constant a in $In_xGa_{1-x}As$ can be determined to a high degree of accuracy as $a(x) = (1 - x)\,a(\text{GaAs}) + x\,a(\text{InAs})$, where $a(\text{GaAs})$ and $a(\text{InAs})$ are the GaAs and InAs lattice constants, respectively. For some material parameters, typically the band gap and valence-band (VB) spin-orbit-splitting energy, experimental measurements reveal slight deviation from the linear variation described by Végard's law, which can be described in terms of a bowing parameter b—for example, the band gap E_g in $\text{In}_x\text{Ga}_{1-x}\text{As}$ can be determined as $E_g(x) = (1 - x)\,E_g(\text{GaAs}) + x\,E_g(\text{InAs}) - b\,x\,(1 - x)$, where b is independent of the alloy composition x and typically takes a small positive value <1 eV. The variation of the properties with alloy composition in conventional semiconductor alloy can then be described theoretically in a straightforward manner using, for example, the virtual crystal approximation in which the group-III atoms in $\text{In}_x\text{Ga}_{1-x}$ As can be treated as though they have the properties of an average "$\text{In}_x\text{Ga}_{1-x}$" atom.

In highly mismatched semiconductor alloys—such as the dilute nitride and bismide alloys we discuss here and in Chapter 10—the situation is markedly different. Highly mismatched semiconductor alloys are typically characterized by the fact that the constituent elements forming the alloy have significantly different sizes (covalent radii) and chemical properties (electronegativities). For example, in the dilute nitride (bismide) alloy $\text{GaN}_x\text{As}_{1-x}$ ($\text{GaBi}_x\text{As}_{1-x}$), the N (Bi) atoms are significantly smaller (larger) and more electronegative (electropositive) than the As atoms they replace. In these cases, the N and Bi atoms act as isovalent impurities when incorporated substitutionally in dilute concentrations, strongly perturbing the electronic structure. As a result, the material properties are found to vary rapidly, and in some cases non-monotonically, with alloy composition, leading to a breakdown of simple descriptions such as the virtual crystal approximation. This is typically characterized by the band gap E_g, which has two general properties not found in conventional semiconductor alloys: (1) a rapid variation of the magnitude of E_g with alloy composition, and (2) a large bowing parameter b which itself depends strongly upon the alloy composition. In a simple sense, the observation in experiment of one or more large, composition-dependent bowing parameters serves as a useful indication that a given semiconductor alloy is highly mismatched and is hence likely to possess a range of unusual material properties. The presence of such behavior in the dilute nitride alloys $\text{GaN}_x\text{As}_{1-x}$ and $\text{GaN}_x\text{P}_{1-x}$ has been known for some time: experimental measurements have revealed the presence due to N incorporation of (1) large, composition-dependent band gap bowing, (2) large and strongly composition-dependent CB effective masses, (3) significantly reduced pressure deformation potentials, and (4) a progression of sharp impurity lines in spectroscopic measurements, lying energtically within the host matrix (GaAs or GaP) band gap.

The presence of strongly composition-dependent band gap bowing in the dilute nitride alloy $\text{GaN}_x\text{As}_{1-x}$ was first theoretically analyzed in detail by Wei and Zunger [18]. Using a series of DFT supercell calculations, they identified two distinct regimes in the variation of the material properties with alloy composition, which serve as an indication of the general nature of a given semiconductor alloy: (1) an impurity-like composition regime, within which the material properties vary abruptly and strongly in response to small changes in composition, and (2) a band-like regime, within which the material properties vary smoothly and gradually in response to changes in composition. A comparative analysis of the band gap bowing in $\text{GaN}_x\text{As}_{1-x}$ and $\text{GaAs}_{1-x}\text{P}_x$ first revealed that the strong, composition-dependent bowing in $\text{GaN}_x\text{As}_{1-x}$ persists to large N compositions >25%, suggesting strong impurity-like behavior over a wide range of alloy compositions. The band gap bowing was calculated to be as large as $b = 15.69$ eV in a small supercell containing 12.5% N ($x = 0.125$), which was found to reduce to 7.6 eV by 25% N ($x = 0.25$). In contrast, such behavior was not calculated for $\text{GaAs}_{1-x}\text{P}_x$, with b calculated to vary from $b = 0.23$ eV at 12.5% P to 0.18 eV by 25% P. This allowed Wei and Zunger to rigorously distinguish between conventional and highly mismatched semiconductor alloys on the basis of the impurity- and band-like regimes described above. In a conventional alloy, substitutional incorporation of a dilute composition of atoms of a different species—for example, P atoms in GaAs to form $\text{GaAs}_{1-x}\text{P}_x$ at low x—leads to impurity-like behavior only over an extremely limited range of alloy compositions, with conventional band-like behavior emerging for compositions as low as $x \approx 1\%$. Distinct to this calculated behavior for $\text{GaAs}_{1-x}\text{P}_x$, Wei and Zunger demonstrated that highly mismatched alloys such as $\text{GaN}_x\text{As}_{1-x}$ can be characterized by the fact that the impurity-like regime extends to much higher alloy compositions.

Further analysis suggests that this persistence of the impurity-like regime is, as outlined above, associated with the large differences in orbital energies and covalent radius between an alloy's constituent elements. That this is the case has been generally known since the seminal work of, among others, Slater and Koster [19] and Hjalmarson et al. [20] on the electronic structure of deep-level impurities in semiconductors— subsequent interest in dilute nitrides and related materials has helped bring these concepts to the forefront of contemporary semiconductor research. As a general point, it is interesting to note that this unusual material behavior—i.e. strong qualitative changes in the material properties—are typically associated with one of several classes of phase transitions (for example, structural transitions between crystal phases or topological transitions in the electronic band structure), while in highly mismatched alloys such changes can be brought about by incorporating a dilute concentration of an impurity atom, suggesting significant potential to control, design, and optimize desirable material properties.

Following this general classification of conventional versus highly mismatched alloys, Wei and Zunger performed a systematic analysis of the large, composition-dependent band gap bowing in GaN_xAs_{1-x}. Specifically, they analyzed the contributions to b associated with N incorporation arising from (1) volume deformation, (2) structural relaxation of the crystal, as well as (3) charge exchange between atoms. Each of these factors contributes to the observed strong reduction of the GaN_xAs_{1-x} band gap with increasing x, associated, respectively, with (1) compression (extension) of the Ga—As (Ga—N) bonds compared to their equilibrium lengths an a binary compound, (2) strain due to local relaxation of the crystal lattice about substitutional N impurities, and (3) a change in the spatial localization of the ground state charge density of VB (bonding) electrons in GaN_xAs_{1-x} as compared to that calculated for unrelaxed GaAs and GaN having the GaN_xAs_{1-x} lattice constant [18]. In this interpretation of the GaN_xAs_{1-x} alloy electronic structure, the local structural relaxation and charge exchange depend directly on the differences in atomic size and electronegativity between. As and N. Using a series of supercell calculations, and again comparing the calculated results for both N and P impurities in GaN_xAs_{1-x} and $GaAs_{1-x}P_x$, they demonstrated that while the volume deformation was comparable between the two alloys, the local structural relaxation and charge exchange were significantly more pronounced in GaN_xAs_{1-x}. The prominence of these effects is then associated with the large ~20% lattice mismatch between GaAs and GaN, as well as the large differences ~4 and 2 eV between the s- and p-like valence orbitals in As and N. The calculated strong contribution of the relaxation of the local crystal structure about isolated N impurities, as well as the observed transfer of valence (conduction) charge from N to As (As to N), confirmed that the large, composition-dependent band gap bowing GaN_xAs_{1-x} is a direct result of the large chemical and size differences between As and N. This fundamental insight into the nature of the GaN_xAs_{1-x} electronic structure has, as we will see throughout the remainder of this section, motivated and informed a rich body of subsequent work on the theory of GaN_xAs_{1-x} and related alloys.

Following this initial study of the electronic structure of ordered GaN_xAs_{1-x} alloys, Bellaiche et al. [21] undertook a more detailed investigation of the electronic structure of GaN_xAs_{1-x} and the related dilute nitride alloy GaN_xP_{1-x}. Through a combination of DFT and EP calculations on supercells containing up to 512 atoms, they calculated, compared, and contrasted the electronic structure of GaN_xAs_{1-x} and GaN_xP_{1-x} in terms of the impact of N incorporation on the CB structure. First, their calculations indicated the presence of impurity states associated with a substitutional N atom, which lie energetically within (and are hence resonant with) the CB in GaAs, but form a bound state in GaP (lying energetically within the GaP band gap). In both GaAs and GaP, the N-related impurity states were found to be highly localized about the N atomic site(s). Later calculations by Kent et al. [22], which we discuss further below, indicate that isolated N impurities are responsible for introducing strong N-mediated coupling of the extended host matrix CB states which creates a set of A_1-symmetric impurity states from a basis formed of a combination of GaAs CB states from the Γ-, L- and X-valleys, in addition to A_1-symmetric states associated directly with N. This leads to the N-derived localized states being formed of an admixture of a significant number of large wave vector host matrix states, with the A_1-like N-localized state associated with an isolated N impurity in GaN_xAs_{1-x} (GaN_xP_{1-x}) calculated to lie approximately 150 meV (30 meV) above (below) the GaAs (GaP) Γ-point CB states. As we describe in further detail throughout the remainder of this section, it is this N-mediated coupling which (1) is responsible for the strong perturbation to the GaAs(P) CB structure caused by N incorporation and (2) forms the basis of simplified theoretical models which have been used to understand and model the properties of dilute nitride alloys and heterostructures. A key success of this theoretical approach has been the explanation of the unusual pressure dependence of the GaN_xAs_{1-x} electronic structure: in the ultra-dilute limit, Kent et al. used a series of EP calculations to confirm that the localized states associated with isolated N impurities—which have intrinsically low-pressure dependence compared to the CB states of the host matrix semiconductor—can be pushed into the band gap under the application of hydrostatic pressures ~2 GPa. This is consistent with a range of experimental measurements and confirms the role of the N-mediated coupling described above in determining the observed low-pressure deformation potentials in GaN_xAs_{1-x} alloys.

Given the presence of N-related localized states in GaN_xP_{1-x} lying energetically within the GaP band gap, incorporation of dilute concentrations of N is found to bring about an abrupt transition in the nature of the band structure: the unperturbed GaP band structure is an indirect gap, with the lowest energy CB lying at the X-point in the Brillouin zone. Incorporation of N gives rise to N-related localized bound states which lie below the GaP X-point CB minimum at Γ, therefore suggesting an abrupt change from an indirect to a direct band gap in GaN_xP_{1-x} even at ultra-dilute N compositions [23–25]. This unusual material behavior is particularly appealing from a technological perspective, as it brings about the opportunity to develop direct-gap III–V alloys which have lattice constants compatible with epitaxial growth on GaP or Si substrates (cf. Section 9.4.3).

Analysis of the bond lengths in $GaAs_{1-x}P_x$ and GaN_xAs_{1-x} alloys further revealed strong local relaxation of the crystal structure about the N atomic site(s), and comparison to calculations performed for isolated P impurities in GaAs again emphasized that these calculated material properties are distinct from those associated with conventional alloys. Following this, Bellaiche et al. investigated the impact of increasing the N composition x on the structural properties of GaN_xAs_{1-x} and GaN_xP_{1-x}, demonstrating the persistence and increase of strong structural relaxation with increasing x. Specifically, these calculations demonstrated the emergence of a binomial distribution of nearest-neighbor bond lengths corresponding, for example, to longer (shorter) Ga–As (Ga–N) bonds in disordered GaN_xAs_{1-x}. These calculations clearly demonstrated the emergence of percolation in GaN_xAs_{1-x} alloys for $x > 0.19 \equiv x_p$, associated with the energetically preferential formation of Ga–N–Ga–N–... nearest-neighbor "chains" throughout the crystal. EP calculations on disordered 512-atom supercells further confirmed the strong deviation of the GaN_xAs_{1-x} band gap from the variation predicted on the basis of Végard's law, throughout the entire composition range $0 \leq x \leq 1$—the electronic structure of GaN_xAs_{1-x} was found to divide roughly into three distinct regions: (1) $x < x_p$ in which N acts as an impurity primarily perturbing the CB structure, (2) $x_p \leq x \leq 0.4$ in which the material properties varying relatively smoothly with N composition, and (3) $x > 0.4$ in which As acts as an impurity primarily perturbing the VB structure. Overall, these calculations indicate highly unusual material behavior consisting of a range of complicated and correlated factors, including the presence of localized impurity states and their strong perturbation of the band structure— leading to strong band mixing and localization effects—as well as structural relaxation, charge transfer, and, ultimately, anion percolation.

Further theoretical analysis of the electronic structure of GaN_xAs_{1-x} and GaN_xP_{1-x} alloys was presented by Kent et al. in Refs. [26] and [22]. In these studies, Kent et al. applied the EP method to investigate the electronic structure of large, disordered GaN_xAs_{1-x} and GaN_xP_{1-x} supercells containing up to 14,000 atoms. Key results of these theoretical calculations include (1) the identification and quantitative analysis of a range of N-related localized states—associated not only with isolated N impurities, but also with N–N pairs (in which two N atoms share a common Ga nearest neighbor) and larger N clusters (in which multiple N atoms are second-nearest neighbors)—features associated with which had previously been observed in spectroscopic measurements, and (2) the identification of a localized-to-delocalized transition both in GaN_xAs_{1-x} and GaN_xP_{1-x}, related to strong, composition-dependent hybridization between the extended states of the host matrix (GaAs or GaP) CB states and highly localized N-related impurity states, and associated with a breakdown of the Bloch character of the alloy CB edge states.

Beginning with (1)—the existence and impact of N-related localized states—the detailed analysis undertaken by Kent et al. distinguished the impact of clustering of substitutional N impurities on the electronic structure of GaAs and GaP as again being distinct from that observed in conventional alloys such as $GaAs_{1-x}P_x$. First, while the formation of P–P pairs and larger clusters of P atoms tends not to form impurity states in $GaAs_{1-x}P_x$—thus leaving the electronic structure relatively unperturbed—as noted above, the formation of N–N pairs and larger clusters of N atoms in GaN_xAs_{1-x} and GaN_xP_{1-x} produces a series of strongly localized impurity states. EP calculations indicate that these localized states typically become more strongly bound as the number of N atoms associated with a given localized perturbation increases, thereby giving rise to the progression of sharp impurity lines observed in spectroscopic measurements. Recalling that the localized states associated with isolated N impurities in GaN_xAs_{1-x} are resonant with the GaAs CB, the presence of N–N pairs and larger N clusters gives rise to more strongly bound localized states lying lower in energy, accounting for the observed progression of impurity lines observed within the GaN_xAs_{1-x} band gap at dilute N compositions.

The calculations of Kent et al. further suggested that these individual N cluster states are sufficiently localized that they tend not to significantly overlap or interact with one. Further analysis highlighted that the N-induced downward shift in energy of the alloy CB edge with increasing x is sufficiently large that it results in strong hybridization between CB edge states and N cluster states—as the former pass through the latter in energy—which, in conjunction with experimental measurements, suggests that N cluster states play a key role in bringing about the observed large CB edge effective masses and anomalously low pressure deformation potentials in GaN_xAs_{1-x} [22,40,44]. These results suggest that some of the more detailed features of the GaN_xAs_{1-x} (and GaN_xP_{1-x}) CB structure then depend on the precise N-related disorder present in a given material sample. It turns out, as we discuss in Sections 9.2.3 and 9.5.1, that this is indeed the case: a quantitative understanding of the dilute nitride CB structure, and in particular the CB edge effective mass and g factor, is possible only when the localized states associated with a full distribution of N clusters are treated explicitly. This again distinguishes dilute nitride alloys from conventional semiconductors, indicating that sophisticated, highly-specialized and computationally demanding theoretical approaches are required in order to understand observed trends in many important material parameters. A detailed review of the quantitative analysis of various N cluster states undertaken by Kent et al. is beyond the scope of this chapter. We refer the reader to Ref. [22] for the relevant details.

Turning our attention to (2)—the existence of a localized-to-delocalized transition—the EP calculations of Kent et al. further quantified the evolution of the alloy CB edge states in GaN_xAs_{1-x} and GaN_xP_{1-x} as the N composition x is increased. At low N composition, the aforementioned strong N-induced hybridization between the extended states of the host matrix CB and N-related localized impurity states leads to alloy CB edge states having significant localized character (corresponding to a loss of the Bloch character that is largely preserved in conventional alloys). This behavior is characteristic of the extended impurity-like composition regime identified above, within which the material properties are strongly influenced by the nature of the localized states present in a given alloy. As the N composition is increased sufficiently to emerge from the impurity-like regime to the band-like regime, the presence and influence of these localized states is diminished, leading to a transition from alloy CB states which have strong localized (N-derived) character to extended alloy states which have significant Bloch character and more closely resemble those present in conventional alloys. We again refer the reader to Ref. [22] for further details and a quantitative discussion of this aspect of the dilute nitride electronic structure.

Having reviewed at a high level the impact of N incorporation on the structural and electronic properties of the prototypical dilute nitride alloys GaN_xAs_{1-x} and GaN_xP_{1-x}, we defer for now further discussion of the impact of N incorporation on the detailed features of the electronic structure. Instead, we turn our attention to simplified models of the dilute nitride CB structure, which have proved highly useful for (1) the interpretation of the general features of the band structure as revealed through experimental measurements, as well as for (2) the theory and modeling of dilute nitride semiconductor devices. The material we discuss in Section 9.2.2 will then informs our further discussion of the dilute nitride electronic structure in Sections 9.2.3 and 9.5, in which we respectively address the impact of N incorporation and N-related alloy disorder on the CB structure and transport properties of GaN_xAs_{1-x} and related alloys.

9.2.2 Band-Anticrossing: Empirical, Tight-Binding, and k·p Models

On the basis of the DFT and EP electronic structure calculations described above, it was established that the localized, impurity-like nature of the states associated with substitutional N atoms in (In)GaAsP alloys was responsible for the strong perturbation of the CB structure brought about by N incorporation. While the interpretation of these atomistic calculations was in general complex, in Ref. [27] Shan et al. demonstrated that the impact of N incorporation on the CB structure in (In)GaN$_x$As$_{1-x}$ could in general be understood in a straightforward, phenomenological manner. Specifically, Shan et al. performed pressure-dependent photo-modulated reflectance (PR) measurements on a series of (In)GaN$_x$As$_{1-x}$ alloys which revealed a number of important trends. First, experiments undertaken at ambient temperature and pressure suggested that N incorporation splits the Γ-point CB minimum into two subbands, clearly identifiable as distinct optical transitions in measured PR spectra. Second, performing PR measurements under applied hydrostatic pressure, Shan et al. observed a strongly non-linear variation of the alloy CB edge energy with applied hydrostatic pressure, a clear deviation from the linear variation which is characteristic of conventional semiconductor alloys. Furthermore, under high applied pressure saturation of the alloy CB edge energy was observed, clearly indicating anticrossing behavior.

On the basis of their experimental measurements, Shan et al. demonstrated that these unusual trends could be explained quantitatively by a simple two-band model of the (In)GaN$_x$As$_{1-x}$ CB structure in which the extended states of the (In)GaAs host matrix undergo a strong anticrossing interaction with an energetically narrow band of localized N-related states that are resonant with the (In)GaAs CB. This is the well-known two-band BAC model, in that the CB structure is obtained by diagonalizing the Hamiltonian [27]

$$H(k) = \begin{pmatrix} E_{\text{N}} & V_{\text{Nc}} \\ V_{\text{Nc}}^* & E_{\text{CB}}^{(0)} + \dfrac{\hbar^2 k^2}{2m_e^*(0)} \end{pmatrix}, \tag{9.1}$$

where $E_{\text{CB}}^{(0)}$ and $m_e^*(0)$ are the CB edge energy and electron effective mass of the (unperturbed) (In)GaAs host matrix, and E_{N} and V_{Nc} are the energy of the N-related localized states, and a matrix element describing a repulsive interaction between the (In)GaAs CB and N-related resonant states. Diagonalizing Equation 9.1 to obtain the (In)GaN$_x$As$_{1-x}$ CB structure demonstrates the splitting of the (In)GaAs CB into two subbands $E_\pm(k)$ given by [27]

$$E_\pm(k) = \frac{E_{\text{CB}}(k) + E_{\text{N}}}{2} \pm \sqrt{\left(\frac{E_{\text{N}} - E_{\text{CB}}(k)}{2}\right)^2 + |V_{\text{Nc}}|^2}, \tag{9.2}$$

where $E_{\text{CB}}(k) = E_{\text{CB}}^{(0)} + \dfrac{\hbar^2 k^2}{2m_e^*(0)}$ describes the dispersion of the unperturbed (In)GaAs CB.

Comparing the CB structure calculated using Equation 9.2 with the results of experimental measurements, Shan et al. demonstrated that this simple model is capable of describing the observed variation of (1) the energy of the optical transitions associated with $E_-(0)$ and $E_+(0)$ with N composition x, and (2) the alloy CB edge energy $E_-(0)$ with applied hydrostatic pressure. In order to obtain a quantitative description, Shan et al. determined on the basis of experimental data the matrix element V_{Nc} describing the N-induced coupling/hybridization of the (In)GaAs CB and N-related localized states depends strongly on the N composition, varying from 0.12 eV at $x = 0.9\%$ to 0.40 eV at $x = 2.3\%$. This composition dependence then describes that the observed strong, N-composition-dependent bowing of (In)GaN$_x$As$_{1-x}$ band gap is associated with an N composition-dependent BAC interaction between the (In)GaAs CB and N-localized resonant states—that is, the strongly N composition-dependent reduction in the alloy CB edge energy $E_-(0)$ is the origin of the observed rapid reduction of the (In)GaN$_x$As$_{1-x}$ band gap with increasing x (cf. Equation (9.2)).

The (In)GaN$_x$As$_{1-x}$ band structure calculated using the BAC model is depicted schematically in Figure 9.1a [28], which demonstrates that the interaction of the (In)GaAs CB (dashed line) with the N-related resonant states (dotted line) splits the (In)GaN$_x$As$_{1-x}$ CB into the two subbands $E_\pm(k)$ described

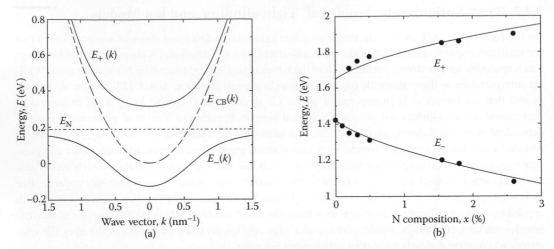

FIGURE 9.1 (a) Band structure of $GaN_{0.01}As_{0.99}$ ($x = 1\%$) calculated using the two-band BAC model of Equation 9.1. The BAC model describes that the GaAs host matrix CB (dashed line) undergoes an N composition-dependent anti-crossing interaction with a set of localized N-related resonant states at energy E_N (dotted line), splitting the CB into two bands $E_{\pm}(k)$ (solid lines). The zero of energy is taken at the GaAs CB edge. (b) Calculated (solid lines) and measured (closed circles) room temperature energies of the optical transitions from the VB edge to the $E_{\pm}$ band edge states (at $k = 0$), as a function of N composition x for a series of GaN_xAs_{1-x} samples. (Adapted from P.J. Klar et al., *Appl. Phys. Lett.*, 76, 3439, 2000.) The excellent agreement between theory and experiment confirms that the BAC model describes the observed strong reduction and composition-dependent bowing of the band gap in terms of the BAC-induced reduction of the alloy CB edge energy $E_-(0)$.

by Equation 9.2 (solid lines). We note that the BAC interaction (1) pushes the alloy CB edge downward in energy, accounting for the strong N-induced reduction of the band gap, and (2) pushes down the dispersion of the lower band at fixed k, leading to a strong N-induced increase in the effective mass (cf. Section 9.2.3). An important point to note regarding the BAC band structure is that the energy $E_-(k)$ of the alloy CB edge approaches E_N asymptotically from below as $k \to \infty$ (cf. Equation 9.2), indicating the presence of an unphysical singularity in the DOS at $E = E_N$. Despite its success in describing the evolution of the main features of the (In)GaN$_x$As$_{1-x}$ band structure, this inability to describe the DOS is an intrinsic limitation of the BAC model, which impairs its ability to provide a comprehensive description of the impact of N incorporation on the CB structure. While this issue is of particular importance for analysis of the transport properties of dilute nitride alloys, it is not of significant consequence for analysis of the optoelectronic properties (where we are generally concerned with carriers lying only in the immediate vicinity of the CB and VB edges in energy). As such, we defer discussion of the DOS until our review of the theory of the impact of N incorporation on carrier transport in Section 9.5.

Turning our attention to the parameterization of the BAC model, we note that the predictive capability of the model is limited due to the approach followed by Shan et al. in Ref. [27]. That this is the case can be understood by noting that the calculated transition energies $E_-(0)$ and $E_+(0)$ associated with the optical features observed in the experiment depend on two independent parameters: (1) the assumed energy E_N of the N-related resonant states, and (2) the magnitude V_{Nc} of the BAC interaction matrix element. In their initial study, Shan et al. determined E_N in (In)GaAs via an extrapolation of the available experimental data for the energy of the N-related states in GaAsP alloys, which for sufficiently large P compositions are bound states lying energetically within the host matrix band gap. For each composition considered, V_{Nc} was then determined by fitting to the measured transition energy associated with the $E_-(0)$ optical transition (which, by Equation 9.1, simultaneously determines $E_+(0)$). From a theoretical perspective, this approach has two significant shortcomings. First, the fitted value of V_{Nc} at a given N composition x depends on the energy E_N assumed for the N-related resonant states, leading to parametric ambiguity in the absence of detailed information regarding E_N and its dependence on composition. Second, the strong composition dependence of V_{Nc} means that—in the absence of experimental data against which to fit V_{Nc}—it is not possible to use Equation 9.1 to predict N composition-dependent trends in the (In)GaN$_x$As$_{1-x}$ CB structure.

In order to overcome these issues, O'Reilly et al. developed an atomistic TB model to study the electronic structure of dilute nitride alloys [29]. This approach has proved useful in elucidating the impact of N incorporation on the electronic structure of GaN$_x$As$_{1-x}$ and related alloys, including quantitative predictions of several key material properties and their dependence on N composition [30], as well as providing detailed information regarding the nature of the localized states associated with substitutional N impurities [31]. A full description of the TB model is beyond the scope of this chapter—we refer the reader to Ref. [29] for full details. In this section, we focus on one aspect of the TB analysis of the GaN$_x$As$_{1-x}$ electronic structure which is relevant to the BAC model. Using the TB method, O'Reilly et al. demonstrated that it is possible to explicitly construct the N-related localized resonant states associated with substitutional N impurities in supercell calculations, thereby allowing the N-related parameters of the BAC model to be computed directly and unambiguously for a given host matrix, and hence delivering predictive power to the BAC model. In the BAC model, the (In)GaN$_x$As$_{1-x}$ alloy CB edge state $|\psi_{CB}^{(1)}\rangle$ is hybridized—i.e. it is formed of a linear combination of the unperturbed (In)GaAs CB edge state $|\psi_{CB}^{(0)}\rangle$ and the localized resonant state $|\psi_N\rangle$ associated with an isolated, substitutional N impurity. As a result, $|\psi_N\rangle$ can be constructed directly via TB calculations carried out on N-free and N-containing supercells as [29,31]

$$|\psi_N\rangle = \frac{|\psi_{CB}^{(1)}\rangle - \langle\psi_{CB}^{(0)}|\psi_{CB}^{(1)}\rangle\,|\psi_{CB}^{(0)}\rangle}{\sqrt{1 - |\langle\psi_{CB}^{(0)}|\psi_{CB}^{(1)}\rangle|^2}}, \tag{9.3}$$

where $|\psi_{CB}^{(0)}\rangle$ and $|\psi_{CB}^{(1)}\rangle$ are, respectively, the CB edge eigenstates of the N-free and N-containing supercells, and $f_{\Gamma,CB} = |\langle\psi_{CB}^{(0)}|\psi_{CB}^{(1)}\rangle|^2$ is the fractional Γ character—that is, the squared modulus of the overlap between the CB edge state $|\psi_{CB}^{(1)}\rangle$ in an N-containing alloy supercell and the unperturbed host matrix CB edge state $|\psi_{CB}^{(0)}\rangle$ [29,31].

Using Equation 9.3 to construct $|\psi_N\rangle$ for a large supercell (containing ~2000 atoms in the presence of a single, isolated substitutional N impurity) then allows E_N and V_{Nc} to be computed directly as $E_N = \langle\psi_N|\hat{H}|\psi_N\rangle$ and $V_{Nc} = \langle\psi_{CB}^{(0)}|\hat{H}|\psi_N\rangle$, where $\hat{H}$ is the full TB Hamiltonian for the N-containing supercell. By applying this approach to a series of ordered GaN$_x$As$_{1-x}$ supercells, O'Reilly et al. directly determined the N composition dependence of the BAC interaction matrix element as $V_{Nc} = \beta\sqrt{x}$, where the coupling parameter β depends on (1) the type of N-related localized state—i.e. whether it is associated with an isolated N impurity, an N–N pair, or larger cluster of N atoms—and (2) the host matrix semiconductor into which N is incorporated.

The calculated values of E_N and β for isolated, substitutional N impurities are summarized in Table 9.1 for a range of III–V host matrix semiconductors [12]. The energies of these N–related localized states are given relative to the host matrix VB edge (as E_N) and relative to the host matrix CB edge (as ΔE_N). Positive (negative) values of E_N indicate that a substitutional N impurity forms resonant (bound) localized states, which lie energetically within the host matrix CB (band gap). Also given in Table 9.1 are the calculated energies E_{NN} of the localized states associated with N–N pairs—in which two N atoms share a group-III (cation) nearest neighbor—and are again given relative to the host matrix VB edge (as E_{NN}) and relative to the host matrix CB edge (as ΔE_{NN}). We note that the values of E_N calculated using the TB method are in quantitative agreement with experiment. In particular, the TB calculations correctly describe that an isolated N impurity forms a resonant (bound) state in GaAs (GaP), lying energetically within the CB (band gap) [23]. Furthermore, we note that $\Delta E_{NN} < \Delta E_N$ for all host matrices considered, indicating that the localized states associated with N–N pairs are more strongly bound than those associated with isolated N impurities. We note also the impact of temperature on the nature of the N-related localized states: given that the N-related (impurity-like) are relatively temperature insensitive, the reduction in the host matrix CB edge energy brought about by an increase in temperature means that, for certain host matrices, an

TABLE 9.1   Energies of the Localized States Associated with Isolated Substitutional N Impurities (E_N) and N–N Pairs(E_{nn}), as well as the Coupling Strength β of the BAC Interaction between $|\psi_N\rangle$ and the Unperturbed Host Matrix CB Edge Calculated at Low Temperature using the TB method for a Range of III–V Semiconductor Host Matrices

Host Matrix	E_g (eV)	E_N (eV)	ΔE_N (meV)	β (eV)	E_{NN} (eV)	ΔE_{NN} (meV)
GaP	2.886 (2.777)	2.306	−580 (−471)	1.737	2.180	−706 (−597)
GaAs	1.519 (1.422)	1.706	187 (284)	2.000	1.486	−33 (64)
GaSb	0.812 (0.727)	0.770	−42 (43)	2.400	0.540	−272 (−187)
InP	1.424 (1.353)	1.903	479 (550)	0.901	1.770	346 (417)
InAs	0.417 (0.354)	1.386	969 (1032)	1.212	1.201	784 (847)
InSb	0.235 (0.174)	0.720	485 (546)	1.970	0.520	285 (346)

Values in parentheses are those computed at room temperature. E_N and E_{NN} are given relative to the unperturbed host matrix VB edge. The band gap E_g of the unperturbed host matrix is provided for reference. ΔE_N (ΔE_{NN}) denotes the difference in energy between E_N (E_{NN}) and the unperturbed host matrix CB edge. As such, positive (negative) values of ΔE_N or ΔE_{NN} indicate that an isolated N impurity or N–N pair forms a resonant (bound) localized state lying energetically within the host matrix CB (band gap).

isolated N impurity forms a bound state at low temperature but a resonant state at room temperature. For example, the calculated negative (positive) value of ΔE_N indicates that an isolated N impurity forms a bound (resonant) state in GaSb at low (room) temperature. A similar trend is observed in GaAs, in which the N–N pair states lie close in energy to the unperturbed host matrix CB edge.

A final important point to note is that the N-related localized states in In-containing compounds lie significantly higher in energy than the host matrix CB edge. As we see when we discuss the impact of alloy disorder on the electronic structure in Section 9.2.3, the presence of N-related states lying close in energy to the CB edge leads to extremely strong perturbation of the electronic structure. As such, it is generally found that the BAC model—which treats only the impact of isolated N impurities on the CB structure—is generally more accurate when applied to In-containing dilute nitride alloys due to the large energy separation between the host matrix CB edge and the localized states associated with N pairs and clusters [12]. It is for this reason that the BAC-based 10-band **k·p** Hamiltonian we describe below is found to be sufficiently detailed to accurately describe the electronic and optical properties of InGaNAs alloys and heterostructures (as we see when we discuss the theory of 1.3-μm dilute nitride semiconductor lasers in Section 9.3).

The success of the BAC model—parameterized explicitly via TB supercell calculations—in describing the main features of the (In)GaN$_x$As$_{1-x}$ CB structure has been explicitly confirmed by comparison of the calculated alloy band edge energies $E_\pm(0)$ to PR measurements of the associated optical transition energies performed on a range of samples having different N compositions [28]. Using the values of E_N and β given for GaN$_x$As$_{1-x}$ in Table 9.1, it is possible to calculate the variation with x of $E_\pm(0)$, which can then be compared to experiment. This theory–experiment comparison is shown in Figure 9.1b, which compares the measured transition energies $E_\pm(0)$ (closed circles) to those calculated using Equation 9.2 with $k = 0$ (solid lines). We note the excellent agreement between theory and experiment, confirming that the BAC model is capable of quantitatively describing the evolution of the main features of the GaN$_x$As$_{1-x}$ band structure with N composition x.

For the purposes of modeling the electronic and optical properties of dilute nitride semiconductor lasers and related photonic devices, it is desirable to develop appropriate **k·p** models which treat the states lying close in energy to the CB and VB edge on an equal footing. For conventional III–V semiconductor alloys an eight-band model—taking as a basis the spin degenerate zone-center Bloch states associated with the lowest energy CB, as well as those of the heavy-hole (HH), light-hole (LH), and spin-split-off-hole (SO) VBs—is typically found to be sufficiently detailed to accurately describe the electronic and optical properties of quantum-confined heterostructures. However, as we have seen, experimental measurements and atomistic calculations suggest that N incorporation introduces localized impurity states lying close in energy to the host matrix CB edge which, due to their strong composition-dependent interactions with the host matrix CB edge, must be treated explicitly in order to understand the impact of N on the band structure. This suggests that the conventional eight-band $\mathbf{k \cdot p}$ model is insufficient to describe the dilute nitride band structure, and hence that an extended $\mathbf{k \cdot p}$ model must be developed in which the N-related localized resonant states are included explicitly in the basis set. Given the success of the BAC model, the natural choice of basis set is then to augment the zone-center Bloch states of the eight-band model for the host matrix with a pair of spin degenerate N-related localized states (having energy E_N). This leads us to a 10-band $\mathbf{k \cdot p}$ Hamiltonian for (In)GaN$_x$As$_{1-x}$ and related alloys.

The TB model introduced by O'Reilly et al. provides a natural framework within which to derive and parameterize the 10-band model. In the interest of brevity, the manner by which this parameterization proceeds—full details of which can be found in Refs. [29] and [32]—is described in more detail in Chapter 10, when we consider a similar extended basis set $\mathbf{k \cdot p}$ Hamiltonian for the band structure of the dilute bismide alloy (In)GaBi$_x$As$_{1-x}$. Essentially, by using the zone-center eigenstates calculated for the host matrix in conjunction with the N-related localized states constructed following Equation 9.3 to construct the 10-band basis set, one can directly evaluate the matrix elements of the full TB Hamiltonian for the N-containing supercell and hence fully parametrize the 10-band model without recourse to post hoc fitting via experimental data. The 10-band $\mathbf{k \cdot p}$ Hamiltonian obtained in this manner is similar to the eight-band model for the host matrix semiconductor, but is distinguished by the addition of (1) virtual crystal (conventional alloy) contributions describing the linear variation of the band edge energies with N composition x, and (2) N composition-dependent BAC interactions $V_{Nc} = \beta\sqrt{x}$ between the host matrix CB edge and the N-related localized states. TB calculations reveal that the N-related states $|\psi_N\rangle$ do not couple to either of the HH, LH, or SO VB edge states [29,32]. The resulting 10-band $\mathbf{k \cdot p}$ Hamiltonian for dilute nitride alloys is then of the form

$$
H = \begin{pmatrix}
E_N & V_{Nc} & 0 & 0 & 0 & 0 & 0 & 0 & 0 & 0 \\
 & E_{CB} & -\sqrt{3}T & \sqrt{2}U & -U & 0 & 0 & 0 & -T^* & -\sqrt{2}T^* \\
 & & E_{HH} & \sqrt{2}S & -S & 0 & 0 & 0 & -R & -\sqrt{2}R \\
 & & & E_{LH} & Q & 0 & T^* & R & 0 & \sqrt{3}S \\
 & & & & E_{SO} & 0 & \sqrt{2}T^* & \sqrt{2}R & -\sqrt{3}S & 0 \\
 & & & & & E_N & V_{Nc} & 0 & 0 & 0 \\
 & & & & & & E_{CB} & -\sqrt{3}T^* & \sqrt{2}U & -U \\
 & & & & & & & E_{HH} & \sqrt{2}S^* & -S^* \\
 & & & & & & & & E_{LH} & Q \\
 & & & & & & & & & E_{SO}
\end{pmatrix}, \tag{9.4}
$$

where we have shown only the upper triangle since H is a Hermitian matrix—the diagonals below can be obtained by Hermitian conjugation. The definition of the basis $|u_1\rangle, ..., |u_{10}\rangle$ of zone-center crystal eigenstates in which Equation 9.4 is written can be found in Ref. [33]. Taking the zero of energy at the unperturbed host matrix VB edge, the matrix elements of Equation 9.4 are defined at N composition x, and in the presence of pseudomorphic strain corresponding to growth along the (001) direction, as [33]

$$
E_{CB} = E_g + (\kappa - \alpha)x + \frac{\hbar^2}{2m_0}s_c\left(k_{\parallel}^2 + k_z^2\right) + \delta E_{CB}^{hy}, \tag{9.5}
$$

$$
E_{HH} = \kappa x - \frac{\hbar^2}{2m_0}\left((\gamma_1 + \gamma_2)k_{\parallel}^2 + (\gamma_1 - 2\gamma_2)k_z^2\right) + \delta E_{VB}^{hy} - \delta E_{VB}^{ax}, \tag{9.6}
$$

$$
E_{LH} = \kappa x - \frac{\hbar^2}{2m_0}\left((\gamma_1 - \gamma_2)k_{\parallel}^2 + (\gamma_1 + 2\gamma_2)k_z^2\right) + \delta E_{VB}^{hy} + \delta E_{VB}^{ax}, \tag{9.7}
$$

$$
E_{SO} = \kappa x - \Delta_{SO} - \frac{\hbar^2}{2m_0}\gamma_1\left(k_{\parallel}^2 + k_z^2\right) + \delta E_{VB}^{hy}, \tag{9.8}
$$

$$Q = -\frac{1}{\sqrt{2}}\frac{\hbar^2}{m_0}\gamma_2 k_\parallel^2 + \sqrt{2}\frac{\hbar^2}{m_0}\gamma_2 k_z^2 - \sqrt{2}\,\delta E_{VB}^{ax}, \tag{9.9}$$

$$R = \frac{\sqrt{3}}{2}\frac{\hbar^2}{m_0}\left(\gamma_{av} k_-^2 - \mu k_+^2\right), \tag{9.10}$$

$$S = \sqrt{\frac{3}{2}}\frac{\hbar^2}{m_0}\gamma_3 k_z k_-, \tag{9.11}$$

$$T_\pm = \frac{P}{\sqrt{6}}k_\pm, \tag{9.12}$$

$$U = \frac{P}{\sqrt{3}}k_z, \tag{9.13}$$

where $k_\parallel = \sqrt{k_x^2 + k_y^2}$, $k_\pm = k_x \pm ik_y$, m_0 is the free electron mass, E_g and Δ_{SO} are the host matrix band gap and VB spin-orbit-splitting energy, P is the Kane interband momentum matrix element, and $\gamma_{1,2,3}$ are the modified Luttinger parameters. γ_{av} and μ are defined in terms of the modified Luttinger parameters as $\gamma_{av} = \frac{1}{2}(\gamma_2 + \gamma_3)$ and $\mu = \frac{1}{2}(\gamma_3 - \gamma_2)$, and the definition of the parameter s_c can be found in Ref. [33]. For brevity, the definitions of the energy shifts δE_{CB}^{hy}, δE_{VB}^{hy}, and δE_{VB}^{ax} to the CB and VB edges, resulting from hydrostatic and axial strain, are provided in Chapter 10. The composition-dependent N-localized state energy E_N of Equation 9.4 is given relative to the host matrix VB edge as $E_N = E_g + \Delta E_N + (\kappa - \gamma)x$ and, as above, the BAC interaction matrix element is $V_{Nc} = \beta\sqrt{x}$.

A detailed discussion of the N-related parameters α, γ, and κ, which describe the aforementioned virtual crystal contributions to the N-induced shifts of the band edge energies, can be found in Ref. [34]. For the calculations based on the 10-band $\mathbf{k}\cdot\mathbf{p}$ model to be presented in Section 9.3, we have used the parameters provided by Tomić et al. in Ref. [33]. The band dispersion calculated in the vicinity of the CB and VB edges in energy using this 10-band model is in excellent agreement with full TB supercell calculations [32,35]. The 10-band model then describes the strong perturbation of the CB structure brought about by N incorporation (due to strong, composition-dependent BAC), while simultaneously describing that N has relatively little impact on the VB structure (described purely in terms of minor shifts in the band edge energies associated with strain and conventional alloy effects).

Having described that the main features of the CB structure of (In)GaNAs(P) alloys can be straightforwardly described by a strongly composition-dependent BAC interaction between the extended states of the host matrix CB edge and highly localized states associated with substitutional N impurities, we now turn our attention to more detailed aspects of the electronic structure of GaN$_x$As$_{1-x}$ alloys, focusing in particular on the impact of N-related alloy disorder. This highlights conditions under which the BAC model can be expected to break down and demonstrates that a quantitative description of the dilute nitride CB structure depends upon a direct treatment of the full range of N-related localized states appearing in realistic, disordered alloys.

9.2.3 Tight-Binding and Alloy Disorder: Linear Combination of Isolated Nitrogen States

As we have described so far in this section, initial investigations of the electronic structure of dilute nitride alloys were primarily based on two complementary approaches: (1) detailed electronic structure calculations using DFT, EP and TB methods, and (2) the empirically motivated BAC model. Detailed comparisons between theory and experiment have demonstrated that the BAC model is generally capable of describing the dependence of the band edge energies in bulk and QW (In)GaN$_x$As$_{1-x}$ samples on N composition and applied hydrostatic pressure. However, despite the successes of these approaches, there are some important experimental data which were not satisfactorily described by either the atomistic supercell or the BAC models described above. First, the BAC model fails to describe the anomalously large values m_e^* and g_e^* of the CB edge electron effective mass and g factor measured in some GaN_xAs_{1-x} samples even for ultra-dilute N compositions $x \approx 0.2\%$ [36–38]. Second, a feature associated with the higher-energy N-derived E_+ state appearing in the BAC model is only observable in spectroscopic measurements on GaN_xAs_{1-x} alloys over a limited range of compositions, appearing for $x > 0.2\%$ and growing in sharpness and prominence with increasing x, before undergoing rapid energy broadening and vanishing for $x > 3\%$ [28,39]. Description of these detailed features of the GaN_xAs_{1-x} electronic structure and, in particular, description of the observed strong, nonmonotonic variation of m_e^* and g_e^* with N composition then serves as a stringent test of theoretical models of the GaN_xAs_{1-x} CB structure.

These previously unexplained aspects of the electronic properties of GaN_xAs_{1-x} alloys motivated further theoretical investigations of the CB structure beyond the BAC model. Of critical importance is to note the fact that the BAC model assumes that all N-related states have the same energy E_N, the energy of the localized state associated with an isolated N impurity (giving rise to the aforementioned singularity in the BAC DOS at the energy of the N-localized states). This is not the case in a real, disordered GaN_xAs_{1-x} alloy: as described for isolated N versus N–N pairs in Section 9.2.2, the localized states associated with different types of N impurities have different energies. This is confirmed by experimental measurements that indicate the presence of a series of N-related levels in the GaN_xAs_{1-x} CB, lying close in energy to the unperturbed GaAs CB edge, suggesting the presence of a broad collection of N-related localized states in a disordered alloy. Such states have also been observed in theoretical calculations based on the EP and TB methods. Two important points then become clear. First, given that the BAC model is capable of describing the dependence of the band edge energies on N composition and applied hydrostatic pressure to a reasonable degree of accuracy, it can be concluded that N-related localized states associated with N–N pairs and larger clusters of N atoms play only a secondary role in determining the variation of the band edge energies (and hence band gap) with x, but play an important role in determining the electron effective mass (and hence the transport properties). Second, any quantitative description of the details of GaN_xAs_{1-x} electronic structure must explicitly take account of the full distribution of N-related localized states, including those associated not only with isolated N impurities, but also with N–N pairs and larger clusters of N atoms.

In Ref. [40], O'Reilly et al. developed an approach that explicitly treats the N-related localized states in realistic, disordered GaN_xAs_{1-x} alloys. This approach proceeds on the basis that the N-related states associated with different N complexes are highly localized about the N atomic sites, suggesting that the N-related states in a disordered alloy can be represented as a LCINS—that is, as a linear combination of the localized states associated with each individual N complex in a given alloy (supercell). Each of these localized states can be constructed explicitly using the TB method (cf. Sec. 9.2.2) allowing their energies and interactions with the unperturbed GaAs CB edge states to be quantified. Taking as a basis these N-related localized states in conjunction with the unperturbed GaAs CB edge state $|\psi_{CB}^{(0)}\rangle$, the LCINS approach then allows a suitable Hamiltonian to be constructed which explicitly treats the interaction between the unperturbed GaAs CB edge and the full distribution of N-related localized states present in a realistic, disordered alloy. More specifically, given the fact that the localized state associated with each N complex is highly localized about the N site(s), one can associate a separate localized state with each isolated N impurity, N–N pair, or larger N cluster. To verify that this is the case, O'Reilly et al. examined the N-related states in a series of ordered, $2M$-atom $Ga_MN_nAs_{M-n}$ supercells ($x = \frac{n}{M}$), in which a total of n substitutional N impurities were placed on a regular grid of sites on the anion sublattice. The N-related states in such a structure were constructed as an LCINS via [40]

$$|\psi_N\rangle = \frac{1}{\sqrt{n}} \sum_{m=1}^{n} |\psi_N^{(m)}\rangle, \qquad (9.14)$$

where $|\psi_N^{(m)}\rangle$ is the N-related localized state associated with an *isolated* substitutional N impurity located at the mth N site. For all bar one of the supercells considered, the calculated overlap of the LCINS state described by Equation 9.14 with the N-related state computed directly for the ordered supercell via

Equation 9.3 was $\geq 94\%$. A smaller overlap ($\approx 70\%$) was obtained only for an 8-atom $Ga_4N_1As_3$ supercell having a large N composition $x = 25\%$, and in which each N atom had an N second-nearest neighbor (manifested as a chain of Ga–N nearest-neighbor pairs due to the Born-von Karman supercell boundary conditions). Furthermore, the energy $E_N = \langle \psi_N | \hat{H} | \psi_N \rangle$ computed for each supercell using the N-related states described by Equations 9.3 and 9.14—in conjunction with the full TB supercell Hamiltonian $\hat{H}$—was found to be in excellent agreement for all supercells considered. These comparisons confirmed that the N-related states in a GaN_xAs_{1-x} alloy supercell can in general be accurately described via an LCINS of the form of Equation 9.14.

In order to investigate the accuracy of the LCINS approach for describing the electronic structure of disordered alloy supercells, O'Reilly et al. next considered a series of disordered, 1000-atom $Ga_{500}N_{n+2p}As_{500-n-2p}$ supercells ($x = \frac{n+2p}{M}$) containing a total of $n + 2p$ substitutional N impurities in the form of n isolated N atoms and p N–N pairs. In general, the GaN_xAs_{1-x} alloy CB edge state is expected to be formed of an admixture of the unperturbed GaAs CB edge state $|\psi_{CB}^{(0)}\rangle$ and the LCINS with which it interacts. On this basis they defined a $(1 + n + 2p) \times (1 + n + 2p)$ Hamiltonian matrix describing the interaction between $|\psi_{CB}^{(0)}\rangle$ and (1) the n localized states associated with isolated N impurities, and (2) the $2p$ localized states associated with N–N pairs. This Hamiltonian matrix, $H_{ij} = \langle \psi_i | \hat{H} | \psi_j \rangle$ where $i, j = 1, ..., n + 2p$, is constructed explicitly using the full TB supercell Hamiltonian. The energies of the LCINS eigenstates are then calculated as the eigenvalues of the generalized eigenvalue equation $H u_i = E_i S u_i$, where $S_{ij} = \langle \psi_i | \psi_j \rangle$ is the overlap matrix between the $1 + n + 2p$ basis states (which, in general, has nonzero off-diagonal elements due to the partial overlap between N-related localized states centered on different lattice sites). The LCINS eigenstates are then computed as [40]

$$|\phi_i\rangle = u_{CB,i}|\psi_{CB}^{(0)}\rangle + \sum_{m=1}^{n} u_{N,i}^{(m)}|\psi_N^{(m)}\rangle + \sum_{m=1}^{n} \left(u_{NN+,ia}^{(m)}|\psi_{NN+}^{(m)}\rangle + u_{NN-,i}^{(m)}|\psi_{NN-}^{(m)}\rangle \right), \tag{9.15}$$

where $|\psi_{NN+}^{(m)}\rangle$ and $|\psi_{NN-}^{(m)}\rangle$ are the localized states associated with isolated N–N pairs which are, respectively, symmetric and antisymmetric about the Ga (cation) nearest neighbor shared by the N atoms forming the pair.

By applying this model to the aforementioned series of disordered $Ga_{500}N_{n+2p}As_{500-n-2p}$ supercells and computing (1) the eigenstate energies E_i, and (2) the fractional Γ character associated with each of the LCINS eigenstates $|\phi_i\rangle$, it was confirmed that the LCINS approach is in excellent agreement with the results of full TB supercell calculations. These detailed comparisons for both ordered and disordered alloys emphatically demonstrate that the GaN_xAs_{1-x} electronic structure can be described straightforwardly and accurately in terms of an extended BAC-like approach in which the interactions between the unperturbed GaAs CB edge state $|\psi_{CB}^{(0)}\rangle$ and the full distribution of localized states associated with different N complexes are treated explicitly via an LCINS.

As described above, the LCINS approach appears to be supercell-dependent—that is, it appears to depend strongly on the assumed distribution of isolated N impurities, N–N pairs, and largers clusters of N atoms in a given supercell—which draws into question its general predictive capability. However, O'Reilly et al. demonstrated that this apparent limitation can be overcome by treating the N-related localized states in disordered alloy supercells containing large numbers ($\sim 10^6 - 10^7$) of atoms and hence large numbers ($\sim 10^4$) of substitutional N impurities, even at ultra-dilute N compositions, with the occurrence of N–N pairs and larger clusters of N atoms determined statistically. This then minimizes the impact of statistical fluctuations on the calculated electronic structure and, in this sense, the LCINS approach is capable of providing an unambiguous description of the electronic structure at any given N composition x. The following approach to analyzing the LCINS electronic structure of a large supercell for a given N composition x provides significant insight into the nature of the CB structure. First, the full distribution of N-related states is considered independently of $|\psi_{CB}^{(0)}\rangle$ in order to determine the energies ϵ_i of the (interacting) N-related localized states centered on different lattice sites. For a supercell containing a total of M substitutional N impurities this consists of diagonalizing an $M \times M$ Hamiltonian matrix $\langle \psi_{N,i} | \hat{H} | \psi_{N,j} \rangle$ $(i, j = 1, ..., M)$ constructed in the basis $\{ | \psi_{N,i} \rangle \}$ of N-related localized states associated with all N complexes present in the supercell. Second, the eigenvectors of this $M \times M$ Hamiltonian are used to construct the N-related LCINS basis states $| \phi_{N,i} \rangle$, as in Equation 9.15 but omitting $| \psi_{CB}^{(0)} \rangle$. Third, the interaction $V_i = \langle \phi_{N,i} | \hat{H} | \psi_{CB}^{(0)} \rangle$ between $| \psi_{CB}^{(0)} \rangle$ and each of the N-related LCINS basis states $| \phi_{N,i} \rangle$ is computed explicitly for each $| \phi_{N,i} \rangle$. Finally, the energies ϵ_i of the N-related LCINS basis states, and their interactions V_i with $| \psi_{CB}^{(0)} \rangle$, are used to generate an $(M+1) \times (M+1)$ Hamiltonian matrix in the basis $\{ | \phi_{N,i} \rangle, | \psi_{CB}^{(0)} \rangle \}$, which can then be diagonalized to obtain the energies E_i and eigenstates $| \phi_i \rangle$ of the large, disordered N-containing supercell. This procedure enables the interactions (1) between the N-related localized states $| \psi_{N,i} \rangle$, and (2) between each N-related localized state $| \psi_{N,i} \rangle$ and $| \psi_{CB}^{(0)} \rangle$, to be quantified explicitly, thereby identifying the contributions of distinct classes of N isolated impurity, N–N pair, and N cluster states to determining the electronic structure of the disordered alloy.

As discussed above, a stringent test of any theoretical model of the GaN_xAs_{1-x} electronic structure is the accuracy with which it is capable of reproducing the measured (1) anomalously large values of m_e^* and g_e^* and their nonmonotonic variation with x, and (2) emergence of an "E_+" feature associated with N-derived states which are resonant with the GaAs CB for $x > 0.2\%$, and its energy broadening and disappearance for $x > 3\%$. We briefly summarize the analysis of these detailed aspects of the electronic structure here, both of which are described quantitatively by the LCINS approach.

The GaN_xAs_{1-x} electron effective mass at N composition x can be computed as [40]

$$m_e^*(x) = \frac{m_e^*(0)\, E_g(x)}{E_g(0)\, f_{\Gamma,CB}(x)\, f_{\Gamma,VB}(x)}, \tag{9.16}$$

where $m_e^*(0)$ and $E_g(0)$ are the GaAs $(x = 0)$ electron effective mass and band gap, $E_g(x)$ is the GaN_xAs_{1-x} band gap at N composition x, and $f_{\Gamma,CB(VB)}(x)$ is the Γ character of the GaN_xAs_{1-x} alloy CB (VB) edge state at N composition x. $f_{\Gamma,CB}(x)$ can be calculated directly by projecting the alloy CB edge eigenstate computed via the LCINS approach onto $| \psi_{CB}^{(0)} \rangle$ (cf. Section 9.2.2), while it can be assumed—in line with detailed supercell calculations—that $f_{\Gamma,VB}(x) = 1 - x$ since the GaN_xAs_{1-x} VB edge eigenstates evolve continuously from those of the unperturbed GaAs VB edge as in a conventional semiconductor alloy [12,29].

Closely related to the effective mass is the Landé effective g factor g_e^* of CB electrons. g_e^* is a key parameter in semiconductor materials, describing the response of the electron spins to externally applied magnetic fields as well as providing useful information regarding the symmetry of the crystal eigenstates at the Γ point. In addition to providing information pertinent to evaluating the potential of specific materials for spintronic applications, analysis of the electron spin properties and g factor therefore provides valuable insight into the material band structure. In particular, g_e^* is extremely sensitive to the separation in energy between zone-center eigenstates, as well as hybridzation between eigenstates caused by, e.g. a reduction of symmetry due to strain [41] or quantum confinement [42] or, in this case, incorporation of N. As a result, calculations of g_e^* also provide a stringent test of theoretical models of the electronic structure [43]. The GaN_xAs_{1-x} electron effective g factor at N composition x can be computed as [12]

$$g_e^*(x) = g_e^*(0) \left(1 - \frac{P^2 f_{\Gamma,CB}(x) f_{\Gamma,VB}(x)}{3} \left(\frac{1}{E_g(x)} - \frac{1}{E_g(x) - \Delta_{SO}(x)} \right) \right), \tag{9.17}$$

where P is the Kane interband momentum matrix element in GaAs, and where $E_g(x)$ and $\Delta_{SO}(x)$ are, respectively, the GaN_xAs_{1-x} band gap and spin-orbit-splitting energy at composition x.

Figure 9.2a and 9.2b [12,44] show, respectively, the variation of m_e^* and g_e^* with x in GaN_xAs_{1-x} calculated using the two-band BAC model of Section 9.2.2 (solid line) and the LCINS method (open circles), compared to experimental measurements (closed circles) [12]. First, we note the large deviation of the measured values of m_e^* and g_e^* from those calculated using the two-band BAC model, even at ultra-dilute

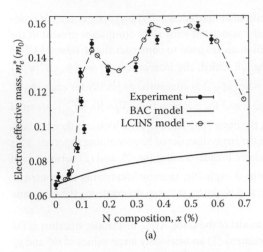
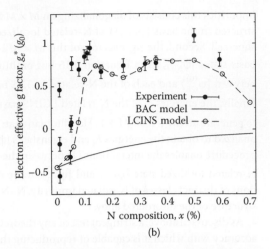

(a)   (b)

FIGURE 9.2 Variation of (a) the electron effective mass m_e^*, and (b) the electron effective g factor g_e^*, as a function of N composition x in GaN$_x$As$_{1-x}$, calculated via Equation 9.16 using the two-band BAC (solid line) and LCINS (open circles, dashed lines) models, compared to experimental measurements (closed circles). The strong influence of N pair and cluster states on the alloy CB edge leads to high values of m_e^*, a change in the sign of g_e^*, and strong non-monotonic variation of both m_e^* and g_e^* with x, none of which is captured by the two-band BAC model. The values of m_e^* and g_e^* calculated using the LCINS approach are in good, quantitative agreement with experiment, verifying the strong impact of N-related alloy disorder on the GaN$_x$As$_{1-x}$ CB structure and emphasizing that a quantitative theoretical description requires direct treatment of N pair and cluster states. (Adapted from E.P. O'Reilly et al., *Semicond. Sci. Technol.*, 24, 033001, 2009; A. Lindsay and E.P. O'Reilly, *Phys. Rev. Lett.*, 93, 196402, 2004.)

N compositions $x < 0.1\%$. Second, we note the excellent quantitative agreement between the measured values of m_e^* and g_e^* and those calculated using the LCINS method. The LCINS approach accurately captures the strong increase in m_e^* between $x = 0.1\%$ and 0.2%, as well as the associated change in the sign of g_e^*, neither of which is captured by the BAC model. This quantitative agreement with experiment highlights that the LCINS method describes the nature and evolution of the CB structure of disordered GaN$_x$As$_{1-x}$ alloys with a sufficient level of accuracy to account for the observed nonmonotonic variation of m_e^* and g_e^* with x. Furthermore, the fractional Γ character $f_{\Gamma,CB}(x)$ of the GaN$_x$As$_{1-x}$ CB eigenstates calculated using the LCINS approach provides insight into the origin of these unusual material trends.

We recall that the LCINS model directly describes the hybridzation of GaAs CB edge eigenstates with a full distribution of N-related localized states, including states which lie below the GaAs CB edge in energy (e.g., the N–N pair states described in Table 9.1). Detailed analysis of LCINS calculations on ultralarge supercells reveals that it is the hybridization between localized states associated with N–N pairs and larger clusters of N atoms that largely brings about the observed large values of m_e^* and g_e^*. As the alloy CB edge moves downward in energy with increasing x it passes through these N pair and cluster states, undergoing strong hybridization when its energy is close to the energy ϵ_i of N-related LCINS states having appreciable (but not necessarily large) V_i. This results in low calculated values $f_{\Gamma,CB} < 50\%$ for the GaN$_x$As$_{1-x}$ CB edge eigenstates across certain ranges of composition, leading to the presence of several distinct maxima in the variation of m_e^* and g_e^* at select N compositions, and thereby accounting for both the large measured values of m_e^* and g_e^* as well as their nonmonotonic variation with x. It should be noted that this is in contrast to the two-band BAC model of GaN$_x$As$_{1-x}$, in which all N-related states are assumed to lie above the unperturbed GaAs CB edge in energy (cf. Table 9.1) and, hence, in which $f_{\Gamma,CB}$ must exceed 50% for all x. These results emphasize that while the evolution of the E_g with x is primarily governed by the interaction of the GaAs CB with localized states associated with isolated N impurities (with N pair and cluster states playing a secondary role), and can hence be described by a simple BAC approach, the evolution of m_e^* and g_e^* is strongly impacted by alloy disorder and can hence only be described accurately via theoretical approaches which explicitly account for the impact of N pair and cluster states on the alloy CB edge eigenstates [12, 40,44]. Additional analysis using this approach has elucidated the variation of m_e^* as a function of applied hydrostatic pressure: the upward movement in energy of the host matrix CB edge in response to applied pressure causes strong hybridization as described above, resulting in strong, nonmonotonic variation of m_e^*, with increasing pressure [40].

Further calculations using the LCINS method have also quantitatively described the observed trends associated with the higher-energy, N-derived "E_+" feature in GaN$_x$As$_{1-x}$, in terms of the evolution with increasing x of the GaAs CB edge Γ character $f_{\Gamma,\text{CB}}$ acquired by N-related localized resonant states [40]. At ultra-dilute N compositions $x < 0.2\%$, the composition-dependent hybridization of the N-related resonant states is sufficiently weak to ensure that the E_+ feature acquires little appreciable GaAs CB edge Γ character, indicating that it is not likely to be resolvable as a distinct, optically active feature in spectroscopic measurements. As the N composition increases, the N-related resonant states hybridize more strongly with $|\psi_{\text{CB}}^{(0)}\rangle$, acquiring larger $f_{\Gamma,\text{CB}}$ and leading to the emergence of the distinct, energetically narrow E_+ feature observed in experiment. As the N composition is further increased above $x \approx 1.4\%$, the repulsive BAC interaction between $|\psi_{\text{CB}}^{(0)}\rangle$ and the N-related resonant states pushes the E_+ feature sufficiently high in energy that it becomes resonant with the large density of CB states associated with the L-minima. The resonance of these localized N-related states with the large density of L-valley states then leads to a strong energetic broadening of the E_+ feature (cf. Section 9.5.1), manifest in the LCINS calculations by a distribution of $f_{\Gamma,\text{CB}}$ over a large number of higher-energy alloy CB states. This redistribution of $f_{\Gamma,\text{CB}}$ brought about by resonance with L-valley states continues with increasing x, leading to the E_+ feature weakening and diminishing in prominence. For N compositions $x > 2.5\%$ this broadening of the E_+ feature becomes sufficiently large as to become indistinguishable as a clear feature in the CB structure. The LCINS approach, therefore, not only accounts for the inability to observe a distinct E_+ feature in spectroscopic measurements at N compositions $x > 3\%$, but also elucidates the mechanism by which this feature vanishes.

We note that it is also possible to calculate the band dispersion of N-containing alloys using the LCINS method. We reserve discussion of the LCINS band dispersion until Section 9.5.1, in which we review the use of Green's function methods in conjunction with the AIM to facilitate quantitative analysis of the CB structure and DOS in disordered GaN$_x$As$_{1-x}$ alloys, accurate calculations of which are essential to predicting the transport properties of N-containing alloys. Overall, the LCINS approach highlights (1) the breakdown of the simple two-band BAC model of the dilute nitride CB structure in the presence of N-related alloy disorder as well as (2) the ability to quantitatively describe the electronic structure of a disordered N-containing alloy by explicitly treating the interaction between the unperturbed GaAs CB edge and the full distribution of N-related localized states. In particular, these results highlight the important role played by short-range alloy disorder in determining the details of the electronic structure. This is a general characteristic of highly-mismatched alloys, clearly distinguishing their properties from those of conventional semiconductors. While we have outlined the main principles and some key results of the LCINS model here, full details of this approach are beyond the scope of this chapter. We refer the reader to Refs. [40] and [44] for further information on the development of the LCINS approach and its application to GaN$_x$As$_{1-x}$ alloys, and to Refs. [24]and [25] for details of its application to GaN$_x$P$_{1-x}$ alloys.

9.3 Theory of GaAs-Based 1.3-µm InGaNAs Quantum Well Lasers

Having focused so far on developing a consistent approach to model the (In)GaN$_x$As$_{1-x}$ band structure, in this section we turn our attention to applying this theory to analyze photonic devices. Specifically, we consider the theory and modeling of 1.3-µm InGaNAs QW lasers grown on GaAs. We begin Section 9.3.1 by outlining a theoretical model for the electronic and optical properties of InGaNAs QW lasers based on the 10-band $\mathbf{k \cdot p}$ model of Section 9.2.2. Next, in Section 9.3.2, we review a series of theoretical calculations of the band structure and optical gain in an In$_{0.36}$Ga$_{0.64}$N$_{0.018}$As$_{0.982}$ single QW laser structure, containing 1.8% N and chosen on the basis of the laser device investigated in Refs. [45] and [33]. Through this analysis we elucidate the impact of N incorporation on the electronic and optical properties of ideal N-containing QWs, allowing us to identify key trends in the material and device parameters for this exemplar 1.3-μm dilute nitride device. Finally, in Section 9.3.3, we discuss the performance of real InGaNAs QW lasers. We focus on the impact of N incorporation on the loss mechanisms present in devices of this type, by reviewing calculations of the temperature dependence of defect-related, radiative, and nonradiative (Auger) recombination in 1.3-μm devices. We first outline how experimental measurements can be undertaken to identify and quantify the recombination pathways present in a QW laser, before comparing the results of theoretical calculations of the recombination rates directly to experiment.

The comparisons to experimental data we review display the predictive capability of the theory, highlighting in particular the ability of the 10-band **k·p** model to describe the impact of N incorporation on the band structure and optical properties. This confirms the validity of the theoretical model described in Section 9.3.1 for use in the design and optimization of future dilute nitride semiconductor lasers and related devices.

9.3.1 Theoretical Model for Dilute Nitride Quantum Well Lasers

Here, we provide a brief outline of a theoretical model that has been developed to study the electronic and optical properties of InGaNAs QW lasers. Full details of this model can be found in Refs. [33,34]. The theoretical model is based on the 10-band **k·p** Hamiltonian presented in Section 9.2.2. The QW band offsets are calculated following the procedure outlined in Ref. [33], while the QW eigenstates are calculated in the envelope function approximation using a numerically efficient plane wave expansion method (similar to that commonly employed in *ab initio* electronic structure calculations). An overview of the plane wave expansion method within the context of **k·p** theory can be found in Ref. [46].

In the plane wave approach, the QW eigenstate $|\psi_n(\mathbf{k}_\parallel, z)\rangle$ at position z and in-plane wave vector $\mathbf{k}_\parallel$, having energy $E_n(\mathbf{k}_\parallel)$, is written as

$$|\psi_n(\mathbf{k}_\parallel, z)\rangle = \frac{1}{\sqrt{L}} \sum_{b=1}^{10} \sum_{m=-M}^{M} a_{nbm}(\mathbf{k}_\parallel) e^{iG_m z} |u_b\rangle, \tag{9.18}$$

where L is the periodicity (length) of the calculational supercell ($-\frac{L}{2} \leq z \leq \frac{L}{2}$), b indexes the bands of the bulk **k·p** Hamiltonian, and m indexes the set of plane wave basis states with which the components of the envelope function associated with $|\psi_n(\mathbf{k}_\parallel, z)\rangle$ are represented. The discrete wave vectors G_m are defined as $G_m = \frac{2m\pi}{L}$, and $|u_b\rangle$ is the zone center Bloch states of the bulk 10-band **k·p** Hamiltonian. We note that this description of the QW eigenstates, which consists of $2M + 1$ independent Fourier components, is formally exact as $M \rightarrow \infty$. However, for most practical calculations a basis set containing ~ 70 plane waves ($M \sim 35$) is sufficient to ensure that the calculated eigenstates have converged with respect to further increases in the basis set size [34].

By substituting Equation 9.18 into the multiband Schrödinger equation it is straightforward to derive analytical expressions for the matrix elements of the QW Hamiltonian in the plane wave basis set defined by the set of wave vectors $\{G_m\}$ [34,46], leading to a reciprocal space Hamiltonian matrix of size $10(2M+1) \times 10(2M+1)$. In this manner, an independent Hamiltonian matrix is obtained at each in-plane wave vector $\mathbf{k}_\parallel$, and diagonalizing these matrices yields the QW band structure $E_n(\mathbf{k}_\parallel)$ and eigenstates $|\psi_n(\mathbf{k}_\parallel, z)\rangle$. We note that, by definition, the eigenstates described by Equation 9.18 satisfy periodic boundary conditions over the calculational supercell, meaning that the plane wave expansion method can readily be used to investigate the electronic and optical properties of superlattices.

In order to facilitate accurate calculation of the optical properties, it is important to include the strong impact of N on the QW eigenstates. For this reason, the QW band structure and envelope functions (calculated in the axial approximation—that is, by setting μ = 0 in Equation 9.10 [47]) are used *directly* to compute the optical properties of a given laser structure. Key to this analysis is the calculation of the optical (momentum) matrix elements. Following Ref. [48], the optical matrix element of an electron–hole pair at in-plane wave vector $\mathbf{k}_\parallel$ is given in terms of the QW eigenstates described by Equation 9.18 as

$$P_{n_c,n_v}\left(\mathbf{k}_\parallel\right) = \frac{m_0}{\hbar} \left\langle \psi_{n_v}\left(\mathbf{k}_\parallel, z\right) \left| \frac{\partial \hat{H}}{\partial \mathbf{k}_\parallel} \cdot \hat{e} \right| \psi_{n_c}\left(\mathbf{k}_\parallel, z\right) \right\rangle, \tag{9.19}$$

where $\hat{e}$ is a unit vector denoting the polarization of the emitted/absorbed photon and n_c (n_v) indexes the conduction (valence) subbands. This approach to calculating the optical matrix elements then explicitly incorporates key band structure effects in the computation of the optical properties, including N-induced BAC and carrier localization in addition to temperature- and injection-dependent carrier spillout from the QW(s), band mixing, and pseudomorphic strain. When discussing optical transition "strengths" below, we refer for simplicity to the quantity $\frac{1}{2m_0}|P_{n_c,n_v}\left(\mathbf{k}_\parallel\right)|^2$, which has units of energy.

Using the optical matrix elements computed in this manner, the calculation of the optical (material) gain spectrum at a given carrier density follows the approach outlined in Ref. [49], which is based upon a transformation of the spontaneous emission (SE) spectra. We note that we are concerned in this section solely with the transverse electric (TE)-polarized optical gain, which is of significantly larger magnitude than the transverse magnetic (TM)-polarized gain in the compressively strained structures under investigation. Comparison of the SE spectra calculated using this approach with temperature-dependent experimental measurements suggests that the spectral broadening for InGaNAs QWs is best described using a hyperbolic secant lineshape of the form $S(\hbar\omega) = \frac{1}{\pi\delta}\text{sech}\left(\frac{E_0 - \hbar\omega}{\delta}\right)$, where δ is the spectral linewidth [33]. Analysis of the SE spectra at threshold for the devices investigated in Ref. [33] demonstrated that incorporation of N leads to significant spectral broadening. This strong broadening of the optical spectra is primarily attributable to the impact of N-related alloy disorder on the CB structure, highlighting the important role played by the full distribution of N-related localized states in determining the details of the material properties (cf. Section 9.2). The determined value $\delta = 18$ meV for the spectral linewidth at room temperature is approximately three times larger than the typical value $\delta = 6.6$ meV (corresponding to an interband relaxation time of approximately 0.1 ps) observed in N-free devices at the same wavelength [33].

As we demonstrate in Sections 9.3.2 and 9.3.3, the results of calculations undertaken using this theoretical model are in good agreement with a range of experimental measurements, confirming the strong impact of N incorporation on the device performance and quantifying key trends in the expected device properties. We note that alternative theoretical approaches have been developed by several other groups. These alternative models are generally based on the 10-band $\mathbf{k}\cdot\mathbf{p}$ model of Section 9.2.2, which is used to calculate the QW band structure and single-particle eigenstates that serve as the starting point for the computation of the optical properties. For example, Koch et al. have performed many-body calculations of the optical gain [50] and carrier recombination [51] in 1.3-μm InGaNAs lasers using the semiconductor Bloch equations [52]. The device properties and performance computed using this detailed approach are largely in accordance with those calculated using that outlined in this chapter, and have been shown to be in good quantitative agreement with experimental measurements [53].

9.3.2 Impact of Nitrogen Incorporation on the Band Structure and Gain of Quantum Wells

In order to quantify the effects of N incorporation on the electronic and optical properties of QW laser structures, we compare the calculated gain characteristics of an $In_{0.36}Ga_{0.64}N_{0.017}As_{0.983}/GaAs$ ($x = 1.7\%$) single QW laser structure to those of an N-free $In_{0.36}Ga_{0.64}As/GaAs$ QW laser structure [33]. We review here the influence of N incorporation on the calculated QW band structure, optical transition strengths, radiative current density, material gain, and differential gain, and identify general trends in the electronic and optical properties of N-containing laser structures. Following this, we review the impact of electrostatic confinement on the optical gain by summarizing the results of calculations in which the QW band structure and eigenstates are computed self-consistently by solving the multiband Schrödinger equation for the laser structure coupled to Poisson's equation for the (assumed) thermalized carrier density associated with injected electrons and holes [34].

The calculated CB, HH, and LH band offsets are $\Delta E_{CB} = 245$ meV, $\Delta E_{HH} = 130$ meV, and $\Delta E_{LH} = 5$ meV in the compressively strained N-free QW. The incorporation of 1.7% N reduces the compressive in-plane strain from 2.51% in the N-free QW to 2.19% in the $In_{0.36}Ga_{0.64}N_{0.017}As_{0.983}$ QW and, when BAC effects are taken into account, this leads to calculated CB, HH, and LH band offsets $\Delta E_{CB} = 354$ meV, $\Delta E_{HH} = 181$, meV, and $\Delta E_{LH} = 7$ meV in the N-containing QW.

The solid lines in Figures 9.3a and 9.3b show, respectively, the CB and VB structures calculated using the 10-band $\mathbf{k}\cdot\mathbf{p}$ Hamiltonian in conjunction with the plane wave expansion method for ideal, 7-nm-thick $In_{0.36}Ga_{0.64}As$/GaAs (dashed lines; N-free) and $In_{0.36}Ga_{0.64}N_{0.017}As_{0.983}$/GaAs (solid lines; $x = 1.7\%$) QWs. The calculated transition energy between the lowest energy bound electron state ($e1$) and the highest energy hole state ($hh1$) is calculated to be 1.124 eV in the N-free QW and 0.962 eV in the N-containing QW. Of this 161 meV reduction in the QW band gap, only 50 meV is due to the N-induced changes in the VB edge energy, while the remaining 111 meV results from the BAC-induced downward shift of the CB edge energy. The impact of N incorporation on the CB structure, described in terms of the BAC interaction (cf. Section 9.2.2), is clearly visible: (1) incorporation of N leads to a strong decrease in the energy of the electron states at the QW zone center, $k_\| = 0$, and (2) at nonzero in-plane wave vector the BAC interaction pushes the CB dispersion downward in energy leading to strong CB nonparabolicity, increased band edge electron effective masses, and a strong increase in the CB DOS. These calculations indicate that the DOS

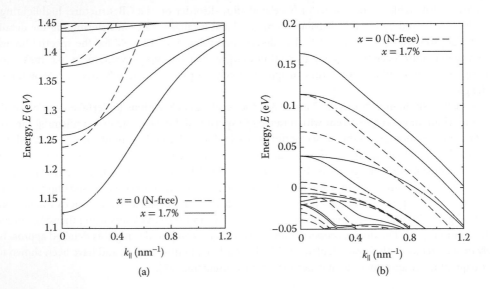

(a)   (b)

FIGURE 9.3 Calculated room temperature (a) CB and (b) VB structure for 7-nm-thick $In_{0.36}Ga_{0.64}As$ (dashed lines; N-free) and $In_{0.36}Ga_{0.64}N_{0.017}As_{0.983}$ (solid lines; $x = 1.7\%$) QWs having GaAs barriers. The zero of energy in each case is taken at the VB edge of the unstrained GaAs barrier. Incorporation of N leads to a strong anticrossing effect and reduction (increase) in the energy (effective mass and DOS) of the CB states, while leaving the VB states relatively unaffected, aside from an upward shift in energy brought about by a reduction of the in-plane compressive strain in the N-containing QW. The primary result of the impact of N on the band structure is a considerable redshift of the QW emission wavelength. The consequences of the N-induced modifications to the band structure and DOS for laser operation are described in detail in the text.

effective mass at $k_\parallel = 0$ for the lowest energy CB increases from $0.046m_0$ in the N-free QW, to $0.060m_0$ in the $In_{0.36}Ga_{0.64}N_{0.017}As_{0.983}$ QW (where m_0 is the free electron mass).

By comparison, the impact of N incorporation on the VB structure is relatively minor. The primary effect on the VB structure is associated with the N-induced reduction in compressive strain in the N-containing QW, which acts to slightly increase the energies of the bound hole states at fixed in-plane wave vector. The ordering of the hole states is calculated to be the same in both the N-free and N-containing QWs: the three highest-energy hole states have predominately HH character at $k_\parallel = 0$, with the ordering of the hole states being *hh*1, *hh*2, *hh*3, and *lh*1. Given that the VB edge DOS is largely unaltered by incorporation (aside from a relatively rigid strain-related energy shift), the impact of N incorporation on the gain characteristics is almost entirely attributable to the N-induced modifications to (1) the CB structure and DOS, (2) the interband optical transition strengths.

The N-induced hybridization between the host matrix CB edge states and N-related localized states leads to a portion of the CB eigenstates consisting of an admixture of N-related localized states which do not couple optically to the *p*-like states at the VB edge [29]. Despite the tendency of N incorporation to enhance the electron confinement—due to (1) the increase in CB offset ΔE_{CB}, and (2) enhanced spatial localization of the electron charge density, in the InGaNAs QW layer—the net effect of the N-induced hybridization is to decrease the interband optical transition strengths compared to those in a conventional N-free structure. Figure 9.4a and 9.4b shows, respectively, the calculated variation of the TE- and TM-polarized optical transition strengths as a function of in-plane wave vector $k_\parallel$, for transitions from the lowest-energy electron states *e*1 to the highest-energy HH- and LH-like states *hh*1 (solid) and *lh*1 (dashed), in the $In_{0.36}Ga_{0.64}N_{0.017}As_{0.983}$/GaAs (N-containing; black lines) and $In_{0.36}Ga_{0.64}As$/GaAs (N-free; gray lines) QW structures. As described above, N-induced hybridization leads to an overall decrease in the optical transition strengths. For example, the optical transition strength $|M_{e1-hh1}|^2$ calculated at $k_\parallel = 0$ for the TE-polarized *e*1–*hh*1 transition in the $In_{0.36}Ga_{0.64}N_{0.017}As_{0.983}$ QW is approximately 70% of that calculated for the N-free structure. It is also interesting to note that the calculated reduction in $|M_{e1-hh1}|^2$ and increase in the CB edge effective mass m_e^* causes the product $|M_{e1-hh1}|^2 \times m_e^*$ to be reduced when N

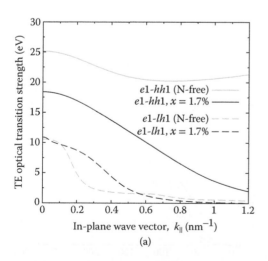
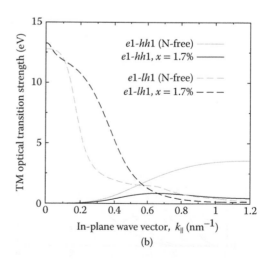

(a) (b)

FIGURE 9.4 Calculated variation of the (a) TE-polarized and (b) TM-polarized optical transition strengths as a function of the in-plane wave vector $k_\parallel$ for the *e*1-*hh*1 (solid lines) and *e*1-*lh*1 (dashed lines) transitions in $In_{0.36}Ga_{0.64}As$/GaAs (N-free; gray lines) and $In_{0.36}Ga_{0.64}N_{0.017}As_{0.983}$/GaAs ($x = 1.7\%$; black lines) QWs. Note the difference in scale between (a) and (b). It can be clearly seen that N-induced hybridization of CB states leads to a significant reduction of the strength of the TE-polarized *e*1-*hh*1 transitions, which account for the majority of the material gain *g* in these compressively strained QWs. This suggests that N incorporation can be expected to reduce the magnitude of *g* compared to that which can be achieved an N-free QW.

is incorporated in the QW. This is expected on the basis of $\mathbf{k} \cdot \mathbf{p}$ theory, in which the dominant contribution to the inverse bulk CB effective mass is directly proportional to $|M_{e1-hh1}|^2$, which describes the coupling of the s-like CB edge states to the p-like VB edge states via the momentum operator, and hence the large increase in the CB DOS associated with N incorporation.

Having investigated the N-induced modifications to the band structure and optical transition strengths we now turn our attention to the calculated gain characteristics, which we again compare for the N-containing and N-free laser structures. The band structure and optical transition strengths presented in Figures 9.3a and 9.3b, and , and in Figures 9.4a and 9.4b were used to calculate the variation of the peak TE-polarized material gain g as a function of carrier density n, and of radiative current density J_{rad}, for temperatures $T = 200$, 300 and 400 K. The solid (dashed) lines in Figures 9.5a and 9.5b denote the calculated variation of g with n and J_{rad} for the laser structure having an N-containing (N-free) $In_{0.36}Ga_{0.64}N_{0.017}As_{0.983}$ ($In_{0.36}Ga_{0.64}As$) QW [54].

Examining first the calculated variation of g with n in Figure 9.5a, we note that incorporation of N leads to an increase in the carrier density n_{tr} at transparency. Further analysis reveals that this behavior arises largely due to two factors: (1) the general reduction of the optical transition strengths due to N-induced hybridization with localized impurity states, and (2) the N-induced increase in the CB edge effective mass m_e^*. The increase in m_e^* is associated with an increase in the DOS at the CB edge, which leads to a reduction in the energy separation ΔF between the electron and hole quasi-Fermi levels at fixed carrier density. As a result of these N-induced changes to the band structure and optical transition strengths, the peak gain in the N-containing structure at fixed n is expected to be reduced compared to that in the N-free structure. This is confirmed by the numerical calculations, which compared to the N-free case show (1) an overall reduction of g, and (2) a saturation of the material gain at a lower level.

Turning our attention to Figure 9.5b, we note that the impact of N incorporation on the calculated variation of the peak material gain with J_{rad} is broadly similar to that observed for the variation of g with n. The calculated reduction of g at fixed J_{rad} in the N-containing structures reflects the fact that, for fixed quasi-Fermi level separation ΔF, the radiative current density J_{rad} in a QW laser is approximately proportional to $m_r^* \times |M_{e1-hh1}|^2$, where m_r^* is the reduced effective mass of a band edge electron–hole pair. Because the effective mass of a hole at the VB edge is larger than that of a CB edge electron, the reduced effective mass m_r^* is determined primarily by m_e^*. Based on our analysis of the band structure and optical transition

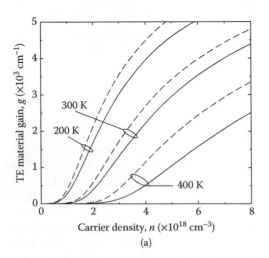
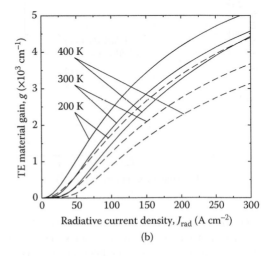

(a)   (b)

FIGURE 9.5 Calculated variation of the TE-polarized material gain g as a function of (a) carrier density n and (b) radiative current density J_{rad}, for $In_{0.36}Ga_{0.64}As/GaAs$ (N-free; dashed lines) and $In_{0.36}Ga_{0.64}N_{0.017}As_{0.983}/GaAs$ ($x = 1.7\%$; solid lines) QWs at temperatures $T = 200$, 300, and 400 K. Incorporation of N brings about a reduction of g at fixed n and J_{rad} and leads to gain saturation at lower g compared to the N-free QW.

strengths above, we conclude that the N-induced increase in m_r^* is insufficient to overcome the corresponding reduction in $|M_{e1-hh1}|^2$, meaning that the product $m_r^* \times |M_{e1-hh1}|^2$ undergoes a moderate reduction when N is incorporated into the QW. The result of this is that $J_{rad} \propto m_r^* \times |M_{e1-hh1}|^2$ is reduced at fixed n compared to that in an N-free QW, thereby accounting for the calculated reduction of the material gain at fixed J_{rad}, primarily in terms of the impact of N on the TE-polarized interband optical transition strengths.

The differential gain $\frac{dg}{dn}$ plays a key role in determining both the threshold carrier density and bandwidth of a directly modulated semiconductor laser. For example, the modulation response frequency ω_r is proportional to the square root of the differential gain with respect to the carrier density, $\omega_r \propto \sqrt{\frac{dg}{dn}}$ [55,56]. Figure 9.6 shows the calculated variation of $\frac{dg}{dn}$ at transparency (squares) and at threshold (circles) in a series of ideal 7-nm-thick $In_yGa_{1-y}N_xAs_{1-x}$ QW structures, calculated using linewidths $\delta = 6.6$ meV (closed symbols, solid lines) and 18 meV (open symbols, dashed lines) [57]. In order to maintain a constant QW ground-state $e1–hh1$ transition energy (emission wavelength) of 0.954 eV (1.3 μm) in the range of N compositions from $x = 0$ to 3%, the In composition must be reduced in order to compensate for the reduction in the QW band gap brought about by N incorporation. The calculated In compositions y required to maintain 1.3-μm emission ranges from 55% in the N-free case to 30% at $x = 3\%$ [57]. For fixed spectral linewidth δ, the calculated differential gain decreases with increasing N composition, with the most rapid decrease observed at low x. As expected, $\frac{dg}{dn}$ also decreases with increasing spectral linewidth. Due to the fact that the measured spectral broadening in InGaNAs QWs increases with increasing x [58], we conclude that the calculated value $\frac{dg}{dn} \sim 0.86 \times 10^{-15}$ cm^{-1} at threshold for $x = 2\%$ is approximately about 40% of that which could be achieved in an ideal InGaAs 1.3-μm QW laser [9,57].

As can be seen from the calculated variation of the peak material gain with carrier density in Figure 9.5a, the initial rapid decrease in $\frac{dg}{dn}$ is due to the strong hybridization between the N-resonant states and the CB edge states, which occurs even for small N compositions x. Figure 9.6 then suggests that an optimized InGaNAs/GaAs QW laser device should contain minimal N, ideally being N-free. However, this is not possible due to the excessively large strain required to achieve 1.3-μm emission from an InGaAs/GaAs QW

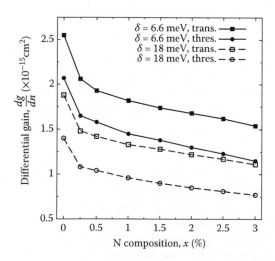

FIGURE 9.6 Calculated variation of the room temperature differential gain $\frac{dg}{dn}$ at transparency (closed squares) and at threshold (closed circles) as a function of N composition x in a series of (In)GaN$_x$As$_{1-x}$/GaAs QW structures, using homogenous linewidths $\delta = 6.6$ meV (solid lines, closed symbols) and $\delta = 18$ meV (dashed lines, open symbols) for the computation of the material gain g. At each N composition x the In composition in the QW has been adjusted to maintain the emission wavelength at 1.3 μm. It can be seen that both N incorporation and N-induced broadening of the gain spectrum lead to a reduction of $\frac{dg}{dn}$ for devices having a fixed 1.3-μm emission wavelength (Adapted from S. Tomić and E. P. O'Reilly, *IEEE Photon. Tech. Lett.*, 15:6, 2003.).

QW [9]. Using the strain-thickness criteria presented in Ref. [59], it is estimated on the basis of the calculated trends for this series of InGaNAs QWs that one needs N and In composition $x > 1.5\%$ and $y < 39\%$ to achieve 1.3-μm emission in a pseudomorphic $In_yGa_{1-y}N_xAs_{1-x}$/GaAs QW, compared to approximately 55% In in an N-free InGaAs structure [9]. The N composition required to achieve 1.3-μm emission can be further reduced by adding more In to narrower QWs and/or by growing tensile strained barrier layers above and below the QW, as demonstrated by Tansu et al., who achieved 1.3-μm emission with $x = 0.5\%$ in a 6-nm-thick QW by utilizing strain compensating tensile strained GaAsP [60,61] and GaAsN [62] barrier layers.

Experimental analysis has led to some controversy as to the origin of the magnitude and temperature sensitivity of the threshold current density J_{th} in 1.3-μm InGaNAs/GaAs QW lasers. Tansu et al. have suggested that the observed large values of J_{th} are associated with hole leakage, which impacts upon the device performance at high temperature [63]. That this might be the case appears consistent with the analysis reviewed above, which demonstrates that the BAC interaction leads to strong electron confinement, but low VB offsets and comparatively weak confinement of holes. Given the large mismatch in the CB and VB offsets in InGaNAs/GaAs QWs, electrostatic confinement of holes by strongly bound and localized electrons can be expected to play an important role in determining the electronic and optical properties of lasers under injection. In order to investigate the potential for hole leakage in InGaNAs laser structures, it is then important to calculate the electron and hole charge densities self-consistently, by solving Poisson's equation to obtain the net electrostatic potential at each injected carrier density n. This task can be accomplished efficiently using a plane wave basis set, since Poisson's equation can be solved analytically in reciprocal space [34].

This self-consistent approach has been applied to theoretically analyze the impact of hole leakage on the gain characteristics of InGaNAs/GaAs(P) QW laser structures. Full details of this self-consistent analysis of the gain characteristics, which are beyond the scope of this chapter, can be found in Ref. [34]. Here we summarize that these temperature- and injection-dependent calculations of the material gain suggest that electrostatic confinement of holes by electrons significantly reduces hole leakage at higher carrier densities, a conclusion which is confirmed by comparing the hole states calculated for structures having GaAs barriers to those having GaAsP barriers (the latter having larger VB offsets and enhanced hole confinement). Taking electrostatic confinement into account leads to a slight increase in the hole quasi-Fermi level at fixed injection, decreasing the quasi-Fermi level separation ΔF and hence increasing the material gain at fixed carrier density. Aside from these minor qualitative changes, the gain characteristics calculated self-consistently for InGaNAs/GaAs(P) QWs are largely the same as those described above. This analysis then suggests that hole leakage does not play a major role in determining the threshold characteristics and performance of 1.3-μm InGaNAs QW lasers, and that other mechanisms must therefore be considered in order to describe the measured magnitude and temperature sensitivity of J_{th}. It is to this question that we turn our attention in Section 9.3.3.

Overall, we conclude that despite the tendency of N to slightly degrade the gain characteristics, 1.3-μm InGaNAs QW lasers offer good (theoretical) performance compared to idealized highly strained InGaAs/-GaAs QWs at the same wavelength. While incorporating N entails trading off some performance compared to equivalent N-free laser structures, utilizing InGaNAs QWs allows to overcome the strain limitations associated with InGaAs QWs, and extends the range of emission wavelengths accessible using a GaAs substrate. Comparing these results for the exemplar GaAs-based InGaNAs laser structure considered above with theoretical analysis of conventional 1.3-μm InP-based InGaAsP and AlInGaAs QW laser structures [33,64–66], we find that InGaNAs QWs offer (1) higher optical gain, and (2) larger differential gain, with the value at threshold $\frac{dg}{dn} \sim 8 \times 10^{-14}$ cm^2 (calculated at $T = 300$ K) being approximately 33% larger than the value $\sim 6 \times 10^{-14}$ cm^2 calculated for an InP-based device having the same compressive strain, optical gain, and QW thickness. We note that this is the case despite the significantly larger spectral linewidth $\delta = 18$ meV in the InGaNAs structure, compared to the value $\delta = 6.6$ meV assumed for the InP-based structure, highlighting the intrinsically superior characteristics of GaAs-based heterostructures. Furthermore, improved optical confinement can be achieved in GaAs-based 1.3-μm InGaNAs laser structures compared to InP-based devices, because of the larger refractive index contrast achievable through use of AlGaAs cladding layers. This reduces the number of QWs required in an optimized edge-emitting device, while also opening the possibility of using InGaNAs alloys to develop GaAs-based 1.3- and 1.55-μm VCSELs. The theoretical analysis reviewed here therefore confirms the potential of dilute nitride alloys for applications in long-wavelength GaAs-based photonics, and in particular for edge- and surface-emitting laser applications in the telecommunication wavelength ranges.

9.3.3 Impact of Nitrogen Incorporation on Carrier Recombination: Theory versus Experiment

While there has been considerable interest in the influence of N incorporation on the electronic structure and optical transitions in InGaNAs/GaAs QWs, there have been comparatively few quantitative investigations of the primary loss mechanisms in lasers based on these material systems. Such studies are of interest since, for example, they allow a direct comparison between the relative importance of radiative and non-radiative recombination pathways in dilute nitride lasers versus their conventional InP-based (N-free) counterparts. Non-radiative recombination processes make a significant contribution to the threshold current in conventional InP-based lasers, leading to strong temperature dependence of the threshold current [3,67]. A number of additional factors can also contribute to this temperature dependence, such as carrier leakage due to poor confinement and thermionic emission [68,69], or intervalence band absorption (IVBA) [64,70]. However, the dominant loss mechanism in 1.3- and 1.55-μm InP-based lasers is generally acknowledged to be non-radiative Auger recombination [67,71]. We therefore begin our discussion of the performance of InGaNAs QW lasers with a description of the quantitative analysis of recombination pathways in 1.3-μm InGaNAs/GaAs lasers, highlighting the importance of defect-related and Auger recombination [45]. We provide an interpretation of the observed device characteristics in terms of the detailed theoretical analysis of the impact of N incorporation on Auger recombination rates in InGaNAs QWs [72]. This is then used as a basis to understand the threshold characteristics measured for a range of InGaNAs QW lasers across the 1.2- to 1.6-μm wavelength range.

Under the assumptions of (1) charge neutrality in the active region, that is, the electron density n being equal to the hole density p, and (2) negligible carrier leakage, the current density J in a semiconductor laser can be written as

$$J = J_{\text{mono}} + J_{\text{rad}} + J_{\text{Auger}} + J_{\text{leak}} = e\left(An + Bn^2 + Cn^3\right) + J_{\text{leak}}, \tag{9.20}$$

where J_{mono}, J_{rad}, and J_{Auger}, respectively refer to the current densities associated with defect-related (monomolecular), radiative, and Auger recombination, e is the magnitude of the electrical charge of an individual carrier (electron or hole), and J_{leak} is the current density associated with leakage of carriers from the active region. In the Boltzmann approximation, the current density associated with each of these recombination mechanisms is proportional to the products of the densities of each type of carrier involved in the process. For example, the radiative component of J—describing SE due to recombining electron–hole pairs—is proportional to np which, under the aforementioned assumption of overall charge neutrality, is equal to n^2. Similarly, the component of J associated with Auger recombination processes—three carrier processes involving two electrons and one hole, or one electron and two holes—is proportional to n^3. The coefficients A, B, and C on the right-hand side of Equation 9.20, which describe the monomolecular, radiative, and Auger recombination rates, respectively, depend strongly on the details of the active region band structure and, in a real device, on the material quality. We note that the assumption of negligible carrier leakage, $J_{\text{leak}} \approx 0$, is found to be generally valid for 1.3-μm InGaNAs laser structures, due to good carrier confinement brought about by the large CB and VB offsets.

Experimental determination of the individual components J_{mono}, J_{rad}, and J_{Auger} of Equation 9.19 relies on the fact that, over a limited range of current densities, Equation 9.20 can be written as $J \propto n^z$, where $1 \leq z \leq 3$, and with the value of z varying depending on whether the dominant contribution to the current density is due to defect-related ($z = 1$), radiative ($z = 2$), or Auger ($z = 3$) recombination [67,73].

Measuring the integrated SE rate $L \propto Bn^2$ then allows the total current density J to be related to L as $J \propto L^{\frac{z}{2}}$, which can be rewritten as $\log(J) \propto z \log(\sqrt{L})$. This enables z to be quantified experimentally for a given laser structure by plotting $\log(J)$ against $\log(\sqrt{L})$, with the power factor z then given as the slope of the log–log plot [45,74]. Based on this approach, it is possible to directly determine the individual components of J described by Equation 9.20. We note that this analysis is typically undertaken at threshold, which is straightforward to identify due to the pinning of the carrier density n, and hence the integrated SE L, in QW lasers. This procedure provides a useful method by which to compare the performance of different laser structures at a given wavelength, since it enables the relative importance of different loss mechanisms to be compared and contrasted for different material systems and laser structures [45]. Full details of this approach to quantifying and analyzing loss mechanisms in semiconductor lasers can be found in Refs. [45] and [74].

Using the temperature-dependent components of the threshold current density J_{th} extracted in this manner, it is possible to determine the recombination coefficients A, B, and C at fixed temperature for a given laser structure using the calculated threshold carrier density n_{th} in Equation 9.20. Using $J_{mono} = eAn_{th}$ allows A to be determined straightforwardly at fixed temperature, with B and C determined similarly using $J_{rad} = eBn_{th}^2$ and $J_{Auger} = eCn_{th}^3$. Repeating this procedure across a range of temperatures enables direct determination of the temperature-dependent recombination coefficients, which can be compared directly to theoretical calculations. This provides quantitative insight into the loss mechanisms present in the laser, as well as their relative importance and dependence on temperature. Comparison of the measured recombination coefficients to theoretical calculations then enables further analysis and interpretation on the basis of the calculated material band structure providing, in the case of dilute nitride alloys, valuable insight into the impact of N incorporation on carrier recombination. An analysis of this type has been undertaken by Fehse et al. in Ref. [45], for a GaAs-based laser structure containing a single $In_{0.36}Ga_{0.64}N_{0.017}As_{0.983}$/GaAs QW (cf. Section 9.3.2). We summarize the results of this analysis here and highlight the ability of the theoretical model outlined in Section 9.3.1 to quantitatively describe trends in real dilute nitride laser devices.

The closed circles in Figure 9.7a show the values of the monomolecular recombination coefficient A extracted from temperature-dependent measurements of current density J_{mono} [45]. However, A cannot readily be computed within the framework of the theoretical models outlined in this chapter—since it is in practice primarily associated with the material quality—making a direct comparison between theory and experiment difficult. Nonetheless, a comparison can be made based on the following simple model: we assume that the defect-related recombination occurs when free carriers having (thermal) velocity v are captured in the InGaNAs material by defects of uniform cross-section σ at a rate $A = \tau_{mono}^{-1}$. Assuming there are n_D such defects per unit volume, the monomolecular recombination rate is given as $\tau_{mono}^{-1} = \sigma v n_D$. The dependence of the free carrier velocity v on temperature is given by the relationship between the kinetic and thermal energy as $\frac{1}{2}mv^2 \propto k_B T$, so that $v \propto \sqrt{T}$. Using this relationship and assuming that σ is independent of T, we conclude that $A = \sigma v n_D \propto \sqrt{T}$. The solid line in Figure 9.7a shows the variation of A with temperature predicted by this simple model, normalized to the measured value of A at $T = 150$ K—that is, $A(T) = A(150 \text{ K})\sqrt{\frac{T}{150 \text{ K}}}$ for $T \geq 150$ K. This simple description of the temperature dependence of A is in good agreement with experiment up to room temperature.

On the other hand, the radiative and Auger recombination coefficients B and C can be computed explicitly from the QW band structure and eigenstates, and therefore compared directly to experiment, allowing the recombination pathways in a given device to be analyzed quantitatively. This provides a stringent test of theoretical models of the electronic and optical properties. The calculation of B is straightforward and consists at each temperature of integrating over the calculated SE spectrum at threshold with respect to the photon energy in order to determine the radiative current density at threshold $J_{rad}^{th} = eBn_{th}^2$, and hence B. Figure 9.7b compares the measured (closed circles) and calculated (open squares) values of B, for temperatures between 150 and 300 K. The calculated values of B are in good overall agreement with experiment: the theoretical model describes well the magnitude of B at low temperature, and the calculated variation of

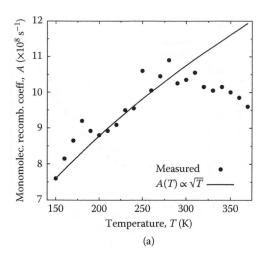
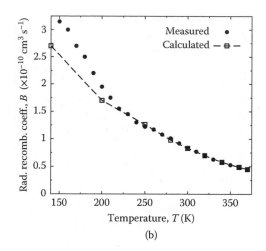

FIGURE 9.7 (a) Measured (at threshold; closed circles) variation of the monomolecular (defect-related) recombination coefficient A at threshold as a function of temperature T for an $In_{0.36}Ga_{0.64}N_{0.017}As_{0.983}$/GaAs ($x = 1.7\%$) 1.3-μm single QW laser structure. The solid black line shows the calculated $A(T)$ based on the simple model described in the text, with the value of A in the theoretical calculation normalized to the measured value at $T = 150$ K. (b) Measured (at threshold; closed circles) and calculated (open squares) variation of the radiative recombination coefficient B as a function of temperature T for the same device as in (a). The calculated values of B—obtained in the Boltzmann approximation, that is, assuming $J_{rad}^{th} = eBn_{th}^2$, where J_{rad}^{th} is calculated directly by integrating over the calculated temperature-dependent SE spectrum at threshold for the device—display the $B(T) \propto T^{-1}$ temperature dependence characteristic of an ideal QW (Adapted from R. Fehse et al., *IEEE J. Sel. Topics Quantum Electron.*, 8:801, 2002).

B with temperature agrees well with experiment. We further note that the calculated values of B between 150 and 350 K follow the $B(T) \propto T^{-1}$ temperature dependence expected for an ideal QW [75].

The calculation of the Auger recombination coefficient C is significantly more complicated due to the large number of matrix elements describing three-body (Coulombic) scattering processes that must be evaluated and integrated over in order to determine the Auger current density $J_{Auger} = eCn^3$. Full details of the calculation of C are beyond the scope of this chapter. We refer the reader to Ref. [72] and references therein for more details on various approaches to the calculation of Auger recombination rates in quantum-confined heterostructures. Figure 9.8a compares the measured (closed circles) and calculated (open squares) values of C, for temperatures between 200 and 300 K. Given the well-known challenges associated with quantitative calculations of Auger recombination rates—the values of which are known to vary by at least an order of magnitude based on the approximations employed in the calculation—we conclude that the calculated values of C can be considered to be in good agreement with experiment. Further analysis of C indicates that the dominant Auger recombination mechanism, accounting for the majority of the calculated value of C, is the so-called CHSH pathway. The CHSH Auger recombination mechanism occurs when an electron in the CB recombines with a hole in the VB, leading to the excitation of a hole from the VB edge to the SO band (cf. Chapter 10). The "hot" SO hole generated by this process dissipates its excess energy and crystal momentum primarily via scattering within the lattice, which increases the lattice temperature and generates significant waste heat. The other notable Auger recombination mechanism is the so-called CHCC pathway, in which an electron in the CB recombining with a hole in the VB leads to the excitation of an electron to a high-energy state in the CB. Given the strong impact of N on the CB structure, it could be expected that N incorporation has significant impact on the Auger recombination rates. However, calculations of C for the InGaNAs/GaAs QW considered in Section 9.3.2 demonstrate that incorporation of 1.7% N leads to a decrease of only ∼ 5% in the total Auger recombination rate compared to that in an equivalent N-free QW [72]. This indicates that the nature of Auger recombination in

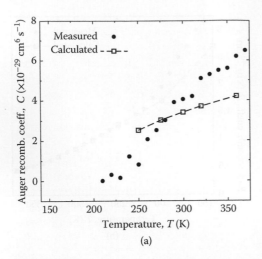
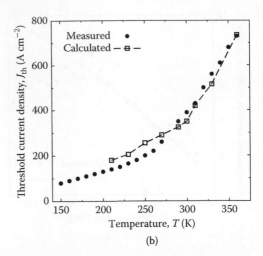

(a)   (b)

FIGURE 9.8 (a) Measured (at threshold; closed circles) and calculated (open squares) variation of the Auger recombination coefficient C as a function of temperature T for the same device as in Figures 9.7a and 9.7b. Despite the difficulties associated with direct calculations of the Auger current density, we note that the measured and calculated values of C are in reasonably good quantitative agreement. (b) Measured (closed circles) and calculated (open squares) threshold current density J_{th} as a function of temperature for the same device as in (a) and Figures 9.7a and 9.7b. J_{th} was computed at each temperature using the calculated values of A, B, and C from Figure 9.7a, 9.7b and 9.8a. We note good, quantitative agreement between theory and experiment across a broad range of temperatures from 200 to 360 K (Adapted from R. Fehse et al., *IEEE J. Sel. Topics Quantum Electron.*, 8, 801, 2002).

GaAs-based InGaNAs laser structures is largely the same as that in conventional InP-based InGaAsP and AlInGaAs structures. We note that typical values of C in InGaAsP/InP 1.3-μm QWs are in the range of 1–8 $\times 10^{29}$ cm^6 s^{-1} [76,77], which are broadly similar to those calculated by Andreev and O'Reilly for a 1.3-μm InGaNAs/GaAs laser structure [72].

Figure 9.8b compares the measured (closed circles) and calculated (open squares) variation of the threshold current density J_{th} with temperature for this In$_{0.36}$Ga$_{0.64}$N$_{0.017}$As$_{0.983}$/GaAs single QW laser structure [45] where, overall, we note good agreement between theory and experiment. Based on the measured and calculated current densities and recombination coefficients, we conclude that defect-related, radiative, and Auger recombination account, respectively, for approximately 55%, 20%, and 25% of J_{th} at room temperature. This suggests that defect-related recombination, as opposed to carrier leakage or Auger recombination, is primarily responsible for the observed large magnitude of J_{th}, as well as the measured variation of J_{th} with temperature.

The strong role played by defect-related recombination in 1.3-μm InGaNAs QW lasers has been confirmed by a systematic study of a range of devices, where experimental measurements indicate that devices having higher J_{th} tend to have higher values of the characteristic temperature[†] T_0 [79]. As described above, the contribution of Auger recombination to J_{th} in InGaNAs devices is broadly similar to that in conventional InGaAsP lasers at the same wavelength. Furthermore, the relatively weak temperature

---

[†] The threshold current density J_{th} of a semiconductor laser at two temperatures T_1 and T_2 can typically be related in the simple form $J_{th}(T_2) = J_{th}(T_1) \exp(\frac{T_2 - T_1}{T_0})$. Assuming that T_2 is greater (less) than T_1, it follows from this relation that changing the temperature by an amount T_0 leads to an increase (decrease) of J_{th} by a factor of e. In this manner, the "characteristic" temperature T_0 describes the temperature dependence of J_{th}, with a large (small) value of T_0 indicating that a large (small) change in temperature is required to bring about a given change in J_{th}. T_0 therefore describes the temperature stability of a given laser device, with a large value of T_0 generally indicating temperature stable operation. For example, InP-based 1.55-μm QW lasers—in which J_{th} is dominated by strongly temperature-dependent Auger recombination—have low characteristic temperatures $T_0 \sim \frac{T}{4}$ [78].

dependence of J^{th}_{mono} suggests that the large values of J_{th} measured in devices having high measured T_0 arise due primarily due to a dominant contribution to J_{th} from defect-related recombination. This strong defect-related recombination is primarily associated with the impact of N incorporation, as well as a degradation in material quality brought about by the nonequilibrium growth conditions required to facilitate N incorporation in the laser active region. We note that similar trends have been observed in 1.3-μm metamorphic (Al)InGaAs QW lasers grown on relaxed InGaAs buffer layers, in which case the strong defect-related contribution to J_{th} is associated with the formation of threading dislocations [9].

Overall, while N incorporation can be used to successfully realize 1.3-μm emission on a GaAs substrate, we conclude that (1) the performance of these devices suffers due to the prevalence of defect-related recombination, and (2) Auger recombination plays an important role in determining the high-temperature performance, as in conventional InP-based devices. The theoretical approach outlined in this and the previous section has provided detailed insight into the impact of N incorporation on the properties and performance of InGaNAs QW lasers and has enabled quantitative analysis of the recombination pathways in dilute nitride alloys and laser structures. This confirms the potential of theory and simulation for use in the design and optimization of dilute nitride semiconductor lasers, both at 1.3-μm and at longer wavelengths. The above analysis also highlights that the main issues to be overcome for the realization of high-performance dilute nitride semiconductor lasers are related to material growth, which must be refined in order to minimize the impact of defect-related losses. Despite these issues, there has been significant success achieved in the development of 1.3-μm InGaNAs/GaAs QW lasers, including realization of low-threshold [80,81] and high-speed [82,83] devices, whose properties and performance are competitive with existing InP-based technologies.

9.4 Alternative Applications of Dilute Nitride Alloys

So far in this chapter, we have focused primarily on the development of InGaNAs alloys to realize GaAs-based semiconductor lasers emitting at the 1.3-μm optical communications wavelength. While the development of 1.3-μm QW lasers on GaAs is the application which attracted a significant share of the initial research effort on dilute nitride alloys, the possibilities brought about by the impact of N incorporation on the band structure of a range of III–V compounds and alloys have stimulated a broader program of research to address related and alternative applications. In this section, we provide an overview of applications of dilute nitride alloys beyond 1.3-μm semiconductor lasers. The applications for which dilute nitride alloys are under investigation are diverse, spanning from the use of InGaNAs(Sb) alloys to develop 1.55-μm semiconductor lasers (Section 9.4.1), through to applications including 1.3- and 1.55-μm SOAs, EAMs, and APDs (Section 9.4.2), and monolithic integration of III–V lasers on Si (Section 9.4.3). An important focus of research interest has been the use of dilute nitride alloys to facilitate the development of 1.3- and 1.55-μm vertical-cavity devices on GaAs, including VCSELs, vertical-external-cavity surface-emitting lasers (VECSELs), and vertical-cavity semiconductor optical amplifiers (VCSOAs) [84]. Such devices have a number of key advantages over conventional InP-based edge-emitting/absorbing devices, including lower fabrication costs, power consumption and noise, as well as enhanced temperature stability. Vertical-cavity devices also open up a range of potential applications in, e.g. on-chip and chip-to-chip optoelectronics, particularly in key components such as optical switches and interconnects. The range of applications described in this section is not exhaustive—for example, there have been significant activity and much success in developing dilute nitride multijunction solar cells and VECSELs, which we do not discuss in detail here—but should give the reader an appreciation of the significant potential to exploit the unique properties of dilute nitride alloys to deliver photonic and photovoltaic devices with enhanced performance and capabilities.

9.4.1 GaAs-Based 1.55-μm Quantum Well Lasers

As described in Section 9.1, a major goal for the III–V semiconductor community is to develop reliable 1.3- and 1.55-μm semiconductor lasers on GaAs substrates to take advantage of the enhanced material properties offered by the GaAs platform (compared to InP), and to exploit technological advantages (such as the ability to utilize vertical-cavity architectures and to monolithically integrate photonic components with high-speed microelectronics). While there has been much success in developing GaAs-based 1.3-μm edge-emitting lasers and VCSELs using InGaNAs alloys, in general it has not been possible to readily and reproducibly grow InGaNAs-based devices having emission wavelengths close to 1.55-μm. This limitation of the InGaNAs material system is primarily associated with the large In and N compositions (~40% and 4%, respectively) required to extend the emission wavelength beyond 1.5 μm. The existence of a narrow window of suitable growth conditions for InGaNAs alloys with emission wavelengths ~1.55 μm means that the growth of InGaNAs alloys with increased In and N compositions is generally associated with significant reductions in material quality and optical efficiency. While some success has been had in developing 1.5-μm InGaNAs broad area lasers [85], there has been less success in developing InGaNAs VCSELs emitting at > 1.5-μm. As such, InGaNAs alloys grown on GaAs are commonly considered as not well suited for the development of 1.55-μm lasers on GaAs.

These limitations of InGaNAs/GaAs alloys have motivated interest in related materials, in order to extend the wavelength range accessible on GaAs beyond 1.3 μm. The approach which has primarily been pursued is to extend the emission wavelength by co-alloying N and Sb in InGaAs to form the InGaNAsSb alloy. During epitaxial growth of this penternary alloy Sb acts as a reactive surfactant enabling N incorporation at higher growth temperatures, and ultimately leading to materials with improved morphology compared to equivalent Sb-free alloys [86]. InGaNAsSb QWs are generally found to have improved material quality and optical efficiency compared to InGaNAs, and have the additional benefits that (1) the window of optimal growth parameters is wider than in InGaNAs, and (2) there is generally a simplification of the alloy growth by which the N composition (and hence emission wavelength) is found to be controllable directly via the total growth rate [87]. As a result, the use of InGaNAsSb alloys to develop GaAs-based 1.55-μm edge-emitting lasers and VCSELs—a review of which can be found in Ref. [87]—has generally been quite successful. For example, Korpijärvi et al. [88] recently demonstrated the first 1.55-μm VECSEL, in which the active region consisted of InGaNAsSb QWs and the laser structure was grown on a GaAs substrate. Recent developments such as this highlight the continuing strong interest in and potential of dilute nitride alloys to facilitate novel long-wavelength devices on GaAs.

One significant benefit of co-alloying N and Sb to form InGaNAsSb is that, contrary to trends in InGaNAs alloys, there is little degradation of the optical efficiency with increasing N composition [87]. As a result, it has been possible to develop InGaNAsSb QW lasers operating across a broad range of wavelengths, from 1.45 to 1.64 μm, which display low J_{th}. For example, a value $J_{th} = 373$ A cm^{-2} was measured for a 1.55-μm edge-emitting device under continuous wave operation at room temperature, which is competitive with the lowest values observed for 1.3-μm InGaNAs devices [80,81,87]. The fact that such low threshold current densities have been achieved in InGaNAsSb devices—despite the impact of (1) significantly higher N compositions than in 1.3-μm InGaNAs devices, and (2) Auger recombination, which is known to increase strongly with increasing wavelength [89]—highlights the benefits of co-alloying N and Sb to extend the wavelength range accessible on GaAs to 1.55 μm and beyond.

However, despite the largely successful development of InGaNAsSb 1.55-μm edge-emitting lasers and VCSELs on GaAs, there are three intrinsic issues associated with this approach that limit the device performance. First, in order to compensate the relatively large compressive strain brought about by co-alloying In and Sb, InGaNAsSb QW laser structures are typically strain compensated by growing the QW(s) between tensile strained GaNAs barrier layers. While this approach has been successful in terms of facilitating the growth of QWs displaying high optical efficiencies, it comes at the cost of associated effects that impact negatively upon the device performance. For example, the tensile strain present in the GaNAs barrier layers pushes the barrier LH band edge upward in energy, leading to a reduction in the VB offset seen by holes in the QW. As such, thermal leakage of holes limits the high-temperature performance of InGaNAsSb QW lasers [90]. Second, while Sb incorporation tends to decrease the defect density in the QW, in a strain-compensated InGaNAsSb QW structure there remains a significant density of N-related defects in the tensile strained GaNAs barriers. Carriers which leak from the QWs then tend to undergo defect-related recombination in the GaNAs layers of the structure. It is possible that these first two issues could be mitigated by, for example, using GaAsP as the barrier material, in order to provide strain compensation without (1) degrading the carrier confinement, or (2) introducing a high density of monomolecular recombination centers [90]. Third, temperature-dependent analysis of the power factor z (cf. Section 9.3.3) suggests that Auger recombination is an important loss mechanism in 1.55-μm InGaNAsSb QW lasers and plays a dominant role in limiting the high-temperature performance, as in conventional InP-based devices [87].

Recalling the analysis of 1.3-μm InGaNAs lasers in Section 9.3.3, we note that the impact of Auger recombination and IVBA is determined largely by the nature of the active region band structure, with the CHSH mechanism being dominant in materials where the magnitude of VB spin-orbit-splitting energy is less than the band gap [3]. The nature of Auger and IVBA losses in semiconductor lasers operating in the 1–2 μm wavelength range is known to be largely independent of whether the device is grown on GaAs or InP, but instead depends upon the details of the VB structure, which are broadly similar both in conventional InGaAsP or AlInGaAs materials and in highly mismatched InGaNAs(Sb) alloys. On this basis, it is not surprising that the performance of InGaNAsSb QW lasers is degraded and limited by Auger recombination. Moreover, the presence of strong Auger recombination in these devices is reflective of the fact that while developing 1.3- and 1.55-μm semiconductor lasers on GaAs substrates is desirable from a technological point of view (due to the significant benefits offered by the GaAs material platform), many of the approaches which have been used to reach this remain encumbered by fundamental issues that need to be addressed in order to overcome the limitations of existing 1- to 2-μm semiconductor laser technologies. These issues, which are associated primarily with Auger recombination and IVBA, account for the low wall-plug efficiencies and poor high-temperature performance of existing GaAs- and InP-based devices. Ultimately, addressing these issues requires a detailed consideration of the band structure of the materials forming the active region. Such considerations have led to a recent surge of interest in dilute bismide alloys—containing bismuth (Bi)—and in particular the GaBiAs alloy, which offers the potential to engineer the band structure to directly mitigate issues associated with Auger recombination and IVBA in GaAs-based 1.55-μm laser structures. An introduction to the theory and simulation of dilute bismide alloys, a class of highly mismatched III–V semiconductor alloys which can, in several ways, be considered naturally complementary to dilute nitride alloys, is provided in Chapter 10.

9.4.2 Semiconductor Optical Amplifiers, Electro-Absorption Modulators, and Avalanche Photodiodes

Based on the strong interest in dilute nitride alloys for applications in semiconductor lasers, several of the measured properties of InGaNAs(Sb) devices have suggested that there is significant potential to use dilute nitrides to deliver new capabilities and improved performance in related devices such as 1.3- and 1.55-μm SOAs, EAMs, and APDs. In this section, we briefly review research on using dilute nitride alloys to develop each of these classes of devices, highlighting the principles motivating this research in terms of the potential benefits of dilute nitride alloys, and summarizing the progress that has been made toward developing devices having improved characteristics compared to conventional InP-based technologies.

In addition to the benefits of developing 1.3- and 1.55-μm optoelectronic devices on GaAs as opposed to InP, for applications in SOAs and EAMs one of the most attractive properties of InGaNAs(Sb) alloys and heterostructures is their reduced sensitivity to temperature. Incorporation of N, which acts as an isovalent impurity, is known to lead to a reduction in the temperature sensitivity of the band gap of bulk materials [91] and QWs [92] compared to that of N-free materials. This reduced temperature dependence of the band gap has its origins in the hybridization of the extended states of the host matrix CB edge with N-related localized states (cf. Section 9.2)—the properties of the latter are largely independent of temperature, so that N-induced hybridization leads to InGaNAs(Sb) alloy CB edge states containing an admixture of N-related localized character and hence having reduced sensitivity to temperature [93]. This property is particularly appealing for the development of SOAs and EAMs: the temperature dependence of the operation wavelength in conventional InP-based devices, as well as the degree of degradation in performance with increasing temperature, mandates the use of external cooling equipment to maintain operational stability which dramatically increases the power consumption of the overall system. The large CB offsets present in dilute nitride QWs compared to those in conventional InGaAsP/InP or AlInGaAs/InP structures lead to reduced thermal leakage of electrons, which can be expected to contribute further to improving high-temperature device performance. The ability to (1) prevent degradation of the optical absorption/amplification at high temperature, and (2) reduce temperature-dependent drift of the operational wavelength make InGaNAs(Sb) structures ideal candidates for the development of GaAs-based 1.3- and 1.55-μm SOAs and EAMs. Such devices additionally offer the potential to achieve uncooled operation, and hence, bring about significant energy savings.

From the perspective of SOA development, further strong motivations for the exploration of dilute nitride alloys include the presence of broad-band optical gain and a significant degree of polarization tunability in InGaNAs(Sb) QWs, both of which can be expected to lead to SOAs displaying enhanced performance [94,95]. These potential benefits of InGaNAs(Sb) alloys, as well as the possibility to grow long-wavelength SOAs on GaAs, have stimulated the development of 1.3-μm InGaNAs SOAs and VCSOAs, as well as 1.55-μm InGaNAsSb SOAs. Initial investigations by Hashimoto et al. [96] on a 1.3-μm traveling-wave InGaNAs/GaAs SOA confirmed the potential of dilute nitride SOAs, by demonstrating significantly enhanced temperature stability compared to InP-based devices. Further studies have developed this device concept further, leading to the demonstration of multiwavelength amplification at 1.3 μm [97], as well as high optical gain and favorable gain dynamics. Following the confirmation of the reduced sensitivity of dilute nitride SOAs to temperature, uncooled operation of a 1.3-μm device at 10 Gbit s^{-1} has been demonstrated [98]. Continued refinement of InGaNAs SOA design and fabrication has since led to further improvements in performance. Recently, Fitsios et al. [99,100] demonstrated error-free operation at 10 Gbit s^{-1} of a 1.3-μm InGaNAs/GaAs SOA displaying high gain (~28 dB), fast gain recovery (~100 ps), and temperature-stable gain (between 20°C and 50°C). The high performance of this device, combined with its reduced cost compared to incumbent 1.3-μm InP-based SOAs, suggests that existing dilute nitride SOAs are already suitable for deployment in, e.g. metro and optical access networks.

Substantial progress has also been made in the development of 1.3-μm InGaNAs/GaAs VCSOAs, from initial theoretical analysis [101] through to investigations of optically pumped devices [102–104] and electrically pumped "HELLISH" devices [105,106], to the recent realization of room temperature gain in an electrically pumped device [107]. These advances clearly illustrate the progress that has been made in developing the materials growth and device engineering of GaAs-based dilute nitride SOAs. The high performance of these devices and potential for on-chip integration further demonstrate that dilute nitride VCSOAs have significant potential for deployment in applications such as optical switches and interconnects. While the main focus of the research effort on GaAs-based dilute nitride SOAs has centered on 1.3-μm devices incorporating InGaNAs QWs, in line with research on dilute nitride QW lasers, there has also been some effort dedicated to using InGaNAsSb QWs to facilitate the development of 1.55-μm GaAs-based SOAs. Dilute nitride SOAs operating at 1.55-μm have also displayed fast gain recovery [108,109] and high temperature stability [109,110]. While these 1.55-μm devices are not as developed as their 1.3-μm counterparts, existing results at 1.55 μm highlight that there is scope for further development, and hence significant potential to develop high-performance GaAs-based (VC)SOAs in this technologically important wavelength range.

By comparison to the ongoing research effort on SOAs, there has been very little development of dilute nitride EAMs. Initial electric-field-dependent measurements of the optical absorption both in 1.3- and 1.55-μm InGaNAs(Sb) QW structures have demonstrated a strong quantum-confined Stark effect (QCSE) [111], suggesting that there is broad scope for the development of GaAs-based EAMs at optical communication wavelengths. Detailed theoretical analysis of a prototypical 1.3-μm InGaNAs structure suggests that dilute nitride EAMs operating at 1.3 μm should offer extinction ratios slightly improved over those expected for N-free devices [112]. These favorable characteristics result primarily from the CB offsets in InGaNAs(Sb) QWs, which are substantially larger than those available in the QWs employed in conventional InP-based EAMs. These large band offsets offer improved carrier confinement, and hence an enhancement of the QCSE. On this basis, reduced thermal leakage of electrons from dilute nitride QWs at higher temperatures is expected to lead to significant temperature stability of the QCSE in such structures. This has been confirmed by experimental measurements undertaken on a series of multi-QW 1.3-μm InGaNAs/GaAs EAMs, in which no appreciable degradation in the QCSE was measured between 25°C and 100°C [113]. These measurements additionally confirmed improved overall high-temperature performance of dilute nitride EAMs compared to InP-based devices, including the expected reduced temperature dependence of the absorption edge wavelength on temperature. Overall, initial studies suggest that dilute nitride alloys are a promising material system for the development of GaAs-based EAMs and, in addition to the benefits associated with growth on GaAs, dilute nitride EAMs should offer high-temperature performance which exceeds that of existing 1.3- and 1.55-μm InP-based devices.

Additional research interest in InGaNAs alloys has been motivated by their potential for applications in APDs. Of key importance in determining the performance of an APD are the electron and hole impact ionization rates[†] α and β. In Ref. [114], Adams identified the potential to use InGaNAs alloys to overcome the limitations associated with the nature of impact ionization in conventional III–V semiconductors by providing a low-noise medium for hole multiplication in APDs. As we discuss in detail in Section 9.5, the strong impact of N incorporation on the CB structure leads to a significant degradation of the electron mobility in InGaNAs alloys. Adams proposed that this N-induced limitation of high-field electron transport, combined with the fact that N incorporation leaves the VB relatively unchanged, should produce a bottleneck for electrons by limiting their ability to accelerate to sufficiently high energies to initiate impact ionization. Hence, incorporation of N can be expected to dramatically reduce α while leaving β relatively unchanged, and thereby leading to low-noise hole multiplication in a device having $k \gg 1$. Such devices are also expected to have reduced band-to-band tuneling due in part to the tendency of N incorporation to introduce intervalley mixing of CB states, which should lead to improved performance in long-wavelength devices [114].

While a favorable ratio $k \ll 1$ of the impact ionization rates can be obtained in Si-based APDs, the operation of these low-noise electron multiplying devices is limited to wavelengths <1.1 μm by the magnitude of the Si band gap. The reduction in band gap brought about by N incorporation in InGaNAs alloys then further adds the potential to realize low-noise APDs operating at 1.3 and 1.55 μm. This has the potential to overcome key limitations associated with existing InP-based technologies in this wavelength range (which must typically incorporate separate absorbing and carrier multiplying regions) by offering improved performance in devices having a single gain medium which offers efficient long-wavelength absorption and low-noise hole multiplication. When combined with the benefits of fabrication on GaAs substrates, including robust fabrication and reduced cost, the development of GaAs-based 1.3- and 1.55-μm APDs represents a promising opportunity to overcome fundamental issues associated with the InP-based devices deployed in existing optical communication networks. However, initial experimental analysis of the impact ionization rates in a series of InGaNAs diode structures has demonstrated that while N incorporation does cause a decrease in α leading to an increase of k by a factor of approximately four, this is accompanied by a reduction of both α and β by close to two orders of magnitude [115]. This strong suppression of the impact ionization rates resulting from N incorporation suggests that dilute nitride alloys may not be suitable for use as a multiplication medium in telecom-wavelength APDs, but may instead have potential for deployment in, for example, GaAs-based heterojunction bipolar transistors. Further investigations are required to ascertain whether or not this is the case, and to definitively demonstrate or rule out the promise of dilute nitrides for applications in long-wavelength APDs.

[†] The nature of avalanche multiplication in an APD depends on the relative values of the electron and hole impact ionization rates α and β, characterized by the ratio $k = \frac{\beta}{\alpha}$. For a low-noise APD it is important that α and β differ significantly in magnitude in order to allow for large, stable electrical gain. Otherwise, in the case where $k \approx 1$—that is, where the electron and hole multiplication rates are roughly equal—the impact ionization process leads to noisy and uncontrollable avalanche breakdown. The latter of these two situations is the case in most III–V semiconductors, in which α and β typically differ by no more than a factor of approximately two in magnitude.

9.4.3 Monolithic Integration of III–V Lasers on Silicon

The ongoing exponential growth of the internet, mobile/cloud computing, and "big data" has created strong demand for new hardware architectures to deliver both increased bandwidth and decreased energy consumption. Key to the continued development of data processing in computing and communications is the need to prevent the formation of bottlenecks by increasing the available bandwidth from chip to server level. Electrical data transmission reaches practical limits at a bandwidth-distance product of approximately 100 Gb m s^{-1} due to the intrinsic limitations of electrical interconnects. A promising approach to overcome these issues is direct integration of optoelectronic devices on silicon (Si) to enable the development of high bandwidth optical interconnects for on-chip and chip-to-chip data transfer. This photonics-based approach has significant advantages over traditional electronics: optical data transfer has been heavily developed for optical communications, leading to interconnects offering significantly improved bandwidth compared to their electronic counterparts, over distances ranging from as low as several meters to long-distance links of the order of thousands of kilometers in length. In addition to offering higher bandwidth, optoelectronic integration on Si has the additional benefit that optical interconnects do not suffer from the losses due to resistive heating that play a significant role in limiting the performance of electrical interconnects.

Compressively strained SiGe/Si alloys and heterostructures have been used to develop many of the components required to realize practical integrated optoelectronic architectures, including photodetectors, filters, waveguides, and switches. However, Si-compatible group-IV materials and heterostructures are generally limited by their inability to efficiently emit/absorb photons—due to their indirect band gaps—meaning that while some progress has been made there are, as yet, no efficient, electrically pumped Si-compatible light sources available to facilitate the widespread adoption of Si-based optoelectronic integrated circuits. Given these limitations, as well as the comparably high optical emission efficiencies offered by direct-gap III–V materials, there has been a growth of interest in integrating III–V optoelectronic devices on Si to deliver monolithically integrated emitters and absorbers. A review of the complete range of approaches being investigated to integrate III–V light sources on Si is beyond the scope of this chapter—we focus here on the potential of dilute nitride alloys to realize electrically pumped lasers sources on Si and refer the reader to Ref. [116] for a recent overview of alternative approaches.

The primary requirements for Si-compatible III–V materials are that the III–V material has (1) a lattice constant close to that of Si, to facilitate the growth of high-quality epitaxial layers and heterostructures, and (2) a direct band gap, to facilitate efficient optical emission/absorption. Of the III–V semiconductor compounds, GaP has the closest lattice constant to Si. This suggests that GaP offers a suitable starting platform for the development of Si-compatible optoelectronic devices. However, GaP also has an indirect band gap, placing severe limitations on its potential for the development of efficient light-emitting devices. Incorporation of a large fraction of As in GaAsP alloys is required to bring about a direct band gap, but this then leads to a large increase in lattice constant which limits the ability to grow direct-gap GaAsP alloys directly on Si. These issues can be overcome by the incorporation of N, which simultaneously lowers the lattice constant and the energy of the Γ-point CB edge. This therefore raises the possibility of using N incorporation to produce direct-gap GaNAsP alloys which can be grown directly on Si with acceptable levels of compressive strain. Analysis of the GaNAsP band structure suggests that As compositions >70% and N compositions ~4% are required to produce suitable direct-gap alloys which are compatible with growth on Si substrates [117].

Direct growth of GaNAsP alloys on Si is a challenging task for epitaxial growth. For example, in addition to the difficulties associated with the growth of high-quality dilute nitride alloys, the mismatch in thermal expansion coefficients between Si and GaNAsP can lead to the development of strain-induced dislocations, and even cracking, during epitaxial growth. Nonetheless, significant progress has been made in developing GaNAsP QW lasers on Si, including the development of novel boron (B) containing alloys to serve as suitable strain compensating barrier layers and mitigate the effects of the aforementioned thermal expansion issues during growth [118]. These developments led to the demonstration in 2011 of the first monolithically integrated III–V laser on Si [119]. The initial single QW device of Ref. [119] demonstrated electrically pumped lasing at a wavelength of 861 nm up to $T = 165$ K, with a threshold current density $J_{th} \sim 1.6$ kA cm^{-2}. Detailed experimental analysis, similar to that described in Section 9.3.3, highlights that GaNAsP/Si lasers offer broadly comparable performance to equivalent test devices grown on GaP [120,121]. These Si-based lasers were shown to suffer from increased defect-related recombination at low temperature compared to the GaP-based devices [116]. However, taking into account the fact that (1) electrically pumped lasing has already been demonstrated using this novel class of semiconductor lasers, and (2) that significant development that has occurred in the epitaxial growth of related dilute nitride alloys and heterostructures suggests that direct-gap GaNAsP laser structures grown directly on Si represent a promising approach to the development of monolithically integrated Si-based light sources suitable for applications in optoelectronic integration. Research on this novel device concept is ongoing, with the aim of realizing room temperature electrically pumped lasing from a monolithically integrated III–V device on Si.

9.5 Theory of the Density of States and Carrier Transport in Dilute Nitride Alloys

Having focused so far in this chapter on the electronic and optical properties of dilute nitride alloys, we turn our attention in this section to the impact of N incorporation on the carrier transport in GaN$_x$As$_{1-x}$. Since N incorporation primarily affects the CB structure while leaving the VB structure virtually unchanged, transport of VB holes in GaN$_x$As$_{1-x}$ is akin to that in GaAs while N impacts extremely strongly on the transport of CB electrons. Carrier transport plays a key role in the operation in several of the classes of dilute nitride–based devices described in Section 9.4, including in particular APDs and multijunction solar cells. As such, developing a detailed understanding of the impact of N incorporation on the carrier transport in GaN$_x$As$_{1-x}$ and related alloys is crucial to facilitate the design and optimization of a range of novel optoelectronic and photovoltaic devices. Investigation of electron transport in dilute nitride alloys has proven to be particularly enlightening from a theoretical perspective, as attempts to describe, e.g., the dramatic reductions in electron mobility observed in experimental measurements, have stimulated detailed considerations of the impact of N incorporation on the CB structure.

The simple BAC model of Section 9.2.2—which proved to be extremely useful and surprisingly accurate for the description of the electronic and optical properties of InGaNAs semiconductor lasers (cf. Section 9.3)—is insufficient to analyze the details of the CB structure and DOS at energies close to N-related localized states. In particular, the unphysical nature of the CB DOS predicted by the BAC model has significant implications for calculations of the transport properties in GaN$_x$As$_{1-x}$ and related alloys. In order to lay the foundations for a detailed understanding of the carrier transport in dilute nitride alloys, we therefore begin our discussion in Section 9.5.1 by reviewing approaches to calculating the CB structure and DOS beyond the BAC model. We discuss the development of Green's function techniques to obtain solutions of the AIM, and review that a self-consistent approach to calculating the AIM Green's function can be used to produce a description of the DOS which is in quantitative agreement with the results of large-scale supercell calculations. We further outline how this approach has been combined with the LCINS description of the electronic structure (cf. Section 9.2.3) to provide significant insight into the impact of N on the nature of the CB dispersion and DOS in realistic, disordered GaN$_x$As$_{1-x}$ alloys.

In Section 9.5.2 we review the theory of electron transport in GaN$_x$As$_{1-x}$ alloys. We begin with simple analytical considerations, which demonstrate that N incorporation places intrinsic limitations on the electron mobility. This is followed by a discussion of resonant scattering of electrons by N-related localized states, highlighting the important role played by N–N pairs and larger clusters of N atoms in determining the transport properties. We review refinements in the theory of carrier transport in dilute nitride alloys—encompassing models based on the Boltzmann transport equation (BTE) as well as Monte Carlo techniques—and show how these approaches (in conjunction with the detailed description of the CB structure described in Sections 9.2 and 9.5.1) have facilitated detailed analysis of electron transport in GaN$_x$As$_{1-x}$, and have provided quantitative agreement with experimental measurements.

9.5.1 The Anderson Impurity Model: Self-Consistent Green's Function Method

That the BAC model provides an unphysical description of the CB structure and DOS can be seen clearly by considering the BAC band structure described in Section 9.2. As the wave vector k increases, the energy $E_-(k)$ of the lower-energy band increases asymptotically toward the N-resonant state energy E_N (cf. Equation 9.2). As a result, the DOS calculated using the BAC model increases strongly with increasing energy above the alloy CB edge $E_-(0)$, leading to a singularity in the DOS at energy $E = E_N$. This clearly unphysical result is associated with the fact that the BAC model intrinsically miscounts the number of N-related states per unit volume in an (assumed infinite) crystal [122,123]. Another problematic feature of the BAC model from the perspective of transport calculations is that it predicts the existence of a gap in the DOS in the energy range $E_N < E < E_+(0)$, even at ultra-dilute N compositions. The presence of such a gap in the DOS is counter to the results of experimental measurements carried out on GaN$_x$As$_{1-x}$ materials, which indicate that the CB DOS is continuous and gapless at energies close to E_N [124,125].

These features of the BAC model, which were not an issue in the calculation of the properties of semiconductor lasers (due to the fact that for analysis of the optical properties we are interested primarily in carriers lying close in energy to the CB and VB edges), have significant implications for calculations of the carrier transport in GaN$_x$As$_{1-x}$ and related alloys. For example, the singularity in the DOS at $E = E_N$ is associated with a breakdown in **k**-selection, suggesting that carriers can acquire extremely large crystal momenta without increasing their energy. Additionally, the presence of a gap in the DOS suggests that electrons in GaN$_x$As$_{1-x}$ are not capable of accelerating out of the lower E_- band, even under the influence of large applied electric fields. Given these shortcomings of the BAC model, it is clear that a quantitative understanding of carrier scattering and transport, particularly under intermediate and high field conditions, must begin with a detailed analysis of the nature of the CB structure and DOS. Such an analysis has been carried out by several authors, who used Green's function techniques to solve the AIM [122,123,126–129]. We summarize these results here and demonstrate by comparison with the results of large-scale supercell diagonalizations of the AIM that an accurate description of the CB structure and DOS, particularly at energies close to N-related localized states, requires a self-consistent approach.

The AIM for dilute nitride alloys describes a set of N-related localized states $|j\rangle$ interacting with the extended CB states $|\mathbf{k}\rangle$ of the unperturbed host matrix semiconductor via the Hamiltonian $\hat{H} = \hat{H}_0 + \hat{V}$. Here, $\hat{H}_0$ is a sum of two terms, describing the energies of the extended and localized states [126,129,130]

$$\hat{H}_0 = \sum_{\mathbf{k}} E_{\mathbf{k}} |\mathbf{k}\rangle\langle\mathbf{k}| + \sum_{j} E_j |j\rangle\langle j|, \tag{9.21}$$

where $E_{\mathbf{k}}$ and E_j are, respectively, the energy of the unperturbed host matrix CB at wave vector $\mathbf{k}$, and of the N-related localized state $|j\rangle$. $\hat{V}$ describes the interaction between the extended and localized states [126,129,130]

$$\hat{V} = \frac{1}{\sqrt{N_c}} \sum_{\mathbf{k},j} \left(V_{\mathbf{k},j}\, e^{i\mathbf{k}\cdot\mathbf{r}_j} |\mathbf{k}\rangle\langle j| + V_{\mathbf{k},j}^*\, e^{-i\mathbf{k}\cdot\mathbf{r}_j} |j\rangle\langle\mathbf{k}| \right), \tag{9.22}$$

where $V_{\mathbf{k},j}$ is a matrix element describing the hybridization of the extended and localized states, $\mathbf{r}_j$ is the position of the substitutional N impurity associated with the localized state $|j\rangle$, and N_c is the total number of primitive unit cells in the system. It is generally assumed that the hybridization parameters $V_{\mathbf{k},j}$ depend only weakly on the wave vector $\mathbf{k}$, so that we can reasonably approximate them as being independent of **k**—that is, $V_{\mathbf{k},j} \approx V_j = \frac{\beta_j}{\sqrt{N_c}}$ [126].

In the case that an N-related localized state $|j\rangle$ lies energetically within the host matrix CB, its interaction with an extended host matrix CB state $|\mathbf{k}\rangle$ results in the creation of a quasi-localized resonance having a finite lifetime [131]. Mathematically, this N-derived, quasi-localized alloy state can be described in terms of a complex-valued energy broadening ΔE_j, corresponding to an eigenstate time dependence described by the phase factor $\exp(\frac{i(E_j+\Delta E_j)t}{\hbar}) = \exp(\frac{i(E_j+\text{Re}\{\Delta E_j\})t}{\hbar}) \times \exp(-\frac{\text{Im}\{\Delta E_j\}t}{\hbar})$. From this, we see that the real and imaginary parts of the complex energy broadening ΔE_j respectively describe an energy shift and homogeneous broadening, with the lifetime of the N-derived quasilocalized resonance given by $\tau_j \sim \frac{\hbar}{\text{Im}\{\Delta E_j\}}$ [128].

Exact extended energy eigenstates of the AIM Hamiltonian $\hat{H}$ can be found via the poles of the Green's function [126]

$$G_{\mathbf{kk}}(E) = \left(E - E_{\mathbf{k}} - \frac{1}{N_c} \sum_j \frac{|V_j|^2}{E - E_j - \Delta E_j(E)} \right)^{-1}, \qquad (9.23)$$

where the complex energy-dependent broadening associated with each localized state $|j\rangle$ is given by

$$\Delta E_j(E) = \frac{|V_j|^2}{N_c} \sum_{\mathbf{k}} G_{\mathbf{kk}}(E), \qquad (9.24)$$

the real and imaginary parts of which then describe, respectively, the energy shift and broadening of the N-related localized states, brought about by their interaction with the extended host matrix CB states $|\mathbf{k}\rangle$ with which they are resonant [129]. The alloy CB DOS projected to extended (Bloch) states, $D(E)$, can be calculated directly from the imaginary part of the extended state Green's function, Equation 9.23, as [122,126]

$$D(E) = -\frac{1}{\pi} \sum_{\mathbf{k}} \text{Im}\{G_{\mathbf{kk}}(E)\}. \qquad (9.25)$$

In practice, Equations 9.24 and 9.25 are evaluated by converting the sum into an integral over the continuous wave vector $\mathbf{k}$ so that, e.g. the DOS is given by

$$D(E) = -\frac{1}{\pi} \text{Im}\left\{ \int_0^\infty G_{\mathbf{kk}}(E) D_0(E_{\mathbf{k}}) \mathrm{d}E_{\mathbf{k}} \right\}, \qquad (9.26)$$

where $D_0(E) = \frac{(2m_0^*)^{2/3}\sqrt{E}}{4\pi^2\hbar^3}$ is the DOS of the unperturbed host matrix CB, (with the zero of energy taken at the host matrix CB edge) [123,129].

In general, Equations 9.23 and 9.24 do not admit an analytical solution due to the energy dependence of the complex localized state broadening $\Delta E_j(E)$. For this reason, several authors have introduced approximations for $\Delta E_j(E)$ in order to produce analytically tractable solutions to Equation 9.23. Typically, this consists of assuming $\Delta E_j(E)$ to be purely complex and independent of energy: $\Delta E_j(E) \approx i\Delta_j$ so that it described uniform (energy-independent) localized state broadening. For dilute nitride alloys this approximation was originally introduced by Wu et al., who used it to solve the AIM in the coherent potential approximation for a BAC-like two-band model in which all of the N-related localized states are assigned the same energy E_N [122]. The real and imaginary parts of the resultant complex energy eigenstates were then interpreted, respectively, in terms of a BAC-like band dispersion and an N-induced energy broadening of the band dispersion at each wave vector $\mathbf{k}$. This approach was further developed by Vaughan and Ridley, who derived the AIM Green's function in the form given by Equation 9.23 and extended the treatment of the band structure to a three-band model including localized states associated with N–N pairs in addition to isolated N impurities [126]. A similar approach has also been developed by Vogiatzis and Rorison, who derived and applied the Matsubara Green's function to calculate the complex energy band structure and localized state lifetimes of the AIM eigenstates for the two- and three-band models both in bulk alloys and in (In)GaN$_x$As$_{1-x}$ QW heterostructures [127,128].

As we describe below, the inclusion of a purely imaginary, energy-independent energy broadening removes the issues associated with the BAC description of the DOS: the singularity at the N-resonant state energy E_N is removed, and the gap in the DOS between E_N and $E_+(0)$ is closed. However, while the closing of the gap in the DOS brought about by taking into account complex energy broadening of the N-related localized states in this manner is qualitatively in agreement with the results of experimental measurements, Seifikar et al. showed by comparison to large-scale supercell calculations that this approximation to the AIM Green's function introduces a number of errors into the description of the CB structure and DOS at energies close to localized states [123]. In order to overcome these issues, it is necessary to obtain a direct numerical solution of Equations 9.23 and 9.24 in order to fully determine the extended state Green's function, by computing the full energy-dependent broadening $\Delta E_j(E)$. These issues were identified by Seifikar et al. [123,129], who demonstrated the need for and developed a self-consistent approach to solving Equations 9.23 and 9.24. The need for a self-consistent Green's function (SCGF) approach was originally identified by Juaho and Wilkins, who studied the general problem of resonant scattering impurities interacting with the extended eigenstates of a parabolic band, a generic version of the problem encountered in dilute nitride and related alloys [132].

The SCGF calculation proceeds as follows: first, an initial guess is generated for the broadening function $\Delta E_j(E)$ using perturbation theory [123]. Second, this broadening is used to compute the energy-dependent extended state Green's function $G_{kk}(E)$ according to Equation 9.23. Third, this Green's function is used to recompute the energy-dependent broadening $\Delta E_j(E)$ according to Equation 9.24. Fourth, this new energy-dependent broadening is used to recompute $G_{kk}(E)$. The third and fourth steps in this procedure are then iterated until $\Delta E_j(E)$ has converged throughout the energy range of interest. The converged complex energy broadening $\Delta E_j(E)$ can then be used to calculate the CB structure and DOS via Equations 9.23 and 9.24.

Having described the motivation for and implementation of the SCGF method, we now summarize the results of applying this approach to study the impact of N incorporation in GaN$_x$As$_{1-x}$ alloys for two models of the CB structure: (1) a BAC-like two-band model in which the N-related localized states are all assumed to have the same energy E_N, and (2) a multiband model based upon the LCINS approach outlined in Section 9.2.3, in which a full distribution of N-related localized states—associated with not only isolated N impurities, but also with N–N pairs and larger clusters of N atoms—are included in the sums in Equations 9.23 and 9.24. We note that the results we summarize here for the SCGF method correspond roughly to the strong scattering regime investigated by Juaho and Wilkins [132], highlighting that substitutional N impurities have a strongly perturbative effect on the CB structure.

We begin by analyzing the CB structure and DOS calculated using the SCGF method for a two-band model of the GaN$_x$As$_{1-x}$ CB. In the two-band model, all N-related localized states are assumed to have energy E_N and the composition dependence of the interaction matrix element is assumed to be of the form $V_j = \beta\sqrt{x}$. As a result, Equation 9.23 can be written for the two-band model as [129]

$$G_{kk}(E) = \left(E - E_k - \frac{\beta^2 x}{E - E_N - \Delta E_N(E)} \right)^{-1}, \tag{9.27}$$

where the complex energy broadening of the N-related localized states is given by

$$\Delta E_N(E) = \beta^2 x \sum_k G_{kk}(E). \tag{9.28}$$

The results of the self-consistent calculation of $G_{kk}(E)$ and $\Delta E_N(E)$ using Equations 9.27 and 9.28 are shown in Figures 9.9a and 9.9b for GaN$_{0.002}$As$_{0.998}$ ($x = 0.2\%$; solid lines). The zero of energy is taken at the CB edge of the unperturbed GaAs host matrix. The initial, energy-independent estimates for the real and imaginary parts of $\Delta E_N(E)$, generated using second-order perturbation theory as outlined in Ref. [123], are

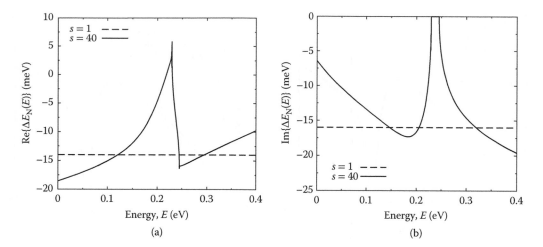

FIGURE 9.9 (a) Real and (b) imaginary parts $\mathrm{Re}\{\Delta E_N(E)\}$ and $\mathrm{Im}\{\Delta E_N(E)\}$ of the complex energy-dependent N-localized state broadening $\Delta E_N(E)$, calculated using the SCGF method for a two-band model of the $GaN_{0.002}As_{0.998}$ ($x = 0.2\%$) CB structure in which all N-related localized states are taken to have the same energy (cf. Equations 9.27 and 9.28). Dashed lines show the initial energy-independent values of $\mathrm{Re}\{\Delta E_N(E)\}$ and $\mathrm{Im}\{\Delta E_N(E)\}$ at the first iteration $s = 1$ of the SCGF calculation, computed using perturbation theory. Solid lines show the converged $\mathrm{Re}\{\Delta E_N(E)\}$ and $\mathrm{Im}\{\Delta E_N(E)\}$ after $s = 40$ iterations. (Adapted from M. Seifikar et al., *J. Phys. Condens. Matter*, 26, 365502, 2014.)

shown in each case using a dashed line. It is found that approximately 40 iterations ($s = 40$) are required to ensure that $\Delta E_N(E)$ is fully converged. However, we note that convergence is much more rapid at energies away from the N-resonant state energy E_N, with as few as two iterations required to ensure convergence. Comparing the converged real and imaginary parts of $\Delta E_N(E)$ to the constant initial values in Figure 9.9a and 9.9b highlights the strong energy dependence of the N-localized state resonant energy and broadening, which are neglected in non-self-consistent approaches. For example, the real part $\mathrm{Re}\{\Delta E_N(E)\}$ of $\Delta E_N(E)$ has a value of approximately -20 meV at the alloy band edge, from which it increases monotonically with increasing energy, changing sign and reaching a value of approximately $+5$ meV at $E = E_N$. $\mathrm{Re}\{\Delta E_N(E)\}$, then decreases strongly in the energy range $E_N < E < E_+(0)$, again changing sign, before once again increasing monotonically with energy for $E > E_+(0)$. It follows from Equations 9.25 and 9.28 that the imaginary part $\mathrm{Im}\{\Delta E_N(E)\}$ of $\Delta E_N(E)$ is proportional to the DOS. We note that the converged self-consistent calculation indicates that $\mathrm{Im}\{\Delta E_N(E)\}$, and hence the DOS, vanishes in the energy range $E_N < E < E_+(0)$. This confirms, as described above, that the DOS calculated using the SCGF method for a two-band model of the GaN_xAs_{1-x} CB contains a gap at energies just above the N-resonant state energy [129].

The opening of this gap in the DOS can be observed in Figure 9.10a, which shows the DOS calculated using the converged $\Delta E_N(E)$ of Equation 9.28 in Equation 9.27 for (1) the first three iterations of the self-consistent calculation ($s = 1, 2,$ and 3), and (2) the fortieth iteration, $s = 40$, by which $\Delta E_N(E)$ has fully converged. The first iteration for the DOS (dashed black line) corresponds to the standard, nonself-consistent Green's function approach [122,126,128], with the broadening calculated using perturbation theory [123]. For reference, we have also included the DOS calculated using the BAC model (solid gray line), which contains the aforementioned singularity at the N-resonant state energy $E_N = 0.23$ eV. For $s = 1$ there is no singularity in the DOS at E_N and no gap in the DOS above E_N. As the self-consistent calculation progresses, the rapid opening up of a gap in the DOS is observed after two ($s = 2$) iterations (depicted using a dash-dotted black line). Comparing these calculations highlights the role of the energy dependence of the complex localized state broadening $\Delta E_N(E)$: the converged DOS is well described by the $s = 1$ (energy independent) broadening at energies away from E_N, while iterating to obtain self-consistency results in large changes to the calculated DOS at energies close to E_N.

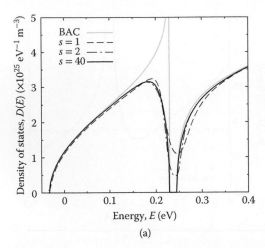 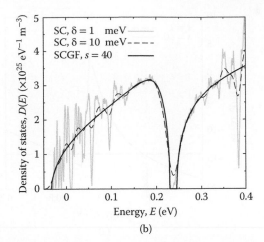

(a)   (b)

FIGURE 9.10   (a) DOS calculated for $GaN_{0.002}As_{0.998}$ ($x = 0.2\%$) using the SCGF method for a two-band model of the CB structure in which all N-related localized states are taken to have the same energy (cf. Equations 9.27 and 9.28). Dashed and dash-dotted black lines show the calculated DOS after $s = 1$ and 2 iterations of the SCGF calculation. The solid black line shows the converged DOS, obtained after $s = 40$ iterations of the SCGF calculation. The solid gray line shows the DOS calculated using the simple two-band BAC model of the CB structure (cf. Section 9.2.2), which is singular at the N-localized state energy $E_N = 0.23$ eV. The opening up of a gap in the DOS above E_N is observed as the SCGF calculation converges, which is not present in the standard, nonself-consistent ($s = 1$) Green's function calculation. (b) Converged DOS calculated using the SCGF model (solid black line, as in (a)), compared to the DOS calculated via a large-scale supercell (SC) diagonalization of the AIM Hamiltonian for the same two-band model of the CB structure. Solid gray and dashed black lines show the SC DOS calculated by assuming in the numerical calculations that the contribution of each $GaN_{0.002}As_{0.998}$ CB state to the DOS is described by a Gaussian lineshape having a standard deviation $\delta = 1$ and 10 meV, respectively. The zero of energy is taken in each case at the unperturbed GaAs CB edge. (Adapted from M. Seifikar et al., *J. Phys. Condens. Matter*, 26, 365502, 2014.)

In order to investigate the importance of including self-consistent energy-dependent broadening to describe the DOS close in energy to N-related localized states, Seifikar et al. performed a direct diagonalization of the AIM Hamiltonian using a large-scale supercell model [123]. Figure 9.10b compares the converged DOS calculated using the SCGF method (solid black line) to that calculated using the supercell method for $GaN_{0.002}As_{0.998}$. The DOS calculated using the supercell approach is shown for Gaussian broadenings $\delta = 1$ and 10 meV associated with each supercell eigenstate (solid gray and dashed black lines, respectively). It was demonstrated in Ref. [123] that assuming a large > 5 meV broadening for the calculation of the supercell DOS leads to results which are in good agreement with the standard, non–self-consistent Green's function method. However, this large broadening obscures the details that are visible in the supercell DOS calculated using a smaller 1 meV broadening, in which case there is a clear indication of a gap in the supercell DOS at energies above E_N. It can be further seen in this case that the DOS calculated using the SCGF and supercell models are in excellent agreement at energies close to E_N. Further detailed comparison of the SCGF and supercell models verifies that the SCGF approach describes the DOS projected to individual GaAs **k** states at energies close to E_N with a high degree of accuracy, elucidating detailed features of the CB structure which are not captured by the standard Green's function approach [129].

Having demonstrated that the SCGF model is capable of fully capturing the physics of the AIM, we now turn our attention to the presence of the gap in the DOS at energies just above E_N. Both the SCGF and supercell models confirm that the presence of this gap is an intrinsic feature of the two-band description of the GaN_xAs_{1-x} CB structure, suggesting that one must go beyond this simple approach in order to obtain a description of the DOS which is in agreement with experiment—i.e. for which there is no gap in the DOS. Seifikar et al. extended the SCGF model to take into account the full distribution of N-related states present in a disordered GaN_xAs_{1-x} alloy using the LCINS approach outlined in Section 9.2.3. In practice, this again consists of solving Equations 9.23 and 9.24 self-consistently, but where the sum in Equation 9.23 runs over all N-related states in the LCINS distribution, and where the broadening described by Equation 9.24 is evaluated directly for each individual N-related localized state $|j\rangle$.

The DOS calculated using this SCGF-LCINS approach is shown in Figure 9.11a (solid line) for $GaN_{0.0036}As_{0.9964}$ ($x = 0.36\%$), where it is compared to the results of a nonself-consistent LCINS Green's function calculation in which a purely imaginary energy-independent broadening $\Delta E_j = i\Delta_N$, with $\Delta_N = 1$ meV, is assumed for all N-related states (dashed line). We note that there are several qualitative differences between the DOS calculated using the SCGF-LCINS and two-band SCGF approaches [129]. First, and most importantly, we note that the inclusion of a full distribution of N-related localized states in the SCGF calculation leads to a closing of the gap in the DOS. This confirms that (1) the presence of a gap in the DOS in the converged SCGF calculation shown in Figure 9.10a (solid line) is an intrinsic property of simple two-band description of the CB structure, and (2) the lack of a gap in the measured DOS in real GaN_xAs_{1-x} materials [124,125] is associated with N-related alloy disorder, which plays a strong role in determining the details of the CB structure (cf. Section 9.2). Second, it is possible to directly identify features associated with distinct classes of N-related localized states in the DOS [129] and optical absorption spectra [133], on the basis of the understanding of these states developed through the LCINS theory. For example, the large dip in the DOS between 0.2 and 0.3 eV originates from hybridization of GaAs extended states with localized states associated with isolated N impurities, while the shoulder in the DOS lying just above the alloy CB edge in energy is associated with N–N pair-related localized states. The quantitative differences between the DOS calculated using these two approaches are significant at energies close to those of the N-related localized states associated with isolated N impurities and N–N pairs, once again highlighting

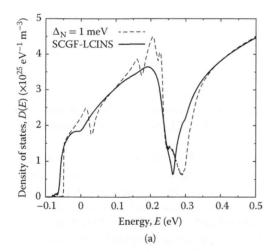 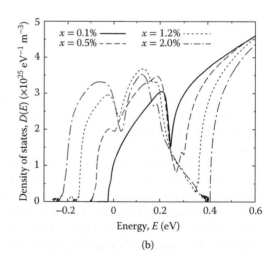

(a)   (b)

FIGURE 9.11 (a) DOS calculated for $GaN_{0.0036}As_{0.9964}$ ($x = 0.36\%$) using the standard, nonself-consistent Green's function method (dashed line)—assuming an energy-independent N-localized state broadening $\Delta E_j = i\Delta_N$ for all localized states, with $\Delta_N = 1$ meV—compared to that calculated using the SCGF method (solid line). In both cases, the CB structure is described by incorporating the full LCINS distribution of N-related localized states (cf. Section 9.2.3 and Equations 9.23 and 9.24). Significant quantitative differences exist between the DOS calculated using these approaches, particularly at energies close to N-related localized states, emphasizing the importance of the energy dependence of ΔE_j (see text). (b) DOS calculated using the full SCGF-LCINS model for GaN_xAs_{1-x} alloys having N compositions $x = 0.1\%$, 0.5%, 1.2%, and 2.0% (solid, dashed, dotted, and dash-dotted lines, respectively). The zero of energy is taken in each case at the unperturbed GaAs CB edge. (Adapted from M. Seifikar et al., *J. Phys. Condens. Matter*, 26, 365502, 2014.)

the importance of the energy-dependent complex localized state broadening in determining the details of the CB structure.

Figure 9.11b shows the DOS calculated using the SCGF-LCINS method for GaN_xAs_{1-x} alloys having N compositions $x = 0.1\%$, 0.5%, 1.2%, and 2.0% (shown using solid, dashed, dotted, and dash-dotted lines, respectively). Despite the detailed features present at each composition due to the effects of N-related alloy disorder on the CB structure, several general trends in the calculated evolution of the DOS with increasing N composition can be understood qualitatively in terms of the BAC description of the band structure. First, the reduction in energy of the alloy CB edge with increasing x is clearly apparent. Second, the dip in the DOS at energies above E_N increases in depth and width as x increases. This is generally consistent with the BAC model, in which the energy $E_-(0)$ $(E_+(0))$ of the lower (higher) energy BAC band at $k = 0$ decreases (increases) with increasing x. In disordered alloys this trend is clearly apparent as the energy at which the increase in the DOS above E_N occurs increases with increasing x, from approximately 0.23 eV in $GaN_{0.001}As_{0.999}$ to 0.41 eV in $GaN_{0.02}As_{0.98}$. This upward shift in energy of the higher-energy N-derived bands leads to an overall reduction of the DOS at energies above E_N, suggesting that a gap is present in the DOS of real, disordered GaN_xAs_{1-x} alloys for N compositions $x > 2\%$. This is to be contrasted with the results of the two-band calculations summarized above where, in the absence of alloy disorder, a gap in the DOS was predicted even at ultra-dilute N compositions $x \sim 0.1\%$. Features in the DOS which are similar to those associated with isolated N impurities are clearly visible at energies close to the unperturbed GaAs CB edge (0 eV), reflecting the impact of hybridization between the extended GaAs CB states and localized states associated with N–N pairs.

The SCGF method can also be used to calculate the broadened GaN_xAs_{1-x} CB dispersion described by the AIM. This is achieved by using the alloy CB eigenstates of the AIM Hamiltonian to compute the DOS projected to individual extended GaAs CB states $|\mathbf{k}\rangle$—i.e. by neglecting the integration (sum) over wave vectors in Equation 9.25 and instead computing the projected DOS separately at each wave vector $\mathbf{k}$ as $-\frac{1}{\pi}\text{Im}\{G_{\mathbf{k}\mathbf{k}}(E)\}$, using the converged complex broadening functions $\Delta E_j(E)$ [129]. The broadened CB structures computed for $GaN_{0.0036}As_{0.9964}$ ($x = 0.36\%$) in this manner are shown in Figure 9.12a and 9.12b. Figure 9.12a shows the CB structure calculated using the two-band model described by Equations 9.27 and 9.28, in which all N-related localized states $|j\rangle$ are assumed to have the same energy E_N. The SCGF CB structure calculated using the two-band model corresponds well to that calculated using the two-band BAC model [129] and demonstrates (1) the strong N-induced energy broadening of the CB states, (2) the breakdown in the extended nature of the states in the lowest-energy alloy CB with increasing $k = |\mathbf{k}|$, as evidenced by the strong reduction in the projected DOS at large wave vectors, and (3) the presence of a gap in the CB DOS—that is, a range of energies at and above E_N for which the projected DOS vanishes at all $\mathbf{k}$ (cf. Figure 9.10a).

Figure 9.12b shows the CB structure calculated using the full SCGF-LCINS method, which takes into account the full LCINS distribution of N-related localized states. Comparing Figure 9.12a and 9.12b we note two important qualitative differences between the two-band and LCINS descriptions of the CB structure. First, the strong impact of localized states associated with N–N pairs can be observed at energies ~ 50 meV above the alloy CB edge (cf. Figure 9.11a)—hybridization of GaAs extended states $|\mathbf{k}\rangle$ with N–N pair-related localized states leads to a strong breakdown of the CB dispersion at energies close to those of the latter states, as evidenced by a strong reduction in the projected DOS. Second, the strong broadening of the CB structure associated with N-related disorder in the LCINS model leads to a closing of the gap in the DOS that is present in the two-band model (cf. Figure 9.11a): the lower and upper CBs are then linked by an energy range across which the project DOS is nonzero, thereby providing a pathway for scattering of electrons to higher-energy CB states (which has important consequences for the electron transport in intermediate and high applied fields). We again note that the CB dispersion obtained from the BAC model provides a reasonable description of the main features of the CB structure, but emphasize that the SCGF method provides much deeper insight into the nature of the GaN_xAs_{1-x} CB eigenstates, particularly given the strong impact of N-related alloy disorder [129].

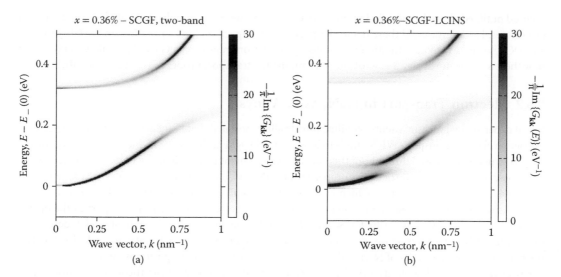

FIGURE 9.12 CB structure of GaN$_{0.0036}$As$_{0.9964}$ ($x = 0.36\%$) calculated using the SCGF method—by projecting the alloy CB states to individual GaAs extended CB states, to obtain the alloy DOS projected to extended host matrix **k** states——$\frac{1}{\pi}$Im$\{G_{\mathbf{kk}}(E)\}$—for (a) a two-band model in which all N-related states have the same energy E_N and (b) the full LCINS distribution of N-related localized states (cf. Section 9.2.3). The two-band SCGF CB structure in (a) corresponds well to the band dispersion calculated using the two-band BAC model (cf. Equation 9.2). The SCGF-LCINS CB structure in (b) shows enhanced N-related broadening associated with alloy disorder, and a strong deviation from the BAC band dispersion due to hybridization of the GaAs CB edge with localized states associated with N–N pairs (lying approximately 50 meV above the alloy CB edge in energy). The zero of energy is taken at the GaN$_{0.0036}$As$_{0.9964}$ CB edge energy $E_-(0)$, calculated using the two-band BAC model. (Adapted from From M. Seifikar et al., *J. Phys. Condens. Matter*, 26, 365502, 2014.)

In summary, we have reviewed how the shortcomings of the BAC approach to modeling the CB structure and DOS in GaN$_x$As$_{1-x}$ and related alloys can be overcome by using a Green's function approach to solve the AIM in the presence of N impurities. In particular, we emphasized that the key physics missing from the BAC model is associated with (1) the complex energy-dependent broadening of N-derived quasilocalized states, consideration of which is necessary to remove unphysical aspects of the description of the DOS based on the BAC model and, (2) N-related alloy disorder, which is accurately captured using the LCINS method and introduces important qualitative and quantitative differences compared to calculations of the CB structure and DOS undertaken using a simplified two-band model. The detailed features of the CB structure and DOS revealed through a combined LCINS and SCGF analysis are of importance for understanding carrier transport in dilute nitride alloys, as they quantitatively capture the strong intrinsic and disorder-induced energy broadening of localized states in real materials, as well as the breakdown in Bloch character associated with the strong hybridization of the extended host matrix CB states with localized impurities.

We note that the theory outlined above is generally applicable to analyzing the band structure of highly mismatched semiconductor alloys in which highly electronegative isovalent impurities strongly perturb the CB, for example in dilute oxide [134,135] or dilute carbide [136,137] alloys. It is further expected that similar approaches could be applied to investigate highly mismatched alloys containing strongly electropositive isovalent impurities—for example, dilute bismide alloys, which will be discussed in the next chapter). However, in dilute bismides and related alloys the strong perturbation of the electronic structure at energies close to the VB edge, combined with the T_2 (as opposed to A_1) symmetry of the localized states associated with isolated impurities, necessitates that spin-orbit coupling be taken into account, which significantly complicates the description of the band structure and which would hence require that the SCGF method outlined above be modified in order to account for the more complicated nature of the p-like VB eigenstates. Having reviewed these detailed aspects of the impact of N incorporation on the CB structure and DOS, we now turn our attention to the theory of electron transport in GaN_xAs_{1-x} alloys, and review how these same features play a key role in determining the transport properties of real materials.

9.5.2 Electron Transport in GaN_xAs_{1-x} Alloys

Initial experimental measurements on dilute nitride alloys revealed that N incorporation has a significant impact on electron transport. While this is to be expected in light of the strong N-induced modifications to the CB structure discussed throughout this chapter, development of sufficiently accurate theoretical models to describe the measured transport properties of GaN_xAs_{1-x} has motivated detailed investigations of the physics of dilute nitride alloys which have significantly enhanced the fundamental understanding of this unusual class of semiconductor materials. Given the importance of electron transport in the operation of many of the devices described in Section 9.4, developing quantitative theoretical models describing the impact of N incorporation on electron transport represents a task which not only seeks to elucidate further aspects of the physics of N-containing alloys, but which is also of significant importance from the perspective of practical applications. Here, we review the development of the theory of electron transport in GaN_xAs_{1-x} alloys, beginning from general considerations that provide fundamental insight into the impact of N incorporation and analytical expressions describing important qualitative trends in the electron mobility, before considering more detailed models and numerical calculations that quantitatively describe the results of experimental measurements.

Experimental measurements have identified two general trends which must be described by theoretical models of the electron transport in GaN_xAs_{1-x}. First, early experiments on GaN_xAs_{1-x} materials revealed that N incorporation results in a strong reduction of the electron mobility μ, from ~ 8500 cm^2 V^{-1} s^{-1} at $x = 0$ (GaAs) to <2000 cm^2 V^{-1} s^{-1}, even at ultra-dilute N compositions $x \sim 0.1\%$ [138,139]. Second, experimental measurements have suggested that while negative differential velocity[†] (NDV) is present in ultra-dilute GaN_xAs_{1-x} alloys [140–142], incorporation of N leads to a significant weakening of this effect compared to that present in GaAs. Additional measurements have demonstrated that NDV in GaN_xAs_{1-x} is strongly suppressed by incorporation of N at slightly higher compositions $x \sim 1\%$ [143], with NDV then being absent in GaN_xAs_{1-x} alloys except for a limited, ultradilute range of N compositions.

These experimental studies clearly demonstrated that N incorporation has a significant impact on the electron transport in GaN_xAs_{1-x} alloys, which must be understood in detail in order to predict and optimize the transport properties of dilute nitride materials and devices. That the electron mobility in dilute nitride alloys is intrinsically limited by N incorporation was explicitly demonstrated by Fahy and O'Reilly, who used an S-matrix approach to derive the scattering cross section σ of N impurities in the ultra-dilute limit [144]. Within the context of the BAC model they obtained the following expression connecting σ and the dependence of the CB edge energy E_- on the N composition x

$$\sigma = \frac{\pi a_0^6}{4} \left(\frac{m_e^*}{2\pi \hbar^2} \right)^2 \left(\frac{dE_-}{dx} \right)^2, \qquad (9.29)$$

[†] The drift velocity v_d of electrons in semiconductors tends to initially increase with the strength F of an applied electric field, displaying conventional ohmic behavior. As F increases it is possible for electrons to be accelerated to energies sufficient to facilitate scattering from high-mobility Γ valley states to states having lower mobilities in higher energy-satellite valleys. This intervalley scattering of electrons results in the existence of a range of F for which further increases in F result in a decrease of v_d—that is, the differential velocity $\frac{dv_d}{dF}$ is negative. In terms of the flow of electrical current through the material, this results in a negative (differential) electrical resistance for a given range of F, which is of use for a variety of applications. For example, NDV arising from intervalley scattering of electrons from the Γ to L valleys facilitates the operation of GaAs-based Gunn diodes.

where m_e^* is the electron effective mass, a_0 is the host matrix lattice constant, and $\frac{\mathrm{d}E_-}{\mathrm{d}x}$ is the derivative of the BAC model CB edge energy with respect to the N composition. One significant piece of information can immediately be gleaned from Equation 9.29: the fact that σ is directly proportional to the square of $\frac{\mathrm{d}E_-}{\mathrm{d}x}$ means that the large, composition-dependent bowing of the CB edge energy present in GaN_xAs_{1-x} and related alloys ensures that the cross section for electron scattering by N impurities is large. For example, a straightforward evaluation of Equation 9.29 for GaN_xAs_{1-x} gives an estimated electron scattering cross section $σ \approx 0.3$ nm^2 for an isolated N impurity in GaAs [144].

It is noteworthy that Equation 9.29 was derived independently of the fact that the scattering impurities are N atoms. This implies that the connection between σ and the dependence of the CB edge energy on the impurity concentration holds generally, regardless of the nature of the localized perturbation or defect which acts as a scattering center for electrons. Fahy and O'Reilly further demonstrated that the Born approximation (which is commonly used to analyze alloy scattering in conventional semiconductor alloys) underestimates σ, and hence the scattering rate due to N impurities, by roughly two orders of magnitude [144]. This stems from the fact that the Born approximation is inadequate for the calculation of *resonant* scattering, such as that associated with substitutional N impurities in GaAs.

Despite the fact that the above expression for σ is strictly valid only in the ultra-dilute limit, it can be extrapolated to higher N compositions with some degree of success. On the basis of Equation 9.29, Fahy and O'Reilly derived the following expression for the electron mobility μ at temperature T and N composition x [144]

$$\mu = \frac{e}{\pi a_0^3 x} \frac{1}{\sqrt{3m_e^* k_B T}} \left(\frac{2\pi \hbar^2}{m_e^*} \right)^2 \left(\frac{\mathrm{d}E_-}{\mathrm{d}x} \right)^{-2}. \tag{9.30}$$

Despite the fact that this simple treatment takes into account scattering due only to isolated N impurities, Equation 9.30 importantly describes the placement of an intrinsic limit on the electron mobility in dilute nitride alloys due to the presence of resonant scattering by N impurities. Interestingly, we deduce from Equation 9.30 that the strong reduction of the GaN_xAs_{1-x} CB edge energy E_- (and hence band gap) with increasing N composition—a major motivating factor for the use of dilute nitride alloys to develop GaAs-based photonic devices—plays a direct role in limiting the electron mobility, since μ is inversely proportional to the square of $\frac{\mathrm{d}E_-}{\mathrm{d}x}$. Using Equation 9.30 to calculate the variation of μ as a function of N composition suggests that the electron mobility is limited to values <1000 cm^2 V^{-1} s^{-1} at $x = 1\%$, an estimation which is consistent with the available experimental data which explicitly confirms the notion of intrinsically limited electron mobility in GaN_xAs_{1-x} and related alloys [144].

Following this initial analysis of μ based on the BAC model, in Ref. [131] Fahy et al. took the first steps toward developing a quantitative theory of the impact of N incorporation on the electron mobility by (1) explicitly deriving the scattering rates associated with N-related localized states, (2) evaluating the total scattering rate at a given N composition x using the full LCINS distribution of N-related localized states (cf. Section 9.2.3), and (3) using this scattering rate in conjunction with the BTE to derive an expression for μ which describes the impact of N-related scattering on the electron mobility.

In the case that an N-related localized state $|j\rangle$ lies energetically within the host matrix CB, we recall that the interaction of this localized state with an extended host matrix CB state results in its broadening into a quasi-localized resonance having a finite lifetime, described mathematically by a complex energy whose imaginary part is related to the lifetime as $\tau_j \sim \frac{\hbar}{\Delta_j}$ [128]. Under the assumption that the energy broadening associated with this resonance can be described using a Lorentzian lineshape [128], Fahy et al. used Fermi's golden rule to derive an expression for the scattering rate $R_j(E)$ between extended CB states $|\mathbf{k}\rangle$ and $|\mathbf{k}'\rangle$ due to the N-related localized state $|j\rangle$ [131]

$$R_j(E) = \frac{2\pi}{\hbar} \frac{|V_j|^4}{(E - E_j')^2 + \Delta_j^2} D_0(E). \tag{9.31}$$

There are two important points to note regarding Equation 9.31. First, the scattering rate is isotropic—that is, it does not depend on $|\mathbf{k} - \mathbf{k}'|$. Second, Equation 9.31 is an analytical continuation of the expression describing the rate of scattering from a nonresonant localized state, which amounts to replacing the localized state energy E_j by the complex energy $E_j' + i\Delta_j$ to encapsulate the energy shift and broadening of $|j\rangle$ (cf. Section 9.5.1).

In order to calculate the total scattering rate $R_N(E)$ due to N impurities Fahy et al. employed the full LCINS distribution of N-related localized states $|j\rangle$ in the independent scattering approximation. The independent scattering approximation assumes that the total scattering rate due to randomly placed localized states is equal to the sum of the scattering rates associated with each individual localized state—that is, $R_N(E) = \sum_j R_j(E)$ [131]. This approximation is justified in practice because (1) by construction, the LCINS basis of N-related localized states interact only with the extended states of the unperturbed CB and not with one another, and (2) due to the low concentrations of large clusters of N atoms at dilute N compositions, it can reasonably be assumed that N-related localized states are not influenced by statistical correlations between their occurrence and location in a randomly disordered alloy.

Figure 9.13 shows the total energy-dependent resonant scattering rate $R_N(E)$ due to N-related localized states in GaN_xAs_{1-x}, computed using the full LCINS distribution (cf. Section 9.2.3) in the independent scattering approximation, for N compositions $x = 0.1\%$, 0.36%, 0.5%, 1.2%, and 2.0% (shown using solid, dashed, dotted, dash-dotted, and short-dash-gapped lines, respectively). Examining $R_N(E)$ in light of Equation 9.31 and the distribution of N impurity states described by the LCINS distributions of Section 9.2.3 we note a crucial trend: the N-related scattering is strongly enhanced for CB electrons having energies close to those of the N-related states in a given alloy. As we will see below, this is borne out by detailed theoretical calculations which are in quantitative agreement with experimental measurements of μ. This general trend is clearly evident for scattering by localized states associated with isolated N atoms, by the presence of a peak in $R_N(E)$ lying approximately 200 meV above the CB edge energy at $x = 0.1\%$, which increases in magnitude and shifts to higher energies as x increases to 2% and more N-related resonant scattering centeres are introduced into the CB. An additional, smaller peak in $R_N(E)$—describing scattering by localized states associated with N–N pairs—can be seen to emerge at energies closer to the CB edge, and increases in magnitude as x increases and more N-N pairs form in a statistically random

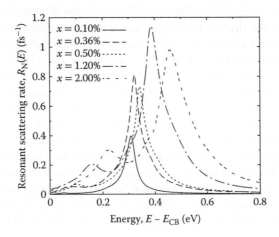

FIGURE 9.13 Total resonant scattering rate $R_N(E)$ associated with N-related localized states in GaN_xAs_{1-x}, calculated using the full LCINS distribution (cf. Section 9.2) in Equation 9.31 for N compositions $x = 0.1\%$, 0.36%, 0.5%, 1.2%, and 2.0% (solid, dashed, dotted, dash-dotted, and short-dash-gapped lines, respectively). The zero of energy is taken in each case at the alloy CB edge energy E_{CB}. A large peak in $R_N(E)$, associated with scattering by isolated N impurities, can be seen in each case. A second smaller peak in $R_N(E)$, associated with scattering by N–N pairs, can also be seen at lower energy. (Adapted from S. Fahy et al., *Phys. Rev. B*, 74, 035203, 2006.)

alloy. This demonstrates, as we see below, that while the number of N–N pairs at dilute N compositions is relatively small, localized states associated with N–N pairs can be expected to have a significant impact on the transport of CB electrons even at low energies. The strong impact of the energy dependence of N-resonant scattering on the transport properties can also be understood qualitatively in terms of the mean free path for electrons, $\lambda(E) = \frac{v_d}{R_N(E)}$, which in applied fields can be as low as 5 nm for certain combinations of N compositions and electron energies [131].

As a point of technical significance, we emphasize that the complex energy broadening $i\Delta_j$ of the localized states in Equation 9.31 is strictly independent of energy. Ideally, calculations of the transport properties of electrons in dilute nitride alloys should incorporate complex energy-dependent broadening of the N-related localized states, as described by the SCGF model of Section 9.5.1 (cf. Equation 9.24). However, these aspects of the band structure are not straightforward to incorporate into conventional models of carrier transport. As a result, published calculations have to date employed energy-independent broadening in the computation of the scattering rates and simplified models of the CB structure which, despite the fact that they omit some of the details described in Section 9.5.1, provide reliable and quantitative descriptions of the measured trends in the electron mobility and NDV in GaN$_x$As$_{1-x}$ alloys. Following an approach of this type, Fahy et al. used the total N-related scattering rate $R_N(E)$ in conjunction with the BTE to derive an expression for μ under the assumptions that (1) the CB dispersion is parabolic, with $E_{\mathbf{k}} = \frac{\hbar^2 |\mathbf{k}|^2}{2m_e^*}$, and (2) the thermal (energy) distribution of electrons in the CB can be described using a Boltzmann (Fermi–Dirac) distribution in the non-degenerate (degenerate) regime [131]. Starting from the BTE in the relaxation time approximation, Fahy et al. calculated the current density J, at carrier density n, associated with electron drift in the presence of an applied electric field of strength F. This was then used to compute the electron mobility as $\mu = \frac{J}{enF}$, resulting in the following expression for μ at temperature T [131]

$$\mu = \frac{4e}{3\pi a_0^3} \sqrt{\frac{2}{\pi m_e^* k_B T}} \left(\frac{2\pi \hbar^2}{m_e^*} \right)^2 \int_0^\infty \frac{\epsilon^3 \, e^{-\epsilon^2}}{M(\epsilon^2 k_B T)} \, d\epsilon, \qquad (9.32)$$

where the zero of energy has been taken at the CB minimum, $\epsilon = \frac{\hbar^2 k^2}{2m_e^* k_B T}$ is a dimensionless measure of the energy of a CB electron, defined as the ratio of the kinetic energy to the thermal energy, and the function $M(E)$ is closely related to the total scattering rate $R_N(E)$ due to N-related localized states [131]

$$M(E) = \frac{1}{N_c} \sum_j \frac{\beta_j^4}{(E - E_j')^2 + \Delta_j^2}. \qquad (9.33)$$

In addition to highlighting the strong influence of N-resonant scattering on μ, Equation 9.32 also reveals the important role played by the difference in energy between N-related localized states and the Fermi level in determining the mobility. The important role played by the Fermi level in this context can lead to unusual trends such as, e.g. an increase in μ with increasing x depending upon the carrier density n. Additionally, the dependence of the Fermi level on T leads to further variations in μ, due to the change in energy between the temperature-dependent extended CB states and the N-related localized states (which are relatively insensitive to changes in temperature). This aspect of the calculation of μ raises interesting points related to the interpretation of measurements of μ, particularly in doped materials, emphasizing the highly unusual nature of the impact of N incorporation on the transport properties. A detailed analysis of this aspect of the calculation of μ is beyond the scope of the current discussion—we refer interested readers to Refs. [131] and [145] for a discussion of this issue relating to bulk GaN$_x$As$_{1-x}$ alloys, as well as in QWs and gated QW heterostructures.

In order to ascertain the predictive capability of this more sophisticated model of μ, it is instructive to compare the results obtained using Equation 9.32 to experiment. Measurements of μ in GaN$_x$As$_{1-x}$

at ultra-dilute N compositions $x < 0.4\%$ have been undertaken by a number of groups, both in bulk [146] and QW samples [138,147]. For example, Young et al. measured extremely low values of 262 and 187 cm^2 V^{-1} s^{-1} for μ in bulk-like thin films having respective N compositions $x = 0.1\%$ and 0.4% and relatively high carrier densities $n \sim 5 \times 10^{18}$ cm^{-3} [146]. Using Equation 9.32, the calculated mobilities at these compositions and carrier densities are 2200 and 385 cm^2 V^{-1} s^{-1} for $x = 0.1\%$ and 0.4%, respectively [131]. We note that the calculated value at $x = 0.1\%$ overestimates μ by close to an order of magnitude. While the calculated value of μ at $x = 0.4\%$ is much closer to experiment, the theoretical value obtained using Equation 9.32 still represents an overestimation of the mobility. This trend is relatively consistent across bulk and QW samples, as well as across different temperatures and doping levels. For example, a value $\mu = 3000$ cm^2 V^{-1} s^{-1} for a QW having $x = 0.12\%$ and an areal carrier density $\sim 1 \times 10^{12}$ cm^{-2}— calculated using the equivalent expression to Equation 9.32 in two dimensions [131]—overestimates the measured value $\mu \sim 600$ cm^2 V^{-1} s^{-1} [138] by a factor of five.

From these comparisons to experimental measurements, it is clear that while N-related scattering plays the dominant role in limiting μ in GaN$_x$As$_{1-x}$, the tendency of Equation 9.32 to overestimate μ highlights that additional scattering mechanisms also play an important role in the electron transport. From a theoretical perspective, this mandates that the models presented above be refined in order to better capture the physics involved in quantitatively determining μ in GaN$_x$As$_{1-x}$. In practice, this is achieved by abandoning the assumed carrier distribution function(s) used to derive Equation 9.32 and instead solving the BTE numerically in order to directly compute the influence of additional scattering mechanisms on the drift current density J. An initial analysis of this type was undertaken by Vaughan and Ridley, who in Ref. [148] developed a ladder method to solve the BTE for a nonparabolic CB. Using the BAC description of the CB structure as the basis for their solution of the BTE, Vaughan and Ridley calculated the dependence of μ in GaN$_x$As$_{1-x}$ on x and T in a model including polar optical, acoustic phonon, and piezoelectric scattering, as well as scattering by ionized, neutral, and N impurities. Despite the level of detail included in this model, the conclusions reached by Vaughan and Ridley are qualitatively similar to those described by Fahy and O'Reilly in Ref. [144] on the basis of Equation 9.30—i.e. that the room temperature mobility in GaN$_x$As$_{1-x}$ is limited to values <2000 cm^2 V^{-1} s^{-1} by scattering associated with isolated N impurities. Quantitatively, the analysis of Vaughan and Ridley suggests a slight reduction of the intrinsic "ceiling" on μ, with the calculated value $\mu \sim 800$ cm^2 V^{-1} s^{-1} at $x = 1\%$ being approximately 80% of that estimated using Equation 9.30. However, regardless of its detailed treatment of a range of scattering mechanisms, this BTE-based model still significantly overestimates μ compared to experimental measurements. This strongly reinforces that the description of the CB structure by the BAC model, which in this context describes N-resonant scattering associated with isolated N impurities only, is insufficient to account for the significant reduction in μ observed in real GaN$_x$As$_{1-x}$ alloys.

In light of these insights, it is clear that a quantitative analysis of the transport properties of GaN$_x$As$_{1-x}$ alloys requires a refined approach which, in addition to resonant scattering by isolated N impurities, also takes into account resonant scattering associated with N pair and cluster states, as well as the influence of additional scattering mechanisms. Such a model was developed in Ref. [149] by Seifikar et al. who solved the BTE numerically in the relaxation time approximation and in the presence of a range of scattering mechanisms including polar optical and acoustic phonon scattering, intervalley scattering between the Γ and L valleys, and resonant scattering by the full LCINS distribution of N-related localized states. Comparing the calculated carrier distributions for the Γ and L valleys as a function of the strength of F of an applied electric field, Seifikar et al. first showed that resonant scattering by N-localized states strongly inhibits acceleration of electrons to higher energies. As a result, intervalley scattering at fixed F is strongly suppressed in GaN$_x$As$_{1-x}$ compared to GaAs. For example, intervalley scattering in room temperature GaAs results in approximately 80% of electrons occupying L valley states at $F = 10$ kV cm^{-1}, while at $x = 0.1\%$ (2.0%) a much stronger applied field $F = 30$ kV cm^{-1} (45 kV cm^{-1}) is required to bring about the same level of electron transfer from Γ to L valley states. That this reduction of the intervalley scattering is brought about by N incorporation was confirmed by a calculation of the average energy of Γ electrons as a function of F, demonstrating explicitly that N-resonant scattering erodes the initial strong increase in electron energy with F calculated for GaAs. The analysis of Seifikar et al. further showed that the strong nonparabolicity of the lowest CB in GaN$_x$As$_{1-x}$ brings about significant increases in the rates of polar optical and acoustic phonon scattering, particularly at energies close to N-related states, suggesting in particular that scattering by polar optical phonons plays an important (but not dominant) role in determining the transport properties of GaN$_x$As$_{1-x}$ alloys [149].

Using the carrier distribution functions obtained by numerical solution of the BTE to compute J, Seifikar et al. calculated the variation of μ ($= \frac{J}{eFn}$) and v_d ($= \frac{J}{en}$) with F for a series of GaN$_x$As$_{1-x}$ alloys having N compositions $x = 0.1 - 2.0\%$. The results of these calculations are summarized in Figures 9.14a and 9.14b. Figure 9.14a shows the calculated variation of μ with F in GaAs (solid gray line) and in GaN$_x$As$_{1-x}$ for $x = 0.1\%, 0.36\%, 0.5\%, 1.2\%$ and 2.0% (solid, dashed, dotted, dash-dotted, and short-dash-gapped lines, respectively) [150,151]. The calculated low-field mobility $\mu = 2000$ cm^2 V^{-1} s^{-1} at $x = 0.1\%$ is relatively close to the measured value of 1200–1400 cm^2 V^{-1} s^{-1} [139], while the calculated value $\mu = 410$ cm^2 V^{-1} s^{-1} at $x = 0.4\%$ is closer to the measured value of 280–360 cm^2 V^{-1} s^{-1} [139]. Given that this model overestimates μ in GaAs—due in part to the fact that the model neglects CB nonparabolicity in the absence of N—these comparisons to experiment suggest that including resonant scattering by the full LCINS distribution of N-localized states correctly describes the impact of N incorporation on the electron transport. That this is the case can also be seen by considering the calculated variation of v_d with F, and the corresponding implications for the presence of NDV in GaN$_x$As$_{1-x}$.

Figure 9.14b shows the calculated variation of v_d with F in GaAs (solid gray line) and in GaN$_x$As$_{1-x}$ for $x = 0.1\%, 0.36\%, 0.5\%, 1.2\%$ and 2.0% (solid, dashed, dotted, dash-dotted, and short-dash-gapped lines, respectively). These calculations suggest that incorporating N results in a strong reduction of the maximum value $v_{d,max}$ of v_d, as well as an increase in the field strength F required to reach $v_{d,max}$. For

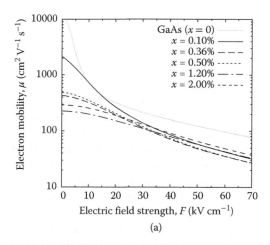
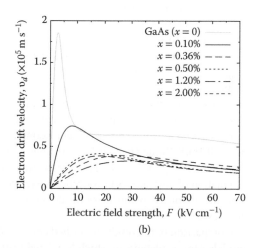

(a) (b)

FIGURE 9.14 (a) Variation of the electron mobility μ in GaN$_x$As$_{1-x}$ as a function of the strength F of an applied electric field, calculated using the total N-related resonant scattering rate of Equation 9.31 in a numerical solution of the BTE, for N compositions $x = 0.1\%, 0.36\%, 0.5\%, 1.2\%$, and 2.0% (solid, dashed, dotted, dash-dotted, and short-dash-gapped lines, respectively). The calculated variation of μ with F for GaAs is shown for reference (solid gray line). A reduction of μ by approximately an order of magnitude is calculated even at the ultra-dilute N composition $x = 0.1\%$, reflecting the strong impact of N-resonant scattering on the electron transport. (b) Variation of the electron drift velocity v_d in GaN$_x$As$_{1-x}$ with F, calculated using the total N-related resonant scattering rate of Equation 9.31 in a numerical solution of the BTE, for N compositions $x = 0.1\%, 0.36\%, 0.5\%, 1.2\%$, and 2.0% (solid, dashed, dotted, dash-dotted, and short-dash-gapped lines, respectively). The calculated variation of v_d with F for GaAs is shown for reference (solid gray line). N incorporation leads to a strong reduction in the maximum value $v_{d,max}$ of v_d. We also note a calculated strong reduction in NDV compared to GaAs, and that appreciable NDV is present only for ultra-dilute N compositions $x < 0.4\%$. (Adapted from M. Seifikar et al., *Phys. Rev. B*, 84, 165216, 2011.)

example, the calculated maximum value $v_{d,\max} \sim 0.5 \times 10^5$ m s^{-1} at $x = 1.2\%$ is approximately one-quarter of that calculated for GaAs, while the field strength required to achieve $v_{d,\max}$ increases from $F = 4$ kV cm^{-1} in GaAs to ~ 30 kV cm^{-1} at $x = 1.2\%$. Additionally, while the variation of v_d with F displays strong NDV above $F = 4$ kV cm^{-1} in GaAs, at $x = 1.2\%$ there is no evidence of NDV, with the calculated $v_d(F)$ saturating for applied fields $F > 30$ kV cm^{-1} [149]. Examining the calculated variation of v_d with F for the other N compositions shown in Figure 9.14b demonstrates that (1) N incorporation leads to a weakening of the NDV effect, as evidenced by a strong reduction in the magnitude of $\frac{dv_d}{dF}$ above $v_{d,\max}$, and (2) NDV is absent except at ultra-dilute N compositions $x < 0.5\%$. We note that these trends are in good agreement with those observed in experiment [140–143] and that the increasing occurrence of N pairs and clusters with increasing x leads to enhanced resonant scattering by N impurities which suppresses NDV by simultaneously limiting electron acceleration and reducing intervalley scattering from Γ to L states [149].

Calculations of the transport properties of GaN$_x$As$_{1-x}$ have also been undertaken by Vogiatzis and Rorison, who used a single-electron Monte Carlo model to analyze μ and v_d as a function of F in the ultra-dilute regime [150,151]. In Ref. [150] they considered a model of the transport including resonant scattering by localized states associated with isolated N impurities only, but which also included polar optical and acoustic phonon scattering, as well as intervalley scattering between Γ and L states. Qualitatively, the conclusions reached on the basis of this detailed analysis are similar to those of Seifikar et al.—i.e. that (1) N-related scattering limits the acceleration of electrons in the presence of an applied electric field, leading to a strong reduction of μ and $v_{d,\max}$, (2) intervalley scattering mediated by polar optical phonons is reduced compared to that present in GaAs, and (3) NDV is present in ultra-dilute GaN$_x$As$_{1-x}$ alloys having $x < 0.4\%$, but with the magnitude of $\frac{dv_d}{dF}$ above $v_{d,\max}$ strongly reduced compared to that in GaAs. Quantitatively, the analysis of Vogiatzis and Rorison suggested that the presence and character of NDV in the ultra-dilute regime is governed by the interplay of polar optical scattering with N-related resonant scattering of electrons to higher-energy N-derived CB states. This is in contrast to the nature of NDV in GaAs, which is understood to arise primarily due to intervalley scattering of electrons from Γ to L valley states. This Monte Carlo analysis was further developed in Ref. [151], in which Vogiatzis and Rorison extended their treatment to include resonant scattering due to localized states associated with N–N pairs and selected larger clusters of N atoms. Using this approach, it was again confirmed that N pairs and clusters play a dominant role in limiting μ and $v_{d,\max}$ in GaN$_x$As$_{1-x}$. However, while the conclusions of these Monte Carlo analyses are qualitatively similar to those arising from studies based on the BTE, it is noteworthy that the electron mobilities calculated by Vogiatzis and Rorison using this extended model are in excellent quantitative agreement with experiment. For example, the calculated values $\mu = 1296$, 738, and 344 cm^2 V^{-1} s^{-1} at $x = 0.1\%$, 0.2%, and 0.4% [151] correspond very well with the measured values $\mu \sim 1200$–1400, 700, and 280–360 cm^2 V^{-1} s^{-1} [139]. These calculations highlight the utility of the Monte Carlo approach for the calculation of the transport properties of semiconductors, and demonstrate the ability of this approach to provide a predictive platform for the design and optimization of carrier transport in dilute nitride materials and devices.

Comparison of the results described above—obtained using the BTE [149] and Monte Carlo [150,151] models—to calculations in which N-related scattering associated only with isolated N impurities is taken into account demonstrates that scattering of electrons by N pair and cluster states (lying close in energy to the CB edge) plays an important role in bringing about the measured strong reduction in μ and suppression of NDV in GaN$_x$As$_{1-x}$. These analyses reinforce the need for a detailed treatment of N-related localized states and their role in electron scattering beyond the BAC model, and further emphasize the utility of the LCINS description of the CB structure to predict the properties of real dilute nitride alloys, in which short-range disorder plays a crucial role (cf. Section 9.2).

To summarize, in this section we have presented an overview of the theory and simulation of electron transport in GaN$_x$As$_{1-x}$. The theoretical models reviewed in this section—beginning from simple descriptions which identify intrinsic limitations on the electron mobility in N-containing alloys and increasing in sophistication to provide quantitative descriptions of measured transport properties—constitute the most complete available treatment of carrier transport in dilute nitride alloys. The results of calculations based on these models not only provide quantitative insight into the complicated nature of electron transport in dilute nitride alloys, but also serve as rigorous tests of models describing the impact of N incorporation on the CB structure in GaN_xAs_{1-x} and related alloys. In particular, the sophisticated nature of the models required to understand measurements of basic transport properties in GaN_xAs_{1-x} alloys emphasizes the unusual characteristics of dilute nitride alloys, and highlights the challenges that N-containing and related classes of highly mismatched alloys present on a theoretical level.

9.6 Summary and Conclusions

In this chapter, we have provided an introduction to the theory and simulation of dilute nitride materials and devices, which we have framed within the context of recent and ongoing research on $(In)GaN_xAs_{1-x}$ and related N-containing alloys. Beginning with an overview of the electronic properties of GaN_xAs_{1-x} and related alloys, we highlighted the potential to exploit the strong N-induced modifications of the band structure of conventional III–V semiconductors to develop a range of photonic and photo-voltaic devices with enhanced performance and capabilities. In particular, we discussed that the extremely rapid reduction in band gap brought about by N incorporation offers the potential to greatly extend the wavelength range accessible via growth on GaAs substrates, which has stimulated significant interest in developing $(In)GaN_xAs_{1-x}$ and related alloys for applications in a range of GaAs-based photonic and pho-tovoltaic technologies. Additionally, we highlighted that experimental measurements have demonstrated that the impurity-like behavior of substitutional N atoms leads to a reduction in the temperature sensitivity of the $(In)GaN_xAs_{1-x}$ band gap. This, combined with the enhanced carrier and optical confinement offered by GaAs heterostructures (compared to conventional InP-based structures), means that $(In)GaN_xAs_{1-x}$ alloys are particularly attractive for the development of GaAs-based 1.3- and 1.55-μm edge-emitting lasers and VCSELs, which have the potential to offer improved performance and new capabilities compared to the incumbent InP-based technologies in this wavelength range.

Beginning in Section 9.2, we highlighted the need for detailed atomistic models to provide a quantitative understanding of the strong impact of N incorporation on the band structure of (In)GaAs(P). We reviewed detailed analysis of the electronic structure of GaN_xAs_{1-x} and GaN_xP_{1-x} alloys based on first principles calculations and on the EP method, which revealed that N acts as an isovalent impurity when incorporated into common III–V compounds and alloys in dilute concentrations and leads to a strong perturbation of the CB structure. We described how a simple two-band BAC model provides a simple phenomenological description of the main features of the $(In)GaN_xAs_{1-x}$ band structure, allowing for simple interpretations of a broad range of experimental measurements. We then reviewed that the TB method can be used to explicitly parameterize the BAC model, placing it on a quantitative footing, and summarized the results of the calculations that describe the impact of isolated substitutional N impurities on the CB structure of a range of III–V compounds. On this basis we outlined the derivation of an extended basis set 10-band **k·p** Hamiltonian for $(In)GaN_xAs_{1-x}$ alloys, which directly accounts for the impact of N incorporation on the CB structure using a BAC approach. We provided a consistent set of band structure parameters for performing calculations on $(In)GaN_xAs_{1-x}$ alloys and heterostructures.

Following this, we discussed models of the dilute nitride band structure beyond the BAC model, focus-ing in particular on the TB-derived LCINS method, which is capable of quantitatively describing the highly unusual properties of dilute nitrides via direct treatment of the full range of N-related localized impurity states present in real, disordered alloys. The results of LCINS calculations were shown to be in excellent quantitative agreement with experimental measurements of the electron effective mass and g factor in GaN_xAs_{1-x} alloys, highlighting the predictive power of the approach and underscoring its use as a plat-form to facilitate detailed analysis of the electronic, optical, and transport properties of real N-containing materials.

Next, in Section 9.3, we provided an overview of the theory and modeling of 1.3-μm InGaNAs QW lasers grown on GaAs. We applied the 10-band $\mathbf{k \cdot p}$ model to the study of InGaNAs/GaAs QW heterostructures and quantified the impact of N on the (1) CB structure and DOS, (2) material gain, and (3) threshold characteristics of a realistic device structure. We showed that while N incorporation compensates the large compressive strain associated with long-wavelength InGaAs/GaAs QWs and readily enables the emission wavelength to be extended to 1.3 μm, this is associated with a slight degradation of the gain characteristics resulting from (1) a reduction in the optical transition strengths due to N-induced hybridization, and (2) a reduction in the quasi-Fermi level separation at fixed injection arising from the N-induced increase of the DOS in the CB. However, these calculations also indicate that InGaNAs single QW laser structures designed to emit at 1.3 μm offer (1) enhanced carrier confinement, and (2) roughly equivalent differential gain than that calculated for optimized InP-based laser structures. This demonstrates that InGaNAs alloys can be used to achieve telecom-wavelength devices on GaAs having low leakage currents, and which offer impressive high-speed/bandwidth performance.

Following this, we considered the impact of N incorporation on the recombination pathways in 1.3 μm InGaNAs lasers. We outlined a method by which experimental measurements of the integrated SE from a QW device can be used to provide detailed insight into the loss mechanisms present in a semiconductor laser, by allowing direct extraction of the defect-related, radiative, and nonradiative (Auger) recombination coefficients. We reviewed a direct comparison between theory and experiment for the temperature dependence of the recombination coefficients, thereby demonstrating that a model based on the 10-band $\mathbf{k \cdot p}$ Hamiltonian is capable of providing a quantitative description of the temperature dependent loss mechanisms in real InGaNAs QW lasers. Experimental measurements show that defect-related recombination accounts for approximately one-half of the threshold current density at room temperature in a 1.3-μm InGaNAs laser, similar to that observed in GaAs-based metamorphic laser structures. This suggests that there is significant potential for improvement in the performance of these devices, through refinement of the material growth in order to reduce the impact of N-related defects.

While good characteristics have been demonstrated by 1.3- and 1.55-μm dilute nitride semiconductor lasers, the advantages of using dilute nitrides—e.g. in GaAs-based long-wavelength VCSEL structures—have not yet been sufficiently pronounced to displace the well-established incumbent technologies based on InP. However, as outlined in this chapter, following a loss of momentum in the development of dilute nitride devices in the mid-2000s, more recently dilute nitrides have made a resurgence due to interest in a wider range of potential applications. In Section 9.4 we provided an overview of a number of alternative and emerging applications of dilute nitride alloys. We reviewed recent progress related to the development of (1) GaAs-based 1.55-μm lasers incorporating GaAs-based InGaNAs(Sb) heterostructures, (2) 1.3-μm SOAs, EAMs, and APDs based on $(In)GaN_xAs_{1-x}$ alloys, and (3) GaNAsP alloys and heterostructures for monolithic integration of direct-gap III–V laser structures on Si. Through these summaries we demonstrated that several classes of dilute nitride devices display impressive characteristics and significant promise for practical applications, and also highlighted areas of ongoing research interest relating to the use of dilute nitride alloys in the development of novel GaAs- and Si-based photonics technologies.

In Section 9.5, we presented an overview of the theory and simulation of electron transport in dilute nitride alloys. We began with a detailed consideration of the CB structure and demonstrated that while a BAC-based approach is sufficient to calculate, for example, the electronic and optical properties to a level required to understand and predict the properties of dilute nitride semiconductor lasers, the BAC model has fundamental limitations which make it unsuitable to analyze carrier transport. We reviewed how, in order to overcome these limitations, the AIM has been adapted and applied to the study of the dilute nitride CB structure. We reviewed theoretical work based on solving the AIM for GaN_xAs_{1-x} alloys using Green's function methods, and showed how a self-consistent model provides a conceptually simple, physically rich, and quantitatively accurate description of the strongly perturbed CB structure. In light of the SCGF model, we highlighted the important role played by alloy disorder in determining the nature of the CB structure and DOS in real GaN_xAs_{1-x} alloys.

Next, we reviewed the development of theoretical models describing the impact of N incorporation on the transport of electrons in GaN_xAs_{1-x}. Beginning with general considerations, we highlighted a generic relationship between the electron scattering cross section and the strong N composition dependence of the CB edge energy, thereby demonstrating that N incorporation places intrinsic limitations on the electron mobility in dilute nitride alloys. We then reviewed the resonant scattering of CB electrons by N-related localized states, highlighting that (1) electrons having energies close to those of N-related states undergo particularly strong impurity scattering, and (2) while N-related scattering plays the dominant role in bringing about the low electron mobilities observed in experimental measurements, alternative scattering mechanisms must be considered in order to develop a quantitative understanding of the electron transport. On this basis, we reviewed detailed numerical calculations of the transport properties based on BTE and Monte Carlo approaches. These calculations again confirm that N-related scattering plays a dominant role in determining the transport properties of real GaN_xAs_{1-x} alloys, and indicate that scattering by N–N pairs and larger clusters of N atoms plays a crucial role in determining the transport properties as the N composition increases above the ultra-dilute regime. This highlights that approaches beyond the BAC model are required to quantitatively understand basic experimental results pertaining to electron transport in GaN_xAs_{1-x}. Overall, the calculations reviewed in Section 9.5 emphasize that (1) the observed low electron mobilities in GaN_xAs_{1-x} bulk materials and heterostructures are an intrinsic property of dilute nitride alloys, and (2) resonant scattering by N-related localized states plays a dominant role in limiting the electron mobility and suppressing NDV in GaN_xAs_{1-x}.

Overall, we conclude that the potential to develop photonic and photovoltaic materials and devices with enhanced characteristics and capabilities is certain to continue to stimulate interest in dilute nitride alloys in the coming years. The theoretical results and models described in this chapter form the basis of a consistent and accurate approach to the theory and simulation of dilute nitride alloys, which continues to find use in the interpretation of experimental studies, as well as in the design and optimization of new materials and structures for novel device applications.

Acknowledgments

The authors dedicate this chapter to the memory of Prof. Naci Balkan (University of Essex, UK), with whom they shared many enjoyable and fruitful years of collaboration. The writing of this chapter was supported by the Engineering and Physical Sciences Research Council, UK (project no. EP/K029665/1), and by Science Foundation Ireland (project no. 15/IA/3082). The authors acknowledge the contributions of the many colleagues with whom they have had the pleasure to collaborate on the study of dilute nitride alloys. In particular, the authors thank Profs. S. J. Sweeney and A. R. Adams (University of Surrey, UK), Prof. S. Tomić (University of Salford, UK), Prof. S. Fahy (University College Cork, Ireland), Prof. H. Riechert (Paul-Drude-Institut für Festkörperelektronik, Germany), and Dr. N. Vogiatzis (Semtech Corporation, UK) for many invaluable discussions on the physics and applications of dilute nitrides, and for their contributions to the research described in this chapter. C.A.B. thanks Ms. W. Xiong (University of Bristol, UK) for her assistance in producing several of the figures contained in this chapter.

References

1. P. S. Zory. *Quantum Well Lasers*. Boston, MA: Academic Press, 1993.
2. E. Kapon. *Semiconductor Lasers*, Vols. I & II. San Diego, CA: Academic Press, 1999.
3. C. A. Broderick, M. Usman, S. J. Sweeney, and E. P. O'Reilly. Band engineering in dilute nitride and bismide semiconductor lasers. *Semicond. Sci. Technol.*, 27:094011, 2012.
4. E. Söderberg, J. S. Gustavsson, P. Modh, A. Larsson, Z. Zhang, J. Berggren, and M. Hammar. High-temperature dynamics, high-speed modulation, and transmission experiments using 1.3-μm InGaAs single-mode VCSELs. *J. Lightwave Technol.*, 25:2791, 2007.

5. D. L. Huffaker, G. Park, Z. Zou, O. B. Shchekin, and D. G. Deppe. 1.3 μm room-temperature GaAs-based quantum-dot laser. *Appl. Phys. Lett.*, 73:2564, 1998.

6. V. M. Ustinov, N. A. Maleev, A. E. Zhukov, A. R. Kovsh, A. Yu. Egorov, A. V. Lunev, B. V. Volovik, I. L. Krestnikov, Yu. G. Musikhin, N. A. Bert, P. S. Kop'ev, Zh. I. Alferov, N. N. Ledentsov, and D. Bimberg. InAs/InGaAs quantum dot structures on GaAs substrates emitting at 1.3 μm. *Appl. Phys. Lett.*, 74:2815, 1999.

7. I. Tångring, S. M. Wang, M. Sadeghi, and A. Larsson. 1.27 μm metamorphic InGaAs quantum well lasers on GaAs substrates. *Electron. Lett.*, 42:691, 2006.

8. S. M. Wang, editor. *Lattice Engineering: Technology and Applications*. Singapore: Pan Stanford Publishing, 2012.

9. S. Bogusevschi, C. A. Broderick, and E. P. O'Reilly. Theory and optimization of 1.3 μm metamorphic quantum well lasers. *IEEE J. Quant. Electron.*, 52:2500111, 2016.

10. M. Kondow, K. Uomi, A. Niwa, T. Kitatani, S. Watahiki, and Y. Yazawa. Gas-source molecular beam epitaxy of GaN$_x$As$_{1-x}$ using a N radical as the N source. *Jpn. J. Appl. Phys.*, 33:L1056, 1996.

11. M. Kondow, T. Kitatani, S. Nakatsuka, M. C. Larson, K. Hakahara, Y. Yazawa, M. Okai, and K. Uomi. GaInNAs: A novel material for long-wavelength semiconductor lasers. *IEEE J. Sel. Top. Quantum Electron.*, 3:719, 1997.

12. E. P. O'Reilly, A. Lindsay, P. J. Klar, A. Polimeni, and M. Capizzi. Trends in the electronic structure of dilute nitride alloys. *Semicond. Sci. Technol.*, 24:033001, 2009.

13. D. A. Livshits, A. Y. Egorov, and H. Riechert. 8 W continuous wave operation of InGaAsN lasers at 1.3 μm. *Electron. Lett.*, 36:1381, 2000.

14. G. Steinle, H. Riechert, and A. Y. Egorov. Monolithic VCSEL with InGaAsN active region emitting at 1.28 μm and CW output power exceeding 500 μW at room temperature. *Electron. Lett.*, 37:632, 2001.

15. H. Riechert, A. Ramakrishnan, and G. Steinle. Development of InGaAsN-based 1.3 μm VCSELs. *Semicond. Sci. Technol.*, 17:892, 2002.

16. S. R. Bank, H. P. Bae, H. B. Yuen, M. A. Wistey, L. L. Goddard, and J. S. Harris. Room-temperature continuous wave 1.55 μm GaInNAsSb laser on GaAs. *Electron. Lett.*, 42:156, 2006.

17. S. R. Bank, M. A. Wistey, L. L. Goddard, H. B. Yuen, V. Lordi, and J. S. Harris. Low-threshold continuous-wave 1.5-μm GaInNAsSb lasers grown on GaAs. *IEEE J. Quant. Electron.*, 40:656, 2004.

18. S.-H. Wei and A. Zunger. Giant and composition-dependent optical bowing coefficient in GaAsN alloys. *Phys. Rev. Lett.*, 76:664, 1996.

19. J. Slater and G. Koster. Simplified LCAO method for the periodic potential problem. *Phys. Rev.*, 94:1498, 1954.

20. H. P. Hjalmarson, P. Vogl, D. J. Wolford, and J. D. Dow. Theory of substitutional deep traps in covalent semiconductors. *Phys. Rev. Lett.*, 44:810, 1980.

21. L. Bellaiche, S.-H. Wei, and A. Zunger. Localization and percolation in semiconductor alloys: GaAsN vs. GaAsP. *Phys. Rev. B*, 54:17568, 1996.

22. P. R. C. Kent and A. Zunger. Theory of electronic structure evolution in GaAsN and GaPN alloys. *Phys. Rev. B*, 64:115208, 2001.

23. J. Wu, W. Walukiewicz, K. M. Yu, J. W. Ager III, E. E. Haller, Y. G. Hong, H. P. Xin, and C. W. Tu. Band anticrossing in GaP$_{1-x}$N$_x$ alloys. *Phys. Rev. B*, 65:241303, 2002.

24. M. Güngerich, P. J. Klar, W. Heimbrodt, G. Weiser, J. F. Geisz, C. Harris, A. Lindsay, and E. P. O'Reilly. Experimental and theoretical investigation of the conduction band edge of GaN$_x$P$_{1-x}$. *Phys. Rev. B*, 74:241202, 2006.

25. C. Harris, A. Lindsay, and E. P. O'Reilly. Evolution of N defect states and optical transitions in ordered and disordered GaP$_{1-x}$N$_x$ alloys. *J. Phys. Condens. Matter*, 20:295211, 2008.

26. P. R. C. Kent and A. Zunger. Evolution of III–V nitride alloy electronic structure: The localized to delocalized transition. *Phys. Rev. Lett.*, 86:2613, 2001.

27. W. Shan, W. Walukiewicz, J. W. Ager III, E. E. Haller, J. F. Geisz, D. H. Friedman, J. M. Olson, and S. R. Kurtz. Band anticrossing in GaInNAs alloys. *Phys. Rev. Lett.*, 82:1221, 1999.

28. P. J. Klar, H. Grüning, W. Heimbrodt, J. Koch, F. Höhnsdorf, W. Stolz, P. M. A. Vicente, and J. Camassel. From N isoelectronic impurities to N-induced bands in the $GaAs_{1-x}N_x$ alloy. *Appl. Phys. Lett.*, 76:3439, 2000.

29. E. P. O'Reilly, A. Lindsay, S. Tomić, and M. Kamal-Saadi. Tight-binding and **k·p** models for the electronic structure of Ga(In)NAs and related alloys. *Semicond. Sci. Technol.*, 17:870, 2002.

30. A. Lindsay and E. P. O'Reilly. Theory of enhanced bandgap non-parabolicity in GaN_xAs_{1-x} and related alloys. *Solid State Commun.*, 112:443, 1999.

31. A. Lindsay and E. P. O'Reilly. Influence of nitrogen resonant states on the electronic structure of GaN_xAs_{1-x}. *Solid State Commun.*, 118:313, 2001.

32. A. Lindsay, S. Tomić, and E. P. O'Reilly. Derivation of a 10-band **k·p** model for dilute nitride semiconductors. *Solid State Commun.*, 47:443, 2003.

33. S. Tomić, E. P. O'Reilly, R. Fehse, S. J. Sweeney, A. R. Adams, A. D. Andreev, S. A. Choulis, T. J. C. Hosea, and H. Riechert. Theoretical and experimental analysis of 1.3-μm InGaNAs/GaAs lasers. *IEEE J. Sel. Top. Quantum Electron.*, 9:1228, 2003.

34. S. B. Healy and E. P. O'Reilly. Influence of electrostatic confinement on optical gain in GaInNAs quantum-well lasers. *IEEE J. Quant. Electron.*, 42:608, 2006.

35. C. A. Broderick, M. Usman, and E. P. O'Reilly. Derivation of 12- and 14-band **k·p** Hamiltonians for dilute bismide and bismide–nitride alloys. *Semicond. Sci. Technol.*, 28:125025, 2013.

36. I. A. Buyanova, G. Pozina, P. N. Hai, W. M. Chen, H. P. Xin, and C. W. Tu. Type I band alignment in the GaN_xAs_{1-x}/GaAs quantum wells. *Phys. Rev. B*, 64:033303, 2000.

37. P. N. Hai, W. M. Chen, I. A. Buyanova, H. P. Xin, and C. W. Tu. Direct determination of electron effective mass in GaNAs/GaAs quantum wells. *Appl. Phys. Lett.*, 77:1843, 2000.

38. F. Masia, A. Polimeni, G. B. H. von Högersthal, M. Bissiri, M. Capizzi, P. J. Klar, and W. Stolz. Early manifestation of localization effects in diluted Ga(AsN). *Appl. Phys. Lett.*, 82:4474, 2003.

39. J. D. Perkins, A. Mascarenhas, Y. Zhang, J. F. Geisz, D. J. Friedman, J. M. Olson, and S. R. Kurtz. Nitrogen-activated transitions, level repulsion, and band gap reduction in $GaAs_{1-x}N_x$ with x <0.03. *Phys. Rev. Lett.*, 82:3312, 1999.

40. E. P. O'Reilly, A. Lindsay, and S. Fahy. Theory of the electronic structure of dilute nitride alloys: Beyond the band-anti-crossing model. *J. Phys. Condens. Matter*, 16:S3257, 2004.

41. G. Hendorfer and J. Schneider. g-factor and effective mass anisotropies in pseudomorphic strained layers. *Semicond. Sci. Technol.*, 6:595, 1991.

42. P. Le Jeune, D. Robart, X. Marie, T. Amand, M. Brousseau, J. Barrau, V. Kalevich, and D. Rodichev. Anisotropy of the electron Landé g-factor in quantum wells. *Semicond. Sci. Technol.*, 12:380, 1997.

43. M. Oestreich, S. Hallstein, and W. W. Rühle. Spin quantum beats in semiconductors. *IEEE J. Sel. Top. Quantum Electron.*, 2:747, 1996.

44. A. Lindsay and E. P. O'Reilly. Unification of the band anticrossing and cluster-state models of dilute nitride semiconductor alloys. *Phys. Rev. Lett.*, 93:196402, 2004.

45. R. Fehse, S. Tomić, A. R. Adams, S. J. Sweeney, E. P. O'Reilly, A. D. Andreev, and H. Riechert. A quantitative study of radiative, Auger, and defect related recombination in 1.3-μm GaInNAs-based quantum-well lasers. *IEEE J. Sel. Top. Quantum Electron.*, 8:801, 2002.

46. M. Ehrhardt and T. Koprucki, editors. *Multiband Effective Mass Approximations: Advanced Mathematical Models and Numerical Techniques*. Berlin: Springer, 2014.

47. A. T. Meney, B. Gonul, and E. P. O'Reilly. Evaluation of various approximations used in the envelope-function method. *Phys. Rev. B*, 50:10893, 1994.

48. F. Szmulowicz. Derivation of a general expression for the momentum matrix elements within the envelope-function approximation. *Phys. Rev. B*, 51:1613, 1995.

49. C.-S. Chang, S. L. Chuang, J. R. Minch, W.-C. W. Fang Y. K. Chen, and T. Tanbun-Ek. Amplified spontaneous emission spectroscopy in strained quantum well lasers. *IEEE J. Sel. Top. Quantum Electron.*, 1:1100, 1995.

50. J. Hader, S. W. Koch, and J. V. Moloney. Microscopic theory of gain and spontaneous emission in GaInNAs laser material. *Solid State Electron.*, 47:513, 2003.

51. J. Hader, J. V. Moloney, and S. W. Koch. Microscopic evaluation of spontaneous emission- and Auger-processes in semiconductor lasers. *IEEE J. Quant. Electron.*, 41:1217, 2005.

52. S. W. Koch and W. W. Chow. *Semiconductor–Laser Fundamentals: Physics of the Gain Materials.* Berlin: Springer, 1999.

53. M. Hofmann, A. Wagner, C. Ellmers, C. Schlichenmeier, S. Schäfer, F. Höhnsdorf, J. Koch, W. Stolz, S. W. Koch, W. W. Rühle, J. Hader, J. V. Moloney, E. P. O'Reilly, B. Borchert, A. Y. Egorov, and H. Riechert. Gain spectra of (GaIn)(NAs) laser diodes for the 1.3-μm-wavelength regime. *Appl. Phys. Lett.*, 78:3009, 2001.

54. S. Tomić and E. P. O'Reilly. Gain characteristics of ideal dilute nitride quantum well lasers. *Physica E*, 13:1102, 2002.

55. E. P. O'Reilly and A. R. Adams. Band-structure engineering in strained semiconductor lasers. *IEEE J. Quant. Electron.*, 30:366, 1994.

56. A. R. Adams. Strained-layer quantum-well lasers. *IEEE J. Sel. Top. Quant. Electron.*, 17:1364, 2011.

57. S. Tomić and E. P. O'Reilly. Optimization of material parameters in 1.3μm InGaAsN–GaAs lasers. *IEEE Photon. Tech. Lett.*, 15:6, 2003.

58. P. J. Klar, H. Grüning, W. Heimbrodt, G. Weiser, J. Koch, K. Volz, W. Stolz, S. W. Koch, S. Tomić, S. A. Choulis, T. J. C. Hosea, E. P. O'Reilly, M. Hofmann, J. Hader, and J. V. Moloney. Interband transitions of quantum wells and device structures containing Ga(N,As) and (Ga,In)(N,As). *Semicond. Sci. Technol.*, 17:830, 2002.

59. E. P. O'Reilly. Valence band engineering in strained-layer structures. *Semicond. Sci. Technol.*, 4:121, 1989.

60. N. Tansu, N. J. Kirsch, and L. J. Mawst. Low-threshold-current-density 1300-nm dilute-nitride quantum well lasers. *Appl. Phys. Lett.*, 81:2523, 2002.

61. N. Tansu and L. J. Mawst. Low-threshold strain-compensated InGaAs(N) ($\lambda = 1.19$-1.31) quantum-well lasers. *IEEE Photon. Technol. Lett.*, 14:444, 2002.

62. N. Tansu, J.-Y. Yeh, and L. J. Mawst. Low-threshold 1317-nm InGaAsN quantum-well lasers with GaAsN barriers. *Appl. Phys. Lett.*, 83:2512, 2003.

63. N. Tansu, J.-Y. Yeh, and L. J. Mawst. Experimental evidence of carrier leakage in InGaAsN quantum-well lasers. *Appl. Phys. Lett.*, 83:2112, 2003.

64. S. Seki, H. Oohashi, H. Sugiura, T. Hirono, and K. Yokoyama. Study on the dominant mechanisms for the temperature sensitivity of threshold current in 1.3-μm InP-based strained-layer quantum-well lasers. *IEEE J. Quant. Electron.*, 32:1478, 1996.

65. J. W. Pan and J. I. Chyi. Theoretical study of the temperature dependence of 1.3-μm AlGaInAs-InP multiple-quantum-well lasers. *IEEE J. Quant. Electron.*, 32:2133, 1996.

66. J. Hader, S. W. Koch, J. V. Moloney, and E. P. O'Reilly. Gain in 1.3 μm materials: InGaNAs and InGaPAs semiconductor quantum-well lasers. *Appl. Phys. Lett.*, 77:630, 2000.

67. A. F. Phillips, S. J. Sweeney, A. R. Adams, and P. J. A. Thijs. The temperature dependence of 1.3- and 1.5-μm compressively strained InGaAs(P) MQW semiconductor lasers. *IEEE J. Sel. Top. Quantum Electron.*, 5:401, 1999.

68. T. R. Chen, B. Chang, L. C. Chiu, K. L. Yu, S. Margalit, and A. Yariv. Carrier leakage and temperature dependence of InGaAsP lasers. *IEEE J. Sel. Top. Quantum Electron.*, 43:217, 1983.

69. J. Barrau, T. Amand, M. Brousseau, R. J. Simes, and L. Goldstein. Induced electrostatic confinement of the electron gas in tensile strained InGaAs/InGaAsP quantum well lasers. *J. Appl. Phys.*, 8:801, 2002.

70. A. R. Adams, M. Asada, Y. Suematsu, and S. Arai. The temperature dependence of the efficiency and threshold current of $In_{1-x}Ga_xAs_yP_{1-y}$ lasers related to intervalence band absorption. *Jpn. J. Appl. Phys.*, 19:L621, 1980.

71. N. K. Dutta and R. J. Nelson. The case for Auger recombination in $In_{1-x}Ga_xAs_yP_{1-y}$. *Appl. Phys. Lett.*, 38:407, 1981.

72. A. D. Andreev and E. P. O'Reilly. Theoretical study of Auger recombination in a GaInNAs 1.3 μm quantum well laser structure. *Appl. Phys. Lett.*, 84:1826, 2004.

73. T. Higashi, S. J. Sweeney, A. F. Phillips, A. R. Adams, E. P. O'Reilly, T. Uchida, and T. Fujii. Experimental analysis of temperature dependence in 1.3μm AlGaInAs-InP strained MQW lasers. *IEEE J. Sel. Top. Quantum Electron.*, 5:413, 1999.

74. S. J. Sweeney. Novel experimental techniques for semiconductor laser characterisation and optimisation. *Phys. Scripta*, T114:152, 2004.

75. A. Haug. Relations between the T_0 values of bulk and quantum well GaAs. *Appl. Phys. B*, 44:151, 1987.

76. G. P. Agrawal and N. K. Dutta. *Long-Wavelength Semiconductor Lasers*. New York, NY: Van Nostrand Reinhold, 1986.

77. J. M. Pikal, C. S. Menoni, H. Temkin, P. Thiagarajan, and G. Y. Robinson. Carrier lifetime and recombination in long-wavelength quantum-well lasers. *IEEE J. Sel. Top. Quantum Electron.*, 5:613, 1999.

78. E. P. O'Reilly, G. Jones, M. Silver, and A. R. Adams. Determination of gain and loss mechanisms in semiconductor lasers using pressure techniques. *Phys. Status Solidi B*, 198:363, 1996.

79. S. J. Sweeney, R. Fehse, A. R. Adams, and H. Riechert. Intrinsic temperature sensitivities of 1.3 μm GaInNAs/GaAs-, InGaAsP/InP- and AlGaInAs/InP-based semiconductor lasers. In *Proceedings of the 16th Annual Meeting of the IEEE Lasers and Electro-Optics Society*, 2003.

80. Y. Q. Wei, M. Sadeghi, S. M. Wang, P. Modh, and A. Larsson. High performance 1.28 μm GaInNAs double quantum well lasers. *Electron. Lett.*, 41:1328, 2005.

81. S. M. Wang, G. Adolfsson, H. Zhao, Y. Q. Wei, J. S. Gustavsson, M. Sadeghi, and A. Larsson. High performance 1.3 μm GaInNAs quantum well lasers on GaAs. *Proc. SPIE*, 6909:690905, 2008.

82. J. S. Gustavsson, Y. Q. Wei, M. Sadeghi, S. M. Wang, and A. Larsson. 10 Gbit/s modulation of 1.3 μm GaInNAs lasers up to 110 °C. *Electron. Lett.*, 42:925, 2006.

83. Y. Q. Wei, J. S. Gustavsson, Å. Haglund, P. Modh, M. Sadeghi, S. M. Wang, and A. Larsson. High-frequency modulation and bandwidth limitations of GaInNAs double-quantum-well lasers. *Appl. Phys. Lett.*, 88:051103, 2006.

84. E. P. O'Reilly, N. Balkan, I. A. Buyanova, X. Marie, and H. Riechert. Dilute nitride and related mismatched semiconductor alloys. *IEE Proc. Optoelectron.*, 151:245, 2004.

85. G. Jaschke, R. Averbeck, L. Geelhaar, and H. Riechert. Low threshold InGaAsN/GaAs lasers beyond 1500 nm. *J. Cryst. Growth*, 278:224, 2005.

86. X. Yang, M. J. Jurkovic, J. B. Heroux, and W. I. Wang. Molecular beam epitaxy growth of InGaAsN: Sb-GaAs quantum wells for long-wavelength semiconductor lasers. *Appl. Phys. Lett.*, 75:178, 1999.

87. S. R. Bank, H. Bae, L. L. Goddard, H. B. Yuen, M. A. Wistey, R. Kudraweic, and J. S. Harris. Recent progress on 1.55-μm dilute-nitride lasers. *IEEE J. Quant. Electron.*, 43:773, 2007.

88. V.-M. Korpijärvi, E. L. Kantola, T. Leinonen, R. Isoaho, and M. Guina. Monolithic GaInNAsSb/GaAs VECSEL operating at 1550 nm. *IEEE J. Sel. Top. Quant. Electron.*, 21:1700705, 2015.

89. M. Silver, E. P. O'Reilly, and A. R. Adams. Determination of the wavelength dependence of Auger recombination in long-wavelength quantum-well semiconductor lasers using hydrostatic pressure. *IEEE J. Quant. Electron.*, 33:1557, 1997.

90. S. R. Bank, L. L. Goddard, M. A. Wistey, H. B. Yuen, and J. S. Harris. On the temperature sensitivity of 1.5-μm GaInNAsSb lasers. *IEEE J. Sel. Top. Quant. Electron.*, 11:1089, 2005.

91. K. Uesugi, I. Suemune, T. Hasegawa, T. Akutagawa, and T. Nakamura. Temperature dependence of band gap energies of GaAsN alloys. *Appl. Phys. Lett.*, 76:1285, 2000.

92. A. Polimeni, M. Capizzi, M. Geddo, M. Fischer, M. Reinhardt, and A. Forchel. Effect of nitrogen on the temperature dependence of the energy gap in $In_xGa_{1-x}As_{1-y}N_y$/GaAs single quantum wells. *Phys. Rev. B*, 63:195320, 2001.

93. I. Suemune, K. Uesugi, and W. Walukiewicz. Role of nitrogen in the reduced temperature dependence of band-gap energy in GaNAs. *Appl. Phys. Lett.*, 77:3021, 2000.

94. D. Alexandropoulos, M. J. Adams, Z. Hatzopoulos, and D. Syvridis. Proposed scheme for polarization insensitive GaInNAs-based semiconductor optical amplifiers. *IEEE J. Quant. Electron.*, 41:817, 2005.

95. X. Sun, N. Vogiatzis, and J. M. Rorison. Theoretical study on dilute nitride 1.3 μm quantum well semiconductor optical amplifiers: Incorporation of N compositional fluctuations. *IEEE J. Quant. Electron.*, 49:811, 2013.

96. J. Hashimoto, K. Koyama, T. Katsuyama, Y. Iguchi, T. Yamada, S. Takagishi, M. Ito, and A. Ishida. 1.3 μm travelling-wave GaInNAs semiconductor optical amplifier. In *Proceedings of Optical Amplifiers and Their Applications*, p. WB3, 2003.

97. J. Pozo, N. Vogiatzis, O. Ansell, P. J. Heard, J. M. Rorison, P. Tuomisto, J. Konttinen, M. Saarinen, C. Peng, J. Viheriälä, T. Leinonen, and M. Pessa. Fabrication and characterization of GaInNAs/GaAs semiconductor optical amplifiers. *Proc. SPIE*, 6997:69970C, 2008.

98. M. Dumitrescu, M. Wolf, K. Schulz, Y. Q. Wei, G. Adolfsson, J. Gustavsson, J. Bengtsson, M. Sadeghi, S. M. Wang, A. Larsson, J. Lim, E. Larkins, P. Melanen, P. Uusimaa, and M. Pessa. 10 Gb/s uncooled dilute-nitride optical transmitters operating at 1.3 μm. In *Proceedings of the Optical Fiber Communication Conference*, p. OWJ7, 2009.

99. D. Fitsios, G. Giannoulis, N. Iliadis, V.-M. Korpijärvi, J. Viheriälä, A. Laakso, S. Dris, M. Spyropoulou, H. Avramopoulos, G. T. Kanellos, N. Pleros, and M. Guina. High gain 1.3-μm GaInNAs SOA with fast-gain dynamics and enhanced temperature stability. *Proc. SPIE*, 8982:898208, 2014.

100. D. Fitsios, G. Giannoulis, V.-M. Korpijärvi, J. Viheriälä, A. Laakso, N. Iliadis, S. Dris, M. Spyropoulou, H. Avramopoulos, G. T. Kanellos, N. Pleros, and M. Guina. High-gain 1.3 μm GaInNAs semiconductor optical amplifier with enhanced temperature stability for all-optical signal processing at 10 Gb/s. *Appl. Optics*, 54:47, 2015.

101. D. Alexandropoulos and M. J. Adams. GaInNAs-based vertical cavity semiconductor optical amplifiers. *J. Phys. Condens. Matter*, 16:S3345, 2004.

102. A. H. Clark, S. Calvez, N. Laurand, R. Macaluso, H. D. Sun, M. D. Dawson, T. Jouhti, J. Kontinnen, and M. Pessa. 1.3 μm GaInNAs optically-pumped vertical cavity semiconductor optical amplifier. *IEEE J. Quant. Electron.*, 9:878, 2004.

103. N. Laurand, S. Calvez, M. D. Dawson, A. C. Bryce, T. Jouhti, J. Konttinen, and M. Pessa. Performance comparison of GaInNAs vertical-cavity semiconductor optical amplifiers. *IEEE J. Quant. Electron.*, 41:642, 2005.

104. N. Laurand, S. Calvez, M. D. Dawson, and A. E. Kelly. Index and gain dynamics of optically pumped GaInNAs vertical-cavity semiconductor optical amplifiers. *Appl. Phys. Lett.*, 87:231115, 2005.

105. N. Laurand, S. Calvez, M. D. Dawson, and A. E. Kelly. Slow-light in a vertical-cavity semiconductor optical amplifier. *Opt. Express*, 9:6858, 2006.

106. F. A. I. Chaqmaqchee and N. Balkan. Gain studies of 1.3-μm dilute nitride HELLISH-VCSOA for optical communications. *Nanoscale Res. Lett.*, 7:526, 2012.

107. S. B. Lisesivdin, N. A. Khan, S. Mazzucato, N. Balkan, M. J. Adams, V.-M. Korpijärvi, M. Guina, G. Mezosi, and M. Sorel. Optical gain in 1.3μm electrically driven dilute nitride VCSOAs. *Nanoscale Res. Lett.*, 9:22, 2014.

108. T. Piwonski, J. Pulka, G. Madden, G. Huyet, J. Houlihan, J. Pozo, N. Vogiatzis, P. Ivanov, J. M. Rorison, P. J. Barrios, and J. A. Gupta. Ultrafast gain and refractive index dynamics in GaInNAsSb semiconductor optical amplifiers. *J. Appl. Phys.*, 106:083104, 2009.

109. G. Giannoulis, V.-M. Korpijärvi, N. Iliadis, J. Mäkelä, J. Viheriälä, D. Apostolopoulos, M. Guina, and H. Avramopoulos. Bringing high-performance GaInNAsSb/GaAs SOAs to true data applications. *IEEE Photon. Tech. Lett.*, 27:1691, 2015.

110. G. Giannoulis, N. Iliadis, D. Apostolopoulos, P. Bakopoulos, H. Avramopoulos, V.-M. Korpijärvi, J. Mäkelä, J. Viheriälä, and M. Guina. 1.55-μm dilute nitride SOAs with low temperature sensitivity

for coolerless on-chip operation. In *Proceedings of the IEEE International Conference on Electronics, Circuits, and Systems*, p. 677, 2015.

111. V. Lordi, H. B. Yuen, S. R. Bank, and J. S. Harris. Quantum-confined Stark effect of GaInNAs(Sb) quantum wells at 1300–1600 nm. *Appl. Phys. Lett.*, 85:902, 2004.

112. T. Fujisawa, M. Arai, and F. Kano. Many-body design of highly strained GaInNAs electroabsorption modulators on GaInNAs ternary substrates. *J. Appl. Phys.*, 107:093107, 2010.

113. J. Hashimoto, K. Koyama, T. Ishizuka, Y. Tsuji, K. Fujii, T. Yamada, and T. Katsuyama. Electroabsorption effect of GaInNAs in waveguiding structure. *Jpn. J. Appl. Phys.*, 48:122403, 2009.

114. A. R. Adams. Band structure engineering to control impact ionisation and related high-field processes. *Electron. Lett.*, 40:1086, 2004.

115. S. L. Tan, W. M. Soong, J. E. Green, M. J. Steer, S. Zhang, L. J. J. Tan, J. S. Ng, I. P. Marko, S. J. Sweeney, A. R. Adams, J. Allam, and J. P. R. David. Experimental evaluation of impact ionization in dilute nitride GaInNAs diodes. *Appl. Phys. Lett.*, 103:102101, 2013.

116. G. Read, I. P. Marko, N. Hossain, and S. J. Sweeney. Physical properties and characteristics of III–V lasers on silicon. *IEEE J. Sel. Top. Quantum Electron.*, 21:1502208, 2015.

117. B. Kunert, K. Volz, and W. Stolz. Properties of laser applications of the GaP-based (GaNAsP)-material system for integration to Si substrates. *Springer Ser. Mat. Sci.*, 105:317, 2008.

118. N. Hossain, T. J. C. Hosea, S. J. Sweeney, S. Liebich, M. Zimprich, K. Volz, B. Kunert, and W. Stolz. Band structure properties of novel $B_x Ga_{1-x}P$ alloys for silicon integration. *J. Appl. Phys.*, 110:063101, 2011.

119. S. Liebich, M. Zimprich, A. Beyer, C. Lange, D. J. Franzbach, S. Chatterjee, N. Hossain, S. J. Sweeney, K. Volz, B. Kunert, and W. Stolz. Laser operation of Ga(NAsP) lattice-matched to (001) silicon substrate. *Appl. Phys. Lett.*, 99:071109, 2011.

120. J. Chamings, A. R. Adams, S. J. Sweeney, B. Kunert, K. Volz, and W. Stolz. Temperature dependence and physical properties of Ga(NAsP)/GaP semiconductor lasers. *Appl. Phys. Lett.*, 93:201108, 2008.

121. J. Chamings, S. Ahmed, A. R. Adams, S. J. Sweeney, V. A. Odnoblyudov, C. W. Tu, B. Kunert, and W. Stolz. Band anti-crossing and carrier recombination in dilute nitride phosphide based lasers and light emitting diodes. *Phys. Status Solidi B*, 246:527, 2009.

122. J. Wu, W. Walukiewicz, and E. E. Haller. Band structure of highly mismatched semiconductor alloys: Coherent potential approximation. *Phys. Rev. B*, 65:233210, 2001.

123. M. Seifikar, E. P. O'Reilly, and S. Fahy. Analysis of band-anticrossing model in GaNAs near localised states. *Phys. Status Solidi B*, 248:1176, 2011.

124. L. Ivanova, H. Eisele, M. P. Vaughan, Ph. Ebert, A. Lenz, R. Timm, O. Schumann, L. Geelhaar, M. Dähne, S. Fahy, H. Riechert, and E. P. O'Reilly. Direct measurement and analysis of the conduction band density of states in diluted $GaAs_{1-x}N_x$ alloys. *Phys. Rev. B*, 82:161201, 2010.

125. M. P. Vaughan, S. Fahy, E. P. O'Reilly, L. Ivanova, H. Eisele, and M. Dähne. Modelling and direct measurement of the density of states in GaAsN. *Phys. Status Solidi B*, 248:1167, 2011.

126. M. P. Vaughan and B. K. Ridley. Electron-nitrogen scattering in dilute nitrides. *Phys. Rev. B*, 75:195205, 2007.

127. N. Vogiatzis and J. M. Rorison. Single impurity Anderson model and band anti-crossing in the $Ga_{1-x}In_x N_y As_{1-y}$ material system. *Phys. Status Solidi A*, 205:120, 2008.

128. N. Vogiatzis and J. M. Rorison. Density of states of dilute nitride systems: Calculation of lifetime broadening. *J. Phys. Condens. Matter*, 21:255801, 2009.

129. M. Seifikar, E. P. O'Reilly, and S. Fahy. Self-consistent Green's function method for dilute nitride conduction band structure. *J. Phys. Condens. Matter*, 26:365502, 2014.

130. P. W. Anderson. Localized magnetic states in metals. *Phys. Rev.*, 124:41, 1961.

131. S. Fahy, A. Lindsay, H. Ouerdane, and E. P. O'Reilly. Alloy scattering of n-type carriers in $GaN_x As_{1-x}$. *Phys. Rev. B*, 74:035203, 2006.

132. A. P. Jauho and Wilkins J. W. Dilute resonant scatterers in a parabolic band: Density of states as a function of scattering strength. *Phys. Rev. B.*, 28:4628, 1983.

133. M. Seifikar, E. P. O'Reilly, and S. Fahy. Optical absorption of dilute nitride alloys using self-consistent Green's function method. *Nanoscale Res. Lett.*, 9:51, 2014.

134. M. Welna, R. Kudraweic, Y. Nabetani, and W. Walukiewicz. Band anticrossing in ZnOSe highly mismatched alloy. *Appl. Phys. Express*, 7:071202, 2014.

135. W. Welna, R. Kudraweic, Y. Nabetani, T. Tanaka, M. Jaquez, O. D. Dubon, K. M. Yu, and W. Walukiewicz. Effects of a semiconductor matrix on the band anticrossing in dilute group II-VI oxides. *Semicond. Sci. Technol.*, 30:085018, 2015.

136. C. A. Stephenson, W. A. O'Brien, M. Qi, M. Penninger, W. F. Schneider, and M. A. Wistey. Band anticrossing in dilute germanium carbides using hybrid density functionals. *J. Electron. Mater.*, 45:2121, 2016.

137. C. A. Stephenson, W. A. O'Brien, M. Penninger, W. F. Schneider, M. Gillet-Kunnath, J. Zajicek, K. M. Yu, R. Kudraweic, R. A. Stillwell, and M. A. Wistey. Band structure of germanium carbides for direct bandgap silicon photonics. *J. Appl. Phys.*, 120:053102, 2016.

138. R. Mouillet, L. de Vaulchier, E. Delporte, Y. Guldner, L. Travers, and J.-C. Harmand. Role of nitrogen in the mobility drop of electrons in modulation-doped GaAsN/AlGaAs heterostructures. *Solid State Commun.*, 126:333, 2003.

139. A. Patanè, G. Allison, L. Eaves, N. V. Kozlova, Q. D. Zhuang, A. Krier, M. Hopkinson, and G. Hill. Electron coherence length and mobility in highly-mismatched III–V-N alloys. *Appl. Phys. Lett.*, 93:252106, 2008.

140. A. Patanè, A. Ignatov, D. Fowler, O. Makarovsky, L. Eaves, L. Geelhaar, and H. Riechert. Hot-electrons and negative differential conductance in GaAs$_{1-x}$N$_x$. *Phys. Rev. B*, 72:033312, 2005.

141. A. Ignatov, A. Patanè, O. Makarovsky, and L. Eaves. Terahertz response of hot electrons in dilute nitride Ga(AsN) alloys. *Appl. Phys. Lett.*, 88:032107, 2006.

142. G. Allison, S. Spasov, A. Patanè, L. Eaves, A. Ignatov, D. K. Maude, M. Hopkinson, and R. Airey. Magnetophonon oscillations in the negative differential conductance of dilute nitride GaAs$_{1-x}$N$_x$ submicron diodes. *Phys. Rev. B*, 75:115325, 2007.

143. Y. Sun, M. P. Vaughan, A. Agarwal, M. Yilmaz, B. Ulug, A. Ulug, N. Balkan, M. Sopanen, O. Reentilä, M. Mattila, C. Fontaine, and A. Arnoult. Inhibition of negative differential resistance in modulation-doped n-type Ga$_x$In$_{1-x}$N$_y$As$_{1-y}$/GaAs quantum wells. *Phys. Rev. B*, 75:205316, 2007.

144. S. Fahy and E. P. O'Reilly. Intrinsic limits on electron mobility in dilute nitride semiconductors. *Appl. Phys. Lett.*, 83:3731, 2003.

145. J. Buckeridge and S. Fahy. Mobility in gated GaN$_x$As$_{1-x}$ heterostructures as a probe of nitrogen-related electronic states. *Phys. Rev. B*, 84:144120, 2011.

146. D. L. Young, J. F. Geisz, and T. J. Coutts. Nitrogen-induced decrease of the electron effective mass in GaAs$_{1-x}$N$_x$ thin films measured by thermomagnetic transport phenomena. *Appl. Phys. Lett.*, 82:1236, 2003.

147. D. Fowler, O. Makarovsky, A. Patanè, L. Eaves, L. Geelhaar, and H. Riechert. Electron conduction in two-dimensional GaAs$_{1-y}$N$_y$ channels. *Phys. Rev. B*, 69:153305, 2003.

148. M. P. Vaughan and B. K. Ridley. Solution of the Boltzmann equation for calculating Hall mobility in GaN$_x$As$_{1-x}$. *Phys. Rev. B*, 72:075211, 2005.

149. M. Seifikar, E. P. O'Reilly, and S. Fahy. Theory of intermediate- and high-field mobility in dilute nitride alloys. *Phys. Rev. B*, 84:165216, 2011.

150. N. Vogiatzis and J. M. Rorison. Negative differential velocity in ultradilute GaAs$_{1-x}$N$_x$ alloys. *J. Appl. Phys.*, 109:083720, 2011.

151. N. Vogiatzis and J. M. Rorison. Electron transport in bulk GaAsN. *Phys. Status Solidi B*, 248:1183, 2011.

10

Dilute Bismide Alloys

Christopher A.
Broderick

Igor P. Marko

Eoin P. O'Reilly

and

Stephen J. Sweeney

10.1 Introduction

Dilute bismide alloys are III–V semiconductor alloys containing small fractions of substitutional bismuth (Bi) atoms. Similar to the incorporation of nitrogen (N) to form the dilute nitride alloys described in Chapter 9, dilute bismide alloys are characterized by the fact that Bi acts as an isovalent impurity when incorporated into, e.g. (In)GaAs, to form the (In)GaBi$_x$As$_{1-x}$ alloy. Similar to the case of dilute nitride alloys, it is the large differences in size (covalent radius) and chemical properties (electronegativity) between Bi atoms and the group-V atoms they replace that brings about this impurity-like behavior, which in practice means that incorporating Bi at dilute concentrations has a significant impact on the material properties. Dilute bismide alloys, like dilute nitrides, are considered to be highly mismatched semiconductor materials. However, while N incorporation primarily affects the conduction band (CB) structure of the material into which it is incorporated, Bi, being significantly larger and more electropositive than N, primarily affects the valence band (VB). Since several general aspects of the physics of dilute bismide alloys are qualitatively similar to those of dilute nitrides, dilute bismide alloys can be considered a naturally complementary material system to the dilute nitrides.

The dilute bismide alloy $GaBi_xAs_{1-x}$ has several novel electronic properties. For example, substitution of As by Bi causes a rapid reduction in the band gap (E_g) with increasing Bi composition x, by up to 90 meV per % Bi at low x [1–6]. This strong Bi-induced decrease in E_g is accompanied by a strong increase in the VB spin-orbit-splitting energy (Δ_{SO}) [5–7], and the changes in both E_g and Δ_{SO} are characterized by strong, composition-dependent bowing, much like the band gap reduction in GaN_xAs_{1-x} [5,8]. These unusual material properties—among others, which we discuss in Section 10.2—have prompted significant interest in the development of dilute bismide alloys for a range of practical applications, including in semiconductor lasers [9–15] and photodiodes [16–19], as well as photovoltaics [20–23], spintronics [7,24,25], and thermoelectrics [26].

The theoretical modeling of dilute bismide alloys is significantly complicated by the impurity-like behavior of the Bi atoms in the alloy, and the resultant strong perturbation of the electronic structure of the host matrix semiconductor. The presence of alloy disorder due to the formation of pairs and larger clusters of Bi atoms in $GaBi_xAs_{1-x}$ has a particularly significant impact on the electronic structure, meaning that any serious attempt to quantify the consequences of Bi incorporation on the material properties must encompass an atomistic viewpoint. In Section 10.2, we provide an overview of the theory of the unusual properties of dilute bismide alloys, as revealed by experimental measurements, and discuss the various atomistic approaches that have been developed to understand the impact of Bi incorporation on the electronic structure. In particular, we focus on an atomistic tight-binding (TB) model we have developed to study the electronic structure of dilute bismide alloys of (In)GaAs(P) and compare the results of calculations undertaken using this approach to a series of experimental measurements of the electronic [5], optical [6,27,28], and spin properties [29] of $GaBi_xAs_{1-x}$ alloys.

While atomistic models provide significant insight into the properties of Bi-containing alloys, they are too computationally expensive to apply directly on the length scales required to model the properties of optoelectronic devices. This motivates the development of continuum models which are suited to calculating the properties of realistically sized Bi-containing device heterostructures. Through a series of detailed atomistic calculations, we review that the main features of the band structure of $GaBi_xAs_{1-x}$ and related alloys can be understood in terms of a band-anticrossing interaction, much like that considered for the dilute nitrides in Chapter 9. However, for $GaBi_xAs_{1-x}$ alloys the situation is the converse of that in GaN_xAs_{1-x}. In GaN_xAs_{1-x}, the band-anticrossing model describes the N composition-dependent coupling between the extended states of the host matrix (GaAs) CB edge and highly localized N-related impurity states which are resonant with the CB. In $GaBi_xAs_{1-x}$, atomistic calculations reveal that a Bi composition-dependent band-anticrossing interaction occurs between the extended states of the host matrix VB edge and highly localized Bi-related impurity states which are resonant with the VB.

Taking into account the p-like nature of the GaAs VB edge eigenstates, as well as the presence of strong spin-orbit-coupling in the VB, it is clear that even the simplest description of the $GaBi_xAs_{1-x}$ VB structure will be more complicated than that of the GaN_xAs_{1-x} CB structure. It is to this issue that we turn our attention in Section 10.3, where we review how atomistic calculations can be used to directly derive and parameterize valence band-anticrossing (VBAC) models for dilute bismide alloys, thereby removing the parametric ambiguity with which phenomenological models of this type are typically associated. Following this approach, we review the derivation of a 12-band $\mathbf{k \cdot p}$ Hamiltonian for the band structure of (In)$GaBi_xAs_{1-x}$ alloys directly from atomistic supercell calculations, and present a general method for calculating the band offsets in (001)-oriented pseudomorphically strained epilayers and quantum wells (QWs).

The practical application which has to date seen the largest research effort and most significant progress has been the development of $GaBi_xAs_{1-x}$ alloys for applications in GaAs-based semiconductor QW lasers [10–12,15,30]. A critical issue with existing InP-based 1.3- and 1.55-μm QW lasers is that their threshold currents and optical (cavity) losses tend to increase strongly with increasing temperature, due largely to a combination of two intrinsic loss mechanisms: non-radiative Auger recombination [31,32] and inter-valence band absorption (IVBA) [33]. $GaBi_xAs_{1-x}$ alloys having Bi compositions $x > 10\%$ have been demonstrated to have a band structure in which $\Delta_{SO} > E_g$, while also possessing a band gap $E_g \approx 0.8$ eV

(1.55 μm) [5,6]. It has therefore been proposed that this large VB spin-orbit-splitting energy could lead to the suppression of the dominant "conduction-hole-spin-hole" (CHSH) Auger recombination process in 1.55-μm semiconductor lasers [9,10,12] in which the energy and crystal momentum of an electron–hole pair recombining across the band gap excite a hole from the VB edge to the spin-split-off (SO) band, as depicted schematically in Figure 10.1.

The CHSH Auger process is a loss mechanism which accounts for the majority of the current at threshold in conventional InP-based devices, and which also strongly degrades the stability and efficiency of the device operation above room temperature [31]. Suppression of the CHSH Auger recombination pathway is therefore expected to bring about highly efficient, temperature-stable laser operation, opening the route to uncooled GaAs-based telecom lasers with significantly reduced power consumption and providing large energy savings compared to the InP-based technologies currently deployed in optical communication networks [9–12]. Using dilute bismide alloys as a route to realizing long-wavelength semiconductor lasers on GaAs substrates should also enable the growth of monolithic telecom-wavelength vertical-cavity surface-emitting lasers (VCSELs), thereby bringing the benefits of (Al)GaAs-based distributed Bragg reflectors (DBRs) to 1.55 μm. This is appealing from a technological perspective, because of the possibility of taking advantage of the enhanced carrier and optical confinement offered by (Al)GaAs-based heterostructures, as well as the potential to monolithically integrate long-wavelength semiconductor lasers with high-speed GaAs-based microelectronics.

Considerable progress has been made in developing epitaxial growth of $GaBi_xAs_{1-x}$ alloys, leading to the demonstration of an optically pumped bulk-like laser by Tominaga et al. in 2010 [34]. Room temperature electrically pumped lasing from a $GaBi_xAs_{1-x}$ QW laser was first demonstrated at $x = 2.1\%$ by Ludewig et al. in 2013 [30], and the Bi composition in such laser structures has since extended up to

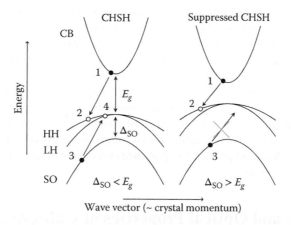

FIGURE 10.1 Left: Schematic illustration of the hot hole-producing CHSH Auger recombination process, in which a CB edge electron (1) recombines with a VB edge hole (2), with the energy released then exciting a VB edge hole (4) to the SO band (3), such that energy and crystal momentum ($\hbar k$) are conserved. The hot hole generated by this process dissipates its excess energy and crystal momentum primarily via scattering within the lattice, which generates phonons, increasing the lattice temperature and generating significant waste heat. The presence of strong, temperature-dependent CHSH Auger recombination in the materials forming the active region of InP-based 1.3- and 1.55-μm semiconductor lasers means that external cooling equipment is required to extract Auger-generated waste heat, in order to maintain operational stability, but at the cost of significantly increasing power consumption. Right: Schematic illustration of suppressed CHSH Auger recombination, achieved for a band structure in which the spin-orbit-splitting energy SO exceeds the band gap: $\Delta_{SO} > E_g$. In this case, the CHSH process is forbidden by conservation of energy, since the energy made available by an electron–hole pair recombining across the band gap is insufficient to excite a VB edge hole to the SO band. The suppression of IVBA proceeds in a similar manner: for $\Delta_{SO} > E_g$ the energy of a photon emitted by an electron–hole pair recombining across the band gap is insufficient to promote (by optical absorption) an electron from the SO band to a hole state at the VB edge, thereby suppressing photon reabsorption in the active region.

$x = 6.5\%$ by Butkutė et al. [35,36]. However, Bi compositions $x > 10\%$ are required to obtain a band structure having $\Delta_{SO} > E_g$ (and hence suppression of Auger recombination and IVBA processes involving the SO band) in GaBi$_x$As$_{1-x}$ alloys. Achieving high-quality GaBi$_x$As$_{1-x}$ layers with $x > 10\%$ is challenging due to the difficulty associated with epitaxial growth of Bi-containing alloys (see, e.g., Refs. [37,38] and references therein). While significant progress has been made in developing GaBi$_x$As$_{1-x}$ materials and devices over the last 5 years, efforts are ongoing to grow heterostructures with sufficiently high Bi compositions to demonstrate lasing at 1.55 μm, as well as suppression of Auger recombination and IVBA, in this new class of semiconductor alloys.

Given the significant potential of dilute bismide alloys for applications in next-generation semiconductor lasers, our focus throughout much of this chapter is on reviewing the theory of dilute bismide lasers. In Section 10.4 we provide an overview of the theory of GaAs-based dilute bismide QW lasers, using a theoretical model based on the 12-band $\mathbf{k}\cdot\mathbf{p}$ Hamiltonian derived in Section 10.3. Using this approach, we review (1) the impact of Bi on the band structure, density of states (DOS), and optical gain, (2) pathways toward device optimization at low Bi compositions, (3) trends in the expected device performance as a function of Bi composition, and (4) the potential of GaBi$_x$As$_{1-x}$ alloys for the development of highly efficient and temperature-stable 1.55-μm lasers. We also review direct comparisons of theoretical calculations of the (1) threshold characteristics, (2) spontaneous emission (SE), and (3) optical gain, to experimental measurements performed on first-generation GaBi$_x$As$_{1-x}$ QW laser devices [15,39].

While the majority of research on dilute bismides has to date centered on GaBi$_x$As$_{1-x}$ alloys and laser applications, the potential of alternative Bi-containing alloys for a range of practical applications has begun to drive an increasing diversity of research into related dilute bismide semiconductors. It is to these emerging directions in dilute bismide research that we turn our attention in Section 10.5. We discuss the significant potential offered by the quaternary alloys In$_y$Ga$_{1-y}$Bi$_x$As$_{1-x}$ and GaBi$_x$N$_y$As$_{1-x-y}$—for respective applications in InP-based mid-infrared photonics and multijunction solar cells—due to the broad flexibility of their band structures, associated with the ability to engineer the band gap, spin-orbit-splitting energy, strain, and band offsets over wide ranges [12,13,40–45]. We also review the recent development of GaBi$_x$As$_{1-x}$/GaN$_y$As$_{1-y}$ type-II QWs, and demonstrate that this novel class of GaAs-based, strain-balanced heterostructures has great potential to deliver emission/absorption across a broad range of near- and mid-infrared wavelengths on, making them of interest for applications in photonics and photovoltaics [46]. Finally, we provide a brief overview of emerging interest in narrow-gap Bi-containing alloys, which are generating strong interest for applications in mid-infrared photonics, as well as in spintronics.

Overall, this chapter should equip the reader with a comprehensive overview of the theory and simulation of dilute bismide alloys and provide an up-to-date overview of ongoing research on this emerging class of semiconductor materials.

10.2 Electronic and Optical Properties of GaBi$_x$As$_{1-x}$ Alloys: Atomistic Theory

In this section, we discuss some of the general properties of dilute bismide alloys, the understanding of which has been strongly informed by theoretical interpretations of experimental measurements. Since Bi acts as an isovalent impurity in (In)GaAs(P) and strongly perturbs the electronic structure, alloy disorder, which is inevitably present in real materials, plays an important role in determining the material properties. This, combined with the associated breakdown of the virtual crystal approximation in Bi-containing alloys, means that atomistic calculations are required in order to provide a quantitative understanding of the impact of Bi incorporation on the electronic structure. As such, we present our discussion in this section within the context of atomistic electronic structure calculations. In Section 10.2.1, we trace how such analyses have contributed strongly to developing the current understanding of this unusual class of semiconductor alloys. Next, in Section 10.2.2, we present and compare to experiment the results of atomistic calculations of a range of material properties for realistic, disordered materials. In Section 10.2.3 we provide a brief outlook for the atomistic theory of dilute bismide alloys, highlighting the tasks and challenges that must be completed and overcome to improve the fundamental understanding of dilute bismides and related emerging classes of highly mismatched semiconductor alloys.

Since $GaBi_xAs_{1-x}$ is the material which has been the subject of most of the research effort on dilute bismide alloys to date, it is to this material that we devote our attention in this section. However, the understanding of the impact of Bi incorporation on the properties of $GaBi_xAs_{1-x}$ alloys has been found to transfer to a range of related materials, so that the theory and material trends we outline in this section can be understood to apply in a general sense to other Bi-containing III–V semiconductor alloys.

10.2.1 Impact of Bi Incorporation on the Electronic Structure: Atomistic Theory

The first atomistic study of the impact of Bi incorporation on the electronic structure of GaAs was undertaken by Janotti et al. [47]. Using a first-principles approach based on density functional theory (DFT), Janotti et al. analyzed the electronic structure of $GaBi_xAs_{1-x}$ alloys, as well as the impact of co-alloying Bi and N in GaAs to form the $GaBi_xN_yAs_{1-x-y}$ quaternary alloy (cf. Section 10.5.2). In Ref. [47] the first detailed calculations for the III–V compound GaBi were also presented. In addition to calculating structural parameters such as the lattice constant and bulk modulus, Janotti et al. presented a calculation of the band structure of bulk GaBi in the zincblende phase. This calculation confirmed the presence of an extremely large spin-orbit-splitting energy in the VB, which first principles calculations indicate to be >2 eV [48]. Furthermore, an unusual (topological) band ordering was identified at the Γ-point in GaBi. As a result of the extremely large spin-orbit coupling, the Γ_{6c} CB edge states were calculated to lie lower in energy than the Γ_{8v} VB edge states, but higher in energy than the Γ_{7v} SO band edge states. As such, GaBi is predicted to be a metallic compound, with its negative band gap $E_g = E(\Gamma_{6c}) - E(\Gamma_{8v}) \simeq -1.5 eV$. While the analysis of Janotti et al. correctly suggested that Bi incorporation leads to a reduction of E_g with increasing x in $GaBi_xAs_{1-x}$, the calculated band gap bowing was approximately one-sixth of that calculated for GaN_xAs_{1-x} using the same approach. However, experimental measurements and subsequent theoretical analyses have revealed that replacing 1% of the As atoms in GaAs by Bi (N) causes a reduction of E_g of up to 90 (150) meV, meaning that while the impact of Bi incorporation is not as strong as that of N incorporation, the effects are broadly comparable in magnitude.

The first major advancement in the understanding of the electronic structure of $GaBi_xAs_{1-x}$ came from the work of Zhang et al. [49]. Using first-principles pseudopotential calculations to investigate the electronic structure as a function of Bi composition in a series of ordered $GaBi_xAs_{1-x}$ supercells, Zhang et al. provided a number of key insights into the impact of Bi incorporation on the GaAs electronic structure and also highlighted the differences between Bi and N incorporation in terms of their relative impact on the electronic structure. First, it was calculated that Bi incorporation causes a significant decrease (increase) in E_g (Δ_{SO}). The calculated E_g data for ordered alloy supercells were shown to be in good agreement with experiment for Bi compositions up to $x = 3.125\%$, the largest dilute composition which could be reliably investigated using the small calculational supercells employed. However, no comment on the precise nature of the strong Bi-induced band gap bowing was provided, and no *direct* attempt was made to elucidate its origin. Second, it was demonstrated that a substitutional Bi atom in GaAs forms impurity states which are (1) highly localized about the Bi atomic site, and (2) resonant with the extended states of the GaAs host matrix VB—i.e. lie below the GaAs VB edge in energy. Third, it was demonstrated that application of hydrostatic pressure decreases the energy of the Bi-related resonant states. As such, while the N-related states in the GaN_xAs_{1-x} CB can be pushed into the band gap under the influence of high hydrostatic pressure [50], the Bi-related states in $GaBi_xAs_{1-x}$ are expected to move deeper into the VB with increasing pressure.

The presence of spin-orbit coupling splits the resonant states associated with an isolated Bi impurity into a doublet (lying close in energy to the top of the GaAs VB) and a singlet (lying deep within the VB, due to the large spin-orbit coupling). Zhang et al. focused on the higher-energy Bi-related doublet, $|\psi_{Bi}\rangle$, and investigated the character and evolution of the these states as a function of Bi composition. Following this approach, it was found that the states $|\psi_{Bi}\rangle$ are formed of a linear combination of host matrix VB states—having wave vectors located along the Δ, Λ, and Σ directions in the Brillouin zone—which are folded back to Γ. This suggests that the Bi-related states, which lie energetically within the VB, form at the expense of a host matrix VB state. This is in contrast to the case of N incorporation in GaAs, in which N is understood to contribute an additional state to the CB. It was shown using a series of ordered $GaBi_xAs_{1-x}$ supercells containing a single substitutional Bi impurity that the calculated energy E_{Bi} of the doublet states $|\psi_{Bi}\rangle$ increases strongly with increasing supercell size (decreasing Bi composition), with the trend stabilizing in large supercells containing >2000 atoms where the impurity limit is approached. A significant number of host matrix states with large wave vectors are then folded back to Γ in these large supercells, and are hence available to contribute to the supercell representation of the states $|\psi_{Bi}\rangle$. As a result, a more accurate representation of the impact of Bi incorporation is obtained in larger supercells. In the impurity limit, Zhang et al. estimated that an isolated substitutional Bi impurity produces resonant doublet states that lie approximately 80 meV below the GaAs VB edge in energy—i.e. lying directly between the GaAs VB and SO band edge states in energy.

While this analysis has played a significant role in informing the current understanding of the fundamental aspects of the $GaBi_xAs_{1-x}$ electronic structure, the charge-patching method used for larger supercells by Zhang et al. is restricted in the size and type of alloys to which it can be applied. For example, such an approach is not well-suited to analyze the properties of randomly disordered alloys at low composition, since it relies on patching the converged charge densities from small supercells into larger supercells: finite-size effects associated with the periodic boundary conditions employed in the underlying small supercell calculation(s) are implicitly retained, which introduces some degree of spurious Bi-Bi interactions in the large supercell calculation(s). In order to overcome these limitations Usman et al. [5] developed a semi-empirical atomistic TB model, based on the sp^3s^* basis set originally introduced by Vogl et al. [51]. Since it employs a localized basis of atomic orbitals, the TB method is well-suited to study the impact of localized impurities on the electronic structure (as we highlighted and reviewed in our discussion of dilute nitride alloys in Chapter 9). Compared to related atomistic approaches, the TB method is computationally cheap and can be used to efficiently study the electronic structure of large alloy supercells containing upward of thousands of atoms. As described in Chapter 9, the TB method has previously been applied with much success to the study of GaN_xAs_{1-x} alloys, where it has provided quantitative predictions of several important material properties in real, disordered alloys (see, e.g., Refs. [50], [52] and [53]).

Using the TB method, in Ref. [5], Usman et al. significantly broadened the scope of the analysis undertaken by Zhang et al. to include detailed investigation of (1) coupling between localized Bi-related states and the extended states of the host matrix VB, (2) the nature of the localized impurity states associated with the formation of pairs and larger clusters of Bi atoms in a realistic, disordered alloy, (3) the impact of alloy disorder on the overall electronic structure, (4) trends in the electronic structure across an extended range of Bi compositions, and (5) the nature of the VB structure of $GaBi_xP_{1-x}$ alloys. The TB method has been highly successful in elucidating the impact of Bi on the electronic and optical properties of $GaBi_xAs_{1-x}$ alloys. For example, the calculated evolution of the band gap, spin-orbit-splitting energy, and electron effective g factor [5,29] are in quantitative agreement with experiment across the full composition range for which experimental data are available (cf. Section 10.2.2). Further investigations based on the TB method have also quantified the origin and consequences of the strong band gap bowing in Bi-containing alloys [42,54], as well as the effect of intrinsic alloy disorder on the optical properties of bulk and QW $GaBi_xAs_{1-x}$ [27,28].

Before describing the TB analysis, it is worthy of note that the prediction by Zhang et al. that Bi forms a resonant state in GaAs, energetically within the VB, is contradictory to the conclusions of Janotti et al. in Ref. [47], whose analysis suggested that a substitutional Bi impurity in GaAs forms a bound state lying approximately 100 meV above the GaAs VB edge in energy. This result of the analysis of Janotti et al. is surprising. It has been known since the 1960s that a substitutional Bi impurity in GaP produces a bound state lying approximately 100 meV above the GaP VB edge in energy [55]. Since the difference in energy between the $3p$ valence orbitals of P and the $6p$ valence orbitals of Bi is larger than that between the latter and the energy of the $4p$ valence orbitals of As, it is expected that Bi is less likely to form a bound impurity state in GaAs than in GaP, with the energy of the Bi-related impurity states in GaAs expected to be lower than those in GaP (relative to the host matrix VB edge) [5]. That Bi forms a bound state in $GaBi_xAs_{1-x}$ is therefore highly unlikely, given the known energy of the Bi-related states in GaP:Bi and the fact that the corresponding states in GaAs:Bi are expected to be at lower energy based on well-established chemical trends.

It has been proposed [4,54] that a band-anticrossing approach, similar to that for the GaN_xAs_{1-x} CB, can be applied to describe the VB structure in $GaBi_xAs_{1-x}$ alloys. However, this interpretation of the VB structure has generated to a degree of controversy [54,56]. In order to confirm the presence of a Bi-induced VBAC interaction, Usman et al. used the TB method to investigate the validity of this approach for the VB structure of ordered $GaBi_xP_{1-x}$ and $GaBi_xAs_{1-x}$ alloys by directly analyzing the Bi-related doublet states $|\psi_{Bi}\rangle$ [5]. For the case of a substitutional Bi atom in GaP the analysis and interpretation are straightforward, since the Bi-related states lie energetically within the GaP band gap. By explicitly constructing the states $|\psi_{Bi}\rangle$ in a series of ordered $GaBi_xP_{1-x}$ supercells (cf. Section 10.3.1), and by directly computing the strength with which these states couple to the GaP VB edge as a function of Bi composition, it was found that (1) in the impurity limit, an isolated, substitutional Bi impurity in GaP forms bound states which are highly localized about the Bi atomic site and lie approximately 120 meV above the GaP VB edge in energy, and (2) the coupling of the states $|\psi_{Bi}\rangle$ to the GaP VB edge states is precisely of the form $V_{Bi} = \beta\sqrt{x}$, as in GaN_xAs_{1-x} and GaN_xP_{1-x} alloys [5]. The calculated Bi bound state energy for GaP:Bi in the impurity limit is in good agreement with the experimentally measured value of Ref. [55], and the TB analysis explicitly confirms the presence of a VBAC interaction in ordered $GaBi_xP_{1-x}$ alloys [5].

In order to examine the validity of the VBAC model for $GaBi_xAs_{1-x}$, Usman et al. repeated the impurity state analysis described above for $GaBi_xP_{1-x}$ [5,54], and reached broadly similar conclusions to those of Zhang et al. [49]: (1) a substitutional Bi atom in $GaBi_xAs_{1-x}$ forms a set of impurity states that are res-onant with the extended states of the GaAs VB, (2) these Bi-related states are highly localized about the Bi atomic site, and (3) the calculated energy E_{Bi} of these Bi-related states varies strongly with supercell size (emphasizing the role of the finite-size effects described above), and converges in the impurity limit, a prediction which is qualitatively the same as, and quantitatively very close to, that of Zhang et al. The difference in the energies E_{Bi} calculated using the first principles pseudopotential and TB methods is likely related in part to the treatment of the supercells: the pseudopotential calculations rely on patching the con-verged charge density from a 64-atom supercell into larger supercells in order to reach the impurity limit, while the TB method treats all supercells directly. This analysis predicts that Bi forms resonant states in $GaBi_xAs_{1-x}$ which lie approximately 180 meV below the GaAs VB edge in energy in the impurity limit. Usman et al. further found that the coupling between the Bi-related resonant states and the GaAs VB edge states in $GaBi_xAs_{1-x}$ varies with Bi composition as $V_{Bi} = \beta\sqrt{x}$ in ordered alloys, explicitly confirming that the alloy VB edge states can be described in terms of a VBAC interaction. Recent experimental studies have provided direct evidence of the presence of Bi-related resonant states lying energetically within the $GaBi_xAs_{1-x}$ VB [57,58]. These detailed measurements support the general conclusions of the theoretical calculations of Refs. [5], [49] and [54], and provide direct experimental support for this interpretation of the VB structure of $GaBi_xAs_{1-x}$ alloys.

On the basis of DFT calculations, Deng et al. [56] suggested that the VBAC approach breaks down at higher Bi compositions, where alloy disorder effects become important and lead to strong modifications of the VB structure resulting from interactions between neighboring Bi atoms. Detailed analysis of this "band broadening" concept by Virkkala et al. [59] showed that a model Hamiltonian considering Bi–Bi interactions is indeed capable of producing a significant band gap with increasing Bi composition in disor-dered $GaBi_xAs_{1-x}$ alloys. However, this approach has two significant shortcomings. First, the predicted decrease in the $GaBi_xAs_{1-x}$ band gap is linear in x, meaning that the strong, composition-dependent bowing of the band gap observed in experimental measurements is not captured by the band broaden-ing interpretation of the band structure. Second, the proposed model requires extraction of supercell- and composition-dependent parameters from DFT calculations in order to define the alloy band gap, which limits predictive power. Furthermore, it has been demonstrated [54,60] that the conclusions of Deng et al. in Ref. [56] were based on a misinterpretation of the VB ordering in the small $GaBi_xAs_{1-x}$ supercells studied. As such, while the strict validity of the VBAC approach must necessarily break down in a disordered alloy, it is noteworthy that the predictions of the model are in quantitative agreement with experimental measurements of E_g and Δ_{SO} in $GaBi_xAs_{1-x}$ across the full range of compositions for which data are available (cf. Section 10.3.2). Detailed investigations suggest that the VBAC model describes well the band edge states in $GaBi_xAs_{1-x}$ alloys, but breaks down for states lying deeper in the VB, particularly in the vicinity of the SO band [29,61]. As we review in Section 10.4, this means that a VBAC approach provides a simple and accurate means by which to describe the properties of semiconductor lasers and related optoelectronic devices, where we are interested only in states lying energetically in the immediate vicinity of the CB and VB edges.

The understanding of the $GaBi_xAs_{1-x}$ electronic structure developed through atomistic calculations and experimental measurements then describes that (1) Bi introduces highly localized impurity states which are resonant with the GaAs VB, (2) the main features of the band edge states, and hence the alloy band structure, can be effectively described in terms of a VBAC interaction between localized Bi-related doublet states (lying between the GaAs VB and SO band edges in energy) and the extended states of the GaAs VB edge, (3) Bi-related singlet states, which lie much deeper (by several eV) within the VB, play a negligible role in determining the details of the electronic structure, and (4) the CB and SO bands in $GaBi_xAs_{1-x}$ are relatively unperturbed by Bi incorporation—their evolution with increasing Bi composition can be described in a conventional manner using, e.g. the virtual crystal approximation. Given the strong impact of alloy disorder on the electronic properties, interpretation of the details of the $GaBi_xAs_{1-x}$ band structure at higher Bi compositions then remains a somewhat open question: detailed and systematic analysis of the relative impact of As–Bi (VBAC-like) and Bi–Bi (band broadening) interactions on the band structure at higher compositions—building upon that already present in the literature—is required to provide definitive insight. However, as we describe throughout this chapter, describing the impact of Bi incorporation via a simple VBAC approach can be used to analyze the VB edge states in $(In)GaBi_xAs_{1-x}$ alloys with a degree of accuracy that is sufficient to describe and predict the optical properties of semiconductor lasers and related devices.

Since our focus in this chapter is on developing theory with a view to the modeling of photonic devices, we have concentrated in this overview on the fundamental features of the dilute bismide electronic structure which are relevant for such applications (primarily the evolution of the band edge energies and character of the band edge eigenstates with Bi composition). We note there have been several additional studies which have either further explored the electronic structure, or which have focused on other aspects of the interesting and unusual properties of dilute bismide alloys. For example, further DFT-based calculations of the electronic structure of $GaBi_xAs_{1-x}$ and related Bi-containing alloys have been carried out by Mbarki et al. [62], and more recently by Polak et al. [63,64]. Calculations of the structural, elastic, and vibrational properties of III-Bi compounds have been undertaken by Ferhat et al. [65,66], and theoretical investigations of the thermo-dynamics and kinetics of the growth of $GaBi_xAs_{1-x}$ alloys have been carried out by Morgan et al. [67,68] and by Punkinnen et al. [69,70]. We direct the reader to the references provided for further information on these and related aspects of the physics and chemistry of Bi-containing compounds and alloys.

10.2.2 Comparison of Atomistic Theory with Experimental Measurements

Having outlined the general features of the electronic structure of $GaBi_xAs_{1-x}$ alloys, we now turn our attention to direct comparisons between theory and experiment for this material system. We reserve detailed discussion of the impact of isolated Bi impurities and quantitative analysis of the VBAC model until Section 10.3.1, and focus here on disordered alloys. The TB method described in Section 10.2.1 provides a particularly useful framework within which to analyze such systems, since it enables direct calculation of the electronic structure in large supercells containing several thousand atoms, thereby allowing alloy disorder effects *at dilute compositions* to be treated in a realistic manner. Theoretical analysis of the electronic structure of disordered GaBi$_x$As$_{1-x}$ alloys and QWs using the TB method—full details of which are beyond the scope of this chapter—can be found in Refs. [5,27–29]. Here, we provide a brief overview of this work, describing some general features of the electronic structure, and comparing the results of TB calculations on disordered alloys directly with experimental measurements of the band gap, spin-orbit-splitting energy, and electron effective g factor.

The electronic structure calculations we review in this section were performed on a series of disordered 4096-atom supercells, which are either free-standing (unstrained) or under compressive pseudomorphic strain. Beginning with a GaAs supercell, a disordered GaBi$_x$As$_{1-x}$ supercell was generated by replacing As atoms by Bi atoms at randomly chosen sites on the anion sublattice. In the free-standing supercells, all atomic positions were allowed to relax freely—using a valence force field model based on the Keating potential [71,72]—with no constraints applied to the atoms at the supercell boundaries. For pseudomorphically strained supercells, a macroscopic pseudomorphic strain was imposed by fixing the positions of the atoms on the exterior boundaries and relaxing the positions of the atoms in the interior of the supercell [29]. For the construction of the supercell TB Hamiltonian, the orbital energies at a given atomic site were computed depending on the overall neighbor environment. Local strain effects were taken into account by scaling the interatomic interaction matrix elements based on the relaxed nearest-neighbor bond length d as $\left(\frac{d_0}{d}\right)^\eta$, where d_0 is the equilibrium bond length in the equivalent binary compound and η is a dimensionless scaling parameter (the value of which depends on the type and symmetry of the interaction). The bond angle dependence of the interaction matrix elements was represented using the two-center integrals of Slater and Koster [73]. Full details of the TB model are described in Ref. [5].

In order to facilitate the TB analysis of the electronic structure of GaBi$_x$As$_{1-x}$ alloys, the fractional Γ character spectrum $G_\Gamma(E)$ serves as a useful tool to probe the nature and origin of the alloy eigenstates. $G_\Gamma(E)$ is calculated in general by projecting a specific choice of host matrix (in this case, GaAs) Γ-point eigenstates onto the full spectrum of zone-center eigenstates for a given alloy supercell. Specifically, using (0) and (1) to respectively denote the unperturbed host matrix and Bi-containing alloy supercell states, $G_\Gamma(E)$ is computed by projecting the unperturbed zone-center eigenstates $\{|\psi_l^{(0)}\rangle\}$ of a $2M$-atom Ga$_M$As$_M$ supercell onto the eigenstates $\{|\psi_k^{(1)}\rangle\}$ computed at Γ for a given Ga$_M$Bi$_L$As$_{M-L}$ alloy supercell

$$G_\Gamma(E) = \sum_k \sum_{l=1}^{g(E_l)} |\langle \psi_k^{(1)} | \psi_l^{(0)} \rangle|^2 \, T\left(E - E_k\right), \tag{10.1}$$

where $x = \frac{L}{M}$ is the Bi composition, E_k is the energy of the zone-center alloy supercell eigenstate $|\psi_k^{(1)}\rangle$, $g(E_l)$ is the degeneracy of the unperturbed host matrix eigenstate having energy E_l at Γ (so that $g(E_l) = 2, 4$, and 2 for the CB, VB, and SO band edge eigenstates, respectively), and where the "top hat" function $T\left(E - E_k\right)$ is defined so that $G_\Gamma(E_k)$ has a value of unity for a doubly (spin) degenerate host matrix Γ state at energy E_k. As has previously been demonstrated—in Refs. [5], [27] and [29]—computing $G_\Gamma(E)$ for a range of band edge states provides (1) a consistent approach with which to analyze the evolution of the electronic structure in both ordered and disordered alloy supercells, and (2) detailed insight into the nature of the zone-center eigenstates in a Bi-containing alloy, by quantifying their origin in terms of those of the unperturbed host matrix semiconductor.

Figure 10.2 shows the calculated distribution of the Γ character associated with the extended GaAs host matrix heavy-hole (HH), light-hole (LH), and SO band edge states over the full spectrum of zone-center alloy states, for a disordered 4096-atom Ga$_{2048}$Bi$_{82}$As$_{1966}$ ($x = 4\%$) supercell under compressive pseudomorphic strain [29]. The top, middle, and bottom panels of Figure 10.2 show, respectively, the $G_\Gamma^{HH}(E_{j,1})$, $G_\Gamma^{LH}(E_{j,1})$ and $G_\Gamma^{SO}(E_{j,1})$ spectra, calculated using the unperturbed GaAs host matrix HH, LH, and SO states

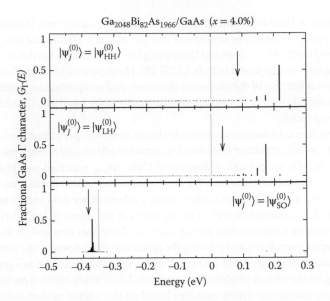

FIGURE 10.2 Black lines: Calculated distribution of the Γ character associated with the GaAs HH (top panel), LH (middle panel), and SO (bottom panel) band edge states over the full spectrum of alloy zone-center valence states in a disordered, 4096-atom $Ga_{2048}Bi_{82}As_{1966}$ ($x = 4\%$) supercell (cf. Equation 10.1). The supercell was placed under compressive pseudomorphic strain corresponding to epitaxial growth on a GaAs substrate, as described in the text. (Adapted from C.A. Broderick et al., *Phys. Rev. B*, 90, 195301, 2014.) Gray lines: Corresponding distribution of the Γ character in a Bi-free $Ga_{2048}As_{2048}$ supercell. The zero of energy is taken at the GaAs VB edge.

$|\psi_{l,0}\rangle = |\psi_{HH,0}\rangle$, $|\psi_{LH,0}\rangle$ and $|\psi_{SO,0}\rangle$ in Equation 10.1. For comparative purposes, the corresponding Γ character of the unperturbed $Ga_{2048}As_{2048}$ host matrix is also shown in Figure 10.2 using gray lines.

Comparing the computed spectra for the host matrix and Bi-containing supercells reveals several important effects of Bi incorporation on the VB structure. First, incorporation of Bi leads to strong hybridization between the GaAs VB edge states and a series of Bi-related states lying at lower energies, within the VB. Second, this hybridization acts to push to the alloy VB edge states upward in energy with increasing Bi composition. In the ordered supercell case, this behavior can be explained directly in terms of a VBAC interaction (cf. Section 10.3.1), and this general behavior is clearly observed in disordered $GaBi_xAs_{1-x}$ alloys across a broad composition range [5]. Third, we note that the Γ character associated with the GaAs HH and LH states is distributed over a large number of VB states in a disordered $GaBi_xAs_{1-x}$ supercell. There are primarily two reasons for this: (1) the localized impurity states associated with substitutional Bi atoms are resonant with a large density of host matrix VB states, which leads to strong energy broadening of the associated quasi-localized resonances (as described for N impurities in Chapter 9) and is reflected by the large number of alloy VB states having small Γ character [54], and (2) the presence of Bi–Bi pairs (in which a single Ga atom has two Bi nearest neighbors) and larger clusters of Bi atoms (having shared Ga nearest neighbors) leads to the formation of localized impurity states which more strongly bind holes—i.e. which lie close to and/or above the GaAs VB edge in energy [5]. This progression of localized Bi-related cluster states leads, in a disordered alloy, to the distribution of a significant fraction of the GaAs Γ character over a range of VB states lying close in energy to the alloy VB edge. This then quantifies quantifying the strong inhomogeneous broadening of the band edge features in $GaBi_xAs_{1-x}$ alloys observed in spectroscopic measurements in terms of the strong impact of Bi-related disorder on the nature of the $GaBi_xAs_{1-x}$ VB [27]. Fourth, in contrast to the character of the alloy VB edge, we note that the feature associated with the Γ character of the GaAs SO band edge states, visible in the bottom panel of Figure 10.2, remains relatively sharp in $GaBi_xAs_{1-x}$. This suggests that the impact of Bi incorporation and alloy disorder on the SO band is less perturbative than it is on the HH- and LH-like states at the VB edge. However, additional calculations indicate that this distribution of the SO-related Γ character persists as the Bi composition is increased, leading to strong energy broadening and a breakdown of the alloy SO band edge at higher Bi compositions [5]. This suggests a breakdown of a conventional band description of the SO states, which has been confirmed by further analysis [29,61]. Finally, as expected for an epilayer under compressive strain, we see that the highest-energy alloy VB state is predominantly HH-like, having >50% GaAs HH Γ character. We additionally note that both the LH and HH band edges in the alloy are strongly inhomogeneously broadened, which is evidenced by the large number of alloy VB states over which the GaAs VB edge Γ character is distributed [27].

Similar analysis of the conduction states suggests that the impact of Bi incorporation on the CB is generally weak, leading to a downward shift in energy and decrease in the CB-related Γ character that is consistent with alloying effects in conventional semiconductor materials. Overall, these calculated trends confirm that Bi incorporation almost exclusively and strongly perturbs the VB structure, while the CB structure is, by comparison, relatively unaffected [5].

On the basis of these supercell calculations, the variation of the band gap and spin-orbit-splitting energy as a function of Bi composition can be extracted for disordered GaBi$_x$As$_{1-x}$ alloys. These values were calculated directly for each supercell on the basis of the calculated $G_\Gamma(E)$ spectra: the band gap was computed as the difference in energy between the (largely GaAs-like) lowest-energy CB state and the alloy VB state having the largest HH or LH GaAs Γ character; the spin–orbit-splitting energy was then computed as the difference in energy between this (typically highest-energy) alloy VB state and a weighted average energy for the alloy SO band (obtained by averaging over the full set of zone-center alloy energies using the value of the SO Γ character as a weight for each contributing eigenstate) [5]. Given the important role played by alloy disorder in determining the details of the VB structure, for each Bi composition considered this analysis was undertaken for five supercells in which the Bi atoms were randomly distributed at different sites on the anion sublattice. The values of E_g and Δ_{SO} were then obtained for that Bi compostion as the average of those calculated using each of the different supercells considered. The results of these calculations are shown in Figure 10.3a, which compares the calculated variation of E_g and Δ_{SO} as a function of Bi composition x in GaBi$_x$As$_{1-x}$ to a range of experimental measurements. We note that the theoretical calculations are in excellent agreement with the experimental measurements, across the full composition range for which data are available. In particular, the TB model accurately captures the strong, composition-dependent bowing of E_g and Δ_{SO} demonstrating—in line with the discussion above—that the observed strong decrease (increase) of E_g (Δ_{SO}) with increasing x is primarily attributable to the impact of Bi incorporation on the VB structure, and results largely from the upward shift in energy of the alloy VB edge. The TB analysis confirms that this behavior originates due to hybridization of the GaAs VB edge states with a full distribution of Bi-related localized states in a disordered alloy [5,54].

The Landé effective g factor of conduction electrons (g_e^*) is a key parameter in semiconductor materials, describing the response of the electron spins to externally applied magnetic fields, as well as providing useful information regarding the symmetry of the electron and hole states at the Γ-point in the Brillouin zone. Therefore, in addition to providing information pertinent to evaluating the potential of specific materials for spintronic applications, analysis of the electron spin properties and g factor gives insight into the material band structure, providing valuable data for material characterization and modeling. In particular, g_e^* is extremely sensitive to the separation in energy between zone-center crystal eigenstates, as well as hybridization between zone-center eigenstates caused by, e.g. a reduction of symmetry due to strain [75] or quantum confinement [76]. As a result, calculations of g_e^* provide a stringent test of theoretical models of the material band structure [77].

Optical excitation of a semiconductor material with circularly polarized light is known to generate a spin-polarized population of electrons in the CB. These electrons can then relax by recombining with holes in the VB and emitting photons which must be circularly polarized in order to conserve angular momentum, with the nature of the circular polarization determined by the orientation of the electron spin. As such, Larmor precession of electron spins in the presence of an externally applied magnetic field leads to beating of the circular polarization of the photoluminescence (PL) generated by exciting a material sample with

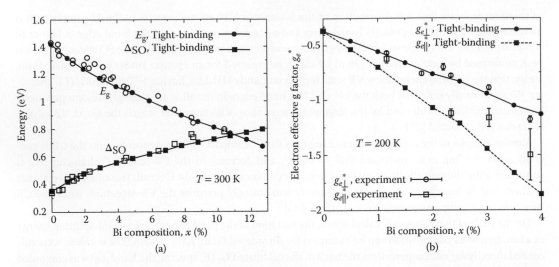

(a)   (b)

FIGURE 10.3 (a) Comparison between theory and experiment for the variation of the band gap (E_g) and spin–orbit-splitting energy (Δ_{SO}) as a function of Bi composition x in GaBi$_x$As$_{1-x}$ at 300 K. Experimental measurements of E_g are depicted by open circles. (The experimental data are taken from S. Tixier et al., *Appl. Phys. Lett.*, 82, 2245, 2003; S. Francoeur et al., *Appl. Phys. Lett.*, 82, 3874, 2003; W. Huang et al., *J. Appl. Phys.*, 98, 053505, 2005; K. Alberi et al., *Phys. Rev. B*, 75, 045203, 2007; Z. Batool et al., *J. Appl. Phys.*, 111, 113108, 2012; Yoshida et al., *Jpn. J. Appl. Phys.*, 42, 371, 2003.) Experimental measurements of Δ_{SO} are depicted by open squares. (The experimental data are taken from K. Alberi et al., *Phys. Rev. B*, 75, 045203, 2007; Z. Batool et al., *J. Appl. Phys.*, 111, 113108, 2012; B. Fluegel et al., *Phys. Rev. Lett.*, 97, 067205, 2006.). Closed circles (squares) show the variation of E_g (Δ_{SO}) calculated using the TB model for a series of disordered, 4096-atom supercells. (Adapted from M. Usman et al., *Phys. Rev. B*, 84, 245202, 2011; C.A. Broderick et al., *Semicond. Sci. Technol.*, 28, 125025, 2013.). (b) Comparison between theory and experiment for the variation of the transverse and longitudinal components $g^*_{\perp,e}$ and $g^*_{\parallel,e}$ of the electron effective g factor g^*_e as a function of Bi composition x in GaBi$_x$As$_{1-x}$ at 200 K. Closed circles (squares) show the variation of $g^*_{\perp,e}$ ($g^*_{\parallel,e}$) calculated using the TB model for a series of disordered, 4096-atom supercells placed under pseudomorphic strain as described in the text. Open circles (squares) show the variation of $g^*_{\perp,e}$ ($g^*_{\parallel,e}$) measured for a series of GaBi$_x$As$_{1-x}$/GaAs epitaxial layers using spin quantum beat spectroscopy. (Adapted from Broderick, C.A. et al., *Phys. Rev. B*, 90, 195301, 2014.)

circularly polarized light [78]. Using time-resolved PL spectroscopy, the Larmor frequency $\omega = \frac{g^*_e \mu_B}{\hbar} B$ of these "quantum beats" can be determined, allowing the relevant component of g^*_e to be determined directly for a given relative orientation of the material sample and externally applied magnetic field [76]. As a result of the long coherence times associated with the beating of the PL polarization, this experimental technique, known as spin quantum beat spectroscopy, enables highly accurate determination of the electron Larmor frequency, and hence the effective g factor g^*_e [77,78]. In order to investigate in further detail the impact of Bi incorporation on the electronic structure, as well as to provide insight into the impact of Bi incorporation on the spin properties of electrons, Broderick et al. [29] developed a TB-based model for the calculation of g^*_e in disordered GaBi$_x$As$_{1-x}$ alloys. Here, we provide a brief summary of the analysis of g^*_e in GaBi$_x$As$_{1-x}$, including comparison to experimental data, full details of which can be found in Ref. [29].

The hybridization between zone-center LH and SO states brought about by the tetragonal deformation and associated reduction of symmetry in pseudomorphically strained layers is known to play a key role in determining the anisotropy of g^*_e in conventional semiconductor alloys, leading to distinct transverse and longitudinal components $g^*_{\perp,e}$ and $g^*_{\parallel,e}$ of the g^*_e tensor [75]. Returning our attention to Figure 10.2, we begin by noting that the calculated distribution of GaAs SO character for this typical pseudomorphically strained GaBi$_x$As$_{1-x}$ supercell highlights that the strong inhomogeneous broadening of the LH-like alloy VB states brought about by alloy disorder minimizes this expected strain-induced LH–SO hybridization—i.e. there is no notable projection of the GaAs SO band edge character onto LH-like states close to the alloy

VB edge—providing the first hint that the unusual VB structure of Bi-containing alloys is likely to have a strong impact on g_e^*.

In the absence of strain, g_e^* is expected to be isotropic, with $g_{\perp,e}^* = g_{\parallel,e}^*$ [75]. Comparative calculations we have undertaken on free-standing (unstrained) supercells demonstrate strain-induced mixing between the LH and SO states in ordered supercells, enabling us to conclude that the strong inhibition of such mixing observed in Figure 10.2 arises due to the disorder-induced energy broadening of the GaAs LH character over a significant number of alloy VB states. In order to investigate the impact of Bi incorporation, pseudo-morphic strain, and Bi-related alloy disorder on the isotropy of g_e^*, we have therefore first calculated $g_{\perp,e}^*$ and $g_{\parallel,e}^*$ as a function of Bi composition x in a series of free-standing supercells. These calculations show that (1) Bi incorporation leads to a rapid, monotonic increase in the magnitude of g_e^* with increasing x, and (2) the reduced local symmetry present in a disordered GaBi$_x$As$_{1-x}$ alloy supercell has a negligible effect on the isotropy of g_e^* [29]. The latter result suggests that the anisotropy of g_e^* observed in experimental measurements performed on GaBi$_x$As$_{1-x}$ epitaxial layers arises primarily due to the reduction in symmetry brought about by the impact of biaxial stress on the band structure, much like in conventional semiconductor materials, but that the LH–SO mixing to which this behavior is attributed in conventional semiconductor alloys [75] is not a significant factor determining the anisotropy of g_e^* in GaBi$_x$As$_{1-x}$ alloys.

In order to determine the anisotropy of g_e^* in real, pseudomorphically strained epitaxial layers, as well as to compare the results of the theoretical calculations to experiment, Broderick et al. repeated this analysis for disordered GaBi$_x$As$_{1-x}$ supercells under the application of a macroscopic strain corresponding to pseudomorphic growth on a GaAs substrate [29] (applied as described above). The results of this analysis are summarized in Figure 10.3b, which shows the calculated variation of $g_{\perp,e}^*$ (closed circles) and $g_{\parallel,e}^*$ (closed squares) with x, as well as experimental measurements of $g_{\perp,e}^*$ (open circles) and $g_{\parallel,e}^*$ (open squares) performed on a series of strained GaBi$_x$As$_{1-x}$ epitaxial layers at a temperature of 200 K [29].

In the presence of pseudomorphic strain, we see that g_e^* is strongly anisotropic. As in the unstrained case, we note that the magnitudes of both $g_{\perp,e}^*$ and $g_{\parallel,e}^*$ increase strongly and monotonically with increasing x. We also calculate that (1) the magnitude of $g_{\perp,e}^*$ exceeds that of $g_{\parallel,e}^*$ for non-zero x, in agreement with experiment, and (2) the magnitude of $g_{\parallel,e}^*$ is close to the isotropic value of g_e^* calculated in the absence of strain [29], both of which are expected for compressively strained epitaxial layers [75]. Based on these comparative calculations, we conclude that while the overall symmetry of g_e^* is largely independent of the effects of alloy disorder, the impact of pseudomorphic strain is crucial for determining the experimentally observed anisotropy of g_e^* in GaBi$_x$As$_{1-x}$ epitaxial layers. We note that the calculated enhancement of $g_{\perp,e}^*$ with increasing x is in excellent agreement with experiment, confirming not only the accuracy of the detailed picture of the GaBi$_x$As$_{1-x}$ VB structure provided by the TB model, but also the strong impact of Bi incorporation on the spin properties of electrons in GaBi$_x$As$_{1-x}$ alloys.

The effects of alloy disorder on the calculated variation of $g_{\perp,e}^*$ and $g_{\parallel,e}^*$ with x can be understood in more detail by considering the distribution of host matrix Γ character shown in Figure 10.2. In a disordered GaBi$_x$As$_{1-x}$ alloy, the VB edge GaAs Γ character is distributed over a large number of alloy VB states, with the details of the distribution determined by the precise short-range disorder present in the alloy. The details of this distribution of the GaAs Γ character onto VB states lying below the alloy VB edge in energy plays a crucial role in determining the calculated variation of $g_{\perp,e}^*$ and $g_{\parallel,e}^*$ with x, since the contribution from a given alloy VB state to a specific component of g_e^* is weighted by its associated host matrix Γ character [29]. Specifically, mixing of GaAs VB edge Γ character into alloy VB states with greater separation in energy from the alloy CB edge results in contributions to g_e^* which act against the much stronger increase in the magnitudes of $g_{\perp,e}^*$ and $g_{\parallel,e}^*$ than would be expected based solely upon the measured variation of E_g and Δ_{SO} with x [24,29]. We recall from Section 10.2.1 that this breakdown of the simple VBAC picture of the VB structure in the vicinity of the SO band [29] suggests that detailed approaches are typically required to understand material properties involving eigenstates that are energetically remote from the CB and VB edges.

Overall, we conclude that the TB model developed in Ref. [5] provides a detailed and quantitative understanding of the electronic structure of GaBi$_x$As$_{1-x}$ and related alloys. The variation of the band gap, spin–orbit-splitting energy, and electron effective g factor as a function of Bi composition calculated using the TB method are in good, quantitative agreement with experiment, and this theoretical approach has delivered significant insight into the impact of Bi incorporation on the GaBi$_x$As$_{1-x}$ band structure.

10.2.3 Outlook for Atomistic Theory

Despite the detailed insight atomistic calculations have provided into the impact of Bi incorporation on the electronic and optical properties of GaBi$_x$As$_{1-x}$ and related alloys, there is still significant potential to develop and apply atomistic methods and models to further analyze a range of fundamental material properties in this emerging class of III-V semiconductors.

For example, based on the observation that the VBAC picture of the band structure—which very accurately describes the evolution of E_g and Δ_{SO} with x (cf. Section 10.3), as well as the optical properties of dilute bismide semiconductor lasers (cf. Section 10.4)—breaks down for states energetically remote from the VB edge, it is possible to conclude that simple, phenomenological approaches are ill equipped to account for the impact of Bi incorporation and alloy disorder on a number of technologically relevant material properties. This is perhaps unsurprising, since the VBAC model treats only the main features of the band structure and does so in a phenomenological manner. However, this conclusion has significant implications for further theoretical analysis of dilute bismide alloys, implying, e.g. that simple models of the band structure are ill equipped to quantitatively describe key processes such as Auger recombination and IVBA, which involve a broad energy range of alloy VB states including, in particular, states lying close to the SO band edge in energy. As was described for dilute nitride alloys in Chapter 9, detailed and specialized theoretical approaches must therefore be developed in order to obtain a quantitative understanding of complicated processes and properties of this nature.

There have to date been no *direct* theoretical investigations on a fundamental level of the impact of Bi incorporation on a number of important material properties and processes including, e.g. (1) carrier lifetimes, (2) radiative and non-radiative recombination rates, (3) carrier transport, and (4) impact ionization. Detailed understandings of these and related properties must be developed in order to inform the development of Bi-containing alloys for applications in a range of novel semiconductor devices.

Previous theoretical analysis has highlighted the strong influence of Bi-related alloy disorder in determining the details of the electronic structure, meaning that further quantitative theoretical analysis of the properties of Bi-containing alloys is likely to require detailed atomistic treatments. This requirement presents the theoretical community with a series of interesting challenges, centered on developing the capability to describe—on an atomic scale—the impact of short-range disorder on a range of fundamental properties and processes in highly mismatched alloys, where dilute concentrations of impurities strongly perturb the electronic structure in a manner that is not compatible with many commonly employed approximations due to the prominence of alloy disorder effects in determining the material properties.

On this basis, further theoretical research on dilute bismides—perhaps informed by related research on dilute nitride alloys—is likely to deliver not only an improved understanding of this interesting class of semiconductors, but also insights and methods that can be applied to a broad range of emerging highly mismatched materials.

10.3 Modeling Dilute Bismide Band Structure: The k·p Method

Having presented an overview of atomistic electronic structure calculations for GaBi$_x$As$_{1-x}$ alloys, in this section, we present a (continuum) **k·p** model of the (In)GaBi$_x$As$_{1-x}$ band structure which is well suited to the simulation of the electronic and optical properties of dilute bismide materials and devices.

In Section 10.3.1, we outline how a VBAC approach can be used to provide a simple yet effective interpretation of the band structure of dilute bismide alloys and describe how detailed atomistic calculations have provided a means to deliver quantitative predictive power to this phenomenological picture of the electronic properties.

In Section 10.3.2 we review the use of TB supercell calculations to derive an extended basis set 12-band **k·p** Hamiltonian for (In)GaBi$_x$As$_{1-x}$ alloys. This rigorous analysis verifies that localized impurity states associated with isolated substitutional Bi atoms play the dominant role in determining the main features of the band structure close in energy to the VB edge. The result of this is—despite the fact that more sophisticated atomistic approaches are typically required to understand the detailed properties of these materials—a VBAC-based approach is sufficient to provide a understanding of the electronic and optical properties of dilute bismide devices such as QW lasers in which only quantitative optically active states lying close in energy to the CB and VB edges are of significant importance.

In Section 10.3.3 we present a general method—based on the 12-band **k·p** Hamiltonian—for calculating the band offsets in pseudomorphically strained (In)GaBi$_x$As$_{1-x}$ epitaxial layers and QWs, such as those which are attracting significant attention for the development of highly efficient GaAs- and InP-based semiconductor lasers. The theory outlined in this section lays the foundation for our analysis of dilute bismide QW lasers in Section 10.4.

10.3.1 Valence Band-Anticrossing

As described in Chapter 9, the strong, composition-dependent band gap bowing observed in the dilute nitride alloy (In)GaN$_x$As$_{1-x}$ is well explained in terms of a band-anticrossing interaction between two levels: one at energy E_{CB} associated with the extended CB edge states of the (In)GaAs host matrix, and the second at energy E_N associated with the highly localized N-related impurity states. In this simple model, the CB edge energy of the N-containing alloy is given by the lower eigenvalue of the two-band Hamiltonian [8]:

$$H = \begin{pmatrix} E_N & V_{Nc} \\ V_{Nc} & E_{CB} \end{pmatrix}, \tag{10.2}$$

where V_{Nc} is the N composition dependent matrix element describing the interaction between E_{CB} and E_N, which varies with N composition x as $V_{Nc} = \beta\sqrt{x}$.

Bismuth, being the heaviest stable group-V element, is significantly larger and more electropositive than As. It is therefore expected that any Bi-related impurity levels in (In)GaAs should lie close to or below the VB edge in energy and, if present, a VBAC interaction will occur between the Bi-related impurity levels and the VB edge states of the host matrix (cf. Section 10.2.1). The presence of such an interaction has been proposed on the basis of experimental measurements [4], in order to account for the observed composition-dependent bowing of the (In)GaBi$_x$As$_{1-x}$ band gap. While detailed analysis of atomistic supercell calculations [5,54,60] directly confirms the validity of this approach to describing the band structure in the case of ordered alloys, we note that the VBAC model introduced in Ref. [4] is in practice strongly limited by parametric ambiguity which removes its predictive capability. For example, one typically has two parameters (the energy of the Bi-related states E_{Bi} and the VBAC coupling strength β) which must be fitted to a single piece of experimental data (the measured alloy band gap at a given Bi composition), meaning that there is insufficient information available with which to provide an unambiguous fit without resorting to limiting assumptions. For example, initial estimation of the energy of the localized states associated with an isolated substitutional Bi impurity in GaAs produced a value $E_{Bi} = -0.4$ eV, which lies approximately 60 meV below the SO band edge [4]. However, first principles pseudopotential and large-supercell TB calculations confirm that these states lie above the SO band edge in energy (cf. Section 10.2.1), suggesting that simple estimates based on experimental data produce a qualitatively different description of the VB structure. Usman et al. overcame this limitation by using atomistic TB calculations to directly evaluate the

VBAC parameters, thereby producing a quantitative VBAC approach that provides a predictive capability which has been verified by comparison to a range of experimental data [15,39,54,79]. Here, we outline the TB analysis of the VBAC interaction, as a prelude to the derivation of a suitable **k·p** Hamiltonian with which to model the band structure of (In)GaBi$_x$As$_{1-x}$ alloys [54] and heterostructures [15,39,79].

To directly analyze the VBAC interaction, we consider the electronic structure of ordered GaBi$_x$As$_{1-x}$ and InBi$_x$As$_{1-x}$ alloys by inserting a single substitutional Bi atom in a series of cubic X$_M$Bi$_1$As$_{M-1}$ (X = Ga, In) supercells containing a total of $2M = 8N^3$ atoms, for $2 \leq N \leq 8$. In the VBAC model the Bi-hybridized (In)GaBi$_x$As$_{1-x}$ VB edge states are an admixture of the unperturbed GaAs VB edge states and the T_2 symmetric localized states associated with isolated Bi impurities. Writing the alloy VB edge states as a linear combination in this manner, we obtain an expression for the Bi-related (doublet) states that interact with the (In)GaAs VB edge [5,42,54]

$$|\psi_{\text{Bi},i}\rangle = \frac{|\psi_i^{(1)}\rangle - \sum_n |\psi_n^{(0)}\rangle\langle\psi_n^{(0)}|\psi_i^{(1)}\rangle}{\sqrt{1 - \sum_n |\langle\psi_n^{(0)}|\psi_i^{(1)}\rangle|^2}}, \tag{10.3}$$

where the indices n and i run respectively over the four-fold degenerate states of the (In)GaAs and (In)GaBi$_x$As$_{1-x}$ VB edge, respectively. The superscripts (0) and (1) refer respectively to the unperturbed host matrix and alloy VB edge eigenstates.

From Equation 10.3 we have calculated the energy of the Bi-related states, $E_{\text{Bi},i} = \langle\psi_{\text{Bi},i}|\hat{H}(x)|\psi_{\text{Bi},i}\rangle$, as well as the coupling strength $V_{\text{Bi},i}$ of the VBAC interaction between $|\psi_{\text{Bi},i}\rangle$ and the host matrix VB edge states, $V_{\text{Bi},i} = \langle\psi_{\text{Bi},i}|\hat{H}(x)|\psi_{\text{VB},i}^{(0)}\rangle$, where $\hat{H}(x)$ is the full TB Hamiltonian for the Bi-containing supercell and $|\psi_{\text{VB},i}^{(0)}\rangle$ is the unperturbed host matrix VB edge state with which $|\psi_{\text{Bi},i}\rangle$ interacts [5]. Doing this shows directly that (1) a substitutional Bi atom in Ga$_M$Bi$_1$As$_{M-1}$ or in In$_M$Bi$_1$As$_{M-1}$ forms a set of four-fold degenerate impurity states (at energy $E_{\text{Bi}} = E_{\text{Bi},i}$), which are highly localized about the Bi site, and (2) the strength of the VBAC interaction between the states $|\psi_{\text{Bi},i}\rangle$ and those of the unperturbed host matrix VB edge varies with Bi composition x as $V_{\text{Bi}} = \beta\sqrt{x}$.

Figure 10.4 shows the calculated values of the difference in energy between the Bi-related states and the unperturbed host matrix VB edge, $\Delta E_{\text{Bi}} = E_{\text{Bi}} - E_{\text{VB}}^{(0)}$, as well as the VBAC coupling parameter β, as a function of supercell size (alloy composition) for a series of ordered X$_M$Bi$_1$As$_{M-1}$ (X = Ga, In) supercells. We calculate a strong variation of ΔE_{Bi} and β with Bi composition at larger x, with the trends stabilizing as x approaches the impurity limit in the larger (>2000 atom) supercells [5]. In line with the first-principles pseudopotential calculations of Zhang et al. [49], we find that this behavior is attributable to the folding of host matrix VB states back to the Γ point as the supercell size is increased, which makes more zone-center supercell states having large **k** components available to construct $|\psi_{\text{Bi},i}\rangle$ in the larger supercells [54]. Therefore, we conclude that one should refer to large supercell calculations when examining the behavior of a substitutional Bi impurity in (In)GaAs alloys. We also note that $\Delta E_{\text{Bi}} < 0$ in both GaBi$_x$As$_{1-x}$ and InBi$_x$As$_{1-x}$, indicating that a substitutional Bi atom forms *resonant* states in (In)GaAs which lie below the unperturbed host matrix VB edge in energy. This is distinct from the case of a substitutional Bi impurity in GaP, where Bi forms *bound* states that lie energetically with the host matrix band gap [5,55].

10.3.2 12-Band k·p Hamiltonian for Dilute Bismide Alloys

In order to derive an appropriate **k·p** Hamiltonian for (In)GaBi$_x$As$_{1-x}$, we began with an eight-band model for the (In)GaAs host matrix [80], which employs a basis of spin-degenerate zone-center Bloch states for the lowest-energy CB, as well as the HH, LH, and SO VBs. By diagonalizing the (In)GaAs TB Hamiltonian at Γ, we obtain these basis states $|u_1\rangle, \ldots, |u_8\rangle$ directly. Then, we use this set of supercell basis states to construct an eight-band parameter set for (In)GaAs directly from the full TB calculations. This gives a

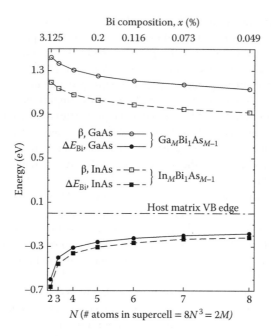

FIGURE 10.4 Calculated difference in energy between the Bi-related localized impurity states and host matrix VB edge, ΔE_{Bi} (closed symbols) and VBAC coupling parameter, β (open symbols) for a substitutional Bi impurity in a series of ordered, cubic $Ga_M Bi_1 As_{M-1}$ (circles) and $In_M Bi_1 As_{M-1}$ (squares) supercells. (Adapted from C.A. Broderick et al., *Semicond. Sci. Technol.*, 28, 125025, 2013; C.A. Broderick et al., *Phys. Stat. Sol. B*, 250, 773, 2013.) Each supercell contains a total of $2M = 8N^3$ atoms (where N is the number of cubic 8-atom unit cells along each of the x-, y- and z-directions), corresponding to Bi compositions ranging from $x = 0.049\%$ (4096 atoms) to $x = 3.125\%$ (64 atoms). The zero of energy is taken in each case at the host matrix VB edge and is denoted by a dash-dotted line at 0 eV.

parameterization of the eight-band Hamiltonian that reproduces the full TB band structure of an $X_M As_M$ (X = Ga, In) supercell with a high degree of accuracy in the vicinity of the zone center [42,54].

Having established this description of the host matrix band structure, we extend the eight-band basis set that describes the host matrix band structure by directly including the four Bi-related localized impurity (doublet) states described by Equation 10.3. The manner in which these Bi-related basis states couple to the host matrix VB edge states is then straightforward to include in the modified **k·p** Hamiltonian: the matrix elements between the localized Bi-related states and the extended VB edge states of the host matrix are precisely those evaluated in Section 10.3.1 as $V_{Bi} = \beta\sqrt{x}$. In this manner, we arrive at a 12-band, extended basis set **k·p** Hamiltonian for (In)GaBi$_x$As$_{1-x}$ alloys in which the impact of Bi on the band structure is described primarily via the VBAC interaction. In addition to the VBAC-induced shift in the VB edge energies, there are also conventional (virtual crystal) contributions to each of the band edge energies which depend linearly upon the Bi composition x. We have again used the TB model to directly quantify these effects, by evaluating the matrix elements $\langle u_b | \hat{H}(x) | u_b \rangle$, where $|u_b\rangle$ are the aforementioned basis states for the host matrix. Following this approach, we find firstly that the Bi-related localized states do not coupled directly to the extended CB edge states of the host matrix. Next, we calculate that the virtual crystal contributions to the variation of the (In)GaAs band edge energies with Bi composition x are given by: $E_{CB}(x) = E_{CB}^{(0)} - \alpha x$, $E_{HH}(x) = E_{LH}(x) = E_{VB}^{(0)} + \kappa x$, and $E_{SO}(x) = E_{SO}^{(0)} - \gamma x$, for the CB, VB, and SO band edges, respectively (where the superscript (0) denotes the band edge energy of the unperturbed host matrix).

We note that the incorporation of Bi then introduces five band structure parameters in addition to those that describe the (In)GaAs host matrix, all of which are required to provide a reliable description of the $GaBi_xAs_{1-x}$ band structure using a VBAC-based approach. (As we see in Section 10.3.3, this number increases to seven in the presence of pseudomorphic strain.) These parameters are (1–3) the virtual crystal contributions to the energies of the CB (α), VB (κ), and SO (γ) band edge energies, (4) the energy of the Bi-related localized states relative to the (In)GaAs VB edge (ΔE_{Bi}), and (5) the VBAC coupling strength (β). This number of free parameters is too large to provide an unambiguous fit to the available experimental band structure data, thereby emphasizing the importance of atomistic calculations in informing not only the understanding of the fundamental properties of Bi-containing alloys, but also the development and parametrization of sufficiently detailed phenomenological models to enable the investigation of heterostructures and devices. The 12-band $\mathbf{k} \cdot \mathbf{p}$ Hamiltonian derived by Broderick et al. using this approach is presented in full in Ref. [54]; the Bi-related band structure parameters derived for $GaBi_xAs_{1-x}$ and $InBi_xAs_{1-x}$ alloys on the basis of the TB analysis are listed in Table 10.1. We note that while this 12-band model contains fewer bands than the 14-band model originally introduced by Alberi et al. in Ref. [4], it ultimately provides a more accurate description of the band structure since, in addition to VBAC interactions, the contributions of strain and virtual crystal effects on the composition dependence of the band edge energies have been explicitly taken into account. Additionally, the 12-band model is fully parametrized on the basis of atomistic supercell calculations, without making recourse to fits to experimental data, thereby providing genuine predictive capability.

Figure 10.5a compares the results of calculations of the $Ga_{32}Bi_1As_{31}$ ($x = 3.125\%$) band structure in the vicinity of the zone center using the 12-band $\mathbf{k} \cdot \mathbf{p}$ (solid lines) and full TB (closed circles) Hamiltonians. Note that since the TB calculation is carried out on a 64-atom supercell, the appropriate values of ΔE_{Bi} and β for the supercell must be used in the $\mathbf{k} \cdot \mathbf{p}$ calculation in order to directly compare the band structures calculated using the two methods (cf. Figure 10.4). We observe excellent agreement between the $\mathbf{k} \cdot \mathbf{p}$ and TB band structures. It can be seen that (1) the VBAC interaction has pushed the alloy VB edge upward in energy, and (2) the CB and SO band edge energies in the alloy are well described by the Bi-induced virtual crystal shifts described above. The VBAC interaction, which describes the strong composition-dependent band gap bowing, accounts for the majority of the observed band gap reduction with increasing x. We also note that the presence of the VBAC interaction is verified by the presence of TB bands that correspond to the lower energy Bi-related ($E_-^{LH/HH}$) bands in the $\mathbf{k} \cdot \mathbf{p}$ calculation. These calculations therefore confirm that the 12-band Hamiltonian provides an accurate description the unusual band structure of $(In)GaBi_xAs_{1-x}$ alloys [42,54].

The main features of the band structure calculated using the 12-band Hamiltonian are in excellent agreement with those obtained from TB calculations performed on realistic disordered alloys across a wide range of Bi compositions, despite having been derived and parameterized for ordered alloys [54]. Figure 10.5b shows the variation of E_g (solid black line) and Δ_{SO} (dashed black line) as a function of Bi composition x at 300 K in $GaBi_xAs_{1-x}$, calculated using the 12-band $\mathbf{k} \cdot \mathbf{p}$ model. Comparing the results of these calculations to a range of experimental data from the literature, we see that the 12-band model produces a highly accurate description of the main features of the $GaBi_xAs_{1-x}$ band structure, and does so across a broad composition range. Overall, this analysis suggests that a suitably constructed and parametrized

TABLE 10.1 Bi-related Band Structure Parameters for the 12-band $\mathbf{k} \cdot \mathbf{p}$ Hamiltonian, Computed Directly Using TB Supercell Calculations for $GaBi_xAs_{1-x}$ and $InBi_xAs_{1-x}$ Alloys as Described in the Text

Alloy	ΔE_{Bi} (eV)	β (eV)	α (eV)	κ (eV)	γ (eV)
$GaBi_xAs_{1-x}$	−0.183	1.13	2.82	1.01	0.55
$InBi_xAs_{1-x}$	−0.217	0.92	2.60	1.03	0.02

Source: C.A. Broderick et al., *Semicond. Sci. Technol.*, 28, 125025, 2013.
The parameters ΔE_{Bi} and β specify the nature of the VBAC interaction, while α, κ, and γ, respectively, describe the Bi-induced virtual crystal (conventional alloy) shifts to the CB, VB, and SO band edge energies.

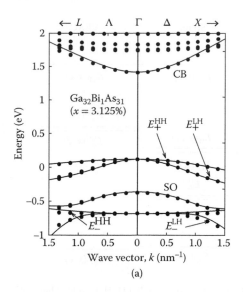
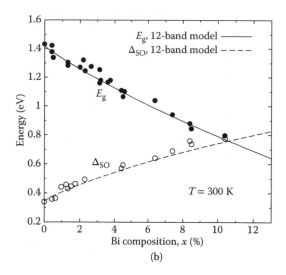

(a) (b)

FIGURE 10.5 (a) Calculated band structure of an ordered, cubic $Ga_{32}Bi_1As_{31}$ ($x = 3.125\%$, 64-atom) supercell, along the Λ- and Δ-directions, close to the center of the Brillouin zone, using the TB (closed circles) and 12-band **k·p** (solid lines) Hamiltonians. (Adapted from Broderick, C.A. et al., *Semicond. Sci. Technol.*, 28, 125025, 2013; C.A. Broderick et al., *Phys. Stat. Sol. B*, 250, 773, 2013.) The zero of energy is taken at the host matrix (GaAs) VB edge. (b) Comparison between theory and experiment for the variation of the band gap (E_g) and spin-orbit-splitting energy (Δ_{SO}) as a function of Bi composition x in $GaBi_xAs_{1-x}$ at 300 K. Experimental measurements of E_g are depicted by closed circles. (The experimental data are taken from S. Tixier et al., *Appl. Phys. Lett.*, 82, 2245, 2003; S. Francoeur et al., *Appl. Phys. Lett.*, 82, 3874, 2003; W. Huang et al., *J. Appl. Phys.*, 98, 053505, 2005; K. Alberi et al., *Phys. Rev. B*, 75, 045203, 2007; Z. Batool et al., *J. Appl. Phys.*, 111, 113108, 2012; J. Yoshida et al., *Jpn. J. Appl. Phys.*, 42, 371, 2003.) Experimental measurements of Δ_{SO} are depicted by open circles. (The experimental data are taken from K. Alberi et al., *Phys. Rev. B*, 75, 045203, 2007; Z. Batool et al., *J. Appl. Phys.*, 111, 113108, 2012; B. Fluegel et al., *Phys. Rev. Lett.*, 97, 067205, 2006.) Solid (dashed) lines show the variation of E_g (Δ_{SO}) calculated using the 12-band **k·p** model. (Adapted from C.A. Broderick et al., *Semicond. Sci. Technol.*, 28, 125025, 2013; C.A. Broderick et al., *Phys. Stat. Sol. B*, 250, 773, 2013; C.A. Broderick et al., *IEEE J. Sel. Top. Quantum Electron.*, 21, 1503313, 2015.)

VBAC model can be used to reliably describe the evolution of the main features of the band structure of (In)$GaBi_xAs_{1-x}$ alloys across a wide range of Bi compositions.

10.3.3 Calculation of Band Offsets in Bi-Containing Quantum Wells

Having considered the theoretical description of the bulk band structure of dilute bismide alloys, we turn our attention now to the calculation of the band offsets in pseudomorphically strained $GaBi_xAs_{1-x}$/(Al)GaAs QWs, as a prelude to our discussion of the electronic and optical properties of GaAs-based dilute bismide QW lasers in Section 10.4. Here, we outline the calculation of the band offsets at arbitrary Bi composition x using the 12-band **k·p** model described in Section 10.3.2. Full details of the analysis upon which this analysis of the QW band offsets is based can be found in Ref. [79].

In order to calculate the strained $GaBi_xAs_{1-x}$ band edge energies we consider the 12-band Hamiltonian including the Bir-Pikus matrix elements describing the impact of pseudomorphic strain on the band edge energies at Γ [81]. The steps in the calculation of the band offsets for a $GaBi_xAs_{1-x}$/GaAs QW are illustrated schematically in Figure 10.6. While we focus here on GaAs-based heterostructures, we note that this method for calculating the QW band offsets is generally applicable to alloys whose band structure can be described using the 12-band model, and so can readily be applied to Bi-containing alloys grown

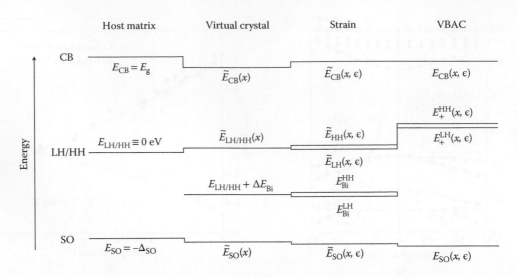

FIGURE 10.6 Schematic illustration of the calculation of the bulk $GaBi_xAs_{1-x}/GaAs$ CB, LH, HH, and SO band offsets, using the 12-band $\mathbf{k \cdot p}$ model of Section 10.3.2 and including the effects of pseudomorphic strain on the band edge energies. (Adapted from Broderick, C.A. et al., *Semicond. Sci. Technol.*, 30, 094009, 2015; G.M.T. Chai et al., *Semicond. Sci. Technol.*, 30, 094015, 2015.) The zero of energy is taken at the VB edge of the unstrained GaAs host matrix. The full details of each step in the calculation are outlined in the text.

using alternative barriers and/or substrates. For barrier materials other than GaAs, the calculation follows straightforwardly from that presented here: one needs simply to calculate the $(In)GaBi_xAs_{1-x}$ band edge energies as outlined below, with the barrier band edge energies then calculated with respect to the (In)GaAs host matrix in the usual way (using, e.g., the model solid theory [82]) in order to determine the band offsets.

The first step in the calculation is shown on the left side of Figure 10.6, in the portion labeled "Host matrix." Here, we begin with a GaAs host matrix (barrier material) and choose the zero of energy to lie at the unstrained VB edge. Second, in the section labeled "Virtual crystal," we include the conventional alloy contributions to the Bi-induced band edge energy shifts for a given Bi composition x as $\widetilde{E}_{CB}(x) = E_{CB} - \alpha x \equiv E_g - \alpha x$, $\widetilde{E}_{LH/HH}(x) = E_{LH/HH} + \kappa x \equiv \kappa x$, and $\widetilde{E}_{SO}(x) = E_{SO} - \gamma x \equiv -\Delta_{SO} - \gamma x$, where, as described above, incorporation of Bi has introduced a set of degenerate resonant impurity levels at energy ΔE_{Bi} relative to the host matrix VB edge.

Third, in the section labeled "Strain," we calculate the strain-induced shifts to the band edge energies, thereby taking account of the fact that an $(In)GaBi_xAs_{1-x}$ epilayer grown on a GaAs substrate will be under compressive pseudomorphic strain. Following the conventions of Krijn [83], we calculate these strain-induced shifts to the CB, HH, LH, and SO band edge energies as $\widetilde{E}_{CB}(x, \epsilon) = \widetilde{E}_{CB}(x) + \delta E_{CB}^{hy}$, $\widetilde{E}_{LH}(x, \epsilon) = \widetilde{E}_{LH}(x) + \delta E_{VB}^{hy} + \delta E_{VB}^{ax}$, $\widetilde{E}_{HH}(x, \epsilon) = \widetilde{E}_{HH}(x) + \delta E_{VB}^{hy} - \delta E_{VB}^{ax}$, and $\widetilde{E}_{SO}(x, \epsilon) = \widetilde{E}_{SO}(x) + \delta E_{VB}^{hy}$, where the energy shifts associated with the hydrostatic and axial components of the strain are $\delta E_{CB}^{hy} = a_c \left(\epsilon_{xx} + \epsilon_{yy} + \epsilon_{zz} \right)$, $\delta E_{VB}^{hy} = a_v \left(\epsilon_{xx} + \epsilon_{yy} + \epsilon_{zz} \right)$, and $\delta E_{VB}^{ax} = -\frac{b}{2} \left(\epsilon_{xx} + \epsilon_{yy} - 2\epsilon_{zz} \right)$ [81]. The non-zero components of the strain tensor given by $\epsilon_{xx} = \epsilon_{yy} = \frac{a_0 - a(x)}{a(x)}$ and $\epsilon_{zz} = -\frac{2C_{12}(x)}{C_{11}(x)}$ [83], where a_0 is the substrate lattice constant, and $a(x)$, $C_{11}(x)$, and $C_{12}(x)$ are the $(In)GaBi_xAs_{1-x}$ lattice and elastic constants.

In order to obtain $a(x)$, $C_{11}(x)$, and $C_{12}(x)$ for $(In)GaBi_xAs_{1-x}$ we interpolate linearly between those of the end-point binary compounds: we use the (Al)GaAs lattice and elastic constants recommended by Vurgaftman et al. [84], and for (In)GaBi we use those calculated *ab initio* by Ferhat and Zaoui [65]. Due to a lack of information in the literature regarding the band edge deformation potentials of (In)GaBi, we take a_c, a_v, and b for $(In)GaBi_xAs_{1-x}$ to be equal to those of the (In)GaAs host matrix, which is expected to be a good approximation at the dilute Bi compositions under consideration.

Due to the *p*-like symmetry of the Bi-related impurity states, there is a lifting of the degeneracy of the LH- and HH-like impurity states in the presence of pseudomorphic strain, resulting from the tetragonal distortion of the crystal lattice. Based on the results of TB calculations on large, ordered supercells, it is found that the energies of these states, E_{Bi}^{LH} and E_{Bi}^{HH}, vary with strain as $E_{Bi}^{LH} = \Delta E_{Bi} + \delta E_{Bi}^{hy} + \delta E_{Bi}^{ax}$ and $E_{Bi}^{HH} = \Delta E_{Bi} + \delta E_{Bi}^{hy} - \delta E_{Bi}^{ax}$, with δE_{Bi}^{hy} and δE_{Bi}^{ax} given by $\delta E_{Bi}^{hy} = a_{Bi} \left(\epsilon_{xx} + \epsilon_{yy} + \epsilon_{zz} \right)$ and $\delta E_{Bi}^{ax} = -\frac{b_{Bi}}{2} \left(\epsilon_{xx} + \epsilon_{yy} - 2\epsilon_{zz} \right)$. The Bi-impurity state hydrostatic and axial deformation potentials—determined using large supercell TB calculations to track the evolution of the Bi-related localized states as functions of hydrostatic and axial strain—are $a_{Bi} = -1.11$ eV and $b_{Bi} = -1.71$ eV, respectively.

Finally, in the section labeled "VBAC," the VBAC interactions between the strained virtual crystal band edges and the Bi-related localized states are included to finally determine the band offsets. Examining the 12-band Hamiltonian [54,79] we see that in bulk (with $k = 0$), the CB is decoupled from the Bi-related states, so that $E_{CB}(x, \epsilon) = \tilde{E}_{CB}(x, \epsilon)$, while the host matrix HH-band couples directly to E_{Bi}^{HH}, giving

$$E_{\pm}^{HH}(x, \epsilon) = \frac{\tilde{E}_{HH}(x, \epsilon) + E_{Bi}^{HH}}{2} \pm \sqrt{\left(\frac{\tilde{E}_{HH}(x, \epsilon) - E_{Bi}^{HH}}{2} \right)^2 + |V_{Bi}|^2} . \tag{10.4}$$

Since the axial component of the pseudomorphic strain couples the host matrix LH and SO bands [83], there exists a second-order strain-induced coupling between the LH-like Bi-related localized states and the host matrix SO band. As a result, the SO band cannot be treated independently and the LH, SO, and lower-energy LH-like VBAC band edges in the strained alloy are obtained as the eigenvalues $E_{\pm}^{LH}(x, \epsilon)$ and $E_{SO}(x, \epsilon)$ of the 3×3 Hermitian matrix

$$\begin{pmatrix} \tilde{E}_{LH}(x, \epsilon) & -\sqrt{2}\,\delta E_{VB}^{ax} & V_{Bi} \\ & \tilde{E}_{SO}(x, \epsilon) & 0 \\ & & E_{Bi}^{LH} \end{pmatrix}, \tag{10.5}$$

which completes the calculation.

On the basis of detailed spectroscopic measurements performed on a series of GaBi$_x$As$_{1-x}$/(Al)GaAs QW laser structures with $x \approx 2\%$ [79], the Bi-related parameters α and β of Table 10.1 have been refined in order to produce an optimized set of parameters for modeling *real* GaBi$_x$As$_{1-x}$ QWs at low x, giving $\alpha = 2.63$ eV and $\beta = 1.45$ eV, with the remaining parameters unchanged from those listed in Table 10.1. These optimized Bi-related band structure parameters have been demonstrated to produce band offsets and transitions energies which are in good, quantitative agreement with those measured for a series of GaBi$_x$As$_{1-x}$/(Al)GaAs QWs. This not only verifies the predictive capability of the 12-band model, but also emphasizes its potential for application to the calculation and analysis of the electronic and optical properties of dilute bismide QWs and related heterostructures.

10.4 GaBi$_x$As$_{1-x}$/(Al)GaAs Quantum Well Lasers

Having focused so far on constructing a consistent approach to model the (In)GaBi$_x$As$_{1-x}$ band structure, in this section we turn our attention to the theory and modeling of dilute bismide semiconductor lasers. In order to address the potential of GaBi$_x$As$_{1-x}$ for the development of long-wavelength GaAs-based lasers, we introduce a theoretical model—introduced by Broderick et al. [15]—describe the impact of Bi on the performance of real and ideal QW laser devices. After outlining the theoretical model in Section 10.4.1, in Sections 10.4.2 and 10.4.3 we analyze the effect of Bi incorporation on the electronic and optical properties of an exemplar GaBi$_x$As$_{1-x}$/(Al)GaAs QW, which has been chosen on the basis of the active region employed in the first electrically pumped dilute bismide laser [30]. Using the theoretical model we review the Bi-induced changes to the band structure, DOS, optical transition matrix elements, and optical gain, and show how the choice of barrier material plays a key role in optimizing the performance of the device at low Bi composition.

Next, in Section 10.4.4, we review gain calculations for a $GaBi_xAs_{1-x}/GaAs$ laser structure with high Bi composition ($x = 13\%$), designed to emit at 1.55 µm and to have $\Delta_{SO} > E_g$ in order to suppress CHSH Auger recombination and IVBA (cf. Figure 10.1). We highlight that the modal and differential gain at this composition is significantly higher than that at lower x due to the combined effects of increased CB offset and compressive strain at larger x. Finally, in Section 10.4.5 we review comparison of the results of these theoretical calculations directly to experimental measurements of the SE and gain [39]. These are the first measurements of SE and optical gain for a dilute bismide semiconductor laser and the only such comparison between theory and experiment to date for this material system.

The theoretical results reviewed in this section elucidate important trends in the properties and performance of $GaBi_xAs_{1-x}$ QW lasers grown on GaAs substrates, and confirm the potential of dilute bismide QW structures for the development of highly efficient 1.55-µm lasers. Comparison to experimental data displays the predictive capability of the theoretical model, highlighting its potential for use in the design and optimization of future dilute bismide devices.

10.4.1 Theoretical Model for Dilute Bismide Quantum Well Lasers

Here, we provide a brief outline of the theoretical model developed to study the electronic and optical properties of dilute bismide QW lasers. Full details of the model can be found in Ref. [15]. The theoretical model is based upon the 12-band $\mathbf{k \cdot p}$ Hamiltonian presented in Section 10.3.2. The QW eigenstates are calculated in the envelope function approximation using a numerically efficient plane wave expansion method, similar to that widely employed in first-principles calculations. An overview of the plane wave expansion method within the context of $\mathbf{k \cdot p}$ theory can be found in Ref. [85].

In the plane wave approach, the QW eigenstate $|\psi_n(k_\|, z)\rangle$ at position z and in-plane wave vector $k_\|$, having energy $E_n(k_\|)$, is written as

$$|\psi_n(k_\|, z)\rangle = \frac{1}{\sqrt{L}} \sum_{b=1}^{12} \sum_{m=-M}^{M} a_{nbm}(k_\|) \, e^{iG_m z} |u_b\rangle, \qquad (10.6)$$

where L is the length of the calculational supercell ($-\frac{L}{2} \leq z \leq \frac{L}{2}$), b indexes the bands of the bulk $\mathbf{k \cdot p}$ Hamiltonian, m indexes the plane waves, and the discrete wave vectors G_m are defined as $G_m = \frac{2m\pi}{L}$. $|u_b\rangle$ are the zone-center Bloch states of the bulk 12-band $\mathbf{k \cdot p}$ Hamiltonian. We note that this description of the QW eigenstates, which consists of $2M + 1$ independent Fourier components, is formally exact as $M \to \infty$. However, for most practical applications, a basis set containing ~ 70 plane waves is sufficient to ensure that the calculated eigenstates have converged with respect to further increases in the basis set size [86].

By substituting Equation 10.6 into the multi-band Schrödinger equation, it is straightforward to derive analytical expressions for the matrix elements of the QW Hamiltonian in the plane wave basis set defined by the set of wave vectors G_m [85,86], leading to a reciprocal space Hamiltonian matrix of size $12(2M + 1) \times 12(2M + 1)$. In this manner, an independent Hamiltonian matrix is obtained at each in-plane wave vector $k_\|$, diagonalization of which yields the QW band structure $E_n(k_\|)$ and eigenstates $|\psi_n(k_\|, z)\rangle$. We note that the eigenstates defined by Equation 10.6 satisfy periodic boundary conditions over the calculational supercell, meaning that a QW will tend to undergo spurious interactions with identical QWs in neighboring supercells. This can readily be mitigated by choosing a sufficiently large supercell length L, and in practice this feature of the method means that it can be used to efficiently investigate the electronic and optical properties of superlattices.

Analysis of the band offsets (cf. Section 10.3.3) suggests that the small CB offset leads to very weak electron confinement in $GaBi_xAs_{1-x}/GaAs$ QWs at low x. As such, hole-induced electrostatic confinement of electron states is likely to play an important role in determining the electronic and optical properties of GaBi$_x$As$_{1-x}$/GaAs QW lasers at low x. For selected laser structures (cf. Section 10.4.3), we therefore also consider self-consistent calculations of the electronic and optical properties, in which the QW band structure and eigenstates are calculated self-consistently by using Poisson's equation in order to determine the net carrier-induced electrostatic potential at each injected carrier density n. For the self-consistent calculations, the coupled multi-band Schrödinger–Poisson system is solved in reciprocal space, using the technique outlined in Ref. [86].

The atomistic supercall calculations reviewed in Section 10.2 clearly demonstrate that Bi-induced hybridization plays an important role in determining the character of the VB edge eigenstates. As such, in order to facilitate accurate calculation of the optical properties of dilute bismide laser structures, it is important to include the impact of Bi incorporation on the QW eigenstates. For this reason, in the theoretical model the QW band structure, envelope functions, and Hamiltonian (calculated in the axial approximation [87]) are used *directly* to compute the optical properties of a given laser structure. Key to this analysis is the calculation of the optical (momentum) transition matrix elements. Following Ref. [88], the transition matrix element describing optical recombination of an electron-hole pair at in-plane wave vector $k_{\parallel}$ is given in terms of the QW eigenstates described by Equation 10.6 as [88]

$$P_{n_c,n_v}\left(k_{\parallel}\right) = \frac{m_0}{\hbar} \left\langle \psi_{n_v}\left(k_{\parallel}, z\right) \left| \frac{\partial \widehat{H}}{\partial k_{\parallel}} \cdot \widehat{e} \right| \psi_{n_c}\left(k_{\parallel}, z\right) \right\rangle, \tag{10.7}$$

where $\widehat{e}$ is a unit vector denoting the polarization of the emitted photon, and n_c (n_v) indexes the conduction (valence) subbands. This approach to calculating the optical matrix elements then explicitly incorporates key band structure effects in the computation of the optical properties including: Bi-induced VB hybridization (VBAC) and localization, temperature- and injection-dependent carrier spillout from the QW(s), band mixing at non-zero in-plane wave vector, and pseudomorphic strain, all of which are necessary to quantitatively describe the SE and optical gain in real devices [15,39]. Note that when discussing optical transition "strengths" below we are referring to the quantity $\frac{1}{2m_0}|P_{n_c,n_v}\left(k_{\parallel}\right)|^2$, which has units of energy.

Using the optical matrix elements computed in this manner, the calculation of the material gain spectrum $g(\hbar\omega)$ follows the approach of Ref. [89], which consists of transforming the corresponding polarization component of the SE rate $r_{\mathrm{spon}}(\hbar\omega)$ at fixed carrier density n according to

$$g(\hbar\omega) = \frac{3\pi^2\hbar c^2}{2n_r^2\omega^2} \left(1 - \exp\left(\frac{\hbar\omega - \Delta F}{k_{\mathrm{B}}T}\right)\right) r_{\mathrm{spon}}(\hbar\omega), \tag{10.8}$$

where $\hbar\omega$ is the photon energy, ΔF is the quasi-Fermi level separation at temperature T in the presence of an injected carrier density n, and n_r is the refractive index of the optical mode in the active region. The SE rate r_{spon} is calculated as described in Refs. [15,89]. Here, we are concerned solely with transverse electric (TE)-polarized gain, which is of larger magnitude than the transverse magnetic (TM)-polarized gain in the compressively strained structures under investigation. As such, r_{spon} is understood to represent only the TE-polarized component of the total SE rate (which gives rise to the leading factor of $\frac{3}{2}$ in Equation 10.8). This approach to calculating the optical gain spectrum has the benefits that (1) it removes the anomalous absorption at energies below the QW band gap that can plague calculations undertaken using an energy and crystal momentum-independent interband relaxation time (homogeneous spectral linewidth), and (2) by definition, $g(\Delta F) = 0$, so that the transparency point on the high-energy side of the gain peak is maintained at the correct, thermodynamically consistent energy. As we will see, when we review the comparison of theory and experiment for the optical gain in Section 10.4.5, this latter characteristic of Equation 10.8 enables a distinct carrier density n in the theoretical calculations to be associated with each current density J at which the gain spectrum is measured, thereby facilitating *direct* comparison of theoretical calculations to experiment.

Based on theory-experiment comparisons, it is found that the spectral broadening of $GaBi_xAs_{1-x}$ QWs is well described using a hyperbolic secant lineshape of the form $S(\hbar\omega) = \frac{1}{\pi\delta} \operatorname{sech}\left(\frac{E_0 - \hbar\omega}{\delta}\right)$, where δ is the (homogeneous) spectral linewidth. We note that this lineshape function was previously found to describe well the SE spectra of 1.3-μm GaInNAs dilute nitride QW lasers [90]. For the calculations in Section 10.4.2 we set $\delta = 6.6$ meV, a typical value for III–V semiconductors such as InGaAs. This enables direct comparison between the calculated SE and gain in Bi-free and Bi-containing structures, and hence quantitative analysis on the impact of Bi incorporation on the optical properties. This value of the spectral linewidth is smaller than that observed in real $GaBi_xAs_{1-x}$ devices by a factor of approximately four [15]. Larger more realistic linewidths are included in several of the calculations presented in Sections 10.4.2 and 10.4.3 by inhomogeneously broadening the calculated spectra [15], and the spectral linewidth is directly analyzed in Section 10.4.5 when review the comparison to experimental measurements of the SE and modal gain spectra [39].

For the calculation of the threshold properties in Sections 10.4.2 and 10.4.3, internal optical (cavity) losses of $\alpha_i = 4$ cm^{-1} are assumed, lattice-matched $Al_{0.4}Ga_{0.6}As$ cladding layers are chosen to form the separate confinement heterostructure (SCH), and an overall cavity length of 1 mm is assumed for all of the devices considered [30]. The calculated facet reflectivity for this structure is $R = 0.35$. The confinement factor Γ of the fundamental (TE-polarized) optical mode was calculated for each laser structure using an effective index approach [91]. It was found for all laser structures considered that Γ is maximized for 150-nm-thick (Al)GaAs barriers. Alternative approaches, based on similar models of the band structure, have been proposed by other authors [14,92]. However, the theoretical model presented here benefits from detailed and unambiguous parameterization based on atomistic calculations, and has been confirmed to provide reliable predictions of key trends in the properties and performance of GaAs-based dilute bismide QW lasers [15], including descriptions of the measured SE and optical gain that are in quantitative agreement with experimental measurements on real devices [39].

10.4.2 Impact of Bi Incorporation on the Band Structure and Gain of Quantum Wells

In order to quantify the impact of Bi incorporation on the electronic and optical properties of laser structures having low Bi compositions, we compare the calculated gain characteristics of a $GaBi_xAs_{1-x}$ QW laser structure at low x with those of an equivalent laser structure containing a GaAs (Bi-free, $x = 0$) QW [15]. We focus here on Bi-containing laser structures having a single 6.9-nm-thick $GaBi_{0.021}As_{0.979}$ ($x = 2.1\%$) QW with $Al_{0.144}Ga_{0.856}As$ barriers [30,79].

The calculated CB and VB offsets are $\Delta E_{CB} = 150$ meV and $\Delta E_{HH/LH} = 65$ meV in the unstrained Bi-free QW [79]. The incorporation of 2.1% Bi introduces a compressive in-plane strain of 0.25% and, when VBAC effects are taken into account, this leads to calculated HH and LH band offsets $\Delta E_{HH} = 224$ meV and $\Delta E_{LH} = 208$ meV, and a CB offset $\Delta E_{CB} = 185$ meV [15,79]. The transition energy between the lowest-energy bound electron state ($e1$) and the highest-energy bound hole state ($hh1$) is calculated to be 1.480 eV in the Bi-free QW, which is reduced to 1.289 eV in the $x = 2.1\%$ QW. Of this 191 meV reduction in the QW band gap, only 31 meV is due to the Bi-induced reduction of the CB edge energy, while the remaining 160 meV results from the VBAC-induced upward shift of the VB edge energy.

Figure 10.7 shows the calculated VB structure and DOS for the $x = 0$ and 2.1% QWs. We focus our analysis of the band structure and DOS on the VB, since Bi incorporation has a comparatively minor impact on the CB (cf. Section 10.2). The zero of energy is taken at the energy of the $hh1$ state (i.e., the QW VB edge) in each case, to facilitate a direct comparison of the DOS. Three factors contribute to differences in the band structure and DOS between the two QWs: (1) the compressive strain in the Bi-containing structure leads to a slight increase in the energy separation between $hh1$ and the highest-energy bound LH state ($lh1$), leading in turn to a reduction in the DOS at energies from 0 to −20 meV, (2) the VBAC interaction pushes the VB dispersion upward in energy at fixed $k_\parallel$, increasing the DOS in the $x = 2.1\%$ QW relative

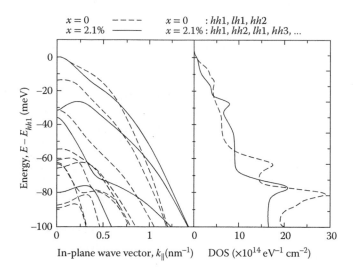

FIGURE 10.7 Calculated band structure (left panel) and density of states (DOS; right panel) in the vicinity of the VB edge for 6.9-nm-thick GaAs (Bi-free; dashed lines) and $GaBi_{0.021}As_{0.979}$ ($x = 2.1\%$; solid lines) QWs with $Al_{0.144}Ga_{0.856}As$ barriers. (Adapted from Broderick, C.A. et al., *IEEE J. Sel. Top. Quantum Electron.*, 21, 1503313, 2015.) Note that the zero of energy has been taken in each case at the energy of the highest bound hole state ($hh1$) to facilitate comparison between the two QWs.

to the Bi-free QW for energies between -20 and -60 meV, and (3) the barrier VB edge lies approximately 60 meV (200 meV) below $hh1$ in energy in the $x = 0$ ($x = 2.1\%$) QW, so that the calculated DOS in the Bi-free structure increases above that of the $x = 2.1\%$ QW for energies between -60 and -100 meV due to the large density of VB states in the AlGaAs barrier. We note that, overall, the calculated impact of Bi incorporation on the VB structure and DOS are relatively minor at low Bi compositions $x \approx 2\%$.

Having considered the impact of Bi incorporation on the QW band structure and DOS, we now turn our attention to the impact of strain and Bi-induced VBAC on the interband optical transition strengths, and then investigate the effect of Bi incorporation on the material gain. Figure 10.8a shows the TE-polarized optical transitions strengths, calculated as a function of $k_\parallel$ for the $e1$-$hh1$ and $e1$-$lh1$ transitions, for the Bi-free (dashed lines) and Bi-containing (solid lines) QWs. Incorporation of Bi reduces the optical transition strength at $k_\parallel = 0$ for transitions between the $e1$ and $hh1/lh1$ states. For example, the $e1$-$hh1$ transition strength at $k_\parallel = 0$ in the Bi-containing QW is approximately 67% of that in the Bi-free QW. The majority of the reduction in the optical transition strengths at $x = 2.1\%$ is due to the VBAC effect, which accounts for all but 1.1% of the calculated 33.4% reduction for the TE-polarized $e1$-$hh1$ transition in the $x = 2.1\%$ QW. This is due to the fact that the hybridization of the bound hole states in the QW with Bi-related localized states leads to a portion of the hole states consisting of an admixture of p-like Bi-related localized states which do not couple optically to the s-like states at the CB edge [54] (cf. Section 10.3.2).

Next, we investigate the variation of the peak material gain as a function of carrier density n, for temperatures $T = 100, 200,$ and 300 K. The results of these calculations are shown in Figure 10.8b for the $x = 2.1\%$ and $x = 0$ laser structures using solid and dashed lines, respectively. Since the material gain at fixed n is inversely proportional to the photon energy, to facilitate a direct comparison of the gain characteristics of the two QW structures—which have different band gaps—all CB states in the $x = 0$ calculation have been shifted downward in energy by 191 meV in order to bring the $e1$-$hh1$ transition energies into coincidence for both QW structures. Therefore, the differences between the material gain for two laser structures shown in Figure 10.8b arise only due to the impact of Bi incorporation on the QW band structure and optical transition strengths. Based on a calculated optical confinement factor $\Gamma = 1.66\%$ for the

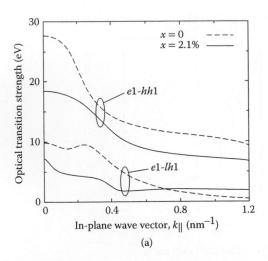
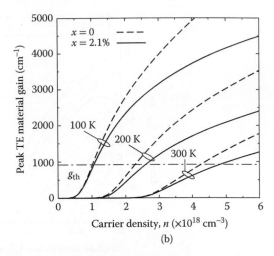

(a)　　　　　　　　　　　　　　　　　　　　　　　　(b)

FIGURE 10.8 (a) Calculated TE-polarized optical transition strengths as a function of in-plane wave vector ($k_\parallel$), for optical transitions between the lowest-energy bound electron and highest-energy bound LH- and HH-like states for 6.9-nm-thick GaAs (Bi-free; dashed lines) and $GaBi_{0.021}As_{0.979}$ (x = 2.1%; solid lines) QWs having $Al_{0.144}Ga_{0.856}As$ barriers. (Adapted from Broderick, C.A. et al., *IEEE J. Sel. Top. Quantum Electron.*, 21, 1503313, 2015.) (b) Calculated variation of the peak TE-polarized material gain as a function of carrier density, at temperatures of 100, 200, and 300 K for the same QWs as in (a). (Adapted from Broderick, C.A. et al., *IEEE J. Sel. Top. Quantum Electron.*, 21, 1503313, 2015.) The horizontal dash-dotted line denotes the calculated threshold material gain, g_{th} = 919 cm^{-1}.

x = 2.1% laser structure, we estimate a threshold material gain g_{th} = 919 cm^{-1} (denoted by a horizontal dash-dotted line in Figure 10.8b).

There is little change in the calculated transparency carrier density, n_{tr}, between the x = 0 and 2.1% laser structures for all temperatures considered. This reflects the small difference between the calculated VB edge DOS for the two QWs (cf. Figure 10.7). However, we find for $n > n_{tr}$ that (1) the gain saturates at a lower level with increasing n at x = 2.1%, and (2) the threshold carrier density, n_{th}, is higher at x = 2.1% than in the Bi-free QW. Both of these features arise due primarily to the general reduction in the optical transition strengths in the Bi-containing QW structure. We also note that while the differential gain at threshold, $\frac{dg}{dn}$, is approximately equal for both the Bi-free and Bi-containing lasers structures at 100 K, $\frac{dg}{dn}$ for the Bi-containing QW decreases with respect to that of the Bi-free QW with increasing temperature. Since the modulation response frequency of a semiconductor laser is proportional to the square root of the differential gain [93,94], these calculations suggest that Bi incorporation can be expected to lead to a degradation of the dynamical performance at low Bi compositions compared to that of an equivalent Bi-free structure.

10.4.3 Optimization of $GaBi_xAs_{1-x}$/(Al)GaAs Laser Structures

Theoretical and experimental analysis has shown that $GaBi_xAs_{1-x}$/GaAs QWs have very small CB offsets at low x, estimated at only 55 meV for a QW with x = 2.1% [79]. Such a low CB offset is detrimental to laser operation, leading to low electron-hole spatial overlap and significant electron leakage from the QW at high temperatures and carrier densities, and is therefore a major factor limiting the material gain at low x. Electron confinement must then be improved by growing $GaBi_xAs_{1-x}$ QWs with $Al_yGa_{1-y}As$ barriers to increase the CB offset, but at the cost of reducing the optical confinement in a laser structure with a fixed Al composition y in in the cladding layers. This leads to a trade-off between the material gain (which increases with increasing y due to improved electron confinement) and the optical confinement (which decreases with increasing y due to a reduced refractive index contrast between the barrier and cladding layers). The theoretical model described above has been used to determine the Al composition

y in the barrier layers required to optimize the performance of a laser structure containing the same 6.9-nm-thick $GaBi_{0.021}As_{0.979}$ QW considered in the exemplar structure of Section 10.4.2. The modal gain of the laser structure, Γg, is the quantity optimized, in order to minimize n_{th}. The gain calculations for this optimization include (1) electrostatic effects, computed self-consistently via coupling the Poisson's equation, and (2) inhomogeneous broadening, in order to produce realistic estimates of the material gain in the laser structure [15], an approach which has been shown to be in excellent agreement with experiment [39,95].

Figure 10.9 shows the calculated variation of n_{th} and g_{th} with the barrier Al composition y for this series of laser structures. Examining first the variation of g_{th} with y (closed circles, solid line) we see that the degradation of the optical confinement when Al is incorporated in the barrier layers leads to an increase in g_{th} with increasing y. This increase in g_{th} is modest at low Al compositions $y < 10\%$, with the optical confinement factor at $y = 10\%$ being 90% of its maximum calculated value $\Gamma = 1.95\%$ at $y = 0$. For $y > 10\%$ Γ degrades rapidly with increasing y, reaching values as low as 1.24% at $y = 25\%$. This results in a large increase in g_{th}, since $g_{th} \propto \Gamma^{-1}$. The net effect of this trend on the laser performance can be seen by considering the variation of n_{th} with y (open circles, dashed line). Beginning with a QW having GaAs barriers ($y = 0$), a large threshold carrier density $n_{th} = 6.3 \times 10^{18}$ cm^{-3} is calculated. When Al is incorporated in the barrier layers n_{th} decreases rapidly with increasing y, reaching a minimum value for $10\% < y < 15\%$. For $y < 10\%$, we therefore conclude that the increase in the material gain at fixed carrier density arising from the enhanced CB offset and electron confinement is sufficient to overcome the associated reduction in Γ, leading to a strong reduction in n_{th}. However, for $y > 15\%$, we find that n_{th} increases rapidly with increasing y, reflecting that any further improvement in the material gain is insufficient to overcome the rapid further degradation of Γ and associated increase in g_{th}. These theoretical calculations therefore suggest that QWs having barrier Al compositions $y = 10\%–15\%$ offer enhanced performance as compared to a QW having GaAs barriers, for $x \approx 2\%$.

While the first electrically pumped $GaBi_xAs_{1-x}$ QW laser had a nominal barrier Al composition $y = 20\%$ [30], more recently improved device performance has been demonstrated following the structural optimization outlined here [96]. Specifically, measurements confirm that choosing $Al_{0.12}Ga_{0.88}As$ as opposed to $Al_{0.20}Ga_{0.80}As$ barriers leads to a decrease of the threshold current density, J_{th}, by a factor of

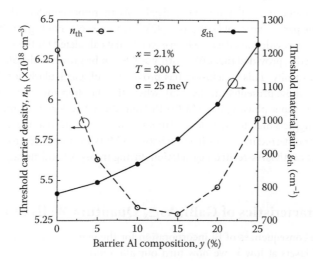

FIGURE 10.9 Calculated variation of the threshold carrier density (n_{th}; open circles, dashed lines) and threshold material gain (g_{th}; closed circles, solid lines) as a function of the barrier Al composition y for a series of 6.9-nm-thick $GaBi_{0.021}As_{0.979}/Al_yGa_{1-y}As$ QWs. (Adapted from Broderick, C.A. et al., *IEEE J. Sel. Top. Quantum Electron.*, 21, 1503313, 2015.) The calculated gain spectra, from which the values of n_{th} were determined, include an inhomogeneous broadening of 25 meV.

TABLE 10.2 Measured Room Temperature Emission Wavelength λ_{meas} and Threshold Current Density J_{th} for a Series of GaBi$_x$As$_{1-x}$/Al$_y$Ga$_{1-y}$As Laser Structures Containing Nominally Identical 6.4-nm-Thick GaBi$_{0.022}$As$_{0.978}$ ($x = 2.2\%$) QWs

QW Material	Barrier Material	λ_{meas} (nm)	J_{th} (kA cm^{-2})
GaBi$_{0.022}$As$_{0.978}$	GaAs	938	7.5
GaBi$_{0.022}$As$_{0.978}$	Al$_{0.12}$Ga$_{0.88}$As	947	1.0–1.1
GaBi$_{0.022}$As$_{0.978}$	Al$_{0.20}$Ga$_{0.80}$As	947	1.5–1.6

Source: Marko et al., *J. Phys. D: Appl. Phys.*, 47, 345103, 2014.
The measured variation of J_{th} with barrier Al composition y closely follows the trend calculated for the threshold carrier density n_{th} (cf. Figure 10.9), verifying that J_{th} can be minimized by varying y to control the trade-off between the carrier and optical confinement.

approximately three. The results of these measurements, which emphasize the importance of AlGaAs barriers for the design and optimization of GaBi$_x$As$_{1-x}$ QW lasers at low x, are summarized in Table 10.2. Full details of this optimization analysis can be found in Refs. [15] and [96]. Examining the experimental data in Table 10.2 we can clearly identify the trends predicted by the theoretical calculations. First, the laser structure with Al-free (GaAs) barriers offers poor material gain leading to high J_{th}. Second, incorporation of 12% Al in the barrier (close to the optimum composition suggested by the theoretical calculations) results in a strong reduction of J_{th} as a consequence of the enhanced modal gain. Finally, increasing the Al composition in the barrier from 12% to 20% brings no further improvement in the laser performance; an increase in J_{th} is observed, in line with the degradation of the optical confinement and model gain predicted by the theoretical calculations.

Since the ultimate goal is to grow GaBi$_x$As$_{1-x}$ lasers with Bi compositions $x > 10\%$, this analysis has been repeated for a series of GaBi$_x$As$_{1-x}$/(Al)GaAs QW laser structures as a function of x. In general, it is found that Al incorporation in the barrier layers is required to enhance the otherwise weak electron confinement for $x < 5\%$. For $x = 6\%$, the GaBi$_x$As$_{1-x}$/GaAs CB offset is sufficiently large that incorporating Al in the barrier layers has a negligible impact on the material gain. We therefore conclude that AlGaAs barriers are required to optimize the performance of GaBi$_x$As$_{1-x}$ QWs with $x < 6\%$. For higher Bi compositions, GaAs barriers should suffice to provide modal gain which exceeds that calculated for the $x = 2.1\%$ laser structure considered above at fixed n, provided that QWs with $x > 6\%$ can be grown with sufficiently high material, structural, and optical quality. This conclusion has proved useful, particularly for the growth of hybrid laser structures such as those presented in Ref. [35], where the cladding and part of the barrier layers are grown by metal-organic vapor phase epitaxy (MOVPE) and the active region by molecular beam epitaxy (MBE). This analysis suggests that, for sufficiently large x, GaAs can be used as the final layer grown before a laser substructure is transferred in either direction between MOVPE and MBE growth chambers, thereby minimizing the impact of AlGaAs-related degradation during transport and storage of partly grown laser structures.

10.4.4 Gain Characteristics of GaBi$_x$As$_{1-x}$ Quantum Wells at $\lambda = 1.55$ μm

Having elucidated the consequences of Bi incorporation for the gain characteristics and performance of real GaBi$_x$As$_{1-x}$ QW lasers at low x, we now turn our attention to an ideal 1.55-μm device at higher Bi composition. We again choose a 6.9-nm-thick GaBi$_x$As$_{1-x}$ QW, in order to compare the calculated modal gain directly to that calculated for the $x = 2.1\%$ QW of Sections 10.4.2 and 10.4.3, and, in light of the analysis described in Section 10.4.3, we choose GaAs as the barrier material. For this 6.9-nm-wide GaBi$_x$As$_{1-x}$/GaAs QW, we calculate that a Bi composition $x = 13\%$ produces an emission wavelength close to 1.55 μm. We note that the bulk band structure of the GaBi$_{0.13}$As$_{0.87}$ QW material has $\Delta_{SO} > E_g$, so that it should be possible to use $GaBi_xAs_{1-x}$ alloys to develop, on a GaAs substrate, a 1.55-μm laser in which the CHSH Auger recombination and IVBA processes are suppressed (cf. Figure 10.1) [12,15].

Incorporation of 13% Bi in the QW produces an in-plane compressive strain of 1.53%, compared to 0.25% in a QW with $x = 2.1\%$. We calculate that the tendency of the VBAC interaction to increase the effective masses of the bound hole states in the QW is overcome by the strain-induced reduction of the in-plane hole effective masses [93], leading to a significant reduction of the DOS in the vicinity of the VB edge at $x = 13\%$ [15]. This strong reduction of the DOS at energies close to the QW VB edge should then enhance the performance of this high Bi composition QW compared to the low x laser structures considered above.

Compared to the value $\Gamma = 1.66\%$ calculated for the optical confinement factor of the $x = 2.1\%$ laser structure of Section 10.4.2, we calculate $\Gamma = 1.26\%$ in the $x = 13\%$ QW. The reduction in Γ results from the longer emission wavelength of the $x = 13\%$ structure and leads to an increase of the material gain at threshold g_{th}. We note that the degradation of the optical confinement a long wavelengths can be mitigated by increasing the barrier thickness. We have not pursued this approach here, as our intention is to analyze only the impact of changing the Bi composition in otherwise identical laser structures. Based on the calculated difference in Γ between the two laser structures we calculate $g_{th} = 1208$ cm^{-1} at $x = 13\%$, which is 32% larger than the value of 919 cm^{-1} calculated for the $x = 2.1\%$ laser structure. However, the calculated CB offset of 245 meV in the $GaBi_{0.13}As_{0.87}$/GaAs QW is 60 meV larger than that calculated for the $GaBi_{0.021}As_{0.979}$/$Al_{0.144}Ga_{0.856}$As QW above, leading to enhanced electron confinement in the 1.55-μm QW without the need to incorporate Al in the barrier layers. The calculated optical transition strength at $k_\parallel = 0$ for the e1–hh1 optical transition in the $x = 13\%$ QW is approximately 90% of that calculated at $x = 2.1\%$, suggesting that the Bi-induced decrease of the optical transition strengths described above is most prominent at low x, but has little additional effect as x is increased. When combined with the favorable impact of strain on the VB structure and DOS at larger values of x, these effects together ensure that the material gain at fixed n is significantly improved in the $x = 13\%$ QW compared to the $x = 2.1\%$ laser structure considered above.

The $x = 13\%$ laser structure has a calculated threshold carrier density $n_{th} = 3.42 \times 10^{18}$ cm^{-3} at 300 K, which is approximately 60% of that calculated for the $x = 2.1\%$ laser structure. Additionally, for the $x = 13\%$ laser structure $\frac{dg}{dn} = 2.76 \times 10^{-16}$ cm^2 for the differential (material) gain at threshold, which is approximately a factor of two larger than the value of 1.32×10^{-16} cm^2 calculated for the $x = 2.1\%$ laser structure. This analysis suggests that the marked improvement in the threshold characteristics of the device at $x = 13\%$ results primarily from the increased compressive strain in the QW. First, the strain-induced reduction in the DOS at the VB edge leads to an increase in the quasi-Fermi level separation at fixed n, meaning that population inversion can be achieved at lower carrier densities. Second, the reduced in-plane hole effective masses ensure that holes occupy VB states over a smaller range of $k_\parallel$ (i.e. a narrow $\mathbf{k}$-space distribution) at $x = 13\%$ than at $x = 2.1\%$, better matching the distribution of electrons in the CB and hence ensuring that more carriers are available to contribute to the lasing mode [97]. These factors ensure that a larger increase in material gain can be obtained for a given change in n at $x = 13\%$, leading to the calculated significant improvement in $\frac{dg}{dn}$.

Previous calculations for InP-based QW structures designed to emit at 1.55 μm suggest that the maximum $\frac{dg}{dn}$ obtainable using optimized quaternary InGaAsP QWs are $\sim 14 \times 10^{-16}$ cm^2 for a multi-QW device containing four QWs [98]. While the calculations here suggest that GaAs-based $GaBi_xAs_{1-x}$ QWs offer slightly lower $\frac{dg}{dn}$ than conventional InP-based devices, we recall that this analysis has been performed on a prototypical 1.55-μm dilute bismide structure in which the waveguide has not been optimized. We therefore conclude that 1.55-μm $GaBi_xAs_{1-x}$ QW lasers have the potential to offer differential gain, and hence dynamical performance, that compares favorably with existing InP-based devices, particularly for structures incorporating multiple $GaBi_xAs_{1-x}$ QWs. We further note that the calculated value of $\frac{dg}{dn}$ for this 1.55-μm dilute bismide device exceeds that calculated for ideal 1.3-μm GaIn(N)As/GaAs QW structures [99].

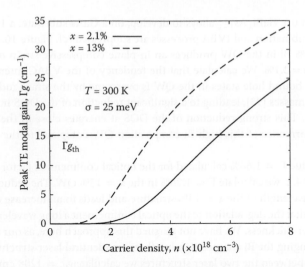

FIGURE 10.10 Calculated variation of the peak TE-polarized modal gain as a function of carrier density at 300 K for the $GaBi_{0.021}As_{0.979}/Al_{0.144}Ga_{0.856}As$ (solid line) and $GaBi_{0.13}As_{0.87}/GaAs$ (dashed line) QWs described in the text. (Adapted from C.A. Broderick et al., *IEEE J. Sel. Top. Quantum Electron.*, 21, 1503313, 2015.) The $x = 2.1\%$ structure has a room temperature emission wavelength ~950 nm, while the $x = 13\%$ structure is designed to have $\Delta_{SO} > E_g$ and to emit at 1.55 μm. The horizontal dash-dotted line denotes the calculated threshold modal gain, $\Gamma g_{th} = 15.2\ \mathrm{cm}^{-1}$.

Figure 10.10 shows the calculated variation of the modal gain as a function of n for the $x = 2.1\%$ (solid line) and $x = 13\%$ (dashed line) laser structures at 300 K. We see that, even in the presence of significant inhomogeneous broadening of the gain spectrum, the enhancement of the material gain described above is sufficient to overcome the reduction in Γ between $x = 2.1\%$ and 13%, leading to significantly improved modal gain at fixed n. Additionally, we note that the differential modal gain at threshold is significantly improved at $x = 13\%$, due to the fact that the calculated increase in the differential material gain at threshold is also sufficient to overcome the reduction in Γ between $x = 2.1\%$ and 13%. We recall that optimization of the waveguide can be undertaken for long-wavelength emission, which can be expected to further improve the performance of QW structures having high Bi compositions over that of low x structures operating at shorter wavelengths.

We therefore conclude overall that the performance of ideal $GaBi_xAs_{1-x}$ QW lasers at high Bi compositions ($x > 10\%$) should be significantly improved compared to that of shorter wavelength devices at lower x. The calculated improvement in the threshold characteristics with increasing x is primarily attributable to the associated increase in compressive strain in the QW when more Bi is incorporated. We also emphasize that the presence of a QW band structure in which $\Delta_{SO} > E_g$ promises to deliver even greater benefit for the laser operation, due to the elimination of the dominant non-radiative CHSH Auger recombination and IVBA losses.

10.4.5 Theory versus Experiment for First-Generation $GaBi_xAs_{1-x}/(Al)GaAs$ Lasers

Having explored general trends in the electronic and optical properties of $GaBi_xAs_{1-x}/(Al)GaAs$ QW lasers, we now compare the results of the theoretical calculations directly to experimental measurements of the SE and gain spectra. The experimental measurements were carried out using the segmented contact method, which allows measurement of the optical absorption, SE, and gain spectra of a multi-section device [100]. The multi-section devices used in this study were fabricated from a wafer on which a single QW $GaBi_xAs_{1-x}/(Al)GaAs$ laser structure was grown. The laser structure is as described in Ref. [30]

and consists of $Al_{0.4}Ga_{0.6}As$ cladding layers, 150-nm-thick AlGaAs barriers and a $GaBi_xAs_{1-x}$ QW with a nominal thickness of 6.4 nm.

Two multisection devices were fabricated in order to carry out the segmented contact measurements: (1) a device fabricated from material located close to the edge of the wafer ("device 1"), and (2) a device fabricated from material located closer to the center of the wafer ("device 2"). Full details of the growth and fabrication of these devices, as well as the theory-experiment comparison to be outlined here, can be found in Ref. [39]. While the barrier has a nominal Al composition of 20%, based on the analysis of Ref. [79] we find that the actual composition is 14.4%—we use the latter in the theoretical calculations, as in Section 10.4.2.

Comparing the measured optical absorption spectra for the two multisection devices, it is found that (1) the optical (cavity) losses are in the range $\alpha_i = 10–15$ cm^{-1} at room temperature, and (2) the absorption edge measured for device 2 is red-shifted by approximately 20 nm compared to the that measured for device 1 [39]. The latter suggests the presence of Bi composition fluctuations across the wafer, with the Bi composition being slightly reduced toward the wafer edge. This is consistent with the measurements of Ref. [79], in which $GaBi_xAs_{1-x}$/(Al)GaAs laser structures having nominally identical $x = 2.2\%$ QWs-displayed variations of approximately 30 meV in the QW band gap (corresponding in theoretical calculations Bi compositional variations of up to $\pm 0.4\%$).

In order to investigate this behavior, we first analyze the measured SE spectra for the two devices. Figure 10.11a shows the SE spectra measured at threshold using the segmented contact method for device 1 (closed circles) and device 2 (closed squares). Both spectra have similar overall shape with the main difference between them being the wavelength of the SE peak, which is approximately 20 nm shorter in device 1. This shifted emission peak is consistent with the observed shift in the absorption edge between the two

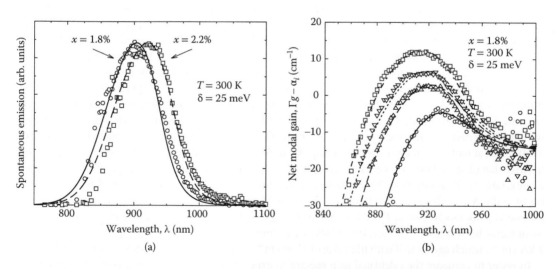

(a)

(b)

FIGURE 10.11 (a) Measured (using the segmented contact method; open symbols) and calculated (lines) SE spectra at threshold for a multisection device fabricated from material located close to the wafer edge (open circles; solid line) and fabricated from material located close to the center of the wafer (closed squares; dashed line). The ~20 nm shift in the SE peak wavelength between the two devices suggests the presence of Bi compositional variations of up to 0.4% across the wafer from which the devices were fabricated. (Adapted from Marko, I.P. et al., *Sci. Rep.*, 6, 28863, 2016.) (b) Measured (using the segmented contact method; closed symbols) and calculated (lines) net modal gain spectra for the device fabricated from material located at the wafer edge, at injected current densities $J = 0.7, 1.4, 2.0,$ and 2.4 kA cm^{-2} (open circles, upright triangles, inverted triangles, and squares, respectively). The carrier densities used in the theoretical calculations (determined as outlined in the text) are $n = 5.12, 7.11, 8.24,$ and 9.38×10^{18} cm^{-3}; the corresponding gain spectra are shown using solid, dashed, dotted, and dash-dotted lines, respectively. (Adapted from Marko, I.P. et al., *Sci. Rep.*, 6, 28863, 2016.)

devices. In order to quantify this variation in Bi composition across the wafer, and to facilitate theoretical analysis of the measured gain spectra for device 1, the theoretical model has been used to analyze the SE spectra shown in Figure 10.11a. The careful parameterization of the 12-band $\mathbf{k}\cdot\mathbf{p}$ model for the GaBi$_x$As$_{1-x}$ band structure described in Section 10.3 enables us to treat the Bi composition x in the QW to be treated as the only free variable in the analysis of the SE emission peak. Through the theoretical calculations we find that the measured SE peak wavelength of 910 nm for device 1 corresponds to a Bi composition $x = 1.8\%$ in the QW (assuming that the QW thickness to be is fixed at its nominal value of 6.4 nm).

Next, by comparing the full calculated SE spectrum for device 1 (shown in Figure 10.11a using a solid line) to the experimental data, we find that the spectral broadening is best described by a hyperbolic secant lineshape with a homogeneous linewidth $\delta = 25$ meV. The theoretical calculation was also carried out at the threshold injection level, which was determined as the carrier density required to produce the TE-polarized threshold material gain $g_{th} = 1325$ cm^{-1} (computed using an effective index calculation to determine $\Gamma = 1.60\%$ for a QW containing 1.8% Bi, and using the measured threshold current density $J_{th} \approx 1.6$ kA cm^{-2} in similar devices [96]). Due to uncertainty in the absolute units and relative intensity of the SE measured for the two devices, the calculated SE spectrum has been normalized to the measured value at the SE peak in order to compare the theoretical and experimental data. We note that the measured and calculated SE spectra are in good overall agreement, with the spectral broadening observed in the experiment well described by a combination of the energy broadening at room temperature of the electron and hole (quasi-Fermi) distribution functions, as well as the hyperbolic secant line broadening.

In order to quantify the variation of the Bi composition across the wafer the SE spectrum for device 2 has also been calculated. The calculation of the SE spectrum in this case proceeds as above, this time with the Bi composition again being the only parameter allowed to vary. The result of this calculation is shown using a dashed line in Figure 10.11a. We find that (1) the measured SE peak at 932 nm is well described in the theoretical calculations, assuming that the QW contains the nominal Bi composition of 2.2%, and (2) the calculated SE spectrum for $x = 2.2\%$ is in good overall agreement with experiment for this device. Overall, these theoretical results suggest that the Bi composition in the QW is close to the nominal value of 2.2% near the center of the wafer, but is slightly reduced, to 1.8%, toward the wafer edges. We recall that this variation in Bi composition is consistent with that determined previously via spectroscopic measurements [79].

The closed symbols in Figure 10.11b show the (TE-polarized) net modal gain spectra, $\Gamma g - \alpha_i$, measured using the segmented contact method for device 1. These gain spectra, which were measured at room temperature for current densities $J = 0.7$, 1.4, 2.0, and 2.4 kA cm^{-2}, are shown using closed circles, upright triangles, inverted triangles and squares, respectively. The measured gain spectra are relatively broad, with a full width at half maximum of approximately 100 meV at a current density of 2 kA cm^{-2}, which is close to twice that observed for an InGaAs/GaAs SQW laser operating at a similar wavelength [101] and is most likely related to the strong Bi-induced inhomogeneous broadening that is characteristic of the optical spectra of GaBi$_x$As$_{1-x}$ alloys [27,28]. Based on the $\alpha_i = 15$ cm^{-1} optical losses measured for device 1, the estimated peak modal gain at $J = 2$ kA cm^{-2} is $\Gamma g_{peak} = 24$ cm^{-1}. Using the calculated optical confinement factor for this laser structure, $\Gamma = 1.60\%$, we estimate a peak material gain $g_{peak} \approx 1500$ cm^{-1} at $J = 2$ kA cm^{-2}, which agrees well with the value of 1560 cm^{-1}, calculated using the theoretical model.

In order to compare the calculated gain spectra to experiment we must determine the carrier density n corresponding to each current density J at which each of the gain spectra depicted by open circles in Figure 10.11b was measured. In order to do so, we recall from Equation 10.8 that the transparency point at which there is zero material/modal gain on the high-energy side of the gain peak is given when the photon energy is equal to the quasi-Fermi level separation at a given level of injection (i.e., $g = 0$ for $\hbar\omega = \Delta F$). We therefore proceed by shifting the measured net modal gain spectra, $\Gamma g - \alpha_i$, upward by the $\alpha_i = 15$ cm^{-1} optical losses in order to obtain the absolute modal gain Γg in the device. Next, we use the transparency points on each of these absolute modal gain spectra to extract the quasi-Fermi level separation ΔF corresponding to each current density J in the experiment. Finally, using the theoretical model we calculate ΔF as a function of the injected carrier density n for the laser structure and, in doing so, determine the value of n which produces the extracted ΔF for each current density J. Following this procedure, we find that the current densities $J = 0.7, 1.4, 2.0$, and 2.4 kA cm^{-2} at which the gain spectra were measured, correspond to quasi-Fermi level separations of 1.375, 1.409, 1.423, and 1.435 eV, which in turn correspond to carrier densities $n = 5.12, 7.11, 8.24$, and 9.38×10^{18} cm^{-3}, respectively, in the theoretical calculations. The TE-polarized component of the SE rate at each of these carrier was then calculated as outlined in Section 10.4.1, from which the theoretical net modal gain spectrum was calculated at each carrier density as $\Gamma g - \alpha_i$, with $\Gamma = 1.60\%$ (calculated) and $\alpha_i = 15$ cm^{-1} (measured), and where g is the TE-polarized material gain spectrum computed using Equation 10.8.

The results of these calculations are shown in Figure 10.11b using, in order of increasing current/carrier density, solid, dashed, dotted, and dash-dotted lines. We note that the theoretical gain spectra shown using solid, dashed and dotted lines include $e1$-$hh1$ optical transitions only while the dash-dotted line, corresponding to the highest current density of 2.4 kA cm^{-2}, also includes $e1$-$lh1$ optical transitions. Including $e1$-$hh1$ transitions only at the highest carrier density was found to underestimate the peak modal gain by ~10%, highlighting that TE-polarized optical recombination involving LH-like states contributes appreciably to the optical gain at higher levels of injection. This conclusion is consistent with further experimental analysis of the SE spectra [39]. Overall, we see that the theoretical spectra are in good, quantitative agreement with experiment: the calculated magnitude of the net modal gain is in excellent agreement with that measured using the segmented contact method across the full investigated range of current densities (confirming in particular the accuracy of the optical transition matrix elements computed within the framework of the 12-band $\mathbf{k \cdot p}$ model), and the shape (in photon energy/wavelength) of the experimental gain spectrum is well reproduced at each current/carrier density by the theoretical model. This analysis has been repeated for a series of multisection and Fabry–Pérot laser devices having $x \approx 2\%$, demonstrating that the theoretical model is generally capable of quantitatively predicting the variation of g_peak with J for this new class of laser structures [39]. Overall, we note that the theoretical model presented in Section 10.4.1, which is directly underpinned by detailed analysis of the impact of Bi on the band structure (facilitated by atomistic supercell calculations), allows calculation of the device properties and performance that are generally in good, quantitative agreement with the available experimental data. The results presented here therefore confirm the predictive capability of the model, suggesting that it can be used as a reliable tool in the analysis, design and optimization of future devices.

10.5 Emerging Directions in Dilute Bismide Research

Having primarily focused so far in this chapter on describing developments in the growth, characterization, theory and applications of GaBi$_x$As$_{1-x}$ alloys grown on GaAs substrates, in this section, we turn our attention to alternative, emerging directions in research on dilute bismide alloys. Sections 10.5.1 and 10.5.2 outline the basic band structure properties of the quaternary Bi-containing alloys In$_y$Ga$_{1-y}$Bi$_x$As$_{1-x}$ and GaBi$_x$N$_y$As$_{1-x-y}$, grown on InP and GaAs substrates, respectively. By extending the $\mathbf{k \cdot p}$ model for GaBi$_x$As$_{1-x}$ alloys outlined in Sections 10.3.2 and 10.3.3, we demonstrate how the incorporation of Bi leads in each case to highly flexible alloy band structures and long emission wavelengths beyond those currently accessible using GaAs or InP substrates. We also briefly describe the development of GaBi$_x$As$_{1-x}$/GaN$_y$As$_{1-y}$ type-II QWs grown on GaAs and demonstrate that this novel class of strain-balanced heterostructures have significant promise for the development of photonic devices operating at and beyond 1.55 μm. GaBi$_x$As$_{1-x}$/GaN$_y$As$_{1-y}$ type-II QWs also have the potential to act as a route to extending the emission wavelength beyond that which has been demonstrated to date in GaBi$_x$As$_{1-x}$/GaAs type-I QWs, which is currently limited by the challenges associated with the epitaxial growth of GaBi$_x$As$_{1-x}$ QWs having sufficiently high optical quality and Bi composition to demonstrate electrically pumped lasing at 1.55 μm. In all three cases, we demonstrate the ability to engineer the band gap, spin-orbit-splitting energy, strain, and band offsets over wide ranges, highlighting the potential to use these new material concepts to deliver enhanced capabilities in near- and mid-infrared photonic devices, as well as (at shorter wavelengths) in photovoltaics.

In addition to these three emerging directions in dilute bismide research, we note that ongoing increase of interest in Bi-containing materials has led to initial demonstrations of a number of additional classes of alloys and heterostructures for a range of potential applications. While it is beyond the scope of this chapter to discuss all of these topics in detail, we briefly list prominent examples in Section 10.5.4 and refer the reader to the references provided therein for further information.

10.5.1 $In_yGa_{1-y}Bi_xAs_{1-x}$/InP Alloys

The vibrational-rotational spectra of many important gases are characterized by strong optical absorption at mid-infrared wavelengths. As such, there is broad scope for practical applications of semiconductor lasers operating in the 2- to 6-μm wavelength range, including: (1) chemical monitoring in industrial processes, (2) detection of environmental pollutants, (3) remote analysis of hazardous substances (including toxic gases and explosives), and (4) detection of biological markers in medical diagnostics. Despite this, mid-infrared semiconductor laser technology is generally less developed than the GaAs- and InP-based technologies that are employed in optical communications at wavelengths <2 μm. Despite the fact that much effort has been dedicated to developing mid-infrared emitters and detectors over the past two decades [102,103], there remains a need to develop new material and device concepts in order to overcome the limitations associated with current technologies. Existing mid-infrared semiconductor lasers typically (1) incorporate Sb-containing materials grown on nonideal and relatively expensive GaSb or InAs substrates, (2) suffer from difficulties associated with the growth of Sb-containing alloys, and/or (3) incorporate complicated heterostructures, such as in quantum (intraband) or interband cascade devices, where the laser structure must be formed of many layer repeats which require a high degree of design optimization and growth control.

In order to overcome these issues, as well as to improve device performance, there has been increasing effort directed toward developing Sb-free devices grown on InP substrates. From a practical perspective, it is desirable to develop InP-based mid-infrared devices, for several reasons. First, materials growth and processing is significantly more advanced for the InP platform than for GaSb or InAs, since the former has experienced heavy technological and commercial development for applications optical in communications. As such, the InP platform offers well-developed, reproducible, and cost-effective device fabrication. Second, from a material physics perspective, InP has several advantages over GaSb, including higher thermal conductivity and lower electrical resistance. Third, growth of InP-based materials stands to benefit from the ready availability of advanced optical components such as low-loss waveguides.

$In_yGa_{1-y}Bi_xAs_{1-x}$/InP alloys offer a promising alternative to existing approaches in this challenging wavelength range, and provide the possibility to develop interband diode lasers incorporating type-I QWs which (1) significantly extend the wavelength range accessible using InP substrates, (2) fully exploit the well-established aspects of the InP material platform [45], and (3) use the Bi-induced modifications to the band structure to suppress the Auger and IVBA loss mechanisms. Full details of our theoretical analysis of the $In_yGa_{1-y}Bi_xAs_{1-x}$ band structure and its suitability for applications in mid-infrared light-emitting devices, can be found in Refs. [13] and [42]. Figure 10.12a shows the variation of the band gap E_g and spin-orbit-splitting energy Δ_{SO} with Bi composition x in $In_{0.53}Ga_{0.47}Bi_xAs_{1-x}$ alloys grown pseudomorphically on InP. Solid (dashed) lines show the variation of E_g (Δ_{SO}) calculated using the 12-band $\mathbf{k \cdot p}$ Hamiltonian described in Section 10.3, while closed (open) circles show the values of E_g (Δ_{SO}) measured using a combination of optical absorption, PL, and photomodulated reflectance spectroscopy [40,42]. We note that (1) the calculated variation of E_g and Δ_{SO} with x is in quantitative agreement with experiment, and (2) the theoretical and experimental data indicate that a band structure in which $\Delta_{SO} > E_g$ can be obtained at low Bi compositions $x > 3.5\%$ in compressively strained $In_{0.53}Ga_{0.47}Bi_xAs_{1-x}$/InP alloys, which is significantly lower than the ~10% required on GaAs (cf. Figure 10.3aa).

In order to evaluate the suitability of $In_yGa_{1-y}Bi_xAs_{1-x}$/InP alloys for applications in mid-infrared devices, we have used the 12-band $\mathbf{k \cdot p}$ model to investigate general trends in the band structure as a function of the Bi and In compositions x and y [44,45]. Figure 10.12b shows a composition space map of the

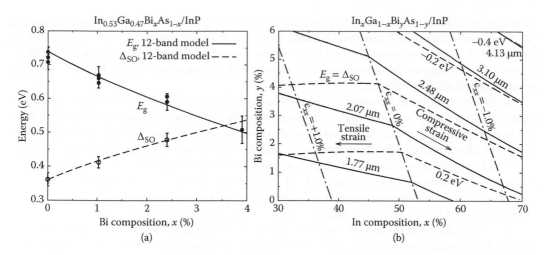

FIGURE 10.12 (a) Variation of the band gap (E_g) and spin-orbit-splitting energy (Δ_{SO}) as a function of Bi composition x, measured for $In_{0.53}Ga_{0.47}Bi_xAs_{1-x}$ alloys grown pseudomorphically on InP (closed and open circles, respectively), compared to calculations undertaken using the 12-band **k·p** Hamiltonian of Section 10.3 (solid and dashed lines, respectively). (Adapted from Marko, I.P. et al., *Appl. Phys. Lett.*, 101, 221108, 2012; C.A. Broderick et al., *Phys. Stat. Sol. B*, 250, 773, 2013.) (b) Calculated variation of E_g, and of the difference between the band gap and spin–orbit-splitting energy, $E_g - \Delta_{SO}$, as a function of Bi and In composition (x and y) for $In_yGa_{1-y}Bi_xAs_{1-x}$/InP alloys (Adapted from Broderick, C.A. et al., In *Proceedings of the 16th International Conference on Numerical Simulation of Optoelectronic Devices*, p. 47 (2016); C.A. Broderick et al., *Theory and design of $In_yGa_{1-y}As_{1-x}Bi_x$ mid-infrared semiconductor lasers: type-I quantum wells for emission beyond 3 μm on InP substrates*, submitted 2017.) Solid and dashed lines denote, respectively, paths in the composition space along which E_g and $E_g - \Delta_{SO}$ are constant. Dash-dotted lines denote paths in the composition space along which strain is constant. We see that $In_yGa_{1-y}Bi_xAs_{1-x}$ alloys can be grown either lattice-matched ($\epsilon_{xx} = 0$), or under compressive ($\epsilon_{xx} < 0$) or tensile ($\epsilon_{xx} > 0$) strain, and that compressively strained alloys enable emission at wavelengths >3 μm at relatively low Bi compositions. Alloys with compositions lying above the $E_g = \Delta_{SO}$ contour are alloys in which $\Delta_{SO} > E_g$, and in which suppression of the nonradiative Auger recombination and IVBA processes involving the SO band can hence be expected.

band gap (E_g), and of the difference between the band gap and spin-orbit-splitting energy ($E_g - \Delta_{SO}$), in pseudomorphically strained $In_yGa_{1-y}Bi_xAs_{1-x}$/InP, calculated using the 12-band model. The dash-dotted lines denote paths in the composition space along which the in-plane strain ϵ_{xx} is constant when the $In_yGa_{1-y}Bi_xAs_{1-x}$ alloy is grown pseudomorphically on an InP substrate and show that the alloy can be grown either lattice-matched ($\epsilon_{xx} = 0$) or under compressive ($\epsilon_{xx} < 0$) or tensile ($\epsilon_{xx} > 0$) strain. Solid lines denote alloy compositions for which E_g is constant. Examining the calculated variation of ϵ_{xx} and E_g with alloy composition, we observe that the band gap of $In_yGa_{1-y}Bi_xAs_{1-x}$ alloys can be varied over an extremely wide range, from ~1.55 to >4 μm, and that this broad spectral coverage can be obtained at modest strains $|\epsilon_{xx}| < 1.5\%$.

The dashed lines in Figure 10.12b denote alloy compositions for which $E_g - \Delta_{SO}$ is constant. Following the $\epsilon_{xx} = 0$ line, we calculate that (1) the amount of Bi required to bring about the $\Delta_{SO} > E_g$ band structure condition (all alloys lying above the $E_g = \Delta_{SO}$ contour) is close to 4% (a prediction in quantitative agreement with the results of a range of experimental measurements [40,43,54]), and (2) the amount of Bi required to achieve $\Delta_{SO} > E_g$ remains constant at approximately 4% in tensile strained alloys less In while, by contrast, it is significantly reduced in compressively strained alloys having higher In compositions. Compared to GaAs, $In_yGa_{1-y}As$/InP alloys have reduced band gaps and comparable spin-orbit-splitting energies, so that significantly less Bi is required to achieve the $\Delta_{SO} > E_g$ band structure condition in $In_yGa_{1-y}Bi_xAs_{1-x}$/InP alloys than in the $GaBi_xAs_{1-x}$/GaAs materials considered previously [13,40,42,43].

Therefore, in addition to providing the possibility of suppressing the CHSH Auger recombination and IVBA loss mechanisms in InP-based mid-infrared semiconductor lasers, the $In_yGa_{1-y}Bi_xAs_{1-x}$/InP material system also has the potential to provide an important proof of principle regarding the use of dilute bismide alloys to suppress the dominant loss mechanisms in GaAs-based semiconductor lasers operating at telecommunication wavelengths.

The difference in band structure between the compressive and tensile strained alloys shown in Figure 10.12b is readily understood: Bi, being larger than the As atoms it replaces, introduces compressive strain so that a significant fraction of In needs to be removed in order to obtain a tensile strained quaternary alloy containing Bi. This removal of In leads to a tensile strained InGaAs host matrix with an increased band gap, as well as a VB structure in which the VBAC coupling is weakened compared to the unstrained or compressively strained cases [13,42]. When these effects are combined, we find that significantly larger Bi compositions are required in tensile strained $In_yGa_{1-y}Bi_xAs_{1-x}$ in order to bring about a given band gap reduction for a fixed magnitude of the strain, as compared to the compressively strained case in which both In and Bi simultaneously contribute to the band gap reduction and compressive strain. Bearing this in mind, and examining Figure 10.12b from the perspective of laser design, we note that the composition region of interest are those lying above the $E_g = \Delta_{SO}$ contour, since it is expected that the Auger and IVBA processes involving the SO band will be suppressed in these alloys, while they are close to resonant for $2 < \lambda < 2.5$ μm, at which compositions $E_g \approx \Delta_{SO}$. Within the reduced composition range of alloys having $\Delta_{SO} > E_g$ we note that the alloy band gap covers a large range of wavelengths >2.5 μm and, significantly, that the 2.5- to >4-μm spectral range can be accessed in alloys having modest compressive strains and Bi compositions.

Our theoretical analysis of $In_yGa_{1-y}Bi_xAs_{1-x}$/InP laser structures [45] suggests that there is broad scope to use this material system to realize compressively strained type-I QWs having emission wavelengths in the 2.5- to 5 μm range. Furthermore, our analysis indicates that the physical characteristics of $In_yGa_{1-y}Bi_xAs_{1-x}$/InP QWs are intrinsically superior to existing GaSb-based heterostructures in the same wavelength range. First, as outlined above, theoretical calculations and experimental measurements have confirmed that $\Delta_{SO} > E_g$ can be achieved in $In_{0.53}Ga_{0.47}Bi_xAs_{1-x}$/InP alloys for Bi compositions as low as $x \approx 4\%$ [13,40,42,43]. This means that it should be possible to suppress the CHSH Auger recombination and IVBA processes and bring about highly efficient, temperature-stable operation in mid-infrared semiconductor lasers. Second, our calculations indicate that large type-I band offsets can readily be engineered in $In_yGa_{1-y}Bi_xAs_{1-x}$ QWs having unstrained ternary $In_{0.53}Ga_{0.47}As$ barriers. This has two significant benefits: (1) the presence of large CB and VB offsets will mitigate carrier leakage, thereby overcoming a key factor limiting the performance of GaSb-based devices at high temperature, and (2) the ability to grow lattice-matched, ternary InGaAs barriers, without the need to incorporate Al or P to enhance the carrier confinement, promises to simplify the growth of these laser structures. We note that (3) also means that it is not necessary to introduce a trade-off between the carrier and optical confinement in these structures (cf. Section 10.4.3). Furthermore, these ternary barrier materials are the same unstrained InGaAs buffer layers upon which $In_yGa_{1-y}Bi_xAs_{1-x}$ epilayers have typically been grown. Overall, this means that $In_yGa_{1-y}Bi_xAs_{1-x}$ QWs can be used to develop mid-infrared lasers in which the remainder of the laser structure is essentially identical to those commonly employed in the well-established 1.3- and 1.55-μm InP-based devices. As such, $In_yGa_{1-y}Bi_xAs_{1-x}$/InP alloys offer the possibility not only to provide type-I InP-based QW diode lasers operating at wavelengths >3 μm, but also promise to deliver several key advantages over competing mid-infrared device concepts.

While there have been some investigations of $In_yGa_{1-y}Bi_xAs_{1-x}$ alloys grown on InP substrates, this material system has to date been the subject of much less research attention than the GaAs-based $GaBi_xAs_{1-x}$ alloys discussed above. Initial growth of $In_yGa_{1-y}Bi_xAs_{1-x}$/InP materials was reported by Feng et al. [104,105], and has since been established by several other groups [106–109]. Enhanced understanding of the incorporation of Bi in InGaAs has led to growth of $In_yGa_{1-y}Bi_xAs_{1-x}$ alloys with Bi compositions as high as 7% [108], and recent experimental and theoretical analysis of a sample containing 5.8% Bi has verified the presence of a $\Delta_{SO} > E_g$ band structure up to and above room temperature [43]. Growth of

$In_yGa_{1-y}Bi_xAs_{1-x}$ alloys has, however, to date been limited to bulk-like epitaxial layers and so, in order to develop this material system for applications in mid-infrared semiconductor lasers, future studies will need to focus on (1) the growth of thin, strained QWs having high optical quality, and (2) providing evidence of reduced CHSH Auger recombination and IVBA in materials and heterostructures having $\Delta_{SO} > E_g$. While research on $In_yGa_{1-y}Bi_xAs_{1-x}$/InP alloys still faces challenges related to the establishment of growth and fabrication of high-quality materials and devices, recent successes in the development of $GaBi_xAs_{1-x}$ alloys, including the demonstration of electrically pumped lasers [30,35,96], suggest that these issues can be overcome with dedicated research effort. We therefore conclude that $In_yGa_{1-y}Bi_xAs_{1-x}$ alloys are a promising candidate for the development of highly efficient, temperature-stable semiconductor lasers, grown on InP substrates and operating in the 3- to 5-μm wavelength range.

10.5.2 $GaBi_xN_yAs_{1-x-y}$/GaAs Alloys

An alternative quaternary Bi-containing alloy which has begun to attract research interest is $GaBi_xN_yAs_{1-x-y}$. Initial theoretical analysis has demonstrated that the band structure of $GaBi_xN_yAs_{1-x-y}$ alloys grown on GaAs substrates is extremely flexible, providing significant potential to develop materials suited to a range of applications in the near- and mid-infrared [12,41,54]. First, since Bi is larger than the As atoms it replaces in GaAs to form $GaBi_xAs_{1-x}$, it introduces compressive strain when grown on a GaAs substrate. Likewise, N atoms, which are smaller than As, introduce tensile strain. As a result, independent control of the Bi and N compositions in $GaBi_xN_yAs_{1-x-y}$ allows precise strain engineering with respect to a GaAs substrate. Second, Bi or N incorporation in GaAs is known to cause a rapid reduction of the band gap with increasing composition. Theoretical calculations [54] suggest that coalloying Bi and N in GaAs to form $GaBi_xN_yAs_{1-x-y}$ enhances these effects and leads to an extremely large reduction in the band gap with increasing Bi and N composition, which is larger than that observed in either $GaBi_xAs_{1-x}$ or GaN_xAs_{1-x} alloys. As a result, the band gaps of $GaBi_xN_yAs_{1-x-y}$ alloys are capable of covering a large spectral range in the near- and mid-infrared, and come with the added benefit of being able to independently control the strain and band gap. These properties make $GaBi_xN_yAs_{1-x-y}$ alloys of particular interest for applications in multi-junction solar cells [21], since they offer the potential to provide optical absorption across an extremely wide range of wavelengths while remaining lattice-matched to either a GaAs or Ge substrate. In addition to the ability to readily control the strain in these alloys, the impact of Bi on the VB structure are retained, meaning that it should also be possible to achieve a band structure in which $\Delta_{SO} > E_g$ in order to suppress Auger and IVBA processes involving the SO band.

Detailed atomistic calculations suggest that the effects of Bi and N on the band structure are effectively decoupled [54], with the modified CB (VB) structure describable primarily in terms of the well-known effects of N-(Bi-)related localized resonant states. On this basis, Broderick et al. derived a 14-band $\mathbf{k \cdot p}$ Hamiltonian to describe the band structure of $GaBi_xN_yAs_{1-x-y}$ alloys [54]. In this model the effects of N and Bi incorporation on the CB and VB structure are described independently in terms of band-anticrossing interactions between N-(Bi-)related localized states and the extended states of the GaAs host matrix CB (VB) edge (cf. Section 10.3.1), with additional virtual crystal and strain-induced modifications to the CB, VB, and SO band edge energies (cf. Section 10.3.2). The fact that N (Bi) primarily affects the CB (VB) then means that the $GaBi_xN_yAs_{1-x-y}$/GaAs band offsets are effectively independently controllable, so that there is significant potential to engineer the properties of QWs and related heterostructures. Incorporating N brings about a large reduction in the CB edge energy, resulting in a large CB offset and a small VB offset with respect to GaAs. The converse is true of Bi incorporation, which leads to a small CB offset and a large VB offset. Combining these effects means that the CB and VB offsets can be engineered relatively independently of one another in $GaBi_xN_yAs_{1-x-y}$, opening up significant potential to develop $GaBi_xN_yAs_{1-x-y}$ QWs providing tunable long-wavelength emission and absorption across a broad spectral range. Full details of our theoretical analysis of the $GaBi_xN_yAs_{1-x-y}$ band structure can be found in Refs. [12], [41] and [54].

Figure 10.13 shows a composition space map of the band gap (E_g), and of the difference between the band gap and spin–orbit-splitting energy ($E_g - \Delta_{SO}$), in pseudomorphically strained $GaBi_xN_yAs_{1-x-y}$/GaAs, calculated using the 14-band **k·p** Hamiltonian described in Ref. [54]. The dashed lines denote paths in the composition space along which the in-plane strain is constant when the $GaBi_xN_yAs_{1-x-y}$ alloy is grown pseudomorphically on a GaAs substrate, and show that the alloy can be grown either lattice-matched ($\epsilon_{xx} = 0$), or under compressive ($\epsilon_{xx} < 0$) or tensile ($\epsilon_{xx} > 0$) strain. Solid lines denote alloy compositions for which E_g is constant. Examining the calculated variation of ϵ_{xx} and E_g with alloy composition we observe that the band gap of $GaBi_xN_yAs_{1-x-y}$ alloys can be varied over an extremely wide range, from that of GaAs (~850 nm) through the near-infrared, and out to wavelengths >3 μm in the mid-infrared. We also note that this broad spectral coverage can be obtained in alloys that are lattice-matched to GaAs, or in alloys having modest tensile or compressive strains with $|\epsilon_{xx}| < 1\%$. The dashed lines in Figure 10.13 denote alloy compositions for which $E_g - \Delta_{SO}$ is constant. The fact that N incorporation brings about a large band gap reduction while leaving the VB structure almost unaltered (apart from minor strain-induced changes) means that co-alloying N and Bi has the potential to bring about an $\Delta_{SO} > E_g$ band structure at lower Bi compositions than in the N-free $GaBi_xAs_{1-x}$ alloy. Following the $\epsilon_{xx} = 0$ line we calculate that the amount of Bi required to bring about the $\Delta_{SO} > E_g$ band structure condition (all alloys lying to the right of the $E_g = \Delta_{SO}$ contour) is close to 7% in the lattice-matched case, which is approximately 30% less than that required in $GaBi_xAs_{1-x}$. In addition to the aforementioned potential of this material system for applications in multijunction solar cells, we also note that the wide composition range for which $\Delta_{SO} > E_g$ indicates that there is also potential to grow $GaBi_xN_yAs_{1-x-y}$ QWs for >2-μm emission, bringing about the possibility of developing GaAs-based mid-infrared diode lasers with suppressed CHSH Auger recombination and IVBA. This also provides an interesting opportunity to develop monolithic GaAs-based VCSELs and related devices operating in the mid-infrared.

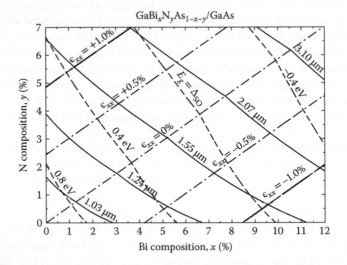

FIGURE 10.13 Calculated variation of the band gap (E_g), and of the difference between the band gap and spin-orbit-splitting energy ($E_g - \Delta_{SO}$), as a function of Bi and N compositions (x and y) for $GaBi_xN_yAs_{1-x-y}$ alloys grown pseudomorphically on GaAs compositions. (Adapted from Broderick, C.A. et al., *Semicond. Sci. Technol.*, 27, 094011, 2012.) Solid and dashed lines denote, respectively, paths in the composition space along which E_g and $E_g - \Delta_{SO}$ are constant. Dash-dotted lines denote paths in the composition space along which strain is constant. We see that $GaBi_xN_yAs_{1-x-y}$ alloys can be grown either lattice-matched ($\epsilon_{xx} = 0$), or under compressive ($\epsilon_{xx} < 0$) or tensile ($\epsilon_{xx} > 0$) strain, and that the band gap can be varied over an extremely wide spectral range in the near- and mid-infrared. Alloys with compositions lying to the right of the $E_g = \Delta_{SO}$ contour are alloys in which $\Delta_{SO} > E_g$, and in which suppression of the non-radiative Auger recombination and IVBA processes involving the SO band can hence be expected.

While there is increasing interest in the development of GaBi$_x$N$_y$As$_{1-x-y}$ alloys for applications in the areas described above, to date there have been limited growth studies of this novel material system. Initial growth of GaBi$_x$N$_y$As$_{1-x-y}$ bulk-like epitaxial layers was undertaken via MBE in the mid-2000s [110,111], while more recently epitaxial layers and QWs have been grown using MOVPE [112]. MOVPE growth has realized co-alloying of Bi and N compositions up to $x = 3.5\%$ and $y = 2.7\%$, but PL has only been observed for samples with compositions up to $x = 1.8\%$ and $y = 1.8\%$. However, the observed PL at these compositions demonstrated an alloy band gap close to 1 eV, confirming that co-alloying of Bi and N can bring about an extreme reduction of the band gap at relatively low compositions [112]. Meanwhile, MBE growth of GaBi$_x$N$_y$As$_{1-x-y}$ alloys has demonstrated electroluminescence at wavelengths out to 1.3 μm [113], and Bi and N incorporations up to $x = 4.5\%$ and $y = 8.0\%$ have been obtained [110]. However, no measurements of the band gap have been performed on these higher composition materials, with the Bi and N compositions having been deduced on the basis of x-ray diffraction measurements. Despite the significant challenges presented by the growth of this highly mismatched quaternary alloy, initial investigations have been promising, and the realization of a 1-eV band gap at such low compositions merits further investigations in order to refine the growth of GaBi$_x$N$_y$As$_{1-x-y}$ alloys to exploit their significant potential for practical applications.

10.5.3 GaBi$_x$As/GaN$_y$As$_{1-y}$ Type-II Quantum Wells on GaAs

In addition to the direct co-alloying Bi and N in GaAs to form GaBi$_x$N$_y$As$_{1-x-y}$, Bi and N incorporation in GaAs opens up additional possibilities for band structure engineering and the development of novel heterostructures. For example, GaN$_x$As$_{1-x}$ (GaBi$_x$As$_{1-x}$) can be used to form tensile (compressively) strained type-I QWs it possible to grow strain-balanced type-II QWs and superlattices in which electrons (holes) are confined in the GaN$_x$As$_{1-x}$ (GaBi$_x$As$_{1-x}$) layer(s) [46,114]. Such structures offer a number of advantages for applications in semiconductor lasers, photodetectors, and solar cells, including the possibility of facilitating optical emission and absorption across a wide range of wavelengths, as well as exploiting a high degree of control over the built-in strain, carrier transport, radiative lifetimes and non-radiative recombination rates [46]. A range of type-II QWs grown on GaAs using different combinations of III-V alloys are currently under investigation for long-wavelength applications. However, strain-balancing of type-II QWs on GaAs has not been possible using the (In)Ga(N)As$_{1-x}$Sb$_x$/In$_y$Ga$_{1-y}$As or GaAs$_{1-x}$Bi$_x$/In$_y$Ga$_{1-y}$As structures that have been investigated recently [115–118], since the constituent alloys are all compressively strained when grown on GaAs. By comparison, GaAs$_{1-x}$Bi$_x$/GaN$_y$As$_{1-y}$ type-II structures are highly engineerable, can be grown with little or no net strain relative to a GaAs substrate, and significant potential for applications in GaAs-based near- and mid-infrared photonic devices [46] and (at shorter wavelengths) in photovoltaics [114].

The band alignment, band offsets, and emission wavelengths achievable using GaAs$_{1-x}$Bi$_x$/GaN$_y$As$_{1-y}$ structures on GaAs are summarized in Figure 10.14a. Dashed lines denote Bi and N compositions x and y for which the CB offset ΔE_{CB} and VB offset ΔE_{VB} are equal in GaAs$_{1-x}$Bi$_x$ and GaN$_y$As$_{1-y}$. Above (below) the lower dashed line ΔE_{CB} is larger (smaller) in GaN$_y$As$_{1-y}$ than in GaAs$_{1-x}$Bi$_x$, while to the right (left) of the upper dashed line ΔE_{VB} is larger (smaller) in GaAs$_{1-x}$Bi$_x$ than in GaN$_y$As$_{1-y}$. This divides the composition space into three regions. Regions A and C correspond to type-I band alignment, with electrons and holes both confined within either the GaN$_y$As$_{1-y}$ (A) or GaAs$_{1-x}$Bi$_x$ (C) layers. Region B corresponds to type-II band alignment, with holes (electrons) confined in the GaAs$_{1-x}$Bi$_x$ (GaN$_y$As$_{1-y}$) layers. Closed circles denote increases of 50 meV in the respective band offsets, beginning from zero at $x = y = 0$ (GaAs). Solid (dash-dotted) lines in region B denote alloy compositions for which the band gap in a bulk (QW) type-II structure—between the GaN$_y$As$_{1-y}$ CB and GaAs$_{1-x}$Bi$_x$ VB—is constant. These calculations demonstrate that type-II GaAs$_{1-x}$Bi$_x$/GaN$_y$As$_{1-y}$ QWs grown on GaAs have the potential to cover an extremely broad spectral range, through the near-infrared to mid-infrared wavelengths in excess of 3 μm [46].

Broderick et al. [46] recently presented the first demonstration of this new class of III-V heterostructures. The black and gray lines in Figure 10.14b show, respectively, the room temperature PL and optical absorption measured for a prototypical type-II GaBi$_x$As$_{1-x}$/GaN$_y$As$_{1-y}$ structure containing five QWs, having Bi and N compositions x = 3.3% and y = 5.2%, and with GaBi$_x$As$_{1-x}$ and GaN$_y$As$_{1-y}$ layer thicknesses of 10.5 and 9.2 nm [46]. This structure was grown using MOVPE and characterized using high-resolution x-ray diffraction measurements, which indicated high structural quality of the strained type-II QWs. Examining Figures 10.14a and 10.14b we note that, despite the intrinsically low optical efficiency of the type-II structure (resulting from the low electron–hole overlap under optical excitation), the measured PL peak at 0.72 eV agrees well with the *e*1-*hh*1 transition energy of 0.74 eV calculated using a carefully parameterized 14-band **k·p** Hamiltonian [54] in conjunction with the plane wave expansion method described in Section 10.4.1 [46]. We note that this band gap corresponds to an emission wavelength ~1.7μm, which demonstrates that GaBi$_x$As$_{1-x}$/GaN$_y$As$_{1-y}$ type-II QWs grown on GaAs can readily provide emission at and beyond 1.55 μm with significantly reduced Bi compositions compared to those required in type-I GaBi$_x$As$_{1-x}$/(Al)GaAs structures (cf. Section 10.4). Indeed, it is promising to observe clear room temperature PL and optical absorption in initial growth of a prototype structure, particularly given the significant scope to optimize the layer ordering and thicknesses to enhance the optical efficiency [46]. Theoretical investigations are ongoing to quantify the electronic and optical properties of this new class of GaAs-based heterostructures, and refinement of epitaxial growth is expected to lead to the development of structures with improved optical efficiency for potential applications in QW solar cells, as well as in semiconductor lasers and photodetectors.

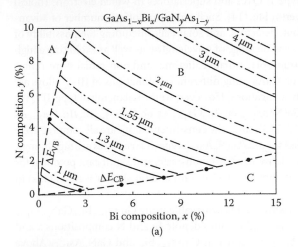

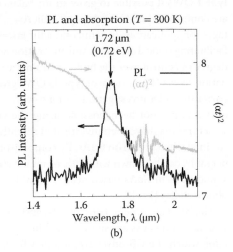

(a)   (b)

FIGURE 10.14 (a) Composition space map showing regions of type-I (A and C) and type-II (B) band alignment, as well as the emission wavelengths accessible using GaAs$_{1-x}$Bi$_x$/GaN$_y$As$_{1-y}$ type-II QWs grown on GaAs. Solid (dash-dotted) lines denote paths in the composition space along which the type-II band gap in bulk (QW) GaAs$_{1-x}$Bi$_x$/GaN$_y$As$_{1-y}$ is constant. Dashed lines denote paths in the composition space along which the CB and VB offsets ΔE_{CB} and ΔE_{VB} are equal in GaAs$_{1-x}$Bi$_x$ and GaN$_y$As$_{1-y}$. Closed circles denote 50 meV increases in ΔE_{CB} and ΔE_{VB} starting from zero at $x = y = 0$. (Adapted from Broderick, C.A. et al., GaAs$_{1-x}$Bi$_x$/GaN$_y$As$_{1-y}$ type-II quantum wells: Novel strain-balanced heterostructures for GaAs-based near- and mid-infrared photonics, *Sci. Rep.*, 7, 46371, 2017) (b) Room temperature photoluminescence (PL) spectrum (black line) and squared product $(\alpha t)^2$ of the optical absorption α and propagation length t (from transmission measurements; gray line) measured for a prototypical GaAs$_{0.967}$Bi$_{0.033}$/GaN$_{0.062}$As$_{0.938}$ type-II QW structure, grown by MOVPE. (Adapted from C.A. Broderick et al., Type-II strain-balanced superlattices on GaAs: novel heterostructure for photonics and photovoltaics, in *Proceedings of the 17th International Conference on Numerical Simulation of Optoelectronic Devices*, 2017)

10.5.4 Further Directions: Quaternary Alloys, Type-II Structures, and Narrow-Gap Materials

Here, we briefly outline additional emerging directions in research on dilute bismide alloys by describing some recent work on materials and heterostructures distinct from, but related to, those discussed above. We refer the reader to the reference provided for further information.

Recently, the quaternary dilute bismide alloy $GaBi_xAs_{1-x-y}P_y$ has been proposed as an alternative material system to $GaBi_xN_yAs_{1-x-y}$ that can provide a 1 eV band gap while remaining lattice matched to a GaAs substrate (cf. Section 10.5.2). In contrast to co-alloying Bi with N to form $GaBi_xN_yAs_{1-x-y}$, incorporation of phosphorus (P), which also compensates the compressive strain brought about by Bi incorporation, tends to increase the band gap at fixed Bi composition. This means that larger Bi compositions will be required to achieve lattice-matched alloys having a given band gap compared to those required in $GaBi_xN_yAs_{1-x-y}$ alloys. However, initial investigations of $GaBi_xP_yAs_{1-x-y}$ have demonstrated that incorporation of a small amount of P significantly enhances Bi incorporation during MOVPE growth [119]. Although it is likely to be related to the P-induced reduction of compressive strain in the quaternary alloy, the precise mechanism by which co-alloying P and Bi enhances Bi incorporation has yet to be definitively determined. Related structures have been grown which include compressively strained $GaBi_xAs_{1-x}$ QWs having tensile strained $GaAs_{1-y}P_y$ barriers, with the first results for this new class of GaAs-based strain-compensated laser structures having been reported recently in Ref. [120]. Investigations of quaternary dilute bismide alloys containing P (or N) are at a relatively early stage, but initial results have been promising and investigations of these novel semiconductor alloys are ongoing.

Further examples of recently emerging directions in dilute bismide research are the growth of type-II and metamorphic heterostructures. In Section 10.5.3 we described that $GaBi_xAs_{1-x}/GaN_yAs_{1-y}$ type-II QWs can be used to achieve long emission wavelengths on GaAs substrates. Recently, GaAs-based $In_yGa_{1-y}As/GaBi_xAs_{1-x}$ type-II QWs have been suggested as an alternative approach to achieving this goal. These heterostructures are similar in principle to the $GaBi_xAs_{1-x}/GaN_yAs_{1-y}$ type-II QWs discussed above, but differ in two important aspects. First, the electron-confining layer(s) of the structure are formed of the conventional alloy $In_yGa_{1-y}As$. The ease and reproducibility with which high-quality $In_yGa_{1-y}As$ strained layers can be grown on GaAs is likely to be of benefit for the realization of electrically pumped lasers and related devices. Second, while the electron-confining GaN_yAs_{1-y} layer(s) in a type-II $GaBi_xAs_{1-x}/GaN_yAs_{1-y}$ structure are tensile strained, the $In_yGa_{1-y}As$ layer(s) in a type-II $In_yGa_{1-y}As/GaBi_xAs_{1-x}$ structure are compressively strained. This means that it is not possible to grow strain-balanced $In_yGa_{1-y}As/GaBi_xAs_{1-x}$ type-II structures on GaAs, which limits the ability to grow large numbers of QWs due to the accumulation of compressive strain in the device active region. This may limit the ability to grow device structure with sufficiently high optical efficiency. MBE growth of $In_yGa_{1-y}As/GaBi_xAs_{1-x}$ type-II structures has recently been established by Pan et al. [115], who reported low temperature PL at a wavelength close to 1.1 μm, thereby providing an initial confirmation of the potential of these structure to overcome the limitations associated with the growth of $GaBi_xAs_{1-x}/GaAs$ type-I QWs having high Bi compositions ($\sim 10\%$).

The first metamorphic heterostructures based on dilute bismide alloys were presented recently by Gu et al. [121], who demonstrated MBE growth of InP-based type-I structures incorporating $Al_yIn_{1-y}As$ metamorphic buffer layers and $InBi_xAs_{1-x}$ QWs. The aim of these structures is the same as that outlined for $In_yGa_{1-y}Bi_xAs_{1-x}/InP$ in Section 10.5.1: to extend the emission wavelength in InP-based QWs beyond 3 μm. Current approaches to obtaining emission at wavelengths >3 μm on InP include InAs metamorphic QWs grown on $Al_yIn_{1-y}As$ metamorphic buffer layers. It is hoped that the strong reduction of the $InBi_xAs_{1-x}$ band gap with increasing Bi composition [54] should allow the wavelength to be extended significantly beyond that obtainable using existing devices [19]. The initial work of Gu et al. demonstrated low temperature PL at wavelengths beyond 3.1 μm, confirming the validity of this approach. However, further effort, is required to quantify and exploit the potential of this class of dilute bismide heterostructure for applications in mid-infrared photonic devices.

Finally, we note that there has also been a steady growth of interest in narrow-gap dilute bismide alloys such as $(In)GaBi_xSb_{1-x}$. These alloys, typically grown on GaSb or InAs substrates, offer the potential to use Bi to reduce the already narrow $(In)GaSb$ band gap, and hence cover an extremely broad range of wavelengths in the mid-infrared and beyond. As such, while investigations of Sb-containing dilute bismide alloys are at a relatively early stage, $(In)GaBi_xSb_{1-x}$ alloys are attracting interest for the development of type-I and -II QW lasers and photodetectors throughout the full mid-infrared spectral range [17,122]. Furthermore, due to their intrinsically large spin-orbit coupling, in addition to the potential of using Bi incorporation to introduce large, tunable Rashba spin-splittings, interest in narrow-gap dilute bismide alloys for applications in spintronics is likely to attract significant attention in the coming years [25].

10.6 Summary and Conclusions

In this chapter we have provided an introduction to the theory and simulation of dilute bismide materials and devices, which we have framed within the context of recent and ongoing research on $GaBi_xAs_{1-x}$ and related Bi-containing alloys. Beginning with a general overview of the electronic properties of dilute bismide alloys, we highlighted the potential to exploit the Bi-induced modifications of the band structure of conventional III–V semiconductors to develop photonic, photovoltaic, and spintronic devices with enhanced performance and capabilities. In particular, we discussed that the incorporation of Bi in $(In)GaAs$ leads to a significant reduction in the material band gap (E_g), which is accompanied by a strong enhancement of the spin-orbit-splitting energy (Δ_{SO}), both of which are characterized by strong composition–dependent bowing. We demonstrated that it is possible, through Bi incorporation, to engineer a band structure in which $\Delta_{SO} > E_g$ at telecommunication wavelengths, which is expected to lead to suppression of the dominant nonradiative Auger recombination and IVBA processes (involving the SO band) that limit the high-temperature performance of existing InP-based 1.55-μm semiconductor lasers. As such, we concluded that $GaBi_xAs_{1-x}$ alloys have the potential to deliver the long sought-after goal of realizing uncooled operation of highly efficient and temperature-stable telecom lasers, which promises to deliver significant energy savings in next-generation optical communications networks. Growth of long-wavelength devices on GaAs also presents the opportunity to exploit vertical-cavity architectures, meaning that dilute bismide alloys have significant potential to extend the benefits of the GaAs platform to telecommunications and mid-infrared wavelengths and, hence, facilitate the development of advanced photonics technologies, with enhanced performance and new capabilities.

Taking into account the fact that Bi incorporated into $(In)GaAs$ acts as an isovalent impurity, in Section 10.2 we highlighted the need for detailed atomistic models to provide a quantitative understanding of the electronic properties. On this basis, we provided a review of the atomistic methods which have been developed for, and applied to, the study of $GaBi_xAs_{1-x}$ and related alloys. Having identified the theoretical methods and models that are capable of describing the impact of Bi on the electronic structure, we then compared the results of atomistic TB calculations for realistic, disordered $GaBi_xAs_{1-x}$ alloys to experiment. We showed that a TB approach is capable of describing, in a quantitative manner, the strong Bi-induced reduction (increase) in the band gap (spin-orbit-splitting energy), across the full composition range for which experimental data are available. We further demonstrated that a TB approach quantitatively predicts the strong Bi-induced increase in the magnitude of the electron effective g factor in pseudomorphically strained $GaBi_xAs_{1-x}/GaAs$ alloys, thereby providing detailed insight into roles played by strain and short-range alloy disorder in determining the details of the strongly perturbed VB structure and confirming the predictive capability of the theory.

In Section 10.3 we turned our attention to deriving a continuum $\mathbf{k \cdot p}$ model which is suited to modeling the electronic and optical properties of dilute bismide heterostructures. Using an atomistic TB model, we reached two significant conclusions. First, we reviewed that the main features of the band structure of $(In)GaBi_xAs_{1-x}$ alloys are well described in terms of a phenomenological Bi composition–dependent VBAC interaction between the extended VB edge states of the host matrix semiconductor and localized

Bi-related impurity states which are resonant with the host matrix VB. Second, we demonstrated how the TB method can be used to directly determine the energies of the Bi-related impurity states, as well as the coupling strength between the Bi-related impurity states and host matrix VB edge states, thereby removing the parametric ambiguity associated with band-anticrossing models parameterized solely with reference to experimental measurements. On this basis, we reviewed the derivation of a 12-band $\mathbf{k} \cdot \mathbf{p}$ Hamiltonian for $(In)GaBi_xAs_{1-x}$ directly from a series of detailed atomistic supercell calculations, provided a consistent set of parameters for performing calculations on real alloys, and used the $\mathbf{k} \cdot \mathbf{p}$ model to explicitly derive expressions for the band edge energies and offsets in pseudomorphically strained $(In)GaBi_xAs_{1-x}$ alloys and QWs, respectively.

Next, in Section 10.4, we discussed the theory and modeling of dilute bismide QW lasers grown on GaAs substrates. We applied the 12-band $\mathbf{k} \cdot \mathbf{p}$ model to the study of $GaBi_xAs_{1-x}/(Al)GaAs$ QW heterostructures and quantified the impact of Bi incorporation on the (1) VB structure and DOS, (2) optical gain, and (3) threshold characteristics of realistic device structures. By considering the incorporation of Al in the barrier layers to improve electron confinement in the QW, we demonstrated the presence of a trade-off between the carrier and optical confinement which allows the design of $GaBi_xAs_{1-x}/(Al)GaAs$ QWs at low Bi compositions $x < 6\%$ to be engineered in order to minimize the threshold current density, a prediction which has been verified experimentally. Repeating this analysis as a function of the QW Bi composition x, we demonstrated that the need to incorporate Al in the barrier layers is removed for $x > 6\%$, which should lead to improved device performance as the Bi composition in GaAs-based laser structures is increased. We extended this analysis to consider the gain characteristics of a $GaBi_xAs_{1-x}$ laser structure designed to emit at 1.55 µm and showed that QWs having higher Bi compositions stand to benefit not only from suppressed Auger recombination and IVBA, but also from the effects of Bi-induced compressive strain, with the latter leading to a favorable VB structure that should, in an ideal device, deliver low threshold current density and high differential gain. Having confirmed the potential of $GaBi_xAs_{1-x}$ QWs for the development of highly efficient and temperature-stable 1.55-µm GaAs-based semiconductor lasers, we then turned our attention to a detailed comparison between theory and experiment for the first generation of electrically pumped $GaBi_xAs_{1-x}$ laser devices. We demonstrated that the theoretical approach outlined in Sections 10.3 and 10.4 is capable of quantitatively describing the SE and optical gain of this new class of GaAs-based semiconductor lasers, confirming its predictive capability and highlighting its potential for use in the design and optimization of future $GaBi_xAs_{1-x}$ lasers and related devices.

Finally, in Section 10.5, we provided an overview of emerging directions in dilute bismide research, including quaternary Bi-containing alloys grown on InP and GaAs substrates, type-II QWs, and narrow-gap materials. We described the potential of $In_yGa_{1-y}Bi_xAs_{1-x}$ and $GaBi_xN_yAs_{1-x-y}$ alloys for respective applications in InP-based mid-infrared photonics, and in multi-junction solar cells. In both cases, we showed that Bi-containing quaternary alloys enable a large degree of control over the band structure, allowing for flexible engineering of the band gap, spin-orbit-splitting energy, band offsets, and strain over wide ranges, thereby making it possible to design new photonic and photovoltaic materials and device structures with enhanced capabilities in the near- and mid-infrared. We described that $GaBi_xAs_{1-x}/GaN_yAs_{1-y}$ type-II structures grown on GaAs offer a route to emission on GaAs at wavelengths out to and beyond 3 µm, in strain-balanced QWs and superlattices. We further highlighted emerging work on $In_yGa_{1-y}As/GaBi_xAs_{1-x}$ type-II QWs and metamorphic $InBi_xAs_{1-x}$ QWs, which, respectively, offer alternative routes to extending the spectral ranges accessible using GaAs and InP substrates. Finally, we described growing interest in narrow-gap $(In)GaBi_xSb_{1-x}$ alloys, which have significant potential for applications across the entirety of the mid-infrared as well as in spintronic devices.

Overall, we conclude that dilute bismide alloys are a rich and rapidly growing area of semiconductor research in which there remain a range of materials physics and device engineering challenges that must be overcome in order to realize the significant potential of this new class of III–V semiconductor materials for practical applications. The theoretical models and calculations reviewed in this chapter constitute the most complete analysis to date of the electronic, optical, and spin properties of dilute bismide alloys, and are expected to serve as a platform to support, interpret, and stimulate ongoing and future research on Bi-containing materials and devices.

Acknowledgments

The writing of this chapter was supported by the Engineering and Physical Sciences Research Council, UK (project nos. EP/K029665/1, EP/H005587/01 and EP/H050787/1), and by Science Foundation Ireland (project no. 15/IA/3082). The research described in this chapter was further supported by the European Commission Seventh Framework Programme project BIANCHO (project no. FP7-257974), by the Irish Research Council (RS/2010/2766), and by Science Foundation Ireland (project no. 10/IN.1/I2994). The authors thank Dr. M. Usman (University of Melbourne, Australia), Prof. K. Volz (Philipps-Universität Marburg, Germany), Prof. A. Krotkus (Center for Physical Sciences and Technology, Lithuania), Prof. J. P. R. David (University of Sheffield, UK), Prof. T. Tiedje (University of Victoria, Canada), Dr. S. Mazzucato (Institut National des Sciences Appliquées, Toulouse, France), Prof. J. M. O. Zide (University of Delaware, USA), and their colleagues, for supplying materials, devices, and experimental data to facilitate aspects of this research, as well as for many useful discussions on the physics of dilute bismide materials and devices. C.A.B. thanks Prof. J. M. Rorison (University of Bristol, UK) for useful feedback on an earlier version of this chapter. The authors also thank their colleagues at their respective institutions for the contributions they have made to the research described in this chapter.

References

1. S. Tixier, M. Adamcyk, T. Tiedje, S. Francoeur, A. Mascarenhas, P. Wei, and F. Schiettekatte. Molecular beam epitaxy growth of GaAs$_{1-x}$Bi$_x$. *Appl. Phys. Lett.*, 82:2245, 2003.
2. S. Francoeur, M. J. Seong, A. Mascarenhas, S. Tixier, M. Adamcyk, and T. Tiedje. Band gap of GaAs$_{1-x}$Bi$_x$, 0 <x < 3.6%. *Appl. Phys. Lett.*, 82:3874, 2003.
3. W. Huang, K. Oe, G. Feng, and M. Yoshimoto. Molecular-beam epitaxy and characteristics of GaN$_y$As$_{1-x}$Bi$_x$. *J. Appl. Phys.*, 98:053505, 2005.
4. K. Alberi, J. Wu, W. Walukiewicz, K. M. Yu, O. D. Dubon, S. P. Watkinsa, C. X. Wang, X. Liu, Y. J. Cho, and J. Furdyna. Valence-band anticrossing in mismatched III-V semiconductor alloys. *Phys. Rev. B*, 75:045203, 2007.
5. M. Usman, C. A. Broderick, A. Lindsay, and E. P. O'Reilly. Tight-binding analysis of the electronic structure of dilute bismide alloys of GaP and GaAs. *Phys. Rev. B*, 84:245202, 2011.
6. Z. Batool, K. Hild, T. J. C. Hosea, X. Lu, T. Tiedje, and S. J. Sweeney. The electronic band structure of GaBiAs/GaAs layers: Influence of strain and band anti-crossing. *J. Appl. Phys.*, 111:113108, 2012.
7. B. Fluegel, S. Francoeur, A. Mascarenhas, S. Tixier, E. C. Young, and T. Tiedje. Giant spin-orbit bowing in GaAs$_{1-x}$Bi$_x$. *Phys. Rev. Lett.*, 97:067205, 2006.
8. W. Shan, W. Walukiewicz, J. W. Ager, E. E. Haller, J. F. Geisz, D. J. Friedman, J. M. Olson, and S. R. Kurtz. Band anticrossing in GaInNAs alloys. *Phys. Rev. Lett.*, 82:1221, 1999.
9. S. J. Sweeney. Light emitting semiconductor device. Patent no. WO2010149978 A1 (filed: June 25, 2010, issued: December 29, 2010).
10. S. J. Sweeney. Bismide-alloys for higher efficiency infrared semiconductor lasers. In *Proceedings of the 22nd IEEE International Semiconductor Laser Conference (ISLC)*, p. 111, 2010.
11. S. J. Sweeney, Z. Batool, K. Hild, S. R. Jin, and T. J. C. Hosea. The potential role of bismide alloys in future photonic devices. In *Proceedings of the 13th International Conference on Transparent Optical Networks*, p. 1, 2011.
12. C. A. Broderick, M. Usman, S. J. Sweeney, and E. P. O'Reilly. Band engineering in dilute nitride and bismide semiconductor lasers. *Semicond. Sci. Technol.*, 27:094011, 2012.
13. S. Sweeney, S. R. Jin. Bismide–nitride alloys: Promising for ecient light emitting devices in the nearand mid-infrared. *J. Appl. Phys.* 113, 043110, 2013.
14. S. Imhof, C. Bückers, A. Thränhardt, J. Hader, J. V. Moloney, and S. W. Koch. Microscopic theory of the optical properties of Ga(AsBi)/GaAs quantum wells. *Semicond. Sci. Technol.*, 23:125009, 2008.

15. C. A. Broderick, P. E. Harnedy, and E. P. O'Reilly. Theory of the electronic and optical properties of dilute bismide quantum well lasers. *IEEE J. Sel. Top. Quantum Electron.*, 21:1503313, 2015.

16. A. Geižutis, V. Pačebutas, R. Butkutė, P. Svidovskya, V. Strazdienė, and A. Krotkus. Growth and characterization of UTC photo-diodes containing $GaAs_{1-x}Bi_x$ absorber layer., *Solid State Electron.*, 99:101, 2014.

17. J. J. Lee, J. D. Kim, and M. Razeghi. Growth and characterization of InSbBi for long wavelength infrared photodetectors. *Appl. Phys. Lett.*, 70:3266, 1997.

18. I. C. Sandall, F. Bastiman, B. White, R. Richards, D. Mendes, J. P. R. David, and C. H. Tan. Demonstration of InAsBi photoresponse beyond 3.5 µm. *Appl. Phys. Lett.*, 104:171109, 2014.

19. Y. Gu, Y. G. Zhang, X. Y. Chen, Y. J. Ma, S. P. Xi, B. Du, and H. Li. Nearly lattice-matched short-wave infrared InGaAsBi photodetectors on InP. *Appl. Phys. Lett.*, 108:032102, 2016.

20. C. J. Hunter, F. Bastiman, A. R. Mohmad, R. Richards, J. S. Ng, S. J. Sweeney, and J. P. R. David. Absorption characteristics of $GaAs_{1-x}Bi_x$/GaAs diodes in the near-infrared. *IEEE Photon. Tech. Lett.*, 24:2191, 2012.

21. S. J. Sweeney, K. Hild, and S. R. Jin. The potential of GaAsBiN for multi-junction solar cells. In *Proceedings of the 39th IEEE Photovoltaic Specialists Conference*, p. 2474, 2013.

22. T. Thomas, A. Mellor, N. P. Hylton, M. Führer, D. Alonso-Álvarez, A. Braun, N. J. Ekins-Daukes, J. P. R. David, and S. J. Sweeney. Requirements for a GaAsBi 1 eV sub-cell in a GaAs-based multi-junction solar cell. *Semicond. Sci. Technol.*, 30:094010, 2015.

23. S. J. Sweeney, and K. Hild. Light receiving device. Patent no. 20160149060 (filed: June 19, 2014, issued: May 26, 2016).

24. S. Mazzucato, T. T. Zhang, H. Carrère, D. Lagarde, P. Boonpeng, A. Arnoult, G. Lacoste, A. Balocchi, T. Amand, C. Fontaine, and X. Marie. Electron spin dynamics and g-factor in GaAsBi. *Appl. Phys. Lett.*, 102:252107, 2013.

25. R. A. Simmons, S. R. Jin, S. J. Sweeney, and S. K. Clowes. Enhancement of the Rashba interaction in GaAs/AlGaAs quantum wells due to the incorporation of bismuth. *Appl. Phys. Lett.*, 107:142401, 2015.

26. P. Dongmo, Y. Zhong, P. Attia, C. Bomberger, R. Cheaito, J. F. Ihlefeld, P. E. Hopkins, and J. M. O. Zide. Enhanced room temperature electronic and thermoelectric properties of the dilute bismuthide InGaBiAs. *J. Appl. Phys.*, 112:093710, 2012.

27. M. Usman, C. A. Broderick, Z. Batool, K. Hild, T. J. C. Hosea, S. J. Sweeney, and E. P. O'Reilly. Impact of alloy disorder on the band structure of compressively strained $GaBi_xAs_{1-x}$. *Phys. Rev. B*, 87:115104, 2013.

28. M. Usman, and E. P. O'Reilly. Atomistic tight-binding study of the electronic structure and interband optical transitions in $GaBi_xAs_{1-x}$/GaAs quantum wells. *Appl. Phys. Lett.*, 104:071103, 2014.

29. C. A. Broderick, S. Mazzucato, H. Carrère, T. Amand, H. Makhloufi, A. Arnoult, C. Fontaine, O. Donmez, A. Erol, M. Usman, E. P. O'Reilly, and X. Marie. Anisotropic electron g factor as a probe of the electronic structure of $GaBi_xAs_{1-x}$/GaAs epilayers. *Phys. Rev. B*, 90:195301, 2014.

30. P. Ludewig, N. Knaub, N. Hossain, S. Reinhard, L. Natterman, I. P. Marko, S. R. Jin, K. Hild, S. Chatterjee, W. Stolz, S. J. Sweeney, and K. Volz. Electrical injection Ga(AsBi)/(AlGa)As single quantum well laser. *Appl. Phys. Lett.*, 102:242115, 2013.

31. S. J. Sweeney, A. F. Phillips, A. R. Adams, E. P. O'Reilly, and P. J. A. Thijs. The effect of temperature dependent processes on the performance of 1.5-µm compressively strained InGaAs(P) MQW semiconductor diode lasers. *IEEE Photon. Tech. Lett.*, 10:1076, 1998.

32. S. J. Sweeney, A. R. Adams, M. Silver, E. P. O'Reilly, J. R. Watling, A. B. Walker, and P. J. A. Thijs. Dependence of threshold current on QW position and on pressure in 1.5-µm InGaAs(P) lasers. *Phys. Stat. Sol. B*, 211:525, 1999.

33. A. R. Adams. Band-structure engineering for low-threshold high-efficiency semiconductor lasers. *Electron. Lett.*, 22:249, 1986.

34. Y. Tominaga, K. Oe, and M. Yoshimoto. Low temperature dependence of oscillation wavelength in $GaAs_xBi_{1-x}$ laser by photo-pumping. *Appl. Phys. Express*, 3:062201, 2010.

35. R. Butkutė, A. Geižutis, V. Pačebutas, B. Čechavičius, V. Bukauskas, R. Kundrotas, P. Ludewig, K. Volz, and A. Krotkus. Multi-quantum well Ga(AsBi)/GaAs laser diodes with more than 6% of bismuth. *Electron. Lett.*, 50:1155, 2014.

36. I. P. Marko, S. R. Jin, K. Hild, Z. Batool, Z. L. Bushell, P. Ludewig, W. Stolz, K. Volz, V. Pačebutas, A. Geižutis, A. Krotkus, and S. J. Sweeney. Properties of hybrid MOVPE/MBE grown GaAsBi/GaAs-based near-infrared emitting quantum well lasers. *Semicond. Sci. Technol.*, 30:094008, 2015.

37. P. Ludewig, L. Natterman, W. Stolz, and K. Volz. MOVPE growth mechanisms of dilute bismide III/V alloys. *Semicond. Sci. Technol.*, 30:094017, 2015.

38. R. D. Richards, F. Bastiman, C. J. Hunter, D. F. Mendes, A. R. Mohmad, J. S. Roberts, and J. P. R. David. Molecular beam epitaxy growth of GaAsBi using As_2 and As_4. *J. Cryst. Growth*, 390:120, 2014.

39. I. P. Marko, C. A. Broderick, S. R. Jin, P. Ludewig, W. Stolz, K. Volz, J. M. Rorison, E. P. O'Reilly, and S. J. Sweeney. Optical gain in GaAsBi/GaAs quantum well diode lasers. *Sci. Rep.*, 6:28863, 2016.

40. I. P. Marko, Z. Batool, K. Hild, S. R. Jin, N. Hossain, T. J. C. Hosea, J. P. Petropoulos, Y. Zhong, P. B. Dongmo, J. M. O. Zide, and S. J. Sweeney. Temperature and Bi-concentration dependence of the bandgap and spin–orbit splitting in InGaBiAs/InP semiconductors for mid-infrared applications. *Appl. Phys. Lett.*, 101:221108, 2012.

41. S. R. Jin, and S. J. Sweeney. InGaAsBi alloys on InP for effcient near-and mid-infrared light emitting devices. *J. Appl. Phys.* 114:213103, 2013.

42. C. A. Broderick, M. Usman, and E. P. O'Reilly. 12-band $\mathbf{k} \cdot \mathbf{p}$ model for dilute bismide alloys of (In)GaAs derived from supercell calculations. *Phys. Stat. Sol. B*, 250:773, 2013.

43. G. M. T. Chai, C. A. Broderick, S. R. Jin, J. P. Petropoulous, Y. Zong, P. B. Dongmo, J. M. O. Zide, E. P. O'Reilly, S. J. Sweeney, and T. J. C. Hosea. Experimental and modelling study of InGaBiAs/InP alloys with up to 5.8% Bi, and with $\Delta_{SO} > E_g$. *Semicond. Sci. Technol.*, 30:094015, 2015.

44. C. A. Broderick, W. Xiong, and J. M. Rorison. Theory of InGaBiAs dilute bismide alloys for highly efficient InP-based mid-infrared semiconductor lasers. In *Proceedings of the 16th International Conference on Numerical Simulation of Optoelectronic Devices*, p. 47, 2016.

45. C. A. Broderick, W. Xiong, S. J. Sweeney, E. P. O'Reilly, J. M. Rorison, Theory and design of $In_yGa_{1-y}As_{1-x}Bi_x$ mid-infrared semiconductor lasers: type-I quantum wells for emission beyond 3 μm on InP substrates. Submitted, 2017.

46. C. A. Broderick, S. R. Jin, I. P. Marko, K. Hild, P. Ludewig, Z. L. Bushell, W. Stolz, J. M. Rorison, E. P. O'Reilly, K. Volz, and S. J. Sweeney. $GaAs_{1-x}Bi_x/GaN_yAs_{1-y}$ type-II quantum wells: Novel strain-balanced heterostructures for GaAs-based near- and mid-infrared photonics. *Sci. Rep.* 7, 46371, 2017.

47. A. Janotti, S. H. Wei, and S. B. Zhang. Theoretical study of the effects of isovalent coalloying of Bi and N in GaAs. *Phys. Rev. B*, 65:115203, 2002.

48. P. Carrier, and S. H. Wei. Calculated spin-orbit splitting of all diamondlike and zinc-blende semiconductors: Effects of $p_{1/2}$ local orbitals and chemical trends. *Phys. Rev. B*, 70:035212, 2004.

49. Y. Zhang, A. Mascarenhas, and L. W. Wang. Similar and dissimilar aspects of III–V semiconductors containing Bi versus N. *Phys. Rev. B*, 71:155201, 2005.

50. E. P. O'Reilly, A. Lindsay, P. J. Klar, A. Polimeni, and M. Capizzi. Trends in the electronic structure of dilute nitride alloys. *Semicond. Sci. Technol.*, 24:033001, 2009.

51. P. Vogl, H. P. Hjalmarson, and J. D. Dow. A semi-empirical tight-binding theory of the electronic structure of semiconductors. *J. Phys. Chem. Solids*, 44:365, 1983.

52. E. P. O'Reilly, A. Lindsay, S. Tomić, and M. Kamal-Saadi. Tight-binding and $\mathbf{k} \cdot \mathbf{p}$ models for the electronic structure of Ga(In)NAs and related alloys. *Semicond. Sci. Technol.*, 17:870, 2002.

53. A. Lindsay, and E. P. O'Reilly. Unification of the band-anticrossing and cluster-state models of dilute nitride semiconductor alloys. *Phys. Rev. Lett.*, 93:196402, 2004.

54. C. A. Broderick, M. Usman, and E. P. O'Reilly. Derivation of 12- and 14-band $\mathbf{k} \cdot \mathbf{p}$ Hamiltonians for dilute bismide and bismide–nitride alloys. *Semicond. Sci. Technol.*, 28:125025, 2013.

55. F. A. Trumbore, M. Gershenzon, and D. G. Thomas. Luminescence due to the isoelectronic substitution of bismuth for phosphorus in gallium phosphide. *Appl. Phys. Lett.*, 9:4, 1966.

56. H. X. Deng, J. Li, S. S. Li, H. Peng, J. B. Xia, L. W. Wang, and S. H. Wei. Band crossing in isovalent semiconductor alloys with large size mismatch: First-principles calculations of the electronic structure of Bi and N incorporated GaAs. *Phys. Rev. B*, 82:193204, 2010.

57. R. S. Joshya, A. J. Ptak, R. France, A. Mascarenhas, and R. N. Kini. Resonant state due to Bi in the dilute bismide alloy $GaAs_{1-x}Bi_x$. *Phys. Rev. B*, 90:165203, 2014.

58. K. Alberi, D. A. Beaton, and A. Mascarenhas. Direction observation of the E_- resonant state in $GaAs_{1-x}Bi_x$. *Phys. Rev. B*, 92:241201, 2015.

59. V. Virkkala, V. Havu, F. Tuomisto, and M. J. Puska. Modeling Bi-induced changes in the electronic structure of $GaAs_{1-x}Bi_x$ alloys. *Phys. Rev. B*, 88:235201, 2013.

60. O. Rubel, A. Bokhanchuk, S. J. Ahmed, and E. Assmann. Unfolding the band structure of disordered solids: From bound states to high-mobility Kane fermions. *Phys. Rev. B*, 90:115202, 2014.

61. J. Hader, J. V. Moloney, O. Rubel, S. C. Badescu, S. Johnson, and S. W. Koch. Microscopic modelling of opto-electronic properties of dilute bismide materials for the mid-IR. *Proc. SPIE*, 9767:976709, 2016.

62. M. Mbarki and A. Rebey. First-principles calculation of the physical properties of $GaAs_{1-x}Bi_x$ alloys. *Semicond. Sci. Technol.*, 26:105020, 2011.

63. R. Kudraweic, J. Kopaczek, M. P. Polak, P. Scharoch, M. Gladysiewicz, R. D. Richards, F. Bastiman, and J. P. R. David. Experimental and theoretical studies of band gap alignment in $GaAs_{1-x}Bi_x$/GaAs quantum wells. *J. Appl. Phys.*, 116:233508, 2014.

64. M. P. Polak and P. Scharoch, and R. Kudraweic. First-principles calculations of bismuth-induced changes in the band structure of dilute Ga–V–Bi and In–V–Bi alloys: Chemical trends versus experimental data. *Semicond. Sci. Technol.*, 30:094001, 2015.

65. M. Ferhat and A. Zaoui. Structural and electronic properties of III–V bismuth compounds. *Phys. Rev. B*, 73:115107, 2006.

66. A. Belabbes, A. Zaoui, and M. Ferhat. Lattice dynamics study of bismuth III–V compounds. *J. Phys.: Condens. Matter*, 20:415221, 2008.

67. H. Jacobsen, B. Puchala, T. F. Keuch, and D. Morgan. Ab initio study of the strain dependent thermodynamics of Bi doping in GaAs. *Phys. Rev. B*, 86:085207, 2012.

68. G. Luo, S. Yang, J. Li, M. Arjmand, I. Szlufarska, A. S. Brown, T. F. Keuch, and D. Morgan. First-principles studies on molecular beam epitaxy growth of $GaAs_{1-x}Bi_x$. *Phys. Rev. B*, 92:035415, 2015.

69. M. P. J. Punkinnen, P. Laukkanen, M. Kuzmin, H. Levämäki, J. Løang, M. Tuominen, M. Yasir, J. Dahl, S. Lu, E. K. Delczeg-Czirjak, L. Vitos, and K. Kokko. Does Bi form clusters in $GaAs_{1-x}Bi_x$ alloys. *Semicond. Sci. Technol.*, 29:115007, 2014.

70. M. P. J. Punkinnen, A. Lahti, P. Laukkanen, M. Kuzmin, M. Tuominen, M. Yasir, J. Dahl, J. Makela, H. L. Zhang, L. Vitos, and K. Kokko. Thermodynamics of pseudobinary $GaAs_{1-x}Bi_x$ ($0 \leq x \leq 1$) alloys studied by different exchange-correlation functionals, special quasi-random structures and Monte Carlo simulations. *Computat. Condens. Matter*, 5:7, 2015

71. P. N. Keating. Effect of invariance requirements on the elastic strain energy of crystals with application to the diamond structure. *Phys. Rev.*, 145:637, 1966.

72. O. L. Lazarenkova, P. von Allmen, F. Oyafuso, S. Lee, and G. Klimeck. Effect of anharmonicity of the strain energy on band offsets in semiconductor nanostructures. *Appl. Phys. Lett.*, 85:4193, 2004.

73. J. C. Slater and G. F. Koster. Simplified LCAO method for the periodic potential problem. *Phys. Rev.*, 94:1498, 1954.

74. J. Yoshida, T. Kita, O. Wada, and K. Oe. Temperature dependence of $GaAs_{1-x}Bi_x$ band gap studied by photo-reflectance spectroscopy. *Jpn. J. Appl. Phys.*, 42:371, 2003.

75. G. Hendorfer and J. Schneider. g-factor and effective mass anisotropies in pseudomorphic strained layers. *Semicond. Sci. Technol.*, 6:595, 1991.

76. P. L. Jeune, D. Robart, X. Marie, T. Amand, M. Brousseau, J. Barrau, V. Kalevich, and D. Rodichev. Anisotropy of the electron Landé g-factor in quantum wells. *Semicond. Sci. Technol.*, 12:380, 1997.

77. M. Oestreich and S. Hallstein, W. W. Rühle. Spin quantum beats in semiconductors. *IEEE J. Sel. Top. Quantum Electron.*, 2:747, 1996.

78. A. P. Heberle, W. W. Rühle, and K. Ploog. Quantum beats of electron Larmor precession in GaAs wells. *Phys. Rev. Lett.*, 72:3887, 1994.

79. C. A. Broderick, P. E. Harnedy, P. Ludewig, Z. L. Bushell, K. Volz, R. J. Manning, and E. P. O'Reilly. Determination of type-I band offsets in $GaBi_xAs_{1-x}$ quantum wells using polarisation-resolved photovoltage spectroscopy and 12-band $\mathbf{k}\cdot\mathbf{p}$ calculations. *Semicond. Sci. Technol.*, 30:094009, 2015.

80. T. B. Bahder. 8-Band $\mathbf{k}\cdot\mathbf{p}$ model of strained zinc-blende crystals. *Phys. Rev. B*, 41:11992, 1990.

81. G. L. Bir, G. E. Pikus. *Symmetry and Strain-Induced Effects in Semiconductors* New York: Wiley, 1974.

82. C. G. Van de Walle. Band lineups and deformation potentials in the model-solid theory. *Phys. Rev. B*, 39:1871, 1989.

83. M. P. C. M. Krijn. Heterojunction band offsets and effective masses in III–V quaternary alloys. *Semicond. Sci. Technol.*, 6:27, 1991.

84. I. Vurgaftman, J. R. Meyer, and L. R. Ram-Mohan. Band parameters for III–V compound semiconductors and their alloys. *J. Appl. Phys.*, 89:5815, 2001.

85. M. Ehrhardt and T. Koprucki, editors. *Multiband Effective Mass Approximations: Advanced Mathematical Models and Numerical Techniques*. Berlin: Springer, 2014.

86. S. B. Healy, and E. P. O'Reilly. Influence of electrostatic confinement on optical gain in GaInNAs quantum-well lasers. *IEEE J. Quant. Electron.*, 42:608, 2006.

87. A. T. Meney, B. Gonul, and E. P. O'Reilly. Evaluation of various approximations used in the envelope-function method. *Phys. Rev. B*, 50:10893, 1994.

88. F. Szmulowicz. Derivation of a general expression for the momentum matrix elements within the envelope-function approximation. *Phys. Rev. B*, 51:1613, 1995.

89. C. S. Chang, S. L. Chuang, J. R. Minch, W. C. W. Fang, Y. K. Chen, and T. Tanbun-Ek. Amplified spontaneous emission spectroscopy in strained quantum-well lasers. *IEEE J. Sel. Top. Quantum Electron.*, 1:1100, 1995.

90. S. Tomić, E. P. O'Reilly, R. Fehse, S. J. Sweeney, A. R. Adams, A. D. Andreev, S. A. Choulis, T. J. C. Hosea, and H. Riechert. Theoretical and experimental analysis of 1.3-μm InGaAsN/GaAs lasers. *IEEE J. Sel. Top. Quantum Electron.*, 9:1228, 2003.

91. K. Kawano and T. Kitoh. *Introduction to Optical Waveguide Analysis: Solving Maxwell's Equation and the Schrödinger Equation*. New York, NY: Wiley, 2001.

92. M. Gladysiewicz, R. Kudrawiec, and M. S. Wartak. 8-Band and 14-band $\mathbf{k}\cdot\mathbf{p}$ modeling of electronic band structure and material gain in Ga(In)AsBi quantum wells grown on GaAs and InP substrates. *J. Appl. Phys.*, 118:055702, 2015.

93. E. P. O'Reilly and A. R. Adams. Band-structure engineering in strained semiconductor lasers. *IEEE J. Quant. Electron.*, 30:366, 1994.

94. S. L. Chuang. *Physics of Photonic Devices*. New York, NY: Wiley, 2009.

95. C. A. Broderick, I. P. Marko, J. M. Rorison, S. J. Sweeney, and E. P. O'Reilly. GaAs-based dilute bismide semiconductor lasers: Theory vs. experiment. In *Proceedings of the 16th International Conference on Numerical Simulation of Optoelectronic Devices*, p. 209, 2016.

96. I. P. Marko, P. Ludewig, Z. L. Bushell, S. R. Jin, K. Hild, Z. Batool, S. Reinhard, L. Natterman, W. Stolz, K. Volz, and S. J. Sweeney. Physical properties and optimization of GaBiAs/(Al)GaAs based near-infrared laser diodes grown by MOVPE with up to 4.4% Bi. *J. Phys. D: Appl. Phys.*, 47:345103, 2014.

97. A. Ghiti, M. Silver, and E. P. O'Reilly. Low threshold current and high differential gain in ideal tensile- and compressive-strained quantum-well lasers. *J. Appl. Phys.*, 71:4626, 1992.

98. M. Silver and E. P. O'Reilly. Optimization of long wavelength InGaAsP strained quantum-well lasers. *IEEE J. Quant. Electron.*, 31:1193, 1995.

99. S. Tomić and E. P. O'Reilly. Optimization of material parameters in 1.3-μm InGaAsN-GaAs lasers. *IEEE Photon. Tech. Lett.*, 15:6, 2003.

100. P. Blood, G. M. Lewis, P. M. Smowton, H. Summers, J. Thomson and J. Lutti. Characterization of semiconductor laser gain media by the segmented contact method. *IEEE J. Sel. Top. Quantum Electron.*, 9:1275, 2003.

101. D. J. Bossert and D. Gallant. Gain, refractive index, and α-parameter in InGaAs–GaAs SQW broad-area lasers. *IEEE Photon. Tech. Lett.*, 8:322, 1996.

102. A. Bauer, K. Rößner, T. Lehnhardt, M. Kamp, S. Höfling, L. Worschech, and A. Forchel. Mid-infrared semiconductor heterostructure lasers for gas sensing applications. *Semicond. Sci. Technol.*, 26:014032, 2011.

103. E. Tournié and A. N. Baranov. Mid-infrared semiconductor lasers: A review. *Semiconduct. Semimet.*, 86:183, 2012.

104. G. Feng, M. Yoshimoto, K. Oe, A. Chayahara, and Y. Horino. New III–V semiconductor InGaAsBi alloy grown by molecular beam epitaxy. *Jpn. J. Appl. Phys.*, 44:L1161, 2005.

105. G. Feng, K. Oe, and M. Yoshimoto. Bismuth containing III–V quaternary alloy InGaAsBi grown by MBE. *Phys. Stat. Sol. A*, 203:2760, 2006.

106. J. P. Petropoulos, Y. Zhong, and J. M. O. Zide. Optical and electrical characterization of InGaAsBi for use as a mid-infrared optoelectronic material. *Appl. Phys. Lett.*, 99:031110, 2011.

107. Y. Zhong, P. B. Dongmo, J. P. Petropoulos, and J. M. O. Zide. Effects of molecular beam epitaxy growth conditions on composition and optical properties of $In_xGa_{1-x}Bi_yAs_{1-y}$. *Appl. Phys. Lett.*, 100:112110, 2012.

108. J. Devenson, V. Pačebutas, R. Butkutė, A. Baranov, and A. Krotkus. Structure and optical properties of InGaAsBi with up to 7% bismuth. *Appl. Phys. Express*, 5:015503, 2012.

109. S. Zhou, M. Qi, L. Ai, A. Xu, and S. Wang. Effects of buffer layer preparation and Bi concentration on InGaAsBi layers grown by gas source molecular beam epitaxy. *Semicond. Sci. Technol.*, 30:125001, 2015.

110. M. Yoshimoto, W. Huang, Y. Takehara, J. Saraie, A. Chayahara, Y. Horino, and K. Oe. New semiconductor GaNAsBi alloy grown by molecular beam epitaxy. *Jpn. J. Appl. Phys.*, 43:L845, 2004.

111. S. Tixier, S. E. Webster, E. C. Young, T. Tiedje, S. Francoeur, A. Mascarenhas, P. Wei, and F. Schiettekatte. Band gaps for the dilute quaternary alloys $GaN_xAs_{1-x-y}Bi_y$ and $Ga_{1-y}In_yN_xAs_{1-x}$. *Appl. Phys. Lett.*, 86:112113, 2005.

112. Z. L. Bushell, P. Ludewig, N. Knaub, Z. Batool, K. Hild, W. Stolz, S. J. Sweeney, and K. Volz. Growth and characterisation of Ga(AsNBi) alloy by metal-organic vapour phase epitaxy. *J. Cryst. Growth*, 396:79, 2014.

113. M. Yoshimoto, W. Huang, G. Feng, Y. Tanaka, and K. Oe. Molecular-beam epitaxy of GaNAsBi layer for temperature-insensitive wavelength emission. *J. Cryst. Growth*, 301–302:975, 2007.

114. J. Hwang, J. D. Phillips. Band structure of strain-balanced GaAsBi/GaAsN superlattices on GaAs. *Phys. Rev. B*, 83:195327, 2011.

115. W. Pan, L. Zhang, L. Zhu, Y. Li, X. Chen, X. Wu, F. Zhang, J. Shao, and S. Wang. Optical properties and band bending in InGaAs/GaAsBi/InGaAs type-II quantum well grown by gas source molecular beam epitaxy. *J. Appl. Phys.*, 120:105702, 2016.

116. J. F. Klem, O. Blum, S. R. Kurtz, I. J. Fritz, and K. D. Choquette. GaAsSb/InGaAs type-II quantum wells for long-wavelength lasers on GaAs substrates. *J. Vac. Sci. Technol. B*, 18:1605, 2000.

117. J. Y. Yeh, L. J. Mawst, A. A. Khandekar, T. F. Kuech, I. Vurgaftman, J. R. Meyer, and N. Tansu. Long wavelength emission of InGaAsN/GaAsSb type II "W" quantum wells. *Appl. Phys. Lett.*, 88:051115, 2006.

118. C. Berger, C. Möller, P. Hens, C. Fuchs, W. Stolz, S. W. Koch, A. R. Perez, J. Hader, and J. V. Moloney. Novel type-II material system for laser applications in the near-infrared regime. *AIP Adv.*, 5:047105, 2015.

119. K. Forghani, Y. Guan, M. Losurdo, G. Luo, D. Morgan, S. E. Babcock, A. S. Brown, L. J. Mawst, and T. F. Kuech. $GaAs_{1-y-z}P_yBi_z$, an alternative reduced band gap alloy system lattice-matched to GaAs. *Appl. Phys. Lett.*, 105:111101, 2014.

120. H. Kim, K. Forghani, Y. Guan, G. Luo, A. Anand, D. Morgan, T. F. Keuch, L. J. Mawst, Z. R. Lingley, B. J. Foran, and Y. Sin. Strain-compensated $GaAs_{1-y-z}P_y/GaAs_{1-x}Bi_x/GaAs_{1-y}P_y$ quantum wells for laser applications. *Semicond. Sci. Technol.*, 30:094011, 2015.
121. Y. Gu, Y. G. Zhang, X. Y. Chen, Y. J. Ma, S. P. Xi, B. Du, W. Y. Ji, and Y. H. Shi. Metamorphic $InAs_{1-x}Bi_x/In_{0.83}Al_{0.17}As$ quantum well structures on InP for mid-infrared emission. *Appl. Phys. Lett.*, 109:122102, 2016.
122. M. K. Rajpalke, W. M. Linhart, M. Birkett, K. M. Yu, D. O. Scanlon, J. Buckeridge, T. S. Jones, M. J. Ashwin, and T. D. Veal. Growth and properties of GaSbBi alloys. *Appl. Phys. Lett.*, 103:142106, 2013.

III

Nanostructures

11

Quantum Wells

Seoung-Hwan Park

and

Doyeol Ahn

11.1 Introduction

A quantum well (QW) is a heterostructure in which one thin-well layer is surrounded by two barrier layers. This layer is so thin that both electrons and holes are quantized. The electronic and the optical properties of quantized states offer new opportunities in developing practical devices, such as QW infrared photo-detectors, quantum cascade lasers, all-optical switches, modulators, and many others [1–4]. Hence, it is very important to obtain eigenvalues and wave functions for the design of the active region in these opto-electronic devices based on QW structures. In this chapter, we review theoretical formalism to calculate eigenvalues and wave functions of (001)-oriented zinc-blende and (0001)-oriented wurtzite QW structures [5–10]. We block diagonalize zinc-blende and wurtzite Luttinger–Kohn 6 × 6 Hamiltonians for the valence bands to two 3 × 3 Hamiltonians, which have analytical solutions for eigenvalues and eigenvectors. We derive several important forms such as interband optical matrix elements and optical gains [11–15]. Also, as a numerical example, we calculate eigenvalues and wave functions for zinc-blende and wurtzite Hamiltonians using a finite-difference method (FDM) [4]. On the basis of this information, we discuss crystal orientation effects on electronic and optical properties of strained zinc-blende and wurtzite QW structures, including the Hamiltonian for nonpolar wurtzite QW structures.

11.2 Band Structures of Bulk Semiconductors

11.2.1 Zinc-Blende Hamiltonian of the (oo1) Orientation

11.2.1.1 6 × 6 Hamiltonian for the Valence Band

The Luttinger–Kohn Hamiltonian for the valence band of the (001)-oriented zinc-blende semiconductor is given by [2,4,5]

$$
H^{\mathrm{LK}}(\vec{k},\bar{\bar{\varepsilon}}) = -
\begin{pmatrix}
P+Q & -S & R & 0 & -S/\sqrt{2} & \sqrt{2}R \\
-S^\dagger & P-Q & 0 & R & -\sqrt{2}Q & \sqrt{3/2}S \\
R^\dagger & 0 & P-Q & 0 & \sqrt{3/2}S^\dagger & \sqrt{2}Q \\
0 & R^+ & S^\dagger & P+Q & -\sqrt{2}R^\dagger & -S^\dagger/\sqrt{2} \\
-S^\dagger/\sqrt{2} & 0 & \sqrt{3/2}S & -\sqrt{2}R & P+\Delta & 0 \\
\sqrt{2}R^\dagger & \sqrt{3/2}S^\dagger & \sqrt{2}Q^\dagger & -S/\sqrt{2} & 0 & P+\Delta
\end{pmatrix},
\tag{11.1}
$$

where

$$
P = P_k + P_\varepsilon, \quad Q = Q_k + Q_\varepsilon,
$$

$$
R = R_k + R_\varepsilon, \quad S = S_k + S_\varepsilon,
$$

$$
P_k = \left(\frac{\hbar^2}{2m_0}\right)\gamma_1(k_x^2 + k_x^2 + k_z^2), \quad Q_k = \left(\frac{\hbar^2}{2m_0}\right)\gamma_2(k_z^2 + k_x^2 - 2k_z^2)
$$

$$
R_k = \left(\frac{\hbar^2}{2m_0}\right)\sqrt{3}[-\gamma_2(k_x^2 - k_x^2) + 2i\gamma_3 k_x k_y], \quad S_k = \left(\frac{\hbar^2}{2m_0}\right)2\sqrt{3}\gamma_3(k_x - ik_y)k_z,
\tag{11.2}
$$

$$
P_\varepsilon = -a_v(\varepsilon_{xx} + \varepsilon_{yy} + \varepsilon_{zz}), \quad Q_\varepsilon = -\frac{b}{2}(\varepsilon_{xx} + \varepsilon_{yy} - 2\varepsilon_{zz}),
$$

$$
R_\varepsilon = \frac{\sqrt{3}}{2}b(\varepsilon_{xx} - \varepsilon_{yy}) - id\varepsilon_{xy}, \quad S_\varepsilon = -d(\varepsilon_{zx} - i\varepsilon_{yz}),
$$

and $\bar{\bar{\varepsilon}} = (\varepsilon_{ij})$ for $i, j = x, y, z$ is a symmetric strain tensor; $\gamma_1, \gamma_2,$ and γ_3 are the Luttinger parameters; $a_v, b,$ and d are the Bir–Pikus deformation potentials; Δ is the spin–orbit split-off energy; m_0 is the free electron mass; $\hbar$ is Planck's constant divided by 2π; and k_i is the wave vector. The superscript $\dagger$ means taking both transpose ($\sim$) and complex conjugate ($*$). In the Hamiltonian H, we restricted ourselves to the biaxial strain case for simplicity, namely,

$$
\varepsilon_{xx} = \varepsilon_{yy} \neq \varepsilon_{zz}, \quad \varepsilon_{ij} = 0 \quad \text{for} \quad i \neq j.
\tag{11.3}
$$

For the case of a strained-layer semiconductor pseudomorphically grown on a (001)-oriented substrate,

$$
\varepsilon_{xx} = \varepsilon_{yy} = \frac{a_s^z - a_1^z}{a_1^z}, \quad \varepsilon_{zz} = -\frac{2C_{12}}{C_{11}}\varepsilon_{xx},
\tag{11.4}
$$

where a_s^z and a_1^z are the lattice constants of the substrate (s) and the layer (l) material, and C_{11} and C_{12} are the stiffness constants for the zinc-blende structure. The bases for the Hamiltonian are

$$
|u_1\rangle = \left|\frac{3}{2}, \frac{3}{2}\right\rangle = -\frac{1}{\sqrt{2}}|(X + iY)\uparrow\rangle,
$$

$$|u_2\rangle = \left|\frac{3}{2}, \frac{1}{2}\right\rangle = \frac{1}{\sqrt{6}}\left|-(X - iY)\downarrow + 2Z\uparrow\right\rangle,$$

$$|u_3\rangle = \left|\frac{3}{2}, -\frac{1}{2}\right\rangle = \frac{1}{\sqrt{6}}\left|(X - iY)\uparrow + 2Z\downarrow\right\rangle,$$

$$|u_4\rangle = \left|\frac{3}{2}, -\frac{3}{2}\right\rangle = \frac{1}{\sqrt{2}}\left|(X - iY)\downarrow\right\rangle, \tag{11.5}$$

$$|u_5\rangle = \left|\frac{1}{2}, \frac{1}{2}\right\rangle = \frac{1}{\sqrt{3}}\left|(X + iY)\downarrow + Z\uparrow\right\rangle,$$

$$|u_6\rangle = \left|\frac{1}{2}, -\frac{1}{2}\right\rangle = \frac{1}{\sqrt{3}}\left|(X - iY)\uparrow - Z\downarrow\right\rangle.$$

11.2.1.2 Block-Diagonalized 3 × 3 Hamiltonian

Under the axial approximation [2], we write the R_k term

$$R_k = -\left(\frac{\hbar^2}{2m_0}\right)\sqrt{3}\left[\frac{\gamma_2 + \gamma_2}{2}(k_x - ik_x)^2 + \frac{\gamma_2 - \gamma_3}{2}(k_x + ik_x)^2\right]$$

$$\cong -\left(\frac{\hbar^2}{2m_0}\right)\sqrt{3}\left[\bar{\gamma}(k_x - ik_x)^2\right], \tag{11.6}$$

where $\bar{\gamma} = \frac{\gamma_2 + \gamma_3}{2}$. In this approximation, we assume $\gamma_2 \cong \gamma_3$ in the R_k term only, whereas we still use γ_2 and γ_3 in the other terms. When we define the angle φ by

$$k_x + ik_x = k_{||}e^{i\varphi}, \tag{11.7}$$

we can write

$$R_k = R_\rho e^{-2i\varphi}, \quad S_k = S_\rho e^{-i\varphi}, \quad R_\rho = -\left(\frac{\hbar^2}{2m_0}\right)\sqrt{3}\bar{\gamma}k_{||}^2, \quad S_\rho = \left(\frac{\hbar^2}{2m_0}\right)2\sqrt{3}\gamma_3 k_{||}k_z, \tag{11.8}$$

where $k_{||} = \sqrt{k_x^2 + k_x^2}$. Then, the 6×6 Hamiltonian can be block-diagonalized into two 3×3 Hamiltonians by using the transformation matrix U

$$H = UH^{LK}U^+$$

$$= -\begin{pmatrix} P+Q & -R_\rho - iS_\rho & -\sqrt{2}R_\rho + i\sqrt{\frac{1}{2}}S_\rho & 0 & 0 & 0 \\ -R_\rho + iS_\rho & P-Q & \sqrt{2}Q + i\sqrt{\frac{3}{2}}S_\rho & 0 & 0 & 0 \\ -\sqrt{2}R_\rho - i\sqrt{\frac{1}{2}}S_\rho & \sqrt{2}Q - i\sqrt{\frac{3}{2}}S_\rho & P+\Delta & 0 & 0 & 0 \\ 0 & 0 & 0 & P+Q & -R_\rho + iS_\rho & -\sqrt{2}R_\rho - i\sqrt{\frac{1}{2}}S_\rho \\ 0 & 0 & 0 & -R_\rho - iS_\rho & P-Q & \sqrt{2}Q - i\sqrt{\frac{3}{2}}S_\rho \\ 0 & 0 & 0 & -\sqrt{2}R_\rho + i\sqrt{\frac{1}{2}}S_\rho & \sqrt{2}Q + i\sqrt{\frac{3}{2}}S_\rho & P+\Delta \end{pmatrix}$$

$$\tag{11.9}$$

and

$$U = \begin{pmatrix} \alpha & 0 & 0 & i\alpha^* & 0 & 0 \\ 0 & -i\beta & -\beta^* & 0 & 0 & 0 \\ 0 & 0 & 0 & 0 & i\beta & -\beta^* \\ \alpha & 0 & 0 & -i\alpha^* & 0 & 0 \\ 0 & i\beta & -\beta^* & 0 & 0 & 0 \\ 0 & 0 & 0 & 0 & -i\beta & -\beta^* \end{pmatrix}, \tag{11.10}$$

where

$$\alpha = \frac{1}{\sqrt{2}} e^{i\frac{3}{2}\varphi}, \quad \beta = \frac{1}{\sqrt{2}} e^{i\frac{1}{2}\varphi}. \tag{11.11}$$

A new basis set is given by using the basis transformation, $|i>= \sum T_{ij}|u_j>$, where $T = U^*$. That is,

$$|1\rangle = -\frac{1}{2}\Big|(X+iY)\uparrow\Big\rangle e^{-i\frac{3}{2}\varphi} - i\frac{1}{2}\Big|(X-iY)\downarrow\Big\rangle e^{-i\frac{3}{2}\varphi}$$

$$|2\rangle = i\frac{1}{2\sqrt{3}}\Big|-(X-iY)\downarrow +2Z\uparrow\Big\rangle e^{-i\frac{1}{2}\varphi} - \frac{1}{2\sqrt{3}}\Big|(X-iY)\uparrow +2Z\downarrow\Big\rangle e^{i\frac{1}{2}\varphi}$$

$$|3\rangle = -i\frac{1}{\sqrt{6}}\Big|(X+iY)\downarrow +Z\uparrow\Big\rangle e^{-i\frac{1}{2}\varphi} - \frac{1}{\sqrt{6}}\Big|(X-iY)\uparrow -Z\downarrow\Big\rangle e^{i\frac{1}{2}\varphi}$$

$$|4\rangle = -\frac{1}{2}\Big|(X+iY)\uparrow\Big\rangle e^{-i\frac{3}{2}\varphi} + i\frac{1}{2}\Big|(X-iY)\downarrow\Big\rangle e^{i\frac{3}{2}\varphi} \tag{11.12}$$

$$|5\rangle = -i\frac{1}{2\sqrt{3}}\Big|-(X-iY)\downarrow +2Z\uparrow\Big\rangle e^{-i\frac{1}{2}\varphi} - \frac{1}{2\sqrt{3}}\Big|(X-iY)\uparrow +2Z\downarrow\Big\rangle e^{i\frac{1}{2}\varphi}$$

$$|6\rangle = i\frac{1}{\sqrt{6}}\Big|(X+iY)\downarrow +Z\uparrow\Big\rangle e^{-i\frac{1}{2}\varphi} - \frac{1}{\sqrt{6}}\Big|(X-iY)\uparrow -Z\downarrow\Big\rangle e^{i\frac{1}{2}\varphi}.$$

11.2.2 Wurtzite Hamiltonian of the (0001) Orientation

11.2.2.1 6 × 6 Hamiltonian for the Valence Band

The c-plane Hamiltonian for the valence band of the (0001)-oriented wurtzite semiconductor in $\{|1\rangle, |2\rangle, |3\rangle, |4\rangle, |5\rangle, |6\rangle\}$ bases is given by [4,6]

$$H(\vec{k}, \bar{\bar{\varepsilon}}) = \begin{pmatrix} F & -K^* & -H^* & 0 & 0 & 0 \\ -K & G & H & 0 & 0 & \Delta \\ -H & H* & \lambda & 0 & \Delta & 0 \\ 0 & 0 & 0 & F & -K & H \\ 0 & 0 & \Delta & -K* & G & -H^* \\ 0 & \Delta & 0 & H^* & -H & \lambda \end{pmatrix}, \tag{11.13}$$

where

$$F = \Delta_1 + \Delta_2 + \lambda + \vartheta,$$

$$G = \Delta_1 - \Delta_2 + \lambda + \vartheta,$$

$$\lambda = \frac{\hbar^2}{2m_0}\Big[A_1 k_z^2 + A_2\left(k_x^2 + k_y^2\right)\Big] + D_1\varepsilon_{zz} + D_2\left(\varepsilon_{xx} + \varepsilon_{yy}\right),$$

$$\vartheta = \frac{\hbar^2}{2m_0}\left[A_3\,k_z^2 + A_4\left(k_x^2 + k_y^2\right)\right] + D_3\varepsilon_{zz} + D_4\left(\varepsilon_{xx} + \varepsilon_{yy}\right), \tag{11.14}$$

$$K = \frac{\hbar^2}{2m_0}A_5\left(k_x + ik_y\right)^2 + D_5\left(\varepsilon_{xx} - \varepsilon_{yy}\right),$$

$$H = \frac{\hbar^2}{2m_0}A_6\left(k_x + ik_y\right)k_z + D_6\left(\varepsilon_{xz} + i\varepsilon_{yz}\right),$$

$$\Delta = \sqrt{2}\Delta_3,$$

and the A_i's are the valence-band effective-mass parameters analogous to the Luttinger parameters for the zinc-blende semiconductors, the D_i's are the deformation potentials for wurtzite semiconductors, Δ_1 is the crystal-field split energy, and Δ_2 and Δ_3 are the spin–orbit interaction energies. Under the cubic approximation [6,10], the following relations hold for the parameters A_i's and D_i's:

$$A_1 - A_2 = -A_3 = 2A_4, \quad A_3 + 4A_5 = \sqrt{2}A_6, \quad D_1 - D_2 = -D_3 = 2D_4, \quad D_3 + 4D_5 = \sqrt{2}D_6. \tag{11.15}$$

Here, we restricted ourselves to the biaxial strain case for simplicity, namely,

$$\varepsilon_{xx} = \varepsilon_{yy} \neq \varepsilon_{zz}, \quad \varepsilon_{ij} = 0 \quad \text{for} \quad i \neq j. \tag{11.16}$$

For the case of a strained-layer semiconductor pseudomorphically grown on a (0001)-oriented substrate,

$$\varepsilon_{xx} = \varepsilon_{yy} = \frac{a_s^w - a_1^w}{a_1^w}, \quad \varepsilon_{zz} = -\frac{2C_{13}}{C_{33}}\varepsilon_{xx}, \tag{11.17}$$

where a_s^w and a_1^w are the lattice constants of the substrate (s) and the layer (l) material, and C_{13} and C_{33} are the stiffness constants for the wurtzite structure. The bases for the Hamiltonian are

$$|u_1\rangle = -\frac{1}{\sqrt{2}}\big|(X + iY)\uparrow\big\rangle,$$

$$|u_2\rangle = \frac{1}{\sqrt{2}}\big|(X - iY)\uparrow\big\rangle,$$

$$|u_3\rangle = \big|Z\uparrow\big\rangle,$$

$$|u_4\rangle = \frac{1}{\sqrt{2}}\big|(X - iY)\downarrow\big\rangle, \tag{11.18}$$

$$|u_5\rangle = -\frac{1}{\sqrt{2}}\big|(X + iY)\downarrow\big\rangle,$$

$$|u_6\rangle = \big|Z\downarrow\big\rangle.$$

11.2.2.2 Block-Diagonalized 3×3 Hamiltonian

When we define the angle φ by

$$k_x + ik_x = k_{||}e^{i\varphi}, \tag{11.19}$$

we can write

$$K = K_t e^{2i\varphi}, \quad H = H_t e^{i\varphi}, \quad K_t = \left(\frac{\hbar^2}{2m_0}\right) A_5 k_{||}^2, \quad H_t = \left(\frac{\hbar^2}{2m_0}\right) A_6 k_{||} k_z. \tag{11.20}$$

Then, the 6×6 Hamiltonian can be block-diagonalized into two 3×3 Hamiltonians following a similar procedure to that of the zinc-blende structure by using the transformation matrix U [4,6,7]

$$H' = UHU^+ = \begin{pmatrix} F & K_t & -iH_t & 0 & 0 & 0 \\ K_t & G & \Delta - iH_t & 0 & 0 & 0 \\ iH_t & \Delta + iH_t & \lambda & 0 & 0 & 0 \\ 0 & 0 & 0 & F & K_t & iH_t \\ 0 & 0 & 0 & K_t & G & \Delta + iH_t \\ 0 & 0 & 0 & -iH_t & \Delta - iH_t & \lambda \end{pmatrix}, \tag{11.21}$$

and

$$U = \begin{pmatrix} \alpha & 0 & 0 & \alpha^* & 0 & 0 \\ 0 & \beta^* & 0 & 0 & \beta & 0 \\ 0 & 0 & \beta & 0 & 0 & \beta^* \\ \alpha & 0 & 0 & -\alpha^* & 0 & 0 \\ 0 & \beta^* & 0 & 0 & -\beta & 0 \\ 0 & 0 & -\beta & 0 & 0 & \beta^* \end{pmatrix} \tag{11.22}$$

where

$$\alpha = \frac{1}{\sqrt{2}} e^{i\left(\frac{3}{4}\pi + \frac{3}{2}\varphi\right)}, \quad \beta = \frac{1}{\sqrt{2}} e^{i\left(\frac{1}{4}\pi + \frac{1}{2}\varphi\right)}. \tag{11.23}$$

Also, a new basis set is given by using the basis transformation, $|i> = \sum T_{ij}|u_j>$, where $T = U^*$. That is,

$$|1\rangle = -\frac{1}{\sqrt{2}}\left|(X + iY)\uparrow\right\rangle \frac{1}{\sqrt{2}} e^{i\left(\frac{3}{4}\pi + \frac{3}{2}\varphi\right)} + \frac{1}{\sqrt{2}}\left|(X - iY)\downarrow\right\rangle \frac{1}{\sqrt{2}} e^{-i\left(\frac{3}{4}\pi + \frac{3}{2}\varphi\right)}$$

$$|2\rangle = \frac{1}{\sqrt{2}}\left|(X - iY)\uparrow\right\rangle \frac{1}{\sqrt{2}} e^{-i\left(\frac{1}{4}\pi + \frac{1}{2}\varphi\right)} - \frac{1}{\sqrt{2}}\left|(X + iY)\downarrow\right\rangle \frac{1}{\sqrt{2}} e^{i\left(\frac{1}{4}\pi + \frac{1}{2}\varphi\right)}$$

$$|3\rangle = \left|Z\uparrow\right\rangle \frac{1}{\sqrt{2}} e^{i\left(\frac{1}{4}\pi + \frac{1}{2}\varphi\right)} + \left|Z\downarrow\right\rangle e^{i\frac{1}{2}\varphi} \frac{1}{\sqrt{2}} e^{-i\left(\frac{1}{4}\pi + \frac{1}{2}\varphi\right)}$$

$$\tag{11.24}$$

$$|4\rangle = -\frac{1}{\sqrt{2}}\left|(X + iY)\uparrow\right\rangle \frac{1}{\sqrt{2}} e^{i\left(\frac{3}{4}\pi + \frac{3}{2}\varphi\right)} - \frac{1}{\sqrt{2}}\left|(X - iY)\downarrow\right\rangle \frac{1}{\sqrt{2}} e^{-i\left(\frac{3}{4}\pi + \frac{3}{2}\varphi\right)}$$

$$|5\rangle = \frac{1}{\sqrt{2}}\left|(X - iY)\uparrow\right\rangle \frac{1}{\sqrt{2}} e^{-i\left(\frac{1}{4}\pi + \frac{1}{2}\varphi\right)} + \frac{1}{\sqrt{2}}\left|(X + iY)\downarrow\right\rangle \frac{1}{\sqrt{2}} e^{i\left(\frac{1}{4}\pi + \frac{1}{2}\varphi\right)}$$

$$|6\rangle = -\left|Z\uparrow\right\rangle \frac{1}{\sqrt{2}} e^{i\left(\frac{1}{4}\pi + \frac{1}{2}\varphi\right)} + \left|Z\downarrow\right\rangle e^{i\frac{1}{2}\varphi} \frac{1}{\sqrt{2}} e^{-i\left(\frac{1}{4}\pi + \frac{1}{2}\varphi\right)}.$$

11.3 Band Structures of Strained-Layer QW

11.3.1 Zinc-Blende Semiconductor

Here, we consider a strained-layer QW structure, assuming that the growth direction is along the z-axis and the strain caused by lattice mismatch is entirely elastically accommodated in the QW.

11.3.1.1 Conduction Band

For the unstrained QW, the effective mass theory for the conduction band is obtained from the dispersion relation

$$E(k) = \frac{\hbar^2 k^2}{2m_e}, \tag{11.25}$$

where the effective mass of the electron in the conduction band is $m_e = m_b$ in the barrier region and $m_e = m_w$ in the QW. The potential for the unstrained QW is given by

$$V_c(z) = \begin{cases} \Delta E_c & |z| > \frac{L_w}{2} \\ 0 & |z| \leq \frac{L_w}{2} \end{cases}, \tag{11.26}$$

where ΔE_c is the conduction-band offset. The energies are all measured from the conduction-band edge of unstrained QW. For a strained QW, the effective mass equation for a single band is

$$\left[-\frac{\hbar^2}{2m_e^c} \frac{\partial^2}{\partial z^2} + \frac{\hbar^2}{2m_e^c} \nabla_t^2 + V(z) + a_c(\varepsilon_{xx} + \varepsilon_{yy} + \varepsilon_{zz}) \right] \psi(\vec{r}) = E\psi(\vec{r}), \tag{11.27}$$

where m_e^c is the effective mass in the conduction band for zinc-blende structure and a_c is the conduction-band deformation potential. In general, the wave function $\psi(\vec{r})$ can be written

$$\psi^{cn}(\vec{r}) = \frac{e^{i(k_x x + k_y y)}}{\sqrt{A}} f(z) |S, \eta > \tag{11.28}$$

and

$$\left[-\frac{\hbar^2}{2m_e^c} \frac{\partial^2}{\partial z^2} + V(z) + a_c(\varepsilon_{xx} + \varepsilon_{yy} + \varepsilon_{zz}) \right] f(z) = \left(E(k_t) - \frac{\hbar^2 k_t^2}{2m_e^c} \right) f(z), \tag{11.29}$$

where η is the electron spin and $|S >$ is the basis function near the zone center in the conduction band. The eigenvalues and eigenfunctions in the conduction band are obtained from the earlier equation. When we ignore the k_t dependence of $f(z)$, Equation 11.29 is solved at $k_t = 0$ for the nth sub-band energy $E_n(0)$ and we have $E_n(k_t) = E_n(0) + \hbar^2 k_t^2 / 2m^*$.

11.3.1.2 Valence Band

For the unstrained QW, a QW potential is given by

$$V_h(z) = \begin{cases} -\Delta E_v & |z| > \frac{L_w}{2} \\ 0 & |z| \leq \frac{L_w}{2} \end{cases}, \tag{11.30}$$

where ΔE_v is the valence-band offset. The eigenvalues and eigenfunctions for the strained QW can be obtained by solving the effective mass equation for a QW potential $V_h(z)$ given earlier:

$$\left[H^{LK}\left(k_x, k_y, k_z = -i\frac{\partial}{\partial z}\right) + V_h(z) \right] \cdot \begin{bmatrix} F_1 \\ F_2 \\ F_3 \\ F_4 \\ F_5 \\ F_6 \end{bmatrix} = E \begin{bmatrix} F_1 \\ F_2 \\ F_3 \\ F_4 \\ F_5 \\ F_6 \end{bmatrix}, \quad (11.31)$$

where H^{LK} is the 6×6 Hamiltonian given by Equation 11.1 and the envelope functions F_1, F_2, F_3, F_4, F_5, and F_6 can be written in the vector form

$$\vec{F}_{\vec{k}}(\vec{r}) = \begin{bmatrix} F_1 \\ F_2 \\ F_3 \\ F_4 \\ F_5 \\ F_6 \end{bmatrix} = \begin{bmatrix} g_{3/2,3/2}(k_x, k_y, z) \\ g_{3/2,1/2}(k_x, k_y, z) \\ g_{3/2,-1/2}(k_x, k_y, z) \\ g_{3/2,03/2}(k_x, k_y, z) \\ g_{1/2,1/2}(k_x, k_y, z) \\ g_{1/2,-1/2}(k_x, k_y, z) \end{bmatrix} \frac{e^{i(k_x x + k_y y)}}{\sqrt{A}}. \quad (11.32)$$

The wave function in component form is expressed as

$$\psi_{\vec{k}}(\vec{r}) = F_1 \left| \frac{3}{2}, \frac{3}{2} \right\rangle + F_2 \left| \frac{3}{2}, \frac{1}{2} \right\rangle + F_3 \left| \frac{3}{2}, -\frac{1}{2} \right\rangle + F_4 \left| \frac{3}{2}, -\frac{3}{2} \right\rangle + F_5 \left| \frac{1}{2}, \frac{1}{2} \right\rangle + F_6 \left| \frac{1}{2}, -\frac{1}{2} \right\rangle$$

$$= \frac{e^{i(k_x x + k_y y)}}{\sqrt{A}} \sum_v g_{3/2,v}(k_x, k_y, z) \left| \frac{3}{2}, v \right\rangle, \quad (11.33)$$

where $v = \frac{3}{2}, \frac{1}{2}, -\frac{1}{2}, -\frac{3}{2}, \frac{1}{2}$, and $-\frac{1}{2}$. Denoting $\vec{k}_t = k_x \hat{x} + k_y \hat{y}$, we can write

$$\left[H^{LK}\left(\vec{k}_t, k_z = -i\frac{\partial}{\partial z}\right) + V_h(z) \right] \cdot \begin{bmatrix} g_{3/2,3/2}(\vec{k}_t, z) \\ g_{3/2,1/2}(\vec{k}_t, z) \\ g_{3/2,-1/2}(\vec{k}_t, z) \\ g_{3/2,03/2}(\vec{k}_t, z) \\ g_{1/2,1/2}(\vec{k}_t, z) \\ g_{1/2,-1/2}(\vec{k}_t, z) \end{bmatrix} = E(\vec{k}_t) \begin{bmatrix} g_{3/2,3/2}(\vec{k}_t, z) \\ g_{3/2,1/2}(\vec{k}_t, z) \\ g_{3/2,-1/2}(\vec{k}_t, z) \\ g_{3/2,03/2}(\vec{k}_t, z) \\ g_{1/2,1/2}(\vec{k}_t, z) \\ g_{1/2,-1/2}(\vec{k}_t, z) \end{bmatrix}. \quad (11.34)$$

However, it is convenient to solve eigenvalues and eigenfunctions with a block-diagonalized 3×3 Hamiltonian for a numerical calculation. Let us look at the upper 3×3 Hamiltonian in Equation 11.9. The wave functions can be written as

$$\psi^U(\vec{k}_t, \vec{r}) = \frac{e^{i\vec{k}_t \cdot \vec{r}_t}}{\sqrt{A}} \left[g^{(1)}(k_z, z) |1\rangle + g^{(2)}(k_z, z) |2\rangle + g^{(3)}(k_z, z) |3\rangle \right]. \quad (11.35)$$

The wave function satisfies the upper Hamiltonian equation

$$
-\begin{pmatrix} P+Q & -R_\rho - iS_\rho & -\sqrt{2}R_\rho + i\sqrt{\frac{1}{2}}S_\rho \\ -R_\rho + iS_\rho & P-Q & \sqrt{2}Q + i\sqrt{\frac{3}{2}}S_\rho \\ -\sqrt{2}R_\rho - i\sqrt{\frac{1}{2}}S_\rho & \sqrt{2}Q - i\sqrt{\frac{3}{2}}S_\rho & P+\Delta \end{pmatrix} \begin{bmatrix} g^{(1)}(\vec{k}_t, z) \\ g^{(2)}(\vec{k}_t, z) \\ g^{(3)}(\vec{k}_t, z) \end{bmatrix}
$$

$$
+\begin{pmatrix} V_h(z) & 0 & 0_\rho \\ 0 & V_h(z) & 0 \\ 0 & 0 & V_h(z) \end{pmatrix} \begin{bmatrix} g^{(1)}(\vec{k}_t, z) \\ g^{(2)}(\vec{k}_t, z) \\ g^{(3)}(\vec{k}_t, z) \end{bmatrix} = E(\vec{k}_t) \begin{bmatrix} g^{(1)}(\vec{k}_t, z) \\ g^{(2)}(\vec{k}_t, z) \\ g^{(3)}(\vec{k}_t, z) \end{bmatrix}. \tag{11.36}
$$

A similar procedure holds for the lower 3×3 Hamiltonian. That is, the wave function is given by

$$
\psi^L(\vec{k}_t, \vec{r}) = \frac{e^{i\vec{k}_t \cdot \vec{r}_t}}{\sqrt{A}} \left[g^{(4)}(k_z, z) \,|\rangle + g^{(5)}(k_z, z) \,|\rangle + g^{(6)}(k_z, z) \,|\rangle \right], \tag{11.37}
$$

which satisfies the following 3×3 Hamiltonian equation

$$
-\begin{pmatrix} P+Q & -R_\rho + iS_\rho & -\sqrt{2}R_\rho - i\sqrt{\frac{1}{2}}S_\rho \\ -R_\rho - iS_\rho & P-Q & \sqrt{2}Q - i\sqrt{\frac{3}{2}}S_\rho \\ -\sqrt{2}R_\rho + i\sqrt{\frac{1}{2}}S_\rho & \sqrt{2}Q + i\sqrt{\frac{3}{2}}S_\rho & P+\Delta \end{pmatrix} \begin{bmatrix} g^{(4)}(\vec{k}_t, z) \\ g^{(5)}(\vec{k}_t, z) \\ g^{(6)}(\vec{k}_t, z) \end{bmatrix}
$$

$$
+\begin{pmatrix} V_h(z) & 0 & 0_\rho \\ 0 & V_h(z) & 0 \\ 0 & 0 & V_h(z) \end{pmatrix} \begin{bmatrix} g^{(1)}(\vec{k}_t, z) \\ g^{(2)}(\vec{k}_t, z) \\ g^{(3)}(\vec{k}_t, z) \end{bmatrix} = E(\vec{k}_t) \begin{bmatrix} g^{(4)}(\vec{k}_t, z) \\ g^{(5)}(\vec{k}_t, z) \\ g^{(6)}(\vec{k}_t, z) \end{bmatrix}. \tag{11.38}
$$

11.3.2 Wurtzite Semiconductor

We consider nitride-based semiconductors as an example of wurtzite materials. We assume that a strained-layer semiconductor is pseudomorphically grown along the (0001) direction (c-axis) on a (0001)-oriented substrate [16]. The wurtzite structure differs from the zinc-blende structure in several aspects. First, in the (0001) wurtzite GaN-based QW structures, there exists a large internal field due to the strain-induced piezoelectric (PZ) and spontaneous (SP) polarizations [17–21]. Second, there are energy splittings in the valence band such as a crystal field splitting and the splitting due to spin–orbit interaction [6].

11.3.2.1 Internal Field

For binary AB compounds such as a GaN wurtzite structure, the sequence of the atomic layers of the constituents A and B is reversed along the [0001] and [000$\bar{1}$] directions. Defining the $+z$ direction with a vector pointing from a Ga atom to the nearest neighbor N atom, the Ga face means that Ga is on the top position of the {0001} bilayer, corresponding to the [0001] polarity. The induced piezoelectric polarization is given by [20]

$$
P_z^{PZ} = e_{31}(\varepsilon_{xx} + \varepsilon_{yy}) + e_{33}\varepsilon_{zz} = 2\left(e_{31} - \frac{C_{13}}{C_{33}}e_{33}\right)\varepsilon_{xx}, \tag{11.39}
$$

where e_{ij} are piezoelectric constants. In the case of AlGaN system, for example, the piezoelectric polarization is negative for tensile strain and positive for compressive strain because $2\left(e_{31} - \frac{C_{13}}{C_{33}}e_{33}\right) < 0$ for the whole range of Al compositions. Also, there exists a spontaneous polarization along the c-axis. That is, the spontaneous polarization for GaN (or AlN) has been found to be negative, meaning that the polarization for Ga-face (or Al-face) is pointing toward the substrate. As a result, in the AlGaN system, the alignment of the piezoelectric polarization and the spontaneous polarization is parallel in the case of tensile strain and antiparallel in the case of compressive strain [18]. In the case of multiple QW (MQW) structure, the internal field F_z is determined from the periodic boundary condition and the difference between the sum of spontaneous and piezoelectric polarizations of the well and barrier layers. The continuity of the displacement vector (D_z) normal to the surface gives

$$D_z = \varepsilon_w F_w + P_w = \varepsilon_b F_b + P_b, \tag{11.40}$$

where the subscripts w and b mean the well region and the barrier region, respectively, $P_w = P_w^{PZ} + P_w^{SP}$, and $P_b = P_b^{PZ} + P_b^{SP}$. We consider the periodic boundary condition that the net voltage drop over one period is zero. That is,

$$F_w L_w + F_b L_b = 0, \tag{11.41}$$

where L_w and L_b are the well width and the barrier width, respectively. Then, we find the electric fields in the well and the barrier:

$$F_w = \frac{L_b}{\varepsilon_b L_w + \varepsilon_w L_b}(P_b - P_w)$$

$$F_b = \frac{-L_w}{\varepsilon_b L_w + \varepsilon_w L_b}(P_b - P_w). \tag{11.42}$$

Hence, the self-consistent solution, which solves the Schrödinger equation and Poisson equation simultaneously, is necessary for the wurtzite QWs.

11.3.2.2 Energy Splitting

At the zone center ($k_x = k_y = k_z = 0$), we can obtain the following doubly degenerate band-edge energies

$$E_1 = \Delta_1 + \Delta_2$$

$$E_2 = \frac{\Delta_1 - \Delta_2}{2} + \sqrt{\left(\frac{\Delta_1 - \Delta_2}{2}\right)^2 + 2\Delta_3^2}$$

$$E_3 = \frac{\Delta_1 - \Delta_2}{2} - \sqrt{\left(\frac{\Delta_1 - \Delta_2}{2}\right)^2 + 2\Delta_3^2}. \tag{11.43}$$

Without the spin–orbit interaction, $\Delta_2 = \Delta_3 = 0$, we have the top two degenerate bands and the lower band as the reference level E_v. That is,

$$E_1 = E_2 = E_v + \Delta_1 = \Delta_1$$

$$E_3 = E_v = 0. \tag{11.44}$$

Here, we set the reference energy E_v to 0. Then, with the spin–orbit interaction, the top valence-band energy is $E_1 = \Delta_1 + \Delta_2$, and the conduction-band edge is given by adding the bandgap energy E_g to E_v:

$$E_c = E_g + \Delta_1 + \Delta_2. \tag{11.45}$$

11.3.2.3 Self-Consistent Calculations with the Screening of Eigenvalues and Eigenfunctions in the Conduction Band and the Valence Band

The total potential profiles for the electrons and the holes are

$$V_c(z) = \begin{cases} E_c + \Delta E_c + |e|\, F_b z - |e|\, \varphi(z) & |z| > \frac{L_w}{2} \\ E_c + |e|\, F_w z - |e|\, \varphi(z) & |z| \le \frac{L_w}{2} \end{cases} \tag{11.46}$$

and

$$V_h(z) = \begin{cases} -\Delta E_v + |e|\, F_b z - |e|\, \varphi(z) & |z| > \frac{L_w}{2} \\ |e|\, F_w z - |e|\, \varphi(z) & |z| \le \frac{L_w}{2} \end{cases}. \tag{11.47}$$

Here, $\varphi(z)$ is the screening potential induced by the charged carriers and satisfies the Poisson equation

$$\frac{d}{dz}\left(\varepsilon(z)\frac{d}{dz}\right)\varphi(z) = -|e|[p(z) - n(z)], \tag{11.48}$$

where $\varepsilon(z)$ is the dielectric constant and we assume that there is no doping in the well and the barrier. The electron and the hole concentrations, $n(z)$ and $p(z)$, are related to the wave functions of the nth conduction sub-band and the mth valence sub-band by

$$n(z) = \frac{kTm_e}{\pi\hbar^2}\sum_n |f_n(z)|^2 \ln\left(1 + e^{[E_{fc} - E_{cn}]/kT}\right) \tag{11.49}$$

and

$$p(z) = \sum_{\sigma = U,L}\sum_m \int dk_{\parallel}\, \frac{k_{\parallel}}{2\pi}\sum_v |g^{\sigma(v)}_{mk_{\parallel}}(z)|^2\, \frac{1}{1 + e^{[E_{fv} - E_{vm}(k_{\parallel})]/kT}}, \tag{11.50}$$

where n and m are the quantized sub-band indices for the conduction and the valence bands, E_{fc} and E_{fv} are the quasi-Fermi levels of the electrons and the holes, respectively, and E_{cn} and $f_n(z)$ are the quantized energy level of the electrons at a band-edge and eigenfunctions, respectively. The eigenvalues and eigenfunctions in the conduction band are obtained by solving

$$\left[-\frac{\hbar^2}{2m^w_{ez}}\frac{\partial^2}{\partial z^2} + V_c(z) + a^w_{ct}(\varepsilon_{xx} + \varepsilon_{yy}) + a^w_{cz}\varepsilon_{zz}\right]f(z) = \left(E(k_{\parallel}) - \frac{\hbar^2 k_{\parallel}^2}{2m^w_{et}}\right)f(z), \tag{11.51}$$

where a^w_{ct} and a^w_{cz} are the conduction-band deformation potentials along the c-axis and perpendicular to the c axis, respectively. Usually, it is convenient to obtain eigenvalues in the conduction band with $E_c = 0$ in Equation 11.46. The term $(E_g + \Delta_1 + \Delta_2)$ can be added when we calculate optical properties such as spontaneous emission coefficient and optical gain.

Also, $E_{vm}(k_{||})$ is the energy for the mth sub-band in the valence band, σ denotes the upper (U) and the lower (L) blocks of the Hamiltonian, $k_{||}$ is the in-plane wave vector, v refers to the new bases for the Hamiltonian, and $g_{mk_{||}}^{\sigma(v)}(z)$ is the envelope function in the valence band.

The eigenvalues and eigenfunctions for the upper Hamiltonian equation of the valence band are obtained by solving

$$
-\begin{pmatrix} F & K_t & -iH_t \\ K_t & G & \Delta - iH_t \\ iH_t & \Delta + iH_t & P + \Delta \end{pmatrix} \begin{bmatrix} g^{(1)}(\vec{k}_t, z) \\ g^{(2)}(\vec{k}_t, z) \\ g^{(3)}(\vec{k}_t, z) \end{bmatrix}
$$

$$
+\begin{pmatrix} V_h(z) & 0 & 0_\rho \\ 0 & V_h(z) & 0 \\ 0 & 0 & V_h(z) \end{pmatrix} \begin{bmatrix} g^{(1)}(\vec{k}_t, z) \\ g^{(2)}(\vec{k}_t, z) \\ g^{(3)}(\vec{k}_t, z) \end{bmatrix} = E(\vec{k}_t) \begin{bmatrix} g^{(1)}(\vec{k}_t, z) \\ g^{(2)}(\vec{k}_t, z) \\ g^{(3)}(\vec{k}_t, z) \end{bmatrix}. \tag{11.52}
$$

A similar procedure holds for the lower 3×3 Hamiltonian, which satisfies the following 3×3 Hamiltonian equation

$$
-\begin{pmatrix} F & K_t & iH_t \\ K_t & G & \Delta + iH_t \\ -iH_t & \Delta - iH_t & P + \Delta \end{pmatrix} \begin{bmatrix} g^{(4)}(\vec{k}_t, z) \\ g^{(5)}(\vec{k}_t, z) \\ g^{(6)}(\vec{k}_t, z) \end{bmatrix}
$$

$$
+\begin{pmatrix} V_h(z) & 0 & 0_\rho \\ 0 & V_h(z) & 0 \\ 0 & 0 & V_h(z) \end{pmatrix} \begin{bmatrix} g^{(1)}(\vec{k}_t, z) \\ g^{(2)}(\vec{k}_t, z) \\ g^{(3)}(\vec{k}_t, z) \end{bmatrix} = E(\vec{k}_t) \begin{bmatrix} g^{(4)}(\vec{k}_t, z) \\ g^{(5)}(\vec{k}_t, z) \\ g^{(6)}(\vec{k}_t, z) \end{bmatrix}. \tag{11.53}
$$

The potential $\varphi(z)$ is obtained by integration

$$
\varphi(z) = -\int_{-L/2}^{z} E(z')dz', \tag{11.54}
$$

where

$$
E(z) = \int_{-L/2}^{z} \frac{1}{\varepsilon(z)} \rho(z')dz'. \tag{11.55}
$$

The procedures for the self-consistent calculations consist of the following steps:

1. Start with the potential profiles V_c and V_h with $\varphi^{(0)}(z) = 0$ in Equations 11.46 and 11.47.
2. Solve the Schrödinger equation (for electrons) and the block-diagonalized Hamiltonian (for holes) with the potential profiles $\varphi^{(n-1)}(z)$ in step (1) to obtain band structures and wave functions.
3. For a given carrier density, obtain the Fermi energies from Equations 11.49 and 11.50 by using the band structures and the charge distribution by using the wave functions.
4. Solve Poisson's equation to find $\varphi^{(n)}(z)$.

   Check if $\varphi^{(n)}(z)$ converges to $\varphi^{(n-1)}(z)$. If not, set $\varphi^{(n)}(z) = w\varphi^{(n)}(z) + (1-w)\varphi^{(n-1)}(z), n = n+1$; then, return to step (2). If yes, the band structures and the wave functions obtained with $\varphi^{(n-1)}(z)$ are solutions. An adjustable parameter w ($0 < w < 1$) is typically set to be 0.5 at low carrier densities. With increasing carrier densities, a smaller value of w is needed for rapid convergence.

11.4 Optical Matrix Elements

11.4.1 Zinc-Blende Structure

The optical momentum matrix elements are defined as

$$|\hat{\varepsilon} \cdot M_{lm}^{\eta\sigma}(k_{||})|^2 = |<\psi_l^{\eta}|\hat{\varepsilon} \cdot \vec{p}|\psi_m^{v\sigma}>|^2, \tag{11.56}$$

which represents the interband transition probability between electrons and hole. Using the expressions given in Equations 11.28 and 11.35 and taking the φ integration of the momentum matrix elements, we obtain the following momentum matrix elements for the upper Hamiltonian.

Transverse electric (TE) polarization ($\hat{\varepsilon} = \hat{x}$ or $\hat{\varepsilon} = \hat{y}$):

$$|M_{\text{TE}}|^2 = |<\psi_l^{\eta}|p_x|\psi_m^{v\sigma}>|^2 = |<\psi_l^{\eta}|p_y|\psi_m^{v\sigma}>|^2$$

$$= \delta_{\vec{k}_t', \vec{k}_t} <f_n|g^{(1)}> <S\uparrow |p_x|1> + <f_n|g^{(2)}> <S\uparrow |p_x|2> + <f_n|g^{(3)}> <S\uparrow |p_x|3>|^2$$

$$= \delta_{\vec{k}_t', \vec{k}_t} |<S\uparrow |p_x|X>|^2| >f_n|g^{(1)}> \left(-\frac{1}{2}e^{-i\frac{3}{2}\varphi}\right)$$

$$+ <f_n|g^{(2)}> \left(-\frac{1}{2\sqrt{3}}e^{i\frac{1}{2}\varphi}\right) + <f_n|g^{(3)}> \left(-\frac{1}{\sqrt{6}}e^{i\frac{1}{2}\varphi}\right)|^2$$

$$= \delta_{\vec{k}_t'\vec{k}_t} |<S\uparrow |p_x|X>|^2 \frac{1}{12}$$

$$\times \left\{ \begin{array}{l} 3 <f_n|g^{(1)}>^2 + [<f_n|g^{(2)}> + \sqrt{2} <f_n|g^{(3)}>]^2 \\ +2\sqrt{3}\cos 2\phi <f_n|g^{(1)}> <f_n|g^{(2)}> +2\sqrt{6}\cos 2\phi <f_n|g^{(1)}> <f_n|g^{(3)}> \end{array} \right\}. \tag{11.57}$$

When we calculate the absorption coefficient, the summation over k_t is given by

$$\frac{1}{V}\sum_{k_t} = \frac{1}{L_z}\int_0^\infty \frac{k_t}{2\pi}dk_t \int_0^{2\pi} \frac{d\varphi}{2\pi}. \tag{11.58}$$

Thus, the term containing the $\cos 2\varphi$ factor does not contribute to the absorption coefficient because the integration over φ vanishes. We obtain

$$|M_{\text{TE}}|^2 = \delta_{\vec{k}_t', \vec{k}_t} |<S\uparrow |p_x|X>|^2 \frac{1}{12} \left\{ 3 <f_n|g^{(1)}>^2 + [<f_n|g^{(2)}> + \sqrt{2} <f_n|g^{(3)}>]^2 \right\}. \tag{11.59}$$

Transverse magnetic (TM) polarization ($\hat{\varepsilon} = \hat{z}$):

$$|M_{\text{TM}}|^2 = |<\psi_l^{\eta}|p_z|\psi_m^{v\sigma}>|^2$$

$$= \delta_{\vec{k}_t', \vec{k}_t} |<f_n|g^{(1)}> <S\uparrow |p_z|1> + <f_n|g^{(2)}> <S\uparrow |p_z|2> + <f_n|g^{(3)}> <S\uparrow |p_z|3>|^2$$

$$= \delta_{\vec{k}_t', \vec{k}_t} |<S\uparrow |p_z|Z>|^2 |<f_n|g^{(2)}> i\frac{1}{\sqrt{3}}e^{-i\frac{1}{2}\varphi} - <f_n|g^{(3)}> i\frac{1}{\sqrt{6}}e^{-i\frac{1}{2}\varphi}|^2$$

$$= \delta_{\vec{k}_t', \vec{k}_t} |<S\uparrow |p_z|Z>|^2 \frac{1}{3} \left| <f_n|g^{(2)}> - <f_n|g^{(3)}> \frac{1}{\sqrt{2}} \right|.$$

$$\tag{11.60}$$

Here,

$$|<S\uparrow|p_x|X>|^2 = |<S\uparrow|p_z|Z>|^2 = \frac{m_0}{2}\left(\frac{m_0}{m_e}-1\right)\frac{(E_g+\Delta)E_g}{(E_g+2\Delta/3)}. \tag{11.61}$$

The momentum matrix elements for the lower Hamiltonian can be also obtained by a similar procedure:

$$|M_{TE}|^2 = \delta_{\vec{k}_t',\vec{k}_t}|<S\downarrow|p_x|X>|^2\frac{1}{12}\left\{3<f_n|g^{(6)}>^2 + [<f_n|g^{(4)}> + \sqrt{2}<f_n|g^{(5)}>]^2\right\} \tag{11.62}$$

and

$$|M_{TM}|^2 = \delta_{\vec{k}_t',\vec{k}_t}|<S\downarrow|p_z|Z>|^2\frac{1}{3}\left|<f_n|g^{(5)}> - <f_n|g^{(4)}>\frac{1}{\sqrt{2}}\right|^2. \tag{11.63}$$

11.4.2 Wurtzite Structure

Similarly to the zinc-blende case, we can obtain the optical momentum matrix elements as follows.

TE polarization ($\hat{\varepsilon} = \hat{x}$ or $\hat{\varepsilon} = \hat{y}$):

$$|M_{TE}|^2 = |<\psi_l^\eta|p_x|\psi_m^{v\sigma}>|^2 = |<\psi_l^\eta|p_y|\psi_m^{v\sigma}>|^2$$

$$= \delta_{\vec{k}_t',\vec{k}_t}|<f_n|g^{(1)}><S\uparrow|p_x|1> + <f_n|g^{(2)}><S\uparrow|p_x|2> + <f_n|g^{(3)}><S\uparrow|p_x|3>|^2$$

$$= \delta_{\vec{k}_t',\vec{k}_t}|<S\uparrow|p_x|X>|^2\left|-\frac{1}{2}<f_n|g^{(1)}>e^{i\left(\frac{3}{4}\pi+\frac{3}{2}\varphi\right)} + <f_n|g^{(2)}>\frac{1}{2}e^{-i\left(\frac{1}{4}\pi+\frac{1}{2}\varphi\right)}\right|^2$$

$$= \delta_{\vec{k}_t',\vec{k}_t}|<S\uparrow|p_x|X>|^2\frac{1}{4}\left|<f_n|g^{(1)}>^2 + <f_n|g^{(2)}>^2 + 2\cos\varphi<f_n|g^{(1)}><f_n|g^{(2)}>\right|. \tag{11.64}$$

Finally,

$$|M_{TE}|^2 = |\delta_{\vec{k}_t',\vec{k}_t}|<S\uparrow|p_x|X>|^2\frac{1}{4}\left\{<f_n|g^{(1)}>^2 + <f_n|g^{(2)}>^2\right\}. \tag{11.65}$$

TM polarization ($\hat{\varepsilon} = \hat{z}$):

$$|M_{TM}|^2 = |<\psi_l^\eta|p_z|\psi_m^{v\sigma}>|^2$$

$$= \delta_{\vec{k}_t',\vec{k}_t}|<f_n|g^{(1)}><S\uparrow|p_z|1> + <f_n|g^{(2)}><S\uparrow|p_z|2> + <f_n|g^{(3)}><S\uparrow|p_z|3>|^2$$

$$= \delta_{\vec{k}_t',\vec{k}_t}<S\uparrow|p_z|Z>^2|<f_n|g^{(3)}>\frac{1}{\sqrt{2}}e^{i\left(\frac{1}{4}\pi+\frac{1}{2}\varphi\right)}|^2$$

$$= \delta_{\vec{k}_t',\vec{k}_t}<S\uparrow|p_z|Z>^2\frac{1}{2}<f_n|g^{(3)}>^2. \tag{11.66}$$

The momentum matrix elements for the lower Hamiltonian also can be obtained by a similar procedure:

$$|M_{TE}|^2 = |\delta_{\vec{k}_t',\vec{k}_t}|<S\downarrow|p_x|X>|^2\frac{1}{4}\left\{<f_n|g^{(4)}>^2 + <f_n|g^{(5)}>^2\right\} \tag{11.67}$$

and

$$|M_{TM}|^2 = \delta_{\vec{k}_t',\vec{k}_t}<S\downarrow|p_z|Z>^2\frac{1}{2}<f_n|g^{(6)}>^2. \tag{11.68}$$

11.5 Band Structures of Bulk Semiconductors with an Arbitrary Crystal Orientation

11.5.1 Zinc-Blende Structure

The Hamiltonian for an arbitrary crystal orientation can be obtained using a rotation matrix

$$U = \begin{pmatrix} \cos\theta\cos\varphi & \cos\theta\sin\varphi & -\sin\theta \\ -\sin\varphi & \cos\varphi & 0 \\ \sin\theta\cos\varphi & \sin\theta\sin\varphi & \cos\theta \end{pmatrix}, \tag{11.69}$$

The rotation with the Euler angles θ and φ transforms physical quantities from the (x, y, z) coordinates to the (x', y', z') coordinates. Figure 11.1 shows configuration of the $[lmn]$-oriented coordinate system to the conventional $[001]$-oriented coordinate system. The relation between the coordinate systems for vectors and tensors is expressed as

$$k_\alpha = U_{i\alpha} k_i',$$
$$\varepsilon_{\alpha\beta} = U_{i\alpha} U_{j\beta} \varepsilon_{ij}', \tag{11.70}$$

where the summation over repeated indices is assumed. Our object is to obtain $P_{k'}$, $Q_{k'}$, $R_{k'}$, and $S_{k'}$ in the (x', y', z') coordinates. First we obtain the 4×4 Hamiltonian for general crystal orientation by using invariant method and then extending these results to obtain the 6×6 Hamiltonian. Here, for simplicity, we consider the case of the QWs with $(11n)$ orientations with $\theta = \arctan(\sqrt{2}/n)$ and $\varphi = \pi/4$. For example, (110) means $\theta = \pi/2$ and $\varphi = \pi/4$. The 4×4 Hamiltonian to describe the interaction between the heavy-hole and light-hole bands along with the Bir–Pikus Hamiltonian for strain can be formally written as

$$H_t^v = H_0^v(\vec{k}) + H_s^v, \tag{11.71}$$

with $H_0^v(\vec{k})$ representing the valence-band Hamiltonian and H_s^v the valence-band strain or Bir–Pikus Hamiltonian. The Luttinger formulation of the most general Hamiltonian for the (001) crystal orientation

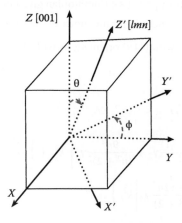

FIGURE 11.1 Configuration of the $[lmn]$-oriented coordinate system to the conventional $[001]$-oriented coordinate system.

is given by [22,23]

$$
H_0^v = \frac{\hbar^2}{2m_0}(\gamma_1 k^2 I_4 - 2\gamma_2[(J_x^2 - 1/3J^2)k_x^2 + (J_y^2 - 1/3J^2)k_y^2 + (J_z^2 - 1/3J^2)k_z^2]
$$

$$
- 4\gamma_3[\{J_xJ_y\}\{k_xk_y\} + \{J_yJ_z\}\{k_yk_z\} + \{J_zJ_x\}\{k_zk_x\}]),
$$

$$
H_s^v = -a_v(\varepsilon_{xx} + \varepsilon_{yy} + \varepsilon_{zz})I_4 + b[(J_x^2 - 1/3J^2)\varepsilon_{xx} + (J_y^2 - 1/3J^2)\varepsilon_{yy} + (J_z^2 - 1/3J^2)\varepsilon_{zz}]
$$

$$
+ 2d/\sqrt{3}[\{J_xJ_y\}\{\varepsilon_{xx}\} + \{J_yJ_z\}\{\varepsilon_{yy}\} + \{J_zJ_x\}\{\varepsilon_{zz}\}]), \tag{11.72}
$$

where the J_i's are the angular momentum matrices for a state with spin 3/2; I_4 is a 4×4 identity matrix; and $\{k_xk_y\} = (k_xk_y + k_yk_x)/2$. The angular momentum matrices are any three matrices, $(J_xJ_yJ_z)$, which satisfy the rules of commutation for angular momentum. These matrices given by Luttinger are as follows:

$$
J_x = \begin{pmatrix} 0 & 0 & \sqrt{3}/2 & 0 \\ 0 & 0 & 1 & \sqrt{3}/2 \\ \sqrt{3}/2 & 1 & 0 & 0 \\ 0 & \sqrt{3}/2 & 0 & 0 \end{pmatrix}, \quad J_y = \begin{pmatrix} 0 & 0 & -i\sqrt{3}/2 & 0 \\ 0 & 0 & i & -i\sqrt{3}/2 \\ i\sqrt{3}/2 & -i & 0 & 0 \\ 0 & i\sqrt{3}/2 & 0 & 0 \end{pmatrix},
$$

$$
J_z = \begin{pmatrix} 3/2 & 0 & 0 & 0 \\ 0 & -1/2 & 0 & 0 \\ 0 & 0 & 1/2 & 0 \\ 0 & 0 & 0 & -3/2 \end{pmatrix}, \quad I_4 = \begin{pmatrix} 1 & 0 & 0 & 0 \\ 0 & 1 & 0 & 0 \\ 0 & 0 & 1 & 0 \\ 0 & 0 & 0 & 1 \end{pmatrix}. \tag{11.73}
$$

We can obtain the (001)-oriented Hamiltonian by substituting Equation 11.73 into 11.72:

$$
H_0^v = \begin{pmatrix} P+Q & R & -S & 0 \\ R^* & P-Q & 0 & S^* \\ -S^* & 0 & P-Q & R \\ 0 & S^* & R^* & P+Q \end{pmatrix}. \tag{11.74}
$$

This is exactly the same from as the upper 4×4 Hamiltonian given in Equation 11.1. However, note that, in this case, the ordering of the basis functions for the choice of the J_i's is $\left|\frac{3}{2}, \frac{3}{2}\right\rangle, \left|\frac{3}{2}, -\frac{1}{2}\right\rangle, \left|\frac{3}{2}, \frac{1}{2}\right\rangle, \left|\frac{3}{2}, -\frac{3}{2}\right\rangle$.

Next, we consider the case of QWs with $(11n)$ orientations with $\theta = \arctan(\sqrt{2}/n)$ and $\varphi = \pi/4$. Then, from Equations 11.69 and 11.70,

$$
P_{k'} = \left(\frac{\hbar^2}{2m_0}\right)\gamma_1(k_{x'}^2 + k_{y'}^2 + k_{z'}^2),
$$

$$
Q_{k'} = \left(\frac{\hbar^2}{2m_0}\right)\left\{\left[\frac{(n^2-4)(n^2-1)}{(n^2+2)^2}\gamma_2 + \frac{9n^2}{(n^2+2)^2}\gamma_3\right]k_{x'}^2 + \left[\frac{(n^2+2)(n^2-1)}{(n^2+2)^2}\gamma_2 + \frac{3}{(n^2+2)}\gamma_3\right]k_{y'}^2 \right.
$$

$$
\left. + \left[-2\frac{(n^2-1)^2}{(n^2+2)^2}\gamma_2 - \frac{6+12n^2}{(n^2+2)}\gamma_3\right]k_{z'}^2\right\},
$$

$$
R_{k'} = \left(\frac{\hbar^2}{2m_0}\right)\left\{\sqrt{3}\left[\frac{(n^2-4)}{(n^2+2)^2}\gamma_2 - \frac{5n^2+n^4}{(n^2+2)^2}\gamma_3\right]k_{x'}^2 + \sqrt{3}\left[\frac{1}{(n^2+2)}\gamma_2 + \frac{1+n^2}{(n^2+2)^2}\gamma_3\right]k_{y'}^2 \right.
$$

$$+ 2\sqrt{3}\left[-\frac{(n^2-1)}{(n^2+2)^2}(\gamma_2-\gamma_3)\right]k_{z'}^2 + i2\sqrt{3}\left[\frac{(n^2\gamma_2+2\gamma_3)}{(n^2+2)}\right]k_{x'}k_{y'}$$

$$+ 2\sqrt{6}\left[i\frac{n}{(n^2+2)}(\gamma_2-\gamma_3)\right]k_{y'}k_{z'} + 6\sqrt{6}\left[\frac{n}{(n^2+2)^2}(\gamma_2-\gamma_3)\right]k_{x'}k_{z'}\Bigg\},$$

$$-S_{k'} = \left(\frac{\hbar^2}{2m_0}\right)\Bigg\{\left[\sqrt{6}\frac{n(4-n^2)}{(n^2+2)^2}(\gamma_2-\gamma_3)\right]k_{x'}^2 + \left[\sqrt{6}\frac{n}{(n^2+2)}(\gamma_3-\gamma_2)\right]k_{y'}^2$$

$$+ \left[2\sqrt{6}\frac{n(n^2-1)}{(n^2+2)^2}(\gamma_2-\gamma_3)\right]k_{z'}^2 + i2\sqrt{6}\left[\frac{n}{(n^2+2)}(\gamma_2-\gamma_3)\right]k_{x'}k_{y'}$$

$$+ i2\sqrt{3}\left[\frac{2\gamma_2+n^2\gamma_3}{(n^2+2)}\right]k_{y'}k_{z'} + \sqrt{3}\left[-\frac{12n^2}{(n^2+2)^2}\gamma_2 - 2\frac{(4-2n^2+n^4)}{(n^2+2)^2}\gamma_3\right]k_{x'}k_{z'}\Bigg\},$$

$$P_{\varepsilon'} = -a_{v1}(\varepsilon_{x'x'} + \varepsilon_{y'y'} + \varepsilon_{z'z'}),$$

$$Q_{\varepsilon'} = -\frac{b(n^2-1)(-\Gamma + 3\sqrt{2}\varepsilon_{x'z'}n + \Gamma n^2) + d\sqrt{3}[\Gamma + \sqrt{2}\varepsilon_{x'z'}n + n^2(2\Gamma - \sqrt{2}\varepsilon_{x'z'}n)]}{(n^2+2)^2},$$

$$R_{\varepsilon'} = -\frac{b\sqrt{3}(-\Gamma + 3\sqrt{2}\varepsilon_{x'z'}n + \Gamma n^2) + d(\Gamma - 3\sqrt{2}\varepsilon_{x'z'}n - \Gamma n^2)]}{(n^2+2)^2},$$

$$-S_{\varepsilon'} = -\frac{b\sqrt{3}n(\sqrt{2}\Gamma - 6\varepsilon_{x'z'}n - \sqrt{2}\Gamma n^2) + d[\sqrt{2}n(n^2-1)\Gamma + \varepsilon_{x'z'}(-4 + 2n^2 - n^4)]}{(n^2+2)^2}, \quad (11.75)$$

where $\Gamma = \varepsilon_{||} - \varepsilon_{z'z'}, \varepsilon_{||} = (a_s - a_e)/a_e$, and a_s and a_e are lattice constants for the substrate and the epilayer materials, respectively. The biaxial strain components for a general crystal orientation are determined from the condition that the layer is grown pseudomorphically and these strain coefficients should minimize the strain energy of the layer simultaneously [24,25]. That is,

$$\varepsilon_{x'x'} = \varepsilon_{y'y'} = \varepsilon_{||}, \quad \varepsilon_{x'y'} = \varepsilon_{y'z'} = 0, \quad \varepsilon_{z'z'} = -2\frac{K_3}{\sqrt{2}K_2}\varepsilon_{||}, \quad \varepsilon_{x'z'} = -2\frac{K_1}{\sqrt{2}K_2}\varepsilon_{||}, \quad (11.76)$$

where

$$K_1 = (C_{11} + 2C_{12})(-C_{11} + C_{12} + 2C_{44})n(n^2-1),$$

$$K_2 = 2C_{11}C_{44} + 2C_{12}C_{44} + 2C_{44}^2 + (C_{11}^2 + C_{11}C_{12} - 2C_{12}^2 + 2C_{11}C_{44} - 4C_{12}C_{44} + C_{11}C_{44}n^2)n^2,$$

$$K_2 = -2\left[C_{11}C_{44} + 3C_{12}C_{44} - 2C_{44}^2 + (C_{11}^2 + C_{11}C_{12} - 2C_{12}^2 - C_{11}C_{44} + C_{12}C_{44}n^2)n^2\right], \quad (11.77)$$

and C_{ij} are the stiffness constants in the strained epilayers. Then, the 4×4 Hamiltonian of $(11n)$-oriented zinc-blende crystal is given by

$$H'_{4\times4} = \begin{pmatrix} P'+Q' & R' & -S' & 0 \\ R'^* & P'-Q' & 0 & S'^* \\ -S'^* & 0 & P'-Q' & R' \\ 0 & S'^* & R'^* & P'+Q' \end{pmatrix}, \quad (11.78)$$

where

$$P' = P_{k'} + P_{\varepsilon'}, \quad Q' = Q_{k'} + Q_{\varepsilon'}, \quad R' = R_{k'} + R_{\varepsilon'}, \quad S' = S_{k'} + S_{\varepsilon'}. \quad (11.79)$$

Off-diagonal strains in zinc-blende structure semiconductors induce a polarization given by [25]

$$P_i^S = 2e_{14}\varepsilon_{jk}, \tag{11.80}$$

where P_i^S is the induced polarization, e_{14} is the piezoelectric constant, and $i, j, k = x, y, z$. Here, i, j, k are in cyclic order. A strained layer with a [001] growth direction has only diagonal strains. Thus, the (001)-oriented layer will not have strain-induced piezoelectric polarization fields. However, strained layers with any other growth direction will have piezoelectric polarization fields. The piezoelectric polarization along the growth direction is important because the electric field in the QW originates from polarization charges at heterojunction interfaces. The electric field induced perpendicular to the growth direction due to the strain can be calculated as

$$E_\perp = -\frac{\vec{P}^S \cdot \hat{u}}{\varepsilon}, \tag{11.81}$$

where the unit vector $\hat{u}$ along the growth direction is given by $\hat{u} = \sqrt{1/(n^2 + 2)}(\hat{i} + \hat{j} + n\hat{k})$. For an [11$n$] growth direction,

$$E_\perp = -\frac{2e_{14}}{\sqrt{n^2 + 2}}(n^2\varepsilon_{xy} + \varepsilon_{yz} + \varepsilon_{zx}), \tag{11.82}$$

where

$$\varepsilon_{xy} = \frac{1}{(n^2 + 2)}\left(-\varepsilon_{x'x'} + \sqrt{2}n\varepsilon_{x'z'} + 2n\varepsilon_{z'z'}\right),$$

$$\varepsilon_{yz} = \frac{1}{2(n^2 + 2)}\left(-2n\varepsilon_{x'x'} - \sqrt{2}(2 - n^2)\varepsilon_{x'z'} + 2n\varepsilon_{z'z'}\right)$$

$$\varepsilon_{zx} = \varepsilon_{yz}. \tag{11.83}$$

It is straightforward to obtain the 6×6 Hamiltonian of (11n)-oriented zinc-blende crystal using components P', Q', R', and S' in the (x', y', z') coordinates. That is, the 6×6 Hamiltonian for the valence-band structure in the (x', y', z') coordinates can be written as

$$H'_{6\times6} = -\begin{pmatrix} P'+Q' & -S' & R' & 0 & -S'/\sqrt{2} & \sqrt{2}R' \\ -S'^\dagger & P'-Q' & 0 & R' & -\sqrt{2}Q' & \sqrt{3/2}S' \\ R'^\dagger & 0 & P'-Q' & 0 & \sqrt{3/2}S'^\dagger & \sqrt{2}Q' \\ 0 & R'^\dagger & S'^\dagger & P'+Q' & -\sqrt{2}R'^\dagger & -S'^\dagger/\sqrt{2} \\ -S'^\dagger/\sqrt{2} & 0 & \sqrt{3/2}S' & -\sqrt{2}R' & P'+\Delta & 0 \\ \sqrt{2}R'^\dagger & \sqrt{3/2}S'^\dagger & \sqrt{2}Q' & -S'/\sqrt{2} & 0 & P'+\Delta \end{pmatrix} \begin{matrix} |3/2, 3/2\rangle' \\ |3/2, 1/2\rangle' \\ |3/2, -1/2\rangle' \\ |3/2, -3/2\rangle' \\ |1/2, 1/2\rangle' \\ |1/2, -1/2\rangle' \end{matrix}.$$

$$\tag{11.84}$$

Here, the basis set in the (x', y', z') coordinates consists of the following basis functions:

$$|u_1\rangle' = \left|\frac{3}{2}, \frac{3}{2}\right\rangle' = -\frac{1}{\sqrt{2}}\left|(X' + iY')\uparrow'\right\rangle,$$

$$|u_2\rangle' = \left|\frac{3}{2}, \frac{1}{2}\right\rangle' = \frac{1}{\sqrt{6}}\left|-(X' - iY')\downarrow' + 2Z'\uparrow'\right\rangle,$$

$$|u_3\rangle' = \left|\frac{3}{2}, -\frac{1}{2}\right\rangle' = \frac{1}{\sqrt{6}}\left|(X' - iY')\uparrow' + 2Z'\downarrow'\right\rangle,$$

$$|u_4\rangle' = \left|\frac{3}{2}, -\frac{3}{2}\right\rangle' = \frac{1}{\sqrt{2}}\left|(X' - iY')\downarrow'\right\rangle, \qquad (11.85)$$

$$|u_5\rangle' = \left|\frac{1}{2}, \frac{1}{2}\right\rangle' = \frac{1}{\sqrt{3}}\left|(X' + iY')\downarrow' + Z'\uparrow'\right\rangle,$$

$$|u_6\rangle' = \left|\frac{1}{2}, -\frac{1}{2}\right\rangle' = \frac{1}{\sqrt{3}}\left|(X' - iY')\uparrow' - Z'\downarrow'\right\rangle.$$

11.5.2 Wurtzite Structure

Analytical expressions for strain components for general crystal orientation can be obtained from the condition that the layer is grown pseudomorphically and the condition that strain energy density should be minimal. Figure 11.2 shows (a) configuration of the coordinate system (x', y', z') in $(hkil)$-oriented crystals and (b) a wurtzite primitive cell. The z-axis corresponds to the c-axis [0001] and the growth axis or the z'-axis is normal to the QW plane $(hkil)$. We define the unit vectors $\hat{x}'$, $\hat{y}'$ and $\hat{z}'$ along the $\vec{x}'$–, $\vec{y}'$-, and $\vec{z}'$-axes and they are related to unit vectors $\hat{x}$, $\hat{y}$, and $\hat{z}\hat{z}'$ along the $\vec{x}$-, $\vec{y}$-, and $\vec{z}$-axes of the original crystal orientation (0001) by the rotation matrix Equation 11.69. We define the hexagonal primitive translational vectors as

$$\vec{\alpha}_i = a_i\hat{x},$$

$$\vec{\beta}_i = -\frac{a_i}{2}\hat{x} + \frac{\sqrt{3}a_i}{2}\hat{y}, \qquad (11.86)$$

$$\vec{\gamma}_i = c_i\hat{z},$$

where a_i and c_i are lattice constants of the hexagonal structure with the subscript i denoting the epilayer (e) and substrate (s), respectively. When the crystal is strained the primitive translation vectors become

$$\vec{\alpha}''_i = a_i\hat{x}'',$$

$$\vec{\beta}''_i = -\frac{a_i}{2}\hat{x}'' + \frac{\sqrt{3}a_i}{2}\hat{y}'', \qquad (11.87)$$

$$\vec{\gamma}''_i = c_i\hat{z}'',$$

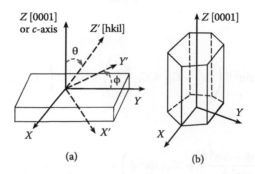

(a)   (b)

FIGURE 11.2 (a) Configuration of the coordinate system (x', y', z') in $(hkil)$-oriented crystals and (b) a wurtzite primitive cell.

with

$$\hat{x}'' = (1 + \varepsilon_{xx})\hat{x} + \varepsilon_{xy}\hat{y} + \varepsilon_{xz}\hat{z},$$

$$\hat{y}'' = \varepsilon_{yx}\hat{x} + (1 + \varepsilon_{yy})\hat{y} + \varepsilon_{yz}\hat{z}, \tag{11.88}$$

$$\hat{z}'' = \varepsilon_{zx}\hat{x} + \varepsilon_{zy}\hat{y} + (1 + \varepsilon_{zz})\hat{z}.$$

When a pseudomorphic interface is formed during the epitaxial growth, the translation vectors of each strained layer must have the same projection onto the growth plane, which is so-called the lattice commensurability constraint. By applying the lattice commensurability constraint on the QW and after some mathematical manipulations, we obtain the following analytical expressions for the strain tensors:

$$\varepsilon_{xx} = \varepsilon_{xx}^{(0)} + \frac{\sin\theta\cos\phi}{\cos\theta}\varepsilon_{xz},$$

$$\varepsilon_{xy} = \frac{\sin\theta\sin\phi}{\cos\theta}\varepsilon_{xz},$$

$$\varepsilon_{yy} = \varepsilon_{xx}^{(0)} + \frac{\sin\theta\sin^2\phi}{\cos\theta\cos\phi}\varepsilon_{xz}, \tag{11.89}$$

$$\varepsilon_{yz} = \frac{\sin\phi}{\cos\phi}\varepsilon_{xz},$$

$$\varepsilon_{zz} = \varepsilon_{zz}^{(0)} + \frac{\cos\theta}{\sin\theta\cos\phi}\varepsilon_{xz},$$

where $\varepsilon_{xx}^{(0)} = \frac{a_s - a_e}{a_e}$ and $\varepsilon_{zz}^{(0)} = \frac{c_s - c_e}{c_e}$. The strain tensor ε_{xz} can be evaluated by minimizing the strain energy density which is given by

$$W = \frac{1}{2}\left[C_{11}\varepsilon_{xx}^2 + C_{11}\varepsilon_{yy}^2 + C_{33}\varepsilon_{12}^2 + 2C_{12}\varepsilon_{xx}\varepsilon_{yy} + 2C_{13}(\varepsilon_{xx}\varepsilon_{zz} + \varepsilon_{yy}\varepsilon_{zz}) + 4C_{44}\varepsilon_{zz}^2\right]. \tag{11.90}$$

By minimizing the strain energy density with respect to ε_{xz}, we get

$$\varepsilon_{xz} = -\frac{N(\theta,\phi)\sin\theta\cos\theta}{D(\theta,\phi)\cos\phi}\varepsilon_{xx}^{(0)}, \tag{11.91}$$

where

$$N(\theta,\phi) = \sin^2\theta\left(C_{11} + C_{12} + C_{13}\frac{\varepsilon_{zz}^{(0)}}{\varepsilon_{xx}^{(0)}}\right) + \cos^2\theta\left(2C_{13} + C_{33}\frac{\varepsilon_{zz}^{(0)}}{\varepsilon_{xx}^{(0)}}\right), \tag{11.92}$$

and

$$D(\theta,\phi) = \sin^4\theta\left(C_{11}\frac{\sin^4\phi + \cos^4\phi}{\cos^2\phi} + 2C_{12}\sin^2\phi\right) + C_{33}\frac{\cos^4\theta}{\cos^2\phi} + 2\left(\frac{C_{13}}{\cos^2\phi} + 2C_{44}\right)\sin^2\theta\cos^2\theta. \tag{11.93}$$

Using Equations 11.70 and 11.89, general expressions for the Hamiltonian for the valence-band structure in (x', y', z') coordinates can be written as

$$\tilde{H}'(\vec{k}', \bar{\bar{\varepsilon}}') = \begin{pmatrix} F' & -K'^* & -H'^* & 0 & 0 & 0 \\ -K' & G' & H' & 0 & 0 & \Delta \\ -H' & H'^* & \lambda' & 0 & \Delta & 0 \\ 0 & 0 & 0 & F' & -K' & H' \\ 0 & 0 & \Delta & -K'^* & G' & -H'^* \\ 0 & \Delta & 0 & H'^* & -H' & \lambda' \end{pmatrix}, \tag{11.94}$$

where

$$F' = \Delta_1 + \Delta_2 + \lambda' + \vartheta',$$

$$G' = \Delta_1 - \Delta_2 + \lambda' + \vartheta',$$

$$\lambda' = \frac{\hbar^2}{2m_0} A_1 \left(\sin^2\theta k_x'^2 - 2\sin\theta\cos\theta k_x' k_z' + \cos^2\theta k_z'^2 \right)$$

$$+ \frac{\hbar^2}{2m_0} A_2 \left(\cos^2\theta k_x'^2 + s\cos\theta\sin\theta k_x' k_z' + \sin^2\theta k_z'^2 + k_y'^2 \right)$$

$$+ D_1 \left(\varepsilon_{zz}^{(0)} - \frac{N(\theta,\phi)\cos^2\theta}{D(\theta,\phi)\cos^2\phi}\varepsilon_{xx}^{(0)} \right) + D_2 \left(2 - \frac{N(\theta,\phi)\sin^2\theta}{D(\theta,\phi)\cos^2\phi} \right) \varepsilon_{xx}^{(0)},$$

$$\vartheta' = \frac{\hbar^2}{2m_0} A_3 \left(\sin^2\theta k_x'^2 - 2\sin\theta\cos\theta k_x' k_z' + \cos^2\theta k_z'^2 \right)$$

$$+ \frac{\hbar^2}{2m_0} A_4 \left(\cos^2\theta k_x'^2 + s\cos\theta\sin\theta k_x' k_z' + \sin^2\theta k_z'^2 + k_y'^2 \right)$$

$$+ D_3 \left(\varepsilon_{zz}^{(0)} - \frac{N(\theta,\phi)\cos^2\theta}{D(\theta,\phi)\cos^2\phi}\varepsilon_{xx}^{(0)} \right) + D_4 \left(2 - \frac{N(\theta,\phi)\sin^2\theta}{D(\theta,\phi)\cos^2\phi} \right) \varepsilon_{xx}^{(0)},$$

$$K' = \frac{\hbar^2}{2m_0} A_5 e^{2i\phi} \left(\cos^2\theta k_x'^2 - k_y'^2 + \sin^2\theta k_z'^2 + 2\sin\theta\cos\theta k_x' k_y' + 2i\cos\theta k_x' k_y' + 2i\sin\theta k_y' k_z' \right)$$

$$- D_5 \frac{N(\theta,\phi)\left(\cos^2\phi - \sin^2\phi - 2i\sin\phi\cos\phi\right)\sin^2\theta}{D(\theta,\phi)\cos^2\phi}\varepsilon_{xx}^{(0)},$$

$$H' = \frac{\hbar^2}{2m_0} A_6 e^{i\phi} \left\{ \sin\theta\cos\theta \left(k_z'^2 - k_x'^2\right) + \cos^2\theta k_x' k_z' - \sin^2\theta k_x' k_z' + i\left(\cos\theta k_y' k_z' - \sin\theta k_x' k_y'\right) \right\}$$

$$- D_6 e^{i\phi} \frac{N(\theta,\phi)\sin\theta\cos\theta}{D(\theta,\phi)\cos^2\phi}\varepsilon_{xx}^{(0)}. \tag{11.95}$$

We note that the old bases $\{|1\rangle, |2\rangle, |3\rangle, |4\rangle, |5\rangle, |6\rangle\}$ are used in the matrix representation of the Hamiltonian in Equation 11.94 and we assumed the hexagonal symmetry in the calculation. The Hamiltonian for nonpolar wurtzite QW structures can be obtained by substituting $\theta = \pi/2$ into the above equations.

11.5.3 Interband Optical-Matrix Elements for QW with an Arbitrary Crystal Orientation

11.5.3.1 (11n)-Oriented Zinc-Blende QW

The hole wave function in a QW can be written as

$$\Psi_m(z'; k_t') = \frac{e^{i\vec{k}_t \cdot \vec{r}_t}}{\sqrt{A}} \sum_{\nu=1}^{6} g_m^{(\nu)}(z'; k_t') |u_\nu\rangle, \tag{11.96}$$

where $g_m^{(\nu)}(z'; k_t')(\nu = 1, 2, 3, 4, 5,$ and $6)$ is the wave function for the mth sub-band in (x', y', z') coordinates, $|\rangle u_\nu\rangle'$ is given by Equation 11.85, and $g_m'^{(i)}$ follows the normalization rules

$$\sum_{\nu=1}^{6} \int dz' \, |g_m^{(\nu)}(z'; \, k_t')|^2 = 1. \tag{11.97}$$

Since we solve one Schrödinger equation for electrons and 6×6 Hamiltonian for holes, we need only the optical matrix elements for two cases: spin up and spin down. The polarization-dependent interband momentum-matrix elements are written as follows.

TE polarization ($\vec{e}' = \cos\varphi'\hat{x}' + \sin\varphi'\hat{y}'$):

$$\left|\vec{e}' \cdot \vec{M}'^{\uparrow}\right|^2 = \left|\cos\varphi' \left\{ -\frac{1}{\sqrt{2}}P_x \left\langle g_m'^{(1)}|\phi_l\right\rangle + \frac{1}{\sqrt{6}}P_x \left\langle g_m'^{(3)}|\phi_l\right\rangle + \frac{1}{\sqrt{3}}P_x \left\langle g_m'^{(6)}|\phi_l\right\rangle \right\} \right.$$
$$\left. + \sin\varphi' \left\{ -i\frac{1}{\sqrt{2}}P_x \left\langle g_m'^{(1)}|\phi_l\right\rangle - i\frac{1}{\sqrt{6}}P_x \left\langle g_m'^{(3)}|\phi_l\right\rangle - i\frac{1}{\sqrt{3}}P_x \left\langle g_m'^{(6)}|\phi_l\right\rangle \right\} \right|^2$$

$$\left|\vec{e}' \cdot \vec{M}'^{\downarrow}\right|^2 = \left|\cos\varphi' \left\{ -\frac{1}{\sqrt{6}}P_x \left\langle g_m'^{(2)}|\phi_l\right\rangle + \frac{1}{\sqrt{2}}P_x \left\langle g_m'^{(4)}|\phi_l\right\rangle + \frac{1}{\sqrt{3}}P_x \left\langle g_m'^{(5)}|\phi_l\right\rangle \right\} \right. \tag{11.98}$$
$$\left. + \sin\varphi' \left\{ -i\frac{1}{\sqrt{6}}P_x \left\langle g_m'^{(2)}|\phi_l\right\rangle - i\frac{1}{\sqrt{2}}P_x \left\langle g_m'^{(4)}|\phi_l\right\rangle + i\frac{1}{\sqrt{3}}P_x \left\langle g_m'^{(5)}|\phi_l\right\rangle \right\} \right|^2.$$

TE polarization ($\vec{e}' = \hat{z}'$):

$$\left|\vec{e}' \vec{M}'^{\uparrow}\right|^2 = \left|-\frac{2}{\sqrt{6}}P_z \left\langle g_m'^{(2)}|\phi_l\right\rangle + \frac{1}{\sqrt{3}}P_z \left\langle g_m'^{(5)}|\phi_l\right\rangle \right|^2$$

$$\left|\vec{e}' \vec{M}'^{\downarrow}\right|^2 = \left|\frac{2}{\sqrt{6}}P_x \left\langle g_m'^{(3)}|\phi_l\right\rangle - \frac{1}{\sqrt{3}}P_x \left\langle g_m'^{(6)}|\phi_l\right\rangle \right|^2. \tag{11.99}$$

11.5.3.2 Wurtzite QW

The optical matrix elements general crystal orientation are given by

$$|\vec{e}' \cdot \vec{M}'^\eta|^2 = \left|\left\langle \Phi_l'^\eta \middle| \vec{e}' \cdot \vec{p}' |\rangle \Psi_m'\right\rangle\right|^2. \tag{11.100}$$

The polarization-dependent interband momentum-matrix elements can be written as follows. TE polarization $(\hat{e}' = \cos\varphi'\hat{x} + \sin\varphi'\hat{y})$:

$$\hat{e}' \cdot \vec{M}'^{\uparrow} = -\cos\varphi' \sin\theta P_z \left\langle \varphi_l' \,\middle|\, g_m'^{(3)} \right\rangle$$

$$-\frac{1}{\sqrt{2}}\cos\varphi'\cos\theta\cos\phi P_x \left\{ (1+i)\left\langle \varphi_l' \,\middle|\, g_m'^{(1)}\right\rangle - (1-i)\left\langle \varphi_l' \,\middle|\, g_m'^{(2)}\right\rangle \right\}$$

$$+\frac{1}{\sqrt{2}}\sin\varphi' P_x \left\{ \sin\phi \left(\left\langle \varphi_l' \,\middle|\, g_m'^{(1)}\right\rangle - \left\langle \varphi_l' \,\middle|\, g_m'^{(2)}\right\rangle\right) - i\cos\phi\left(\left\langle \varphi_l' \,\middle|\, g_m'^{(1)}\right\rangle + \left\langle \varphi_l' \,\middle|\, g_m'^{(2)}\right\rangle\right)\right\},$$

$$\hat{e}' \cdot \vec{M}'^{\downarrow} = -\cos\varphi'\sin\theta P_z\left\langle \varphi_l' \,\middle|\, g_m'^{(6)}\right\rangle$$

$$+\frac{1}{\sqrt{2}}\cos\varphi'\cos\theta\cos\phi P_x\left\{ (1-i)\left\langle \varphi_l' \,\middle|\, g_m'^{(4)}\right\rangle - (1+i)\left\langle \varphi_l' \,\middle|\, g_m'^{(5)}\right\rangle\right\}$$

$$-\frac{1}{\sqrt{2}}\sin\varphi' P_x\left\{ \sin\phi\left(\left\langle \varphi_l' \,\middle|\, g_m'^{(4)}\right\rangle - \left\langle \varphi_l' \,\middle|\, g_m'^{(5)}\right\rangle\right) + i\cos\phi\left(\left\langle \varphi_l' \,\middle|\, g_m'^{(4)}\right\rangle + \left\langle \varphi_l' \,\middle|\, g_m'^{(5)}\right\rangle\right)\right\}.$$

$$(11.101)$$

TM polarization $(\hat{e}' = \hat{z}')$:

$$\hat{e}' \cdot \vec{M}'^{\uparrow} = \cos\theta P_z\left\langle \varphi_l' \,\middle|\, g_m'^{(3)}\right\rangle + \frac{1}{\sqrt{2}}\sin\theta\cos\phi\left\{ -\left\langle \varphi'_l \,\middle|\, g_m'^{(1)}\right\rangle + \left\langle \varphi_l' \,\middle|\, g_m'^{(2)}\right\rangle\right\}$$

$$-\frac{i}{\sqrt{2}}\sin\theta\sin\phi\left\{\left\langle \varphi_l' \,\middle|\, g_m'^{(1)}\right\rangle + \left\langle \varphi_l' \,\middle|\, g_m'^{(2)}\right\rangle\right\},$$

$$\hat{e}' \cdot \vec{M}'^{\downarrow} = \cos\theta P_z\left\langle \varphi_l' \,\middle|\, g_m'^{(6)}\right\rangle + \frac{1}{\sqrt{2}}\sin\theta\cos\phi\left\{\left\langle \varphi_l' \,\middle|\, g_m'^{(4)}\right\rangle - \left\langle \varphi_l' \,\middle|\, g_m'^{(5)}\right\rangle\right\}$$

$$-\frac{i}{\sqrt{2}}\sin\theta\sin\phi\left\{\left\langle \varphi_l' \,\middle|\, g_m'^{(4)}\right\rangle + \left\langle \varphi_l' \,\middle|\, g_m'^{(5)}\right\rangle\right\},$$

$$(11.102)$$

11.6 Optical Gain Model with Many-Body Effects

The optical gain spectra are calculated using the non-Markovian gain model with many-body effects [11,13]. The many-body effects include the plasma screening, band-gap renormalization (BGR), and the excitonic or the Coulomb enhancement (CE) of the interband transition probability [14]. The simplest non-Markovian quantum kinetics is Gaussian line-shape function, which is connected with memory effects in the system–reservoir interaction [13]. The optical gain with many-body effects including the effects of anisotropy on the valence-band dispersion is given by

$$g(\omega) = \sqrt{\frac{\mu_o}{\varepsilon}}2\left(\frac{e^2}{m_o^2\omega}\right)\int\limits_0^{2\pi}d\varphi_0\int\limits_0^{\infty}dk_{||}\frac{2k_{||}}{(2\pi)^2 L_w}\left|M_{nm}(k_{||},\varphi_0)\right|^2 [f_l^c(k_{||},\varphi_0) - f_m^v(k_{||},\varphi_0)]L(\omega,k_{||},\varphi_0),$$

$$(11.103)$$

where ω is the angular frequency, μ_o is the vacuum permeability, ε is the dielectric constant, e is the charge of an electron, m_o is the free electron mass, $k_{||}$ is the magnitude of the in-plane wave vector in the QW plane, L_w is the well width, $\left|M_{nm}(k_{||},\varphi_0)\right|^2$ is the momentum matrix element in the strained QW, f_l^c and f_m^v are the Fermi functions for occupation probability by the electrons in the conduction sub-band states and the valence sub-band states, respectively, and the indices l and m denote the electron states in conduction sub-band and heavy hole (light hole) sub-band states, respectively. Also, $E_{lm}(\hbar\omega, k_{||}, \varphi_0) = E_l^c(k_{||}, \varphi_0) - E_m^v(k_{||}, \varphi_0) + E_g + \Delta E_{SX} + \Delta E_{CH} - \hbar\omega$ is the renormalized transition energy between electrons and holes, where E_g is the band gap of the material, and ΔE_{SX} and ΔE_{CH} are the screened exchange and the Coulomb-hole contributions [14] to the BGR, respectively. The factor $Q(\hbar\omega, k_{||}, \varphi_0)$ accounts for the excitonic or CE of the interband transition probability [14,15]. The Gaussian line-shape function $L(E_{lm}(\hbar\omega, k_{||}, \varphi_0)$ renormalized with many-body effects is given by

$$\mathrm{Re}[L(E_{lm}(\hbar\omega, k_{||}, \varphi_0))] = \sqrt{\frac{\pi\tau_{in}(\hbar\omega, k_{||}, \varphi_0)\tau_c}{2\hbar^2}} \exp\left(-\frac{\tau_{in}(\hbar\omega, k_{||}, \varphi_0)\tau_c}{2\hbar^2} E_{lm}^2(\hbar\omega, k_{||}, \varphi_0)\right)$$

$$\mathrm{Im}[L(E_{lm}(\hbar\omega, k_{||}, \varphi_0))] = \frac{\tau_c}{\hbar} \int_0^\infty dt \exp\left(-\frac{\tau_c}{2\tau_{in}(\hbar\omega, k_{||}, \varphi_0)}t^2\right) \sin\left(\frac{\tau_c E_{lm}(\hbar\omega, k_{||}, \varphi_0)}{\hbar}t\right). \quad (11.104)$$

The correlation time τ_c is related to the non-Markovian enhancement of optical gain [13] and is assumed to be constant. The τ_{in} and the τ_c used in the calculation are 25 and 10 fs, respectively. The BGR is given as a summation of the Coulomb-hole self-energy and the screened-exchange shift [14]. The φ_0 dependence of the BGR is very small and neglected for simplicity. The Coulomb-hole contribution to the BGR is written as

$$\Delta E_{CH} = -2E_R a_o \lambda_s \ln\left(1 + \sqrt{\frac{32\pi N L_w}{C\lambda_s^3 a_o}}\right), \quad (11.105)$$

where N is the carrier density, λ_s is the inverse screening length, and C is a constant usually taken between 1 and 4. The Rydberg constant E_R and the exciton Bohr radius a_o are given by

$$\Delta E_R(eV) = 13.6\frac{m_o/m_r}{(\varepsilon/\varepsilon_0)^2} \quad (11.106)$$

and

$$a_o(\text{Å}) = 0.53\frac{\varepsilon/\varepsilon_0}{m_o/m_r}, \quad (11.107)$$

where m_r is the reduced electron–hole mass defined by $1/m_r = 1/m_e + 1/m_h$. The exchange contribution to the BGR is given by

$$\Delta E_{SX} = -\frac{2E_R a_o}{\lambda_s} \int_0^\infty dk_{||} k_{||} \frac{1 + \frac{C\lambda_s a_o k_{||}^2}{32\pi N L_w}}{1 + \frac{k_{||}}{\lambda_s} + \frac{Ca_o k_{||}^3}{32\pi N L_w}} \left[f_n^c(k_{||}) + 1 - f_m^v(k_{||})\right]. \quad (11.108)$$

The factor $1/(1 - Q(k_{||}, \hbar\omega))$ represents the CE in the Padé approximation. Here, the factor $Q(k_{||}, \hbar\omega)$ is given by [11,14]

$$Q(k_{||}, \hbar\omega) = i\frac{E_R a_o}{\pi\lambda_s \left|M_{nm}(k_{||})\right|^2} \int_0^\infty dk'_{||} k'_{||} \left|M_{nm}(k_{||})\right| [f_n^c(k'_{||}) - f_m^v(k'_{||})]\Xi(E_{lm}(\hbar\omega, k_{||}))\Theta(k_{||}, k'_{||}), \quad (11.109)$$

where

$$\Theta(k_{||}, k'_{||}) = \int_0^\infty d\theta \frac{1 + \frac{C\lambda_s a_o q_{||}^2}{32\pi N L_w}}{1 + \frac{q_{||}}{\lambda_s} + \frac{C a_o q_{||}^3}{32\pi N L_w}} \quad (11.110)$$

and

$$q = \left|k_{||} - k'_{||}\right|. \quad (11.111)$$

The spontaneous emission coefficient $g_{SP}(\omega)$ can be obtained by replacing $[f_l^c(k_{||}, \varphi_0) - f_m^v(k_{||}, \varphi_0)]$ in Equation 11.103 by $f_l^c(k_{||}, \varphi_0)(1 - f_m^v(k_{||}, \varphi_0))$.

$$g_{SP}(\omega) = \sqrt{\frac{\mu_o}{\varepsilon}} 2\left(\frac{e^2}{m_o^2 \omega}\right) \int_0^{2\pi} d\varphi_0 \int_0^\infty dk_{||} \frac{2k_{||}}{(2\pi)^2 L_w} \left|M_{nm}(k_{||}, \varphi_0)\right|^2 f_l^c(k_{||}, \varphi_0)(1 - f_m^v(k_{||}, \varphi_0))L(\omega, k_{||}, \varphi_0).$$

$$(11.112)$$

The spontaneous emission rate $r_{spon}(E)$ can be obtained from our calculated spontaneous emission spectrum $g_{SP}(E)$ by using

$$r_{spon}(E) = \left(\frac{4n^2}{\hbar\lambda^2}\right) g_{SP}(E),$$

where n is the refractive index of the QW and $E = 2\pi c\hbar/\lambda$ with c being the speed of light.

11.7 Numerical Example

As a numerical example, we calculate valence-band structure, optical matrix element, and optical gain as a function of crystal angle for zinc-blende and wurtzite GaN/AlGaN QW structures. For zinc-blende case, we consider the compressively strained GaN well by assuming that QW structures are grown on AlGaN substrate. On the other hand, in the case of the wurtzite semiconductor, we assume that QW structures are grown on GaN substrate. The material parameters for zinc-blende GaN and AlN used in the calculation were taken from Refs. [9,26–32] and references therein. Also, the material parameters for wurtzite GaN and AlN used in the calculation were taken from Refs. [33–35] and references therein. All parameters used in the calculation are summarized in Table 11.1.

11.7.1 Zinc-Blende Structure

Figure 11.3 shows the valence-band structures along $k_{x'}$ and $k_{y'}$ of (a) (001)-, (b) (111)-, and (c) (110)-oriented zinc-blende GaN/Al$_x$Ga$_{1-x}$N QWs (L_w = 2.5 nm). The sub-bands are labeled HH$_i$ and LS$_i$, where i denotes the sub-band level. Here SL is the acronym for "splitoff-hole-light-hole," and the first letter denotes the dominant component for the wave function. The crystal angles for QWs with (11n) orientations

TABLE 11.1 Parameters Used in the Calculation

Parameters	Wurtzite			Zinc-Blende	
	GaN	AlN		GaN	AlN
Lattice constant (Å)					
a	3.1892	3.112	a	4.460	4.342
c	5.185	4.982			
Energy parameter					
E_g(eV)	3.44	6.16	E_g(eV)	3.1	4.9
$\Delta_1 = \Delta_{cr}$(meV)	22.0	−58.5	Δ (meV)	11.0	11.0
$\Delta_2 = \Delta/3$ (meV)	15.0	20.4			
$\Delta_3 = \Delta_2$(meV)	5.0	6.8			
Conduction-band effective masses					
m_{ez}^w/m_0	0.2	0.3	m_e^c/m_0	0.13	0.21
$m_{et}^w = m_{ez}^w$					
Valence-band effective-mass parameters					
A_1	−6.56	−3.95	γ_1	3.06	2.42
A_2	−0.91	−0.27	γ_2	0.91	0.58
A_3	−3.13	−1.95	γ_3	1.03	0.71
Deformation potentials					
a_c	−4.6	−4.5	a_c	−2.77	−6.8
D_1	−1.7	−2.89	a_v	3.63	2.3
D_2	6.3	4.89	b	−2.67	−1.5
D_3	−4.0	−3.34	d	−4.62	−4.5
Dielectric constant					
ε	10.0	8.5	ε	10.69	8.5
Elastic stiffness constant					
C_{11}	39.0	39.8	C_{11}	29.6	30.4
C_{12}	14.5	14.0	C_{12}	15.4	15.2
C_{13}	10.6	12.7			
C_{33}	39.8	38.2			
C_{44}	10.5	9.6	C_{44}	20.0	19.9
C_{66}	12.3	12.9			
Piezoelectric constant					
$d_{31}(\times 10^{-12}$ m/V)	−1.7	−2.0	e_{14}(C/m^2)	−1.11	−0.526
Spontaneous polarization (C/m^2)					
P_{SP}	−0.029	−0.081			

($\varphi = \pi/4$) are obtained by using the relation $\theta = \tan^{-1}(\sqrt{2}/n)$. For example, crystal angles for (111) and (110) are $\theta \approx 55°$ and $\theta \approx 90°$, respectively. The valence-band structures of the self-consistent model are calculated at a carrier density of $N_{2D} = 10 \times 10^2$ cm^{-2}. The strain-induced piezoelectric field in the GaN well for the (111) orientation is about 1.33 MV/cm. The sub-band structures for different orientations are apparently different. For the (001) and the (111) orientations, the sub-band structures are nearly spherically symmetric. On the other hand, the valence-band structure of the (110)-oriented QW shows anisotropy. In particular, the effective mass of the first sub-band along $k_{y'}$ is observed to be greatly reduced for the (110)-oriented structure. A larger energy spacing between the first two sub-bands (HH1 and SL1) and higher sub-bands are observed for the (001)-oriented structure. The increase in the sub-band energy spacing will reduce the carrier population in the higher sub-bands. However, the energy spacing is gradually reduced with increasing polar angle θ.

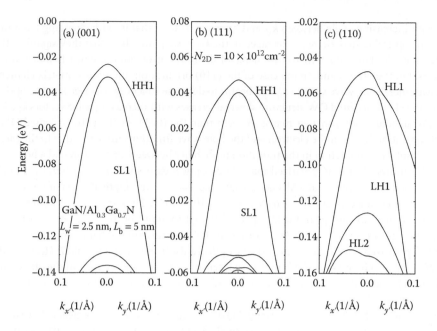

FIGURE 11.3 Valence-band structures along $k_{x'}$ and $k_{y'}$ of (a) (001)-, (b) (111)-, and (c) (110)-oriented zinc-blende GaN/Al$_x$Ga$_{1-x}$N QWs (L_w = 2.5 nm). The sub-bands are labeled HH$_i$ and LS$_i$, where i denotes the sub-band level.

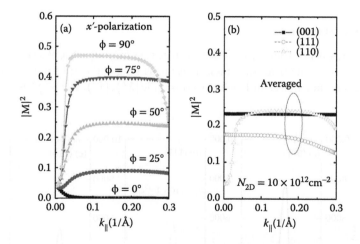

FIGURE 11.4 (a) x'-Polarized normalized optical matrix element as a function of $k_{||}$ for various values of φ for (110)-oriented zinc-blende GaN/Al$_x$Ga$_{1-x}$N quantum wells (QWs) (L_w = 2.5 nm) and (b) averaged optical matrix element as a function of $k_{||}$ for (001)-, (111)-, and (110)-oriented zinc-blende GaN/Al$_x$ Ga$_{1-x}$N QWs (L_w = 2.5 nm).

Figure 11.4 shows the x'-polarized normalized optical matrix element as a function of $k_{||}$ for various values of φ for (110)-oriented zinc-blende GaN/Al$_x$Ga$_{1-x}$N QWs (L_w = 2.5 nm) and (b) the averaged optical matrix element as a function of $k_{||}$ for (001)-, (111)-, and (110)-oriented zinc-blende GaN/Al$_x$Ga$_{1-x}$N QWs (L_w = 2.5 nm). The angle φ is defined by $\tan\phi = k_{y'}/k_{x'}$ and $k_{||} = \sqrt{k_{x'}^2 + k_{y'}^2}$. Here, φ = 0° and 90° mean that optical matrix elements are plotted along $k_{x'}$ and $k_{y'}$. The optical matrix elements of the self-consistent model are calculated at a carrier density of $N_{2D} = 10 \times 10^{12}$ cm^{-2}. The optical matrix element increases with increasing angle between $k_{x'}$ and $k_{y'}$ wave vectors, that is, with changing from the x' to y' direction. The optical matrix elements averaged in the $k_{x'} - k_{y'}$ plane show that they significantly depend on the crystal angle. The optical matrix elements of the (001) orientation are nearly independent of the wave vectors. On the other hand, in the case of the (110) orientation, the optical matrix element rapidly increases with increasing wave vectors $k_{||}$ and begins to decrease in a range of large $k_{||}$. The optical matrix elements of the (111)-oriented QW structure slowly decreases with increasing wave vectors $k_{||}$. The (111)-oriented QW structure has a smaller matrix element than the (001)- or the (110)-oriented QW because the spatial separation between the electron and the hole wave functions increases due to the piezoelectric field, which results in a reduction in the transition probability between an electron and a hole.

Figure 11.5 shows x'- and y'-polarized many-body optical gain spectra of (a) (001)-, (b) (111)-, and (c) (110)-oriented zinc-blende GaN/Al$_x$Ga$_{1-x}$N QWs (L_w = 2.5 nm). Optical gains of the self-consistent model are calculated at a carrier density of $N_{2D} = 10 \times 10^{12}$ cm^{-2}. They are also averaged in the $k_{x'} - k_{y'}$ plane because of their anisotropy in the QW plane. The optical gain spectra have peaks corresponding to C1-HH1 and C1-HL1 transitions. The (111)-oriented QW structure shows that y'-polarized optical gain is larger than x'-polarized optical gain. The in-plane optical anisotropy ρ is about -0.17. Here, ρ is defined as $(I_{x'} - I_{y'})/(I_{x'} + I_{y'})$, where $I_i (i = x', y')$ means an intensity of peak optical gain. On the other hand, in the case of the (111)-oriented QW structure, y'-polarized optical gain is shown to be similar to x'-polarized optical gain with $\rho \approx 0$.

For a given carrier density, the (111)-oriented QW structure has a smaller optical gain than the (001)- or the (110)-oriented QW. This can be explained by the fact that the (111)-oriented QW has much smaller matrix elements than the (001)- or the (110)-oriented QW structures, as shown in Figure 11.4b. This is because the spatial separation between the electron and the hole wave functions is increased due to the piezoelectric field, which results in a reduction in the transition probability between electrons and holes.

11.7.2 Wurtzite Structure

Figure 11.6 shows the valence-band structures of wurtzite GaN/Al$_x$Ga$_{1-x}$N QWs (L_w = 2.5 nm) with (a) $\theta = 0°$ (c-plane), (b) 30°, and (c) 90° (a-plane). Here, the naming of the sub-bands follows the dominant composition of the wave function at the Γ point in terms of the $|X'\rangle, |Y'\rangle$, and $|Z'\rangle$ bases. The components

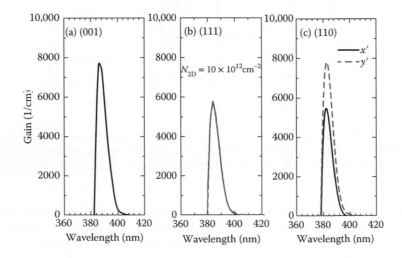

FIGURE 11.5 x'- and y'-polarized many-body optical gain spectra of (a) (001)-, (b) (111)-, and (c) (110)-oriented zinc-blende GaN/Al$_x$Ga$_{1-x}$N quantum wells (QWs) (L_w = 2.5 nm). Optical gains of the self-consistent model are calculated at a carrier density of N$_{2D}$ 10×10^{12} cm^{-2}.

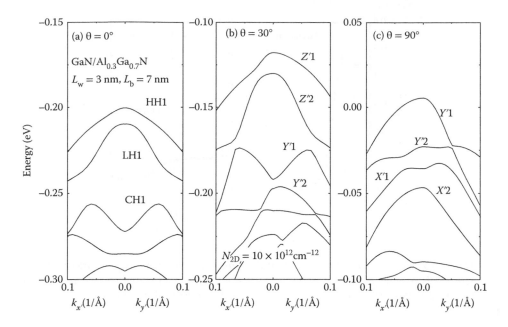

FIGURE 11.6 Valence-band structures of wurtzite GaN/Al$_x$Ga$_{1-x}$N quantum wells (QWs) (L_w = 2.5 nm) with (a) θ = 0° (*c*-plane), (b) 30°, and (c) 90° (*a*-plane).

$P_m^{i(=X',Y',Z')}$ of each wave function are given by

$$P_m^{X'} = \left\langle g_m'^{(3)} \mid g_m'^{(3)} \right\rangle + \left\langle g_m'^{(6)} \mid g_m'^{(6)} \right\rangle$$

$$P_m^{Y'} = \frac{\left\langle g_m'^{(1)} + g_m'^{(2)} \mid g_m'^{(1)} + g_m'^{(2)} \right\rangle + \left\langle g_m'^{(4)} + g_m'^{(5)} \mid g_m'^{(4)} + g_m'^{(5)} \right\rangle}{2}$$

$$P_m^{Z'} = \frac{\left\langle g_m'^{(2)} - g_m'^{(1)} \mid g_m'^{(2)} - g_m'^{(1)} \right\rangle + \left\langle g_m'^{(4)} - g_m'^{(5)} \mid g_m'^{(4)} - g_m'^{(5)} \right\rangle}{2}. \tag{11.113}$$

The valence-band structures of QW structures with θ = 30° and 90° show anisotropy in the QW plane, unlike the (0001)-oriented structure (*c*-plane). The effective mass of the topmost valence band along $k_{y'}$ is smaller than that along $k_{x'}$. Also, the hole effective mass (1.24 m_o for *a*-plane) of the topmost valence band along $k_{y'}$ is lower than that (1.56 m_o) of the *c*-plane. However, the hole effective mass (1.39 m_o) averaged in the $k_{x'} - k_{y'}$ QW plane is slightly larger than that along $k_{y'}$. Here, to estimate the magnitude of the hole effective mass, we considered a parabolic band fitted to the lowest sub-band of the exact band structure. The effective mass is determined so that, for a given carrier density and the quasi-Fermi level for holes, the carrier density and the quasi-Fermi level agree with those of the exact band structure. Hence, the effective mass of the fitted parabolic band reflects an averaged density of states.

Figure 11.7 shows the y'-polarized normalized optical matrix element as a function of $k_{\parallel}$ for various values of φ for wurtzite *a*-plane GaN/Al$_x$Ga$_{1-x}$N QWs (L_w = 2.5 nm) and (b) averaged optical matrix element as a function of $k_{\parallel}$ for QW structures with crystal angles of (a) θ = 0° (*c*-plane), (b) 30°, and (c) 90° (*a*-plane). The optical matrix elements of the self-consistent model are calculated at a carrier density of N_{2D} = 10 × 10^{12} cm^{-2}. The optical matrix element rapidly decreases with increasing $k_{\parallel}$ and greatly depends on angle φ between $k_{x'}$ and $k_{y'}$ wave vectors. In particular, in the case with φ = 90°, the decrease in the optical matrix element is observed to be much more rapid, compared to the other case. The optical

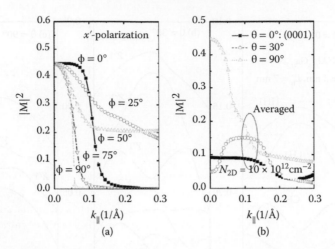

FIGURE 11.7 (a) y'-Polarized normalized optical matrix element as a function of $k_{||}$ for various values of φ for wurtzite *a*-plane GaN/Al$_x$Ga$_{1-x}$N quantum wells (QWs) ($L_w = 2.5$ nm) and (b) averaged optical matrix element as a function of $k_{||}$ for QW structures with crystal angles of $\theta = 0°$ (*c*-plane), 30°, and 90° (*a*-plane).

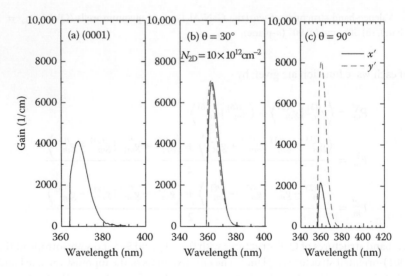

FIGURE 11.8 x'- and y'-polarized, many-body optical gain spectra for quantum well (QW) structures with crystal angles of $\theta = 0°$ (*c*-plane), 30°, and 90° (*a*-plane).

matrix elements averaged in the $k_{x'} - k_{y'}$ plane show that they significantly depend on the crystal angle, as observed for the zinc-blende case of Figure 11.3. The optical matrix element rapidly increases with increasing crystal angle. This can be explained by the fact that the internal field is reduced for the QW structure with larger crystal angle. In particular, in the QW structure with $\theta = 90°$, the internal field is disappeared and the transition probability between an electron and a hole is significantly enhanced due to the reduction in the spatial separation between the electron and the hole wave functions.

Figure 11.8 shows x'- and y'-polarized many-body optical gain spectra for QW structures with crystal angles of (a) $\theta = 0°$ (*c*-plane), (b) 30°, and (c) 90° (*a*-plane). Optical gains of the self-consistent model are calculated at a carrier density of $N_{2D} = 10 \times 10^{12}$ cm^{-2}. They are also averaged in the $k_{x'} - k_{y'}$ plane because of their anisotropy in the QW plane. The (0001)-oriented QW structure shows that y'-polarized optical gain is the same as x'-polarized optical gain because of its isotropy in the QW plane. The QW structure with $\theta = 30°$ also shows that x'- and y'-polarized optical gains are similar to each other. On the other hand, the a-plane QW structure shows that y'-polarized optical gain is much larger than x'-polarized optical gain. The in-plane optical anisotropy ρ is about -0.58, which is much larger than that observed for the zinc-blende QW structures.

11.8 Summary

In summary, we reviewed theoretical formalism to obtain eigenvalues and wave functions of (001)-oriented zinc-blende and (0001)-oriented wurtzite QW structures. In addition, we reviewed crystal orientation effects on electronic and optical properties of strained zinc-blende and wurtzite QW structures. As a numerical example, we calculated valence-band structures, optical matrix elements, and optical gains as a function of crystal orientation for zinc-blende and wurtzite AlGaN/GaN QW structures. These results can be used for a design of QW-based optoelectronic devices.

References

1. Corzine, S. W., R. H. Yan, and L. A. Coldren. 1993. Optical gain in III-V bulk and quantum well semiconductors. In *Quantum Well Lasers*, edited by P. S. Zory, Jr. San Diego, CA: Academic Press.
2. Chuang, S. L. 1995. *Physics of Optoelectronic Devices*. New York, NY: Wiley.
3. Piprek, J. 2003. *Semiconductor Optoelectronic Devices: Introduction to Physics and Simulation*. San Diego, CA: Academic Press.
4. Ahn, D. and S.-H. Park. 2011. *Engineering Quantum Mechanics*. Hoboken, NJ: John Wiley & Sons.
5. Chao, C. Y.-P. and S. L. Chuang. 1992. Spin–orbit-coupling effects on the valence-band structure of strained semiconductor quantum wells. *Phys. Rev. B* **46**: 4110–4122.
6. Chuang, S. L. and C. S. Chang. 1996. The k.p method for strained wurtzite semiconductors. *Phys. Rev. B* **54**: 2491–2504.
7. Park, S.-H. and S. L. Chuang. 1999. Crystal-orientation effects on the piezoelectric field and electronic properties of strained wurtzite semiconductors. *Phys. Rev. B* **59**: 4725–4737.
8. Takeuchi, T., H. Amano, and I. Akasaki. 2000. Theoretical study of orientation dependence of piezoelectric effects in wurtzite strained GaInN/GaN heterostructures and quantum wells. *Jpn. J. Appl. Phys.* **39**: 413–416.
9. Park, S.-H. and S. L. Chuang. 2000. Comparison of zinc-blende and wurtzite GaN semiconductors with spontaneous polarization and piezoelectric field effects. *J. Appl. Phys.* **87**: 353–364.
10. Bir, G. L. and G. E. Pikus, 1974. *Symmetry and Strain-Induced Effects in Semiconductors*. New York, NY: Wiley.
11. Park, S.-H., S. L. Chuang, and D. Ahn. 2000. Intraband relaxation time effects on non-Markovian gain with many-body effects and comparison with experiment. *Semicond. Sci. Technol.* **15**: 203–208.
12. Chuang, S. L. 1996. Optical gain of strained wurtzite GaN quantum-well lasers. *IEEE J. Quan. Electron.* **32**: 1791–1800.
13. Ahn, D. 1997. Theory of non-Markovian optical gain in quantum-well lasers. *Prog. Quant. Electron.* **21**: 249–287.
14. Chow, W. W., S. W. Koch, and M. Sergent III. 1994. *Semiconductor-Laser Physics*. Berlin: Springer.
15. Haug, H. and S. W. Koch. 1993. *Quantum Theory of the Optical and Electronic Properties of Semiconductors*. Singapore: World Scientific.
16. Nakamura, S. and G. Fasol. 1997. *The Blue Laser Diode*. Berlin: Springer.
17. Martin, G., A. Botchkarev, A. Rockett, and H. Morkoç. 1996. Valence-band discontinuities of wurtzite GaN, AlN, and InN heterojunctions measured by x-ray photoemission spectroscopy. *Appl. Phys. Lett.* **68**: 2541–2543.

18. Bernardini, F., V. Fiorentini, and D. Vanderbilt. 1997. Spontaneous polarization and piezoelectric constants of III–V nitrides. *Phys. Rev. B.* **56**: 10024–10027.
19. Park, S.-H. and S. L. Chuang. 1998. Piezoelectric effects on electrical and optical properties of wurtzite GaN/AlGaN quantum well lasers. *Appl. Phys. Lett.* **72**: 3103–3105.
20. Ambacher, O., J. Smart, J. R. Shealy, N. G. Weimann, K. Chu, M. Murphy, W. J. Schaff, L. F. Eastman, R. Dimitrov, L. Wittmer, M. Stutzmann, W. Rieger, and J. Hilsenbeck. 1999. Two-dimensional electron gases induced by spontaneous and piezoelectric polarization charges in N- and Ga-face AlGaN/GaN heterostructures. *J. Appl. Phys.* **85**: 3222–3233.
21. Park, S.-H. 2002. Crystal orientation effects on electronic properties of wurtzite InGaN/GaN quantum wells. *J. Appl. Phys.* **91**: 9904–9908.
22. Henderson, R. H. and E. Towe. 1995. Strain and crystallographic orientation effects on interband optical matrix elements and band gaps of [II/]-oriented III–V epilayers. *J. Appl. Phys.* **78**: 2447–2455.
23. Henderson, R. H. and E. Towe. 1996. Effective mass theory for III–V semiconductors on arbitrary (hkl) surfaces. *J. Appl. Phys.* **79**: 2029–2037.
24. Bykhovski, A., B. Gelmont, and S. Shur. 1993. Strain and charge distribution in GaN-AIN-GaN semiconductor–insulator–semiconductor structure for arbitrary growth orientation. *Appl. Phys. Lett.* **63**: 2243–2245.
25. Smith, D. L. and C. Mailhiot. 1998. Piezoelectric effects in strained-layer superlattices. *J. Appl. Phys.* **63**: 2717–2719.
26. Wright, A. F. and J. S. Nelson. 1995. Consistent structural properties for AlN, GaN, and InN. *Phys. Rev. B* **51**: 7866.
27. Rubio, A., J. L. Corkill, M. L. Cohen, E. L. Shirley, and S. G. Louie. 1993. Quasiparticle band structure of AlN and GaN. *Phys. Rev. B.* **48**: 11810.
28. Hellwege, K. H. and O. Madelung (Eds.). 1982. *Physics of Group IV Elements and III–V Compounds*, Landolt-Börnstein, New Series, Group III, Vol. 17, Part A. Berlin: Springer-Verlag.
29. Fan, W. J., M. F. Li, T. C. Chong, and J. B. Xia. 1996. Electronic properties of zinc-blende GaN, AlN, and their alloys $Ga_{1-x}Al_xN$. *J. Appl. Phys.* **79**: 18896.
30. Meney, A. T. and E. P. O'Reilly. 1995. Theory of optical gain in ideal GaN heterostructure lasers. *Appl. Phys. Lett.* **67**: 3013.
31. Kim, K., W. R. L. Lambrecht, and B. Segall. 1996. Elastic constants and related properties of tetrahedrally bonded BN, AlN, GaN, and InN. *Phys. Rev. B* **53**: 16310.
32. Van de Walle, C. G. and J. Neugebauer. 1997. Small valence-band offsets at GaN/InGaN heterojunctions. *Appl. Phys. Lett.* **70**: 2577.
33. Vurgaftman, I. and J. R. Meyer. 2003. Band parameters for nitrogen-containing semiconductors. *J. Appl. Phys.* **94**: 3675–3696.
34. Park, S.-H. 2011. Optical gain characteristics of non-polar Al-rich AlGaN/AlN quantum well structures. *J. Appl. Phys.* **110**: 063105.
35. Park, S.-H., Y.-T. Moon, D.-S. Han, J. S. Park, M.-S. Oh, and D. Ahn. 2011. Light emission enhancement in blue InGaAlN/InGaN quantum well structures. *Appl. Phys. Lett.* **99**: 181101.

12

Nanowires

12.1 Introduction: Semiconductor Nanowires

During the past years, much research effort has been dedicated to explore the properties of semiconductor nanowires (NWs). Correspondingly, remarkable progress has been achieved in design, growth, characterization, and theoretical description of these structures. Many different applications have been suggested, ranging from energy harvesting and storage (Wallentin et al. 2013) to single-photon emitters (Heiss et al. 2013) and general lighting, with some of them poised to enter the commercial market in the near future.

Free-standing semiconductor NWs, sometimes also referred to as *nanorods*, are grown using molecular-beam epitaxy (MBE) or metal-organic chemical vapor deposition (MOCVD) or produced from planar layers by lithography. They have diameters of a few tens to hundreds of nanometers and lengths of a few hundred nanometers to a few micrometers and exhibit some unique properties in comparison to planar semiconductor quantum wells (QWs) or embedded quantum dots (QDs). If the diameter of a NW is small enough such that quantum confinement along the transverse direction is ensured, the NW is called a *quantum wire*. A particular advantage of the NW geometry is that it facilitates a much better strain relaxation, resulting in a better overall material quality. Additionally, the large surface-to-volume ratio of NWs represents an advantage for light emitting and detecting devices.

If heterostructures such as QWs or QDs are incorporated in NWs, the free side facets additionally facilitate elastic relaxation, allowing one to design heterostructures from materials that have a large lattice mismatch with respect to the NW material, which could not be produced in a planar structure. Typical heterostructures in NWs are of either axial or radial character, as depicted in Figure 12.1. The active layer of the heterostructure is then either perpendicular to the growth direction (axial heterostructure) or parallel to it (radial heterostructure). Thickness and material composition of axial and radial layers can be controlled during the growth process, so that the electronic properties of the NW can be tailored to suit specific applications.

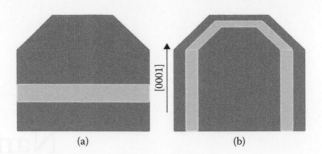

FIGURE 12.1   Cross-section of an axial (a) and a radial (b) heterostructure in a hexagonal nanowire (NW). The active layer is light gray.

Moreover, surface states at the side facets of NWs give rise to additional electrostatic potentials whose impact on the electronic properties needs to be understood. The simulation of semiconductor surfaces is a very challenging field and very few sophisticated approaches exist to model a semiconductor–vacuum interface.

Most III–V semiconductor NWs also show a high density of stacking faults. In some cases, the switching in the crystal structure along the length of an NW can be controlled. This allows one to investigate material properties of semiconductors in crystal structures that do not occur otherwise in the bulk. For example, whereas bulk GaAs exhibits a zinc-blende (ZB) crystal structure, GaAs NWs with the hexagonal wurtzite (WZ) structure have often been reported (Hoang et al. 2009; Heiss et al. 2011; Ahtapodov et al. 2012). Since the electronic properties of ZB and WZ GaAs are different, both phases have to be addressed in a simulation of an NW, with a particular focus on the treatment of their interface.

Finally, the photonic properties of NWs are altered by their peculiar geometry. In particular, the light absorption and extraction efficiencies depend strongly on the polarization of light with respect to the NW axis and it is mandatory to take this effect into account when correlating the degree of polarization of optical transitions with possible valence-band mixing in the NW.

12.2 Basic Charge Confining Mechanisms in NWs

In the following, we provide a brief overview of mechanisms that influence the localization and the energy levels of carriers in NWs. We compare these mechanisms to those that occur in planar layers or QDs and discuss specific features of NWs.

12.2.1 Bulk Electronic Properties: Band Offsets and Work Functions

Similar to planar heterostructures and QDs (see Chapters 11 and 13), the electronic properties of axial or radial semiconductor NW heterostructures are determined by the band offsets between the conduction and valence bands, E_c and E_v, of the materials involved. If we neglect surface states in a first step, the potential barrier arising at the interface between the NW and the surrounding vacuum is best described by the work function of the NW material. For the hole state, the following consideration yields a good description of the confining potential: The minimum energy to create a hole in the valence band is the sum of bandgap (to excite an electron from the valence to the conduction band) and the electron affinity. Therefore, we consider the sum of bandgap and electron affinity as a potential barrier for the valence band at the side facet of the NW. However, as the work function and the electron affinity of most semiconductors is on the order of a few eV and typically much larger than bandgaps and band offsets within an NW heterostructure, a good approximation can already be achieved by assuming an infinite potential at the semiconductor–vacuum interface. This simplification is in particular reasonable for the electron and hole states closest to the conduction and valence-band edges, respectively (Levine 1965).

12.2.2 Approaches to Compute Elastic Properties of Nanostructures

The electronic properties of a semiconductor nanostructure are significantly modified if the system is subject to elastic relaxation. The impact of strain on the electronic properties is commonly taken into account via deformation potentials. For example, within the ZB structure, the lowest conduction band is modified via the deformation potential a_c by the additional contribution $a_c \cdot \text{Tr}(\varepsilon)$. In a WZ system, two deformation potentials a_1 and a_2 are employed such that the conduction band is modified by the term $a_2 \cdot (\varepsilon_{xx} + \varepsilon_{yy}) + a_1 \cdot \varepsilon_{zz}$. The strain-induced modifications of the valence bands are more complex. As an example, the impact of strain on the three highest valence bands can be computed using a six-band Pikus–Bir Hamiltonian (Ghosh et al. 2002). Chapter 1 provides a more detailed discussion on how strain is considered in the simulation of electronic properties of semiconductor nanostructures.

NWs exhibit quite specific strain profiles as the free side facets facilitate elastic relaxation. This feature allows not only the NW itself to relax toward the bulk lattice constants but also a much better elastic relaxation of axial and radial heterostructures incorporated within the wire. The possibility to relax even a large lattice mismatch is a particular strength of NWs in comparison to planar heterostructures, where large differences in the lattice constants lead to defects and crack formation.

To shed some light on the specific elastic properties of semiconductor NWs and their impact on the electronic properties, we discuss different approaches to compute the elastic properties of an NW in the following.

12.2.2.1 Valence Force Field Model

The most complete description of the elastic properties of an NW is achieved using an atomistic model. However, accurate models rely on *ab initio* methods, which are computationally highly expensive thus limiting the simulation domain to a small number of atoms (Yang et al. 2008; Xiang et al. 2008). Given the typical dimensions of semiconductor NWs in the order of tens to hundreds of nanometers, it is clear that such models are not well suited to describe NWs of the commonly observed dimensions.

An alternative approach is the computationally less expensive valence force field (VFF) model. Within the VFF model (Keating 1966), the atomic coordinates r_i of the individual atoms of the NW are the degrees of freedom. For an NW with a ZB crystal structure, the potential energy of the crystal to be minimized can be written as (Keating 1966):

$$E = \sum \sum \frac{3\alpha}{8r_{ij0}^2} \left(r_{ij}^2 - r_{ij0}^2 \right)^2 + \sum \sum \frac{3\beta}{8r_{ij0} \cdot r_{ik0}} \left(r_{ij} \cdot r_{ik} - r_{ij0} \cdot r_{ik0} \right)^2, \qquad (12.1)$$

where the equilibrium bond angles satisfy $r_{ij0} \cdot r_{ik0} = -|r_{ij0}||r_{ik0}|/3$ and N indicates summation over the nearest neighbors. $r_{ij} = r_i - r_j$ is the distance between atoms i and j and r_{ij0} denotes the value of r_{ij} in a bulk, strain-free crystal. The parameters α and β are coupling constants between neighboring atoms.

However, the basic VFF model is inaccurate for hexagonal crystal structures and thus requires improvements (Rücker and Methfessel 1995; Große and Neugebauer 2001), and it is questionable whether this model is an appropriate choice for the description of semiconductor NWs (Singh et al. 2011). Moreover, the computational effort of the VFF scales nearly linearly with the number of atoms involved. For typical NWs, even the VFF becomes computationally highly expensive if the super cell spans over a large segment of the NW.

12.2.2.2 Linear Continuum Elasticity Theory

As a computationally inexpensive alternative to the atomistic VFF model, the elastic properties of an NW can also be computed using a continuum-based approach. Here, the underlying atomistic lattice is ignored and the strain tensor $\varepsilon_{ij}(r)$ is expressed in a continuous manner. The elastic energy to be minimized is then (Landau and Lifshitz 1959) given as follows:

$$E = \frac{1}{2} \int \sum C_{ijkl}(r)\varepsilon_{ij}(r)\varepsilon_{kl}(r)dV, \tag{12.2}$$

where $C_{ijkl}(r)$ denotes the elastic constants of the materials involved and $i, j, k, l = x, y, z$. Note that all information on the underlying crystal structure is contained in the elastic constants. Within the ZB lattice, only the constants $C_{11} = C_{iiii} = C_{xxxx} = C_{yyyy} = C_{zzzz}$, $C_{12} = C_{iijj}$ ($i \neq j$), and $C_{44} = C_{ijij}$ ($i \neq j$) are nonzero. The advantage of the continuum elasticity model is of course its computational efficiency, as the discretization of the super cell is not bound to atomistic coordinates. However, the atomistic nature of the system under consideration is fully neglected.

12.2.3 Elastic Properties of NW Heterostructures

In this section, we discuss strain distributions in axial and radial NW heterostructures, as well as their impact on the electronic properties. We also discuss the specific advantages of axial and radial NW heterostructures in terms of elastic relaxation.

12.2.3.1 Radial NWs Heterostructures

In radial semiconductor NW heterostructures—typically referred to as core–shell NW—a core material is surrounded by one or more shells consisting of different materials or having different doping densities. The elastic properties of core and shells are controlled by their respective thicknesses, which finally allows the control of the electronic properties. In a simplified picture, a core–shell NW can be approximated as a cylinder with a lattice mismatched shell. The strain inside such a cylindrical core–shell NW is best described within cylindrical coordinates and the nonzero components of the strain tensor are (Menendez et al. 2011)

$$\varepsilon_{rr} = \varepsilon_{\theta\theta} = -\varepsilon_0 \left(R_s^2 - R_c^2 \right) \left[\frac{(\nu+1)(1-2\nu)}{(1-\gamma)(1-2\nu)R_c^2 - (1-2\nu+\gamma)R_s^2} + \frac{\nu}{\left(R_s^2 - R_c^2 \right) + \gamma R_c^2} \right]$$

$$\varepsilon_{zz} = \varepsilon_0 \left[\frac{\left(R_s^2 - R_c^2 \right)}{\left(R_s^2 - R_c^2 \right) + \gamma R_c^2} \right] \tag{12.3}$$

in the core and

$$\varepsilon_{rr}(r) = \frac{\varepsilon_0 (\nu+1) \gamma R_c^2}{(1-\gamma)(1-2\nu)R_c^2 - (1-2\nu+\gamma)R_s^2} \left[(1-2\nu) - \frac{R_s^2}{r^2} \right] + \frac{\nu\gamma R_c^2 \varepsilon_0}{\left(R_s^2 - R_c^2 \right) + \gamma R_c^2}$$

$$\varepsilon_{\theta\theta}(r) = \frac{\varepsilon_0 (\nu+1) \gamma R_c^2}{(1-\gamma)(1-2\nu)R_c^2 - (1-2\nu+\gamma)R_s^2} \left[(1-2\nu) + \frac{R_s^2}{r^2} \right] + \frac{\nu\gamma R_c^2 \varepsilon_0}{\left(R_s^2 - R_c^2 \right) + \gamma R_c^2}$$

$$\varepsilon_{zz} = \frac{\gamma R_c^2 \varepsilon_0}{\left(R_s^2 - R_c^2 \right) + \gamma R_c^2} \tag{12.4}$$

in the shell, where r is the radial position. $\varepsilon_0 = \left(a_s - a_c \right) / a_c$ is the initial misfit for core and shell materials with lattice constants a_c and a_s, respectively. R_c and R_s denote core and shell radii. For the Poisson ratio ν, we use the Voigt average (Hirth and Lothe 1968) and consider this ratio as being constant inside both the core and the shell. Within a cubic crystal, this ratio reads

$\nu^{c,s} = C_{11}^{c,s} + 4C_{12}^{c,s} - 2C_{44}^{c,s}/4C_{11}^{c,s} + 6C_{12}^{c,s} + 2C_{44}^{c,s}$ where the $C_{ij}^{c,s}$ are the elastic constants of the core and the shell.

$\gamma = E_{core}/E_{shell}$ with E denoting Young's modulus. In the case of an NW grown along the [111] axis, this modulus can be determined from the stress–strain relation as (Hirth and Lothe 1968)

$$E_{111} = 3\frac{C_{44}\left(C_{11} + 2C_{12}\right)}{C_{44} + C_{11} + 2C_{12}}.$$

All components of the strain tensor are constants inside the core and depend only on the elastic constants of the materials involved as well as on the ratio between core and shell radii. Within the shell, ε_{zz} is constant. In contrast, $\varepsilon_{rr}(r)$ and $\varepsilon_{\theta\theta}(r)$ are functions of the radial distance from the NW core, and converge for large radii, as shown in Figure 12.2 for a Ge–Si core–shell NW with a core radius of 10 nm and a shell radius of 20 nm. The ratio between the radii of the core and the shell determines the strain state of the NW and is a property that can be easily controlled during the growth process.

For more realistic NW geometries that cannot be described using analytical models, the elastic properties can be computed only numerically using continuum-based or atomistic approaches. A comparison between the analytic solution for the cylindrical NW and a numerical solution using the finite-elements method for hexagonal, [011]-grown Ge–Si NWs can be found in Singh et al. (2011), where the authors report a very good agreement with the analytic model for the ε_{zz} component.

Discrepancies are observed in ε_{rr} and $\varepsilon_{\theta\theta}$ inside the shell toward the corners of the realistic, hexagonal NW, whereas a very reasonable agreement is reported in the vicinity of the core. Note, however, that ε_{rr} and $\varepsilon_{\theta\theta}$ are not uniform within the hexagonal core anymore, as derived analytically for a cylindrical NW. The agreement in ε_{zz} is very good with uniform values in both core and shell (Singh et al. 2011).

A comparative study between a continuum elasticity model and the atomistic VFF model is presented by Grönquist et al. (2009) for GaAs/GaP core–shell NWs with a total diameter of 27.8 nm. The agreement between VFF and continuum elasticity model is very good throughout the NW cross-section and deviations occur only at the free ends of the finite wires. For typical semiconductor NWs, elasticity theory within a continuum picture can thus be considered as a reasonable and computationally inexpensive model.

The elastic relaxation within the model NW has a significant influence on its electronic properties. $\mathrm{Tr}(\varepsilon) > 0$ within the shell leads to a reduction of both the conduction and the valence-band energies. Within the core, the hydrostatic strain is negative and the energies of the conduction band minimum and valence band maximum increase. The impact of strain on the conduction band is larger, such that the bandgap of the core material (GaAs) increases and a red shift of luminescence from the core is predicted. As the magnitude of the strain depends on the thicknesses of the core and the shell, bandgap engineering can be applied in core–shell NWs by a variation of the core and shell thicknesses. An example for such bandgap engineering is presented in Liu et al. (2014), where the bandgap of an InAs shell around an InP core is tuned by up to 210 meV solely by a variation of the core diameter.

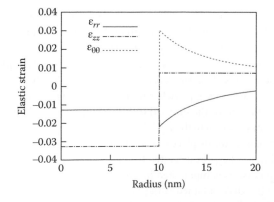

FIGURE 12.2 Nonzero strain components of a Ge–Si core–shell nanowire (NW) in cylindrical coordinates. The radius of the Ge core is 10 nm.

12.2.3.2 Axial NW Heterostructures

Axial NW heterostructures consist, similar to planar heterostructures, of thin layers separated by the matrix material (barriers). In contrast to their planar counterparts, the elastic strain caused by the mismatch between the insertion and the matrix crystal lattices relaxes at the side facets. This relaxation only partially eliminates the strain in the wells and gives rise to a complicated inhomogeneous strain distribution that depends on the relation between the NW diameter, the thickness of the insertion, the barrier thickness, and also the thickness of the whole stack of the insertions. The elastic relaxation facilitated by the free side facets is of particular importance for $In_xGa_{1-x}N$ disks in GaN NWs, as InN and GaN exhibit a lattice mismatch of approximately 10%, making it difficult to produce planar $In_xGa_{1-x}N/GaN$ heterostructures with large In content and high crystal quality.

The difference in lattice parameters between the matrix and the inserted layer gives rise to the strain $\varepsilon_{zz}^0 = (c - c_0)/c_0$, $\varepsilon_{rr}^0 = \varepsilon_{\theta\theta}^0 = (a - a_0)/a_0$, which itself does not cause stress and is called the intrinsic strain (the terms self-strain and eigenstrain are also used). a (a_0) and c (c_0) denote the layer (matrix) lattice constants. The elastic strain ε and the stress σ result from the coherency of the crystal lattices of the layer and the matrix. In a planar heterostructure, the layer is clamped by the matrix. Therefore, the components of the total strain in the plane $\varepsilon_{rr}^T = \varepsilon_{rr}^0 + \varepsilon_{rr}$ and $\varepsilon_{\theta\theta}^T = \varepsilon_{\theta\theta}^0 + \varepsilon_{\theta\theta}$ are equal to zero. The whole structure can freely expand in the direction normal to the layer, so that the stress $\sigma_{zz} = 0$. Using Hooke's law, these conditions lead to the stress in the layer plane $\sigma_{rr} = \sigma_{\theta\theta} = -\sigma^0$, where $\sigma^0 = (C_{11} + C_{12} - 2C_{13}^2/C_{33})\varepsilon_{rr}^0$. In NWs, this stress is relaxed at the free side facets. The solution of the above problem for the NW is, therefore, the sum of two terms: The solution in a planar structure described earlier and the image field due to stress $\sigma_{rr}^{im} = \sigma^0$ applied at the interface of the layer to satisfy the boundary conditions $\sigma_{rr} = 0$, $\sigma_{rz} = 0$ at the side facets. The elastic problem can be solved analytically for a circular cylinder shape of the NW and an arbitrary intrinsic strain distribution $\varepsilon^0(z)$ (Kaganer and Belov 2012). In this case, the image displacements are written as

$$u_r^{im}(x, \varsigma) = -\frac{R\sigma^0}{c_{44}} \int_0^\infty \left[\frac{1}{1 + k_1} \frac{I_1(qx/v_1)}{I_1(q/v_1)} - \frac{1}{1 + k_2} \frac{I_1(qx/v_2)}{I_1(q/v_2)} \right] \frac{g(q)}{D(q)} \cos q\varsigma dq,$$

$$u_z^{im}(x, \varsigma) = -\frac{R\sigma^0}{c_{44}} \int_0^\infty \left[\frac{k_1 v_1}{1 + k_1} \frac{I_0(qx/v_1)}{I_1(q/v_1)} - \frac{k_2 v_2}{1 + k_2} \frac{I_0(qx/v_2)}{I_1(q/v_2)} \right] \frac{g(q)}{D(q)} \sin q\varsigma dq,$$

$$D(q) = \frac{c_{11} - c_{12}}{c_{44}} \frac{k_2 - k_1}{(1 + k_1)(1 + k_2)} - q \left(v_1 \frac{I_0(q/v_1)}{I_1(q/v_1)} - v_2 \frac{I_0(q/v_2)}{I_1(q/v_2)} \right). \tag{12.5}$$

I_0 and I_1 are modified Bessel functions, $\varsigma = z/R$, and $x = r/R$ with R being the radius of the cylinder. The constants k_1, k_2, v_1, v_2 are defined by the elastic moduli of the material, which are $k_1 = 0.234$, $k_2 = 4.264$, $v_1 = 1.589$, $v_2 = 0.623$ for GaN. The function $g(z)$ describes the strain distribution and $g(q)$ its Fourier transform. For a homogeneous disk of thickness $2d$, it is given by $g(q) = 2sin(qd/R)/(\pi q)$.

Figure 12.3 illustrates the strain state in the disk center of a cylindrical NW with an inserted disk. In the limit of large diameters, the strain state approaches to the one of a planar structure. When the diameter is much smaller than the disk thickness, the inserted disk is strain free. As shown in Figure 12.3, the strain changes sign when going from very thin to thick NWs since both the insertion and the matrix are distorted. The insert in Figure 12.3 visualizes the displacements calculated by Equation 12.5 for the cylinder diameter corresponding to the minimum of ε_{zz} (exaggerated for better visibility).

The strain in the inserted disk is compensated by an opposite strain in the matrix. The displacements well above and well below the disk obtained using Equation 12.5 show that the total displacement due to transverse intrinsic strain is zero. In other words, the strain averaged over the whole cylinder is zero. This result has important consequences for multi-QW NW heterostructures since each QW experiences

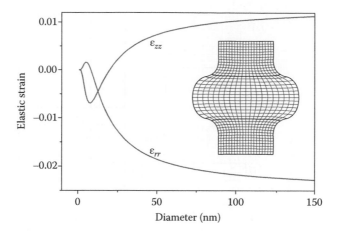

FIGURE 12.3 Elastic strain at the center of a 5-nm-thick $In_{0.25}Ga_{0.75}N$ disk inserted in a GaN cylinder as a function of the diameter of the cylinder. The inset shows the displacements (magnified for better visibility) for a cylinder diameter of 8 nm corresponding to the minimum of ε_{zz}.

strain produced by the other QWs. Variation of the barrier thickness and the NW diameter are therefore additional degrees of freedom for strain engineering of NW heterostructures (Wölz et al. 2013).

Figure 12.4 shows three strain states of lattice-mismatched heterostructures. In the case of a planar heterostructure (cf. Figure 12.4a), the Poisson effect gives rise to a homogeneous strain in the layer. In a multilayer NW heterostructure, the lateral elastic relaxation occurs in both materials. Figure 12.4b shows the displacement (center) and the out-of-plane strain (right) that were obtained for an axial superlattice using Equation 12.5. The strain is nonuniformly distributed. In the center of the cylinder, shown by the red line, the barrier crystal is laterally expanded. The QW is still under lateral compression, but to a smaller extent than in the planar case of Figure 12.4a. Going outwards, the strain distribution remains qualitatively similar to the one in the center up to about 0.8R, shown by the green line. Toward the surface, the strain profile is more complex as shown by the blue line at 0.95R.

This complicated strain state can be approximated by the strain state shown in Figure 12.4c, where it is assumed that the in-plane lattice spacing is constant everywhere in the superlattice. The strain can be determined by the condition that the average strain on the side surface is zero (Wölz et al. 2011, 2012). This relaxation reduces the strain in the QW by the factor $d_{barrier}/(d_{barrier} + d_{well})$ compared to the case of a planar heterostructure in Figure 12.4a. This lateral relaxation approximation is a much simpler description of the dominant strain states in the superlattice compared to the full consideration of the inhomogeneous strain resulting in a configuration as shown in Figure 12.4b. A systematic comparison of the full calculation based on Equation 12.5 with this approximation allows one to establish criteria for its validity (Wölz et al. 2013). First, the height of the whole stack of QWs should be larger than the NW diameter. Second, to achieve a large strain in QWs, the NWs should be sufficiently thick.

For more complicated geometries or material compositions, the strain can be determined by finite element calculations. For example, Figure 12.5 shows the strain distribution in an axial $In_xGa_{1-x}N/GaN$ NW heterostructure containing two $In_xGa_{1-x}N$ insertions (Knelangen et al. 2011). Each insertion consists of a 2-nm-thick $In_xGa_{1-x}N$ layer with an In content of 19% and a 9-nm-thick layer with an In content of 39%. They are embedded in a GaN NW with a diameter of 30 nm and separated from the side facets by a 7-nm-thick GaN shell. The distance between the two disks is 3 nm. The magnitude of the strain increases toward the side facets, whereas elastic relaxation occurs near the center of the NW. Nevertheless, significant strain persists at the side facets of the NW.

A study on single $In_xGa_{1-x}N$ disks in a GaN NW that extend throughout the whole NW section without a GaN shell also reports persisting strains (Böcklin et al. 2010). An $In_{0.4}Ga_{0.6}N$ layer with a thickness

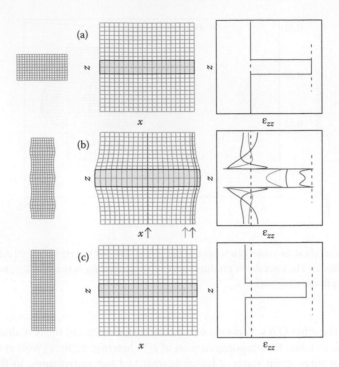

FIGURE 12.4 Strain states of lattice-mismatched heterostructures: sketch of the entire structure (left), atomic displacement (center), and out-of-plane strain ε_{zz} (right) near one quantum well (QW). (a) Planar pseudomorphic epitaxial QW growth, the substrate remains strain free. (b) Axial nanowire (NW) superlattice, nonuniform strain profile calculated with a cylindrical model. The red, green, and blue curves represent the displacement and strain at the radial positions indicated by the arrows (NW center, 80% and 95% of NW radius). (c) Axial NW superlattice, lateral relaxation estimate, QW and barrier assume a common average in-plane lattice parameter. (Reprinted with permission from Wölz et al. (2013). Copyright [2013] **American Chemical Society**.)

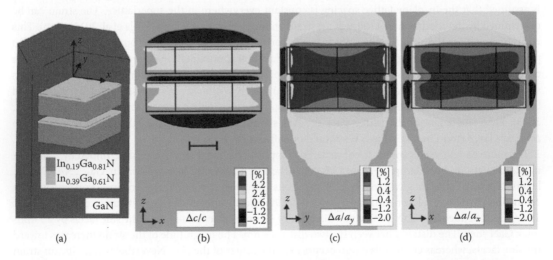

FIGURE 12.5 (a) Perspective view of the GaN nanowire (NW) with two $In_xGa_{1-x}N$ disks. The diagonal components of the strain tensor, ε_{zz}, ε_{yy}, and ε_{xx}, are shown in (b), (c), and (d), respectively. (Reproduced with permission from Knelangen, M. et al., *Nanotechnology*, 2011. © IOP Publishing. All rights reserved.)

of 3 nm within a hexagonal GaN NW with a diameter of 50 nm is highly strained due to the large In content and experiences a significant lateral extension. In the center of the layer, the magnitude of the strain approaches the value predicted in the biaxial limit. As a consequence, the bandgap is significantly lowered at the NW surface, inducing a localization of electrons and holes at the corners and at the side facets, respectively.

12.2.4 Piezoelectric and Spontaneous Polarization

Both GaN and InN exhibit pyroelectric and piezoelectric polarization. Planar heterostructures are commonly grown along the polar [0001] axis, resulting in large internal electrostatic fields in the $In_xGa_{1-x}N$ QWs. These fields lead to a spatial separation of electrons and holes and thus to a considerable reduction of the radiative recombination rate. The use of axial GaN NW heterostructures promises to resolve some of these problems (Kikuchi et al. 2004; Kim et al. 2004; Li and Waag 2012). As discussed in the previous section, the free side facets facilitate the elastic relaxation of strain in the $In_xGa_{1-x}N$ disk, so that higher In contents can be achieved (Björk et al. 2002; Ertekin et al. 2005; Kaganer and Belov 2012). Along with the elastic relaxation, a significant reduction of the built-in piezoelectric potential and hence the quantum confined Stark effect (QCSE) (Miller et al. 1984) is expected. In fact, some researchers report evidence of a vanishing QCSE (Lin et al. 2010; Armitage and Tsubaki 2010; Nguyen et al. 2011; Bardoux et al. 2009), whereas others observe a significant QCSE (Wölz et al. 2013). Typically, the polarization P is computed directly using the strain tensor. For WZ crystals, it reads (Nye 1985):

$$P(r) = \begin{pmatrix} 2e_{15}\varepsilon_{13} \\ 2e_{15}\varepsilon_{23} \\ e_{31}\left(\varepsilon_{11} + \varepsilon_{22}\right) + e_{33}\varepsilon_{33} + P_{sp} \end{pmatrix}, \tag{12.6}$$

where $e_{ij} = e_{ij}(r)$ are the piezoelectric constants and $P_{sp}(r)$ is the spontaneous polarization is the spontaneous polarization: WZ crystals exhibit a charge separation even in the absence of an electric field. The polarization potential $V(r)$ is computed solving the Poisson equation.

$$\rho_p(r) = \varepsilon_0 \nabla \cdot \left[\varepsilon_r(r)\nabla V(r)\right] \quad \text{with} \quad \rho_p(r) = -\nabla \cdot P(r). \tag{12.7}$$

ε_0 and ε_r denote the vacuum and the relative dielectric constants, respectively.

The resulting electrostatic potential $V(r)$ is used in the model to compute the electronic properties as a potential contribution added to both the conduction and the valence bands.

For the system considered by Böcklin et al. (2010), accounting for the polarization potential leads to a confinement of the hole ground state at the center of the NW, whereas the electron remains at the edges. Depending on size and material composition of NW and disk, both strain and the polarization potential can therefore dominate the localization of carriers in a NW.

For an $In_xGa_{1-x}N$ disk in a circular cylinder GaN NW, the Poisson equation can be solved using Equation 12.5 and the electrostatic potential is expressed in similar integrals (Kaganer et al. 2016). Figure 12.6 displays the electrostatic potential V along the axis of the cylinder ($r = 0$) for an $In_{0.25}Ga_{0.75}N$ disk of thickness 5 nm inserted into GaN cylinders of different diameters. The potential for a film (i.e., a cylinder of an infinite diameter) is included for comparison. Already for very thick NWs with a diameter of 150 nm, the electrostatic potential of the disk along the central NW axis is visibly reduced and this reduction increases for decreasing diameters. However, even for a very thin NW with a diameter of 15 nm, a potential difference of approximately 300 meV persists.

Although the polarization potential is significantly reduced compared to the one of planar heterostructures of similar growth direction, thickness, and In content, a very small NW diameter ($d < 30$ nm) is required to achieve a substantial reduction of the piezoelectric potential for the typical disk thicknesses of only a few nanometers. Even in such systems, a nonvanishing built-in potential remains. The magnitude of

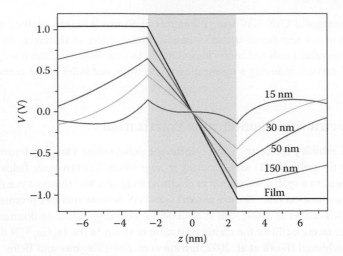

FIGURE 12.6 The electrostatic potential V along the axis of the cylinder ($r = 0$) for a 5-nm-thick $In_{0.25}Ga_{0.75}N$ disk inserted into GaN cylinders of different diameters, as indicated in the figure. The interior of the disk is highlighted. (Reproduced with permission from Kaganer, V. M. et al., *Nanotechnology*, 2016. ©IOP Publishing. All rights reserved.)

the polarization potential decreases with larger aspect ratio and the extrema of the potential shift from the central axis of the NW toward the side facets (Kaganer et al. 2016). The polarization potential therefore not only affects the transition energies but also the confinement of electrons and holes and thus the recombination rates. Strain relaxation in axial NW heterostructures represents an additional degree of freedom to control the emission wavelength and the electron–hole overlap in comparison to their planar counterparts. However, electrostatic potentials remain and exert an important influence on the electronic properties of the system.

12.3 Doping and the Influence of Surfaces

The electrical conductivity of semiconductor materials can be tuned over many orders of magnitude by the intentional incorporation of impurities (Schubert 2005). The ability to dope semiconductors in a controlled fashion is a key feature of all semiconductor-based technology (Sze 1985; Seeger 1991). For a theoretical description, doping poses a formidable problem, as dopants occupy random positions in the host lattice. Established solid-state simulation methods exploiting the periodicity of the system cannot be employed (Shklovskii and Efros 1984). For device simulations, the discrete nature of the dopant positions is therefore commonly ignored altogether, and a homogeneous, continuous distribution of charge is assumed (Keyes 1975).

In semiconductor NWs, intentional or unintentional doping is of particular importance due to the vicinity of free surfaces. In the case of an $In_xGa_{1-x}N$ based NW, the residual doping concentration of O and Si atoms is $\rho_D = 10^{16} - 10^{17}$ cm^{-3}. These donors transfer their extra electrons to surface states, leaving the donor ionized and the NW depleted at all temperatures, as long as the radius of the NW does not exceed the typical thickness of the depletion region in the bulk material. In the following, we discuss concepts to simulate the impact of doping on the electronic structure of semiconductor NWs.

12.3.1 Continuous, Homogeneous Doping-Related Background Charge

In a simple picture, the potential arising from ionized donors is treated as an electrostatic potential obtained from a homogeneous background charge density by solving the Poisson equation. In the case of a semiconductor NW with doping density ρ_D and with a cylindrical shape, the surface potential V arising from a homogeneous background charge density is

$$V(r) = E_F - \frac{e \cdot \rho_D}{4\varepsilon_0 \cdot \varepsilon_r} R^2 \left[1 - \left(\frac{r}{R} \right)^2 \right], \tag{12.8}$$

with R being the radius of the NW and E_F being the surface Fermi level pinning. For a cylindrical GaN NW with $R = 40$ nm and a doping density of 10^{17} cm^{-3}, the potential difference between the center and the side facets of the NW is 80 mV. If the residual doping consists of Si and O, which is typical for GaN, the potential is positive at the side facets, that is, attractive for holes.

Figure 12.7 shows the electron (red) and hole (blue) ground-state-charge density in an axial In$_x$Ga$_{1-x}$N/GaN NW heterostructure for various values of x and accounting for the influence of a surface potential, computed using an eight-band **k·p** model (see Chapter 1 for details). It can be seen that the surface potential gives rise to a behavior specific to NWs. As long as the thickness of the In$_x$Ga$_{1-x}$N disk is small compared to the NW diameter, the strain-induced piezoelectric potential exhibits extrema at the top and bottom surface of the In$_x$Ga$_{1-x}$N disk along the central axis of the NW (see discussion in Section 12.2.4). For the hole state, this potential is attractive at the bottom center of the disk. A surface potential arising from an unintentional background doping, on the contrary, localizes hole states at the side facets of the NW. The localization of the hole state therefore depends on the magnitude of surface and polarization potentials: For large In contents, the polarization potential increases and induces a localization of holes along the central axis of the NW. For small In contents, the localization of the hole state is dominated by the surface potential and the hole is confined near the side facets of the NW (Marquardt et al. 2013).

12.3.2 Random Dopant Fluctuations: The Discrete Nature of Dopants in an NW

In a typical axial In$_x$Ga$_{1-x}$N/GaN NW heterostructure, the number of dopants is commonly not large enough to justify the picture of a homogeneous doping-related background charge (Corfdir et al. 2014). With a diameter of 80 nm, a doping concentration of 10^{17} cm^{-3} corresponds on average to only 8.3 charges within a segment of 20 nm length. Therefore, individual, randomly distributed dopants have to be considered in NWs. Of course, the ideal approach to deal with influences of individual atoms is an atomistic picture, such as empirical tight binding or pseudopotential methods or density functional theory (for more details on these models see Chapter 1). At the same time, experimentally relevant semiconductor

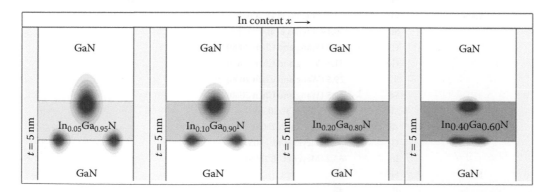

FIGURE 12.7 Side view of the electron (red) and hole (blue) ground-state charge density in an axial In$_x$Ga$_{1-x}$N/GaN nanowire (NW) heterostructure for $x = 0.05, 0.1, 0.2$, and 0.4 (left to right). The NW diameter and disk thickness are not to scale. (Reprinted with permission from Marquardt et al. (2013). Copyright 2013 American Chemical Society.)

NWs exhibit typical diameters of a few tens to hundreds of nanometers. Therefore, a reasonable simulation domain commonly contains millions of atoms. Random dopant fluctuations (RDFs) induce variations of the electronic potential landscape between individual NWs. A meaningful study of the impact of RDFs on the electronic structure of NWs will thus require a statistical evaluation of these variations so that a correspondingly large number of simulations needs to be performed. Within an atomistic picture, this represents a massive computational effort. However, individual dopants can also be included within the computationally inexpensive continuum approaches, for example, effective mass or multiband $\mathbf{k} \cdot \mathbf{p}$ models (see Chapter 1 for more details) under certain conditions.

If an impurity can be considered a hydrogenic dopant, its main influence on the electronic structure is a long-range Coulomb potential. These shallow donors and acceptors induce only a small deformation to the host lattice and require only small energies (less than 3 $k_B T$ ~75 meV at room temperature) to be ionized. Some examples for shallow donors and acceptors in different host materials are listed in Table 12.1.

A shallow impurity can be modeled as a point charge. The respective Coulomb potential obtained from the Poisson equation is then simply added to the potential landscape within the simulation cell. Note that a shallow impurity in a binary material is not necessarily a shallow impurity in a ternary system. For example, O is a shallow donor in GaN but its behavior in $Al_xGa_{1-x}N$ alloys is still a matter of controversy. Some studies report a shallow–deep transition (Park and Chadi 1997; McCluskey et al. 1998), whereas others claim O to remain a shallow donor in $Al_xGa_{1-x}N$ even for large Al contents (Kakanakova-Georgieva et al. 2013). For deep dopants, the model of a simple point charge for an ionized dopant becomes inaccurate.

Figure 12.8a shows an example of a Coulomb potential resulting from RDFs in an axial $In_xGa_{1-x}N$/GaN NW heterostructure. The respective electron (red) and hole (blue) ground-state charge densities are shown in the center of Figure 12.8, computed using an eight-band $\mathbf{k} \cdot \mathbf{p}$ model (Chuang and Chang 1996). To investigate the variations of electron and hole wave functions as well as binding energies in an ensemble of otherwise identical NWs, one can perform a statistical study by employing a larger number of random donor distributions. The electron–hole ground-state recombination energy resulting from such a study is shown in Figure 12.8c. Additionally, the energy values obtained by assuming a continuous, homogeneous

TABLE 12.1   Activation Energies of Shallow Impurities in Different Host Materials

Host Material	Donor	$E_c - E_d$ (meV)	Acceptor	$E_a - E_v$ (meV)
GaAs	S	6	Mg	28
	Se	6	Zn	31
	Te	30	Cd	35
	Si	58	Si	26
GaN	O	33.21 (Freitas et al. 2003)		
	Si	30.19 (Freitas et al. 2003)		
	C	34.0 (Wang and Chen 2000)		
	Ge	31.1 (Wang and Chen 2000)		
	S	29.5 (Wang and Chen 2000)		
	Se	29.5 (Wang and Chen 2000)		
AlN	Si	60 (Zeisel et al. 2000)		
ZnO	Al	53 (Meyer et al. 2005)		
	Ga	54.5 (Meyer et al. 2005)		
	In	63.2 (Meyer et al. 2005)		
Si	Sb	39	B	45
	P	45	Al	67
	As	54	Ga	73

Source: If not indicated differently, taken from Whitaker (2005). Freitas, J. A. et al., *Phys. Stat. Solidi B,* 240, 330, 2003; Wangm, H. and A.-B. Chen, *J. Appl. Phys.,* 87, 7859, 2000; Zeisel, R. et al. *Phys. Rev. B,* 61, R16283, 2000; Meyer, B. K. et al., *Semicond. Sci. Technol.,* 20, S62, 2005.

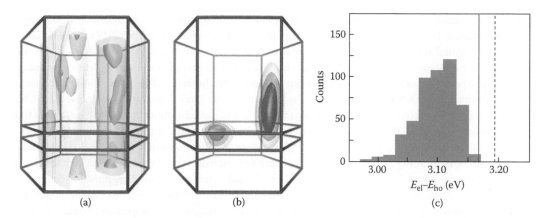

FIGURE 12.8 (a) Coulomb potential isosurfaces for a random distribution of nine individual dopants in an axial $In_xGa_{1-x}N$/GaN nanowire (NW) heterostructure. The active layer is marked in the lower part of the NW. (b) Electron (red) and hole (blue) ground-state charge density. (c) Histogram of the electron–hole ground-state recombination energy for an ensemble of 500 different random dopant distributions with an average of 8.3 charges in the vicinity of the active layer. For the dimensions of the NW segment under consideration, this number corresponds to a doping density of 10^{17} cm^{-3}.

doping-related background charge (solid red line) as well as the one of an ideal, undoped NW (dashed black line) are shown. It is seen that both of these energies are larger than the ones obtained assuming RDFs, which arises from the fact that RDFs induce much deeper local potentials than the surface potential arising from a homogeneous doping-related background charge. The ground-state recombination energy of the undoped NW represents an upper limit for the other two models (Marquardt et al. 2015).

In contrast to shallow impurities, an atomistic treatment as described in Chapter 1 is necessary for deep donors or acceptors, that is, impurities that exhibit a large chemical shift or induce a large deformation to the host lattice.

12.3.3 Dielectrically Enhanced Ionization Energies of Dopants

In the previous sections, we have neglected the mismatch in dielectric constants at the surface of the NWs. Still, the discontinuity of the dielectric constant at the surface of NWs strongly affects the properties of dopants in these nanostructures. In particular, it has been demonstrated that the Coulomb interaction is enhanced in nanostructures surrounded by a barrier material with a lower dielectric constant (Keldysh 1979). This so-called dielectric confinement is a consequence of the electrostatic potential that is set up by the presence of a charged particle in a nanostructure and that can be calculated using the image charge method (Kumagai and Takagahara 1989).

Whereas the mismatch in dielectric constants at the interface of heterostructures such as GaN (Al,Ga)N or GaAs/(Al,Ga)As is small and induces only minor changes in the binding energy of excitons and in the ionization energy of dopants (Andreani and Pasquarello 1990), the mismatch is maximum for NWs in air or vacuum, and the increase in the dopant ionization energies due to the dielectric confinement is considerable (Shik 1993; Diarra et al. 2007). Although quantum confinement leads as well to increased ionization energies for impurities in NWs with a radius smaller than the Bohr radius of donors in the bulk, a_B (Bryant 1984), the dielectric confinement in semiconductor NWs in vacuum sets in already for NW radii smaller than $5a_B$ (Diarra et al. 2007). This finding has been verified experimentally by Björk et al. (2009) who measured a significant increase in resistivity in *n*-doped Si NWs with a diameter smaller than 15 nm.

The impact of image charges on donors or excitons in low dimensional systems has been studied using envelope function calculations (Kumagai and Takagahara 1989), tight binding methods (Allan et al. 1995; Diarra et al. 2007), or density functional calculations (Chan et al. 2008). All of these simulations predict an increase in ionization energy for donors located at the center of a nanostructure as a result of the dielectric confinement (Shik 1993; Diarra et al. 2007; Chan et al. 2008; Corfdir and Lefebvre 2012a). The situation is more complicated for donors located off the axis of NWs. Early effective-mass simulations neglecting the dielectric confinement have predicted that the infinite potential barrier at the surface leads to a change in the symmetry of the wave function of the donor ground-state electron and to a monotonic decrease in donor ionization energy when getting closer to the surface (Levine 1965; Satpathy 1983). In particular, the binding energy of an electron to a donor located at the surface of a semi-infinite layer is one quarter of that of a bulk donor (Levine 1965). However, accounting for the combination of the surface potential barrier with the image charges, one obtains that the ionization energy of a donor in a thin slab bounded by vacuum exhibits a nonmonotonic dependence on the donor site: It first increases and then decreases when the donor moves from the center to the surface of the slab (Corfdir and Lefebvre 2012a). This position-dependent binding energy of electrons to donors has been used to explain the inhomogeneous broadening of the photoluminescence lines related to donor bound excitons in unintentionally doped and strain free GaN NWs (Corfdir et al. 2014). However, using scanning tunneling microscopy, Wijnheimer et al. measured a gradual increase in the ionization of Si over the last 1.2 nm below the (110) surface of a thick GaAs layer (Wijnheijmer et al. 2009). This increase, not reproduced by the effective mass theory, was tentatively explained by the authors in terms of broken symmetry at the surface.

The increase in dopant ionization energies in thin NWs due to the dielectric confinement leads to a decrease in conductivity, which may hinder the realization of efficient electronic and optoelectronic devices based on these structures (Björk et al. 2009). In addition, theoretical and experimental studies have shown that the deactivation of dopants is not homogeneous across the NW section (Wijnheijmer et al. 2009). This finding, together with the nonuniform distribution of dopants in NWs discussed above and the modification of the dielectric environment by electrical contacts or gates (Diarra et al. 2008), renders the simulation of NW-based devices challenging.

12.4 Crystal-Phase Bandgap Engineering

An additional peculiarity of III–V NWs with respect to their planar counterpart is their pronounced polytypism, that is, the crystal structure along the NW length alternates between WZ and ZB. Although III-nitride NWs usually grow within the WZ structure with only a few stacking faults, III-arsenides and III-phosphides NWs show frequent alternation between ZB and WZ segments of various lengths along their axis (Figure 12.9a; Algra et al. 2008; Bao et al. 2008; Caroff et al. 2009; Jacopin et al. 2011; Heiss et al. 2011; Graham et al. 2013). This polytypism is usually regarded as detrimental for the electronic properties of NWs: Alternation between WZ and ZB crystal structures efficiently scatters electrons (Konar et al. 2011) and leads to a significant increase in the resistivity of NWs (Thelander et al. 2011). In addition, the presence of stacking faults along the length of NWs leads to localization of electrons and holes, increasing the ionization energies of dopants in these nanostructures (Corfdir et al. 2009).

Although considerable effort has been devoted to the elimination of stacking defects in NWs, some degree of control on the switching between WZ and ZB during the NW growth has been demonstrated (Algra et al. 2008; Caroff et al. 2009). This control opens the possibility of realizing crystal-phase bandgap engineering in NWs, where carrier confinement is obtained by alternation of the crystal structure and not by the composition (Algra et al. 2008; Caroff et al. 2009; Akopian et al. 2010; Corfdir et al. 2013).

Such crystal-phase quantum structures exhibit atomically flat interfaces (Bolinsson et al. 2011) and are free of alloy disorder. Thanks to the high quality of their interfaces, crystal-phase quantum structures exhibit a sub-meV photoluminescence line width at cryogenic temperatures (Figure 12.9b; Akopian et al. 2010; Jacopin et al. 2011; Graham et al. 2013) and they can be of interest for quantum optics applications (Akopian et al. 2010; Castelletto et al. 2014).

The band alignment between the WZ and the ZB phases of a III–V semiconductor is usually considered to be of type II, with the electron and the hole localized in the ZB and WZ phases, respectively

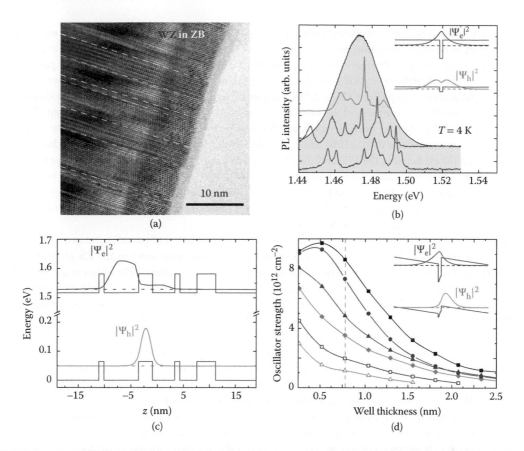

FIGURE 12.9 (a) High-resolution transmission electron micrograph of a polytype GaAs nanowire (NW). Wurtzite (WZ) segments are highlighted in red. (Adapted with permission from Corfdir et al. (2013). Copyright 2013 American Chemical Society.) (b) Photoluminescence spectra at 4 K from single NWs (red, blue, green) and from an ensemble (gray) of polytype GaAs NWs. The signal between 1.46 and 1.50 eV arises from recombination at crystal-phase quantum disks. The inset depicts schematically the band profile for a zinc-blende (ZB) inclusion in WZ GaAs (blue). The energy and the wave function of the electron (red) and the hole (green) are shown with solid and dashed lines, respectively. (Adapted with permission from Graham et al. (2013). Copyright 2013 by the American Physical Society.) (c) Band profiles and electron and hole energy and wave function for the NW segment shown in (a). (Adapted with permission from Corfdir et al. (2013). Copyright 2013 American Chemical Society.) (d) Oscillator strength for an exciton in a QW of ZB GaN in WZ GaN as a function of the well thickness and in the presence of built-in electric fields. The lines from top to bottom correspond to built-in electric fields of 0, 0.1, 0.5, 1, 2, and 3 MV/cm, respectively. The built-in electric fields along the confinement axis lead to a spatial separation of the electron and hole wave functions (inset). (Adapted with permission from Corfdir and Lefebvre (2012a). Copyright 2013, American Institute of Physics.)

(Figure 12.9b; Murayama and Nakayama 1994). *Ab initio* calculations have predicted small band offsets and small bandgap differences between the WZ and the ZB phases (Bechstedt and Belabbes 2013). Therefore, although long radiative lifetimes have been calculated for thick crystal-phase QWs in NWs (Zhang et al. 2010), the electron and hole wave functions for thin ZB insertions in the WZ show significant spreading into the WZ and the ZB, respectively. Their squared overlap is large (Corfdir and Lefebvre 2012b; Corfdir et al. 2013) and thin crystal-phase QWs and QDs in NWs act as efficient light emitters, despite their type-II band alignment.

Effective-mass calculations have been used to calculate the photoluminescence energy associated to single crystal-phase QWs and to polytype semiconductor NW segments (Bao et al. 2008; Jacopin et al. 2011; Heiss et al. 2011; Corfdir and Lefebvre 2012b; Lähnemann et al. 2012; Corfdir et al. 2013). For instance, such calculations have been combined with photoluminescence and transmission electron microscopy experiments to obtain information on the band structure of WZ GaAs (Figure 12.9c; Corfdir et al. 2013). However, the conclusions obtained from those studies remain only qualitative since many parameters required for the calculations (e.g., the conduction and valence-band offsets) are poorly known. The situation is particularly complex for GaAs NWs. Despite a large number of experimental and theoretical studies, there is still no agreement on the value of the bandgap of WZ GaAs; values larger and smaller than the bandgap of ZB GaAs have been reported, and the symmetry of the lowest energy conduction band of this material remains under debate (Martelli et al. 2015). Note also that, due to the small conduction and valence-band offsets between the WZ and the ZB phases of a semiconductor, the energy computed for an exciton in crystal-phase quantum structures of a thickness of a few monolayers is not sensitive to the exact values of the band offsets. As shown in Lähnemann et al. (2012), the photoluminescence energy for three-to-five-monolayers-thick crystal-phase QWs in GaN NWs can in fact be reproduced using either a type-I or a type-II band alignment between WZ and ZB GaN.

The connection of bands at the interface of different crystal phases is of particular importance for the modeling of polytype NWs. In the case of GaAs, the lowest conduction band in the ZB phase has a Γ_6 symmetry and is well separated in energy from any remote conduction bands (Belabbes et al. 2012). The equivalent of the Γ_6 band in the WZ structure is the Γ_7 band. However, compounds with a WZ structure show an additional band with a Γ_8 symmetry that results from the zone folding due to doubling of the WZ unit cell along the [0001] direction when compared to the ZB unit cell. For WZ GaAs, recent studies (De and Pryor 2010; Heiss et al. 2011; Belabbes et al. 2012; Graham et al. 2013) suggest that this Γ_8 band is very close in energy to the Γ_7 band. It is, however, not clear yet which of these two bands exhibits the lowest energy. This point is, however, essential since (1) these two bands are expected to exhibit very different effective masses and a weak coupling with the fundamental Γ_7 and Γ_9 valence bands, and (2) confinement in crystal-phase quantum disks could give rise to some conduction band mixing.

Finally, although strain-free compounds with a ZB structure do not exhibit any spontaneous polarization, WZ III–V semiconductors are polar materials. Differential phase contrast microscopy and photoluminescence spectroscopy on polytype NWs have shown that the spontaneous polarization in WZ GaAs and GaN is aligned along the c-axis of the WZ and has a strength equal to 2.7×10^{-3} and -2.2×10^{-2} C/m^2, respectively (Bauer et al. 2014; Lähnemann et al. 2012). Since the axis of most WZ NWs is parallel to the c direction, the spontaneous polarization field in polytype NWs shows discontinuities at the interfaces between WZ and ZB segments. These discontinuities give rise to built-in electric fields parallel to the axis of polytype NWs with an intensity on the order of MV/cm in WZ/ZB GaN quantum structures (Corfdir and Lefebvre 2012b; Lähnemann et al. 2012). The resulting quantum confined Stark effect strongly affects the optical properties of thick crystal-phase QWs. The photoluminescence transitions are redshifted and may be centered even at energies below the band gap of ZB GaN, as observed for ZB insertions thicker than six monolayers (Jacopin et al. 2011), and the oscillator strength of the exciton is decreased (Figure 12.9d; Corfdir and Lefebvre 2012b).

12.5 Optical Anisotropy of NWs—The Antenna Effect

Semiconductors in the form of NWs not only show altered electrical characteristics but also strongly modified optical properties. As shown by Wang et al. (2001), the electric field at the sidewalls of NWs with a diameter much smaller than the wavelength is discontinuous (continuous) for light polarized perpendicular (parallel) to the NW axis. For an NW considered as an infinite dielectric cylinder, the electric field is $E_i = 2\kappa_0 / \left(\kappa_r + \kappa_0 \right) E_0$ for light polarized perpendicular to the NW axis, where E_0 is the electric field outside the NW and κ_0 and κ_r are the dielectric constants of vacuum and of the NW, respectively. Since $\kappa_r \approx 10$ in III–V semiconductors, this *antenna effect* leads to significant polarization anisotropies of interband transitions in NWs. In particular, the antenna effect may affect the emission intensity ratio between different interband transitions. For instance, while heavy holes have dipoles oriented perpendicular to the axis of GaAs NWs, light holes exhibit a large on-axis dipole. Optical transitions related to light–hole excitons are thus promoted by the antenna effect (Spirkoska et al. 2012). Complete reversal of the polarization anisotropy with respect to the bulk has even been predicted for WZ NWs with crystal field and spin–orbit splittings such that the A and B valence bands are close in energy (Efros and Lambrecht 2014).

The situation is more complicated for NWs with diameters on the order of the wavelength, as they support guided modes (Ruppin 2002). The latter gives rise to oscillations in the degree of polarization of the NW luminescence with varying wavelength or NW diameter (Giblin et al. 2009; Corfdir et al. 2015). NW ensembles usually exhibit a broad diameter distribution, and the polarization response may vary from one NW to another (Corfdir et al. 2015).

The polarization properties of interband transitions have been commonly used to extract the strain state of planar structures. Comparison between the measured optical anisotropy and the one computed using **k·p** calculations has made it possible to estimate the strain state of group-III-nitrides layers grown along a nonpolar or a semipolar axis (Ghosh et al. 2002; Funato et al. 2013). The analysis is more complicated for NWs since the combination of strain, dielectric contrast and possible surface-induced wave function distortion (see Section 12.3.3) may result in a complex polarization anisotropy of the interband transitions in NWs. For instance, the degree of polarization of single InP/InAs NWs exhibiting valence-band mixing and dispersed on a carbon grid had to be calculated using detailed finite difference time domain calculations (Anufriev et al. 2015).

Acknowledgments

The authors would like to thank Oliver Brandt and Lutz Geelhaar as well as present and former colleagues at Paul-Drude-Institut für Festkörperelektronik, Berlin, Germany, and at the Cavendish Laboratory, Cambridge, UK, for stimulating and fruitful discussions. We also thank Lutz Schrottke for critical reading of the manuscript. Pierre Corfdir acknowledges funding from the Fonds National Suisse de la Recherche Scientifique through project 161032.

References

Ahtapodov, L., J. Todorovic, P. Olk, T. Mjåland, P. Slåttnes, D. L. Dheeraj, A. T. J. Van Helvoort, B. O. Fimland, and H. Weman. 2012. A story told by a single nanowire: Optical properties of wurtzite GaAs. *Nano Lett.* 12:6090.

Akopian, N., G. Patriarche, L. Liu, J.-C. Harmand, and V. Zwiller. 2010. Crystal phase quantum dots. *Nano Lett.* 10:1198.

Algra, R. E., M. A. Verheijen, M. T. Borgström, L.-F. Feiner, G. Immink, W. J. P. van Enckevort, E. Vlieg, and E. P. A. M. Bakkers. 2008. Twinning superlattices in indium phosphide nanowires. *Nature* 456:369.

Allan, G., C. Delerue, M. Lannoo, and E. Martin. 1995. Hydrogenic impurity levels, dielectric constant, and Coulomb charging effects in silicon crystallites. *Phys. Rev. B* 52:11982.

Andreani, L. C., and A. Pasquarello. 1990. Accurate theory of excitons in GaAs-Ga$_{1-x}$Al$_x$As quantum wells. *Phys. Rev. B* 42:8928.

Anufriev, R., J.-B. Barakat, G. Patriarche, X. Letartre, C. Bru-Chevallier, J.-C. Harmand, M. Gendry, and N. Chauvin. 2015. Optical polarization properties of InAs/InP quantum dot and quantum rod nanowires. *Nanotechnology* 26:395701.

Armitage, R., and K. Tsubaki. 2010. Multicolour luminescence from InGaN quantum wells grown over GaN nanowire arrays by molecular-beam epitaxy. *Nanotechnology* 21:195202.

Bao, J., D. C. Bell, F. Capasso, J. B. Wagner, T. Mårtensson, J. Trägårdh, and L. Samuelson. 2008. Optical properties of rotationally twinned InP nanowire heterostructures. *Nano Lett.* 8:836.

Bardoux, R., A. Kaneta, M. Funato, Y. Kawakami, A. Kikuchi, and K. Kishino. 2009. Positive binding energy of a biexciton confined in a localization center formed in a single In$_x$Ga$_{1-x}$N/GaN quantum disk. *Phys. Rev. B* 79:155307.

Bauer, B., J. Hubmann, M. Lohr, E. Reiger, D. Bougeard, and J. Zweck. 2014. Direct detection of spontaneous polarization in wurtzite GaAs nanowires. *Appl. Phys. Lett.* 104:211902.

Bechstedt, F., and A. Belabbes. 2013. Structure, energetic, and electronic states of III–V compound types. *J. Phys. Condens. Matter* 25:273201.

Belabbes, A., C. Panse, J. Furthmüller, and F. Bechstedt. Electronic bands of III–V semiconductor polytypes and their alignment. 2012. *Phys. Rev. B* 86:075208.

Björk, M. T., B. J. Ohlsson, T. Sass, A. I. Persson, C. Thelander, M. H. Magnusson, K. Deppert, L. R. Wallenberg, and L. Samuelson. 2002. One-dimensional heterostructures in semiconductor whiskers. *Appl. Phys. Lett.* 80:1058.

Björk, M. T., H. Schmid, J. Knoch, H. Riel, and W. Riess. 2009. Donor deactivation in silicon nanostructures. *Nat. Nanotechnol.* 4:103.

Böcklin, C., R. G. Veprek, S. Steiger, and B. Witzigmann. 2010. Computational study of an InGaN/GaN nanocolumn light-emitting diode. *Phys. Rev. B* 81:155306.

Bolinsson, J., P. Caroff, B. Mandl, and K. A. Dick. 2011. Wurtzite-zincblende superlattices in InAs nanowires using a supply interruption method. *Nanotechnology* 22:265606.

Bryant, G. W. 1984. Hydrogenic impurity states in quantum-well wires. *Phys. Rev. B* 29:6632.

Caroff, P. K., A. Dick, J. Johansson, M. E. Messing, K. Deppert, and L. Samuelson. 2009. Controlled polytypic and twin-plane superlattices in III–V nanowires. *Nat. Nanotechnol.* 4:50.

Castelletto, S., Z. Bodrog, A. P. Magyar, A. Gentle, A. Gali, and I. Aharonovich. 2014. Quantum-confined single photon emission at room temperature from SiC tetrapods. *Nanoscale* 6:10027.

Chan, T. L., M. L. Tiago, E. Kaxiras, and J. R. Chelikowsky. 2008. Size limits on doping phosphorus into silicon nanocrystals. *Nano Lett.* 8:596.

Chuang, S. L., and C. S. Chang. 1996. **k·p** method for strained wurtzite semiconductors. *Phys. Rev. B* 54, 2491.

Corfdir, P., P. Lefebvre, J. Ristić, J.-D. Ganière, and B. Deveaud-Plédran. 2009. Electron localization by a donor in the vicinity of a basal stacking fault in GaN. *Phys. Rev. B* 80:153309.

Corfdir, P., and P. Lefebvre. 2012a. Role of the dielectric mismatch on the properties of donors in semiconductor nanostructures bounded by air. *J. Appl. Phys.* 112:106104.

Corfdir, P., and P. Lefebvre. 2012b. Importance of excitonic effects and the question of internal electric fields in stacking faults and crystal phase quantum discs: The model-case of GaN. *J. Appl. Phys.* 112:053512.

Corfdir, P., B. Van Hattem, E. Uccelli, S. Conesa-Boj, P. Lefebvre, A. Fontcuberta i Morral, and R. T. Phillips. 2013. Three-dimensional magneto-photoluminescence as a probe of the electronic properties of crystal-phase quantum disks in GaAs nanowires. *Nano Lett.* 13:5303.

Corfdir, P., J. K. Zettler, C. Hauswald, S. Férnandez-Garrido, O. Brandt, and P. Lefebvre. 2014. Sub-meV linewidth in GaN nanowire ensembles: Absence of surface excitons due to the field ionization of donors. *Phys. Rev. B* 90:205301.

Corfdir, P., F. Feix, J. K. Zettler, S. Fernández-Garrido, and O. Brandt. 2015. Importance of the dielectric contrast for the polarization of excitonic transitions in single GaN nanowires. *New J. Phys.* 17:033040.

De, A., and C. Pryor. 2010. Predicted band structures of III–V semiconductors in the wurtzite phase. *Phys. Rev. B* 81:155210.

Diarra, M., Y.-M. Niquet, C. Delerue, and G. Allan. 2007. Ionization energy of donor and acceptor impurities in semiconductor nanowires: Importance of dielectric confinement. *Phys. Rev. B* 75:045301.

Diarra, M., C. Delerue, Y.-M. Niquet, and G. Allan. 2008. Screening and polaronic effects induced by a metallic gate and a surrounding gate oxide on donor and acceptor impurities in silicon nanowires. *J. Appl. Phys.* 103:073703.

Efros, A. L., and W. R. L. Lambrecht. 2014. Theory of light emission polarization reversal in zinc-blende and wurtzite nanowires. *Phys. Rev. B* 89:035304.

Ertekin, E., P. A. Greaney, D. C. Chrzan, and T. D. Sands. 2005. Equilibrium limits of coherency in strained nanowire heterostructures. *J. Appl. Phys.* 97:114325.

Freitas, J. A., W. J. Moore, B. V, Shanabrook, G. C. B. Braga, D. D. Koleske, S. K. Lee, S. S. Park, and J. Y. Han. 2003. Shallow donors in GaN. *Phys. Stat. Solidi B* 240:330.

Funato, M., K. Matsuda, R. G. Banal, R. Ishii, and Y. Kawakami. 2013. Strong optical polarization in nonpolar (1–100) $Al_xGa_{1-x}N/AlN$ quantum wells. *Phys. Rev. B* 87:041306(R).

Ghosh, S., P. Waltereit, O. Brandt, H. T. Grahn, and K. H. Ploog. 2002. Electronic band structure of wurtzite GaN under biaxial strain in the M plane investigated with photoreflectance spectroscopy. *Phys. Rev. B* 65: 075202.

Giblin, J., V. Protasenko, and M. Kuno. 2009. Wavelength sensitivity of single nanowire excitation polarization anisotropies explained through a generalized treatment of their linear absorption. *ACS Nano* 3:1979.

Graham, A. M., P. Corfdir, M. Heiss, S. Conesa-Boj, E. Uccelli, A. Fontcuberta, I. Morral, and R. T. Phillips. 2013. Exciton localization mechanisms in wurtzite/zinc-blende GaAs nanowires. *Phys. Rev. B* 87:125304.

Grönquist, J., N. Søndergaard, F. Boxberg, T. Guhr, S. Åberg, and H. Q. Xu. 2009. Strain in semiconductor core–shell nanowires. *J. Appl. Phys.* 106:053508.

Große, F., and J. Neugebauer. 2001. Limits and accuracy of valence force field models for $In_xGa_{1-x}N$ alloys. *Phys. Rev. B* 63:085207.

Heiss, M., S. Conesa-Boj, J. Ren, H. H. Tseng, A. Gali, A. Rudolph, E. Uccelli, F. Peiró, J. R. Morante, D. Schuh, E. Reiger, E. Kaxiras, J. Arbiol, and A. Fontcuberta i Morral. 2011. Direct correlation of crystal structure and optical properties in wurtzite/zinc-blende GaAs nanowire heterostructures. *Phys. Rev. B* 83:045303.

Heiss, M., Y. Fontana, A. Gustafsson, G. Wüst, C. Magen, D. D. O'Regan, J. W. Luo, B. Ketterer, S. Conesa-Boj, A. V. Kuhlmann, J. Houel, E. Russo-Averchi, J. R. Morante, M. Cantoni, N. Marzari, J. Arbiol, A. Zunger, R. J. Warburton, and A. Fontcuberta i Morral. 2013. Self-assembled quantum dots in a nanowire system for quantum photonics. *Nat. Materials* 12:439.

Hirth, J.P., and J. Lothe. 1968. *Theory of Dislocations*. New York: McGraw-Hill.

Hoang, T. B., A. F. Moses, H. L. Zhou, D. L. Dheeraj, B. O. Fimland, and H. Weman. 2009. Observation of free exciton photoluminescence emission from single wurtzite GaAs nanowires. *Appl. Phys. Lett.* 94:133105.

Jacopin, G., L. Rigutti, L. Largeau, F. Fortuna, F. Furtmayr, F. H. Julien, M. Eickhoff, and M. Tchernycheva. 2011. Optical properties of wurtzite/zincblende heterostructures in GaN nanowires. *J. Appl. Phys.* 110:064313.

Kaganer, V. M., and A. Yu. Belov. 2012. Strain and x-ray diffraction from axial nanowire heterostructures. *Phys. Rev. B* 85:125402.

Kaganer, V. M., O. Marquardt, and O. Brandt. 2016. Piezoelectric potential in axial (In,Ga)N/GaN nanowire heterostructures. *Nanotechnology* 27(16):165201.

Kakanakova-Georgieva, A., D. Nilsson, X. T. Trinh, U. Forsberg, N. T. Son, and E. Janzén. 2013. The complex impact of silicon and oxygen on the n-type conductivity of high-Al-content AlGaN. *Appl. Phys. Lett.* 102:132113.

Keating, P. N. 1966. Effect of invariance requirements on the elastic strain energy of crystals with application to the diamond structure. *Phys. Rev.* 145:637.

Keldysh, L. 1979. Coulomb interaction in thin semiconductor and semimetal films. *JETP Lett.* 29:658.

Keyes, W. R. 1975. The effect of randomness in the distribution of impurity atoms on FET thresholds. *Appl. Phys.* 8:251.

Kikuchi, A., M. Kawai, M. Tada, and K. Kishino. 2004. InGaN/GaN multiple quantum disk nanocolumn light-emitting diodes grown on (111) Si substrate. *Jpn. J. Appl. Phys.* 43:1524.

Kim, H., Y. Cho, H. Lee, S. Kim, S. Ryu, D. Kim, T. Kang, and K. Chung. 2004. High-brightness light emitting diodes using dislocation-free indium gallium nitride/gallium nitride multiquantum-well nanorod arrays. *Nano Lett.* 4:1059.

Knelangen, M., M. Hanke, E. Luna, L. Schrottke, O. Brandt, and A. Trampert. 2011. Monodisperse (In,Ga)N insertions in catalyst-free-grown GaN(0001) nanowires. *Nanotechnology* 22:365703.

Konar, A., T. Fang, N. Sun, and D. Jena. 2011. Charged basal stacking fault scattering in nitride semiconductors. *Appl. Phys. Lett.* 98:022109.

Kumagai, M., and T. Takagahara. 1989. Excitonic and nonlinear-optical properties of dielectric quantum-well structures. *Phys. Rev. B* 40:12359.

Lähnemann, J., O. Brandt, U. Jahn, C. Pfüller, C. Roder, P. Dogan, F. Grosse, A. Belabbes, F. Bechstedt, A. Trampert, and L. Geelhaar. 2012. Direct experimental determination of the spontaneous polarization of GaN. *Phys. Rev. B* 86:081302(R).

Landau, L. D., and E. M. Lifshitz. 1959. *Theory of Elasticity*. London: Pergamon.

Levine, J. 1965. Nodal hydrogenic wave functions of donors on semiconductor surfaces. *Phys. Rev.* 140:A586.

Li, S., and A. Waag. 2012. GaN-based nanorods for solid state lighting. *J. Appl. Phys.* 111:071101.

Lin, H. W., Y. J. Lu, H. Y. Chen, H. M. Lee, and S. Gwo. 2010. InGaN/GaN nanorod array white light-emitting diode. *Appl. Phys. Lett.* 97:073101.

Liu, P., H. Huang, X. Liu, M. Bai, D. Zhao, Z. Tang, X. Huang, J.-Y. Kim, and J. Guo. 2014. Core–shell nanowire diode based on strain-engineered bandgap. *Phys. Stat. Solidi A* 212:617.

Marquardt, O., C. Hauswald, M. Wölz, L. Geelhaar, and O. Brandt. 2013. Luminous efficiency of axial $In_xGa_{1-x}N$/GaN nanowire heterostructures: Interplay of polarization and surface potentials. *Nano Lett.* 13:3298.

Marquardt, O., L. Geelhaar, and O. Brandt. 2015. Impact of random dopant fluctuations on the electronic properties of $In_xGa_{1-x}N$/GaN axial nanowire heterostructures. *Nano Lett.* 15:4289.

Martelli, F., G. Priante, and S. Rubini. 2015. Photoluminescence of GaAs nanowires at an energy larger than the ZB band-gap: Dependence on growth parameters. *Semicond. Sci. Tech.* 30:055020.

McCluskey, M. D., N. M. Johnson, C. G. Van de Walle, D. P. Bour, M. Kneissl, and W. Walukiewicz. 1998. Metastability of oxygen donors in AlGaN. *Phys. Rev. Lett.* 80:4008.

Menéndez, J., R. Singh, and J. Drucker. 2011. Theory of strain effects on the Raman spectrum of Si–Ge core–shell nanowires. *Ann. Phys.* 523:145.

Meyer, B. K., J. Sann, D. M. Hofmann, C. Neumann, and A. Zeuner. 2005. Shallow donors and acceptors in ZnO. *Semicond. Sci. Technol.* 20:S62.

Miller, D. A. B., D. S. Chemla, T. C. Damen, A. C. Gossard, W. Wiegmann, T. H. Wood, and C. A. Burrus. 1984. Band-edge electroabsorption in quantum well structures: The quantum-confined stark effect. *Phys. Rev. Lett.* 53:2173.

Murayama, M., and T. Nakayama. 1994. Chemical trend of band offsets at wurtzite/zinc-blende hetero-crystalline semiconductor interfaces. *Phys. Rev. B* 49:4710–4724.

Nguyen, H. P. T., S. Zhang, K. Cui, X. Han, S. Fathololoumi, M. Couillard, G. A. Botto, and Z. Mi. 2011. p-Type modulation doped InGaN/GaN dot-in-a-wire white-light-emitting diodes monolithically grown on Si(111). *Nano Lett.* 11:1919.

Nye, J. F. 1985. *Physical Properties of Crystals—Their Representation by Tensors and Matrices*. Clarendon: Oxford.

Park, C. H., and D. J. Chadi. 1997. Stability of deep donor and acceptor centers in GaN, AlN, and BN. *Phys. Rev. B* 55:12995.

Rücker, H., and M. Methfessel. 1995. Anharmonic Keating model for group-IV semiconductors with application to the lattice dynamics in alloys of Si, Ge, and C. *Phys. Rev. B* 52:11059.

Ruppin, R. 2002. Extinction by a circular cylinder in an absorbing medium. *Optics Commun.* 211:335.

Satpathy, S. 1983. Eigenstates of Wannier excitons near a semiconductor surface. *Phys. Rev. B* 28:4585.

Schubert, E. F. 2005. *Doping in III–V Semiconductors*. Cambridge: Cambridge University Press.

Seeger, K. 1991. *Semiconductor Physics*. Berlin: Springer.

Shik, A. 1993. Excitons and impurity centers in thin wires and in porous silicon. *J. Appl. Phys.* 74:2951.

Shklovskii, B. I., and A. L. Efros. 1984. *Electronic Properties of Doped Semiconductors*. Berlin: Springer.

Singh, R., E. J. Dailey, J. Drucker, and J. Menéndez. 2011. Raman scattering from Ge–Si core–shell nanowires: Validity of analytical strain models. *J. Appl. Phys.* 110:124305.

Spirkoska, D., A. L. Efros, W. R. L. Lambrecht, T. Cheiwchanchamnangij, A. Fontcuberta i Morral, and G. Abstreiter. 2012. Valence band structure of polytypic zinc-blende/wurtzite GaAs nanowires probed by polarization-dependent photoluminescence. *Phys. Rev. B* 85:045309.

Sze, S. M. 1985. *Semiconductor Devices: Physics and Technology*. New York: Wiley.

Thelander, C., P. Caroff, S. Plissard, A. W. Dey, and K. A. Dick. 2011. Effects of crystal phase mixing on the electrical properties of InAs nanowires. *Nano Lett.* 11:2424.

Wallentin, J., N. Anttu, D. Asoli, M. Huffman, I. Åberg, M. H. Magnusson, G. Siefer, P. Fuss-Kailuweit, F. Dimroth, B. Witzigmann, H. Q. Xu, L. Samuelson, K. Deppert, and M. T. Borgström. 2013. InP nanowire array solar cells achieving 13.8% efficiency by exceeding the ray optics limit. *Science* 339:1057.

Wang, H., and A.-B. Chen. 2000. Calculation of shallow donor levels in GaN. *J. Appl. Phys.* 87:7859.

Wang, J., M. S. Gudiksen, X. Duan, Y. Cui, and C. M. Lieber. 2001. Highly polarized photoluminescence and photodetection from single indium phosphide nanowires. *Science* 293:1455.

Whitaker, J. C. 2005. *Microelectronics*. Boca Raton, FL: Taylor & Francis.

Wijnheijmer, A. P., J. K. Garleff, K. Teichmann, M. Wenderoth, S. Loth, R. G. Ulbrich, P. A. Maksym, M. Roy, and P. M. Koenraad. 2009. Enhanced donor binding energy close to a semiconductor surface. *Phys. Rev. Lett.* 102:166101.

Wölz, M., V. M. Kaganer, O. Brandt, L. Geelhaar, and H. Riechert. 2011. Analyzing the growth of $In_xGa_{1-x}N$/GaN superlattices in self-induced GaN nanowires by x-ray diffraction. *Appl. Phys. Lett.* 98:261907.

Wölz, M., V. M. Kaganer, O. Brandt, L. Geelhaar, and H. Riechert. 2012. Erratum: Analyzing the growth of $In_xGa_{1-x}N$/GaN superlattices in self-induced GaN nanowires by x-ray diffraction. *Appl. Phys. Lett.* 100:179902.

Wölz, M., M. Ramsteiner, V. M. Kaganer, O. Brandt, L. Geelhaar, and H. Riechert. 2013. Strain engineering of nanowire multi-quantum well demonstrated by Raman spectroscopy. *Nano Lett.* 13:4053.

Xiang, H. J., S.-H. Wei, J. L. F. Da Silva, and J. Li. 2008. Strain relaxation and band-gap tunability in ternary $In_xGa_{1-x}N$ nanowires. *Phys. Rev. B* 77:193301.

Yang, L., R. N. Musin, X.-Q. Wang, and M. Y. Chou. 2008. Quantum confinement effect in Si/Ge core–shell nanowires: First-principles calculations. *Phys. Rev. B* 77:195325.

Zeisel, R., M. W. Bayerl, S. T. B. Goennenwein, R. Dimitrov, O. Ambacher, M. S. Brandt, and M. Stutzmann. 2000. DX-behavior of Si in AlN. *Phys. Rev. B* 61:R16283.

Zhang, L., J.-W. Luo, A. Zunger, N. Akopian, V. Zwiller, and J.-C. Harmand. 2010. Wide InP nanowires with wurtzite/zinc-blende superlattice segments are type-II whereas narrower nanowires become type-I: An atomistic pseudopotential calculation. *Nano Lett.* 10:4055.

13

Quantum Dots

Stanko Tomić

and

Nenad Vukmirović

13.1 Introduction

Semiconductor nanocrystals or quantum dots (QDs) are the subject of intensive research due to a number of novel properties, which make them attractive for both fundamental studies and technological applications [1–6]. QDs are of particular interest for solar cell applications due to their ability to increase efficiency via the generation of multiexcitons from a single photon [7–9]. QDs can be synthesized with a high degree of control using colloidal chemistry [10,11]. Much research effort has been directed toward studying QDs grown from more than one semiconductor, e.g., core/shell heterostructures [12–14]. Such core/shell nanostructures provide means to control the optical properties by tuning the electron–hole wave function overlap, which is affected by the alignment of the conduction band (CB) and valence band (VB) edges, as well as the QD shape and size [15–17]. In addition, such core/shell structures can provide for type-II alignments with staggered CB and VB edges so the lowest energy states for electrons and holes lie in different spatial regions, leading to charge separation between the carriers. Such staggered band alignments have several useful physical consequences, including longer radiative recombination times for more efficient charge extraction in photovoltaic applications [18,19], optical gaps that can be made smaller than the bulk values of constituent materials [12,20,21] and control of the electron–hole wave function overlap that determines the exchange interaction energy [22]. Charge separation in type-II QDs can also be used to increase the repulsion between like-sign charges in biexciton states [23,24], leading to the possibility of lasing in the single exciton regime [6,25,26].

To determine the energetics of many-body states in QDs both the confinement potential and many-body interactions between the carriers need to be taken into account. Many-body interactions lead to Coulomb (charge) and Fermi (spin) correlation. Coulomb correlation arises from the electrostatic interaction of charge carriers in the many-body complex, while spin correlation occurs due to the fermionic character of charge carriers (i.e., the Pauli exclusion principle) [27]. Correlated many-body states may be calculated with the configuration interaction (CI) method, which can be used in the framework of continuum or atomistic descriptions of single-particle (SP) states. [28–38].

Colloidal QDs are usually embedded or dispersed in media [39] of lower dielectric constant than the semiconductor itself—this dielectric confinement leads to a modification of the Coulomb interaction, which can be described using classical image charge theory. While atomistic calculations [40] showed that dielectric confinement significantly affects the charging energies of QDs, in single-material spherical QDs the similar electron and hole charge distributions lead to a weakened dielectric confinement effect [41] on exciton states, which mainly increases the binding energy [28,42]. It is, therefore, natural to wonder if the optical properties of spherical type-II core/shell QDs can be significantly affected by the dielectric environment.

The effect of dielectric confinement and many-electron correlation means that the single-particle picture is not good enough to faithfully predict exciton energetics or wave functions in colloidal QDs. The proper treatment of charges requires a many-electron description that goes beyond mean-field theories, and the CI method is the most appropriate approach. However, the full configuration method becomes progressively computationally expensive as the number of states increases. Luckily, however, the interpretation of physical experiments often requires detailed knowledge of just a few excitons of particular symmetry. This allows us to evaluate the effect of correlation on exciton energies and dipole matrix elements for arbitrary QD geometries.

In this chapter, we review the theoretical methods for the description of the electronic structure and transport properties of colloidal nanocrystal quantum dots (NQD). The chapter is organized as follows: after introducing the core/shell morphology in Section 13.2, in Section 13.3, we review the widely used **k·p** theory for electronic structure of semiconductor nanostructures, both the 8-band and 14-band version, taking into account effects like band-mixing, strain, piezoelectric field, and spin–orbit interaction. In Section 13.4, we outline the plane wave (PW) basis set implementation of the **k·p** Hamiltonian and necessary modifications for QDs, i.e., three-dimensional (3D) confined systems. In Section 13.5, we discuss the effect of dielectric confinement, polarization, and self-polarization due to large contrast and spatial variation of the dielectric constant in colloidal NQD. In Section 13.6, we derive the CI Hamiltonian for the description of many-electron states in colloidal NQD. In Section 13.7, we present the basic theory of the electron transport through NQD networks. In Section 13.8, we present our results on the ground state exciton absorption edge, radiative times of the ground state exciton, and correlation energies of the ground state exciton, as well as electronic charge transport in NQD solids.

13.2 The Core/Shell QD Structure

The core/shell QD structures are comprised of a single-crystal "core" region enveloped by a second "shell" region of a different semiconductor material. The shell region usually possesses a larger band gap than the core region. In such a configuration, with appropriate materials combinations and alloys, such QD can exhibit a type-II band alignment, thus offering an extra degree of freedom in *quantum engineering* of nanostructures for specific purposes. The addition of the shell has two main effects. First, the shell serves to passivate dangling bonds at the surface of the dot (surface ligands), which arise from the lack of coordination in atoms at the dot surface, vastly reducing the number of trap states near the surface. This reduces the effect of the nonradiative processes on device operations. Second, the energy offsets forming the type-II heterostructure result in confinement of the charge carriers into separate regions of the QD: electrons are confined to the core, while holes are confined to the shell region, or vice versa, depending on the relative band alignment between the two regions. Such staggered band alignments have several useful physical consequences, including longer radiative recombination times for more efficient charge extraction in photovoltaic applications and optical gaps that can be made smaller than the bulk values of the constituent materials. To illustrate the problem and introduce basic parameters, in Figure 13.1 we show

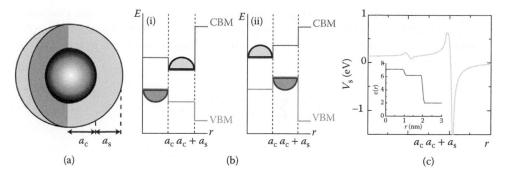

FIGURE 13.1 (a) Cutaway view of a spherical core/shell quantum dot (QD) characterized by core radius a_c and shell thickness a_s; (b) conduction band minimum (CBM) and the valence band maximum (VBM) corresponding to type-II alignment, in (i) an h/e CdTe/CdSe QD and (ii) an e/h CdSe/CdTe QD.

(a) a schematic of a spherical core/shell QD characterized by core radius a_c and shell thickness a_s, the staggered alignments of the CB minimum (CBM) and VB maximum (VBM) in type-II structures where (b,i) holes (h) are confined in the core and electrons (e) are confined in the shell, as in CdTe/CdSe QDs and (b,ii) electrons are confined in the core and holes are confined in the shell, as in CdSe/CdTe QDs.

13.3 The k·p Method for SP Electronic Structure

The quantum mechanical description of electrons in any material requires detailed knowledge of their wave functions, $\psi_{n,\mathbf{k}}(\mathbf{r})$, which are found by solving the Schrödinger equation (in the single-electron approximation):

$$H_0 \psi_{n,\mathbf{k}}(\mathbf{r}) = E_{n,\mathbf{k}} \psi_{n,\mathbf{k}}(\mathbf{r}). \tag{13.1}$$

The Hamiltonian in Equation 13.1, $H_0 = p^2/2m_0 + V(\mathbf{r})$, is the function of the quantum mechanical momentum operator, $\mathbf{p} = -i\hbar\nabla$, and the crystal potential experienced by electrons, $V(\mathbf{r}) = V(\mathbf{r}+\mathbf{R})$, which is a periodic function, with the periodicity of the crystal lattice, $\mathbf{R}$, and m_0 is the free electron mass. According to Bloch's theorem, the solutions to this Schrödinger equation can be written as $\psi_{n,\mathbf{k}}(\mathbf{r}) = e^{i\mathbf{k}\cdot\mathbf{r}} u_{n,\mathbf{k}}(\mathbf{r})$ where $\mathbf{k}$ is the electron wave vector, n is the band index, and $u_{n,\mathbf{k}}$ is the *cell-periodic* function, with the same periodicity as the crystal lattice. The cell-periodic function, $u_{n,\mathbf{k}}$, satisfies the equation

$$H_{\mathbf{k}} u_{n,\mathbf{k}} = E_{n,\mathbf{k}} u_{n,\mathbf{k}}, \tag{13.2}$$

where the Hamiltonian

$$H_{\mathbf{k}} = H_0 + H_{\mathbf{k}}' = \frac{p^2}{2m_0} + V + \frac{\hbar^2 k^2}{2m_0} + \frac{\hbar\mathbf{k}\cdot\mathbf{p}}{m_0} \tag{13.3}$$

is given as a sum of two terms: the unperturbed, H_0, which in fact equals the exact Hamiltonian at $\mathbf{k} = 0$ (i.e., at the Γ point in the Brillouin zone) and the "perturbation," $H_{\mathbf{k}}'$. The Hamiltonian given by Equation 13.3 is called the k·p Hamiltonian [43–45]. If the eigenvalues are nondegenerate, the first-order energy correction is given by $E'_{n,\mathbf{k}} \approx \langle u_{n,0} | H_{\mathbf{k}}' | u_{n,0}\rangle$ and there is no correction (to first order, in the absence of nondiagonal matrix elements) in the eigenfunctions. The second-order correction arises from nondiagonal terms; the energy correction is given by $E''_{n,\mathbf{k}} \approx \sum_{n'\neq n} |\langle u_{n,0}|H_{\mathbf{k}}'|u_{n',0}\rangle|^2/(E_{n,0}-E_{n',0})$. The "perturbation" term $H_{\mathbf{k}}'$ gets progressively smaller as $\mathbf{k}$ approaches zero. Therefore, the k·p perturbation theory is most accurate for small values of **k**. However, if enough terms are included in the perturbative expansion, then the theory can in fact be reasonably accurate for any value of **k** in the entire Brillouin zone [46,47]. For a band n, with an extremum at $\mathbf{k} = 0$, and with no spin–orbit coupling, the result of **k·p** perturbation theory is (to the lowest nontrivial order):

$$u_{n,\mathbf{k}} = u_{n,0} + \frac{\hbar}{m_0} \sum_{n' \neq n} \frac{\langle u_{n,0} | \mathbf{k} \cdot \mathbf{p} | u_{n',0} \rangle}{E_{n,0} - E_{n',0}} u_{n',0} \tag{13.4}$$

$$E_{n,\mathbf{k}} = E_{n,0} + \frac{\hbar^2 k^2}{2m_0} + \frac{\hbar^2}{m_0^2} \sum_{n' \neq n} \frac{|\langle u_{n,0} | \mathbf{k} \cdot \mathbf{p} | u_{n',0} \rangle|^2}{E_{n,0} - E_{n',0}}. \tag{13.5}$$

The parameters required to do these calculations, the band edge energies, $E_{n,0}$, and the optical matrix elements, $\langle u_{n,0} | \mathbf{p} | u_{n',0} \rangle$, are typically inferred from experimental data or detailed atomistic-based theories.

A particular strength of the **k·p** theory is a straightforward inclusion of the spin–orbit interaction and of the strain effects on the band structure via deformation potential theory [48].

Relativistic effects in the **k·p** method are included also perturbatively via the spin–orbit (SO) interaction Hamiltonian, $H_{\mathrm{SO}} = \frac{\hbar}{4m_0^2 c^2} \left[\nabla V(\mathbf{r}) \times \mathbf{p} \right] \cdot \vec{\sigma}$, where $\vec{\sigma} = (\sigma_x, \sigma_y, \sigma_z)$ is a vector consisting of the three Pauli spin matrices. Finally, the **k·p** Hamiltonian becomes

$$H_{\mathbf{k}} = \frac{p^2}{2m_0} + V + \frac{\hbar^2 k^2}{2m_0} + \frac{\hbar \mathbf{k} \cdot \mathbf{p}}{m_0} + \frac{\hbar}{4m_0^2 c^2} (\vec{\sigma} \times \nabla V) \cdot (\hbar \mathbf{k} + \mathbf{p}). \tag{13.6}$$

In a strained system, the lattice constants are stretched or compressed [48]. This affects the SP potential experienced by electrons and consequently yields an additional term in the **k·p** Hamiltonian. Taking into account only the effects up to first order in strain tensor ϵ, the strained **k·p** Hamiltonian becomes

$$H_{\mathbf{k}} = \frac{p^2}{2m_0} + V + \frac{\hbar^2 k^2}{2m_0} + \frac{\hbar \mathbf{k} \cdot \mathbf{p}}{m_0} + \frac{1}{4m_0^2 c^2} (\vec{\sigma} \times \nabla V) \cdot (\mathbf{k} + \mathbf{p}) + \mathbf{D}^\epsilon \cdot \epsilon, \tag{13.7}$$

where $\mathbf{D}^\epsilon$ is the deformation potential operator. This Hamiltonian can be subjected to the same sort of perturbation theory analysis as above [49].

The most widely used version of the **k·p** Hamiltonian is the 8-band model [50], in which the operator given in Equation 13.7 is represented in the basis of s-bonding and p-antibonding states around the energy gap:

$$|u_1\rangle = |\tfrac{1}{2}, +\tfrac{1}{2}\rangle = |s;\uparrow\rangle \qquad\qquad |u_5\rangle = |\tfrac{1}{2}, -\tfrac{1}{2}\rangle = -|s;\downarrow\rangle$$

$$|u_2\rangle = |\tfrac{3}{2}, +\tfrac{3}{2}\rangle = \frac{i}{\sqrt{2}}[|x;\uparrow\rangle + i|y;\uparrow\rangle] \qquad |u_6\rangle = |\tfrac{3}{2}, -\tfrac{3}{2}\rangle = -\frac{i}{\sqrt{2}}[|x;\downarrow\rangle - i|y;\downarrow\rangle]$$

$$|u_3\rangle = |\tfrac{3}{2}, +\tfrac{1}{2}\rangle = \frac{i}{\sqrt{6}}[|x;\downarrow\rangle + i|y;\downarrow\rangle - 2|z;\uparrow\rangle] \quad |u_7\rangle = |\tfrac{3}{2}, -\tfrac{1}{2}\rangle = +\frac{i}{\sqrt{6}}[|x;\uparrow\rangle - i|y;\uparrow\rangle + 2|z;\downarrow\rangle]$$

$$|u_4\rangle = |\tfrac{1}{2}, +\tfrac{1}{2}\rangle = \frac{i}{\sqrt{3}}[|x;\downarrow\rangle + i|y;\downarrow\rangle + |z;\uparrow\rangle] \quad |u_8\rangle = |\tfrac{1}{2}, -\tfrac{1}{2}\rangle = +\frac{i}{\sqrt{3}}[|x;\uparrow\rangle - i|y;\uparrow\rangle - |z;\downarrow\rangle]$$

Occasionally, the 14-band **k·p** Hamiltonian is also used, in which one exploits the basis of p-bonding, s-bonding, and p-antibonding states around the energy gap [51]:

$$|u_1\rangle = |\tfrac{3}{2},+\tfrac{3}{2}\rangle = \tfrac{i}{\sqrt{2}}|(x_a+iy_a)\uparrow\rangle \qquad |u_8\rangle = |\tfrac{3}{2},-\tfrac{3}{2}\rangle = -\tfrac{i}{\sqrt{2}}|(x_a-iy_a)\downarrow\rangle$$

$$|u_2\rangle = |\tfrac{3}{2},+\tfrac{1}{2}\rangle = \tfrac{i}{\sqrt{6}}[|(x_a+iy_a)\downarrow\rangle - 2|z_a\uparrow\rangle] \qquad |u_9\rangle = |\tfrac{3}{2},-\tfrac{1}{2}\rangle = +\tfrac{i}{\sqrt{6}}[|(x_a-iy_a)\uparrow\rangle + 2|z_a\downarrow\rangle]$$

$$|u_3\rangle = |\tfrac{1}{2},+\tfrac{1}{2}\rangle = \tfrac{i}{\sqrt{3}}[|(x_a+iy_a)\downarrow\rangle + |z_a\uparrow\rangle] \qquad |u_{10}\rangle = |\tfrac{1}{2},-\tfrac{1}{2}\rangle = +\tfrac{i}{\sqrt{3}}[|(x_a-iy_a)\uparrow\rangle - |z_a\downarrow\rangle]$$

$$|u_4\rangle = |\tfrac{1}{2},+\tfrac{1}{2}\rangle = |s_a\uparrow\rangle \qquad |u_{11}\rangle = |\tfrac{1}{2},-\tfrac{1}{2}\rangle = -|s_a\downarrow\rangle$$

$$|u_5\rangle = |\tfrac{3}{2},+\tfrac{3}{2}\rangle = \tfrac{i}{\sqrt{2}}|(x_b+iy_b)\uparrow\rangle \qquad |u_{12}\rangle = |\tfrac{3}{2},-\tfrac{3}{2}\rangle = -\tfrac{i}{\sqrt{2}}|(x_b-iy_b)\downarrow\rangle$$

$$|u_6\rangle = |\tfrac{3}{2},+\tfrac{1}{2}\rangle = \tfrac{i}{\sqrt{6}}[|(x_b+iy_b)\downarrow\rangle - 2|z_b\uparrow\rangle] \qquad |u_{13}\rangle = |\tfrac{3}{2},-\tfrac{1}{2}\rangle = +\tfrac{i}{\sqrt{6}}[|(x_b-iy_b)\uparrow\rangle + 2|z_b\downarrow\rangle]$$

$$|u_7\rangle = |\tfrac{1}{2},+\tfrac{1}{2}\rangle = \tfrac{i}{\sqrt{3}}[|(x_b+iy_b)\downarrow\rangle + |z_b\uparrow\rangle] \qquad |u_{14}\rangle = |\tfrac{1}{2},-\tfrac{1}{2}\rangle = +\tfrac{i}{\sqrt{3}}[|(x_b-iy_b)\uparrow\rangle - |z_b\downarrow\rangle]$$

13.4 PW Implementation of the k·p Hamiltonian and Coulomb Integrals for QD Structures

The QD as a three-dimensional (3D) object breaks the translational symmetry of the bulk material along all three Cartesian directions implying operator replacement $k_\nu \rightarrow -i\partial/\partial\nu$ in Equation 13.6, where $\nu = (x, y, z)$. To solve the multiband system of Schrödinger equations, Equation 13.6, the PW methodology is employed as an expansion method [33,52–56]. In the PW representation, the eigenvalues (E_i) and the coefficients [$A_{n,\mathbf{k}}^{(i)}$] of the ith eigenvector [$\psi_n^{(i)}(\mathbf{r}) = \sum_{\mathbf{k}} A_{n,\mathbf{k}}^{(i)} e^{i\mathbf{k}\mathbf{r}}$], are linked by the relation

$$\sum_{n,\mathbf{k}'} h_{m,n}(\mathbf{k}, \mathbf{k}') A_{n,\mathbf{k}'}^{(i)} = E_i A_{m,\mathbf{k}}^{(i)}, \tag{13.8}$$

where $h_{m,n}(\mathbf{k}, \mathbf{k}')$ are the Hamiltonian matrix elements in the PW basis, and $m, n \in \{1,...,8\}$ or $m, n \in \{1,...,14\}$ are the band indices in the k·p Hamiltonian. All the elements in the Hamiltonian matrix, Equation 13.8, can be expressed as a linear combination of different kinetic and strain related terms and their convolution with the characteristic function of the QD shape, $\chi_{qd}(\mathbf{k})$ [33,57]. The whole k-space is discretized by embedding the QD in a rectangular box of dimensions L_x, L_y, and L_z and volume $\Omega = L_x \times L_y \times L_z$ and choosing the k-vectors in the form $\mathbf{k} = 2\pi(n_x/L_x, n_y/L_y, n_z/L_z)$, where n_x, n_y, and n_z are integers. Maximal absolute values of these integers control the accuracy of the method.

Having determined PW expansion coefficients, $A_{n,\mathbf{k}}^{(i)}$, from Equation 13.8, we can use the single-particle wave functions, $\psi_n^{(i)}(\mathbf{r})$, to calculate the Coulomb integrals, V_{ijkl}, relevant for many-electron processes that occur in QDs. The Coulomb integral among states i, j, k, and l is defined as

$$V_{ijkl} = \sum_{b=1}^{N_b} \sum_{b'=1}^{N_b} \int_\Omega d^3\mathbf{r} \int_\Omega d^3\mathbf{r}' \psi_b^{(i)}(\mathbf{r})^* \psi_b^{(j)}(\mathbf{r}) V(|\mathbf{r}-\mathbf{r}'|) \psi_{b'}^{(k)}(\mathbf{r}')^* \psi_{b'}^{(l)}(\mathbf{r}'), \tag{13.9}$$

where $V(\mathbf{u}) = e^2/4\pi\epsilon\mathbf{u}$, with ϵ being the static dielectric constant, b is the band index, and N_b is the number of bands, i.e., 8 or 14 in the k·p Hamiltonian. The integral 13.9 can be rewritten as

$$V_{ijkl} = \int_\Omega d^3\mathbf{r} \int_\Omega d^3\mathbf{r}' B_{ij}(\mathbf{r}) V(|\mathbf{r}-\mathbf{r}'|) B_{kl}(\mathbf{r}'), \tag{13.10}$$

where we introduce

$$B_{ij}(\mathbf{r}) = \sum_{b=1}^{N_b} \psi_b^{(i)}(\mathbf{r})^* \psi_b^{(j)}(\mathbf{r}). \tag{13.11}$$

The integral 13.10 can be calculated by performing a six-dimensional integration in real space, which is, however, numerically very demanding, especially when a large number of integrals has to be calculated, as in the CI approach [33,58]. One of the advantages of the PW method is that relevant physical quantities can be expressed in terms of the coefficients in the PW expansion, i.e., by the Fourier transform (FT). The last statement holds in principle when the evaluation of Coulomb integrals is concerned, since the FT of the Coulomb potential is fully analytical. This can be shown by the following set of transformations leading to Equation 13.16. Defining the PW expansion of $B_{ij}(\mathbf{r})$

$$B_{ij}(\mathbf{r}) = \sum_{\mathbf{k}} B_{ij}(\mathbf{k})e^{i\mathbf{k}\cdot\mathbf{r}} \tag{13.12}$$

and putting the last expression into Equation 13.10 one obtains

$$V_{ijkl} = \sum_{\mathbf{k}} B_{ij}(\mathbf{k}) \sum_{\mathbf{k}'} B_{kl}(\mathbf{k}') \int_{\Omega} d^3\mathbf{r} \int_{\Omega} d^3\mathbf{r}'\, e^{i\mathbf{k}\cdot\mathbf{r}} V(|\mathbf{r} - \mathbf{r}'|)e^{i\mathbf{k}'\cdot\mathbf{r}'}. \tag{13.13}$$

Using the properties of convolution and Parseval's theorem in inverse space, the $B_{ij}(\mathbf{k})$ can be expressed in terms in the coefficients in the PW expansion of the envelope functions as

$$B_{ij}(\mathbf{k}) = \sum_{b=1}^{N_b} \sum_{\mathbf{q}} A_{b,\mathbf{q}}^{(i)*} A_{b,\mathbf{q}+\mathbf{k}}^{(j)}. \tag{13.14}$$

By introducing the approximation changing the domain of integration in one of the integrals in Equation 13.13 from Ω to the whole space (which is valid when Ω is large enough) one gets after the replacement of variables from $\mathbf{r}$ and $\mathbf{r}'$, to $\mathbf{r}$ and $\mathbf{u} = \mathbf{r} - \mathbf{r}'$

$$V_{ijkl} = \sum_{\mathbf{k}} B_{ij}(\mathbf{k}) \sum_{\mathbf{k}'} B_{kl}(\mathbf{k}')[\int_{\Omega} d^3\mathbf{r}\, e^{i\mathbf{k}\cdot\mathbf{r}} e^{i\mathbf{k}'\cdot\mathbf{r}}][\int d^3\mathbf{u}\, V(|\mathbf{u}|)e^{-i\mathbf{k}'\cdot\mathbf{u}}]. \tag{13.15}$$

Using the relations

$$\int d^3\mathbf{u}\, e^{-i\mathbf{k}'\mathbf{u}} V(|\mathbf{u}|) = \frac{e^2}{\epsilon k'^2}$$

and

$$\frac{1}{\Omega} \int_{\Omega} d^3\mathbf{r}\, e^{i(\mathbf{k}+\mathbf{k}')\mathbf{r}} = \delta_{\mathbf{k}+\mathbf{k}',0}$$

one obtains

$$V_{ijkl} = \Omega \sum_{\substack{\mathbf{k} \\ \mathbf{k}\neq 0}} B_{ij}(\mathbf{k}) B_{kl}(-\mathbf{k}) \frac{e^2}{\epsilon k^2}. \tag{13.16}$$

13.5 The Effect of Dielectric Confinement

For colloidal QDs, the dielectric constant ϵ of the QD material is typically much larger than that of the surrounding medium, and the spatial variation $\epsilon = \epsilon(\mathbf{r})$ cannot be ignored. Such dielectric contrast means that any free charge in the QD induces polarization charge in the QD and its surroundings. As a consequence, the Coulomb energy for the electron–hole pair for a system with a spatially varying dielectric constant is

$$V(\mathbf{r}_e, \mathbf{r}_h) = V_c(\mathbf{r}_e, \mathbf{r}_h) + V_s(\mathbf{r}_e) + V_s(\mathbf{r}_h), \tag{13.17}$$

where V_c is the interparticle Coulomb potential and V_s is the self-polarization potential. The potential, $V_c(\mathbf{r}_e, \mathbf{r}_h)$, encompasses both the "direct" interparticle Coulomb interaction and the interface polarization potential due to the interaction between a real particle and the induced charge of the other particle. The self-polarization potential $V_s(\mathbf{r})$ arises from the interaction of a particle and its own induced charge.

The interparticle Coulomb potential in real space:

$$V_c(\mathbf{r}_e, \mathbf{r}_h) = \frac{1}{\epsilon(\mathbf{r}_e, \mathbf{r}_h)} \frac{e^2}{4\pi\epsilon_0 |\mathbf{r}_e - \mathbf{r}_h|}, \tag{13.18}$$

under the assumption, $\epsilon(\mathbf{r}_e, \mathbf{r}_h) \approx \epsilon(\mathbf{r}_e - \mathbf{r}_h) = \epsilon(\mathbf{r})$ [59], can be recast in the form of product of two functions

$$V_c(\mathbf{r}) = \frac{1}{\epsilon(\mathbf{r})} \frac{e^2}{4\pi\epsilon_0 |\mathbf{r}|} = \frac{1}{\epsilon(\mathbf{r})} U(\mathbf{r}), \tag{13.19}$$

where $U(\mathbf{r}) = e^2/4\pi\epsilon_0 \mathbf{r}$ is the bare Coulomb potential due to charge e, with the well-known FT in the inverse space: $\mathcal{F}[U(\mathbf{r})] = \tilde{U}(\mathbf{k}) = e^2/\epsilon_0 k^2$. Considering stepwise uniform regions where $\epsilon(r) = \epsilon_m = $ const., and m is the index of the region, and spherical QD shape, the general expression, Equation 13.18, acquires the simple analytical form:

$$\tilde{V}_c(\mathbf{k}) = \frac{e^2}{\epsilon_0 k^2} \left[\frac{1}{\epsilon_c} + \left(\frac{1}{\epsilon_s} - \frac{1}{\epsilon_c} \right) \cos(kR_c) + \left(\frac{1}{\epsilon_{\text{coll.}}} - \frac{1}{\epsilon_s} \right) \cos(kR_s) \right], \tag{13.20}$$

where ϵ_c, ϵ_s, and $\epsilon_{\text{coll.}}$ are the dielectric constants of the QD's core, shell, and surrounding colloid respectively, and R_c and R_s are the radius of core and outer radius of the shell, respectively.

Calculation of the self-polarization potential for the arbitrary QD shape in the inverse space is a bit involved. Alternatively, for several characteristic QD shapes, like spherical QD [60], spherical core/shell QD [61,62], elliptical QD [63], elliptical core/shell QD [64–66], and cuboidal QD [67], there are available expressions for $V_s(\mathbf{r})$ in real space. Here, we list a general expression for the self-polarization potential of spherical QD with an arbitrary number, m, of core/shells:

$$V_s^m(r) = \frac{e^2}{8\pi\epsilon_0 \epsilon_m} \sum_{l=0}^{\infty} \frac{1}{1 - p_{m,l} q_{m,l}} \left[p_{m,l} r^{2l} + p_{m,l} q_{m,l} r^{-1} + q_{m,l} r^{-2(l+1)} \right], \tag{13.21}$$

where $p_{m,l}$ and $q_{m,l}$ are recursive coefficients that depend on the dielectric constants, ϵ_m, and radii, R_m, in the region m, of the core/shell structures (see Ref. [61]). Such expressions are then simple to FT into $\tilde{V}_s(\mathbf{k})$ and are used when calculating the Coulomb matrix elements, V_{ijkl} for the CI Hamiltonian; see Section 13.6. The expression for the contribution of the direct Coulomb, Equation 13.20, and self-polarization energy, Equation 13.21, to the total Coulomb integral reads as

$$V_{ijkl}^{(c)} = \Omega \sum_{\substack{\mathbf{k} \\ \mathbf{k} \neq 0}} B_{ij}(\mathbf{k}) B_{kl}(-\mathbf{k}) \tilde{V}_c(\mathbf{k}) \tag{13.22}$$

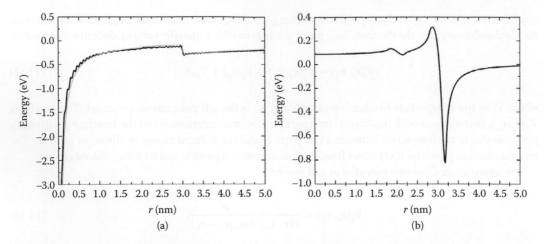

FIGURE 13.2 Variation of the interparticle Coulomb potential (a), $V_c(\mathbf{r})$, and self-polarization potential (b), $V_s(\mathbf{r})$, of CdSe/CdTe core shell quantum dot (QD) with $\epsilon_c = 7.1$, $\epsilon_s = 6.2$, $\epsilon_{coll.} = 2$, and $R_c = 2$ nm and $R_s = 3$ nm. In (a), the tiny bright line represents analytic $V_c(r)$ obtained by Equation 13.19, while solid thick line represents the same $V_c(r)$ recovered from the analytical expression in the inverse space, Equation 13.20.

and

$$V^{(s)}_{ijkl} = \Omega \sum_{\mathbf{k}} [B_{ij}(\mathbf{k})\delta_{kl} + \delta_{ij}B_{kl}(\mathbf{k})]\tilde{V}_s(\mathbf{k}). \tag{13.23}$$

In colloidal core/shell QDs, the self-polarization potential is characterized by a small peak and well near the core/shell interface due to the small dielectric mismatch between the core and shell materials. However, a much larger peak just inside $r = R_c + R_s$ and a deep well slightly outside the QD are due to the far greater dielectric mismatch of the shell and matrix material. In Figure 13.2, we present the radial variation of $V_c(\mathbf{r})$ and $V_s(\mathbf{r})$ of CdSe/CdTe core shell QD with $\epsilon_c = 7.1$, $\epsilon_s = 6.2$, $\epsilon_{coll.} = 2$, and $R_c = 2$ nm and $R_s = 3$ nm.

13.6 The Configuration Interaction Hamiltonian for NQD

In this section, we derive the Hamiltonian describing interacting electrons and holes in QDs. This Hamiltonian can be used to calculate neutral and charged excitonic and multiexcitonic states that are populated upon photoexcitation of the QD.

We start with a general Hamiltonian describing interacting electrons in the solid

$$H = H_s + H_{int}, \tag{13.24}$$

where the noninteracting part of the Hamiltonian is

$$H_s = \int d^3x\, \psi^\dagger(x) \left[-\frac{\hbar^2}{2m_0}\nabla^2 + V(x) \right] \psi(x)\, d^3x \tag{13.25}$$

and the interaction part is

$$H_{int} = \frac{1}{2} \int d^3x\, d^3y\, \psi^\dagger(x)\, \psi^\dagger(y)\, V_{int}(x,y)\, \psi(y)\, \psi(x). \tag{13.26}$$

In previous equations, $\psi(x)$ is the fermionic field operator, $\psi^\dagger(x)$ is its adjoint operator, V is the potential of atomic cores in the solid, and $V_{int}(x,y)$ is the interaction potential between the particles at x and y, which for the bare Coulomb interaction reads $V_{int}(x,y) = \frac{e^2}{4\pi\epsilon_0|x-y|}$.

Next, we express the field operator in terms of annihilation operators as

$$\psi(x) = \sum_{v \in VB} a_v \phi_v(x) + \sum_{c \in CB} a_c \phi_c(x), \tag{13.27}$$

where $\phi_c(x)$ and $\phi_v(x)$ are the SP wave function for states in the CB and VB, while a_c and a_v are the corresponding electron annihilation operators. The operator a_c acts on arbitrary many-body state of the system by destroying an electron in state c and a_v acts by destroying an electron in state v. Their adjoint operators $a_c^\dagger$ and $a_v^\dagger$ act by creating an electron in the corresponding state. After substitution of Equation 13.27 into Equations 13.25 and 13.26, one arrives at the expressions

$$H_s = \sum_{j_1 j_2} a_{j_1}^\dagger a_{j_2} \int d^3x\, \phi_{j_1}^*(x) \left[-\frac{\hbar^2}{2m_0}\nabla^2 + V(x) \right] \phi_{j_2}(x) \tag{13.28}$$

$$H_{int} = \frac{1}{2} \sum_{j_1 j_2 j_3 j_4} a_{j_1}^\dagger a_{j_2}^\dagger a_{j_3} a_{j_4} \int d^3x\, d^3y\, \phi_{j_1}^*(x) \phi_{j_4}(x) V_{int}(x,y) \phi_{j_2}^*(y) \phi_{j_3}(y), \tag{13.29}$$

i.e.,

$$H = \sum_{j_1 j_2} v_{j_1 j_2} a_{j_1}^\dagger a_{j_2} + \frac{1}{2} \sum_{j_1 j_2 j_3 j_4} V_{j_1 j_4 j_2 j_3} a_{j_1}^\dagger a_{j_2}^\dagger a_{j_3} a_{j_4}, \tag{13.30}$$

where

$$v_{j_1 j_2} = \int d^3x\, \phi_{j_1}^*(x) \left[-\frac{\hbar^2}{2m_0}\nabla^2 + V(x) \right] \phi_{j_2}(x) \tag{13.31}$$

$$V_{j_1 j_4 j_2 j_3} = \int d^3x\, d^3y\, \phi_{j_1}^*(x) \phi_{j_4}(x) V_{int}(x,y) \phi_{j_2}^*(y) \phi_{j_3}(y). \tag{13.32}$$

The summations in Equation 13.30 should in principle include all SP states in the valence and the CB. However, we are interested mainly in the lowest energy excitations of the system where electrons at the bottom of the CB and holes at the top of the VB are created. To study these excitations, it is more convenient to use electron and hole creation and annihilation operators $e_j(e_j^\dagger)$ and $h_j(h_j^\dagger)$ defined as $e_j = a_j$, when $j \in CB$ and $h_j = a_j^\dagger$ when $j \in VB$. These operators satisfy the following fermionic commutation rules: $\{e_i, e_j^\dagger\} = \delta_{ij}$, $\{h_i, h_j^\dagger\} = \delta_{ij}$, $\{e_i, e_j\} = 0$, $\{h_i, h_j\} = 0$, $\{e_i^\dagger, h_j\} = 0$, $\{e_i, h_j^\dagger\} = 0$, where $\{a, b\} = ab + ba$ denotes the anticommutator of two operators. The H_s term then reads

$$H_s = \sum_{j_1 j_2 \in CB} v_{j_1 j_2} e_{j_1}^\dagger e_{j_2} - \sum_{j_1 j_2 \in VB} v_{j_2 j_1} h_{j_1}^\dagger h_{j_2} + \sum_{j \in VB} v_{jj}, \tag{13.33}$$

where the last term contributes an irrelevant constant shift and can be omitted from further consideration. To obtain the H_{int} term, we first introduce an approximation to consider only the terms in H_{int} that contain the same number of electron creation and annihilation operators, as well as the same number of hole creation and annihilation operators. The other terms that do not conserve the number of pair excitations give rise to the polarization of the electronic orbitals, and their main effects can be taken into account by a dielectric constant in the Coulomb potential, which can be inserted into the final result [68].

Under such an assumption the H_{int} term consists of three sets of terms $H_{\text{int}} = H_{\text{int}}^{(1)} + H_{\text{int}}^{(2)} + H_{\text{int}}^{(3)}$. The first set of these terms (when $j_1, j_2, j_3, j_4 \in$ CB) reads:

$$H_{\text{int}}^{(1)} = \frac{1}{2} \sum_{j_1 j_2 j_3 j_4} V_{j_1 j_4 j_2 j_3} e_{j_1}^\dagger e_{j_2}^\dagger e_{j_3} e_{j_4}. \tag{13.34}$$

The second set of these terms $H_{\text{int}}^{(2)}$ that is obtained when $j_1, j_2, j_3, j_4 \in$ VB contains the term $h_{j_1} h_{j_2} h_{j_3}^\dagger h_{j_4}^\dagger$. By exploiting the fermionic commutation relations, this term can be rewritten as

$$h_{j_1} h_{j_2} h_{j_3}^\dagger h_{j_4}^\dagger = \delta_{j_2 j_3} \delta_{j_1 j_4} - \delta_{j_1 j_3} \delta_{j_2 j_4} - \delta_{j_2 j_3} h_{j_4}^\dagger h_{j_1} - \delta_{j_1 j_4} h_{j_3}^\dagger h_{j_2} + \delta_{j_1 j_3} h_{j_4}^\dagger h_{j_2} + \delta_{j_2 j_4} h_{j_3}^\dagger h_{j_1} + h_{j_3}^\dagger h_{j_4}^\dagger h_{j_1} h_{j_2}. \tag{13.35}$$

The $H_{\text{int}}^{(2)}$ term is then given as

$$\begin{aligned}
H_{\text{int}}^{(2)} = {} & \frac{1}{2} \sum_{j_1 j_2} V_{j_1 j_1 j_2 j_2} - \frac{1}{2} \sum_{j_1 j_2} V_{j_1 j_2 j_2 j_1} - \frac{1}{2} \sum_{j_1 j_2 j_4} V_{j_1 j_4 j_2 j_2} h_{j_4}^\dagger h_{j_1} - \frac{1}{2} \sum_{j_1 j_2 j_3} V_{j_1 j_1 j_2 j_3} h_{j_3}^\dagger h_{j_2} \\
& + \frac{1}{2} \sum_{j_1 j_2 j_4} V_{j_1 j_4 j_2 j_1} h_{j_4}^\dagger h_{j_2} + \frac{1}{2} \sum_{j_1 j_2 j_3} V_{j_1 j_2 j_2 j_3} h_{j_3}^\dagger h_{j_1} + \frac{1}{2} \sum_{j_1 j_2 j_3 j_4} V_{j_1 j_4 j_2 j_3} h_{j_3}^\dagger h_{j_4}^\dagger h_{j_1} h_{j_2}.
\end{aligned} \tag{13.36}$$

In Equation 13.36, the first and the second terms describe the Coulomb direct and exchange interactions between the carriers in fully occupied VB. These terms lead to an irrelevant shift of the total energy and can be omitted from the Hamiltonian. The third and the fourth terms describe the direct Coulomb interaction of holes with fully occupied VB, while the fifth and the sixth terms describe the exchange interaction of holes with fully occupied VB. The third, fourth, fifth, and sixth terms are of the same mathematical form as the second term in H_s [Equation 13.33] and can be considered to originate from the effective potential that describes the interaction of holes with occupied VB. Consequently, these terms can be incorporated into H_s. Finally, the last term in Equation 13.36 describes the hole–hole interaction.

The third set of terms, $H_{\text{int}}^{(3)}$, includes the terms that contain two electron and two hole operators. Among these terms, the terms that contain $h_{j_1} e_{j_2}^\dagger e_{j_3} h_{j_4}^\dagger$ and $e_{j_1}^\dagger h_{j_2} h_{j_3}^\dagger e_{j_4}$ contribute equally to $H_{\text{int}}^{(3)}$. The same holds for the terms that contain $h_{j_1} e_{j_2}^\dagger h_{j_3}^\dagger e_{j_4}$ and $e_{j_1}^\dagger h_{j_2} e_{j_3} h_{j_4}^\dagger$. By exploiting the identities

$$h_{j_1} e_{j_2}^\dagger e_{j_3} h_{j_4}^\dagger = e_{j_2}^\dagger e_{j_3} h_{j_1} h_{j_4}^\dagger = \delta_{j_1 j_4} e_{j_2}^\dagger e_{j_3} + e_{j_2}^\dagger h_{j_4}^\dagger e_{j_3} h_{j_1} \tag{13.37}$$

and

$$h_{j_1} e_{j_2}^\dagger h_{j_3}^\dagger e_{j_4} = -e_{j_2}^\dagger e_{j_4} h_{j_1} h_{j_3}^\dagger = -\delta_{j_1 j_3} e_{j_2}^\dagger e_{j_4} - e_{j_2}^\dagger h_{j_3}^\dagger e_{j_4} h_{j_1} \tag{13.38}$$

one obtains

$$\begin{aligned}
H_{\text{int}}^{(3)} = {} & \sum_{j_1 j_2 j_3 j_4} V_{j_1 j_4 j_2 j_3} e_{j_2}^\dagger h_{j_4}^\dagger e_{j_3} h_{j_1} - \sum_{j_1 j_2 j_3 j_4} V_{j_1 j_4 j_2 j_3} e_{j_2}^\dagger h_{j_4}^\dagger e_{j_3} h_{j_1} \\
& + \sum_{j_1 j_2 j_3} V_{j_1 j_1 j_2 j_3} e_{j_2}^\dagger e_{j_3} - \sum_{j_1 j_2 j_4} V_{j_1 j_4 j_2 j_1} e_{j_2}^\dagger e_{j_4}.
\end{aligned} \tag{13.39}$$

The last two terms in Equation 13.39 describe direct and exchange Coulomb interaction of electrons with fully occupied VB. In the same way as similar terms for holes these can be incorporated into H_s. One thus obtains the Hamiltonian

$$
\begin{aligned}
H = & \sum_{j_1 j_2 \in CB} \tilde{v}_{j_1 j_2} e_{j_1}^\dagger e_{j_2} - \sum_{j_1 j_2 \in VB} \tilde{v}_{j_2 j_1} h_{j_1}^\dagger h_{j_2} \\
& + \frac{1}{2} \sum_{ijkl} V_{iljk} e_i^\dagger e_j^\dagger e_k e_l + \frac{1}{2} \sum_{ijkl} V_{likj} h_i^\dagger h_j^\dagger h_k h_l \\
& - \sum_{ijkl} e_i^\dagger h_j^\dagger h_k e_l \left(V_{ilkj} - V_{ijkl} \right).
\end{aligned}
\tag{13.40}
$$

So far, we have not specified the SP states $\phi_c(x)$ and $\phi_v(x)$. If we choose these states to be eigenstates of the effective SP Hamiltonian $\tilde{v}_{ij}$, Equation 13.40 reduces to

$$
\begin{aligned}
H = & \sum_{j \in CB} \varepsilon_j e_j^\dagger e_j - \sum_{j \in VB} \varepsilon_j h_j^\dagger h_j \\
& + \frac{1}{2} \sum_{ijkl} V_{iljk} e_i^\dagger e_j^\dagger e_k e_l + \frac{1}{2} \sum_{ijkl} V_{likj} h_i^\dagger h_j^\dagger h_k h_l \\
& - \sum_{ijkl} e_i^\dagger h_j^\dagger h_k e_l \left(V_{ilkj} - V_{ijkl} \right).
\end{aligned}
\tag{13.41}
$$

To obtain the energies of exciton states, one needs to diagonalize the Hamiltonian given in Equation 13.41. Since all terms in the Hamiltonian contain only the operators that conserve the number of electrons and the number of holes, one can separately find exciton states (one electron and one hole), biexciton states (two electrons and two holes), negative trions (two electrons and one hole), etc. For example, to find the exciton states one has to diagonalize the Hamiltonian in the Hilbert space spanned by the vectors $e_a^\dagger h_b^\dagger |G\rangle$, where $|G\rangle$ is the ground state of the system (fully occupied VB and empty CB), $a \in CB$ and $b \in VB$. Each element of the Hamiltonian matrix depends on Coulomb integrals defined in Equation 13.32. When the **k·p** method is used to evaluate the SP wave functions, these are expressed as

$$
\phi_j(x) = \sum_b \psi_j^{(b)}(x) u_b(x),
\tag{13.42}
$$

where b is the band index, $u_b(x)$ is the Bloch function of underlying crystal, and $\psi_j^{(b)}$ is the slowly varying envelope function. The Coulomb integral (Equation 13.32) then reads

$$
\begin{aligned}
V_{j_1 j_4 j_2 j_3} = & \sum_{b_1 b_2 b_3 b_4} \int d^3 x \, d^3 y \, \psi_{b_1}^{(j_1)}(x)^* \psi_{b_4}^{(j_4)}(x) V_{int}(x, y) \psi_{b_2}^{(j_2)}(y)^* \psi_{b_3}^{(j_3)}(y) \times \\
& \times u_{b_1}^*(x) u_{b_4}(x) u_{b_2}^*(y) u_{b_3}(y).
\end{aligned}
\tag{13.43}
$$

Next, we exploit the fact that for the slowly varying function $f(\mathbf{u})$ and rapidly varying periodic function $g(\mathbf{u})$, the following approximation holds

$$
\int d\mathbf{u} f(\mathbf{u}) g(\mathbf{u}) \approx \int d\mathbf{u} f(\mathbf{u}) \langle g(\mathbf{u}) \rangle_T,
\tag{13.44}
$$

where $\langle g(\mathbf{u}) \rangle_T$ is the average of the function $g(\mathbf{u})$ over its period. In Equation 13.43, the terms in the first line define a slowly varying function, while the terms in the second line define a rapidly varying function whose average over the period is

$$\left\langle u^*_{b_1}(x)\, u_{b_4}(x)\, u^*_{b_2}(y)\, u_{b_3}(y) \right\rangle_T = \delta_{b_1 b_4} \delta_{b_2 b_3}, \tag{13.45}$$

due to orthonormality of the Bloch functions. Consequently, the Coulomb integral 13.43 can be expressed only in terms of envelope functions as

$$V_{j_1 j_4 j_2 j_3} = \sum_{b_1 b_2} \int d^3 x\, d^3 y\, \psi^{(j_1)}_{b_1}(x)^* \, \psi^{(j_4)}_{b_1}(x)\, V_{\text{int}}(x,y)\, \psi^{(j_2)}_{b_2}(y)^* \, \psi^{(j_3)}_{b_2}(y). \tag{13.46}$$

These integrals can be evaluated in the PW representation using the expression given by Equation 13.16 in Section 13.4. More details on the implementation of a numerically efficient scheme for evaluation of Coulomb integrals when the PW basis is used to represent the envelope functions is given in Ref. [33].

13.7 Transport Properties of QD Nanocrystal Arrays

In this section, we discuss electrical transport properties of colloidal QD nanocrystal arrays and we derive the formula for the charge carrier hopping probability from one dot to another.

QDs obtained by colloidal chemistry typically contain ligand molecules that passivate the surface and eliminate unwanted surface trap states. However, the ligands also increase the separation between the neighboring dots and consequently decrease the wave function overlap between the electron wave functions in neighboring dots (and their electronic coupling). The presence of ligands, therefore, significantly inhibits charge transport between neighboring dots. The approaches where short ligands are used enable the improvement of electrical transport between the dots.

To engineer the transport in QD arrays, it is highly desirable to have a theoretical approach for calculating the electrical transport properties of these systems. Since electronic coupling between the dots is weak, it can be considered a perturbation that leads to occasional carrier hopping from one dot to another. To evaluate the hopping rate of charge carrier between the dots, we model the system using the following Hamiltonian

$$H = H_1 + H_2 + V. \tag{13.47}$$

In the previous equation, H_1 is the Hamiltonian of the first dot, H_2 is the Hamiltonian of the second dot, and V is the part of the Hamiltonian that describes the electronic coupling between two dots. For notational simplicity, we will assume that each of the dots accommodates only one electronic energy level that is coupled to all vibrational modes of that dot. The Hamiltonian H_1 is then given as

$$H_1 = \varepsilon_1 a^\dagger_1 a_1 + \sum_s \hbar\omega_{1s} b^\dagger_{1s} b_{1s} + \sum_s g_{1s} \hbar\omega_{1s} a^\dagger_1 a_1 \left(b^\dagger_{1s} + b_{1s} \right). \tag{13.48}$$

In the previous equation, a_1 is the electron annihilation operator that annihilates an electron in state of energy ε_1, g_{1s} is the dimensionless coupling constant describing the coupling of electron in dot 1 to the phonon mode s of dot 1, and $\hbar\omega_{1s}$ is the energy of that phonon mode, while b_{1s} is the corresponding phonon annihilation operator. Electronic operators satisfy fermionic commutation relations, while phonon operators satisfy bosonic commutation relations. Analogously, the Hamiltonian H_2 reads

$$H_2 = \varepsilon_2 a_2^\dagger a_2 + \sum_s \hbar\omega_{2s} b_{2s}^\dagger b_{2s} + \sum_s g_{2s}\hbar\omega_{2s} a_2^\dagger a_2 \left(b_{2s}^\dagger + b_{2s} \right), \tag{13.49}$$

where the notation is analogous to that in the Hamiltonian H_1. The last term in Equation 13.47 reads

$$V = -t \left(a_1^\dagger a_2 + a_2^\dagger a_1 \right), \tag{13.50}$$

where t is the electronic coupling parameter between the two electronic states in dots 1 and 2.

We are interested in determining the transition rate for the process in which an electron that is initially in dot 1 makes a transition to dot 2. Possible initial states for this process are $|i\rangle = a_1^\dagger |i_{\mathrm{ph}}\rangle$, where $|i_{\mathrm{ph}}\rangle$ denotes a state of the phonon subsystem. The final state is given as $|f\rangle = a_2^\dagger |f_{\mathrm{ph}}\rangle$, where $|f_{\mathrm{ph}}\rangle$ is the final state of the phonon subsystem. The transition rate is then given by Fermi's Golden Rule rule expression

$$W = \frac{2\pi}{\hbar} \sum_i p_i \sum_f \left| V_{if} \right|^2 \delta\left(E_i - E_f \right), \tag{13.51}$$

where p_i is the probability that the initial state is $|i\rangle$ $V_{if} = \langle i|V|f\rangle$ is the matrix element of the operator V, while E_i and E_f are energies of the unperturbed system in the initial and final states. We will transform this conventional expression to a form that is more convenient for our analysis. Using the identities

$$\delta\left(E_i - E_f \right) = \frac{1}{2\pi\hbar} \int_{-\infty}^{\infty} dt \exp\left[\frac{i}{\hbar}\left(E_i - E_f \right) t \right], \tag{13.52}$$

$$\left| V_{if} \right|^2 = \langle i|V|f\rangle \langle f|V|i\rangle, \tag{13.53}$$

and

$$\langle i|V|f\rangle \exp\left[\frac{i}{\hbar}\left(E_i - E_f \right) t \right] = \left\langle i \left| \exp\left[\frac{i}{\hbar} H_0 t \right] V \exp\left[-\frac{i}{\hbar} H_0 t \right] \right| f \right\rangle \tag{13.54}$$

where $H_0 = H_1 + H_2$, we arrive at the expression for transition rates that is convenient for our analysis

$$W = \frac{1}{\hbar^2} \int_{-\infty}^{\infty} dt \sum_i p_i \sum_f \langle i|V_I(t)|f\rangle \langle f|V_I(0)|i\rangle, \tag{13.55}$$

where $V_I(t) = \exp\left[\frac{i}{\hbar} H_0 t \right] V \exp\left[-\frac{i}{\hbar} H_0 t \right]$.

The Hamiltonians H_1 and H_2 can be diagonalized exactly using the unitary transformation

$$U = \exp\left[a_1^\dagger a_1 \sum_s g_{1s}\left(b_{1s}^\dagger - b_{1s} \right) \right] \exp\left[a_2^\dagger a_2 \sum_s g_{2s}\left(b_{2s}^\dagger - b_{2s} \right) \right]. \tag{13.56}$$

This transformation acts on electron and phonon operators as

$$Ua_i U^\dagger = a_i \exp\left[\sum_s g_{is}\left(b_{is}^\dagger - b_{is}\right)\right] \tag{13.57}$$

and

$$Ub_{is}U^\dagger = b_{is} - g_{is}a_i^\dagger a_i. \tag{13.58}$$

The transformed Hamiltonian then reads

$$\tilde{H} = UHU^\dagger = \tilde{H}_1 + \tilde{H}_2 + \tilde{V}, \tag{13.59}$$

with

$$\tilde{H}_1 = \left(\varepsilon_1 - \sum_s g_{1s}^2 \hbar\omega_s\right)a_1^\dagger a_1 + \sum_s \hbar\omega_{1s}b_{1s}^\dagger b_{1s}, \tag{13.60}$$

$$\tilde{H}_2 = \left(\varepsilon_2 - \sum_s g_{2s}^2 \hbar\omega_s\right)a_2^\dagger a_2 + \sum_s \hbar\omega_{2s}b_{2s}^\dagger b_{2s}, \tag{13.61}$$

and

$$\tilde{V} = -t\left[a_1^\dagger a_2 \exp\left[-\sum_s g_{1s}\left(b_{1s} - b_{1s}^\dagger\right)\right]\exp\left[\sum_s g_{2s}\left(b_{2s} - b_{2s}^\dagger\right)\right] + c.c.\right]. \tag{13.62}$$

To obtain the expression for transition probability given by Equation 13.55, we first express the matrix elements as

$$\langle i|V_I(t)|f\rangle = \langle\Psi_i|\exp\left(\frac{i}{\hbar}\tilde{H}_0 t\right)\tilde{V}\exp\left(-\frac{i}{\hbar}\tilde{H}_0 t\right)|\Psi_f\rangle. \tag{13.63}$$

In the last expression $|\Psi_i\rangle = U|i\rangle$ and $|\Psi_f\rangle = U|f\rangle$. By noting that $\tilde{H}_0 = \tilde{H}_0^e + \tilde{H}_0^{ph}$ and $\exp\left[-\frac{i}{\hbar}\tilde{H}_0^e t\right]|\Psi_f\rangle = \exp\left[-\frac{i}{\hbar}\varepsilon_f t\right]|\Psi_f\rangle$, using Equations 13.62 and 13.55 one arrives at

$$W = \frac{t^2}{\hbar^2}\int_{-\infty}^{\infty} du \exp\left[\frac{i}{\hbar}\left(\varepsilon_i - \varepsilon_f\right)u\right]\sum_{i_{ph}} p_{i_{ph}}$$

$$\times \sum_{f_{ph}}\langle i_{ph}|\exp\left(\frac{i}{\hbar}\tilde{H}_0^{ph}u\right)\exp\left[-\sum_s g_{1s}\left(b_{1s} - b_{1s}^\dagger\right)\right]\exp\left[\sum_s g_{2s}\left(b_{2s} - b_{2s}^\dagger\right)\right]\exp\left(-\frac{i}{\hbar}\tilde{H}_0^{ph}u\right)|f_{ph}\rangle$$

$$\times \langle f_{ph}|\exp\left[\sum_s g_{1s}\left(b_{1s} - b_{1s}^\dagger\right)\right]\exp\left[-\sum_s g_{2s}\left(b_{2s} - b_{2s}^\dagger\right)\right]|i_{ph}\rangle. \tag{13.64}$$

Taking into account that $\sum_{f_{ph}} |f_{ph}\rangle\langle f_{ph}| = 1$ and $\sum_{i_{ph}} P_{i_{ph}} \langle i_{ph}|X|i_{ph}\rangle = \langle X\rangle_{ph}$, where $\langle X\rangle_{ph}$ denotes the expectation value of the operator X, we obtain

$$W = \frac{t^2}{\hbar^2} \int_{-\infty}^{\infty} du \exp\left[\frac{i}{\hbar}(\varepsilon_i - \varepsilon_f) u\right]$$

$$\times \left\langle \exp\left\{-\sum_s g_{1s}\left[b_{1s}(u) - b_{1s}^\dagger(u)\right]\right\} \exp\left\{\sum_s g_{1s}\left[b_{1s} - b_{1s}^\dagger\right]\right\}\right\rangle \qquad (13.65)$$

$$\times \left\langle \exp\left\{\sum_s g_{2s}\left[b_{2s}(u) - b_{2s}^\dagger(u)\right]\right\} \exp\left\{-\sum_s g_{2s}\left[b_{2s} - b_{2s}^\dagger\right]\right\}\right\rangle.$$

Next, we exploit the identity

$$\left\langle \exp\left\{-g_{is}\left[b_{is}(u) - b_{is}^\dagger(u)\right]\right\} \exp\left\{g_{is}\left[b_{is} - b_{is}^\dagger\right]\right\}\right\rangle =$$
$$\exp\left\{-g_{is}^2\left[2n_{is} + 1 - (n_{is} + 1)e^{-i\omega_{is}u} - n_{is}e^{i\omega_{is}u}\right]\right\}, \qquad (13.66)$$

where g_{is}, n_{is}, and $\hbar\omega_{is}$ are, respectively, the electron–phonon coupling constant, the phonon occupation number and the phonon energy for the sth phonon mode at site i. We finally obtain the transition rate as [69–71]

$$W = \frac{t^2}{\hbar^2} \int_{-\infty}^{\infty} du \exp\left[\frac{i}{\hbar}(\varepsilon_i - \varepsilon_f) u\right] \times$$
$$\prod_s \exp\left\{-g_{1s}^2\left[2n_{1s} + 1 - (n_{1s} + 1)e^{-i\omega_{1s}u} - n_{1s}e^{i\omega_{1s}u}\right]\right\} \times \qquad (13.67)$$
$$\prod_s \exp\left\{-g_{2s}^2\left[2n_{2s} + 1 - (n_{2s} + 1)e^{-i\omega_{2s}u} - n_{2s}e^{i\omega_{2s}u}\right]\right\}.$$

Equation 13.67 can be used directly to calculate the transition rate from one dot to another. In some cases, a simpler Marcus formula is often used instead of Equation 13.67. In what follows, we introduce additional approximations that lead to the reduction of Equation 13.67 to the Marcus formula. However, we alert the reader that the conditions for the validity of the Marcus formula should be carefully checked before that formula is used.

A detailed analysis of the exponential terms in Equation 13.67 leads to the conclusion that when electron–phonon interaction is strong ($\sum_s g_{is}^2 \gg 1$) exponential terms decay very quickly when $|\omega_{is}u| \gg 1$. Therefore, only the terms with $|\omega_{is}u| \ll 1$ contribute significantly to the integral. By keeping terms up to the quadratic in the Taylor expansion $e^{i\omega_{is}t} = 1 + i\omega_{is}t - \frac{1}{2}\omega_{is}^2 t^2$, one obtains

$$\prod_s \exp\left\{-g_{1s}^2\left[2n_{1s} + 1 - (n_{1s} + 1)e^{-i\omega_{1s}u} - n_{1s}e^{i\omega_{1s}u}\right]\right\} \approx$$
$$\prod_s \exp\left(i\omega_{1s}g_{1s}^2 u\right) \exp\left[-\frac{1}{2}g_{1s}^2\omega_{1s}^2 u^2 (2n_{1s} + 1)\right]. \qquad (13.68)$$

When the temperature is high, $k_B T \gg \hbar\omega_{is}$, the $(2n_{is} + 1)$ term reduces to $\frac{2k_B T}{\hbar\omega_{is}}$. Using that approximation one obtains

$$W \approx \frac{t^2}{\hbar^2} \int_{-\infty}^{\infty} du \exp\left[\frac{i}{\hbar}(\varepsilon_f - \varepsilon_i + \lambda_1 + \lambda_2) u\right] \exp\left[-\frac{k_B T}{\hbar^2}(\lambda_1 + \lambda_2) u^2\right], \qquad (13.69)$$

where $\lambda_i = \sum_s \hbar\omega_{is}\, g_{is}^2$. The solution of the last integral gives the Marcus formula

$$W \approx \frac{t^2}{\hbar}\sqrt{\frac{\pi}{k_B T \lambda}}\exp\left[-\frac{\left(\lambda + \varepsilon_f - \varepsilon_i\right)^2}{4 k_B T \lambda}\right], \tag{13.70}$$

where $\lambda = \lambda_1 + \lambda_2$.

13.8 Practical Examples: Cd-Based Chalcogenide QDs

13.8.1 Comparison with Experiment: Absorption Edge Wavelength of CdSe/CdTe Type II CQDs

In this section, we apply the theoretical approach described in previous sections to calculate the absorption wavelength in CdSe/CdTe core/shell QDs. We analyze the influence of CdSe/CdTe core/shell colloidal QD (CQD) morphology on the variation of the first exciton peak in the absorption spectra, which corresponds to excitation of the $1S_{1/2}^e 1S_{3/2}^h$ state. As a first step, we validate our theoretical methodology, Sections 13.3–13.6, based on the combination of **k·p** and CI Hamiltonians, against available experimental and theoretical results on CdSe and CdTe core-only CQD, as they are the constituent materials of our core/shell structure. Figure 13.3 shows the variation of the $1S_{1/2}^e 1S_{3/2}^h$ exciton energy as a function of CdSe and CdTe QD radius, a_c, respectively. Solid symbols represent experimental results from several independent measurements from different groups and the line is an empirical inverse polynomial fitting curve, $E_X^{1S_{1/2}1S_{3/2}}(a_c) = E_g^{bulk} + (Aa_c^2 + Ba_c + C)^{-1}$, to this experimental data proposed by de Mello Donega in Ref. [72], where A, B, and C are fitting parameters. Results of $1S_{1/2}^e 1S_{3/2}^h$ exciton energies predicted by our calculation are shown on the same figures and exhibit excellent agreement with the experimental measurements [73–82].

Having validated the methodology for core-only CQD structures, we now employ our method to describe CdSe/CdTe core/shell type-II CQD structures. The remaining question when modeling the system of core/shell structures is the parametrization of the valence band offset (VBO) between core and shell materials, in our case CdSe and CdTe, respectively. From our previous analysis [58] of the variation of $1S_{1/2}^e 1S_{3/2}^h$ and $1S_{1/2}^e 2S_{3/2}^h$ excitonic energies with the shell thickness in CdTe/CdSe core/shell CQD, we have estimated the $\text{VBO}_{CdTe/CdSe} = 0.4$ eV. We have assumed the same VBO for inverse structures, i.e., for CdSe/CdTe core/shell CQD considered here.

The $1S_{1/2}^{(e)} nS_{3/2}^{(h)}$ ($n = 1, 2$) states are the two lowest-energy excitons observed in the absorption spectra of colloidal CdTe/CdSe NCs [23,83,84], making them the most important for understanding the near band-edge absorption characteristics of such nanoparticles. Figure 13.3 shows the $1S_{1/2}^{(e)} 1S_{3/2}^{(h)}$ and $1S_{1/2}^{(e)} 2S_{3/2}^{(h)}$ exciton energies (solid lines) calculated by the CI as a function of shell thickness for CdTe/CdSe QDs with (a) $a_c = 1.7$ nm, (b) $a_c = 1.72$ nm, (c) $a_c = 1.75$ nm, and (d) $a_c = 1.95$ nm. Dashed lines show upper and lower limits on the exciton energies resulting from an uncertainty of 1 monolayer (ML) in the displayed core radii ($\sim \pm 0.3$ nm). Filled circles show exciton energies taken from the first and second absorption peak positions in absorption spectra measured by (a) Gong et al., [84], (b) Ma et al., [85], (c) Cai et al. [86], and (d) Oron et al. [23]. We see good quantitative agreement between the calculated exciton energies and the experimental data, with the data lying in the channels defined by an uncertainty of ± 1 ML width in the core size. It should be noted that the results of Oron et al. [23] were obtained on zinc-blende NC structures, in addition to those of Cai et al. [86] The papers by Gong et al. [84] and Ma et al. [85] do not explicitly state the crystal structures of the core/shell nanoparticles, although Ma et al. [85] note that their core/shell NCs gave very similar absorption and photoluminescence spectra to those of Cai and coworkers [86]. Our calculations accurately reproduce the 0.25 eV energy separation between the $1S_{1/2}^{(e)} 1S_{3/2}^{(h)}$ and $1S_{1/2}^{(e)} 2S_{3/2}^{(h)}$ excitons that is nearly independent of shell thickness [84]. This constant energy separation is

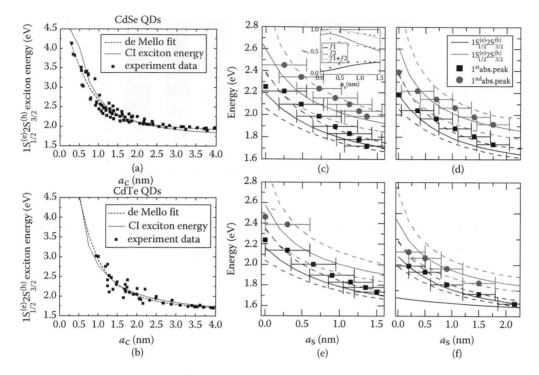

FIGURE 13.3 Variation of the $1S^e_{1/2}1S^h_{3/2}$ exciton energy as a function of colloquial quantum dot (CQD) core radius, a_c, in (a) CdSe and (b) CdTe CQDs: experimental results (square symbols), de Mello Donega's empirical fit to the experimental data (dotted line) [72], and theoretical model (solid line). Calculated (lines) energies of the $1S^{(e)}_{1/2}nS^{(h)}_{3/2}$ ($n = 1, 2$) excitons for CdTe/CdSe QDs with shell thickness a_s and core radius of (c) $a_c = 1.7$ nm, (d) $a_c = 1.72$ nm, (e) $a_c = 1.75$ nm, and (f) $a_c = 1.95$ nm; experimental data taken from Refs. [84], [85],[86], and [23] are shown as filled symbols. Error bars represent an uncertainty of 1 mono-layer (ML) ($\approx \pm 0.3$ nm) in the shell thickness. Dashed lines show upper and lower limits on the exciton energies resulting from an uncertainty of 1 ML in the nominal core radii.

characteristic of changing electron confinement but approximately constant hole confinement in the h/e heterostructure. We also find good agreement between the oscillator strength obtained by Gong et al. [84] from the absorption spectra and our calculations. Calculating the oscillator strength f_n of the $1S^{(e)}_{1/2}nS^{(h)}_{3/2}$ ($n = 1, 2$) excitons as $f_n = 2P^2_X/m_0E_X$ we find that $f_1 + f_2 \sim$ constant (inset Figure 13.3), confirming the validity of relevant excitonic wave functions too.

In contrast to core-only CQDs, the size dependence of $1S^e_{1/2}1S^h_{3/2}$ exciton energies in core/shell CQDs is more complex and it is not possible to capture its trend by a simple polynomial fitting curve. Figure 13.5 shows the shell thickness, a_s, dependence of the $1S^e_{1/2}1S^h_{3/2}$ absorption wavelengths for CdSe/CdTe core/shell CQDs for different core radii, a_c, ranging from a_c=1.5 nm to 2.5 nm. Also shown in Figure 13.5 are experimental data for a CdSe core of 1.7-nm radius, as measured by transmission electron microscopy, and different CdTe shell thicknesses [87,88]. Good agreement is seen between the calculated energy of the $1S^e_{1/2}1S^h_{3/2}$ exciton and the spectral position observed for the first absorption peak, further validating our theoretical approach.

To illustrate our theoretical model outlined in Section 13.6, in Figure 13.4, we show the excitonic spectra of an $a_c = 2$ nm CdSe core-only QD ($a_s = 0$ case) and a set of CdSe/CdTe core/shell QDs with shell thicknesses varying from $a_s = 0.5$ to 3.0 nm. By changing the shell thickness from $a_s = 0$ (no shell) to $a_s = 3$ nm the excitonic ground state energy, E_{X0}, is changed from 2.30 eV to 1.49 eV. It is also interesting to note that for certain shell thicknesses, i.e., between $a_s = 0.5$ and 1 nm, the character of the ground state exciton changes from $1S_{1/2}1S_{3/2}$ to $1S_{1/2}1P_{3/2}$. It can be observed that the density of excitonic states

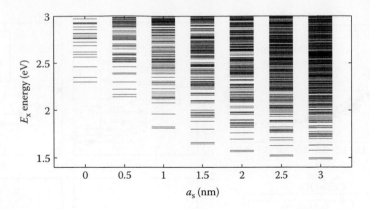

FIGURE 13.4 Excitonic spectra of CdSe and CdSe/CdTe quantum dots (QDs) as a function of the shell thickness, a_s.

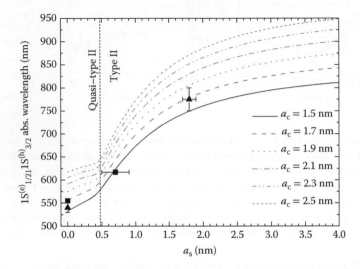

FIGURE 13.5 Core radius, a_c, and shell thickness, a_s, dependence of the $1S^e_{1/2}1S^h_{3/2}$ exciton peak in the absorption spectra of CdSe/CdTe core/shell quantum dots (QDs). Also shown for comparison are experimental data from Ref. [87] (squares) and Ref. [88] (triangles).

increases with shell thickness: this suggests that phononic gaps are more likely to be found in core-only or core/shell structures with thin shells [16].

In Figure 13.5, we can clearly distinguish between two trends in the $1S^e_{1/2}1S^h_{3/2}$ absorption wavelengths in the region of shell thicknesses <0.5 nm and for shell thicknesses >0.5 nm. For the QDs with shells of thickness approximately or less than 0.5 nm, we can see that (1) the $1S^e_{1/2}1S^h_{3/2}$ absorption wavelengths are approximately linearly dependent on a_s, suggesting strong influence of the dielectric confinement on $1S^e_{1/2}1S^h_{3/2}$ excitons and their wavelengths in this region and (2) as the core size, a_c, increases the gradient of the $1S^e_{1/2}1S^h_{3/2}$ absorption wavelength versus a_s curves decreases. We explain such nonmonotonic behavior in terms of the changing localization regime of the $1s^h_{3/2}$ SP hole, in the correlated $1S^e_{1/2}1S^h_{3/2}$ exciton, as the QD dimensions change [58]. The lower gradient of the curves for shell widths <0.5 nm is due to the fact that the $1s^h_{3/2}$ hole is in the delocalized regime, i.e., its probability density is spread over the whole heterostructure and its energy is mainly determined by the global confinement provided by the QD potential well of radius $a_c + a_s$. In this regime, the size dependence of the hole confinement is closer to that of one confined in a core-only QD. For the QDs with shells of width greater than 0.5 nm, we can see that the $1S^e_{1/2}1S^h_{3/2}$ absorption wavelengths again acquire a trend that can be described by a quadratic polynomial function. As a_s increases, for a particular core size, the $1s^h_{3/2}$ SP hole localizes in the shell fully so the absorption wavelength is again more strongly affected by the effect of shell thickness on the confinement energy of $1s^h_{3/2}$. This behavior is closer to the SCR in a core/shell heterostructure. Maps of the absorption wavelength against QD core/shell dimensions as depicted in Figure 13.5 can only be obtained numerically and should be of use to experimentalists studying such systems.

13.8.2 Radiative Lifetimes in CdSe/CdTe CQDs

Next, we analyze the radiative lifetimes in core/shell CdSe/CdTe CQDs [17]. In order to assess the variation of the radiative times, $\tau_{rad.}$, with the core size, a_c, and the shell thickness, a_s, we use the following expression:

$$\frac{1}{\tau_{rad.}} = \frac{1}{3}\frac{F^2\bar{n}e^2}{\pi\epsilon_0 c^3\hbar^4}E_X\frac{1}{d_X}\sum_{i=1}^{d_X}|\mathbf{P}_{X_i}|^2, \tag{13.71}$$

where $\bar{n}$ is the refractive index of the colloidal material, e is the electron charge, ϵ_0 is the permittivity of free space, m_0 is the electron rest mass, c is the speed of light, and $\hbar$ is the reduced Planck constant. In the expression above, $|\mathbf{P}_X|$ is the modulus of excitonic dipole matrix element, given in (eVÅ), and E_X is the excitonic energy, both obtained from the CI calculation, and d_X is the integer number representing the degree of degeneracy of a particular excitonic state. For the $1S^e_{1/2}1S^h_{3/2}$ exciton considered below the degeneracy is $d_X = 8$. The expression for dielectric screening of spherical core/shell QDs is given as $F = 9\epsilon_s\epsilon_{coll.}/(\epsilon_s\epsilon_a + 2\epsilon_{coll.}\epsilon_b)$, where $\epsilon_a = \epsilon_c(3 - 2\Omega_s/\Omega_{qd}) + 2\epsilon_s\Omega_s/\Omega_{qd}$, $\epsilon_b = \epsilon_c\Omega_s/\Omega_{qd} + \epsilon_s(3 - \Omega_s/\Omega_{qd})$, and Ω_{qd} and Ω_s are the volume of QD and shell, respectively [89].

Figure 13.6 shows the mean radiative lifetime of the $1S^e_{1/2}1S^h_{3/2}$ exciton in CdSe/CdTe CQDs as a function of shell thickness a_s for several different core radii. The most noticeable feature is the sudden increase in $\tau_{rad.}$ at around $a_s \sim 0.5$ nm. In comparison to core-only CQDs, in which the value of $1S^e_{1/2}1S^h_{3/2}$ excitonic radiative time τ_{rad} changes relatively weakly with the CQD size [35,90], in core/shell structures it is

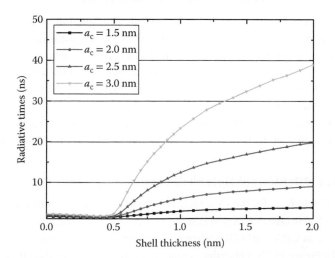

FIGURE 13.6 Plot of the $1S^e_{1/2}1S^h_{3/2}$ exciton radiative lifetime as a function of the shell thickness for different cores: $a_c = 1.5, 2.0, 2.5,$ and 3.0 nm in CdSe/CdTe colloquial quantum dots (CQDs).

possible to increase this radiative time over one order of magnitude with suitable change in the core and shell sizes. Several observations regarding the trend in τ_{rad} in CdSe/CdTe core/shell type II CQDs can be made here: (1) again as in the case of the absorption wavelength, shown in Figure 13.5, we can distinguish clearly between two different regions of the shell thicknesses $a_s < 0.5$ nm and $a_s > 0.5$ nm. In the region of $a_s < 0.5$ nm, the whole CQD system behaves similar to core-only CQDs, since both electron and hole correlated charge densities are either largely confined in the core region or just starting to delocalize over the whole CQD structure. In this region, the optical dipole matrix element, $|\mathbf{P}_X|$, is strong while the magnitude of both $|\mathbf{P}_X|$ and E_X changes very little with the overall sizes of QDs, which explains the almost constant τ_{rad}. (2) As shell thickness increases, and in particular beyond $a_s > 1$ nm, the hole states become strongly confined in the shell region. Here, dielectric confinement does not have enough strength to overcome the confinement imposed by the type II aligned VB edge of CdTe, so that the exciton reaches the SCR. Consequently, the electron–hole wave function overlap is dramatically reduced and E_X is mainly now determined by the variation of the $1s_{3/2}^{(h)}$ hole confinement with a_s. For $a_s > 1$ nm, τ_{rad} continues to increase, but more slowly, reaching values that are about one order of magnitude greater than those for core-only CQDs. For all shell thicknesses, a monotonic trend to larger τ_{rad} with increasing core size is also observed. It is interesting to observe that for CdSe/CdTe type II CQDs, the value of τ_{rad} increases over "only" one order of magnitude, while in other systems, such as epitaxially grown type II InAs/GaAs/GaAsSb structure, the radiative times can increase over three orders of magnitude [91,92]. This difference in behavior is attributed to the absence of significant dielectric confinement in epitaxially grown QDs, compared to CQDs.

13.8.3 Correlation Energy

The correlation energy of the exciton can be defined as

$$E_{corr} = E_{X,CI} - E_X, \tag{13.72}$$

where E_X is the exciton energy calculated according to first-order perturbation theory inside the strong confinement approximation for the exciton wave function [42,61,83,93], and $E_{X,CI}$ is the excitonic energy calculated using the CI method.

13.8.3.1 CdTe/CdSe QD: Effect of Electron Shell Localization

In Figure 13.7a, we present E_{corr} for the $1S_{1/2}^{(e)}1S_{3/2}^{(h)}$ exciton as a function of core radius for fixed shell $a_s = 2$ nm CdTe/CdSe QDs. We see that in the presence of dielectric confinement $|E_{corr}| \lesssim 20$ meV and that E_{corr} exhibits at least one minimum as a function of a_c in the $\epsilon = $ const. and $\epsilon = \epsilon(r)$ case. $|E_{corr}|$ is up to four times greater in the presence of dielectric confinement, $\epsilon = \epsilon(r)$, compared to the $\epsilon = $ const. case. This result highlights the importance of a proper treatment of the dielectric environment in such nanostructures. The minimum in E_{corr} for the $1S_{1/2}^{(e)}1S_{3/2}^{(h)}$ exciton, Figure 13.7a, is a consequence of two competing effects: proximity of the self-polarization potential peak, which tends to reduce the electron–hole separation and the effect of the type-II confinement profile, which tends to separate the carriers as a_c increases.

13.8.3.2 CdSe/CdTe QD: Effect of Hole Shell Localization

In Figure 13.7b, we present E_{corr} for the $1S_{1/2}^{(e)}1S_{3/2}^{(h)}$ exciton as a function of shell thickness for an $a_c = 3.5$ nm fixed core in CdSe/CdTe QDs. We observe the largest size correlation energy of the considered excitons for the $1S_{1/2}^{(e)}1S_{3/2}^{(h)}$ state in CdSe/CdTe QDs when $\epsilon = \epsilon(r)$, with E_{corr} reaching -62 meV for an $a_c = 3.5$ nm, $a_s = 0.6$ nm QD, Figure 13.7b. This value is more than six times larger than the corresponding value for the $\epsilon = $ const. case, highlighting the particularly strong effect of the dielectric environment on this exciton. It can be observed from Figure 13.7b that the effect of dielectric mismatch on E_{corr} is strongest

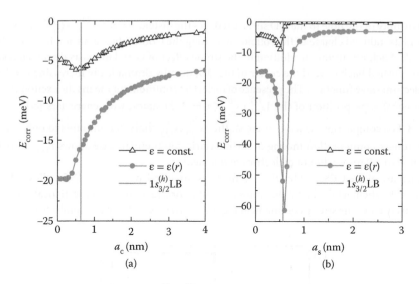

FIGURE 13.7 Correlation energy, E_{corr}, of $1S_{1/2}^{(e)}1S_{3/2}^{(h)}$ excitons in: (a) $a_s = 2.0$ nm CdTe/CdSe quantum dots (QDs), and (b) $a_c = 3.5$ nm CdSe/CdTe QDs. Cases of $\epsilon = $ const. and $\epsilon = \epsilon(r)$ are represented by open and solid symbols, respectively. Vertical line points to the localization/delocalization boundary (LB) of $1s_{3/2}$ hole state.

in the vicinity of the $1s_{3/2}^{(h)}$ localization boundary (LB), i.e., once the SP hole starts to move significantly toward the shell region and becomes delocalized over the whole QD.

$1S_{1/2}^{(e)}1S_{3/2}^{(h)}$ *exciton:* In the case of a core-only CdSe QD correlation causes both carriers, correlated electron and correlated hole, to move closest toward the center of the QD compared with their strongest SP character. This is purely a result of the direct interparticle Coulomb interaction. Introduction of the self-polarization potential, i.e., $\epsilon = \epsilon(r)$, further exaggerates this move of the radial probability densities (RPD) of both carriers away from the QD surface in the core-only CdSe QD, Figure 13.8a. This effect increases localization of both carriers near the center of QD, increasing overlap between them and giving correlation energies $E_{corr} = -18$ meV. For the $1S_{1/2}^{(e)}1S_{3/2}^{(h)}$ exciton in the CdSe QD, the shift in RPD is mainly due to an increase in $2s_{1/2}^{(e)}1s_{3/2}^{(h)}$ character.

To assess the effect of dielectric confinement on the correlated carriers in the CdSe QD, we consider the expectation value of the $1s$ electron (hole) radial coordinate, denoted $\langle r_{e(h)} \rangle$. When $\epsilon = $ const. (no self-polarization) we find $\langle r_h \rangle = 1.55$ nm compared to $\langle r_h \rangle = 1.44$ nm when $\epsilon = \epsilon(r)$ for the $1s_{3/2}^{(h)}$ state. In contrast, the effect of dielectric confinement moves the $1s_{1/2}^{(e)}$ electron from $\langle r_e \rangle = 2.01$ nm to $\langle r_e \rangle = 1.89$. Although the SP $1s_{1/2}^{(e)}$ RPD has significantly greater overlap with the repulsive peak in self-polarization potential near the QD surface than the SP $1s_{3/2}^{(h)}$ hole RPD, the correlated electron is shifted by dielectric confinement by almost the same distance as the correlated hole. These results reflect the larger sensitivity of the correlated hole wave function to the dielectric environment compared to the electron in the CdSe core-only QD.

In Figure 13.8b, we see that the introduction of a thin CdTe shell allows the uncorrelated hole to start to localize nearer the QD surface (at $r = a_c + a_s$), dramatically reducing its overlap with the uncorrelated electron. However, the effect of correlation is strong enough to pull the hole back toward the center, mainly due to the addition of the $1s_{1/2}^{(e)}2s_{3/2}^{(h)}$ electron–hole pair (EHP) character to the exciton wave function. We see that the introduction of dielectric confinement exaggerates this move of the carriers further away from the QD surface compared to the $\epsilon = $ const. case [58]. The effect of dielectric confinement is particularly strong in this case because $\langle r_h \rangle$ for the $1s_{3/2}^{(h)}$ state is close to the value of QD's outermost radius, $a_c + a_s$.

The close proximity of the hole to the QD surface reduces the distance $\xi = \langle r_h^{QD} \rangle - \langle r_h^{ind.} \rangle$ between the hole in the QD and its induced charge in the colloid, increasing the Coulombic repulsion between them which scales as $\propto 1/\xi$. Such repulsion, in addition to the strong effect of correlations as discussed above, causes the hole to be pushed back toward the center of the QD, thereby dramatically increasing overlap with the correlated electron wave function. The presence of dielectric confinement means the exciton wave function is an almost equal superposition of the $1s_{1/2}^{(e)}ns_{3/2}^{(h)}$ ($n = 1, 2$) states, with characters $|c_1|^2 = 0.449$ and $|c_2|^2 = 0.458$. For comparison, when $\epsilon = $ const. the $1s_{1/2}^{(e)}2s_{3/2}^{(h)}$ character amounts to only $|c_2|^2 = 0.019$. The much stronger configuration mixing in the dielectric confinement case allows E_{corr} to reach ~ -62 meV, compared to -9 meV without dielectric confinement.

Further increase of the CdSe/CdTe QD shell thickness to $a_s = 1$ nm, Figure 13.8c, allows the SP hole to fully localize in the shell, while the SP electron stays in the core, reaching the type-II localization limit. The carriers effectively enter the strong confinement regime (SCR) in which the Coulomb effects are overridden

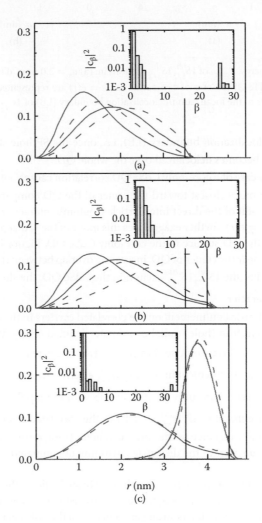

FIGURE 13.8 Solid (dashed) lines represent radial probability densities (RPDs) of the $1S_{1/2}^{(e)}1S_{3/2}^{(h)}$ exciton (single particle $1s_{1/2}^{(e)}1s_{3/2}^{(h)}$ electron–hole pair [EHP]) of CdSe/CdTe quantum dots (QDs) with $a_c = 3.5$ nm and (a) $a_s = 0$, (b) $a_s = 0.5$ nm, and (c) $a_s = 1$ nm. Vertical lines denote the boundaries between the core, shell, and external medium. Insets are bar charts of single EHP characters, $|c_\beta|^2$, in the exciton.

by the effects of the type-II spatial confinement. In the SCR the charge density of a correlated excitonic state is very similar to the charge density of the uncorrelated product of SP states, $\rho_X^{e(h)} \simeq \rho_{SP}^{e(h)}$, and the effect of correlations is lost. Again, E_{corr} is only nonzero when the hole is delocalized; once it localizes in the shell, the effect of VBM confinement overrides the interparticle Coulomb attraction.

Overall, we find that dielectric confinement affects the correlated hole density more than the correlated electron density for two reasons. First, the larger effective mass and deeper potential well experienced by SP hole states compared to electron states allows the former to localize more fully in the shell, closer to the peak in $V_s(r)$ at $r \simeq a_c + a_s$. Second, the smaller energy spacing between the hole SP basis states (i.e., the larger density of hole states) compared to electron SP basis states means that the resulting correlated hole density has more "degrees of freedom" to adjust to the effects of dielectric confinement.

13.8.4 Absorption Spectra of Equivalent CdTe and CdTe/CdSe QD

In Figure 13.9, we compare the exciton dipole spectrum of (a) an $a_c = 2$ nm, $a_s = 1$ nm CdTe/CdSe QD and (b) an "equivalent" $a_c = 3.8$ nm CdTe QD calculated using the CI Hamiltonian. The radius of the CdTe core-only QD is chosen such that the absorption wavelength of its ground-state exciton is the same as that of the CdTe/CdSe core/shell QD. The energy gap between the $1S_{1/2}^{(e)}1S_{3/2}^{(h)}$ and $1S_{1/2}^{(e)}2S_{3/2}^{(h)}$ excitons is increased from ~ 0.1 eV in the core-only QD to ~ 0.18 eV in the core/shell structure. The size of the ground-state exciton optical dipole matrix element is reduced by about 30%, from $0.15P_0^2$ (where P_0 is the bulk optical dipole matrix element) in the core-only QD to $0.11P_0^2$ in the core/shell QD due to reduced electron–hole overlap and electron delocalization (the $a_c = 2$ nm, $a_s = 1$ nm QD lies in the quasi-type-II regime). As expected, the CdTe QD shows slightly stronger absorption than its CdTe/CdSe QD counterpart (see Figure 13.9c), mainly due to the better overall overlap between electron and hole states in core-only QDs compared to type-II structures. However, we note that type-II QDs overall have superior absorption

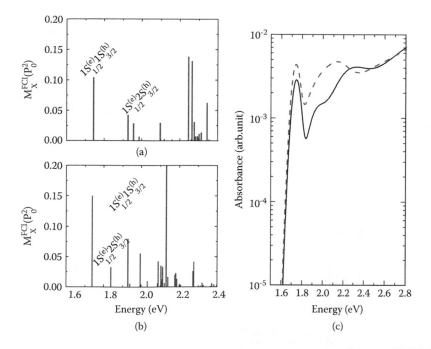

FIGURE 13.9 The exciton dipole spectrum of (a) an $a_c = 2$ nm, $a_s = 1$ nm CdTe/CdSe quantum dot (QD) and (b) an $a_c = 3.8$ nm CdTe QD. (c) The absorption spectra corresponding to the dipole spectra shown in (a) and (b) are shown as solid and dashed lines, respectively.

properties compared with core-only QDs for the important application area of QD-sensitized solar cells. For example the type-II band alignment allows the band-edge absorption to be red-shifted compared to the core-only QD—this is often desirable because the optimum energy for exploitation of the solar spectrum is ~1.35 eV. Type-II QDs also allow greater absorption ranges to be achieved compared to core-only QDs since the absorption edges are not limited by the energy gap of the underlying bulk materials [94].

13.8.5 Charge Carrier Mobility in CdSe QD Arrays

In this section, we evaluate the charge carrier mobility in QD arrays. Good charge carrier mobility is essential for applications of QDs in solar cells since photogenerated carriers have to reach the contacts to form the current in external circuit.

We consider a three-dimensional lattice of CdSe QDs and calculate the temperature dependence of electron mobility in such a system. For hopping transport in a cubic lattice of QDs, the mobility is related to hopping rate between two neighboring dots via $\mu = \frac{ea^2}{k_B T} W$, where a is the distance between centers of two neighboring dots (lattice constant of a QD supercrystal) and W is the transition rate between the dots that can be evaluated using either Equation 13.67 or 13.70 from Section 13.7. If all dots are identical the mobility reads

$$\mu = \int_{-\infty}^{\infty} dt \frac{ea^2 t^2}{\hbar^2 k_B T} \prod_s \exp\left\{-2g_s^2 \left[2n_s + 1 - (n_s + 1)e^{-i\omega_s t} - n_s e^{i\omega_s t}\right]\right\} \tag{13.73}$$

and takes the form $\mu = \int_{-\infty}^{\infty} dt\, h(t)$. By exploiting the fact that $h(t) = h(-t)^*$, the last integral reduces to $\mu = \int_0^{\infty} dt\, f(t)$, where $f(t) = 2 \operatorname{Re} h(t)$.

We consider two different dots. For dot 1, the diameter is $D = 3.6$ nm, relevant phonon energies and electron–phonon coupling constants are $\hbar\omega_s(1 - 6) = \{3.78, 8.10, 1.46, 2.87, 24.0, 24.0\}$ meV, $g_{1-6}\hbar\omega_s(1 - 6) = \{2.82, 0.14, 0.78, 0.52, 20.33, 3.52\}$ meV, the supercrystal lattice constant is $a = 5.0$ nm, and the electronic coupling parameter is $t = 4.0$ meV. For dot 2, the relevant parameters are $D = 7.0$ nm, $\hbar\omega_s(1 - 6) = \{1.95, 4.16, 0.75, 1.48, 24.0, 24.0\}$ meV, $g_{1-6}\hbar\omega_s(1 - 6) = \{1.06, 0.14, 0.40, 0.27, 14.58, 2.52\}$ meV, the supercrystal lattice constant is $a = 8.0$nm, the electronic coupling parameter is $t = 2.5$meV.

One might expect that it is difficult to numerically integrate the function $f(t)$ because of oscillatory terms of the form $\exp\left(i\omega_s t\right)$. However, in the strong electron–phonon coupling regime this function is smooth and decays with t very quickly. Its form for dot 1 at temperatures of $T = 200$ K and $T = 300$ K is presented in Figure 13.10.

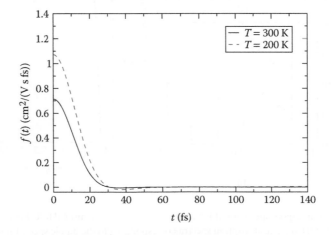

FIGURE 13.10 The function $f(t)$ for dot 1 at temperatures of $T = 200$ K and $T = 300$ K.

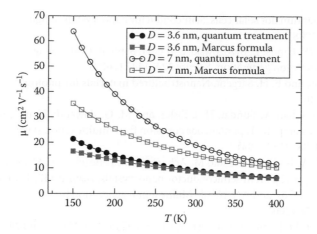

FIGURE 13.11 The dependence of mobility, μ, on temperature, T, for dots 1 and 2 with diameters $D = 3.6$ and 7 nm, respectively, calculated using Equations 13.70 and 13.67.

Temperature dependence of the mobility for dots 1 and 2 is presented in Figure 13.11. It has been checked that the assumption of strong electron–phonon coupling is valid in this temperature range [95]. The mobility was calculated using both the fully quantum treatment from Equation 13.67 and the Marcus formula given by Equation 13.70. We find that for this system in the relevant temperature range the mobility decreases with increasing temperature. This temperature dependence is often observed in the case of band transport. However, one should note that the charge transport in our case takes place by carrier hopping between neighboring dots. This suggests that the measurement of temperature dependence of mobility is not sufficient to identify the charge carrier transport regime. By comparing the results obtained using fully quantum treatment and Marcus formula, we find that, as expected, the agreement between these approaches is the best at high temperatures.

13.9 Conclusion

In conclusion, we have reviewed the methodology for calculation of SP, (multi)excitonic states, and charge transport in colloidal QDs. The methodology was then applied to analyze the optical absorption wavelength, radiative lifetimes, and correlation energy in CdSe/CdTe core/shell QDs, as well as charge carrier mobility in CdSe QD arrays.

Acknowledgments

Stanko Tomić acknowledges the EPSRC UK grant "Enhanced multiple exciton generation in colloidal quantum dots" (EP/K008587/1) for financial support. Nenad Vukmirović is supported by the Ministry of Education, Science, and Technological Development of the Republic of Serbia under project ON171017, and by the European Commission under H2020 project VI-SEEM, Grant No. 675121. We would also like to acknowledge the contribution of the COST Action MP1406 "MultiscaleSolar," and we wish to thank David J. Binks, Jacek M. Miloszewski, Nikola Prodanović, and Edward J. Tyrrell for useful discussions.

References

1. B. N. Pal, Y. Ghosh, S. Brovelli, R. Laocharoensuk, V. I. Klimov, J. A. Hollingsworth, and H. Htoon. Giant CdSe/CdS core/shell nanocrystal quantum dots as efficient electroluminescent materials: Strong influence of shell thickness on light-emitting diode performance. *Nano Lett.* **12**, 331 (2012).

2. P. V. Kamat. Quantum dot solar cells. Semiconductor nanocrystals as light harvesters. *J. Phys. Chem. C* **112**, 18737 (2008).

3. V. Sukhovatkin, S. Hinds, L. Brzozowski, and E. H. Sargent. Colloidal quantum-dot photodetectors exploiting multiexciton generation. *Science* **324**, 1542 (2009).

4. G. Konstantatos, and E. H. Sargent. Nanostructured materials for photon detection *Nat. Nanotechnol.* **5**, 391 (2010).

5. J.-M. Caruge, Y. Chan, V. Sundar, H. J. Eisler, and M. G. Bawendi. Transient photoluminescence and simultaneous amplified spontaneous emission from multiexciton states in CdSe quantum dots. *Phys. Rev. B* **70**, 085316 (2004).

6. V. I. Klimov, S. A. Ivanov, J. Nanda, M. Achermannn, I. Bezel, J. A. McGuire, and A. Piryatinski. Single-exciton optical gain in semiconductor nanocrystals. *Nature* **447**, 441 (2007).

7. A. Shabaev, A. L. Efros, and A. J. Nozik. Multiexciton generation by a single photon in nanocrystals. *Nano Lett.* **6**, 2856 (2006).

8. O. E. Semonin, J. M. Luther, S. Choi, H.-Y. Chen, J. Gao, A. J. Nozik, and M. C. Beard. Peak external photocurrent quantum efficiency exceeding 100 in a quantum dot solar cell. *Science* **334**, 1530 (2011).

9. M. T. Trinh, L. Polak, J. M. Schins, A. J. Houtepen, R. Vaxenburg, G. I. Maikov, G. Grinbom, A. G. Midgett, J. M. Luther, M. C. Beard, A. J. Nozik, M. Bonn, E. Lifshitz, and L. D. A. Siebbeles. Anomalous independence of multiple exciton generation on different group IV–VI quantum dot architectures. *Nano Lett.* **11**, 1623 (2011).

10. S. Ithurria, M. D. Tessier, B. Mahler, R. P. S. M. Lobo, B. Dubertret, and A. L. Efros. Colloidal nanoplatelets with two-dimensional electronic structure. *Nat. Mater.* **10**, 936 (2011).

11. J. Jasieniak, M. Califano, and S. E. Watkins. Size-dependent valence and conduction band-edge energies of semiconductor nanocrystals. *ACS Nano* **5**, 5888 (2011).

12. S. Kim, B. Fisher, H.-J. Eisler, and M. Bawendi. Type-II quantum dots: CdTe/CdSe(core/shell) and CdSe/ZnTe(core/shell) heterostructures. *J. Am. Chem. Soc.* **125**, 11466 (2003).

13. J. J. Li, J. M. Tsay, X. Michalet, and S. Weiss. Wavefunction engineering: From quantum wells to near-infrared type-II colloidal quantum dots synthesized by layer-by-layer colloidal epitaxy. *Chem. Phys.* **318**, 82 (2005).

14. R. Xie, U. Kolb, J. Li, T. Basch, and A. Mews. Synthesis and characterization of highly luminescent CdSe-Core CdS/Zn$_{0.5}$Cd$_{0.5}$S/ZnS multishell nanocrystals. *J. Am. Chem. Soc.* **127**, 7480 (2005).

15. U. Aeberhard, R. Vaxenburg, E. Lifshitz, and S. Tomić. Fluorescence of colloidal PbSe/PbS QDs in NIR luminescent solar concentrators. *Phys. Chem. Chem. Phys.* **14**, 16223 (2012).

16. C. T. Smith, E. J. Tyrrell, M. A. Leontiadou, J. Miloszewski, T. Walsh, M. Cadirci, R. Page, P. O'Brien, D. Binks, and S. Tomić. Energy structure of CdSe/CdTe type {II}colloidal quantum dots: Do phonon bottle necks remain for thick shells? *Sol. Energ. Mater.* **158**, 160–167 (2016),

17. M. A. Leontiadou, E. J. Tyrrell, C. T. Smith, D. Espinobarro-Velazquez, R. Page, P. O'Brien, J. Miloszewski, T. Walsh, D. Binks, and S. Tomić. Influence of elevated radiative lifetime on efficiency of CdSe/CdTe type {II}colloidal quantum dot based solar cells. *Sol. Energ. Mater.* **159**, 657–663 (2017).

18. S. Kumar, M. Jones, S. Lo, and G. D. Scholes. Nanorod heterostructures showing photoinduced charge separation. *Small* **3**, 1633 (2007).

19. H. Zhong, Y. Zhou, Y. Yang, C. Yang, and Y. Li. Synthesis of Type II CdTe–CdSe nanocrystal heterostructured multiple-branched rods and their photovoltaic applications. *J. Phys. Chem. C* **111**, 6538 (2007).

20. S. Itzhakov, H. Shen, S. Buhbut, H. Lin, and D. Oron. Type-II quantum-dot-sensitized solar cell spanning the visible and near-infrared spectrum. *J. Phys. Chem. C* **117**, 22203 (2013).

21. N. McElroy, R. Page, D. Espinbarro-Valazquez, E. Lewis, S. Haigh, P. O'Brien, and D. Binks. Comparison of solar cells sensitised by CdTe/CdSe and CdSe/CdTe core/shell colloidal quantum dots with and without a CdS outer layer. *Thin Solid Films* **560**, 65 (2014).

22. S. Brovelli, R. D. Schaller, S. A. Crooker, F. Garcia-Santamaria, Y. Chen, R. Viswanatha, J. A. Hollingsworth, H. Htoon, and V. I. Klimov. Nano-engineered electron–hole exchange interaction controls exciton dynamics in core–shell semiconductor nanocrystals. *Nat. Commun.* **2**, 280 (2011).

23. D. Oron, M. Kazes, and U. Banin. Multiexcitons in type-II colloidal semiconductor quantum dots. *Phys. Rev. B* **75**, 035330 (2007).

24. P. G. McDonald, E. J. Tyrrell, J. Shumway, J. M. Smith, and I. Galbraith. Tuning biexciton binding and antibinding in core/shell quantum dots. *Phys. Rev. B* **86**, 125310 (2012).

25. S. A. Ivanov, J. Nanda, A. Piryatinski, M. Achermann, L. P. Balet, I. V. Bezel, P. O. Anikeeva, S. Tretiak, and V. I. Klimov. Light amplification using inverted core/shell nanocrystals: Towards lasing in the single-exciton regime. *J. Phys. Chem. B* **108**, 10625 (2004).

26. J. Nanda, S. A. Ivanov, M. Achermann, I. Bezel, A. Piryatinski, and V. I. Klimov. Light amplification in the single-exciton regime using exciton–exciton repulsion in type-II nanocrystal quantum dots. *J. Phys. Chem. C* **111**, 15382 (2007).

27. A. Szabo, and N. S. Ostlund. *Modern Quantum Chemistry: Introduction to Advanced Electronic Structure Theory* (New York, NY: Dover Publications, 1982).

28. V. A. Fonoberov, E. P. Pokatilov, and A. A. Balandin. Exciton states and optical transitions in colloidal CdS quantum dots: Shape and dielectric mismatch effects. *Phys. Rev. B* **66**, 085310 (2002).

29. E. Menéndez-Proupin, and C. Trallero-Giner. Electric-field and exciton structure in CdSe nanocrystals. *Phys. Rev. B* **69**, 125336 (2004).

30. N. Vukmirović, Ž. Gačević, Z. Ikonić, D. Indjin, P. Harrison, and V. Milanović. Electric-field and exciton structure in CdSe nanocrystals. *Semicond. Sci. Technol.* **21**, 1098 (2006).

31. N. Vukmirović, Z. Ikonić, I. Savić, D. Indjin, and P. Harrison. A microscopic model of electron transport in quantum dot infrared photodetectors. *J. Appl. Phys.* **100**, 074502 (2006).

32. N. Vukmirović, D. Indjin, Z. Ikonić, and P. Harrison. Origin of detection wavelength tuning in quantum dots-in-a-well infrared photodetectors. *Appl. Phys. Lett.* **88**, 251107 (2006).

33. N. Vukmirović, and S. Tomić. Plane wave methodology for single quantum dot electronic structure calculations. *J. Appl. Phys.* **103**, 103718 (2008).

34. N. Vukmirović, Z. Ikonić, V. D. Jovanović, D. Indjin, and P. Harrison. Optically pumped intersublevel mid-infrared lasers based on InAs/GaAs quantum dots. *IEEE J. Quant. Electron.* **41**, 1361 (2005).

35. M. Califano, A. Franceschetti, and A. Zunger. Lifetime and polarization of the radiative decay of excitons, biexcitons, and trions in CdSe nanocrystal quantum dots. *Phys. Rev. B* **75**, 115401 (2007).

36. S. Tomić, and N. Vukmirović. Excitonic and biexcitonic properties of single GaN quantum dots modeled by 8-band $\mathbf{k} \cdot \mathbf{p}$ theory and configuration-interaction method. *Phys. Rev. B* **79**, 245330 (2009).

37. M. Korkusinski, O. Voznyy, and P. Hawrylak. Fine structure and size dependence of exciton and biexciton optical spectra in CdSe nanocrystals. *Phys. Rev. B* **82**, 245304 (2010).

38. G. Allan, and C. Delerue. Tight-binding calculations of the optical properties of HgTe nanocrystals. *Phys. Rev. B* **86**, 165437 (2012).

39. A. L. Efros, and M. Rosen. The electronic structure of semiconductor nanocrystals. *Ann. Rev. Mater. Sci.* **30**, 475 (2000).

40. A. Franceschetti, A. Williamson, and A. Zunger. Addition spectra of quantum dots: The role of dielectric mismatch. *J. Phys. Chem. B* **104**, 3398 (2000).

41. J. I. Climente, M. Royo, J. L. Movilla, and J. Planelles. Strong configuration mixing due to dielectric confinement in semiconductor nanorods. *Phys. Rev. B* **79**, 161301 (2009).

42. L. E. Brus. A simple model for the ionization potential, electron affinity, and aqueous redox potentials of small semiconductor crystallites. *J. Chem. Phys.* **80**, 4403 (1984).

43. J. M. Luttinger, and W. Kohn. Motion of electrons and holes in perturbed periodic fields. *Phys. Rev.* **97**, 869 (1955).

44. E. O. Kane. Band structure of indium antimonide. *J. Phys. Chem. Solids* **1**, 249 (1957).

45. E. O. Kane. The $\mathbf{k} \cdot \mathbf{p}$ method. *Semiconduct. Semimet.* **1**, 75 (1966).

46. M. Cardona, and F. H. Pollak. Energy-band structure of germanium and silicon: The k · p method. *Phys. Rev.* **142**, 530 (1966).

47. D. Rideau, M. Feraille, L. Ciampolini, M. Minondo, C. Tavernier, H. Jaouen, and A. Ghetti. Strained Si, Ge, and $Si1-xGex$ alloys modeled with a first-principles-optimized full-zone k · p method. *Phys. Rev. B* **74**, 195208 (2006).

48. G. L. Bir, and G. E. Pikus. *Symmetry and Strain-Induced Effects in Semiconductors* (New York, NY: Wiley, 1974).

49. J. P. Loehr. *Physics of Strained Quantum Well Lasers* (New York, NY: Kluwer Academic Publishers, 1998).

50. C. R. Pidgeon, and R. N. Brown. Interband magneto-absorption and Faraday rotation in InSb. *Phys. Rev.* **146**, 575 (1966).

51. S. Tomić, and N. Vukmirović. Symmetry reduction in multiband Hamiltonians for semiconductor quantum dots: The role of interfaces and higher energy bands. *J. Appl. Phys.* **110**, 053710 (2011).

52. M. A. Cusack, P. R. Briddon, and M. Jaros. Electronic structure of InAs/GaAs self-assembled quantum dots. *Phys. Rev. B* **54**, R2300 (1996).

53. A. D. Andreev, and E. P. O'Reilly. Strain distributions in quantum dots of arbitrary shape. *Phys. Rev. B* **62**, 15851 (2000).

54. N. Vukmirović, D. Indjin, V. D. Jovanović, Z. Ikonić, and P. Harrison. Symmetry of $k · p$ Hamiltonian in pyramidal InAs/GaAs quantum dots: Application to the calculation of electronic structure. *Phys. Rev. B* **72**, 075356 (2005).

55. S. Tomić, A. G. Sunderland, and I. J. Bush. Parallel multi-band $k · p$ code for electronic structure of zinc blend semiconductor quantum dots. *J. Mater. Chem.* **16**, 1963 (2006).

56. N. Vukmirović, Z. Ikonić, D. Indjin, and P. Harrison. Symmetry-based calculation of single-particle states and intraband absorption in hexagonal GaN/AlN quantum dot superlattices. *J. Phys. Condens. Matter* **18**, 6249 (2006).

57. A. D. Andreev, J. R. Downes, D. A. Faux, and E. P. O'Reilly. Strain distributions in quantum dots of arbitrary shape. *J. Appl. Phys.* **86**, 297 (1999).

58. E. J. Tyrrell, and S. Tomić. Effect of correlation and dielectric confinement on $1s_{1/2}^{(e)} ns_{3/2}^{(h)}$ excitons in CdTe/CdSe and CdSe/CdTe type-II quantum dots. *J. Phys. Chem. C* **119**, 12720 (2015).

59. A. Franceschetti, H. Fu, L. W. Wang, and A. Zunger. Many-body pseudopotential theory of excitons in InP and CdSe quantum dots. *Phys. Rev. B* **60**, 1819 (1999).

60. Y. Z. Hu, M. Lindberg, and S. W. Koch. Theory of optically excited intrinsic semiconductor quantum dots. *Phys. Rev. B* **42**, 1713 (1990).

61. P. G. Bolcatto, and C. R. Proetto. Partially confined excitons in semiconductor nanocrystals with a finite size dielectric interface. *J. Phys. Condens. Matter* **13**, 319 (2001).

62. J. Movilla, and J. Planelles. Image charges in spherical quantum dots with an off-centered impurity: Algorithm and numerical results. *Comput. Phys. Comm.* **170**, 144 (2005).

63. C. G. Koay, J. E. Sarlls, and E. zarslan. Three-dimensional analytical magnetic resonance imaging phantom in the Fourier domain. *Magn. Reson. Med.* **58**, 430 (2007).

64. C. Xue, and S. Deng. Three-layer dielectric models for generalized coulomb potential calculation in ellipsoidal geometry. *Phys. Rev. E* **83**, 056709 (2011).

65. S. Deng. A robust numerical method for self-polarization energy of spherical quantum dots with finite confinement barriers. *Comput. Phys. Commun.* **181**, 787 (2010).

66. A. Sihvola, and I. Lindell. Polarizability and effective permittivity of layered and continuously inhomogeneous dielectric ellipsoids. *J. Electromagnet. Wave.* **4**, 1 (1990).

67. T. Takagahara. Effects of dielectric confinement and electron–hole exchange interaction on excitonic states in semiconductor quantum dots. *Phys. Rev. B* **47**, 4569 (1993).

68. E. Hanamura, and H. Haug. Condensation effects of excitons. *Phys. Rep.* **33**, 209 (1977).

69. G. Nan, X. Yang, L. Wang, Z. Shuai, and Y. Zhao. Nuclear tunneling effects of charge transport in rubrene, tetracene, and pentacene. *Phys. Rev. B* **79**, 115203 (2009).

70. S. H. Lin, C. H. Chang, K. K. Liang, R. Chang, Y. J. Shiu, J. M. Zhang, T.-S. Yang, M. Hayashi, and F. C. Hsu. *Ultrafast Dynamics and Spectroscopy of Bacterial Photosynthetic Reaction Centers* (New York, NY: John Wiley & Sons, Inc., 2002), pp. 1–88.

71. I.-H. Chu, M. Radulaski, N. Vukmirovic, H.-P. Cheng, and L.-W. Wang. Charge transport in a quantum dot supercrystal. *J. Phys. Chem. C* **115**, 21409 (2011).

72. C. de Mello Donegá, and R. Koole. Size dependence of the spontaneous emission rate and absorption cross section of CdSe and CdTe quantum dots. *J. Phys. Chem. C* **113**, 6511 (2009).

73. W. W. Yu, L. Qu, W. Guo, and X. Peng. Experimental determination of the extinction coefficient of CdTe, CdSe, and CdS nanocrystals. *Chem. Mater.* **15**, 2854 (2003), doi:10.1021/cm034081k.

74. T. Rajh, O. I. Micic, and A. J. Nozik. Synthesis and characterization of surface-modified colloidal cadmium telluride quantum dots. *J. Phys. Chem.* **97**, 11999 (1993), doi:10.1021/j100148a026.

75. C. B. Murray, D. J. Norris, and M. G. Bawendi. Synthesis and characterization of nearly monodisperse CdE (e = sulfur, selenium, tellurium) semiconductor nanocrystallites. *J. Am. Chem. Soc.* **115**, 8706 (1993), doi:10.1021/ja00072a025.

76. X. Peng, J. Wickham, and A. P. Alivisatos. Kinetics of II–VI and III–V colloidal semiconductor nanocrystal growth: "Focusing" of size distributions. *J. Am. Chem. Soc.* **120**, 5343 (1998), doi:10.1021/ja9805425.

77. A. L. Rogach, A. Kornowski, M. Gao, A. Eychmuller, and H. Weller. Synthesis and characterization of a size series of extremely small thiol-stabilized CdSe nanocrystals. *J. Phys. Chem. B* **103**, 3065 (1999), doi:10.1021/jp984833b.

78. V. N. Soloviev, A. Eichhofer, D. Fenske, and U. Banin. Molecular limit of a bulk semiconductor: Size dependence of the band gap in CdSe cluster molecules. *J. Am. Chem. Soc.* **122**, 2673 (2000), doi:10.1021/ja9940367.

79. F. V. Mikulec, M. Kuno, M. Bennati, D. A. Hall, R. G. Griffin, and M. G. Bawendi. Organometallic synthesis and spectroscopic characterization of manganese-doped CdSe nanocrystals. *J. Am. Chem. Soc.* **122**, 2532 (2000), doi:10.1021/ja991249n.

80. Y. Masumoto, and K. Sonobe. Size-dependent energy levels of CdTe quantum dots. *Phys. Rev. B* **56**, 9734 (1997).

81. D. V. Talapin, S. Haubold, A. L. Rogach, A. Kornowski, M. Haase, and H. Weller. A novel organometallic synthesis of highly luminescent CdTe nanocrystals. *J. Phys. Chem. B* **105**, 2260 (2001), doi:10.1021/jp003177o.

82. P. Dagtepe, V. Chikan, J. Jasinski, and V. J. Leppert. Quantized growth of CdTe quantum dots; observation of magic-sized CdTe quantum dots. *J. Phys. Chem. C* **111**, 14977 (2007).

83. E. J. Tyrrell, and J. M. Smith. Effective mass modeling of excitons in type-II quantum dot heterostructures. *Phys. Rev. B* **84**, 165328 (2011).

84. K. Gong, Y. Zeng, and D. F. Kelley. Extinction coefficients, oscillator strengths, and radiative lifetimes of CdSe, CdTe, and CdTe/CdSe nanocrystals. *J. Phys. Chem. C* **117**, 20268 (2013).

85. X. Ma, A. Mews, and T. Kipp. Determination of electronic energy levels in type-II CdTe-core/CdSe-shell and CdSe-core/CdTe-shell nanocrystals by cyclic voltammetry and optical spectroscopy. *J. Phys. Chem. C* **117**, 16698 (2013).

86. X. Cai, H. Mirafzal, K. Nguyen, V. Leppert, and D. F. Kelley. Spectroscopy of CdTe/CdSe type-II nanostructures: Morphology, lattice mismatch, and band-bowing effects. *J. Phys. Chem. C* **116**, 8118 (2012).

87. E. A. Lewis, R. C. Page, D. J. Binks, T. J. Pennycook, P. O'Brien, and S. J. Haigh. Probing the core–shell–shell structure of CdSe/CdTe/CdS type II quantum dots for solar cell applications. *J. Phys.* **522**, 012069 (2014).

88. C.-Y. Chen, C.-T. Cheng, C.-W. Lai, Y.-H. Hu, P.-T. Chou, Y.-H. Chou, and H.-T. Chiu. Type-II CdSe/CdTe/ZnTe (core/shell/shell) quantum dots with cascade band edges: The separation of electron (at CdSe) and hole (at ZnTe) by the CdTe layer. *Small* **1**, 1215 (2005).

89. B. D. Geyter, Y. Justo, I. Moreels, K. Lambert, P. F. Smet, D. V. Thourhout, A. J. Houtepen, D. Grodzinska, C. de Mello Donega, A. Meijerink, D. Vanmaekelbergh, and Z. Hens. The different nature of band edge absorption and emission in colloidal PbSe /CdSe core/shell quantum dots. *ACS Nano* **5**, 58 (2011), PMID: 21189031.

90. D. Espinobarro-Velazquez, M. A. Leontiadou, R. C. Page, M. Califano, P. O'Brien, and D. J. Binks. Effect of chloride passivation on recombination dynamics in CdTe colloidal quantum dots. *Chem. Phys. Chem.* **16**, 1239 (2015).

91. K. Nishikawa, Y. Takeda, T. Motohiro, D. Sato, J. Ota, N. Miyashita, and Y. Okada. Extremely long carrier lifetime over 200ns in GaAs wall-inserted type II InAs quantum dots. *Appl. Phys. Lett.* **100**, 113105 (2012).

92. S. Tomić. Effect of Sb induced type II alignment on dynamical processes in InAs/GaAs/GaAsSb quantum dots: Implication to solar cell design. *Appl. Phys. Lett.* **103**, 072112 (2013).

93. L. E. Brus. A simple model for the ionization potential, electron affinity, and aqueous redox potentials of small semiconductor crystallites. *J. Chem. Phys.* **79**, 5566 (1983).

94. S. Tomić, J. M. Miloszewski, E. J. Tyrrell, and D. J. Binks. Design of core/shell colloidal quantum dots for meg solar cells. *IEEE J. Photovolt.* **6**, 179 (2016).

95. N. Prodanović, N. Vukmirović, Z. Ikonić, P. Harrison, and D. Indjin. Synthesis and characterization of surface-modified colloidal cadmium telluride quantum dots. *J. Phys. Chem. Lett.* **5**, 1335 (2014).

IV

Light-Emitting Diodes (LEDs)

VI

Light-Emitting Diodes (LEDs)

14

Light-Emitting Diode Fundamentals

Sergey Yu. Karpov

14.1 Introduction

Due to the breakthroughs in technological advances in the development of III-nitride semiconductors made in the early 1990s, visible light-emitting diodes (LEDs) now occupy the dominant sector of the optoelectronics market. The emerging LED production is also stimulated by ever-increasing demands from solid-state lighting, which is expected to not only reduce world energy consumption but also change human conceptions of light quality and ways to utilize it. However, because of the constantly increasing capacity of production equipment and, in particular, of growth reactors, the cost of research and development in the LED industry continues to increase at a very high rate, making comprehensive investigations virtually impossible for all but the biggest universities and industrial players. In this situation, modeling and simulation are especially important, saving time, money, and manpower at the research and development stage.

Simulation of LEDs is essentially a multiscale and multidisciplinary problem. The relevant specific sizes range from a few nanometers for the widths of LED active regions to a few centimeters for the LED lamp dimensions, that is, about seven orders of magnitude. This means that, electromechanical, electrical, thermal, and optical phenomena involved in LED operations are often considered to be self-consistent and, in the case of state-of-the-art LEDs, in three-dimensional (3D) formulation. In practice, this makes straightforward LED simulations unacceptably time- and computer resource-consuming. The commonly utilized approach to resolving this problem and making simulation a helpful research and engineering tool is to split the task into independent but interrelated subtasks. This approach correlates three stages of LED fabrication: heterostructure growth, chip fabrication, and lamp assembly. We choose the LED structure, chip, and lamp as the major objects of investigation within the above-mentioned subtasks.

This chapter focuses on fundamental mechanisms underlying the LED operation that should be included in the simulation models. LED heterostructures, chips, and lamps are considered separately, with a discussion on links between them. Here, inorganic LEDs will be considered only when the organic LED models require special modifications.

14.2 LED Family

There is no wonder that most LED applications are related to various kinds of lighting, and thus utilize the visible optical spectrum. The visible spectral range is shared by phosphide LEDs with $(Al_xGa_{1-x})_{0.51}In_{0.49}P$ active regions (emission wavelength $\lambda = 555-625$ nm), usually lattice matched to the GaAs substrates and nitride LEDs with InGaN-based active regions ($\lambda = 400-600$ nm). Beyond the visible spectral range, the most important region is the ultraviolet (UV) region where UVA ($\lambda = 315-400$ nm), UVB ($\lambda = 280-315$ nm), and UVC ($\lambda = 100-280$ nm) diapasons are commonly distinguished. The active regions of UV LEDs are normally based on AlGaN compounds, with the emission wavelength decreasing with the Al content. These LEDs are used for resin curing, sterilization of water, air, and food, and in medical and biotechnology applications. Infrared (IR) LEDs are situated at the other siide of the visible spectral range, with the active region made up of AlGaInAs and GaInAsSb compounds. They are used in niche applications such as IR lighting, remote control, and gas sensing with mid-IR LEDs ($\lambda = 1.8-5.0\,\mu m$).

LED efficiency is the most important characteristic of the devices (see Section 14.3.4). It is also simultaneously a measure for the maturity of their fabrication technology. Figure 14.1 summarizes the maximum external quantum efficiency (EQE) achieved to date in different spectral ranges by LEDs from various industrial companies and leading research centers. There are two distinct EQE maxima: one of ~77%–82% has been reported for InGaN-based LEDs ($\lambda = 410-440$ nm), whereas another of ~71% has been attained by red AlGaInP ($\lambda = 625$ nm) and IR AlGaAs ($\lambda = 850$ nm) devices. Between them, that is, in the spectral range of 500–600 nm, a so-called "green gap" is seen where EQE drops to ~10%–30% in both nitride and phosphide LEDs. Electron leakage into p-type layers has been proved to be responsible for the efficiency

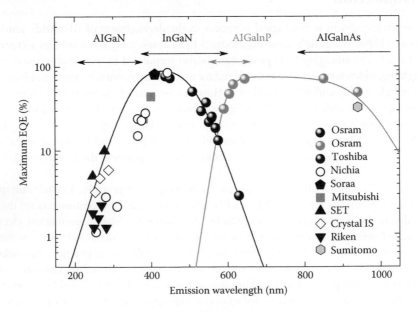

FIGURE 14.1 Maximum external quantum efficiencies (EQEs) of light-emitting diodes (LEDs) as a function of their emission wavelength (symbols are data compiled from numerous sources, lines are drawn for eye). Arrows indicate the ranges of utilizing various III–V compounds for the active region fabrication.

reduction in phosphide LED (Altieri et al., 2005), whereas the nature of the "green gap" in nitride LEDs is not yet completely understood.

Figure 14.1 shows that EQE of AlGaN-based ultraviolet (UV) LEDs decrease dramatically with the Al content in the alloy (at shorter λ), which is attributed to (1) insufficiently high light extraction efficiency partly related to strong transverse-magnetic polarization of the emitted light and absence of highly reflective metallic contacts in this spectral range, (2) problems of p-doping of AlGaN alloys with high Al content, and (3) higher defect density in the grown materials, as compared to Al-free nitride compounds.

Efficiency of IR LEDs suffers primarily from nonradiative Auger recombination inherent in narrow-bandgap semiconductors. In particular, EQE of mid-IR GaInAsSb LEDs emitting in the spectral range of 2.0–4.5 µm (not shown in Figure 14.1) does not exceed ~0.5%–3.0% at room temperature.

To date, as shown earlier, nitride and phosphide LEDs have attained the best performance in terms of emission efficiency. Therefore, simulation examples given in this chapter focus primarily on these two material systems.

14.3 Carrier Transport and Recombination in LED Heterostructures

Any LED structure consists of a stack of epitaxial layers grown in a certain sequence on a substrate of choice. At an arbitrary orientation of heterostructure growth surface, the epilayer interfaces do not always correspond to the main low-index facets of the grown crystals. Therefore, it is convenient to distinguish between the epitaxial coordinate system (ECS) $x'y'z'$, with the x' and y' axes lying in the interface plane and the z'-axis being normal to this plane, and the crystal coordinate system (CCS) xyz, with the axes corresponding to the symmetry axes of the crystal (see Figure 14.2). While ECS is suitable for analysis of carrier transport across the LED structure, CCS enables easy description of anisotropic properties of the crystalline materials (Nye, 1964).

Any crystal of wurtzite symmetry has elastic properties isotropic in the plane normal to its hexagonal C-axes. If the CCS z-axis corresponds to the C-axis of the crystal, the other x- and y-axes can be chosen rather arbitrarily. It is convenient to choose the x-axis of CCS and the x'-axis of ECS to coincide with each other and to be normal to both z- and z'-axes (Romanov et al., 2006). Then the y'- or y-axis orthogonal to both x'- and z'- or x- and z-axes can be unambiguously determined. In ECS defined in such a way, the in-plane lattice constant mismatch can be characterized by two parameters corresponding to the lattice

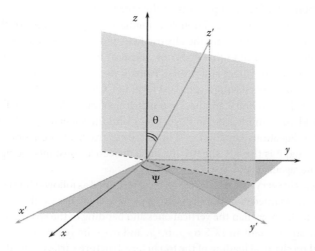

FIGURE 14.2 Crystal coordinate system (CCS) xyz, epitaxial coordinate system (ECS) $x'y'z'$, and specific angles between their axes.

constant variation along the x'- and y'-axes (Romanov et al., 2006):

$$\eta_x = 1 - a_S/a_E \quad \text{and} \quad \eta_y = 1 - [(a_S/a_E)^2 \cos^2 \theta + (c_S/c_E)^2 \sin^2 \theta]^{1/2}, \tag{14.1}$$

where a_E and c_E are the lattice constant of the epitaxial layer considered, a_S and c_S are the lattice constants of the substrate or a thick template layer on which the LED structure is coherently grown, and θ is the polar angle between the z'- and z-axes (see Figure 14.2). This angle can be expressed via four-digit Miller indices $[klmh]$ $(m = -k-l)$ characterizing the orientation of the LED structure interfaces (Gil, 2009):

$$\cos \theta = \frac{h}{[h^2 + \frac{4}{3}(c_E/a_E)^2(k^2 + l^2 + kl)]^{1/2}} \tag{14.2}$$

In the case of zinc-blende materials having no in-plane isotropy of elastic properties, the x'-axis normal to both z'- and z-axes does not coincide with the x-axis of CCS but lies in the xy-plane (Figure 14.2). Then the angles θ and ψ (see Figure 14.2) can be expressed via three-digit Miller indexes $[klm]$ as follows:

$$\cos \theta = \frac{m}{(k^2 + l^2 + m^2)^{1/2}}, \quad \cos \psi = \frac{l}{(k^2 + l^2)^{1/2}} \tag{14.3}$$

The lattice mismatch in zinc-blende crystals is described by the only parameter η_x (see Equation 14.1), where a_E and a_S denote the lattice constants of the epilayer and substrate/template, respectively.

Generally, epitaxial layers may have the lattice constants highly mismatched with those of substrate/template layer. The mismatch produces elastic strain affecting the band structure of the material, its electric polarization, and the band offsets at the structure interfaces. Therefore, determination of strain distribution in an LED structure is the primary task in analysis of its operation.

14.3.1 Electromechanical Coupling

Under isothermal conditions, the equations for elastic strain, stress, and electric polarization in every epitaxial layer of the LED structure can be written in matrix form (Nye, 1964):

$$\boldsymbol{\sigma} = \hat{C}\boldsymbol{\varepsilon} - \hat{e}^T \mathbf{E} \ \text{(a)}, \quad \boldsymbol{\varepsilon} = \hat{S}\boldsymbol{\sigma} + \hat{d}^T \mathbf{E} \ \text{(b)}, \quad \mathbf{D} = \mathbf{P}_{sp} + \hat{d}\boldsymbol{\sigma} + \kappa_0 \hat{\kappa} \mathbf{E} \ \text{(c)}. \tag{14.4}$$

Here, $\boldsymbol{\sigma}$ is the vector built up of six components of the stress tensor (σ_{xx}, σ_{yy}, σ_{zz}, σ_{yz}, σ_{xz}, and σ_{xy}), $\boldsymbol{\varepsilon}$ is the vector consisting of the diagonal and doubled nondiagonal components of the strain tensor (u_{xx}, u_{yy}, u_{zz}, $2u_{yz}$, $2u_{xz}$, and $2u_{xy}$), $\mathbf{E}$ and $\mathbf{D}$ are the electric field and induction vectors, respectively, and $\mathbf{P}_{sp}$ is the spontaneous electric polarization that does not vanish in wurtzite crystals. The first and second of Equations 14.4 correspond to the direct and reverse Hook laws, where $\hat{C}$ and $\hat{S}$ are the 6 × 6 matrices of stiffness and compliance constants having a standard form in CCS (Nye, 1964); $\hat{C} \cdot \hat{S} = \hat{S} \cdot \hat{C} = \hat{I}$, where $\hat{I}$ is the 6 × 6 unity matrix. Here, $\hat{d}$ and $\hat{e}$ are the 3 × 6 matrices of piezoelectric coefficients ($\hat{e} = \hat{d} \cdot \hat{C}$ and $\hat{d} = \hat{e} \cdot \hat{S}$) having a standard form in CCS too (Nye, 1964), and the superscript "T" denotes transposition of a matrix. The third of Equations Equation 14.4 links the electric field, induction, and stress vectors; κ_0 is the electric constant; and $\hat{\kappa}$ is the 3 × 3 matrix of static dielectric constants, being diagonal in CCS for both zinc-blende and hexagonal crystals.

The strain components in every epitaxial layer can be determined as follows (Romanov et al., 2006). First, the stress components are assumed to be uniform within the layer, which is valid, if the lateral dimensions of an LED structure are much larger than the vertical ones and bending of the structure is negligible. Second, the components of the stress vector in ECS $\sigma_{z'z'}$, $\sigma_{y'z'}$, and $\sigma_{x'z'}$ are regarded as vanishing according to the absence of external mechanical loading of the top heterostructure surface. Third, the shear component $\sigma_{x'y'}$ is also equal to zero due to uniformity of the strain distribution in the x'- and y'-directions. Then the strain vector contains only two components, $\sigma_{x'x'}$ and $\sigma_{y'y'}$. To determine them, the respective components of the strain vector $\varepsilon_{x'x'}$ and $\varepsilon_{y'y'}$ should be expressed through $\sigma_{x'x'}$ and $\sigma_{y'y'}$ using Equation 14.4 in ECS. These components are known from geometrical consideration, that is, $\varepsilon_{x'x'} = -\eta_x$ and $\varepsilon_{y'y'} = -\eta_y$, which enables the determination of two nonzero components $\sigma_{x'x'}$ and $\sigma_{y'y'}$. As soon as they are found other components of the strain vector ε in ECS can be calculated using Equation 14.4 and then transformed into CCS in a standard way.

In the simplest case of wurtzite crystals, the above procedure provides the analytical solution, which is formulated as follows, neglecting the electric field contribution to the stress components:

$$\sigma_{x'x'} = \sigma_1 = -\frac{S'_{22}\eta_x - S'_{12}\eta_y}{S'_{11}S'_{22} - (S'_{12})^2}, \quad \sigma_{y'y'} = \sigma_2 = -\frac{S'_{11}\eta_y - S'_{12}\eta_x}{S'_{11}S'_{22} - (S'_{12})^2}, \tag{14.5}$$

where $S'_{11} = S_{11}$, $S'_{12} = S_{12}\cos^2\theta + S_{13}\sin^2\theta$ and $S'_{22} = S_{11}\cos^2\theta + S_{33}\sin^2\theta + \frac{1}{4}(2S_{13} + S_{44} - S_{11} - S_{33})\sin^2 2\theta$.

Then the strain components in CCS can be obtained from Equation 14.5:

$$\varepsilon_{xx} = -\eta_a; \quad \varepsilon_{xz} = \varepsilon_{xy} = 0$$

$$\varepsilon_{yy} = S_{12}\sigma_1 + (S_{11}\cos^2\theta + S_{13}\sin^2\theta)\sigma_2$$

$$\varepsilon_{zz} = S_{13}\sigma_1 + (S_{13}\cos^2\theta + S_{33}\sin^2\theta)\sigma_2 \tag{14.6}$$

$$\varepsilon_{yz} = -\frac{1}{2}S_{44}\sigma_2\sin 2\theta.$$

Nonzero stress components in zinc-blende crystals depend generally on both angles θ and ψ. Using the above approach, one can obtain the strain components in CCS for three practically important orientations of the LED structure interfaces, that is,

- for (001) orientation:

$$\varepsilon_{xx} = \varepsilon_{yy} = -\eta_x, \quad \varepsilon_{zz} = -\frac{2S_{12}}{S_{11} + S_{12}}\eta_x, \quad \varepsilon_{xz} = \varepsilon_{xy} = \varepsilon_{yz} = 0; \tag{14.7}$$

- for (111) orientation:

$$\varepsilon_{xx} = \varepsilon_{yy} = \varepsilon_{zz} = -\eta_x\frac{S_{11} + 2S_{12}}{S_{11} + 2S_{12} + \frac{1}{4}S_{44}}, \quad \varepsilon_{xz} = \varepsilon_{xy} = \varepsilon_{yz} = \eta_x\frac{\frac{1}{2}S_{44}}{S_{11} + 2S_{12} + \frac{1}{4}S_{44}}; \tag{14.8}$$

- for (0lm) orientations: strain components can be found from Equation 14.6, assuming $S_{33} = S_{11}$, $S_{13} = S_{12}$, and finding the angle θ from Equation 14.3; in particular, $\theta = 45°$ for the (011) orientation.

The obtained strain components in CCS can then be used for calculating strain-dependent band structure and electric polarization of every epitaxial layer. In wurtzite crystals, the components of the total electric polarization vector **P** in ECS are

$$P_{x'} = 0$$

$$P_{y'} = e_{15}\varepsilon_{yz} \cdot \cos\theta - [P_{sp} + e_{31}(\varepsilon_{xx} + \varepsilon_{yy}) + e_{33}\varepsilon_{zz}] \cdot \sin\theta \tag{14.9}$$

$$P_{z'} = [P_{sp} + e_{31}(\varepsilon_{xx} + \varepsilon_{yy}) + e_{33}\varepsilon_{zz}] \cdot \cos\theta + e_{15}\varepsilon_{yz} \cdot \sin\theta.$$

In zinc-blende crystals,

$$P_{x'} = P_{y'} = P_{z'} = 0, \qquad\qquad \text{for (001) orientation}$$

$$P_{x'} = P_{y'} = 0 \text{ and } P_{z'} = \sqrt{3}e_{14}\varepsilon_{yz}, \qquad \text{for (111) orientation} \qquad (14.10)$$

$$P_{x'} = e_{14}\varepsilon_{yz} \text{ and } P_{y'} = P_{z'} = 0, \qquad \text{for } (0lm) \text{ orientation.}$$

The difference in the z'-component of the electric polarization vector $\Delta P_{z'}$ of neighboring layers determines the polarization charge accumulated at their interface. Figure 14.3a shows $\Delta P_{z'}$ at the interfaces between the $In_{0.2}Ga_{0.8}N$, $Al_{0.2}Ga_{0.8}N$, $In_{0.17}Al_{0.83}N$, and GaN layer as a function of the C-axis inclination angle θ (the sign of $\Delta P_{z'}$ corresponds to the sign of the interface charge). To calculate $\Delta P_{z'}$ for the nitride alloys, Vegard's rule was used for estimation of the compliance constants S_{ij}. The application of Vegard's rule to the stiffness constants C_{ij} had proved to provide the results graphically indistinguishable from those shown in Figure 14.3a. The $\Delta P_{z'}$ values plotted in Figure 14.3a refine the similar results reported by Romanov et al. 2006, as the nondiagonal component of the strain tensor u_{yz} was erroneously used in this study instead of ε_{yz}.

An alternative approach to analysis of electromechanical coupling in nitride compounds is based on minimization of the elastic energy in every epitaxial layer (Park and Chuang, 1999). This approach provides incorrect results as the predicted stress does not correspond to unloaded free surface of the LED structure and its interfaces.

Equation 14.10 relevant to zinc-blende semiconductors shows, in particular, that only the (111)-orientation possesses polarization charges at the LED structure interfaces.

14.3.2 Transport Equations

In the hierarchy of transport models, the drift-diffusion one (DDM) was first successfully applied to electron and hole transport in semiconductor devices. Introduced in the mid-twentieth century (Van Roosbroeck, 1950), DDM is commonly recognized as the best compromise between the predictability of the theory and the cost of its numerical implementation.

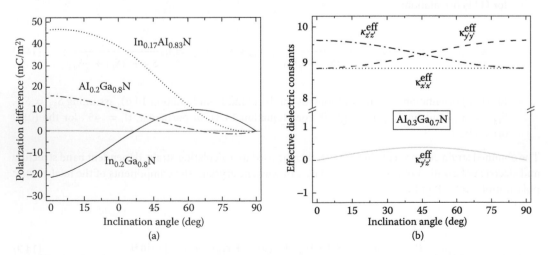

FIGURE 14.3 Difference between the electric polarization of III-nitride alloys and GaN (a) and components of the dielectric constant matrix of $Al_{0.3}Ga_{0.7}N$ (b) in the epitaxial coordinate system (ECS) as a function of inclination angle θ.

Within DDM, the electron n and hole p concentrations can be found from the continuity equations solved in ECS:

$$\frac{\partial n}{\partial t} + \nabla \cdot \mathbf{J_n} = G - R, \quad \frac{\partial p}{\partial t} + \nabla \cdot \mathbf{J_p} = G - R, \tag{14.11}$$

where G is the carrier generation rate, for example, due to light absorption, R is the electron/hole recombination rate, and $\mathbf{J_n}$ and $\mathbf{J_p}$ are the electron and hole fluxes, respectively, related to their partial current densities $\mathbf{j_n}$ and $\mathbf{j_p}$ as follows: $\mathbf{j_n} = -q\mathbf{J_n}, \mathbf{j_p} = q\mathbf{J_p}$; q is the elementary charge.

The most compact definition of the carrier fluxes $\mathbf{J_n}$ and $\mathbf{J_p}$ is based on linear nonequilibrium thermodynamics, involving electron (F_n) and hole (F_p) quasi-Fermi levels measured in electron volts (eV):

$$\mathbf{J_n} = -\mu_n n \nabla F_n, \quad \mathbf{J_p} = -\mu_p p \nabla F_p. \tag{14.12}$$

Here, scalar electron (μ_n) and hole (μ_p) mobilities are used as there is no distinct evidence in the literature for noticeable anisotropy of carrier electric conductivities in various LED structures. The expression 14.12 for the carrier fluxes account for all the forces driving the carrier movement if the electron and hole concentrations are defined as

$$n = N_C \cdot \mathcal{F}_n \left(\frac{F_n - E_C + q\phi}{kT} \right), \quad p = \sum_s N_s \cdot \mathcal{F}_s \left(\frac{E_s - F_p - q\phi}{kT} \right). \tag{14.13}$$

Here, E_C is the conduction band energy determined from the known electron affinity of every epitaxial layer in the LED structure, E_s is the energy of the sth valence sub-band obtained from E_C and the energy gap of the sub-band, ϕ is the electric potential, k is the Boltzmann constant, T is temperature, N_C is the effective density of states in the conduction band, and N_s is effective density of states in the s-th valence sub-band (heavy-hole, light-hole, or split-off hole one), that is,

$$N_C = 2m^n_{xx}(m^n_{zz})^{1/2} \left(\frac{kT}{2\pi\hbar^2} \right)^{3/2}, \quad N_s = 2m^s_{xx}(m^s_{zz})^{1/2} \left(\frac{kT}{2\pi\hbar^2} \right)^{3/2}, \tag{14.14}$$

where $\hbar$ is Planck's constant $(m^n_{xx})^{-1}, (m^n_{zz})^{-1}$, and $(m^s_{xx})^{-1}, (m^s_{zz})^{-1}$ are the components of the inverse effective mass tensor of electrons and holes in the s-th valence sub-band ($s = hh, lh$, or so) in CCS, respectively ($m^n_{xx} = m^n_{zz} = m_n$ and $m^s_{xx} = m^s_{zz} = m_s$ in cubic crystals within isotropic approximation). Integrals $\mathcal{F}_n$ and $\mathcal{F}_s$ in Equation 14.13 are

$$\mathcal{F}_\nu(\zeta) = \int_0^\infty \frac{g_\nu(x)}{1 + \exp(x - \zeta)}, \quad (\nu = n, s), \tag{14.15}$$

where $g_\nu(x)$ is the density of states in the ν-th band, which is expressed as a function of carrier kinetic energy normalized by kT. For a simple parabolic band, $g_\nu(x) = x^{1/2}$ and the integral 14.15 is reduced to the standard Fermi integral of the 1/2-order. In a general case, Equation 14.15 enables accounting for the band nonparabolicity or a more complex energy spectrum typical for holes.

Electric potential distribution in the LED structure can be found from the Poisson equation solved in ECS. In wurtzite heterostructures of arbitrary orientation, the Poisson equation should account for anisotropic properties of the materials:

$$\kappa_0 \nabla \cdot (\hat{\kappa}^{eff} \nabla \phi) = \nabla \cdot \mathbf{P} - \rho, \quad \hat{\kappa}^{eff} = \hat{\kappa} - \frac{(\hat{e}\hat{S}\hat{e}^T)}{\kappa_0}, \tag{14.16}$$

where $\rho = q(N_D^+ + p - N_A^- - n)$ is the charge density, and N_D^+ and N_A^- are the concentrations of ionized donors and acceptors, respectively. The effective dielectric constant $\hat{\kappa}^{eff}$ is a diagonal matrix in CCS. To obtain its components in the ECS, the standard transformation should be applied. The second term in the expression for $\hat{\kappa}^{\text{eff}}$ originates from the elastic stress produced by the electric field (see Equation 14.4) and the back contribution of the stress to the electric polarization.

The dielectric constant of cubic crystals κ is a scalar as well as the effective dielectric constant $\kappa^{\text{eff}} = \kappa - S_{44}e_{14}^2/\kappa_0$. Therefore, the Poisson equation for the cubic heterostructures has the same form in both CCS and ECS. As a rule, $\kappa \gg S_{44}e_{14}^2/\kappa_0$, which allows neglecting the stress contribution to the dielectric constants. In the case of wurtzite semiconductors, the stress contribution to κ_{zz}^{eff} of ~7%–8% is not negligible, changing remarkably the anisotropy degree of the dielectric constant. Variation of $\hat{\kappa}^{\text{eff}}$ with the *C*-axis inclination angle (Figure 14.3b) may exceed ~10% and should be accounted for in the analysis of nonpolar and semipolar LEDs; in the latter case, the nondiagonal $\kappa_{y'z'}^{\text{eff}}$ component of the effective dielectric constant should also be considered.

Concentrations of ionized donors and acceptors in Equation 14.16 depend locally on the quasi-Fermi level positions relative to the band edges, that is,

$$N_D^+ = \frac{N_D}{1 + g_D \exp\left(\frac{F_n - E_C + E_D + q\phi}{kT}\right)}, \quad N_A^- = \frac{N_A}{1 + g_A \exp\left(\frac{E_V + E_A - F_p - q\phi}{kT}\right)}, \tag{14.17}$$

Here N_D and N_A are total donor and acceptor concentrations, respectively, E_D and E_A are the activation energies of the impurities, and g_D and g_A are their degeneracy factors. These equations assume free carriers and those trapped by donors and acceptors to be in thermodynamic equilibrium with each other.

Equations 14.11 through 14.17 form the simplest DDM for the carrier transport in LED structures. More elaborate models (Lundstrom, 2000; Jüngel, 2009; Jacoboni, 2010; Querlioz and Dollfus, 2010), both semiclassical and quantum mechanical, are not discussed here as most of the practically important results have been obtained with DDM allowing clear and straightforward interpretation.

Figure 14.4 presents band diagrams of single-quantum well (SQW) LED structures consisting of a thick n-GaN contact layer, an undoped 15 nm i-GaN interlayer, an undoped 3-nm In$_{0.17}$Ga$_{0.83}$N SQW, a an 8-nm i-GaN spacer where Mg was assumed to diffuse to from the neighboring 20-nm p-Al$_{0.1}$Ga$_{0.9}$N electron-blocking layer (EBL), and a 150 nm p-GaN contact layer. The above structures differ in crystal orientation only, demonstrating orientation-dependent band diagrams primarily controlled by the interface polarization charges. The Ga- and N-polar structures (Figure 14.4a and b) form QWs with strong built-in polarization fields and asymmetric barriers, whereas nonpolar (Figure 14.4c) and semipolar (Figure 14.4d) ones possess a rather weak electric field inside the well, and nearly symmetric barriers. The QW asymmetry varies with the forward bias applied and produces a strong, by tens of nanometers, bias-dependent blue shift of the emission wavelength of the Ga- and N-polar structures. In contrast, the emission wavelengths of the nonpolar and semipolar structures are predicted to be stable within ~3–5 nm under the bias/current variation. It is also interesting that GaN claddings of the InGaN SQW in N-polar structure form rather high natural barriers, preventing electrons and holes from leakage to the p- and n-layers of the structure, respectively (Figure 14.4b).

14.3.3 Recombination Models

Various recombination channels should be accounted for in LED simulations. The first one is the nonradiative Shockley–Read–Hall (SRH) recombination mediated by defects. The respective recombination rate is

$$R^{\text{SRH}} = \frac{np}{\tau_p n (1 + \xi_n) + \tau_n p (1 + \xi_p)} \cdot \left[1 - \exp\left(-\frac{F_n - F_p}{kT}\right)\right];$$

$$\xi_n = \exp\left(\frac{E_t - F_n}{kT}\right), \quad \xi_p = \exp\left(\frac{F_p - E_t}{kT}\right). \tag{14.18}$$

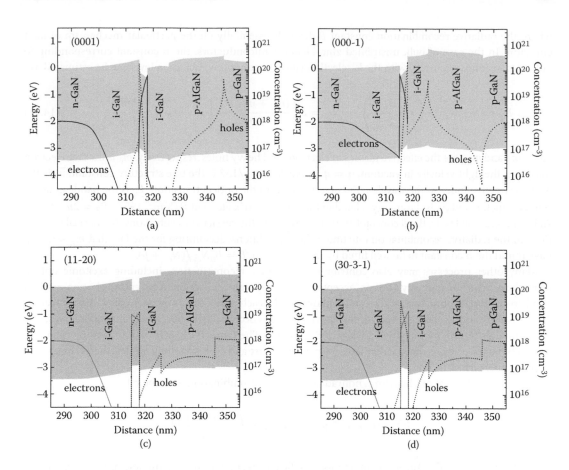

FIGURE 14.4 Band diagrams (gray shading indicates the bandgap) and distributions of electron (solid lines) and hole (dotted lines) concentrations in Ga-polar (a), N-polar (b), nonpolar (c), and semipolar (d) InGaN-based SQW light-emitting diode (LED) structures at the current density of 30 A/cm^2.

Here, E_t is the dominant energy level of a defect responsible for SRH recombination, τ_n and τ_p are the electron and hole life times. One can see from Equation 14.18 that ξ_n, $\xi_p \ll 1$, if the energy E_t is located deep in the materials bandgap.

In the case of recombination via point defects, $\tau_{n,p} = (c_{n,p} V_{n,p} N_t)^{-1}$, where $c_{n,p}$ is the electron/hole trapping cross-section by the defects, $V_{n,p}$ is the carrier thermal velocity, and N_t is the defect concentration in the material. Dislocations threading through the LED active region are inherent defects in III-nitride epitaxial structures because of lack of native substrates. Here, the electron/hole life times can be estimated by the expression (Karpov and Makarov, 2002)

$$\tau_{n,p} = \frac{1}{4\pi D_{n,p} N_d} \left\{ \ln \left(\frac{1}{\pi r_c^2 N_d} \right) + \frac{2D_{n,p}}{r_c V_{n,p}} - \frac{3}{2} \right\}, \tag{14.19}$$

where $D_{n,p}$ is the carrier diffusivity, N_d is the dislocation density, and r_c is the dislocation core radius equal to the lattice constant a in the order of magnitude.

Another recombination channel is the radiative one. The rate of radiative recombination between free carriers in the conduction and valence bands can be calculated as

$$R^{\text{rad}} = Bnp \cdot \left[1 - \exp \left(-\frac{F_n - F_p}{kT} \right) \right], \tag{14.20}$$

where the radiative recombination constant B depends essentially on the particular mechanism of light emission. In the case of bulk unstrained zinc-blende semiconductors, the B constant corresponding to optical transitions occurring with the **k**-selection rule at a rather low excitation level can be estimated as

$$B_0 = (N_{cv}\tau_B)^{-1} \cdot \left(1 + \frac{3kT}{2E_g}\right); \quad N_{cv} = 2\left[\frac{(m_n + m_{hh})kT}{2\pi\hbar^2}\right]^{3/2}, \quad \tau_B = \frac{3\hbar m_0 c^2}{2\alpha\bar{n}_r E_g E_P} \quad (14.21)$$

Here, m_n and m_{hh} are the effective masses of electrons and heavy holes, respectively, m_0 is the free electron mass, c is the light velocity in vacuum, $\alpha = q^2/(4\pi\varepsilon_0\hbar c) \approx 1/137$ is the fine structure constant, $\bar{n}_r$ is the group refractive index at the wavelength corresponding to the optical transition, E_g is the bandgap of the semiconductor, and E_P is the energy related to the Kane's matrix element P (Kane, 1957): $E_P = 2m_0 P^2/\hbar^2$. At high excitation levels, the B constant starts to depend on the carrier concentrations. Numerical calculations of the radiative recombination rate under these conditions, accounting for the fact that $m_n \ll m_{hh}$ have found the B constant to be well fitted by the expression: $B = B_0 N_{cv}/(N_{cv} + p)$.

Many other processes may also contribute to radiative recombination, including excitonic optical transitions, band-to-impurity (donor, acceptor, or a deep center), donor-to-acceptor, phonon-assisted band-to-band recombination, and so on (Pankove, 1971), each providing its own specific radiative recombination constant. In addition, each of the above processes occurs differently in bulk materials, QWs, and quantum dots (QDs), multiplying a variety of B constants. Therefore, the choice of dominant mechanisms should be based on experimental data in every particular case and corresponding models should be applied in the framework of DDM.

The third important mechanism is nonradiative Auger recombination with the rate

$$R^A = (C_n n + C_p p)\, np \cdot \left[1 - \exp\left(-\frac{F_n - F_p}{kT}\right)\right], \quad (14.22)$$

where C_n and C_p are the Auger recombination coefficients related to the microscopic processes involving two electrons and a hole and two holes and an electron, respectively. In bulk unstrained zinc-blende semiconductors, the C_n coefficient corresponding to the process involving two electrons and a heavy hole, that is, the CHCC[†] process, is given by the equation (Gelmont and Sokolova, 1982; Abakumov et al., 1991)

$$C_n = \frac{16}{\sqrt{\pi}}\chi\frac{\nu_D}{N_C N_{hh}}\left(\frac{kT}{E_g}\right)^{3/2}\frac{\langle\varepsilon_c\rangle}{E_g}\exp\left(-E_{th}/kT\right);$$

$$E_{th} = \frac{m_n}{m_{hh}}E_g\frac{(2E_g + \Delta)(3E_g + 2\Delta)}{(E_g + \Delta)(3E_g + \Delta)}, \quad \nu_D = \frac{E_D}{\hbar}, \quad (14.23)$$

$$\chi = \left(\frac{E_g + \Delta}{3E_g + 2\Delta}\right)^{3/2}\left(\frac{3E_g + \Delta}{2E_g + \Delta}\right)^{1/2}, \quad E_D = \frac{m_n q^4}{32\pi^2(\kappa_0\kappa)^2\hbar^2}.$$

[†] To distinguish between various microscopic Auger processes, the following four-letter notation is commonly accepted. First two letters indicate the initial and final electronic states of the recombining particle, e.g., an electron in the conduction band (C) and a hole in the heavy hole sub-band (H). Last two letters indicate the initial and final states of the third particle to which the excess energy released in course of recombination is transferred. Therefore, the CHCC notation means that an electron in the conduction band and a heavy hole recombine with each other, transferring the excess energy to one more electron (C), which is scattered within the same conduction band (C). In turn, the CHHS notation means that an electron in the conduction band and a heavy hole recombine, transferring the excess energy to another heavy hole (H), which is scattered into the split-off hole sub-band (S).

Here, N_C and N_{hh} are the effective densities of states in the conduction band and heavy-hole sub-band, respectively (see Equation 14.14), E_D is the ionization energy of a shallow donor, $\langle \varepsilon_c \rangle$ is the mean electron energy in the conduction band, and Δ is the spin–orbital splitting of the valence band. The energy $\langle \varepsilon_c \rangle = \frac{3}{2} kT$ in the case of nondegenerated electrons and $\langle \varepsilon_c \rangle = \frac{3}{5} \varepsilon_F \propto n^{2/3}$ for degenerated electrons, and ε_F is the Fermi energy. In the latter case, Auger recombination depends even more on the electron concentration than in the case of nondegenerated carriers.

The CHCC Auger process has a threshold accounted for by the exponential factor in Equation 14.23, with E_{th} being the threshold energy. The existence of the threshold originates from simultaneous conservation of the momentum and energy during recombination. As follows from Equation 14.23, the rate of the CHCC Auger recombination decreases dramatically with the material bandgap; therefore, this nonradiative process is especially important in narrow-bandgap semiconductors.

Among the processes involving two holes and an electron in bulk cubic semiconductors, the most important is the CHHS one, where the high-energy hole transfers from the heavy-hole to the split-off hole sub-band. At $E_g - \Delta \gg kT$, the recombination coefficient associated with this process has an analytical representation (Abakumov et al., 1991):

$$
C_p = \frac{216}{\sqrt{\pi}} \chi \frac{\nu_D}{N_{hh} N_{so}} \left(\frac{kT}{E_{th}} \right)^{1/2} \left(\frac{kT}{E_g} \right)^3 \exp\left(\frac{-E_{th}}{kT} \right);
$$

$$
E_{th} = \frac{m_{so}}{m_{hh}} E_g \frac{E_g(E_g - \Delta)}{(E_g + \Delta)(3E_g - 2\Delta)}, \quad \chi = \left(\frac{m_{hh}}{m_{so}} \right)^{1/2} \left(\frac{E_g + \Delta}{E_g} \right)^2.
$$

(14.24)

Equation 14.24 shows that the energy threshold of the CHHS process disappears under resonance conditions $\Delta \approx E_g$; near the resonance, the threshold energy is relatively small, resulting in a considerable increase in the recombination rate.

Similarly to the case of radiative recombination, there is a variety of microscopic Auger recombination processes, involving, in addition to free carriers, those trapped by shallow donors, acceptors, and deep centers (Abakumov et al., 1991), or injected into QWs and QDs (Zegrya and Kharchenko, 1992; Dyakonov and Kachorovskii, 1994; Zegrya and Samosvat, 2007) as well as phonons (Kioupakis et al., 2011). In all these cases, the momentum conservation is violated in one or more directions, leading to the disappearance of the Auger recombination threshold and, as a result, to an increase in the nonradiative recombination rate. Unfortunately, not all of the mechanisms allow even simplified analytical treatment to derive the recombination coefficient; moreover, the theory for some important mechanisms is not yet completely developed (Abakumov et al., 1991). Therefore, only experimental observations may provide guidelines for the choice of dominant Auger mechanisms, in particular materials and device structures, and fitting of the C_n and C_p coefficient to the available data instead of using a rigorous theory may be applied in some cases.

The final nonradiative mechanism of carrier losses to be considered is that their surface recombination is occurs mainly at free surfaces of the LED active region. The surface recombination is accounted for via the flux J_s of carriers arriving at the free surface to recombine nonradiatively. The flux may be calculated as

$$
J_s = \frac{np}{n/S_p + p/S_n} \cdot \left[1 - \exp\left(-\frac{F_n - F_p}{kT} \right) \right],
$$

(14.25)

where S_n and S_p are the surface recombination velocities associated with n-type and p-type materials, respectively. Frequently the carrier losses localized at the active region surface are accounted for by the effective recombination rate distributed over the whole active region with the area $\mathcal{A}$ and perimeter $\mathcal{P}$, which can be calculated by Equation 14.18 with $\xi_n = \xi_p = 0$ and $\tau_{n,p} = \mathcal{A}/(\mathcal{P} S_{n,p})$ (Royo et al., 2002). Such a simplified model is valid if the carrier concentration variation in the active region is much lower than their mean concentrations. In other cases, a more elaborate approach should be applied (see Section 14.4.2).

14.3.4 Efficiency of Light Emission

A number of parameters characterize the efficiency of LED structures. One is the injection efficiency η_{inj}, which accounts for the losses of minority carriers arriving at the contact electrodes without recombination inside the LED structure. The injection efficiency is defined as the ratio of the recombination current $I_{rec} = q \int_V R(\mathbf{r}) \, d^3\mathbf{r}$ to the total current I flowing through the LED (here, the integration is performed over the whole LED volume V). Another important parameter is the internal quantum efficiency (IQE) η_i, defined as the ratio of the radiative recombination current $I_{rad} = q \int_V R^{rad}(\mathbf{r}) \, d^3\mathbf{r}$ to I_{rec}. EQE η_e characterizes the number of photons outgoing from the LED per one electron–hole pair passing through the LED contact electrodes. Generally, $\eta_e = \eta_{ext}\eta_{inj}\eta_i$, where η_{ext} is the efficiency of light extraction (LEE) from the LED die. Finally, the wall–plug efficiency (WPE) η_w is the ratio of the output optical power to the electrical power consumed by LED during its operation. The ratio η_w/η_e is sometimes regarded as the so-called electric efficiency. However, this ratio may be greater than unity at low currents and, therefore, use of this parameter appears to be somewhat misleading.

Both blue/green InGaN-based and red/amber AlGaInP-based LEDs exhibit rather strong EQE reduction at high operating currents but the dominant mechanisms responsible for the reduction are different in nitride and phosphide devices. In AlGaInP LEDs, the EQE reduction with current has been primarily attributed to electron leakage into the p-side of the LED structure because of insufficiently high barriers inherent in the AlGaInP materials system (Royo et al., 2002; Altieri et al., 2005). The leakage is found to be strongly increased by elevated temperature, with the activation energy nearly equal to the difference between the electron energy levels involved in optical transitions and the edge of the barrier conduction band (Altieri et al., 2005). This experimental fact was the major argument for attributing the EQE reduction with current to the electron leakage.

Blue/green nitride LEDs operate at the nonequilibrium carrier density in the active region, at least an order of magnitude higher than that in phosphide LEDs (compare Figures 14.4 and 14.5a). Therefore, the EQE reduction (droop) with current is associated here with Auger recombination (Shen et al., 2007; Bula-shevich and Karpov, 2008; Laubsch et al., 2009); see also various reviews (Piprek, 2010; Avrutin et al., 2013; Verzellesi et al., 2013; Cho et al., 2013) for more detailed discussion on this and alternative mechanisms and experimental studies (Iveland et al., 2013; Galler et al., 2013; Binder et al., 2013) directly demonstrating the importance of Auger recombination. Although particular microscopic mechanisms of Auger recombination in nitride LEDs are not reliably identified, the concept of enlarged recombination volume used in the LED structure design (see, e.g., the review by Weisbuch et al. 2015) and aimed at suppression of the nonradiative carrier losses due to lowering the carrier concentration in the active region works well in practice.

Figure 14.5a shows the band diagrams and nonequilibrium carrier concentrations in a red ($\lambda = 622$ nm) LED structure grown on an n-GaAs substrate and consisting of an undoped $5 \times (3.5$ nm $Ga_{0.51}In_{0.49}P/7$ nm $(Al_{0.8}Ga_{0.2})_{0.51}In_{0.49}P)$ multiple-quantum well (MQW) active region sandwiched between n- and p-$Al_{0.51}In_{0.49}P$ confinement layers, both 100 nm thick; a thick p-GaP contact layer completes the structure. As one can see from a comparison of Figures 14.5a and 14.4, the electron/hole concentration in the active region of the red LED is ~1–2 orders of magnitude lower than in InGaN-based LEDs at the same operating current density of 30 A/cm^2. This originates from different properties of the phosphide and nitride semiconductors: conduction and valence band offsets, carrier mobilities, and recombination constants. The band alignment in the MQW active region is rather flat, providing a uniform distribution of the recombination rate among all the QWs.

The injection efficiency of the red LED structure plotted in Figure 14.5b is strongly dependent on temperature. It differs from unity rather weakly at 300 K, and is considerably reduced at 340 and 370 K, indicating increasing carrier losses via leakage. The electron leakage is predicted to dominate over the hole one at all current densities considered. IQE of the red LED structure varies weakly with temperature (Figure 14.5c) due to temperature-dependent recombination constants. The IQE dependence on current

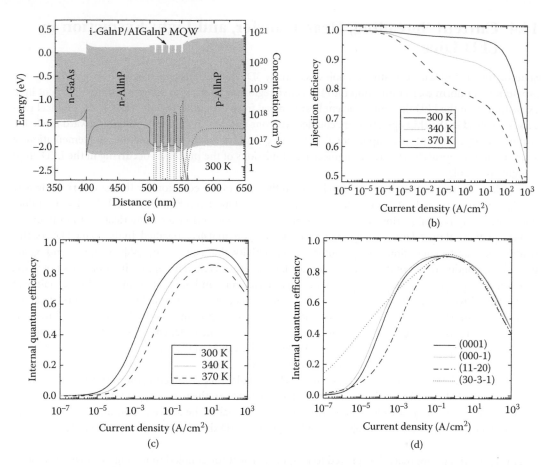

FIGURE 14.5 Band diagram (gray shading indicates the bandgap) and distributions of electron (solid lines) and hole (dotted lines) concentrations in a red MQW light-emitting diodes (LEDs) structure at the current density of 30 A/cm^2 (a). Injection efficiency (b) and IQE (c) of the structure as a function of current density at three different temperatures. IQEs of the InGaN-based SQW LED structures shown in Figure 14.4 are also given versus current density for comparison (d).

has a dome-like shape, reflecting competition between the radiative and nonradiative recombination channels. In particular, the IQE droop at high currents is related to nonradiative carrier losses caused by Auger recombination. Because IQE starts to decline at much higher current density than IE does, degradation of the overall red LED efficiency with temperature should be primarily attributed to the carrier leakage.

Figure 14.5d displays current-dependent IQEs of the nitride SQW LED structures discussed in Section 14.3.2. Simulation of the structures was intentionally carried out with the same recombination constants in order to emphasize the role of the crystal polarity. The computed injection efficiency is equal to unity for all the structures in the whole range of the current density variation. Therefore, the IQE droop clearly seen in Figure 14.5d should be attributed entirely to Auger recombination. As the carrier concentrations in the active region of InGaN-based LEDs considerably exceed those in the red LED structure, the IQE droop in nitride emitters starts at lower current densities (compare Figure 14.5c and d).

The efficiency droop is also observed in UV LEDs with AlGaN active regions (Mickevičius et al., 2015). The nature of the droop here, however, remains unclear and requires further investigation.

14.4 Current Spreading, Heat Transfer, and Light Extraction in LED Dice

State-of-the-art LEDs utilize rather complex essentially 3D chip designs aimed at achieving various goals: providing a uniform carrier injection into the active regions at minimum series resistance, efficient light extraction from the LED dice, and efficient heat removal from the active region. Most of the physical processes involved in the chip operation are coupled with one another so that their straightforward modeling requires a large amount of computational time and computer resources. Therefore, implementation of approximation approaches to create engineering simulations of the processes occurring in the LED chips is highly desirable.

A rather effective approach can be developed on the basis of the equation for the total current density $\mathbf{j} = \mathbf{j_n} + \mathbf{j_p}$; under steady-state conditions, this equation follows from Equation 14.11: $\nabla \cdot \mathbf{j} = 0$. It does not contain any recombination terms and enables much easier and faster numerical solution than full DDM equations, if the relationship between $\mathbf{j}$ and electric potential ϕ is specified. In the contact layers, this relationship can be approximated by a unipolar conductivity, that is, $\mathbf{j} = -\sigma_{n,p} \nabla \phi$, where σ_n and σ_p are the electric conductivities of the n- and p-contact layers, respectively. This approximation does not work, however, in the LED active region. On the other hand, electrons and holes are transferred here mainly in the vertical direction due to dominant vertical components of the electric field and gradients of carrier concentrations. Therefore, the carrier transport in the active region may be well approximated by a 1D model. The 1D solution to the DDM equations provides a nonlinear dependence of the vertical current density j_v on the p-n junction bias $U_b = F_n - F_p$, where the values of the quasi-Fermi levels F_n and F_p are taken at the n- and p-sides of the active region, respectively. Thus, the $j_v(U_b)$ dependence obtained may be used as a nonlinear boundary condition in 3D chip simulations, connecting the vertical component of the local current density $\mathbf{j}$ with the electric potential difference on the n-side (ϕ_n) and p-side (ϕ_p) borders of the active region: $\mathbf{n} \cdot \mathbf{j} = j_v(\phi_p - \phi_n)$, where $\mathbf{n}$ is the normal to the active region directed from the p- to n-contact layer. Solution to the above problem provides 3D distribution of the current density in an LED die.

This approximate approach has proved to be quite efficient for coupled modeling of LEDs with rather complex chip designs (Bogdanov et al., 2008); it can also be easily refined by using additional experimental input (López and Margalith, 2008). Comparison of theoretical results obtained by the above approach with available observations has demonstrated its good predictability (Chernyakov et al., 2013).

As it was already mentioned, the carrier transport in the LED active region occurs primarily in the vertical direction. Therefore, considering the lateral electron and hole transfer as a perturbation, one can also account for the lateral redistribution of carriers due to ambipolar diffusion and their surface recombination at the free surface of the active region.

To illustrate, Figure 14.6 shows simulated 2D current density distribution over the active region and some characteristics of a vertical red LED die shaped in such a way as to increase its light extraction efficiency (LEE) using inclined side walls of the die (see Figure 14.7a). The LED structure presented in Figure 14.5a has been chosen for the simulations, whereas modeling of current spreading, heat transfer, and light extraction was carried out self-consistently in 3D approximation. Due to the vertical chip geometry, current crowding (see Section 14.4.1 for a more detailed discussion) is rather weak in the chip (Figure 14.6a), resulting in a nearly linear relationship between the mean current density and current (Figure 14.6b). Simulated current–voltage and light–current characteristics of the red LED are shown in Figure 14.6c, whereas EQE and WPE variation with current is given in Figure 14.6d. At relatively low currents, WPE exceeds EQE (see Figure 14.6d), which is typical for LEDs operating at the forward voltage $V_f < E_{ph}/q$, where E_{ph} is the mean energy of emitted photons. The decrease of EQE and WPE at high currents is caused by the thermally enhanced electron leakage, as discussed in Section 14.3.4. One can see from Figure 14.6b that the mean temperature in the LED die increases superlinearly with the operating current.

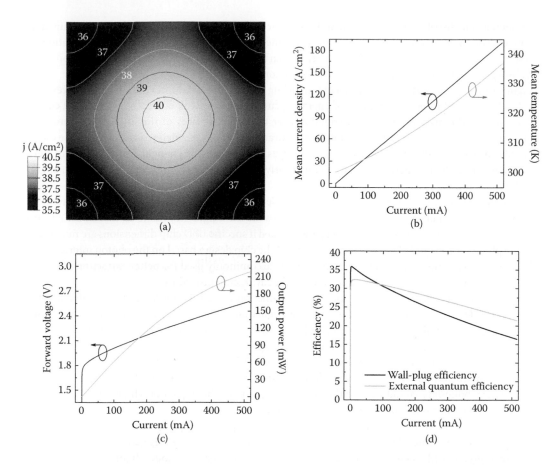

FIGURE 14.6 Two-dimensional (2D) vertical current density distribution in the active region at the current of 100 mA (a) and mean current density and temperature (b), forward voltage and output optical power (c), and external quantum efficiency (EQE) and wall–plug efficiency (WPE) (d) as a function of operating current of the red light-emitting diode (LED) chip shown in Figure 14.7a.

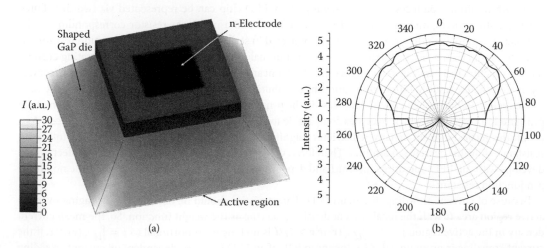

FIGURE 14.7 Near-field emission intensity distribution over the outer surfaces of red light-emitting diode (LED) die with shaped side walls (a) and far-field emission pattern of the LED (b).

As soon as the distribution of electron and hole concentrations in the active region is determined, the radiative recombination rate in each point can be obtained (Bogdanov et al., 2008). Using the radiation pattern specific for the active region band structure, one can use the ray tracing to simulate photon propagation and absorption in the LED die, as well as their extraction efficiency.

Figure 14.7 shows results of the ray-tracing modeling of the shaped red LED die considered previously. LEE has been found to be independent of operating current due to weak current crowding and equal to 35.2%. Near-field emission intensity distribution over the outer surfaces of the LED die (Figure 14.7a) demonstrates that most of the emitted photons are extracted through the inclined side walls of the die and primarily through the regions adjacent to the die edges. As a result, the far-field emission pattern differs substantially from the Lambertian one, providing nearly constant intensity in a wide observation angle of ~100° (Figure 14.7b).

Ray tracing is quite suitable for accounting properties of the photon ensemble in LED dice. However, its applicability is rather limited as the approach ignores the wave nature of photons, that is, the effects of their diffraction and interference. Ray tracing is warranted if specific LED chip dimensions are much larger than the photon wavelength. However, any elements of the chip design based on the photon interference or diffraction, such as photonic crystals or textured surface commonly used for better extraction of photons from LEDs, cannot be considered to be accurate in the ray tracing procedure.

An alternative approach frequently applied for estimation of LEE is the direct finite-difference time domain (FDTD) simulations, where the Maxwell's equations are solved for the electromagnetic field induced by a dipole imitating the act of radiative recombination of an electron and a hole. The FDTD method is quite suitable, in particular, for analysis of light diffraction at a textured surface of the LED dice. Despite the accounting for photon wave nature, the FDTD approach has also some limitations related to (1) ignoring the ensemble properties of electrons/holes and, hence, of emitted photons and (2) using some artificial boundary conditions for the Maxwell's equations, the impact of which on the accuracy of theoretical predictions is still uncertain (Zhmakin, 2011).

To summarize, ray tracing and FDTD simulation are complementary approaches to LEE calculation rather than competing ones. The choice between them depends on the LED chip geometry and parameters desirable to predict LEE value, emission pattern in the near- and far-field zones, light absorption in LED die, emission polarization, and so on.

14.4.1 Current Crowding

Current crowding is the effect of the current density localization inside the LED die around the streamlines with minimum path resistance. Generally, every LED chip can be represented via two distributed linear resistors corresponding to n- and p-contact layers and a nonlinear resistor corresponding to the LED active region, including p-n junction, all connected in series. The nonlinear p-n junction resistor has a nearly exponential resistance dependence on the internal p-n junction bias. At a low operating current, the p-n junction resistance is much higher than both linear ones and the lateral current flow in the contact layers dominates, resulting in a rather uniform distribution of the vertical current density over the LED active region. At higher currents, the increase in the p-n junction bias leads to a dramatic reduction of the p-n junction resistance, which becomes comparable to or lower than both contact layer resistances. As a result, the vertical current density redistributes significantly in such a way as to minimize the total path resistance along the current streamlines. Therefore, current crowding is essentially a nonlinear effect, which develops with the LED operating current I and depends on temperature primarily through the nonlinear p-n junction resistance.

Because of current crowding, the main LED characteristics should be obtained by averaging over the active region area a with the local current density j_v serving as the weight function. So, the mean current density in the active region $\bar{j} = I^{-1} \int_a j_v^2(\mathbf{r}) \, d^2\mathbf{r} > I/a$ is no longer proportional to $I = \int_a j_v(\mathbf{r}) \, d^2\mathbf{r}$, if the current crowding is pronounced. Also the mean IQE of an LED becomes dependent on current crowding (see simulation results obtained by Bogdanov et al. (2010) and experimental data by Malyutenko et al.

(2010)): $\overline{\eta_i} = I^{-1} \int_a \eta_i(j_v) \cdot j_v(\mathbf{r}) \, d^2\mathbf{r}$, where $\eta_i(j_v)$ is the IQE of LED structure dependent on the local current density j_v.

Another important effect of current crowding found by simulations (Bogdanov et al., 2010) and reported experimentally by Laubsch et al. (2010) is the current localization under metallic electrodes deposited on the emitting surface of LEDs. Depending largely on the operating current, the localization results in LEE decreasing with current, enhancing the overall efficiency droop. The effect was found to be much stronger than the IQE reduction caused by current crowding (Bogdanov et al., 2010). Avoiding metallic electrodes on the emitting surface by the use of advance chip designs (see, for instance, Laubsch et al. 2010) improves LEE substantially.

Generally, current crowding is especially pronounced in lateral LED chips utilizing nonconductive sapphire substrate and having one-side electrode access to both n- and p-contact layers. This is because of a relatively thin n-contact layer providing dominant lateral current flow inside the chip. In a vertical LED die similar to that shown in Figure 14.7a, the current crowding is weaker but, nevertheless, still tangible (see Figure 14.6a) despite a relatively large distance, ~200 µm, between the n- and p-electrodes, which looks at first glance to be quite sufficient for the complete homogenization of the current density distribution inside the die.

14.4.2 Surface Recombination

There are three main parameters affecting the surface recombination impact on the LED efficiency: the recombination velocity S, the ambipolar carrier diffusion coefficient D_a in the active region, and the carrier differential life time τ_d generally dependent on the carrier concentration in the active region. The latter two parameters determine the carrier diffusion length $L_d = (D_a \tau_d)^{1/2}$, which characterizes the width of the region adjacent to the free surface of the active region where contribution of surface recombination to the carrier losses is significant.

Red/amber AlGaInP LEDs operate normally at a reduced, ~10^{17}–10^{18} cm^{-3}, carrier concentration in the active region (Figure 14.5a), providing a relatively long τ_d. The carrier diffusivity in the active region enhanced in undoped QWs may be rather large. As a result, L_d may approach ~10–25 µm in such devices. Therefore, surface recombination results in a remarkable rise to the carrier losses already at the LED chip dimensions as small as ~150–250 µm (Royo et al., 2002).

The carrier concentration in the active regions of blue and green InGaN-based LEDs is ~1–2 orders of magnitude higher (Figure 14.4), shortening significantly the lifetime τ_d; the carrier diffusivity here is also about an order of magnitude smaller. Therefore, L_d does not normally exceed a few micrometers in nitride LEDs. On the other hand, the free active region surface is normally located close to the regions of strong current crowding in the III-nitride LED chips. Estimates show that surface recombination may be valuable in blue/green LEDs at low currents corresponding to the maximum of their efficiency. Surface recombination is expected to be especially pronounced in deep-UV LEDs because of (1) a high Al content in the active region normally increasing the surface recombination velocity, (2) a weaker carrier localization resulting in their higher diffusivity in the active region, and (3) smaller LED die dimensions usually utilized in practice.

14.5 Phosphor-Converted White Light Emission

An important sector of LED industry is the production of white light sources for solid-state lighting. Such sources are based on emulation of sunlight by a number of LEDs emitting in different spectral ranges or by using a partial light conversion by one or more phosphors. As it has been shown by Bulashevich et al. (2015), the phosphor-converted LEDs (PC-LEDs) provide a higher efficacy at a better color rendition, as compared to white light sources utilizing LEDs only. In addition, fabrication of PC-LEDs represents the main industrial approach to the white light production. Therefore, this section will focus on simulation models necessary for the analysis of light conversion by phosphors pumped with LEDs.

14.5.1 White Light Characteristics

The fundamentals of color perception and white light characterization are described in detail by Judd and Wyszecki (1975). Here, the main white light characteristics will be briefly discussed in view of discussion on their further modeling.

Starting with J.C. Maxwell's studies in the mid-nineteenth century, it became customary to decompose the light color into three components. According to CIE (1986), three color coordinates, X, Y, and Z, can be brought in by the convolution of a light source spectral power density emission spectrum $S_\lambda(\lambda)$ with three tabulated color matching functions, $\bar{x}(\lambda)$, $\bar{y}(\lambda)$, and $\bar{z}(\lambda)$. Then the relative chromatic coordinates $x = X/(X + Y + Z)$ and $y = Y/(X + Y + Z)$ project the emission spectrum into the CIE1931 color chart, which is represented by points in the (x, y) space, all the possible colors that can be encountered in practice.

A natural source producing white light is a black body heated up to a certain temperature. Under the temperature variation, chromatic coordinates of the black-body radiation (BBR) form in the CIE1931 diagram, a locus nearly corresponding to white light. In practice, some deviations of the white-light chromatic coordinates x_w and y_w from the BBR locus are admissible. In this case, a certain projection of the (x_w, y_w) point to the BBR locus produces the so-called correlated color temperature (CCT), which is one of the most important characteristics of white light.

Another important characteristic is the luminous efficacy of radiation (LER) obtained by convolution of the spectral power density spectrum $S_\lambda(\lambda)$ with the human eye sensitivity function $\Phi(\lambda)$ (Judd and Wyszecki, 1975), accounting statistically for the perception of human eye: LER $= \int S_\lambda(\lambda)\,\Phi(\lambda)\,d\lambda$. The efficacy of a white light source is the product of LER and total WPE of the source, which depends on WPEs of individual LEDs and the power fractions of the emission spectra supplied by these LEDs into the total spectrum $S_\lambda(\lambda)$ (Bulashevich et al., 2015). If a phosphor emission is used for white light generation, its effective WPE depends on the WPE of the pumping LED, the Stokes shift caused by down conversion of the photon energy, and quantum yield of the phosphor emission (Bulashevich et al., 2015).

Ability of a light source to reproduce the colors of illuminated objects accurately is characterized by the color rendering index (CRI). The CRI accounts for the differences between the chromatic coordinates of light reflected from a number of standard color samples at their illuminating by the light source studied and by one or another standard illuminant. Detailed procedure of the CRI evaluation is recommended by CIE (1986). The CRI is reduced from the maximum value of 100 by every chromatic coordinate difference obtained from the selected color samples.

14.5.2 Light Conversion by Phosphor

Conversion medium in an LED lamp typically consists of a silicone layer with embedded particles of one or more phosphors deposited on the surface of pumping LED. Therefore, the light absorption and scattering by the phosphor particle, as well as emission of the down-converted light and its scattering by the particles, occurs in the medium. Here, spectral ray tracing is the approach most suitable for simulations of light conversion by phosphors. Assuming the phosphor particles to be randomly and uniformly distributed in silicone, one can derive the light conversion to depend on scattering properties of an individual particle and total volume fraction f_{ph} of the particles in the conversion medium.

Rigorous Mie theory (Van de Hulst, 1957) is currently the best model for analysis of light scattering by individual phosphor particles. Assuming spherical particle shape, the theory predicts the wavelength-dependent scattering and absorption cross-sections necessary for ray tracing simulations, as well as the scattering pattern in the far-field zone. For particles of realistic size and monochromatic light, the scattering patterns exhibit strong interference oscillations (black lines in Figure 14.8), which should be absent in

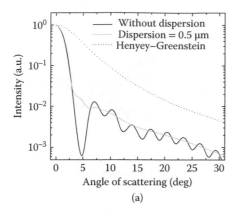 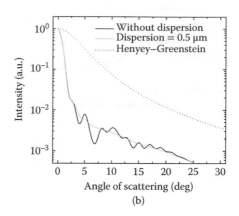

FIGURE 14.8 Scattering patterns of YAG:Ce^{3+} particle with the radius of 4 μm obtained by the Mie theory with and without account of particle size dispersion and by Henyey–Greenstein approximation for the wavelength of 600 nm (a) and 400 nm (b).

practice in view of (1) dispersion of the particle sizes, (2) deviation of the particle shapes from the spherical one, and (3) a nonmonochromatic spectra of the pumping LED and phosphor. Accounting for the size dispersion of phosphor particles is already sufficient to avoid the oscillations in the scattering pattern (gray lines in Figure 14.8). It is interesting that the Henyey–Greenstein scattering functions (Henyey and Greenstein, 1941) commonly used for empirical fitting of the scattering pattern do not work well in the case of realistic phosphor particles, if the asymmetry parameter g of the function is correctly estimated as the first moment of the scattering pattern predicted by the Mie theory (dotted lines in Figure 14.8). Being fitted to the computed pattern, the Henyey–Greenstein scattering function at g ~0.98–0.99 is capable of predicting the enhanced forward scattering seen in Figure 14.8 but fails to reproduce the wings of the scattering pattern predicted by the Mie theory. So, there is an unlikely alternative to the use of Mie theory for modeling the light interaction with the phosphor particles.

The absorption/scattering cross-sections and scattering pattern of individual phosphor particles involved in spectral ray tracing allow simulating the total emission spectrum in the far-field zone, estimating the color characteristics of white light, that is, chromatic coordinates x and y, CCT, and CRI, and their dependence on the observation angle. The latter dependence is quite important as the overall white light LED performance requires uniform color characteristics in a wide range of the observation of angle variation. As soon as the angle dependence of the total emission spectrum is obtained, integration over the angle enables calculation of LER and then of efficacy of the white light source.

One of the approaches aimed at improving angle uniformity of color characteristics is based on the fact that the scattering cross-section of phosphor particles depends very critically on the particle size. Figure 14.9 shows the simulated angle dependence of CCT and CRI of the white light source comprising of a blue (450 nm) LED and a yellow YAG:Ce^{3+} phosphor with the emission spectrum peaked at 560 nm and its full width at half maximum of 120 nm. Simulations were carried out for a square-shaped LED chip uniformly coated by a 230-nm silicone layer with 8% of its volume filled by the phosphor particles. One can see from Figure 14.9 that the angle uniformity of CCT and CRI is strongly dependent on the particle size. Among the chosen values of the mean particle radius, only that of 2 μm provides simultaneously the angle-uniform CCT and CRI.

The above results have been obtained by ray tracing performed in 50 spectral intervals from 380 to 730 nm, using 10^8 rays. Even at that, the simulated color characteristics in the far-field zone are rather noisy because of insufficient ray statistics. Therefore, parallel computing is normally required for the spectral ray tracing in order to accelerate simulations.

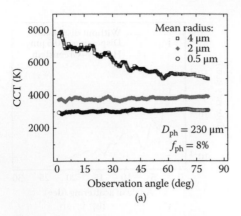
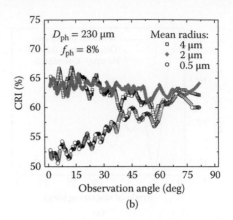

FIGURE 14.9 Angle-dependent correlated color temperature (CCT) (a) and color rendering index (CRI) (b) of a white light source consisting of a blue light-emitting diode (LED) and a YAG:Ce^{3+} yellow phosphor with the particles of various sizes. The conversion media thickness $D_{ph} = 230$ μm is assumed in the ray-tracing simulations.

14.6 Concluding Remarks

This chapter presents a general frame and a strategy for multiscale and multidisciplinary simulations of state-of-the-art LEDs from heterostructure to lamp. The basic models and the most important mechanisms underlying the simulations are reviewed with the focus on those that were not discussed in detail previously. It should be noted that many aspects of simulations are just outlined in the chapter and should be adopted to specific subjects of the study. In particular, the radiative and nonradiative recombination mechanisms require more detailed description that accounts for both direct and phonon-assisted processes, as well as the variety of active region designs (quasi-bulk, QWs, or QDs). Another issue beyond the scope of the chapter is the proper choice of materials properties, which are known to often be highly inaccurate, and its impact on the simulation results. Finally, the role of a number of material factors important for LED operation, such as high-density threading dislocations and compositional fluctuations in InGaN and other alloys, are still in the early stages of study. A better understanding of these factors may result in a reconsideration of the basic carrier transport and recombination models in the future.

References

Abakumov VN, Perel VI, Yassievich IN (1991) Nonradiative recombination in semiconductors (Ch. 11). In: *Modern Problems in Condensed Matter Science*, V. M. Agranovich and A. A. Maradudin (Eds.), vol. 33. Amsterdam: North-Holland.

Altieri P, Jaeger A, Windisch R, Linder N, Stauss P, Oberschmid R, Streubel K (2005) Internal quantum efficiency of high-brightness AlGaInP light-emitting devices. *J. Appl. Phys.* 98: 086101.

Avrutin V, Hafiz S-D-A, Zhang F, Özgür Ü, Morkoç H (2013) InGaN light-emitting diodes: Efficiency-limiting processes at high injection. *J. Vac. Sci. Technol. A* 31: 050809.

Binder M, Nirschl A, Zeisel R, Hager T, Lugauer H-J, Sabathil M, Bougeard D, Wagner J, Galler B (2013) Identification of nnp and npp Auger recombination as significant contributor to the efficiency droop in (GaIn)N quantum wells by visualization of hot carriers in photoluminescence. *Appl. Phys. Lett.* 103: 071108.

Bogdanov MV, Bulashevich KA, Evstratov IY, Zhmakin AI, Karpov SY (2008) Coupled modeling of current spreading, thermal effects, and light extraction in III-nitride light-emitting diodes. *Semicond. Sci. Technol.* 23: 125023.

Bogdanov MV, Bulashevich KA, Khokhlev OV, Evstratov IY, Ramm MS, Karpov SY (2010) Current crowding effect on light extraction efficiency of thin-film LEDs. *Phys. Status Solidi C* 7: 2124–2126.

Bulashevich KA, Karpov SY (2008) Is Auger recombination responsible for the efficiency rollover in III-nitride light-emitting diodes? *Phys. Status Solidi C* 5: 2066–2069.

Bulashevich KA, Kulik AV, Karpov SY (2015) Optimal ways of colour mixing for high-quality white-light LED sources. *Phys. Status Solidi A* 212: 914–919.

Chernyakov AE, Bulashevich KA, Karpov SY, Zakgeim AL (2013) Experimental and theoretical study of electrical, thermal, and optical characteristics of InGaN/GaN high-power flip-chip LEDs. *Phys. Status Solidi A* 210: 466–469.

Cho J, Schubert EF, Kim JK (2013) Efficiency droop in light-emitting diodes: Challenges and countermeasures. *Laser Photonics Rev.* 7: 408–421.

CIE Publ. No. 15.2 (1986) Colorimetry. *Official Recommendations of the International Commission on Illumination.* Vienna: Central Bureau of the CIE.

Dyakonov MI, Kachorovskii VY (1994) Nonthreshold Auger recombination in quantum wells. *Phys. Rev. B*, 49: 17130–17138.

Galler B, Lugauer H-J, Binder M, Hollweck R, Folwill Y, Nirschl A, Gomez-Iglesias A, Hahn B, Wagner J, Sabathil M (2013) Experimental determination of the dominant type of Auger recombination in InGaN quantum wells. *Appl. Phys. Express* 6: 112101.

Gelmont B, Sokolova Z (1982) Auger recombination in direct-gap N-type semiconductors. *Sov. Phys. Semicond.* 16: 1067–1068.

Gil B (2009) Wurtzitic semiconductors heterostructures grown on (hk.l) oriented substrates: The interplay between spontaneous and piezoelectric polarization fields, elastic energy and the modification of quantum confined Stark effect. *Proc. SPIE* 7216: 7216OE.

Henyey LG, Greenstein JL (1941) Diffuse radiation in the galaxy. *Astrophys. J.* 93: 70–83.

Iveland J, Martinelli L, Peretti J, Speck JS, Weisbuch C (2013) Direct measurement of Auger electrons emitted from a semiconductor light-emitting diode under electrical injection: Identification of the dominant mechanism for efficiency droop. *Phys. Rev. Lett.* 110: 177406.

Jacoboni C (2010) *Theory of Electron Transport in Semiconductors.* Berlin: Springer.

Judd DB, Wyszecki G (1975) *Color in Business, Science and Industry.* New York, NY: Wiley.

Jüngel A (2009) *Transport Equations for Semiconductors.* Berlin: Springer.

Kane EO (1957) Band structure of indium antimonide. *J. Phys. Chem Solids* 1: 249–261.

Karpov SY, Makarov YN (2002) Dislocation effect on light emission efficiency in gallium nitride. *Appl. Phys. Lett.* 81: 4721–4723.

Kioupakis E, Rinke P, Delaney KT, Van de Walle CG (2011) Indirect Auger recombination as a cause of efficiency droop in nitride light-emitting diodes. *Appl. Phys. Lett.* 98: 161107.

Laubsch A, Sabathil M, Bergbauer W, Strassburg M, Lugauer H, Peter M, Lutgen S, Linder N, Streubel K, Hader J, Moloney JV, Pasenow B, Koch SW (2009) On the origin of IQE-"droop" in InGaN LEDs. *Phys. Status Solidi C* 6: S913–S916.

Laubsch A, Sabathil M, Baur J, Peter M, Hahn B (2010) High-power and high-efficiency InGaN-based light emitters. *IEEE Trans. Electron Devices* 57: 79–87.

López T, Margalith T (2008) Electro-thermal modelling of high power light emitting diodes based on experimental device characterisation. www.comsol.de/paper/5004.

Lundstrom M (2000) *Fundamentals of Carrier Transport.* Cambridge: Cambridge University Press.

Malyutenko VK, Bolgov SS, Podoltsev AD (2010) Current crowding effect on the ideality factor and efficiency droop in blue lateral InGaN/GaN light emitting diodes. *Appl. Phys. Lett.* 97: 251110.

Mickevičius J, Tamulaitis G, Jurkevičius J, Shur MS, Shatalov M, Yang J, Gaska R (2015) Efficiency droop and carrier transport in AlGaN epilayers and heterostructures. *Phys. Status Solidi B* 252: 961–964.

Nye JF (1964) *Physical Properties of Crystals: Their Representation by Tensors and Matrices* (Ch. 10). Oxford: Clarendon Press.

Pankove JI (1971) *Optical Processes in Semiconductors*. New York, NY: Dover Publications.

Park S-H, Chuang S-L (1999) Crystal-orientation effects on the piezoelectric field and electronic properties of strained wurtzite semiconductors. *Phys. Rev. B* 59: 4725–4737.

Piprek J (2010) Efficiency droop in nitride-based light-emitting diodes. *Phys. Status Solidi A* 207: 2217–2225.

Querlioz D, Dollfus P (2010) *The Wigner Monte Carlo Method for Nanoelectronic Devices*. London: ISTE Ltd. & John Wiley & Sons.

Romanov AE, Baker TJ, Nakamura S, Speck JS (2006) Strain-induced polarization in wurtzite III-nitride semipolar layers. *J. Appl. Phys.* 100: 023522.

Royo P, Stanley RP, Ilegems M, Streubel K, Gulden KH (2002) Experimental determination of the internal quantum efficiency of AlGaInP microcavity light-emitting diodes *J. Appl. Phys.* 91: 2563–2568.

Shen YC, Müller GO, Watanabe S, Gardner NF, Munkholm A, Krames MR (2007) Auger recombination in InGaN measured by photoluminescence. *Appl. Phys. Lett.* 91: 141101.

Van de Hulst HC (1957) *Light Scattering by Small Particles*. New York, NY: Dover Publications.

Van Roosbroeck W (1950) Theory of flow of electrons and holes in germanium and other semiconductors. *Bell System Tech. J.* 29: 560–607.

Verzellesi G, Saguatti D, Meneghini M, Bertazzi F, Goano M, Meneghesso G, Zanoni E (2013) Efficiency droop in InGaN/GaN blue light-emitting diodes: Physical mechanisms and remedies. *J. Appl. Phys.* 114: 071101.

Weisbuch C, Piccardo M, Martinelli L, Iveland J, Peretti J, Speck JS (2015) The efficiency challenge of nitride light-emitting diodes for lighting. *Phys. Status Solidi A* 212: 899–913.

Zegrya GG, Kharchenko VA (1992) New mechanism of Auger recombination of nonequilibrium current carriers in semiconductor heterostructures. *Sov. Phys. JETP* 74: 173–181.

Zegrya GG, Samosvat DM (2007) Mechanisms of Auger recombination in semiconducting quantum dots. *J. Exp. Theor. Phys.* 104: 951–965.

Zhmakin AI (2011) Enhancement of light extraction from light emitting diodes. *Phys. Rep.* 498: 189–241.

Organic Light-Emitting Diodes

Pascal Kordt

Peter Bobbert

Reinder Coehoorn

Falk May

Christian Lennartz

and

Denis Andrienko

15.1 Introduction

The 2014 Nobel Prize in Physics was awarded "for the invention of efficient blue light-emitting diodes (LEDs) which has enabled bright and energy-saving white light sources" [1–3], setting a clear target to mankind: energy-efficient and environmental-friendly light sources [4]. The first *organic* LED (OLED) was reported in 1987 by a team at Kodak [5]. This publication, cited to date more than 10,000 times, stipulated the entire field of organic electronics. Shortly afterward, a polymer LED (PLED) was demonstrated [6], paving the way for flexible lighting applications [7]. Nowadays, OLEDs are successfully used in displays of mobile phones and televisions: In 2008, Samsung announced a flexible display that was only 50 μm thick [8], about half the thickness of a sheet of paper. A prototype of an OLED display for the automotive market was presented recently by Continental [9]: In OLED displays, black pixels are completely switched off, allowing the driver's eye to adapt better to the darkness. Contrary to liquid crystal displays (LCDs), OLED screens do not require backlight illumination, yielding exceptionally good contrast ratios and reduced power consumption. OLED displays also provide viewing angles and response times superior to LCDs and are, in general, thinner and lighter. Last but not least, many organic materials can be printed from solution, enabling cost-effective large-scale manufacturing on mechanically flexible films.

The prime challenge in OLED development is an improvement of device operation lifetimes. In many cases, the steady decrease of luminescence efficiency of OLEDs under continuous operation is compensated by a gradual rise in bias voltages, or by using larger pixel sizes, which can then be operated at lower voltages and luminances. In particular, blue phosphorescent devices, with their high energy of emitted photons and long-lived excited states, are prone to rapid material degradation. The underlying mechanisms of degradation, discussed in Section 15.12.5, are unfortunately not fully understood.

Designing new organic materials is crucial for tuning OLED properties, performance, and stability. One, therefore, needs to provide rigorous links between device characteristics and the chemical composition of the layers, or structure–property relationships. This can be done with the help of cheminformatics tools [10], i.e., by correlating available experimental data to underlying chemical structures. However, the number of experimental samples is normally limited to a few hundred since synthesis, device optimization, and characterization are often costly and time consuming. A substantial extension of the training set is therefore experimentally not feasible to achieve, again motivating the development of computer simulations techniques capable of predicting device characteristics *in silico* [11]. In this chapter, we aim to provide an overview of theoretical models and simulation approaches developed to study charge and exciton transport in organic semiconductors, as well as to predict and eventually optimize current–voltage characteristics, electroluminescence efficiency, and lifetimes of OLEDs.

15.2 Working Principles of an OLED

We start by reviewing the elementary processes taking place in an OLED. The simplest two-layer OLED consists of hole and (luminescent) electron transporting layers (ETLs), which are sandwiched between two electrodes, as shown in Figure 15.1a. When a voltage is applied to the electrodes, holes are injected from the anode into the hole transporting layer (HTL) and electrons from the cathode into the ETL. The external field forces electrons and holes to drift toward the interface between these layers, where they recombine and emit light. Energetically, such a heterojunction is designed to facilitate the hole injection from HTL to ETL as well as to block electron diffusion into the opposite direction.

To fine-tune device properties such as luminescence, driving voltage, light outcoupling, and lifetime, more layers are added to a two-layer OLED: A typical phosphorescent OLED is shown in Figure 15.1b. Here, when a voltage is applied to the electrodes, electrons are injected from the reflective metal cathode (Al) and holes from the semitransparent anode (indium tin oxide or ITO). The cathode has a low work function, while the anode has a high one. This energy difference is compensated by the externally applied voltage, which forces charge carriers to drift into the emission layer. In the emission layer, holes and electrons form excitons, predominantly on the emitter (guest) molecules. Finally, radiative decay of an exciton leads to light emission. In the next sections, we briefly review the functionality of all layers.

15.2.1 Electrodes

The outermost layers, metallic or metal-oxide electrodes, inject electrons and holes into organic layers. A common material for the semitransparent anode is ITO [12]. Its low ionization potential (IP) ensures

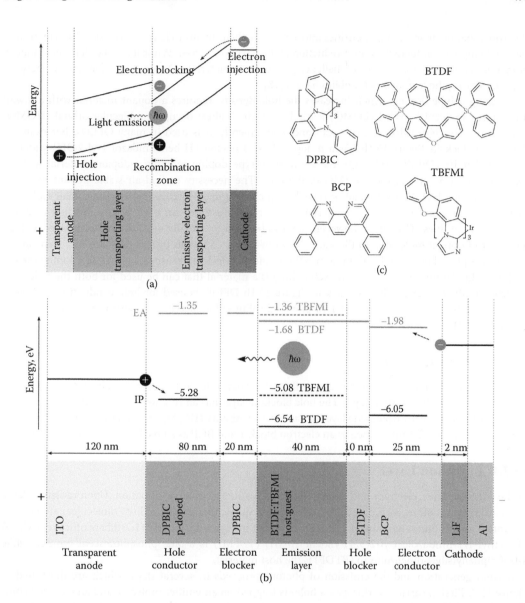

FIGURE 15.1 (a) Schematic of a two-layer electroluminescent organic light-emitting diode (OLED) and its energy level diagram under applied voltage. (b) Structure of a blue phosphorescent OLED with a transparent anode (indium tin oxide or ITO), a reflective cathode (Al), emission layer, and hole/electron conduction/blocking layers. Indicated energy levels are without the applied bias. (c) Compounds used in the OLED (b). (Adapted from P. Kordt et al., *Adv. Funct. Mater.*, 25, 1955–1971, 2015.)

efficient hole injection into the organic layer. For efficient electron injection a low work function cathode is required. Since such materials are often air-unstable, normally bilayered structures are used [13], e.g., Al/LiF [14,15] or Ag/MgAg [16,17].

15.2.2 Charge Transport Layers

The role of electron and hole transport layers (HTLs) is to provide an Ohmic contact to the electrodes (barrierless injection) and to help control light outcoupling. These layers are often doped. By inserting electron donating or accepting impurities into a material, their intrinsic charge carrier density is increased, leading to higher conductivities and a reduction of the injection barrier. While it is a standard technique in inorganic semiconductors, it can be challenging in organic materials: Dopant molecules create energetic traps, hindering the formation of mobile charges [18].

Realizing *p*-type doping, which increases the hole density, requires a dopant material with a lowest unoccupied molecular orbital (LUMO) level close to the highest occupied molecular orbital (HOMO) of the host. This is feasible in most cases and, consequently, it is used in many OLEDs. For example, the OLED stack in Figure 15.1b has a *p*-doped Tris[(3-phenyl-1H-benzimidazol-1-yl-2(3H)-ylidene)-1,2-phenylene]Ir (DPBIC) HTL. On the contrary, *n*-type doping requires a dopant material with an HOMO level near or above the LUMO of the host. The necessity to find air-stable dopant materials with such a high HOMO level makes *n*-type doping challenging. Moreover, when using small molecules such as O_2, Br_2, and I_2 as dopant materials, these can diffuse into an organic host, leading to undesired doping profiles. This effect, however, can also be used to actually dope the adjacent layer with an interlayer of dopant molecules. In the stack shown in Figure 15.1b this technique is applied: Here an LiF interlayer is placed between the aluminum cathode and the electron transport layer (ETL) doping the latter [19]. Alq_3 (aluminum-tris(8-hydroxchinolin)) is a material that can be used for both the hole- and electron-conducting layers. In the stack in Figure 15.1b DPBIC is used for hole conduction and bathocuproine-4,7-diphenyl-2,9-dimethyl-1,10-phenanthroline (BCP) serves as an electron-conducting layer [14,16,17,20,21].

15.2.3 Blocking Layers

Hole/electron blocking layers suppress charge flow to the opposite electrodes, enhancing their recombination probability in the emission layer. The hole blocking layer has an IP lower than the IP of the emission layer. Similarly, an electron blocking layer has an electron affinity (EA) higher than the EA of the emission layer. In Figure 15.1, DPBIC is used as an electron blocker and BCP as a hole blocker [14,16,17,20,21].

15.2.4 Emission Layer

In the emission layer, electrons and holes recombine, leading to exciton formation. Upon radiative decay of the excitons, photons are emitted. To avoid exciton quenching, emitters (host molecules) are embedded into a matrix (guest molecules). In the stack shown in Figure 15.1b, Tris[(1,2-dibenzofurane-4-ylene)(3-methyl-1/1-imidazole-1-yl-2(3/1)-ylidene)]Ir(III) (TBFMI) is a blue phosphorescent emitter, while 8-bis(triphenylsilyl)-dibenzofuran (BTDF) is the host material.

Exciton generation and the emission of photons proceeds in several steps, which are illustrated in Figure 15.2. First, a carrier (in this case a hole) is trapped on an emitter molecule and an oxidized complex is formed. Driven by Coulomb forces, a carrier of opposite charge (electron) moves on host molecules toward the trapped carrier. When the electron reaches a host molecule' neighboring oxidized emitter, a charge transfer (CT) state is formed, as shown in Figure 15.2b. In a CT state, the short-range exchange interaction leads to an energy splitting of singlet (S) and triplet (T) states. The triplet consists of three triplet substates, which differ from one another by their relative spin orientations. Statistically, one obtains a population ratio of 1:3 between singlet and triplet substates. In a final step, the electron moves in a very rapid process directly to the emitter molecule, forming an excited emitter, see Figure 15.2c. This process may occur via a singlet or a triplet path, depending on the initial spin orientation of the electron–hole pair. The two spins of an electron and a hole are then coupled to four new combined states: one singlet state, with total spin momentum 0, and one triplet state, with the total spin momentum 1. In a statistical limit, all four substates are populated with an equal probability.

An exciton formed on an emitter molecule can decay radiatively by fluorescence or phosphorescence. In *fluorescence*, which is the most common radiative decay pathway for organic molecules, the radiative decay occurs from the excited singlet state, S_1, to the ground state. If triplet states are not harvested, the internal quantum efficiency of an OLED is limited to 25%. *Phosphorescence*, conversely, allows harvesting of

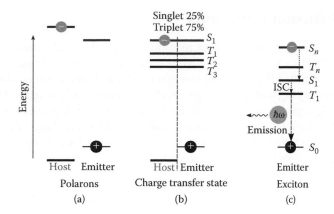

FIGURE 15.2 Exciton formation in the emission layer of a phosphorescent organic light-emitting diode (OLED). (a) Electron moves toward a hole which is trapped on the emitter molecule. (b) Charge transfer state is formed with the electron on a host and hole on a neighboring emitter molecule. Spins of hole and electron are already correlated to one singlet and three triplet states. (c) Exciton is formed on the emitter molecule with a statistical probability to be in a 25% singlet and 75% triplet state. Singlet state undergoes fast intersystem crossing (ISC) into a triplet state which then emits light.

triplet states. One can use compounds with large spin–orbit couplings, such as organometallic complexes, to achieve this. The large spin–orbit coupling capacitates a spin flip, (see Figure 15.2c), leading to ISC, i.e., a transition between singlet and triplet states. Thus, triplets can also decay radiatively ($T_1 \rightarrow S_0$), yielding an internal quantum efficiency of 100%.

Another way of harvesting triplet states is via a *thermally activated delayed fluorescence* (TADF) [22]. TADF can be realized in organic molecules even with small spin–orbit coupling. In TADF triplets are converted to singlets thermally (reverse ISC) and a radiative $S_1 \rightarrow S_0$ decay (fluorescence) takes place. TADF requires small energy differences between T_1 and S_1 levels $\Delta E_{ST} \sim k_B T$. To achieve this, a large spatial HOMO–LUMO separation is required. This, however, competes with the requirement of large oscillator strengths of fluorescence, i.e., a large transition dipole moment [23].

Several processes in an OLED can lead to efficiency losses. In a phosphorescent OLED, these are *triplet–triplet quenching*, where two triplets are annihilated as a result of the interaction of neighboring excited emitter molecules, and *triplet–polaron quenching*, i.e., an interaction between electronic and vibrational modes, energy transfer to polarons, and dissociation into free carriers [12,24]. These processes are discussed in more detail in Sections 15.12.3 and 15.12.5.

A comprehensive OLED model should incorporate all the aforementioned elementary processes: charge injection from electrodes, charge trapping, and transport to the emission layer, long-range electron–hole interactions, formation of CT and excited states, exciton–exciton and exciton–polaron interactions, and radiative decay of excitons. The ultimate goal of such a model is to predict the current–voltage–luminescence characteristics of the multilayered OLED structure. Since the typical thickness of an organic layer of an OLED stack ranges from 10 to 100 nm, it is computationally prohibitive to use only first-principles methods for OLED modeling. State-of-the-art OLED simulations employ either continuous models, such as drift–diffusion equations (see Section 15.3), or solve the master equation for charge/exciton occupation probabilities (see Section 15.4). Both approaches use models of different complexity. For example, if the target is to optimize the composition of the stack made of well-characterized organic layers, drift–diffusion equations and lattice models with phenomenologically fitted parameters are used. If one also needs to retain the link to the underlying chemical composition of layers, off-lattice models based on first-principles parametrizations are required. In all cases, we are dealing with a typical multiscale problem, which requires the development of scale-bridging techniques (see Section 15.11). In the next sections, we review these approaches, paying special attention to limitations, parametrizations, and scale-bridging issues.

15.3 Drift–Diffusion Equations

We start with the most coarse model: On a macroscopic level, the drift and diffusion of electrons, holes, and excitons in an OLED stack can be described by the corresponding densities, denoted here as n, p, and s, respectively. The drift–diffusion model assumes that local charge densities, charge mobilities, μ_n, μ_p, diffusion constants, D_e, D_n, and field strength, F, all vary continuously in space. The set of equations describing the time-dependent spatial distribution of charge and exciton densities is then based on the respective conservation laws. In what follows, we describe a model where charges drift-diffuse in a given density of states (DOS) $g(E)$ (see also Section 15.7), can be trapped, have density- and field-dependent mobilities, and recombine either radiatively or nonradiatively.

15.3.1 Charges

Charge conservation leads to two continuity equations, for electrons and holes

$$e\frac{\partial n}{\partial t} = \vec{\nabla} \cdot \vec{J}_n - eR(n,p) - e\frac{\partial n_t}{\partial t}, \qquad e\frac{\partial p}{\partial t} = -\vec{\nabla} \cdot \vec{J}_p - eR(n,p) - e\frac{\partial p_t}{\partial t}, \tag{15.1}$$

where $R(n,p)$ is the recombination rate (see Section 15.3.2), n_t and p_t are densities of trapped charges, and e is the electron charge. The current equations for electron and holes drift–diffusion in an electrostatic potential ψ read [25,26]

$$\vec{J}_n = -en\mu_n\vec{\nabla}\psi + eD_n\vec{\nabla}n, \qquad \vec{J}_p = -ep\mu_p\vec{\nabla}\psi - eD_p\vec{\nabla}p. \tag{15.2}$$

Note that only mobile charges contribute to the current. Summing up electron and hole currents yields the total current in the device, $\vec{J} = \vec{J}_n + \vec{J}_p$. The electrostatic potential ψ is related to the electron and hole densities via the Poisson equation

$$\varepsilon_0\varepsilon_r\Delta\psi = e\left(n - p + n_t - p_t\right), \tag{15.3}$$

where ε_0 is the vacuum and ε_r is the relative permittivity. The densities of trapped charges obey the phenomenological rate equations

$$\frac{\partial n_t}{\partial t} = r_{c,n}n(N_t - n_t) - r_{e,n}n_t, \qquad \frac{\partial p_t}{\partial t} = r_{c,p}p(N_t - p_t) - r_{e,p}p_t, \tag{15.4}$$

where it is assumed that traps occupy two levels (one for holes and one for electrons). Here N_t is the trap density, r_e is the escape rate, and r_c is the capture rates of electrons (n) and holes (p), respectively.

Since two electrons or holes cannot occupy the same energy level at the same time, the occupation probabilities of energy levels follow the Fermi–Dirac statistics [27]

$$f(E, E_F) = \left[1 + \exp\left(\frac{E - E_F}{k_B T}\right)\right]^{-1}, \tag{15.5}$$

where $f(E, E_F)$ is the occupation probability of a level with energy E, k_B is the Boltzmann constant, and T is the temperature. The carrier density is then related to the quasi-Fermi level, E_F, as

$$p\left(E_F\right) = \frac{N}{V}\int_{-\infty}^{\infty} g(E)f(E, E_F)dE, \tag{15.6}$$

where $g(E)$ is the DOS (see Section 15.7), N is the number of holes, and V is the box volume. A similar relation holds also for electrons. Fermi–Dirac statistics implies that the diffusion coefficient and mobility in Equation 15.1 are related via the generalized Einstein relation [28]

$$D = \frac{p\mu}{e} \left(\frac{\partial p}{\partial E_F} \right)^{-1}.$$

(15.7)

In order to solve the drift–diffusion equations one needs to know the dependence of mobility on charge density, electric field, and temperature, $\mu(p, n, \vec{F}, T)$. These can be obtained either from experiments or Monte Carlo simulations, as discussed in Section 15.11.

15.3.2 Recombination

The recombination rate $R(n, p)$ includes two loss mechanisms: Shockley–Read–Hall (SRH), or trap-assisted recombination, and bimolecular recombination. In SRH recombination, electrons are trapped in low-energy states and recombine with free holes. The macroscopic rate for this process reads [29]

$$R_{\mathrm{SRH}} = \frac{C_n C_p N_t (np - n_i p_i)}{C_n(n + n_i) + C_p(p + p_i)}.$$

(15.8)

Here C_n and C_p are the capture coefficients for electrons and holes, respectively, N_t is the trap density for electrons, n and p are the electron and hole densities, and n_i and p_i are the intrinsic electron and hole densities.

Bimolecular recombination is often modeled using the Langevin rate

$$R_L = \gamma(np - n_i p_i), \qquad \gamma = \frac{q}{\varepsilon_0 \varepsilon_r} \left(\mu_n + \mu_p \right).$$

(15.9)

Here γ is the recombination constant and μ_n and μ_p are electron and hole mobilities. The intrinsic electron and hole carrier densities are often neglected [29,30], simplifying the expression to

$$R_L \approx \gamma np.$$

(15.10)

There are several assumptions on which the Langevin rate is based and which might not hold in the case of organic semiconductors. Its validity can be verified by performing kinetic Monte Carlo (KMC) simulations on lattices, as discussed in Section 15.6.2.

The overall recombination rate, $R_{\mathrm{SRH}} + R_L$, enters the continuity Equation 15.1. Experimentally, the prevailing recombination mechanism can be probed by means of the classical Shockley diode equation with an ideality factor η that differs depending on the dominant mechanism [31,32].

15.3.3 Excitons

The generation, transport, and decay of excitons can be described with a phenomenological equation for the population of excitons S_i, where i denotes the exciton type (triplet, singlet) [26]

$$\frac{\partial S_i}{\partial t} = G_i R(n, p) + \vec{\nabla} \cdot \vec{J}_i - \left(k_i^{(r)} + k_i^{(n)} \right) S_i - k_i^{(a)} f S_i^2 + \sum_{j=S,T} (k_{ji} S_j - k_{ij} S_i) - k_{\mathrm{TPQ}}(n + p).$$

(15.11)

Here G_i is the exciton generation efficiency: For singlet excitons $G_S = 0.25$, while for triplet excitons $G_T = 0.75$. $k^{(r)}$ and $k^{(n)}$ are the radiative and nonradiative (position-dependent) decay rates, and $k^{(a)}$ is the annihilation rate. The factor f is 0.5 if only one triplet is lost or 1 if both are lost. The radiative decay rate can be calculated using the dipole emission model [26,33]. The exciton energy transfer rate, k_{ij}, describes the conversion of triplet excitons to singlets and vice versa. k_{TPQ} is the triplet–polaron quenching rate [24] (assumed here to be the same for electrons and holes). Since excitons are charge-neutral, their transport is purely diffusive, i.e., $\vec{J_i} = D_i \vec{\nabla} S_i$. Note that the exciton dissociation into an electron and a hole has been neglected in Equation 15.11.

15.3.4 Boundary Conditions and Numerical Solution

Equations 15.1 through 15.11 are complemented by the boundary conditions for the electrostatic potential, ψ, by setting the potential difference at the boundaries to $\psi_{eff} = V_{app} - V_{int}$, where V_{app} is the applied potential and V_{int} is the built-in potential, defined as the difference of the materials' work functions.

The set of Equations 15.1 through 15.11 is normally solved using iterative schemes, until self-consistency for the electrostatic potential, density, and current is reached. The Gummel iteration method [34] with a discretization according to a scheme proposed by Scharfetter and Gummel [35] can be used to solve linearized equations. This method is less sensitive to the initial guess than a Newton algorithm and thus is the method of choice despite its slower convergence in terms of iteration steps [36].

15.3.5 Limitations

Field, temperature, and charge density dependencies of the mobility entering drift–diffusion equations are normally parametrized in equilibrium or under stationary conditions. These dependencies cannot in general be used to describe nonequilibrium, time-dependent processes. For example, charge-carrier relaxation makes the mobility effectively time dependent, as evident from, e.g., impedance and dark-injection studies (see Section 15.12.2). Moreover, it has been shown that the current density in the device can be spatially inhomogeneous in all three dimensions (filamentary structures) [37–39], which then questions the applicability of mean-field descriptions. Last, some processes, e.g., exciton–electron interactions or energetic barriers between organic layers, are difficult to incorporate into drift–diffusion equations. The master equation, which we review in the next section, allows us to consider the individual rates between CT and other events. In the subsequent section, we describe the Monte Carlo technique that also allows us to include excitonic processes.

15.4 Master Equation

In inorganic, crystalline semiconductors charges are delocalized and energy eigenvalues of the electronic Hamiltonian form smooth bands. Consequently, one deals with band-like charge transport, where mobility decreases with increasing temperature. In contrast, in amorphous organic semiconductors, molecules are weakly bound by van der Waals forces and, consequently, intermolecular electronic couplings are small. As a result, excess electrons are localized on single molecules or their fragments. Moreover, orientational and positional *energetic* disorder helps further localize charged excitations. In this situation, charge transport can be modeled as thermally activated charge *hopping* between neighboring molecules. Mathematically, this is described by a Poisson process: The probability of an event to happen depends only on the time interval (no memory) and events do not occur simultaneously. The corresponding master equation, which describes the time evolution of a system with a discrete set of states i, then reads

$$\frac{dP_i(t)}{dt} = \sum_{j \neq i} \left[w_{ij} P_i(t) - w_{ji} P_j(t) \right], \tag{15.12}$$

where $P_i(t)$ is the probability of finding the system in state i at time t and w_{ij} is the transition rate from state i to j. In most situations, the dimension of the state space is too large to use direct numerical differential equation solvers to solve for P_i. In special cases, e.g., for single charge carrier transport or transport of many carriers in a mean-field approximation, this equation can be rewritten in terms of site occupation probabilities, p_i.

15.4.1 Single-Carrier Charge Transport

If only one charge carrier is present in the system, the state of the system is fully determined by the position of the charge or the index of the molecule which this charge occupies. In this situation, the occupation probability of the system state, P_i, is equivalent to the site (molecule) occupation probability, p_i. The rates for transitions between states are then given by CT rates, ω_{ij} (see Section 15.6), and the master equation can be rewritten in terms of site occupation probabilities

$$\frac{dp_i(t)}{dt} = \sum_{j \neq i} \left[\omega_{ij} p_i(t) - \omega_{ji} p_j(t) \right].$$ (15.13)

Equation 15.13 is a set of linear equations of size N, where N is the number of molecules in the system. It can be solved using standard numerical differential equation solvers [40]. In special cases, e.g., for one-dimensional charge transport, it is even possible to obtain an analytic solution [41–43]. Alternatively, one can use the KMC algorithm, which is discussed in Section 15.5.

15.4.2 Finite Charge-Carrier Densities

In experimentally relevant conditions, charge densities have finite values. The system state is now given by a vector of indices of occupied molecules and the number of states increases dramatically. It is, however, still possible to reformulate the master Equation 15.12 in terms of site occupation probabilities by using a mean field approximation [44,45]. The master equation then reads

$$\frac{dp_i}{dt} = \sum_{j \neq i} \left[\omega_{ij} p_i \left(1 - p_j \right) - \omega_{ji} p_j \left(1 - p_i \right) \right].$$ (15.14)

The resulting equation is no longer linear in p_i and thus requires more involved numerical solvers. In most cases, where one deals with many carriers the KMC method (see Section 15.5) becomes more practical: It does not rely on the mean field approximation but solves the original master equation, and it can easily be extended to other processes, such as exciton transport, triplet–triplet annihilation, etc.

15.4.3 Mobility and Diffusion Constant

The stationary solution of the master equation for a system in an external field $\vec{F}$ allows to evaluate both charge carrier mobility and diffusion constant. For single-carrier transport, the mobility tensor $\hat{\mu} = \vec{v} \otimes \vec{F}/F^2$ reads

$$\mu_{\alpha\beta} = F^{-2} \sum_{ij} \omega_{ij} p_i r_{ij,\alpha} F_\beta,$$ (15.15)

where $r_{ij,\alpha} = r_{i,\alpha} - r_{j,\alpha}$ and $\vec{v}$ is the average velocity of a charge carrier [45]. For finite charge carrier density $\rho = N/V$, where N is the number of carriers and V is the volume of the box, Equation 15.15 takes the form

$$\mu_{\alpha\beta} = \frac{1}{\rho F^2 V} \sum_{ij} \omega_{ij} p_i \left(1 - p_j \right) r_{ij,\alpha} F_\beta.$$ (15.16)

In a similar fashion, occupation probabilities can be used to calculate charge and current distributions in the system [45]. Alternatively, one can directly analyze charge trajectories of KMC simulations and evaluate mobility tensor as

$$\hat{\mu} = \frac{\langle \vec{v} \rangle \otimes \vec{F}}{F^2}, \quad \text{with} \quad \langle \vec{v} \rangle = \frac{\Delta \vec{r}}{\Delta t}.$$ (15.17)

Here $\vec{F}$ is the external electric field, $\vec{r}$ is the charge position, and Δt is the simulation time. The diffusion tensor can also be calculated directly from the charge trajectory generated without the applied external field [46]

$$D_{\alpha\beta} = \frac{\langle \Delta r_\alpha \Delta r_\beta \rangle}{2\tau}, \tag{15.18}$$

where $\Delta r_\alpha(t) = r_\alpha(t + \tau) - r_\alpha(\tau)$, or can be obtained from charge mobility with the help of the generalized Einstein relation, Equation 15.7.

15.5 Kinetic Monte Carlo

As we saw in the previous section, rewriting the master Equation 15.12 to a form tractable for large systems (i.e., in terms of site occupation probabilities) requires certain approximations. Moreover, this step becomes less and less straightforward if events other than CT are included, e.g., electron–hole recombination, exciton splitting, exciton decay, or transfer. For these reasons, a different way of solving the master equation becomes more practical. Here, system dynamics, or trajectories in the phase space, or Markov chains, are generated explicitly by using the so-called KMC method.

KMC, also known as dynamic Monte Carlo, Gillespie algorithm, residence-time algorithm, n-fold way, or the Botz–Kalos–Lebowitz algorithm, was initially developed by Doob [47,48]. The time update as it is used here was first introduced by Young and Elcock [49]. Bortz, Kalos, and Lebowitz developed the same algorithm independently and applied it to the Ising model [50]. Gillespie provided a physics-based derivation of the algorithm [51,52], which was then improved in terms of computational efficiency by Fichthorn [53] and Jansen [54]. The version we describe here is known as a variable step size method (VSSM) [54].

15.5.1 Variable Step Size Method

The VSSM allows us to group certain events and treat the groups and events in the group hierarchically. In the case of charge transport with multiple charge carriers, e.g., a two-level approach can be used. First, a charge is selected with the probability proportional to its escape rate ω_i, that is the sum of the rates of all possible moves of this charge away from the occupied site

$$\omega_i = \sum_{j=1}^{m} \omega_{ij}, \tag{15.19}$$

where m is the coordination number, i.e., the number of neighbors to which the site is connected, and ω_{ij} is the rate for all possible moves to connected sites. Afterward, the destination site is chosen with a probability ω_{ij}/ω_i. Since we are dealing with a Poisson process, waiting times are exponentially distributed with a parameter λ that is the inverse of the sum of all escape rates

$$\Delta t \sim \exp(\lambda), \ \lambda = \left(\sum_i \omega_i \right)^{-1}. \tag{15.20}$$

In order to reproduce this distribution, the time in the VSSM algorithm is updated after each move with Δt drawn from an exponential distribution. In practice, this is achieved by drawing a random number u from a uniform distribution on the interval $(0, 1]$ and setting Δt to

$$\Delta t = -\lambda \ln(u), \qquad u \sim \mathcal{U}((0, 1]). \tag{15.21}$$

15.5.2 Forbidden Events

The VSSM algorithm can be adapted to efficiently treat forbidden events [54]. In the case of charge transport, e.g., each site (molecule) can be occupied by one charge carrier at a time (the Pauli exclusion principle). As a result, charge carriers obey Fermi–Dirac occupation statistics (see Section 15.3). Hence, if site j is already occupied, all incoming rates, $\omega_{i \to j}$ should be set to zero. This is computationally inefficient, since all rates (as well as escape rates) must be updated after every event. A much more efficient way is to keep the forbidden event in the event list. Once this event is attempted, i.e., a carrier attempts to hop to an occupied site, this destination site is marked as forbidden and another destination is selected from the remaining possibilities. If all surrounding sites are occupied, the algorithm switches to the level above, i.e., a different charge is selected and the previous one is added to a temporary list of forbidden events. The time is updated regardless of whether or not the event is forbidden (the charge moved or not). One can show that this strategy results in the same statistics as *a priori* removing all forbidden events from the event list [54].

15.5.3 Efficiency and Parallelization

Developing an efficient KMC code for charge/exciton transport is a rather challenging task. Indeed, if only the Pauli exclusion principle is taken into account, a single charge transfer event does not affect the rates, and only the (local) list of forbidden events should be updated. Hence, all rates can be precomputed before the KMC simulation is performed. If (long-range) Coulomb interactions are included, every CT event changes all rates. In other words, energy differences between two states of the system depend now on the relative positions of all charges in the system. To avoid updating rates at every KMC step, one can use scale-separation techniques: Since mesoscopic charge densities evolve on time scales much longer than typical CT times, electrostatic fields created by far-off charges can be treated in a mean-field way, e.g., by solving the Poisson equation. In this case, one can update local rates at every step, while all rates are updated only every 100–1000 steps [55].

Even the local rate updates can become computationally costly for systems with many charge carriers. In this situation, the binary tree search algorithm helps improve computational efficiency by providing a faster way of identifying which events and escape rates should be updated. By using this algorithm one can gain a factor of 10 speed-up for systems with 10% occupied sites [55]. Another alternative is to use the next reaction method [56], an improved version of the first reaction method, where a dependency graph is used to determine which rates to update.

Apart from trivial parallelization, where multiple copies of the system with different initial conditions are run in parallel, Monte Carlo schemes can be parallelized by dividing the simulation volume into subvolumes treated as individual simulations [57–60]. An efficient parallel implementation on graphics processing units (GPUs) has also been demonstrated [61].

15.6 Rates

To parametrize the master Equation 15.12, we need to evaluate the rates of all elementary processes of interest. For example, for charge transport, CT rates must be evaluated for all neighboring molecular pairs. To make this computationally feasible, we often rely on simplified theoretical treatments of transfer reactions. In this section, we review several of such theories as well as introduce the corresponding rate expressions.

15.6.1 Charge Transfer

The simplest expression for the CT rate, proposed by Miller and Abrahams [62], is the rate of thermally activated, barrierless tunneling between two localized electronic states. The corresponding rate is proportional to the Boltzmann prefactor of the free energy difference between the initial and final states, ΔE_{ij}, and the electronic overlap between the states, which decays exponentially with the molecular separation

$$
\omega_{ij} = \begin{cases} \omega_0 \exp\left(-2\gamma r_{ij}\right) \exp\left(-\dfrac{\Delta E_{ij}}{k_B T}\right) & \Delta E_{ij} > 0, \\ 1, & \Delta E_{ij} \leq 0, \end{cases}
\tag{15.22}
$$

where $1/\gamma$ is the localization length of the charge, r_{ij} is the distance between the two molecules, and ω_0 is the attempt frequency. The simplicity and intuitiveness of this rate justified its use in the early stages of understanding charge transport in organic semiconductors [63].

Environmental effects, as well as molecular reorganization upon charging/discharging (i.e., the energetic barrier between the initial and final states), have been accounted for in the so-called Marcus CT rate [64,65]

$$
\omega_{ij} = \frac{2\pi}{\hbar} \frac{J_{ij}^2}{\sqrt{4\pi\lambda_{ij}k_B T}} \exp\left[-\frac{\left(\Delta E_{ij} - \lambda_{ij}\right)^2}{4\lambda_{ij}k_B T}\right].
\tag{15.23}
$$

Again, $\Delta E_{ij} = E_i - E_j$ is the driving force or the free energy difference of the final and initial states of the CT reaction [66]. An accurate evaluation of these energies using polarizable force-fields is discussed in Section 15.7. J_{ij} is the electronic coupling element. They are intimately related to the overlap of the diabatic electronic states, as discussed in Section 15.9. Finally, λ_{ij} is the sum of the external and internal reorganization energies, discussed in more detail in Section 15.8.

The main issue with the classical Marcus rate is that the intramolecular vibrational modes promoting the CT reaction are energetically comparable to the C–C bond stretching mode at room temperature, $\hbar\omega_{CC} \sim 0.2\,\mathrm{eV} \gg k_B T \sim 0.025\,\mathrm{eV}$. Therefore, these modes should be treated quantum mechanically. For a common set of intramolecular high-frequency (quantum-mechanical) and an outer sphere low-frequency (classical) vibrational coordinates, a mixed quantum-classical multichannel generalization of the Marcus formula is readily available [66,67]. A generalization for the bimolecular electron transfer rate with independent sets of coordinates for donor and acceptor has also been proposed [45].

Another rate expression has been proposed by Weiss and Dorsey [68–70]. In the low temperature limit, $\hbar\nu_c/k_B T \gg 1$, it reads

$$
\omega_{ij}(\epsilon) = \frac{J_{ij}^2}{\hbar^2 \nu_c} \left(\frac{\hbar\nu_c}{2\pi k_B T}\right)^{1-2\alpha} \frac{\left|\Gamma\left(\alpha + i\dfrac{\Delta E_{ij}}{2\pi k_B T}\right)\right|^2}{\Gamma(2\alpha)} \exp\left(\frac{\Delta E_{ij}}{2 k_B T} - \frac{\left|\Delta E_{ij}\right|}{\hbar\nu_c}\right).
\tag{15.24}
$$

Here $\Gamma(z)$ is the gamma function and ν_c is the characteristic frequency, or the largest frequency in the Ohmic bath, which is related to the reorganization energy, λ, by $\lambda = 2\alpha\hbar\nu_c$. The Kondo parameter, α, describes the coupling strength between the charge and the heat bath [71]. In the high temperature limit, the Weiss–Dorsey rate simplifies to the Marcus rate. Indications that the Weiss–Dorsey rate is better suited for describing charge transport, especially at low temperatures and high fields, have recently been reported [72].

15.6.2 Electron–Hole Recombination

The microscopic process of electron–hole recombination is straightforward: Whenever a hole and an electron occupy the same site (molecule), they recombine with a certain rate, w_r. Radiative recombination leads to a photon emission which is the basis of OLED functionality.

On a macroscopic level, electron–hole recombination is traditionally described by the Langevin equation (see Section 15.3.2), with a prefactor proportional to the sum of electron and hole mobilities. Extensive KMC simulations were performed to verify this dependency [73]. An excellent agreement was found if the electron and hole mobilities are extracted from simulations with both carrier types, which was attributed to the change in mobilities upon inclusion of two carrier types due to their Coulomb attraction. Deviations at high charge concentrations ($>10^{-3}$ carriers per molecule) were also observed: Here, the average electron–hole distance becomes smaller than the thermal capture radius, $r_c = e^2/4\pi\epsilon\epsilon_0 k_B T$, violating assumptions used in the derivation of the Langevin formula.

15.6.3 Energy Transfer

We now briefly review rate expressions used to describe electronic excitation (energy) transfer (EET). EET is also known as a resonant energy transfer (RET).

A theory explaining the mechanism of EET was first proposed by Förster [74]. The electronic interaction promoting EET relies on a coupling of the donor and acceptor molecules via Coulombic interaction. Similar to CT, Förster theory relies on Fermi's Golden Rule with the electronic coupling between donor and acceptor treated perturbatively. Additional assumptions are that the system equilibrates after the electronic excitation of the donor on a time scale much faster than that of EET and that coupling to the bath (given by the absorption line shape) is much greater that the electronic coupling between donor and acceptor. Energy conservation in the weak coupling limit results in a coupling element, which is proportional to the overlap of the donor fluorescence spectrum with the acceptor absorption spectrum. The spectral overlap includes nuclear overlap (in the form of Franck–Condon factors), which depends on the spectral line shapes and thus provides the temperature dependence of the EET rate. Under these assumptions, the EET rate between emitters i (donor) and j (acceptor) takes a simple form

$$\omega_{ij}^F = \frac{1}{\tau_{r,i}} \left(\frac{R_{F,ij}}{r_{ij}} \right)^6, \tag{15.25}$$

where $\tau_{r,i}$ is the radiative life time of the emitter (donor), r_{ij} is the separation between donor and acceptor, and $R_{F,ij}$ is the Förster radius for transfer from donor to acceptor.

An EET process can occur even when Coulomb-term-mediated electronic transitions are forbidden. Dexter provided a derivation for the case when the Coulombic interaction is negligible and EET is due to the exchange part [75]. This rate decays exponentially with intermolecular separation

$$\omega_{ij}^D = k_D \exp\left(-2\gamma r_{ij} \right), \tag{15.26}$$

where $1/\gamma$ is the wavefunction decay length and k_D is the Dexter prefactor proportional to the exchange integral.

Note that since the exchange integral is a quantum mechanical correction to the Coulombic repulsion, the total EET rate is always a sum of the two rates. However, due to the exponential decay of the Dexter coupling, the electronic interaction that mediates EET at separations greater than 5 Å is invariably Coulombic.

The importance of exchange and other short-range interactions, as well as higher multipole contributions to the Coulombic interaction, has been examined fairly extensively, helping improve the accuracy of rate expressions 15.25 and 15.26. An overview of these extensions can be found in Refs. [76–80].

15.7 Density of States

In the previous section, we saw that CT rates depend on the energies of the initial and final states of the CT complex. In an amorphous organic semiconductor, every molecule is embedded in a unique environment and therefore every molecule has its unique set of energy levels, EA, and IP. The set of energy levels available for an excess charge is termed the DOS. In this section, we discuss several methods of evaluation of the DOS of an amorphous organic semiconductor.

15.7.1 Gaussian Disorder

The simplest approach to model the DOS of organic semiconductor is to assume a Gaussian distribution of energy levels

$$g(E) = \frac{1}{\sqrt{2\pi\sigma^2}} \exp\left[-\frac{(E - \bar{E})^2}{2\sigma^2}\right]. \tag{15.27}$$

This phenomenological expression, first proposed by Bässler [63] and termed the Gaussian disorder model (GDM), is motivated as follows: Randomly oriented dipole moments in an amorphous material interact with one another, thus influencing the energy levels of neighboring molecules. As a result of the central limit theorem, this leads to approximately Gaussian distributed energy levels. The width of the distribution, σ, is called the *energetic disorder* and can be extracted from temperature-dependent mobility measurements [81, 82] or from simulations, as discussed in Section 15.11.1.

The original model was extended to the correlated disorder model (CDM) by accounting for spatial correlations that arise from the dipole interaction of neighboring molecules [83]. For randomly oriented dipoles, $\vec{p}_j$, the the electrostatic energy of a site i is given by [84]

$$E_i = -\frac{q}{4\varepsilon} \sum_{j \neq i} \frac{q\vec{p}_j \cdot (\vec{r}_j - \vec{r}_i)}{|\vec{r}_j - \vec{r}_i|^3}, \tag{15.28}$$

where ε is the dielectric permittivity and q is the charge. In the case of equal absolute values, $|\vec{p}_j| = p$, and dipoles fixed on a cubic lattice with lattice constant a, the sum can be evaluated using Ewald summation [85,86], yielding a Gaussian DOS with energetic disorder $\sigma = 2.35\,q\,p/\varepsilon\,a^2$ [83,87]. The spatial energy correlation function

$$\kappa(r) = \frac{\mathbb{E}\left[\left(E(\vec{r}_i) - \bar{E}\right)\left(E(\vec{r}_j) - \bar{E}\right)\right]}{\sigma^2} \tag{15.29}$$

is then given by $\kappa(r) \approx 0.74\frac{a}{r}$ [88]. Here $r = |\vec{r}_i - \vec{r}_j|$ is the distance between two molecules, $\bar{E}$ is the mean of the energy distribution, and $\mathbb{E}\,[\cdot]$ is the expectation value. Note that the spatial correlation function of this simple model system depends only on the lattice spacing a, limiting its applicability to realistic morphologies [89], as discussed in Section 15.11.1.

The Gaussian DOS is at the heart of the family of GDM, where Miller–Abrahams rates, Equation 15.22, with the Gaussian DOS are assumed for a charge hopping on a cubic lattice. Extensive KMC simulations of these models helped parametrize the mobility as a function of temperature, field, and field carrier density [63]. It was extended by including the important influence of charge carrier density in the extended Gaussian disorder model (EGDM) [90] without spatial correlation of site energies. The mobility expression in the EGDM at finite can be written as a product of three functions,

$$\mu(T, F, \rho) = \mu_0(T)g(T, \rho)f(T, F), \tag{15.30}$$

where

$$\mu_0(T) = 1.8 \times 10^{-9}\mu_0 \exp\left[-C\hat{\sigma}^2\right], \qquad g(T, \rho) = \exp\left[\frac{1}{2}\left(\hat{\sigma}^2 - \hat{\sigma}\right)\left(2\rho a^3\right)^{\delta}\right], \tag{15.31}$$

$$f(T, F) = \exp\left[0.44\left(\hat{\sigma}^{3/2} - 2.2\right)\left(\sqrt{1 + 0.8\hat{F}^2} - 1\right)\right], \qquad \delta = 2\frac{\ln\left(\hat{\sigma}^2 - \hat{\sigma}\right) - \ln(\ln 4)}{\hat{\sigma}^2},$$

with $C = 0.42$, $\hat{F} = eaF/\sigma$ and $\hat{\sigma} = \sigma/k_B T$. A similar parametrization, termed the extended correlated disorder model (ECDM), also exists for the spatially correlated DOS [91].

15.7.2 Perturbative Approach

While Gaussian disorder models have been successful in describing many properties of organic semi-conductors, they do not provide a direct link to the material morphology or chemical composition. A perturbative approach allows us to evaluate site energies in atomistically resolved (see Section 15.10), large-scale morphologies. Furthermore, it can also be used to parametrize Gaussian disorder models and perform large-scale simulations (see Section 15.11.2).

For every molecular pair, our quantity of interest is the energy difference $\Delta E_{ij} = E_i - E_j$, which is the energy separation between the minima of the diabatic potential energy surfaces (PES). In systems with weak intermolecular couplings, one can treat interactions with the environment perturbatively: The inter-molecular electrostatic and induction contributions are then given by the first- and second-order terms in the expansion of the interaction energy [92].

The total energy energy of an ion embedded in a molecular environment thus includes an internal con-tribution E_i^{int}, i.e., the electron affinities for electrons and IPs for holes of *isolated* molecules. These can vary from one molecular pair to another because of different energy levels for different types of molecules, or different conformers of the same molecule. Correspondingly, the *external* contribution is due to the electrostatic and induction interactions, E_i^{el} and E_i^{ind}, of a charged molecule with the environment. These interaction energies are determined by the electrical charge distribution and the polarizability distribu-tion, respectively, in the environment of the charged molecule. Overall, the total energy of molecule i is then given by

$$E_i = E_i^{int} + E_i^{el} + E_i^{ind}. \tag{15.32}$$

Most difficult to evaluate is the interaction with the environment. This is for two reasons: First, the under-lying interactions are long-ranged and thus large system sizes are needed to converge the values of site energies. Second, special summation techniques are needed in order to evaluate interactions of an ion with a neutral *periodic* environment [93–96]. Also note that the perturbative evaluation of site energies relies on accurate molecular representations in terms of distributed multipoles and polarizabilities and is computationally demanding, in spite of being classical [97].

15.7.3 Hybrid Approaches

Using Equation 15.32, the site energies can be evaluated by solving a microscopic analogue of the Poisson equation. The self-consistent solution is normally achieved using iterative schemes for induced multipoles and Ewald summation techniques for static multipoles, which is computationally demanding. The role of hybrid schemes is to reduce computational cost. Here, one relies on the fact that the fields created by the far-off charges can be treated in a mean-field way, i.e., obtained by solving the Poisson equation. Nearest-neighbor interactions (i.e., interactions between molecular pairs within a certain cutoff distance) are still evaluated explicitly. This scheme also allows us to add metallic electrodes as image charges and has been used extensively to simulate multilayered OLED devices [98–100]. To improve computational efficiency, in this approach all induction interactions are taken into account effectively, by rescaling interaction energies by the (effective) dielectric constant of the medium.

15.8 Reorganization Energy

The reorganization energy, responsible for the energy barrier between diabatic states as well as for the broadening of energy levels of the electron detachment/attachment spectra, has two contributions. The *internal* reorganization energy is a measure for how much the geometry of the CT complex adapts while the charge is transferred. It can be estimated based on four points on the diabatic PES [45,101]

$$\lambda_{ij}^{int} = E_i^{nC} - E_i^{nN} + E_j^{cN} - E_j^{cC}. \tag{15.33}$$

Here, small letters denote the state and capital the geometry of a molecule, e.g., E_i^{nC} is the internal energy of the molecule i in the geometry of its charged state. Treatments that do not approximate the PES in terms of a single shared normal mode are also available [45,66,102].

An additional contribution to the overall reorganization energy results from the rearrangement of the environment in which the CT takes place. In a classical case, this outer-sphere reorganization energy, λ^{out}, contributes to the exponent in the rate expression in the same way as its internal counterpart. Assuming that CT is significantly slower than electronic polarization but much faster than the nuclear rearrangement of the environment, λ^{out} can be evaluated from the electric displacement fields created by the CT complex [66], provided that the Pekar factor is known. Alternatively, one can use polarizable force-fields [103] or quantum mechanics/molecular mechanics (QM/MM) methods [104]. It also turns out that the classical Marcus expression for the outersphere reorganization energy (inversely proportional to the molecular separation) can predict negative values of λ^{out} for small intramolecular separations, which are unphysical and hence should be used with care [45].

15.9 Electronic Coupling Elements

Electronic coupling elements, or transfer integrals , J_{ij}, entering the CT rate (Equations 15.23 and 15.24) are off-diagonal matrix elements, $J_{ij} = \langle \psi_i | \hat{H}^{el} | \psi_j \rangle$, of the electronic Hamiltonian, $\hat{H}^{el} = \hat{T}^{el} + \hat{V}^{el-el} + \hat{V}^{nuc-el}$, based on a diabatic (noninteracting) state, ψ_i [66]. A number of approaches can be used to evaluate electronic coupling elements. Their efficiency and accuracy depend on how the diabatic states and Hamiltonian are constructed, as well as how the matrix projection is performed.

Diabatic states are often approximated by the HOMO of monomers for hole transport, or the LUMO for electron transport ("frozen core" approximation) [105–107]. An approximate diabatic basis can also be constructed using constrained density functional theory [108].

The dimer Hamiltonian can be constructed using semiempirical methods, e.g., the Zerner's Intermediate Neglect of Differential Overlap (ZINDO) Hamiltonian [101,109–111]. This approach does not require a self-consistent evaluation of the dimer Hamiltonian and is therefore computationally very efficient [110]. One can also employ density functional theory (DFT) and use either the fully converged Hamiltonian or only the initial guess [105]. Another way to improve the efficiency is to reduce the number of orbitals for which electronic couplings are calculated. A detailed comparison of accuracy and efficiency of different approaches can be found in Refs. [105,112,113].

For (approximately) spherically shaped molecules the logarithm of the squared electronic coupling, $\log(J_{ij}^2)$, decays linearly as a function of the intermolecular separation r (at least for large distances), justifying the functional form of the Miller–Abrahams rate (see Equation 15.22). If molecular pairs are extracted from the respective dimers in the realistic atomistic morphology, $\log(J_{ij}^2)$ is often Gaussian distributed, with a mean and variance that depend on the molecular separation [55,114]. This observation can be used to parametrize coarse-grained (stochastic) charge transport models, as discussed in Section 15.11.1.

Since electronic couplings are related to the overlap of electronic orbitals participating in charge transport, they are very sensitive to the relative positions and orientations of molecules. Hence, an amorphous morphology of an organic material should be generated as precisely as possible. The corresponding methods are covered in the next section.

15.10 Morphology

In the previous sections, we explained how charge/exciton transport and recombination can be mapped on a series of events with specified rates. We have also emphasized that rates are very sensitive to both local and global molecular ordering. For computer-based predictions of material properties it is therefore important to simulate the material morphology as realistically as possible. In OLEDs, we usually deal with *amorphous* molecular arrangements—crystalline films are often less efficient in OLEDs because of the resulting large exciton diffusion distances, leading to enhanced exciton loss due to quenching processes. Amorphous films have a well-defined local structure, which depends on molecular interactions and processing conditions. In this section, we describe how amorphous morphologies of 10^4–10^5 molecules can be simulated using atomistic force fields or coarse-grained models. These systems can then be used to study small-scale charge transport [45,115–119], or to parameterize mesoscale models, as discussed in Section 15.11.1.

15.10.1 Classical Force Fields

The role of classical force fields in simulations of organic semiconductors is two-fold. First, they are used to generate atomistically resolved morphologies of molecular assemblies. Second, they are employed to evaluate the solid-state electrostatic and induction contributions to site energies (see Section 15.7). In both cases, these classical molecular representations should be appropriately parametrized.

For site energy calculations, it is important to evaluate electrostatic and induction contributions as accurately as possible. The corresponding parametrization is rather straightforward: A perturbative expansion of the intermolecular interaction energies is based on distributed multipoles (electrostatic interaction) and distributed polarizabilities (induction interaction) [92,97,120–122]. Van der Waals interactions are normally ignored since we are interested in free energy differences between charged and neutral states of the system. To improve computational efficiency, one can also use machine-learning techniques to devise simple structure–property relations for, e.g., atomic multipoles of molecular conformers [123].

Polarizable force fields based on distributed multipole expansions are still computationally prohibitive for simulating molecular arrangements of large systems. Moreover, parametrizations of effective *pairwise* potentials (partial charges and Lennard–Jones parameters), which mimic many-body van der Waals and induction interaction energies, are a nontrivial task, which largely relies on experimental input [124,125]. Standard forces fields, e.g., those suitable for biosystems, are often not transferable to organic molecules with large π-conjugated subsystems. A representative example is the comparison of the Williams 99 and the optimized potential for liquid simulations (OPLS) force fields for the amorphous mesophase of Alq$_3$ [126], which predict fairly different densities, radial distribution functions, and glass transition temperatures.

That being said, the development of new force fields for organics is currently limited to refitting of partial charges using the Merz–Singh–Kollman [127] scheme or the CHarges from ELectrostatic potentials using a grid-based method (CHELPG) [128] scheme, and a parametrization of missing bonded interactions from first-principles scans (see Figure 15.3). The remaining parameters are often taken from standard force fields, such as OPLS [124]. Reparametrized force fields are then verified against experimentally available densities and glass transition temperatures.

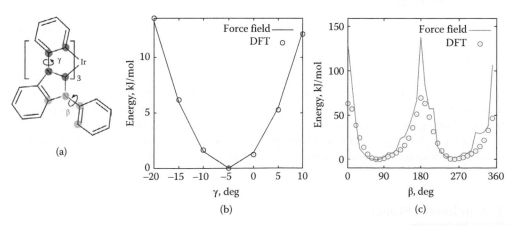

FIGURE 15.3 Potential energy scans of two dihedral angles of the DPBIC molecule, the chemical structure of which together with the definition of dihedral angles is shown in the inset (a). Symbols show DFT energies (B3LYP/6-311g(d,p)) and lines indicate the force-field energies after adjusting the force-field parameters. (Reproduced from P. Kordt et al., *Adv. Funct. Mater.*, 25, 1955–1971, 2015. With permission.)

15.10.2 Morphology Simulations

With the force field at hand, one can, in principle, simulate amorphous molecular assemblies. For systems with periodic boundaries in all directions this is done by first equilibrating the system above the glass transition temperature and then quenching it to room temperature in the *NPT* ensemble using molecular dynamics [126,129,130].

For thin slabs (2D-periodic systems) one can directly deposit molecules either using molecular dynamics [94] or Monte Carlo [131] techniques. In both cases the dynamics of deposition is not realistic. In other words, if surface diffusion plays an important role (e.g., in the case of guest aggregation in the host) these techniques cannot be applied in a straightforward manner, since the length scales and time scales required of host aggregation cannot be reached by atomistic molecular dynamics simulations. It is, however, possible to explore the fact that certain parts of molecules evolve on much slower time scales and larger-length scales and to combine several coherently moving atoms, connected via stiff degrees of freedom (e.g., bonds) into a single interaction site, as it is done, e.g., in the united atom force-fields with hydrogens incorporated into heavier atoms [132]. By doing this, we reduce the number of degrees of freedom to be propagated and, more importantly, obtain a much smoother potential energy landscape in terms of the *coarse-grained* degrees of freedom [133] (softer interaction potentials, less friction), allowing one to simulate 10–100 longer times and system sizes [132].

In order to perform correct statistical sampling of the coarse-grained degrees of freedom, the potential of mean force should be used as the interaction potential [134], which is inherently a many-body potential. To reduce computational cost, this potential is represented as the sum of a few functions, i.e., projected onto the basis functions of the force field. The accuracy of the coarse-grained model thus becomes sensitive to the way the projection is performed as well as the number of basis functions that are used to represent the coarse-grained force-field [135]. Existing projection schemes try to reproduce various pair distribution functions (structure-based coarse-graining [136–139]), to match the forces [134,140,141], to minimize the information loss in terms of relative entropy [142], or to use liquid state theory [143]. An extensive overview of such coarse-graining techniques is provided in Ref. [144]. Often, the accuracy of the coarse-grained model can be improved by explicitly incorporating information about macroscopic properties of the system, i.e., its equation of state or the symmetry of its mesophase [145–147]. Moreover, one can further reduce the number of degrees of freedom by introducing anisotropic interaction potentials. Eventually, atomistic details can be reintroduced into coarse-grained morphologies [148].

15.11 Scale Bridging

Now that we have seen how an OLED device can be modeled on different time and length scales, we eventually would like to transfer our knowledge about the system between the scales. For example, drift–diffusion equations (Section 15.3) can be used to calculate current–voltage curves of a device (micrometer scale). However, they require expressions for charge carrier mobility, the diffusion constant and, in the case recombination is taken into account, the recombination rate, all as a function of external field, carrier density, and temperature. Microscopic simulations (see Sections 15.4–15.10) can provide this information, at the same time retaining the link to the molecular structure. However, they are becoming computationally too demanding for high charge densities and large system sizes required to parametrize these dependencies. Our aim here is to provide several strategies, which can be used to link different time and length scales.

15.11.1 Stochastic Models

In an OLED, charges are inhomogeneously distributed [149] and charge density variations span several orders of magnitude. To cover the required density range in simulations, one needs to deal with relatively large systems—this quickly becomes computationally demanding if all rates are evaluated from first principles, as described in Section 15.6. To remedy the situation, one can devise a phenomenological algorithm to parametrize the master equation. In this section we outline how such an algorithm can be constructed for charge transport simulations.

We first note that the master equation (Section 15.4) is completely determined by the event rates. In the case of charge transport, these rates depend on site energy differences, electronic coupling elements, and reorganization energies. In order to evaluate observables of interest (e.g., charge mobility), one additionally needs site positions. Hence, the algorithm should be able to reproduce (statistically) several distribution and correlation functions.

Let us start with molecular positions. In an amorphous solid, both positions and orientations are completely defined by a set of (many-body) spatial correlation functions. For (approximately) spherically shaped molecules, the pair correlation function, or radial distribution function, $g(r)$, which quantifies the density of molecules at a separation r, contains the most relevant structural information. To reproduce this function approximately, one can use "thinning of a Poisson process" [114,150]. More accurate coarse-graining techniques, such as iterative Boltzmann inversion [138,151] or inverse Monte Carlo [136,137,152], allow a (numerically) exact reproduction of the radial distribution function [55]. These methods optimize a pair interaction potential, $U(r)$, in a way that the corresponding $g(r)$ is reproduced. They rely on the Henderson theorem [153], which states that there is a unique correspondence between $U(r)$ and $g(r)$. An illustration of this algorithm for an amorphous layer of DPBIC is presented in Figure 15.4a. The approach can also be applied to nonspherical molecules, by using several interacting sites per molecule [135].

The second ingredient of the stochastic model is the connectivity: In the atomistic model only molecules within a certain cutoff distance are used for calculating CT rates. The rest of the rates are set to zero. This is justified by the fact that electronic coupling elements decrease roughly exponentially with molecular separation [55], see, e.g., Equation 15.22. The distance that determines whether or not two molecules are connected is given by their two closest atoms. Since this information is not present in the coarse-grained model, the resulting probability of two sites to be connected, as a function of their center-of-mass separation, is given by the corresponding probability extracted from the reference data. Figure 15.4b shows this probability for an amorphous DPBIC layer.

As explained in Section 15.7, the stochastic generation of site energies should reproduce both their distribution function (DOS) and their spatial correlation. This can be achieved by mixing in site energy contributions of neighboring sites [89]. Figure 15.4c shows the spatial correlation function for an amorphous DPBIC layer.

Electronic coupling elements can also be generated using appropriate distributions. These distributions are, however, separation dependent: The logarithm of squared transfer integrals, $\log J^2$ (which is often Gaussian distributed), depends on molecular separation. For DPBIC, the distance dependence of the mean and the standard deviation is shown in Figure 15.4d. In the stochastic model, transfer integrals are then drawn from such distant-dependent distributions.

With all necessary rate ingredients, one can now validate the model, e.g., by evaluating the distribution of rates, see Figure 15.4e, or by directly comparing charge carrier mobilities, as shown in Figure 15.4f. Note that charge transport in systems with large energetic disorder has pronounced finite size-effects [130]. Therefore, similar system sizes should be used to compare stochastic and reference simulations.

Since stochastic models are computationally significantly less demanding, they can serve as an intermediate step between atomistic and macroscopic (drift–diffusion) descriptions, as discussed in the next section.

15.11.2 Parametrization of Gaussian Disorder Models

As mentioned in Section 15.3, macroscopic OLED modeling requires the charge mobility as a function of external field, temperature, and carrier density, $\mu(T, F, \rho)$. Analytic expressions of these dependencies can be provided by the EGDM and ECDM, as discussed in Section 15.7.1. These generic expressions include

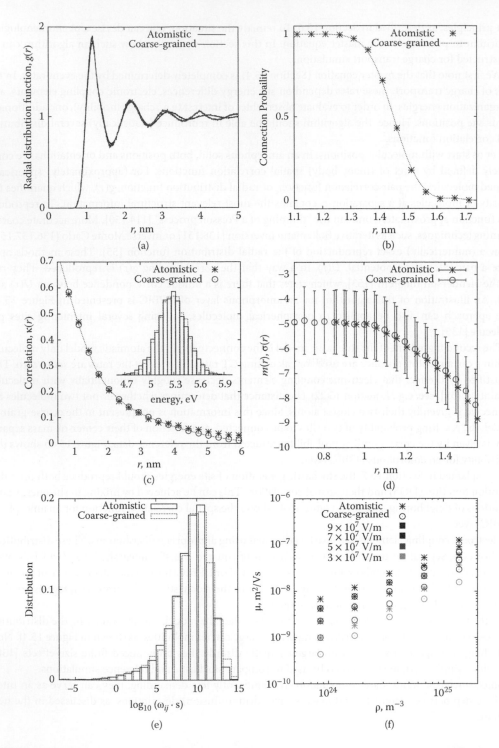

FIGURE 15.4 Comparison of hole states and hole mobilities in the atomistic reference system of DPBIC and in a stochastic model. (a) Radial distribution function, $g(r)$, where r is the center-of-mass distance. (b) Probability for sites to be connected. (c) Spatial correlation function, $\kappa(r)$, as defined in Equation 15.29; inset: site energy distribution with a mean (ionization potential) of 5.28 eV and an energetic disorder of $\sigma = 0.176$ eV. (d) Mean m and width σ of a distribution of the logarithm of squared electronic couplings, $\log_{10}(J^2/\text{eV}^2)$, for molecules at a fixed separation r. (e) Rate distributions. (f) Mobility as a function of hole density at different electric fields. (Reproduced from P. Kordt et al., *Adv. Funct. Mater.*, 25, 1955–1971, 2015, with permission.)

several material-specific parameters. In this section, we describe how to determine these parameters from simulations of small systems.

Both EGDM and ECDM depend parametrically on the lattice constant a, the energetic disorder σ (see Section 15.7), and a prefactor μ_0, which is related to the temperature-dependent mobility at zero field and charge density by $\mu_0(T) = 1.8 \times 10^{-9}\mu_0 \exp(-0.42\hat{\sigma}^2)$ for the EGDM (see Section 15.7.1) or $\mu_0(T) = 1.0 \times 10^{-9}\mu_0 \exp(-0.29\hat{\sigma}^2)$ in the case of ECDM.

In principle, both a and σ can be evaluated in a relatively small system: a as the mean distance between neighboring molecules and σ as the width of the DOS that results from perturbative energy calculations. μ_0 can be extracted from charge transport simulations performed at different temperatures. This approach, however, does not lead to reliable parameterizations [55,114]. Indeed, a multidimensional fit of simulated mobilities to the EGDM or ECDM expressions, for a wide range of temperatures, charge densities, and external fields, yields a very different set of parameters. A comparison of these two approaches for amorphous dicyanovinyl-substituted quaterthiophe (DCV4T) and DPBIC is given in Table 15.1. One can see, e.g., that the EGDM underestimates the energetic disorder, while the ECDM overestimates it. In both cases, spatial site energy correlations are responsible for this discrepancy: EGDM does not include correlations and compensates for higher mobility values by reducing the energetic disorder σ. On the other hand, ECDM overestimates spatial correlations and compensates this by reducing the lattice constant [89]. The discrepancy between microscopic values and fits to EGDM and ECDM teaches us that parameters of these models do not have a clear physical interpretation. Nevertheless, they still provide reasonable parametrizations and can eventually be used in conjunction with drift–diffusion equations; see Section 15.12.1. Figure 15.5 compares EGDM and ECDM fits to microscopic simulations for an amorphous mesophase of DCV4T.

On a technical side, the stochastic models described in Section 15.11.1 become very useful to perform the fits. They help cover the required range of charge carrier densities and reduce finite-size effects. Note, however, that finite-size effects in systems with small charge carrier densities and large disorder are so large that the actual value of mobility is overestimated by several orders of magnitude [130]. In this case, one needs to use the extrapolated mobility values [114].

15.11.3 Tabulated Mobilities

Fitting the results of KMC simulations to the parametrizations provided by the EGDM or ECDM imposes a constraint on the functional form of $\mu(\rho, \vec{F}, T)$. To avoid this, one can tabulate the mobility in a wide range of charge densities, temperatures, and electric fields. This tabulated function can then be used directly in the drift–diffusion equations solver [149]. The tabulation is computationally feasible only with the help of a stochastic model—otherwise it is not possible to reach the necessary system sizes and to span the wide density regime. Before using the tabulated function, it has to be interpolated and smoothed to ensure numeric stability [149]. Figure 15.6 shows the tabulated and smoothed mobility for amorphous DPBIC, which is eventually used to evaluate current–voltage characteristics of a DPBIC film; see Section 15.12.1.

TABLE 15.1 Lattice Spacing, Energetic Disorder, and Mobility at Zero Field and Density Extracted From a Microscopic System and From Fitting Simulated Hole Mobilities to Extend Gaussian Disorder Model (EGDM) and Extended Correlated Disorder Model (ECDM) for Amorphous Phases of (a) dicyanovinyl-substituted quaterthiophe (DCV4T) and (b) DPBIC

(a) DCV4T	a [nm]	σ [eV]	$\mu_0(300\,K)$ [m^2/Vs]	(b) DPBIC	a [nm]	σ [eV]	$\mu_0(300\,K)$ [m^2/Vs]
Microscopic	0.86	0.253	2.0×10^{-21}	Microscopic	1.06	0.176	3.4×10^{-12}
EGDM	1.79	0.232	2.1×10^{-21}	EGDM	1.67	0.134	2.1×10^{-11}
ECDM	0.34	0.302	3.3×10^{-22}	ECDM	0.44	0.211	1.8×10^{-13}

Source: DCV4T values: Reprinted from P. Kordt et al., *J. Chem. Theory Comput.*, 10, 2508–2513, 2014. With permission; DPBIC values: Reproduced from P. Kordt et al., *Adv. Funct. Mater.*, 25, 1955–1971, 2015. With permission.

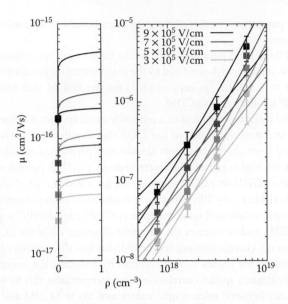

FIGURE 15.5 Parametrization of lattice models from simulated dicyanovinyl-substituted quaterthiophe (DCV4T) hole mobilities, μ, for different hole densities, ρ. Symbols are the simulated values for four different external fields, solid lines are the fit to the extended Gaussian disorder model (EGDM), and dashed lines are the fit to the extended correlated disorder model (ECDM). An extrapolation has been used to obtain non-dispersive values (i.e., without finite-size effects) in the limit of zero density. (Reprinted from P. Kordt et al., *J. Chem. Theory. Comput.*, 10, 2508–2513, 2014. With permission.)

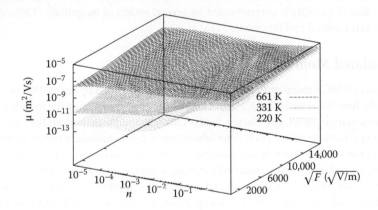

FIGURE 15.6 Hole mobilities for amorphous DPBIC as a function of the charge concentration n (number of charge carriers per site) after tabulation from simulations, smoothing, and interpolation. (Reprinted from P. Kordt et al., *Phys. Chem. Chem. Phys.*, 17, 22778–22783, 2015. With permission.) Published by the PCCP Owner Societies.

15.12 Case Studies

So far we have described various methodological developments and simulation approaches, which can be used to simulate multilayered OLED structures. In the following sections, we show how these methods can be used to simulate steady-state current–voltage characteristics of an OLED, perform impedance spectroscopy simulations, estimate OLED efficiency, and study electroluminescence of a white OLED, as well as to gain insight into OLED stability.

15.12.1 Current–Voltage Characteristics

We start by showing how the steady-state current–voltage characteristics of a single layer device can be simulated starting from the chemical structure of an organic semiconductor. As an example, we use a thin layer of DPBIC, a hole-conducting material, which is sandwiched between an ITO and aluminum electrode.

After parametrizing the DPBIC force field [55], amorphous boxes of 4000 DPBIC molecules are simulated by molecular dynamics (MD) simulations, as described in Section 15.10.2. The DOS of an amorphous solid state is then evaluated for holes using the perturbative scheme (see Section 15.7.2), yielding a mean value of 5.28 eV and an energetic disorder of $\sigma = 0.176$ eV. Density functional theory calculations (B3LYP/6-311g(d,p), see also Section 15.8) yield a hole reorganization energy of $\lambda = 0.068$ eV.

The reference system of 4000 molecules is further used to parametrize a stochastic algorithm (see Section 15.11.1) and to generate larger systems of 40,000 sites (details can be found in Ref. [149]). KMC simulations in large systems allow us to tabulate charge mobility as a function of charge density, electric field, and temperature. The interpolated and smoothed tabulated function, depicted in Figure 15.6, is then used to solve the drift–diffusion equations, as described in Section 15.3. For the electrode IPs we use average values of experimental reports: 4.73 eV for ITO [154–157] and 4.16 eV for aluminum [158]. These values, together with the DPBIC IPs, provide the value of the injection barrier, which is required to solve the Poisson Equation 15.3.

Without further microscopic calculations, it is now possible to solve the drift–diffusion equations for film thicknesses of 203, 257, and 314 nm and temperatures of 233, 293, and 313 K, corresponding to different experiments. Figure 15.7 shows a comparison of simulated and experimental current–voltage curves in these situations. The agreement is remarkable for higher temperatures (293 and 313 K). A possible reason for the larger differences at low temperatures are nonequilibrium processes, such as charge relaxation, which are not accounted for in our approach that relies on mobility parametrizations under stationary conditions. Another possibility is that the Marcus rate expression, Equation 15.23, is no longer valid at these temperatures, since it is derived assuming a classical promoting mode and is valid only for high enough temperatures.

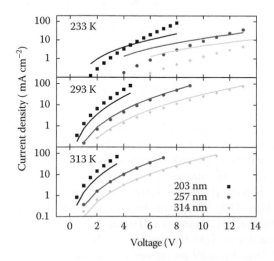

FIGURE 15.7 Current–voltage characteristics for DPBIC films of different thickness sandwiched between an indium tin oxide (ITO) anode and an aluminum cathode, measured at different temperatures. Theoretic predictions (lines) were obtained using parameter-free coupling of microscopic hole mobility data to drift–diffusion equations. Symbols are experimental results. (Reprinted from P. Kordt et al., *Phys. Chem. Chem. Phys.*, 17, 22778–22783, 2015. With permission.)

15.12.2 Impedance Spectroscopy

In addition to studies of steady-state current–voltage characteristics, valuable insight into the functioning of organic devices can be obtained from impedance spectroscopy studies. In particular, impedance spectroscopy can provide information about carrier relaxation in the DOS, can help distinguish between different trapping regimes in organic-semiconductor devices [36], and can be used to determine the width σ of the Gaussian DOS [159].

In impedance spectroscopy a dc bias V is applied over a device and, in addition to that, a small ac component $\Delta V(t) = \Delta V \exp(2\pi i f t)$ is added, where f is the frequency. The impedance $Z = Z' + iZ''$ is defined as the zero-amplitude limit of the ratio of $\Delta V(t)$ and the response $\Delta I \exp\left[2\pi i (f + \phi)t\right]$ in the current, with ϕ a phase difference. Of particular interest is the capacitance–voltage, C–V, characteristic, with the capacitance given by $C = -Z''/2\pi f |Z|^2$.

Applying KMC simulations to extract the small response $\Delta I(t)$ is extremely cumbersome because of the noise present in such simulations. It has been shown, however, that for single-carrier organic devices the current–voltage characteristics obtained by solving the master equation are practically the same as those obtained from KMC simulations [160]. The influence of small perturbations can be rather easily evaluated using the master equation. Within the framework of the time-dependent master equation, the small ac component of the voltage in a single-carrier device leads to a time-dependent probability, $p_i(t)$, of the occupation of a site i by a charge, obeying Equation 15.14:

$$\frac{dp_i}{dt} = \sum_{j \neq i} \left[\omega_{ji} p_j (1 - p_i) - \omega_{ij} p_i (1 - p_j) \right] \equiv g_i(\vec{p}), \tag{15.34}$$

where $\vec{p}$ is the vector of occupational probabilities of all sites.

Using a perturbative approach, first the steady-state solution, $\vec{p}_0$, for $dp_i/dt = 0$ at the applied static voltage, V, has to be evaluated. The procedure for doing so has been described in Ref. [99]. Sheets of sites representing the electrodes are introduced at either side of a simulation box representing the device. An additional small-amplitude ac voltage with frequency f induces a small change, $\vec{\Delta p}$. Linearizing, we write $\vec{p}(t) \approx \vec{p}_0 + \exp\left(2\pi i f t\right) \vec{\Delta p}$ and $\vec{g}(\vec{p}) \approx \vec{g}(\vec{p}_0) + \exp(2\pi i f t)\left[\Delta V \partial \vec{g}/\partial V + \hat{J}\vec{\Delta p}\right]$, with the matrix elements of the Jacobian, $\hat{J}$, given by $J_{ij} = \partial g_i/\partial p_j\big|_{\vec{p}_0}$. Substituting these expressions into Equation 15.34 and linearizing leads to the equation

$$(2\pi i f \hat{I} + \hat{J})\vec{\Delta p} = -\Delta V \frac{\partial \vec{g}}{\partial V}, \tag{15.35}$$

with $\hat{I}$ denoting the identity matrix. Equation 15.35 can be solved for $\vec{\Delta p}$ with standard techniques, and from this the current, ΔI, and the capacitance, C, are readily obtained.

As an example, we consider two hole-only devices with the structure glass, ITO (100 nm), poly(3,4-ethylenedioxythiophene) polystyrene sulfonate (PEDOT:PSS) (100 nm), light-emitting polymer (LEP), Pd (100 nm). The LEP consists of polyfluorene with 7.5 mol% copolymerized triarylamine units for hole transport; see the Figure 15.8a inset. The LEP-layer thicknesses are $L = 97$ and 121 nm for the two devices and their areas are $A = 9 \times 10^{-6}$ m^2. According to EGDM modeling studies of the current density–voltage, J–V, characteristics of these devices [161,162] no injection barrier is present at the anode (PEDOT:PSS) and injection barriers of 1.65 and 1.90 eV are present at the cathode (Pd) for the $L = 97$ and $L = 121$ nm device, respectively. These modeling studies gave best fits for the J–V characteristics with σ = 0.13 eV.

We apply the above method to calculate the C–V characteristics of these two devices. As in the EGDM studies [161,162] we assume an uncorrelated Gaussian DOS with standard deviation σ. The use of Marcus rates would require knowledge of the reorganization energy λ_{ij}, which is not available here. We therefore assume Miller–Abrahams nearest-neighbor hopping with rates given by Equation 15.22. In the absence

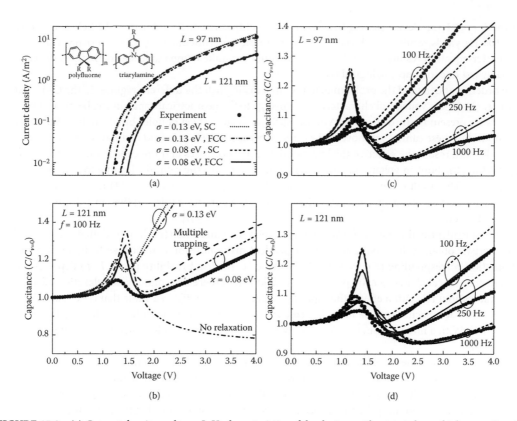

FIGURE 15.8 (a) Current density–voltage, *J–V*, characteristics of the devices with organic layer thicknesses $L = 97$ and $L = 121$ nm at $T = 295$ K. Dots: measurements [163]. Curves: solutions of the steady-state master equation with σ = 0.13 and 0.08 eV for simple cubic (SC) and face-centered cubic (FCC) lattices. Inset: the used hole-transporting copolymer. (b) Capacitance, *C*, normalized to its value at $V = 0$, as a function of *V* at a frequency $f = 100$ Hz and temperature $T = 295$ K for the $L = 121$ nm device. Dots: measurements [164]. Dash–dot–dotted curve: drift–diffusion calculation with an extended Gaussian disorder model (EGDM) mobility function for σ = 0.13 eV, neglecting relaxation. Long-dashed curve: multiple-trapping result, which includes relaxation [164]. Other curves: Solutions of the time-dependent master equation for σ = 0.13 and 0.08 eV and for SC and FCC lattices. (c) and (d) *C–V* characteristics at different frequencies for the $L = 97$ nm (c) and $L = 121$ (d) device. Dots: Measurements [164]. Curves: Solutions of the time-dependent master equation. (Reprinted with permission from M. Mesta, J. Cottaar, R. Coehoorn, and P. A. Bobbert, Study of charge-carrier relaxation in a disordered organic semiconductor by simulating impedance spectroscopy. *Appl. Phys. Lett.*, 104(21), 213301, 2014. Copyright [2014], American Institute of Physics.)

of any morphological information we assume a regular lattice of hole-transporting sites. To investigate a possible influence of morphology we investigate simple cubic (SC) as well as face-centered cubic (FCC) lattices with a lattice constant $a = 1.19$ nm for the SC lattice and $a = 1.88$ nm for the FCC lattice, in accordance with the known density 1.8×10^{26} m^{-3} of hole-transporting units. The simulation boxes have dimensions $L \times L_y \times L_z$, with $L_y = L_z = 50a$ and periodic boundary conditions in the *y*- and *z*-directions, yielding a sufficient lateral averaging. For further details we refer to Ref. [159].

As expected from the EGDM modeling [161,162] it can be observed in Figure 15.8a that the experimental *J–V* characteristics (dots) at room temperature ($T = 295$ K) of the two devices are very well described by the solution of the steady-state master equation for σ = 0.13 eV, both for the SC (dotted curve) and the FCC (dash-dotted curve) lattice (a prefactor in the hopping rates was adjusted in both cases to obtain an optimal fit). However, this is not at all true for the *C–V* characteristics. It is seen in Figure 15.8b that for the $L = 121$-nm device solving the time-dependent master equation for $\sigma = 0.13$ eV at a frequency $f = 100$ Hz yields results that deviate strongly from the experimental C–V characteristic. The fact that the master-equation results for the SC and FCC lattice are quite comparable shows that this deviation is probably not due to a morphological issue.

In order to understand the problem better we first distinguish the different regimes in the C–V characteristics. (1) At low voltage all characteristics converge to the geometrical capacitance, because almost no carriers are present in the device. (2) With increasing voltage, a sheet of holes builds up by diffusion from the anode, but these cannot yet move to the cathode because the electric field is still directed from cathode to anode. As a result, the effective thickness of the device decreases and the capacitance rises. (3) When approaching the built-in voltage V_{bi} these holes start to move to the cathode, leading to a decrease of the capacitance. The result is a peak in the C–V curve before V_{bi} is reached [165]. (4) In the regime beyond V_{bi} the C–V curve rises again. Here relaxation effects play a dominant role and therefore this is the regime we want to focus on.

In order to identify the effects of relaxation we display in Figure 15.8b (dash–dot–dotted curve) the C–V characteristic obtained by solving the time-dependent drift–diffusion equation with the EGDM mobility function corresponding to $\sigma = 0.13$ eV. In this case, the local mobility $\mu(x; \rho, F, T)$ depends on the instantaneous local charge density ρ and electric field F, and therefore contains no relaxation effects. It is seen that without relaxation effects the capacitance decreases after V_{bi} to a value that is even smaller than the geometrical capacitance. The long-dashed curve is the result of a multiple-trapping model for relaxation [164]. With a fitted conduction-level energy $E_c = -0.75\sigma$, this model leads to a fair agreement with experiment.

Since solving the time-dependent master equation for $\sigma = 0.13$ eV apparently overestimates relaxation effects and since such effects decrease with decreasing σ we solved the time-dependent master equation for lower values of σ. With a value of $\sigma = 0.08$ eV we find a very satisfactory agreement with the experimental C–V characteristics, not only for the device and frequency considered in Figure 15.8b, but also for both devices and all considered frequencies; see Figure 15.8c and 15.8d. The only clear disagreement is in the peak, which is more pronounced in the calculations than in the experiment. This may be partially explained by lateral variations in V_{bi} of the devices [164]. The dashed (SC) and full (FCC) curves in Figure 15.8a are the corresponding J–V curves obtained by solving the steady-state master equation. It is observed that for high voltages the experimental J–V curves are very well described, but significant deviations occur at low voltages around V_{bi}.

The analysis brings up the question why there is such an apparent discrepancy between the description of steady-state and time-dependent charge transport. A possible explanation is that in steady-state transport the low-energy tail of the DOS is important, represented by a relatively large σ, while in time-dependent transport relaxing carriers probe a larger part of the DOS, represented by a smaller σ. This would mean that the shape of the DOS is more complicated than a single Gaussian. It would also explain the difference in the description of the J–V curves in Figure 15.8c. At low voltage, when carriers only occupy the low energy tail of the DOS, $\sigma = 0.13$ eV gives a better description, while at higher voltage, when the DOS is filled up further, $\sigma = 0.08$ eV provides an excellent description, which is even slightly better than with $\sigma = 0.13$ eV. A value of $\sigma = 0.13$ eV at low voltage could also partially explain the lower peak in the C–V curve as compared to the calculations with $\sigma = 0.08$ eV in Figure 15.8b through d. We note that the position of the peak could be improved by adapting the used built-in voltages V_{bi}. These voltages were obtained from an EGDM fit of the J–V characteristics with $\sigma = 0.13$ eV [161,162], but should be optimized again in a fit with $\sigma = 0.08$ eV. In addition, the EGDM neglects spatial correlations of site energies. This can lead to discrepancies when analyzing an experimental system with correlations in terms of a model without correlations [89].

The present conclusion that $\sigma = 0.08$ eV should be used to describe carrier relaxation in the considered devices is fully in agreement with the conclusion that dark-injection experiments on the same devices, which also probe carrier relaxation, can be described by solving a time-dependent master equation with the same value of σ [163].

15.12.3 Efficiency

In OLEDs, electrical power is converted to a radiant flux (radiant energy emitted per unit time), Φ_e. The power efficiency, sometimes called the wall-plug efficiency, is given by

$$\eta_{power} = \frac{\Phi_e}{IV} = \frac{\int_0^\infty \Phi_{e,\lambda,OLED}(\lambda)d\lambda}{IV}, \quad (15.36)$$

where I is the current, V is the applied voltage, and $\Phi_{e,\lambda,OLED}$ is the total optical power that is emitted externally per unit wavelength λ. The power efficiency is generally limited by inevitable Ohmic losses in the electrodes and in the organic charge transport layers, and sometimes also by Ohmic losses due to the presence of internal organic–organic energy barriers outside the emissive layer. When judging the efficiency of the conversion process in the emissive layer, one therefore often focuses on a complementary quantity, the external quantum efficiency η_{EQE} (EQE), that is defined as the total number of externally emitted photons per charge carrier that has passed the device:

$$\eta_{EQE} = \frac{e}{I} \int_0^\infty \Phi_{e,\lambda,OLED}(\lambda) \frac{\lambda}{ch} d\lambda, \quad (15.37)$$

where e is the fundamental charge, c is the speed of light, and h is Planck's constant. Due to full or partial internal reflection of light, not all photons that are internally generated will escape from the microcavity that is formed by the OLED layer structure. It is therefore useful to introduce an additional quantity, the internal quantum efficiency (IQE) η_{IQE}, which is defined as the ratio of the total number of photons generated within the device and the number of electrons injected. The IQE is not directly measurable, but may be derived from the EQE using the expression

$$\eta_{IQE} = \frac{\eta_{EQE}}{\eta_{out}}, \quad (15.38)$$

where η_{out} is the light-outcoupling efficiency. For emission from a specific position in a planar OLED microcavity, under a specific angle and for a specific wavelength, the (s and p) polarization-dependent emitted light intensity may be obtained from optical simulations [166–169]. The light-outcoupling efficiency is thus an effective value, which is determined by averaging over the entire emission profile and the entire emission spectrum and which is sensitive to the precise angular dependence of the emission from the dye molecules. Application of advanced emission profile reconstruction techniques [170,171] and a measurement of the emitter orientation distribution [172] are required to determine η_{out} for a specific case with high precision. In the absence of such information, one often assumes that for well-designed phosphorescent OLEDs with a (glass | ITO | organic semiconductor | Al) layer structure and with a random emitter orientation η_{out} is approximately 0.2. Larger values, up to ~0.25–0.30, are possible by optimizing all layer thicknesses [173]. Several methods, including the use of a roughened external glass surface or the use of internal high-refractive index scattering layers, have been developed to enhance the light-outcoupling efficiency to values above 0.5 [174,175].

Recently, much progress has been made in advanced molecular-scale KMC calculations of the IQE [176–178]. We focus in this section on applications of KMC simulations to phosphorescent OLEDs based on a small concentration of metal-organic emitter molecules in a matrix material. In general, the IQE may be expressed as [174]

$$\eta_{IQE} = \eta_{rec}\eta_{ST}q_{eff}, \quad (15.39)$$

with η_{rec} the recombination efficiency, defined as the fraction of injected charges which contributes to exciton formation, η_{ST} the singlet–triplet factor, defined as the fraction of generated excitons which is quantum-mechanically allowed to decay radiatively, and q_{eff} the effective radiative decay efficiency, defined as the fraction of such excitons which actually decays radiatively. In phosphorescent OLEDs based on heavy metal-organic molecules, strong spin–orbit interaction gives rise to triplet states with some mixed-in singlet-character, so that also triplets are emissive and $\eta_{ST} = 1$. The recombination efficiency can be close to unity by making use of appropriate electron and hole blocking layers. A highly effective radiative decay efficiency may be obtained, first, by using emissive dye materials with a large radiative decay rate, Γ_{rad}, and a small nonradiative decay rate, Γ_{nr}. In the absence of other loss processes, the IQE is then equal to $\Gamma_{rad}/(\Gamma_{rad} + \Gamma_{nr}) \equiv \eta_{PL}$, the photoluminescence (PL) efficiency. Nonradiative decay is a result of the nuclear motion, so that the energy of the molecule in its excitonically excited state can be equienergetic with a highly vibrationally excited excitonic ground state [179]. Second, the matrix material and the adjacent blocking material should have a triplet energy level significantly larger than the dye triplet level so that the triplet excitons stay confined to the dye sites. These design rules are already relevant to the IQE at small current densities. Experimentally, the IQE is found to depend on the current density, J. At large J, η_{IQE} decreases with increasing J. For some devices, η_{IQE} is found to show a broad maximum before the decrease ("roll-off") sets in. A practical measure is the current density J_{90} at which the IQE has decreased to 90% of the maximum value. For efficient phosphorescent OLEDs, maximum reported values of J_{90} are approximately 300 A/m^2 [180]. In commercial white OLEDs for lighting conditions, operated at high luminance levels, the efficiency loss due to roll-off can be indeed of the order of 10%.

Understanding the roll-off is not only important as a first step toward enhancing the efficiency. Loss processes that limit the IQE at high J can also trigger local degradation processes with a certain probability, as will be discussed in Section 15.12.5. By building "virtual OLEDs" in which the interplay of all charge transport and excitonic processes is included mechanistically using KMC simulations, the functioning of OLEDs can be studied with subnanosecond time and molecular-scale spatial resolution. The first demonstration of the feasibility of such an approach was presented by van Eersel et al. [176]. We discuss their simulation results for OLEDs based on the green-emitting metal-organic molecule tris[2-phenylpyridine]iridium (Ir(ppy)$_3$) and the red phosphorescent dye platinum octaethylporphyrin (PtOEP).

The simulations were based on the three-dimensional (3D) KMC code Bumblebee (http:/simbeyond .com): For a detailed discussion of the model used, we refer to Refs. [176,177,181,182]. Ref. [176] also gives motivations for the parameter values used and provides analyses of the sensitivity of the simulation results to the parameter values. Briefly, the OLEDs were modeled as a collection of molecular sites on a simple cubic lattice. For each type of molecule, the site energies for electrons and holes were taken randomly from a Gaussian DOS with an average energy as given in Figure 15.9a. Charges were assumed to hop with a rate as described within the Miller–Abrahams formalism Equation 15.22. The hopping attempt frequency ω_0 and the wavefunction decay length $\lambda = 1/\gamma$ were taken equal for all pairs of sites. The simulations included the Coulomb interactions between all charge carrier pairs and with image charges in the metallic electrodes. In a natural way, the formation of space-charge layers near the injecting and organic–organic interfaces, and the resulting "band bending," was thus included. Cottaar et al. have demonstrated that in energetically disordered materials, as they are used in OLEDs, explicitly taking the individual 3D Coulomb interactions into account is important for properly treating charge accumulation near internal interfaces [183]. Instantaneous ISC was assumed, so that only triplet excitons were considered. Exciton generation and dissociation were treated in the same way as hops of electrons and holes, but including the triplet exciton binding energy. Radiative and nonradiative triplet exciton decay was included, as well as exciton transfer between the dye molecules, leading to exciton diffusion. The transfer rate was expressed as a sum of Förster-type and Dexter-type contributions, Equations 15.25 and 15.26, where $\tau = (\Gamma_{rad} + \Gamma_{nr})^{-1}$ is the effective decay time, R_F is the Förster radius for diffusion, and k_D is the Dexter prefactor. The two bimolecular loss processes that potentially contribute to the IQE roll-off, triplet–polaron quenching (TPQ) and triplet–triplet annihilation (TTA), were both included in a parameter-free manner, viz. by assuming an infinite (zero) rate when an exciton and a polaron or two excitons, respectively, are present on nearest-neighbor (more distant) sites. Table 15.2 gives an overview of the used parameter values.

Figure 15.10a and b shows a comparison of the calculated and experimental $J-V$ and IQE roll-off curves, respectively, for a temperature of 300 K. The slope of the $J-V$ curves is well described, but the absolute

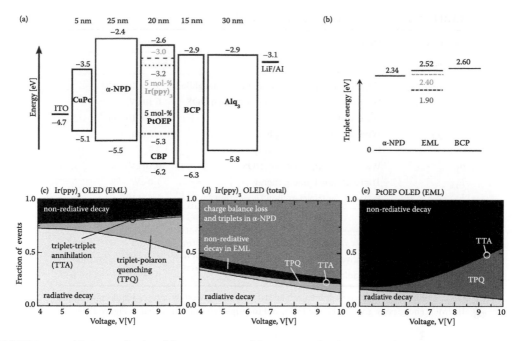

FIGURE 15.9 (a) Energy level and layer structure of the green and red organic light-emitting diodes (OLEDs) studied in Ref. [176]. The phosphorescent emissive layer (EML) is sandwiched in between materials facilitating hole and electron injection, transport and blocking: CuPc (copper phthalocyanine), α-NPD (4,4′-bis[N-(1-naphthyl)-N-phenyl-amino] biphenyl), BCP (2,9-dimethyl, 4,7-diphenyl, 1,10-phenanthroline), and Alq₃ (tris [8-hydroxyquinoline] aluminum). (b) Triplet energies for the materials used in the EML (solid line: CBP; dashed line: Ir(ppy)₃; dotted line: PtOEP) and the layers adjacent to the EML. (c)–(e) Contribution of the various exciton decay processes in the EML of the Ir(ppy)₃ device (b), in the entire Ir(ppy)₃ device (c), and in the EML of the PtOEP device (d). The figures show that even above 6 V only a small fraction of the efficiency loss (less than 2% (0.5%) for the Ir(ppy)₃ (PtOEP) devices) is due to triplet–triplet annihilation (TTA). (Reprinted with permission from H. van Eersel, P. A. Bobbert, R. A. J. Janssen, and R. Coehoorn. Monte Carlo study of efficiency roll-off of phosphorescent organic light-emitting diodes: Evidence for dominant role of triplet-polaron quenching. *Appl. Phys. Lett.*, 105(14):143303, 2014. Copyright [2014], American Institute of Physics.)

value of the current density is somewhat overestimated. As argued in Ref. [176], this might be related to an underestimation of the HOMO–LUMO gap, which was taken to be equal to the optical gap. This often-used approach neglects the exciton binding energy, which can be around 1 eV [184]. Such a correction would horizontally shift the *J–V* curves by 1 eV, giving rise to a significant reduction of the discrepancy. The roll-off curve would not be affected by such a correction. We note that (as mentioned above) the triplet exciton binding energy was included when calculating the rates of exciton generation and dissociation processes. For Ir(ppy)₃, the simulation results agree within the error margin with experiment, whereas for PtOEP the roll-off is slightly underestimated at high current densities. A sensitivity analysis was carried out to find out which uncertainties in the choice of parameter values have the largest impact. It was found, e.g., that the *J–V* characteristics are quite strongly determined by the energy-level differences at interfaces and between the host and guest states in the emissive layer (EML). The sensitivity to the hopping attempt frequency was found to be relatively small.

The simulations provided detailed views on the cause of the roll-off, as shown in Figure 15.9c through e. In both devices, most of the emission was found to occur near the anode-side of the EML. This may be understood from Figure 15.9a, from which the guest molecules are expected to give rise to stronger hole trapping than electron trapping. As a result, the effective electron mobility is larger than the hole mobility. In the emissive layer, most of the IQE loss was found to be due to TPQ, and only at high voltages

TABLE 15.2   Overview of the Simulation Parameters

Parameter	Description	Value	
Common			
ω_0	Hopping attempt frequency to the first neighbor	$3.3 \times 10^{10}\ s^{-1}$	
σ	Width of the electron and hole Gaussian DOS	0.10 eV	
N_t	Site density	$1.0 \times 10^{27}\ m^{-3}$	
$\lambda \equiv 1/\gamma$	Wavefunction decay length	0.3 nm	
ε_r	Relative dielectric permittivity	3.5	
σ_T	Width of the triplet exciton DOS	0.10 eV	
$E_{T,b}$	Triplet exciton binding energy	1.0 eV	
k_D	Prefactor for triplet exciton Dexter transfer	$1.6 \times 10^{10}\ s^{-1}$	
Material-specific		Ir(ppy)$_3$	PtOEP
R_F	Förster radius for triplet exciton diffusion	1.5 nm	1.5 nm
Γ_{rad}	Radiative decay rate	0.816 µs^{-1}	0.1 µs^{-1}
Γ_{nr}	Nonradiative decay rate	0.249 µs^{-1}	0.525 µs^{-1}

Note: The highest occupied molecular orbital (HOMO) and lowest unoccupied molecular orbital (LUMO) energies are given in Figure 15.9a, and the triplet energies are given in Figure 15.9b.

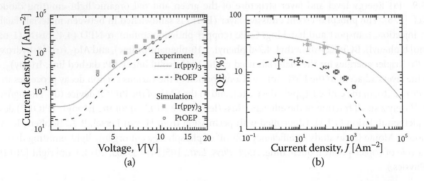

FIGURE 15.10 (a) Simulated and experimental J–V characteristics. (b) Simulated internal quantum energy (IQE) and experimental IQE (assuming 20% outcoupling efficiency) as a function of the current–density. Experimental data from Ref. [185]. (From N. C. Giebink and S. R. Forrest, *Phys. Rev. B.*, 77, 235215, 2008; Reprinted with permission from H. van Eersel et al., Monte Carlo study of efficiency roll-off of phosphorescent organic light-emitting diodes: Evidence for dominant role of triplet–polaron quenching. *Appl. Phys. Lett.*, 105(14):143303, 2014. Copyright [2014], American Institute of Physics.)

was a small TTA contribution was found (see Figure 15.9c). For the Ir(ppy)$_3$ devices, the overall loss was found to be determined mostly by a nonideal recombination efficiency due to imperfect electron blocking and by triplet exciton diffusion to the α-NPD layer (see Figure 15.9d). The finding of electron loss to the α-NPD layer is consistent with the observation of some blue emission from that layer [185]. Due to subsequent TPQ, triplet transfer to the α-NPD layer also gives rise to a loss. From simulations, improving the devices by introducing perfect electron blocking was predicted to give rise to a 10% increase of the IQE at small voltages, from ∼ 34% to ∼ 44%. The IQE at small voltages was found to become equal to the PL efficiency assumed (77%) when the triplet transfer to the HTL also was eliminated. For the red devices, at low voltages no significant electron and triplet loss to the α-NPD layer was found, as may be understood from the lower LUMO energy and the smaller triplet energy of PtOEP. The IQE is then close to the PL efficiency assumed (16.4%).

15.12.4 Electroluminescence of a White OLED

The KMC simulations described in Section 15.5 can be employed to model all molecular-scale electronic processes that finally lead to electroluminescence of an OLED: injection, transport, and recombination of electrons and holes as well as diffusion and radiative decay of excitons. We consider here the white multilayer OLED stack of Figure 15.11a, which was studied experimentally and by KMC simulations in Ref. [181]. It concerns a so-called "hybrid" OLED, which combines red and green phosphorescent emission with blue fluorescent emission. Phosphorescent emission can be very efficient because of the harvesting of both singlet and triplet excitons (see Section 15.2.4). However, since stable blue phosphorescent emitters with long-term stability are to date unavailable, many commercial white OLEDs make use of blue fluorescent emission, despite the fact that then only singlet excitons are harvested.

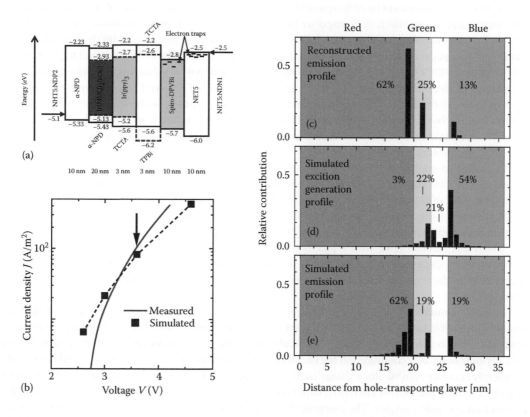

FIGURE 15.11 (a) Stack structure of the investigated white organic light-emitting diode (OLED) with the different layers and their thicknesses. The highest occupied molecular orbital (HOMO) and lowest unoccupied molecular orbital (LUMO) energy levels as used in the kinetic Monte Carlo (KMC) simulations are indicated by solid lines for host materials and dashed lines for guest materials. The horizontal arrows indicate the work functions used for the doped hole- and electron-injecting layers. Electron traps were assumed in the 2,2′,7,7′-tetrakis(2,2-diphenylvinyl)spiro-9,9′-bifluorene (spiro-DPVBi) and NET5 layers with concentrations of 10^{-3} and 5×10^{-3}, and characteristic energies $k_B T_0 = 0.2$ and 0.12 eV, respectively, of their exponential density of states (DOS). (b) Measured and simulated room-temperature current density-voltage, J–V characteristics. All other results presented are for the bias voltage $V = 3.6$ V, indicated by the arrow. The relative error in the simulated current density is estimated to be about 10%. (c) Emission profile reconstructed from the measured angle- and polarization-dependent emission spectrum [181]. The percentages of red, green, and blue emission are indicated. (d) Exciton generation profile as obtained from the KMC simulations. (e) Emission profile as obtained from the KMC simulations, after taking into account exciton diffusion and the radiative decay probabilities of the emitters. The error in the simulated profiles is about 1%. Reprinted by permission from Macmillan Publishers Ltd. *Nat. Mater.* [181] Copyright (2013).

The working principle of the OLED of Figure 15.11a is the following. Holes reach the light-emitting layers of the stack after being injected by an ITO layer into a 4 mol% p-doped injection layer of NHT5:NDP2 and transported through a hole-transporting and electron-blocking layer of α-NPD (N,N'-di(naphthalen-1-yl)-N,N'-diphenyl-benzidine). Electrons reach the emitting layers after being injected by an aluminum cathode into a 4 mol% n-doped electron-injection of NET5:NDN1 and transported through an electron-transporting and hole-blocking layer of NET5 (materials supplied by Novaled). Blue light is generated in a 10-nm-thick fluorescent layer of spiro-DPVBi (2,2',7,7'-tetrakis(2,2-diphenylvinyl)spiro-9,9'-bi fluorene) adjacent to the NET5 layer. Green light is generated in a 3-nm-thick layer of TCTA (4,4',4''-tris(N-carbazoyl)-triphenylamine) doped with 8 mol% of the green phosphorescent dye Ir(ppy)$_3$ (fac-tris(2-phenylpyridyl)iridium). Red light is generated in a 20-nm-thick α-NPD layer doped with 5 mol% of the red phosphorescent dye Ir(MDQ)$_2$(acac) ((acelylacetonate)bis(2-methyldibenzo[f,h]quinoxinalate)iridium).

The green phosphorescent layer is separated from the blue fluorescent layer by a thin (3-nm) interlayer consisting of a mixture of the hole-transporter TCTA with 33 mol% of the electron-transporter TPBi (1,3,5-tris(N-phenylbenzimidazol-2-yl)benzene). This interlayer has several purposes [186]. It should block the transfer of singlet excitons from the blue to the green layer and of triplet excitons from the green to the blue layer (spiro-DPVBi has a triplet energy lower than that of Ir(ppy)$_3$). Also, this interlayer should allow the passage of electrons from the blue to the green layer (by the TPBi) and of holes from the green to the blue layer (by the TCTA). The red phosphorescent layer is purposely in direct contact with the green phosphorescent layer, allowing triplet excitons formed on the phosphorescent dye in the green layer to diffuse to the phosphorescent dye in the red layer. This diffusion is an important process in establishing the right color balance.

The measured room-temperature current density–voltage characteristic of the OLED is shown in Figure 15.11b. The arrow indicates the bias voltage of 3.6 V for most of the reported results. The CIE 1931 color point of the perpendicularly emitted light at this bias was measured to be $[x, y] = [0.47, 0.45]$, which corresponds to warm-white emission. The EQE, i.e., the fraction of emitted photons per injected electron–hole pair, is measured to be $5 \pm 1\%$ [181]. Figure 15.11c shows the color-resolved emission profile, which was reconstructed with a precision on the order of a nanometer from the measured angle- and polarization-dependent emission spectra [170,181]. It is observed that in the blue layer, the emission occurs close to the interface with the interlayer and in the red layer close to the interface with the green layer.

The KMC simulations of the charge dynamics were carried out in the following way. Point sites are arranged on a cubic lattice with a lattice constant $a = 1$ nm, the typical intermolecular distance of the used molecular semiconductors represent the molecules in the stack. A simulation box of $50 \times 50 \times 56$ sites with periodic boundary conditions in the lateral (x- and y-) directions was used, which turned out to yield sufficiently accurate results. Since charge transport in various small-molecule materials was found to be described well by the ECDM [161,191], correlated disorder caused by random dipoles was assumed for the electron and hole energies. The energetic disorder was taken to be $\sigma = 0.1$ eV, corresponding to the value found for hole transport in α-NPD [161]. With this value the charge transport in all materials in the stack is expected to be reasonably described. Red and green emitting guests were introduced according to the known concentration of the emitters, with appropriately adapted energy levels. Electron traps were introduced in the layers in which electron transport is important: the blue fluorescent layer and the ETL. They were modeled with an exponential DOS of trap energies [192–194] with a concentration c_{trap} and a characteristic energy $k_B T_0$.

Nearest-neighbor hopping of charges on the lattice using the Miller–Abrahams rate (see Equation 15.22) was assumed. The energy differences in the hopping rates contain, apart from the random site energies, an electrostatic contribution due to the bias applied to the OLED and the Coulomb energy due to all present charges. The doped injection layers were treated as metallic, injecting and collecting charges with an energy according to their work function, indicated by the arrows in Figure 15.11a. Exciton generation was assumed to occur by hopping of an electron to a site where a hole resides, or vice versa, and was assumed to be always an energetically downward process.

Table 15.3 gives the parameters of the stack materials used in the KMC simulations of the charge dynamics. The parameters were determined from charge-transport and spectroscopic studies of the various materials [181]. Exciton diffusion within the green and red layers and from the green to the red layer was included in the simulations. Since the red and green emitters trap electrons as well as holes (see the energy level scheme in Figure 15.11a), almost all excitons in the red and green layers are generated on the emitters. The diffusion of excitons among the emitters was described by Förster transfer (see Equation 15.25), made possible by the spin–singlet character that is mixed into the exciton wave function by the spin-orbit coupling of the heavy iridium atoms. Apart from being transferred, excitons can decay radiatively with a rate $\Gamma_{\mathrm{rad},i} = 1/\tau_{\mathrm{r},i}$ or nonradiatively with a rate $\Gamma_{\mathrm{nr},i} = 1/\tau_{\mathrm{nr},i}$. These rates are related to the radiative decay probabilities η_r by $\eta_{\mathrm{r},i} = \Gamma_{\mathrm{rad},i}/(\Gamma_{\mathrm{rad},i} + \Gamma_{\mathrm{nr},i})$. We took $\eta_r = 0.84$ and 0.76 for the red and green phosphorescent emitters in their respective hosts [169]. Other parameters used in the exciton dynamics are given in Table 15.4. Only exciton transfer from green to red was taken into account. Transfer from red to green should be negligible.

TABLE 15.3 Highest Occupied Molecular Orbital (HOMO) and Lowest Unoccupied Molecular Orbital (LUMO) Energies, Room Temperature Hole- and Electron-Mobilities $\mu_{0,\mathrm{h}}$ and $\mu_{0,\mathrm{e}}$ at Low Field and Low Carrier Density, Electron-Trap Concentration c_{trap}, and Trap Temperature T_0 of the Exponential Trap Density of States (DOS) in the Different Layers of the Stack, as Used in the Kinetic Monte Carlo (KMC) Simulations

Material	E_{HOMO} (eV)	E_{LUMO} (eV)	$\mu_{0,\mathrm{h}}$ (m^2/Vs)	$\mu_{0,\mathrm{e}}$ (m^2/Vs)	c_{trap}	T_0 (K)
NHT5:NDP2	−5.10					
α-NPD	−5.43	−2.33	6×10^{-9}	6×10^{-10}		
Ir(MDQ)$_2$(acac)	−5.13	−2.93	6×10^{-9}	6×10^{-10}		
TCTA	−5.60	−2.20	2×10^{-8}	2×10^{-9}		
Ir(ppy)$_3$	−5.20	−2.70	2×10^{-8}	2×10^{-9}		
TCTA	−5.60	−2.20	2×10^{-8}	2×10^{-9}		
TPBi	−6.20	−2.60	2×10^{-8}	2×10^{-9}		
Spiro-DPVBi	−5.70	−2.80	6×10^{-9}	8×10^{-9}	0.001	2350
NET5	−6.00	−2.50	1.5×10^{-11}	1.5×10^{-10}	0.005	1400
NET5:NDN1		−2.50				

Source: M. Mesta et al., *Nat. Mater.*, 12, 652–658, 2013.

Note: Spiro-DPVBi, 2,2′,7,7′-tetrakis(2,2-diphenylvinyl)Spiro-9,9′-bifluorene; TCTA, 4,4′,4″-tris(N-carbazoyl)-triphenylamine; α-NPD, N,N′-di(naphthalen-1-yl)-N,N′-diphenyl-benzidine; MDQ, acelylacetonate)bis(2-methyldibenzo[f,h]quinoxinalate)iridium; TPBI, 1,3,5-tris(N-phenylbenzimidazol-2-yl)benzene. NHT5:NDP2 and NET5:NDN1 are doped hole and electron transporter materials supplied by Novaled.

TABLE 15.4 Radiative and Nonradiative Exciton Decay Rates Γ_{rad} and Γ_{nr}, Förster Radii R_F for Exciton Transfer between Phosphorescent Emitter Molecules and Triplet Energies E_T

Material	Γ_{rad} (μs^{-1})	Γ_{nr} (μs^{-1})	R_F (nm)	E_T (eV)
Ir(MDQ)$_2$(acac)	0.588	0.112	1.5	2.0
Ir(ppy)$_3$	0.816	0.249	1.5	2.4

Note: For the green emitter Ir(ppy)$_3$ Γ_{rad} and Γ_{nr} were taken from Ref. [187]. For the red emitter Ir(MDQ)$_2$(acac), Γ_{rad} was obtained from Ref. [188], and Γ_{nr} from η_r is given in Ref. [169] and the relation $\eta_r = \Gamma_{\mathrm{rad}}/(\Gamma_{\mathrm{rad}} + \Gamma_{\mathrm{nr}})$. The value $R_F = 1.5$ nm for transfer between equal emitter molecules is a typical value given in Ref. [189]. According to the estimate in Ref. [190] we took $R_{F,\mathrm{GR}} = 3.5$ nm for the transfer from a green to a red emitter. The triplet energies were taken from Ref. [186].

For each exciton generated in the red or green layer a separate simulation of its dye-to-dye diffusion and final radiative or nonradiative decay was performed. This diffusion was assumed to proceed independently from all other processes, which means that exciton quenching processes were neglected. Excitons generated on host sites in the red and green layers (a small fraction) were assumed to transfer instantaneously to an emitter in their neighborhood. Diffusion of excitons generated in the blue layer was not accounted for because the diffusion length of these excitons is short and because their transfer to the green layer is blocked by the interlayer. We assumed that in the blue fluorescent layer singlet and triplet excitons are generated in a quantum-statistical ratio of 1:3 and that the triplet excitons are lost. For the radiative decay probability of singlet excitons in the blue layer we took $\eta_r = 0.35$ [195]. Excitons generated in the interlayer were assumed to be lost by nonradiative decay or emission outside the visible spectrum.

The J–V characteristic following from the KMC simulations is given in Figure 15.11b. Considering the various simplifications that were made, the agreement with the measured characteristic is fair. The underestimation of J at high voltage could be due to a heating effect, while the overestimation of J at low voltage could result from a systematic underestimation of the LUMO energies due to neglect of the exciton binding energies. At the bias voltage of 3.6 V for most of the reported results, the experimental and simulated current density agree quite well.

Figure 15.11d and e present the simulated exciton generation profile and emission profile, respectively, at 3.6 V. The effect of exciton transfer from green to red is clearly observable and is very important for the color balance of this OLED. The simulated emission profile is in fair agreement with the reconstructed emission profile from Figure 15.11c. The broadening of the simulated emission profile over a few nanometers in the red layer found in the simulations is not seen in the reconstructed emission profile but this could be due to the limited resolution of the reconstruction procedure [170,181]. The total percentages of emission in the red, green, and blue agree quite well with the reconstructed emission profile.

The results presented in this section show that KMC simulations of all molecular-scale electronic processes leading to electroluminescence in quite complicated multilayer OLEDs with commercial relevance are feasible. The parameters in the present study were almost all obtained from experimental studies, but there is no obstacle for obtaining these from first-principles computational studies. Complete *in silico* studies of the functioning of commercial OLEDs therefore seem to be within reach.

15.12.5 Degradation

Developing an improved understanding of the mechanisms that limit the operational lifetime of OLEDs is of key importance toward the further adoption of OLED technology for display, lighting, and signage applications. Experimentally, given fixed current density and ambient temperature conditions, the luminance is often observed to decrease with time in an approximately exponential or stretched-exponential manner. For commercial white OLEDs for lighting applications, the time at which the luminance has dropped to 70% of the initial value (the so-called LT_{70} lifetime) can today be as large as 10,000 hours or more at a luminance of 8000 cd/m^2. However, long lifetimes are in practice often realized by making a trade-off with device efficiency (e.g., when using hybrid OLEDs, see Section 15.12.4) or production cost and ease of manufacturing (e.g., when using multiply vertically stacked OLEDs). Examples of other aspects of OLED reliability during prolonged operation are the voltage stability (at a fixed current density), the color point stability, the stability at high ambient temperatures, and the stability under prolonged exposure to the ambient atmosphere [196,197]. The latter issue, leading to so-called black-spot formation, has been mitigated by the development of improved encapsulation technologies [198]. Recently, an excellent overview of the degradation mechanisms and reactions in OLEDs has been given by Scholz et al. [199].

In this section, we focus on the use of molecular-scale OLED device modeling as a means to elucidate the role of various possible intrinsic degradation processes, i.e., processes that are caused by excitons (including those due to absorption of internally emitted photons), charges, and fields in the opto-electronically active organic semiconducting layer. When setting up an extension of the KMC simulations described in Sections 15.5 and 15.12.3, in order to include degradation, first an inventory should be made of the processes which for a specific system are expected to give rise to degradation. Formally, a distinction should be made between (1) monomolecular degradation processes, which occur when the state of a single-specific molecule is modified, e.g., due to the presence of a polaron or an exciton or due to a local electric field, and (2) bimolecular degradation processes, which occur upon an interaction between charges and/or excitons on two different molecules, e.g., exciton–exciton annihilation or exciton–polaron quenching.

Including these two types of degradation processes requires a rate and a probability, respectively, as well as a description of the resulting changes of the KMC parameter values. Monomolecular degradation has been observed, e.g., in OLEDs containing the electron transport material Alq_3, in which the presence of holes gives rise to a reduction of the emission by the formation of fluorescence quenchers [200]. Degradation due to the presence of singlet excitons has been reported for 4,4′-bis(N-carbazolyl)biphenyl (CBP), used as a matrix material in the EML of green-emitting phosphorescent OLEDs [201]. The degradation products can act as nonradiative recombination centers and trap states. Degradation due to singlet excitons has also been reported for OLEDs based on the fluorescent emitter material spiro-DPVBi [202]. An example of a bimolecular process which can be accompanied by degradation is triplet–polaron quenching (TPQ). When TPQ is due to excitation of the polaron by the triplet exciton, followed by nonradiative decay of the excited polaron due to internal conversion, the locally dissipated energy may, with a certain probability, lead to a chemical change of the molecule. Such a mechanism was found for a blue phosphorescent OLED using 4,4′-bis(3-methylcarbazol-9-yl)-2,2′-biphenyl (mCBP) as the host material in the EML [203]. The mCBP defect sites were argued to act as deep charge traps and the dissociation products were argued to damage the guest (emitter) so that it becomes a nonradiative center and a luminescence quencher.

A KMC degradation study starts in general with running a simulation without degradation until dynamic equilibrium is achieved under the operational conditions (voltage, temperature) of interest. Subsequently, the simulations are continued while degradation is switched on. Monomolecular degradation due to a charge or an exciton on a sensitive site is included as a new possible process which competes with all other possible processes. Degradation that accompanies a bimolecular process such as TPQ is included by branching the end result of that process, so that a defect site is formed with a certain probability. We note that in KMC simulations only the primary event needs to be described. All subsequent effects (e.g., a shift of the recombination zone to a less favorable position due to a changed mobility balance, or an efficiency loss due to TPQ at charges residing on defect molecules which act as traps [204]) will follow "automatically" from the simulations. KMC lifetime simulations are necessarily strongly accelerated, as practically realistic simulated times are usually at most of the order of 1 ms, 6–11 orders of magnitude smaller than actual lifetimes in the range of 1–100,000 hours. This can be accomplished by assuming an enhanced value of the degradation rate (for monomolecular degradation) or the degradation probability (for degradation accompanying a bimolecular process). As in the case of experimental OLED lifetime tests, it is also possible to accelerate the simulations by carrying them out for high current densities and/or elevated temperatures, followed by extrapolation to application-relevant operational conditions.

A first demonstration of the feasibility of 3D KMC OLED lifetime simulations was presented in Ref. [177]. The simulations were carried out for a symmetric OLED with an energy level structure as shown in Figure 15.12a, for a temperature of 300 K. The mixed-matrix emissive layer contains equal concentrations of the HTL and ETL material as a host, and 4 mol% of emitter molecules (guest). The hole and electron blocking is excellent, so that the recombination efficiency is 100%. The simulation parameters are as given in Table 15.2, with the following exceptions: $\varepsilon_r = 3$, $\sigma_T = 0$ eV (no triplet energy disorder), and $\Gamma_{rad} = 0.544 \ \mu s^{-1}$ and $\Gamma_{nr} = 0.181 \ \mu s^{-1}$ (values typical for the orange-red emitter $Ir(MDQ)_2(acac)$ in an α-NPD matrix [205]). As in Section 15.12.3, TPQ and TTA were treated in a parameter-free manner as instantaneous nearest-neighbor processes.

Van Eersel et al. [176] showed that for these symmetric devices and with a dye trap depth $\Delta = 0.2 - 0.3$ eV the emission profile at small voltages is quite uniform across the emissive layer. This is illustrated by Figure 15.12b, which shows the emission profile at 3 V for the case studied ($\Delta = 0.2$ eV). On the one hand,

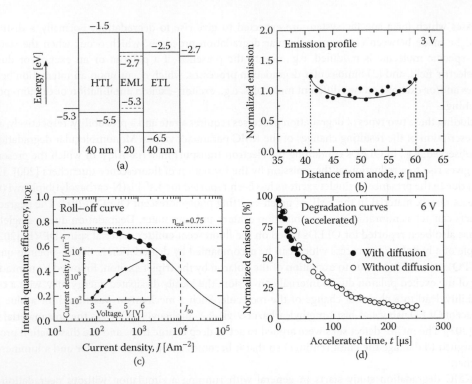

FIGURE 15.12 (a) Organic layer and energy level structure of the organic light-emitting diodes (OLEDs) considered in the kinetic Monte Carlo (KMC) degradation simulation studies. Dashed lines: Emitter energy levels (4 mol%). (b) Emission profile at 3 V. The full curve is a guide-to-the-eye. (c) Current density dependence of the internal quantum efficiency. The full curve is a fit using Equation 15.41 with $m = 0.80$ and $J_{50} = 10$ kA/m². The vertical dashed line indicates the current density at which the IQE has dropped to 50% (J_{50}) of the value at small current densities. (d) Time-dependence of the normalized emission at 6 V as obtained from a KMC degradation study which assumes that (1) the degradation occurs on triplet-excited dye sites upon triplet-polaron quenching processes (TPQ-t, see the main text) resulting from the displacement of a polaron on a neighbor site to the dye site, and which assumes that (2) each quenching is followed by a degradation process upon which the involved dye molecule becomes nonemissive. The simulations were carried out with and without exciton diffusion (closed and open symbols, respectively). The dashed curve is a stretched-exponential fit to the simulation results without diffusion (see the main text). (Reproduced from R. Coehoorn et al., *Adv. Funct. Mater.*, 25, 2024–2037, 2015. With permission.)

this choice avoids a large Ohmic loss due to deep trapping and a large overvoltage due to the enhanced built-in voltage (for large Δ). On the other hand, it also avoids a large loss due to strong roll-off caused by emission from thin zones near the blocking layer interfaces (for small Δ). The optimum value of Δ will depend on the application goal (see Ref. [176]) and is sensitive to the detailed mechanisms of the TPQ and TTA processes. Figure 15.12c shows the calculated IQE roll-off curve before degradation. The J–V curve is given in the inset. At small voltages, the IQE is equal to the assumed radiative decay (PL) efficiency ($\eta_{rad} = \Gamma_{rad}/(\Gamma_{rad} + \Gamma_{nr}) = 0.75$). The J_{90} current density, defined in Section 15.12.3, is approximately 640 A/m². This is larger than the largest experimental value obtained so far ($\sim$ 300 A/m², see Section 15.12.3). The full curve gives a fit through the data points, discussed below (see Equation 15.41). For the nearest-neighbor TPQ and TTA mechanisms considered, the roll-off is almost completely due to TPQ.

Figure 15.12d shows the dependence of the normalized emission on the simulated time, at constant voltage conditions (6 V), for a degradation scenario in which, upon a TPQ process, the polaron involved is displaced to the site at which the triplet exciton resides (a "TPQ-t process," see Figure 2 in Ref. [177]), which then becomes nonemissive with a degradation probability $p_{degr} = 1$. The simulations thus employ the largest possible acceleration factor. All other parameters are kept identical. The current density was found to remain essentially unchanged during the degradation process. Furthermore, it was found that choosing a smaller degradation probability does not significantly change the results, apart from changing the time scale. That indicates that the lifetime is still much larger than all other relevant time scales. Simulations including exciton diffusion are computationally more expensive, and were stopped when a 50% emission reduction had been obtained. Simulations without diffusion were continued until a reduction to only about 10% of the initial emission was reached. An approximate description of the decay is given by the stretched exponential curve, shown in the figure, with the form $I(t) = I(0)\exp[-(t/\tau_{\mathrm{sim,acc}})^\beta]$, with a simulated accelerated (1/e) lifetime $\tau_{\mathrm{sim,acc}} \cong 80$ µs and a stretching exponent $\beta = 0.91$. The actual (1/e) lifetime as predicted from the simulations is given by

$$\tau_{\mathrm{sim}} = \frac{\tau_{\mathrm{sim,acc}}}{p_{\mathrm{degr}}}. \tag{15.40}$$

Conversely, using Equation 15.40 the value of p_{degr} could be deduced from a degradation simulation and a measurement of the lifetime. Such an analysis led in Ref. [177] to an estimated order-of-magnitude value of $p_{\mathrm{degr}} \sim 10^{-8}$ when assuming that the simulations discussed above are relevant to state-of-the-art white OLEDs.

Within a refined approach, the probability that a TPQ process gives rise to degradation could be treated stochastically, e.g., by treating it as a thermally activated process with an activation energy with a Gaussian distribution. Such an approach is expected to give rise to a smaller value of the stretching exponent β. Experimentally, values of β around 0.5 have been observed. The effect is so far generally explained in a rather phenomenological manner [206]. It should be noted that the decay can also become more stretched-exponential like for OLEDs with imbalanced electron and hole mobilities, resulting in a highly nonuniform emission profile. If this picture is correct, lifetime studies could also, albeit indirectly, provide information about the shape of the emission profile.

We envisage that, using Equation 15.40, lifetime predictions can be obtained from KMC lifetime simulations if the parameter p_{degr} (or its distribution) can be determined from a few well-chosen combined experimental and KMC calibration studies. Subsequently, KMC-based lifetime predictions can be obtained for other measurement conditions (e.g., current density and temperature), device architectures, and other dye concentrations in the same host. So far, studies which could validate this view have not been carried out. As a first step, KMC simulations were carried out of the iridium dye concentration dependence of the lifetime, for otherwise identical simulation parameters [177]. Figure 15.13a shows the accelerated LT_{90} lifetime obtained from the simulations (large open circles), and results from simulations in which exciton diffusion was switched off (small open circles). For iridium dye concentrations above about 7 mol%, exciton diffusion is found to yield a significant decrease of the lifetime, up to a factor ~4. The exciton diffusion length is then larger than the average distance to a degraded site, so that a large fraction of the excitons which have been generated on nondegraded sites is lost due to diffusion and subsequent nonradiative decay on degraded sites. Interestingly, the IQE obtained when exciton diffusion was switched off was found to be slightly reduced for all systems studied, namely by 1%–2%. The reduction cannot be due to switching off the transfer of triplets from matrix to guest sites, as due to the large energy gap of the matrix materials all excitons are generated directly on the dye sites. We surmise that the reduction is due to switching off the possibility that excitons diffuse to molecular sites in regions with a slightly smaller average polaron density, in which the loss due to TPQ is reduced. This effect may for the highest iridium dye concentrations considered be compensated in part by an increase of the IQE due to a smaller exciton diffusion contribution to the TPQ loss. For the devices and the parameter values employed, the latter effect was found to be almost negligible for iridium dye concentrations around and below 8 mol% [177].

The simulation data shown in Figure 15.13a may be analyzed more quantitatively with the help of a useful relationship which has been established between the lifetime and the IQE roll-off [177]. It is based on a model which assumes uniform electron, hole, and exciton densities in the emissive layer. If

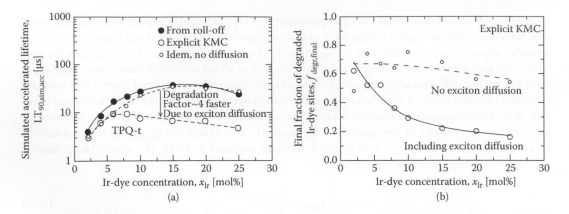

(a)   (b)

FIGURE 15.13 (a) Ir-dye concentration dependence of the LT_{90} lifetime as obtained from accelerated KMC lifetime simulations for the OLEDs shown in Figure 15.12a, at 6 V, for the TPQ-t degradation scenario discussed in the main text (large open circles) and when switching off exciton diffusion (small open circles). The closed circles show the accelerated lifetime which is expected from the simulated roll-off using Equation 15.42. The curves are guides-to-the-eye. The predicted actual LT_{90} lifetime is equal to $LT_{90,sim,acc}/p_{degr}$, with p_{degr} the probability that upon a TPQ event degradation takes place. Note that due to a variation of the current density with the Ir-dye concentration (see the main text) the lifetime as obtained using Equation 15.42 is not proportional to that concentration. (b) Ir-dye concentration dependence of the fraction of degraded Ir-dye molecules in the $t \to \infty$ limit, as obtained from the same KMC simulations including and without exciton diffusion. (Reproduced from R. Coehoorn et al., *Adv. Funct. Mater.*, 25, 2024–2037, 2015. With permission.)

the charge carrier mobility is charge carrier concentration (c) dependent and proportional to c^b, with b a positive exponent which increases with an increasing width of the polaron DOS, the IQE roll-off curve is given by

$$\eta_{IQE} = \frac{\eta_{rad}}{1 + \left(\frac{J}{J_{50}}\right)^m}, \tag{15.41}$$

with η_{rad} the PL efficiency (0.75 for the case studied in this section), J_{50} the current density at which the IQE has dropped to 50% of its low-voltage value, and $m = (1 + b)/(2 + b)$. In the absence of exciton diffusion, the simulated accelerated (1/e) lifetime is then given by

$$\tau_{sim,acc} = edn_{dye}J_{50}^m/J^{m+1}, \tag{15.42}$$

where d is the EML layer thickness and n_{dye} is the dye molecule volume density. From this formalism, a current density acceleration exponent $m + 1 = (3 + 2b)/(2 + b)$ in the range 1.5–2 is expected, as is indeed often observed for phosphorescent OLEDs. The linear n_{dye} dependence is due to the linear increase with increasing dye concentration of the probability that a dye is still emissive after a certain period of operation. The closed circles in Figure 15.13 show the simulated accelerated LT_{90} lifetime, obtained using Equation 15.42 under the assumption of exponential decay ($\beta = 1$), so that $LT_{90,sim,acc} = -\ln(0.9) \times \tau_{sim,acc} \cong 0.105 \times \tau_{sim,acc}$. These predictions from the roll-off curves agree quite well with the explicit KMC simulation results, obtained when exciton diffusion is switched off (small open circles). We note that at the constant voltage (6-V) condition employed, the current density shows a weak but nonnegligible nonmonotonic dependence on the Ir-dye concentration. It shows a broad minimum around a concentration of approximately 10 mol%, and is ~15% larger for the 2 and 25 mol% systems. Within the concentration range studied, the transport shows a cross-over from a low-concentration guest–host–guest hopping regime, in which the guest molecule states act as traps, to a high-concentration regime, in which the transport is predominantly due to direct guest–guest hopping (see, e.g., Figure 15.5a in Ref. [100]). The lifetime at a fixed voltage, as obtained from Equation 15.42, is therefore not proportional to the Ir-dye concentration.

Various optical and chemical analytical techniques have been used to investigate the degradation mechanisms of OLEDs and to quantify the concentrations of degraded molecules (see Ref. [199] and references therein). In future studies, it would be useful to compare such experimental results with the results of KMC simulations. Interestingly, the simulations for the model systems discussed in this section revealed that, to a good approximation, the fraction of degraded Ir-dye molecules increases with time as

$$f_{\text{degr}}(t) = f_{\text{degr,final}} \left[1 - \exp\left(-\left(t/\tau_{\text{sim,acc}} \right)^{\beta} \right) \right], \tag{15.43}$$

with $f_{\text{degr,final}}$ the final ($t \to \infty$) fraction of degraded Ir-dye molecules.

The values of $\tau_{\text{sim,acc}}$ and β are, within the numerical uncertainty, equal to the values describing the luminance decay. Figure 15.13b shows the Ir-dye concentration dependence of $f_{\text{degr,final}}$ as obtained from KMC simulations with and without exciton diffusion. The figure shows that in the absence of diffusion, only 60%–70% of the sites have degraded in the $t = \infty$ limit. This result indicates that, on a significant fraction (30%–40%) of the sites, excitons are either never formed, or that excitons on those sites are well protected against TPQ due to a position of those sites well outside the somewhat filamentary electron and hole current density pathways. The first explanation is consistent with the finding that in systems with a monomodal Gaussian DOS exciton generation preferentially takes place on sites with a low-lying electron or hole state [73]. When exciton diffusion is included, $f_{\text{degr,final}}$ is found to decrease significantly with increasing Ir-dye concentration, to only approximately 0.16 for 25 mol% systems. This is consistent with the view that, due to the energetic disorder, the average polaron density and the local polaron diffusivity will be quite nonuniform [99,207], so that in the case of strong exciton diffusion, degradation will occur predominantly on the relatively small fraction of sites that are located in a region with a large average polaron density and diffusivity.

We emphasize that the degradation scenario assumed in the case study discussed in this section was only chosen for the purpose of giving a demonstration of the feasibility of KMC lifetime simulations. Including monomolecular or other bimolecular scenarios or including refinements (e.g., a degradation probability distribution, conversion of the degraded molecules to polaron trap sites, or an extension of the TPQ interaction range so that the role of the degraded molecule as an exciton quencher is enhanced) is straightforward. It also will be useful to develop analytical models as discussed above for other degradation scenarios and to extend these to a more realistic nonuniform emission from the EML, in which the lifetime becomes position-dependent, so that the emission decay becomes more stretched-exponential like.

15.13 Outlook

In this chapter, we have reviewed multiscale techniques used to simulate organic LEDs and demonstrated the feasibility of full 3D OLED modeling. As an outlook, we would like to mention areas where substantial method development is still required in order to achieve a parameter-free modeling of realistic devices. Refined studies that aim at developing a final view on the detailed performance of specific devices should consider the following: (1) first-principles evaluations of charge injection rates, (2) explicit treatment of the induction interaction when solving the master equation, (3) quantitative treatment of excited states embedded in a heterogeneous polarizable molecular environment, (4) more quantitative descriptions of charge–exciton and exciton–exciton interactions, and (5) descriptions of TTA and TPQ as longer-range Förster and Dexter-type interactions. Advancements in all these directions are absolutely vital for devising accurate structure–property relationships for organic semiconductors used in OLEDs.

Acknowledgments

This work was supported in part by the Federal Ministry of Education and Research BMBF grants MEDOS (FKZ 03EK3503B), MESOMERIE (FKZ 13N10723), and InterPhase (FKZ 13N13661). The project received funding from the NMP-20-2014—"Widening materials models" program under Grant Agreement No. 646259 (MOSTOPHOS). The project was also supported by NanoNextNL, a nanotechnology program of the Dutch Ministry of Economic Affairs. Part of the work of one of the authors (RC) was carried out at the Philips Research Laboratories (Eindhoven, The Netherlands).

References

1. I. Akasaki. Nobel lecture: Fascinated journeys into blue light. *Rev. Mod. Phys.*, 87(4):1119–1131, 2015.
2. H. Amano. Nobel lecture: Growth of GaN on sapphire via low-temperature de-posited buffer layer and realization of p-type GaN by Mg doping followed by low-energy electron beam irradiation. *Rev. Mod. Phys.*, 87(4):1133–1138, 2015.
3. S. Nakamura. Nobel lecture: Background story of the invention of efficient blue InGaN light emitting diodes. *Rev. Mod. Phys.*, 87(4):1139–1151, 2015.
4. S. R. Forrest. The path to ubiquitous and low-cost organic electronic appliances on plastic. *Nature*, 428(6986):911–918, 2004.
5. C. W. Tang and S. A. VanSlyke. Organic electroluminescent diodes. *Appl. Phys. Lett.*, 51(12):913–915, 1987.
6. R. Friend, J. Burroughes, and D. Bradley. Electroluminescent devices, Patent WO9013148 (A1). Cambridge Research and Innovation Limited; Cambridge Capital Management Limited; Lynxvale Limited, 1990.
7. R. Coehoorn, V. van Elsbergen, and C. Verschuren. High efficiency OLEDs for lighting applications. In E. Cantatore, editor, *Applications of Organic and Printed Electronics, Integrated Circuits and Systems*, pages 83–100. New York, NY: Springer, 2013.
8. T. Otani. Samsung shows new 4 flexible AMOLED that is so thin (0.05 mm) it "aps" in the wind textbar OLED-Info. 2008. https://www.oled-info.com/samsung-shows-new-4-flexible-amoled-so-thin-005mm-it-flaps-wind
9. R. Mertens. Continental shows a prototype dual-screen flexible AMOLED display for the automotive market—OLED-Info, 2014. https://www.oled-info.com/continental-shows-prototype-dual-screen-flexible-amoled-display-automotive-market
10. R. Olivares-Amaya, C. Amador-Bedolla, J. Hachmann, S. Atahan-Evrenk, R. S. Sanchez-Carrera, L. Vogt, and A. Aspuru-Guzik. Accelerated computational discovery of high-performance materials for organic photovoltaics by means of cheminformatics. *Energy Environ. Sci.*, 4(12):4849–4861, 2011.
11. P. Deglmann, A. Schaefer, and C. Lennartz. Application of quantum calculations in the chemical industry—An overview. *Int. J. Quantum Chem.*, 115(3):107–136, 2015. WOS:000346654700001.
12. H. Yersin. Triplet emitters for OLED applications. Mechanisms of exciton trapping and control of emission properties. In *Transition Metal and Rare Earth Compounds*, Yersin, Hartmut (Ed.) No. 241 in Topics in Current Chemistry, Hartmut Yersin (Ed.), pp. 1–26. Berlin: Springer, 2004.
13. L. Duan, K. Xie, and Y. Qiu. Review paper: Progress on efficient cathodes for organic light-emitting diodes. *J. Soc. Inf. Disp.*, 19(6):453–461, 2011.
14. A. Fukase, K. Luan Thanh Dao, and J. Kido. High-efficiency organic electroluminescent devices using iridium complex emitter and arylamine-containing polymer buffer layer. *Polymer. Adv. Tech.*, 13(8):601–604, 2002.
15. X. Zhou, D. S. Qin, M. Pfeiffer, J. Blochwitz-Nimoth, A. Werner, J. Drechsel, B. Maen-nig, K. Leo, M. Bold, P. Erk, and H. Hartmann. High-efficiency electrophosphorescent organic light-emitting diodes with double light-emitting layers. *Appl. Phys. Lett.*, 81(21):4070–4072, 2002.

16. C. Adachi, M. A. Baldo, S. R. Forrest, S. Lamansky, M. E. Thompson, and R. C. Kwong. High-efficiency red electrophosphorescence devices. *Appl. Phys. Lett.*, 78(11):1622–1624, 2001.

17. Y. Kawamura, S. Yanagida, and S. R. Forrest. Energy transfer in polymer electrophosphorescent light emitting devices with single and multiple doped luminescent layers. *J. Appl. Phys.*, 92(1):87–93, 2002.

18. B. A. Gregg, S.-G. Chen, and R. A. Cormier. Coulomb forces and doping in organic semiconductors. *Chem. Mater.*, 16(23):4586–4599, 2004.

19. K. Walzer, B. Maennig, M. Pfeiffer, and K. Leo. Highly efficient organic devices based on electrically doped transport layers. *Chem. Rev.*, 107(4):1233–1271, 2007.

20. M. A. Baldo, S. Lamansky, P. E. Burrows, M. E. Thompson, and S. R. Forrest. Very high-efficiency green organic light-emitting devices based on electrophosphorescence. *Appl. Phys. Lett.*, 75(1):4–6, 1999.

21. Md. K. Nazeeruddin, R. Humphry-Baker, D. Berner, S. Rivier, L. Zuppiroli, and M. Graetzel. Highly phosphorescence iridium complexes and their application in organic light-emitting devices. *J. Am. Chem. Soc.*, 125(29):8790–8797, 2003.

22. Q. Peng, W. Li, S. Zhang, P. Chen, F. Li, and Y. Ma. Evidence of the reverse intersystem crossing in intra-molecular charge transfer fluorescence based organic light-emitting devices through magneto-electroluminescence measurements. *Adv. Opt. Mater.*, 1(5):362–366, 2013.

23. S. Hirata, Y. Sakai, K. Masui, H. Tanaka, S. Youn Lee, H. Nomura, N. Nakamura, M. Yasumatsu, H. Nakanotani, Q. Zhang, K. Shizu, H. Miyazaki, and C. Adachi. Highly efficient blue electroluminescence based on thermally activated delayed fluorescence. *Nat. Mater.*, 14(3):330–336, 2015.

24. S. Reineke, K. Walzer, and K. Leo. Triplet–exciton quenching in organic phosphorescent light-emitting diodes with Ir-based emitters. *Phys. Rev. B*, 75(12):125328, 2007.

25. E. Knapp, R. Husermann, H. U. Schwarzenbach, and B. Ruhstaller. Numerical simulation of charge transport in disordered organic semiconductor devices. *J. Appl. Phys.*, 108(5):054504, 2010.

26. B. Ruhstaller, E. Knapp, B. Perucco, N. Reinke, D. Rezzonico, and F. Mueller. Advanced numerical simulation of organic light-emitting devices. In O. Sergiyenko, editor, *Optoelectronic Devices and Properties*. InTech, 2011. Rijeka, Croatia: DOI: 10.5772/14626.

27. G. Kaniadakis and P. Quarati. Kinetic equation for classical particles obeying an exclusion principle. *Phys. Rev. E*, 48(6):4263–4270, 1993.

28. Y. Roichman and N. Tessler. Generalized Einstein relation for disordered semiconductors-implications for device performance. *Appl. Phys. Lett.*, 80(11):1948–1950, 2002.

29. G. Lakhwani, A. Rao, and R. H. Friend. Bimolecular recombination in organic photovoltaics. *Annu. Rev. Phys. Chem.*, 65(1):557–581, 2014.

30. B. Perucco, N. A. Reinke, D. Rezzonico, E. Knapp, S. Harkema, and B. Ruhstaller. On the exciton profile in OLEDs-seamless optical and electrical modeling. *Org. Electron.*, 13(10):1827–1835, 2012.

31. G.-J. A. H. Wetzelaer and P. W. M. Blom. Diffusion-driven currents in organic-semiconductor diodes. *NPG Asia Mater.*, 6(7):e110, 2014.

32. J.-H. Lee, S. Lee, S.-J. Yoo, K.-H. Kim, and J.-J. Kim. Langevin and trap-assisted recombination in phosphorescent organic light emitting diodes. *Adv. Funct. Mater.*, 24(29):4681–4688, 2014.

33. R. R. Chance, A. Prock, and R. Silbey. Molecular fluorescence and energy transfer near interfaces. In I. Prigogine and S. A. Rice, editors, *Advances in Chemical Physics*, pp. 1–65. Hoboken, NJ: John Wiley & Sons, 1978.

34. H. K. Gummel. A self-consistent iterative scheme for one-dimensional steady state transistor calculations. *IEEE Trans. Electron Dev.*, 11(10):455–465, 1964.

35. D. L. Scharfetter and H. K. Gummel. Large-signal analysis of a silicon Read diode oscillator. *IEEE Trans. Electron Dev.*, 16(1):64–77, 1969.

36. E. Knapp and B. Ruhstaller. Numerical impedance analysis for organic semiconductors with exponential distribution of localized states. *Appl. Phys. Lett.*, 99(9), 093304, 2011.

37. E. Tuti, I. Batisti, and D. Berner. Injection and strong current channeling in organic disordered media. *Phys. Rev. B*, 70(16):161202, 2004.

38. K. D. Meisel, W. F. Pasveer, J. Cottaar, C. Tanase, R. Coehoorn, P. A. Bobbert, P. W. M. Blom, D. M. de Leeuw, and M. A. J. Michels. Charge-carrier mobilities in disordered semiconducting polymers: Effects of carrier density and electric field. *Phys. Status Solidi C*, 3(2):267–270, 2006.

39. N. Rappaport, Y. Preezant, and N. Tessler. Spatially dispersive transport: A mesoscopic phenomenon in disordered organic semiconductors. *Phys. Rev. B*, 76(23):235323, 2007.

40. J. Honerkamp. *Stochastische Dynamische Systeme : Konzepte, Numerische Methoden, Datenanalysen*. Weinheim: Wiley-VCH, 1990.

41. B. Derrida. Velocity and diffusion constant of a periodic one-dimensional hopping model. *J. Stat. Phys.*, 31(3):433–450, 1983.

42. K. Seki and M. Tachiya. Electric field dependence of charge mobility in energetically disordered materials: Polaron aspects. *Phys. Rev. B*, 65:014305, 2001.

43. S. D. Baranovskii. Theoretical description of charge transport in disordered organic semiconductors. *Phys. Status Solidi B*, 251(3):487–525, 2014.

44. J. Cottaar and P. A. Bobbert. Calculating charge-carrier mobilities in disordered semiconducting polymers: Mean field and beyond. *Phys. Rev. B*, 74(11):115204, 2006.

45. V. Rühle, A. Lukyanov, F. May, M. Schrader, T. Vehoff, J. Kirkpatrick, B. Baumeier, and D. Andrienko. Microscopic simulations of charge transport in disordered organic semiconductors. *J. Chem. Theory Comput.*, 7(10):3335–3345, 2011.

46. M. Heidernaetsch, M. Bauer, and G. Radons. Characterizing N-dimensional anisotropic Brownian motion by the distribution of diffusivities. *J. Chem. Phys.*, 139(18):184105, 2013.

47. J. L. Doob. Topics in the theory of Markoff chains. *Trans. Am. Math. Soc.*, 52(1):37–64, 1942.

48. J. L. Doob. Markoff chains—Denumerable case. *Trans. Am. Math. Soc.*, 58(3):455, 1945.

49. W M. Young and E. W. Elcock. Monte Carlo studies of vacancy migration in binary ordered alloys: I. *Proc. Phys. Soc.*, 89(3):735–746, 1966.

50. A. B. Bortz, M. H. Kalos, and J. L. Lebowitz. A new algorithm for Monte Carlo simulation of Ising spin systems. *J. Comput. Phys.*, 17(1):10–18, 1975.

51. D. T. Gillespie. A general method for numerically simulating the stochastic time evolution of coupled chemical reactions. *J. Comput. Phys.*, 22(4):403–434, 1976.

52. D. T. Gillespie. Exact stochastic simulation of coupled chemical reactions. *J. Phys. Chem.*, 81(25):2340–2361, 1977.

53. K. A. Fichthorn and W. H. Weinberg. Theoretical foundations of dynamical Monte Carlo simulations. *J. Chem. Phys.*, 95(2):1090, 1991.

54. A. P. J. Jansen. Monte Carlo simulations of chemical reactions on a surface with time-dependent reaction-rate constants. *Comput. Phys. Commun.*, 86(1–2):1–12, 1995.

55. P. Kordt, J. J. M. van der Holst, M. Al Helwi, W. Kowalsky, F. May, A. Badinski, C. Lennartz, and D. Andrienko. Modeling of organic light emitting diodes: From molecular to device properties. *Adv. Funct. Mater.*, 25(13):1955–1971, 2015.

56. M. A. Gibson and J. Bruck. Efficient exact stochastic simulation of chemical systems with many species and many channels. *J. Phys. Chem. A.*, 104(9):1876–1889, 2000.

57. B. D. Lubachevsky. Efficient parallel simulations of dynamic Ising spin systems. *J. Comput. Phys.*, 75(1):103–122, 1988.

58. Y. Shim and J. G. Amar. Semirigorous synchronous sublattice algorithm for parallel kinetic Monte Carlo simulations of thin film growth. *Phys. Rev. B*, 71(12), 125432, 2005.

59. M. Merrick and K. A. Fichthorn. Synchronous relaxation algorithm for parallel kinetic Monte Carlo simulations of thin film growth. *Phys. Rev. E*, 75(1), 011606, 2007.

60. E. Martnez, J. Marian, M. H. Kalos, and J. M. Perlado. Synchronous parallel kinetic Monte Carlo for continuum diffusion-reaction systems. *J. Comput. Phys.*, 227(8):3804–3823, 2008.

61. N. J. van der Kaap and L. J. A. Koster. Massively parallel kinetic Monte Carlo simulations of charge carrier transport in organic semiconductors. *J. Comput. Phys.*, 307:321–332, 2016.

62. A. Miller and E. Abrahams. Impurity conduction at low concentrations. *Phys. Rev.*, 120(3):745–755, 1960.

63. H. Baessler. Charge transport in disordered organic photoconductors: A Monte Carlo simulation study. *Phys. Status Solidi B*, 175(1):15–56, 1993.

64. R. A. Marcus. Electron transfer reactions in chemistry. Theory and experiment. *Rev. Mod. Phys.*, 65(3):599–610, 1993.

65. G. R. Hutchison, M. A. Ratner, and T. J. Marks. Hopping transport in conductive heterocyclic oligomers: Reorganization energies and substituent effects. *J. Am. Chem. Soc.*, 127(7):2339–2350, 2005.

66. V. May and O. Kuehn. *Charge and Energy Transfer Dynamics in Molecular Systems*, 3rd ed. Weinheim: Wiley-VCH, 2011.

67. M. Bixon and J. Jortner. Electron transfer from isolated molecules to biomolecules. In I. Prigogine and S. A. Rice, editors, *Advances in Chemical Physics*, pp. 35–202. Hoboken, NJ: John Wiley & Sons, 2007.

68. R. Egger, C. H. Mak, and U. Weiss. Quantum rates for nonadiabatic electron transfer. *J. Chem. Phys.*, 100(4):2651–2660, 1994.

69. M. P. A. Fisher and A. T. Dorsey. Dissipative quantum tunneling in a biased double-well system at finite temperatures. *Phys. Rev. Lett.*, 54(15):1609–1612, 1985.

70. H. Grabert and U. Weiss. Quantum tunneling rates for asymmetric double-well systems with Ohmic dissipation. *Phys. Rev. Lett.*, 54(15):1605–1608, 1985.

71. A. J. Leggett, S. Chakravarty, A. T. Dorsey, M. P. A. Fisher, A. Garg, and W. Zwerger. Dynamics of the dissipative two-state system. *Rev. Mod. Phys.*, 59(1):1–85, 1987.

72. K. Asadi, A. J. Kronemeijer, T. Cramer, L. J. A. Koster, P. W. M. Blom, and D. M. de Leeuw. Polaron hopping mediated by nuclear tunnelling in semiconducting polymers at high carrier density. *Nat. Commun.*, 4:1710, 2013.

73. J. J. M. van der Holst, F. W. A. van Oost, R. Coehoorn, and P. A. Bobbert. Electron–hole recombination in disordered organic semiconductors: Validity of the Langevin formula. *Phys. Rev. B*, 80(23):235202-1–235202-8, 2009.

74. T. H. Foerster. Zwischenmolekulare Energiewanderung und Fluoreszenz. *Ann. Phys.*, 437(1–2):55–75, 1948.

75. D. L. Dexter. A theory of sensitized luminescence in solids. *J. Chem. Phys.*, 21(5):836–850, 1953.

76. G. D. Scholes and K. P. Ghiggino. Rate expressions for excitation transfer I. Radiationless transition theory perspective. *J. Chem. Phys.*, 101(2):1251–1261, 1994.

77. G. D. Scholes, R. D. Harcourt, and K. P. Ghiggino. Rate expressions for excitation transfer. III. An ab initio study of electronic factors in excitation transfer and exciton resonance interactions. *J. Chem. Phys.*, 102(24):9574–9581, 1995.

78. R. D. Harcourt, G. D. Scholes, and K. P. Ghiggino. Rate expressions for excitation transfer. II. Electronic considerations of direct and through configuration exciton resonance interactions. *J. Chem. Phys.*, 101(12):10521–10525, 1994.

79. S. Speiser. Photophysics and mechanisms of intramolecular electronic energy transfer in bichromophoric molecular systems: Solution and supersonic jet studies. *Chem. Rev.*, 96(6):1953–1976, 1996.

80. R. F. Fink, J. Pfister, H. Mei Zhao, and B. Engels. Assessment of quantum chemical methods and basis sets for excitation energy transfer. *Chem. Phys.*, 346(1–3):275–285, 2008. WOS:000256142000032.

81. C. Tanase, P. W. M. Blom, D. M. de Leeuw, and E. J. Meijer. Charge carrier density dependence of the hole mobility in poly(p-phenylene vinylene). *Phys. Status Solidi A*, 201(6):1236–1245, 2004.

82. F. Laquai and D. Hertel. Influence of hole transport units on the efficiency of polymer light emitting diodes. *Appl. Phys. Lett.*, 90(14):142109, 2007.

83. S. V. Novikov, D. H. Dunlap, V. M. Kenkre, P. E. Parris, and A. V. Vannikov. Essential role of correlations in governing charge transport in disordered organic materials. *Phys. Rev. Lett.*, 81(20):4472–4475, 1998.

84. D. H. Dunlap, P. E. Parris, and V. M. Kenkre. Charge-dipole model for the universal field dependence of mobilities in molecularly doped polymers. *Phys. Rev. Lett.*, 77(3):542–545, 1996.

85. S. W. de Leeuw, J. W. Perram, and E. R. Smith. Simulation of electrostatic systems in periodic boundary conditions. I. Lattice sums and dielectric constants. *Proc. Royal Soc. A: Math. Phys. Eng. Sci.*, 373(1752):27–56, 1980.

86. P. P. Ewald. Die Berechnung optischer und elektrostatischer Gitterpotentiale. *Ann. Phys.*, 369(3):253–287, 1921.

87. R. H. Young. Dipolar lattice model of disorder in random media analytical evaluation of the Gaussian disorder model. *Philos. Mag. B*, 72(4):435–457, 1995.

88. S. V. Novikov and A. V. Vannikov. Cluster structure in the distribution of the electrostatic potential in a lattice of randomly oriented dipoles. *J. Phys. Chem.*, 99(40):14573–14576, 1995.

89. P. Kordt and D. Andrienko. Modeling of spatially correlated energetic disorder in organic semiconductors. *J. Chem. Theory Comput.*, 12(1):36–40, 2016.

90. W. F. Pasveer, J. Cottaar, C. Tanase, R. Coehoorn, P. A. Bobbert, P. W. M. Blom, D. M. de Leeuw, and M. A. J. Michels. Unified description of charge-carrier mobilities in disordered semiconducting polymers. *Phys. Rev. Lett.*, 94(20), 2005.

91. M. Bouhassoune, S. L. M. van Mensfoort, P. A. Bobbert, and R. Coehoorn. Carrier-density and field-dependent charge-carrier mobility in organic semiconductors with correlated Gaussian disorder. *Org. Electron.*, 10(3):437–445, 2009.

92. A. J. Stone. *The Theory of Intermolecular Forces*. Oxford: Clarendon Press, 1997.

93. C. Poelking and D. Andrienko. Long-range embedding of molecular ions and excitations in a polarizable molecular environment. *J. Chem. Theory Comput.*, 12(9):4516–4523, 2016.

94. C. Poelking, M. Tietze, C. Elschner, S. Olthof, D. Hertel, B. Baumeier, F. Wrthner, K. Meerholz, K. Leo, and D. Andrienko. Impact of mesoscale order on open-circuit voltage in organic solar cells. *Nat. Mater.*, 14(4):434–439, 2014.

95. C. Poelking and D. Andrienko. Design rules for organic donor-acceptor heterojunctions: Pathway for charge splitting and detrapping. *J. Am. Chem. Soc.*, 137(19):6320–6326, 2015.

96. D. Andrienko. Simulations of Morphology and Charge Transport in Supramolecular Organic Materials. In *Supramolecular Materials for Opto-Electronics*, Norbert Koch (Ed.). Cambridge, UK: The Royal Society of Chemistry, 2014.

97. B. T. Thole. Molecular polarizabilities calculated with a modified dipole interaction. *Chem. Phys.*, 59(3):341–350, 1981.

98. S. L. M. van Mensfoort and R. Coehoorn. Effect of Gaussian disorder on the voltage dependence of the current density in sandwich-type devices based on organic semiconductors. *Phys. Rev. B*, 78(8):085207, 2008.

99. J. J. M. van der Holst, M. A. Uijttewaal, B. Ramachandhran, R. Coehoorn, P. A. Bobbert, G. A. de Wijs, and R. A. de Groot. Modeling and analysis of the three-dimensional current density in sandwich-type single-carrier devices of disordered organic semiconductors. *Phys. Rev. B*, 79(8):085203, 2009.

100. R. Coehoorn and P. A. Bobbert. Effects of Gaussian disorder on charge carrier transport and recombination in organic semiconductors. *Phys. Status Solidi A*, 209(12):2354–2377, 2012.

101. J.-L. Bredas, D. Beljonne, V. Coropceanu, and J. Cornil. Charge-transfer and energy-transfer processes in pi-conjugated oligomers and polymers: A molecular picture. *Chem. Rev.*, 104(11):4971–5004, 2004.

102. J. R. Reimers. A practical method for the use of curvilinear coordinates in calculations of normal-mode-projected displacements and Duschinsky rotation matrices for large molecules. *J. Chem. Phys.*, 115(20):9103–9109, 2001.

103. D. P. McMahon and A. Troisi. Evaluation of the external reorganization energy of polyacenes. *J. Phys. Chem. Lett.*, 1(6):941–946, 2010.

104. J. E. Norton and J.-L. Bredas. Polarization energies in oligoacene semiconductor crystals. *J. Am. Chem. Soc.*, 130(37):12377–12384, 2008.

105. B. Baumeier, J. Kirkpatrick, and D. Andrienko. Density-functional based determination of inter-molecular charge transfer properties for large-scale morphologies. *Phys. Chem. Chem. Phys.*, 12(36):11103, 2010.

106. J. Huang and M. Kertesz. Validation of intermolecular transfer integral and bandwidth calculations for organic molecular materials. *J. Chem. Phys.*, 122(23):234707, 2005.

107. E. F. Valeev, V. Coropceanu, D. A. da Silva Filho, S. Salman, and J.-L. Bredas. Effect of electronic polarization on charge-transport parameters in molecular organic semiconductors. *J. Am. Chem. Soc.*, 128(30):9882–9886, 2006.

108. T. Van Voorhis, T. Kowalczyk, B. Kaduk, L.-P. Wang, C.-L. Cheng, and Q. Wu. The diabatic picture of electron transfer, reaction barriers, and molecular dynamics. *Annu. Rev. Phys. Chem.*, 61(1):149–170, 2010.

109. J. Ridley and M. Zerner. An intermediate neglect of differential overlap technique for spectroscopy: Pyrrole and the azines. *Theor. Chim. Acta*, 32(2):111–134, 1973.

110. J. Kirkpatrick. An approximate method for calculating transfer integrals based on the ZINDO Hamiltonian. *Int. J. Quantum Chem.*, 108(1):51–56, 2008.

111. V. Coropceanu, J. Cornil, D. A. da Silva Filho, Y. Olivier, R. Silbey, and J.-L. Bredas. Charge transport in organic semiconductors. *Chem. Rev.*, 107(4):926–952, 2007.

112. A. Kubas, F. Hoffmann, A. Heck, H. Oberhofer, M. Elstner, and J. Blumberger. Electronic couplings for molecular charge transfer: Bench-marking CDFT, FODFT, and FODFTB against high-level ab initio calculations. *J. Chem. Phys.*, 140(10):104105, 2014.

113. F. Gajdos, S. Valner, F. Hoffmann, J. Spencer, M. Breuer, A. Kubas, M. Dupuis, and J. Blumberger. Ultrafast estimation of electronic couplings for electron transfer between pi-conjugated organic molecules. *J. Chem. Theory Comput.*, 10(10):4653–4660, 2014.

114. P. Kordt, O. Stenzel, B. Baumeier, V. Schmidt, and D. Andrienko. Parametrization of extended Gaussian disorder models from microscopic charge transport simulations. *J. Chem. Theory Comput.*, 10(6):2508–2513, 2014.

115. S. Athanasopoulos, J. Kirkpatrick, D. Martnez, J. M. Frost, C. M. Foden, A. B. Walker, and J. Nelson. Predictive study of charge transport in disordered semiconducting polymers. *Nano Lett.*, 7(6):1785–1788, 2007.

116. J. Kirkpatrick, V. Marcon, J. Nelson, K. Kremer, and D. Andrienko. Charge mobility of discotic mesophases: A multiscale quantum and classical study. *Phys. Rev. Lett.*, 98(22):227402, 2007.

117. J. Nelson, J. J. Kwiatkowski, J. Kirkpatrick, and J. M. Frost. Modeling charge transport in organic photovoltaic materials. *Acc. Chem. Res.*, 42(11):1768–1778, 2009.

118. Y. Nagata and C. Lennartz. Atomistic simulation on charge mobility of amorphous tris(8-hydroxyquinoline) aluminum (Alq[sub 3]): Origin of Poole–Frenkel-type behavior. *J. Chem. Phys.*, 129(3):034709, 2008.

119. A. Fuchs, T. Steinbrecher, M. S. Mommer, Y. Nagata, M. Elstner, and C. Lennartz. Molecular origin of differences in hole and electron mobility in amorphous Alq3—A multiscale simulation study. *Phys. Chem. Chem. Phys.*, 14(12):4259–4270, 2012.

120. A. J. Stone. Distributed polarizabilities. *Mol. Phys.*, 56(5):1065–1082, 1985.

121. A. J. Stone. Distributed multipole analysis—Stability for large basis sets. *J. Chem. Theory Comput.*, 1(6):1128–1132, 2005.

122. A. J. Misquitta and A. J. Stone. Distributed polarizabilities obtained using a constrained density-fitting algorithm. *J. Chem. Phys.*, 124(2):024111, 2006.

123. T. Bereau, D. Andrienko, and O. Anatole von Lilienfeld. Transferable atomic multipole machine learning models for small organic molecules. *J. Chem. Theory Comput.*, 11(7):3225–3233, 2015.

124. W. L. Jorgensen and J. Tirado-Rives. The OPLS [optimized potentials for liquid simulations] potential functions for proteins, energy minimizations for crystals of cyclic peptides and crambin. *J. Am. Chem. Soc.*, 110(6):1657–1666, 1988.

125. W. L. Jorgensen, D. S. Maxwell, and J. Tirado-Rives. Development and testing of the OPLS all-atom force field on conformational energetics and properties of organic liquids. *J. Am. Chem. Soc.*, 118(45):11225–11236, 1996.

126. A. Lukyanov, C. Lennartz, and D. Andrienko. Amorphous films of tris(8-hydroxyquinolinato)aluminium: Force-field, morphology, and charge transport. *Phys. Status Solidi (A)*, 206(12):2737–2742, 2009.

127. B. H. Besler, K. M. Merz, and P. A. Kollman. Atomic charges derived from semiempirical methods. *J. Comput. Chem.*, 11(4):431–439, 1990.

128. C. M. Breneman and K. B. Wiberg. Determining atom-centered monopoles from molecular electrostatic potentials. The need for high sampling density in formamide conformational analysis. *J. Comput. Chem.*, 11(3):361–373, 1990.

129. F. May, M. Al-Helwi, B. Baumeier, W. Kowalsky, E. Fuchs, C. Lennartz, and D. Andrienko. Design rules for charge-transport efficient host materials for phosphorescent organic light-emitting diodes. *J. Am. Chem. Soc.*, 134(33):13818–13822, 2012. WOS:000307699000042.

130. A. Lukyanov and D. Andrienko. Extracting nondispersive charge carrier mobilities of organic semiconductors from simulations of small systems. *Phys. Rev. B*, 82(19):193202, 2010.

131. T. Neumann, D. Danilov, C. Lennartz, and W. Wenzel. Modeling disordered morphologies in organic semiconductors. *J. Comput. Chem.*, 34(31):2716–2725, 2013.

132. M. Moral, W.-J. Son, J. C. Sancho-Garca, Y. Olivier, and L. Muccioli. Cost-effective force field tailored for solid-phase simulations of OLED materials. *J. Chem. Theory Comput.*, 11(7):3383–3392, 2015.

133. O. Anatole von Lilienfeld, I. Tavernelli, U. Rothlisberger, and D. Sebastiani. Optimization of effective atom centered potentials for London dispersion forces in density functional theory. *Phys. Rev. Lett.*, 93(15):153004, 2004.

134. W. G. Noid, J.-W. Chu, G. S. Ayton, V. Krishna, S. Izvekov, G. A. Voth, A. Das, and H. C. Andersen. The multiscale coarse graining method. 1. A rigorous bridge between atomistic and coarse-grained models. *J. Chem. Phys.*, 128:244114, 2008.

135. V. Rühle, C. Junghans, A. Lukyanov, K. Kremer, and D. Andrienko. Versatile object-oriented toolkit for coarse-graining applications. *J. Chem. Theory Comput.*, 5(12):3211–3223, 2009.

136. A. K. Soper. Empirical potential Monte Carlo simulation of fluid structure. *Chem. Phys.*, 202(2–3):295–306, 1996.

137. A. P. Lyubartsev and A. Laaksonen. Calculation of effective interaction potentials from radial distribution functions: A reverse Monte Carlo approach. *Phys. Rev. E*, 52(4):3730–3737, 1995.

138. D. Reith, M. Pütz, and F. Müller-Plathe. Deriving effective mesoscale potentials from atomistic simulations. *J. Comput. Chem.*, 24(13):1624–1636, 2003.

139. M. Jochum, D. Andrienko, K. Kremer, and C. Peter. Structure-based coarse-graining in liquid slabs. *J. Chem. Phys.*, 137(6):064102, 2012.

140. F. Ercolessi and J. B. Adams. Interatomic potentials from first-principles calculations: The force-matching method. *EPL*, 26(8):583, 1994.

141. S. Izvekov, M. Parrinello, C. J. Burnham, and G. A. Voth. Effective force fields for condensed phase systems from ab initio molecular dynamics simulation: A new method for force-matching. *J. Chem. Phys.*, 120(23):10896–10913, 2004.

142. M. Scott Shell. The relative entropy is fundamental to multiscale and inverse thermo-dynamic problems. *J. Chem. Phys.*, 129(14):144108, 2008.

143. W. G. Noid, J.-W. Chu, G. S. Ayton, and G. A. Voth. Multiscale coarse-graining and structural correlations: Connections to liquid-state theory. *J. Phys. Chem. B*, 111(16):4116–4127, 2007.

144. W. G. Noid. Perspective: Coarse-grained models for biomolecular systems. *J. Chem. Phys.*, 139(9):090901, 2013.

145. K. Ch. Daoulas and M. Müller. Comparison of simulations of lipid membranes with membranes of block copolymers. In W. Peter Meier and W. Knoll, editors, *Polymer Membranes/Biomembranes*, No. 224 in Advances in Polymer Science, pp. 43–85. Berlin: Springer, 2010.

146. P. Gemnden, C. Poelking, K. Kremer, D. Andrienko, and K. Ch. Daoulas. Nematic ordering, conjugation, and density of states of soluble polymeric semiconductors. *Macromolecules*, 46(14):5762–5774, 2013.

147. C. Poelking, K. Daoulas, A. Troisi, and D. Andrienko. Morphology and charge transport in P3ht: A theorist's perspective. In S. Ludwigs, editor, *P3HT Revisited—From Molecular Scale to Solar Cell Devices*, No. 265 in Advances in Polymer Science, pp. 139–180. Berlin: Springer, 2014. doi:10.1007/12-2014-277.

148. P. Gemnden, C. Poelking, K. Kremer, K. Daoulas, and D. Andrienko. Effect of mesoscale ordering on the density of states of polymeric semiconductors. *Macromol. Rapid Commun.*, 36(11):1047–1053, 2015.

149. P, Kordt, S. Stodtmann, A. Badinski, M. Al Helwi, C. Lennartz, and D. Andrienko. Parameter-free continuous drift–diffusion models of amorphous organic semiconductors. *Phys. Chem. Chem. Phys.*, 17(35):22778–22783, 2015.

150. B. Baumeier, O. Stenzel, C. Poelking, D. Andrienko, and V. Schmidt. Stochastic modeling of molecular charge transport networks. *Phys. Rev. B*, 86(18):184202, 2012.

151. W. Tschoep, K. Kremer, J. Batoulis, T. Buerger, and O. Hahn. Simulation of polymer melts. I. Coarse-graining procedure for polycarbonates. *Acta Polym.*, 49(2–3):61–74, 1998.

152. T. Murtola, A. Bunker, I. Vattulainen, M. Deserno, and M. Karttunen. Multiscale modeling of emergent materials: Biological and soft matter. *Phys. Chem. Chem. Phys.*, 11(12):1869–1892, 2009.

153. R. L. Henderson. A uniqueness theorem for fluid pair correlation functions. *Phys. Lett. A*, 49(3):197–198, 1974.

154. W.-K. Oh, S. Qamar Hussain, Y.-J. Lee, Y. Lee, S. Ahn, and J. Yi. Study on the ITO work function and hole injection barrier at the interface of ITO/a-Si:H(p) in amorphous/crystalline silicon heterojunction solar cells. *Mater. Res. Bull.*, 47(10):3032–3035, 2012.

155. R. Schlaf, H. Murata, and Z. H. Kafafi. Work function measurements on indium tin oxide films. *J. Electron Spectrosc. Relat. Phenom.*, 120(1–3):149–154, 2001.

156. G. M. Wu, H. H. Lin, and H. C. Lu. Work function and valence band structure of tin-doped indium oxide thin films for OLEDs. *Vacuum*, 82(12):1371–1374, 2008.

157. C. Hyung Kim, C. Deuck Bae, K. Hee Ryu, B. Ki Lee, and H. Jung Shin. Local work function measurements on various inorganic materials using Kelvin probe force spectroscopy. *Solid State Phenom.*, 124–126:607–610, 2007.

158. D. R. Lide. *CRC Handbook of Chemistry and Physics: A Ready-Reference Book of Chemical and Physical Data*, 79th ed. Boca Raton, FL: CRC Press, 1998.

159. M. Mesta, J. Cottaar, R. Coehoorn, and P. A. Bobbert. Study of charge-carrier relaxation in a disordered organic semiconductor by simulating impedance spectroscopy. *Appl. Phys. Lett.*, 104(21):213301-1–213301-4, 2014.

160. J. J. M. van der Holst, F. W. A. van Oost, R. Coehoorn, and P. A. Bobbert. Monte Carlo study of charge transport in organic sandwich-type single-carrier devices: Effects of Coulomb interactions. *Phys. Rev. B*, 83(8):085206, 2011.

161. S. L. M. van Mensfoort, S. I. E. Vulto, R. A. J. Janssen, and R. Coehoorn. Hole transport in polyfluorene-based sandwich-type devices: Quantitative analysis of the role of energetic disorder. *Phys. Rev. B*, 78(8):085208, 2008.

162. R. J. de Vries, S. L. M. van Mensfoort, R. A. J. Janssen, and R. Coehoorn. Relation between the built-in voltage in organic light-emitting diodes and the zero-field voltage as measured by electroabsorption. *Phys. Rev. B*, 81(12):125203, 2010.

163. M. Mesta, C. Schaefer, J. de Groot, J. Cottaar, R. Coehoorn, and P. A. Bobbert. Charge-carrier relaxation in disordered organic semiconductors studied by dark injection: Experiment and modeling. *Phys. Rev. B*, 88(17):174204, 2013.

164. W. C. Germs, J. J. M. van der Holst, S. L. M. van Mensfoort, P. A. Bobbert, and R. Coehoorn. Modeling of the transient mobility in disordered organic semiconductors with a Gaussian density of states. *Phys. Rev. B*, 84(16):165210, 2011.

165. S. L. M. van Mensfoort and R. Coehoorn. Determination of injection barriers in organic semiconductor devices from capacitance measurements. *Phys. Rev. Lett.*, 100(8):086802, 2008.

166. K. A. Neyts. Simulation of light emission from thin-film microcavities. *J. Opt. Soc. Am. A*, 15(4):962–971, 1998.

167. V. Bulovi, V. B. Khalfin, G. Gu, P. E. Burrows, D. Z. Garbuzov, and S. R. Forrest. Weak microcavity effects in organic light-emitting devices. *Phys. Rev. B*, 58(7):3730–3740, 1998.

168. Z. B. Wang, M. G. Helander, X. F. Xu, D. P. Puzzo, J. Qiu, M. T. Greiner, and Z. H. Lu. Optical design of organic light emitting diodes. *J. Appl. Phys.*, 109(5):053107, 2011.

169. M. Furno, R. Meerheim, S. Hofmann, B. Luessem, and K. Leo. Efficiency and rate of spontaneous emission in organic electroluminescent devices. *Phys. Rev. B*, 85(11):115205, 2012.

170. S. L. M. van Mensfoort, M. Carvelli, M. Megens, D. Wehenkel, M. Bartyzel, H. Greiner, R. A. J. Janssen, and R. Coehoorn. Measuring the light emission profile in organic light-emitting diodes with nanometre spatial resolution. *Nat. Photon.*, 4(0):329–335, 2010.

171. M. Flaemmich, D. Michaelis, and N. Danz. Accessing OLED emitter properties by radiation pattern analyses. *Org. Electron.*, 12(1):83–91, 2011.

172. J. Frischeisen, D. Yokoyama, C. Adachi, and W. Bruetting. Determination of molecular dipole orientation in doped fluorescent organic thin films by photoluminescence measurements. *Appl. Phys. Lett.*, 96(7):073302, 2010.

173. S.-Y. Kim and J.-J. Kim. Outcoupling efficiency of organic light emitting diodes and the effect of ITO thickness. *Org. Electron.*, 11(6):1010–1015, 2010.

174. W. Bruetting, J. Frischeisen, T. D. Schmidt, B. J. Scholz, and C. Mayr. Device efficiency of organic light-emitting diodes: Progress by improved light outcoupling. *Phys. Status Solidi A*, 210(1):44–65, 2013.

175. M. C. Gather and S. Reineke. Recent advances in light outcoupling from white organic light-emitting diodes. *J. Photon. Energy*, 5(1):057607–057607, 2015.

176. H. van Eersel, P. A. Bobbert, R. A. J. Janssen, and R. Coehoorn. Monte Carlo study of efficiency roll-off off phosphorescent organic light-emitting diodes: Evidence for dominant role of triplet–polaron quenching. *Appl. Phys. Lett.*, 105(14):143303, 2014.

177. R. Coehoorn, H. van Eersel, P. Bobbert, and R. Janssen. Kinetic Monte Carlo study of the sensitivity of OLED efficiency and lifetime to materials parameters. *Adv. Funct. Mater.*, 25(13):2024–2037, 2015.

178. Y. Shen and N. C. Giebink. Monte Carlo simulations of nanoscale electrical in-homogeneity in organic light-emitting diodes and its impact on their efficiency and lifetime. *Phys. Rev. Appl.*, 4(5):054017, 2015.

179. G. Lanzani. *The Photophysics behind Photovoltaics and Photonics*. Weinheim: Wiley-VCH, 2012. OCLC: ocn768072884.

180. C. Murawski, K. Leo, and M. C. Gather. Efficiency roll-off in organic light-emitting diodes. *Adv. Mater.*, 25(47):6801–6827, 2013.

181. M. Mesta, M. Carvelli, R. J. de Vries, H. van Eersel, J. J. M. van der Holst, M. Schober, M. Furno, B. Luessem, K. Leo, P. Loebl, R. Coehoorn, and P. A. Bobbert. Molecular-scale simulation of electroluminescence in a multilayer white organic light-emitting diode. *Nat. Mater.*, 12(7):652–658, 2013.

182. H. van Eersel, P. A. Bobbert, and R. Coehoorn. Kinetic Monte Carlo study of triplet–triplet annihilation in organic phosphorescent emitters. *J. Appl. Phys.*, 117(11):115502, 2015.

183. J. Cottaar, R. Coehoorn, and P. A. Bobbert. Modeling of charge transport across disordered organic heterojunctions. *Org. Electron.*, 13(4):667–672, 2012.

184. H. Yoshida and K. Yoshizaki. Electron affinities of organic materials used for organic light-emitting diodes: A low-energy inverse photoemission study. *Org. Electron.*, 20:24–30, 2015.

185. N. C. Giebink and S. R. Forrest. Quantum efficiency roll-off at high brightness in fluorescent and phosphorescent organic light emitting diodes. *Phys. Rev. B*, 77(23):235215, 2008.

186. G. Schwartz, S. Reineke, T. Conrad Rosenow, K. Walzer, and K. Leo. Triplet harvesting in hybrid white organic light-emitting diodes. *Adv. Funct. Mater.*, 19(9):1319–1333, 2009.

187. S. Mladenovski, S. Reineke, and K. Neyts. Measurement and simulation of exciton decay times in organic light-emitting devices with different layer structures. *Opt. Lett.*, 34(9):1375–1377, 2009.

188. K.-C. Tang, K. Lin Liu, and I.-C. Chen. Rapid intersystem crossing in highly phosphorescent iridium complexes. *Chem. Phys. Lett.*, 386(4–6):437–441, 2004.

189. Y. Kawamura, J. Brooks, J. J. Brown, H. Sasabe, and C. Adachi. Intermolecular interaction and a concentration-quenching mechanism of phosphorescent Ir(III) complexes in a solid film. *Phys. Rev. Lett.*, 96(1):017404, 2006.

190. F. S. Steinbacher, R. Krause, A. Hunze, and A. Winnacker. Triplet exciton transfer mechanism between phosphorescent organic dye molecules. *Phys. Stat. Sol. (A)*, 209(2):340–346, 2012.

191. S. L. M. van Mensfoort, R. J. de Vries, V. Shabro, H. P. Loebl, R. A. J. Janssen, and R. Coehoorn. Electron transport in the organic small-molecule material BAlq—The role of correlated disorder and traps. *Org. Electron.*, 11(8):1408–1413, 2010.

192. M. M. Mandoc, B. de Boer, G. Paasch, and P. W. M. Blom. Trap-limited electron transport in disordered semiconducting polymers. *Phys. Rev. B*, 75(19):193202, 2007.

193. F. May, B. Baumeier, C. Lennartz, and D. Andrienko. Can lattice models predict the density of states of amorphous organic semiconductors? *Phys. Rev. Lett.*, 109(13):136401, 2012.

194. S. Olthof, S. Mehraeen, S. K. Mohapatra, S. Barlow, V. Coropceanu, J.-L. Bredas, S. R. Marder, and A. Kahn. Ultralow doping in organic semiconductors: Evidence of trap filling. *Phys. Rev. Lett.*, 109(17):176601, 2012.

195. G. Li, C. H. Kim, Z. Zhou, J. Shinar, K. Okumoto, and Y. Shirota. Combinatorial study of exciplex formation at the interface between two wide band gap organic semiconductors. *Appl. Phys. Lett.*, 88(25):253505, 2006.

196. H. Aziz and Z. D. Popovic. Degradation phenomena in small-molecule organic light-emitting devices. *Chem. Mater.*, 16(23):4522–4532, 2004.

197. F. So and D. Kondakov. Degradation mechanisms in small-molecule and polymer organic light-emitting diodes. *Adv. Mater.*, 22(34):3762–3777, 2010.

198. J.-S. Park, H. Chae, H. Kyoon Chung, and S. I. Lee. Thin film encapsulation for flexible AM-OLED: A review. *Semicond. Sci. Technol.*, 26(3):034001, 2011.

199. S. Scholz, D. Kondakov, B. Luessem, and K. Leo. Degradation mechanisms and reactions in organic light-emitting devices. *Chem. Rev.*, 115(16):8449–8503, 2015.

200. H. Aziz, Z. D. Popovic, N.-X. Hu, A.-M. Hor, and G. Xu. Degradation mechanism of small molecule-based organic light-emitting devices. *Science*, 283(5409):1900–1902, 1999.

201. D. Y. Kondakov, W. C. Lenhart, and W. F. Nichols. Operational degradation of organic light-emitting diodes: Mechanism and identification of chemical products. *J. Appl. Phys.*, 101(2):024512, 2007.

202. R. Seifert, S. Scholz, B. Luessem, and K. Leo. Comparison of ultraviolet- and charge-induced degradation phenomena in blue fluorescent organic light emitting diodes. *Appl. Phys. Lett.*, 97(1):013308, 2010.

203. N. C. Giebink, B. W. D'Andrade, M. S. Weaver, P. B. Mackenzie, J. J. Brown, M. E. Thompson, and S. R. Forrest. Intrinsic luminance loss in phosphorescent small-molecule organic light emitting devices due to bimolecular annihilation reactions. *J. Appl. Phys.*, 103(4):044509, 2008.

204. T. D. Schmidt, L. Jaeger, Y. Noguchi, H. Ishii, and W. Bruetting. Analyzing degradation effects of organic light-emitting diodes via transient optical and electrical measurements. *J. Appl. Phys.*, 117(21):215502, 2015.

205. T. D. Schmidt, D. S. Setz, M. Flaemmich, J. Frischeisen, D. Michaelis, B. C. Krummacher, N. Danz, and W. Bruetting. Evidence for non-isotropic emitter orientation in a red phosphorescent organic light-emitting diode and its implications for determining the emitter's radiative quantum efficiency. *Appl. Phys. Lett.*, 99(16):163302, 2011.

206. C. Fry, B. Racine, D. Vaufrey, H. Doyeux, and S. Cin. Physical mechanism responsible for the stretched exponential decay behavior of aging organic light-emitting diodes. *Appl. Phys. Lett.*, 87(21):213502, 2005.

207. A. Mass, R. Coehoorn, and P. A. Bobbert. Universal size-dependent conductance fluctuations in disordered organic semiconductors. *Phys. Rev. Lett.*, 113(11), 2014.

16

Tunnel-Junction Light-Emitting Diodes

Yen-Kuang Kuo

Jih-Yuan Chang

Ya-Hsuan Shih

Fang-Ming Chen

and

Miao-Chan Tsai

16.1 Introduction

The use of p-n tunnel junctions for applications in GaN-based electronic and optoelectronic devices has become increasingly attractive. The use of tunnel junctions has conceptually been desired for the reuse of carriers for coupled active regions, enabling high quantum efficiencies and improved vertical transport (Ozden et al., 2001). Concepts of the tunnel field-effect transistors, multiple active region light-emitting diodes (LEDs) and laser diodes, high-conductivity hole injection layers, and multijunction solar cells may be realized with the III-nitride material system providing that a sufficiently low-resistivity tunnel junction can be obtained. However, the feasibility of forming low-resistivity tunnel junctions in the wide-bandgap III-nitride devices has been a challenge due to the high hole and electron concentrations required for band alignment. Substantial effort is still required to resolve this critical issue.

16.1.1 Polarization-Assisted Tunneling Effect

The probability of interband tunneling across a potential barrier is mainly governed by the tunneling barrier height determined by the bandgap and the tunneling barrier thickness (Simon et al., 2009). For wide-bandgap III-nitrides, tunneling is difficult due to the high barrier height and the difficulty in achieving degenerate p-type impurity doping (Simon et al., 2009; Krishnamoorthy et al., 2010). However, the strong polarization-induced electric field available in the III-nitrides and other highly polar semiconductor materials provides a new design approach for tunneling structures. In hetero-structures of highly polar materials, the polarization-induced sheet charges can create significantly high electric fields resulting in large band bending over a small distance. The insufficient band-bending for interband tunneling from doping-induced built-in electric field (E_{bi}) can thus be supplemented by the polarization-induced electric field (E_p) assuming that the two electric fields are oriented in the same direction. The probability of interband tunneling can therefore be increased.

Recently, GaN-based tunnel junctions, which are supplemented by the polarization-induced electric field, have been demonstrated (Simon et al., 2009; Krishnamoorthy et al., 2010, 2011, 2013; Schubert, 2010a; Grundmann and Mishra, 2007). The use of large spontaneous and piezoelectric polarization fields present in the III-nitrides enables the formation of thin depletion region and relaxes the strict requirement for degenerate doping. Structures including the GaN/AlN/GaN, AlN/GaN/AlN, and GaN/InGaN/GaN have been investigated theoretically and experimentally (Simon et al., 2009; Krishnamoorthy et al., 2010; Schubert, 2010a). In particular, tunnel-junction structures using an intermediate InGaN layer have been developed due to the advantages of low bandgap and large piezoelectric polarization. It was shown by estimation of tunneling probability using the Wentzel–Kramers–Brillouin (WKB) approximation that the optimal performance, with n-doping $N_D = 5 \times 10^{19}$ cm^{-3} and p-doping $N_A = 1 \times 10^{19}$ cm^{-3}, would occur when the indium composition in InGaN is greater than 30% and the thickness is larger than 3 nm, which was achieved on the N-polar GaN with molecular beam epitaxy (MBE) growth method (Grundmann and Mishra, 2007; Krishnamoorthy et al., 2011, 2013). In many of the envisioned applications in optoelectronics, an indium composition of greater than approximately 20% in InGaN is not favorable due to the high optical absorption caused by the InGaN and the problem in crystal growth for the layers subsequent to the tunnel junction. Moreover, the doping levels are limited by the Si-induced roughening and strain, ionization efficiency, and solubility of Mg dopants. In this chapter, design and analysis for the low-resistivity tunnel junctions, which are with adequate indium composition and doping density for a typical Ga-polar InGaN multiple-quantum well (MQW) LED structure, are introduced.

16.1.2 Efficiency Droop in III-Nitride LEDs

For the application in solid-state lighting, the development of high-efficiency and high-power III-nitride LEDs is required. In this demand, the major issue that needs to be resolved might be the problem of efficiency droop, i.e., the reduction in illumination efficiency at high current density. It is believed that heating is not the major cause for efficiency droop due to the fact that the efficiency droop occurs in both pulsed and continuous wave (CW) conditions and the droop becomes severe with the decrease of ambient temperature (Laubsch et al., 2009; Kim et al., 2007). To date, the physical origin of efficiency droop remains debatable even though numerous possible physical mechanisms, such as the carrier delocalization (Monemar and Sernelius, 2007; Chichibu et al., 2006), Auger recombination (Shen et al., 2007; Delaney et al., 2009), insufficient hole injection (Rozhansky and Zakheim, 2006), carrier leakage (Shim et al., 2012; Lin et al., 2012), and polarization effect (Kim et al., 2007; Kuo et al., 2009), have been proposed and demonstrated to be related to the efficiency droop in GaN-based LEDs. Even though continuous efforts have been made, an overall solution for the efficiency droop of III-nitride LEDs is still lacking. In this chapter, instead of trying to promote the quantum efficiency at high current density, another approach in circumventing the issue of efficiency droop is proposed, which is based on the use of GaN-based tunnel junctions. Specifically, in this approach it is suggested to operate the III-nitride LEDs within the range of high efficiency, i.e., to operate the LEDs at low current density, with the insertion of a low-loss tunnel junction between two active regions, which is beneficial in reducing the carrier density in each active region and mitigating the problem of efficiency droop.

16.2 Physical Models and Simulation Parameters

16.2.1 Bulk Band Structure

The characteristics of III-nitride tunnel junction and tunnel-junction LEDs constructed along the c-axis are numerically studied using a self-consistent simulation program Advanced Physical Models of Semiconductor Devices (APSYS). APSYS employs the 6×6 **k·p** model, which was developed for the strained wurtzite semiconductor by Chuang and Chang (1996), to calculate the energy-band structures of the non-active bulk regions. The unstrained bandgap energies of the $In_xGa_{1-x}N$ and $Al_yGa_{1-y}N$ ternary alloys can be expressed by the following formula (Vurgaftman et al., 2001)

$$E_g(In_xGa_{1-x}N) = x \cdot E_g(GaN) + (1-x) \cdot E_g(InN) - x \cdot (1-x) \cdot B(InGaN), \tag{16.1}$$

$$E_g(Al_yGa_{1-y}N) = y \cdot E_g(GaN) + (1-y) \cdot E_g(AlN) - y \cdot (1-y) \cdot B(AlGaN), \tag{16.2}$$

where $E_g(GaN)$, $E_g(InN)$, and $E_g(AlN)$ are the bandgap energies of GaN, InN, and AlN, which have values of 3.42, 0.64, and 6.0 eV at 300 K, respectively (Nepal et al., 2005; Wu et al., 2003). $B(InGaN)$ and $B(AlGaN)$ are the bandgap bowing parameters of InGaN and AlGaN, respectively. The bandgap bowing parameters of InGaN and AlGaN are assumed to be 2.1 and 1.0 eV, respectively (Gorczyca et al., 2011a, b). The band-offset ratio is assumed to be 0.7/0.3 for the III-nitride material systems.

16.2.2 MQW Model: Approximation of Effective Mass

A model of effective mass for the computation of density of states (DOS) and for a simplified MQW model is needed because, in general, the dispersion of bulk valence bands is nonparabolic and anisotropic with strong mixing or anticrossing behavior in the direction perpendicular to the *c*-axis. Based on the method of Chuang and Chang (1996), an analytical model of effective mass is implemented for the valence band.

Within a range of small k, a situation when the valence band is lightly populated by holes, the following effective masses hold:

$$m_{hh}^z = -m_0(A_1 + A_3)^{-1}, \tag{16.3}$$

$$m_{lh}^z = -m_0\left[A_1 + \left(\frac{E_2^0 - \lambda_e}{E_2^0 - E_3^0}\right)A_3\right]^{-1}, \tag{16.4}$$

$$m_{ch}^z = -m_0\left[A_1 + \left(\frac{E_3^0 - \lambda_e}{E_3^0 - E_2^0}\right)A_3\right]^{-1}, \tag{16.5}$$

$$m_{hh}^t = -m_0(A_2 + A_4)^{-1}, \tag{16.6}$$

$$m_{lh}^t = -m_0\left[A_2 + \left(\frac{E_2^0 - \lambda_e}{E_2^0 - E_3^0}\right)A_4\right]^{-1}, \tag{16.7}$$

$$m_{ch}^t = -m_0\left[A_2 + \left(\frac{E_3^0 - \lambda_e}{E_3^0 - E_2^0}\right)A_4\right]^{-1}, \tag{16.8}$$

where E_i^0 ($i = 1, 2, 3$) are the valence-band edge (values) at $k = 0$.

For a large range of k, i.e., a situation when the valence band is heavily populated by holes, the following effective masses are valid:

$$m_{hh}^z = -m_0(A_1 + A_3)^{-1}, \tag{16.9}$$

$$m_{lh}^z = -m_0(A_1 + A_3)^{-1}, \tag{16.10}$$

$$m_{ch}^z = -m_0A_1^{-1}, \tag{16.11}$$

$$m_{hh}^t = -m_0(A_2 + A_4 - A_5)^{-1}, \tag{16.12}$$

$$m_{lh}^t = -m_0(A_2 + A_4 - A_5)^{-1}, \tag{16.13}$$

$$m_{ch}^t = -m_0A_2^{-1}. \tag{16.14}$$

In the present simulation, a compromise of the above two models is used, i.e., to average them. Note that the choice of model is based on how heavily the valence bands are populated by holes. Other parameters of band structure for the binary nitride wurtzite semiconductors employed in the simulation are listed in Table 16.1 (Wu, 2009; Vurgaftman and Meyer, 2007).

16.2.3 Incomplete Ionization of Impurities

The degree of ionization is described by the occupancies f_D and f_A. Shallow impurities are assumed to be in equilibrium with the local carriers and therefore the occupancy of the shallow impurities can be described by

$$f_D = \frac{1}{1 + g_d^{-1} \exp[(E_D - E_{fn})/kT]}, \quad (16.15)$$

$$f_A = \frac{1}{1 + g_a \exp[(E_A - E_{fp})/kT]}, \quad (16.16)$$

where the subscripts D and A are used to denote shallow donors and acceptors, respectively. The degeneracy levels are $g_d = 2$ and $g_a = 4$ (Tiwari, 1992).

TABLE 16.1 Material Parameters of the Binary Semiconductors GaN, AlN, and InN (Wu, 2009; Vurgaftman and Meyer, 2007)

Parameter	Symbol (Unit)	GaN	AlN	InN
Electron effective mass (c-axis)	m_c^z (m$_0$)	0.21	0.33	0.068
Electron effective mass (transverse)	m_c^t (m$_0$)	0.19	0.32	0.065
Hole effective mass parameter	A_1	−7.21	−3.86	−8.21
	A_2	−0.44	−0.25	−0.68
	A_3	6.68	3.58	7.57
	A_4	−3.46	−1.32	−5.23
	A_5	−3.40	−1.47	−5.11
	A_6	−4.90	−1.64	−5.96
Spin–orbit split energy	Δ_{so}(eV)	0.017	0.019	0.005
Crystal-field split energy	Δ_{cr}(eV)	0.010	−0.169	0.040
Lattice constant a	a_0(Å)	3.189	3.112	3.533
Lattice constant c	c_0(Å)	5.185	4.982	5.693
Elastic stiffness constant	C_{11} (GPa)	390	396	223
Elastic stiffness constant	C_{12} (GPa)	145	137	115
Elastic stiffness constant	C_{13} (GPa)	106	108	92
Elastic stiffness constant	C_{33} (GPa)	398	373	224
Elastic stiffness constant	C_{44} (GPa)	105	116	48
Hydrost. deform. potential (c-axis)	a_z (eV)	−7.1	−3.4	−4.2
Hydrost. deform. potential (transverse)	a_t (eV)	−9.9	−11.8	−4.2
Shear deform. potential	D1 (eV)	−3.6	−2.9	−3.6
	D2 (eV)	1.7	4.9	1.7
	D3 (eV)	5.2	9.4	5.2
	D4 (eV)	−2.7	−4.0	−2.7
	D5 (eV)	−2.8	−3.3	−2.8
	D6 (eV)	−4.3	−2.7	−4.3
Dielectric constant	ε	8.9	8.5	10.5

Note: $\Delta_{cr} = \Delta_1$, $\Delta_{so} = 3\Delta_2 = 3\Delta_3$.

In the simulation, the acceptor activation energy (E_A) of Mg doped in InGaN is scaled linearly from 60 meV (InN) (Khan et al., 2007) to 166 meV (GaN) (Kumakura et al., 2003). The acceptor activation energy of Mg doped in AlGaN is scaled linearly from 166 meV (GaN) (Kumakura et al., 2003) to 510 meV (AlN) (Nam et al., 2003).

16.2.4 Electronic Polarization and Interface Charges

In typical III-nitride semiconductor devices, the device structure is usually constructed of ternary AlGaN and InGaN alloys. In order to consider the influences of internal polarization in tunnel junction and tunnel-junction LEDs, the method developed by Fiorentini et al. (2002) is employed to estimate the internal polarization in Ga-face configuration, which is represented by fixed surface charges at hetero-interfaces (Fiorentini et al., 2002). The composition of the compound determines the net polarization charges that remain at each hetero-interface. The nonlinear model described in the papers by Fiorentini et al. (2002) and Vurgaftman and Meyer (2003) is used here. The spontaneous polarization P_{sp} (Cm^{-2}) is calculated as

$$P_{sp}(Al_xGa_{1-x}N) = -0.090 \cdot x - 0.034 \cdot (1-x) + 0.019 \cdot x \cdot (1-x), \tag{16.17}$$

$$P_{sp}(In_xGa_{1-x}N) = -0.042 \cdot x - 0.034 \cdot (1-x) + 0.038 \cdot x \cdot (1-x). \tag{16.18}$$

For binary compounds, the piezoelectric polarization P_{pz} (Cm^{-2}) is given as the nonlinear function of the transverse strain ε_{xx} by

$$P_{pz}(GaN) = -0.918 \cdot \varepsilon_{xx} + 9.541 \cdot \varepsilon_{xx}^2, \tag{16.19}$$

$$P_{pz}(InN) = -1.373 \cdot \varepsilon_{xx} + 7.559 \cdot \varepsilon_{xx}^2, \tag{16.20}$$

$$P_{pz}(AlN) = -1.808 \cdot \varepsilon_{xx} - 7.888 \cdot \varepsilon_{xx}^2 \; (\varepsilon_{xx} > 0), \tag{16.21}$$

$$P_{pz}(AlN) = -1.808 \cdot \varepsilon_{xx} + 5.624 \cdot \varepsilon_{xx}^2 \; (\varepsilon_{xx} < 0). \tag{16.22}$$

The piezoelectric polarization of a specific ternary compound is linearly interpolated by the piezoelectric polarizations of the corresponding binary compounds.

The built-in electric fields of III-nitride devices obtained from experimental measurements are usually smaller than the values obtained from theoretical calculation, ranging from 20% to 80% (Piprek, 2007), due mainly to partial compensation of the polarization field by fixed defect and interface charges. In this study, the polarization-induced surface charge densities are assumed to be 50% of the values determined from theoretical calculation.

16.2.5 Interband Tunneling Model

Direct interband tunneling in a p-n junction between the valence and conduction bands in an external applied electric field is commonly referred to as a tunnel junction, or an Esaki junction, or a Zener tunneling diode under reverse bias. The probability of tunneling within the WKB approximation, following the derivation of band-to-band tunneling in the papers by Kane (1960), Duke (1969), and Moll (1964), is calculated as follows:

$$D = \exp(-2J) = P_0 \exp\left(\frac{-E_\perp}{E}\right), \tag{16.23}$$

where

$$J\left(E_{\parallel}\right) = \int_{x_2}^{x_1} \left| \left(\frac{2m^*}{\hbar^2}\right) \frac{\left[\left(E_g/2\right)^2 - \left(\varepsilon_c\right)^2\right]}{E_g} + E_{\perp} \right|^{1/2} dx, \tag{16.24}$$

$$P_0 = \exp\left[\frac{\pi(m^*)^{1/2}(E_g)^{3/2}}{2(2)^{1/2}qF\hbar}\right] = \exp\left(-\frac{E_g}{4\underline{E}}\right) \tag{16.25}$$

$$\underline{E} = \frac{(2)^{1/2}qF\hbar}{2\pi(m^*)^{1/2}(E_g)^{1/2}}, \tag{}$$

$$m^* = \frac{2m_c m_v}{(m_c + m_v)}. \tag{16.26}$$

$E_{\parallel}$ and $E_{\perp}$ are the electron kinetic energies in the directions along and perpendicular to the tunneling direction, respectively. P_0 is the tunneling probability with a momentum of zero perpendicular to the x-direction. The electrons that can tunnel through the bandgap barrier are those electrons with perpendicular momentum near zero if $\underline{E}$ is small. $\underline{E}$ is a measure of significance of perpendicular momentum range and m^* is the effective tunneling mass.

16.3 III-Nitride Tunnel Junction

For the applications in Ga-polarity InGaN-based LEDs, with the use of InGaN as the barrier material between the heavily doped n- and p-type GaN layers, the vectors of the polarization and normal built-in fields have the same direction for an n-on-p tunnel diode, which helps the generation of the electric field required for tunneling. Furthermore, the reduced barrier height of InGaN also promotes the tunneling probability of carriers since the bandgap energy of InGaN is relatively low when compared to GaN. Figure 16.1 shows the schematic diagram of n-GaN/p-GaN and n-GaN/InGaN/p-GaN tunnel-junction structures. The illustrations for the directions of built-in field and polarization field are also indicated in Figure 16.1 for better reference. Here, let us begin with the n-GaN/InGaN/p-GaN tunnel-junction structure, which is composed of a heavily doped n-GaN layer, a 7-nm-thick InGaN barrier layer, and a heavily doped p-GaN layer (Krishnamoorthy et al., 2011), and calculate the tunneling current as a function of reverse bias voltage, as shown in Figure 16.2. In the simulation, the dopant concentrations of n- and p-type GaN layers are assumed to be identical, while the ionization energies for the computation of incomplete ionization of donors and acceptors are quite distinct, making the electron and hole concentrations also different. It is observed in Figure 16.2a that the tunneling current increases dramatically with the increase of indium composition. It is required that the Fermi levels are in the valence band and the conduction band

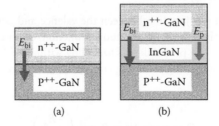

(a)   (b)

FIGURE 16.1 Schematic diagrams of (a) n-GaN/p-GaN and (b) n-GaN/InGaN/p-GaN tunnel-junction structures.

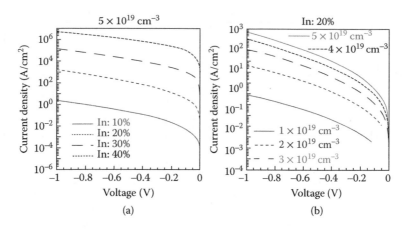

FIGURE 16.2 Electrical characteristics of the n-i-p tunnel diode in the reversed bias regime under various conditions of (a) indium composition and (b) acceptor and donor densities.

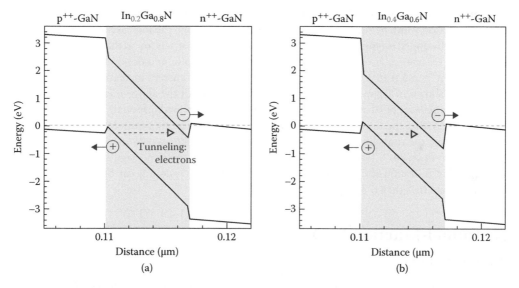

FIGURE 16.3 Energy band diagrams of (a) n-GaN/In$_{0.2}$Ga$_{0.8}$N/p-GaN and (b) n-GaN/In$_{0.4}$Ga$_{0.6}$N/p-GaN tunnel diodes at equilibrium.

on the p- and n-side, respectively, so that the carriers can tunnel between the bands as in a Zener diode. The improved tunneling characteristics with the increase of indium composition are thus ascribed to the promotion of the polarization field and the reduction of potential barrier height, as shown in Figure 16.3. However, it is noteworthy that when the composition of indium is high, the crystalline layer tends to relax and the film quality may deteriorate due to the large lattice mismatch and thermal mismatch between the InN and GaN. Under these circumstances, the strain-related polarization field reduces and the defect density increases, which is harmful for the applications in optoelectronics since the high defect density may act as the nonradiative recombination centers (Cherns et al., 2001). Moreover, for many applications in optoelectronics, it is usually required that the tunnel junction is with a reduced percentage of indium due to the problem of optical absorption. In the present study, the indium composition is fixed at 20% for subsequent simulations.

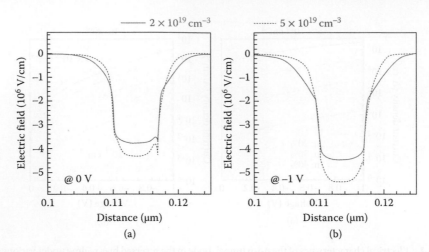

FIGURE 16.4   Electric fields of n-GaN/In$_{0.2}$Ga$_{0.8}$N/p-GaN tunnel diode with various acceptor and donor densities (a) at equilibrium and (b) at −1 V.

Another approach to improve the performance of tunnel junction is to increase the built-in field and reduce the depletion width by supporting more space charges through the increase of acceptor and donor densities of heavily doped p-type and n-type GaN layers. In Figure 16.4, it is found that the total electric field of the In$_{0.2}$Ga$_{0.8}$N layer can be effectively enhanced when the acceptor and donor densities are increased from 2×10^{19} cm^{-3} to 5×10^{19} cm^{-3}. Consequently, the tunneling current increases with the increase of doping level, as shown in Figure 16.2b. Note that there is a limit for the above-mentioned improvement because the doping level of III-nitrides cannot be unlimitedly increased, especially for p-type doping. It is also noteworthy that the tunneling characteristics can be significantly promoted by increasing the Si doping level, while the tunneling current cannot be appreciably increased by increasing the Mg doping (Krishnamoorthy et al., 2013; Tsai et al., 2013). The findings provide another design approach that might be applicable for the realization of III-nitride tunnel junctions since heavy doping can be more easily achieved for the n-type materials than the p-type materials.

16.4 Tunnel-Junction LED

16.4.1 Single LED

Characterization of the blue single LED structure under study, which is used as a reference for the following investigations in stacking LEDs with tunnel junctions, is introduced first in this section. The LED structure is composed of a 3-μm-thick n-GaN layer (1×10^{19} cm^{-3}), four pairs of QWs with 3-nm-thick In$_{0.2}$Ga$_{0.8}$N wells and 6-nm-thick GaN barriers, a 20-nm-thick p-Al$_{0.15}$Ga$_{0.85}$N electron-blocking layer (EBL) (1×10^{19} cm^{-3}), and 100-nm-thick p-GaN layer (1×10^{19} cm^{-3}). The layer structure of the reference blue single LED is listed in Table 16.2. In the simulation, to focus on the topic explored and to eliminate the issue of current crowing, ideal ohmic contacts are assumed to cover the full top and bottom surfaces of the simulated vertical LED structure. The Shockley–Read–Hall (SRH) recombination lifetime and Auger recombination coefficient are set to be 50 ns and 2×10^{-30} cm^6/s, respectively (Yoshida et al., 2010; Piprek and Li, 2013). It is also assumed that the light extraction efficiency is 70%.

Figure 16.5 shows the simulated electrical and optical characteristics of the blue single LED. In the energy band diagram, as shown in Figure 16.5a, obvious band deformation caused by the polarization field is observed. In particular, the energy bands of QWs and barriers are severely sloped, forming triangular-shaped band profiles, which impact the transport, injection, and confinement of carriers in the MQW active region (Kim et al., 2007; Kuo et al., 2009). The polarization-induced sloped well results in spatial

TABLE 16.2 Layer Structure of the Reference Blue Single Light-Emitting Diode (LED) under Study

Parameter (unit)	d(nm)	$N_{dop}(1/cm^3)$
p-GaN	100	1×10^{19}
p-Al$_{0.15}$Ga$_{0.85}$N (EBL)	20	1×10^{19}
i-GaN (barrier)	6	–
i-In$_{0.2}$Ga$_{0.8}$N (well)	3	–
i-GaN (barrier)	6	–
i-In$_{0.2}$Ga$_{0.8}$N (well)	3	–
i-GaN (barrier)	6	–
i-In$_{0.2}$Ga$_{0.8}$N (well)	3	–
i-GaN (barrier)	6	–
i-In$_{0.2}$Ga$_{0.8}$N (well)	3	–
i-GaN (barrier)	6	–
n-GaN	3000	1×10^{19}

Note: d, layer thickness; N_{dop}, dopant density.

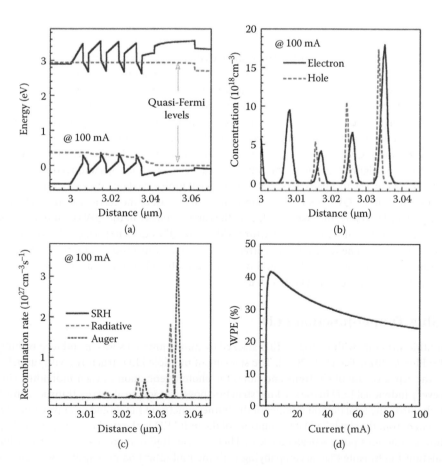

FIGURE 16.5 (a) Energy band diagram, (b) carrier concentrations, (c) recombination rates at 100 mA, and (d) wall-plug efficiency (WPE) of the reference blue single light-emitting diode (LED).

separation of electron and hole wave functions in the well, which is believed to be one of the key issues that deteriorate the radiative recombination. Moreover, the electron leakage current is negligibly small, which indicates that the capability of carrier confinement of the MQW active region is superior. As shown in Figure 16.5b, the distribution of carriers in the QWs is quite nonuniform. Most of the carriers accumulate in the QWs close to the p-side. Similar nonuniform carrier distribution in the InGaN LED has also been reported, which is attributed to the poor injection and transportation of holes (David et al., 2008). Under these circumstances, the illumination efficiency of the blue single LED is severely limited by the huge transition loss of Auger recombination. The droop of wall-plug efficiency (WPE), defined as $(WPE_{max} - WPE_{100mA})/WPE_{max}$ where WPE_{max} is the maximum WPE and WPE_{100mA} is the WPE at 100 mA, is as large as 42%, as shown in Figure 16.5c and d. Note that the WPE is referred to the energy conversion efficiency with which the LED converts the electrical power into optical power, as defined below:

$$WPE = \frac{\text{Light output power}}{\text{Current} \times \text{Voltage}}. \tag{16.27}$$

The influence of Auger recombination on the efficiency of LEDs is typically analyzed based on the reduced rate equation. Assuming that the injection efficiency is 100%, the internal quantum efficiency (IQE), which is directly related to the light output power, can be expressed as

$$IQE = \frac{B \cdot n^2}{A \cdot n + B \cdot n^2 + C \cdot n^3}, \tag{16.28}$$

where A, B, C, and n are the SRH coefficient, radiative coefficient, Auger coefficient, and carrier density, respectively. According to Equation 16.28, the Auger recombination dominates the total recombination rate at high carrier density because it scales with the cubic power of the free carrier density, which is believed to be one of the key factors contributing to the efficiency droop (Shen et al., 2007). If the carrier density in the active region can be reduced while the total carriers confined within the active region remains unchanged, the efficiency at high current may be improved due to the suppression of Auger recombination. It might be helpful to use more QWs in the active region because the carrier density per QW may be reduced under this situation. Figure 16.6 shows the simulated carrier concentrations and recombination rates of the single LED structure with 8 pairs and 12 pairs of QWs. It is observed that the carrier distribution of the LED structures with more QWs is still quite nonuniform and most of the carriers accumulate in the QWs close to the p-side. Therefore, the goal of reducing the carrier density in the QWs cannot be satisfactorily achieved. Under this circumstance, the structure with more QWs does not benefit from the suppressed Auger recombination and hence, similar radiative-Auger relations are observed for the 4-, 8-, and 12-QW LED structures as shown in Figure 16.6c and d. As a result, the efficiency characteristics do not exhibit much difference when the number of QWs changes.

16.4.2 Blue Tunnel-Junction LED

The monolithic structure with multiple LEDs stacked by tunnel junctions was introduced recently (Akyol et al., 2013; Piprek, 2014; Tsai et al., 2014). The stacking of multiple LED structures with tunnel junctions allows for the repeated use of electrons and holes for photon generation in each individual single LED. Ideally, if every individual LED in the stacking structure possesses identical carrier transport and recombination mechanism, the total photon number of the tunnel-junction LED should be approximately N times to that obtained from the single LED if the number of the unit LED in the stacking structure is N, provided that the tunnel junctions possess negligible loss. Therefore, an IQE of more than 100% is possible for the tunnel-junction LEDs. Note that, accompanying with the multiplication of output power, bias and input power multiply with the number of unit LED in the stacking structure simultaneously. Hence, the WPE of tunnel-junction LEDs remains below 100% (Piprek, 2014).

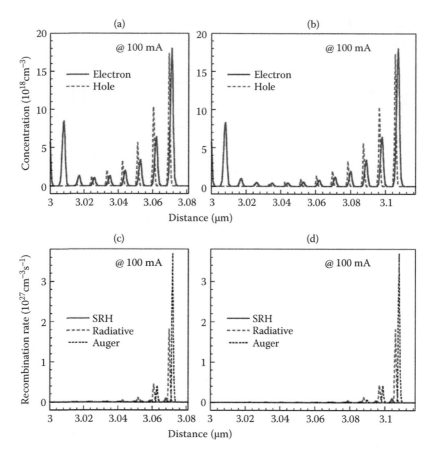

FIGURE 16.6 (a) and (b) Carrier concentrations and (c) and (d) recombination rates of the blue single light-emitting diode (LED) with 8 pairs and 12 pairs of QWs at 100 mA.

The simulated energy band diagram and radiative recombination rate at 100 mA for the blue tunnel-junction LEDs are depicted in Figure 16.7. Note that, in this figure, the n-type layer is located in the left-hand side of each unit LED. In the simulation, the n-GaN/In$_{0.2}$Ga$_{0.8}$N/p-GaN tunnel junction with a dopant concentration of 5×10^{19} cm^{-3} is utilized to ensure good tunneling characteristics, as shown in Figure 16.2c. When the tunnel-junction LED is turned on, electrons are injected and then transported in the conduction band of n-type layer of the first stacked LED. In the meantime, in the valance band of the p-type layer, electrons are transported and then transferred into the conduction band of the second stacked LED, which is regarded as holes moving in the opposite direction. The electrons and holes then recombine in the MQW active region of the first stacked LED to generate photons. The carrier transport and photon generation repeat in subsequent stacked LEDs. The emission profiles of all unit LEDs are almost identical, as shown in Figure 16.7.

Figure 16.8 shows the light output power as a function of current (L–I) and current as a function of voltage (I–V) characteristics of the blue single LED and tunnel-junction LEDs. It is observed that the light output power and operation voltage of the stacked LEDs are almost equivalent to the light output power and operation voltage of the blue single LED multiplied by the number of unit LED in the stacked LED structure. Under these circumstances, the WPE of the tunnel-junction LEDs would not vary too much comparing to that of the single LED. As mentioned previously, the approach of stacking LEDs with tunnel junctions is not intended to promote the quantum efficiency at high current density; instead, it finds a way to operate the LED devices at low current density that possesses relatively high IQE. Figure 16.9 shows the

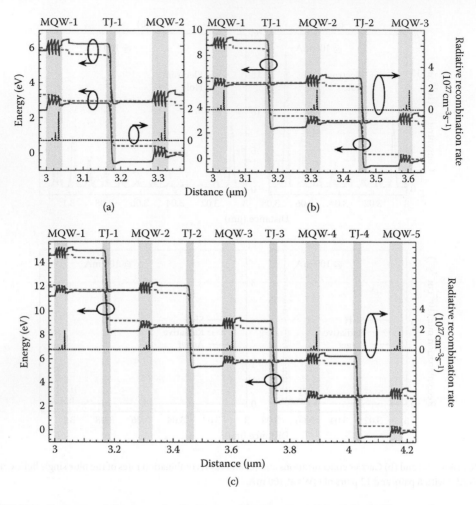

FIGURE 16.7 Energy band diagram and radiative recombination rate at 100 mA for the blue tunnel-junction light-emitting diodes (LEDs) with (a) two unit LEDs, (b) three unit LEDs, and (c) five unit LEDs.

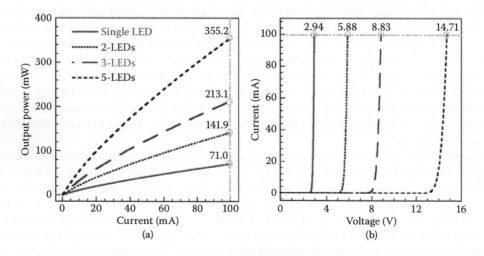

FIGURE 16.8 L–I and I–V characteristics of the blue single light-emitting diode (LED) and tunnel-junction LEDs.

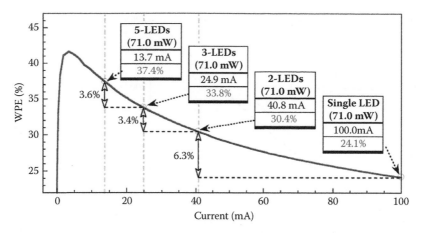

FIGURE 16.9 WPE of the blue light-emitting diode (LED) structures under study.

WPE of the blue LED structures under study. Only one curve is presented in this figure since all the WPE profiles are quite similar. For the single LED structure, the output power is 71 mW at 100 mA. For the tunnel-junction LEDs with two and three stacked unit LEDs, in order to obtain the same output power, the operation currents are 40.8 and 20.9 mA, and the WPE are 30.4% and 33.8%, respectively. As for the stacked structure of five unit LEDs, it can be operated at a relatively low current of 13.7 mA, which has a WPE of 37.4%. Consequently, if the tunnel-junction LED is stacked by more unit LEDs, the input power required to obtain a specific output power can be markedly reduced because of the relatively high WPE, as shown in Figure 16.10.

16.4.3 Green Tunnel-Junction LED

The concept of stacking identical LED structures with tunnel junctions is also investigated for green InGaN LEDs in this chapter to probe its effectiveness. The green single LED structure under study is identical to the blue LED except that the indium composition of InGaN wells is increased to 29%. Moreover, considering the enlarged lattice mismatch between the wells and barriers and the limited solubility of indium in GaN, the SRH recombination lifetime is reduced to 20 ns in the simulation. Figure 16.11 shows the energy band diagram, carrier concentrations, and recombination rates at 100 mA and WPE of the green single LED. The electrical and optical characteristics of the green single LED are quite similar to those of the blue single LED shown in Figure 16.5. However, since the indium composition of the InGaN wells in the green single LED is higher than that in the blue single LED, deeper QWs are observed in the band profile. Due to the deeper wells, the feature of nonuniform carrier distribution becomes much more severe, in which only the QW closest to the p-side contributes to interband transitions, as shown in Figure 16.11b and c. Under these circumstances, the droop of WPE in the green LED is more severe than that in the blue LED (48% versus 42%).

Since the holes almost accumulate in the last QW closest to the p-side, similar WPE is obtained no matter how many QWs are utilized in the green single LED under study, as shown in Figure 16.12a. Moreover, it was reported that the defect-related SRH recombination is an important interband transition loss that slashes the low current efficiency (Schubert et al., 2007). The WPEs of the green single LED structures with various SRH lifetimes are plotted in Figure 16.12b. The peak efficiency increases with the increase of SRH lifetime, which is similar to the results reported in the published literature (Schubert et al., 2007). The WPE droop deteriorates when the SRH lifetime increases, due mainly to the marked variation of peak efficiency. Since the high current efficiency remains almost unchanged, the high-power applications cannot benefit much from the improved film quality in the single LED. However, it is not the case for the tunnel-junction LED because the design concept for tunnel-junction LEDs is to operate the LEDs at low current

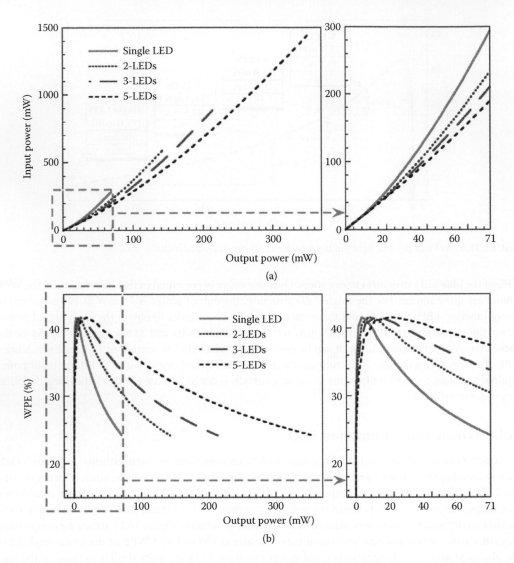

FIGURE 16.10 Required input power and corresponding wall-plug efficiency (WPE) of the blue single light-emitting diode (LED) and tunnel-junction LEDs as a function of the light output power.

density. Figure 16.13 shows the WPE of the green single LED and tunnel-junction LEDs when the SRH lifetimes are 20 and 50 ns. It is observed that the WPE of the tunnel-junction LED with 5 unit LEDs in the stacked structure is almost twice that of the single LED, especially when the SRH lifetime is high. The design strategy of tunnel-junction LED is thus applicable to the green LEDs. Since the efficiency of green LEDs is difficult to upgrade due to the nature of deeper InGaN wells, the idea of tunnel-junction LED may be of great benefit in improving the physical and optical performance.

16.5 Conclusion

In this chapter, several design approaches of the GaN-based polarization-assisted tunnel junctions are explored. With appropriate design via polarization engineering, low-resistivity and excellent tunneling characteristics can be achieved in the structures with practical indium composition and doping density. Note that the tunneling currents of the tunnel-junction structures obtained from experiments typically

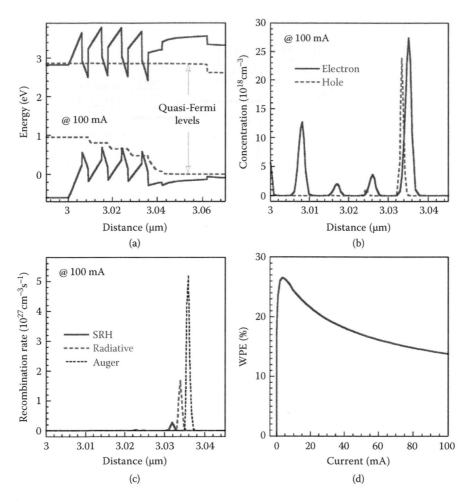

FIGURE 16.11 (a) Energy band diagram, (b) carrier concentrations, (c) recombination rates at 100 mA, and (d) wall-plug efficiency (WPE) of the green single light-emitting diode (LED).

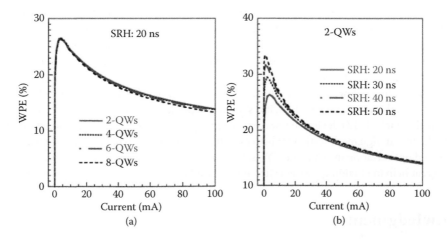

FIGURE 16.12 Wall-plug efficiency (WPE) of the green single light-emitting diode (LED) structures with (a) various pairs of QWs and (b) various Shockley–Read–Hall (SRH) lifetimes.

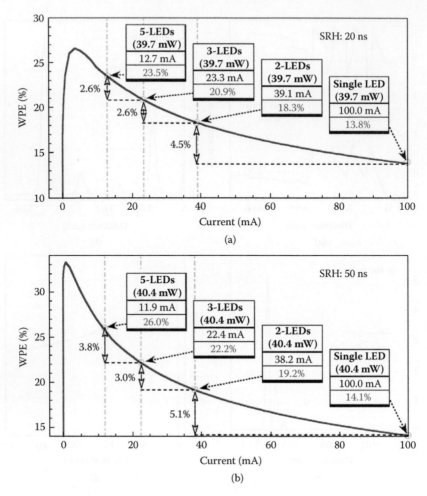

FIGURE 16.13 Wall-plug efficiency (WPE) of the green single light-emitting diode (LED) and tunnel-junction LEDs when the Shockley–Read–Hall (SRH) lifetimes are (a) 20 ns and (b) 50 ns.

exceed the values obtained from theoretical calculation. The additional tunneling current may be owing to the defect and impurity-related mid-gap electronic states (Schubert, 2010b), which are not considered in the present study. Hence, the utilization of tunnel junction in nitride-based applications is quite feasible, which enables the development of high-efficiency GaN-based devices, such as the multijunction solar cells, multijunction optoelectronic devices, and multijunction electronic devices. Specifically, in the present chapter, the application of low-resistivity tunnel junction in both the blue and green tunnel-junction LEDs is explored. Through the stacking of LEDs, it is shown that marked improvement in WPE can be achieved by operating the stacked LED devices at low current density. Furthermore, the tunnel-junction LED structure allows further design modification for better electrical and optical characteristics, such as the removal of the EBL and optimization in number of QWs. It is our belief that the method introduced in this chapter may be of great help in obtaining cost-effective and high-efficiency LED devices for solid-state lighting.

Acknowledgments

This work is supported by the Ministry of Science and Technology of Taiwan under grants NSC-99-2119-M-018-002-MY3 and NSC-102-2112-M-018-004-MY3.

References

Akyol F, Krishnamoorthy S, and Rajan S (2013) Tunneling-based carrier regeneration in cascaded GaN light emitting diodes to overcome efficiency droop. *Applied Physics Letters.* 103:081107-1–081107-4.

Cherns D, Henley SJ, and Ponce FA (2001) Edge and screw dislocations as nonradiative centers in InGaN/GaN quantum well luminescence. *Applied Physics Letters.* 78:2691–2693.

Chichibu SF et al. (2006) Origin of defect-insensitive emission probability in In-containing (Al,In,Ga)N alloy semiconductors. *Nature Materials.* 5:810–816.

Chuang SL and Chang CS (1996) k·p method for strained wurtzite semiconductors. *Physical Review B.* 54:2491–2504.

David A et al. (2008) Carrier distribution in (0001) InGaN/GaN multiple quantum well light-emitting diodes. *Applied Physics Letters.* 92: 053502-1–053502-3.

Delaney KT, Rinke P, and Van de Walle CG (2009) Auger recombination rates in nitrides from first principles. *Applied Physics Letters.* 94:191109-1–191109-3.

Duke CB (1969) *Tunneling in Solids.* New York, NY: Academic Press.

Fiorentini V, Bernardini F, and Ambacher O (2002) Evidence for nonlinear macroscopic polarization in III–V nitride alloy heterostructures. *Applied Physics Letters.* 80:1204–1206.

Gorczyca I, Suski T, Christensen NE, and Svane A (2011a) Band gap bowing in quaternary nitride semiconducting alloys. *Applied Physics Letters.* 98:241905-1–241905-3.

Gorczyca I, Suski T, Christensen NE, and Svane A (2011b) Size effects in band gap bowing in nitride semiconducting alloys. *Physical Review B.* 83:153301-1–153301-4.

Grundmann MJ and Mishra UK (2007) Multi-color light emitting diode using polarization-induced tunnel junctions. *Physica Status Solidi C.* 4:2830–2833.

Kane EO (1960) Zener tunneling in semiconductors. *Journal of Physics and Chemistry of Solids.* 12:181–188.

Khan N, Nepal N, Sedhain A, Lin JY, and Jiang HX (2007) Mg acceptor level in InN epilayers probed by photoluminescence. *Applied Physics Letters.* 91:012101-1–012101-3.

Kim MH et al. (2007) Origin of efficiency droop in GaN-based light-emitting diodes. *Applied Physics Letters.* 91:183507-1–183507-3.

Krishnamoorthy S, Akyol F, Park PS, and Rajan S (2013) Low resistance GaN/InGaN/GaN tunnel junctions. *Applied Physics Letters.* 102:113503-1–113503-5.

Krishnamoorthy S, Nath DN, Akyol F, Park PS, Esposto M, and Rajan S (2010) Polarization-engineered GaN/InGaN/GaN tunnel diodes. *Applied Physics Letters.* 97:203502-1–203502-3.

Krishnamoorthy S, Park PS, and Rajan S (2011) Demonstration of forward inter-band tunneling in GaN by polarization engineering. *Applied Physics Letters.* 99:233504-1–233504-3.

Kumakura K, Makimoto T, and Kobayashi N (2003) Mg-acceptor activation mechanism and transport characteristics in p-type InGaN grown by metal organic vapor phase epitaxy. *Journal of Applied Physics.* 93:3370–3375.

Kuo YK, Chang JY, Tsai MC, and Yen SH (2009) Advantages of blue InGaN multiple-quantum well light-emitting diodes with InGaN barriers. *Applied Physics Letters.* 95:011116-1–011116-3.

Laubsch A et al. (2009) On the origin of IQE-"droop" in InGaN LEDs. *Physica Status Solidi C.* 6:S913–S916.

Lin GB, Meyaard D, Cho J, Schubert EF, Shim H, and Sone C (2012) Analytic model for the efficiency droop in semiconductors with asymmetric carrier-transport properties based on drift-induced reduction of injection efficiency. *Applied Physics Letters.* 100:161106-1–161106-4.

Moll JL (1964) *Physics of Semiconductors.* New York, NY: McGraw-Hill.

Monemar B and Sernelius BE (2007) Defect related issues in the "current roll-off" in InGaN based light emitting diodes. *Applied Physics Letters.* 91:181103-1–181103-3.

Nam KB, Nakarmi ML, Li J, Lin JY, and Jiang HX (2003) Mg acceptor level in AlN probed by deep ultraviolet photoluminescence. *Applied Physics Letters.* 83:878–880.

Nepal N, Li J, Nakarmi ML, Lin JY, and Jiang HX (2005) Temperature and compositional dependence of the energy band gap of AlGaN alloys. *Applied Physics Letters.* 87:242104-1–242104-3.

Ozden I, Makarona E, Nurmikko AV, Takeuchi T, and Krames M (2001) A dual-wavelength indium gallium nitride quantum well light emitting diode. *Applied Physics Letters*. 79:2532–2534.

Piprek J, Li ZM, Farrell R, DenBaars SP, and Nakamura S (2007) Electronic properties of InGaN/GaN vertically-cavity lasers, In: *Nitride Semiconductor Devices: Principles and Simulation*, Piprek J (ed.), pp. 423–445. Weinheim: John Wiley & Sons.

Piprek J (2014) Origin of InGaN/GaN light-emitting diode efficiency improvements using tunnel-junction-cascaded active regions. *Applied Physics Letters*. 104:051118-1–051118-4.

Piprek J and Li SZM (2013) Origin of InGaN light-emitting diode efficiency improvements using chirped AlGaN multiquantum barriers. *Applied Physics Letters*. 102:023510-1–023510-4.

Rozhansky IV and Zakheim DA (2006) Analysis of the causes of the decrease in the electroluminescence efficiency of AlGaInN light-emitting-diode heterostructures at high pumping density. *Semiconductors*. 40:839–845.

Schubert MF (2010a) Interband tunnel junctions for wurtzite III-nitride semiconductors based on heterointerface polarization charges. *Physical Review B*. 81:035303-1–035303-8.

Schubert MF (2010b) Polarization-charge tunnel junctions for ultraviolet light emitters without p-type contact. *Applied Physics Letters*. 96:031102-1–031102-3.

Schubert MF et al. (2007) Effect of dislocation density on efficiency droop in GaInN/GaN light-emitting diodes. *Applied Physics Letters*. 91:231114-1–231114-3.

Shen YC et al (2007) Auger recombination in InGaN measured by photoluminescence. *Applied Physics Letters*. 91:141101-1–141101-3.

Shim JI et al. (2012) Efficiency droop in AlGaInP and GaInN light-emitting diodes. *Applied Physics Letters*. 100:111106-1–111106-4.

Simon J et al. (2009) Polarization-induced Zener tunnel junctions in wide-band-gap heterostructures. *Physical Review Letters*. 103:026801-1–026801-4.

Tiwari S (1992) *Compound Semiconductor Device Physics*. London: Academic Press.

Tsai MC, Leung B, Hsu TC, and Kuo YK (2013) Low resistivity GaN-based polarization-induced tunnel junctions. *Journal of Lightwave Technology*. 31:3575–3581.

Tsai MC, Leung B, Hsu TC, and Kuo YK (2014) Tandem structure for efficiency improvement in GaN based light-emitting diodes. *Journal of Lightwave Technology*. 32:1801–1806.

Vurgaftman I and Meyer JR (2003) Band parameters for nitrogen-containing semiconductors. *Journal of Applied Physics*. 94:3675–3696.

Vurgaftman I and Meyer JR (2007) Electron band structure parameters. In: *Nitride Semiconductor Devices: Principles and Simulation*, Piprek J (ed.), pp. 13–48. Weinheim: John Wiley & Sons.

Vurgaftman I, Meyer JR, and Ram-Mohan LR (2001) Band parameters for III–V compound semiconductors and their alloys. *Journal of Applied Physics*. 89:5815–5875.

Wu J (2009) When group-III nitrides go infrared: New properties and perspectives, *Journal of Applied Physics*. 106:011101-1–011101-27.

Wu J et al. (2003) Temperature dependence of the fundamental band gap of InN. *Journal of Applied Physics*. 94:4457–4460.

Yoshida H, Kuwabara M, Yamashita Y, Uchiyama K, and Kan H (2010) Radiative and nonradiative recombination in an ultraviolet GaN/AlGaN multiple-quantum-well laser diode. *Applied Physics Letters*. 96:211122-1–211122-3.

17

Quantum Disk Nanowire Light-Emitting Diodes

Fabio Sacconi

17.1 Introduction

Due to their properties, nanowire (NW) based light-emitting diode (LED) structures offer very interesting advantages, such as defect-free material, due to small footprint on the substrate, which allows higher LED efficiency, and three-dimensional (3D) geometry features, such as lateral strain relaxation, which can also be exploited to increase efficiency (Ristic et al., 2005). Group III nitride semiconductors have attracted much attention for quite a long time especially for their light-emitting device applications. A new approach for reaching exceptionally high efficiencies of LEDs for the whole visible spectrum is therefore based on nanostructured InGaN emitters, where nanorods, which have been shown to be defect-free, serve as active light-emitting structures. Light-emitting nanorod arrays based on InGaN as a material for the active region have been investigated (Kim et al., 2004); in principle they could cover the whole visible spectrum from blue to red and be incorporated into a single device for phosphor-free white light emission. By controlling indium content in the InGaN quantum disk (QD) embedded in the NW, nanorod emitters are expected to exhibit efficient green, yellow, and red emission, allowing coverage of the full visible spectrum with GaN-based LEDs. This could overcome a major drawback of today's red–green–blue (RGB) light sources, that is, the lack of an efficient green emitter due to the efficiency gap existing between the blue InGaN and the red InGaAlP materials system (Piprek, 2010).

In this chapter, we discuss several simulation approaches for the numerical study of the transport and optoelectronic properties of GaN NW diode structures with an embedded InGaN QD. Nanostructure-based emerging electronic and optoelectronic devices, such as LEDs or photodetectors, need a fully

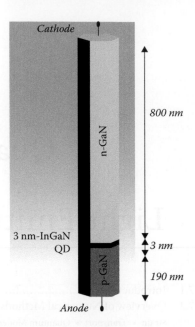

FIGURE 17.1　Nanowire (NW) quantum dot (QD) light-emitting diode (LED) model structure.

quantum mechanical or atomistic approaches to obtain a correct description of structural, electrical, and optical properties. Usually, the active region of a device, which needs such an advanced investigation, is small compared to the overall simulation domain. The computational cost of more accurate quantum mechanical models, however, makes their application to the whole device unfeasible. It is, therefore, necessary to implement a multiscale simulation approach, which couples the semiclassical models describing the bulk regions of the device to the quantum mechanical or atomistic models, acting only on the nanostructured regions of the device.

We see in the rest of the chapter how an integrated multiscale and multiphysics simulation environment (see Auf der Maur et al., 2011, 2013; Auf der Maur, 2015) may be capable of coupling different models on different scales, ranging from macroscopical to atomistic representations.

In the following, we first describe briefly the different physical models that need to be applied for an adequate description of these devices. Then we show the main results obtained from these numerical methods in the simulation of QD-based LEDs. If not stated otherwise, in the rest of the text we refer to the model NW LED structure with a single QD described in Figure 17.1. This model structure is a GaN NW usually around 1 μm high with a hexagonal base and a radius between 25 nm and several hundreds of nanometers. The embedded QD, which constitutes the LED active region, is a 3-nm-thick disk made of InGaN alloy, with In composition varying between 10% and 40%.

17.2　Overview of Numerical Methods

17.2.1　Strain

Strain has a critical influence on the behavior of heterostructures due to its effect on the band energies and the strain-induced piezoelectric polarization. The latter is particularly important in nitride-based devices. A straightforward approach for the calculation of strain in lattice mismatched heterostructures is the one based on the linear elasticity theory of solids. Usually, in this approach it is assumed that pseudomorphic interfaces are present between different materials (Povolotskyi and Di Carlo, 2006). One also assumes small deformations, such that the strain is a linear function of deformation and that Hooke's law, which linearly relates stress to strain, can be used. The strain and deformation field are found by minimizing the elastic energy of the system. As a result, one obtains the strain tensor in any point of the structure, the shape deformation, and the piezoelectric polarization, which to the first order depends linearly on strain. Self-consistent electromechanical simulations can be carried out by including the converse piezoelectric effect.

This continuum elastostatic model described here has several limitations.

In particular, it does not contain any information on internal strain, which can be very important, particularly in wurtzite crystals. A more accurate description of strain can be obtained by means of the valence force field (VFF) method, which offers the best compromise between accuracy and computational effort in the study of semiconductor nanostructures. One of the implementations is based on the Keating model (Keating, 1966), for diamond structures, as generalized by Camacho and Niquet (2010) to treat nonideal wurtzite structures. As VFF is computationally more intensive than the continuous media model, a possibly convenient implementation is the one that allows coupling of the two models in a multiscale simulation (see Auf der Maur et al., 2011, 2013).

17.2.2 Transport

Usually transport of electrons and holes in LED structures is treated in a semiclassical picture based on the drift–diffusion model (Sze, 1981). The particle fluxes are written in terms of the electrochemical potentials. The carrier statistics are given by Boltzmann or Fermi–Dirac statistics, assuming as usual local equilibrium. The conduction and valence band edges and effective masses may be obtained from bulk $\mathbf{k} \cdot \mathbf{p}$ calculations; in this way, the local corrections due to strain can be easily included.

To account for quantum effects, a correction to this purely classical approach is needed. Several methods can be applied, such as the density gradient quantum correction model (Ancona, 1990) up to the more computationally demanding nonequilibrium Green's function (NEGF) (Datta, 2000).

A self-consistent approach in many cases may be a good compromise. Some applications of self-consistent techniques will be described in the following.

17.2.3 Quantum Models

Quantum mechanical models are used for the calculation of eigenstates of confined particles in nano-structures, either based on the envelope function approximation (EFA) or on atomistic approaches. In the former, the Hamiltonian of the system is constructed in the framework of single-band and multiband $\mathbf{k} \cdot \mathbf{p}$ theory (Chuang, 1995; Chuang and Chang, 1996). The single particle wave functions are expanded in terms of bulk Bloch functions (usually taken at the zone center $k = 0$ for direct band gap materials) of the constituent materials, leading to a system of equations for the envelope functions. The atomistic approaches are typically based on the empirical tight binding (ETB) method (Di Carlo, 2003). In this method, the electronic states are written as a linear combination of atomic orbitals (LCAO). In this case, an atomistic structure describing the heterostructure studied has to be generated according to the macroscopic device description and crystallographic orientation. The solution of the eigenvalue problems resulting from the EFA and ETB models provides the energy spectrum, the particle densities, and the probabilities of optical transitions. The particle densities are calculated by populating the electron and hole states according to the expectation value of the corresponding electrochemical potential. The particle densities may then be fed back to the Poisson/drift–diffusion model for a self-consistent Schrödinger–Poisson/drift–diffusion calculation.

17.3 Strain Maps

It is interesting to study the behavior of the strain tensor in an GaN NW embedding an InGaN QD. As previously discussed, strain induced by the lattice mismatch in a heterostructure can be calculated, for example, by means of an elasticity model. In the case of the QD NW LED, strain obtained from this numerical model is due to the lattice mismatch between the material of the NW or nanocolumn and the material that constitutes the QD. It has been shown (Hugues et al., 2013) that a GaN NW grown bottom-up becomes strain free after around 100 nm. Thus, we can expect that a typical GaN NW, with a height of 1 μm or more, is completely relaxed beyond the region surrounding the QD. This justifies the choice of the NW material (GaN in this case) as the substrate reference material for the strain calculations. This means that the components of the strain tensor as resulting from the minimization of the elastic energy in the system composed by the whole heterostructure will tend to vanish in the substrate material.

If we look at the calculated in-plane strain tensor component ε_{xx} in an NW, we clearly see a strong compressive strain inside the InGaN QD (Sacconi et al., 2010, 2012). In fact, the InN lattice constant is higher than that of the reference GaN lattice material; thus, a negative in-plane strain is determined in the pseudomorphic InGaN/GaN heterostructure. Besides, due to 3D effects, a tensile strain is present in the GaN barrier, close to the interface with the quantum well, while it vanishes in the bulk of the GaN NW. A peculiar feature of the NW structure is the strong reduction of the in-plane strain close to the column surface. The strain relaxation due to the surface boundary is visible in Figure 17.2 where ε_{xx} is shown, along the z-axis (growth direction), for two different radial positions: in the center of the QD ($r = 0$) and close to the surface boundary ($r = 25$). Strain is clearly higher in the center of the column, while it tends to vanish at the surface boundary. Piezoelectric polarization is obtained from the strain field and exhibits a similar behavior: polarization is higher at the center of the QD and tends to decrease toward the lateral surface of the NW. Piezoelectric polarization (together with spontaneous or pyroelectric polarization) has a direct effect on the band profiles and therefore on the quantum and optical properties.

It is interesting to analyze the dependence of the ε_{xx} strain tensor component in the QD on the column size. It is found (Sacconi et al., 2012) that, by increasing the NW size, the (tensile) strain in the barrier regions outside the QD tends to vanish, while the compressive strain slightly increases inside the QD. Moreover, it can clearly be seen that the effect of strain relaxation is more evident for columns with a radius lower than 50 nm and it tends to vanish for very large columns.

The dependence on QD thickness of the ε_{xx} strain tensor component in the NW is shown in Figure 17.3, for an $In_{0.1}GaN$ QD in an NW with radius 25 nm. The value of ε_{xx} is shown along the z-axis, close to the surface boundary. It can be noted that by increasing QD thickness from 2 to 5 nm, the compressive strain slightly decreases inside the QD, while the tensile strain in the GaN surrounding the QD slightly increases. On the other hand, by comparing strain behavior in the center and close to the lateral surface (Figure 17.3)

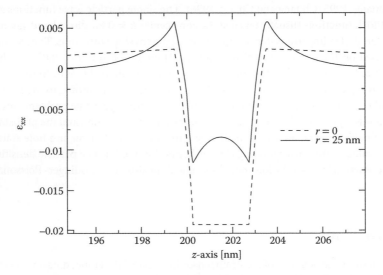

FIGURE 17.2 ε_{xx} strain tensor component along the z-axis, respectively, at the center of the column (dashed line) and close to the column surface (solid line), for an $In_{0.2}GaN$ quantum dot (QD) with $r = 25$ nm.

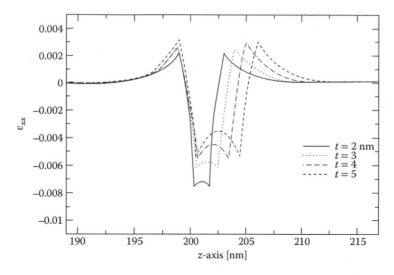

FIGURE 17.3 ε_{xx} strain tensor component along the z-axis, close to the lateral surface, for an $In_{0.1}GaN$ quantum dot (QD) thickness from 2 to 5 nm.

of the NW, the strain relaxation is clearly noticed, which is more evident for thicker QDs (50% reduction for 5-nm-thick QD). We see in the following how these peculiar strain features of the NW structure can be put in relation with the emission properties of the QD LED.

17.4 Transport Properties

Transport in NW QD LED may be studied in a first approach with a classical model. In this way, the IV characteristic of the forward-biased p-i-n LED diode is calculated by imposing ohmic boundary conditions to the contact regions at the two sides of the NW (anode and cathode) and by solving the Poisson/drift–diffusion model, taking care to include the effects of polarization (both spontaneous and piezoelectric), by means of an appropriate polarization vector P in the current equation. Figure 17.4 shows the conduction and valence band profiles and the electron and hole densities along the z-axis in the center of the NW, as obtained from these classical calculations. The dependence of the calculated IV characteristics on the In molar fraction in the InGaN QD is shown in Figure 17.5 for a column with radius $r = 25$ nm (Sacconi et al., 2010). As expected, the threshold voltage decreases from 3.2 to around 2.2 V with an increase of In concentration from $x = 0.1$ to $x = 0.4$, since the energy gap of the alloy material in the QD is decreasing. In addition, from the IV characteristics for a 20% In composition in the QD we obtain an increase of the output current by about two orders of magnitude for increasing values of the column size, from 25 to 200 nm. In the following, we see how a self-consistent calculation can be used to couple these classical results with a quantum model for the QD.

17.5 Quantum Calculations: EFA Models

For the calculations of the eigenvalues of a confined system, an approach based on EFA is usually a good trade-off between computational load and precision of the results. A solution based on the **k·p** model allows us to take into account the electronic structure details close to the valley minima, which is considered a reasonable approximation, provided that particle energies are not too high, as could be the case for hot electrons. The most complete model, the so-called 8 × 8 **k·p** model, takes into account the three valence bands: heavy-hole (HH), light-hole (LH), and crystal-field split-off (CH), together with the first conduction band (CB).

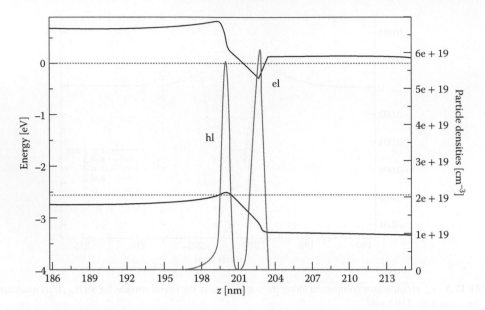

FIGURE 17.4   Band profiles and electron and hole densities from classical calculations.

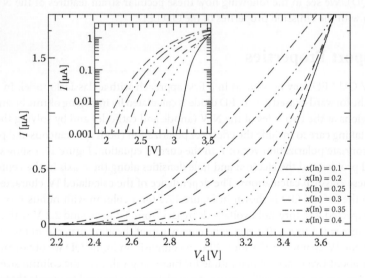

FIGURE 17.5   IV characteristics of the p-i-n diode with radius $r = 25$ nm for several values of In concentration in the InGaN quantum dot (QD).

However, when the energy gap of the material is wide enough, as is the case of GaN and its alloys with AlN and InN (for low InN contents), the interaction between conduction and valence bands is so weak that it can be neglected. Thus, one can apply separately a single band (effective mass) model for CB and a 6×6 $\mathbf{k}\cdot\mathbf{p}$ model for the three VBs, thus reducing the computational time needed without significant loss of accuracy in calculations.

To perform quantum calculations with the EFA model for an LED device, usually the drift–diffusion model is first solved, in order to apply a bias ramp to the diode until the desired operation conditions are obtained. Then a Schrödinger solver is applied for the solution of the eigenvalue problem, restricted

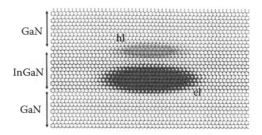

FIGURE 17.6 Conduction and valence band tight-binding states in a 14-nm-wide $In_{0.1}GaN$ quantum dot (QD).

to the active region of the device, comprising the QD and part of the GaN barrier regions. From the calculated conduction and valence ground states, the transition energies between the first electron and hole states can be obtained, for several values of In content (Sacconi et al., 2012). It is found that the transition energy decreases with higher In content mainly due to a lower energy gap in the QD. The transition energy dependence on geometrical and material parameters will be analyzed in more detail later in this chapter.

If we express the conduction and valence band ground states by means of the isosurface of probability density for each quantum state, it is evident that in both cases the quantum states are confined in the QD region (see also Figure 17.6). However, a spatial separation is clearly present between the electron and hole states due to the polarization fields that cause the so-called quantum confined stark effect (QCSE), contrary to what usually happens with GaAs-based structures.

In the following, we see how these results can be corrected by a self-consistent solution of Schrödinger and Poisson/drift–diffusion equations.

17.6 Quantum Calculations: Empirical Tight-Binding Approach

As we have seen before, a more accurate approach for the study of quantum properties of nanostructures is based on atomistic methods. It has been shown recently that this approach is able to provide a description of a realistic InGaN alloy features, contributing to an explanation of the decreasing emission efficiency on nitride-based quantum well (QW) LEDs (green droop) (Auf der Maur et al., 2016). We therefore present the results obtained when ETB calculations are performed to find confined electronic states in an InGaN/GaN QD NW.

In a first example, the simulation model is applied to an GaN NW with an $In_{0.1}GaN$ QD about 14 nm wide and with a thickness of 3 nm. First, an atomistic structure comprising around 77,000 atoms is generated, based on the finite element (FE) grid used for the discretization of continuous partial differential equation (PDE) models (Auf der Maur et al., 2011, 2013). This association with FE allows projection to the atomic positions the displacements obtained from strain calculation in the heterostructure with the PDE-based elasticity model.

Then eigenstates and eigenfunctions of the system are calculated on the strained atomistic system with an ETB model based on an $sp^3d^5s^*$ parameterization (Jancu et al., 2002).

The calculation is restricted to the atomistic structure, which describes the materials included in the LED active region consisting of the InGaN QD (see Figure 17.6) and an appropriate portion of the GaN NW surrounding the QD. As in the EFA **k·p** case, here a Poisson/drift–diffusion calculation also is performed earlier on the NW model (that is its FE grid). Then the potential profile obtained at a given bias, which includes polarization fields, is projected on the atomistic representation to be used for ETB calculations. This is generally accomplished with a shifting of the on-site energies of the ETB Hamiltonian matrix (Auf der Maur et al., 2011, 2013).

In Figure 17.6, we show the first hole and electron quantum states, calculated at an applied bias of $V_d = 3.5$ V. As in the previous EFA results, it is clearly seen that the polarization-induced electric field leads to a spatial separation of electron and hole states.

We see in the following how a multiscale simulation can be performed where both the atomistic (ETB) and the continuous (EFA) models are applied together for a better description of the QD electronic structure.

As a second example, we see an ETB calculation based on a random alloy approach.

Given a ternary alloy in the form $A_x B_{1-x} C$, such as $In_x Ga_{1-x} N$, there are two ways to model it. The first is virtual crystal approximation (VCA), where we consider the alloy as a fictitious material whose properties are a weighted average of the properties of the binary components AC and BC (e.g. InN and GaN), in function of the molar fraction x, according to Vegard's law. The second is the random alloy approach, which consists of studying an ensemble of stochastic realizations of structures where we substitute the anion according to a probability given by the concentration of anion A in the alloy.

Let us consider then a QD NW with a diameter of 10.7 nm with a 2-nm-thick QD. The total number of atoms in the atomistic structure is around 51,000.

A set of atomistic structures of the GaN/InGaN/GaN QD region, including the InGaN QD and two 2-nm-thick GaN barrier regions, are created on the GaN substrate lattice. Random alloy is generated in the InGaN region by including In atoms with a probability given by the chosen In concentration. Then, as an initial guess, macroscopic strain is calculated and the displacements obtained are projected on the atom positions. VFF relaxation is applied on this structure until the atoms' equilibrium positions are found. Finally, ETB calculation is performed on the relaxed structure, taking into account the appropriate scaling of ETB parameters according to the bond length obtained by the relaxation procedure.

Electron and hole ground states have been calculated. Figure 17.7 shows the transition energy E_t for increasing In concentration in the InGaN QD. The average value of E_t for the set of calculations is plotted, together with the range between the minimum and maximum values. Random alloy results are compared with those obtained with ETB for the same model structure, but with a VCA approach and just a macroscopic strain. It can be seen that there is a significant discrepancy between the two models. In fact, they differ by about 50 meV for $x(\text{In}) = 0.1$ and the difference gets larger with increasing In content. These results show that VCA may not be adequate to describe nanostructures based on InGaN alloys due to the

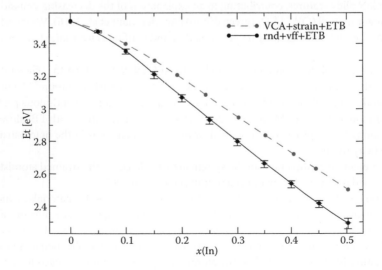

FIGURE 17.7 Dependence of transition energy on In concentration in the quantum dot (QD): Virtual crystal approximation and random alloy model.

great difference in the lattice properties of its component. The random alloy approach, with an atomistic treatment of strain with VFF, may be required to study these devices.

17.7 Self-Consistent Calculations

As discussed previously, a correction to a purely classical description of transport in the NW QD LEDs is needed to account for quantum effects. In this section, we describe a convenient procedure to take into account quantum charge densities of electrons and holes, obtained from an EFA model as described earlier, in the transport calculations of an NW QD LED. The solution to this issue described here consists of a self-consistent coupling of the Schrödinger equation in the EFA framework and the Poisson/drift–diffusion equations. This approach presents the advantage of a lower computational load with respect to more accurate methods for the calculation of quantum transport (e.g., NEGF; see Datta, 2000) while providing all the information about scattering features included in the drift–diffusion model. Thus, even if several approximations must be adopted, such as the assumption of quasiequilibrium of Fermi levels in the quantum regions, the results obtained may be considered reasonably accurate and have been validated by several experimental benchmarks. For more details on the method applied, see, e.g., Auf der Maur et al. (2011).

Let us see now an example of the application of a self-consistent model to an NW QD LED device. It is convenient to perform first purely classical drift–diffusion calculations to ramp the diode voltage up to a desired operating point (e.g., just above its threshold voltage V_{th}). In this way, the LED is biased at a point where the first quantized levels in the QD are populated; this guarantees that the calculated particle quantum density resulting from quantum EFA calculations is not vanishing and thus can be considered comparable with the classical density of the particle resulting from the Fermi level position in the QD. Then, the drift–diffusion model is coupled with a Schrödinger solver for the self-consistent solution of the eigenvalue problem restricted to the active region of the device. The quantum mechanical electron and hole densities are fed back into the Poisson/drift–diffusion equations for a self-consistent Schrödinger–Poisson/drift–diffusion calculation. A self-consistent loop may be implemented by using a simple predictor–corrector scheme that assumes that the quantum density n_q varies with the potentials as the classical density n_{cl} (for more details see Sacconi et al., 2012). As the electron and hole states in the system are calculated applying closed boundary conditions, the quantum densities near the interface between quantum mechanical and classical simulation domains suffer from artificial behavior. To obtain a continuous transition from the purely classical densities far away from the QD to the quantum mechanical densities, one may define an embracing region with an extension of a few nanometers where a linear mixing of classical and quantum density can be applied (Sacconi et al., 2012). Figure 17.8 shows the conduction and valence band profiles along the z-axis in the center of the NW, together with the particle densities obtained after the self-consistent cycle.

By comparing the self-consistent electron and hole charge densities of Figure 17.8 with those of Figure 17.4, obtained with purely classical calculations, the effect of quantum correction is evident. The charge density distribution is no longer peaked at the heterointerface between the QD and the GaN barrier material, as determined by the band bending in classical calculations. Instead, it is moved toward the center of the QD, accordingly with the behavior of the wave functions of the particle confined states.

The distribution of the electron quantum densities on a xy-plane orthogonal to the growth direction is shown in Figure 17.9. The planar quantum confinement toward the center of the QD is clearly visible for the electrons and it is found even more pronounced for holes, due to their larger effective mass. This effect is even more evident when we compare the classical and self-consistent electron density in the QD region along the y-axis on a slice on a xy-plane orthogonal to the growth direction (Sacconi et al., 2012). As for the classical results, for both conduction and valence band a band bending forms close to the lateral surface due to the strain relaxation. A lower strain determines a lower band-gap at the surface than in the center of the NW. This in turn determines a higher value of classical electron and hole densities close to the surface.

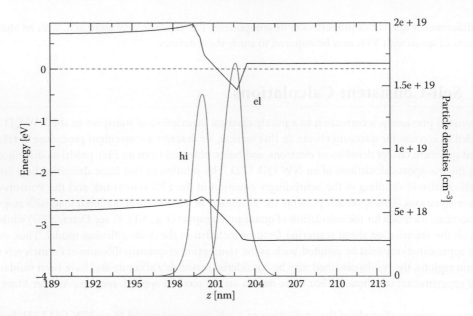

FIGURE 17.8 Self-consistent band profiles and densities along growth direction z in the quantum dot (QD).

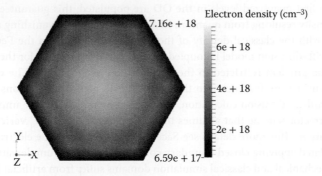

FIGURE 17.9 Electron quantum densities in the xy-plane in the middle of the quantum dot (QD).

On the other hand, the self-consistent densities show the effect of a lateral quantum confinement, so that the carriers arc mainly confined to the center region of the QD, even if for electrons this behavior is slightly more complicated.

Due to the behavior of both hole and electron density, we can expect a very different qualitative result when comparing the radiative recombination rate $R = Bnp$ in the active region, that is, the QD, calculated with classical and quantum model. In fact, it is found that from the classical results the emission appears to originate from a region very close to the surface of the column; on the other hand, when quantum mechanical particle densities are used the radiative recombination is mainly concentrated at the center of the QD due to the spatial confinement of the carriers. In Figures 17.10 and 17.11, we show a comparison of the calculated current density through the NW QD, in both cases. Current flow lines in the self-consistent case (Figure 17.11) indicate that the confinement of the quantum states leads to a current crowding at the center of the NW. In the purely classical calculation (Figure 17.10), on the contrary, flow lines focus toward the lateral surfaces, where the classical densities are higher. This result can explain the difference in the IV characteristics obtained with the two approaches (Sacconi et al., 2012). In fact, it turns out that the self-consistent results show a lower threshold voltage and an integrated current value at the contacts of around

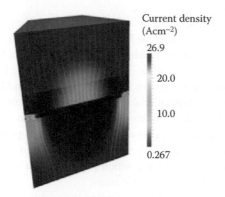

FIGURE 17.10   Classical calculations: Current flow is focused to quantum dot (QD) edges.

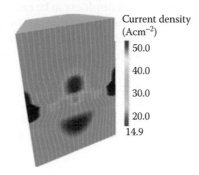

FIGURE 17.11   Self-consistent calculations: Current flow is focused to quantum dot (QD) center.

10 times higher with respect to the classical outcome. This discrepancy can be attributed to the increased recombination rate in the QD active region. In fact, on the one hand, the classical particle densities show higher peak values, but on the other hand, the average value of densities and thus of recombination is higher after quantum correction, which leads to a higher level of output current.

17.7.1 Multiscale Simulation

We now consider an example of a multiscale simulation applied to a NW QD LED (Auf der Maur et al., 2011). Here, a self-consistent classical/quantum calculation is again performed, but this time the information on quantum density of electrons and holes is obtained by an atomistic model. In this case, the device model is a p-i-n LED made of an AlGaN NW with an embedded GaN QD. The simulation is performed as follows.

First, the strain in the NW is calculated based on continuum elasticity theory. This result is then used as a guess for an atomistic VFF method applied for the relaxation of the atomistic structure generated according to the FEM model of the GaN device active region.

Then the electronic transport in the NW LED is simulated based on the self-consistent solution of the Schrödinger/Poisson/drift–diffusion equations using both the continuous media EFA model and the atomistic ETB approach.

To perform the transport calculation, first the drift–diffusion model is applied to get a bias where the diode is nearly in flat band condition. At this bias point the actual multiscale simulation is performed by concurrently solving the Poisson/drift–diffusion and the quantum mechanical models. For the latter both a $\mathbf{k} \cdot \mathbf{p}$ 6×6 model and an ETB model (using an $sp^3 d^5 s^*$ parameterization) are used. Since the hole states result is very dense for this device structure, a large number of states are needed to obtain a reasonably converged quantum hole density. Moreover, the ETB calculation is computationally much more intensive than an EFA calculation. Therefore, one possible choice is to calculate only the first three electron states using ETB and the higher ones with EFA, while EFA alone is used for the hole states.

This approach is a good compromise between the accuracy of full-band ETB calculations of the particle quantum states and an affordable computational load. In fact, it turns out that the hole state energies calculated by EFA and ETB are very similar and thus we can rely on the less computationally demanding EFA model for them. On the other hand, the higher electron states are less populated and therefore they have less impact on the properties of the system. We limit then our ETB approach to the first more populated electron states. This can be considered a decomposition of the problem in energy space since EFA and ETB are used to calculate energetically different particle states.

The coupling between the different models constituting the multiscale system is obtained by a concurrent solution implemented by applying a self-consistent cycle, similar to that shown earlier.

This coupling can be viewed as a quantum correction of the local density of states (LDOS) in the classical expression for the particle density, where the local densities can be easily calculated based on the envelope functions. In addition to the purely continuous case, here we have also the particle density obtained by ETB, which needs to be projected onto the FE mesh that has been used for the discretization of the continuous media models. This projection may be obtained by using an exponentially decaying function centered on each atomic site.

The results of these self-consistent calculations show also in this case that the confinement of the quantum states leads to a current crowding toward the center of the NW (Auf der Maur et al., 2011). We can conclude that a multiscale approach is in principle able to describe with higher accuracy the confinement effects that are critical for the NW LED devices while keeping the simulation domain for the quantum mechanical models reasonably small.

17.8 Modeling of Surface States

Surface states are expected to be present on the surface of GaN NWs due to defects and dangling bonds (Van de Walle and Segev, 2007).

The effect of surface states can be taken into account in a numerical simulation through the implementation in the drift–diffusion model of an acceptor trap model so that the trap is negatively charged if not occupied by a hole. Following the results obtained in Calarco et al. (2005) we can place the trap at an energy value of 0.6 eV below the conduction band, assuming a trap charge density of 1×10^{12} cm^{-2}. If we apply this model to a GaN NW we obtain a conduction band profile that shows a pinning of the Fermi level at the surface at around 0.6 eV below the conduction band. As is shown in Calarco et al. (2005), when a doping is present the pinning of the Fermi level determines a depletion region that, in this case, covers the whole column size. In this way, we have fitted the surface trap model to the experimental value of the Fermi level pinning in a GaN NW. In the following, we assume that, as a first approximation, the same model holds for the InGaN/GaN QD NW structure.

To investigate the influence of surface states on the electronic properties of the QD NW LED, a Shockley–Read–Hall (SRH) surface recombination model has to be associated to the traps. In this way, an SRH model for nonradiative recombination takes into account the surface traps, with a given density and recombination time, and their contribution to the recombination part of the current (Sacconi et al., 2012). Now, if one performs classical drift–diffusion calculations, taking into account the SRH surface recombination component, results (Sacconi et al., 2012) show that the presence of the surface states induces a large recombination current in the diode, which provides an output current around 60 times larger than in the case where surface states are neglected.

As is reported in Sacconi et al. (2012), the flow lines focus close to the surface, where there is the peak of the classical particle densities, which, thus, largely increases the surface recombination contribution.

On the other hand, when self-consistent quantum densities are calculated, this effect almost vanishes, and the self-consistent current at the contacts is just 20% higher than in the case without surface recombination. In fact, it is found that the self-consistent current flow lines tend again to crowd inside the QD since contribution from surface states is now limited by the quantum confinement effect.

17.9 Sensitivity to Geometrical and Material Parameters

We discuss in the following the dependence of the transition energies and thus of the optical emission from the QD on the NW geometrical and material parameters.

The optical emission spectra from spontaneous recombination may be calculated in the following way (Chuang, 2009):

$$P(\hbar\omega) = \sum_{i,j}^{\infty} n \frac{1}{2\pi^2} \frac{\omega_{ij}^2 e^2}{m^2 c^3} \left| M_{i,j} e \right|^2 f_i(E_i) \left[1 - f_j(E_j) \right] \frac{\Gamma/2}{\left(\hbar\omega_{ij} - \hbar\omega \right)^2 + (\Gamma/2)^2} d\Omega,$$

where f_i and f_j are the Fermi distributions and $M_{i,j}$ is the optical matrix element between the states i and j. The matrix elements may be obtained with different quantum models, such as ETB and EFA **k·p**, and provide the probability of optical transitions between each couple of states. The sum is extended to all the quantum states calculated for conduction and valence band. In this way, the optical transition probability is weighted with the occupation functions for each couple of particles, calculated according to the Fermi distribution at each state energy. The result is the total emitted power for transitions in a given energy range, given the particle population of the considered quantum states.

Usually, we are interested in the emission resulting from the spontaneous recombination of electrons in conduction band and holes in valence band. From the simulation point of view, this means that the device conduction band has to be populated with an optical or electrical pumping. In the following, the latter method has been performed, that is, a ramp bias has been applied until the LED is in conduction regime and the conduction and valence band are populated. At this point, a quantum model, such as **k·p**, is applied for the calculation of quantum states. Following that, the optical emission spectra are calculated based on the previous steps.

Figure 17.12 shows the optical spectra obtained for a GaN column with a radius of 25 nm and a 3-nm-thick InGaN QD (Sacconi et al., 2010). The results show a decrease of the peak emission energy with the increase of In concentration in the InGaN QD from 10% to 40%, indicating that an emission wavelength range between 400 and 680 nm can be covered in principle by these devices.

Generally speaking, in NW LEDs, similarly to QWs LED structures, emission energy decreases with higher In content due to a lower energy gap in the QD and a higher QCSE. The variation of the QCSE is caused by the increase of polarization fields with lattice mismatch between InGaN and GaN (piezoelectric polarization) and with In content (pyroelectric polarization). Moreover, it can be seen (Sacconi et al., 2012) that the emission power tends to decrease with higher In concentration due to the increase of spatial separation of the carriers caused by higher strain-induced QCSE. On the other hand, the calculated optical spectra for a fixed In composition and variable column size for the LED considered earlier show an increase in the value of peak emission energy of around 60 meV when the NW width is reduced from 400 to 50 nm. This result is a combination of several effects related to strain relaxation close to the NW boundaries and quantum confinement in the QD; we discuss this point in detail in the following paragraph.

As for the effect of InGaN QD thickness on the optical emission spectra, the behavior is again similar to what occurs in a QW LED. For example, the emission peak in a 50-nm-wide QD increases by 150 meV when the thickness of the QD decreases from 5 to 2 nm. This is mainly due to the higher quantum confinement in the thinner QD; moreover, for the same polarization field, the QCSE is lower for a smaller

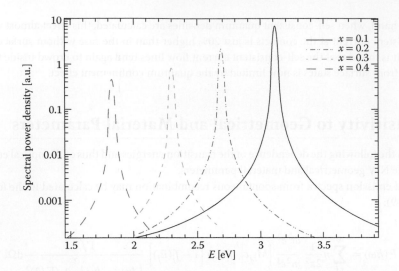

FIGURE 17.12   Optical emission spectra for a column with a radius of 25 nm and several In concentrations.

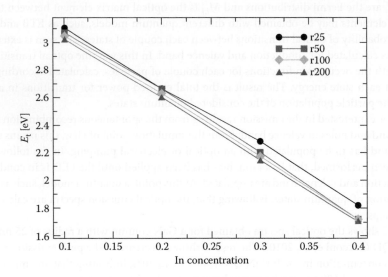

FIGURE 17.13   Dependence of emission energy on In concentration for several column widths.

thickness, and thus QCSE-induced red-shift is reduced. This also determines an increased emission power due to the reduced spatial separation of electron and hole quantum states.

Let us see now in more detail how the transition energies depend on the geometry parameters of the QD NW, in particular, how the lateral scaling of the NW affects the emission properties.

Figure 17.13 shows the dependence of emission energy on In concentration in an InGaN QD NW for several values of the column radius (Sacconi et al., 2012). It can be seen that, for all the column widths, the emission energy has a linear dependence on the In molar fraction. On the other hand, Figure 17.13 shows that the transition energy increases slightly with decreasing width, with a larger slope for radii lower than 50 nm. The effect gets larger with increased In concentration.

Calculations performed in a study by Sacconi et al. (2012) by applying a full self-consistent quantum/drift–diffusion coupling shows an emission energy increase of around 64 meV, from 2.440 to 2.504 eV, when the column radius decreases from 100 to 25 nm. These results show a trend in qualitative agreement with experiments (Ramesh et al., 2010), even if it must be noted that a comparison of calculations with experimental results is quite difficult because the results depend critically on the geometrical and structure details, as well as on the particle density distribution in the NW.

To point out the role of confinement and of strain distribution on this lateral size effect, the dependence of emission energy on NW geometry has been studied in Sacconi et al. (2012) for several structures, beginning with simulations for undoped structures at equilibrium. In Figure 17.14, the result of a 1D calculation for a QW structure, which can be assumed as a model of a NW with an infinite radius, is compared to the result for a 50-nm-wide NW. It is clear that the geometry of the NW increases the transition energy for a given value of In molar fraction. Moreover, calculations for a very wide column (10 μm) yield results very close to the thin film case. This result can be put in relation with the polarization-induced electric field obtained in the QD shown in Figure 17.15. It can be seen that the electric field gets lower in the NW QD

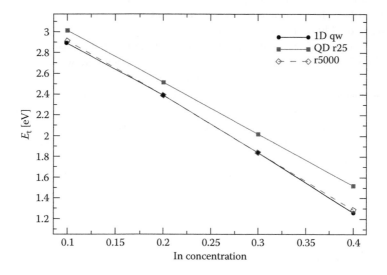

FIGURE 17.14 Dependence of emission energy on In content for undoped structures.

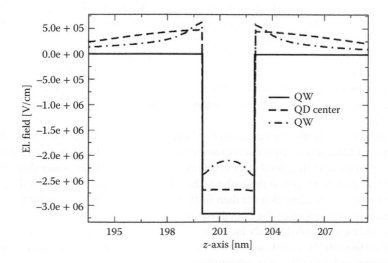

FIGURE 17.15 Electric field along the z-axis in an undoped nanocolumn.

with respect to the QW value due to the 3D strain effects, causing lower QCSE and therefore higher transition energy. Furthermore, the electric field in the QD is even more reduced close to the lateral surface due to the strain relaxation effect. The situation is anyway less clear in the doped structures, where screening effects become appreciable. Moreover, it is found that if polarization and strain are not taken into account, the lateral confinement has a significant effect only when the NW radius is reduced under 10 nm (Sacconi et al., 2012).

From these considerations, it seems reasonable to conclude that the main effect of the lateral scaling of the NW is the reduction of the strain inside the QD due to its relaxation at the NW lateral surfaces. In fact, the smaller the NW size, the more pronounced is this reduction of the strain. A lower strain implies less piezoelectric polarization and thus a lower value of QCSE, which causes a blue shift in emission energies. In addition, a smaller size induces a stronger quantum confinement, whose effect on transition energy would be, however, significant only at an extremely reduced size. Anyway, as we have seen earlier, an important role of quantum confinement is to keep particles far from the lateral surfaces, thus limiting the direct influence of NW surface effects, such as strain relaxation and surface states, on the behavior of current in the QD and thus on the LED emission properties.

We have seen the effects of scaling down the NW geometry; on the other hand, one can wonder what happens when the NW size is scaled up. It has been found experimentally (Kawakami et al., 2010) that for a 500-nm-wide NW, emission seems to occur at the strain relaxed region close to the lateral boundary. From simulations performed on an NW model with this size (Sacconi et al., 2012) it turns out that higher recombination rates are obtained for the regions close to the lateral surface, suggesting that for very wide NWs the confinement of quantum states on the NW lateral plane is not sufficient to focus the recombination at the center of the QD. Thus, emission takes place mainly in the lateral regions, where the strain relaxation induces a lower energy gap. However, it has been found from simulations that already for 200-nm-wide NWs the quantum confinement begins to affect the charge distribution, focusing the emission at the center of the QD.

17.10 Conclusions

After having discussed the several aspects of NW QD behavior, we can now underline the main beneficial features provided by NW geometry compared with planar LED structure.

First of all, the NW geometry can be grown almost without defects, yielding in principle high-emission efficiency. NWs also offer the possibility of covering a wide emission spectrum by means of band-gap engineering through variations of alloy concentration in the QD.

Another peculiar NW feature is the relaxation of strain induced in the QD, which occurs at the lateral surface of the NW. This is part of a 3D effect, which determines a lower strain in the QD active region, with respect to an analogous QW in a planar LED structure. This in turns causes less QCSE and higher overlap of quantum wave functions, increasing recombination rate, and internal quantum efficiency, with respect to the planar case.

A lower strain at lateral boundaries determines a lower energy gap with respect to the rest of the QD. This may affect the LED efficiency by reducing the recombination rate. However, as we have shown earlier, the quantum confinement on the lateral plane is sufficient, at least for NW size lower than 200 nm, to focus particle densities in a limited region at the center of the QD. The resulting current crowding in the QD and the increase of radiative recombination are then further advantages of NW geometry versus planar LED. It must be noted also that very often the embedded QD obtained by the NW growth is surrounded by the NW material (yielding a so-called InGaN inclusion in a GaN NW). In this case, the active region is even less affected by surface effects, such as surface states, while the confinement is increased and thereby also the current focusing and the radiative recombination.

For all of these features, one can expect NW QD structures to be very promising for the design of future high-efficiency LED devices.

References

M. G. Ancona, Density-gradient theory analysis of electron distributions in heterostructures, *Superlattices Microstruct.*, 7(2), 119–130 (1990).

M. Auf der Maur, Multiscale approaches for the simulation of InGaN/GaN LEDs, *J. Comput. Electron.*, 14(2), 398–408 (June 2015).

M. Auf der Maur et al., The multiscale paradigm in electronic device simulation, *IEEE Trans. Electron. Devices*, 58(5), 1425–1432 (May 2011).

M. Auf der Maur, A. Pecchia, G. Penazzi, F. Sacconi, and A. Di Carlo, Coupling atomistic and continuous media models for electronic device simulation, *J. Comput. Electron.*, 12(4), 553–562 (2013).

M. Auf der Maur, A. Pecchia, G. Penazzi, W. Rodrigues, and A. Di Carlo, Efficiency drop in green InGaN/GaN light emitting diodes: The role of random alloy fluctuations. *Phys. Rev. Lett.*, 116, 027401 (2016).

R. Calarco et al., Size-dependent photoconductivity in MBE-grown GaN-nanowires. *Nano Lett.*, 5, 981–984 (2005).

D. Camacho and Y. M. Niquet, Application of Keating's valence force field model to non-ideal wurtzite materials. *Phys. E*, 42, 1361–1364 (2010).

S. L. Chuang, *Physics of Optoelectronic Devices*, 1st ed. Wiley Series in Pure and Applied Optics. New York, NY: Wiley (1995).

S. L. Chuang, *Physics of Photonic Devices*, 2nd ed. New York, NY: Wiley (2009).

S. L. Chuang and C. Chang, k•p method for strained wurtzite semiconductors, *Phys. Rev. B*, 54, 2491–2504 (1996).

S. Datta, Nanoscale device simulation: The Green's Function Method, *Superlattices Microstruct.*, 28, 253–278 (2000).

A. Di Carlo, Microscopic theory of nanostructured semiconductor devices: Beyond the envelope-function approximation, *Semicond. Sci. Technol.*, 18, 1 (2003).

M. Hugues et al., Strain evolution in GaN nanowires: From free-surface objects to coalesced templates, *J. Appl. Phys.*, 114, 084307 (2013).

J.-M. Jancu et al., Transferable tight-binding parametrization for the group-III nitrides. *Appl. Phys. Lett.*, 81, 4838, (2002).

Y. Kawakami et al., Optical properties of InGaN/GaN nanopillars fabricated by postgrowth chemically assisted ion beam etching, *J. Appl. Phys.*, 107(2), 023522-1–023522-7 (January 2010).

P. N. Keating, Effect of invariance requirements on the elastic strain energy of crystals with application to the diamond structure. *Phys. Rev.*, 145(2), 26 (1966).

H. Kim et al., High-brightness light emitting diodes using dislocation-free indium gallium nitride/gallium nitride multiquantum-well nanorod arrays, *Nano Lett.*, 4(6), 1059 (2004).

J. Piprek, Efficiency drop in nitride-based light-emitting diodes, *Phys. Stat. Sol. A*, 207(10), 2217–2225 (October 2010).

M. Povolotskyi and A. Di Carlo, Elasticity theory of pseudomorphic heterostructures grown on substrates of arbitrary thickness, *J. Appl. Phys.*, 100, 063514 (2006).

V. Ramesh, A. Kikuchi, K. Kishino, M. Funato, and Y. Kawakami, Strain relaxation effect by nanotexturing InGaN/GaN multiple quantum well, *J. Appl. Phys.*, 107(11), 114303-1–114303-6 (June 2010).

J. Ristic et al., Carrier-confinement effects in nanocolumnar GaN/AlGaN quantum disks grown by molecular-beam epitaxy, *Phys. Rev. B*, 72, 085330 (2005).

F. Sacconi, M. Auf der Maur, and A. Di Carlo, Optoelectronic properties of nanocolumn InGaN/GaN LEDs, *IEEE Trans. Electron. Dev.*, 59(11), (November 2012).

F. Sacconi, G. Penazzi, A. Pecchia, M. Auf der Maur and A. Di Carlo, Optoelectronic and transport properties of nanocolumnar InGaN/GaN quantum disk LEDs, *Proc. SPIE 7597, Phys. Sim. Optoelectron. Dev.*, XVIII, 75970D (February 25, 2010); doi:10.1117/12.840533.

F. Sacconi, M. Auf der Maur, A. Pecchia, M. Lopez, and A. Di Carlo, Optoelectronic properties of nanocolumnar InGaN/GaN quantum disk LEDs, *Phys. Status Solidi C*, 9(5), 1315–1319 (2012); doi:10.1002/pssc.201100205.

S. M. Sze, *Physics of Semiconductor Devices*, 2nd ed. Hoboken, NJ: Wiley (1981).

C. Van de Walle and D. Segev, Microscopic origins of surface states on nitride surfaces, *J. Appl. Phys.*, 101, 081704 (2007).

18

Influence of Random InGaN Alloy Fluctuations on GaN-Based Light-Emitting Diodes

Chen-Kuo Wu

Tsung-Jui Yang

and

Yuh-Renn Wu

18.1 Introduction to Random Alloy Distribution

Unlike the GaAs-based light-emitting diodes (LEDs), the carrier transport and recombination mechanism in nitride-based LEDs are relatively more complicated. As we know, nitride-based wurtzite structures have a strong spontaneous polarization field inside the material. In addition, when the InGaN quantum well (QW) is grown on the GaN buffer layer, it also suffers extra strain, which induces the piezoelectric polarization. Therefore, the polarization field difference at the interface will induce the quantum confined stark effect (QCSE) and increase the radiative lifetime. The lack of native substrate also makes the crystal quality of the GaN and InGaN layer relatively poor, where a typical 10^8 cm^{-2} dislocation density can be observed in the InGaN QW, which is expected to affect device performance significantly. The efficiency droop, where the light output power increases sublinearly with the drive current, has also been an issue in limiting LED efficiency. There are many studies focused on determining the origin of droop, such as electron overflow [1–3], Auger recombination [4–9], or defects [10–14]. The defect or phonon-assisted Auger recombination is also proposed to explain the larger Auger coefficient. Recently, the direct measurement of Auger electrons [7,15] and a clear correlation between the Auger current and droop seem to indicate that the Auger process might be the main cause of droop. In addition, as mentioned, the QCSE will increase the radiative lifetime and the contribution of nonradiative Auger recombination or carrier delocalization toward droop will become more significant.

Despite these issues, the nitride-based blue LED still has very high internal quantum efficiency (IQE) even under this high dislocation density. The 80% peak IQE of blue LEDs still can be easily achieved in today's commercial LEDs. The reason for this high quantum efficiency has been attributed to the carrier localization effects, where carriers are localized at a local potential minimum so that it will not diffuse into the dislocation center for nonradiative recombination. The spectrum broadening effect is also much stronger compared to GaAs-based LEDs. This is attributed to the random distributed localized state. The origin of carrier localization has been an issue whether it is from indium clustering effect, random alloy fluctuation, or even charged dislocation line-induced potential barrier, etc. More and more evidence shows that random alloy fluctuation should be the main reason for this effect. First, this localization and spectrum broadening is still observed in GaN-negative substrate, which could exclude the effect of a charged dislocation-induced potential barrier. In addition, the three-dimensional (3D) atom probe tomography (APT) data [16–24] show that the indium composition distribution is naturally disordered. Figure 18.1a shows the randomly fluctuated indium distribution in the lateral direction from the APT data [17]. The in-plane map reveals that there are some high indium composition locations corresponding to the red regions and relatively low indium composition sites scattered in the 2D map. Figure 18.1b also indicates that the average indium composition along the growth direction is not uniform and decreases from the middle of the QW to the interface of the InGaN/GaN. Therefore, the QW structure possesses an indium distribution closer to a Gaussian shape rather than the ideal "top-hat" function. A similar idea of the influence of Gaussian shape QW on the IQE has been discussed by Hader et al. [6]. However, their analyses were one dimensional (1D) and lateral fluctuations in indium composition were not considered.

Many simulation tasks [22,24] suggested that carrier localization, induced by these fluctuations, has a strong influence on the broadening of the light emission spectrum. As discussed in Refs. [6,22,23], due to

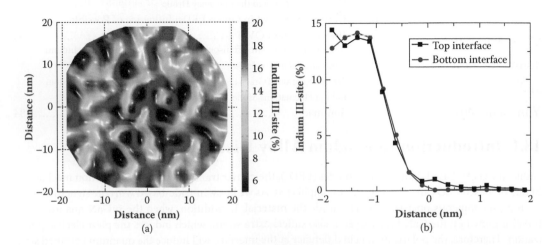

FIGURE 18.1 (a) The in-plane indium distribution of the quantum well (QW). (b) The average indium composition along the growth direction of one 3-nm QW. (From D. A. Browne et al., *Journal of Applied Physics*, 117, 185703, 2015.)

the indium fluctuation, the electrons and holes injected into InGaN QWs will localize in the high indium regions because of the deep localized potential. Additionally, the localized energy states originating from quantum confinement will vary due to fluctuations in indium composition and the QW width, thus broadening the emission spectrum [22,24]. In addition, the carrier localization will increase the local carrier density and the strong screening of the polarization field might be achieved at a smaller current density. The high local carrier density might increase the influence on Auger recombination or defect-assisted recombination [11] at a relatively smaller current density. In addition, the local polarization value induced by composition fluctuation will also influence polarization potential in the QW [25].

Besides these tasks that study the influence on radiative and nonradiative recombination due to the localization effect, there are a few other efforts that analyze the impact on carrier transport by considering random alloy potential fluctuation. In the past, many simulation studies on multiple QW LEDs have been performed under the assumption of a uniform composition and thus potential in the QWs. However, the simulations reveal a large deviation from the experimental data, especially in the InGaN MQWs cases. Due to the strong GaN barrier induced by the piezoelectric charges, a much higher applied bias is usually obtained in the simulation than in the experimental data. The deviation of simulation to experimental results becomes very significant, especially in MQW cases where many triangular shapes of GaN barriers exist in the system. Some studies reduced the theoretical polarization value to fit the experimental data. However, these assumptions might lead to more problems since they might improve electron–hole overlap and result in no droop or droop at a much larger bias, or need a much larger Auger coefficient to reproduce the experimentally observed droop behavior. Some used the tunneling effect to explain this phenomenon. However, the tunneling effect is significant only when a smaller polarization field was employed since the potential barrier induced by the 100% theoretical polarization filed could be very high. If the simulation only tries to fit the current versus IQE and neglects the fitting of I–V curve, usually a higher voltage is obtained in simulation and a conclusion of overflow or overshot effects are often obtained. This is due to that the extra-voltage lowered down polarization barrier induced by polarization field and potential barrier in electron blocking layer and make carrier to overflow much easier. However, this extra voltage is not observed in experimental result in a good commercial LED. Therefore, the conclusion would be wrong since it is based on a non-existence factor.

As mentioned earlier, the influence of random alloy fluctuation on the transport is seldom considered due to the need of large amounts of computational power. Atomic simulation of carrier transport with the 3D random indium fluctuation, especially in MQWs at current computation power, is also impossible. Even modeling a small area of InGaN QW (10 nm × 10 nm × 3 nm) will require a few months calculation with a super computer. Therefore, some quasi-classical simulation work is needed to approach these issues. Our past work [23,24,26] indicated that the results obtained by including the indium fluctuation into the 2D and 3D Poisson, drift–diffusion, and Schrödinger equation simulation model could be closer to experimental data, without the necessity for the assumptions mentioned earlier. The randomly fluctuated alloy composition of ternary epilayers will result in irregular energy bandgap and piezoelectric potential distribution, which will make carrier percolation and confinement more complicated compared to the traditional 1D assumption.

Figure 18.2 shows a sketch of a carrier percolation in the random alloy system. The red regions refer to the high potential region and the blue regions refer to the low potential area. As the sketch shows, the potential distribution is fluctuated either in the QW region or in the electron-blocking layer (EBL) region, where carriers are much easier to percolate through smaller barrier sites and localize in the local low-energy areas. While the large number of localized carriers accumulate in QWs, the local polarization charges will be screened and the effect of Auger recombination will be enhanced at the same time. In addition, the inherent complex potential distribution might provide carriers a path to avoid flow into defect-related regions formed by threading dislocations (TDs) in GaN LEDs. Our past research [23,24] has shown that the calculated turn-on voltages and IQE performances can be better predicted without any parameter reductions by considering random alloy fluctuations. A more reasonable Auger coefficient is used in the model [15], and a broad emission spectrum is also observed due to different localized states produced by alloy fluctuations. Moreover, the emission spectrum shift can be modeled well in the paper [24]. In this chapter, we will introduce how we use the traditional 3D Poisson and drift–diffusion solver to address these issues.

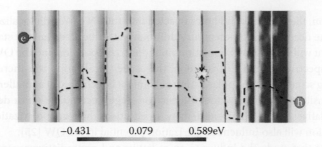

−0.431   0.079   0.589eV

FIGURE 18.2 The sketch illustrates the carrier percolation in the random alloy system. The red regions refer to the high potential region, and the blue regions are low potential area.

18.2 Methodology in Random Alloy Modeling

To model the 3D structure properly, we need to use a full 3D model to examine the device performances of the different structures. The 3D finite element method (FEM) Poisson, drift–diffusion, and Schrödinger solver developed by our lab (named the 3D-DDCC) is mainly used for solving the 3D carrier transport issue of the semiconductor. This program can correlate the electrical characteristic with optical problems and give a self-consistent solution. Our lab has tested and verified the model [24,26–28] and the model has been further developed to apply in the 3D transport simulation. In addition, 3D drift-diffusion charge control (3D-DDCC) program is a very versatile software, in that we can easily use inserted functions in special cases to model the structures or the physics more precisely. In this chapter, the externally inserted functions have the decisive position in the simulation. The details of the functions will be introduced later.

Figure 18.3 shows the full diagram of the simulation flow chart in this modeling process. First, the Gmsh program is used to construct the 3D mesh structure [29]. Afterwards, we need to use the random number

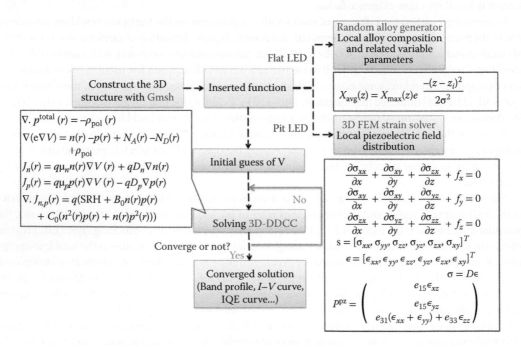

FIGURE 18.3 The full diagram of the simulation flow chart in this modeling process.

generator to generate the random alloy composition which is utilized in the examination of carrier transport in the random alloy system. After obtaining the indium map, we assign the indium composition in each node of the mesh element and all physical parameters change with the indium composition. To obtain a proper consideration of the piezoelectric polarization under the random alloy conditions, the 3D FEM-based strain solver can be used to account for the piezoelectric polarization and calculate the polarization charge. The 3D-DDCC developed by our lab is then used to solve the transport equations with the appropriate input parameters generated by the external modules until we get the converged solutions. Finally, we can obtain the physical profiles such as the potential distributions, current–voltage (*I*–*V*) curve, and IQE curve.

18.2.1 Method for Generating the Random Alloy Distribution

As shown in Figure 18.4a, first the In(Al) and Ga atoms are randomly assigned by the random number generator according to the average indium composition and are aligned to the cation lattice site. The lattice site size is decided by the atom density. In addition, according to the APT data, the average alloy composition of each lateral plane along the *z*-direction, $\chi_{avg}(z)$, is like a Gaussian shape distribution as shown in Figure 18.5a and b, especially when the QW is very thin [6,23,24,30]. Note that the adenosine triphosphate (ATP) resolution in the depth is about one monolayer and lateral resolution is around 2 nm. Therefore, we can obtain the average indium composition of the *i*th QW by the following equation:

$$\chi_{avg}(z) = \chi_{max} e^{\frac{-(z-z_i)^2}{2\sigma_s^2}}, \tag{18.1}$$

where χ_{max} is the peak average composition of the epilayer, z_i is at the middle of the *i*th QW, and σ_s is the half width of the Gaussian broadening coefficient. The value of σ_s is around the half width of the epilayer in the *z*-direction. When atoms are assigned according to Equation 18.1, we need to choose a volume size to get the average local alloy composition. The volume size we use to calculate the local composition is around 2 nm × 2 nm × 0.667 nm, which is close to the APT resolution. Then the alloy composition is weighted at this volume region with Gaussian shape weighting criteria.

As shown in Figure 18.6, the local alloy composition extracted on volumes as mentioned before is following a binomial distribution, which is similar to Refs. [16,18–22].

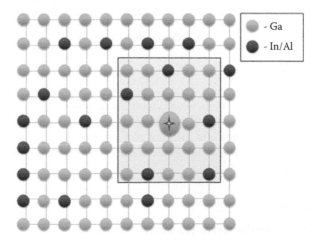

FIGURE 18.4 The concept of the alloy generator. The In, Al, or Ga atom at each lattice site (all with N atoms) is assigned by the random number. Then we choose a proper volume to average the local alloy composition. (From C.-K. Wu et al., *Journal of Computational Electronics*, 14, 416–424, 2015.)

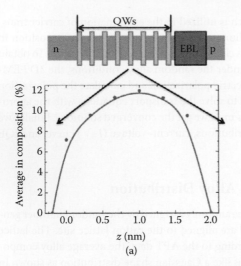
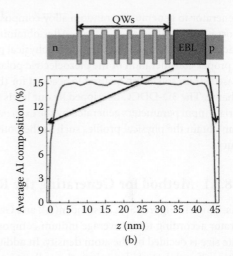

FIGURE 18.5 (a) The average alloy composition distributes along the 2-nm quantum well. (b) The average alloy composition distributes along the 45-nm electron blocking layer (EBL). (From C.-K. Wu et al., *Journal of Computational Electronics*, 14, 416–424, 2015.)

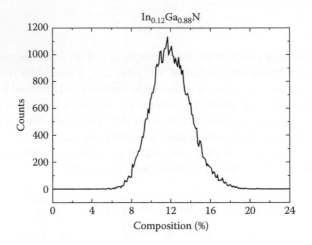

FIGURE 18.6 The counting distribution of certain alloy compositions of the $In_{0.12}Ga_{0.88}N$ quantum well (QW) [26].

After assigning the alloy landscape as shown in Figure 18.7, we combine the mesh file of the structures with the alloy maps and produce the local input parameters simultaneously. The material parameters of each mesh node are assigned according to the local alloy composition map $x(r)$. All parameters (e.g., bandgap [31], effective mass [32,33]) in each node are calculated locally with the indium composition map $x(r)$ by our in-house 3D simulation solver.

18.2.2 Strain Modeling

18.2.3 Calculation of Strain with 3D FEM Method

To understand how the polarization is induced in the random alloy fluctuation condition, we need to do a 3D finite element strain analysis. We need to solve the displacement field by the FEM after considering the incorporation of incoherent lattice constants.

Composition distribution

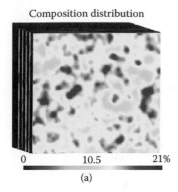

0   10.5   21%

(a)

Conduction band potential

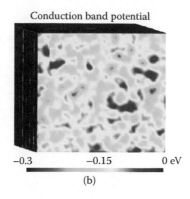

−0.3   −0.15   0 eV

(b)

FIGURE 18.7   (a) One random alloy distribution of the n-i-n InGaN quantum wells (QW). (b) The section view of the fluctuated conduction band potential, which is corresponds to the random alloy distribution, in the n-i-n InGaN quantum well case at 0 V bias. (From C.-K. Wu et al., *Journal of Computational Electronics*, 14, 416–424, 2015.)

18.2.3.1 Equilibrium Equations

When the element is located in the x,y,z coordinates and some regions are constrained, the element deforms under the balanced force. We can describe the deformation of a point $x (= [x, y, z]^T)$ in three components of its displacement:

$$u = [u, v, w]^T. \tag{18.2}$$

The force per unit volume in the vector form can be expressed as:

$$f = [f_x, f_y, f_z]^T. \tag{18.3}$$

In Figure 18.8, we can find the stresses and the body force distributed on the elemental volume dV. Here, we describe the stress by six components as

$$\boldsymbol{\sigma} = [\sigma_{xx}, \sigma_{yy}, \sigma_{zz}, \sigma_{yz}, \sigma_{xz}, \sigma_{xy}]^T, \tag{18.4}$$

where σ_x, σ_y, σ_z are the normal stresses and σ_{yz}, σ_{xz}, σ_{xy} are the shear stresses. Under the equilibrium conditions, the total forces along each direction are zero in each elemental volume. In Figure 18.8, we can pay attention to the x-axis and multiply the stresses to the corresponding areas and list the following equation:

$$\frac{\partial \sigma_{xx}}{\partial x} dx(dydz) + \frac{\partial \sigma_{xy}}{\partial y} dy(dxdz) + \frac{\partial \sigma_{zx}}{\partial z} dz(dxdy) + f_x dV = 0. \tag{18.5}$$

Knowing the $dV = dxdydz$, Equation 18.5 can be simplified as the following:

$$\frac{\partial \sigma_{xx}}{\partial x} + \frac{\partial \sigma_{xy}}{\partial y} + \frac{\partial \sigma_{zx}}{\partial z} + f_x = 0. \tag{18.6}$$

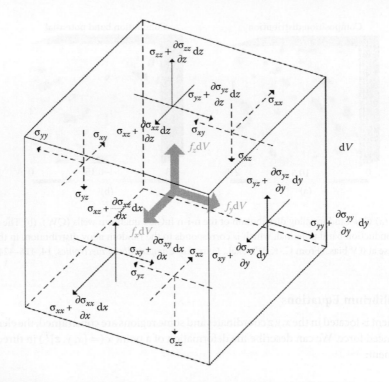

FIGURE 18.8 Equilibrium condition of an elemental volume. (From T. R. Chandrupatla and A. D. Belegundu. *Introduction to Finite Elements in Engineering*. Pearson Education International, 2002.)

After considering the total force along x-, y-, and z-directions, we can list the total equilibrium equations as follows:

$$\frac{\partial \sigma_{xx}}{\partial x} + \frac{\partial \sigma_{xy}}{\partial y} + \frac{\partial \sigma_{zx}}{\partial z} + f_x = 0 \tag{18.7}$$

$$\frac{\partial \sigma_{xy}}{\partial x} + \frac{\partial \sigma_{yy}}{\partial y} + \frac{\partial \sigma_{yz}}{\partial z} + f_y = 0 \tag{18.8}$$

$$\frac{\partial \sigma_{zx}}{\partial x} + \frac{\partial \sigma_{yz}}{\partial y} + \frac{\partial \sigma_{zz}}{\partial z} + f_z = 0. \tag{18.9}$$

18.2.3.2 Stress–Strain–Displacement Relations

In Equations 18.2 and 18.4, we know the form of the stresses and displacement. The strains have the corresponding form as following:

$$\boldsymbol{\epsilon} = [\epsilon_{xx}, \epsilon_{yy}, \epsilon_{zz}, \epsilon_{yz}, \epsilon_{xz}, \epsilon_{xy}]^T, \tag{18.10}$$

where ϵ_{xx}, ϵ_{yy}, and ϵ_{zz} are normal strains and ϵ_{yz}, ϵ_{xy}, and ϵ_{xz} are shear strains. We can also write the strain–displacement relation as

$$\boldsymbol{\epsilon} = \left[\frac{\partial u}{\partial x}, \frac{\partial v}{\partial y}, \frac{\partial w}{\partial z}, \frac{\partial v}{\partial z} + \frac{\partial w}{\partial y}, \frac{\partial u}{\partial z} + \frac{\partial w}{\partial x}, \frac{\partial u}{\partial y} + \frac{\partial v}{\partial x}\right]^T. \tag{18.11}$$

These strain–displacement relations hold under small deformations.

In our research, we assume that the linearly electric stress–strain relations come from the generalized Hooke's law.

$$\boldsymbol{\sigma} = \mathbf{D}\boldsymbol{\epsilon}. \tag{18.12}$$

$\mathbf{D}$ is the material matrix. In our research, we focused on the wurtzite structure, AlN, GaN, and InN, and $\mathbf{D}$ is given by

$$\mathbf{D} = \begin{pmatrix} c_{11} & c_{12} & c_{13} & 0 & 0 & 0 \\ c_{12} & c_{11} & c_{13} & 0 & 0 & 0 \\ c_{13} & c_{13} & c_{33} & 0 & 0 & 0 \\ 0 & 0 & 0 & c_{44} & 0 & 0 \\ 0 & 0 & 0 & 0 & c_{44} & 0 \\ 0 & 0 & 0 & 0 & 0 & c_{66} \end{pmatrix}. \tag{18.13}$$

18.2.4 Calculation of Piezoelectric Polarization and the Implementation Method

Due to the random alloy fluctuation, each element has its own spontaneous polarization $P^{\mathrm{sp}}(r)$ and piezoelectric polarization $P^{\mathrm{pz}}(r)$ depending on the local alloy composition. After calculating the strain energy, we can further analyze the piezoelectric polarization field distribution using the following equation:

$$P^{\mathrm{pz}} = \begin{pmatrix} 0 & 0 & 0 & 0 & e_{15} & 0 \\ 0 & 0 & 0 & e_{15} & 0 & 0 \\ e_{31} & e_{31} & e_{33} & 0 & 0 & 0 \end{pmatrix} \begin{pmatrix} \epsilon_{xx} \\ \epsilon_{yy} \\ \epsilon_{zz} \\ \epsilon_{yz} \\ \epsilon_{xz} \\ \epsilon_{xy} \end{pmatrix} = \begin{pmatrix} e_{15}\epsilon_{xz} \\ e_{15}\epsilon_{yz} \\ e_{31}(\epsilon_{xx} + \epsilon_{yy}) + e_{33}\epsilon_{zz} \end{pmatrix}. \tag{18.14}$$

The e_{15}, e_{31}, and e_{33} are the piezoelectric coefficients, which are listed in Table 18.1.

After obtaining the piezoelectric polarization, the total polarization of each element can be calculated by

$$P^{\mathrm{total}}(r) = P^{\mathrm{sp}}(r) + P^{\mathrm{pz}}(r), \tag{18.15}$$

where P^{sp} is the spontaneous polarization of InGaN alloy where the value can be found in Ref. [35]. After calculating the polarization $P^{\mathrm{total}}(r)$, the polarization charge, $\rho^{\mathrm{pol}}(r)$, induced at each element can be calculated by

$$\nabla \cdot P^{\mathrm{total}}(r) = -\rho_{\mathrm{pol}}(r). \tag{18.16}$$

The induced fixed polarization charges at different positions will be put into the 3D Poisson and drift–diffusion solver to obtain the potential inside the devices.

TABLE 18.1   Piezoelectric Coefficients

	$e_{33}(\mathrm{cm}^{-2})$	$e_{31}(\mathrm{cm}^{-2})$	$e_{15}(\mathrm{cm}^{-2})$
GaN	0.73	−0.49	−0.4
InN	0.73	−0.49	−0.4

18.2.5 Modeling of Carrier Transport under the Random Alloy Fluctuation with Drift–Diffusion Solver

To study the 3D carrier transport simulation under the random alloy fluctuation, we need to apply our 3D FEM-based Poisson and drift–diffusion solver. Although the 3D Poisson and drift–diffusion solver cannot describe the tunneling effect, which can estimate the device performance more accurately, it has been proved that it is a suitable model to describe the carrier transport behavior with proper physical parameters, and a solver that yields an acceptable calculation time and required computation memory [24,26–28]. The solver is based on solving the following equations:

$$\nabla \cdot P^{\text{total}}(r) = -\rho_{\text{pol}}(r) \tag{18.17}$$

$$\nabla(\epsilon \nabla V(r)) = n(r) - p(r) + N_A(r) - N_D(r) + \rho_{\text{pol}}(r), \tag{18.18}$$

$$J_n(r) = q\mu_n n(r)\nabla V(r) + qD_n \nabla n(r), \tag{18.19}$$

$$J_p(r) = q\mu_p p(r)\nabla V(r) - qD_p \nabla p(r), \tag{18.20}$$

$$\nabla \cdot J_{n,p}(r) = q(\text{SRH} + B_0 n(r)p(r) + C_0(n^2(r)p(r) + n(r)p^2(r))), \tag{18.21}$$

$$\text{SRH} = \frac{n(r)p(r) - n_i^2}{\tau_n\left(p(r) + n_i e^{\frac{E_i - E_t}{k_B T}}\right) + \tau_p\left(n(r) + n_i e^{\frac{E_t - E_i}{k_B T}}\right)}, \tag{18.22}$$

where V is the potential, and ϵ is the static dielectric constant. N_A^- and N_D^+ are the doping density. n and p are the free carrier concentration of the electron and hole. $\rho_{\text{pol}}(r)$ is the local analytic polarization charge which varies with local indium composition or the calculated results by the 3D FEM elastic strain solver. q is 1.6×10^{-19}C. $\mu_{n,p}$ are electron and hole mobility. $D_{n,p}$ are the coefficients of diffusion, and $J_{n,p}(r)$ are the electron and hole current, respectively. The Shockley–Read–Hall (SRH) is the defect-assisted nonradiative recombination, where τ_n and τ_p are the nonradiative carrier lifetime depending on the crystal quality. E_t is the trap energy level located at the midgap. E_i and n_i are the intrinsic energy level and intrinsic carrier density, respectively. k_B is the Boltzmann constant and T is the temperature (300 K here). B_0 is the radiative recombination coefficient. C_0 is the Auger recombination coefficient. We solved those equations until we obtained a converged solution.

18.2.6 Schrödinger Equation and Emission Rate

To analyze the emission properties of the LED, we need to solve the time-independent Schrödinger equation:

$$H\Psi\left(x, y, z\right) = E\Psi\left(x, y, z\right), \tag{18.23}$$

where H is the Hamiltonian operator, which is a second-order differential operator. The time-independent Schrödinger equation can be also expressed as

$$\left[\frac{-\hbar}{2m^*}\nabla^2 + E_{c,v}\left(x, y, z\right)\right]\Psi\left(x, y, z\right) = E\Psi\left(x, y, z\right). \tag{18.24}$$

$E_{c,v}$ is the conduction and valence band potential calculated by the 3D-DDCC. Ψ is the carrier wavefunction. m^* is the effective mass and $\hbar$ is the Planck constant divided by 2π. Since the Schrödinger equation is an eigenvalue problem, we obtain the eigenenergy E corresponding to the eigenwave function.

The formula we used to calculate the spontaneous emission rate [36] can be expressed as

$$
R_{\text{sp}}(\hbar\omega) = \frac{e^2 n_r \hbar\omega}{m_0^2 \varepsilon_0 c^3 \hbar^2} \frac{1}{V} \sum_{i,j} |\mathbf{a} \cdot \mathbf{p_{i,j}}|^2 \times \frac{1}{\sigma_b \sqrt{2\pi}} \exp\left[\frac{-\left(E_{i,j} - \hbar\omega\right)^2}{2\sigma_b^2} \right]
$$

$$
\times f_e\left(E_i^e\right) f_h\left(E_j^h\right) \ (\text{cm}^{-3}\text{s}^{-1}\text{eV}^{-1}), \tag{18.25}
$$

where n_r is the refractive index and c is the light velocity. The $|\mathbf{a} \cdot \mathbf{p_{i,j}}|$ is the momentum matrix element, which includes the overlapping of the electron and hole wave function. E_i^e and E_j^h are electron and hole eigenenergy states, respectively. $E_{i,j}$ is the effective bandgap and is equal to $E_i^e - E_j^h$. f_e and f_h are the Fermi–Dirac distribution of electrons and holes, respectively. The equations are shown below:

$$
f_e\left(E_i^e\right) = \frac{1}{1 + \exp\left(\left(E_i - E_{fn}\right)/k_B T\right)} \tag{18.26}
$$

$$
f_h\left(E_j^h\right) = \frac{1}{1 + \exp\left(\left(E_{fp} - E_j\right)/k_B T\right)}. \tag{18.27}
$$

The $|\mathbf{a} \cdot \mathbf{p_{i,j}}|$ is the momentum matrix element term where electron and hole overlapping is taken into account in both the localization by fluctuations. The effect of QCSE is already accounted for in solving the Poisson equation. A Gaussian broadening σ_b was used in the modeling. The value of σ_b used in modeling was 10 meV, which is slightly smaller than $k_B T$ to limit the spectrum broadening due to σ_b. Therefore, the calculated emission spectrum broadening will be mainly determined by the different energy levels of localized states.

18.3 Unipolar Transport for Random Alloy System

In this section, the percolation transport study of the random alloy system will be presented. To understand how the piezoelectric field affects the transport, we investigate the pure electron transport in the n-GaN/i-InGaN/n-GaN QW structures and the simulation results will also be compared to the experimental work from Browne et al. [17,37]. The 3D numerical model considering random alloy fluctuations will be applied.

18.3.1 Electron Transport in n-GaN/i-InGaN/n-GaN Structures

The existence of piezoelectric polarization and spontaneous polarization has been confirmed by many studies [31,38,39]. As shown in Figure 18.9, if the InGaN QW is grown in between the GaN, it will induce a huge electric field (>1 MV/cm) in the QW and cause the potential bending as shown in Figure 18.9b. If the quantum barrier (QB) thickness is 10 nm, the induced potential barrier peak could be more than 1 eV without electron screening. Theoretically, under this large electric field, we need to apply a strong bias to overcome this barrier for carrier to go through the multiple quantum wells (MQWs). For the single QW case, the influence of the polarization-induced barrier is not significant, especially as it will be screened by carriers from the n-GaN cap layer. However, under the MQW condition, it could cause a huge resistance for carrier transport. For the n-i-n structure, there is no influence of radiative or nonraditive recombination. Therefore, it is a good platform to test how alloy fluctuation and the polarization-induced potential barrier affect the carrier transport. In this section, in order to understand the carrier transport mechanism across the InGaN QWs, we show that we must study how carriers overcome the piezoelectric barriers before we start to investigate the LED structures.

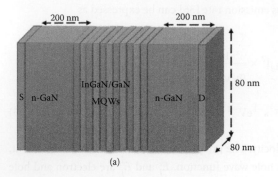
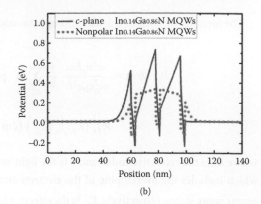

(a)

(b)

FIGURE 18.9 (a) The modeled device paradigm of n-GaN/i-InGaN/n-GaN quantum well structures for modeling the pure electron transport. (b) The potential distribution of $In_{0.14}Ga_{0.86}N$ c-plane and nonpolar piezoelectric barriers.

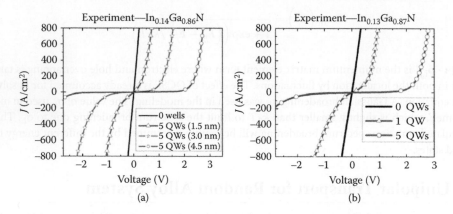

FIGURE 18.10 (a) The experimental I–V curves of c-plane $In_{0.14}Ga_{0.86}N$ with different thicknesses of quantum wells (QWs). (b) The experimental I–V curves of c-plane $In_{0.13}Ga_{0.87}N$ with different numbers of QWs. (From D. A. Browne et al., *Journal of Applied Physics*, 117, 185703, 2015; D. A. Browne et al., Investigation of electron transport through InGaN quantum well structures. In *14th Electronic Materials Conference*, Santa Barbara, CA, June 25–27, 2014.)

The traditional c-plane MQWs have a larger degree of piezoelectric barriers than nonpolar (m-plane) MQWs, as shown in Figure 18.9b. That means the performance of different traditional c-plane MQW structures has a higher correlation with their level of piezoelectric barriers. Browne et. al. [17,37] have found a clear rectifying I–V curve in c-plane MQWs which is related to the QW thicknesses and numbers, where the equivalent piezoelectric barriers will increase with increasing the QW thicknesses and numbers. Figure 18.10 shows that the higher piezoelectric barriers in the devices will increase the driving voltages at the same current density. However, if the QW model is used to study the vertical transport by considering either a full theoretical polarization value or a zero polarization value (nonpolar MQWs), there is a large deviation of predicted driving voltages between experimental work at the same current density, as shown in Figure 18.11.

Besides, it should not be Auger-assisted hot carriers overflow since no Auger recombination could occur in an n-i-n structure and an investigation of the n-i-n structure can be regarded as one that focuses on the carrier transport without any recombination mechanism. The percolation transport through the random alloy system might be a possible reason. A similar effect is also observed experimentally in n-GaN/

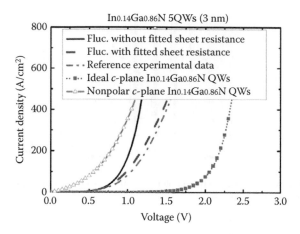

FIGURE 18.11 The *I–V* curves of five experimental and simulated 3-nm quantum wells (QWs) in the n-i-n system. (From D. A. Browne et al., *Journal of Applied Physics*, 117, 185703, 2015; D. A. Browne et al., Investigation of electron transport through InGaN quantum well structures. In *14th Electronic Materials Conference*, Santa Barbara, CA, June 25–27, 2014.)

TABLE 18.2   Detailed Parameters of Each Epilayer for n-GaN/i-InGaN/n-GaN Structures

Epilayer	n-GaN	i-InGaN/GaN
μ_e (cm^2/Vs)	200	600
Doping (1/cm^3)	5.0×10^{18}	1.0×10^{17}
E_a (meV)	25	–

i-AlGaN/n-GaN EBL cases [40]. To verify this, we apply the 3D program by considering the random alloy fluctuation in the InGaN QW.

The structures modeled consiste of two n-type doping sides (5×10^{18} cm^{-3}) that sandwich an intrinsic MQW region. The thicknesses of the QWs are 1.5, 3, and 4.5 nm. The barriers in all the cases are 10 nm. The numbers of QWs are 1 and 5. The configuration of the n-i-n structures is shown in Figure 18.9a. In order to present the nanoscale random alloy fluctuations, a small mesh element size is needed. Therefore, we need to limit our chip to 80 nm × 80 nm in the lateral direction due to computational limitations. The random alloy distribution is based on the rule mentioned in the previous section. The average indium composition is 14% according to Refs. [17,37]. The fluctuation range is around 8%–21%, which follows the binomial distribution. We have simulation results of the ideal In$_{0.14}$Ga$_{0.86}$N QW cases for comparison. The physical parameters are listed in Table 18.2.

One of the alloy distributions is shown in Figure 18.7a, and the corresponding fluctuated conduction band potential at 0 V is shown in Figure 18.7b, where a local composition site will induce a reversed trend of potential distribution. Moreover, the lower indium regions will induce relatively low polarization barriers and the electrons could percolate through the relatively lower barrier, as plotted in Figure 18.12.

Hence, the fluctuated case performs smaller turn-on voltages than the uniform cases, as shown in Figure 18.11. Compared with the experimental data [17,37], the turned-on voltage of the *I–V* curve with indium fluctuations is close to the experimental results. Since the experimental result has additional sheet resistance not modeled by our vertical transport study due to the size limitation, with the fitted sheet resistance, the calculated results can further match the experimental results. Figure 18.13 shows a comparison of

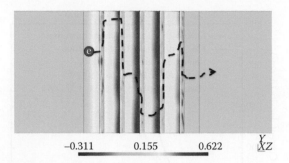

FIGURE 18.12 The side view of conduction band potential in the n-i-n InGaN quantum wells and the scheme of the carrier transport along the fluctuated barriers.

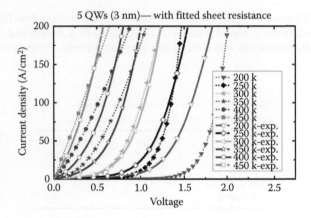

FIGURE 18.13 The *I–V* curves of the experimental and calculated work at various temperatures. (From D. A. Browne et al., Investigation of electron transport through InGaN quantum well structures. In *14th Electronic Materials Conference*, Santa Barbara, CA, June 25–27, 2014.)

the experimental work with the calculated outcomes at different temperatures. At low temperature, the calculations do not fit well. The result reveals that a small portion of tunneling current might dominate at the low temperature rather than the thermionic emission current in the random alloy system because the tunneling current should have a weaker temperature dependence. Consequently, the deviation between predicted current behavior and experimental results might be due to the lack of tunneling process in our transport model.

On the other hand, Figure 18.14 shows the increasing turn-on voltage with increasing the thickness or number of QWs. As we know, for the same polarization electric field, a thicker QW will cause a much larger potential band bending, making it harder for carriers to go across the junction. The turn-on voltage of a single QW is very low because it has only one barrier to be overcome by applying the bias voltage. Even though the thickness of the QW increases to 4.5 nm, the turn-on voltage is still less than 0.5 V. When the number of QWs increases to five, the positive turn-on voltages for 1.5, 3, and 4.5 nm at 20 A/cm^2 current density are 0.10, 0.68, and 1.73 V, respectively.

In addition, there is an asymmetric *I–V* behavior induced by the polarization field, which needs a larger negative voltage to reduce the barriers at reverse bias range. However, since the thermionic current is lower at the reverse bias region, the tunneling current might play a more important role, which requires a program that can handle tunneling transport to model this problem in the future.

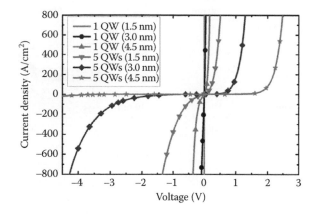

FIGURE 18.14 The *I–V* curves of various thicknesses and numbers of the fluctuated quantum wells (QWs).

18.4 Modeling Result of MQW InGaN LED and the Comparison to the Traditional Model

From the unipolar transport, we find that the polarization-induced barrier does play an important role in limiting the carrier transport behavior. However, without using the random alloy fluctuation, it is hard to approach the experimental *I–V* behavior. In this section, we focus on how indium fluctuation affects the carrier transport in InGaN LEDs.

As we know, InGaN QW LED has become a popular technology in solid-state lighting. Due to the strong lattice mismatch between InN and GaN layers, the self-formed random indium fluctuation has played an important role in influencing LED electrical and optical properties. In this section, we first study the influence of nanoscale indium fluctuation on the emission spectrum and then the relation of carrier transport and radiative efficiency in LEDs. The randomly generated indium fluctuation in the QW in this simulation will be used again in the study with 3D modeling.

18.4.1 Simulation Structure

As we mentioned previously, we divide our study into two parts: the influence of indium fluctuation on (1) emission spectrum and (2) carrier transport. In these two parts, we use different simulation structures. In part (1), since the 3D eigenvalue solver requires huge computer memory and calculating time, the simulation structure is relatively small. The simulation domain for a double QW LED was 30 nm × 30 nm, as shown in Figure 18.15a. The detail parameters setting can be found in Table 18.3. On the other hand, the simulation structure is an 80 nm × 80 nm 6-pair QW LED and included a 40-nm AlGaN EBL in part (2), as shown in Figure 18.15b. The detailed setting can also be found in Table 18.4. We considered a MQW with a 100-nm p-doped GaN layer and a 200-nm n-doped GaN layer, and the QW and barrier width are 3 and 10 nm, respectively. We focus pm a 450-nm MQW blue LED ($In_{0.17}Ga_{0.83}N$). In the uniform QW case, the indium composition is uniformly chosen to be 17%. In our case, as we mentioned previously, we assigned the indium fluctuation in each QW randomly. The maximum local indium composition, which is determined by the random number generator, was around 18%–19%.

A constant nonradiative lifetime was assumed to be 5.0×10^{-8} s, and the radiative recombination coefficient B_0 was assumed to be 2.0×10^{-11} cm^3/s [5,15,41]. Note that recombination rate is decided by B_0 $n(r) \cdot p(r)$ where the electron–hole overlap term is in the term of $n(r) \cdot p(r)$. In all cases, a 100% theoretical polarization value was applied [38]. The EBL thickness was 40 nm with 15% Al content AlGaN. For studying the carrier transport, we solved the 3D Poisson and drift–diffusion solver with the classical particle model. A larger area can be used in the modeling because we do not need to solve for the Schrödinger equation for the eigenstates for transport properties, which needs a much longer computing

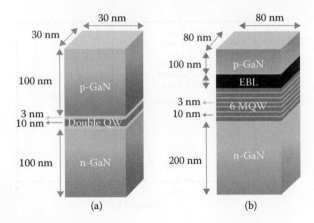

FIGURE 18.15 (a) The simulation structure in analyzing the emission spectrum. (b) The simulation structure in analyzing the carrier transport.

TABLE 18.3 The Material Parameter Settings of the Structure in Analyzing the Emission Spectrum

Unit	Thickness (nm)	Doping (1/cm³)	E_a (meV)	Impurity (1/cm³)	e Mobility (cm²/V·s)	h Mobility (cm²/V·s)
p-GaN	100	2.0×10^{19}	170	0.0	200	5
InGaN	3	0.0	0.0	1.0×10^{17}	600	10
GaN	10	0.0	0.0	1.0×10^{17}	200	10
InGaN	3	0.0	0.0	1.0×10^{17}	600	10
n-GaN	100	5.0×10^{18}	25	0.0	200	10

Note: The unit of thickness is *nm*. The unit of doping and impurity is 1/cm³. The unit of activation energy (E_a) is meV. The unit of electron and hole mobility is cm²/V·s.

time. We modeled 10 different fluctuation maps and took the averaged *I–V* and IQEs. Due to memory limitation, we cannot model the whole area of LED. Therefore, we only modeled the limited area in the p-i-n region, as shown in Figure 18.15b. Therefore, the sheet resistance in our simulation was not included due to the memory limitation. In general, the sheet resistance will cause voltage drops in the p-GaN and n-GaN layers so the calculated voltage across the active QW region is smaller than the externally applied bias in experiments. In addition, the real device structures also suffer current crowding effects, which will lead to a more serious droop effect than the result we obtain here.

18.4.2 The Influence of Indium Fluctuation on the Emission Spectrum

In this section, we discuss the influence of indium fluctuation on the emission spectrum. First, as we mentioned previously, the simulation structure we used in this section is a double QW LED and the simulation domain was 30 nm × 30 nm due to computer memory limitations required for 3D simulation, especially for the eigenvalue solver. To avoid the result being limited by the restricted area, we ran a set of different random cases and took the average results. First, the 3D Poisson and drift–diffusion solver developed by our lab was used to obtain a converged band potential at a fixed current density of 20 A/cm². Second, we solved the 3D Schrödinger equation with the calculated potential and obtained the confined eigenstate $E_{i,j}$.

Figure 18.16 shows the effects of randomly generated fluctuations in the QW for five different random maps. A total of 22 different random cases were run and averaged (dashed line in Figure 18.16). As we can see, the emission spectra broadened when the indium fluctuations were included in the simulation. The main reason is that the indium fluctuations form indium-rich regions in the QW (Figure 18.7a and b) with

TABLE 18.4 The Material Parameter Settings of the Structure in Analyzing the Carrier Transport

Unit	Thickness (nm)	Doping (1/cm^3)	E_a (meV)	Impurity (1/cm^3)	e Mobility (cm^2/V · s)	h Mobility (cm^2/V · s)
p-GaN	100	2.0×10^{19}	170	0.0	200	5
EBL	40	2.0×10^{19}	200	0.0	200	5
p-GaN	10	1.0×10^{18}	170	0.0	200	5
InGaN	3	0.0	0.0	1.0×10^{17}	600	10
GaN	10	0.0	0.0	1.0×10^{17}	200	10
InGaN	3	0.0	0.0	1.0×10^{17}	600	10
GaN	10	0.0	0.0	1.0×10^{17}	200	10
InGaN	3	0.0	0.0	1.0×10^{17}	600	10
GaN	10	0.0	0.0	1.0×10^{17}	200	10
InGaN	3	0.0	0.0	1.0×10^{17}	600	10
GaN	10	0.0	0.0	1.0×10^{17}	200	10
InGaN	3	0.0	0.0	1.0×10^{17}	600	10
GaN	10	0.0	0.0	1.0×10^{17}	200	10
InGaN	3	0.0	0.0	1.0×10^{17}	600	10
n-GaN	10	5.0×10^{18}	25	0.0	200	10
n-GaN	200	5.0×10^{18}	25	0.0	200	10

Note: The unit of thickness is *nm*. The unit of doping and impurity is 1/cm^3. The unit of activation energy (E_a) is meV. The unit of electron and hole mobility is (cm^2/V·s).

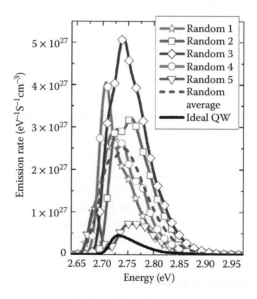

FIGURE 18.16 The calculated emission spectra when including the indium fluctuation. The dashed line is the average result of 22 random cases.

different bandgaps and confined energy levels. However, local peaks are observed in some cases (the random 2 and random 3 cases in Figure 18.16). This might be the result of the small sampling volume used in the simulation. In real devices with much larger area, these different local peaks merge into the overall spectral broadening. Figure 18.17b shows the calculated full width half maximum (FWHM) of the emission spectrum is around 80–120 meV at room temperature, which is close to the experimental results.

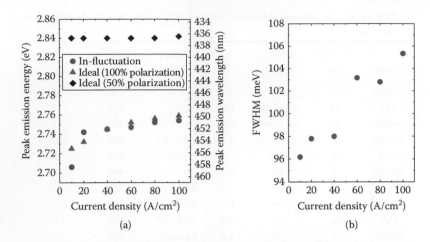

FIGURE 18.17   (a) The position of emission peak shift at different current density. (b) The full width half maximum (FWHM) of the emission spectrum. (From T.-J. Yang et al., *Journal of Applied Physics*, 116, 113104, 2014.)

Homogeneous broadening mechanisms usually will not lead to such large linewidths. If we compare the calculated intensity to the uniform QW case, the emission is stronger due to three factors occurring in the high In content regions where carriers localize in high indium region (deep potential): (1) there is less electron–hole separation; (2) there is more screening of electric field at a given current; and (3) there is an increased QW occupancy factor due to carrier localization. Figure 18.17a shows the blue shift of the spectrum when increasing the current density. A 40 meV ($\sim$ 7 nm) blue shift is observed when the current density increases from 10 to 100 A/cm^2, which is close to most experimental observations. Comparing with simulations of ideal QWs, assuming 50% polarization only leads to a very small blue shift due to a very reduced QCSE and the emission peak shifts to much shorter wavelength, while assuming a 100% polarization gives a reasonable shift, while, however, leading to very large forward bias (see the discussion in Section 18.4.3.1).

18.4.3 The Influence of Indium Fluctuation on the Carrier Transport and the Efficiency Droop

To model the vertical carrier transport in the 3D indium fluctuation cases, we used LEDs with six InGaN QWs with GaN barriers and included an AlGaN EBL, as shown in Figure 18.15b. We considered an MQW with a 100-nm p-doped GaN layer and a 200-nm n-doped GaN layer. The QW and barrier widths are 3 and 10 nm, respectively. We focus on 450-nm MQW blue LEDs (In$_{0.17}$Ga$_{0.83}$N). In the ideal QW case, the indium composition was uniformly 17%. In our case, we assigned the indium fluctuation in each QW randomly as described above. The maximum indium composition, which is determined by the random number generator, was around 18%–19%.

18.4.3.1 The Simulation Result of *c*-Plane LED

The simulation result is shown in Figure 18.18a. This is the calculated conduction band potential landscape. The indium-rich regions correspond to regions with a lower potential. The carriers will localize in the relatively lower potential regions because carriers inherently tend to stay in low-energy regions, as shown in Figure 18.18b. Consequently, the radiative recombination rate increases in the indium-rich region because of the carrier screening of the polarization fields, as shown in Figure 18.18c.

18.4.3.2 The *I–V* Curve

Next, we focused on the comparison of transport between the QW including indium fluctuation and uniform QW. Figure 18.19a shows the *I–V* curves. The forward voltage V_f used here is for a current density of

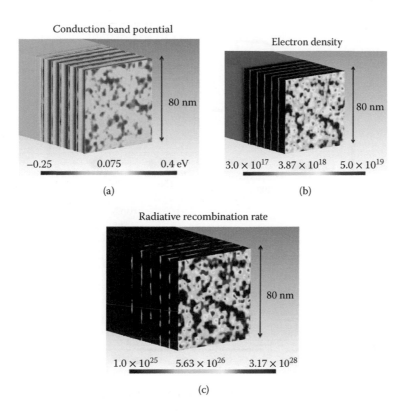

FIGURE 18.18 (a) The conduction band potential at 3.1 V. (b) The electron density at 3.1 V. (c) The radiative recombination rate at 3.1 V. (From T.-J. Yang et al., *Journal of Applied Physics*, 116, 113104, 2014.)

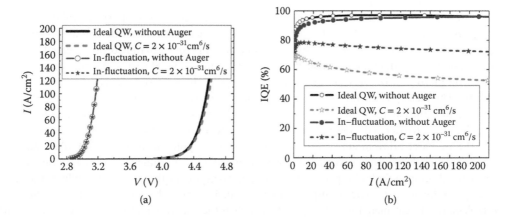

FIGURE 18.19 (a) The influence of indium fluctuation on the *I–V* curves. (b) The internal quantum efficiency (IQE) curves. (From T.-J. Yang et al., *Journal of Applied Physics*, 116, 113104, 2014.)

20 A/cm^2. For commercial blue light LEDs, V_f is around 2.8–3.0 V. For the ideal QW simulation (with uniform indium composition and 100% theoretical polarization value), V_f is almost 4.4 V, which is far larger than experimental results. This is also observed in most commercial simulation software with the same parameters set, which was also discussed in our twodimensional (2D) modeling results [23]. Some suggest including tunneling in the model to get a smaller V_f. But our calculation shows that using Wentzel–Kramers–Brillouin (WKB) tunneling between ideal QWs and 100% polarization cannot push V_f

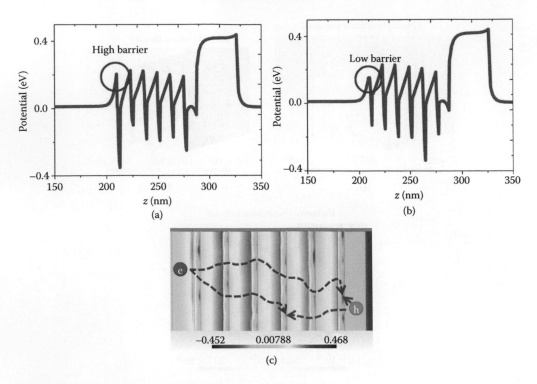

FIGURE 18.20 (a) The high barrier in higher indium region in the first quantum well (QW). (b) The low barrier in low indium region in the first QW. (c) The illustration of carrier transport in the device.

lower than 3.7 V. On the other hand, when including indium fluctuations, we calculated a V_f of 3.05 V even without considering tunneling.

The reason why the V_f shifts to a reasonable value is that the indium composition will directly affect the bandgap potential in the QW due to different strength of the polarization field that in turn affects the bandgap potential in the QW. As we can see for the first QW in Figure 18.20a and b, the potential at the high indium region is lower than the potential at the lower indium region. However, due to the strong lattice mismatch, the piezo polarization charge will induce a triangular shape barrier between the GaN barrier and InGaN QW. The piezo polarization-inducing barrier height also depends on the indium composition. As shown in picture, the barrier at the high indium region is high due to the large polarization difference. On the other hand, the barrier at the low indium side will be reduced. For carriers, they are much easier to flow through the low barrier (low indium side) and localize in the low potential region (high indium side). This is the main reason why the I–V curve shifts to a value that matches experimental data more closely. Because the carriers will inherently find a percolation path and transport in the device, the indium fluctuation will strongly influence the carrier transport. Figure 18.20c illustrates the carrier transport in the device. Nevertheless, the V_f calculated is still slightly higher than the experimental result. Recent studies show that QW LEDs without V-pit or GaN substrate LED with very low dislocation density have a V_f around 3.2–3.4 eV, which is close to our prediction because the influence of V-pit has not yet been included in our simulation. In addition, the indium fluctuations of the piezoelectric field–induced GaN barrier might enhance the tunneling since the induced barrier height will be smaller at the lower indium composition site. This might further lower the V_f.

18.4.3.3 The Droop Behavior of IQE

Turning to the dependence of IQE on current, there are several reasons proposed for origins of droop, such as overflow [1–3], Auger recombination [4–9], and defects [10–14]. For the idealized uniform QW without considering the Auger effect and with EBL, the efficiency peak occurs at 80 A/cm² and at a very large bias ($\sim$ 4.5V). (Figure 18.19b) (considering only vertical transport and neglecting the influence of sheet resistance). Although the drift process becomes dominant at this high bias, the electrons are blocked by the EBL, and the overflow is weak. Even including the Auger recombination, the droop effect only occurs at 4.0 V, which is very large compared to commercial data sheets. Note that this large bias condition across the junction should not occur in the real device application. If we observe the most recent experimental results, we can find that the droop occurs at very low current density ($\sim$10 A/cm^2) and the applied bias is typically lower than 3.1 V. In the past, most researchers focused on the current density and disregarded the voltage except when discussing the wall plug efficiency. However, here the voltage plays a key role in understanding physical processes in LEDs.

When Auger recombination is excluded from the simulation, there is no droop even when the current is $\sim$200 A/cm^2 (Figure 18.19b). This is due to the blocking action of EBL. With In fluctuations included, a small droop occurs when the current density is above 400 A/cm^2 and the applied voltage above 3.5 V is larger than V_{BI} and ΔE_c, which is unphysical as the resistive voltage drop should only occur in n and p-layer surrounding the depletion region. The biased voltage larger than built-in voltage V_{BI} including the ΔE_c should not be exist in the depletion region. Such voltages are very different from observed ones since the experimental droop typically occurs at 2.8–3.0 V from commercial data sheets (Figure 18.24).

At 3.0 V, the bias is smaller than the built-in voltage, V_{BI} (3.3 V in our case), and the carrier transport is mainly dominated by the diffusion process, as we can see in Figure 18.22a. The potential in the p-region

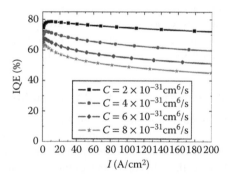

FIGURE 18.21   The internal quantum efficiency (IQE) curves with different Auger coefficients.

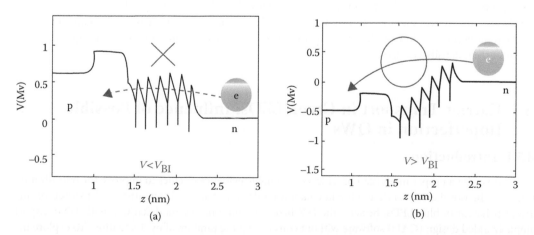

FIGURE 18.22   (a) When $V < V_{BI}$, it is mainly dominated by the diffusion process. It is harder for the carrier to overflow or overshoot the active region. (b) When $V > V_{BI}$, the carrier start to overflow or overshoot the device. However, the droop of a commercial grade light-emitting diode (LED) occurs at $V << V_{BI}$.

including the extra barrier height from the AlGaN EBL is much higher than the n-region and forms a large barrier to the carriers. Therefore, it is harder for carriers to spill over the barrier especially with the existence of the AlGaN barrier. Even with the polar optical phonon absorption process to gain the electron energy (~90 meV), it is still hard to overcome the 0.3–0.4 eV barrier. On the other hand, when the voltage exceeds the V_{BI}, as we can see in Figure 18.22b, the band structure reverses. In this situation, the carrier transport turns from the diffusion process to the drift process. The carriers start to overflow the device. Therefore, the simulation only sees the droop caused by overflow at a voltage higher than the built-in voltage (>3.3 V), if the Auger recombination is not considered. Therefore, the droop caused by the overflow only happens when the bias is larger than the built-in voltage or even larger bias when the AlGaN EBL is added.

Since the actual droop occurs at a very low bias in the experiments, we need to consider other factors that might cause droop effect at low bias. When the Auger recombination is included, the result shows that the droop occurs at 3.0 V, which matches closely to the experimental data. Consequently, we conclude that the droop effect is mainly dominated by the Auger recombination rate or other reasons such as carrier density dependent nonradiative recombination by defects.

As we can see in Figure 18.19a and b, when Auger recombination is included, droop occurs in both cases of ideal QW or QW incorporating In fluctuations. However, in the former case the bias voltage is again an unphysical 4.0 V, while in the latter case it is 3.0 V, which matches experimental data. Consequently, we conclude that the droop effect is mainly dominated by Auger recombination, enhanced by the effect of indium fluctuations. In addition, it should be pointed that due to the memory limitation, we calculated the vertical carrier transport within a limited area. In a real device, current crowding [42] issues can make local current density much higher than expected, making the droop effect worse, as discussed in earlier studies [43,44]. Figure 18.21 shows the IQE curves with the different Auger recombination coefficients, C. With larger values of C, the IQE curves show earlier droop onset, lower IQE peak, and more severe droop effects.

18.4.3.4 The Comparison of IQE between the Ideal QW and In-Fluctuation QW

In Figure 18.19b, the indium fluctuation case shows a good IQE performance with the higher IQE peak value and lower droop effect. Due to the carrier localization, the QCSE will be screened by the localized carrier charge. As a result, the better electron–hole overlapping will lead to a higher efficiency peak. On the other hand, in the ideal QW case, the IQE curve reaches the peak value when the applied bias is over 4.0 V, which is much larger than V_{BI} (3.3V). Therefore, the carrier leakage starts to affect the droop effect so the IQE curve shows a severe droop effect in the ideal QW case.

In conclusion, we prove that by considering the indium fluctuation in MQW LEDs, the electrical and optical properties are much closer to the commercial blue light LEDs. The droop behavior might be dominated by the Auger recombination at the lower current density since the voltage is much smaller than V_{BI}, and the ratio of the overflow mechanism might increase only when the bias is close to or larger than V_{BI} and the current density gets larger.

18.5 Carrier Transport in Green LEDs: Influence of Possible Imperfection in QWs

18.5.1 Introduction

Simulations with a proper model can provide some suitable evidence or clues to improve the optimization of devices. Before the 3D random alloy fluctuation model was proposed, scientists had focused on the current behavior of blue LEDs, because the I–V behavior could not be modeled well with 1D transport computer-aided design (CAD) software without considering the random alloy fluctuation. To explore the carrier transport and recombination mechanism in green LEDs, we conduct a preliminary examination of the electric property of green p-n LEDs.

18.5.2 Carrier Transport in Green LEDs Considering Indium Fluctuations Only

Initially, we constructed a green emitting p-n LEDs with five pairs of uniform 3-nm/10-nm $In_{0.27}Ga_{0.73}N/$ GaN MQWs (517-nm green emission) and a fluctuated case for comparison. Both the cases are composed of 295-nm n-doped layer with 5×10^{18} cm^{-3} doping density and a 100-nm p-type layer with 2×10^{19} cm^{-3} doping density. Figure 18.23a shows the epilayers of all the simulated structure here. The chip size for the 3D case is 250 nm × 40 nm. The average composition for the 3D fluctuated case is 21%, as shown in Figure 18.23b, where the maximum composition of the fluctuated case is around 27%. Table 18.5 lists the input parameters of all the cases in this section. Figure 18.24 shows the calculated *I–V* curves of green LEDs. As expected, the 1D vertical transport model with a 100% theoretical polarization value failed to describe the experimental observation because of strong polarization fields. Even when the ideal polarization is reduced to 50%, the calculated result still shows a large difference with the extracted experimental values. However, the simulated green emitting LED with random alloy fluctuations also performs a larger driving voltage compared to the experimental result. Obviously, carriers still could not percolate through the fluctuated piezoelectric barriers in the simulation because the average compositions of QWs are so high that the piezoelectric fields will result in high potential barriers. The assumption of average 21% fluctuated QWs with 27% maximum composition might be far removed from the real devices. Our past study also shows that a much larger period of composition fluctuation will further reduce the turn-on voltage [23]. Therefore, it is worth taking further investigation.

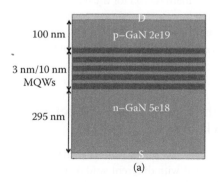
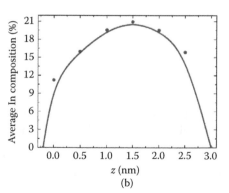

FIGURE 18.23 (a) The structure of the simulated green light-emitting diode (LED). (b) The average composition of the quantum well along the growth direction in the 3D fluctuated case.

TABLE 18.5 Detailed Parameters of Each Epilayer for Green Emission LEDs

Epilayer	n-GaN	i-InGaN/GaN	p-GaN
μ_h (cm²/Vs)	10	10	5.0
μ_e (cm²/Vs)	200	600	200
Doping (1/cm³)	5.0×10^{18}	1.0×10^{17}	2.0×10^{19}
E_a (meV)	25	–	170
τ_n (s^{-1})	5.0×10^{-8}	5.0×10^{-8}	5.0×10^{-8}
τ_p (s^{-1})	5.0×10^{-8}	5.0×10^{-8}	5.0×10^{-8}
B_0(cm³/s)	2.0×10^{-11}	2.0×10^{-11}	2.0×10^{-11}
C_0(cm⁶/s)	2.0×10^{-31}	2.0×10^{-31}	2.0×10^{-31}

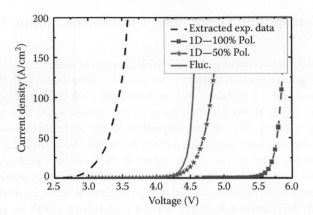

FIGURE 18.24 The $I–V$ curves of the several models and the extracted experimental data. (From C.-H. Lu et al., *Journal of Alloys and Compounds*, 555, 250–254, 2013.)

18.5.3 Imperfection of QWs in Green LEDs

Some studies [47,48] have shown that the QWs are not perfect. The studies in Refs. [47,48] further indicated that large-scale well-width fluctuation exists in the commercial green LED samples. Inconsistent growth temperature between GaN layer and $In_xGa_{1-x}N$ might be the main reason for the interrupted void regions among the QWs. Especially when the indium composition increases, the difference of the required growth temperature with GaN will be larger. To further understand how the imperfect QWs affect the electrical property of green p-n LEDs, we apply the 3D model in examining the imperfection of QW in the green LED device.

18.5.4 3D Examination of Green LEDs with Imperfect QW Model and Random Indium Fluctuations Model

To examine the carrier transport in the void structures, we construct a 3D structure with random alloy fluctuations that is much closer to the real physics. The structure is with different void densities. The distance of void is 30 and 40 nm. The input parameters are listed in Table 18.5. Due to the huge computation time in the 3D calculation, the calculation area is limited to 250 nm × 40 nm in the lateral direction. All the epilayers are the same as mentioned previously. The side view and section view composition distributions are shown in Figure 18.25. Compared to the normal fluctuated QWs, the map with imperfect QWs will have some void regions in the QW region. The void regions are also not aligned with one another according to experimental observations [47,48]. Figure 18.26a shows that the 3D model with different void lengths can better model the V_f of the experimental data. The conduction band potential at 3.05 V of 3D fluctuation with 40-nm l_{void} shows that the piezoelectric barriers are smaller in the void region, which is shown in Figure 18.26b. In addition the piezoelectric fields near the normal fluctuated QW region have the insufficient impact on the void region. Hence, the carriers would prefer to percolate through the void regions into the active area, which might be closer to the real devices.

Figure 18.27 shows the calculated vector profile of both electron and hole currents. The results indicate that the void regions dominate the current flow path for both electron and hole currents. While carriers percolate into the void regions, other carriers might prefer to flow into the low potential areas locatied at the stripe QWs. As a result, the smaller V_fs in imperfect QW cases are attributed to an alternative percolation path provided by the void regions.

To conclude this section, we applied the 3D model to show that the void regions in the green emission LEDs indeed affect the carrier injection. The imperfect regions provide an additional injection path for the

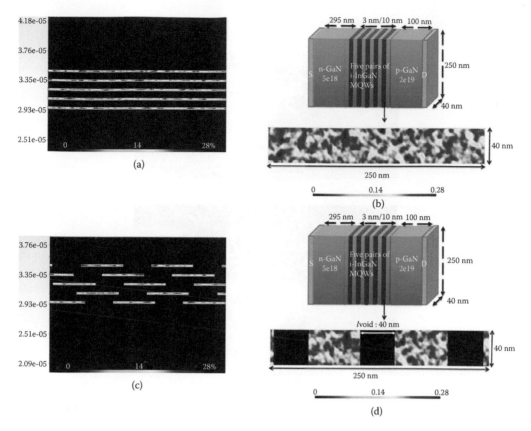

FIGURE 18.25 (a) The side view of the composition map with normal fluctuated QWs. (b) The section view of the composition map with normal fluctuated QWs. (c) The side view of the composition map with imperfect fluctuated QWs. (d) The section view of the composition map with imperfect fluctuated QWs.

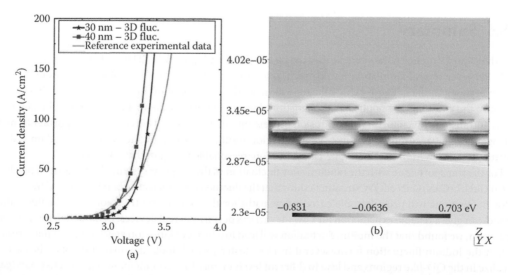

FIGURE 18.26 (a) The I–V curves of the different widths of the void region taking into consideration alloy fluctuations [45]. (b) The conduction band potential of the 40 nm case, where the location of the void regions is consistent, is at 3.06 V, which is near the turn-on point at 20 A/cm^2.

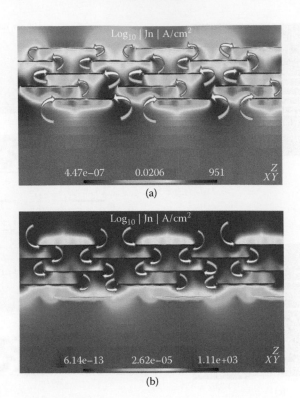

FIGURE 18.27 (a) The calculated vector profile of electron current. (b) The calculated vector profile of hole current.

carrier percolation. Although more reports of microscope observations in green emitting QWs are needed to prove that the structure of green LEDs are commonly imperfect, this preliminary simulated survey can still provide some hints for future modeling of the physical properties of green LEDs.

18.6 Summary

For the sake of simplicity, the simulation of the InGaN/GaN multiple QW LED usually assumes the uniform QW. However typical simulation results, such as the extremely high forward voltage (V_f), were far from the experimental data. These extra voltages will make the interpretation of carrier transport inaccurate. Because the extra voltage applied in the junction does not appear in the real device and will change the current transport mechanism from diffusion process into drift process. Therefore, it is important to make voltage fitting at accurate as possible. The simulation results show that by including the random indium fluctuation distribution in the QW, we can get a more reasonable fitting in the I–V curve.

For the transport issue with the random alloy fluctuation in the unipolar system, the simulation results of n-i-n GaN/InGaN/GaN MQW structures show that the fluctuated piezoelectric barrier induced by random alloy fluctuations will provide some current flow paths for the percolation transport, and the fluctuation model can give a reliable explanation for how the carriers perform in real devices. In modeling the emission spectrum, we found that the indium fluctuation will lead to a broader emission spectrum. The main reason is that the indium fluctuation forms several In-rich clusters, which look like quantum dots. The carriers localize in the QD-like regions and lead to different levels of quantum confinement. The effective bandgaps at each region vary from the different indium composition and the confined size of the indium clusters, which are the two important roles to affecting the discrete energy of quantum confinement. In addition, the emission strength will also be enhanced due to the better electron–hole localization effects.

We also analyzed the influence of indium fluctuations on the carrier transport. The simulation result of the I–V curve shifts to a value that is close to the experimental data. It indicates that the indium fluctuation does strongly affect the transport in the device. The main reason is that the low indium composition region will reduce the large potential barrier induced by the QCSE. For the uniform QW structure, the large V_f mainly comes from the strong band bending, which makes carriers hard to travel across the quantum barrier. For the indium fluctuation case, the reduction of the band bending at low indium regions provides a leaky way for carriers to flow through. The tunneling effect is not included in the indium fluctuation model but this will be a future project and may push the I–V curve to an even lower value. Furthermore, the IQE peaks occur at about 3.0 V, which also matches up to the experimental observation. Based on these simulation results, we tried to explain the cause of the droop effect. At a low bias condition, the droop effect is mainly dominated by the Auger recombination since the voltage is much lower than V_{BI}. When the bias gets higher and reaches the built-in voltage, the ratio of the droop influenced by carrier leakage will increase since the carrier transport turns from diffusion process into drift process at the reverse band condition.

Finally, we analyzed the green LED. We found that green emitting LEDs are difficult to use the traditional model to describe their physical property. Even for simulations that consider nanoscale alloy disorders, the calculated result differs from the experimental result. According to some reports, the green LEDs might exist in large-scale fluctuations, such as imperfect QWs in devices. Hence, we applied 3D models to examine the performance of the LEDs with imperfect QWs. The results show that the void region in the QWs provide an alternative path for carrier percolation. Thus, the IV of our simulation cases may approach the experimental observation more closely. In the future, we can model the green emission LEDs more elaborately based on the imperfect QW model.

References

1. H.-Y. Ryu, D.-S. Shin, and J.-I. Shim. Analysis of efficiency droop in nitride light-emitting diodes by the reduced effective volume of InGaN active material. *Applied Physics Letters*, 100(13):131109, 2012.

2. X. Ni, X. Li, J. Lee, S. Liu, V. Avrutin, A. Matulionis, U. Ozgur, and H. Morkoc. Pivotal role of ballistic and quasi-ballistic electrons on led efficiency. *Superlattices and Microstructures*, 48:133–153, 2010.

3. F. Akyol, D. N. Nath, S. Krishnamoorthy, P. S. Park, and S. Rajan. Suppression of electron overflow and efficiency droop in n-polar GaN green light emitting diodes. *Applied Physics Letters*, 100(11):111118, 2012.

4. F. Bertazzi, M. Goano, and E. Bellotti. A numerical study of Auger recombination in bulk InGaN. *Applied Physics Letters*, 97(23):231118, 2010.

5. K. T. Delaney, P. Rinke, and C. G. V. de Walle. Auger recombination rates in nitrides from first principles. *Applied Physics Letters*, 94(19):191109, 2009.

6. J. Hader, J. V. Moloney, B. Pasenow, S. W. Koch, M. Sabathil, N. Linder, and S. Lutgen. On the importance of radiative and Auger losses in GaN-based quantum wells. *Applied Physics Letters*, 92(26):261103, 2008.

7. J. Iveland, L. Martinelli, J. Peretti, J. S. Speck, and C. Weisbuch. Direct measurement of Auger electrons emitted from a semiconductor light-emitting diode under electrical injection: Identification of the dominant mechanism for efficiency droop. *Physical Review Letters*, 110:177406, April 2013.

8. H.-Y. Ryu, H.-S. Kim, and J.-I. Shim. Rate equation analysis of efficiency droop in InGaN light-emitting diodes. *Applied Physics Letters*, 95(8):081114, 2009.

9. Y. C. Shen, G. O. Mueller, S. Watanabe, N. F. Gardner, A. Munkholm, and M. R. Krames. Auger recombination in InGaN measured by photoluminescence. *Applied Physics Letters*, 91(14):141101, 2007.

10. J. Hader, J. V. Moloney, and S. W. Koch. Density-activated defect recombination as a possible explanation for the efficiency droop in GaN-based diodes. *Applied Physics Letters*, 96(22):221106, 2010.

11. J. Hader, J. V. Moloney, and S. W. Koch. Temperature-dependence of the internal efficiency droop in GaN-based diodes. *Applied Physics Letters*, 99(18):181127, 2011.

12. N. Okada, H. Kashihara, K. Sugimoto, Y. Yamada, and K. Tadatomo. Controlling potential barrier height by changing V-shaped pit size and the effect on optical and electrical properties for InGaN/GaN based light-emitting diodes. *Journal of Applied Physics*, 117(2):025708, 2015.

13. J. Smalc-Koziorowska, E. Grzanka, R. Czernecki, D. Schiavon, and M. Leszczyski. Elimination of trench defects and V-pits from InGaN/GaN structures. *Applied Physics Letters*, 106(10):101905, 2015.

14. H. Wang, X. Wang, Q. Tan, and X. Zeng. V-defects formation and optical properties of InGaN/-GaN multiple quantum well LED grown on patterned sapphire substrate. *Materials Science in Semiconductor Processing*, 29:112–116, 2015.

15. E. Kioupakis, P. Rinke, K. T. Delaney, and C. G. Van de Walle. Indirect Auger re-combination as a cause of efficiency droop in nitride light-emitting diodes. *Applied Physics Letters*, 98(16):161107, 2011.

16. S. E. Bennett, D. W. Saxey, M. J. Kappers, J. S. Barnard, C. J. Humphreys, G. D. Smith, and R. A. Oliver. Atom probe tomography assessment of the impact of electron beam exposure on InxGa1−xN/GaN quantum wells. *Applied Physics Letters*, 99(2):021906, 2011.

17. D. A. Browne, B. Mazumder, Y.-R. Wu, and J. S. Speck. Electron transport in unipolar InGaN/GaN multiple quantum well structures grown by NH$_3$ molecular beam epitaxy. *Journal of Applied Physics*, 117(13):185703, 2015.

18. M. J. Galtrey, R. A. Oliver, M. J. Kappers, C. J. Humphreys, P. H. Clifton, D. Larson, D. W. Saxey, and A. Cerezo. Three-dimensional atom probe analysis of green and blue-emitting InxGa1−xN/GaN multiple quantum well structures. *Journal of Applied Physics*, 104(1):013524, 2008.

19. M. J. Galtrey, R. A. Oliver, M. J. Kappers, C. J. Humphreys, D. J. Stokes, P. H. Clifton, and A. Cerezo. Three-dimensional atom probe studies of an InxGa1−xN/GaN multiple quantum well structure: Assessment of possible indium clustering. *Applied Physics Letters*, 90(6):061903, 2007.

20. J. R. Riley, T. Detchprohm, C. Wetzel, and L. J. Lauhon. On the reliable analysis of indium mole fraction within InxGa1−xN quantum wells using atom probe tomography. *Applied Physics Letters*, 104(15):152102, 2014.

21. R. Shivaraman, Y. Kawaguchi, S. Tanaka, S. DenBaars, S. Nakamura, and J. Speck. Comparative analysis of 2021 and 2021 semipolar GaN light emitting diodes using atom probe tomography. *Applied Physics Letters*, 102(25):251104, 2013.

22. D. Watson-Parris, M. J. Godfrey, P. Dawson, R. A. Oliver, M. J. Galtrey, M. J. Kappers, and C. J. Humphreys. Carrier localization mechanisms in InxGa1−xN/GaN quantum wells. *Physical Review B*, 83:115321, March 2011.

23. Y.-R. Wu, R. Shivaraman, K.-C. Wang, and J. S. Speck. Analyzing the physical properties of InGaN multiple quantum well light emitting diodes from nano scale structure. *Applied Physics Letters*, 101(8):083505, 2012.

24. T.-J. Yang, R. Shivaraman, J. S. Speck, and Y.-R. Wu. The influence of random indium alloy fluctuations in indium gallium nitride quantum wells on the device behavior. *Journal of Applied Physics*, 116(11):113104, 2014.

25. M. A. Caro, S. Schulz, and E. P. O'Reilly. Theory of local electric polarization and its relation to internal strain: Impact on polarization potential and electronic properties of group-III nitrides. *Physical Review B*, 88:214103, December 2013.

26. C.-K. Wu, C.-K. Li, and Y.-R. Wu. Percolation transport study in nitride based LED by considering the random alloy fluctuation. *Journal of Computational Electronics*, 14(2):416–424, 2015.

27. C.-K. Li, H.-C. Yang, T.-C. Hsu, Y.-J. Shen, A.-S. Liu, and Y.-R. Wu. Three dimensional numerical study on the efficiency of a core-shell InGaN/GaN multiple quantum well nanowire light-emitting diodes. *Journal of Applied Physics*, 113(18):183104, 2013.

28. C.-K. Li, P.-C. Yeh, J.-W. Yu, L.-H. Peng, and Y.-R. Wu. Scaling performance of Ga_2O_3/GaN nanowire field effect transistor. *Journal of Applied Physics*, 114(16):163706, 2013.

29. C. Geuzaine and J.-F. Remacle. Gmsh: A 3-D finite element mesh generator with built-in pre- and post-processing facilities. *International Journal for Numerical Methods in Engineering*, 79:1309–1331, 2009.

30. M. Sabathil, A. Laubsch, and N. Linder. Self-consistent modeling of resonant PL in InGaN SQW LED-structure. *SPIE Proceedings*, 6486:64860V–648609, 2007.

31. I. Vurgaftman, J. R. Meyer, and R. L. R. Mohan. Band parameters for III–V compound semiconductors and their alloys. *Journal of Applied Physics*, 89(11):5815–5875, 2001.

32. S. L. Chuang and C. S. Chang. k p method for strained wurtzite semiconductors. *Physical Review B*, 54:2491–2504, July 1996.

33. J. Wu. When group-III nitrides go infrared: New properties and perspectives. *Journal of Applied Physics*, 106(1):011101, 2009.

34. T. R. Chandrupatla., et al. *Introduction to Finite Elements in Engineering*. Vol. 2. Upper Saddle River, NJ: Prentice Hall, 2002.

35. A. Romanov, T. Baker, S. Nakamura, and J. Speck. Strain-induced polarization in wurtzite III-nitride semipolar layers. *Journal of Applied Physics*, 100(2):023522, 2006.

36. Y.-R. Wu, Y.-Y. Lin, H.-H. Huang, and J. Singh. Electronic and optical properties of InGaN quantum dot based light emitters for solid state lighting. *Journal of Applied Physics*, 105(1):013117, 2009.

37. D. A. Browne, B. Mazumder, Y.-R. Wu, and J. S. Speck. Investigation of electron transport through InGaN quantum well structures. In *14th Electronic Materials Conference*, Santa Barbara, CA, June 25–27, 2014.

38. O. Ambacher, J. Majewski, C. Miskys, A. Link, M. Hermann, M. Eickhoff, M. Stuzmann, F. Bernardini, V. Fiorentini, V. Tilak, B. Schaff, and L. F. Eastman. Pyroelectric properties of Al(In)GaN/GaN hetero- and quantum well structures. *Journal of Physics: Condensed Matter*, 14:3399–3434, 2002.

39. O. Ambacher, J. Smart, J. R. Shealy, N. G. Weimann, K. Chu, M. Murohy, W. J. Schaff, L. F. Eastman, R. Dimitrov, L. Wittmer, M. Stutzmann, W. Rieger, and J. Hilsenbeck. Two-dimensional electron gases induced by spontaneous and piezo-electric polarization charges in N- and Ga-face AlGaN/GaN heterostructures. *Journal of Applied Physics*, 85(6):3222–3233, March 1999.

40. D. N. Nath, Z. C. Yang, C.-Y. Lee, P. S. Park, Y.-R. Wu, and S. Rajan. Unipolar vertical transport in GaN/AlGaN/GaN heterostructures. *Applied Physics Letters*, 103(2):022102, 2013.

41. J. Piprek. Efficiency droop in nitride-based light-emitting diodes. *Physica Status Solidi (A)*, 207(10):2217–2225, 2010.

42. H.-Y. Ryu and J.-I. Shim. Effect of current spreading on the efficiency droop of InGaN light-emitting diodes. *Optics Express*, 19(4):2886–2894, February 2011.

43. M. Calciati, M. Goano, F. Bertazzi, M. Vallone, X. Zhou, G. Ghione, M. Meneghini, G. Meneghesso, E. Zanoni, E. Bellotti, G. Verzellesi, D. Zhu, and C. Humphreys. Correlating electroluminescence characterization and physics-based models of InGaN/GaN LEDs: Pitfalls and open issues. *AIP Advances*, 4(6):067118, 2014.

44. C. Kang Li, M. Rosmeulen, E. Simoen, and Y.-R. Wu. Study on the optimization for current spreading effect of lateral GaN/InGaN LEDs. *IEEE Transactions on Electron Devices*, 61(2):511–517, February 2014.

45. C.-H. Lu, Y.-C. Li, Y.-H. Chen, S.-C. Tsai, Y.-L. Lai, Y.-L. Li, and C.-P. Liu. Out-put power enhancement of InGaN/GaN based green light-emitting diodes with high-density ultra-small In-rich quantum dots. *Journal of Alloys and Compounds*, 555:250–254, 2013.

46. W. Lv, L. Wang, L. Wang, Y. Xing, D. Yang, Z. Hao, and Y. Luo. InGaN quantum dot green light-emitting diodes with negligible blue shift of electroluminescence peak wavelength. *Applied Physics Express*, 7(2):025203, 2014.

47. N. Van der Laak, R. Oliver, M. Kappers, and C. Humphreys. Characterization of InGaN quantum wells with gross fluctuations in width. *Journal of Applied Physics*, 102(1):013513, 2007.

48. N. K. van der Laak, R. A. Oliver, M. J. Kappers, and C. J. Humphreys. Role of gross well-width fluctuations in bright, green-emitting single InGaN/GaN quantum well structures. *Applied Physics Letters*, 90(12):121911–121911, 2007.

19

Superluminescent Light-Emitting Diodes

Nicolai Matuschek
and
Marcus Duelk

19.1 Introduction

More than 30 years ago, the concept of a superluminescent light-emitting diode (SLED) was proposed for the first time (Kaminow and Marcuse 1983). It can be briefly described as an edge-emitting semiconductor light source that operates in the so-called superluminescence regime, also known as the amplified spontaneous emission (ASE) regime. This means that the gain medium is pumped to a level beyond transparency but below the threshold for starting lasing activity. As a consequence of this particular operating range, SLEDs combine some aspects of the electro-optical performance from standard light-emitting diodes (LEDs) and laser diodes (LDs). They are able to produce high-output powers with high brightness similar to high-power LDs. On the other hand, they show a broadband emission spectrum similar to LEDs. With respect to their coherence properties, this translates into high spatial coherence similar to LDs and low temporal coherence like LEDs. A more detailed comparison is given in Rossetti et al. (2012).

Due to these intermediate characteristics, SLEDs are preferred light sources for many applications, including fiber-optic gyroscopes (FOGs) (Burns et al. 1983), fiber-optic current sensors (FOCSs) (Bohnert et al. 2002), optical coherence tomography (OCT) (Schmitt 1999; Drexler and Fujimoto 2008), structural health monitoring with optical fiber sensors (Wild and Hinckley 2009), speckle-free illumination (Rossetti et al. 2012), metrology systems (Dufour et al. 2005), or optical test equipment for fiber-optic networks (Senior 2009). In order to meet the different requirements of these applications, broadband SLEDs at various emission wavelengths with different spectral bandwidths and shapes are used. SLEDs can be designed for wavelengths ranging from 390 nm up to 2700 nm. They can be realized in GaN (390–570 nm), GaAs (570–1150 nm), InP (1150–2000 nm), or GaSb (2000–2700 nm). Figure 19.1 shows a selection of typical ASE spectra obtained for various commercially available SLED modules operating at center wavelengths from around 400 to 1550 nm (Exalos 2017).

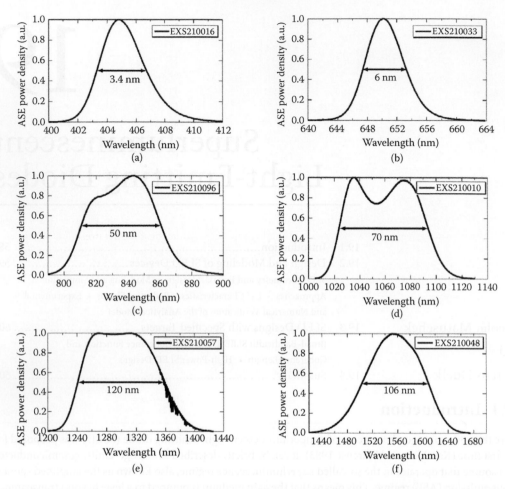

FIGURE 19.1 Measured amplified spontaneous emission (ASE) spectra (on a linear scale) obtained for six different superluminescent light-emitting diode (SLED) modules operating at relevant wavelength ranges from about 400 to 1550 nm. The full-width at half maximum (FWHM) or 3-dB bandwidth is given in the diagrams. The Gaussian-shaped spectra shown in (a), (b), and (f) are based on a single transition epitaxial design, whereas the spectra with two humps shown in (c), (d), and (e) are based on a multitransition epitaxial design, as explained in Section 19.3.1.

Traditionally, the vast majority of SLEDs are employed in systems with light emission in the near infrared from 750 to 1600 nm (see Figure 19.1c through f). Over the last decade, the emerging interest in speckle-free red–green–blue (RGB) applications forced the development of SLEDs operating in the visible wavelength range (see Figure 19.1a and b). Based on the technology for red LDs, which has been well established over the last 30 years, the demonstration of a red SLED realized in GaAs was a straightforward task (Semenov et al. 1993a). In contrast, the development of blue and green SLEDs has been confronted with a couple of intrinsic technological difficulties (Rossetti et al. 2012). One of them is the choice of the substrate material, which can be GaN, SiC, sapphire, or others. For SLEDs and LDs, high-quality substrate materials with good thermal conductivity and low dislocation densities are considered to be of utmost importance. Therefore, past development activities have focused on the growth of the epitaxial layer structure on free-standing (mainly *c*-plane) GaN substrates. The first GaN-based SLEDs emitting in the blue violet spectral region could be demonstrated just a few years ago (Feltin et al. 2009; Rossetti et al. 2010), whereas research and development for the demonstration of green SLEDs is still ongoing. Very recently, the first long-wavelength SLED realized in GaSb for optical sensing at 2.4 μm has been demonstrated (Wootten et al. 2014).

19.2 Design and Modeling of SLED Devices

19.2.1 SLED Geometry and Epitaxial Design

Typically, SLEDs are realized as epitaxially grown layers on a substrate in an index-guided ridge-waveguide geometry similar to Fabry–Pérot-type LDs. The latter consists of a straight waveguide with optical dielectric coatings applied at the facets of the chip, typically a high-reflective (HR) at the back facet and a partial-reflective (PR) coating at the front facet (see, e.g., Macleod 2010). The straight waveguide in combination with reflecting facets defines a resonant cavity that allows for the build-up of longitudinal cavity modes and laser oscillations. In contrast, several tricks are applied to SLEDs for the suppression of resonant cavity modes in order to minimize modulations in the ASE spectrum (so-called spectral ripples) or, even worse, to prevent the starting of lasing. The most popular and very efficient way to achieve this is the tilting of the waveguide by a few degrees with respect to the normal of the facets (Alphonse et al. 1988) in combination with the application of antireflection (AR) coatings (Macleod 2010) at the facet sides. The residual net modal reflectivity of such a design can be calculated using the analytical model presented in Marcuse (1989). Exemplary reflectivity calculations as a function of the tilt angle are given in Matuschek and Duelk (2013). Other methods are based on the incorporation of an absorber section at the back-facet side of the waveguide (Patterson et al. 1994; Kwong et al. 2008), the use of a bent (Semenov et al. 1993b), or tapered (Koyama et al. 1993) waveguide structure, or a combination of various methods (see, e.g., Lee et al. 1973; Nagai et al. 1989; Semenov et al. 1993a; Middlemast et al. 1997).

The ASE process occurs in the active region along the SLEDs waveguide structure. The most common approach to realize a positive material gain is based on a single-quantum well (SQW) or multiquantum well (MQW) active region design (see, e.g., Chuang 1995). For some applications, a bulk layer approach might be advantageous, as it is the case, for example, for SLEDs that require a low polarization extinction ratio (PER),[†] that is, the amount of transverse electric (TE)- and transverse magnetic (TM)-polarized output power should be as equal as possible (Heo et al. 2011). Quantum-dot epi structures for SLEDs have been proposed (Sun et al. 1999) and demonstrated a few times as well. However, this technology has various severe limitations, like a limited wavelength range (so far 1050–1200 nm), low differential gain requiring long SLED chips and large drive currents, manufacturing challenges, and others (Rossetti et al. 2008).

Figure 19.2 shows the schematic front view of a typical SLED structure for an SQW epitaxial layer design. Generally, quantum well (QW) active-region designs allow for the use of strained layers. Compressively strained QW layers are required for SLED designs showing a high PER greater than ~10 dB, which is desired for many applications. Maximum strain values up to 3% (with respect to the lattice constant of the substrate material) may be realized in practice. Higher strain values may lead to the formation of dislocations and are therefore detrimental to achieving a good long-term reliability. In addition to the relative strain of an individual layer, the total integrated net strain for the entire active region must be taken into account too. Hence, strain compensation by incorporation of tensile strained barrier layers is a useful option for strain relaxation over the active region (Tansu and Mawst 2001).

The QW and barrier layers are sandwiched between (undoped or partially doped) waveguide layers that are surrounded by highly doped n- and p-cladding layers. The high refractive index of the active-region and waveguide layers leads to the confinement of the optical mode in vertical direction. On the other hand, the ridge-waveguide geometry ensures the lateral mode confinement. As a result, the epitaxial layer stack realized in a ridge-waveguide geometry allows for the build-up of transverse optical modes with a lateral extension roughly given by the ridge width and a vertical extension as given by the thickness of the waveguide layers. For almost all applications, single-mode operation on the fundamental optical mode is preferred. Hence, care has to be taken to avoid the occurrence of higher-order modes or leaky substrate

---

[†] The PER is defined as $\mathrm{PER} = 10 \cdot \lg\left(\frac{P_{\mathrm{TE}}}{P_{\mathrm{TM}}}\right)$ dB, where P_{TE} and P_{TM} is the SLED's total output power from the front facet in TE and TM polarization, respectively.

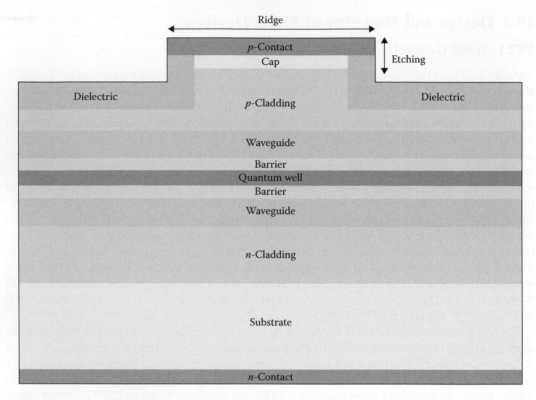

FIGURE 19.2 Basic layer structure of a single-quantum well (SQW) epitaxy. The QW and barrier layers are sandwiched between waveguide layers followed by the cladding layers on both sides. The ridge-waveguide geometry ensures the lateral guiding of the optical mode, whereas the vertical guiding is a consequence of the refractive-index profile defined by the material composition of the different layers.

modes. The latter are particularly a problem for GaN-based epi structures in the visible due to the high refractive index of the substrate material (Laino et al. 2007; Matuschek and Duelk 2013). Moreover, some output power might be lost into higher-order lateral modes if such modes are supported by the lateral waveguide (see, e.g., Coldren et al. 2012). The number of supported lateral modes, and thus, the corresponding power loss scales with the width of the ridge waveguide. Therefore, the stripe width is typically chosen to be smaller than 10 μm.

As mentioned earlier, the appropriate choice of semiconductor materials depends primarily on the wavelength range of interest, and thus, on the bandgap energy of available semiconductor compound material. In Table 19.1, the substrate materials and the most commonly used compound materials for the functional layers of the epitaxial structure, as shown in Figure 19.2, are summarized for the different wavelength regimes.

19.2.2 Modeling and Simulation Approaches

A couple of different approaches exist for the modeling and simulation of SLEDs. Generally, the analytical insight into the electro-optical performance decreases with the complexity of the set of coupled equations taken under consideration. On the other hand, the accuracy increases when more effects are taken into account.

Commercially available packages (Synopsis 2017; Crosslight Software 2017; Photon Design 2017) allow for full three-dimensional (3D) simulations. They are based on the coupling of solvers for the various problems involved in the simulation of an SLED (so-called multiphysics treatment) (Li and Li 2010). Typically,

TABLE 19.1 Materials Used for the Epitaxial Layers of an SLED Grown on a Substrate Depending on the Wavelength Range of Interest

Wavelength Range	390–570 nm	570–1150 nm	1150–2000 nm	2000–2700 nm
Substrate	GaN	GaAs	InP	GaSb
Cladding	$Al_xGa_{1-x}N$	$Al_xGa_{1-x}As$ $Al_xGa_yIn_{1-x-y}P$	InP $Al_xIn_{1-x}As$	$Al_xGa_{1-x}As_{1-y}Sb_y$
Waveguide	GaN	$Al_xGa_{1-x}As$ $Al_xGa_yIn_{1-x-y}P$	$Al_xGa_yIn_{1-x-y}As$ $In_xGa_{1-x}As_{1-y}P_y$	$Al_xGa_{1-x}As_{1-y}Sb_y$
Active region (QWs + barriers)	$In_xGa_{1-x}N$	$Al_xGa_yIn_{1-x-y}P$ $Al_xGa_yIn_{1-x-y}As$ $In_xGa_{1-x}As_{1-y}P_y$	$Al_xGa_yIn_{1-x-y}As$ $In_xGa_{1-x}As_{1-y}P_y$	$Al_xGa_{1-x}As_{1-y}Sb_y$ $In_xGa_{1-x}As_{1-y}Sb_y$

Note: The general definition of the quaternary materials includes the special case that the materials may reduce to ternary or binary materials if the material fractions x and/or y are equal to 0 or 1.

the quantum-mechanical problem is described using the **k·p** method and the optical modes are found from a Helmholtz equation (Chuang 1995). These equations are coupled to the semiconductor equations, for example, as Poisson equation and drift-diffusion equation and to a heat-transfer equation if the thermal properties are also taken into account (Loeser and Witzigmann 2008). All equations are then solved simultaneously and self-consistently for each iteration step, yielding a solution for all local and global variables of the system.

In analytical approaches both the physical dimensionality of the problem and the number of coupled equations are reduced. They are normally based on a traveling-wave equation for the electric field (Park and Li 2006) or the optical power (or photon density) (Matuschek and Duelk 2013; Milani et al. 2015) and on rate equations for the carrier density. The material gain is either calculated from quantum-mechanical equations or using simplified semianalytical equations assuming a logarithmic or linear gain dependence on the carrier density (Coldren et al. 2012). In order to calculate the net modal gain, both the optical confinement factor and the internal loss coefficient are required. The first can be calculated from the overlap integral of the optical modes with the active layers. The latter can be used, for instance, as a free fitting parameter or it can be extracted from an *inverse efficiency versus chip length* plot, which is obtained from the *L–I* characteristics of LDs measured for various chip lengths. The LD design for such measurements differs from the SLED design only in the straight ridge waveguide and its facets are typically uncoated.

19.2.3 *L–I* Characteristics of Reflecting SLEDs

SLEDs operate in the ASE regime, that is, spontaneously emitted photons induce the process of stimulated emission of photons while traveling along the waveguide. Hence, the number of forward- and backward-propagating photons grows exponentially in direction to both facets of the chip. On the other hand, the SLED's *L–I* characteristic follows to good approximation a power law with chip length–dependent exponent because of the logarithmic gain dependence on current density as shown in Matuschek and Duelk (2013). The analytic derivation of this power law was based on the assumption of an ideal SLED without any residual reflectivity at both facet sides.

Here, we want to extend this approach to the general case of SLEDs with arbitrary effective modal reflectivities at the chip's front- and back-facet sides. This derivation includes special cases of standard SLEDs and of so-called reflecting SLEDs (R-SLEDs). Typical SLEDs feature a tilted ridge-waveguide geometry with tilt angles smaller than 15° and high-quality AR coatings deposited at both facet sides in order to avoid any resonant cavity effect. In contrast, R-SLEDs consist of a tilted waveguide section in direction to the front facet with deposited AR coating and a straight waveguide section toward the chip's back facet with a deposited HR coating (Matuschek and Duelk 2014), as shown in Figure 19.3. As explained below in more detail, R-SLEDs are able to produce much higher output powers due to the double-pass geometry

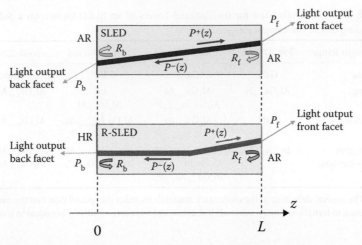

FIGURE 19.3 Schematic top view of an SLED and an R-SLED with effective modal reflectivities R_f and R_b. The power is exponentially amplified in both directions along the (partially) tilted ridge-waveguide structure, where P^+ denotes the power distribution for forward-propagating photons and P^- for backward-propagating photons, respectively. P_f and P_b are the powers, which are coupled out and can be measured at the front- and back-facet sides of the SLED or R-SLED. The z-axis is defined by the propagation direction of photons along the ridge-waveguide structure, so that the single-path length is slightly greater than the vertical distance between the facets due to the tilt.

compared to standard SLEDs under same operating conditions. However, the drawback is that such devices are much more susceptible to the appearance of ripples in the ASE spectrum and/or starting of undesired lasing operation.

In the following, the longitudinal power distribution for two counter-propagating waves is derived under the condition of steady-state continuous wave (CW) operation. Based on a one-dimensional (1D) traveling-wave amplifier approach similar to (Marcuse 1983; Matuschek et al. 2008), the differential equations

$$\frac{dP^+(z)}{dz} = \left(g_{\text{mod}} - \alpha_i\right) \cdot P^+(z) + \frac{n_{\text{sp}}\hbar\omega v_{\text{gr}}}{2L} g_{\text{mod}}, \tag{19.1}$$

$$-\frac{dP^-(z)}{dz} = \left(g_{\text{mod}} - \alpha_i\right) \cdot P^-(z) + \frac{n_{\text{sp}}\hbar\omega v_{\text{gr}}}{2L} g_{\text{mod}}, \tag{19.2}$$

determine the power distribution $P^{\pm}(z)$ for forward- and backward-propagating photons of the same longitudinal mode with angular frequency ω. $g_{\text{mod}} = \Gamma g_{\text{mat}}$ is the modal gain with g_{mat} being the material gain and Γ the mode confinement factor. α_i is the internal loss coefficient, n_{sp} the population inversion factor, L the effective chip length of the tilted ridge-waveguide structure, and $v_{\text{gr}} = c/n_{\text{gr}}$ the group velocity with the speed of light in vacuum, c, and the effective group index, n_{gr}. The independent variable (z-coordinate) is defined in the range $z \in [0; L]$, where the boundaries are defined by the chip's front and back facet (see Figure 19.3). The second term on the right-hand side describes the generation of light by spontaneously emitted photons, whereas the first term describes the stimulated amplification process by already existing photons.

It should be noted that, generally, there is one set of equations for each longitudinal mode. Here, we have omitted an index for the number of the longitudinal mode. We solve the equations for the mode closest to the gain maximum. Finally, the total power values are obtained by summing over an effective number of longitudinal modes, which is proportional to the full-width at half maximum (FWHM) or 3-dB bandwidth of the emitted ASE power spectrum (Matuschek and Duelk 2013). As explained in detail in Marcuse (1983), *it is not necessary to treat this ASE process as a coherent signal since it distributes itself continuously over a relatively wide band of wavelengths with random phases between adjacent wavelength components.* Hence, a description using powers instead of fields with definite amplitude and phase is sufficient.

Equations 19.1 and 19.2 can be easily solved if we assume that the gain is homogeneous in longitudinal direction, that is, we neglect any gain saturation effects. This approximation is justified for low-power SLEDs but might be questionable for SLEDs and R-SLEDs operating at high injection currents, and thus, at high-output powers, as will be shown at the end of this section. Then the general solution is given by

$$P^+(z) = P_0^+ \cdot G(z) + a \cdot \{G(z) - 1\}, \tag{19.3}$$

$$P^-(z) = P_L^- \cdot G(L-z) + a \cdot \{G(L-z) - 1\}, \tag{19.4}$$

with

$$a = \frac{n_{sp}\hbar\omega v_{gr}}{2L} \frac{g_{mod}}{g_{mod} - \alpha_i}, \tag{19.5}$$

where we have introduced the gain amplification factor

$$G(z) = \exp\left\{(g_{mod} - \alpha_i)z\right\}. \tag{19.6}$$

The subscripts 0 and L denote power values at positions $z = 0$ and $z = L$ inside the chip.

Equations 19.3 and 19.4 are coupled by the boundary conditions

$$P_L^- = R_f \cdot P_L^+, \tag{19.7}$$

$$P_0^+ = R_b \cdot P_0^-, \tag{19.8}$$

where R_f and R_b are the effective modal reflectivities at the front and back sides, respectively. Using these boundary conditions, we can derive the expressions

$$P_L^+ = a \cdot \frac{1 + G(L)R_b}{1 - G^2(L)R_bR_f} \cdot (G(L) - 1), \tag{19.9}$$

$$P_0^- = a \cdot \frac{1 + G(L)R_f}{1 - G^2(L)R_bR_f} \cdot (G(L) - 1), \tag{19.10}$$

for the light powers that hit the facets from inside the chip. It should be noted that Equations 19.9 and 19.10 are symmetric with respect to an exchange of the chip's front- and back-facet sides.

The power coupled out of the chip is obtained by multiplying (Equations 19.9 and 19.10) with the effective transmittance of the front- and back-facet coatings, T_f and T_b.[†] Thus, taking the values of all variables at the spectral position of the gain maximum, which we indicate by adding a subscript index 0 at each

---

[†] Generally, the inequality $R_{f/b} + T_{f/b} < 1$ holds because of slight absorption that might occur in the applied dielectric coatings. Moreover, for tilted SLED sections light that is not reflected back at the facet into the waveguide but elsewhere, is lost and not covered by the reflectivities $R_{f/b}$.

relevant variable and summing over the number of effectively contributing modes finally leads to

$$P_f = \tilde{P} \cdot \frac{1 + G_0(L)R_b}{1 - G_0(2L)R_bR_f} \cdot (G_0(L) - 1) \cdot T_f, \tag{19.11}$$

$$P_b = \tilde{P} \cdot \frac{1 + G_0(L)R_f}{1 - G_0(2L)R_bR_f} \cdot (G_0(L) - 1) \cdot T_b. \tag{19.12}$$

The prefactor $\tilde{P}$ is given by

$$
\begin{aligned}
\tilde{P} &= n_{\text{eff}}a_0 = n_{\text{eff}} \cdot \frac{n_{\text{sp}}\hbar\omega_0 v_{\text{gr},0}}{2L} \frac{g_{\text{mod},0}}{g_{\text{mod},0} - \alpha_{i,0}} \\
&= \Delta\lambda_{\text{3dB}} \cdot \frac{n_{\text{sp}}hc^2}{\lambda_0^3} \frac{g_{\text{mod},0}}{g_{\text{mod},0} - \alpha_{i,0}},
\end{aligned} \tag{19.13}
$$

where we have estimated the effective mode number, n_{eff}, by the 3-dB bandwidth of the ASE output spectrum, $\Delta\lambda_{\text{3dB}}$, and the longitudinal mode spacing, $\lambda_0^2/(2n_{\text{gr},0}L)$. Strictly speaking, the replacement of the sum by an effective mode number requires that the dispersion of the effective modal facet reflectivities and coating transmittance can be neglected over the relevant spectral range.

Equations 19.11 and 19.12 together with Equation 19.13 represent the general solution for the front- and back-facet output power of an SLED with arbitrary front- and back-side reflectivity. They are particularly useful for analyzing differences observed in the output power from both facets caused by residual facet reflections. In Matuschek et al. (2008), very similar expressions were derived and a short discussion of the consequences on the front-to-back output power ratio was given. In the following, we apply our general results to two special cases of an SLED, namely an ideal standard SLED and an ideal R-SLED.

19.2.3.1 Ideal Standard SLED

An SLED device is assumed where both facets do not have any residual reflection. This condition is nearly fulfilled for real devices with sufficiently great tilt angle and AR-coated facets. In this limit, we can set $R_f = R_b = 0$ and $T_f = T_b = 1$ and obtain

$$P_f = P_b = \tilde{P} \cdot (G_0(L) - 1) = \Delta\lambda_{\text{FWHM}} \cdot \frac{n_{\text{sp}}hc^2}{\lambda_0^3} \frac{g_{\text{mod},0}}{g_{\text{mod},0} - \alpha_{i,0}} \cdot \left(e^{(g_{\text{mod},0} - \alpha_{i,0}) \cdot L} - 1\right). \tag{19.14}$$

Inserting a logarithmic gain model in Equation 19.14 yields a power law for the SLED's L–I characteristic with chip length–dependent exponent. The consequences of this power law have been discussed and experimentally proven in detail in Matuschek and Duelk (2013) and are not repeated here.

19.2.3.2 Ideal R-SLED

An SLED device is assumed with a perfect front facet having zero residual reflection, similar to an ideal standard SLED, but with a back facet being a perfect reflector that reflects all incident light back into the waveguide. For real devices, this condition is fairly well realized with effective modal reflectivity values of about 90% or even higher for the HR coating. Moreover, the waveguide at the back facet must be straight or perpendicular with respect to the facet in order to avoid any geometrical reduction of the effective modal reflectivity. In the limit of an ideal R-SLED we can set $R_f = T_b = 0$ and $R_b = T_f = 1$ and obtain

$$P_f = \tilde{P} \cdot (G_0(2L) - 1) = \Delta\lambda_{\text{FWHM}} \cdot \frac{n_{\text{sp}}hc^2}{\lambda_0^3} \frac{g_{\text{mod},0}}{g_{\text{mod},0} - \alpha_{i,0}} \cdot \left(e^{2 \cdot (g_{\text{mod},0} - \alpha_{i,0}) \cdot L} - 1\right) \tag{19.15}$$

for the total power out of the front facet. It should be obvious that under those assumptions no power will exit the back facet of the chip, that is, $P_b = 0$.

Comparing Equation 19.15 with Equation 19.14, it follows directly that with respect to output power from the front facet an R-SLED of length L behaves like a standard SLED of twice the length $2L$. This result seems intuitive since the amplification process for photons is extended over a distance, which corresponds to one round-trip in the chip with perfect reflection at the back facet. However, it is important to realize that with respect to current injection the chip is still a short chip of length L. This means that an R-SLED of length L emits the same power out of the front facet as a standard SLED of length $2L$ at same current density values J or, in other words, at half values of the injection current I. Also, under these conditions the average carrier densities are approximately the same and, thus, the emitted ASE output spectra with their corresponding FWHM bandwidth. Consequently, the prefactor of Equation 19.13 has the same value in both cases.

19.2.4 Experimental and Numerical Verification of the Analytical Model

In this section, we want to verify the results derived from the analytical treatment by comparing them with experimental results as well as results obtained from a full 3D simulation of the SLED and R-SLED under consideration. The simulation software is based on a finite-element discretization of the device. Using the **k·p** method, the quantum mechanical system is described by a 6×6 Luttinger–Kohn Hamiltonian and the eigenvalue problem is solved as well as a Helmholtz equation for the optical modes (Chuang 1995). These equations are coupled to the semiconductor equations and thermal equations as described earlier. All equations are then solved self-consistently on a discrete mesh.

As an object for our study, we use SLED and R-SLED chips from the same epi wafer with light emission in the wavelength region around 860 nm. All chips are operated at a heat-sink temperature of 25°C. The MQW epi structure of these chips is based on (Al)GaAs/(In)GaAs layers grown on a GaAs substrate (see Table 19.1). For the simulation, the effective modal reflectivity values have not been set to the ideal values 0 and 1 but to values of 10^{-9} for the AR-coated facets with tilted waveguide section and 0.95 for the back facet with straight waveguide section. These values are quite realistic for well-designed SLED and R-SLED structures. The simulator has been calibrated by adjusting the L–I characteristics and ASE output spectra for short standard SLEDs with chip lengths ranging from 500 to 850 μm.

Figure 19.4 shows the results obtained for the 850-μm-long SLED chip after the calibration procedure. Obviously, the measured L–I curve is almost exactly reproduced by the simulation. Moreover, the detailed shape of the measured ASE spectra is also simulated very accurately for different levels of injection currents. The quality of the analytical model is shown by the dotted L–I characteristic, which is obtained from Equation 19.14 by following the procedure described in Matuschek and Duelk (2013). The center wavelength and 3-dB bandwidth have been taken from the ASE spectrum at an injection current of 100 mA. The modal gain as a function of the current density has been derived from the full 3D simulation by averaging over the z-direction. Apparently, the output power at high-injection currents is somewhat overestimated by the analytical model because the relatively simple model does not take into account self-heating and gain saturation effects of the chip.

After successful calibration, we are able to compare the electro-optical properties of a 1750-μm standard SLED with an 840-μm R-SLED. The chip length ratio is 2.08, which is close to the ideal value of two. Moreover, the back-facet coating of the R-SLED chip is not a real HR coating but only a PR coating with a reflectance value of about 20%. Figure 19.5 shows the measured and simulated front-facet output power as a function of the current density for both types of SLEDs. For our considerations the effective width of current injection is not relevant. Therefore, we scale the current values only by the chip length so that the density values are defined by $J = I/L$. As can be seen, the agreement between all curves is excellent. Only the measured power of the R-SLED (dash-dotted line) is slightly smaller because of the nonideal reflectance of the PR coating. First of all, this proves that an R-SLED behaves like a standard SLED of twice the length

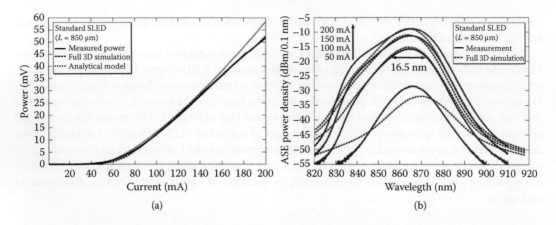

(a)   (b)

FIGURE 19.4 (a) *L–I* characteristics and (b) ASE output spectra (on an absolute logarithmic scale) for an 850-μm-long standard SLED chip. Solid lines represent measurement data and dashed lines simulation results obtained from a full 3D simulation after the calibration procedure. The dotted *L–I* characteristic follows from the analytical model according to Equation 19.14, as explained in the text. The ASE spectra are shown for four different injection currents: 50, 100, 150, and 200 mA. At 100 mA the 3-dB spectral bandwidth is about 16.5 nm. Note that all simulated ASE spectra have been shifted slightly by the same amount in wavelength so that they coincide at their maximum values with the measured spectra. This allows a better comparison of the detailed behavior of the spectral shapes.

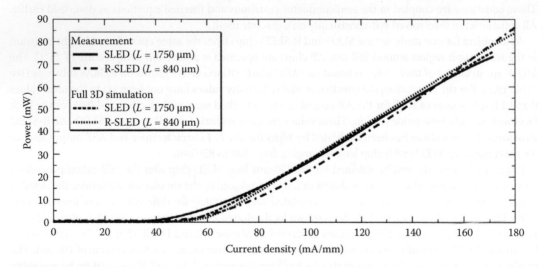

FIGURE 19.5 Measured (solid and dash–dotted line) and simulated (dashed and dotted line) front-facet output power of an SLED chip with length 1750 μm and an R-SLED chip with length 840 μm as function of current density, as described in the text. All curves agree almost perfectly.

at the same current density. Second, experimental results can be accurately reproduced and extrapolated once the simulator is well calibrated.

This is further demonstrated by analyzing the ASE output spectra. Figure 19.6a shows measured ASE spectra and Figure 19.6b shows simulated ASE spectra for both chips at two different levels of current density. In the first case, the current density with values slightly below 60 mA/mm is just above the threshold for ASE operation, which is why the output power is low. In the second case, the current density is beyond 170 mA/mm, resulting in the chips operating at high front-facet output powers of about 75 mW. The simulated curves coincide more or less over the entire spectral range for both current densities, whereas a

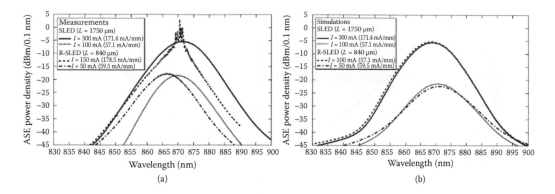

FIGURE 19.6 (a) Measured and (b) simulated ASE output spectra of an SLED chip with length 1750 μm and an R-SLED chip with length 840 μm at two different levels for the current density. The dotted and dash-dotted lines correspond to a current density slightly below 60 mA/mm and the solid and dashed lines to a value slightly above 170 mA/mm. In contrast to Figure 19.4b, no additional wavelength shift has been applied to the ASE spectra.

small wavelength shift can be observed for the measured curves. This is explained by the fact that even nominally identical chips from the same epi wafer have slightly different center wavelengths of the emitted ASE spectra. Furthermore, the thermal heat sinking toward the submount might not be identical for all chips. A more important difference, which can be observed for the measurements at the high current densities, is that the ASE spectrum of the R-SLED is somewhat narrower and shows strong periodic modulations (spectral ripples) in the center part. The periodicity corresponds to the longitudinal mode spacing defined by the length of the linear cavity. This is a clear indication of a resonant cavity effect (Fabry–Pérot modulation) caused by the high back-side reflectivity and demonstrates the tendency of an R-SLED to start lasing at high levels of current injection. In Section 19.3.2, the influence of spectral ripples on the coherence function will be discussed.

To gain further insight into the physical properties of an SLED and R-SLED, respectively, we analyze the longitudinal distribution of the optical intensity and the carrier density along the z-axis inside the chip based on the results obtained from the full 3D simulation (see Figure 19.7). The intensity profiles have been extracted at the center of the optical mode, and the maximum value at the front facet has been normalized to unity. The carrier density profiles have been extracted at the center of one of the QWs. The z-axis has been defined so that the back facet of the R-SLED coincides with the middle of the standard SLED and the front facets almost coincide, too. With this definition, all chip properties of the standard SLED are symmetric with respect to the origin. As shown in the upper plot, the shapes of the optical intensity profiles are quite similar for the same current density. However, as can be clearly seen, for low current densities (dotted and dash–dotted lines) the growth of the optical intensity in direction to the front facet is almost exponential, whereas the optical intensity saturates for higher current densities (solid and dashed lines) and, hence, higher optical output powers. This gain saturation behavior due to high optical powers is directly linked to the longitudinal carrier density distribution as shown in the lower plot. At low current densities, the carrier density is almost constant along the chip with a value of about 3.3×10^{-18} cm^{-3}. In contrast, at high current densities the carrier density decreases strongly by more than 40% from the origin toward the front facet with a terminal value of about 2.8×10^{-18} cm^{-3}. Low values of the carrier density mean low material gain, and thus, low gain amplification of the optical power.

Our investigations can be summarized as follows:

- At same current densities an R-SLED behaves very similar like a standard SLED with twice the length.
- An important difference is that R-SLEDs show ripples in the ASE spectrum and may have the tendency to start lasing operation.
- For long SLED chips and for R-SLEDs at high-output powers, gain saturation effects are not negligible anymore.

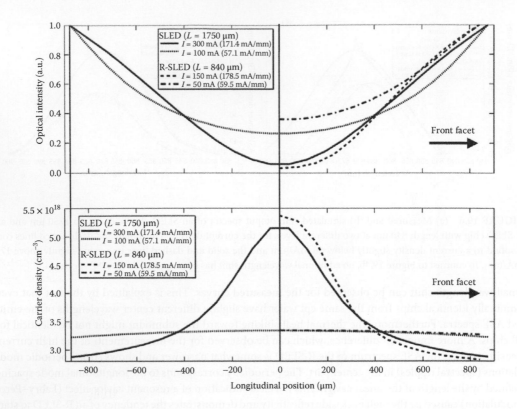

FIGURE 19.7 Longitudinal distribution of the optical intensity (upper plot) and carrier density (lower plot) along the z-axis inside the chip of an SLED chip with length 1750 μm and an R-SLED chip with length 840 μm at two different levels for the current density. The dotted and dash-dotted lines correspond to a current density slightly below 60 mA/mm and the solid and dashed lines to a value slightly above 170 mA/mm.

19.3 SLED Designs with Specified Targets

Commercial SLEDs have to fulfill target specifications for the electro-optical performance (output power, wavelength, bandwidth, spectral shape, ripple, PER, etc.) that are defined by the application. Also, their long-term reliability has to meet certain lifetime requirements (e.g., 10% power drop over 5000 hours of operation or 50% power drop over 100,000 hours of operation) that are dependent on the application and the intended use of operation. The shape and width of the ASE spectrum is directly linked to the coherence function, which is the autocorrelation function of an SLED. This is of particular importance for applications like OCT, as explained below. It is well known that increasing the output power is, in general, accompanied by a reduction in the ASE spectral bandwidth. Hence, methods are required to overcome this limitation.

19.3.1 Broad-Bandwidth SLED Designs

Generation of light in an SLED is based on the recombination from an electronic state in the conduction band with a hole state in the valence band (VB). In a QW, the bound states build a discrete set of sub-bands. In the simplest case, the SLED's epitaxial structure is based on an SQW active-region design or an MQW structure consisting of uncoupled identical QWs. Such designs can be reduced to the discussion of the quantized SQW states. Here, we restrict our discussion to the VB for heavy holes because ASE spectra of SLEDs consisting of compressively strained QWs are dominated by transitions to the heavy-hole sub-bands.

19.3.1.1 SLEDs with Bell-Shaped (Gaussian) ASE Spectra

The spectral shape of the ASE spectra is mainly determined by the fundamental sub-band transition from the first electronic state to the first heavy-hole state $(e_1 \rightarrow hh_1)$, as long as the carrier density is sufficiently low, so that the population of other sub-band states can be neglected (Chen et al. 1990). Generally, single transition designs yield bell-shaped (Gaussian) ASE spectra as shown in Figure 19.1a, b, and f. Such spectra are preferred with respect to the avoidance of side lobes in the near optical path length difference (OPD) range of the coherence function (see Section 19.3.2). An increasing carrier density in the QW leads at first to a rising population of high-energy states in the fundamental sub-bands, and thus, to a broadening of the spectrum. If other sub-band states start to become significantly populated, an additional hump tends to appear on the short-wavelength side (as shown in Figure 19.1c). Hence, with increasing population of the sub-bands, the spectral appearance changes continuously from being bell shaped in direction to an M-shaped double-humped spectrum (see Figure 19.1d).

19.3.1.2 SLEDs with M-Shaped (Double-Humped) ASE Spectra

Generally, all approaches for the design of SLED structures showing an ASE spectrum with ultrawide emission bandwidth are based on a multitransition design. This means that more than just one optical transition contributes to the emission spectrum. One has to distinguish between two principally different approaches. The first one is based on the SQW (or MQW) epitaxial design, as just discussed, in which, in addition to the fundamental transition, other transitions fulfilling the quantum-mechanical selection rules are utilized to contribute to the generation of light. If properly designed, such QW structures allow for the enormous enlargement of the spectral bandwidth (Semenov et al. 1993b; Kondo et al. 1992). The second approach is based on the use of nonidentical (so-called chirped) QWs (Lin et al. 2004). Chirping means that at least one QW of an MQW epitaxial layer structure is differently designed compared to the other QW(s), yielding optical transitions at different wavelengths. Chirping can be achieved by a different QW thickness and/or different material compositions for the QW and/or the barrier layers. The basic idea of utilizing chirped QWs is demonstrated in Figure 19.8 for a chirped double-quantum well (DQW) structure. Due to the greater band-gap energy and thinner QW thickness of the right QW, the fundamental transition results in photons with higher energy (shorter wavelength) compared to the left QW. Thus, an ultrawide bandwidth design may be achieved if the spectral separation chosen between both fundamental transitions is sufficiently large.

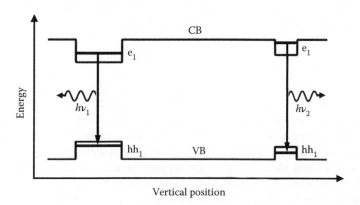

FIGURE 19.8 Energy-band diagram for two uncoupled chirped quantum wells (QWs). The horizontal lines in the wells indicate the first bound electronic state in the conduction band (CB), e_1, and the first bound hole state in the valence band (VB) for each QW separately. For the VB we restrict ourselves to the heavy hole band with its states hh_1. The arrows indicate the fundamental transition from the CB to the VB for both QWs. The electron–hole recombination results in the emission of a long-wavelength photon with energy $h\nu_1$ from the left QW and a short-wavelength photon with energy $h\nu_2$ from the right QW.

As mentioned earlier, ultrawide ASE spectra tend to exhibit a double-humped spectral shape. Such spectra show their maximum spectral bandwidth when both humps are well balanced, i.e., for approximately the same peak strength (the so-called flat-top condition). For a given chip design and a given heat-sink temperature, this situation can be achieved only over a limited range of operation currents. This makes such designs less flexible compared to single-transition designs. Moreover, the non-Gaussian spectral shape is unfavorable with respect to the occurrence of side lobes in the coherence function as mentioned earlier. The stronger the deviation from a Gaussian shape, the higher the side lobes are in the coherence function.

Generally, the width of the ASE spectrum broadens with the energy separation between allowed transitions to the heavy-hole VB. As a consequence of a greater spectral separation, first of all, the dip between both humps obvious in the ASE spectra at flat-top condition increases yielding a negative impact on the coherence function. And second, the SLED has to be driven at higher currents for operation at flat-top condition. The reason is that higher carrier densities in the QWs are required in order to lift the energetically more distant transitions on the short-wavelength side. This might be used as a positive side effect or if it is undesired, this effect can be compensated for by using chips with shorter active segment length. Nevertheless, it is obvious that the approach described here is limited to finding the best compromise between spectral bandwidth and the depth of the spectral dip.

19.3.1.3 Flattening of M-Shaped ASE Spectra

The M-shape approach can be optimized by a flattening of the spectral dip (see Figure 19.1e). This requires the use of at least one additional transition between QW states yielding photons with a wavelength somewhere around the spectral position of the dip. The additional transition fills up the dip between both peaks in the ASE spectrum. It can be realized in different ways, as discussed in the following.

The first possibility is the use of a chirped MQW active region design with at least one of the chirped QWs having its fundamental transition in the spectral range of the dip (Lin et al. 1996). For MQW structures consisting of a large number of QWs an elaborated approach is based on the continuous chirp of the fundamental transition wavelength for each QW over a wide spectral range.

Another possibility is the utilization of coupled QWs, where the coupling is caused by the finite potential walls between the QWs. For a coupled DQW active region design, the coupling leads to a splitting of the fundamental sub-band states, where the coupling strength, and thus, the splitting can be adjusted by the thickness of the barrier layers between the QWs and/or the well depth. The latter can be adjusted by the material composition of the barrier layers. If properly designed, the transition wavelength of the short-wavelength photon appears in the spectral range of the dip.

A third possibility to reduce or even remove the dip is the combination of two or more SLEDs with spectrally shifted ASE spectra. The SLEDs are designed so that the spectral peak of one SLED coincides with the dip of the other SLED. The combined spectrum is flatter and potentially broader if the SLEDs are appropriately designed. The flattening effect is demonstrated by the example of the M-shaped ASE spectrum for the 1050-nm SLED shown in Figure 19.1d. We assume that a second SLED has been designed with similar spectrum just shifted by 24 nm to shorter wavelengths. For simplicity reasons we take the original spectrum (the dashed curve in Figure 19.9) and shift it accordingly in order to obtain the dotted curve. Now, we assume that the output spectra are combined via a dispersion-free 50:50 coupler. This corresponds to a simple linear addition of both spectra with the same weight, resulting in the solid curve shown in Figure 19.9. Obviously, the original dip at around 1050 nm is removed and the resulting spectrum looks rather bell shaped. In the following section, the impact of the SLED combination on the coherence function is discussed.

19.3.2 Coherence Function and Coherence Length

As mentioned earlier, for OCT applications, the coherence function is of major interest. It allows the extraction of the coherence length, which determines the theoretical limit for the axial resolution of an

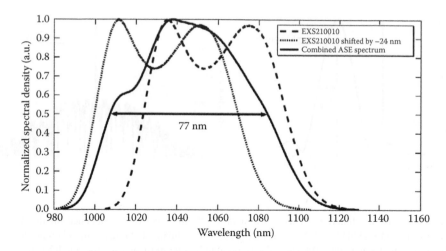

FIGURE 19.9   The dashed line shows the same ASE spectrum as plotted in Figure 19.1d for module EXS210010. The dotted curve shows the same spectrum shifted by 24 nm to shorter wavelengths. The solid curve is obtained as the linear superposition of both spectra. The FWHM bandwidth of the combined spectrum is 7 nm greater compared to the width of each single spectrum.

OCT imaging system (Duelk and Hsu 2015). For a broadband light source, the coherence length can be written as

$$l_{coh} = \gamma \cdot \frac{2 \cdot \ln(2)}{\pi} \cdot \frac{\lambda_{3dB}^2}{\Delta\lambda_{3dB}}, \tag{19.16}$$

with λ_{3dB} being the 3-dB center wavelength and $\Delta\lambda_{3dB}$ the 3-dB bandwidth of the ASE spectrum. The factor γ depends on the shape of the ASE spectrum. For an ideal Gaussian shape, the factor is unity, $\gamma = 1$, for other shapes, $\gamma > 1$. For SLEDs with double-humped ASE spectra operating at flat-top condition, a value of $\gamma \approx 1.19$ is typically used. This yields a penalty of roughly 20% in coherence length caused by the deviation from an ideal Gaussian shape. As an example, for the ASE spectrum shown in Figure 19.1e, we calculate for the coherence length $l_{coh} \approx 7.4$ μm using Equation 19.16 with $\lambda_{3dB} = 1300$ nm, $\Delta\lambda_{3dB} = 120$ nm, and assuming a flat-top shape with $\gamma = 1.19$.

The coherence length can be directly determined from the coherence function, too. The latter is equivalent to the SLED's autocorrelation function, and thus, according to the Wiener–Khinchin theorem, the Fourier transform of the ASE spectrum (Schmitt 1999). Figure 19.10 shows the coherence function as a function of the OPD obtained for the ASE spectrum just discussed. The coherence function is symmetric with respect to the OPD. Its drop-off from the maximum value at zero OPD to 50% defines the coherence length, that is, the coherence length is the half-width at half maximum (HWHM). From the coherence function shown in Figure 19.10a near zero, we extract a coherence length of 7.9 μm. This value is close to the value estimated from Equation 19.16. Moreover, side lobes are visible at an OPD of 20 μm with a side-lobe suppression ratio (SLSR) of about 10 dB. The side lobes are caused by the non-Gaussian flat-top shape of the ASE spectrum. For applications like OCT, those side lobes may generate imaging artifacts if they are not suppressed by an additional windowing function in the OCT signal processing (Duelk and Hsu 2015). Therefore, SLED designs with bell-shaped ASE spectra are preferred for such applications. Figure 19.11 shows the coherence functions, which correspond to the ASE spectra plotted in Figure 19.9 for module EXS210010 and the combined SLED spectrum. Obviously, going from the M-shaped to the rather bell-shaped ASE spectrum leads to a strong reduction of the side lobes with an increase of the SLSR from 7 to 13 dB. As a side effect, the coherence length decreases by almost 1 μm from 8.6 to 7.7 μm due to the broadening of the combined SLED spectrum by 7 nm.

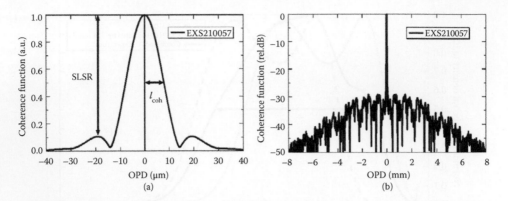

FIGURE 19.10 Coherence function versus optical path length difference (OPD) obtained for the ASE spectrum shown in Figure 19.1e for SLED EXS210057. (a) Linear plot over a small OPD range. The coherence length defined by the half-width at half maximum (HWHM) value is found to be 7.9 μm (in air) and the side-lobe suppression ratio (SLSR) roughly 10 dB. (b) Logarithmic plot over the full OPD range.

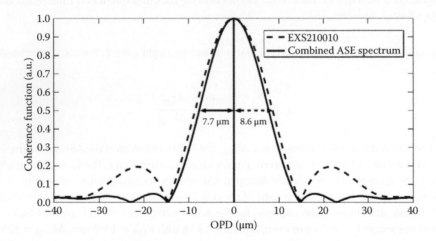

FIGURE 19.11 Coherence function in the near OPD range obtained for the ASE spectra plotted in Figure 19.9 for module EXS210010 (dashed curve) and the combined SLED spectrum (solid curve). From the dashed curve an SLSR of 7 dB and a coherence length of 8.6 μm is extracted, whereas the values found from the solid curve are 13 dB and 7.7 μm, respectively.

Periodic modulations in the ASE spectrum caused by the build-up of a resonant cavity, as discussed in Section 19.2.4, may show up as secondary coherence subpeaks at an OPD of a few millimeters (see, e.g., Figure 18.1 in Duelk and Hsu 2015). Figure 19.10b is free of such secondary coherence peaks as the ASE spectrum is smooth and free of spectral ripple. In order to avoid OCT imaging artifacts (so-called "ghost lines"), a secondary peak suppression ratio (SPSR) of at least 25 dB is needed. The SPSR requirement defines the tolerance for the maximum allowed spectral ripples in the ASE spectrum. As a consequence, this might cause R-SLED designs to be impractical for OCT applications.

19.3.3 High-Power SLED Designs

In addition to the requirements for the ASE output spectrum, SLEDs should typically deliver a minimum amount of output power at a given injection current. A variety of methods exist for increasing the SLED's front-facet output power, and thus achieve a high-power design. In the following, we describe possible approaches and discuss their advantages and drawbacks.

19.3.3.1 Increasing the Injection Current

The most obvious approach is just to increase the injection current. This is possible as long as the point for the thermal roll-over is not reached. But one should take into account that the increasing carrier density may have a strong influence on the shape of the ASE output spectrum. As discussed in Section 19.3.1, ASE spectra resulting from a single optical transition will typically remain bell shaped with broader bandwidth. However, for M-shaped ASE spectra, which may be well balanced for a lower injection current, as shown in Figure 19.1d and e, the low-wavelength peak will strongly grow with increasing injection current. This leads to an unfavorably strong reduction of the spectral bandwidth compared to the preferred flat-top condition.

19.3.3.2 Increasing the Chip Length

The output power can be increased for a given injection current by increasing the active segment length, that is, the length of the gain medium. This method works for operating currents sufficiently above the ASE threshold level because the latter increases with increasing chip length. Longer chips lead to a reduction of both the thermal resistance and the series resistance leading to lower junction temperatures for SLED chips running in CW mode of operation. Moreover, the carrier density is reduced resulting in the opposite effect on the ASE spectra, as explained in Section 19.3.3.1.

19.3.3.3 Using an R-SLED Design

As discussed in Section 19.2.3, a very efficient way to design an ultimate high-power SLED structure is the R-SLED approach. This allows for very high-output powers using relatively short chips. The main drawback of this approach is the enhancement of spectral ripples or even worse the tendency to start lasing due to the build-up of a resonant cavity. This may lead to an increase of the noise level in general and the appearance of strong secondary coherence subpeaks. Hence, the R-SLED design might be the optimum choice for insensitive applications with respect to resonant-cavity effects. For other applications, this approach might not work or is not recommended.

19.3.3.4 High-Power SLEDs by Epitaxial Design

The epitaxial design of the active-region layers has a strong influence on the SLED's *L–I* characteristic. Typically, it is easier to design SLEDs delivering higher-output powers by increasing the number of QWs and/or by increasing the thickness of the QW layers. Modifying the epitaxial design in this way has diverse and complicated effects on the expected electro-optical performance. For example, the injection current required for reaching ASE threshold increases in general with increasing total thickness of the active region. Moreover, the carrier density is reduced, having a similar effect on the ASE spectrum as described earlier in Section 19.3.3.2 for an extended chip length.

Therefore, the approach that has to be chosen for the realization of a specific SLED design depends on the whole set of target specification. The expected *L–I* characteristics and ASE spectra for a given active region design finally determine the appropriate chip length and injection current for device operation. There are no simple and strictly valid rules for finding the optimal SLED design. However, it can be approached using the discussion presented in this section as a guideline.

19.4 Summary

SLEDs are employed over a wide range of different applications with emitted ASE spectra covering the wavelength range from the visible to the near infrared. The band-gap energy of available semiconductor materials determines the appropriate choice of the substrate and epitaxial layer materials for the design of a specific SLED structure. Most SLEDs are realized in a tilted ridge-waveguide geometry similar to a Fabry–Pérot LD, which allows for the formation of optical modes. The target specifications mainly on the output power and the ASE spectrum determine the active-region design. SQW and MQW structures are most commonly used to achieve material gain.

The performance of SLED structures can be simulated like other opto-electronic devices using full 3D simulation tools. Such complex tools couple the different problems involved in the physical device description (multiphysics treatment). The simulation results are rather accurate at the expense of analytical insight. Analytical approaches based on traveling-wave equations describing the longitudinal propagation of light reduce the dimensionality of the problem and the number of coupled equations. They are particularly useful to derive an analytical expression for the SLED's $L–I$ characteristic. We have presented a simple model that includes the influence of arbitrarily strong modal reflectivities at both facets on the output power. Using our analytical results, we have compared the performance of standard SLEDs with the new class of so-called R-SLEDs. Furthermore, our model has been verified by a comparison against experimental data and results obtained from full 3D simulations.

The design of SLEDs with specific target requirements has been discussed. In particular, the appearance of single-peaked (bell-shaped) and double-humped (M-shaped) ASE spectra for broad bandwidth SLED structures and how it affects the corresponding coherence functions has been analyzed in detail. Various methods for boosting the output power have been illustrated with their implications particularly on the ASE spectra. These findings can be finally used as a guideline for the design of a specific broad-bandwidth high-power SLED structure.

References

Alphonse, G. A., D. B. Gilbert, M. G. Harvey, et al. 1988. High-power superluminescent diodes. *IEEE J. Quant. Electron.* 24, no. 12: 2454–2457.

Bohnert, K., P. Gabus, J. Nehring, et al. 2002. Temperature and vibration insensitive fiber-optic current sensor. *IEEE J. Lightwave Technol.* 20, no. 2: 267–275.

Burns, W. K., C. Chen, and R. P. Moeller. 1983. Fiber-optic gyroscopes with broad-band sources. *IEEE J. Lightwave Technol.* 1, no. 1: 98–105.

Chen, T. R., L. Eng, Y. H. Zhuang et al. 1990. Quantum well superluminescent diode with very wide emission spectrum. *Appl. Phys. Lett.* 56, no. 14: 1345–1346.

Chuang, S. L. 1995. *Physics of Optoelectronic Devices.* New York: John Wiley & Sons.

Coldren, L. A., S. W. Corzine, and M. L. Masanovic. 2012. *Diode Lasers and Photonic Integrated Circuits,* 2nd ed. Hoboken, NJ: John Wiley & Sons.

CrosslightSoftware2017

Crosslight Software, Inc. Crosslight Software, 2017. http://www.crosslight.com/

Drexler, W. and J. G. Fujimoto, eds. 2008. *Optical Coherence Tomography: Technology and Applications,* 2nd ed. Berlin: Springer-Verlag.

Duelk, M. and K. Hsu. 2015. SLEDs and swept source laser technology for OCT. In *Optical Coherence Tomography: Technology and Applications,* 2nd ed., Drexler, W. and J. G. Fujimoto (eds.), 527–562. Berlin: Springer-Verlag.

Dufour, M. L., G. Lamouche, V. Detalle, et al. 2005. Low coherence interferometry—An advanced technique for optical metrology in industry. *Insight–Non-Destructive Testing and Condition Monitoring* 47, no. 4: 216–219.

Exalos, A. G. EXALOS: High-performance broadband SLEDs, swept sources and sub-systems, 2017. http://www.exalos.com/sled-modules/

Feltin, E., A. Castiglia, G. Cosendey, et al. 2009. Broadband blue superluminescent light-emitting diodes based on GaN. *Appl. Phys. Lett.* 95: 081107.

Heo, D., I.-K. Yun, J.-S. Lee, et al. 2011. High-power SLD-based BLS module for WDM-PON applications. *J. Korean Phys. Soc.* 58, no. 3: 429–433.

Kaminow, I. P. and D. Marcuse. 1983. Superluminescent LED with efficient coupling to optical waveguide. U.S. Patent 4 376 946.

Kondo, S., H. Yasaka, Y. Noguchi, et al. 1992. Very wide spectrum multiquantum well superluminiscent diode at 1.5 μm. *Electron. Lett.* 28, no. 2: 132–134.

Koyama, F., K. Y. Liou, A. G. Dentai, et al. 1993. Multiple-quantum-well GaInAs/GaInAsP tapered broad-area amplifiers with monolithically integrated waveguide lens for high-power applications. *IEEE J. Photonics Technol. Lett.* 5, no. 8: 916–919.

Kwong, N. S. K., K.-Y. Lau, and N. Bar-Chaim. 2008. High-power, high-efficiency GaAlas superluminescent diodes with integrated absorber for lasing suppression. *IEEE J. Quant. Electron.* 25, no. 4: 696–704.

Laino, V., F. Roemer, B. Witzigmann, et al. 2007. Substrate modes of (Al,In)GaN semiconductor laser diodes on SiC and GaN substrates. *IEEE J. Quant. Electron.* 43, no. 1: 16–24.

Lee, T. P., C. A. Burrus, and B. I. Miller. 1973. A stripe-geometry double-heterostructure amplified-spontaneous-emission (superluminescent) diode. *IEEE J. Quant. Electron.* 9, no. 8: 820–828.

Li, Z. Q. and Z. M. S. Li. 2010. Comprehensive modeling of superluminescent light-emitting diodes. *IEEE J. Quant. Electron.* 46, no. 4: 454–461.

Lin, C.-F., Y.-S. Su, C.-H. Wu, et al. 1996. Broad-band superluminescent diodes fabricated on a substrate with asymmetric dual quantum wells. *IEEE J. Photonics Technol. Lett.* 8, no. 11: 1456–1458.

Lin, C.-F., Y.-S. Su, C.-H. Wu, et al. 2004. Influence of separate confinement heterostructure on emission bandwidth of InGaAsP superluminescent diodes/semiconductor optical amplifiers with nonidentical multiple quantum wells. *IEEE J. Photonics Technol. Lett.* 16, no. 6: 1441–1443.

Loeser, M. and B. Witzigmann. 2008. Multi-dimensional electro-opto-thermal modeling of broadband optical devices. *IEEE J. Quant. Electron.* 44, no. 6: 505–514.

Macleod, H. A. 2010. *Thin-Film Optical Filters*. Boca Raton, FL: Taylor & Francis.

Marcuse, D. 1983. Computer model of an injection laser amplifier. *IEEE J. Quant. Electron.* 19, no. 1: 63–73.

Marcuse, D. 1989. Reflection loss of laser mode from tilted end mirror. *J. Lightwave Technol.* 7, no. 2: 336–339.

Matuschek, N. and M. Duelk. 2013. Modeling and simulation of superluminescent light emitting diodes (SLEDs). *IEEE J. Sel. Top. Quant. Electron.* 19, no. 5: 7800307.

Matuschek, N., T. Pliska, and N. Lichtenstein. 2008. Properties of pump-laser modules exposed to polarization-dependent and wavelength-selective feedback from fiber Bragg gratings. *IEEE J. Quant. Electron.* 44, no. 3: 262–274.

Matuschek, N. and M. Duelk. 2014. Modeling and simulation of reflecting SLEDs. *14th International Conference on Numerical Simulation of Optoelectronic Devices* (NUSOD '14), Postdeadline Poster MPD43, Mallorca, Spain.

Middlemast, I., J. Sarma, and S. Yunus. 1997. High power tapered superluminescent diodes using novel etched deflectors. *Electron. Lett.* 33, no. 10: 903–904.

Milani, N. M., V. Mohadesi, and A. Asgari. 2015. A novel theoretical model for broadband blue InGaN/GaN superluminescent light emitting diodes. *J. Appl. Phys.* 117: 054502.

Nagai H., Y. Noguchi, and S. Sudo. 1989. High-power, high-efficiency 1.3 μm superluminescent diode with buried bent absorbing guide structure. *Appl. Phys. Lett.* 54, no. 18: 1719–1721.

Park, J. and X. Li. 2006. Theoretical and numerical analysis of superluminescent diodes. *IEEE J. Lightwave Technol.* 24, no. 6: 2473–2480.

Patterson, B. D., J. E. Epler, B. Graf, et al. 1994. A superluminescent diode at 1.3 μm with very low spectral modulation. *IEEE J. Quant. Electron.* 30, no. 3: 703–712.

Photon Design. Photon Design - Your source of photonics CAD tools, 2017. https://www.photond.com/

Rossetti, M., J. Dorsaz, R. Rezzonico, et al. 2010. High power blue-violet superluminescent light emitting diodes with InGaN quantum wells. *Appl. Phys. Express.* 3: 061002.

Rossetti, M., L. H. Li, A. Fiore, et al. 2008. Quantum dot superluminescent diodes. In *Handbook of Self Assembled Semiconductor Nanostructures for Novel Devices in Photonics and Electronics*, ed. Henini, M., 565–599. Oxford: Elsevier.

Rossetti, M., J. Napierala, N. Matuschek, et al. 2012. Superluminescent light emitting diodes—The best out of two worlds. *SPIE Photonics West, MOEMS and Miniaturized Systems XI*, 8252–8206.

Schmitt, J. M., 1999. Optical Coherence Tomography (OCT): A Review. *IEEE J. Sel. Top. Quant. Electron.* 5, no. 4: 1205–1215.

Semenov, A. T., V. R. Shidlovski, S. A. Safin, et al. 1993a. Superluminescent diodes for visible (670 nm) spectral range based on AlGaInP/GaInP heterostructures with tapered grounded absorber. *Electon. Lett.* 29, no. 6: 530–532.

Semenov, A. T., V. R. Shidlovski, and S. A. Safin. 1993b. Wide-spectrum SQW superluminescent diodes at 0.8 μm with bent optical waveguide. *Electon. Lett.* 29, no. 10: 854–856.

Senior, J. M. 2009. *Optical Fiber Communications Principles and Practice*. Harlow: Pearson Education Limited.

Sun, Z. Z., D. Ding, Q. Gong, et al. 1999. Quantum-dot superluminescent diode: A proposal for an ultra-wide output spectrum. *Opt. Quant. Electron.* 31: 1235–1246.

Synopsis, Inc. Synopsys, 2017. http://www.synopsys.com/

Tansu, N. and L. J. Mawst. 2001. High-performance strain-compensated InGaAs–GaAsP–GaAs (λ = 1.17 μm) quantum-well diode lasers. *IEEE J. Photonics Technol. Lett.* 13, no. 3: 179–181.

Wild, G. and S. Hinckley. 2009. Distributed optical fibre smart sensors for structural health monitoring: A smart transducer interface module. *5th International Conference Intelligent Sensors, Sensor Networks and Information Processing*, Melbourne, Australia, 373–378.

Wootten, M. B., J. Tan, Y. J. Chien, et al. 2014. Broadband 2.4 μm superluminescent GaInAsSb/AlGaAsSb quantum well diodes for optical sensing of biomolecules. *Semicond. Sci. Technol.* 29, no. 11: 115014.

V

Semiconductor Optical Amplifiers (SOAs)

V

Semiconductor Optical Amplifiers (SOAs)

20

Semiconductor Optical Amplifier Fundamentals

Michael J. Connelly

20.1 Introduction

The rapid growth in the development of optical networks requires small, inexpensive, and easy-to-integrate optical amplifiers for use as basic amplifiers (power boosters, in-line amplifiers to compensate for fiber loss, and optical receiver preamplifiers) and also as optoelectronic signal processing devices such as wavelength converters, optical switches, intensity and phase modulators, logic gates, and dispersion compensators.

There are two main classes of optical amplifier: optical fiber amplifiers (OFAs) and semiconductor optical amplifiers (SOAs). OFAs are optical amplifiers that use optical fiber as the gain medium, which in most cases is a glass fiber doped with rare-earth ions such as erbium (for operation in the 1.55-μm telecommunications band). The active dopant is supplied with energy by an external pump laser. OFAs are relatively large devices, with advantages such as wide optical bandwidth (tens of nm), high gain (>20 dB), high saturation output power (>20 dBm), defined as the output signal power at which the gain is half that of the unsaturated gain, low-noise figure, and low polarization sensitivity. When an OFA is used to amplify an optical data signal, its slow gain dynamics (the lifetime of the excited energy levels are typically in the tens of ms range) are a significant advantage as the amplifier only experiences the average power of the data signal. This results in low intersymbol interference and, when used to amplify wavelength division multiplexed signals, low interchannel cross-talk. However, OFA's slow dynamics preclude its use in optical signal processing applications.

To achieve optical gain, an SOA uses an electrically pumped semiconductor material, as is the case for a semiconductor laser, such that a population inversion occurs between the material conduction and valence bands. An incoming light wave is amplified when the resulting stimulated emission exceeds losses due to stimulated absorption and other material or structural losses. By appropriate choice of the gain material and its bandgap energy, SOAs can be designed to operate in the wavelength region of choice, typically the 1.3- and 1.55-μm optical communication windows (Connelly, 2002a; Dutta and Wang, 2013). Various SOA designs have been shown to achieve gain, noise figure, saturation output power, and optical bandwidths comparable to those for OFAs; however, SOAs can exhibit significant polarization sensitivity. The main advantages of SOAs over fiber amplifiers are their small size and compatibility with photonic integrated circuit (PIC) technology. SOAs have much faster dynamics than OFAs, which can lead to data signal distortion and interchannel cross-talk when operated in the gain-saturated regime. However, high-speed dynamics and nonlinearities can be exploited to realize all-optical signal processing functions.

In this chapter, the basic principles and types of SOA are reviewed followed by descriptions and implementations of relatively simple models that are used to gain insight into important SOA static, dynamic, and nonlinear behavior. The models described are (1) quasi-analytic static model, (2) bulk SOA static model including amplified spontaneous emission (ASE), (3) time-domain model including ASE, which is used to simulate pattern effects and a cross-gain modulation (XGM)-based wavelength converter, (4) pulse amplification analytic model, which is also used to simulate a cross-phase modulation (XPM)-based wavelength converter, and (5) a four-wave mixing (FWM) analytic model. Further chapters describe more detailed descriptions and models of particular types of SOA.

20.2 Basic Principles

The principle of operation of a traveling-wave SOA is shown in Figure 20.1. The amplifier consists of an electrically pumped active waveguide. An incoming light wave is amplified as it propagates through the waveguide. Antireflection (AR) coatings are used to suppress the end facet reflections, which are approximately 32% in uncoated devices. Through the use of AR coatings and optimized waveguide designs (such as using a tilted waveguide geometry), reflectivities as low as 10^{-5} or less can be achieved. The models described in this chapter assume zero facet reflectivities. The input signal experiences a single-pass power gain $G = \exp(gL)$, where g is the gain coefficient at the signal wavelength and L the amplifier length. Although SOA waveguides are designed to be single mode, they support two orthogonal polarization modes, the transverse electric (TE) and transverse magnetic (TM) modes.

The amplification process adds broadband noise (spontaneous emission) to the propagating signal, caused by spontaneous recombination of conduction band electrons to valence band holes. This noise is subsequently amplified, leading to ASE. In the linear operating region, where the gain is constant and the ASE statistics are not affected by the input signal, the output ASE power spectral density (W/Hz) in a single polarization state at the signal energy E_s is given by

$$\rho_{ASE} = n_{sp}E_s(G-1), \tag{20.1}$$

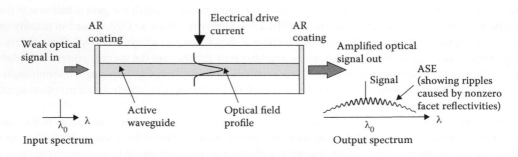

FIGURE 20.1   Semiconductor optical amplifier (SOA) basic structure.

where n_{sp} is the population inversion factor having a maximum possible value of 1. The amplifier noise factor F is defined as the ratio of the amplifier input to output signal-to-noise ratios (SNRs). F quantifies the degradation of the SNR, as determined in terms of the signal and noise levels in the photocurrent of an ideal photodetector placed in the signal path, due to the insertion of the amplifier in a signal transmission system (Baney et al., 2000; Kweon, 2002). The ideal photodetector responsivity $R = e/E$ (having a quantum efficiency equal to 1), where e and E are the electronic charge and light wave photon energy, respectively. Assuming that the input signal is shot-noise limited, the input SNR is equal to $P_s/(2E_s B_e)$, where P_s is the input signal power and B_e the detector electrical bandwidth. In practical receivers, the photodetector is preceded by an optical bandpass filter of bandwidth B_o that passes the signal and greatly reduces the ASE. The average signal and total ASE powers after the optical filter are $P_s G$ and $2\rho_{ASE} B_o$, respectively, with corresponding photodetector shot-noise current variances (A^2), after electrical filtering, of $\sigma_{sig}^2 = 2eRGP_s B_e$ and $\sigma_{sp}^2 = 2eR\rho_{ASE} B_o B_e$ (Olsson, 1989; Kweon, 2002). Since the photodetector is a square-law type detector, the signal also beats with the ASE giving rise to a signal-spontaneous beat noise current variance $\sigma_{s-sp}^2 = 4R^2 GP_s \rho_{ASE} B_e$. The beating of the ASE with itself leads to spontaneous–spontaneous beat noise having a current variance $\sigma_{sp-sp}^2 = 2R^2 \rho_{ASE}^2 B_e (2B_o - B_e)$. The output SNR, SNR_{out}, is the ratio of the square of the signal photocurrent $(RGP_s)^2$ to the sum of the noise variances. If the optical filter bandwidth is small enough such that σ_{sp-sp}^2 and σ_{sp}^2 are negligible compared to the other two noise sources (the signal-spontaneous beat noise limit), then

$$SNR_{out} = \frac{(RGP_s)^2}{4R^2 GP_s \rho_{ASE} B_e + 2eRGP_s B_e}. \tag{20.2}$$

The noise factor is then

$$F = \frac{2\rho_{ASE}}{GE_s} + \frac{1}{G}. \tag{20.3}$$

The noise figure (NF) is F expressed in decibel units, $NF = 10 \log_{10} F$ dB. For large gains, $F \approx 2n_{sp}$. Since the maximum value of n_{sp} is 1, the minimum noise figure possible is 3 dB. Usually, the shot-noise term in Equation 20.3 is much smaller than the signal-spontaneous beat noise term so

$$F = \frac{2\rho_{ASE}}{GE_s}. \tag{20.4}$$

Nonzero reflectivities lead to multiple passes of the signal and ASE, with constructive interference occurring at discrete optical frequencies ν_q that are integer multiples r of the ratio of the speed of light in the amplifier c/n_{eff}, where n_{eff} is the waveguide effective index (defined as the ratio of the propagation constant in the waveguide to the free space propagation constant), to the round-trip distance $2L$. As the latter quantity is much greater than the wavelength, r is very large. The frequency spacing $\Delta\nu$ between adjacent resonances is $\Delta\nu = c/(2n_{eff}L)$. For a typical SOA with $L = 1$ mm and $n_{eff} = 3.5$ and operating in the 1.55-μm region, the wavelength spacing is 1.2 nm. When the effect of nonzero reflectivities is significant, the SOA is termed a Fabry–Pérot SOA (FP-SOA) because of its similarity to a Fabry–Pérot resonator. The signal gain of an FP-SOA at optical frequency ν is given by

$$G_{FP}(\nu) = \frac{(1 - R_1)(1 - R_2)G}{(1 - \sqrt{R_1 R_2}G)^2 + 4\sqrt{R_1 R_2}G \sin^2[\pi(\nu - \nu_r)/\Delta\nu]}, \tag{20.5}$$

where R_1 and R_2 are the SOA input and output facet reflectivities, respectively. FP resonances cause undesirable ripples in the signal gain and ASE spectrums (Figure 20.1). If the single-pass gain is high enough such that the denominator in Equation 20.5 approaches 0, G_{FP} will become very large and the SOA will begin to oscillate, at which point it behaves as a laser and not as an amplifier.

A reflective semiconductor optical amplifier (RSOA) can be formed by applying an AR coating to the input facet and a highly reflective coating to the opposite facet. RSOAs have applications such as modulators in passive optical networks (Lee et al., 2005).

20.2.1 Typical Bulk SOA Structure

A bulk material SOA operating in the 1550-nm region is shown in Figure 20.2 (Connelly, 2007). The active region is sandwiched between two separate confinement heterostructure (SCH) layers, which have a lower refractive index than the active region, and hence confine the light. The p–n junctions formed by the p-type and n-type InP layers act as current blocks, thereby providing good confinement of the injected carriers in the active region. An important SOA geometrical parameter is the optical confinement factor Γ defined as the fraction of the transverse (to the propagation direction) optical intensity overlapping with the active region. In general, Γ is polarization dependent, so the TE and TM confinement factors Γ_{TE} and Γ_{TM} are unequal. They are only equal in an unstrained bulk material SOA having a square cross-section waveguide. Such SOAs are difficult to fabricate as they require optical confinement in both transverse directions. Most commercial SOAs use rectangular cross-section waveguides and consequently, $\Gamma_{TE} \neq \Gamma_{TM}$. Confinement factors can be determined using formulas (Chuang, 2009) or commercial mode solvers. If $\Gamma_{TE} \neq \Gamma_{TM}$ and the material gain is polarization independent, then the SOA gain will be polarization dependent. In Figure 20.2, the introduction of tensile strain between the active region and SCH layers is used to increase the TM to TE material gain ratio, to compensate for the higher Γ_{TE} value caused by the waveguide asymmetry, and thereby reduce polarization sensitivity. The tapered regions act as mode expanders that couple light from the active waveguide to an underlying passive waveguide, thereby simplifying coupling to external optical fibers.

20.2.2 Quantum Well, Dot and Dash SOAs

SOAs that use bulk materials require high transparency current densities (at which the SOA has unity gain). The active layer in quantum well (QW) SOAs have much smaller thicknesses (~5–10 nm) than in conventional bulk structures (~100 nm). In bulk material, the injected carriers can move in three dimensions whereas in a QW they are confined to two dimensions. The enhanced quantum effects result in a significantly different band structure and material gain compared to bulk material. The small thickness of the QW allows many QWs to be stacked to form a multiple quantum well (MQW) active region. MQW SOAs have a number of advantages compared to bulk SOAs. Because of the small active region volume compared to bulk SOAs, a reduced injection current is sufficient to create a large carrier density in the

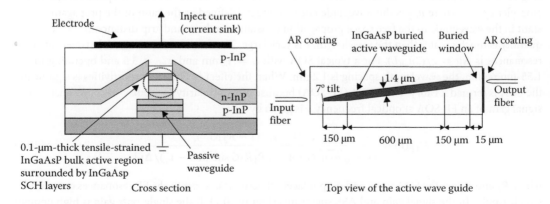

FIGURE 20.2 Typical SOA structure (Connelly, 2007). The buried windows and 7° tilted waveguide (with respect to the end facets) further reduce the effective facet reflectivity.

active region resulting in a broader gain spectrum and shorter carrier lifetime (related to the average time for conduction band electrons to recombine with valence band holes). Shorter carrier lifetimes result in shorter gain recovery times, which is of particular importance in reducing pattern effects when the SOA is used to amplify high-speed data. Furthermore, the loss coefficient in MQW active regions is significantly smaller than that in bulk devices, which leads to an improvement in noise performance. The polarization sensitivity of QW structures can be reduced by using strained QWs. Strain-induced band structure modifications also result in a reduction in loss mechanisms such as Auger recombination and intervalence band absorption (IVBA). The linewidth enhancement factor (LEF) is also reduced, which results in amplified pulses experiencing less spectral broadening, leading to superior high-speed performance as compared to bulk and unstrained QW SOAs. Bulk and QW SOAs have typical gain recovery lifetimes in the range of hundreds of picoseconds; so, when used to amplify data signals, significant pattern effects can be present at baud rates in the region of tens of Gbaud/s. Pattern effects also induce nonlinear phase noise through the self-phase modulation (SPM) effect, which can be a more serious performance degradation factor when advanced modulation formats are used. The amount of SPM is directly related to the LEF.

A quantum dot (QD) is a semiconductor nanostructure with dimensions that typically range from 2 to 10 nm, which confines the motion of injected carriers to all three spatial directions (Akiyama et al., 2007). Compared to bulk and QW SOAs, QD SOAs have been shown to have improved gain bandwidth, noise figure, and saturation output power as well as enhanced nonlinear effects such as FWM. Polarization insensitive QD SOAs have been realized using optimized QD shapes along with close stacking of the QD layers. Of particular significance is the ultrafast gain recovery lifetime, typically of the order of a few picoseconds, a factor of ten lower than that for bulk and QW SOAs so that QD SOAs are capable of amplifying ultrafast data signals with little or no pattern effects (Ben-Ezra et al., 2005). QD SOA LEFs can be much lower than for bulk and QW SOAs and so are particularly suitable for amplifying very high baud rate advanced modulation data signals.

Quantum dash (QDash) SOAs are of interest as an alternative to QD SOAs, since they have some dot-like properties and can more easily be made to operate in the 1.55-μm region, although they also have been shown to have longer gain recovery times in the range of hundreds of picoseconds (Lelarge et al., 2007).

20.3 Quasi-Analytic Static Model

As the input signal power to an SOA is increased, the gain reduces as the electrically pumped conduction band carriers (electrons) in the active material are depleted by stimulated recombination with holes in the valence band. To determine the factors that influence SOA gain at high input powers, a simple traveling-wave-based model can be used (Connelly, 2002a). The carrier density n dependent material gain g_m per unit length at the signal wavelength is assumed to be a linear function $g_m = a(n - n_t)$, where a is the differential of the material gain with respect to n and n_t is the transparency carrier density. The carrier density rate equation is

$$\frac{dn}{dt} = \frac{\eta I}{eV} - \frac{n}{\tau_c} - \frac{\Gamma a(n - n_t)P}{AE_s},$$ (20.6)

where I is the bias current and E_s the signal photon energy. The injection efficiency η is the fraction of the SOA current entering the active region. The active region cross-section area, $A = dW$, where d and W are the active waveguide thickness and width, respectively, and the active region volume $V = AL$. τ_c is the interband carrier lifetime due to nonradiative and radiative spontaneous recombination processes including trap, Auger, and spontaneous emission. The propagation of the signal power P is described by the traveling-wave equation

$$\frac{dP}{dz} = \left[\Gamma a(n - n_t) - \alpha\right] P,$$ (20.7)

where z is the propagation direction (along the amplifier axis) measured from the input and α is the loss coefficient. Under steady-state conditions, the time differential in Equation 20.6 is zero, from which

$$n = \left(\frac{\tau_c \eta I}{eV} \right) \frac{P_{sat}}{P + P_{sat}} + n_t \frac{P}{P + P_{sat}}. \tag{20.8}$$

The saturation power P_{sat} is defined as

$$P_{sat} = \frac{AE_s}{\Gamma a \tau_c}, \tag{20.9}$$

Defining the spatially dependent normalized power $p(z) = P(z)/P_{sat}$, the unsaturated carrier density $n_0 = \tau_c \eta I/(eV)$ and the unsaturated gain coefficient $g_0 = \Gamma a (n_0 - n_t)$, Equation 20.8 can be written as

$$n(z) = \frac{n_0}{1 + p} + \frac{n_t p}{1 + p}. \tag{20.10}$$

Inserting Equation 20.10 into Equation 20.7 gives

$$\frac{dp}{dz} = \left(\frac{g_0}{1 + p} - \alpha \right) p. \tag{20.11}$$

The unsaturated gain $G_0 = \exp[(g_0 - \alpha)L]$. Equation 20.11 can be solved numerically, using, e.g., the Runge–Kutta method, with the boundary condition $p(0) = P_{in}/P_{sat}$, where P_{in} is the input signal power. The amplifier gain G is calculated as the ratio of the output power $P_{out} = p(L)P_{sat}$ to the input power. The calculated gain versus output power is shown in Figure 20.3a for various values of G_0 with $\exp(\alpha L) = 10$. The carrier density is calculated using Equation 20.10. For a given SOA, G_0 corresponds to a particular value of bias current. The normalized saturation output power $p_{o,sat}$ is defined as the normalized output power at which the amplifier gain is half the unsaturated gain. From Figure 20.3a, it can be seen that $p_{o,sat} \approx -3.8$ dB for the 20 dB unsaturated gain curve and almost independent of G_0. The saturation output power $P_{o,sat} = p_{o,sat}P_{sat}$, which from the form of Equation 20.9 can be increased by reducing a and τ_c. In real SOAs, these two parameters are not constants but for a given structure and material depend on wavelength and carrier density.

Consider a square cross-section bulk SOA with parameters $W = d = 0.4$ μm, $L = 600$ μm, $\Gamma = 0.45$, $\tau_c = 0.5$ ns, $a = 1 \times 10^{-20}$ m^2, $n_t = 2.5 \times 10^{24}$ m^{-3}, and $\exp(\alpha L) = 10$. If the unsaturated gain is 20 dB at a bias current of 150 mA and assuming 100% injection efficiency, then $n_o = 4.9 \times 10^{24}$ m^{-3}, $g_0 = 1.1 \times 10^4$ m^{-1} and for a signal wavelength of 1550 nm, $P_{sat} = 9.6$ dBm and so $P_{o,sat} = 5.8$ dBm. Figure 20.3b shows the signal power and carrier density distributions for various values of normalized input power. At input powers $\ll P_{sat}$, the carrier density is uniform throughout the amplifier. Models that include ASE show that below saturation, the carrier density is not uniform but is symmetrical with minima at the SOA ends and a maximum at the center. As the level of saturation increases, the carrier distribution becomes less uniform, until the amplifier is highly saturated, at which the carrier density throughout the amplifier approaches its transparency value. As Figure 20.3b shows, the spatial distribution of the carrier density in a SOA can vary greatly. A further disadvantage of this model is that the spontaneous recombination term in Equation 20.4 is assumed to be a linear function of n. In practice, spontaneous recombination is more accurately modeled as a polynomial function of carrier density, so τ_c is actually carrier dependent.

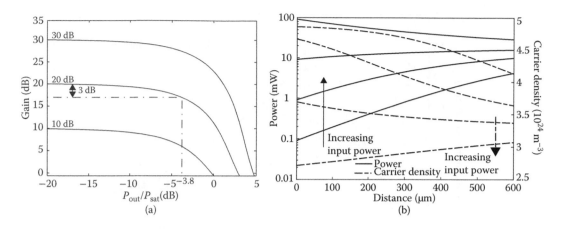

FIGURE 20.3 (a) Gain characteristics for various unsaturated gains. (b) Carrier density and power distributions for normalized input powers of 0.01, 0.1, 1, and 10. The unsaturated gain is 20 dB.

20.4 Bulk SOA Static Model Including ASE

The simple model described above illustrates SOA operational principles, in particular, saturation effects, but it is of limited use in a real SOA analysis. Most importantly, it does not include gain coefficient wavelength dependency or ASE, which contributes to gain saturation and determines the amplifier noise performance. In this section, a modified version of the wideband models for an unstrained bulk SOA that includes ASE (Connelly, 2001, 2007) is considered. The SOA has the same active region geometry and confinement factor as used in the quasi-analytic static model. The bulk material is unstrained, and thereby polarization independent. In the mathematical formulation, Γ_{TE} and Γ_{TM} are assumed to be different but are assumed to be equal in the simulations.

20.4.1 Material Gain

The material gain g_m and spontaneous gain g_{sp} (used to determine the additive spontaneous emission) of the unstrained bulk material are calculated using methods detailed in Jones and O'Reilly (1993) and Connelly (2007). Typical gain spectra are shown in Figure 20.4. The net gain coefficient, at energy E and polarization p (TE or TM), $g_p(n, E) = \Gamma_p g_{m,p} - \alpha_p$ and the loss coefficient $\alpha_p(n) = \alpha_0 + \Gamma_p \alpha_1 n$, where α_0 is the waveguide scattering loss and α_1 the IVBA coefficient loss coefficient (Suematsu and Adams, 1994), with values of 3000 m^{-1} and 7500×10^{-24} m^2, respectively (Connelly, 2001).

20.4.2 Traveling-Wave and Carrier Density Rate Equations

A traveling-wave approach is used to determine signal and the TE and TM ASE photon rates. Transverse variations in the optical field are not considered, and as such, the model has a one-dimensional spatial dependency. The traveling-wave equation for the forward propagating signal photon rate N_s is

$$\frac{dN_s}{dz} = g_s N_s, \tag{20.12}$$

where g_s is the net gain coefficient at the signal polarization and photon energy E_s. The boundary condition at the input $N_s(z = 0) = P_{in}/E_s$, where P_{in} is the input signal power. For a given carrier density spatial

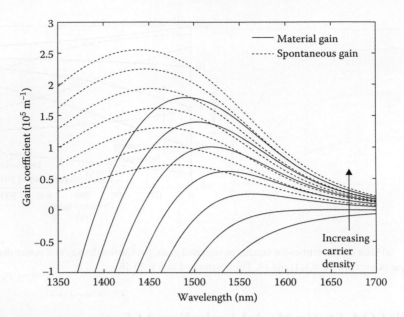

FIGURE 20.4 Unstrained bulk material gain and spontaneous emission gain spectra. The carrier density ranges from 2 to 5×10^{24} m^{-3} in increments of 0.5×10^{24} m^{-3}. The model takes into account intraband relaxation effects and carrier density–dependent bandgap shrinkage. The bandgap wavelength at zero carrier density is 1.55 μm.

distribution, Equation 20.12 has the solution

$$N_s(z) = \frac{P_{in}}{E_s} \exp \left(\int_0^z g_s \, dz \right), \tag{20.13}$$

which can be calculated using numerical integration. To model ASE, a spectral slicing scheme is used in which the ASE is split into M spectral slices, each of energy width ΔE. The traveling-wave equations for the forward (+) and backward (−) propagating kth spectral slice $N_{p,k}^{\pm}$ with polarization p and centered at energy E_k are given by

$$\frac{dN_{p,k}^{\pm}}{dz} = \pm g_{p,k} N_{p,k}^{\pm} \pm \Gamma_p g_{sp,p,k} \Delta E, \tag{20.14}$$

with boundary conditions $N_{p,k}^{+}(z = 0) = 0$ and $N_{p,k}^{-}(z = L) = 0$. The last term on the right-hand side (RHS) of Equation 20.14 is the additive spontaneous emission into the kth spectral slice. For a given carrier density spatial distribution, Equation 20.14 has the solution

$$N_{p,k}^{+}(z) = \exp \left(- \int_0^z g_{p,k} \, dz \right) \int_0^z \left[\exp \left(\int_0^z g_{p,k} \, dz \right) \Gamma_p g_{sp,p,k} \Delta E \right] dz \tag{20.15}$$

$$N_{p,k}^{-}(z) = \exp \left(- \int_z^L g_{p,k} \, dz \right) \int_z^L \left[\exp \left(\int_z^L g_{p,k} \, dz \right) \Gamma_p g_{sp,p,k} \Delta E \right] dz, \tag{20.16}$$

which can be calculated using numerical integration. The carrier density rate equation is

$$\frac{dn}{dt} = \frac{\eta I}{eV} - R(n) - \frac{1}{dW}\left[\Gamma_s g_{m,p} N_s + \sum_{p=TE,TM}\sum_{k=1}^{M}\Gamma_p g_{m,p,k}(N_{p,k}^+ + N_{p,k}^-)\right]. \qquad (20.17)$$

The first term on the RHS of Equation 20.17 is the carrier pumping due to the injected bias current. The injection efficiency η is assumed to be equal to 1. Γ_s is the confinement factor of the signal polarization. The nonstimulated emission recombination rate $R(n)$ is given by

$$R(n) = A_{tr}n + R_{rad}(n) + Cn^3, \qquad (20.18)$$

where A_{tr} and C are the trap and Auger coefficients, respectively. Trap recombination is mainly caused by Shockley–Read–Hall recombination at defects in the material (Fukuda, 1991). The spontaneous radiative recombination rate $R_{rad}(n)$ is commonly assumed to be a bimolecular process approximated by $R_{rad} = Bn^2$, where the bimolecular recombination coefficient B is often taken to be a fitting parameter; however, it can be directly calculated from the material gain (Chuang, 2009). For the material under consideration, $B \approx 2.8 \times 10^{-16} - 0.24 \times 10^{-40}n$ m^{3}s^{-1}. Auger recombination involves three carriers: an electron and hole, which recombine in a band-to-band transition and give off the resulting energy to another electron or hole (Chuang, 2009). Because SOAs usually operate with high carrier (electron) densities, the hole density is almost equal to the carrier density so the Auger recombination rate is proportional to n^3. The Auger coefficient for a particular material or structure can be calculated from first principles, but this is difficult and there is often a high degree of uncertainty in values quoted in the literature. In most models, typical values are used or it is treated as a fitting parameter. Here, C is taken to be equal to 3×10^{-41} m^{6}s^{-1}. At the high carrier densities present in SOAs and if the defect level is low, trap-induced recombination is often assumed to be negligible compared to the other two processes and can be ignored, as is the case here. The third term on the RHS of Equation 20.17 is the amplified signal-induced carrier density depletion rate. $\Gamma_s N_s$ is the photon rate in the active region, which is multiplied by $g_{m,p}$ to obtain the conduction band to valence band carrier recombination rate per unit length. Division by the active region cross-section area results in the carrier density depletion rate. The fourth term on the RHS of Equation 20.17 is the ASE-induced carrier density depletion rate, which is obtained in a similar fashion to that for the amplified signal.

20.4.3 Algorithm and Simulations

Under static conditions, the time differential in Equation 20.17 is equal to zero; the aim of the numerical algorithm is to determine the signal, ASE photon rates, and the carrier density such that this is the case (Connelly, 2007). The model equations cannot be solved analytically, thereby requiring a numerical solution (Connelly, 2001, 2007). Fifty spatial points are used for numerical integration and 150 spectral slices covering a range of 1250–1650 nm. First, the material gain and spontaneous gain are calculated for the spectral slice energies and particular values of carrier density within the expected range. The bimolecular recombination coefficient is calculated for the same carrier density values. Initially, the carrier density in the amplifier is set to some reasonable value (e.g., 3×10^{24} m^{-3}). The signal intensity and ASE photon rates are then estimated using Equations 20.13, 20.15, and 20.16. An updated value for n at each spatial section is then estimated using Equation 20.17 with the time derivative set to zero, and subsequently the gain and spontaneous recombination rate throughout the amplifier, using interpolation of the full calculations, followed by the signal and ASE photon rates. This process is continued until the values of the ASE photon rates converge to a suitable tolerance (0.1%). Amplifier characteristics such as gain, ASE spectrum, and noise figure can then be calculated. The polarization-dependent ASE output spectral density (W/Hz) at E_k is given by

$$\rho_{p,k} = \frac{h E_k}{\Delta E} N_{p,k}^+ (z = L).$$ (20.19)

In the signal-spontaneous beat noise limit, the noise figure spectrum is given by

$$NF_{p,k} = 10 \log_{10} \left(\frac{2\rho_{p,k}}{E_k G_{p,k}} \right),$$ (20.20)

where $G_{p,k}$ is the amplifier gain at E_k (Baney et al., 2000; Olsson, 1989). The maximum gain achievable is mainly limited by ASE. The small-signal gain and ASE spectra are shown in Figure 20.5 for bias currents ranging from 40 to 100 mA. The gain and ASE spectrum peaks shift to shorter wavelengths as the bias current increases, an effect which is present in real SOAs, and they both saturate at high bias currents. The 3-dB optical gain bandwidth at a bias of 100 mA is 30.6 nm. The corresponding ASE spectrum has a 3-dB bandwidth of 34.4 nm; in general, the ASE spectral width is a reasonably good measure of the amplifier gain bandwidth. Figure 20.6a shows the small-signal gain and noise figure bias current dependency. The gain begins to saturate at bias currents at which the ASE saturates; the noise figure reaching a limiting value of approximately 11.7 dB. The ASE and carrier density spatial distributions are shown on Figure 20.6b for no input signal. The latter has a symmetrical profile with a maximum at the center, in contrast to the quasi-analytic model, which does not include ASE. The gain, output signal, and ASE powers versus input power characteristics are shown in Figure 20.7a for a bias current of 100 mA at which the unsaturated gain is 23.3 dB and $P_{\text{o,sat}} = 4.9$ dBm. The reduction in ASE power is clearly seen as the amplifier is driven into saturation by the input signal. As the input power level is increased, the amplified signal begins to compete for the available gain and the carrier density distribution becomes more asymmetrical, as shown in Figure 20.7b for a moderate saturation level.

The model can be used to investigate the effects of different material and geometrical parameters on SOA performance and is thereby useful in SOA design.

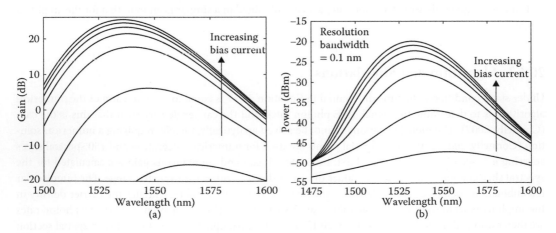

FIGURE 20.5 (a) Gain spectra and (b) amplified spontaneous emission (ASE) spectra (summed over both polarizations) in the absence of an input signal for bias currents of 40–100 mA in increments of 10 mA.

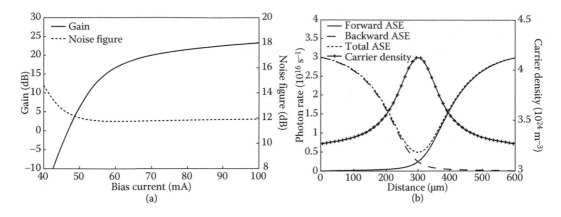

FIGURE 20.6 Small-signal (a) gain and noise figure versus bias current at 1550 nm, (b) ASE and carrier density spatial distributions. The bias current and signal wavelength are 100 mA and 1550 nm, respectively.

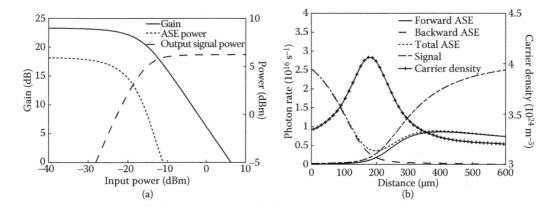

FIGURE 20.7 (a) Gain, output signal and ASE powers versus input power. (b) ASE and carrier density spatial distributions for an input power of −15 dBm, at which the gain has been reduced from its unsaturated value of 23.3–20 dB. The bias current and signal wavelength are 100 mA and 1550 nm, respectively.

20.5 Time-Domain Model Including ASE

The most common application of SOAs is for modulated signal amplification, so it is of interest to model time-domain behavior such as pattern effects, XGM, SPM, and XPM (Obermann et al., 1998; Durhuus et al., 1992; Asghari et al., 1997; Occhi et al., 2003). By including the time differential term in Equation 20.17, the static model described above can be used to model SOA dynamics having timescales as short as hundreds of picoseconds (Connelly, 2002b). The carrier density is initialized to some suitable value and the initial values of the signal and ASE photon rates are set to zero. The carrier density at the next time step is determined by solving Equation 20.17 using the modified Euler method, which is a second-order technique (Press et al., 2007) and for a given time step is significantly more accurate than simply replacing the time differential by a first-order finite difference (Euler method). Nonetheless, an appropriate time-step size must be chosen to prevent stability issues and ensure accuracy. Assuming that the propagation time through the amplifier is much less than the time variation of the optical input and current, the ASE and signal photon rates can be determined using Equations 20.13, 20.15, and 20.16. The new values of the photon rates are then used in Equation 20.17 to determine the carrier density at the next time step until the time span of interest is completed.

The carrier density rate Equation 20.17 can be expressed as

$$\frac{dn}{dt} = \frac{\eta I}{eV} - n\left(\frac{1}{\tau_c} + \frac{1}{\tau_s}\right), \tag{20.21}$$

where the inverse of the nonstimulated emission carrier lifetime τ_c is given by

$$\frac{1}{\tau_c} = \frac{R(n)}{n}, \tag{20.22}$$

and the inverse of the stimulated recombination carrier lifetime τ_s is given by

$$\frac{1}{\tau_s} = \frac{1}{n\,dW}\left[\Gamma_s g_{m,p}N_s + \sum_{p=\text{TE,TM}}\sum_{k=1}^{M}\Gamma_p g_{m,p,k}(N_{p,k}^+ + N_{p,k}^-)\right]. \tag{20.23}$$

The total effective carrier lifetime τ_{eff} is

$$\tau_{\text{eff}} = \left(\frac{1}{\tau_c} + \frac{1}{\tau_s}\right)^{-1}, \tag{20.24}$$

which depends on the amplifier operating point and input optical power and also has a strong spatial dependency. The smaller the τ_{eff}, the faster the gain recovery time. Figure 20.8 shows the output power, spatially averaged effective carrier lifetime $\tau_{\text{eff,m}}$, and instantaneous gain for an amplified 5-Gb/s non-return-to-zero (NRZ) data stream with high and low powers of 0.1 mW and 0.1 μW, respectively. The bias current, unsaturated gain, and signal wavelength are 100 mA, 23.3 dB, and 1550 nm, respectively. The amplified data distortion is due to the dynamic changes of the carrier and photon distributions causing pattern effects on a timescale approximately given by $\tau_{\text{eff,m}}$, which for the simulations shown in Figure 20.8 can vary between 370 and 850 ps. At higher bit rates the severity of pattern effects increases and consequently their effect on system performance.

20.5.1 XGM Wavelength Converter

XGM occurs when an intensity-modulated input data signal (the pump) causes dynamic changes in the carrier density and thereby the gain experienced by a copropagating or counter-propagating light wave (the probe) at a different wavelength to the data signal. Carrier density changes also cause changes in the active region refractive index and thereby the phase of both the pump and probe leading to SPM and XPM. XGM can be used to realize wavelength converters in which data is copied from an input pump data signal to an input continuous wave (CW) probe. An XGM-based 2.5-Gb/s wavelength converter with copropagating pump and probe light waves is shown in Figure 20.9. When the pump power is high, the gain experienced by the probe is reduced and vice versa, hence the inverted pump data is transferred to the probe. The above time-domain model is used to simulate the converter, as shown in Figure 20.9. The performance degrades as the bit rate increases, as the dynamic gain is unable to follow the variations in the input pump power. Because of their superior dynamics, compared to bulk or QW SOAs, QD SOAs can be used as XGM wavelength converters at much higher bit rates.

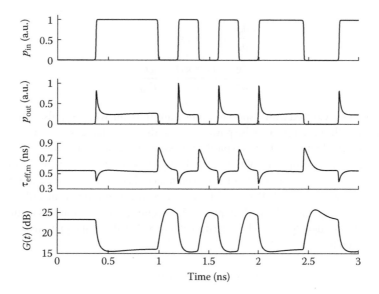

FIGURE 20.8 Simulated SOA output power, spatially averaged effective carrier lifetime, and instantaneous gain for an input NRZ 5-Gb/s data sequence. The data high- and low-level powers are 0.1 mW and 0.1 μW, respectively. The bias current, unsaturated gain, and signal wavelength are 100 mA, 23.3 dB, and 1550 nm, respectively. The simulation time step is 2.5 ps.

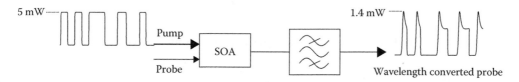

FIGURE 20.9 Cross-gain modulation (XGM)-based wavelength converter and simulated wavelength converted probe output. The bandpass filter is used to select the probe signal. The pump input is a 30-dB extinction ratio NRZ 2.5-Gb/s data sequence with a peak power and wavelength of 5 mW and 1550 nm, respectively. The input probe power and wavelength are 10 μW and 1560 nm, respectively. The bias current is 100 mA.

20.6 Pulse Amplification Analytic Model

Because SOAs have wide optical bandwidths and fast gain recovery lifetimes, they can be used to amplify optical pulses with full width at half-maximum (FWHM) pulse widths in the picosecond range. By exploiting SOA nonlinearities using materials with intrinsically faster gain recovery times, it is possible to amplify femtosecond pulses. Many pulse amplification models are described in the literature; here a simple model (Agrawal and Olsson, 1989), which does not include ASE, is described. The model is applicable to pulse widths greater than the intraband relaxation time (typically <0.1 ps). The amplifier is assumed to be polarization independent. The optical field can be written as $A(t) = \sqrt{P(t)} \exp[j\varphi(t)]$, where P and φ are the temporal power and phase. The gain coefficient g at the signal wavelength is modeled as $g = \Gamma a (n - n_{\mathrm{tr}})$. The power, phase, and gain coefficient dynamics are determined by solving

$$\frac{\partial P}{\partial z} = (g - \alpha)P \tag{20.25}$$

$$\frac{\partial \varphi}{\partial z} = -\frac{1}{2}\alpha_H g \tag{20.26}$$

$$\frac{\partial g}{\partial \tau} = \frac{(g - g_0)}{\tau_c} - \frac{gP}{E_{sat}}. \tag{20.27}$$

τ is the pulse local time measured with respect to a time frame moving with the pulse. The temporal dependence of g, induced by the propagating pulse power, leads to dynamic changes in the pulse phase, i.e., SPM. τ_c is the interband carrier lifetime, which is typically the order of hundreds of picoseconds in bulk and QW SOAs, and α_H is the LEF—a proportionality factor relating phase changes to changes in the gain (Henry, 1982). α_H is often taken to be a constant, but for a given material is carrier density and wavelength dependent. The saturation energy $E_{sat} = E_s Wd/(\Gamma a)$. To determine the evolution of the pulse shape and phase, a numerical solution of Equations 20.25 through 20.27 is required. However, if it is assumed that $\alpha << g$ then Equations 20.23 through 20.25 can be solved to obtain a closed-set of equations for the output pulse power and phase given by

$$P_{out}(\tau) = G(\tau)P_{in}(\tau) \tag{20.28}$$

$$\varphi_{out}(\tau) = \varphi_{in}(\tau) - \frac{\alpha_H}{2}h(\tau), \tag{20.29}$$

where the instantaneous gain $G(\tau) = \exp[h(\tau)]$. If the input FWHM pulse width $\tau_p << \tau_c$, then the integrated gain $h(\tau)$ is given by

$$h(\tau) = -\ln\left\{1 - \left(1 - \frac{1}{G_0}\right)\exp\left[-\frac{U_{in}(\tau)}{E_{sat}}\right]\right\}, \tag{20.30}$$

where $G_0 = \exp(g_0 L)$ is the unsaturated gain, g_0 is the unsaturated gain coefficient, and

$$U_{in}(\tau) = \int_{-\infty}^{\tau} P_{in}(\tau')\,d\tau'. \tag{20.31}$$

The output pulse frequency chirp, defined as the time derivative of its instantaneous phase divided by 2π, is given by

$$\Delta\nu_{out}(\tau) = \Delta\nu_{in}(\tau) + \frac{\alpha(G_0 - 1)}{4\pi G_0}\frac{P_{out}(\tau)}{E_{sat}}\exp\left[\frac{-U_{in}(\tau)}{E_{sat}}\right], \tag{20.32}$$

where $\Delta\nu_{in}$ is the input pulse chirp. The model is applicable to pulse widths of the order of picoseconds to tens of picoseconds. For pulse widths >1 ps, nonlinear intraband processes such as carrier heating (CH) and spectral hole burning (SHB) that have characteristic time constants in the subpicosecond range are important; so, models that take such effects into account are required (Hong et al., 1994; Mecozzi and Mork, 1997). Once the output pulse power and phase are known, its power spectrum can be calculated. To show the effects of an SOA on an amplified pulse consider an unchirped zero phase Gaussian input pulse,

$$P_{in}(\tau) = \frac{E_{in}}{\tau_0\sqrt{\pi}}\exp\left[-\left(\frac{\tau}{\tau_0}\right)^2\right], \tag{20.33}$$

where E_{in} is the pulse energy. τ_o is related to τ_p by $\tau_p \approx 1.665\tau_0$. Simulated amplified pulse shape, chirp, and spectrum are shown in Figure 20.10. The amplified pulse is asymmetric because the leading edge of

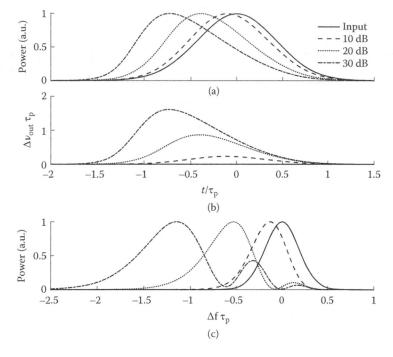

FIGURE 20.10 Amplified pulse; (a) power, (b) chirp, and (c) spectrum. Δf is the frequency deviation from the input light wave center frequency. The input is a zero-chirp Gaussian pulse with $E_{in}/E_{sat} = 0.1$. The parameter is the unsaturated gain and $\alpha_H = 5$.

the pulse experiences a larger gain than the trailing edge. The amplified pulse is chirped and its spectrum is broader than the input spectrum. At high gains, the amplified pulse spectrum exhibits a multipeak structure caused by SPM-induced frequency chirp.

20.6.1 SOA XPM Mach–Zehnder Wavelength Converter

XGM-based wavelength converters suffer from the disadvantage that high-input pump powers are required to obtain the large SOA gain modulation depth necessary to achieve a large converted signal extinction ratio. The associated high-carrier depletion rates lead to a significant increase in the gain recovery lifetime, which limits the data rates at which such converters can operate. To overcome the trade-off between speed and modulation depth SOA XPM can be used (Asghari et al., 1997). XPM occurs when an intensity-modulated input data signal (the pump) causes dynamic changes in the carrier density and, consequently, the active region refractive index and thereby the phase of a copropagating or counter-propagating probe signal at a different wavelength to the pump. In an SOA XPM-based wavelength converter, the probe phase changes are converted into amplitude changes by incorporating one or two SOAs in an interferometric structure such as the Mach–Zehnder interferometer (MZI) shown in Figure 20.11. If sufficient phase shift is achieved with a low-gain modulation index, and the interferometer visibility is high, then the XPM converter performance will be superior to XGM-based converters. An optical delay line is used to adjust the relative delay between the modulated input pump pulses reaching each SOA. The interferometer output probe power, assuming that destructive interference occurs between the recombining probe light waves when their relative phases are equal, is given by

$$p(t) = \frac{P_{probe}}{4}\left[G_1(t) + G_2(t) - 2\sqrt{G_1(t)G_2(t)}\cos\left(\varphi_1(t) - \varphi_2(t)\right)\right], \tag{20.34}$$

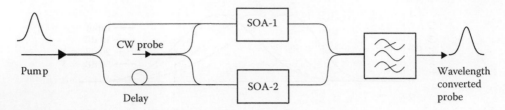

FIGURE 20.11   SOA Mach–Zehnder interferometer (MZI) wavelength converter.

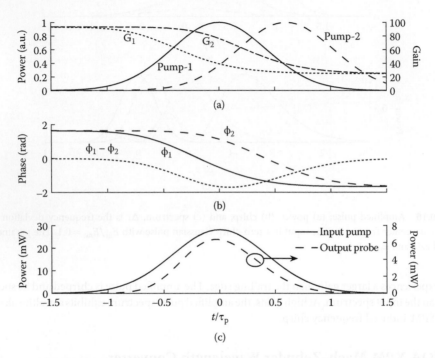

FIGURE 20.12   SOA MZI wavelength converter simulations. (a) SOA input pump powers and dynamic gains, (b) induced SOA probe phase shifts and difference, (c) comparison between the input pump and converted probe pulse shapes.

where P_{probe} is the input CW probe power, $G_1(t)$, $\varphi_1(t)$, $G_2(t)$, and $\varphi_2(t)$ are the gain and phase changes of the SOA in the upper and lower interferometer arms, respectively. The couplers have splitting and coupling ratios of 3 dB. The pulse amplification model described above can also be used to model XPM in the case where the input pump pulse width is much less than the intraband lifetime and assuming that the gain has fully recovered since the previous pulse. Including the effects of relatively slow interband processes requires a numerical solution. In the following simulations, an MZI with two identical SOAs having gains of 20 dB and $\alpha_H = 5$ is considered. The input pump is a zero-chirp Gaussian pulse with $E_{\text{in}}/E_{\text{sat}} = 0.03$. The CW input probe power is -10 dBm. The delay between the pump pulses entering SOA-1 and SOA-2 is equal to 0.6 of the input pump pulse width. These settings were chosen to obtain a converted probe pulse shape similar to the input pump pulse, as shown in Figure 20.12. The input pump pulse causes dynamic gain saturation and associated probe phase shifts. In the unsaturated state, the phase difference between the probe light waves in the interferometer arms is zero, leading to total destructive interference and thereby a zero output probe signal. The arrival of the pump pulses in the SOAs creates a window in which the magnitude of the phase difference increases, leading to constructive interference and thereby a nonzero converted probe signal pulse.

20.7 FWM Analytic Model

FWM is a *coherent* nonlinear process that can occur in an SOA between two copolarized input optical fields, a strong pump and a weaker probe at angular frequencies ω_0 and $\omega_0 - \Omega$, where Ω is the pump and probe detuning Agrawal (1988). Beating between the optical fields results in the establishment of gain and refractive index pulsations in the SOA at the detuning frequency. In most practical applications of FWM, in particular wavelength conversion, the detuning is much larger than the inverse of the interband carrier lifetime (typically in the nanosecond range) and the pulsations are caused by fast intraband processes, in particular CH and SHB. The gain and index pulsations result in nondegenerate FWM, in which energy from the pump and probe is coupled into a new copropagating field at frequency $\omega_0 + \Omega$, called the conjugate, so termed because its dynamic phase is the opposite of the signal phase.

SHB is caused by the injected pump signal creating a hole in the *intraband* carrier distribution. This modulates the occupation probability of carriers within a band leading to fast gain modulation. CH is caused by stimulated emission and free carrier absorption. Stimulated emission subtracts carriers that are cooler than average while free carrier absorption moves carriers to higher-energy levels in the band. The resulting increase in temperature decreases the gain. Two characteristic times are associated with CH: the carrier–phonon scattering time τ_1, which is the average time carriers require to cool down to the semiconductor lattice temperature, and the carrier–carrier scattering time τ_2, which is the average time taken by the carrier population to reach a heated equilibrium from the initial nonheated equilibrium. The latter time constant is also associated with SHB. The characteristic times are of the order of tens to hundreds of femtoseconds, so FWM-based wavelength converters can operate with detuning frequencies as high as hundreds of GHz. A comprehensive analysis of FWM in SOAs requires a numerical solution to a set of coupled-mode equations Agrawal (1988). The electric field in the SOA can be expressed as

$$E(z, t) = E_1 \exp\left\{j[k_1 z - (\omega_0 - \Omega)t]\right\} + E_0 \exp\left[j(k_0 z - \omega_0 t)\right] + E_2 \exp\left\{j[k_2 z - (\omega_0 + \Omega)t]\right\}, \quad (20.35)$$

where the copropagating and copolarized pump, probe, and conjugate field amplitudes are E_0, E_1, and E_2, respectively, with corresponding frequencies of ω_0, $\omega_0 - \Omega$, and $\omega_0 + \Omega$, and propagation coefficients k_0, k_1, and k_2. If it is assumed that the conjugate power is small relative to the copropagating input pump and probe powers, then an analytic model can be used to predict the SOA FWM characteristics (Mecozzi et al., 1995). In this model, the FWM conversion efficiency η defined as the ratio of the output conjugate to input probe powers is given by

$$\eta = G\left|G_1\right|^2. \quad (20.36)$$

G is the saturated amplifier gain, which can be calculated using Equation 20.11, where p is taken to be the sum of the pump and probe powers. The unsaturated gain $G_0 = \exp[(g_0 - \alpha)L]$, where g_0 and α are the unsaturated gain and loss coefficients, respectively.

$$G_1 = \frac{-1 + j\alpha_H}{\alpha_H} \exp\left[-\frac{1}{2}\sigma F_{cd}(\Omega)\right] \sin\left[\frac{\alpha_H}{2}\sigma F_{cd}(\Omega)\right] - \frac{1}{2}\varepsilon_{sh} P_{sat} H_{sh}(\Omega) \sigma F_{sh} - \frac{1}{2}\varepsilon_{ch} P_{sat} H_{ch}(\Omega) \sigma F_{ch}$$

$$(20.37)$$

$$F_{cd}(\Omega) = \frac{1}{1 - j\Omega\tau_c\zeta}\left[\ln\left(\frac{1 + GP_T/P_{sat} - j\Omega\tau_c}{1 + P_T/P_{sat} - j\Omega\tau_c}\right) + \zeta\ln\left(\frac{G_0}{G}\right)\right] \quad (20.38)$$

$$F_{ch} = -\frac{1}{\zeta}\left[\frac{P_T}{P_{sat}}(G - 1) - \ln\left(\frac{G_0}{G}\right)\right] \quad (20.39)$$

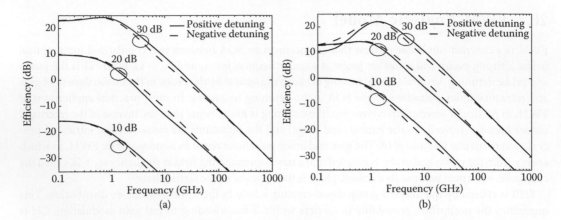

FIGURE 20.13 Four-wave mixing (FWM) efficiency versus absolute value of frequency detuning for input pump and probe powers of (a) 200 and 20 μm, and (b) 20 and 2 μm, respectively.

$$F_{sh} = \ln\left(\frac{G_0}{G}\right) \tag{20.40}$$

$$S(0) = P_{sat}\left(\frac{1-\zeta}{\zeta}\right)\frac{1-(G/G_0)^\zeta}{G-(G/G_0)^\zeta} \tag{20.41}$$

$$\sigma = \frac{P_0}{P_0 + P_1}. \tag{20.42}$$

ε_{ch} and ε_{sh} are parameters characterizing the strengths of CH and SHB processes, respectively, and $\zeta = \alpha/g_0$. P_0 and P_1 are the input pump and probe powers, respectively, and the total input power $P_T = P_0 + P_1$. The Fourier transforms H_{ch} and H_{sh} of the nonlinear gain responses due to CH and SHB, respectively, are given by

$$H_{ch}(\Omega) = \frac{1}{(1-j\Omega\tau_1)(1-j\Omega\tau_2)} \tag{20.43}$$

$$H_{sh}(\Omega) = \frac{1}{1-j\Omega\tau_2}. \tag{20.44}$$

The spatial dependence of the saturated gain G can be determined using Equation 20.9, where p is taken to be the sum of the normalized pump and probe powers. Figure 20.13 shows the calculated efficiency for different values of amplifier unsaturated gain and input pump/probe powers with model parameters: $P_{sat} = 10$ mW, $\alpha_H = 4.0$, $\tau_c = 0.25$ ns, $\tau_1 = 750$ ps, $\tau_2 = 15$ ps, $\varepsilon_{sh} = 10.0$ W^{-1}, $\varepsilon_{ch} = 2.5$ W^{-1}, and $\exp(\alpha L) = 5$ dB. The efficiency characteristics are asymmetric and show a more peaked structure as the amplifier gain is increased. The efficiency can be greater than unity for frequencies up to hundreds of GHz, i.e., frequency conversion with gain, so FWM-based SOA wavelength converters are of particular importance in wavelength division multiplexed systems with wide wavelength channel spacing.

20.8 Conclusion

SOAs have many uses in optical networks, not only as basic amplifiers but also as fundamental components in all-optical signal processing applications. Although SOAs have been investigated for many years, the development of new SOA structures and incorporation in PICs, advanced materials such as QDs, and exploitation of ultrafast nonlinearities make SOA research and development an exciting field of study.

The successful advancement of SOA technology requires the development of accurate models, which can be a significant challenge requiring a deep knowledge and understanding of many diverse areas such as photonic materials, device design, optoelectronic nonlinearities, noise modeling, and communication systems design. This chapter has given an overview of SOA basics and used relatively simple models to explain important SOA characteristics and applications. Further chapters consider more advanced models applied to particular types of SOAs.

References

Agrawal, A. P. 1988. Population pulsations and nondegenerate four-wave mixing in semiconductor lasers and amplifiers. *J. Opt. Soc. Am. B* 5:147–159.

Agrawal, A. P. and Olsson, A. 1989. Self-phase modulation and spectral broadening of optical pulses in semiconductor laser amplifiers. *IEEE J. Quantum Electron.* 25:2297–2306.

Akiyama, T., Sugawara, M. and Arakawa, Y. 2007. Quantum-dot semiconductor optical amplifiers. *Proc. IEEE* 95:1757–1766.

Asghari, M., White, I. H. and Penty, R. V. 1997. Wavelength conversion using semiconductor optical amplifiers. *J. Lightwave Technol.* 15: 1181–1190.

Baney, D. M., Gallion, P. and Tucker, R. S. 2000. Theory and measurement techniques for the noise figure of optical amplifiers. *Opt. Fiber Technol.* 6:122–154.

Ben-Ezra, Y., Haridim, M. and Lembrikov, B. I. 2005. Theoretical analysis of gain-recovery time and chirp in QD-SOA. *IEEE Photonics Technol. Lett.* 17:1803–1805.

Chuang, S. L. 2009. *Physics of Optoelectronic Devices*. New York, NY: Wiley.

Connelly, M. J. 2001. Wide-band semiconductor optical amplifier steady-state numerical model. *IEEE J. Quant. Electron.* 37:439–447.

Connelly, M. J. 2002a. *Semiconductor Optical Amplifiers*. Boston, MA: Springer.

Connelly, M. J. 2002b. Wideband dynamic numerical model of a tapered buried ridge stripe semiconductor optical amplifier gate. *IEE Proc. Circ. Dev. Syst.* 149:173–178.

Connelly, M. J. 2007. Wide-band steady-state numerical model and parameter extraction of a tensile-strained bulk semiconductor optical amplifier. *IEEE J. Quant. Electron.* 43:47–56.

Durhuus, T., Mikkelsen, B. and Stubkjaer, K. E. 1992. Detailed dynamic model for semiconductor optical amplifiers and their crosstalk and intermodulation distortion. *J. Lightwave Technol.* 10:1056–1065.

Dutta, N. K. and Wang, Q. 2013. *Semiconductor Optical Amplifiers*. Singapore: World Scientific Publishing.

Fukuda, M. 1991. *Reliability and Degradation of Semiconductor Lasers and LEDs*. Boston, MA: Artech House.

Henry, C. H. 1982. Theory of the linewidth of semiconductor lasers. *IEEE J. Quant. Electron.* 18:259–264.

Hong, M. Y., Chang, Y. H., Dienes, A., Heritage, J. P. and Delfyett, P. J. 1994. Subpicosecond pulse amplification in semiconductor laser amplifiers: Theory and experiment. *IEEE J. Quant. Electron.* 30:1122–1131.

Jones, G. and O'Reilly, E. P. 1993. Improved performance of long-wavelength strained bulk-like semiconductor lasers. *IEEE J. Quant. Electron.* 29:1344–1345.

Kweon, G. 2002. Noise figure of optical amplifiers. *J. Korean Phys. Soc.* 41:617–628.

Lee, W., Park, M. Y., Cho, S. H., Lee, J., Kim, B. W., Jeong, G. and Kim, B. W. 2005. Bidirectional WDM-PON based on gain-saturated reflective semiconductor optical amplifiers. *IEEE Photon. Technol. Lett.* 17:2460–2462.

Lelarge, F., Dagens, B., Renaudier, J., Brenot, R., Accard, A., van Dijk, F., Make, D., Le Gouezigou, O., Provost, J. G., Poingt, F. and Landreau, J. 2007. Recent advances on InAs/InP quantum dash based semiconductor lasers and optical amplifiers operating at 1.55 μm. *IEEE J. Sel. Top. Quantum Electron.* 13:111–124.

Mecozzi, A. and Mork, J. 1997. Saturation induced by picosecond pulses in semiconductor optical amplifiers. *J. Opt. Soc. Am. B* 14:761–770.

Mecozzi, A., Scotti, S., D'Ottavi, A., Iannone, E. and Spano, P. 1995. Four-wave mixing in traveling-wave semiconductor optical amplifiers. *IEEE J. Quant. Electron.* 31:689–699.

Obermann, K., Kindt, S., Breuer, D. and Petermann, K. 1998. Performance analysis of wavelength converters based on cross-gain modulation in semiconductor-optical amplifiers. *J. Lightwave Technol.* 16:78–85.

Occhi, L., Schares, L. and Guekos, G. 2003. Phase modeling based on the α-factor in bulk semiconductor optical amplifiers. *IEEE J. Sel. Top. Quantum Electron.* 9:788–797.

Olsson, N. A. 1989. Lightwave systems with optical amplifiers. *J. Lightwave Technol.* 7:1071–1082.

Press, W. H., Teukolsky, S. A., Vetterling, W. T. and Flannery, B. P. 2007. *Numerical Recipes: The Art of Scientific Computing*, 3rd ed. New York, NY: Cambridge University Press.

Suematsu, Y. and Adams, A. R. 1994. *Handbook of Semiconductor Lasers and Photonic Integrated Circuits.* London: Chapman & Hall.

21

Traveling-Wave and Reflective Semiconductor Optical Amplifiers

Angelina Totović

and

Dejan Gvozdić

21.1 Introduction

Eras in modern human history have often been labeled by their most significant and widely adopted technological achievements. It is no wonder that the last several decades proudly bear the name "The Information Age." Information, in its many forms, is an invaluable asset and its dissemination, collection, storage, and analysis have been the primary focus of researchers in numerous fields. Ever-increasing hunger for information exchange has driven supporting technologies to new levels. The transition from electric to optical domain has proved to be one of the most significant milestones of modern communications. According to TeleGeography and Huawei Marine Networks, as of early 2017, it has been estimated that close to 1.1 million kilometers of submarine optical cables exist, each bearing several optical fibers, and approximately 99% of all international telecommunications traffic is carried over this infrastructure. Starting from the overloaded backbone links, optical technology was gradually implemented in the lower levels of network hierarchy, with the ultimate goal of reaching individual users in the foreseeable future. Decreasing attenuation, eliminating electromagnetic interference, reducing the power consumption, and diminishing the possibility of eavesdropping are just some of the advantages the new technology has realized (Ramaswami, 2002). With advantages also come challenges, as we are gravitating toward increased efficiency, which includes increased capacity over larger distances with decreased signal degradation and power consumption, and aggregation of fixed and mobile networks into a single network architecture (Carapellese et al., 2014). Aside from information exchange, information processing and storage in the optical domain remain in focus as we pursue all-optical solutions to mitigate unnecessary electro-optic and opto-electric conversion (Ramaswami, 2002; Sygletos et al., 2008). Many steps have already been taken, both in the physical layer, where new materials and production technologies significantly influenced performance, and in the higher levels of abstraction, where new modulation formats and protocols have been introduced (Ramaswami, 2002; Gladisch et al., 2006; Sygletos et al., 2008). Devices that are simple, but can perform multiple functions, have always been highly regarded. One example is the semiconductor optical amplifier (SOA)—a photonic device that under adequate conditions can amplify the input optical signal. SOAs have been extensively studied and perfected over the past three decades. Similar to laser diodes, they are compact and easy to produce and integrate with other photonic devices on a single chip (Xing et al., 2004). Wide amplification spectrum makes them suitable for use in wavelength-division multiplexed (WDM) optical networks (Mecozzi and Wiesenfeld, 2001; Zimmerman and Spiekman, 2004; Kani, 2010). SOAs consume relatively little power (Xing et al., 2004; Sygletos et al., 2008; Koenig et al., 2014), they are transparent to the optical signal modulation format (Schmuck et al., 2013; Koenig et al., 2014), and they can be designed to be polarization insensitive (Mathur and Dapkus, 1992; Carlo et al., 1998; Michie et al., 2006).

Aside from being used as standalone amplifiers, SOAs can be used for numerous other important functions (Olsson, 1989; Mecozzi and Wiesenfeld, 2001), such as switching (Stabile and Williams, 2011; Figueiredo et al., 2015), modulation/remodulation (Totović et al., 2011; Pham, 2014), wavelength conversion (Joergensen et al., 1997; Dailey and Koch, 2009), signal regeneration (Sygletos et al., 2008), all-optical flip-flops (Pitris et al., 2015), and all-optical random access memory (RAM) cells (Vyrsokinos et al., 2014), to name just a few.

21.2 General Structure and Operation Principles of TW- and R-SOA

Although SOAs come in many different forms, they all share several common features. Each SOA consists of an active region and a cladding and can have either highly reflective or antireflective facets, or a combination of the two, as shown in Figure 21.1. The two most commonly used SOA structures are the traveling-wave (TW) SOA and the reflective (R) SOA.

In the TW-SOA, both facets are coated with an antireflective layer, and the optical signal travels only once through the active region. In other words, input and output ports of TW-SOA are on opposite sides. This is preferable configuration for using SOA as a standalone amplifier, whether as a booster, in-line, or preamplifier (Olsson, 1989; Xing et al., 2004). On the other hand, the R-SOA has one antireflective and one highly reflective facet, which causes the signal to be reflected and travel through the active region twice. Both input and output ports are on the same side, namely at the antireflective facet. The rear facet can be made semitransparent, rather than highly reflective, which is usually the case if the interface between the cleaved semiconductor facet and air is responsible for reflection. This may be a useful feature, since now two output signals can be used, one at the rear and one at the front facet. An SOA with both facets highly reflective is the Fabry–Pérot (FP) SOA (Adams et al., 1985; Thylén, 1988). Due to the feedback loop provided by the highly reflective facets, the FP-SOA closely resembles a laser diode, although it is not typically used to generate coherent signal by itself (Yariv and Yeh, 2007). Nevertheless, its spectrum does exhibit pronounced resonances and antiresonances, unlike TW- and R-SOAs, where ripples in spectrum are much more subtle and are the result of residual reflectivity of the antireflective facets (Yariv and Yeh, 2007).

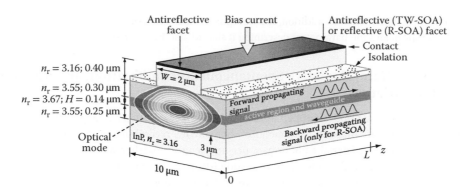

FIGURE 21.1 Schematic representation of the semiconductor optical amplifier (SOA), with relevant geometrical parameters and signals denoted. The active region is based on the unstrained bulk $In_{0.53}Ga_{0.47}As$ and designed to be polarization insensitive. (R-SOA, reflective SOA; TW-SOA, traveling-wave SOA.)

The active region is responsible for the signal amplification through the mechanism of stimulated emission, where the electric bias current is used to achieve population inversion. The purpose of the built-in waveguide is to confine the optical signal that propagates along the SOA and prevent its leaking into the cladding. The output signal of any SOA should ideally be an amplified replica of the input signal, although, in practice, a certain amount of noise is always superimposed, due to the always-present spontaneous emission. Aside from the photons initially generated through the process of spontaneous emission, noise propagation leads to its amplification, which finally results in amplified spontaneous emission (ASE) noise. Although the noise is typically regarded as the unwanted by-product of the device operation, there are numerous examples of using the noise either to generate or deliberately modify the optical signal (Yamatoya and Koyama, 2004; de Oliveira Ribeiro et al., 2005; Kang et al., 2006; Gebrewold et al., 2015), or enhance the overall SOA performance (Valiente et al., 1996).

SOAs can also be classified based on the material used for the active region, where each material has advantages and disadvantages. The simplest choice is bulk semiconductor (Michie et al., 2006; Connelly, 2007; Mazzucato et al., 2015; Totović et al., 2015). However, due to the several drawbacks of bulk SOA, such as high recovery time (Zilkie et al., 2007), and broad ASE spectrum (Totović et al., 2013), other solutions have emerged.

Depending on the desired improvement, different materials with a higher degree of confinement can be used (Zilkie et al., 2007), starting from multiple quantum well (MQW) (Nagarajan et al., 1992; Keating et al., 1999; Qin et al., 2012), and continuing to quantum dash (QDash) (Gioannini, 2004; Reithmaier et al., 2005; Qasaimesh, 2013) and quantum dot (QD) materials (Berg et al., 2001; Qasaimesh, 2003; Kim et al., 2009). SOAs with the active region based on these materials usually exhibit better performance in one or more aspects, at the expense of increased complexity and production cost. For example, QD SOA has fast recovery time, which is crucial for ultrafast signal processing, but due to the dot size distribution, it has moderate material gain and wide ASE spectrum, with full-width at half-maximum (FWHM) reaching double the values reported for MQW and QDash SOAs (Zilkie et al., 2007). On the other hand, QDash SOA has high differential gain and requires much lower driving currents in comparison with other active region types and outperforms bulk SOA in terms of time needed for gain recovery (Reithmaier et al., 2005).

21.3 Modeling of Material and Structural Parameters

Before proceeding to the SOA analysis, it is necessary to understand the physical principles responsible for its operation and quantify them through the parameters typically used for description of photon–electron interaction in the active media. In addition, the waveguide should be optimized to provide polarization-insensitive operation, and the parameters describing it should also be studied.

21.3.1 Material Gain, Radiative Spontaneous Recombination, and Refractive Index Variation

Some of the most important optical properties of an SOA's active layer, determining its functional performance and parameters, are material gain, radiative spontaneous recombination (spontaneous emission) rate, and refractive index. Material gain, or gain per unit length, together with several other structural parameters, is responsible for defining the transmission (device) gain, i.e., the ratio of powers of the output and input optical signals, which is a figure of merit for the amplifier's capability to amplify the input optical signal by stimulated emission of photons (Connelly, 2002). Radiative spontaneous recombination rate and its spectral distribution represent the defining parameters that sculpt ASE spectrum and the SOA's noise (Silver et al., 2000; Connelly, 2002), whereas the refractive index variation directly affects the SOA linewidth enhancement factor (LEF) (Wang et al., 2007), further influencing the SOA's performance in cross-phase modulation (XPM) systems and optical switching.

In SOAs, stimulated emission has interband character, meaning that it occurs between the conduction band (CB) and the valence band (VB) of the semiconductor. Since these electronic bands consist of dispersed energy levels with different density per unit energy and occupation probability, the material gain significantly differs from gain in other solid-state or gas optically active media, which have a few (usually three or four) energy levels involved in the process of stimulated emission and its provisioning. The spreading of energy levels in the bands leads to a broad spectrum of material gain, which increases with the increase of injected carrier density or, equivalently, the current density. The radiative stimulated recombination rate is generally proportional to material gain (Chuang, 1995; Coldren et al., 2012), similar to the radiative spontaneous recombination rate (Chuang, 1995; Coldren et al., 2012). The material gain, as well as the radiative spontaneous recombination rate and the refractive index change, is, generally speaking, the function of three quantities: the joint density of states, electron/hole occupation probability, and a constant corresponding to a single optical transition between the two energy levels, one in CB and the other in VB.

The model of material gain, and the other two parameters, depends on the degree of carrier confinement in the active region of the SOA and the semiconductor strain introduced by the confinement. However, the models may differ with respect to degree of accuracy, i.e., approximations adopted in the model. The common framework for calculating optical properties of semiconductor bulk and quantum confined structures is the envelope function approximation (Meney et al., 1994). Within this framework, a relatively simple, so-called "two-band" model is used for material gain calculation (Chuang, 1995). It is based on the assumption of parabolic band dependence $E(k)$, where E is the energy and k is the wave vector of electron (hole) in the conduction (valence) band. Each band is treated separately by the effective mass Schrödinger equation (Chuang, 1995). More sophisticated multiband envelope function models account for the VB mixing effect, which essentially means that the character of sub-bands is a mixture of heavy hole (HH), light hole (LH), and split-off (SO) band. This effect occurs as a result of quantum confinement and becomes more prominent as the degree of confinement increases.

The most common multiband envelope function models are the Luttinger–Kohn (LK) 4×4 (Chuang, 1991) and 6×6 models (Chao and Chuang, 1992). The former model (LK 4×4) accounts for HH and LH band mixing, whereas the latter (LK 6×6) includes SO band, in addition to HH and LH bands. A more advanced multiband approach is based on the 8×8 $\mathbf{k} \cdot \mathbf{p}$ model (Meney et al., 1994; Liu and Chuang, 2002; Gvozdić and Ekenberg, 2006), which comprises all three VBs and the CB. In addition to nonparabolicity of VB, this model accounts for nonparabolicity of CB as well. However, its fundamental drawback is the occurrence of the spurious solutions, which can be avoided to some extent by various techniques (Foreman, 1997; Cartoixa et al., 2003; Kolokolov et al., 2003; Veprek et al., 2007).

21.3.1.1 Two-Band Model

Although relatively rough, the two-band model might be used in case of bulk, and even other types of quantum confined structures, but with modest accuracy. It is suitable for the unstrained bulk active region since the optical matrix element is isotropic in this case (Chuang, 1995), which means that material gain is the same for any direction of the amplified light polarization. Contrarily, in case of quantum confined semiconductor structures, material gain is polarization dependent. Nevertheless, the two-band model can still be implemented even in the case of quantum wells. If the optical dipole matrix element is averaged over the azimuthal angle corresponding to the plane of the quantum well, it is possible to derive its angular dependence with respect to the confinement direction of the quantum well (Chuang, 1995).

Here, the detailed expressions for material gain g, spontaneous emission rate per unit energy r_{sp} and refractive index variation Δn_r are presented for the unstrained bulk active region, according to the two-band model (Chuang, 1995). They are all functions of photon energy $\hbar\omega$ and carrier density n, which is included in expressions through the carrier-dependent quasi-Fermi levels:

$$g(n, \hbar\omega) = C_g \frac{M_b^2}{4\pi^2} \sum_{i=HH,LH,SO} \left(\frac{2\mu_i}{\hbar^2}\right)^{3/2} \int_{E_{g,i}}^{\infty} \frac{\gamma}{\pi} \frac{\sqrt{x - E_{g,i}}}{(x - \hbar\omega)^2 + \gamma^2} \left(f_{FD}^c - f_{FD}^v\right) dx, \tag{21.1}$$

$$r_{sp}(n, \hbar\omega) = \frac{8\pi n_r^2 (\hbar\omega)^2}{h^3 c^2} C_g \frac{M_b^2}{4\pi^2} \sum_{i=HH,LH,SO} \left(\frac{2\mu_i}{\hbar^2}\right)^{3/2} \int_{E_{g,i}}^{\infty} \frac{\gamma}{\pi} \frac{\sqrt{x - E_{g,i}}}{(x - \hbar\omega)^2 + \gamma^2} f_{FD}^c \cdot \left(1 - f_{FD}^v\right) dx, \tag{21.2}$$

$$\Delta n_r(n, \hbar\omega) = \frac{q^2 \hbar^2}{n_r \varepsilon_0 m_0^2} \frac{M_b^2}{2\pi^2} \sum_{i=HH,LH,SO} \left(\frac{2\mu_i}{\hbar^2}\right)^{3/2} \int_{E_{g,i}}^{\infty} \frac{(x - \hbar\omega)\sqrt{x - E_{g,i}}}{x(x + \hbar\omega)\left[(x - \hbar\omega)^2 + \gamma^2\right]} \left(f_{FD}^v - f_{FD}^c\right) dx. \tag{21.3}$$

In previous equations, $C_g = \pi q^2/(n_r c \varepsilon_0 m_0^2 \omega)$, where q is the absolute electron charge, m_0 is the free-electron mass, n_r is the refractive index of the active region, c is the speed of light in vacuum, ε_0 is the vacuum permittivity, $\omega = 2\pi\nu$ is the angular photon frequency, with ν being its frequency, and $\hbar = h/(2\pi)$ is the reduced Planck's constant, with h being the Planck's constant. In Equations 21.1 through 21.3, $M_b^2 = m_0 E_p/6$ stands for the bulk momentum matrix element squared, where E_p is the interband matrix element, $\mu_i = (m_e^{-1} + m_i^{-1})^{-1}$ and $E_{g,i}$ are the reduced effective mass and the energy gap between the CB bottom E_c and the VB top $E_{v,i}$, respectively, where i stands for the HH, LH, and SO band. With $f_{FD}^c(E) = f_{FD}^c[E_c + \mu_i/m_e \cdot (x - E_{g,i})]$ and $f_{FD}^v(E) = f_{FD}^v[E_{v,i} - \mu_i/m_i \cdot (x - E_{g,i})]$ we denote the Fermi–Dirac distribution $f_{FD}^{(j)}(E) = \{1 + \exp[(E - E_f^{(j)}/k_B T)]\}^{-1}$ for the CB ($j = c$) and the VB ($j = v$), characterized by the corresponding quasi-Fermi levels $E_f^{(j)}$, where k_B is the Boltzmann constant and T is the temperature. The half linewidth of the Lorentzian function is denoted by γ.

21.3.1.2 Multiband Model

The LK 4 × 4 and 6 × 6 Hamiltonians are usually implemented in the analysis of quantum confined structures. However, this approach can be very useful in the calculation of gain of strained bulk material (Connelly, 2007), which is polarization dependent in this case. As shown by Connelly (2007), and Mazzucato et al. (2015), eigenvectors of multiband Hamiltonian can be used to find the components of the optical matrix elements along the directions corresponding to polarizations of incoming light.

Totović et al. (2013) present an analysis of the material gain, radiative spontaneous recombination rate spectrum, and refractive index variation in MQWs based on the 8 × 8 **k·p** method (Liu and Chuang, 2002; Gvozdić and Ekenberg, 2006). The analysis is done for an MQW active region consisting of six coupled 0.13% tensile-strained $In_{0.516}Ga_{0.484}As$ quantum wells, which are strain-compensated by the $In_{0.9}Ga_{0.1}As_{0.3}P_{0.7}$ 0.26% compressively strained barriers. The well thickness is $L_w = 19$ nm, while the barriers between the wells are $L_b = 10$ nm thick, making the total thickness of the structure $L_z = 6L_w + 5L_b = 164$ nm. The electronic band structure calculation accounts for Burt–Foreman Hermitianization (Foreman, 1993) and biaxial strain generated by lattice-mismatched growth of the well-barrier layers. By using an appropriate basis set, the 8×8 Hamiltonian is decoupled into two 4×4 Hamiltonians (H_U and H_L) corresponding to the upper (U) and lower (L) block Hamiltonians (Liu and Chuang, 2002). In calculating the band structure, the finite difference method (FDM) is employed. For determining g, r_{sp}, and Δn_r, the following relations are used (Chuang, 1995; Liu et al., 2001; Liu and Chuang, 2002):

$$g(n, \hbar\omega) = \frac{q^2\pi}{n_r c\varepsilon_0 m_0^2 \omega L_z} \sum_{\eta=U,L} \sum_{\sigma=U,L} \sum_{n,m} \int \left|\mathbf{e} \cdot \mathbf{M}_{nm}^{\eta\sigma}\right|^2 \frac{\gamma}{\pi} \frac{f_{\eta n}^c - f_{\sigma m}^v}{\left(E_{nm}^{\eta\sigma} - \hbar\omega\right)^2 + \gamma^2} \frac{k_t dk_t}{2\pi}, \tag{21.4}$$

$$r_{sp}(n, \hbar\omega) = \frac{q^2 n_r \omega}{\pi\hbar c^3 \varepsilon_0 m_0^2 L_z} \sum_{\eta=U,L} \sum_{\sigma=U,L} \sum_{n,m} \int \left|M_{sp}\right|^2 \frac{\gamma}{\pi} \frac{f_{\eta n}^c \cdot \left(1 - f_{\sigma m}^v\right)}{\left(E_{nm}^{\eta\sigma} - \hbar\omega\right)^2 + \gamma^2} \frac{k_t dk_t}{2\pi}, \tag{21.5}$$

$$\Delta n_r(n, \hbar\omega) = \frac{q^2\hbar^2}{n_r \varepsilon_0 m_0^2 L_z} \sum_{\eta=U,L} \sum_{\sigma=U,L} \sum_{n,m} \int \left|\mathbf{e} \cdot \mathbf{M}_{nm}^{\eta\sigma}\right|^2 \frac{E_{nm}^{\eta\sigma} - \hbar\omega}{E_{nm}^{\eta\sigma} \cdot \left(E_{nm}^{\eta\sigma} + \hbar\omega\right)} \frac{f_{\eta n}^v - f_{\sigma m}^c}{\left(E_{nm}^{\eta\sigma} - \hbar\omega\right)^2 + \gamma^2} \frac{k_t dk_t}{2\pi}, \tag{21.6}$$

where η and σ stand for one of the two 4×4 Hamiltonians (H_U or H_L) and corresponding to the nth conduction and the mth valence sub-band, respectively, and k_t is the in-plane wave vector, while $f_{\eta n}^c(k_t)$ and $f_{\sigma m}^v(k_t)$ are the Fermi–Dirac distributions for CB and VB, respectively. In these relations, $E_{nm}^{\eta\sigma} = E_{nm}^{\eta\sigma}(k_t)$ is the energy difference between the eigenenergy of the nth conduction sub-band corresponding to the η-Hamiltonian and mth valence sub-band of the σ-Hamiltonian, whereas $\mathbf{M}_{nm}^{\eta\sigma} = \mathbf{M}_{nm}^{\eta\sigma}(k_t)$ is the corresponding momentum matrix element vector (Liu and Chuang, 2002). Depending on the unit polarization vector $\mathbf{e}$, the components of the $\mathbf{M}_{nm}^{\eta\sigma}$ vector correspond to the transverse-electric (TE) mode ($\mathbf{e} = \mathbf{x}$ or $\mathbf{z}$) or the transverse-magnetic (TM) mode ($\mathbf{e} = \mathbf{y}$). In the calculation of r_{sp}, the average matrix element for the three polarization directions is used, i.e., $|M_{sp}(k_t)|^2 = (2|\mathbf{M}_{TE}(k_t)|^2 + |\mathbf{M}_{TM}(k_t)|^2)/3$.

The spectral dependencies of material gain, radiative spontaneous recombination rate, and refractive index variation are shown in Figure 21.2. In Figure 21.2a through c, the results are given for the bulk unstrained $In_{0.53}Ga_{0.47}As$ active region, calculated using the two-band model and the parameters listed in Table 21.1. In Figure 21.2d through f, the same quantities, calculated using the multiband model, are given for an MQW active region consisting of six coupled 0.13% tensile-strained $In_{0.516}Ga_{0.484}As$ quantum wells, which are strain-compensated by the $In_{0.9}Ga_{0.1}As_{0.3}P_{0.7}$ 0.26% compressively strained barriers. The spectral dependencies for the MQW active region are given only for TE polarization and are determined based on the parameters given in Table 21.2.

In addition to bulk and QWs, the SOA's active region often comprises self-assembled QDs. The calculation of the optical properties of QDs can be based on sophisticated $\mathbf{k\cdot p}$ models, such as 8×8 $\mathbf{k\cdot p}$ model (Grundmann et al., 1995; Stier et al., 1999). However, due to the statistical nature of the QD size, very precise calculation is usually meaningless. Therefore, some simple methods can also be sufficiently effective as in the two-band model (Kim and Chuang, 2006; Kim et al., 2008).

21.3.2 Waveguide Design and Polarization Insensitivity

In addition to high transmission gain, high optical output power, low-noise, and low-energy consumption, in many SOA applications polarization insensitivity of modal gain is required. This demand can be difficult to fulfill since polarization dependence of modal gain is caused by SOA's waveguide geometry, through confinement factor, and material choice, through the optical transitions. Therefore, careful analysis and design of the waveguide structure and material gain are needed (Labukhin and Li, 2006). Numerical

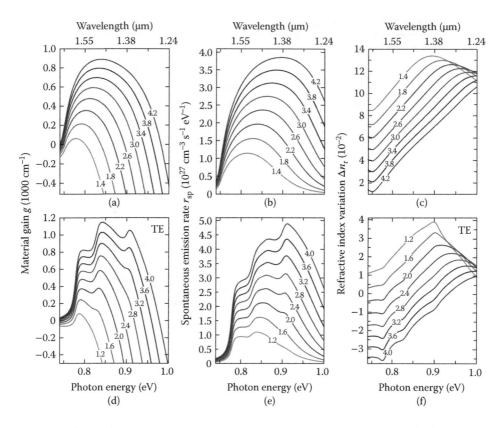

FIGURE 21.2 Spectral dependency of (a, d) material gain, (b, e) radiative spontaneous recombination rate per unit energy, and (c, f) refractive index variation for different carrier densities n (10^{18} cm^{-3}) for (a) through (c) bulk and (d) through (f) multiple quantum well (MQW) active region.

TABLE 21.1 Material Parameters Used for Band Calculation for Bulk SOA

Symbol	Quantity	In$_{0.53}$Ga$_{0.47}$As	In$_{0.76}$Ga$_{0.24}$As$_{0.52}$P$_{0.48}$
E_g	Energy gap CB–VB (HH, LH)	748 meV	1001 meV
$E_{g,SO}$	Energy gap CB–VB (SO)	1115 meV	1246 meV
Δ_{SO}	SO energy	367 meV	245.3 meV
E_P	Interband matrix element	24.93 eV	23.26 eV
m_e^*	Electron effective mass	0.0389 m_0	0.0541 m_0
m_{HH}^*	HH effective mass	0.3410 m_0	0.4069 m_0
m_{LH}^*	LH effective mass	0.0567 m_0	0.0889 m_0
m_{SO}^*	SO effective mass	0.1393 m_0	0.1709 m_0

SOA, semiconductor optical amplifier; LH, light hole; HH, heavy hole; SO, split-off; CB, conduction band; VB, valence band.

modeling of the active region's susceptibility, the real and imaginary parts of which correspond to refractive index and material gain, in combination with the calculation of the waveguide propagation constant and confinement factor is an inevitable and important step in the quest for the optimal, polarization-insensitive design of SOA. Waveguide polarization sensitivity comes from significant disproportion of its cross-section dimensions, where the width is usually one order of magnitude larger than its thickness, raising the difference in TE and TM wave propagation and the corresponding confinement factors. Moreover, as shown in

TABLE 21.2   Material Parameters Used for Band Calculation for MQW SOA

Symbol	Quantity	$In_{0.516}Ga_{0.484}As$	$In_{0.9}Ga_{0.1}As_{0.3}P_{0.7}$
E_g	Energy gap	765.2 meV	1107.3 meV
$E_{v,av}$	Average VB position	−6.6689 eV	−6.8986 eV
Δ_{SO}	SO energy	366.3 meV	189.2 meV
E_p	Interband matrix element	25.03 eV	21.91 eV
m_e^*	Electron effective mass	$0.0395\,m_0$	$0.0617\,m_0$
γ_1	Luttinger parameters	10.1534	6.0906
γ_2		3.6114	1.9545
γ_3		4.5580	2.6785
c_{11}	Elastic stiffness constants	1020.7 GPa	996.79 GPa
c_{12}		507.48 GPa	536.03 GPa
a_c	CB deformation potential	−6.091 eV	−5.941 eV
a_v	VB deformation potential	−1.077 eV	−0.802 eV
b	Shear deformation	−1.897 eV	−1.918 eV

SOA, semiconductor optical amplifier; MQW, multiple quantum well; SO, split-off; CB, conduction band; VB, valence band.

Section 21.3.1, quantum-confined and stressed bulk semiconductors in active region exhibit polarization-dependent dipole optical matrix elements, leading to additional variation of TE and TM modes. An unstrained bulk material is polarization isotropic and in this case square waveguide cross-section may provide SOA polarization insensitivity (Connelly, 2001; Michie et al., 2006). Unfortunately, such waveguide design is technologically very demanding (Connelly, 2001; Michie et al., 2006). Moreover, it leads to large far-field divergence and consequently to poor coupling efficiency from the SOA to optical fiber (Connelly, 2002). This can be overcome by tapering the active waveguide near the amplifier facets (Connelly, 2002; Michie et al., 2006).

Nevertheless, there are other solutions for polarization-insensitive operation of SOA proposed in the literature. The majority of these proposals concern MQW SOA and rely on any of the following concepts: MQWs with tensile barriers (Magari et al., 1991, 1994), tensile-strained QWs (Ito et al., 1998; Carrère et al., 2010), tensile-strained QWs with compressive barriers (Godefroy et al., 1995; Zhang and Ruden, 1999), alternation of tensile and compressive QWs (Mathur and Dapkus, 1992; Joma et al., 1993; Tiemeijer et al., 1993; Tishinin et al., 1997; Silver et al., 2000), and the delta-strained QW (Carlo et al., 1998; Cho and Choi, 2001). The concept based on MQWs with tensile barriers (Magari et al., 1991, 1994) essentially relies on considerable increase of barrier refractive index for TM mode in comparison with TE mode, as a consequence of LH band contribution, which modifies the bandgap and shifts the wavelength toward the longer wavelength side due to the tensile strain. The approach based on tensile-strained MQWs with compressive barriers, proposed by Zhang and Ruden (1999), involves the barrier width optimization such that the HH sub-bands are grouped tightly and the LH sub-bands are widely separated in energy. The uppermost valence sub-bands, which have large occupation probability and strongly contribute to the gain, consist of a single LH sub-band and a group of coupled HH sub-bands, giving rise to balanced gains for the TE and TM polarizations. The method of alternation of separate tensile and compressive QWs (Silver et al., 2000) requires thick QWs (around 14 nm) and with high strain (around −0.8%). Due to the thick tensile layers, the difference in density of states of the two types of wells leads to significant charge redistribution within the active region. The space charge modifies the band profile and forces holes to move from the tensile well to the compressive well, increasing TE gain. In order to compensate for hole redistribution, the number of tensile wells can be increased so that they outnumber the compressive wells.

Otherwise, the thickness of the compressive wells can be increased to minimize the charge redistribution effects.

In unstrained MQWs, HH band sub-bands are usually higher than those of LH band, resulting in higher TE gain than the TM one. In delta-strained QWs (Cho and Choi, 2001), the delta layer introduces larger VB discontinuity for HH bands than LH bands, and the quantized energy levels for HH bands experience a shift downward, in contrast to those corresponding to LH bands, which shift upward. As a consequence, the TM transition strength for the transition between the first conduction sub-band and the top LH-like valence sub-band is much larger than the TE transition strength.

In addition to polarization insensitivity, SOA has to provide as high as possible transmission gain. Totović et al. (2013), presented a case study, where both design criteria have been implemented: high device gain and polarization insensitivity. The study analyzes two types of SOA waveguides, with active regions described in Section 21.3.1: unstrained bulk and tensile-strained MQWs, which are strain compensated by compressively strained barriers. In order to satisfy the first design criterion, the upper limit of the confinement factor is set to approximately 30%, thereby preventing the strong influence of ASE on SOA saturation (de Valicourt, 2012). Due to the gain isotropy, polarization-insensitive operation of bulk SOA requires equal confinement factors for both polarizations ($\Gamma_{TE} = \Gamma_{TM}$) (Connelly, 2001; Labukhin and Li, 2006; Michie et al., 2006). This requires optimization of the waveguide structure since confinement factor Γ_{TE} is usually larger than Γ_{TM} (Connelly, 2002; Labukhin and Li, 2006; Michie et al., 2006). Instead of using the unstrained, it is possible to implement a tensile-strained bulk active region, which leads to the polarization and gain anisotropy (Michie et al., 2006; Connelly, 2007). If the difference between the confinement factors is sufficiently small, a carefully chosen amount of tensile strain (Michie et al., 2006; Connelly, 2007) may compensate for the difference in the confinement factors Γ_{TE} and Γ_{TM} without the waveguide optimization. This can provide an efficient way for polarization-insensitive operation of bulk SOAs.

Similarly, tensile-strained MQWs in the active region, for which TM gain dominates over TE gain, may compensate for the difference in confinement factors Γ_{TE} and Γ_{TM}, leading to balanced modal gains for both polarizations, $\Gamma_{TE}g_{TE} = \Gamma_{TM}g_{TM}$, in the wavelength range of interest. The design of the MQW active region and the waveguide as a whole can be performed by a self-consistent iterative procedure involving calculation of the confinement factors and material gain, which ultimately, should provide optimized well/barrier dimensions, strain, number of wells, and waveguide cladding. The self-consistent procedure becomes important if the calculation of the confinement factor accounts for the refractive index variation of the active region, associated with variation of the material gain, and caused by carrier injection into the MQW structure (Totović et al., 2013). For the MQW active region from Section 21.3.1, the procedure leads to the confinement factor ratio $\Gamma_{TM}/\Gamma_{TE} = 0.806$, with $\Gamma_{TE} = 22\%$. The same procedure is applied in the optimization of the waveguide for unstrained bulk SOA. The optimization is performed for 1.55 µm and an average carrier density of 2×10^{18} cm^{-3}, for both bulk and MQW SOAs. The modal gains for TE and TM polarization of the optimized MQW active region are compared in Figure 21.3 for a range of photon energies and carrier densities.

It should be brought to attention that the temperature variation can significantly affect the refractive index and consequently the confinement factor. If efficient heat sink is not available for SOA, temperature effects should be accounted for in the design of the waveguide. More details on temperature effects will be given in Section 21.4.3.4.

The proposed design of the MQW waveguide is similar to those described by Wünstel et al. (1996) and Wolfson (2001) in both optical confinement and layer thicknesses. A rigorous treatment of the confinement factor should take into account the variation of the refractive index imaginary part, i.e., gain, in the active region (Visser et al., 1997). However, in this case study, the focus was set only on the variation of the real part of refractive index Δn_r due to the carrier injection. Although for other carrier densities and photon energies confinement factors somewhat detune from the optimal values and ratio, they partially compensate for the material gain ratio detuning, toning down the difference in modal gains for the TE and TM modes.

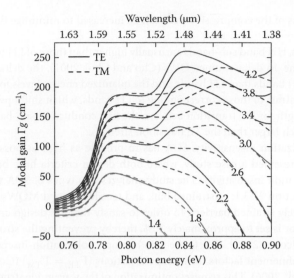

FIGURE 21.3 Spectral dependency of modal gain for polarization insensitive multiple quantum well (MQW) active region for the transverse-electric (TE) (solid lines) and transverse-magnetic (TM) (dashed lines) modes, for different carrier densities n (10^{18} cm^{-3}).

Distribution of the transversal components of the electric $\mathbf{E}(x, y)$ and magnetic $\mathbf{H}(x, y)$ fields is determined by a system of Helmholtz equations (Kawano and Kitoh, 2001):

$$\nabla_\perp^2 \mathbf{E}(x, y) + \left[(n_r + \Delta n_r)^2 k_0^2 - \beta^2 \right] \mathbf{E}(x, y) = 0, \tag{21.7}$$

$$(n_r + \Delta n_r)^2 \, \nabla_\perp \left[\frac{1}{(n_r + \Delta n_r)^2} \nabla_\perp \mathbf{H}(x, y) \right] + \left[(n_r + \Delta n_r)^2 k_0^2 - \beta^2 \right] \mathbf{H}(x, y) = 0, \tag{21.8}$$

where $\Delta n_r = \Delta n_r(n, \hbar\omega)$ is the refractive index variation, which is nonzero only in the active region of SOA, k_0 is the wave vector, $\beta = n_{\text{eff}} k_0$ is the wave propagation constant, and n_{eff} is the effective index of refraction, also dependent on n and $\hbar\omega$. Equations are solved using the finite element method (FEM), with all boundaries set to the Neumann boundary conditions, except for the top ridge, which is set to the Dirichlet boundary condition.

Confinement factors for TE and TM polarizations are defined as (Kawano and Kitoh, 2001)

$$\Gamma_{\text{TE}} = \int_{\text{a.r.}} \left| \mathbf{E}(x, y) \right|^2 dxdy \Big/ \int_{-\infty}^{+\infty} \left| \mathbf{E}(x, y) \right|^2 dxdy, \tag{21.9}$$

$$\Gamma_{\text{TM}} = \frac{1}{n_{\text{a.r.}}^2} \int_{\text{a.r.}} \left| \mathbf{H}(x, y) \right|^2 dxdy \Big/ \sum_i \frac{1}{n_i^2} \int_i \left| \mathbf{H}(x, y) \right|^2 dxdy, \tag{21.10}$$

where a.r. stands for the active region and i stands for each subdomain of the waveguide, including the active region. It should be noted that the background part of the refractive index $n_{r,i}$, with excluded Δn_r for the active region, is derived from Adachi's interpolation formulas (Adachi, 1989), and it is consequently a function of photon energy, i.e., $n_{r,i} = n_{r,i}(\hbar\omega)$. Since this dependence has abrupt changes, quantities shown in Figure 21.4 cannot exhibit smooth behavior.

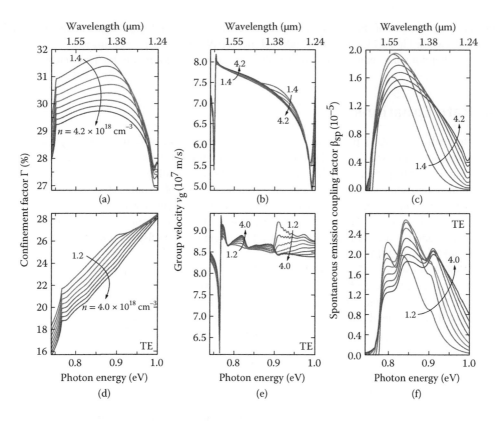

FIGURE 21.4 Spectral dependency of (a, d) confinement factor, (b, e) group velocity, and (c, f) spontaneous emission coupling factor for different carrier densities n (10^{18} cm^{-3}) for (a) through (c) bulk and (d) through (f) multiple quantum well (MQW) active region.

Another important parameter for ASE noise modeling in SOAs is the spontaneous emission coupling factor $\beta_{sp}(n, \hbar\omega)$, which is defined as the spontaneous emission rate coupled to the one optical mode, relative to the total spontaneous emission rate, $R_{sp}(n) = \int r_{sp}(n, \hbar\omega)d(\hbar\omega)$, and can be calculated according to the following expression (Coldren et al., 2012):

$$\beta_{sp} = \frac{\Gamma v_g}{V} \frac{g n_{sp}}{R_{sp}}, \tag{21.11}$$

where V is the active region volume and Γ and g are the averaged values of confinement factor and material gain over polarizations, respectively, obtained in the same way as the average optical matrix element for radiative recombination rate. Here, $n_{sp}(n, \hbar\omega)$ is the population inversion factor defined by (Coldren et al., 2012)

$$n_{sp} = \left[1 - \exp\left(\frac{\hbar\omega - \Delta E_f}{k_B T}\right)\right]^{-1}, \tag{21.12}$$

where $\Delta E_f = E_f^{(c)} - E_f^{(v)}$ is the difference between the quasi-Fermi levels of CB and VB. In order to calculate β_{sp}, in addition to confinement factor, it is necessary to calculate the group velocity of the light $v_g(n, \hbar\omega)$:

$$v_g = \frac{\partial \omega}{\partial \beta} = \frac{1}{\hbar} \cdot \left[\frac{\partial \beta}{\partial(\hbar\omega)}\right]^{-1}. \tag{21.13}$$

Spectral dependencies of the confinement factor, group velocity, and spontaneous emission coupling factor are shown in Figure 21.4, for both bulk and the MQW active region, optimized to be polarization insensitive. Therefore, the quantities for the MQW active region are given only for TE polarization.

21.4 Rate Equations

In order to analyze the SOA's performance, it is first necessary to identify the main physical processes underlying its operation, which are related to the interaction mechanisms between photons and carriers. These are then typically presented in the form of a coupled system of differential equations denoted as the rate equation system. Optical signal can be described either in terms of complex electric field or in terms of photon density and phase. On the other hand, carriers are modeled either via carrier density (Dailey and Koch, 2007; Totović et al., 2013) or material gain (Shtaif et al., 1998), or, in some cases, using the integrated material gain along the direction of signal propagation (Agrawal and Olsson, 1989; Cassioli et al., 2000; Antonelli and Mecozzi, 2013).

For TW-SOA, it usually suffices to analyze only the forward-propagating optical signal, with respect to the longitudinal axis. However, for R-SOA, both forward and backward-propagating optical signals need to be accounted for. Although the model's complexity increases when counterpropagating optical signals are analyzed, it is often good practice to develop the generalized model since it can easily be customized to account for any SOA type by simply changing the power reflectivity coefficient of the front and/or rear facet.

21.4.1 Basic Rate Equations

During the decades of SOA study, a variety of models of different complexity emerged. Some of them are based on signal power (Agrawal and Olsson, 1989), intensity (Adams et al., 1985), or photon density analysis (Totović et al., 2014), which may or may not include a separate phase equation, whereas the others treat the optical signal using the complex electric field envelope (Melo and Petermann, 2008; Schrenk, 2011; Antonelli and Mecozzi, 2013). In its basic form, the SOA model describes the interaction between the carrier density $n(z, t)$ and the two counterpropagating optical signals, forward (+) and backward (−), with respect to the longitudinal z-axis, all of which are dependent on both time t and the z-coordinate. For SOA operation, it is necessary to achieve population inversion, which is done using the electric (bias) current I. When the bias current is low, carriers are mainly drained by spontaneous recombination, which can be nonradiative, yielding phonons (Ghafouri-Shiraz, 2004), or radiative, which produces noncoherent spectrally wide optical signal, regarded as noise. As the current increases beyond the transparency value, enough carriers exist in the active region to make the stimulated emission the dominant process and the input signal can be amplified during propagation. For the simplest case of bulk SOA, carrier dynamics can be described as follows (Coldren et al., 2012; Totović et al., 2013):

$$\frac{dn}{dt} = \frac{I}{qV} - \left[An + R_{sp}(n) + Cn^3\right] - R_{st}(n).$$
(21.14)

The first term on the right-hand side (RHS) of Equation 21.14 describes the rate at which the carriers are "pumped" into the active region, where q is the elementary charge, and $V = WHL$ is the active region volume, dependent on the active region width W, height H, and length L. The second term accounts for the loss of carriers through spontaneous recombination, where $R_{sp}(n)$ is the total radiative spontaneous recombination rate, and A and C are the Shockley–Read–Hall and Auger coefficients, respectively, related to nonradiative spontaneous recombination (Coldren et al., 2012). In some instances, carrier loss due to radiative and nonradiative spontaneous recombination is modeled as n/τ, where τ is the carrier lifetime, which is treated either as constant (Thylén, 1988; Shtaif et al., 1998; Totović et al., 2016) or dependent on carrier density (Melo and Petermann, 2008; Xia and Ghafouri-Shiraz, 2016). Finally, the last term on the RHS of Equation 21.14, $R_{st}(n)$, accounts for the loss of carriers due to stimulated emission, either through the amplification of the signal alone or the signal and noise combined.

21.4.1.1 Electric Field Envelope Model

Each optical signal can be represented by its electric field, which can be separated into the carrier signal and the slowly varying envelope along the propagation direction. The complex envelope can then be analyzed either as is (Loudon et al., 2005), or, more commonly, it can be normalized such that its squared magnitude represents photon density (Connelly, 2007; Totović et al., 2011), power (Agrawal and Olsson, 1989; Cassioli et al., 2000), or intensity (Henry, 1982). Regardless of the choice for normalization, the system of equations remains the same since the underlying spatiotemporal dependence is unmodified by the normalization constants. The evolution of electric field envelopes $E_{\pm}(z, t)$, normalized such that their square magnitude represents photon densities, $|E_{\pm}(z, t)|^2 = S_{\pm}(z, t)$, propagating forward (+) and backward (−) along the longitudinal z-axis, is governed by the following system of equations (Agrawal and Olsson, 1989):

$$\pm \frac{\partial E_{\pm}}{\partial z} + \frac{1}{v_g} \frac{\partial E_{\pm}}{\partial t} = \frac{1}{2} \left(\Gamma g - \alpha_i \right) E_{\pm}. \tag{21.15}$$

Here, v_g is the group velocity, Γ stands for the optical confinement factor, which determines the portion of the signal propagating through the active region relative to the total photon density, and g is the material gain of the active region, all given at the signal frequency ω. These quantities can be derived from the models given in Section 21.3, and in Figures 21.2 and 21.4, by interpolation at the signal central energy $\hbar\omega$. With $\alpha_i = K_0 + \Gamma K_1 n$ active region loss is denoted (Connelly, 2001), where K_0 and K_1 stand for the carrier-independent and carrier-dependent loss coefficients, respectively, which take into account the intrinsic material loss and the free-carrier and inter-VB absorption (Dailey and Koch, 2009; Schrenk, 2011). It should be noted that, in general, all parameters appearing in Equation 21.15, namely v_g, Γ, g, and α_i, depend on carrier density n, and implicitly on t and z. This dependence notation is omitted from Equation 21.15 and the equations that will follow for compactness and clarity. Nevertheless, it is common to treat v_g and/or Γ as constants (Jin et al., 2003; Totović et al., 2014), which simplifies the model at the expense of accuracy.

Input optical signal, given by its normalized envelope E_0, is assumed to be entering the device at the front facet ($z = 0$), with the amplitude reflection coefficient r_1, and transmission coefficient t_1, as shown in Figure 21.5. The front facet is usually coated with the antireflective layer, and it is a common practice to neglect any residual reflectivity and assume $r_1 = 0$ and $t_1 = 1$, especially if the analytical or semianalytical solution is sought (Antonelli and Mecozzi, 2013; Totović et al., 2014). On the other hand, the rear facet ($z = L$) has the amplitude reflection coefficient r_2, which can be neglected for TW-SOA, but needs to be accounted for in the case of an R-SOA. If the cleaved facet is used as a high-reflecting one, the amplitude reflection coefficient can be determined as $r_2 = (1 - n_r)/(1 + n_r)$ under the assumption of TE polarization,

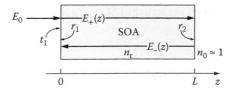

FIGURE 21.5 Semiconductor optical amplifier (SOA) structure with relevant optical signals, refractive indices, and reflection and transmission coefficients denoted.

where n_r is the refractive index of the active region. Boundary conditions according to Figure 21.5 read (Adams et al., 1985, Jin et al., 2003)

$$E_+(0) = t_1 E_0 + r_1 E_-(0),$$
$$E_-(L) = r_2 E_+(L). \tag{21.16}$$

It is common practice to introduce the LEF α in Equation 21.15, which accounts for the phase modulation due to variation of real and imaginary parts of susceptibility χ (Henry, 1982; Agrawal and Olsson, 1989; Dailey and Koch, 2009; Antonelli and Mecozzi, 2013). This variation can be induced by injection of carriers, α_N, and carrier heating (CH), α_{CH} (Occhi et al., 2003). The total phase change will be the sum of these two contributions since they can be treated as uncorrelated. Carrier-induced LEF is defined as (Osinski and Buus, 1987; Occhi et al., 2003)

$$\alpha_N = -\frac{\partial Re\,\{\chi\}/\partial n}{\partial Im\,\{\chi\}/\partial n} = -2k_0 \frac{\partial n_r/\partial n}{\partial g/\partial n}, \tag{21.17}$$

where $k_0 = 2\pi\nu/c$ is the wave vector. For a small change in carrier density, which usually is the case, the derivatives in Equation 21.17 can be replaced with finite differences, which leads to (Henry, 1982; Osinski and Buus, 1987)

$$\alpha_N = -2k_0 \frac{\Delta n_r}{\Delta g}, \tag{21.18}$$

where Δn_r and Δg are the differential change in refractive index and material gain, respectively, induced by carrier density variation (Henry, 1982; Osinski and Buus, 1987; Agrawal, 1990). The deviation Δn_r is a result of several different mechanisms, including dipole band-to-band transition (Dailey and Koch, 2009; Qin et al., 2012), band-filling (Burstein–Moss effect), bandgap shrinkage, and free-carrier absorption (plasma effect) (Bennett et al., 1990). Band-filling is related to the decrease in absorption for photon energies slightly above nominal bandgap, caused by low density of states in CB that are easily filled, and is mostly pronounced for heavily doped semiconductors (Moss et al., 1973). Bandgap shrinkage is related to large concentrations of electrons at the bottom of CB, which repel one another due to Coulomb force, leading to the lowering of the energy of CB (Bennett et al., 1990). Finally, free-carrier absorption describes transition of free carriers to the higher-energy level due to photon absorption, and the change in refractive index can be described using the Drude model (Moss et al., 1973; Dailey and Koch, 2009; Qin et al., 2012). These four effects are assumed to be independent, and the total refractive index deviation can be found as a sum of their respective contributions.

Another contribution to LEF, caused by carrier temperature variation, is defined as (Occhi et al., 2003)

$$\alpha_{CH} = -2k_0 \frac{\partial n_r/\partial T}{\partial g/\partial T}. \tag{21.19}$$

However, this contribution is often neglected in SOA models, and LEF is assumed to be $\alpha \approx \alpha_N$. Equation 21.15 now becomes

$$\pm\frac{\partial E_\pm}{\partial z} + \frac{1}{v_g}\frac{\partial E_\pm}{\partial t} = \frac{1}{2}\left[\Gamma g\,(1 - i\alpha) - \alpha_i\right] E_\pm. \tag{21.20}$$

It should be stressed that, depending on the sign convention adopted for the α-parameter, there may be variation in the literature in the sign prefixing α in Equation 21.20.

Having defined the equations which govern spatiotemporal electric field envelope distribution of the counterpropagating optical signals, it is possible to devise the stimulated emission rate in Equation 21.14, which is proportional to the total photon density in the amplifier's active region, S_{Σ}:

$$R_{st}(n) = v_g g S_{\Sigma} = v_g g |E_+ + E_-|^2. \tag{21.21}$$

It can be seen from Equation 21.21 that the interference between the two counterpropagating optical signals will cause the spatial grating in the carrier density distribution. However, characteristic grating length will be of the order of the signal's wavelength λ, and it will be easily washed out by the diffusion (Yacomotti et al., 2004; Serrat and Masoller, 2006; Totović et al., 2013). This justifies neglecting of all high-frequency terms, keeping only the sum of the squared electric field magnitudes, i.e., photon densities, $S_{\Sigma} = |E_+|^2 + |E_-|^2 = S_+ + S_-$. Depending on the model complexity, S_{Σ} can also include ASE noise.

21.4.1.2 Photon Density Model

By expressing the electric field envelope in terms of photon densities, $S_{\pm}$, and phase, $\varphi_{\pm}$:

$$E_{\pm}(z, t) = \sqrt{S_{\pm}(z, t)} \exp\left[i\varphi_{\pm}(z, t)\right], \tag{21.22}$$

Equation 21.20 can be reduced to the system of coupled TW equations for forward and backward propagation directions, written with respect to photon densities (Qin et al., 2012):

$$\pm \frac{\partial S_{\pm}}{\partial z} + \frac{1}{v_g} \frac{\partial S_{\pm}}{\partial t} = \left(\Gamma g - \alpha_i\right) S_{\pm}, \tag{21.23}$$

and phase (Agrawal and Olsson, 1989):

$$\pm \frac{\partial \varphi_{\pm}}{\partial z} + \frac{1}{v_g} \frac{\partial \varphi_{\pm}}{\partial t} = -\frac{1}{2}\Gamma g \alpha. \tag{21.24}$$

Using the definition of LEF, Equation 21.24 can be written in yet another form that emphasizes the underlying cause of phase variation (Agrawal, 1990; Dailey and Koch, 2007; Qin et al., 2012):

$$\pm \frac{\partial \varphi_{\pm}}{\partial z} + \frac{1}{v_g} \frac{\partial \varphi_{\pm}}{\partial t} = k_0 \Gamma \Delta n_r. \tag{21.25}$$

If the signal is assumed to preserve its coherence during propagation, both Equations 21.23 and 21.25 are required; otherwise, it is sufficient to use only Equation 21.23 in modeling an SOA. Appending the phase equation to the model is important especially if the input signal carries information encoded in its phase, which is the case for many advanced modulation formats, such as phase shift keying (PSK), quadrature amplitude modulation (QAM), and quadrature phase shift keying (QPSK), to name a few (Schmuck et al., 2013; Koenig et al., 2014).

Assuming that the signal entering the device at the front facet ($z = 0$) is $E_0 = \sqrt{S_0}\exp(i\varphi_0)$, boundary conditions given by Equation 21.16 can now be rewritten using Equation 21.22. Additionally, amplitude reflection and transmission coefficients can be written as functions of the power reflectivity coefficients R_1 and R_2, corresponding to front and rear facets, respectively. The input signal does not change the propagation direction during transmission through the front facet, so no phase change is present, and $t_1 = \sqrt{(1 - R_1)}$ (Adams et al., 1985). On the other hand, during reflection at any facet, the signal being reflected does change the propagation direction, and a phase change of π needs to be accounted for. This can also be inferred from the definition of the amplitude reflection coefficient r_2 for the cleaved facet, which is negative. Finally, amplitude reflection coefficients can be written as $r_1 = \sqrt{R_1} \exp(i\pi)$ and $r_2 = \sqrt{R_2} \exp(i\pi)$. Substituting amplitude reflection and transmission coefficients in Equation 21.16, and expressing electric fields in terms of photon densities and phase, yields:

$$\sqrt{S_+(0)} \exp[i\varphi_+(0)] = \sqrt{1-R_1}\sqrt{S_0}\exp(i\varphi_0) + \sqrt{R_1}\exp(i\pi)\sqrt{S_-(0)}\exp[i\varphi_-(0)],$$
$$\sqrt{S_-(L)} \exp[i\varphi_-(L)] = \sqrt{R_2}\exp(i\pi)\sqrt{S_+(L)}\exp[i\varphi_+(L)]. \tag{21.26}$$

By multiplying previous set of equations with their corresponding complex conjugates, the boundary conditions for photon densities can be derived:

$$S_+(0) = (1-R_1)S_0 + R_1 S_-(0) - 2\sqrt{(1-R_1)S_0}\sqrt{R_1 S_-(0)}\cos[\varphi_0 - \varphi_-(0)],$$
$$S_-(L) = R_2 S_+(L). \tag{21.27}$$

Equating the arguments of the left-hand side (LHS) and RHS of Equation 21.26, the following boundary conditions for phases can be found (Totović et al., 2013):

$$\varphi_+(0) = \arctan\left[\frac{\sqrt{(1-R_1)S_0}\sin\varphi_0 - \sqrt{R_1 S_-(0)}\sin\varphi_-(0)}{\sqrt{(1-R_1)S_0}\cos\varphi_0 - \sqrt{R_1 S_-(0)}\cos\varphi_-(0)}\right], \tag{21.28}$$

$$\varphi_-(L) = \varphi_+(L) + \pi.$$

Typically, the input optical signal is given in terms of power rather than photon density. For the input optical power P_0, the photon density can be determined as $S_0 = \Gamma P_0/(\hbar\omega v_g WH)$, where $\hbar\omega$ is the signal energy.

21.4.1.3 Signal Spectral Dependence

Most of the parameters figuring in Equations 21.23 and 21.25 do exhibit spectral dependence, as shown in Figures 21.2 and 21.4. When working with photon densities, signal is assumed to be infinitely spectrally narrow and centered at the energy corresponding to the maximum spectral photon density, $\hbar\omega_0$. This assumption justifies the usage of a simplified model where the parameter values are given at the signal central energy, as is the case with Equation 21.23. In practice, signal will always have a finite spectral width, which is not necessarily narrow and cannot always be neglected. This is especially important if SOAs are used in WDM systems, where multiple signals at different wavelengths travel through the active region. In order to correctly model the signal propagation in this case, Equation 21.23 is rewritten in terms of spectral photon density, $s_\pm(\hbar\omega, z, t)$, for forward and backward propagation (Totović et al., 2013):

$$\pm\frac{\partial s_\pm}{\partial z} + \frac{1}{v_g}\frac{\partial s_\pm}{\partial t} = (\Gamma g - \alpha_i)s_\pm. \tag{21.29}$$

It should be stressed that now v_g, Γ, g, and α_i are spectrally dependent. Additionally, two more changes to the model are required. The first is related to the carriers consumed by stimulated emission in Equation 21.14, which can be determined using

$$R_{st}(n) = \int_0^\infty v_g g s_\Sigma d(\hbar\omega), \tag{21.30}$$

where $s_\Sigma = s_+ + s_-$. The second is related to the boundary conditions given in Equation 21.27, which are now written with respect to $s_\pm$ instead of $S_\pm$. Finally, if no reflection coating is used for the R-SOA's rear facet, the reflection coefficient R_2 also becomes spectrally dependent through $n_r(\hbar\omega)$.

21.4.2 Amplified Spontaneous Emission

Spontaneous recombination can usually be viewed as an undesirable side effect of SOA operation, from several viewpoints. First, the carriers that should be reserved for signal amplification are indefinitely lost, meaning that the higher bias currents are required to achieve the same material gain. Second, the random nature of spontaneous recombination induces fluctuations in carrier density, which in turn causes the material gain and refractive index to change. This leads to fluctuations in both intensity and phase of the amplified signal (Ghafouri-Shiraz, 2004). Finally, radiative spontaneous recombination generates wideband signal with random phase, polarization, and spatial orientation. Once these photons are generated, they become impossible to separate from the signal, so they travel jointly through the active region and become amplified. Even if the signal is filtered at the SOA's output, a certain amount of noise will always be present in the narrow band around the signal central energy (Yariv and Yeh, 2007). An additional drawback is that the noise will have higher optical power when the input signal is vague since more carriers will be available for spontaneous recombination (Talli and Adams, 2003; Totović et al., 2013; Koenig et al., 2014). As the signal becomes more amplified, fewer carriers remain, and the ASE contribution becomes less important in terms of power.

The process of spontaneous emission is at its core stochastic, and, within the framework of the semiclassical approach, it can be modeled statistically in terms of its probability, similarly to the shot noise present in electronic devices (Ghafouri-Shiraz, 2004). The nature of ASE noise led to the development of numerous models of different degree of complexity, which can be roughly divided into two categories: deterministic, which do not require any random generators (Talli and Adams, 2003; Totović et al., 2013), and stochastic, which require random sources with certain probability density functions (Marcuse, 1984; Cassioli et al., 2000; Park et al., 2005; Melo and Petermann, 2008; Totović et al., 2011). In the semiclassical framework, ASE is usually treated as white Gaussian noise process, or as Poisson process (Park et al., 2005), whereas in the quantum mechanical formalism the noise is treated using the set of appropriate quantum noise operators, including spontaneous emission process, internal absorption, and vacuum field fluctuations (Shtaif et al., 1998). It has been shown by Shtaif et al. (1998) that the semiclassical results are essentially the same as the ones obtained using quantum description, with the exception of the shot noise term, which needs to be explicitly added to the semiclassical intensity. However, for sufficiently small bandwidths, the shot noise term can be ignored (Shtaif et al., 1998).

21.4.2.1 Deterministic Approach

It has already been pointed out that signal and noise travel together, which would justify modification of the photon density equation, either Equation 21.23 or Equation 21.29, to include another term modeling the rate with which the spontaneous photons are generated (Park et al., 2005; Dailey and Koch, 2007; Zhou et al., 2007; Totović et al., 2015). However, from the standpoint of detailed device analysis, it may be useful to separate noise and signal equations, allowing them to be coupled through the carrier rate equation since these two processes are uncorrelated (Ghafouri-Shiraz, 2004). ASE is known to have wide spectrum; therefore, it is more appropriate to use spectral photon densities, $a_\pm(\hbar\omega, z, t)$, rather than photon densities, for its description. The system of equations for two counterpropagating ASE signals reads (Talli and Adams, 2003; Melo and Petermann, 2008; Totović et al., 2013):

$$\pm\frac{\partial a_\pm}{\partial z} + \frac{1}{v_g}\frac{\partial a_\pm}{\partial t} = \left(\Gamma g - \alpha_i\right) a_\pm + \frac{1}{2v_g}\Gamma\beta_{sp}r_{sp}, \tag{21.31}$$

where r_{sp} is the radiative spontaneous recombination rate per unit energy, given by Equation 21.2 or Equation 21.5, depending on the active region type, and β_{sp} is the spontaneous emission coupling factor given by Equation 21.11. Boundary conditions for noise are $a_+(0) = R_1 a_-(0)$ and $a_-(L) = R_2 a_+(L)$. Phase is not analyzed since it is random and can be averaged to zero. In other words, under the usual operating conditions, i.e., if no resonant cavity is present, ASE does not exhibit coherent behavior.

In order to have the complete model, it is necessary to include ASE together with the signal when calculating total spectral photon density, $s_\Sigma = s_+ + s_- + a_+ + a_-$, which further influences $R_{st}(n)$, given by Equation 21.30.

It should be noted that many other deterministic approaches for ASE analysis exist. The choice depends on the desired model complexity and the tradeoff between accuracy and computational resource consumption. These approaches include the photon statistic master equation method (Mukai and Yamamoto, 1982), field beating (Olsson, 1989), the equivalent circuit model (Berglind and Gillner, 1994; Ghafouri-Shiraz, 2004), the semiclassical wave theory model (Donati and Giuliani, 1997).

21.4.2.2 Resonant Properties of the SOA Cavity

In the previous pages, during the discussion of facet's power reflectivities and boundary conditions, it was assumed that all antireflective facets have zero reflectivity. Although this assumption can be helpful for model simplification, it can also mask some of the processes present in SOAs (Thylén, 1988; Zhou et al., 2007). Residual reflectivity, which always has some finite value, provides a feedback loop that acts as a filter, giving rise to resonances and antiresonances (Adams et al., 1985; Schrenk, 2011). This effect has been reported in both experimental (Olsson, 1989) and theoretical SOA analyses (Zhou et al., 2007; Totović et al., 2013), mostly related to ASE noise. An example of the output spectrum of the signal and noise combined, with visible resonant footprint of the cavity, calculated using the model presented by

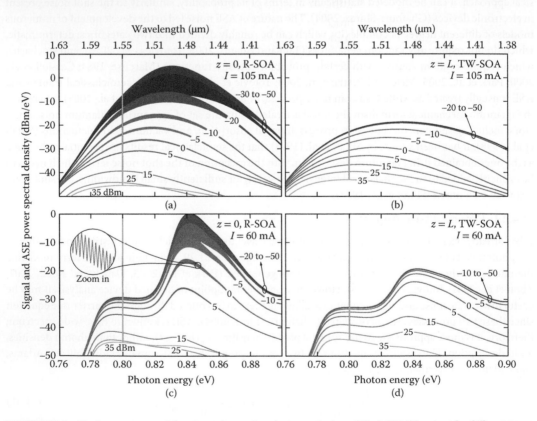

FIGURE 21.6 Total power spectral density at the semiconductor optical amplifier's (SOA's) output for different input optical powers in the case of (a, c) reflective (R) SOA at $z = 0$, and (b, d) traveling wave (TW) SOA at $z = L$, for (a, b) bulk and (c, d) multiple quantum well (MQW) active region.

Totović et al. (2013), is given in Figure 21.6. It can be seen that R-SOA, Figure 21.6a and c, exhibits more pronounced resonances and antiresonances in comparison with TW-SOA, Figure 21.6b and d, which is a consequence of the highly reflective rear facet. The ASE noise spectrum profoundly depends on the optical power of the input signal and resonant pattern washes out as the input optical power increases. This is to be expected since the strong signal easily depletes the carriers and suppresses the noise, consequently reducing the effect of multiple round trips within the cavity on its spectrum. The difference between the output spectrums for distinct active region types, bulk, Figure 21.6a and b, and MQW, Figure 21.6c and d, is a result of different spectrums of radiative spontaneous emission rate and material gain, given in Figure 21.2.

In order to develop the model capable of capturing ripples in the ASE output spectrum, two approaches can be used. One would require an equation for ASE phase evolution in addition to Equation 21.31, which describes ASE spectral photon density. This would enable us to account for interference, and subsequently filtering property of the SOA cavity. However, due to the random nature of the ASE noise phase, this approach would require a random number generator (Park et al., 2005) and could essentially be categorized as a stochastic model. Another approach is based on spectrum slicing and analysis of photon density corresponding to each resonance, or mode, followed by photon redistribution over the energy range between the two antiresonances according to the cavity transfer function. This has been shown to be equivalent to the analysis of the electric field or the spectral photon density and phase (Adams et al., 1985). The phase equation will be included implicitly during the assessment of resonant and antiresonant frequencies of the mth cavity mode, ω_m^r, and ω_m^a, respectively. These can be derived by equating the phase accumulation during the cavity round trip to $2m\pi$ for resonant, or $(2m-1)\pi$ for antiresonant frequencies, where m is an integer. The single-pass phase shift, Φ, will be the same for either propagation direction and can be found from Equation 21.25 as

$$\Phi = \int_0^L \frac{\partial \varphi_+}{\partial z} dz = \int_0^L \left(-\frac{1}{v_g} \frac{\partial \varphi_+}{\partial t} + k_0 \Gamma \Delta n_r \right) dz = \frac{\omega}{v_g} L + k_0 \int_0^L \Gamma \Delta n_r dz. \tag{21.32}$$

Here, $\omega/v_g = n_{\text{eff},0} k_0 = \beta_0$ is the wave propagation constant for zero carrier density, where $n_{\text{eff},0}$ is the effective index of refraction for zero carrier density or the background index of refraction. Resonant and antiresonant frequencies are calculated from

$$2\frac{\omega}{c} \left(n_{\text{eff},0} L + \int_0^L \Gamma \Delta n_r dz \right) = \begin{cases} 2m\pi, & \text{for } \omega_m^r \\ (2m-1)\pi, & \text{for } \omega_m^a \end{cases}. \tag{21.33}$$

Since both Γ and Δn_r depend on photon energy, $\hbar\omega$, resonant and antiresonant frequencies cannot be expressed in a closed analytical form. Additionally, we can conclude that ω_m^r and ω_m^a depend on carrier density through the second term in the LHS of Equation 21.33 since both Γ and Δn_r depend on n. This implies that the spectral output of either TW- or R-SOA will not be static; rather the frequencies corresponding to resonances and antiresonances will be shifted in the presence of the optical signal (Schrenk, 2011; Totović et al., 2013). However, comparing to the first term in the LHS of Equation 21.33, the contribution of the second term is usually small and can be justifiably neglected.

By integrating Equation 21.31 over the energy between each of the two antiresonances, ω_m^a and ω_{m+1}^a, a system of equations can be developed, written with respect to noise photon densities of the mth mode, $A_\pm^m(z, t)$, for forward (+) and backward (−) propagation directions (Schrenk, 2011; Totović et al., 2013):

$$\pm \frac{\partial A_\pm^m}{\partial z} + \frac{1}{v_g} \frac{\partial A_\pm^m}{\partial t} = \left(\Gamma_m g_m - \alpha_i^m \right) A_\pm^m + \frac{1}{2v_g^m} \Gamma_m \beta_{\text{sp}}^m R_{\text{sp}}^m, \tag{21.34}$$

where the index m denotes the parameter values corresponding to the mth cavity resonance, and R_{sp}^m stands for the fraction of radiative spontaneous recombination rate injected into the mth mode. This approach, relying on separating amplification and filtering properties of the amplifier cavity, reduces time, memory, and processing resource consumption during simulation.

The filtering function of the amplifier is given in the form of an Airy function corresponding to the FP cavity (Adams et al., 1985):

$$G_m(\hbar\omega) = \frac{(1 - R_1)(1 - R_2) G_s^m}{\left(1 - G_s^m \sqrt{R_1 R_2}\right)^2 + 4 G_s^m \sqrt{R_1 R_2} \sin^2 \Phi(\hbar\omega)}, \tag{21.35}$$

where Φ is the single-pass phase shift, defined by Equation 21.32, and G_s^m stands for the single-pass gain at resonant energy $\hbar\omega_m^r$:

$$G_s^m = \exp\left[\int_0^L \left(\Gamma_m g_m - \alpha_i^m\right) dz\right]. \tag{21.36}$$

The transmittance given by Equation 21.35 accounts for the signal amplification, so care should be exercised when using it with Equation 21.34. Prior to photon filtering, the transmittance needs to be normalized to unity over the mth mode energy range since the noise amplification has already been accounted for using Equation 21.34, and only photon redistribution is required. The normalized filtering function $T_m(\hbar\omega)$ can be found as

$$T_m(\hbar\omega) = \frac{G_m(\hbar\omega)}{\left(\hbar\omega_{m+1}^a - \hbar\omega_m^a\right)^{-1} \int_{\hbar\omega_m^a}^{\hbar\omega_{m+1}^a} G_m(\hbar\omega) \, d(\hbar\omega)}, \tag{21.37}$$

which is equivalent to

$$T_m(\hbar\omega) = \frac{G_m(\hbar\omega)}{\pi^{-1} \int_0^\pi G_m(\Phi) \, d\Phi} = \frac{\sqrt{1 + \gamma_m}}{1 + \gamma_m \sin^2 \Phi}, \tag{21.38}$$

where the γ-parameter is defined as follows:

$$\gamma_m = \frac{4 G_s^m \sqrt{R_1 R_2}}{\left(1 - G_s^m \sqrt{R_1 R_2}\right)^2}. \tag{21.39}$$

The validity of the previous model can be easily checked by analysis of the corner case where any of R_1 or R_2 is equal to zero. This would yield $\gamma_m = 0$, and $T_m = 1$, which corresponds to the case of uniformly distributed photons. This result is to be expected since for zero reflectivity no feedback loop exists.

Finally, the spectral photon densities of the noise at the SOA's output can be found using (Totović et al., 2013)

$$a_-(0, \hbar\omega) = \sum_m \left[\left(\hbar\omega_{m+1}^a - \hbar\omega_m^a\right)^{-1} A_-^m(0) T_m(\hbar\omega)\right],$$

$$a_+(L, \hbar\omega) = \sum_m \left[\left(\hbar\omega_{m+1}^a - \hbar\omega_m^a\right)^{-1} A_+^m(L) T_m(\hbar\omega)\right], \tag{21.40}$$

where the summation is done over the modes m from 1 to the total number of accounted modes M.

In order to increase the efficiency of numerical calculations related to noise, an approach based on grouping of several adjacent modes into clusters was proposed by Totović et al. (2013). Although this degrades the accuracy to a certain point, the small intermodal space between resonances, which is of the order of $\Delta\nu = 70$ GHz, or $\Delta\lambda = 0.54$ nm for the active region length of 800 μm, ensures that no significant changes for any spectrally dependent parameter exist within the cluster. The proposed optimal number of modes within one cluster is 5, based on the fact that material gain and stimulated emission rate do not change more than 2% for bulk and 3% for MQW active region, within the cluster's range of frequencies, for the most part of the spectrum. In addition, the grouping of modes need not be uniform. In other words, clusters can comprise a lower number of modes in the frequency range close to the signal central frequency and maximum spontaneous emission rate and that number can be increased approaching the edges of spontaneous emission spectrum.

21.4.2.3 Stochastic Approach

Implementing any stochastic process relies on random number generators, or sources, meaning that only numerical analysis is possible in this case. Nonetheless, this approach can be very useful for investigating statistical properties of noise and signals. As both amplitude and phase of noise signal are random, it is suitable to use the equation written with respect to the normalized electric field envelope Equation 21.20, modified such that it includes a noise generator:

$$\pm\frac{\partial E_{\pm}}{\partial z} + \frac{1}{v_g}\frac{\partial E_{\pm}}{\partial t} = \frac{1}{2}\left[\Gamma g\left(1 - i\alpha\right) - \alpha_i\right]E_{\pm} + \mu_{\pm}. \tag{21.41}$$

Here, the complex term $\mu_{\pm}(z,t)$ stands for the random noise source, also known as the Langevin noise source (Henry, 1986; Coldren et al., 2012), which can be modeled as the Poisson or Gaussian phase-independent spatially uncorrelated white noise process (D'Ottavi et al., 1995; Cassioli et al., 2000; Park et al., 2005), based on the fluctuation-dissipation theorem (Shtaif et al., 1998). This implies that the mean value of $\mu_{\pm}(z,t)$ is zero, whereas the autocorrelation function satisfies the following condition:

$$\left\langle\left\langle\mu_{\pm}\left(z,t\right)\mu_{\pm}^{*}\left(z-z',t-t'\right)\right\rangle\right\rangle = \frac{1}{v_g^2}\Gamma\beta_{sp}R_{sp}L\delta\left(z'\right)\delta\left(t'\right). \tag{21.42}$$

Recalling that Equation 21.20 does not include spectrally dependent parameters, it can be seen that the noise source defined in this manner has infinite bandwidth and consequently infinite power due to neglected spectral dependence of material gain (Cassioli et al., 2000). In practice, material gain will have finite bandwidth and will limit the noise power to a finite value. This effect can be included by passing $\mu_{\pm}(z,t)$ through the bandpass filter, which will result in

$$\left\langle\left\langle\mu_{F\pm}\left(z,t\right)\mu_{F\pm}^{*}\left(z-z',t-t'\right)\right\rangle\right\rangle = \frac{1}{v_g^2}\Gamma\beta_{sp}R_{sp}LB_F\delta\left(z'\right), \tag{21.43}$$

where $\mu_{F\pm}(z,t)$ is the band-limited noise source and B_F is the noise equivalent bandwidth (D'Ottavi et al., 1995; Cassioli et al., 2000). If the segmentation along the spatial coordinate is small enough, such that the carrier density can be considered constant, in numerical simulations the Dirac delta function can be replaced with the inverse step size Δz (Marcuse, 1984):

$$\left\langle\left\langle\mu_{F\pm}\left(z,t\right)\mu_{F\pm}^{*}\left(z-z',t-t'\right)\right\rangle\right\rangle = \frac{1}{v_g^2}\Gamma\beta_{sp}R_{sp}LB_F\frac{1}{\Delta z}. \tag{21.44}$$

Now the noise source can be expressed as follows (Marcuse, 1984; Cassioli et al., 2000; Melo and Petermann, 2008):

$$\mu_{F\pm}(z,t) = \frac{1}{v_g}\sqrt{\Gamma\beta_{sp}R_{sp}B_F\frac{L}{\Delta z}}x_e,$$ (21.45)

where x_e is the complex Gaussian random variable. According to the sampling theorem, the sampling period, which is equivalent to the simulation time-step Δt, needs to be less than or equal to $1/(2B_F)$. The complex Gaussian random variable can be expressed in terms of its real and imaginary parts, which finally gives

$$\mu_{F\pm}(z,t) = \sqrt{\frac{1}{2v_g}\Gamma\beta_{sp}R_{sp}\frac{L}{v_g}\frac{1}{\Delta t\Delta z}}\frac{x_1 + ix_2}{\sqrt{2}},$$ (21.46)

where x_1 and x_2 are independent and identically distributed numerically generated Gaussian random variables with zero mean and unit variance (Melo and Petermann, 2008). Similarly, x_e can be expressed in terms of magnitude and phase, also distributed according to the Gaussian statistics, with zero mean and unit variance.

21.4.2.4 Noise Figure

The usual method of quantifying the noise is by determining the noise figure (NF). It can be shown (Simon et al., 1989) that the intrinsic NF of an SOA can be obtained using

$$NF_{dB} = 10\log_{10}\left[\frac{1}{G} + 2\frac{(G_s - 1)(1 + R_1 G_s)}{G_s(1 - R_1)}n_{sp}\frac{\Gamma g}{\Gamma g - \alpha_i}\right],$$ (21.47)

where G stands for the device (transmission) gain and G_s for the single pass gain, defined by Equation 21.36.

21.4.3 Extended Rate Equation Model

Previous discussion on SOA modeling covered the most important aspects of the photon–carrier interaction dynamics. Nevertheless, many more upgrades to the model are possible, and the choice of the model complexity mainly depends on the intended SOA usage. Whenever possible, it is good practice to simplify the model and exclude the effects that bear little to no improvement to the results, in order to reduce resource consumption during computation. Some of the upgrades are listed in the following pages, but many more do exist, e.g., separate analysis of polarization components (Pillai et al., 2006; Melo and Petermann, 2008), inclusion of nonlinearities, etc.

21.4.3.1 Carrier Diffusion

Although often omitted from consideration, carrier diffusion plays an important role in SOA modeling. It provides the means for washing out the spatial grating in carrier density distribution caused by the interference between the electric fields of the counterpropagating optical signals (Yacomotti et al., 2004; Serrat and Masoller, 2006; Totović et al., 2013). In this manner, its effects are implicitly accounted for by neglecting the high frequency terms in carrier density spatial distribution and replacing $|E_+ + E_-|^2$ with $|E_+|^2 + |E_-|^2$. This approach, however, has its limitations. It has been shown that only "fast" spatial grating, coming from counterpropagating fields of the same mode, can be neglected, whereas the "slow" one, resulting from interference effects of the different longitudinal modes, remains to a certain degree (Yacomotti et al., 2004; Serrat and Masoller, 2006). In order to encompass all possible cases, the SOA model can be generalized by modifying the carrier density rate Equation 21.14 to include the term describing diffusion (Agrawal and Olsson, 1989; Serrat and Masoller, 2006):

$$\frac{\partial n}{\partial t} = D\frac{\partial^2 n}{\partial z^2} + \frac{I}{qV} - \left[An + R_{sp}(n) + Cn^3\right] - R_{st}(n),\tag{21.48}$$

where D stands for the diffusion coefficient and $R_{st}(n)$ includes the square magnitude of the sum of both signal and noise counterpropagating electric fields.

21.4.3.2 Nonlinear Gain Suppression

The two most important effects that can influence material gain on a short-term scale are spectral hole burning (SHB) (Ahn and Chuang, 1990) and CH (Willatzen et al., 1991). The former represents formation of frequency selective dip in the gain spectrum as a consequence of stimulated emission, whereas the latter describes the fact that the temperature of carriers and lattice may differ. Both can be phenomenologically accounted for by introducing the nonlinear gain suppression factor, ε, which causes gain reduction proportional to the photon density in the active region (Willatzen et al., 1991; Totović et al., 2013). Instead of using the so-called linear gain g, which is dependent only on photon energy and carrier density, equations should be modified such that nonlinear gain is used $g_{NL} = g/(1 + \varepsilon S_\Sigma)$, which also accounts for gain dependence on photon density. The complete set of equations for wideband SOA modeling can now be written in the following form (Totović et al., 2013):

$$\frac{dn}{dt} = \frac{I}{qV} - \left(An + R_{sp} + Cn^3\right) - \frac{R_{st}}{1 + \varepsilon S_\Sigma},\tag{21.49}$$

$$\pm\frac{\partial s_\pm}{\partial z} + \frac{1}{v_g}\frac{\partial s_\pm}{\partial t} = \left(\frac{\Gamma g}{1 + \varepsilon S_\Sigma} - \alpha_i\right)s_\pm,\tag{21.50}$$

$$\pm\frac{\partial \varphi_\pm}{\partial z} + \frac{1}{v_g}\frac{\partial \varphi_\pm}{\partial t} = \frac{k_0\Gamma\Delta n_r}{1 + \varepsilon S_\Sigma},\tag{21.51}$$

$$\pm\frac{\partial a_\pm}{\partial z} + \frac{1}{v_g}\frac{\partial a_\pm}{\partial t} = \left(\frac{\Gamma g}{1 + \varepsilon S_\Sigma} - \alpha_i\right)a_\pm + \frac{1}{2v_g}\Gamma\beta_{sp}r_{sp}.\tag{21.52}$$

Presence of the nonlinear gain suppression also modifies the position of resonant and antiresonant frequencies, given by Equation 21.33, and the new condition can be derived from Equation 21.51:

$$2k_0\left(n_{eff,0}L + \int_0^L \frac{\Gamma\Delta n_r}{1 + \varepsilon S_\Sigma}dz\right) = \begin{cases} 2m\pi, & \text{for } \omega_m^r \\ (2m-1)\pi, & \text{for } \omega_m^a \end{cases}.\tag{21.53}$$

21.4.3.3 Carrier Transport Model for MQW and QD SOAs

As previously discussed, materials with higher levels of confinement can outperform bulk SOAs in many aspects. The most common choice of active regions aside from bulk are the ones based on MQWs or QDs. Due to the different carrier dynamics, these devices usually require a more elaborate model of carrier rate equations than the one introduced for bulk SOA.

The MQW region in lasers and SOAs is usually embedded into the separate confinement heterostructure (SCH) region. Injected carriers from the outer edges of the SCH region diffuse across the region, leading to the subsequent capture and emission of carriers by the quantum wells.

Figure 21.7 shows a typical SCH MQW active region. Carrier transport effects in MQW SOA can be modeled by rate equations written with respect to the carrier density in the barrier (continuum) states,

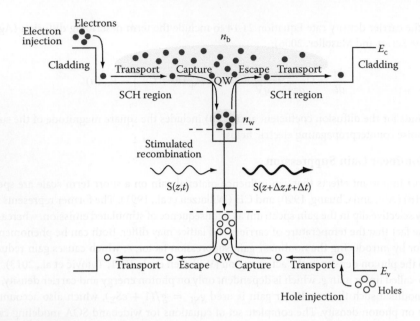

FIGURE 21.7 Band diagram of the multiple quantum well (MQW) semiconductor optical amplifier's (SOA's) active region and schematics showing carrier injection, transport, capture, escape and signal amplification by stimulated emission. (SCH, separate confinement heterostructure; QW, quantum well.)

including the SCH region and active layers, n_b, and the carrier density in the bound states of the well region, n_w, similar to SCH MQW lasers (Nagarajan et al., 1992; Keating et al., 1999; Totović et al., 2012). This model is referred to as the reservoir model (Nagarajan et al., 1992; Keating et al., 1999) and is equivalent to models that incorporate additional effects such as diffusive transport. Coupling of the carrier density in the barrier states above the MQWs to the carrier density in the MQWs is modeled by two terms representing carrier capture and escape into and from the wells, respectively:

$$\frac{dn_b}{dt} = \eta_{inj} \frac{I}{qV_b} - \frac{n_b}{\tau_b} - \frac{n_b}{\tau_{bw}} + \frac{n_w}{\tau_{wb}} \frac{V_w}{V_b}, \tag{21.54}$$

$$\frac{dn_w}{dt} = \frac{n_b}{\tau_{bw}} \frac{V_b}{V_w} - \frac{n_w}{\tau_w} - \frac{n_w}{\tau_{wb}} - \frac{R_{st}}{1 + \varepsilon S_\Sigma}. \tag{21.55}$$

Here, η_{inj} is the injection efficiency, V_b is the volume of the SCH and active region, V_w is the volume of the well region, τ_b and τ_w are the carrier recombination lifetimes in the barrier and in the well region, respectively, τ_{bw} is the effective carrier diffusion across the SCH region and capture time by the wells, and τ_{wb} is the thermionic emission and carrier diffusion time from the well to the barrier states.

The TW equations for the counterpropagating optical photon densities retain the same form as in the case of the bulk SOA, with the exception of material gain and internal loss of the active region, which now dominantly depend on carrier density in the bound states of the well, n_w:

$$\pm \frac{\partial S_\pm}{\partial z} + \frac{1}{v_g} \frac{\partial S_\pm}{\partial t} = \left(\frac{\Gamma g}{1 + \varepsilon S_\Sigma} - \alpha_i \right) S_\pm. \tag{21.56}$$

A more detailed model of the MQW active region's carrier dynamics may comprise additional rate equations for the reservoir model. For example, in addition to barrier states, well states may be divided into the excited state (ES) and the ground state (GS) (Qin et al., 2012). An even more elaborate approach is required for asymmetric MQW structures, where the tunneling effect of carriers through the barriers needs to be accounted for (Lysak et al., 2005, 2006).

The basic rate equations for QD SOA are essentially the same as for the detailed model of the MQW active region (Qin et al., 2012). The discrete energy levels in the well of QDs include the GS level and the ES level, which is doubly degenerated. The populations of these two levels are described by separate carrier densities, n_G and n_E, respectively, which are normalized with respect to the total dot volume V_D. Dots are interconnected by the wetting layer (WL), described by the carrier density, n_W, which is normalized to the WL volume, V_W. It can be assumed that the carriers are injected directly from the contacts into the WL and the barrier dynamics is thus ignored in the model (Berg et al., 2001; Qasaimesh, 2003). The rate equations describing the carrier dynamics read

$$\frac{dn_W}{dt} = \frac{I}{qV_W} + \frac{n_E}{\tau_e^E}\frac{V_D}{V_W}f_W' - \frac{n_W}{\tau_c}f_E' - \frac{n_W}{\tau_{sp}}, \tag{21.57}$$

$$\frac{dn_E}{dt} = \frac{n_W}{\tau_c}\frac{V_D}{V_W}f_E' + \frac{n_G}{\tau_e^G}f_E' - \frac{n_E}{\tau_e^E}f_W' - \frac{n_E}{\tau_0}f_G' - \frac{n_E}{\tau_{sp}}, \tag{21.58}$$

$$\frac{dn_G}{dt} = \frac{n_E}{\tau_0}f_G' - \frac{n_G}{\tau_e^G}f_E' - \frac{n_G}{\tau_{sp}} - \frac{R_{st}}{1+\varepsilon S_\Sigma}. \tag{21.59}$$

Here, τ_e^E is the escape time of carriers from the ES level to WL, τ_c is the capture time of carriers from the WL to the ES level, τ_{sp} is the spontaneous recombination time, which is assumed to be identical for all levels, τ_e^G is the excitation time of carriers from the GS level to the ES level, and τ_0 is the intradot relaxation time. In Equations 21.57 through 21.59, $f_{W,E,G}' = 1 - f_{W,E,G}$ denote the probabilities of finding an empty carrier state at the WL band edge, the ES and GS levels, respectively, which are closely related to the carrier densities of the corresponding levels (Berg et al., 2001). Finally, material gain is now dependent on carrier density in the GS state. A similar model can be implemented in the case of active regions based on QDashes.

Last, when QDs are embedded into the QW region, it is necessary to extend the model with an additional rate equation that deals with the QW dynamics. Each additional state in QDs also requires a separate rate equation (Kim et al., 2009).

21.4.3.4 Carrier Heating

During the discussion on effects that cause material gain to behave nonlinearly, CH was introduced. Moreover, CH is responsible for phase variation, through Equation 21.19. Depending on the SOA regime of operation, CH might be accounted for phenomenologically, using the nonlinear gain suppression factor, ε, and LEF, α_{CH}, or, in the case of ultrafast applications, a more elaborate model can be used. The temperature dynamics in SOAs can be described using the carrier temperature rate equation (Dailey and Koch, 2007, 2009; Xia and Ghafouri-Shiraz, 2015):

$$\frac{dT}{dt} = \frac{1}{\partial U/\partial T}\left(\frac{dU}{dt} - \frac{\partial U}{\partial n}\frac{dn}{dt}\right) - \frac{T - T_0}{\tau}, \tag{21.60}$$

where T is the carrier temperature, T_0 is the lattice temperature, U is the total carrier plasma energy density, and τ is the electron–phonon interaction time. In this model, electron and hole plasmas are assumed to be equal in both temperature, T, and density, n (Dailey and Koch, 2007). The rate of energy density change can be found to be (Xia and Ghafouri-Shiraz, 2015)

$$\frac{dU}{dt} = -v_g \sum_l \left(\hbar\omega_l - E_g\right) g_l \left(S_+^l + S_-^l\right) + v_g \Gamma K_1 n \sum_l \hbar\omega_l \left(S_+^l + S_-^l\right)$$
$$- v_g \sum_m \left(\hbar\omega_m - E_g\right) g_m \left(A_+^m + A_-^m\right) + v_g \Gamma K_1 n \sum_m \hbar\omega_m \left(A_+^m + A_-^m\right), \tag{21.61}$$

where summations are done over each mode l of the two counterpropagating signal photon densities $S_{\pm}^l$, and each mode m of the two counterpropagating ASE noise photon densities $A_{\pm}^m$. The free-carrier and intravalence absorption is captured in ΓK_1. The remaining derivatives, namely, $\partial U / \partial T$ and $\partial U / \partial n$, can be determined from the energy density distribution function (Xia and Ghafouri-Shiraz, 2015), given as (Dailey and Koch, 2009; Coldren et al., 2012)

$$U = \frac{2}{\sqrt{\pi}} k_B T \left(N_c F_{3/2}^c + N_v F_{3/2}^v \right), \tag{21.62}$$

where N_c and N_v stand for the effective densities of states for CB and VB, respectively, and $F_{3/2}^c$ and $F_{3/2}^v$ are the Fermi–Dirac integrals of order 3/2 for CB and VB, respectively. Having determined the rate of temperature change, it is possible to account for its variation in calculating the material gain, radiative spontaneous recombination rate, and variation of index of refraction, which are all temperature dependent through the Fermi–Dirac functions, Equations 21.1 through 21.3, or Equations 21.4 through 21.6, depending on the active region type. The analysis by Xia and Ghafouri-Shiraz (2015) confirms that the difference in carrier density and material gain caused by CH does influence the signal amplification in the picosecond regime of SOA operation.

21.4.3.5 Distributed Bias Current

It is common practice to assume that the driving current of SOA gets instantaneously uniformly distributed along the active region. In practice, this lumped electrode model is not always appropriate. Electric current does require a finite amount of time to travel from the contact to the edges of the electrode, and modeling it as a traveling microwave (TMW) might be more suitable in some cases, especially when the time delay is a significant fraction of the modulation period (Tauber et al., 1994; Mørk et al., 1999; Totović et al., 2015). This distributed nature of the current can have important implications on the results when SOAs are directly modulated at high bitrates (Tauber et al., 1994; Wu et al., 1995; Liljeberg and Bowers, 1997). Moreover, the microwave will be attenuated during the propagation, and the loss can sometimes be very high at typical modulation frequencies (Tauber et al., 1994). An additional point worth noting is that the microwave can be reflected at the end of an electrode, depending on the load impedance. When the microwave reflection coefficient differs from zero, two counterpropagating microwaves will exist (Totović et al., 2015):

$$I(z, t) = \bar{I} + \Delta I_F(z) \exp \left[i \left(2\pi f t - \beta_e z \right) \right] + \Delta I_B(z) \exp \left[i \left(2\pi f t + \beta_e z \right) \right]. \tag{21.63}$$

Here $\bar{I}$ is the stationary value of bias current, ΔI_F and ΔI_B are the small-signal bias current amplitudes for forward (F) and backward (B) propagation with respect to z-axis, which are defined by the spatially variable voltage across the electrode, f is the modulation frequency, $\beta_e = 2\pi f / v_e$ is the microwave propagation constant, $v_e = c / n_e$ is the microwave velocity, and n_e is the effective electric refractive index. Bias current will induce carrier density change through the carrier rate Equation 21.49, which will lead to change in all carrier-dependent parameters.

21.5 Overview of Steady-State and Dynamic TW- and R-SOA Models

Both TW- and R-SOA have found their application niches within the optical networks and photonic circuits. In order to make the optimal choice regarding the SOA material and geometric properties, as well as operating conditions, it is necessary to analyze their performance, in both the steady-state and dynamic regimes. This can be done either experimentally or theoretically, using any of the models presented in this section. Essentially, two borderline approaches can be used depending on required precision, available time, and computational resources. The first one is largely exploited in commercially available software for optical network analysis and relies on a significant number of approximations that either allow for analytical or semianalytical solutions or a relatively simple numerical approach using efficient numerical methods. The second one includes a detailed and comprehensive analysis and is mostly used in SOA design and optimization.

Modeling of steady-state and dynamic properties of SOAs essentially reduces to efficient and accurate self-consistent solving of coupled rate equations written with respect to the counterpropagating signals and noise, and carrier density, with imposed boundary and initial conditions. In the case of TW-SOA, the equation system usually comprises only the forward-propagating TW equation, whereas in the case of R-SOA both forward and backward TW equations are needed. The number of rate equations describing the carrier density depends on the material used for the active region and anticipated model accuracy. A self-consistent approach is required since the carrier depletion in the active region is caused by the stimulated amplification of the input optical signal, and this in turn modifies the material gain and, consequently, further amplification of the signal. If the carrier depletion is weak and the carrier density is approximately uniformly distributed along the active region, the TW equation for the forward-propagating signal (Equation 21.23) can be analytically solved. In the case of the continuous wave (CW), i.e., stationary input optical signal $S_{in} = S_+(z = 0)$, the output photon density $S_{out} = S_+(z = L)$ is given by

$$S_{out} = G_s S_{in} = S_{in} \exp\left[\left(\Gamma g - \alpha_i\right) L\right], \tag{21.64}$$

where G_s is the single pass gain. An analogous relation can be written for the electric field, where the factor 1/2 should be accounted for in the argument of exponential function: $\exp[(\Gamma g - \alpha_i)L/2])$, or $G_s^{1/2}$. In the case of the time-dependent (nonstationary) input optical signal, the solution of Equation 21.23 for the forward-propagating photon density reads

$$S_{out}(t) = S_{in}\left(t - \frac{L}{v_g}\right) \exp\left[\left(\Gamma g - \alpha_i\right) L\right]. \tag{21.65}$$

Here, a reference frame traveling with the signal is used, and therefore the temporal dependence of the input signal, $S_{in}(t)$, is replaced with $S_{in}(t - L/v_g)$. It can be seen that in both cases the output photon density is proportional to G_s.

The above relations can be generalized and may also be applied for an SOA section in which the carrier density is uniformly distributed. In this case, S_{in} and S_{out} represent the section's input and output photon densities, $S_+^{k-1} = S_+(z_{k-1})$ and $S_+^k = S_+(z_k)$ for forward propagation, and $S_-^k = S_-(z_k)$ and $S_-^{k-1} = S_-(z_{k-1})$ for backward propagation, respectively, as shown in Figure 21.8. The decrease of the section length makes the assumption of the uniform carrier distribution more justified. This fact represents the basis for the development of various methods, which can provide efficient stationary and time-domain modeling of TW- and R-SOA.

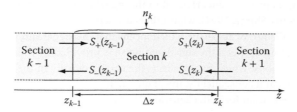

FIGURE 21.8 Schematic representation of the semiconductor optical amplifier (SOA) section with the relevant carrier and photon densities.

21.5.1 Steady-State Models

One of the oldest methods used in modeling of SOAs, optical waveguides, and semiconductor lasers is the transfer matrix method (TMM) (Chu and Ghafouri-Shiraz, 1994), which has long been regarded as an efficient and powerful numerical tool. The robustness and simplicity of its implementation on computer platforms are the major reasons for its success and popularity in using it to analyze complex photonics structures. The method assumes that the entire amplifier is divided into p sections, each with a length of $\Delta z = L/p$, where L is the amplifier length. Each section is labeled such that the kth section extends from $(k-1)\Delta z$ to $k\Delta z$. The value of Δz is small enough such that the carrier density, material gain, and modal gain can be assumed as uniform along each section. If only the forward-propagating wave is of interest, as in TW-SOA, the output photon densities of subsequent sections, S_+^{k-1} and S_+^k, satisfy the relation that is a generalized form of Equation 21.64

$$S_+^k = G_k S_+^{k-1} = S_+^{k-1} \exp\left[\left(\Gamma g_k - \alpha_{i,k}\right)\Delta z\right], \tag{21.66}$$

where g_k and $\alpha_{i,k}$ are the material gain and active region loss corresponding to section k, respectively, and G_k stands for the single pass gain of the kth section. In the case of bidirectional propagation (forward and backward), as in R-SOA or FP-SOA, the matrix relating $(k-1)$th and the kth section may be written in the following form, based on Equation 21.66:

$$\begin{bmatrix} S_+^k \\ S_-^k \end{bmatrix} = \begin{bmatrix} G_k & 0 \\ 0 & G_k^{-1} \end{bmatrix} \begin{bmatrix} S_+^{k-1} \\ S_-^{k-1} \end{bmatrix}. \tag{21.67}$$

Same method can be applied when the TW equation is written with respect to the electric field envelope (Equation 21.15) instead of photon density, intensity, or power. In this case, the transfer matrix may comprise the information about phase (Chu and Ghafouri-Shiraz, 1994):

$$\begin{bmatrix} E_+^k \\ E_-^k \end{bmatrix} = \begin{bmatrix} G_k^{1/2} \exp\left(i\beta\Delta z\right) & 0 \\ 0 & G_k^{-1/2} \exp\left(i\beta\Delta z\right) \end{bmatrix} \begin{bmatrix} E_+^{k-1} \\ E_-^{k-1} \end{bmatrix}. \tag{21.68}$$

By using input parameters, namely, bias current I and injected optical power P_0, or electric field $E_0 = S_0^{1/2} \exp(i\varphi_0)$, along with boundary conditions and transfer matrices, the carrier density for each section can be found from the Equation 21.14. In the case of R-SOA, the carrier rate equation will be converted into a p-dimensional system of transcendental equations with respect to carrier densities corresponding to all sections (Totović et al., 2014). Once the carrier density in each section is known, the transfer matrices can be used in the evaluation of the photon density or the corresponding electric field envelope in each of the SOA sections.

In this consideration, it is assumed that the photon density, or the corresponding envelope of the electric field, represents the signal. Most of the steady-state models do not account for the noise contribution to the signal, assuming that its phase is uncorrelated to the signal's phase. Therefore, some models neglect the noise (Chu and Ghafouri-Shiraz, 1994), while others use separate TW equations for the signal and the ASE (Jin et al., 2003; de Valicourt et al., 2010), which means that the signal and noise are generally treated separately. However, coupling between the signal and noise is provided by the carrier rate equation that accounts for both contributions, which finally affects the carrier density in the active region and all carrier dependent parameters, and consequently, the photon density. In the case of steady-state analysis, it is possible to solve the TW equation for ASE (Equation 21.31) on domain Δz of section k, and use it in the TMM. The solution is given by

$$a_+^k = G_k a_+^{k-1} + \sigma_k \frac{G_k - 1}{\ln G_k}, \tag{21.69}$$

where a_+^{k-1} and a_+^k represent ASE spectral photon densities at the input and the output of the kth section, respectively, whereas $\sigma_k = 1/(2v_g)\Gamma\beta_{sp}r_{sp}^k\Delta z$ is the contribution of the radiative spontaneous recombination rate spectral density in section k. If the ASE spectrum is uniform over an optical noise bandwidth B_N, the photon density corresponding to ASE can be obtained by multiplying Equation 21.69 by B_N. Otherwise, the total ASE photon density can be found by integrating Equation 21.69 over the ASE bandwidth. The above result can be generalized for both propagation directions (de Valicourt et al., 2010) and written in the following matrix form:

$$\begin{bmatrix} a_+^k \\ a_-^k \end{bmatrix} = \begin{bmatrix} G_k & 0 \\ 0 & G_k^{-1} \end{bmatrix} \begin{bmatrix} a_+^{k-1} \\ a_-^{k-1} \end{bmatrix} + \frac{\sigma_k}{\ln G_k} \begin{bmatrix} G_k - 1 \\ G_k^{-1} - 1 \end{bmatrix}. \tag{21.70}$$

This matrix form does not strictly fit into the standard transfer matrix form due to additional vector representing the ASE contribution of the kth section. However, a considerable number of time-domain multisection methods accounts for ASE by using the form defined by Equation 21.70 (Durhuus et al., 1992; Davis and O'Dowd, 1994; Kim et al., 1999; Occhi et al., 2003; Park et al., 2005; Mathlouthi et al., 2006; Morel and Sharaiha, 2009). Since these methods are not essentially different from TMM, they can be referred to as TMMs.

Although TMM is simple for implementation, it requires a significant number of matrix multiplications, increasing linearly with the number of sections. A large number of multiplications may cause error accumulation and inaccuracy in the evaluation of photon density or the corresponding electric field envelope. The TW equations can be solved by implementation of the FDM rather than the TMM (Connelly, 2001, 2002). Forward differences are used for forward-propagating TWs, and backward differences for backward-propagating TWs. The carrier rate equation in each mesh point is used for determining the carrier density value. Since the self-consistent method implementation is required, the process of calculation of carrier and photon densities relies on iterative procedure, with adjustment of at least one variable in each iteration step, usually the carrier density. However, the carrier density variation along the amplifier is rather small even for a wide range of the signal input powers (Totović et al., 2013). On the other hand, the photon density varies significantly, up to several orders of magnitude. Therefore, the control of the iteration process by the carrier density adjustments, as proposed by Connelly (2001, 2002), makes the process very sensitive since the small variation of the carrier density leads to a large variation of the photon density. It is shown by Jin et al. (2003) that iterations based on photon density may provide faster convergence and increased stability of the iteration process.

21.5.2 Dynamic Models

Due to its simplicity and popularity, TMM has been modified and implemented in the development of the large-signal dynamic model (Davis and O'Dowd, 1994; Kim et al., 1999). The model is bidirectional, which makes it suitable for both TW- and R-SOA time-domain simulations. Similar multisection time-domain models, based on the time-dependent version of Equation 21.67 or Equations 21.68 and 21.70, can also be found in the literature (Occhi et al., 2003; Park et al., 2005; Mathlouthi et al., 2006; Kim et al., 2009). It has been shown by Park et al. (2005) that only the "fullwave" model, which accounts for the intensity and phase of the superimposed electric fields of the signal and noise, provides accurate results. The model based solely on photon densities of signal and noise (or their optical powers) overestimates the influence of the noise on the signal (Park et al., 2005). On the other hand, the model in which the signal is represented by the electric field, and spontaneous emission noise by the photon density, underestimates the effect of noise on the signal (Park et al., 2005).

Another time-domain multisection, wideband, and bidirectional method has been derived using the inverse Fourier transform of the frequency-domain propagation equation (Durhuus et al., 1992). It is essentially a type of TMM, comprising complex time delay. Due to included wideband spectral dependence, it is suitable for simulation of multichannel amplification. A similar method proposed by Morel and Sharaiha

(2009) is implemented in a commercially available software, Advanced Design System from Keysight Technologies (www.keysight.com), after conversion of transfer matrices, including complex time delay, into an equivalent SOA circuit. Another recently developed equivalent circuit model, more precisely transmission line model (TLM), was presented by Xia and Ghafouri-Shiraz (2016). A multisection wideband time-domain method based on the finite-impulse response (FIR) filter scheme has been proposed by Toptchiyski et al. (1999) and Runge et al. (2010). The method is based on two computational steps in each section. The first step includes calculation of local gain with its linear and nonlinear components, as well as the local nonlinear phase change. Then, the propagation equations for the forward- and backward-traveling fields are solved by accounting for the spectral profile of the gain through the implementation of an FIR filter. The FDM can also be applied in time-domain simulations of SOAs. A comprehensive improved finite-difference beam propagation model (IFD-BPM) (Razaghi et al., 2009) has been used for modeling of temporal and spectral properties of copropagating and counterpropagating picosecond optical pulses with different wavelengths. The aforementioned model includes the following effects: interband gain and refractive index dynamics, CH, SHB, two-photon absorption (TPA), ultrafast nonlinear refraction (UNR), gain dispersion, gain peak shift with carrier density variation, and group velocity dispersion (GVD). Due to numerous linear and nonlinear effects, the forward and backward TW equations become essentially nonlinear propagation equations, and are named modified nonlinear Schrödinger equations. Solving these equations is based on coordinate transformation, which can be successfully done by trapezoidal integration and central difference technique. The same method has been also used in the development of the R-SOA pulse propagation model (Connelly, 2012). An interesting method for modeling of SOA time-domain response and four-wave mixing (FWM) was proposed by Mecozzi and Mørk (1997) and Cassioli et al. (2000). The method assumes a relatively small variation of carrier density along the amplifier and consequently a small gain variation. It is based on the analytical integration of material gain over the entire SOA length. This eliminates the spatial coordinate dependence and allows for the amplifier dynamics to be described by solving a set of ordinary differential equations for the complex gain. The ASE noise is modeled by an equivalent noise source with appropriate statistical properties. When the internal loss is included, the method's complexity increases. The SOA simulator is implemented in the Simulink® software, based on the MATLAB® engine and routines for numerical calculations, because of its capability to deal with time-domain analysis of dynamical systems. A similar approach, based on the delayed differential equation, has been used in the derivation of R-SOA's time-domain transmission function, which is then implemented in the semianalytical analysis of FWM in R-SOA (Antonelli and Mecozzi, 2013). The fundamental condition for the method implementation is that the round-trip time of the pulse is small compared to its duration. An improved version of this, the so-called reduced model, is proposed by Dúill and Barry (2015), where the improvement is achieved by the inclusion of the internal losses.

If the dynamic effects of interest occur on a much larger time scale than the time required for the signal to travel along the active region, a simplification to the dynamic model can be introduced by assuming instantaneous propagation of signals across the SOA length (Connelly, 2014, 2015). This assumption significantly simplifies the model by reducing the TW equations to their steady-state form, which can then be solved using the corresponding finite differences for each propagation direction. In this manner, spatial distributions of photon and carrier densities are calculated at each point in time, under the quasi-steady-state conditions. This approach might be helpful as a basis for the simple TW- and R-SOA models; however, a detailed model does require accounting for the finite time needed for wave propagation. In this case, the system of coupled partial differential TW equations may be solved numerically using a first-order upwind scheme based on the FDM (Totović et al., 2011) in order to obtain a full spatiotemporal distribution of all relevant quantities.

21.5.3 Case Study: Steady-State Wideband Self-Consistent Numerical Model

In order to fully understand the mechanisms underlying SOA operation, the model needs to be designed such that it accounts for spectral dependencies of all relevant parameters discussed in Section 21.3 and given in Figures 21.2 through 21.4. These models, typically referred to as wideband, are required to operate not with photon densities, but rather their spectral distributions. Additionally, it is good practice to generalize the model such that it supports any facet reflectivity. This makes the model transparent to the SOA type and it can be further used to analyze TW-, FP-, or R-SOA simply by adjusting the facet reflectivities to the desired value. For TW-SOA, the values can be chosen to be either zero, or, if higher precision is required, some small, but finite value, in the range of 10^{-3}–10^{-5}. In the case of FP- or R-SOA, the reflectivity of a highly reflective facet is usually chosen to be equal to the reflectivity of the interface between cleaved semiconductor and air, $R_2 = (n_r-1)^2/(n_r+1)^2$, but it can be set to another value if reflective layers are used.

The full system of equations for a steady-state wideband model can be derived from Equations 21.49 through 21.52 by setting the time derivatives of carrier density and spectral photon densities to zero:

$$\frac{dn}{dt} = \frac{I}{qV} - \left(An + R_{sp} + Cn^3\right) - \frac{R_{st}}{1 + \varepsilon S_\Sigma} = 0, \tag{21.71}$$

$$\pm \frac{ds_\pm}{dz} = \left(\frac{\Gamma g}{1 + \varepsilon S_\Sigma} - \alpha_i\right) s_\pm, \tag{21.72}$$

$$\pm \frac{d\varphi_\pm}{dz} = \beta_0 + k_0 \frac{\Gamma \Delta n_r}{1 + \varepsilon S_\Sigma}, \tag{21.73}$$

$$\pm \frac{da_\pm}{dz} = \left(\frac{\Gamma g}{1 + \varepsilon S_\Sigma} - \alpha_i\right) a_\pm + \frac{1}{2v_g}\Gamma \beta_{sp} r_{sp}, \tag{21.74}$$

where

$$S_\Sigma\left(n\right) = \int_0^\infty \left(s_+ + s_- + a_+ + a_-\right) d\left(\hbar\omega\right), \tag{21.75}$$

$$R_{st}\left(n\right) = \int_0^\infty v_g g \left(s_+ + s_- + a_+ + a_-\right) d\left(\hbar\omega\right). \tag{21.76}$$

Boundary conditions for the signal can be derived from Equations 21.27 and 21.28 and are given by the following set of equations at the front facet:

$$s_+\left(0\right) = \left(1 - R_1\right) s_0 + R_1 s_-\left(0\right) - 2\sqrt{\left(1 - R_1\right) s_0}\sqrt{R_1 s_-\left(0\right)} \cos\left[\varphi_0 - \varphi_-\left(0\right)\right], \tag{21.77}$$

$$\varphi_+\left(0\right) = \arctan\left[\frac{\sqrt{\left(1 - R_1\right) s_0} \sin\varphi_0 - \sqrt{R_1 s_-\left(0\right)} \sin\varphi_-\left(0\right)}{\sqrt{\left(1 - R_1\right) s_0} \cos\varphi_0 - \sqrt{R_1 s_-\left(0\right)} \cos\varphi_-\left(0\right)}\right], \tag{21.78}$$

and by $s_-(L) = R_2 s_+(L)$, and $\varphi_-(L) = \varphi_+(L) + \pi$ at the rear facet. The input signal is assumed to have Gaussian power spectral distribution $\sigma_G(\hbar\omega)$, with FWHM of 0.1 nm, which is a good approximation of the signal generated by the distributed feedback (DFB) laser. Given the input optical power P_0, input signal spectral density can be found by multiplying the input signal photon density, S_0, by the Gaussian function normalized to unity with respect to energy:

$$s_0\left(\hbar\omega\right) = \frac{\Gamma\left(\hbar\omega\right) P_0}{\hbar\omega v_g\left(\hbar\omega\right) WH}\sigma_G\left(\hbar\omega\right). \tag{21.79}$$

In analysis of ASE noise, phase is not included explicitly, and therefore no interference exists at either facet, giving $a_+(0) = R_1 a_-(0)$ and $a_-(L) = R_2 a_+(L)$.

In order to account for the resonant properties of the SOA cavity, the procedure described in Section 21.4.2 for deriving Equation 21.34 from Equation 21.31 can be employed, and Equation 21.74 can be transformed into a system of steady-state equations written with respect to the noise photon densities corresponding to each mode m:

$$\pm \frac{dA_\pm^m}{dz} = \left(\Gamma_m g_m - \alpha_i^m \right) A_\pm^m + \frac{1}{2 v_g^m} \Gamma_m \beta_{sp}^m R_{sp}^m, \tag{21.80}$$

with boundary conditions $A_+^m(0) = R_1 A_-^m(0)$ and $A_-^m(L) = R_2 A_+^m(L)$. The system given by Equation 21.80 represents the steady-state form of Equation 21.34. Introducing the mode photon densities $A_\pm^m$ also implies that modification of Equations 21.75 and 21.76 is required, where the integral over energies for ASE noise now becomes a sum over modes m:

$$S_\Sigma(n) = \int_0^\infty \left(s_+ + s_- \right) d(\hbar\omega) + \sum_m \left(A_+^m + A_-^m \right), \tag{21.81}$$

$$R_{st}(n) = \int_0^\infty v_g g \left(s_+ + s_- \right) d(\hbar\omega) + \sum_m v_g^m g_m \left(A_+^m + A_-^m \right). \tag{21.82}$$

In this manner, signal amplification will be accounted for by TW Equation 21.80 and the photons will be subsequently redistributed within the mode frequency range using the filtering function (Equation 21.38). The resonant and antiresonant frequencies are found by equating the noise phase accumulation to $2m\pi$ and $(2m-1)\pi$, respectively, the same as in Equation 21.33:

$$2k_0 \left(n_{eff,0} L + \int_0^L \frac{\Gamma \Delta n_r}{1 + \varepsilon S_\Sigma} dz \right) = \left\{ \begin{array}{ll} 2m\pi, & \text{for } \omega_m^r \\ (2m-1)\pi, & \text{for } \omega_m^a \end{array} \right. . \tag{21.83}$$

In the pursuit of increased efficiency, several adjacent modes can be grouped into clusters of $2l+1$ mode, spanning between two antiresonances, from $\hbar\omega_{m-l}^a$ to $\hbar\omega_{m+l+1}^a$, where l stands for a small nonnegative integer. All photons corresponding to one cluster are now treated with a single pair of photon densities $A_\pm^m$, centered at the mth mode resonance, $\hbar\omega_m^r$, where the values of all spectrally dependent parameters in Equations 21.80 through 21.82 are given at $\hbar\omega_m^r$. In this manner, the number of equation pairs (comprising one equation for each propagation direction) required for modeling the ASE noise is reduced from the number of modes N to the number of clusters $M = N/(2l+1)$. This method effectively reduces the time and resource consumption at the cost of a modest decrease in accuracy. The error in calculating the photon density corresponding to a cluster will be pronounced in the portions of the spectrum where the ASE noise photon count reaches high values. Since the photon count depends on both the radiative spontaneous recombination rate and the material gain among other parameters, this error can be alleviated by nonuniform clustering of modes; the clusters can comprise only one mode in the vicinity of the signal central frequency, where the material gain and stimulated emission rate are high and the number of modes within the cluster can be increased further away in the spectrum.

21.5.3.1 Implementation of the Self-Consistent Method

As discussed earlier, the system of equations describing SOA needs to be solved in a self-consistent manner due to the coupling between the signal and noise though the carrier rate equation. Numerical solving of the system requires discretization of the space-spectrum domain into a two-dimensional (2D) mesh, where each point represents a unique pair of position along the longitudinal axis z_i and energy $\hbar\omega_j$. The size of the spectral domain should be chosen such that it covers the region where material gain is positive since in the remaining regions spontaneously emitted photons will be attenuated during propagation and their contribution to the ASE spectrum can be neglected. Step size along the spectral axis should be smaller than the difference between two adjacent resonances of the cavity, which can roughly be estimated from Equation 21.83 as $\Delta(\hbar\omega) = \pi\hbar c/(n_{\mathrm{eff},0}L)$, whereas for the longitudinal axis no particular constraints exist and the smaller step size simply enables a more accurate spatial distribution calculation. The choice of 1001 equidistant points z_i along the longitudinal axis, where i spans from 0 ($z = 0$) to 1000 ($z = L$), provides a quasi-continuous spatial domain, with the step length equal to $\Delta z = L/1000$. An example of such 2D mesh is given in Figure 21.9.

The derivatives over the z-coordinate in Equations 21.72 and 21.73 and Equation 21.80 can now be replaced with finite differences. Depending on the propagation direction of signal and ASE noise, forward or backward, appropriate finite differences are used, forward in the former, and backward in the latter case. This approach allows for successive calculation of the variable values in the next point in space based on the values from the previous one. For forward propagation, the equations can be written in the following generalized form:

$$f_+ \left(z_{i+1}\right) = f_+ \left(z_i\right) + F_+ \left(z_i\right) \Delta z, \tag{21.84}$$

where f_+ stands for any forward-propagating variable, s_+, φ_+, or A_+^m, whereas F_+ denotes the RHS of the corresponding equations, evaluated at the ith point in space, z_i. For backward propagation, using the same generalized notation, the equations can be reduced to

$$f_- \left(z_{i-1}\right) = f_- \left(z_i\right) + F_- \left(z_i\right) \Delta z, \tag{21.85}$$

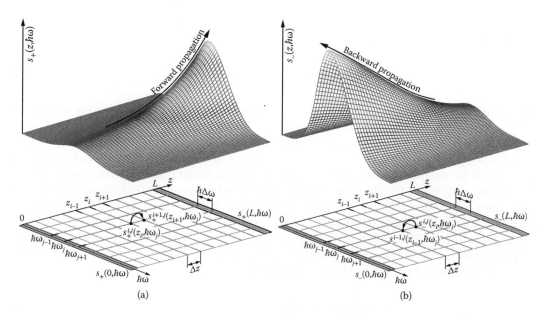

FIGURE 21.9 Illustration of the two-dimensional space-spectrum mesh used in the wideband steady-state self-consistent numerical method for modeling semiconductor optical amplifiers (SOAs). The example includes spectral photon density distribution for (a) forward and (b) backward propagations.

where f_- stands for any backward-propagating variable, s_-, φ_-, or A_-^m, whereas F_- denotes the RHS of the corresponding equations, evaluated at the ith point in space, z_i. The coupling between Equations 21.84 and 21.85 is provided by the boundary conditions at $z = 0$ and $z = L$. Carrier density in each point in space can be found as a solution of the transcendental Equation 21.71, which is then followed by calculation of resonant and antiresonant frequencies, using Equation 21.83.

The system given by Equations 21.71 through 21.73, 21.80, and 21.83 represents the basis of one iteration step in the self-consistent numerical method (SCNM). Starting from the arbitrarily chosen carrier density distribution, signal and noise for both propagation directions can be determined and subsequently used for calculating the new carrier distribution. Assuming that the newly calculated carrier density distribution converges to the steady-state distribution, repeating this process should give improved results with each iteration step. The process can be stopped when the difference between the values calculated within two consecutive iteration steps reaches the tolerance point. Although this simplified algorithm is useful for conceptualization of iterative procedure, in practice it suffers from severe instability, and rarely converges. The instability is caused by significant changes in photon spectral densities, up to several orders of magnitude, for subtle changes in carrier density. An approach typically used to alleviate this issue relies on self-consisting tuning of the carrier density (Connelly, 2001) or photon density (Jin et al., 2003) by its averaging over the current and previous iteration step, with proper weight coefficients. Although these approaches have shown to be successful in modeling TW-SOA, in the case of an R-SOA they still diverge for high driving currents and high input signal optical powers. Choosing to simultaneously tune all variables of interest, namely $s_\pm$, $A_\pm^m$, and n, instead of just one, does increase stability, but not sufficiently enough to provide a reliable algorithm. The main issue comes from the interplay between counterpropagating signals and noise, where the backward-propagating ones are the delayed replicas of their forward-propagating counterparts. This "echoing," present whenever either facet reflectivity is nonzero, destabilizes the algorithm and raises the need for restraining the variables through the fading memory of previous iteration steps. Instead of averaging the variables over the current and previous iteration steps, the generalized algorithm assumes inclusion of k previous iteration steps, where $k \geq 1$. In this manner, sudden changes are prevented, and the system steadily converges toward the solution. The contribution of each iteration step can be adjusted using the set of weight coefficients w_i, with a sum of unity, which ponder the results of the ith iteration step during the averaging process. The updated (averaged) iteration variable that is passed to the next iteration step, x_{i+1}, can be found as

$$x_{i+1} = w_i x_i + w_{i-1} x_{i-1} + \dots + w_{i-k} x_{i-k}. \tag{21.86}$$

The optimal number of iteration steps included will depend on the device's properties, mainly on the front and rear facet reflection coefficients. It has been shown by Totović et al. (2013) that $k = 3$ previous iteration steps, in addition to the current one, suffice to provide a stable algorithm, with a set of weight coefficients $\{w_i, w_{i-1}, w_{i-2}, w_{i-3}\} = \{0.4, 0.3, 0.2, 0.1\}$. The choice of weight coefficients depends on desired robustness of the algorithm and the convergence speed. By using higher pondering values for the current iteration step relative to the previous ones, the iteration process becomes more time efficient at the cost of an increased risk for divergence. Inclusion of several previous iteration steps is required in modeling FP- and R-SOA but can be relaxed to $k = 1$ for TW-SOA.

Prior to entering the iteration procedure, the maximum tolerable relative error δ_{max} needs to be defined. The error δ between the two consecutive iteration steps is calculated as a maximum value of relative errors among all monitored variables, for each point in the mesh $(z_i, \hbar\omega_j)$:

$$\delta = \max \left\{ \delta\left(n\right), \delta\left(s_\pm\right), \delta\left(A_\pm^m\right), \delta\left(\hbar\omega_m^r\right) \right\}. \tag{21.87}$$

The iteration procedure is stopped once the value of the current relative error δ drops below the defined tolerance δ_{max}. The meticulousness with which the iteration error δ is calculated during each iteration step ensures that all results consistently need to converge in order for the iteration procedure to end. Setting δ_{max} to 10^{-3} imposes very strict convergence requirements for all variables.

Aside from the input and control parameters, i.e., I, P_0, weight coefficients, and δ_{max}, SCNM requires a set of initial guess values (IGVs) for all variables. Since the spectral photon density of the input signal, s_0, depends on the confinement factor as shown in Equation 21.79, its IGV is calculated for an arbitrary carrier density between the transparency value n_{tr} and the maximum allowed value, n_{max}, governed by the bias current. The transparency value of the carrier density is determined by equating material gain to zero, whereas the maximum allowed value can be found from Equation 21.71, under the assumption of zero photon density in the active region:

$$\frac{I}{qV} - \left[An_{max} + R_{sp}\left(n_{max}\right) + Cn_{max}^3\right] = 0. \tag{21.88}$$

The IGV of spectral photon density for the forward-propagating signal, s_+, is then set to the input signal spectral photon density s_0, whereas the IGV for s_- is set either to zero, for TW-SOA, or to s_0, for FP- and R-SOA. The IGV for phase in both propagation directions, $\varphi_\pm$, is chosen to be zero. Before setting IGVs for ASE noise, it is necessary to determine the number of analyzed modes and the type of clustering. For a device length between 500 and 1000 μm, the intermodal space is of the order of 0.5 nm, and $N = 600$ modes covers the portion of the spectrum where ASE is expected to be pronounced. For the 100 modes around the signal central frequency, clusters of 1 mode have been chosen, whereas in the rest of the spectrum, clusters consist of 5 modes. This gives a total of $M = 200$ clusters to be analyzed. The ASE noise is usually orders of magnitude lower than the signal, therefore IGVs for all clusters $A_\pm^m$ are set to zero. The IGV for carrier density is calculated based on the IGVs for signal and noise photon densities by solving the transcendental Equation 21.71 with respect to n at each point z_i along the discretized longitudinal axis. Using Equation 21.83, IGVs for resonant and antiresonant frequencies are determined. Finally, using the appropriate expressions depending on the active region type, bulk or MQW, all material parameters dependent on carrier density can be determined. These include material gain (Equation 21.1 or Equation 21.4), radiative spontaneous recombination rate (Equation 21.2 or Equation 21.5), refractive index variation (Equation 21.3 or Equation 21.6), confinement factor (Equation 21.9), group velocity (Equation 21.13), and spontaneous emission coupling factor (Equation 21.11). Once all IGVs are defined, the iteration procedure can be initialized.

As discussed earlier, the iteration procedure essentially relies on repeating one iteration step and updating variable values according to the current and k previous iteration steps using Equation 21.86. It should be noted that until the number of completed iteration steps reaches k, corresponding IGVs are used as a replacement for the missing previous iteration values. The iteration step starts with solving Equations 21.72, 21.73, and 21.80, for forward propagation and determining s_+, φ_+, and A_+^m, based on the initial conditions at $z = 0$. This is followed by adjusting the variable values by averaging them over previous iteration steps and applying boundary conditions at $z = L$. Next, using Equations 21.72, 21.73, and 21.80, backward propagation is simulated and s_-, φ_-, and A_-^m are determined, and subsequently adjusted by averaging their values over k previous iteration steps. The total photon density in the active region, S_Σ, and the rate of stimulated emission, R_{st}, are found from Equations 21.81 and 21.82, respectively, which gives the basis for determining carrier density distribution using Equation 21.71. After adjusting n by averaging it over previous iteration steps, all parameters dependent on carrier density can be calculated. Finally, at the end of the iteration step, the maximum relative error between two consecutive iteration steps is calculated using Equation 21.87 and compared with the tolerance δ_{max}. If the tolerance is satisfied, the iteration process has successfully converged to the solution; otherwise the iteration process continues. The algorithm described for SCNM is summarized in Figure 21.10.

When the iteration process successfully converges to the steady-state solution, normalized filtering function of the cavity $T_m(\hbar\omega)$ can be found from Equation 21.38, and the ASE noise spectral photon

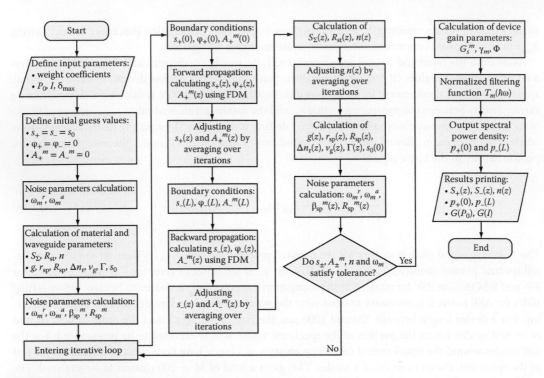

FIGURE 21.10 Flowchart of the steady-state wideband self-consistent numerical method (SCNM). FDM, finite difference method.

densities determined using

$$a_+ (L, \hbar\omega) = \sum_{m=1}^{M} \left[\left(\hbar\omega^a_{m+l+1} - \hbar\omega^a_{m-l} \right)^{-1} A_+^m (L) \, T_m (\hbar\omega) \right], \qquad (21.89)$$

$$a_- (0, \hbar\omega) = \sum_{m=1}^{M} \left[\left(\hbar\omega^a_{m+l+1} - \hbar\omega^a_{m-l} \right)^{-1} A_-^m (0) \, T_m (\hbar\omega) \right]. \qquad (21.90)$$

Finally, the output spectral power densities of signal and noise combined are calculated as

$$p_+ (L, \hbar\omega) = \left[1 - R_2 (\hbar\omega) \right] \left[s_+ (L, \hbar\omega) + a_+ (L, \hbar\omega) \right] v_g (\hbar\omega) \, WH\hbar\omega \Big/ \Gamma (\hbar\omega), \qquad (21.91)$$

$$p_- (0, \hbar\omega) = \left[1 - R_1 (\hbar\omega) \right] \left[s_- (0, \hbar\omega) + a_- (0, \hbar\omega) \right] v_g (\hbar\omega) \, WH\hbar\omega \Big/ \Gamma (\hbar\omega), \qquad (21.92)$$

where the signal spectral distribution is calculated using Equations 21.84 and 21.85, whereas the noise spectral distribution is determined from Equations 21.89 and 21.90. It should be noted that, if reflectivities of both facets are chosen to be zero, which is an idealized case of TW-SOA, the previous algorithm can be simplified. After determining and adjusting the forward-propagating variables, the following step, related to backward propagation, can be reduced to calculation and adjusting of only A_-^m since the backward-propagating signal will not exist.

One of the most valuable figures of merit for SOA characterization is the device (transmission) gain G. Strictly speaking, it is defined as the ratio of the signal's output and input photon density, or their corresponding powers, i.e., $G = 10 \log_{10}(P_{out}/P_0)$. However, as the signal output photon density cannot be separated from ASE noise in the signal's wavelength range, the output photon density usually includes a small fraction of ASE, which has a negligible effect on G, as long as the input optical power P_0 is sufficiently high (above -40 dBm). For TW-SOA, the output photon density is recorded at the rear facet ($z = L$), $P_{out} = \int p_+ (L, \hbar\omega) d(\hbar\omega)$, whereas for the R-SOA, at the front facet ($z = 0$), $P_{out} = \int p_-(0, \hbar\omega)d(\hbar\omega)$. For FP-SOA, either facet can be regarded as the output one, depending on system configuration. In the case when the SOA model bundles up signal and the whole ASE noise spectrum into one quantity, the above definition of device gain G is meaningful for somewhat larger input powers, which can be recognized by the relatively flat dependence of G versus P_0. For insufficient input powers P_0, the device gain tends to increase more rapidly as the input optical power decreases, ultimately reaching the infinite value for the zero-input signal.

21.5.3.2 Results and Discussion

Based on the wideband self-consistent numerical model presented in earlier section, a full analysis of TW- and R-SOA is carried out under the steady-state operation regime. The list of parameters used in the analysis is given in Table 21.3. The remaining parameters, which are both carrier and energy dependent, including g, r_{sp}, Δn_r, Γ, v_g, and β_{sp}, are determined using the models presented in Section 21.3 and are given in Figures 21.2 and 21.4. Both bulk and MQW-based active regions are optimized to be polarization insensitive, as shown in Figure 21.3, so only TE polarization is analyzed.

21.5.3.2.1 Steady-State Signal and Carrier Densities Spatial Distribution

Figures 21.11 and 21.12 show spatial distribution of carrier densities, $n(z)$, and photon densities of forward and backward-propagating signal and ASE noise combined, $\int[s_+(z, \hbar\omega) + a_+(z, \hbar\omega)]d(\hbar\omega)$, and $\int[s_-(z, \hbar\omega) + a_-(z, \hbar\omega)]d(\hbar\omega)$, respectively, for different input optical powers P_0. In Figure 21.11, the distributions are shown for the active region based on the unstrained bulk $In_{0.53}Ga_{0.47}As$, with the parameters listed in Table 21.1, whereas in Figure 21.12, the results are given for MQW consisting of six coupled 0.13% tensile-strained $In_{0.516}Ga_{0.484}As$ quantum wells, which are strain-compensated by the $In_{0.9}Ga_{0.1}As_{0.3}P_{0.7}$ 0.26% compressively strained barriers, with the parameters given in Table 21.2. In both active region types, carrier dynamics is treated using Equation 21.71. Although this is a somewhat simplified model for MQW SOA, its usage is justified when the volumes of the well and barrier regions do not differ significantly. Namely, the carrier recombination time in the barrier, τ_b, figuring in the full system of carrier rate equations describing MQW active region, Equations 21.54 and 21.55, is of the order of nanoseconds, unlike the capture time by the wells τ_{bw}, which has the value between 25 and 55 ps (Keating et al., 1999). On the other hand, it has been shown by Tsai et al. (1995) that the escape time from the wells, τ_{wb}, is one to two orders of magnitude higher in comparison with τ_{bw}. Assuming that the volumes of the well and barrier regions, V_w and V_b, respectively, are similar, the most significant contribution to dn_b/dt comes from the first and the third terms on the RHS of Equation 21.54, whereas the second and fourth terms can be neglected. In

TABLE 21.3 SOA Material and Geometric Parameters

Symbol	Quantity	Value
W	Active region width	2 μm
H	Bulk active region height	140 nm
L_w	MQW well thickness	19 nm
L_b	MQW barrier thickness	10 nm
H	MQW active region height for six QWs	114 nm
L	Active region length	600 μm
K_0	Carrier independent loss coefficient	62 cm^{-1}
K_1	Carrier dependent loss coefficient	7.5×10^{-17} cm^2
A	Shockley–Read–Hall coefficient	1.1×10^8 s^{-1}
C	Auger coefficient	5.82×10^{-29} cm^6/s
ε	Nonlinear gain suppression factor	1.5×10^{-17} cm^3
$\hbar\omega_0$	Signal central energy	0.8 eV
R_1	Antireflective facet power reflectivity	5×10^{-5}

SOA, semiconductor optical amplifier; MQW, multiple quantum well.

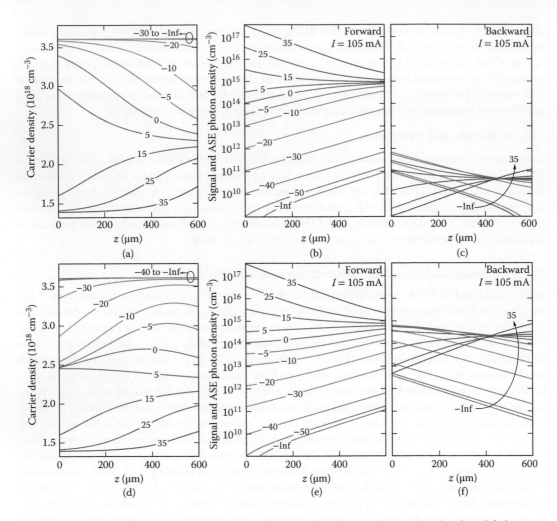

FIGURE 21.11 Spatial distribution of (a), (d) carrier density, and total photon density (signal and amplified spontaneous emission [ASE]) for (b), (e) forward, and (c), (f) backward propagation for different input signal optical powers (given in dBm) in the case of (a) through (c) traveling-wave semiconductor optical amplifier (TW-SOA), and (d) through (f) reflective (R)-SOA, both with the active region based on the unstrained bulk material.

this manner, the generation term $\eta_{inj}I/(qV_b)$ directly influences the carrier density in the barrier states, n_b, which is further reflected in the first term of Equation 21.55. This effect becomes even more pronounced if the capture time is short and the steady-state value of n_b is quickly reached.

It can be seen that for low input optical powers, up to −30 dBm, the carrier density has values close to the maximum allowed by the available bias current, n_{max}, due to the low consumption by vague optical signals. As P_0 increases, it dumps the carrier density overall and leads to its nonuniform spatial distribution caused by the intensified consumption of carriers. For both SOA types, forward-propagating photon densities have similar values and spatial distributions, as shown in Figure 21.11b and e. The signals are exponentially amplified for low to moderate input optical powers, up to 5 dBm. A further increase of P_0 leads to significant carrier depletion close to the front facet, Figure 21.11a and d, and consequently exponential signal attenuation. The backward-propagating photon densities, Figure 21.11c and f, are orders of magnitude lower for TW-SOA than for R-SOA, which is expected due to very low residual reflectivity of the rear facet. However, spatial distributions are qualitatively similar, and resemble the reversed forward-propagating ones, exponentially amplified for low P_0, and attenuated as P_0 increases. The existence of strong backward-propagating signal in the case of R-SOA leads to a different distribution of $n(z)$

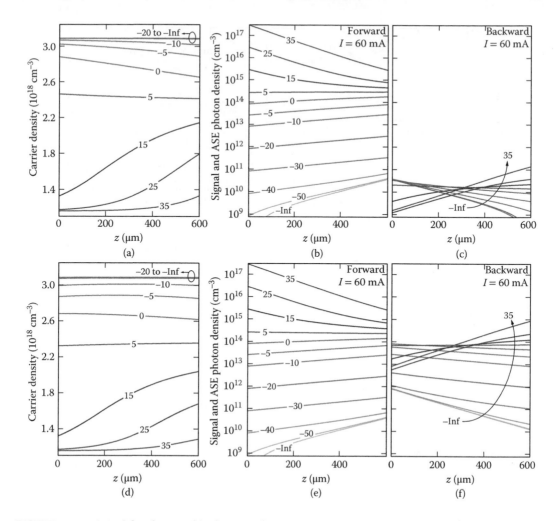

FIGURE 21.12 Spatial distribution of (a, d) carrier density, and total photon density (signal and amplified spontaneous emission [ASE]) for (b, e) forward and (c, f) backward propagation for different input signal optical powers (given in dBm) in the case of (a) through (c) traveling-wave semiconductor optical amplifier (TW-SOA), and (d) through (f) reflective (R)-SOA, both with the active region based on the tensile-strained quantum wells, which are strain-compensated by the compressively strained barriers.

for moderate P_0, between -20 and 5 dBm, in comparison to TW-SOA. The carrier density is more severely depleted closer to the front facet, by the amplified $S_-(z)$, as shown in Figure 21.11d, and has a maximum close to the rear facet, caused by the interplay of the counterpropagating signals.

Spatial distributions of carrier and photon densities for MQW-based SOAs, as shown in Figure 21.12, resemble the ones shown for bulk SOAs, as shown in Figure 21.11, and same qualitative analysis can be applied. The main difference comes from lower carrier density, and material gain, due to the polarization-insensitive design and lower bias current. This leads to the more uniform carrier spatial distribution, as shown in Figure 21.12c and d, and somewhat lower device gain, which can be inferred from the values of the output photon densities, at $z = L$ for TW-SOA, and $z = 0$ for R-SOA.

21.5.3.2.2 Steady-State Device Gain

One of the most valuable figures-of-merit for SOA is its steady-state device (transmission) gain. As defined in earlier section, it is the ratio of the output and input optical powers, or photon densities. Figures 21.13 and 21.14 show the dependence of device gain on operating conditions, i.e., bias current I, and input optical power P_0, for bulk, and MQW active region types, respectively, for both TW- and R-SOA.

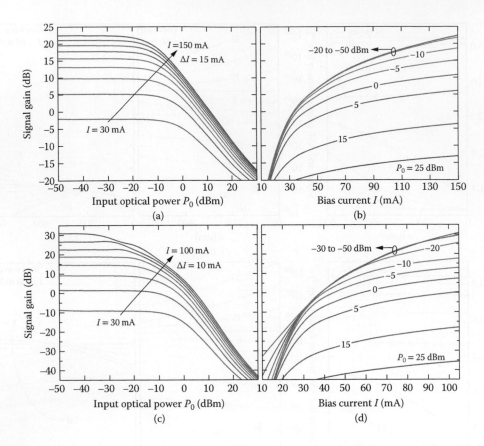

FIGURE 21.13 Device (transmission) gain for bulk active region versus (a), (c) input optical power, and (b), (d) bias current, for (a), (b) traveling-wave semiconductor optical amplifier (TW-SOA), and (c), (d) reflective (R)-SOA.

Figure 21.13 shows that the increase in bias current leads to the higher gain in both SOA types since more carriers are available to provide signal amplification, whereas the increase in P_0 leads to the gain decrease due to excessive carrier consumption. For low to moderate P_0, up to approximately -10 dBm, the device gain does not change significantly with input power and is limited by the length of the active region that is responsible for amplification. Additionally, R-SOA exhibits higher gain, Figure 21.13c and d in comparison with TW-SOA, Figure 21.13a and b, for bias currents beyond 50 mA. In contrast, for the currents below 50 mA, the gain is lower in R-SOA compared with TW-SOA. This comes from double signal propagation through the active region in the case of R-SOA, which emphasizes either amplification, when carrier density is high, or attenuation, when n is low. The maximum operating current of SOA is limited either by maximum bias current density, which provides long-term stable operation, or by the threshold for stimulated emission. In the latter case, the threshold material gain g_{th} can be determined from (Coldren et al., 2012)

$$\Gamma g_{th} - \left(K_0 + \Gamma K_1 n_{th}\right) = \frac{1}{2L} \ln \frac{1}{R_1 R_2}. \tag{21.93}$$

Having the value for g_{th}, it is possible to determine n_{th}, and using Equation 21.71, the corresponding threshold current value. Since the product of the two reflectivities is higher for R-SOA than for TW-SOA, the g_{th} will be lower, and, therefore, the current threshold value will also be lower.

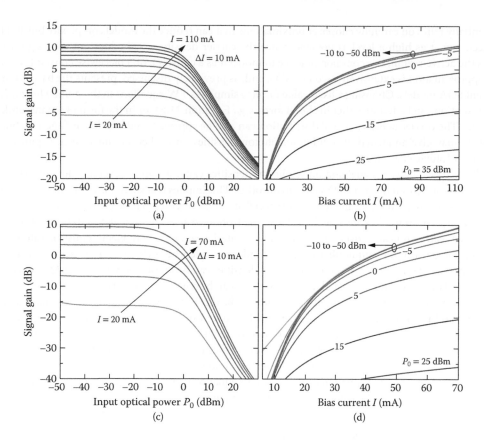

FIGURE 21.14 Device (transmission) gain for the multiple quantum well (MQW) active region versus (a), (c) input optical power, and (b), (d) bias current, for (a), (b) traveling-wave semiconductor optical amplifier (TW-SOA), and (c), (d) reflective (R)-SOA.

An interesting feature in the R-SOA's gain dependence on input optical power can be observed for high bias currents, as shown in Figure 21.13c. As the saturation regime of operation approaches, when the gain begins to decrease, an overshoot in device gain is visible. This effect has already been reported (Schrenk, 2011; Totović et al., 2013) and can be explained by the trade-off between a sudden decrease in the spontaneous emission and ASE with the decrease of carrier density and the corresponding relatively slow decrease in the material gain at the signal wavelength (Totović et al., 2013).

Figure 21.14 shows that MQW SOA provides somewhat lower gain in comparison with bulk, for the same operating conditions. This can be attributed to the lower modal gain, coming from lower confinement factor at the signal central frequency, as shown in Figure 21.4, which is a consequence of the polarization-insensitive design. Nevertheless, qualitatively, the two active regions give similar results, where the device gain increases with bias current increase. For low input optical powers, gain does not vary with P_0, and as the input optical power increases and carrier consumption becomes prominent, the device enters the saturation regime, and gain begins to decrease.

21.5.4 Case Study: Steady-State Semianalytical Model

When simulation time and computational resources are the priority, another approach can be used in SOA analysis. The main underlying idea of the semianalytical approach is to harvest the insightfulness of analytical methods and the versatility of numerical methods to reach a fast and computationally efficient model.

It essentially relies on carefully chosen approximations that can reduce the model complexity but still keep the accuracy at as high a level as possible. Since the carrier density usually does not vary significantly across the active region, it is common practice to model it as a constant, or a piecewise constant function. This approximation is extensively used in TMM, and, as previously discussed, presents the basis for many different SOA models. Choosing a fixed value for $n(z)$ simplifies TW equations to the point where an analytical solution can be found, either in closed form, e.g., Equation 21.64, or in the form of a transcendental equation. The carrier density's spatial distribution can also be approximated by a linear function, but this approach may give an analytical solution for photon densities only when nonlinear gain suppression is neglected (Totović et al., 2014).

Unlike carrier density, signal and noise photon densities can vary up to several orders of magnitude during the propagation from one SOA facet to the other. This implies that small variations in n can build up to significant changes in signal and noise, and special attention is required in selecting the adequate carrier density value. For example, choosing the value at the front or rear facet, $n_0 = n(z = 0)$ or $n_L = n(z = L)$, respectively, can significantly over- or underestimate material gain and spontaneous emission rate and lead to noticeable error in the output photon density, and, consequently, device gain. In order to alleviate this discrepancy, an average value of these two carrier densities can be chosen, $\bar{n} = (n_0 + n_L)/2$, and thus the error caused by neglecting the spatial dependence of n can be partially compensated.

The requirements for increased model efficiency can be met by yet another model approximation, where the signal and noise spectral dependences are neglected and the total spontaneous emission and signal are assigned to a single wavelength corresponding to signal peak, providing model analysis in terms of photon densities rather than spectral photon densities. Since no spectral analysis is present, TW equations written with respect to signal and noise can be wrapped up into a single equation that accounts for both spontaneous and stimulated emission of photons. This equation can be derived as a sum of Equations 21.72 and 21.74, integrated over the whole spectrum, resulting in

$$\pm\frac{dS_\pm}{dz} = \left[\frac{\Gamma g}{1 + \varepsilon\left(S_+ + S_-\right)} - \alpha_i\right]S_\pm + \frac{1}{2v_g}\Gamma\beta_{sp}R_{sp}, \tag{21.94}$$

where $S_\pm$ now stands for the photon density of the signal and noise combined, at the signal central frequency ω_0. All parameters that are dependent on carrier density and photon energy in Equation 21.94, namely, Γ, g, α_i, v_g, and β_{sp}, are now replaced with the corresponding numerical dependencies on n, interpolated from the full spectral and carrier dependent model presented in Figures 21.2 and 21.4, for the energy of $\hbar\omega_0$. Additionally, for parameters that do not vary significantly with n, i.e., Γ, α_i, v_g, and β_{sp}, as shown in Figure 21.4, a fixed value can be used, interpolated for an average carrier density, n_{av}, between the transparency, n_{tr}, and maximum allowed value, n_{max}, given by Equation 21.88. The carrier density at transparency will depend only on material gain, and not on operating conditions, whereas the maximum allowed value for carrier density n_{max} will depend on the value of bias current. In order to choose the adequate value of n_{av} for parameter interpolation, it is useful to determine n_{max} for maximum allowed bias current I_{max}. The value of I_{max} can be calculated as $I_{max} = J_{max}WL$, for a cross-section area defined by the active region's width W and length L, given in Table 21.3, where J_{max} is the maximum allowed bias current density for the long-term stable operation, which is in the range of 25–35 kA/cm². Under the listed conditions, the chosen values for bulk and MQW-SOA are $n_{av} = 2.5 \times 10^{18}$ cm⁻³ and $n_{av} = 2.1 \times 10^{18}$ cm⁻³, respectively. The interpolated values of Γ, v_g, β_{sp}, and additional parameters are listed in Table 21.4 for bulk and MQW at the signal central energy of $\hbar\omega_0 = 0.8$ eV.

If a simpler model is used, particularly if an analytical solution is sought, it proves to be impractical to use numerical dependencies of g and R_{sp} on n (Figure 21.2). Rather, numerical results can be fitted to analytical functions, which further simplifies calculation. The most common choice for $g(n)$ dependence is a linear one, $g(n) = \alpha(n - n_{tr})$, where α is the differential gain (Thylén, 1988; Shtaif et al., 1998; Mørk et al., 1999). This approximation is justified if the bias current is low enough to prevent a significant increase in carrier

TABLE 21.4 Interpolated Parameter Values for Bulk and Multiple Quantum Well (MQW) Active Region at the Signal Central Energy $\hbar\omega_0 = 0.8$ eV and for the Average Carrier Density of $n_{av} = 2.5 \times 10^{18}$ cm^{-3} for Bulk and $n_{av} = 2.1 \times 10^{18}$ cm^{-3} for MQW

Symbol	Quantity	Bulk	MQW
Γ	Optical confinement factor	30.44%	21.62%
v_g	Group velocity	7.76×10^7 m/s	8.73×10^7 m/s
k_0	Wave vector	4.05×10^6 m^{-1}	4.05×10^6 m^{-1}
n_r	Index of refraction	3.660	3.597
Δn_r	Variation of index of refraction	0.071	7.44×10^{-4}
$n_{eff,0}$	Effective index of refraction for zero carrier density	3.405	3.291
n_{eff}	Effective index of refraction	3.490	3.291
β_0	Wave propagation constant for zero carrier density	1.38×10^7 m^{-1}	1.33×10^7 m^{-1}
R_2	Highly reflective facet power reflectivity	0.3258	0.3192
n_{sp}	Inversion factor	1.035	1.055
β_{sp}	Spontaneous emission coupling factor	1.842×10^{-5}	1.869×10^{-5}

All values given are for transverse electric (TE) polarization.

TABLE 21.5 Bulk and Multiple Quantum Well (MQW) Active Region Fitting Parameters at the Signal Central Energy $\hbar\omega_0 = 0.8$ eV for Material Gain and Spontaneous Emission Rate to the two-Parameter Logarithmic and Second-Degree Polynomial Function of Carrier Density, Respectively

Symbol	Quantity	Bulk	MQW
g_0	Material gain fitting parameter	742.182 cm^{-1}	669.205 cm^{-1}
n_{tr}	Carrier density at transparency	1.38×10^{18} cm^{-3}	1.15×10^{18} cm^{-3}
B_0	Spontaneous emission fitting parameter	-8.375×10^{25} cm^{-3}/s	-1.351×10^{26} cm^{-3}/s
B_1	Spontaneous emission fitting parameter	1.344×10^8 s^{-1}	1.975×10^8 s^{-1}
B_2	Spontaneous emission fitting parameter	2.790×10^{-11} cm^3/s	2.146×10^{-11} cm^3/s

density. However, a more suitable model relies on a two-parameter (g_0, n_{tr}) logarithmic dependence (Coldren et al., 2012), which provides gain saturation with carrier density increase, $g(n) = g_0 \ln(n/n_{tr})$, where g_0 is the material gain fitting parameter and n_{tr} is the carrier density at transparency. It should be noted that, although the two-parameter logarithmic gain does provide reasonably good results, for more linear-like dependencies of gain on carrier density, such as for bulk materials, another model may be employed, which includes a third parameter, n_s. This model, $g(n) = g_0 \ln[(n + n_s)/(n_{tr} + n_s)]$, prevents infinite gain value for zero carrier densities, and in the limiting case of $n_s \to 0$ reduces to the purely logarithmic model, whereas for $n_s \to \infty$ reduces to purely linear model (Coldren et al., 2012). The model used in this case study is a two-parameter logarithmic gain since it fits well on numeric results.

The radiative spontaneous recombination rate is usually modeled via the quadratic function (Chu and Ghafouri-Shiraz, 1994; Melo and Petermann, 2008). Although this function fits actual dependence quite well, a more detailed model would include a second-degree polynomial function, $R_{sp}(n) = B_0 + B_1 n + B_2 n^2$, where $B_0, B_1,$ and B_2 are the spontaneous emission rate fitting parameters. Table 21.5 shows a list of fitting parameters at wavelength $\lambda = 1.55$ μm ($\hbar\omega_0 = 0.8$ eV), for active regions based on the unstrained bulk In$_{0.53}$Ga$_{0.47}$As, and the MQW consisting of six coupled 0.13% tensile-strained In$_{0.516}$Ga$_{0.484}$As quantum wells, which are strain-compensated by the In$_{0.9}$Ga$_{0.1}$As$_{0.3}$P$_{0.7}$ 0.26% compressively strained barriers.

In a steady-state analysis, the input signal carries no information encoded in its amplitude or phase, so the phase equation may be omitted from the model, in attempting to reduce the model's complexity. This simplification is additionally supported by the fact that the carrier diffusion will wash out the spatial grating in the carrier density produced by the interference between two counterpropagating optical fields.

For all antireflective facets, residual reflectivity can be neglected and equated with zero. Since the noise power is usually several orders of magnitude lower than the signal power, zero reflectivity implies that in the case of TW-SOA, the signal can be assumed to travel only once through the active region in the forward direction. On the other hand, in the case of R-SOA, the signal will pass through the active region twice, forward and backward, as a consequence of reflection at the rear facet. In other words, no multiple reflections are present, and the cavity does not exhibit resonant behavior. The boundary conditions for the signal at the front and rear facets are $S_+(0) = S_0$ and $S_-(L) = R_2 S_+(L)$, respectively, where S_0 is the input signal photon density. Depending on the SOA type, reflectivity of the rear facet can have either some finite value, for R-SOA, or it can be zero, for TW-SOA, which essentially simplifies the boundary condition to $S_-(L) = 0$.

The steady-state carrier density rate equation can be derived from Equations 21.71, 21.75, and 21.76, under the previously listed assumptions:

$$\frac{I}{qV} - \left(An + R_{sp} + Cn^3\right) - \frac{v_g g \left(S_+ + S_-\right)}{1 + \varepsilon \left(S_+ + S_-\right)} = 0. \tag{21.95}$$

This equation can be used for R-SOA modeling as is, whereas in the case of TW-SOA, it can be simplified by neglecting the backward-propagating noise contribution and setting the photon density of the backward-propagating signal to zero ($S_- = 0$). The system of coupled equations given by Equations 21.94 and 21.95 cannot be solved analytically unless the approximation regarding the carrier density's spatial distribution is introduced. Different approximations require different analytical treatments and implementations, which are further discussed.

21.5.4.1 The Fundamentals of the Semianalytical Models

The starting point of semianalytical treatment is replacement of the carrier density spatial distribution with a fixed value, equal to the arithmetic mean of the carrier density values at the edges of the SOA structure, or the edges of its sections, namely, $n(z) = \bar{n} = (n_0 + n_L)/2$. These carrier densities are deduced by solving the carrier rate Equation 21.95, where photon density distributions are expressed as functions of the fixed carrier density $\bar{n}$. In other words, once the analytical dependence of $S_\pm(\bar{n}, z) = S_\pm(n_0, n_L, z)$ is determined, it can be substituted in Equation 21.95 at $z = 0$ and $z = L$, which results in a system of two coupled transcendental equations written with respect to n_0 and n_L. The difference between these two carrier density equations comes from the spontaneous emission term $R_{sp}(n)$ and material gain $g(n)$, which for $z = 0$ depend on n_0, and for $z = L$ depend on n_L. Solving these equations provides the numerical value of $\bar{n}$, which is then returned to $S_\pm(\bar{n}, z)$, and the output photon density can be evaluated.

The same method can be applied to an SOA section of length $\Delta z = L/p$, given in Figure 21.15, where p is the number of analyzed sections, and the method can be generalized to the piecewise constant carrier density spatial distribution. The advantage of this approach over the TMM is the choice of the kth section carrier density value, $\bar{n}_k$, as the mean value between the carrier densities at the section interfaces, $\bar{n}_k = (n_{k-1} + n_k)/2$. In this manner, a higher degree of accuracy is provided for a smaller number of sections, compared to the choice of carrier density at the edge of a section.

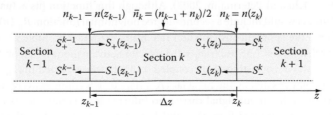

FIGURE 21.15 Schematic representation of the kth section in the segmented semiconductor optical amplifier structure with the relevant carrier and photon densities.

Assuming that the carrier density has a known, fixed value $\bar{n}_k$ at the kth section of an R-SOA, Equation 21.94 needs to be solved analytically for both propagation directions, with initial conditions imposed by the photon densities entering the kth section, which can be determined by solving Equation 21.94 at section $k-1$ for forward, and $k+1$ for backward propagation. However, the analytical solution of Equation 21.94 cannot be obtained for R-SOA without additional approximations, due to the nonlinear coupling of S_+ and S_-. In the case of TW-SOA, only the forward-propagating photon density S_+ is included in Equation 21.94, providing an analytical solution and the corresponding initial conditions for all sections along TW-SOA.

21.5.4.2 Approximate Solutions of Counterpropagating Photon Density Distributions and Implementation of the Semianalytical Algorithm

In order to derive the photon density spatial distributions, S_+ and S_-, corresponding to R-SOA, the system of Equation 21.94 needs to be decoupled, which can be done in two ways.

First, the nonlinear gain suppression can be neglected by setting $\varepsilon = 0$, which results in a linear decoupled system with respect to S_+ and S_-:

$$\pm \frac{dS_\pm}{dz} = \left(\Gamma g - \alpha_i\right) S_\pm + \frac{1}{2v_g} \Gamma \beta_{sp} R_{sp}. \tag{21.96}$$

This assumption greatly simplifies the system, and, for the bias currents and input optical powers that are not too high, it does not degrade the model accuracy significantly. The solution of Equation 21.96 for each section can be found in closed analytical form, corresponding to the matrix form (Equation 21.70) used in description of noise by TMM:

$$S_+^k = S_+^{k-1} G_k + \sigma_{sp}^k \left(G_k - 1\right)/\ln G_k, \tag{21.97}$$

$$S_-^{k-1} = S_-^k G_k + \sigma_{sp}^k \left(G_k - 1\right)/\ln G_k, \tag{21.98}$$

where $G_k = G(\bar{n}_k) = \exp\{[\Gamma g(\bar{n}_k) - \alpha_i(\bar{n}_k)]\Delta z\}$ is the single-pass gain of the kth section and $\sigma_{sp}^k = \sigma_{sp}(\bar{n}_k) = 1/(2v_g)\Gamma \beta_{sp} R_{sp}(\bar{n}_k)\Delta z$ is the single-pass generated spontaneous emission within the kth section. Due to the cascading nature of sections, the output photon density of the $(k-1)$th section, $S_+^{k-1} = S_+(z_{k-1})$, is the input photon density of the kth one for forward propagation, whereas for backward propagation, the output photon density of the $(k+1)$th section, $S_-^k = S_-(z_k)$, is the input photon density of the kth one, as shown in Figure 21.15. This chaining, along with the form of Equations 21.97 and 21.98, suggests that the photon densities at section interfaces are calculated by recursion. The base cases in a recursive method for forward and backward propagation are defined by the boundary conditions at the front ($k=1$) and rear ($k=p$) facets, respectively, $S_+^0 = S_0$ and $S_-^p = R_2 S_+^p$. The forward-propagating signal exiting the kth section will depend on carrier densities of all preceding sections, $S_+^k = S_+(z_k, \bar{n}_1, \ldots, \bar{n}_k) = S_+(z_k, n_0, \ldots, n_k)$, whereas the backward-propagating signal will depend on carrier densities in all sections, $S_-^k = S_-(z_k, \bar{n}_1, \ldots, \bar{n}_p) = S_-(z_k, n_0, \ldots, n_p)$, due to the imposed initial condition at the rear facet, which assumes reflection of $S_+(z_p, n_0, \ldots, n_p)$. These analytical expressions can now be returned to the carrier rate Equation 21.95 at each section interface located at z_k

$$f_k = 0 = \frac{I}{qV} - \left[An_k + R_{sp}\left(n_k\right) + Cn_k^3\right] - v_g g\left(n_k\right) \left[S_+\left(z_k, n_0, \ldots, n_k\right) + S_-\left(z_k, n_0, \ldots, n_p\right)\right], \tag{21.99}$$

which forms a system of $p+1$ transcendental equations, all of which are dependent on $p+1$ carrier densities, from n_0 to n_p. This system cannot be solved analytically, so a numerical procedure is required. An example of such a procedure, given by Totović et al. (2014), relies on numerically assisted elimination of variables, where, in an iterative approach, the range of carrier densities in which the solution n_k is expected to be found is narrowed simultaneously for all k from 0 to p. The initial range for carrier densities at all interfaces is bounded by the carrier density at transparency, n_{tr}, and the maximum carrier density, n_{max}, limited by the available current (Equation 21.88). In each iteration step, the corresponding range of each n_k is reduced to one fifth of its previous size by the appropriate algorithm, still including the solution n_k. After a predefined number of iteration steps, the values for n_k can be determined by evaluating the RHS of the system (Equation 21.99) for the carrier densities in the narrowed ranges, giving a function $f_k(n)$, and subsequent interpolation of the function $f_k(n)$ in search for n_k which returns 0, i.e., the LHS value of Equation 21.99. Usually, four iteration steps suffice for the error coming from neglected gain suppression to prevail over the error coming from the iteration process. Having the values for carrier densities at all interfaces, it is possible to evaluate the photon densities at each z_k using Equations 21.97 and 21.98, and subsequently determine the device gain. Since this model treats the propagation of the signal and ASE noise combined through Equation 21.96, or more generally Equation 21.94, the device gain is defined as the ratio of the output signal power, along with the total ASE noise, and the input signal power. This definition implies that in the limiting case of zero input signal, the device gain will asymptotically approach infinity due to division by zero.

Another approach that can be used for the decoupling of Equation 21.94 with respect to $S_\pm(z,n)$ is to assume that the opposite-propagating photon density is spatially independent, i.e., $S_\mp(z,n) = S_\mp^{cst}(n)$. Unlike the carrier density, which varies modestly across the active region, photon density exhibits exponential increase or decrease from one section edge to the other, so a choice of geometric mean is a more suitable one, which gives $S_\mp^{cst,k}(n) = [S_\mp^{k-1}(n)S_\mp^k(n)]^{1/2}$. The term $1 + \varepsilon(S_+ + S_-)$ in Equation 21.94, which is responsible for equation coupling, can now be replaced with $1 + \varepsilon(S_\pm + S_\mp^{cst})$, or, more generally, with a sum of the spatially independent, and spatially dependent functions, $\theta_\pm + \varepsilon S_\pm$, where $\theta_\pm = 1 + \varepsilon S_\mp^{cst}$. In the case of TW-SOA, no backward-propagating signal exists and $\theta_+ = 1$. Equation 21.94 now reads

$$\pm \frac{dS_\pm}{dz} = \left(\frac{\Gamma g}{\theta_\pm + \varepsilon S_\pm} - \alpha_i \right) S_\pm + \frac{1}{2v_g} \Gamma \beta_{sp} R_{sp}, \tag{21.100}$$

and its solution for the kth section can be found in the implicit form from the transcendental equation

$$2\alpha_i^k \Delta z = \left(\frac{1}{T_\pm^k} - 1 \right) \ln \left| \frac{\theta_\pm^k + \varepsilon S_\pm^{out,k} - (1 - T_\pm^k)/\mu_\pm^k}{\theta_\pm^k + \varepsilon S_\pm^{in,k} - (1 - T_\pm^k)/\mu_\pm^k} \right| - \left(\frac{1}{T_\pm^k} + 1 \right) \ln \left| \frac{\theta_\pm^k + \varepsilon S_\pm^{out,k} - (1 + T_\pm^k)/\mu_\pm^k}{\theta_\pm^k + \varepsilon S_\pm^{in,k} - (1 + T_\pm^k)/\mu_\pm^k} \right|, \tag{21.101}$$

where the auxiliary parameters $\mu_\pm^k$ and $T_\pm^k$ are defined as follows:

$$\mu_\pm^k = \frac{2\alpha_i^k}{\Gamma g_k + \alpha_i^k \theta_\pm^k + \varepsilon \Gamma \beta_{sp} R_{sp}^k / (2v_g)}, \tag{21.102}$$

$$\left(T_\pm^k \right)^2 = 1 - \left(\mu_\pm^k \right)^2 \Gamma g_k \theta_\pm^k / \alpha_i^k. \tag{21.103}$$

In Equation 21.101, $S_\pm^{in,k}$ and $S_\pm^{out,k}$ denote the input and output photon densities into and from the kth section, respectively. For forward propagation, $S_+^{in,k} = S_+^{k-1}$ and $S_+^{out,k} = S_+^k$, whereas in the case of backward propagation, $S_-^{in,k} = S_-^k$ and $S_-^{out,k} = S_-^{k-1}$, according to Figure 21.15. Equation 21.101 is solved with respect to $S_\pm^{out,k}$ recursively, using the previously determined $S_\pm^{in,k}$, with the base cases in a recursive method defined by the boundary conditions in sections $k = 1$ and $k = p$, for forward and backward propagation, respectively. Just as in the previous model, where the nonlinear gain suppression was neglected, the solutions of Equation 21.101 will depend either on carrier densities in all preceding sections for forward propagation, $S_+^{out,k} = S_+(z_k, \bar{n}_1, \dots, \bar{n}_k) = S_+(z_k, n_0, \dots, n_k)$, or on all carrier densities in the case of backward propagation, $S_-^{out,k} = S_-(z_{k-1}, \bar{n}_1, \dots, \bar{n}_p) = S_-(z_{k-1}, n_0, \dots, n_p)$. Substituting these dependencies into the carrier rate Equation 21.95 at each section interface, z_k, gives the system of $p + 1$ transcendental equations:

$$f_k = 0 = \frac{I}{qV} - \left[An_k + R_{sp}\left(n_k\right) + Cn_k^3 \right] - \frac{v_g g\left(n_k\right)\left[S_+\left(z_k, n_0, ..., n_k\right) + S_-\left(z_k, n_0, ..., n_p\right)\right]}{1 + \varepsilon\left[S_+\left(z_k, n_0, ..., n_k\right) + S_-\left(z_k, n_0, ..., n_p\right)\right]}. \quad (21.104)$$

Unlike the previous model, for $\varepsilon = 0$, where the photon density dependencies on n_k could be expressed in closed analytical form, the solutions are now given in implicit form (Equation 21.101), which essentially requires a numerical method. Moreover, in order to employ a numerical method, the values for carrier densities in all sections need to be known. This raises the need for a self-consistent numerical approach in solving the system given by Equations 21.101 and 21.104. The numerical method should comprise two distinct steps. First, the IGVs need to be generated for carrier and photon densities at all section interfaces, which is done using the model for $\varepsilon = 0$. Next, an iterative procedure is used, where carrier and photon densities are alternately calculated, and carrier densities are subsequently updated based on the values from the current and previous iteration steps until the predefined tolerance is satisfied. The maximum relative error between two consecutive iteration steps, δ, is calculated as the maximum relative error for all monitored variables, at all interfaces, i.e., $\delta = \max\{\delta(n_k), \delta(S_+^k), \delta(S_-^k)\}$, and the tolerance is chosen to be $\delta_{max} = 10^{-6}$. The updating of n_k is done using a set of weight coefficients with a sum of unity for the current and previous iteration steps, $\{w_i, w_{i-1}\} = \{0.1, 0.9\}$. The weight coefficients are chosen such that they give balance between the stability and speed of convergence, and usually no more than 100 iteration steps are required for the procedure to reach a self-consistent solution. It should be noted that due to the nature of the algorithm, the system of equations given by Equation 21.104 is not coupled since $S_+(z_k, n_0, \ldots, n_k)$ and $S_-(z_k, n_0, \ldots, n_p)$ are calculated for the carrier values from the previous iteration step and can be treated as parameters. In other words, the current iteration variable n_k is included in Equation 21.104 through the spontaneous emission terms and material gain.

It has been shown by Totović et al. (2014) that depending on the number of chosen sections and the level of approximation, the semianalytical model can be up to two orders of magnitude faster in comparison to the numerical model based on SCNM. Meanwhile, the maximum absolute mismatch in the device gain calculated by the numerical and semianalytical models does not exceed 1.6 dB even for only one section, and neglected nonlinear gain suppression, whereas this mismatch drops to a maximum of 0.11 dB for three sections with included gain suppression for a bulk R-SOA.

21.5.5 Case Study: Dynamic Propagation Model

When the input signal carries information encoded either in its amplitude, or phase, or both, a dynamic approach in SOA modeling is required. The same model needs to be applied for modulation or remodulation purposes, when the bias current is time dependent. The most general dynamic model, suitable for both large- and small-signal modulation, includes spatiotemporal and spectral dependency of both forward- and backward-propagating signals and noise, given by Equations 21.50 through 21.52, and, depending on the active region type, one or several equations describing spatiotemporal carrier density distribution. For bulk SOA, carrier density evolution can be described by Equation 21.49, for the MQW active region by Equations 21.54 and 21.55, and for the QD-SOA by Equations 21.57 through 21.59. Due to the ampleness and complexity of three-dimensional (3D) mesh (time-space-spectrum) implementation in the numerical algorithms, whenever the input signal is spectrally narrow, it is common practice to simplify the model and perform the analysis for photon densities and not their spectral distributions. Similar to Section 21.5.4, material gain and spontaneous emission rate can be approximated with the two-parameter logarithmic and second-order polynomial function, respectively, $g(n) = g_0 \ln(n/n_{tr})$ and $R_{sp}(n) = B_0 + B_1 n + B_2 n^2$, where the fitting parameters are given in Table 21.5 for the bulk and MQW active region, at the signal central wavelength of 1.55 μm. All other parameters, which are dependent on photon energy and carrier density, can be approximated by the corresponding fixed values given in Table 21.4, interpolated for the photon energy of $\hbar\omega_0 = 0.8$ eV and an average carrier density n_{av}, which depends on the active region type. Under these assumptions, the signal can be described by the equation similar to Equation 21.23, with the addition of nonlinear gain suppression:

$$\pm\frac{\partial S_\pm}{\partial z} + \frac{1}{v_g}\frac{\partial S_\pm}{\partial t} = \left(\frac{\Gamma g}{1 + \varepsilon S_\Sigma} - \alpha_i\right) S_\pm. \tag{21.105}$$

The phase evolution for both propagating directions is given by Equation 21.51:

$$\pm\frac{\partial \varphi_\pm}{\partial z} + \frac{1}{v_g}\frac{\partial \varphi_\pm}{\partial t} = \frac{k_0 \Gamma \Delta n_r}{1 + \varepsilon S_\Sigma}. \tag{21.106}$$

ASE noise is treated with a separate equation, resembling Equation 21.105, with the addition of the spontaneous emission rate contribution:

$$\pm\frac{\partial A_\pm}{\partial z} + \frac{1}{v_g}\frac{\partial A_\pm}{\partial t} = \left(\frac{\Gamma g}{1 + \varepsilon S_\Sigma} - \alpha_i\right) A_\pm + \frac{1}{2v_g}\Gamma \beta_{sp} R_{sp}. \tag{21.107}$$

It should be noted that by using Equation 21.107, spectral dependence of ASE is neglected and all noise photons are treated at the single frequency. Boundary conditions are defined by the input signal, described by its photon density $S_0(t)$ and phase $\varphi_0(t)$, and front and rear facet reflectivities, R_1 and R_2, respectively. For antireflective facets, zero reflectivity can be assumed, whereas for the highly reflective ones, the reflectivity coefficient can be calculated as $R_2 = (n_r - 1)^2/(n_r + 1)^2$, if no reflective layer is placed. At the front facet, which is antireflective, the boundary conditions read $S_+(0) = S_0, \varphi_+(0) = \varphi_0, A_+(0) = 0$, and for the rear facet, which can be either highly- or antireflective, $S_-(L) = R_2 S_+(L), \varphi_-(L) = \varphi_+(L) + \pi$, and $A_-(L) = R_2 A_+(L)$. In the case of TW-SOA, rear facet reflectivity R_2 will be zero, and no backward-propagating signal will exist. However, since the noise is generated within the SOA, with equal probability of photon traveling in any direction, backward-propagating noise will exist, both for TW- and R-SOA.

Carrier density distribution is described by Equation 21.49, where the current I can be time dependent:

$$\frac{dn}{dt} = \frac{I}{qV} - \left(An + R_{sp} + Cn^3\right) - \frac{v_g g S_\Sigma}{1 + \varepsilon S_\Sigma}. \tag{21.108}$$

In Equations 21.105 through 21.108, S_Σ stands for the total photon density in the SOA's active region, including both signal and noise, $S_\Sigma = S_+ + S_- + A_+ + A_-$.

The system given by Equations 21.105 through 21.108 cannot be analytically solved, and several numerical approaches are described in Section 21.5.2. Although TMM is widely used due to its simplicity, coarse segmentation along the longitudinal axis might lead to an error in calculating the photon density distribution. Moreover, this error can be accumulated during matrix multiplication and can be even more pronounced when two counterpropagating signals exist.

21.5.5.1 Upwind Scheme Numerical Implementation

In order to mitigate potential errors, a first-order upwind scheme numerical method based on FDM can be developed, where the space and time axis are treated as quasi-continuous through fine segmentation. After choosing the number of points along the longitudinal axis, P, which define the section length, $\Delta z = L/(P - 1)$, segmentation of the temporal axis is done, such that $\Delta t \leq \Delta z/v_g$, according to the Courant–Friedrichs–Lewy stability condition. Depending on the length of the analyzed time interval, T, the number of points along the temporal axis is determined as $Q = \lceil T/\Delta t \rceil + 1$. All derivatives in Equations 21.105 through 21.108 are replaced with finite differences, such that for temporal derivatives forward differences are used:

$$\frac{df}{dt} \approx \frac{f\left(t_{j+1}\right) - f\left(t_j\right)}{\Delta t}, \tag{21.109}$$

where f is any time-dependent variable, i.e., $S_\pm, \varphi_\pm, A_\pm, n$, and for spatial derivatives, the choice is made based on the propagation direction. For forward propagation

$$\frac{df_+}{dz} \approx \frac{f_+\left(z_i\right) - f_+\left(z_{i-1}\right)}{\Delta z}, \tag{21.110}$$

where f_+ denotes any of the following variables S_+, φ_+, A_+, whereas for backward propagation

$$\frac{df_-}{dz} \approx \frac{f_-\left(z_{i+1}\right) - f_-\left(z_i\right)}{\Delta z}, \tag{21.111}$$

where f_- denotes any of the following variables: S_-, φ_-, A_-. Using Equations 21.109 through 21.111, a set of coupled partial differential equations describing signal and noise propagation along the SOA, Equations 21.105 through 21.107, can be transformed into the set of coupled linear algebraic equations, for forward

$$f_+\left(z_i, t_{j+1}\right) = f_+\left(z_i, t_j\right) + \Delta t v_g \left[F_+\left(z_i, t_j\right) - \frac{f_+\left(z_i, t_j\right) - f_+\left(z_{i-1}, t_j\right)}{\Delta z} \right], \tag{21.112}$$

and backward propagation

$$f_-\left(z_i, t_{j+1}\right) = f_-\left(z_i, t_j\right) + \Delta t v_g \left[F_-\left(z_i, t_j\right) + \frac{f_-\left(z_{i+1}, t_j\right) - f_-\left(z_i, t_j\right)}{\Delta z} \right], \tag{21.113}$$

where $f_\pm$ stands for any variable $S_\pm, \varphi_\pm$, or $A_\pm$, whereas $F_\pm$ stands for the RHS of the corresponding equation evaluated at point (z_i, t_j) in the 2D spatiotemporal mesh. For carrier density equation, the transformation yields

$$n\left(z_i, t_{j+1}\right) = n\left(z_i, t_j\right) + \Delta t F\left(z_i, t_j\right), \tag{21.114}$$

where F stands for the RHS of Equation 21.108, evaluated at point (z_i, t_j) in the 2D spatiotemporal mesh. Forward differences for temporal derivatives enable calculation of all variables in the next point in time, t_{j+1}, based on the values from the current one t_j, as shown in Equations 21.112 through 21.114, and illustrated in Figure 21.16.

The numerical implementation of the discretized system (Equations 21.112 through 21.114) essentially requires two distinctive iteration processes: one for the temporal and another for spatial evolution. For every point j along the temporal axis, spatial distribution of carrier density for the next temporal point $(j + 1)$ is determined using Equation 21.114, which is followed by simulation of forward and subsequently

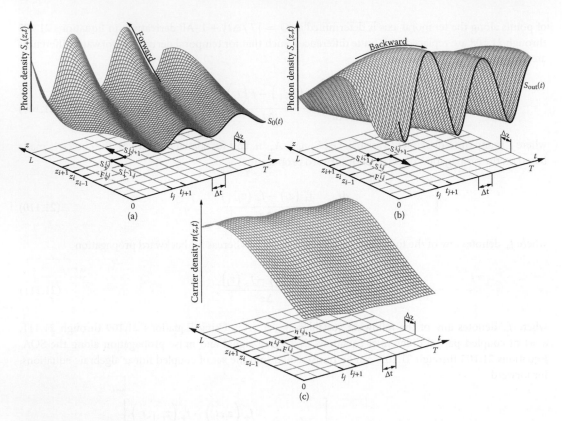

FIGURE 21.16 Illustration of two-dimensional (2D) spatiotemporal mesh with denoted relevant points for calculation of (a) forward- and (b) backward-propagating signal photon densities by upwind finite differences scheme, and (c) carrier densities by the forward finite differences method. The same upwind scheme can be applied to the phase, or noise photon density calculation. The example includes spatiotemporal distribution for (a) forward signal, (b) backward signal, and (c) carrier density.

backward propagation of signal and noise along the SOA's active region using Equations 21.112 and 21.113. This process is repeated until the end of the temporal axis is reached, incrementing j by one, up to Q. Since the results in the $(j + 1)$th point in time are dependent on the values from the previous jth point, initial conditions for spatial distributions of all variables are required (for $j = 0$), in order to start the iteration process over time by calculating the values in the point $j = 1$. Initial values will depend on the assumed SOA state before the initialization of the simulation. If no input signal is present before $t = 0$, it can be assumed that both $S_{\pm}(z, 0)$ and $\varphi_{\pm}(z, 0)$ are zero. The noise will exist regardless of the signal's presence, so stationary distributions can be assumed for $A_{\pm}(z, 0)$. Based on these values, using Equation 21.108, stationary distribution for carrier density can also be determined and associated with $n(z, 0)$. Within the nested loop, for a fixed j, system (Equations 21.112 through 21.114) is solved for every point z_i, from 1 to P for forward-propagation, and P $-$ 1 to 0 for backward propagation. The algorithm is summarized in the flowchart given in Figure 21.17.

21.5.6 Case Study: TMW Small-Signal Model

A very valuable figure-of-merit in using SOA for modulation purposes is its electro-optical (E/O) small-signal -3dB modulation bandwidth Ω_{3dB}. It is defined as the frequency of the modulation signal, i.e., electric current, for which the output SOA small signal, i.e., photon density or optical power, drops

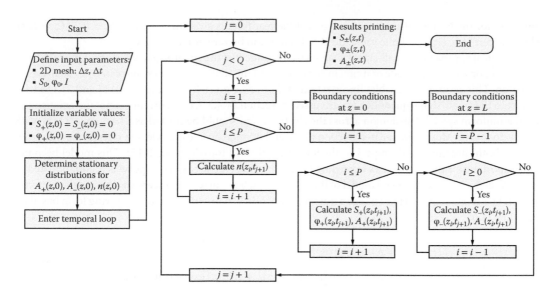

FIGURE 21.17 Flowchart of the upwind finite differences algorithm for the calculation of carrier and photon density spatiotemporal distributions, for both signal and noise.

to the half of its zero-frequency value. Knowing the influence of different structural and material parameters, as well as operating conditions, on Ω_{3dB}, provides the means for optimization of the modulation response and increasing the bandwidth. Under the assumption that the input signal is spectrally narrow, the analysis is based on photon densities for forward and backward propagation, which include both signal and noise. The TW equations have the form similar to Equation 21.105, with the addition of the term describing spontaneous emission, as in Equation 21.107:

$$\pm\frac{\partial S_\pm}{\partial z} + \frac{1}{v_g}\frac{\partial S_\pm}{\partial t} = \left(\frac{\Gamma g}{1+\varepsilon S_\Sigma} - \alpha_i\right)S_\pm + \frac{1}{2v_g}\Gamma\beta_{sp}R_{sp}, \tag{21.115}$$

where $S_\Sigma = S_+ + S_-$. Since the intensity modulation is of interest, the analysis of the signal phase is not necessary. The carrier rate equation has the same form as Equation 21.108:

$$\frac{dn}{dt} = \frac{J}{qH} - \left(An + R_{sp} + Cn^3\right) - \frac{v_g g S_\Sigma}{1+\varepsilon S_\Sigma}, \tag{21.116}$$

where current density $J = I/(WL)$ is used instead of current I. This choice stems from the fact that modulation current I depends on device length, and the current density proves to be a more suitable parameter for benchmarking the performance of SOAs with different active region lengths. Generally, bias current density will consist of the stationary component and a small signal, which will, through Equation 21.116, lead to the carrier density modulation, and consequently to modulation of all carrier-dependent parameters, namely, g, R_{sp}, α_i, and $S_\pm$. From this point onward, all stationary values will be denoted by an overline, whereas all small-signals will be prefixed with Δ.

For the stationary value of bias current density, $\bar{J}$, all temporal derivatives will diminish from Equations 21.115 and 21.116, and the system will reduce to its steady-state form, given by Equations 21.94 and 21.95. In order to determine the steady-state values of $\bar{n}$ and $\bar{S}_\pm$, any of the previously discussed methods for steady-state analysis, given in Sections 21.5.1, 21.5.3, and 21.5.4, can be used. It is assumed that the CW input signal enters the device at the front facet, giving $S_+(z=0,t) = \bar{S}_+(z=0) = S_0$. The boundary condition for the backward-propagating signal at the rear facet is given by $S_-(z = L, t) = R_2 S_+(z = L, t)$, where R_2 can have either a finite value, for R-SOA, or it can be zero, for TW-SOA.

The typical approach in small-signal analysis is to assume that the small-signal values are approximately lower by an order of magnitude or more, in comparison to the stationary values, and can be treated as perturbation. This assumption enables linearization of the system given by Equations 21.115 and 21.116 with respect to the small-signal quantities. As discussed in Section 21.4.3.5, bias current cannot instantaneously reach a uniform spatial distribution across the electrode and should be treated as a harmonic TMW with finite velocity, v_e. Moreover, due to reflection at the end of microstrip electrode, the microwave will propagate in both directions with respect to the longitudinal axis (Totović et al., 2015). This model becomes very important for the frequencies exceeding $v_e/(2\pi L)$, when an entire electric pulse is accommodated in the SOA's active region. A general model of the small-signal bias current density with sinusoidal waveform, defined by the spatially variable voltage across the electrode, can be written in the following form based on Equation 21.63:

$$J(z, t) = \bar{J} + \Delta J(z) \exp(i\Omega t)$$
$$= \bar{J} + \Delta J_F(z) \exp\left[i\left(\Omega t - \beta_e z\right)\right] + \Delta J_B(z) \exp\left[i\left(\Omega t + \beta_e z\right)\right], \tag{21.117}$$

where $\Delta J(z)$ is the total small-signal bias current density spatial distribution, comprising both forward (F)- and backward (B)-propagating microwaves, $\Omega = 2\pi f$ is the angular frequency, with f being the frequency of the TMW, and β_e is the microwave propagation constant.

21.5.6.1 Derivation of the Small-Signal Model

Modulation of the bias current density leads to the modulation of carrier density n, and all carrier-dependent parameters, g, R_{sp}, α_i, and $S_\pm$, commonly denoted ξ, which will also have sinusoidal form

$$\xi(z, t) = \bar{\xi}(z) + \Delta\xi(z) \exp(i\Omega t)$$
$$= \bar{\xi}(z) + \Delta\xi_F(z) \exp\left[i\left(\Omega t - \beta_e z\right)\right] + \Delta\xi_B(z) \exp\left[i\left(\Omega t + \beta_e z\right)\right], \tag{21.118}$$

where $\Delta\xi(z)$ is the total small-signal spatial distribution of the corresponding quantities. For known dependencies of material gain, and spontaneous emission rate, on carrier density, $g(n) = g_0 \ln(n/n_{tr})$, and $R_{sp}(n) = B_0 + B_1 n + B_2 n^2$, respectively, it is possible to express small-signal values of $\Delta g^{F/B}$ and $\Delta R_{sp}^{F/B}$ via $\Delta n^{F/B}$ using the first derivatives of the corresponding functions, evaluated for the steady-state carrier density. This gives

$$\Delta g^{F/B} = \left(dg/dn\right)\Big|_{n=\bar{n}} \Delta n^{F/B} = \left(g_0/\bar{n}\right) \Delta n^{F/B}, \tag{21.119}$$

$$\Delta R_{sp}^{F/B} = \left(dR_{sp}/dn\right)\Big|_{n=\bar{n}} \Delta n^{F/B} = \left(B_1 + 2B_2\bar{n}\right) \Delta n^{F/B}. \tag{21.120}$$

Substituting Equations 21.117 through 21.120 into Equation 21.116, and separating the stationary, and small signals, gives a system comprising a steady-state rate equation, given by Equation 21.95, and small-signal equations for forward and backward TMW propagation. In order to linearize the small-signal equations, two approximations are required. The first one is related to the term describing the stimulated emission in Equation 21.116, namely, $v_g g S_\Sigma/(1 + \varepsilon S_\Sigma)$, which is nonlinear due to the presence of the small-signal photon densities both in numerator and denominator. The function

$$\left(1 + \varepsilon S_\Sigma\right)^{-1} = \left[1 + \varepsilon\bar{S}_\Sigma + \varepsilon\left(\Delta S_+^F + \Delta S_-^F\right) e^{i(\Omega t - \beta_e z)} + \varepsilon\left(\Delta S_+^B + \Delta S_-^B\right) e^{i(\Omega t + \beta_e z)}\right]^{-1} \tag{21.121}$$

can be rephrased as

$$\left(1 + \varepsilon S_\Sigma\right)^{-1} = \frac{1}{1 + \varepsilon \bar{S}_\Sigma} \left[1 + \frac{\varepsilon \left(\Delta S_+^F + \Delta S_-^F\right)}{1 + \varepsilon \bar{S}_\Sigma} e^{i(\Omega t - \beta_e z)} + \frac{\varepsilon \left(\Delta S_+^B + \Delta S_-^B\right)}{1 + \varepsilon \bar{S}_\Sigma} e^{i(\Omega t + \beta_e z)}\right]^{-1}. \quad (21.122)$$

The second term in the product of the RHS of Equation 21.122 can be treated as $(1 + x)^{-1}$, where $x \ll 1$, and therefore approximated with $1 - x$, which leads to

$$\left(1 + \varepsilon S_\Sigma\right)^{-1} \approx \frac{1}{1 + \varepsilon \bar{S}_\Sigma} - \frac{\varepsilon \left(\Delta S_+^F + \Delta S_-^F\right)}{\left(1 + \varepsilon \bar{S}_\Sigma\right)^2} e^{i(\Omega t - \beta_e z)} - \frac{\varepsilon \left(\Delta S_+^B + \Delta S_-^B\right)}{\left(1 + \varepsilon \bar{S}_\Sigma\right)^2} e^{i(\Omega t + \beta_e z)}. \quad (21.123)$$

The following step in linearization is applying the first-order approximation, i.e., neglecting of all small-signal terms of the order higher than one. This gives the following system of small-signal rate equations for forward and backward TMW propagation

$$i\Omega \Delta n^{F/B} = \frac{\Delta J^{F/B}}{qH} - \left(A + B_1 + 2B_2\bar{n} + 3C\bar{n}^2\right)\Delta n^{F/B} - \left[\frac{g_0}{\bar{n}} \frac{v_g \bar{S}_\Sigma}{1 + \varepsilon \bar{S}_\Sigma} \Delta n^{F/B} + v_g \frac{g_{eff}}{\Gamma} \left(\Delta S_+^{F/B} + \Delta S_-^{F/B}\right)\right], \quad (21.124)$$

where $g_{eff} = \Gamma \bar{g}/(1 + \varepsilon \bar{S}_\Sigma)^2$. The previous system can be solved with respect to the small-signal carrier densities resulting from forward (F)- and backward (B)-propagating microwaves:

$$\Delta n^{F/B} = \frac{\Delta J^{F/B}/(qH) - v_g g_{eff} \left(\Delta S_+^{F/B} + \Delta S_-^{F/B}\right)/\Gamma}{A + B_1 + 2B_2\bar{n} + 3C\bar{n}^2 + \frac{g_0}{\bar{n}} \frac{v_g \bar{S}_\Sigma}{1 + \varepsilon \bar{S}_\Sigma} + i\Omega}. \quad (21.125)$$

Substituting Equations 21.118 through 21.120 and Equation 21.125 into Equation 21.115, followed by the linearization process, leads to the system of two steady-state TW equations given by Equation 21.94, and four small-signal TW equations written with respect to the small-signal photon densities propagating in any of the two directions (denoted by subscript sign $\Delta S_\pm$) and resulting from any of the microwave propagation directions (denoted by the superscript $\Delta S^{F/B}$):

$$\pm \frac{d\Delta S_\pm^F}{dz} = \frac{\Gamma \gamma_\pm}{qHv_g} \Delta J_F - \left(\varepsilon \bar{S}_\mp + \gamma_\pm\right) g_{eff}\Delta S_\mp^F + \left[g_{eff}\left(1 + \varepsilon \bar{S}_\mp - \gamma_\pm\right) - \bar{\alpha}_i - i\left(\frac{\Omega}{v_g} \mp \beta_e\right)\right]\Delta S_\pm^F, \quad (21.126)$$

$$\pm \frac{d\Delta S_\pm^B}{dz} = \frac{\Gamma \gamma_\pm}{qHv_g} \Delta J_B - \left(\varepsilon \bar{S}_\mp + \gamma_\pm\right) g_{eff}\Delta S_\mp^B + \left[g_{eff}\left(1 + \varepsilon \bar{S}_\mp - \gamma_\pm\right) - \bar{\alpha}_i - i\left(\frac{\Omega}{v_g} \pm \beta_e\right)\right]\Delta S_\pm^B, \quad (21.127)$$

where

$$\gamma_\pm = \frac{v_g \left(\frac{g_0/\bar{n}}{1 + \varepsilon \bar{S}_\Sigma} - K_1\right)\bar{S}_\pm + \frac{1}{2}\beta_{sp}\left(B_1 + 2B_2\bar{n}\right)}{A + B_1 + 2B_2\bar{n} + 3C\bar{n}^2 + \frac{g_0}{\bar{n}} \frac{v_g \bar{S}_\Sigma}{1 + \varepsilon \bar{S}_\Sigma} + i\Omega}. \quad (21.128)$$

The γ-parameter is dimensionless, complex, and spatially and frequency dependent. It is responsible for frequency dependence of small-signal photon density distributions and their sensitivity to the frequency change for all other parameters fixed. In the case of zero frequency, $\gamma_\pm$ becomes purely real, whereas for high frequencies, such as the ones close to the modulation bandwidth and beyond, its imaginary part becomes dominant over the real one.

Since Equations 21.126 and 21.127 are derived from Equation 21.115, the boundary conditions for small-signal equations are inherited from the ones corresponding to Equation 21.115. The input signal is assumed to be CW, so $\Delta S_+^{F/B}(0) = 0$. At the rear facet, the boundary conditions are conditioned by the power reflectivity coefficient, $\Delta S_-^{F/B}(L) = R_2 \Delta S_+^{F/B}(L)$.

21.5.6.2 Numerical Implementation

In order to determine the -3dB bandwidth, $f_{3dB} = \Omega_{3dB}/(2\pi)$, it is necessary to calculate the output small-signal photon density for a range of modulation frequencies, $\Delta S_{out}(\Omega)$, and determine the one for which the condition $|\Delta S_{out}(\Omega_{3dB})/\Delta S_{out}(0)| = 1/2$ is satisfied. In the case of TW-SOA, the output signal will be recorded at the rear facet, $\Delta S_{out} = \Delta S_+(L) = \Delta S_+^F(L)\exp(-i\beta_e L) + \Delta S_+^B(L)\exp(i\beta_e L)$, whereas for R-SOA, at the front facet, $\Delta S_{out} = \Delta S_-(0) = \Delta S_-^F(0) + \Delta S_-^B(0)$. These signals can be found by solving the boundary value problem given by Equations 21.126 and 21.127. In the case of an R-SOA, this system comprises four coupled differential equations of the first order, with functional coefficients, and can be solved using any of the available numerical algorithms for boundary value problems, e.g., the FDM that implements the three-stage Lobatto IIIA formula (Hairer and Wanner, 1996). For a TW-SOA, backward-propagating signals can be neglected due to zero power reflectivity, and very low contribution of the spontaneous emission, so the system reduces to two noncoupled equations which can be solved with numerical integration.

Prior to solving system of Equations 21.126 and 21.127, it is necessary to calculate spatial distributions of the steady-state variables, $\bar{n}$ and $\bar{S}_\pm$, and auxiliary parameters, g_{eff} and $\gamma_\pm$. As previously discussed, steady-state values can be calculated by any of the models presented in earlier sections.

After the output small-signal variables are calculated, the -3dB bandwidth can be determined by interpolating the function $|\Delta S_{out}(\Omega)/\Delta S_{out}(0)|$ in search of the value Ω_{3dB} which gives $1/2$ as a solution. For rare, special cases, Ω_{3dB} might be found in analytical form. In order to do this, usually a significant number of approximations is required, and the resulting analytical expression might be very complex (Antonelli et al., 2015), or the expression has a simpler form, but is valid under limited conditions.

21.5.6.3 Results and Discussion

Implementing the algorithm described in the previous pages provides the dependence of -3dB bandwidth, f_{3dB}, on the operating conditions, given in Figure 21.18.

The analysis is done for the microwave velocity of $v_e = 4.6 \times 10^7$ m/s (Tauber et al., 1994), and under the assumption that the load and characteristic impedances of the transmission line, Z_L and Z_C, respectively, are matched, which results in the microwave reflection coefficient $\Gamma_L = (Z_L - Z_C)/(Z_L + Z_C) = 0$. The small signal bias current density, ΔJ, is assumed to be equal to $0.1\bar{J}$.

Comparing the two SOA types, TW-SOA, Figure 21.18a and b, and R-SOA, Figure 21.18c and d, it can be concluded that, not only does the R-SOA provide higher bandwidths, but it does so for a much wider range of input optical powers. Both SOA types exhibit an increase in f_{3dB} with the increase of steady-state bias current density, but the behavior with P_0 variation differs significantly. In the case of TW-SOA, the bandwidth generally increases with the increase of P_0, reaching its maximum for deep saturation, and input optical powers between 20 and 25 dBm. Contrarily, R-SOA shows two maxima, providing the choice between two operation regimes, one for the low input optical powers and high device gain, and one for the deep saturation. It should be noted that -3dB bandwidth significantly depends on device length (Totović et al., 2015), which can be optimized to achieve maximum values of f_{3dB}. Aside from changes in values of f_{3dB} with the length variation, the prominence of low- and high-power maxima in R-SOA can also be

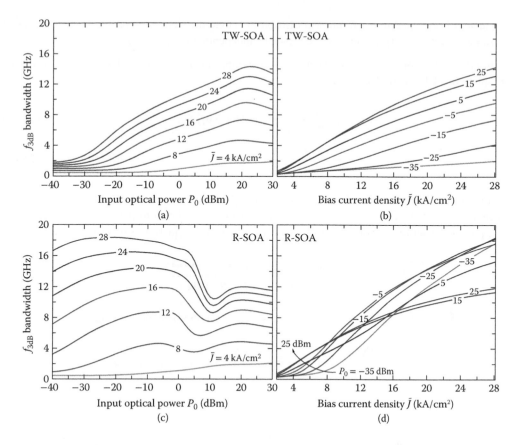

FIGURE 21.18 Bandwidth dependence on (a, c) the continuous wave (CW) input optical power P_0 (dBm) and (b, d) the bias current density $\bar{J}$ (kA/cm^2) for (a), (b) traveling-wave semiconductor optical amplifier (TW-SOA), and (c), (d) reflective (R)-SOA in the case of $L = 800$ μm.

interchanged. It should be noted that calculated bandwidth is intrinsic, and that inclusion of parasitic effects of the supporting circuit may modify the modulation response.

21.5.7 Case Study: Small-Signal Model for Transparent SOA

Transparent SOA has been repeatedly analyzed as a characteristic example of operating conditions under which the analytical expression for -3dB E/O bandwidth may be obtained (Mørk et al., 1999; Antonelli et al., 2015; Totović et al., 2016). By transparency, it is assumed that the steady-state modal gain is equal to the loss of the amplifier cavity, $\Gamma\bar{g} = \bar{\alpha}_i$, and therefore the steady-state photon density remains unchanged during its propagation. However, even under these circumstances, an analytical expression cannot be obtained without a number of approximations which simplify the system.

21.5.7.1 Framework for the Small-Signal Analysis

In this case study, the expressions will be derived under the following assumptions (Totović et al., 2016):

1. Nonlinear gain suppression is neglected, $\varepsilon\bar{S}_\Sigma \ll 1$.
2. Contribution of ASE noise to the signal is neglected, $\beta_{sp} = 0$.
3. Carrier loss due to the spontaneous emission is modeled via carrier lifetime, τ_s.

4. Current is assumed to reach uniform distribution along the electrode instantaneously, i.e., small-signal current density is assumed to be spatially independent, $\Delta J(z) = \Delta J$.

Within this framework, the system of equations governing signal and carrier dynamics can be derived from Equations 21.115 and 21.116, and reads

$$\pm \frac{\partial S_\pm}{\partial z} + \frac{1}{v_g} \frac{\partial S_\pm}{\partial t} = \left(\Gamma g - \alpha_i\right) S_\pm, \tag{21.129}$$

$$\frac{dn}{dt} = \frac{J}{qH} - \frac{n}{\tau_s} - v_g g S_\Sigma. \tag{21.130}$$

Assuming that the bias current has the general form given by Equation 21.117, $J(z,t) = J(t) = \bar{J} + \Delta J \exp(i\Omega t)$, the system (Equations 21.129 and 21.130) can be decoupled to the steady-state and small-signal equations. After linearization, small-signal carrier density has the form similar to Equation 21.125:

$$\Delta n = \frac{\Delta J/(qH) - v_g \bar{g}\left(\Delta S_+ + \Delta S_-\right)}{1/\tau_s + v_g g_0/\bar{n} \cdot \bar{S}_\Sigma + i\Omega}, \tag{21.131}$$

and the system of equations written with respect to small-signal photon densities reads

$$\pm \frac{d\Delta S_\pm}{dz} = \frac{\Gamma \gamma_\pm}{qH v_g} \Delta J + \left(\Gamma \bar{g} - \bar{\alpha}_i - i\frac{\Omega}{v_g}\right) \Delta S_\pm - \Gamma \bar{g} \gamma_\pm \left(\Delta S_\pm + \Delta S_\mp\right), \tag{21.132}$$

where

$$\gamma_\pm = \frac{v_g \left(g_0/\bar{n} - K_1\right)}{1/\tau_s + v_g g_0/\bar{n} \cdot \bar{S}_\Sigma + i\Omega} \bar{S}_\pm \tag{21.133}$$

has a meaning similar to the γ-parameter defined by Equation 21.128. Although the system given by Equation 21.132 is analytically solvable, the complexity of the solution undermines its benefit, and prevents us from determining the analytical solution for Ω_{3dB}.

21.5.7.2 Employing the Transparency Condition

Choosing the transparency as the operating regime defines the steady-state values of carrier and photon densities. From the condition $\Gamma \bar{g} = \bar{\alpha}_i$, the value for carrier density can be determined:

$$\bar{n} = -\frac{g_0}{K_1} W_L \left[-\frac{K_1 n_{tr}}{g_0} \exp\left(\frac{K_0}{\Gamma g_0}\right) \right], \tag{21.134}$$

where $W_L(x)$ stands for the Lambert W (product-logarithm) function. Introducing the transparency condition in the steady-state form of Equation 21.129 results in $dS_\pm/dz = 0$, meaning that the steady-state photon density is not spatially dependent and is defined by the boundary conditions. For both TW- and R-SOA, the forward-propagating steady-state photon density $\bar{S}_+$ will be equal to the input photon density S_0, whereas the backward-propagating photon density will either be zero, for TW-SOA, or $\bar{S}_- = R_2 \bar{S}_+ = R_2 S_0$, for R-SOA. Substituting $\bar{S}_\pm$ in the steady-state form of Equation 21.130 gives correlation between the stationary bias current density and the input photon density, for which the transparency condition is met:

$$S_0 = \frac{1}{v_g g_0 \ln\left(\bar{n}/n_{tr}\right)\left(1 + R_2\right)} \left(\frac{\bar{J}}{qH} - \frac{\bar{n}}{\tau_s}\right). \tag{21.135}$$

Aside from the maximum value of bias current density, dictated by the long-term stable SOA operation, Equation 21.135 gives a limit for the minimal value of $\bar{J}_{\min} = qH\bar{n}/\tau_s$ since the input photon density cannot be negative.

Introducing the transparency condition in Equation 21.132 gives

$$\pm \frac{d\Delta S_\pm}{dz} = \frac{\Gamma \gamma_\pm}{qHv_g} \Delta J - i\frac{\Omega}{v_g}\Delta S_\pm - \Gamma \bar{g}\gamma_\pm \left(\Delta S_\pm + \Delta S_\mp\right). \tag{21.136}$$

Moreover, having a fixed value for $\bar{n}$ reduces the γ-parameter to a spatially independent value, meaning that Equation 21.136 is the system of differential equations of the first order with fixed coefficients. The output small-signal photon density of TW-SOA has the following form

$$\Delta S_+ (L) = \frac{\Gamma}{qHv_g}\Delta J \frac{\gamma_+}{\Gamma \bar{g}\gamma_+ + i\Omega/v_g} \left\{1 - \exp\left[-\left(\Gamma \bar{g}\gamma_+ + i\frac{\Omega}{v_g}\right)L\right]\right\}, \tag{21.137}$$

whereas the output small-signal photon density of an R-SOA reads

$$\Delta S_- (0) = 2R_2 \frac{\Gamma \gamma_+}{qHv_g}\Delta J \times \left\{\frac{1 + 3R_2}{2}\Gamma \bar{g}\gamma_+ + i\frac{\Omega}{v_g} + \sqrt{\left(\frac{1 - R_2}{2}\Gamma \bar{g}\gamma_+\right)^2 + \left[(1 + R_2)\Gamma \bar{g}\gamma_+ + i\frac{\Omega}{v_g}\right]i\frac{\Omega}{v_g}}\right.$$

$$\left. \times \coth\left[L\sqrt{\left(\frac{1 - R_2}{2}\Gamma \bar{g}\gamma_+\right)^2 + \left[(1 + R_2)\Gamma \bar{g}\gamma_+ + i\frac{\Omega}{v_g}\right]i\frac{\Omega}{v_g}}\right]\right\}^{-1}. $$

$$\tag{21.138}$$

21.5.7.3 Analytical Formulae for Modulation Bandwidth

As previously discussed, the γ-parameter influences the SOA's small-signal modulation response through its dependence on Ω. For zero frequency, $\gamma_\pm$ becomes purely real, whereas for high frequencies, close to Ω_{3dB} and beyond, it can be approximated with purely imaginary parameter

$$\gamma_\pm|_{\Omega \to \Omega_{3dB}} \approx v_g \left(g_0/\bar{n} - K_1\right) \bar{S}_\pm \big/ (i\Omega). \tag{21.139}$$

In addition, during the derivation of the analytical expression for Ω_{3dB}, all trigonometric functions that are encountered, namely $\sin(x)$ for $x \to 0$ in case of TW-SOA and $\coth(x)$ for $x \to 0$ in the case of R-SOA, are approximated by the first term of their respective Maclaurin series expansion, $\sin(x) \approx x$ and $\coth(x) \approx 1/x$, respectively. This leads to the simpler form of Equations 21.137 and 21.138 in the vicinity of Ω_{3dB}. For TW-SOA, the output small-signal photon density is

$$|\Delta S_+ (L)|\big|_{\Omega \to \Omega_{3dB}} \approx \frac{\Gamma \Delta J}{qH}\left(\frac{g_0}{\bar{n}} - K_1\right)L\frac{S_0}{\Omega}, \tag{21.140}$$

and for R-SOA

$$|\Delta S_- (0)|\big|_{\Omega \to \Omega_{3dB}} \approx 2R_2\frac{\Gamma \Delta J}{qH}\left(\frac{g_0}{\bar{n}} - K_1\right)L\frac{S_0}{\Omega}. \tag{21.141}$$

Finally, from the condition $|\Delta S_{out}(\Omega_{3dB})/\Delta S_{out}(0)| = 1/2$, the -3dB bandwidth can be found, under the transparency regime. For TW-SOA, it is given by

$$\Omega_{3dB} = \Gamma g_0 \ln \frac{\bar{n}}{n_{tr}} L v_g \left(\frac{g_0}{\bar{n}} - K_1 \right) S_0 \left\{ 1 + \coth \left[\Gamma g_0 \ln \frac{\bar{n}}{n_{tr}} \frac{L}{2} \frac{v_g \left(g_0/\bar{n} - K_1 \right) S_0}{1/\tau_s + v_g g_0/\bar{n} \cdot S_0} \right] \right\}, \qquad (21.142)$$

and for R-SOA it has the form

$$\Omega_{3dB} = 2 \left[\frac{1}{\tau_s} + v_g \frac{g_0}{\bar{n}} \left(1 + R_2 \right) S_0 \right] + \left(1 + 3R_2 \right) \Gamma g_0 \ln \frac{\bar{n}}{n_{tr}} L v_g \left(\frac{g_0}{\bar{n}} - K_1 \right) S_0. \qquad (21.143)$$

21.6 Conclusion

SOAs have been extensively studied ever since WDM networks became widely adopted. In order to enable massive deployment of optical access networks, optical technologies, including SOAs, have expanded toward the lower levels of network hierarchy, aiming to relieve first/last mile bottleneck. Due to their versatility, SOAs are the key components in many of these levels. R-SOA has recently become one of the most promising candidates for the next-generation wavelength division multiplexed passive optical networks (WDM-PONs), representing the new Fiber-to-the-Home paradigm. Moreover, it is on its way to become a crucial component in Worldwide Interoperability for Microwave Access (WiMAX) and Wireless-Fidelity (Wi-Fi) Radio-over-Fiber (RoF) access network architecture. TW- and R-SOAs are devices of choice for many nonlinear applications, including those related to FWM, wavelength conversion, and all optical signal processing in general.

The choice of the SOA type for a specific purpose does not necessarily require better overall performance but usually better performance in one or several aspects. Therefore, both TW- and R-SOA have found their applications. However, optimization of the devices themselves, as well as the system as a whole, should always be performed in order for the SOAs to reach their full potential. The optimization of the device includes the choice of material, device geometry, and operating conditions, whereas the system analysis involves many more aspects that need to be addressed. This process requires an elaborate, but efficient SOA model, which accounts for all relevant effects. In Section 21.5, various models have been presented for both the steady-state and dynamic operating regimes. All models are built starting from the rate equations describing SOA dynamics, which can include multiple levels of approximation, as discussed in Section 21.4. The basic model recognizes only the essential processes present in the SOAs, i.e., carrier injection, stimulated and spontaneous recombination, and signal amplification. More realistic models comprise additional effects, such as wideband ASE noise, carrier transport, nonlinear gain suppression, temperature effects, and distributed bias current, all of which are discussed in detail. Moreover, in order for the model to accurately represent the SOA's performance, a detailed analysis of the optical properties of materials used for the active region and the waveguide geometry should be performed, which can be done using the models presented in Section 21.3. This includes calculation of the full spectral profile of material gain, radiative spontaneous recombination rate, refractive index variation, and optical confinement factor, among other parameters.

For SOA optimization in the steady-state operating regime, the wideband model, based on the self-consistent iterative method, given in Section 21.5.3, can provide detailed results, but does consume a significant amount of computational resources. This method is mainly designed for the characterization and optimization of the device itself. On the other hand, for quick assessment of the SOA's performance, especially in more complex network architectures, the semi-analytical model, presented in Section 21.5.4, proves to be time and resource efficient, at the expense of somewhat decreased precision. In the dynamic operating regime, the upwind scheme finite difference model can be used, as shown in Section 21.5.5, which is transparent to the modulation type, and designed to support even advanced modulation formats. For SOAs used in direct E/O modulation, −3dB bandwidth is a good indicator of the SOAs' performance, and can be determined and optimized based on the detailed numerical model given in Section 21.5.6. When resources are limited, and the SOA operates in the transparency regime, simple analytical solutions, derived in Section 21.5.7, can be used for the same purpose.

The overview of various steady-state and dynamic models, as well as the case studies presented, provides insight into the methods for numerically accurate and efficient simulations of standalone and system-embedded SOAs. These methods can help in the development of computational tools which can be used for the optimization of SOA design based on the dependence of its performance on technological parameters and operating conditions. It should be noted that advanced SOA models, which deal with nonlinear effects, are usually built upon the models which simulate standard SOA amplification regime.

References

Adachi S (1989) Optical dispersion relations for GaP, GaAs, GaSb, InP, InAs, InSb, $Al_xGa_{1-x}As$, and $In_{1-x}Ga_xAs_yP_{1-y}$. *Journal of Applied Physics.* 66:6030–6040.

Adams MJ, Collins JV, Henning ID (1985) Analysis of semiconductor laser optical amplifiers. *IEEE Proceedings Journal—Optoelectronics.* 132:58–63.

Agrawal GP (1990) Effect of gain and index nonlinearities on single-mode dynamics in semiconductor lasers. *IEEE Journal of Quantum Electronics.* 26:1901–1909.

Agrawal GP, Olsson NA (1989) Self-phase modulation and spectral broadening of optical pulses in semiconductor laser amplifiers. *IEEE Journal of Quantum Electronics.* 25:2297–2306.

Ahn D, Chuang SL (1990) Optical gain and gain suppression of quantum-well lasers with valence band mixing. *IEEE Journal of Quantum Electronics.* 26:13–24.

Antonelli C, Mecozzi A (2013) Reduced model for the nonlinear response of reflective semiconductor optical amplifiers. *IEEE Photonics Technology Letters.* 25:2243–2246.

Antonelli C, Mecozzi A, Zhefeng H, Santagiustina M (2015) Analytic study of the modulation response of reflective semiconductor optical amplifiers. *Journal of Lightwave Technology.* 33:4367–4376.

Bennett BR, Soref RA, del Alamo JA (1990) Carrier-induced change in refractive index of InP, GaAs and InGaAsP. *IEEE Journal of Quantum Electronics.* 26:113–122.

Berg TW, Bischoff S, Magnusdottir I, Mørk J (2001) Ultrafast gain recovery and modulation limitations in self-assembled quantum-dot devices. *IEEE Photonics Technology Letters.* 13:541–543.

Berglind E, Gillner L (1994) Optical quantum noise treated with classical electrical network theory. *IEEE Journal of Quantum Electronics.* 30:846–853.

Carapellese N, Tornatore M, Pattavina A (2014) Energy-efficient baseband unit placement in a fixed/mobile converged WDM aggregation network. *IEEE Journal on Selected Areas in Communications.* 32:1542–1551.

Carlo AD, Reale A, Tocca L, Lugli P (1998) Polarization-independent-strained semiconductor optical amplifiers: A tight-binding study. *IEEE Journal of Quantum Electronics.* 34:1730–1739.

Carrère H, Truong VG, Marie X, Brenot R, De Valicourt G, Lelarge F, Amand T (2010) Large optical bandwidth and polarization insensitive semiconductor optical amplifiers using strained InGaAsP quantum wells. *Applied Physics Letters.* 97:121101.

Cartoixa X, Ting DZY, McGill TC (2003) Numerical spurious solutions in the effective mass approximation. *Journal of Applied Physics.* 93:3974–3981.

Cassioli D, Scotti S, Mecozzi A (2000) A time-domain computer simulator of the nonlinear response of semiconductor optical amplifiers. *IEEE Journal of Quantum Electronics.* 36:1072–1080.

Chao CYP, Chuang SL (1992) Spin-orbit-coupling effects on the valence-band structure of strained semiconductor quantum wells. *Physical Review B.* 46:4110–4122.

Cho YS, Choi WY (2001) Analysis and optimization of polarization-insensitive semiconductor optical amplifiers with delta-strained quantum wells. *IEEE Journal of Quantum Electronics.* 37:574–579.

Chu CYJ, Ghafouri-Shiraz H (1994) Analysis of gain and saturation characteristics of a semiconductor laser optical amplifier using transfer matrices. *Journal of Lightwave Technology.* 12:1378–1386.

Chuang SL (1991) Efficient band-structure calculations of strained quantum wells. *Physical Review B.* 43:9649–9661.

Chuang SL (1995) *Physics of Optoelectronic Devices.* New York, NY: Wiley-Interscience.

Coldren LA, Corzine SW, Mašanović ML (2012) *Diode Lasers and Photonic Integrated Circuits.* New York, NY: Wiley-Interscience.

Connelly MJ (2001) Wideband semiconductor optical amplifier steady-state numerical model. *IEEE Journal of Quantum Electronics.* 37:439–447.

Connelly MJ (2002) *Semiconductor Optical Amplifiers.* Norwell, MA: Kluwer.

Connelly MJ (2007) Wide-band steady-state numerical model and parameter extraction of a tensile-strained bulk semiconductor optical amplifier. *IEEE Journal of Quantum Electronics.* 43:47–56.

Connelly MJ (2012) Reflective semiconductor optical amplifier pulse propagation model. *IEEE Photonics Technology Letters.* 24:95–97.

Connelly MJ (2014) Reflective semiconductor optical amplifier modulator dynamic model. *14th International Conference on Numerical Simulation of Optoelectronic Devices (NUSOD).* 2014:41–42.

Connelly MJ (2015) Reflective semiconductor optical amplifier electrode voltage based phase shifter model. *International Conference on Numerical Simulation of Optoelectronic Devices (NUSOD).* 2015:39–40.

D'Ottavi A, Iannone E, Mecozzi A, Scotti S, Spano P, Dall'Ara R, Eckner J, Guekos G (1995) Efficiency and noise performance of wavelength converters based on FWM in semiconductor optical amplifiers. *IEEE Photonics Technology Letters.* 7:357–359.

Dailey JM, Koch TL (2007) Impact of carrier heating on SOA transmission dynamics for wavelength conversion. *IEEE Photonics Technology Letters.* 19:1078–1080.

Dailey JM, Koch TL (2009) Simple rules for optimizing asymmetries in SOA-based Mach–Zehnder wavelength converters. *Journal of Lightwave Technology.* 27:1480–1488.

Davis MG, O'Dowd RF (1994) A transfer matrix method based large-signal dynamic model for multielectrode DFB lasers. *IEEE Journal of Quantum Electronics.* 30:2458–2466.

de Oliveira Ribeiro R, Pontes MJ, Giraldi MTMR, Carvalho MCR (2005) Characterisation of all-optical wavelength conversion by cross-gain modulation of ASE on a SOA. International Conference on Microwave and Optoelectronics (SBMO/IEEE MTT-S) 2005:218–221.

de Valicourt G (2012) Next generation of optical access network based on reflective-SOA, Ch. 1. In *Selected Topics on Optical Amplifiers in Present Scenario,* edited by S. K. Garai. Rijeka, Croatia: InTech.

de Valicourt G, Violas MA, Wake D, Van Dijk F, Ware C, Enard A, Make D, Liu Z, Lamponi M, Duan GH, Brenot R (2010) Radio-over-fiber access network architecture based on new optimized RSOA devices with large modulation bandwidth and high linearity. *IEEE Transactions on Microwave Theory and Techniques.* 58:3248–3258.

Donati S, Giuliani G (1997) Noise in an optical amplifier: Formulation of a new semiclassical model. *IEEE Journal of Quantum Electronics.* 33:1481–1488.

Dúill SPÓ, Barry LP (2015) Improved reduced models for single-pass and reflective semiconductor optical amplifiers. *Optics Communications.* 334:170–173.

Durhuus T, Mikkelsen B, Stubkjaer KE (1992) Detailed dynamic model for semiconductor optical amplifiers and their crosstalk and intermodulation distortion. *Journal of Lightwave Technology.* 10:1056–1065.

Figueiredo RC, Ribeiro NS, Gallep CM, Conforti E (2015) Frequency- and time-domain simulations of semiconductor optical amplifiers using equivalent circuit modeling. *Optical Engineering.* 54:114107.

Foreman BA (1993) Effective-mass Hamiltonian and boundary conditions for the valence bands of semiconductor microstructures. *Physical Review B.* 48:4964–4967.

Foreman BA (1997) Elimination of spurious solutions from eight-band k.p theory. *Physical Review B.* 56:12748–12751.

Gebrewold SA, Bonjour R, Barbet S, Maho A, Brenot R, Chanclou P, Brunero M, Marazzi L, Parolari P, Totovic A, Gvozdic D, Hillerkuss D, Hafner C, Leuthold J (2015) Self-seeded RSOA-fiber cavity lasers vs. ASE spectrum-sliced or externally seeded transmitters—A comparative study. *Applied Sciences.* 5:1922–1941.

Ghafouri-Shiraz H (2004) *The Principles of Semiconductor Laser Diodes and Amplifiers.* London Imperial College Press.

Gioannini M (2004) Numerical modeling of the emission characteristics of semiconductor quantum dash materials for lasers and optical amplifiers. *IEEE Journal of Quantum Electronics.* 40:364–373.

Gladisch A, Braun RP, Breuer D, Ehrhardt A, Foisel HM, Jaeger M, Leppla R, Schneiders M, Vorbeck S, Weiershausen W, Westphal F (2006) Evolution of terrestrial optical system and core network architecture. *Proceedings of the IEEE.* 94:869–891.

Godefroy A, Le Corre A, Clerot F, Salaan S, Loualiche S, Simon JC, Henry L, Vaudry C, Keromnes JC, Joulie G, Lamouler P (1995) 1.55-μm polarization-insensitive optical amplifier with strain-balanced superlattice active layer. *IEEE Photonics Technology Letters.* 7:473–475.

Grundmann M, Stier O, Bimberg D (1995) InAs/GaAs pyramidal quantum dots: Strain distribution, optical phonons, and electronic structure. *Physical Review B.* 52:11969–11981.

Gvozdić DM, Ekenberg U (2006) Superefficient electric-field-induced spin-orbit splitting in strained p-type quantum wells. *Europhysics Letters.* 73:927–933.

Hairer E, Wanner G (1996) *Solving Ordinary Differential Equations II: Stiff and Differential-Algebraic Problems.* Berlin: Springer-Verlag.

Henry CH (1982) Theory of the linewidth of semiconductor lasers. *IEEE Journal of Quantum Electronics.* 18:259–264.

Henry CH (1986) Theory of spontaneous emission noise in open resonators and its application to lasers and optical amplifiers. *Journal of Lightwave Technology.* 4:288–297.

Ito T, Yoshimoto N, Magari K, Sugiura H (1998) Wide-band polarization-independent tensile-strained InGaAs MQW-SOA gate. *IEEE Photonics Technology Letters.* 10:657–659.

Jin CY, Guo WH, Huang YZ, Yu LJ (2003) Photon iterative numerical technique for steady-state simulation of gain-clamped semiconductor optical amplifiers. *IEEE Proceedings—Optoelectronics.* 150:503–507.

Joergensen C, Danielsen SL, Stubkjaer KE, Schilling M, Daub K, Doussiere P, Pommerau F, Hansen PB, Poulsen HN, Kloch A, Vaa M, Mikkelsen B, Lach E, Laube G, Idler W, Wunstel K (1997) All-optical wavelength conversion at bit rates above 10 Gb/s using semiconductor optical amplifiers. *IEEE Journal of Selected Topics in Quantum Electronics.* 3:1168–1180.

Joma M, Horikawa H, Xu CQ, Yamada K, Katoh Y, Kamijoh T (1993) Polarization insensitive semiconductor laser amplifier with tensile strained InGaAsP/InGaAsP multiple quantum well structure. *Applied Physics Letters.* 62:121–122.

Kang JM, Lee SH, Kwon HC, Han SK (2006) WDM-PON with broadcasting function using direct ASE modulation of reflective SOA. *European Conference on Optical Communications (ECOC).* 2006: 24–28.

Kani J-I (2010) Enabling technologies for future scalable and flexible WDM-PON and WDM/TDM-PON systems. *IEEE Journal of Selected Topics in Quantum Electronics.* 16:1290–1297.

Kawano K, Kitoh T (2001) *Introduction to Optical Waveguide Analysis: Solving Maxwell's Equations and the Schrödinger Equation.* New York, NY: Wiley-Interscience.

Keating T, Jin X, Chuang SL, Hess K (1999) Temperature dependence of electrical and optical modulation responses of quantum-well lasers. *IEEE Journal of Quantum Electronics.* 35:1526–1534.

Kim J, Chuang SL (2006) Theoretical and experimental study of optical gain, refractive index change, and linewidth enhancement factor of p-doped quantum-dot lasers. *IEEE Journal of Quantum Electronics.* 42:942–952.

Kim J, Laemmlin M, Meuer C, Bimberg D, Eisenstein G (2008) Static gain saturation model of quantum-dot semiconductor optical amplifiers. *IEEE Journal of Quantum Electronics.* 44:658–665.

Kim J, Laemmlin M, Meuer C, Bimberg D, Eisenstein G (2009) Theoretical and experimental study of high-speed small-signal cross-gain modulation of quantum-dot semiconductor optical amplifiers. *IEEE Journal of Quantum Electronics*. 45:1658–1666.

Kim Y, Lee H, Kim S, Ko J, Jeong J (1999) Analysis of frequency chirping and extinction ratio of optical phase conjugate signals by four-wave mixing in SOA's. *IEEE Journal of Selected Topics in Quantum Electronics*. 5:873–879.

Koenig S, Bonk R, Schmuck H, Poehlmann W, Pfeiffer Th, Koos C, Freude W, Leuthold J (2014) Amplification of advanced modulation formats with a semiconductor optical amplifier cascade. *Optics Express*. 22:17854–17871.

Kolokolov KI, Li J, Ning CZ (2003) k·p Hamiltonian without spurious-state solution. *Physical Review B*. 68:161308.

Labukhin D, Li X (2006) Polarization insensitive asymmetric ridge waveguide design for semiconductor optical amplifiers and super luminescent light-emitting diodes. *IEEE Journal of Quantum Electronics*. 42:1137–1143.

Liljeberg T, Bowers JE (1997) Velocity mismatch limits in semiconductor lasers and amplifiers. *IEEE Conference Proceedings: 10th Annual Meeting Lasers and Electro-Optics Society Annual Meeting (LEOS '97)*. 1:341–342.

Liu G, Chuang SL (2002) Modeling of Sb-based type-II quantum cascade lasers. *Physical Review B*. 65:165220.

Liu G, Jin X, Chuang SL (2001) Measurement of linewidth enhancement factor of semiconductor lasers using an injection-locking technique. *IEEE Photonics Technology Letters*. 13:430–432.

Loudon R, Ramoo D, Adams MJ (2005) Theory of spontaneous emission noise in multisection semiconductor lasers. *Journal of Lightwave Technology*. 23:2491–2504.

Lysak VV, Kawaguchi H, Sukhoivanov IA, Katayama T, Shulika AV (2005) Ultrafast gain dynamics in asymmetrical multiple quantum-well semiconductor optical amplifiers. *IEEE Journal of Quantum Electronics*. 41:797–807.

Lysak VV, Sukhoivanov IA, Shulika OV, Safonov IM, Lee YT (2006) Carrier tunneling in complex asymmetrical multiple-quantum-well semiconductor optical amplifiers. *IEEE Photonics Technology Letters*. 18:1362–1364.

Magari K, Okamoto M, Noguchi Y (1991) 1.55-μm polarization insensitive high-gain tensile-strained-barrier MQW optical amplifier. *IEEE Photonics Technology Letters*. 3:998–1000.

Magari K, Okamoto M, Suzuki Y, Sato K, Noguchi Y, Mikami O (1994) Polarization-insensitive optical amplifier with tensile-strained-barrier MQW structure. *IEEE Journal of Quantum Electronics*. 30:695–701.

Marcuse D (1984) Computer simulation of laser photon fluctuations: Theory of single-cavity laser. *IEEE Journal of Quantum Electronics*. 20:1139–1148.

Mathlouthi W, Lemieux P, Salsi M, Vannucci A, Bononi A, Rusch LA (2006) Fast and efficient dynamic WDM semiconductor optical amplifier model. *Journal of Lightwave Technology*. 24:4353–4365.

Mathur A, Dapkus PD (1992) Polarization insensitive strained quantum well gain medium for lasers and optical amplifiers. *Applied Physics Letters*. 61:2845–2847.

Mazzucato S, Carrère H, Marie X, Amand T, Achouche M, Caillaud C, Brenot R (2015) Gain, amplified spontaneous emission and noise figure of bulk InGaAs/InGaAsP/InP semiconductor optical amplifiers. *IET Optoelectronics*. 9:52–60.

Mecozzi A, Mørk J (1997) Saturation effects in nondegenerate four-wave mixing between short optical pulses in semiconductor laser amplifiers. *IEEE Journal of Selected Topics in Quantum Electronics*. 3:1190–1207.

Mecozzi A, Wiesenfeld JM (2001) The roles of semiconductor optical amplifiers in optical networks. *Optics & Photonics News*. 12:36–42.

Melo AM, Petermann K (2008) On the amplified spontaneous emission noise modeling of semiconductor optical amplifiers. *Optics Communications*. 281:4598–4605.

Meney AT, Gonul B, O'Reilly EP (1994) Evaluation of various approximations used in the envelope-function method. *Physical Review B.* 50:10893–10904.

Michie C, Kelly AE, McGeough J, Armstrong I, Andonovic I, Tombling C (2006) Polarization-insensitive SOAs using strained bulk active regions. *Journal of Lightwave Technology.* 24:3920–3927.

Morel P, Sharaiha A (2009) Wideband time-domain transfer matrix model equivalent circuit for short pulse propagation in semiconductor optical amplifiers. *IEEE Journal of Quantum Electronics.* 45:103–116.

Mørk J, Mecozzi A, Eisenstein G (1999) The modulation response of a semiconductor laser amplifier. *IEEE Journal of Selected Topics in Quantum Electronics.* 5:851–860.

Moss TS, Burrell GJ, Ellis B (1973) *Semiconductor Opto-Electronics.* London: Butterworth & Co.

Mukai T, Yamamoto Y (1982) Noise in an AlGaAs semiconductor laser amplifier. *IEEE Transactions on Microwave Theory and Techniques.* 30:410–421.

Nagarajan R, Ishikawa M, Fukushima T, Geels RS, Bowers JE (1992) High speed quantum-well lasers and carrier transport effects. *IEEE Journal of Quantum Electronics.* 28:1990–2008.

Occhi L, Schares L, Guekos G (2003) Phase modeling based on the α-factor in bulk semiconductor optical amplifiers. *IEEE Journal of Selected Topics in Quantum Electronics.* 9:788–797.

Olsson NA (1989) Lightwave systems with optical amplifiers. *Journal of Lightwave Technology.* 7:1071–1082.

Osinski M, Buus J (1987) Linewidth broadening factor in semiconductor lasers—An overview. *IEEE Journal of Quantum Electronics.* 23:9–29.

Park J, Li X, Huang WP (2005) Comparative study of mixed frequency-time-domain models of semiconductor laser optical amplifiers. *IEE Proceedings—Optoelectronics.* 152:151–159.

Pham QT (2014) Highly effective crosstalk mitigation method using counter-propagation in semiconductor optical amplifier for remodulation WDM-PONs. *Journal of Photonics.* 2014:610967.

Pillai BSG, Premaratne M, Abramson D, Lee KL, Nirmalathas A, Lim C, Shinada S, Wada N, Miyazaki T. (2006) Analytical characterization of optical pulse propagation in polarization-sensitive semiconductor optical amplifiers. *IEEE Journal of Quantum Electronics.* 42:1062–1077.

Pitris S, Vagionas C, Kanellos GT, Pleros N, Kisacik R, Tekin T, Broeke R (2015) Monolithically integrated all-optical SOA-based SR Flip-Flop on InP platform. *International Conference on Photonics in Switching (PS).* 2015:208–210.

Qasaimesh O (2003) Optical gain and saturation characteristics of quantum-dot semiconductor optical amplifier. *IEEE Journal of Quantum Electronics.* 39:794–800.

Qasaimesh O (2013) Broadband gain-clamped linear quantum dash optical amplifiers. *Optical and Quantum Electronics.* 45:1277–1286.

Qin C, Huang X, Zhang X (2012) Theoretical investigation on gain recovery dynamics in step quantum well semiconductor optical amplifiers. *Journal of the Optical Society of America B.* 29:607–613.

Ramaswami R (2002) Optical fiber communication: From transmission to networking. *IEEE Communications Magazine.* 40:138–147.

Razaghi M, Ahmadi V, Connelly MJ (2009) Comprehensive finite-difference time-dependent beam propagation model of counterpropagating picosecond pulses in a semiconductor optical amplifier. *Journal of Lightwave Technology.* 27:3162–3174.

Reithmaier JP, Somers A, Deubert S, Schwertberger R, Kaiser W, Forchel A, Calligaro M, Resneau P, Parillaud O, Bansropun S, Krakowski M, Alizon R, Hadass D, Bilenca A, Dery H, Mikhelashvili V, Eisenstein G, Gioannini M, Montrosset I, Berg TW, van der Poel M, Mørk, J, Tromborg B (2005) InP based lasers and optical amplifiers with wire-/dot-like active regions. *Journal of Physics D: Applied Physics.* 38:2088–2102.

Runge P, Elschner R, Petermann K (2010) Time-domain modeling of ultralong semiconductor optical amplifiers. *IEEE Journal of Quantum Electronics.* 46:484–491.

Schmuck H, Bonk R, Poehlmann W, Haslach C, Kuebart W, Karnick D, Meyer J, Fritzsche D, Weis E, Becker J, Freude W, Pfeiffer T (2013) Demonstration of SOA-assisted open metro-access infrastructure for heterogeneous services. *39th European Conference and Exhibition on Optical Communication (ECOC).* 2013:1–3.

Schrenk B (2011) Characterization and Design of Multifunction Photonic Devices for Next Generation Fiber-to-the-Home Optical Network Units. PhD diss., Universitat Politècnica de Catalunya.

Serrat C, Masoller C (2006) Modeling spatial effects in multi-longitudinal-mode semiconductor lasers. *Physical Review A*. 73:043812.

Shtaif M, Tromborg B, Eisenstein G (1998) Noise spectra of semiconductor optical amplifiers: Relation between semiclassical and quantum descriptions. *IEEE Journal of Quantum Electronics*. 34:869–878.

Silver M, Phillips AF, Adams AR, Greene PD, Collar AJ (2000) Design and ASE characteristics of 1550-nm polarization-insensitive semiconductor optical amplifiers containing tensile and compressive wells. *IEEE Journal of Quantum Electronics*. 36:118–122.

Simon JC, Doussiere P, Pophillat L, Fernier B (1989) Gain and noise characteristics of a 1.5 μm near-travelling-wave semiconductor laser amplifier. *Electronics Letters*. 25:434–436.

Stabile R, Williams KA (2011) Photonic integrated semiconductor optical amplifier switch circuits. In *Advances in Optical Amplifiers*, edited by P. Urquhart, 205–230. Rijeka, Croatia: InTech.

Stier O, Grundmann M, Bimberg D (1999) Electronic and optical properties of strained quantum dots modeled by 8-band k.p theory. *Physical Review B*. 59:5688–5701.

Sygletos S, Tomkos I, Leuthold J (2008) Technological challenges on the road toward transparent networking. *Journal of Optical Networking*. 7:321–350.

Talli G, Adams MJ (2003) Amplified spontaneous emission in semiconductor optical amplifiers: Modelling and experiments. *Optics Communications*. 218:161–166.

Tauber DA, Spickermann R, Nagarajan R, Reynolds T, Holmes AL, Bowers JE (1994) Inherent bandwidth limits in semiconductor lasers due to distributed microwave effects. *Applied Physics Letters*. 64:1610–1612.

Thylén L (1988) Amplified spontaneous emission and gain characteristics of Fabry-Perot and traveling wave type semiconductor laser amplifiers. *IEEE Journal of Quantum Electronics*. 24:1532–1537.

Tiemeijer LF, Thijs PJA, Van Dongen T, Slootweg RWM, van der Heijden JMM, Binsma JJM, Krijn MPCM (1993) Polarization insensitive multiple quantum well laser amplifiers for the 1300 nm window. *Applied Physics Letters*. 62:826–828.

Tishinin D, Uppal K, Kim I, Dapkus PD (1997) 1.3-μm polarization insensitive amplifiers with integrated-mode transformers. *IEEE Photonics Technology Letters*. 9:1337–1339.

Toptchiyski G, Kindt S, Petermann K, Hilliger E, Diez S, Weber HG (1999) Time-domain modeling of semiconductor optical amplifiers for OTDM applications. *Journal of Lightwave Technology*. 17:2577–2583.

Totović AR, Crnjanski JV, Krstić MM, Gvozdić DM (2011) Application of multi-quantum well RSOA in remodulation of 100 Gb/s downstream RZ signal for 10 Gb/s upstream transmission. *19th Telecommunications Forum (TELFOR)*. 2011:840–843.

Totović AR, Crnjanski JV, Krstić MM, Gvozdić DM (2012) Modelling of carrier dynamics in multi-quantum well semiconductor optical amplifiers. *Physica Scripta*. 2012:014032.

Totović AR, Crnjanski JV, Krstić MM, Gvozdić DM (2014) An analytical solution for stationary distribution of photon density in traveling-wave and reflective SOAs. *Physica Scripta*. 2014:014013.

Totović AR, Crnjanski JV, Krstić MM, Gvozdić DM (2014) An efficient semi-analytical method for modeling of traveling-wave and reflective SOAs. *Journal of Lightwave Technology*. 32:2106–2112.

Totović AR, Crnjanski JV, Krstić MM, Gvozdić DM (2015) Numerical study of the small-signal modulation bandwidth of reflective and traveling-wave SOAs. *Journal of Lightwave Technology*. 33:2758–2764.

Totović AR, Crnjanski JV, Krstić MM, Mašanović ML, Gvozdić DM (2013) A self-consistent numerical method for calculation of steady-state characteristics of traveling-wave and reflective SOAs. *IEEE Journal of Selected Topics in Quantum Electronics*. 19:3000411.

Totović AR, Levajac VG, Gvozdić DM (2016) Electro-optical modulation bandwidth analysis for traveling-wave and reflective semiconductor optical amplifiers in transparency operating regime. *Optical and Quantum Electronics*. 48:262.

Tsai CY, Tsai CY, Lo YH, Spencer RM, Eastman LF (1995) Nonlinear gain coefficients in semiconductor quantum-well lasers: Effects of carrier diffusion, capture, and escape. *IEEE Journal of Selected Topics in Quantum Electronics*. 1:316–330.

Valiente I, Lablonde L, Simon JC, Billès L (1996) Effects of amplified spontaneous emission on gain recovery dynamics of semiconductor optical amplifiers. In *Optical Amplifiers and Their Applications*, edited by R. Jopson, K. Stubkjaer, and M. Suyama, Vol. 5, *OSA Trends in Optics and Photonics Series*. Washington, DC: Optical Society of America.

Veprek RG, Steiger S, Witzigmann B (2007) Ellipticity and the spurious solution problem of k · p envelope equations. *Physical Review B*. 76:165320.

Visser TD, Blok H, Demeulenaere B, Lenstra D (1997) Confinement factors and gain in optical amplifiers. *IEEE Journal of Quantum Electronics*. 33:1763–1766.

Vyrsokinos K, Vagionas C, Fitsios D, Miliou A (2014) Frequency and time domain analysis of all optical memories based on SOA and SOA-MZI switches. *IEEE Optical Interconnects Conference*. 2014: 57–58.

Wang J, Maitra A, Poulton CG, Freude W, Leuthold J (2007) Temporal dynamics of the alpha factor in semiconductor optical amplifiers. *Journal of Lightwave Technology*. 25:891–900.

Willatzen M, Uskov A, Mørk J, Olesen H, Tromborg B, Jauho AP (1991) Nonlinear gain suppression in semiconductor lasers due to carrier heating. *IEEE Photonics Technology Letters*. 3:606–609.

Wolfson D. All-optical signal processing and regeneration. PhD diss., Research Center COM, Technical University of Denmark, 2001.

Wu B, Georges JB, Cutrer DM, Lau KY (1995) On distributed microwave effects in semiconductor lasers and their practical implications. *Applied Physics Letters*. 67:467–469.

Wünstel K, Laube G, Idler W, Daub K, Lach E, Dütting K, Klenk M, Schilling M (1996) High speed and polarization insensitive wavelength converters by MQW optimization, In *Conference Proceedings of IPR'96*, paper IMG6, Boston, MA.

Xia M, Ghafouri-Shiraz H (2015) Theoretical analysis of carrier heating effect in semiconductor optical amplifiers. *Optical and Quantum Electronics*. 47:2141–2153.

Xia M, Ghafouri-Shiraz H (2016) A novel transmission line model for quantum well semiconductor optical amplifiers. *Optical and Quantum Electronics*. 48:52.

Xing W, Yikai S, Xiang L, Leuthold J, Chandrasekhar S (2004) 10-Gb/s RZ-DPSK transmitter using a saturated SOA as a power booster and limiting amplifier. *IEEE Photonics Technology Letters*. 16:1582–1584.

Yacomotti AM, Furfaro L, Hachair X, Pedaci F, Giudici M, Tredicce J, Javaloyes J, Balle S, Viktorov EA, Mandel P (2004) Dynamics of multimode semiconductor lasers. *Physical Review A*. 69:053816.

Yamatoya T, Koyama F (2004) Optical preamplifier using optical modulation of amplified spontaneous emission in saturated semiconductor optical amplifier. *Journal of Lightwave Technology*. 22:1290–1295.

Yariv A, Yeh P. *Photonics: Optical Electronics in Modern Communications*. New York, NY: Oxford University Press, 2007.

Zhang Y, Ruden PP (1999) 1.3 µm polarization-insensitive optical amplifier structure based on coupled quantum wells. *IEEE Journal of Quantum Electronics*. 35:1509–1514.

Zhou E, Zhang X, Huang D (2007) Analysis on dynamic characteristics of semiconductor optical amplifiers with certain facet reflection based on detailed wideband model. *Optics Express*. 15:9096–9106.

Zilkie AJ, Meier J, Mojahedi M, Poole PJ, Barrios P, Poitras D, Rotter TJ, Yang C, Stintz A, Malloy KJ, Smith PWE, Aitchison JS (2007) Carrier dynamics of quantum-dot, quantum-dash, and quantum-well semiconductor optical amplifiers operating at 1.55 µm. *IEEE Journal of Quantum Electronics*. 43:982–991.

Zimmerman DR, Spiekman LH (2004) Amplifiers for the masses: EDFA, EDWA, and SOA amplets for metro and access applications. *Journal of Lightwave Technology*. 22:63–70.

22

Tapered Semiconductor Optical Amplifiers

José-Manuel G. Tijero

Antonio Pérez-Serrano

Gonzalo del Pozo

and

Ignacio Esquivias

22.1 Introduction

In recent years, the pervasive extension of semiconductor-based laser sources has been reaching application domains, such as lidar, material processing, metrology, and frequency doubling, where other laser sources are still prevalent because of the concurrent requirements of high power, high spectral purity, and high-beam quality. Two slightly different semiconductor devices are already in use for some of these applications: tapered laser diodes (Walpole, 1996; Wenzel et al., 2003; Sumpf et al., 2009) and master-oscillator power-amplifiers (MOPAs) (O'Brien et al., 1993; Spreemann et al., 2009). A tapered MOPA consists of either a distributed Bragg reflector laser or a distributed feedback laser acting as the master oscillator (MO) and a tapered semiconductor optical amplifier (SOA) acting as the power amplifier. Both hybrid and integrated MOPAs are of great interest. In a hybrid MOPA, the MO and the SOA are coupled through additional optics (Schwertfeger et al., 2011) with the advantage of separate fabrication of the components but at the expense of a more complex set-up. The integrated MOPAs have the advantages of reduced total size but the drawback of possible optical coupling between sections (Vilera et al., 2015).

The schematic of a typical tapered SOA is shown in Figure 22.1a. It is composed of a straight and narrow index-guided section used to filter the input beam and convert it into a single spatial mode, and a gain-guided tapered section where the beam is amplified while preserving its shape. The tapered SOA of an integrated MOPA does not include an index-guided section since the MO provides a single spatial mode to the tapered section. Cross-sectional views of the index and gain-guided sections are shown in Figure 22.1b. The index-guided section is typically a ridge waveguide (RW) structure in which the etched regions (filled with an isolator) provide lateral optical confinement. In the tapered gain-guided section, the injection area is defined by a proton-implanted region.

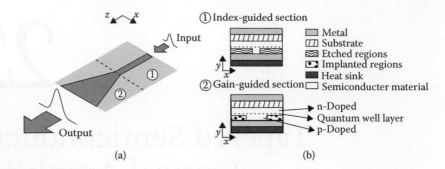

FIGURE 22.1   (a) Schematic planar view of a typical tapered SOA, consisting of a straight and narrow index-guided section and a gain-guided tapered section. (b) Cross-sectional views of the index-guided section (1) and the gain-guided section (2).

The characteristics of the output beam of a tapered SOA are very similar to those of tapered lasers which have been studied in detail by simulations and experiments (Williams et al., 1999; Sujecki et al., 2003; Borruel et al., 2004) and are reviewed in Volume 2 of this book (Chapter 28) (Esquivias et al., 2017). In brief, the single spatial mode of the RW section is launched into the tapered region where it suffers simultaneously diffraction, gain guiding, and index guiding (or antiguiding) due to the carrier and temperature dependencies of the semiconductor refractive index. The full taper angle α_{tap} is usually designed to match the free diffraction angle at $1/e^2$ (Borruel et al., 2008), and therefore it depends on the RW width and the lateral index step. At low power, the beam is strongly astigmatic: the virtual source in the lateral axis (x-axis) is located approximately at a distance from the output facet given by the taper section length divided by the effective index n_{eff}, while the virtual source in the vertical axis (y-axis) is at the output facet. The far-field (FF) pattern is almost Gaussian in both axes and much narrower in the lateral axis. However, when the output power increases, the beam quality degrades due to the combined effects of carrier and thermal lensing, as shown in Section 22.4.

The optimization of the tapered SOA's geometry and epitaxial design is a key point to improve the performance of these devices. The complex interaction between the semiconductor media and the optical field makes it extremely difficult to advance meaningful predictions of the behavior of a specific design by means of simple analytical calculations. Such difficulties together with the complex and costly fabrication process make necessary the use of complete numerical simulation tools to help the analysis and design of tapered SOAs to be used in hybrid or integrated MOPAs. Different numerical models, with different degrees of complexity, have been proposed and used to predict the behavior of tapered SOAs. In the case of the straight SOAs used in communications, the main emphasis has been placed on analyzing the dynamic response and the role of the amplified spontaneous emission in the noise characteristics (Razaghi et al., 2009; Connelly, 2001, and references therein). In contrast, in the case of high-power tapered SOAs the main performance characteristics are the maximum output power and the beam properties, and therefore steady-state models including thermal equations have been employed (Lang et al., 1993; Lai and Lin, 1998; Tijero et al., 2015).

In Section 22.2, we present a review of the different modeling approaches that can be applied to tapered SOAs; in Section 22.3, we explain the implementation of our numerical model; and in Section 22.4 we present as an example the results of the simulations of a 1.5-μm tapered SOA, including the analysis of the influence of some selected design parameters on the device performance. The Section 22.5 summary concludes the chapter.

22.2 Modeling Approaches

There are many options for the modeling approaches needed to simulate tapered SOAs. The proper choice depends on the degree of numerical complexity and on the performance characteristics of interest. In this section, we review the most common approaches, explain the main differences between them, and refer the interested reader to the most important references for a detailed accounts. We will also use references corresponding to tapered lasers and laser diodes in general, as most of the modeling approaches for high-power SOAs are equally appropriate for high-power semiconductor lasers.

One of the main characteristics of tapered SOAs and lasers compared with other semiconductor devices is the nonuniformity of the photon, carrier, and temperature profiles along the longitudinal (z-axis) and the lateral directions. This leads to the need to solve the coupled equations in at least two dimensions and to use and using some approximation to solve the equations in the vertical direction separately.

22.2.1 Carrier Transport Models

The main role of a carrier transport model is to provide the 2D carrier density distribution along the active region, which is needed to calculate the main material parameters affecting the optical wave propagation (optical gain and losses and refractive index variation), as well as to calculate the local heat sources required by the thermal model. But, in turn, the knowledge of the local photon density and the temperature distribution is needed for solving the electrical equations, and therefore iterative schemes are usually applied (see Section 22.3). The external input is the applied voltage (V_0), rather than the total current (I), which is the usual input in experiments. In the RW section, the current flow spreads out of the nominal contact width. Furthermore, the local current density depends not only on the applied voltage but also on the local temperature and stimulated recombination, and hence on the local photon density, which is strongly dependent on the lateral and longitudinal position. In consequence, the current density J is not uniform, but depends on the position in the x–z plane, $J = J(x, z)$.

Most of the published tapered SOA models make use of the unipolar approximation assuming that the electron (n) and hole (p) densities in the active layer are identical (Lang et al., 1993; Lai and Lin, 1998; Spreemann et al., 2009). Under this approximation, which is also very common in laser diode models, the carrier density in the active layer $N(x, z)$ can be calculated using the simple rate equation:

$$\frac{dN(x, z)}{dt} = \frac{J(x, z)}{q d_{act}} - R(N) - R_{stim}(N, S) + D_{eff}\nabla^2_{x,z}N(x, z), \tag{22.1}$$

where q is the elementary charge, d_{act} the active layer thickness, $R(N)$ the carrier recombination rate including nonradiative and spontaneous recombination, and $R_{stim}(N, S)$ the stimulated recombination rate, which depends on the local photon density $S(x, z)$. Finally, finally, $D_{eff}\nabla^2_{x,z}N(x, z)$ accounts for carrier diffusion in the longitudinal and lateral directions and is perhaps the most important consideration. In this expression, D_{eff} is an effective diffusion coefficient accounting for both the carrier diffusion and the current spreading. The local current density $J(x, z)$ can be expressed in terms of the applied voltage, the contact (and/or p-layer) resistivity, and the junction voltage $V_J(x, z)$, which in turn depends on the carrier density. Details on this formulation can be found in Lai and Lin (1998).

The unipolar approximation is a simple way to reduce the complexity of the carrier transport model and provides very reasonable results, but the p–n junction is essentially a bipolar device, thus requiring the solution of the complete semiconductor drift–diffusion equations. These equations can be expressed in terms of three unknown variables, the electrostatic potential (ϕ) and the electron (n) and hole (p) concentrations. In the steady state, the equations are (Selberherr, 1984)

$$\nabla\left(\varepsilon_S\nabla\varphi\right) + q\left(p - n + C_i\right) = 0 \tag{22.2}$$

$$\nabla\mathbf{j_n} - q\left(R(n, p) + R_{stim}(n, p, S)\right) = 0 \tag{22.3}$$

$$\nabla\mathbf{j_p} + q\left(R(n, p) + R_{stim}(n, p, S)\right) = 0, \tag{22.4}$$

where ε_S is the static dielectric constant, C_i is the charged impurity density (ionized donor density minus ionized acceptor density), and j_n (j_p) is the electron (hole) current density. The current densities can be expressed as a function of the local variables ϕ, n, p, and temperature (T) using as local parameters the electron and hole mobilities and thermal diffusivities.

The carrier transport models often consider the capture and escape of carriers into and out of the quantum wells (QWs) (Nagarajan et al., 1992). This effect is well known in high-speed lasers, and it also influences the steady-state emission properties. When the carrier capture time is not negligible, there is an accumulation of carriers in the confinement layers that can result in a reduction of the efficiency (Borruel et al., 2003). In the case of unipolar models using rate equations, the carrier capture/escape processes are modeled by solving a second equation for the unconfined carriers. In the case of bipolar models of laser diodes using the complete semiconductor equations, different approaches have been proposed (Borruel et al., 2003; Grote et al., 2005), but as far as we know the importance of these effects in the modeling of SOAs has not been analyzed.

Bipolar carrier transport models require also appropriate boundary conditions at the metal–semiconductor contacts (Selberherr, 1984) and at the abrupt heterojunctions, where thermionic emission or tunneling-assisted thermionic emission should be considered instead of the standard drift–diffusion equations.

22.2.2 Optical Models

All optical models start from Maxwell equations and a description of the material susceptibility in the frequency or time domain, leading to optical wave equations, which have to be solved numerically. There are two main types of models: those solving the equations in the time domain, usually known as traveling-wave models (TWM), and frequency domain models. TWM provide dynamic solutions as well as spectral properties, and are appropriate for pulse propagation, beam filamentation, transient phenomena, and multifrequency devices. Frequency domain models are much simpler to implement and are useful for steady-state studies where the overall behavior of the device can be described in terms of a single frequency.

The beam propagation method (BPM) (Van Roey et al., 1981) is the most commonly used frequency domain method for the field propagation in waveguide optoelectronic and fiber devices, thanks to its low computational time demand. BPM calculates the propagation along the longitudinal direction of the steady-state transverse field, verifying the Helmholtz equation:

$$\nabla^2 \vec{E}(x, y, z) + \tilde{n}^2(x, y, z)k_0^2\vec{E}(x, y, z) = 0, \tag{22.5}$$

where $\vec{E}(x, y, z)$ is the complex optical field vector, $\tilde{n}(x, y, z)$ is the complex refractive index, and k_0 is the wavenumber in vacuum.

In order to obtain a solution for the Helmholtz equation, the scalar approximation is commonly used in the study of tapered lasers and amplifiers (Sujecki et al., 2003; Lang et al., 1993). This approximation ignores the vectorial nature of the field, assuming continuity of the field and its derivatives through the dielectric interfaces. The approximation implies that the optical polarization remains unchanged along the device. The effective refractive index approximation is used to decrease the computational complexity of the BPM (Buus, 1982). It consists of solving Equation 22.5 for the vertical axis at every lateral position, taking into account the existence of etched regions and considering only a real index vertical profile (passive approximation). This leads to an effective index lateral profile and a vertical distribution of the optical mode $f(y)$. The effective index profile is used for the beam propagation and the function $f(y)$ is usually assumed to be independent of the longitudinal and lateral positions. In this way, the original 3D problem becomes a 2D problem in the x–z plane.

Dynamical TWMs naturally incorporate the spatial effects and the multifrequency character of the field, with the geometry of the device being incorporated through the boundary conditions for the waves. Although TWMs have long been applied to study laser physics (Lugiato and Narducci, 1985), application to semiconductor media requires a proper description in the time domain of the spectral dependence of the gain and the index changes due to the injected carriers. TWMs are based on making the slowly varying amplitude approximation or paraxial approximation of the electric fields. Under this approximation, the equations for the electric field of the optical wave can be written as

$$-\frac{i}{2k_0}\frac{\partial^2 E_\pm}{\partial x^2} \pm \frac{\partial E_\pm}{\partial z} + \frac{1}{v_g}\frac{\partial E_\pm}{\partial t} = f(E_+, E_-, P_+, P_-), \tag{22.6}$$

where $E_\pm$ are the complex counter propagating electric fields, v_g is the group velocity, and the function f contains different terms describing the relationship of the field with the dielectric polarization ($P_\pm$) that includes gain, internal losses, Bragg couplings, etc. Several approaches have been used to solve Equation 22.6, including finite-difference time-domain (Fischer et al., 1996) and Fourier split-step methods (Spreemann et al., 2009; Pérez-Serrano et al., 2013). Although TWMs have been applied mainly for describing laser dynamics, they have also been applied to tapered SOAs (Balsamo et al., 1996) and for the analysis of coupled cavity effects in tapered MOPAs (Spreemann et al., 2009; Pérez-Serrano et al., 2013).

22.2.3 Thermal Models

The temperature distribution along the complete device is a relevant variable in the simulation of a tapered SOA due to the dependence of the gain and refractive index on the local temperature. The 3D steady-state heat equation is expressed as

$$\nabla(\kappa \nabla T) + w(x, y, z) = 0, \tag{22.7}$$

where κ is the temperature-dependent thermal conductivity and $w(x, y, z)$ the local density of heat sources. This term usually includes Joule, nonradiative recombination, and free-carrier absorption heat sources. A complete description, based on a rigorous thermodynamic approach (Wachutka, 1990), can be found in Bandelow et al. (2005). Two heat sources are usually not included in most of the laser and amplifier models due to the difficulties in the quantification of their local distribution. These are the sources of heat arising from the energy lost by the scattered stimulated photons and by the spontaneously emitted photons that do not leave the device. After a photon recycling process, their energy ends up in heat transferred to the lattice somewhere in the device. To account for this effect, as well as for other uncertainties in the expressions and parameters for the local heat sources, an additional excess power P_{exc} shall be included (Borruel et al., 2002). The amount of this power is calculated from the energy conservation expression:

$$P_{exc} = IV_0 - (P_{out} - P_{in}) - \int w(x, y, z)\mathrm{d}V, \tag{22.8}$$

where P_{out} and P_{in} are the output and input powers, respectively, and the integration extends over the entire device volume. After determining P_{exc}, the corresponding heat source w_{exc} is either distributed uniformly over the whole device or weighted according to the different absorption properties of the epitaxial layers.

Thermal equations should be solved taking into account the boundary conditions imposed by the thermal properties and the geometry of the mount and submount and the heatsink temperature.

22.2.4 Material Properties

22.2.4.1 Material Susceptibility

There are different options to describe the interaction between the gain medium and the optical field. The simplest approach, which is usually employed in rate equation models, is to consider a logarithmic or linear dependence of the material gain on the carrier density, together with a constant value for the linewidth enhancement factor (LEF) (Lai and Lin, 1998). The gain spectra can be taken as Lorentzian (Spreemann et al., 2009), although the most sophisticated models include a complete solution of the band dispersion equations to calculate the material gain as a function of the local carrier densities, the wavelength, and the temperature. A complete microscopic 8×8 $\mathbf{k \cdot p}$ model for the semiconductor band structure implies a large computational effort and even a more simplified 4×4 valence band mixing model results in a high computation time when repeated locally. Additionally, most of the input parameters for the complete models are not well known. For these reasons, some modeling approaches perform initial complete or semicomplete band structure calculations and either (1) use a parabolic band model previously fitted to the results of the complete model (Borruel et al., 2004), (2) use curve fits to a more sophisticated model (Mariojouls et al., 2000), or (3) use precalculated look-up tables (Koch et al., 2005).

The refractive index's dependence on the wavelength, carrier density, and temperature can also be calculated using complete microscopic approaches (Koch et al., 2005) and then again using look-up tables to decrease the computational cost. A simpler approach is to neglect the wavelength dependence of the index and to use simple expressions for the carrier and temperature dependencies, i.e.,

$$\delta \tilde{n} = \delta \tilde{n}_T + \delta \tilde{n}_N, \tag{22.9}$$

where $\delta \tilde{n}$ is the variation of the active layer index with respect to its reference value at the heatsink temperature and without injection, $\delta \tilde{n}_T$ is the thermally induced variation, and $\delta \tilde{n}_N$ is the carrier contribution to the variation. $\delta \tilde{n}_T$ is usually considered a linear function of the local temperature increase while $\delta \tilde{n}_N$ can be considered either as dependent on the carrier density (Connelly, 2001; Borruel et al., 2004) or it is calculated from the gain variations using a constant LEF (Lang et al., 1993; Mariojouls et al., 2000).

Full space-time models require a good description of the material complex susceptibility, which has been done in different ways: at microscopic level (Gehrig and Hess, 2001), by using multiple Lorentzians fitted either to experiments or to microscopic calculations (Moloney et al., 1997), and by a convolution integral (Javaloyes and Balle, 2010).

22.2.4.2 Carrier Recombination

The unipolar models typically use a polynomial approximation for the carrier recombination rate in Equation 22.1, which can be given by

$$R(n) = An + Bn^2 + Cn^3, \tag{22.10}$$

where A, B, and C are the coefficients corresponding to nonradiative trap recombination, spontaneous recombination, and Auger recombination, respectively. It is important to remark that even in this simple description, the dependence of the coefficients on temperature, especially the dependence of C, should be taken into account. Some other polynomial approaches have also been employed (Connelly, 2001). The more complete bipolar models use the complete Shockley–Read–Hall (SRH) recombination expression for the nonradiative trap recombination and the electron and hole Auger recombination rates and calculate the spontaneous recombination rate from the complete energy band description by using the same approach used for the optical gain.

22.2.4.3 Internal Losses Coefficient

The internal absorption losses have a dramatic effect on the amplifier performance, and therefore the coefficient accounting for them, α_{in}, is an extremely important parameter for the simulation. It can be taken as a constant value to be estimated from experiments in broad area lasers, or it can be considered the result of two contributions: the scattering losses α_{scat}, and the losses due to different mechanism of photon absorption by free carriers, the so-called free-carrier absorption losses α_{fc}. In the approaches that solve the electrical equations in the vertical direction (Borruel et al., 2004; Grote et al., 2005), α_{fc} is locally calculated taking into account the carrier and photon distribution by

$$\alpha_{fc}(x, z) = \int\limits_{L_y} \left[\kappa_n n(x, y, z) + \kappa_p p(x, y, z) \right] S(x, y, z) dy, \tag{22.11}$$

where L_y is the thickness of the epilayer structure, and $\kappa_n(\kappa_p)$ is the electron (hole) free carrier absorption cross section. In most materials, the main contribution to α_{fc} arises from the hole term due to the mechanism of intervalence band absorption.

The most sophisticated amplifier models, especially those solving the complete electro-thermal-optical semiconductor equations, require many additional material parameters such as carrier mobilities, bandgap narrowing coefficients, energy band parameters, thermoelectric powers, thermal conductivities, etc.

22.3 Model Implementation

22.3.1 Overview of Solution Procedure

In this section, we describe the implementation of HAROLD 4.0, the model and code that we have developed for the simulation of tapered SOAs. Our approach is based on a previous model for tapered lasers (Borruel et al., 2004; Sujecki et al., 2003). The original algorithm has been modified slightly and a part of it has been rewritten in order to consider a single pass amplifier instead of a resonant cavity. HAROLD 4.0 solves self-consistently the steady-state partial differential equations describing the electrical, thermal, and optical behavior of a tapered SOA. The simulator includes a 3D electrical solver for the Poisson and continuity equations (Equations 22.2 through 22.4) coupled to a 3D thermal solver for the heat-flow equation (Equation 22.7). The thermal solver uses the local heat sources provided by the electrical solution. On the optical side, a wide-angle finite-difference beam propagation method (WA-FDBPM) (Hadley, 1992), under the effective index approximation, is used to propagate the optical field in the x–z plane. The optical solution is not fully 3D, as the vertical direction and the x–z plane are solved separately, and therefore the complete model is considered as quasi-3D.

Due to the large number of variables and iteration loops involved in the model, an important point is how to initialize the solution. In order to provide a good initial guess to the quasi-3D algorithm, a 1D laser simulator (Harold, 2001) is used for the initialization of the electrical, thermal, and optical variables. Although the 1D laser simulator was not designed for amplifiers, it provides useful initial guess values for the electrothermal variables as well as some inputs for the quasi-3D algorithm such as the applied voltage V_0 and the optical field modal profile in the vertical direction $f(y)$. The 1D simulator requires the electrical, thermal, and optical properties of all the device materials. The user introduces a target bias current and the 1D simulator solves the electrothermal and optical equations for an equivalent broad area laser having the same electrical injection area as the tapered SOA. The 1D simulator provides V_0, as well as an initialization for the vertical distribution of the electrostatic potential, the electron and hole concentrations, and the temperature. The power, wavelength, and lateral profile of the input optical field are provided by the user as inputs for the algorithm. The input optical field is symmetric and therefore only one half of the device is considered in the simulations.

The main flow of the quasi-3D algorithm is shown in Figure 22.2. It includes the following steps: (1) the 1D simulator provides V_0 and $f(y)$, and initializes all unknown 2D variables according to the initial 1D solution for the target current; (2) the initial lateral optical field profile at the first longitudinal slice ($z_i = 1$) is either set by the user or calculated as the fundamental mode of the waveguide in the lateral direction; (3) A 2D electrical solver is applied to the first slice in the $x - y$ plane at $z_i = 1$. It provides a 2D map of the electrical variables (n, p, and φ) and of the Joule (w_{joule}), the nonradiative (w_{nr}), and the free-carrier (w_{fc}) heat sources. It also calculates the complex effective refractive index along the lateral axis

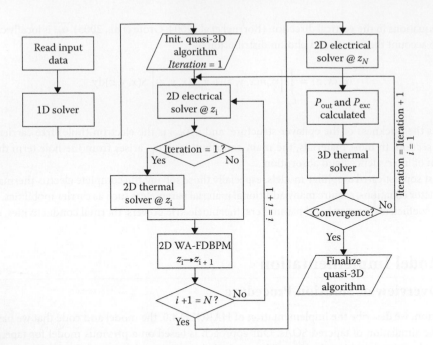

FIGURE 22.2　Main flow of the quasi-3D algorithm HAROLD 4.0 (see text for details).

at $z_i = 1$, thus providing the gain and refractive index variation due to the carriers. (4) The calculated heat sources are used as inputs for a 2D thermal solver applied to the x–y plane at $z_i = 1$. It should be noted that the thermal and electrical meshes are different and therefore some interpolation between the meshes is used. The 2D thermal solver provides a map of the temperature in the x–y plane and the thermally induced refractive index variation. This 2D thermal solver is only applied in the first iteration in order to initialize the thermal solution. (5) The WA-FDBPM is used to propagate the optical field from $z_i = 1$ to $z_i = 2$ using a fine mesh in the x–z plane. It uses the gain and refractive index changes given by the electrothermal solvers in the previous step and takes into account the geometry of the device. (6) Steps (3)–(5) are repeated until the algorithm arrives to the output facet at $z_i = N$. At this point, the output power P_{out} is known and the excess power P_{exc} can be calculated. Then, a 3D thermal solver is applied allowing heat flow between the z-slices. (7) The process described in (3), (5), and (6) is repeated until convergence is reached and a stable optical field, as well as stable temperature, carrier, and current distributions are found.

In the following subsections, the solvers for the electrical, thermal, and optical equations are described in more detail.

22.3.2 Solver for Electrical Equations

The electrical equations are solved in the volume defined by the total thickness of the epitaxial layers, the amplifier length, and the chip width. The input for the equations is taken from material databases according to the epilayer structure. Etched and implanted regions are simulated by assuming negligible carrier mobility. The electrical equations are solved together with two capture–escape continuity equations for electrons and holes at each QW, taking into account the nonequilibrium condition between the confined carriers in the QWs and the unconfined carriers in the barriers. Nonradiative (SRH and Auger) and spontaneous recombination terms are included along the complete device, in addition to the stimulated recombination in the active region. The model also includes free carrier absorption and band gap renormalization with standard dependencies and appropriate parameters.

The material gain and spontaneous recombination rate in the QWs are calculated using the expressions in Coldren and Corzine (1995), considering simple parabolic bands and Lorentzian broadening to decrease the computational effort. The parameters for the gain calculations (energy levels, effective masses, and intraband relaxation time) were previously fitted to yield similar results to those of a more sophisticated gain model that takes into account the valence band mixing. The local material gain (or absorption) and carrier-induced refractive index changes are calculated at the wavelength of the input power for each point of the mesh, taking into account the local temperature and carrier concentrations. They are subsequently used as inputs for the optical solver.

The equations are discretized using a finite difference approach. Nonuniform, separate meshes are taken for the vertical and lateral directions. The denser regions in the vertical and lateral directions correspond to the QW and to the border of the implanted/etched areas, respectively. The lateral mesh depends on the longitudinal position and typically contains around 40 mesh points. We chose 100 μm for the x–y slice separation in the longitudinal direction, after making sure that the results in our structures were similar when decreasing the longitudinal mesh size. The core of the numerical procedure is a Newton–Raphson algorithm, which solves the coupled Equations 22.2 through 22.4 in 1D along the vertical direction. The algorithm considers at each mesh point the influence of the next neighbors in the lateral and longitudinal directions. The 1D solver scans iteratively the lateral direction until convergence, providing a 2D map of the electrical variables at the z-slice i, z_i. The electrical variables are used to calculate the heat sources needed by the thermal solver, while the computed gain (or absorption) and the carrier-induced refractive index changes will be used by the optical solver to propagate the electric field to the slice z_{i+1}. In the resolution of the z_{i+1} slice, the values of the electrical variables previously obtained for z_i and z_{i+2} are considered in order to account for the longitudinal current flow. Finally, when convergence of the whole algorithm is reached, a complete 3D map of the electrical variables (electrostatic potential and electron and hole densities), consistent with the photon and temperature distributions, is produced.

22.3.3 Solver for Optical Equations

As mentioned above, the shape of the field at the amplifier input ($z_i = 1$) is either provided by the user or obtained by solving the Helmholtz equation in the lateral direction making use of the effective index approximation. Then, the WA-FDBPM is applied to propagate the field along a longitudinal optical mesh, which is much denser than the electrothermal mesh. The method uses the modal complex index along the lateral axis provided by the electrothermal solution of the initial slice. The WA-FDBPM uses the (1,1) Padé approximant, allowing the propagation of the field a few degrees around the longitudinal direction (1°–6°). The propagation continues until reaching the next slice of the electrothermal mesh, where the electrical and thermal solvers update the modal complex index. In this way, the tapered section is considered a succession of rectangular slices with increasing width.

To complete the optical solution, perfectly matched layers (PML) (Huang et al., 1996) are used as boundary conditions. The PML method defines a lossy region to which the Helmholtz equation is mapped by an anisotropic complex transformation. The PML attenuates the lateral traveling waves ensuring that the field at the borders of the device in the lateral direction vanishes. Due to the lateral symmetry of the optical field, a zero derivative is applied as the boundary condition at $x = 0$.

22.3.4 Solver for Thermal Equations

The heat flow equation (Equation 22.7) is solved in a larger region and using a different mesh than those of the electrical solver. In addition to the epilayer, the substrate, the metal layers, and the heatsink are also considered in the thermal solution. The boundary conditions for the heat-flow equation in the case of a p-down mounted device are negligible heat flow at the n-metal and lateral chip external interfaces and uniform temperature at the bottom of the heatsink. Etched regions are thermally simulated by introducing the thermal conductivity of the planarization isolator.

As mentioned in Section 22.3.1, during the first iteration of the Quasi-3D algorithm, the 2D heat flow equation is solved at each x–y slice with the local heat sources obtained from the electrical solver. The 2D heat flow equation is discretized with a finite difference method, and the system of equations is solved by using a stabilized biconjugate gradient method. The calculated temperature distribution replaces the initialization values provided by the 1D laser simulator. Once the electro-optical solution process has reached, the output facet and the output power P_{out} has been calculated, the excess power P_{exc} is calculated using Equation 22.8, and the excess power heat source is distributed along the complete device volume. Then, a 3D thermal solver is applied. The 3D thermal solver is based on scanning the 2D solver along the longitudinal direction allowing the longitudinal heat flow between slices until reaching convergence. Successive iterations of the complete algorithm use this 3D thermal solver as indicated in Figure 22.2.

22.4 Simulation Example: 1.5-μm InGaAsP/InP Tapered Amplifier

As an illustrative example, in this section we analyze a typical 1.5-μm InGaAsP/InP tapered amplifier with the HAROLD 4.0 simulation tool. After a summary of the most relevant material, and geometrical and simulation parameters, we present the simulation results of a reference device and based on these results, we illustrate the potential of the simulation tools by analyzing the effect on the device performance of two important design parameters: the taper angle α and the confinement factor Γ. The analysis is presented here only for the sake of illustration, so we will present an overview of some interesting effects without a detailed analysis.

22.4.1 Device Geometry and Simulation Parameters

As a reference device, we have considered an SOA with an InGaAsP/InP multi-QW epitaxial structure similar to that of the MOPA reported in Faugeron et al. (2015). It features an asymmetric cladding structure especially conceived for shifting the broad vertical profile of the optical mode to the n-side of the diode, thus minimizing the overlapping with the highly absorbing p-doped layers. Table 22.1 provides the geometrical parameters of the device as well as a brief summary of the most influential material and device parameters used in the simulation. The taper angle has been selected so as to fit the calculated free diffraction angle assuming an index step $\Delta n_{eff} = 7 \times 10^{-3}$. We have considered a p-side down mounting configuration as more appropriate for the heat management required in a power device. For the Auger recombination

TABLE 22.1 Geometrical Parameters of the Reference SOA and Summary of the Most Relevant Material and Device Parameters Used in the Simulation

Symbol	Parameter	Value	Units
L_{RW}	Length of the ridge waveguide (RW) section	1	mm
W_{RW}	Width of the RW section	3.5	μm
L_{Tap}	Length of the taper section	2	mm
α_{tap}	Taper angle	7	°
Δn_{eff}	Effective index step of the RW section	7×10^{-3}	–
	Mounting configuration (p-up/p-down)	p-Down	–
T_{HS}	Heatsink temperature	18	°C
P_{in}	Input power	20	mW
Γ	Confinement factor	0.037	
α_{scat}	Scattering losses coefficient	0.5	cm^{-1}
$C_n(C_p)$	Electron (hole) Auger recombination coefficient	$8(8) \times 10^{-29}$	cm^{6}s^{-1}
k_e (k_h)	Electron (hole) free carrier absorption coefficient	$1(30) \times 10^{-18}$	cm^2
n_I	Differential refractive index coefficient	10^{-1}	cm$^{3/2}$

coefficients and the electron and hole free carrier absorption coefficients, we have used judicious values in the range of published values (Joindot and Beylat, 1993; Piprek et al., 2000). We have accounted for the dependence of the refractive index on the carrier density by means of a square root function with a coefficient n_1 (Borruel et al., 2004) such that it results in a value of about 5 for the LEF at the operating wavelength and carrier densities. The value used for the internal scattering losses together with the free-carrier absorption losses calculated using Equation 22.9 yield a total internal loss of approximately 8 cm^{-1} for a broad area laser with the same epitaxial structure.

22.4.2 Analysis of the Reference SOA

Figure 22.3 shows the simulated power–current (P–I) characteristics of the reference SOA together with the evolution of the electrical-to-optical conversion efficiency ($\eta_{E\text{-}O}$). For injection currents above a transparency level ($I \sim 1.3$ A) the device actually amplifies with an increasing conversion efficiency. The output power shows an approximately linear increase in the current range $4 < I < 9.5$ A. For higher injection levels, both the output power and the conversion efficiency start a saturation process.

The simulation provides the means to make a more in-depth analysis revealing details of the behavior of relevant variables inside the SOA. As an example, Figure 22.4 shows 3D plots of several magnitudes in the QW region for a relatively low injection current ($I = 4$ A). In the RW section, due to the index guiding, the profile of the photon density (Figure 22.4a) is narrow and stable, and shows an increase of the maximum along the longitudinal direction revealing the optical gain. The maximum drops and the profile broadens when the tapered region starts, and from there up to the output facet, the photon profile evolves smoothly. In this evolution, the maximum photon density increases and so does the profile width, following the evolution of the width of the tapered section. For this relatively low injection level, the electron density profile (Figure 22.4b) is almost flat inside the tapered section. In the RW section, the electron density profile is broader than the width of the section due to the current spreading (not shown). The maximum electron density in this section is lower than the density in the tapered section and decreases slightly along the longitudinal direction. Both effects are a consequence of a higher stimulated recombination ratio in the RW section. The hole density (not shown) is slightly different but shows a similar profile. The mutual dependence between carrier and photon densities is much more pronounced at high-injection levels as explained below. The temperature profile in the lateral direction (Figure 22.4c) reminds the injection profile showing higher values in the injected region and a fast variation at the edge of the region. In the RW section, the lateral profile shows a small dip at the cavity center, whereas in the tapered section, the temperature decreases smoothly from a maximum at the axis to a minimum at the edges of the injected region. In the noninjected regions, the lateral evolution is smooth and shows a slow decrease toward the edges of the chip. In the longitudinal direction, the temperature increases smoothly from the RW section to the output facet. This increase is slow in the RW section and faster in the taper section (see also Figure 22.5b). At

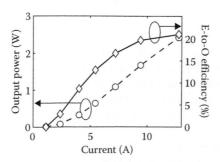

FIGURE 22.3 Power–current characteristics (circles) and electrical-to-optical conversion efficiency (diamonds) of the reference SOA.

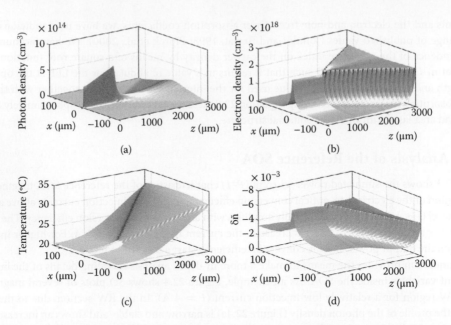

FIGURE 22.4 3D maps of the photon (a) and the averaged electron densities (b), the temperature (c), and the carrier and thermally induced refractive index change (d) in the active region of the reference SOA for an injection $I = 4$ A. The little "ripples" appearing at the edges of the tapered section in some plots are artifacts due to the discretization in the longitudinal direction.

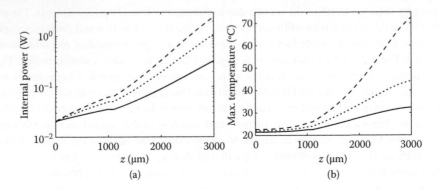

FIGURE 22.5 Evolution along the longitudinal direction of the internal optical power (a) and the maximum QW temperature (b), for different current values. $I = 4$ A (solid), $I = 7.2$ A (dotted), and $I = 12.7$ A (dashed).

this relatively low-injection level, the evolution of the longitudinal temperature profiles are similar in the injected and in the noninjected regions. Finally, Figure 22.4d shows a 3D map of the QW refractive index variation given by Equation 22.9. This variation is a balance between the positive thermal contribution and the negative contribution of the excess carrier density. Therefore the profile is similar to the electron density profile (Figure 22.4b) smoothly modified by the temperature profile (Figure 22.4c). This index variation is the way by which the density of carriers and the local temperature influence the photon density and, hence, the beam profile.

More insight into the evolution of the optical power along the SOA can be provided by plots such as the one in Figure 22.5a, where the internal optical power at each x–y slice is plotted along the longitudinal direction in a logarithmic scale. The evolution of the internal optical power can be described in terms of an effective optical gain g_{eff} by $P(z + \Delta z) = P(z)e^{g_{\text{eff}}\Delta z}$ and therefore g_{eff} is proportional to the slope of the curves in Figure 22.5a. In this description, g_{eff} corresponds to the difference between the modal gain and all the losses occurring in the beam propagation. For the low-injection level considered previously (4 A, solid line), the effective gain is relatively low and slightly decreasing in the RW section and much higher in the tapered section. This can be explained from the maps of photon and carrier densities in Figure 22.4a and b: for approximately the same carrier density in both sections, the photon density is much higher in the RW section and therefore the gain shows an increasing saturation effect which is absent in the tapered section where the photon density is much lower. In a transition region, just at the beginning of the tapered section, there is a dramatic drop of the gain corresponding to an increase of the losses due to the change in the guiding mechanism from index guiding to gain guiding. This drop is due to the poor overlapping of the mode and the injected region at the beginning of the tapered section since the external part of the mode expands to unpumped and, therefore, absorbing regions. In this device, in which the taper angle (and therefore the angle of the gain region) almost matches the free diffraction angle, the gain recovers soon after the transition. For higher-injection levels, Figure 22.5a shows a qualitatively similar behavior of the gain. In both sections, the gain is higher as corresponding to a higher carrier density and evidences some saturation at the end of each section corresponding to a high photon density in these regions.

To complete the overview picture of the longitudinal evolution of internal variables, Figure 22.5b shows the evolution of the maximum QW temperature for the same low-, intermediate-, and high-injection levels of Figure 22.5a. As expected, the QW temperature evolves similarly for the three injection levels, showing a faster increase in the taper section for the highest injection level.

More local details of the behavior of the photon and carrier densities when the injection increases are provided by Figure 22.6. Figure 22.6a shows the electron and photon density profiles along the lateral direction at the output facet of the device. The mutual dependence between carrier and photons pointed above for low injection is much more apparent at high-injection levels and can be explained as follows: The density of carriers injected at these levels is high on average. In comparison with the outer regions, the number of photons generated by stimulated recombination around the axis ($x = 0$) (where the optical mode reaches the maximum) is higher, and therefore the carrier density locally decreases. This is the well-known spatial hole burning (SHB) effect. Due to its negative dependence on the carrier density, the refractive index locally increases in the places more affected by the SHB effect, giving rise to a more abrupt index profile. In turn, the new index profile contributes to a higher confinement of the optical mode in the axial region in a feedback loop in which the local temperature also plays a role through the thermal dependence of the refractive index. As a consequence, the smooth profile of the optical beam at low injection undergoes a slight degradation when the injection increases as illustrated in Figure 22.6b, where the FF profiles at the three injection levels have been compared.

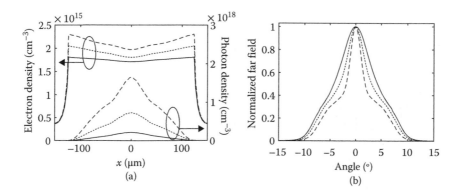

FIGURE 22.6 (a) Photon and electron density lateral profiles at the output facet for three injection levels. (b) The corresponding far field profiles. $I = 4$ A (solid), $I = 7.2$ A (dotted), and $I = 12.7$ A (dashed).

22.4.3 Effect of the Confinement Factor and Taper Angle

The confinement factor and the taper angle are two critical design parameters affecting the performance of a tapered SOA. In order to illustrate the capabilities of the simulation tools for the analysis of the influence of these parameters, we have compared the reference SOA ($\Gamma = 3.7\%$ and $\alpha = 7°$) with two devices: (1) an SOA which epitaxial design has been carefully modified so as to result in a confinement factor $\Gamma = 5.6\%$ without increasing the mode overlapping with the p-doped cladding (high Γ-SOA) and (2) a device with $\alpha = 3°$ and the same RW and tapered section lengths as the reference SOA (3°-SOA). Figure 22.7 shows a comparison of the P–I characteristics, and the conversion efficiency of the three devices. In comparison with the reference SOA, the high Γ-SOA has a lower transparency current, a higher slope of the P–I characteristics, and a higher conversion efficiency. Therefore, at the highest current its output power and conversion efficiency are significantly higher. This can be qualitatively understood just taking into account the higher mode overlapping with the active region. The 3°-SOA shows also a lower transparency current, a higher slope of the P–I characteristics, and a higher conversion efficiency. In this case, the explanation is the higher current density corresponding to the same current as a consequence of the lower area of the device.

Figure 22.8 shows the effective optical gain in the three devices at the highest injection level of around 13 A. Note that the output power of each device is different (see Figure 22.7a). In the RW section, the effective gain of the high Γ-SOA is only slightly higher than the effective gain of the reference SOA in spite of the higher value of Γ. The reason for this is the material gain saturation due to the high photon density in this section that makes the gain almost independent of Γ. The effective gain of the 3°-SOA in this RW section reaches higher values corresponding to the higher carrier injection. In the tapered section, the gain of the high Γ-SOA is clearly superior to that of the reference SOA as can be expected from its higher value of Γ. The difference is more apparent at the beginning of the section, where the photon densities are low, than at the end, where the photon densities (and hence the saturation) become high. The most revealing difference between the gain of the 3°-SOA and the other devices is in the transition region at the beginning of the taper section where the mismatch between the free diffraction angle and the taper angle in the 3°-SOA results in high losses evidenced as a low effective gain in this region.

The improved performance of the high Γ-SOA and the 3°-SOA in terms of output power and conversion efficiency is, however, at the expense of a much faster beam degradation as revealed by the evolution of the FF width and the beam propagation factor (M^2) plotted in Figure 22.9 (see [Esquivias et al., 2017] Chapter 28, Volume 2 of this book). In the case of the high Γ-SOA, the beam quality as revealed by M^2 progressively worsens, reaching high values at relatively low currents. In the case of the 3°-SOA, the degradation takes place at intermediate values of the current whereas the reference SOA keeps a relatively low value of M^2 up to a high-injection level.

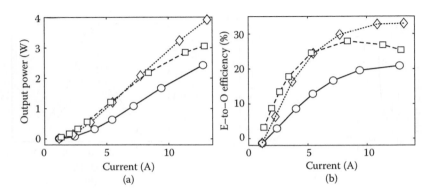

FIGURE 22.7 (a) P–I characteristics of the reference SOA (circles), the high Γ-SOA (diamonds), and the 3°-SOA (squares). (b) Their corresponding electrical-to-optical conversion efficiencies.

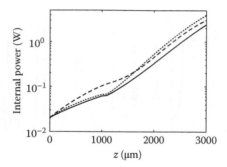

FIGURE 22.8 Evolution along the longitudinal direction of the internal optical power of the three devices at the highest injection ($I = 13$ A), showing the different effective gain (slope of the curves). Reference SOA (solid), high Γ-SOA (dotted), and 3°-SOA (dashed).

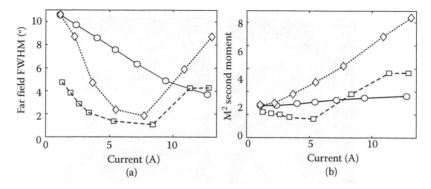

FIGURE 22.9 (a) Far field full width at half-maximum (FWHM) and (b) M^2 dependence on current for the three devices considered. Reference SOA (circles), high Γ-SOA (diamonds), and 3°-SOA (squares).

As mentioned above, more detailed information of the mechanism underlying beam degradation is brought by the plots in Figure 22.10a where the electron and photon density profiles at the output facet of the three devices are plotted for the highest injection level. The mutual dependence between carriers and photons described in Section 22.4.2 for the reference SOA is much more apparent for the high Γ- and the 3°-SOAs and results in deep holes in the axial region of the electron density profiles together with the corresponding narrow peaks in the photon density. As a consequence, the beam quality severely degrades, as revealed, for example, by the FF profiles plotted in Figure 22.10c and d in comparison with the profile of the reference SOA (Figure 22.10b).

The self-focusing is another manifestation of the beam degradation due to the interplay between carriers and photons. Figure 22.11 shows a comparison between the evolution of the beam profiles in the reference (Figure 22.11a and b) and the 3°-SOA (Figure 22.11c and d) for low- (Figure 22.11a and c) and high injection conditions (Figure 22.11b and d). In these plots, the photon density in each slice perpendicular to the longitudinal axis has been normalized to its maximum value in order to visualize the focusing. The white dashed lines show the borders of the injected region. In contrast with the virtually absent focusing of the beam in the reference SOA, Figure 22.11d shows a strong self-focusing in the 3°-SOA at high current. As a consequence, the slightly increasing astigmatism in the reference SOA when the injection increases (not shown) becomes abrupt and then drops down to negative values in the 3°-SOA.

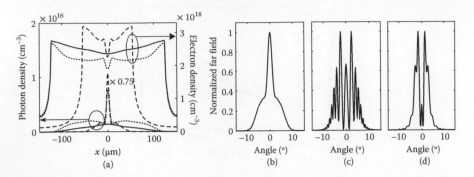

FIGURE 22.10 (a) Photon and electron density lateral profiles at the output facet at the highest injection level ($I = 13A$) for the three devices considered. Reference SOA (solid), high Γ-SOA (dotted), and 3°-SOA (dashed). For clarity, the photon density values of the 3°-SOA device (dashed) have been multiplied by 0.75. (b) through (d) Normalized far field profiles for the reference SOA, the high Γ-SOA, and the 3°-SOA, respectively.

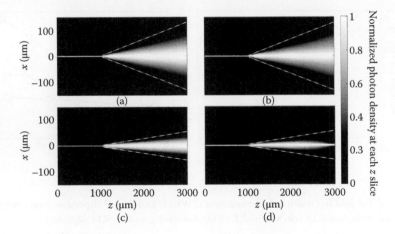

FIGURE 22.11 Maps of the normalized photon density of the reference SOA (top) and the 3°-SOA (bottom) under low-injection (left) and high-injection conditions (right). The white dashed lines show the borders of the injected region. For each value of the longitudinal coordinate z, the photon density was normalized to its maximum along the corresponding lateral x-axis.

22.5 Summary

We have presented an overview of the current state of the art in the modeling of high-power tapered SOAs. We have described the most relevant models used for the simulation of carrier transport, optical field, material properties, and temperature distribution in this type of devices. The particular geometry of tapered amplifiers makes difficult the reduction of the dimensionality of the problem, requiring relatively complex numerical models. The implementation of a steady-state quasi-3D model has been explained in detail, with emphasis on the interaction between the electrical, thermal, and optical solvers. For illustration, the model has been applied to a 1.5-μm tapered SOA, showing the relation between internal carrier and photon distributions and the external measurable characteristics. The influence of the optical confinement factor and the taper angle in the output power and beam quality has also been analyzed.

The main limitation of the steady-state single-frequency models is that they cannot be used to analyze the noise properties of the SOAs, as well as other complex nonlinear phenomena as cross-talk in wavelength division multiplexing applications. Two- or quasi-3D TWMs, coupled to bipolar carrier and thermal equations are the natural evolution of steady-state models to account for dynamical and spectral issues.

Acknowledgments

This work was supported by the European Commission through the FP7-Space project BRITESPACE under grant agreement no. 313200, by Ministerio de Economía y Competitividad of Spain under projects RANGER (TEC2012-38864-C03-02) and COMBINA (TEC2015-65212-C3-2-P), and by the Comunidad de Madrid under program SINFOTON-CM (S2013/MIT-2790). A. Pérez-Serrano acknowledges support from Ayudas a la Formación Posdoctoral 2013 program (FPDI-2013-15740). The authors acknowledge the contribution of L. Borruel, H. Odriozola, S. Sujecki, and E. C. Larkins to the development of the initial model and code for the simulation of tapered lasers.

References

Balsamo, S., Sartori, F., and Montrosset, I. 1996. Dynamic beam propagation method for flared semiconductor power amplifiers. *IEEE J. Quantum Electron.* 2(2):378–384.

Bandelow, U., Gajewski, H., and Hünlich, R. 2005. Fabry–Perot lasers: Thermodynamics-based modeling. In *Optoelectronic Devices. Advanced Simulation and Analysis*. Ed. J. Piprek. 63–85. New York, NY: Springer.

Borruel, L., Arias, J., Romero, B. et al. 2003. Incorporation of carrier capture and escape processes into a self-consistent cw model for Quantum Well lasers. *Microelectron. J.* 34:675–677.

Borruel, L., Odriozola, H., Tijero, J. M. G. et al. 2008. Design strategies to increase the brightness of gain guided tapered lasers. *Opt. Quantum Electron.* 40(2–4):175–189.

Borruel, L., Sujecki, S., Esquivias, I. et al. 2002. Self-consistent electrical, thermal, and optical model of high-brightness tapered lasers. *Proc. SPIE* 4646:355–366.

Borruel, L., Sujecki, S., Moreno, P. et al. 2004. Quasi-3-D simulation of high-brightness tapered lasers. *IEEE J. Quantum Electron.* 40:463–472.

Buus, J. 1982. The effective index method and its application to semiconductor lasers. *IEEE J. Quantum Electron.* 18(7):1083–1089.

Coldren, L. A. and Corzine, S. W. 1995. *Diode Lasers and Photonic Integrated Circuits*, New York, NY: John Wiley & Sons.

Connelly, M. J. 2001. Wideband semiconductor optical amplifier steady-state numerical model. *IEEE J. Quantum Electron.* 37(3):439–447.

Esquivias, I., Pérez-Serrano, A., and Tijero, J. M. G. 2017. High brightness tapered lasers. In: *Handbook of Optoelectronic Device Modeling and Simulation*, vol 2, (Piprek, J., ed) pp.59–80. Boca Raton, FL: Taylor & Francis.

Faugeron, M., Vilera, M., Krakowski, M. et al. 2015. High power three-section integrated master oscillator power amplifier at 1.5 μm. *IEEE Photon. Technol. Lett.* 27:1449–1452.

Fischer, I., Hess, O., Elsäßer, W. et al. 1996. Complex spatio-temporal dynamics in the near-field of a broad-area semiconductor laser. *Europhys. Lett.* 35(8):579–584.

Gehrig, E., and Hess, O. 2001. Spatio-temporal dynamics of light amplification and amplified spontaneous emission in high-power tapered semiconductor laser amplifiers. *IEEE J. Quantum Electron.* 37(10):1345–1355.

Grote, B., Heller, E. K., Scarmozzino, R. et al. 2005. Fabry–Perot lasers: Temperature and many-body effects. In *Optoelectronic Devices. Advanced Simulation and Analysis*. Ed. J. Piprek. 63–85. New York, NY: Springer.

Hadley, G. R. 1992. Wide-angle beam propagation using Padé approximant operators. *Opt. Lett.* 17(20):1426–1428.

Harold™ 2001. *3.0 Reference Manual*. Oxford: Photon Design.

Huang, W. P., Xu, C. L., Lui, W. et al. 1996. The perfectly matched layer (PML) boundary condition for the beam propagation method. *IEEE Photon. Technol. Lett.* 8(5):649–651.

Javaloyes, J. and Balle, S. 2010. Quasiequilibrium time-domain susceptibility of semiconductor quantum wells. *Phys. Rev. A* 81:062505.

Joindot, I. and Beylat, J. L. 1993. Intervalence band absorption coefficient measurements in bulk layer, strained and unstrained multiquantum well 1.55 μm semiconductor lasers. *Electron. Lett.* 29: 604–606.

Koch, S. W., Hader, J., Thränhardt, A. et al. 2005. Gain and absorption: Many-body effects. In *Optoelectronic Devices. Advanced Simulation and Analysis*. Ed. J. Piprek. 63–85. New York, NY: Springer.

Lai, J.-W. and Lin, C.-F. 1998. Carrier diffusion effect in tapered semiconductor-laser amplifier. *IEEE J. Quantum Electron.* 34(7):1247–1256.

Lang, R. J., Hardy, A., Parke, R. et al. 1993. Numerical analysis of flared semiconductor laser amplifiers. *IEEE J. Quantum Electron.* 29:2044–2051.

Lugiato, L. A. and Narducci, L. M. 1985. Single-mode and multimode instabilities in lasers and related optical systems. *Phys. Rev. A* 32(3):1576–1587.

Mariojouls, S., Margott, S., Schmitt, A. et al. 2000. Modeling of the performance of high-brightness tapered lasers. *Proc. SPIE* 3944:395–406.

Moloney, J. V., Indik, R. A., and Ning, C. Z. 1997. Full space-time simulation of high-brightness semiconductor lasers. *IEEE Photon. Technol. Lett.* 9:731–733.

Nagarajan, R., Ishikawa, M., Fukushima, T. et al. 1992. High speed quantum-well lasers and carrier transport effects. *IEEE J. Quantum Electron.* 28:1990–2008.

O'Brien, S., Welch, D. F., Parke, R. A. et al. 1993. Operating characteristics of a high-power monolithically integrated flared amplifier master-oscillator power-amplifier. *IEEE J. Quantum Electron.* 29:2052–2057.

Pérez-Serrano, A., Javaloyes, J., and Balle, S. 2013. Spectral delay algebraic equation approach to broad area laser diodes. *IEEE J. Sel. Top. Quant. Electron.* 19(5):1502808.

Piprek, J., Abraham, P., and Bowers, J. E. 2000. Self-consistent analysis of high-temperature effects on strained-layer multiquantum-well InGaAsP–InP lasers. *IEEE J. Quantum Electron.* 36:366–374.

Razaghi, M., Ahmadi, A., and Connelly, M. J. 2009. Comprehensive finite-difference time-dependent beam propagation model of counter propagating picosecond pulses in a semiconductor optical amplifier. *IEEE /OSA J. Lightwave Technol.* 27(15):3162-3174.

Schwertfeger, S., Klehr, A., Hoffmann, T. et al. 2011. Picosecond pulses with 50 W peak power and reduced ASE background from an all-semiconductor MOPA system. *Appl. Phys. B* 103:603–607.

Selberherr, S. 1984. *Analysis and Simulation of Semiconductor Devices*, Vienna: Springer-Verlag.

Spreemann, M., Lichtner, M., Radziunas, M. et al. 2009. Measurement and simulation of distributed-feedback tapered master-oscillator power-amplifiers. *IEEE J. Quant. Electron.* 45:609–616.

Sujecki, S., Borruel, L., Wykes, J. et al. 2003. Nonlinear properties of tapered laser cavities. *IEEE J. Sel. Top. Quant. Electron.* 9:823–834.

Sumpf, B., Hasler, K.-H., Adamiec, P. et al. 2009. High-brightness quantum well tapered lasers. *IEEE J. Sel. Top. Quant. Electron.* 15(3):1009–1020.

Tijero, J. M. G., Borruel, L., Vilera, M. et al. 2015. Analysis of the performance of tapered semiconductor optical amplifiers: Role of the taper angle. *Opt. Quant. Electron.* 47(6):1437–1442.

Van Roey, J., van der Donk, J., and Lagasse, P. E. 1981. Beam-propagation method: Analysis and assessment. *J. Opt. Soc. Am.* 71:803–810.

Vilera, M., Pérez-Serrano, A., Tijero, J. M. G. et al. 2015. Emission characteristics of a 1.5 μm all semiconductor tapered master oscillator power amplifier. *IEEE Photon. J.* 7(2):1500709.

Wachutka, G. K. 1990. Rigorous thermodynamic treatment of heat generation and conduction in semiconductor device modeling. *IEEE Trans. Comput. Aided Des. Integr. Circuits Syst.* 9:1141–1149.

Walpole, J. N. 1996. Semiconductor amplifiers and lasers with tapered gain regions. *Opt. Quant. Electron.* 28:623–645.

Wenzel, H., Sumpf, B., and Erbert, G. 2003. High-brightness diode lasers. *C. R. Phys.* 4:649–661.

Williams, K. A., Penty, R. V., White, I. H. et al. 1999. Design of high-brightness tapered laser arrays. *IEEE J. Sel. Top. Quant. Electron.* 5:822–831.

23

Quantum-Dot Semiconductor Optical Amplifiers

Benjamin Lingnau

and

Kathy Lüdge

23.1 Introduction

The rapidly increasing need for telecommunications and data-streaming applications within our society still demands a deeper understanding of the physical processes behind this emerging technology. In the last decades, there have been many attempts to optimize the performance of optoelectronic devices by using innovative nanostructured semiconductor gain materials [1,2]. A lot of these innovative devices already have found their way into photonic applications [3,4]. In particular, the demand for faster and more energy-efficient data transfer set the trend to replace the well-established cable-based data transmission with energy efficient optical technologies [5]. As one example, optical amplifiers are needed in optical networks to raise the signal power level and compensate for the inevitable optical losses in glass–fiber connections. Semiconductor-based optical amplifiers (SOAs) are cheap, relatively easy to fabricate, and consume only little energy within an electric circuit. They are structurally similar to semiconductor laser devices, with the difference lying in the absence of an optical cavity (see Chapter 20 SOA Fundamentals).

In this chapter, we describe the modeling of optical amplifiers that contain semiconductor quantum dots (QDs) as active media. Those nearly zero-dimensional QD structures gained a lot of attention as the material of choice for highly energy-efficient and small-footprint optoelectronic devices [6]. The scattering mechanisms and the unique electronic structure of semiconductor QDs have been found to make such devices prime candidates for the implementation of next-generation optoelectronic applications and novel high-speed data transmission schemes. QDs are the final step in miniaturization of the optically active gain material and thus in the confinement of charge carriers below the de Broglie wavelength of electrons (a few nanometers). In addition to the modeling of the quantum-dot semiconductor optical amplifiers (QDSOAs) as active media described within this chapter, other QD-based optical devices such as QD laser and QD light-emitting diodes are treated separately in a different section of this book. Nevertheless, our micro-scopically based material model for the QDs can equally be used as the building block in laser devices [7,8].

Our main focus will be the characterization of both the performance and the special nonlinear features of QDSOAs that are induced by the low dimensional and thus highly confined QD gain medium. While linear amplification with a low noise figure is required, to ensure low distortion of the input signal for data transmission [9–11], nonlinear optical applications, such as four-wave-mixing [12–15] and cross-gain modulation [16–18] for wavelength conversion, as well as regenerative amplification, require a nonlinear response of the optical amplifier. QDSOAs generally have a high gain bandwidth due to the growth-induced inhomogeneity in the dot size and material composition. These fluctuations lead to an inhomogeneous broadening of the localized QDs states and consequently allow for a broadband amplification. Addition-ally, due to the coupling to a charge-carrier reservoir with charge-carrier scattering times in the picosecond range, ultrafast gain recovery [19–24] and nonlinear signal processing are possible [15,16,25]. The com-parably slow dephasing time of the microscopic interband polarization in the localized QD states [26–28] allows the possibility to directly observe quantum-mechanical effects, such as Rabi oscillations [29–33] or self-induced transparency [34,35]. This could potentially open up new applications in the signal processing of ultra-short, ultra-strong optical pulses.

The optical amplifier devices we consider in this work are single-pass devices, where the optical signal coupled into one side of the waveguide structure is ideally passing exactly once across the device. Thus, the edge-emitting QDSOAs are longer than typical laser devices, in order to provide a long enough interaction time between optical signal and the active medium. The model used for the numeric simulations is based on a traveling-wave approach for the pulse propagation, using Maxwell's wave equation in the slowly vary-ing amplitude and rotating wave approximation. The polarization of the active medium is modeled using microscopic considerations for the charge-carrier dynamics, i.e., the light–matter interaction is modeled on the basis of Maxwell–Bloch equations, taking into account the microscopic polarization of the QD medium.

The structure of the chapter is as follows. First, a delay-differential equation model for an efficient and accurate description of the electric field propagation through the QDSOA is presented (Section 23.2). Subsequently, the charge-carrier dynamics is described in the framework of microscopically calculated nonlinear charge-carrier scattering rates before our approach for a quantitative modeling of the amplified spontaneous emission (ASE) inside the QDSOA is discussed. In Section 23.3, we present exemplary results on the performance of the QDSOA, focusing on the unique properties of semiconductor QDs as active medium. After a static characterization of a chosen sample device, we discuss the ultrafast gain dynamics of QDSOAs, identifying the main timescales that determine its gain recovery. We then present results on the coherent propagation of strong optical pulses. The long dephasing time of QDs leads to strong signa-tures of the coherent interaction, leading to modifications in the pulse shape and the gain dynamics due to the appearance of Rabi oscillations. A brief conclusion is provided.

23.2 QD Semiconductor Optical Amplifier Model

QDSOAs differ from conventional devices in the choice of active medium. The charge-carrier dynamics in QDs can be very complex due to the localization of electrons within the QDs embedded in the sur-rounding quantum-well or bulk material. In this chapter, we, therefore, aim to describe the charge-carrier dynamics and derive a set of coupled differential equations for the electronic states involved. The strong nonlinearity of the resulting equations leads to important differences in the performance of QDSOAs when compared to conventional devices. It also opens up the possibility of novel applications, based on their ultrafast gain recovery for linear applications, or their broad gain spectrum for nonlinear or wavelength conversion applications.

In this section, we derive a QDSOA model, in which we focus on the unique dynamics of the QD gain medium. The wave equation used for the propagation of the light will be similar to conventional SOA models (see Chapter 20 on SOA fundamentals in this book). In other works, models without spatial resolution have been employed for the description of semiconductor amplifiers [17,36–38]. As soon as strong spatial inhomogeneities arise, such models are, however, bound to fail. On the other hand, numerically solving the partial differential equation for the electric field propagation along the waveguide axis is computationally very expensive[39]. We, therefore, develop a delay-differential equation model that combines spatial resolution and reasonable computation time. The light–matter interaction will be described within a Maxwell–Bloch framework, which couples the light propagation dynamics to the QD active medium.

In the following, we derive the electric field propagation equations and the material equations describing the active medium. Furthermore, we derive Boltzmann-like equations that govern the charge-carrier scattering dynamics within a quantum-dot-in-a-well system. Spontaneous emission will be included in a phenomenological way to yield stochastic differential equations.

23.2.1 Electric Field Propagation: Delay-Differential-Equation Model

In order to calculate the dynamics of the QDSOA device, we must solve the wave equation of the optical field inside the waveguide. Here we concentrate on narrow-area ridge waveguide structures where the propagation along the longitudinal axis (labeled z) dominates, and the transverse mode profile is assumed to be constant. The electric field is thus governed by the one-dimensional wave equation

$$\frac{\partial^2}{\partial t^2}\mathcal{E}(\mathbf{r}, t) - c_0{}^2 \frac{\partial^2}{\partial z^2}\mathcal{E}(\mathbf{r}, t) = -\frac{1}{\varepsilon_0}\frac{\partial^2}{\partial t^2}\mathbf{P}(\mathbf{r}, t), \qquad (23.1)$$

where $\mathcal{E}$ and $\mathbf{P}$ denote the real electric field and polarization, respectively, and c_0 and ε_0 are the vacuum speed of light and dielectric constant, respectively. The field quantities are expanded in terms of plane waves,

$$\mathcal{E}(z, t) = \frac{1}{2}\left[E^+(z, t)e^{ikz} + E^-(z, t)e^{-ikz} \right] e^{-i\omega t} + \text{c.c.}, \qquad (23.2)$$

where we have introduced the slowly varying field amplitudes $E^\pm$, describing the forward (+) and backward (−) propagating electric field. The wave number is given by k, and ω is the optical frequency of the reference frame. From Equation 23.1, we can derive the propagation equations for the electric field amplitudes within the slowly varying envelope approximation (SVEA), neglecting all but the lowest order of derivatives. The SVEA limits the description to field envelopes that change on time scales and lengths much larger than the respective optical period and wavelength, respectively. The propagation equation for the slowly varying field envelope is then written as

$$\left(\frac{\partial}{\partial t} \pm v_g \frac{\partial}{\partial z}\right) E_\pm(z, t) = \frac{i\omega\Gamma}{2\varepsilon_{bg}\varepsilon_0} P_\pm(z, t) =: S_\pm(z, t), \qquad (23.3)$$

with the group velocity $v_g \equiv \frac{c_0}{\sqrt{\varepsilon_{bg}}}$, with the background permittivity ε_{bg}. $P_\pm(z, t)$ is the macroscopic slowly varying polarization amplitude. The transverse optical confinement factor Γ is introduced phenomenologically to the above equation. It describes the overlap of the electric field mode with the active medium, integrated over the transverse (x and y) coordinates. We summarize the right-hand side in a general source term $S_\pm(z, t)$.

The numerical solution of the above partial differential equation using a finite-difference method requires a spatial discretization into very fine sections of length $v_g dt$, where dt is the numerical time-step, in order to ensure numerical stability [41]. For common device lengths on the order of 1 mm, a high number of spatial discretization points is needed. The numerical integration, therefore, becomes very expensive in terms of computation time and memory requirements [42,43]. A more elegant approach is the formulation of the problem as a delay-differential equation system [40,44], which we do in the following.

The partial differential equation Equation 23.3 can be formally solved by integrating along its characteristic lines, given by

$$z = \pm v_g t + \text{const.} \tag{23.4}$$

Expanding the total derivative $\frac{d}{dt} = \frac{\partial z}{\partial t}\frac{\partial}{\partial z} + \frac{\partial}{\partial t}$, we can thus write

$$\frac{d}{dt}E_\pm(z,t) = \left[\pm v_g\frac{\partial}{\partial z} + \frac{\partial}{\partial t}\right]E_\pm(z,t) = S_\pm(z,t). \tag{23.5}$$

Now we describe the optical amplifier of length ℓ by a number of Z sections along the propagation axis, such that the distance between two discretization points is given by $\Delta z := \ell/Z$. Integrating Equation 23.5 over the time interval $\Delta t = \frac{\Delta z}{v_g}$ thus yields

$$E_\pm(z,t) = E_\pm(z \mp \Delta z, t - \Delta t) + \int_{-\Delta t}^{0} \left[\frac{d}{dt}E_\pm\left(z \pm v_g\tau, t + \tau\right)\right]d\tau'$$

$$= E_\pm(z \mp \Delta z, t - \Delta t) + \int_{-\Delta t}^{0} S_\pm\left(z \pm v_g\tau, t + \tau\right)d\tau$$

$$\approx E_\pm(z \mp \Delta z, t - \Delta t) + \frac{\Delta t}{2}\left[S_\pm(z,t) + S_\pm(z \mp \Delta z, t - \Delta t)\right]. \tag{23.6}$$

The integral over the source term was approximated by its values at the end points of the integration interval. This approximation is valid for negligible change of $S_\pm$ along the integration path, i.e., for a sufficiently small space discretization step. The electric field at time t now depends on the values of $E_\pm$, $S_\pm$ at time $t - \Delta t$, which introduces a time delay into the equations.

The electric field in each of the spatial sections along the amplifier device thus couples to the time-delayed electric field in the neighboring sections, with the time Δt describing the time needed for the electric field propagation along the length of one section [40,44]. The resulting discretization scheme is illustrated in Figure 23.1. The advantage of this approach is the decoupling of the time and space discretization steps, ensuring numerical stability even for $\Delta z \gg v_g dt$. The temporal dynamics at each point within the amplifier device is usually much faster than the characteristic timescale over which a propagating pulse changes its shape. Thus, a reduced number of spatial discretization points can be chosen compared to the finite-difference method, which significantly improves the simulation efficiency. The implementation of the delay-differential equations, however, means that the history of the electric fields as well as the source terms must be saved over a time interval Δt.

23.2.2 Quantum-Dot-in-a-Well Material Equations

The amplifier active medium is composed of an ensemble of of semiconductor quantum dots. For InAs quantum dots grown in the Stranski–Krastanov (SK) mode by molecular beam epitaxy (MBE), their areal density is in the order of $2 - 5\times10^{10}$ cm^{-2} [45]. The considered QDs are assumed to have two optically active localized states. These are denoted by GS and ES, for the ground and first excited state, respectively, which are located within the bands of the surrounding quantum well, as sketched in Figure 23.2. The resulting energy structure is sketched in Figure 23.3.

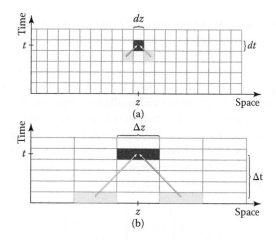

(a)

(b)

FIGURE 23.1 Electric field propagation schemes: (a) Traditional time-domain traveling-wave integration scheme. The device is discretized into spatial sections with width dz, usually related to the numerical time step dt via the electric field group velocity $dz = v_g dt$. The field $E(z, t)$ is then determined from the field in neighboring discretization points at the previous time step $E(z \pm dz, t - dt)$. (b) The discretization into fewer spatial sections separated by Δz leads to a coupling to neighboring points with a time delay Δt. The time evolution in each section is still calculated with a time step dt, separating space and time discretization steps. (After J. Javaloyes and S. Balle, *Opt. Express*, 20, 8496–8502, 2012.)

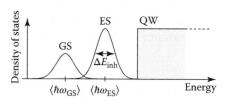

FIGURE 23.2 Sketch of the conduction band density of states of a quantum-dot-in-a-well material system. The QD transitions are assumed to consist of a ground-state (GS) and excited-state (ES) transition, which are inhomogeneously broadened with a full-width-at-half maximum (FWHM) of ΔE_{inh}. The two-dimensional quantum-well (QW) states act as a charge-carrier reservoir for the QD states.

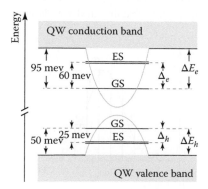

FIGURE 23.3 Energy scheme of the considered dot-in-a-well (DWELL) structure. The QD ground states (GS) lie ΔE_b below the quantum-well (QW) band edge, with an energy spacing of Δ_b between GS and first excited state (ES).

Due to fluctuations in QD size and composition, their emission spectrum is inhomogeneously broadened around the central emission energy of ≈ 950 meV for InAs/InGaAs QDs. Typical inhomogeneous linewidths lie around 30 meV—higher than their homogeneous linewidth [27,46]. An accurate description of the QD medium, therefore, requires a separate treatment of quantum dots in dependence of their respective transition energy. Both the interaction with the optical field as well as the charge-carrier scattering dynamics are influenced by the transition and localization energies. A natural way to deal with the inhomogeneous broadening is, therefore, to distribute the whole QD ensemble into a number of subgroups, characterized by their optical transition energy. This procedure is illustrated in Figure 23.4. We label each subgroup by an index j and its respective mean transition energy $\hbar\omega^j$. We introduce the probability mass function $f(j)$ denoting the fraction of QDs within the jth subgroup. Following a Gaussian distribution with an FWHM of ΔE_{inh}, the distribution function $f(j)$ is then given by

$$f(j) = \frac{1}{\mathcal{N}} \exp\left(-4\ln 2 \left[\frac{\hbar\omega^j - \langle\hbar\omega\rangle}{\Delta E_{\text{inh}}}\right]^2\right), \tag{23.7}$$

with the normalization constant $\mathcal{N}$ chosen such that $\sum_j f(j) \overset{!}{=} 1$, calculated numerically. The inhomogeneous broadening of the optical spectrum is given by the sum of the individual single-particle state broadenings of electrons and holes (subscripts e and h, respectively),

$$\Delta E_{\text{inh}} = \Delta\varepsilon_e + \Delta\varepsilon_h, \tag{23.8}$$

where $\Delta\varepsilon_b$, $b \in \{e, h\}$ is the corresponding electron and hole state broadening. Only the total broadening ΔE_{inh} is experimentally readily accessible, e.g., by measurements of the QD luminescence spectra [47]. For the individual state broadening, we assume widths proportional to the localization energy of the given state:

$$\Delta\varepsilon_b = \Delta E_{\text{inh}} \frac{\Delta E_b}{\Delta E_e + \Delta E_h}. \tag{23.9}$$

We now proceed by deriving the dynamic equation for each of the QD subgroups within each amplifier section. In addition to the QD subgroup index j, we introduce the state index $m \in \{\text{GS}, \text{ES}\}$, distinguishing the charge carriers in the GS and first ES. The material dynamics are described within the Maxwell–Bloch approach [48], characterizing the QD active medium by their occupation probabilities ρ_m^j and the microscopic polarization amplitudes p_m^j. The microscopic polarization is induced by the electric field and describes the transition probability under emission of a photon, and thus couples the

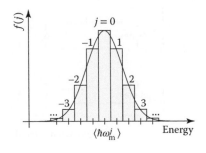

FIGURE 23.4 Illustration of the QD subgroups to model inhomogeneous broadening. The QDs are distributed into j_{max} subgroups, assumed to follow a Gaussian distribution around the mean transition energy $\langle\hbar\omega_m\rangle$. The probability mass function $f(j)$ gives the probability for a QD to be found in the jth subgroup.

charge-carrier equations to the light field. Similar to the electric field, we expand p_m^j in terms of forward and backward propagating amplitudes,

$$p_m^j(z, t) = \left[p_{m,+}^j(z, t)e^{ikz} + p_{m,-}^j(z, t)e^{-ikz} \right] e^{-i\omega t}. \tag{23.10}$$

We assume a coupling of the individual polarization amplitudes $p_{m,\pm}^j$ only to the corresponding co-propagating part of the electric field. This is the rotating wave approximation, neglecting terms oscillating with $e^{\pm 2ikz}$ and $e^{\pm 2i\omega t}$, which would appear in the coupling of counter-propagating polarization and field terms.

The Maxwell–Bloch equations of the QD active medium polarization amplitudes and occupation probabilities then read [49,50]

$$\frac{d}{dt} p_{m,\pm}^j(z, t) = - \left[i(\omega_m^j - \omega) + \frac{1}{\tau_2} \right] p_{m,\pm}^j(z, t) - i\frac{\mu_m}{2\hbar} \left(\rho_{e,m}^j(z, t) + \rho_{h,m}^j(z, t) - 1 \right) E_\pm(z, t), \tag{23.11}$$

$$\frac{d}{dt} \rho_{b,m}^j(z, t) = \frac{1}{\hbar} \text{Im} \left[p_{m,+}^j(z, t)\mu_m^* E_+^*(z, t) \right] + \frac{1}{\hbar} \text{Im} \left[p_{m,-}^j(z, t)\mu_m^* E_-^*(z, t) \right]$$

$$- W_m \rho_{e,m}^j(z, t)\rho_{h,m}^j(z, t) + \frac{\partial}{\partial t} \rho_{b,m}^j(z, t) \Big|_{\text{sc}}. \tag{23.12}$$

Here, we again denote the localized QD states by $m \in \{\text{GS}, \text{ES}\}$ and their subgroup index by j. The above equations describe the micoscopic polarization amplitudes in the reference frame of ω, whereas ω_m^j denotes the transition frequency of the respective optical transition. The polarization decays with the time constant τ_2. This dephasing time describes the loss of the quantum-mechanical coherence between different QDs due to scattering processes with other electrons or phonons. In semiconductors, these processes are usually in the order of a few tens of femtoseconds, due to the high density of charge carriers as potential scattering partners. The polarization is driven by the electric field $E_\pm(z, t)$ and the carrier inversion, with the interaction strength given by the transition dipole moment μ_m.

The QD charge-carrier dynamics is given by stimulated recombination induced by the microscopic polarization, and bimolecular spontaneous recombination with a rate W_m. The term $\frac{\partial}{\partial t} \rho_{b,m}^j(z, t) \Big|_{\text{sc}}$ denotes charge-carrier scattering contributions, which we will address in the following section.

The above equations couple to the macroscopic light field via the macroscopic slowly varying polarization amplitudes $P_\pm(z, t)$. These are given by the macroscopic sum over the individual microscopic contributions:

$$P_\pm(z, t) = \frac{2N^{\text{QD}}}{h^{\text{QW}}} 2 \sum_{j,m} v_m f(j)\mu_m^* p_{m,\pm}^j(z, t), \tag{23.13}$$

where N^{QD} is the areal QD density per quantum-well layer and h^{QW} the layer height. The sum $\sum_{j,m} v_m f(j)$ gives the sum over all QD subgroups and confined states, with v_m the degeneracy (excluding spin) of the mth state and $f(j)$ the distribution function as defined in Equation 23.7. The macroscopic polarization couples back to the electric field equation as a source term, as shown in Equation 23.3:

$$\frac{d}{dt} E_\pm(z, t) = S_\pm(z, t) = \frac{i\omega\Gamma}{2\varepsilon_{\text{bg}}\varepsilon_0} P_\pm(z, t)$$

$$= \frac{2N^{\text{QD}}}{h^{\text{QW}}} \frac{i\omega\Gamma}{\varepsilon_{\text{bg}}\varepsilon_0} \sum_{j,m} v_m f(j)\mu_m^* p_{m,\pm}^j(z, t). \tag{23.14}$$

It is tempting to identify a small-signal gain $g(z, t, \omega)$, as used in traditional rate equation models, from an earlier equation:

$$g(z, t, \omega) = \frac{i\omega\Gamma}{2\varepsilon_{bg}\varepsilon_0} \frac{P_{\pm}(z, t)}{E_{\pm}(z, t)}. \tag{23.15}$$

However, the independent dynamics of the polarization would lead to a strongly varying (and even diverging) value of $g(z, t, \omega)$ in time. Nevertheless, in a static limit, by setting $\frac{\partial}{\partial t}p_m^j = 0$, the adiabatic small-signal gain can be calculated from Equation 23.11, which yields

$$g(z, t, \omega) = \frac{\hbar\omega\Gamma}{\varepsilon_0\varepsilon_{bg}h^{QW}} 2N^{QD} \sum_{j,m} v_m f(j) \frac{\tau_2|\mu_m|^2}{2\hbar^2} \frac{\left[\rho_{e,m}^j(z, t) + \rho_{h,m}^j(z, t) - 1\right]}{1 + [\tau_2(\omega - \omega_m^j)]^2}. \tag{23.16}$$

This expression, however, neglects the dynamics of the polarization, assuming it to follow the incident electric field instantaneously. For coherent interactions, such as Rabi oscillations discussed later in this chapter, the above equation, therefore, cannot be applied. Furthermore, Equation 23.16 is strictly valid only for a monochromatic electric field at a given frequency ω. When simulating broadband amplification, such as optical pulses, the full polarization dynamics must, therefore, be taken into account. For slowly varying and near-monochromatic electric field envelopes, such as encountered in lasers, the adiabatic elimination of the polarization equations is a valid and widely used approximation, and Equation 23.16 can be used to calculate the optical gain.

The Maxwell–Bloch equations only describe the optically active QD states. In order to describe the carrier dynamics in the surrounding charge-carrier reservoir, we model the quantum-well charge-carrier density w_b by a rate equation,

$$\frac{d}{dt}w_b(z, t) = \eta\frac{J}{e_0} - A^S\sqrt{w_e(z, t)w_h(z, t)} - B^S w_e(z, t)w_h(z, t) + \frac{\partial}{\partial t}w_b(z, t)\Big|_{sc}, \tag{23.17}$$

with the pump current density J, the electron charge $-e_0$, and a pump current efficiency η accounting for losses in the surrounding semiconductor layers, as well as imperfect carrier injection into the active region, which is not modeled explicitly. Linear and bimolecular recombination rates in the reservoir are given by A^S and B^S, respectively, which are important to describe carrier losses over a wide range of currents. We neglect Auger recombination, as with typical values ($C = 10^{-28}$ cm$^6 \cdot$ s^{-1}) the losses are found to be dominated by A^S and B^S in the current ranges considered in this chapter. The term $\frac{\partial}{\partial t}w_b(z, t)\Big|_{sc}$ accounts for scattering of charge carriers into the QD states, which is derived in the next section. The above equations are defined for each space discretization point along the amplifier device, allowing for a spatially inhomogeneous distribution of the charge-carrier distribution, as encountered in long amplifier devices [51,52].

23.2.3 QD Charge-Carrier Scattering

QDs differ from conventional quantum-well or bulk gain media in the presence of localized states and thus a strongly modified density of states. Electrically injected charge carriers reach the active region in high-energy states near the bulk material band edges. The optically active QD states must, therefore, be populated by means of scattering processes. While the charge-carrier distributions within the quasi-continuous bulk and quantum-well bands quickly thermalize due to electron–phonon interaction [53,54], an effective scattering of carriers into the QD states by scattering with phonons is limited due to the existence of the "phonon bottleneck"[55]. Instead, at elevated charge-carrier densities, Coulomb-mediated carrier-carrier scattering provides an ultrafast scattering mechanism between the QD states and the surrounding semiconductor material [56]. In the following, we derive expressions for the Coulomb scattering mechanism.

The starting point for calculating the carrier-carrier scattering is the many-body Hamiltonian in second quantization [49,57],

$$
H_{sys} = H_{kin} + H_C = \sum_{\substack{a \\ s}} \varepsilon_a a_{as}^\dagger a_{as} + \frac{1}{2} \sum_{\substack{abcd \\ ss'}} W_{abcd}\, a_{as}^\dagger a_{bs'}^\dagger a_{cs'} a_{ds}\,, \tag{23.18}
$$

where a_x, $a_x^\dagger$ are the electron annihilation and creation operators in the state x with the energy ε_x, respectively. The Hamiltonian consists of the kinetic (free-carrier) contribution H_{kin}, and the Coulomb-interaction Hamiltonian H_C, which includes the many-body interaction between the charge carriers in the semiconductor. In the sums, the labels a, b, c, d denote all possible electronic states, with s, s' denoting their spins. The occupation probability for a given state (ν, s) is given by

$$
\rho_{\nu s} = \langle a_{\nu s}^\dagger a_{\nu s} \rangle. \tag{23.19}
$$

The screened Coulomb interaction matrix element is given by

$$
W_{abcd} = \iint d^3 r\, d^3 r'\; \phi_a^*(r)\phi_b^*(r') \frac{e_0^2}{4\pi\varepsilon_0\varepsilon_{bg}} \frac{e^{-\kappa|r-r'|}}{|r-r'|}\phi_c(r')\phi_d(r)\,, \tag{23.20}
$$

with the single-particle wave functions $\phi_x(r)$, approximated as harmonic oscillator wave functions, which was shown to yield surprisingly accurate results [58]. The vacuum and background permittivity are given by ε_0 and ε_{bg}, respectively, and $-e_0$ is the electron charge. The screening wave number κ describes the screening of the Coulomb interaction potential by the surrounding charge-carrier plasma, which can be calculated in a self-consistent way [57,59,60]. Here, we implement the screening wave number in the static quasi-equilibrium limit as

$$
\kappa = \frac{e_0^2}{2\varepsilon_0\varepsilon_{bg}} \sum_b b\frac{\partial w_b}{\partial E_{F,b}^{eq}} = \frac{e_0^2}{2\varepsilon_0\varepsilon_{bg}} \sum_b D_b f(E_{b,0}^{QW}, E_{F,b}^{eq}, k_B T^{eq})\,, \tag{23.21}
$$

with the quasi-Fermi level $E_{F,b}^{eq}$, quasi-equilibrium temperature T^{eq}, and 2D density of states D_b of the corresponding electron and hole plasma. Within the quasi-equilibrium approximation, the screening can be expressed by the occupation probability at the band edge, $E_{b,0}^{QW}$, increasing the screening of the Coulomb interaction with increasing charge-carrier density. The screening becomes very important at elevated charge-carrier densities where the unscreened Coulomb potential would greatly overestimate the interaction between the charge carriers.

An exact numeric solution to Heisenberg's equation of motion within the given problem is generally not possible. The Coulomb interaction Hamiltonian couples the dynamic evolution of n-operator expectation values to $n+2$-operator expectation values, leading to an infinite number of coupled differential equations. It is, therefore, necessary to apply further approximations. We truncate this infinite chain of dynamic equations by factorizing six-operator expectation values into factors of two-operator expectation values, i.e., occupation probabilities. The Coulomb matrix elements thus enter the dynamic equations in up to second order. First-order contributions lead to renormalization effects of the single-particle energies and polarization amplitudes due to the Coulomb interaction with surrounding charge carriers [59,61]. These effects do not lead to a net change in the charge-carrier distribution and are therefore neglected.

By applying the Markov approximation to the resulting equations of motion, the explicit time dependence of four-operator expectation values is discarded. Instead, a quasi-static dependence of these expectation values on the occupation probabilities is assumed, by setting their time derivative to zero in the adiabatic limit. The scattering processes are thus assumed to follow any changes in the charge-carrier distribution instantaneously. The scattering contribution to the charge-carrier states then yields a Boltzmann-type equation:

$$\frac{\partial}{\partial t}\rho_{v\sigma}\Big|_{sc} = \frac{2\pi}{\hbar}\sum_{\substack{bcd\\s'}}\mathrm{Re}\left[W_{vbcd}\left(W_{vbcd}^* - W_{vbdc}^*\right)\right]\delta(\varepsilon_v + \varepsilon_b - \varepsilon_c - \varepsilon_d)$$

$$\times\left[(1-\rho_{v\sigma})(1-\rho_{bs'})\rho_{cs'}\rho_{d\sigma} - \rho_{v\sigma}\rho_{bs'}(1-\rho_{cs'})(1-\rho_{d\sigma})\right], \tag{23.22}$$

which describes the Coulomb scattering in the second-order Born–Markov approximation [23,62–64]. The summation terms in Equation 23.22 describe the simultaneous scattering between states $d \leftrightarrow v$ and $c \leftrightarrow b$. The delta function ensures energy conservation, such that the total energy of the final states equals that of the initial states. The Coulomb scattering is thus revealed to be of Auger-type, requiring the simultaneous scattering of two electrons, with no net change in the total charge-carrier energy. The individual scattering processes are proportional to the occupation probabilities in the initial states c, d and proportional to the probability to find vacant final states v, b, accounting for Pauli blocking. The second term in the sum in Equation 23.22 describes the reverse scattering processes, with electrons scattering out of the states v, b.

Equation 23.22 can be written in the form of a Boltzmann equation for the occupation probability $\rho(t)$ of any given state in the system,

$$\frac{\partial}{\partial t}\rho(t)\Big|_{sc} = S^{\mathrm{in}}[1-\rho(t)] - S^{\mathrm{out}}\rho(t), \tag{23.23}$$

combining the summation terms into an in-scattering rate S^{in} and a corresponding out-scattering rate S^{out}. These rates are calculated from the sums in Equation 23.22, which include all possible individual scattering processes. An exact treatment of the scattering dynamics within the second-order Born–Markov approximation would still require the dynamic tracking of the complete charge-carrier distribution, making numerical treatment difficult. Fortunately, the given QD-quantum-well system allows the distinction between qualitatively different scattering processes in order to break up the sums in Equation 23.22 into different parts which can be handled more easily.

For the dot-in-a-well (DWELL) structures considered in this chapter, two general classes of charge-carrier scattering processes can be distinguished: the capture of a quantum-well electron into a confined QD state, and the intradot electron relaxation, each with their respective inverse escape processes. This is illustrated in Figure 23.5a and b, respectively. The accompanying Auger-electron can involve either quantum-well states only, or transitions between quantum-well and other QD states. Note that depending on the involved energy differences, not all of these scattering channels are possible. For example, in the depicted case of the intra-dot relaxation in Figure 23.5b, the Auger transition in the valence band from the quantum well to the GS is not possible, as it would violate energy conservation. The possible scattering processes contributing to the total scattering rate thus strongly depend on the exact energy scheme of the QD-quantum-well system. Note that throughout this work impact ionization and Auger-assisted recombination, i.e., the direct scattering between conduction and valence bands, is not considered.

Following the earlier discussion, the scattering dynamics of the localized QD states are rewritten in the electron–hole picture as

$$\frac{\partial\rho_{b,\mathrm{GS}}}{\partial t}\Big|_{sc} = S_{b,\mathrm{GS}}^{\mathrm{cap,in}}(\{\rho_{\mathrm{QW}}\})(1-\rho_{b,\mathrm{GS}}) - S_{b,\mathrm{GS}}^{\mathrm{cap,out}}(\{\rho_{\mathrm{QW}}\})\rho_{b,\mathrm{GS}}$$

$$+ S_{b,\mathrm{GS}}^{\mathrm{rel,in}}(\{\rho_{\mathrm{QW}}\})\rho_{b,\mathrm{ES}}(1-\rho_{b,\mathrm{GS}}) - S_{b,\mathrm{GS}}^{\mathrm{rel,out}}(\{\rho_{\mathrm{QW}}\})(1-\rho_{b,\mathrm{ES}})\rho_{b,\mathrm{GS}}, \tag{23.24}$$

$$\frac{\partial\rho_{b,\mathrm{ES}}}{\partial t}\Big|_{sc} = S_{b,\mathrm{ES}}^{\mathrm{cap,in}}(\{\rho_{\mathrm{QW}}\})(1-\rho_{b,\mathrm{ES}}) - S_{b,\mathrm{ES}}^{\mathrm{cap,out}}(\{\rho_{\mathrm{QW}}\})\rho_{b,\mathrm{ES}}$$

$$+ S_{b,\mathrm{ES}}^{\mathrm{rel,in}}(\{\rho_{\mathrm{QW}}\})\rho_{b,\mathrm{GS}}(1-\rho_{b,\mathrm{ES}}) - S_{b,\mathrm{ES}}^{\mathrm{rel,out}}(\{\rho_{\mathrm{QW}}\})(1-\rho_{b,\mathrm{GS}})\rho_{b,\mathrm{ES}}. \tag{23.25}$$

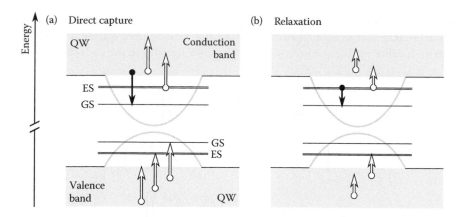

FIGURE 23.5 Possible scattering channels in quantum-dot (QD) quantum-well (QW) systems. (a) Direct capture into the QD ground state (GS). (b) Intradot relaxation from excited state (ES) to GS. The electron scattering process considered is shown by the black arrow, with the white arrows denoting the possible Auger processes. For all processes shown, the corresponding reverse scattering is also possible. Not shown is the direct capture into the QD ES, analogous to (a).

Here, ρ_b denotes either electron or hole occupation probabilities for $b \in \{e, h\}$. The scattering rates $S^{\mathrm{cap,in}}$ denote the direct capture of quantum-well electrons into the QD states, $S^{\mathrm{rel,in}}$ the intradot relaxation between the QD states, with S^{out} the scattering rate of the respective reverse processes. We can relate the relaxation processes of the ES to the corresponding GS terms,

$$S_{b,\mathrm{ES}}^{\mathrm{rel,in}}(\{\rho_{\mathrm{QW}}\}) = -\frac{1}{2} S_{b,\mathrm{GS}}^{\mathrm{rel,out}}(\{\rho_{\mathrm{QW}}\}), \tag{23.26}$$

$$S_{b,\mathrm{ES}}^{\mathrm{rel,out}}(\{\rho_{\mathrm{QW}}\}) = -\frac{1}{2} S_{b,\mathrm{GS}}^{\mathrm{rel,in}}(\{\rho_{\mathrm{QW}}\}), \tag{23.27}$$

with a factor $\frac{1}{2}$ compared to the GS contribution, due to the twofold degeneracy of the ES. All scattering rates in the above equations depend on the whole quantum-well distribution in both bands, denoted by $\{\rho_{\mathrm{QW}}\}$.

Thus far the derived scattering expressions only describe the dynamics of QD states and their interaction with the quantum-well charge carriers. The dynamics of quantum-well carriers can in principle be expressed by Equation 23.22 as well. However, this would require resolving all quantum-well states and tracking their population distribution in time, which greatly increases the dimensionality of the system state. This problem can be resolved by assuming a specific distribution of the carrier population within the quantum well.

The intraband scattering between quantum-well states is typically in the order of $\approx 100\,\mathrm{fs}$ [65–69]. As long as this scattering is faster than the charge-carrier exchange between the quantum well and QDs, the quantum well can be assumed to be in quasi-equilibrium with good accuracy:

$$\rho_{b,\mathrm{QW}}(\varepsilon_{b,k}^{\mathrm{2D}}) \approx f(\varepsilon_{b,k}^{\mathrm{2D}}, E_{\mathrm{F},b}^{\mathrm{eq}}, T^{\mathrm{eq}}) \equiv \left[1 + \exp\left(\frac{\varepsilon_{b,k}^{\mathrm{2D}} - E_{\mathrm{F},b}^{\mathrm{eq}}}{k_{\mathrm{B}} T^{\mathrm{eq}}} \right) \right]^{-1}, \tag{23.28}$$

with the corresponding single-particle energies $\varepsilon_{b,k}^{\mathrm{2D}}$ and the quasi-Fermi level $E_{\mathrm{F},b}^{\mathrm{eq}}$. From this quasi-Fermi distribution the 2D-charge-carrier density w_b in the quantum wells can be calculated by taking the density of states in the quantum well as

$$\mathcal{D}_b(E) = D_b \Theta(E - E_{b,0}^{\mathrm{QW}}) = \frac{m_b^*}{\pi \hbar^2} \Theta(E - E_{b,0}^{\mathrm{QW}}), \tag{23.29}$$

under the assumption that the quantum-well sub-band spacing is large enough that only the lowest sub-band needs to be taken into account. The energy $E_{b,0}^{QW}$ is the corresponding quantum-well band edge and Θ is the Heaviside function. The quantum-well charge-carrier density can then be written as

$$
\begin{aligned}
w_b &= \frac{2}{A_{\mathrm{act}}} \sum_{k^{2D}} \left[1 + \exp \left(\frac{\varepsilon_{b,k}^{2D} - E_{F,b}^{eq}}{k_B T^{eq}} \right) \right]^{-1} \\
&= \int_{-\infty}^{\infty} d\varepsilon_{b,k}^{2D} \, D_b(\varepsilon_{b,k}^{2D}) \left[1 + \exp \left(\frac{\varepsilon_{b,k}^{2D} - E_{F,b}^{eq}}{k_B T^{eq}} \right) \right]^{-1} \\
&= D_b k_B T^{eq} \log \left[1 + \exp \left(\frac{E_{F,b}^{eq} - E_{b,0}^{QW}}{k_B T^{eq}} \right) \right],
\end{aligned}
\tag{23.30}
$$

where the sum over all quantum-well k-states was expressed as the integral over the charge-carrier energy. A_{act} is the active region in-plane area, with the factor 2 accounting for spin degeneracy. By inverting the above expression, the quasi-Fermi level $E_{F,b}^{eq}$ can be expressed in terms of the charge-carrier density in the quantum well,

$$
E_{F,b}^{eq} = E_{b,0}^{QW} + k_B T^{eq} \log \left[\exp \left(\frac{w_b}{D_b k_B T^{eq}} \right) - 1 \right].
\tag{23.31}
$$

Thus, the quantum-well charge-carrier population can be expressed as a function of the carrier density and the quasi-equilibrium temperature:

$$
\rho_{b,QW}(\varepsilon_{b,k}^{2D}) \equiv \rho_{b,QW}(\varepsilon_{b,k}^{2D}, w_b, T^{eq}) = \left[1 + \exp \left(\frac{\varepsilon_{b,k}^{2D} - E_{F,b}^{eq}(w_b, T^{eq})}{k_B T^{eq}} \right) \right]^{-1}.
\tag{23.32}
$$

By entering this relation into the expressions for the scattering rates Equation 23.22, also the individual scattering rates can be expressed as functions of only the 2D charge-carrier densities w_b and their quasi-equilibrium temperature T^{eq}, eliminating the need to keep track of the microscopic carrier population distribution.

Furthermore, it is now possible to relate the in- and out-scattering rates of a given scattering process to each other [70,71]. The out-scattering contribution in Equation 23.22 is equivalent to the in-scattering contribution under the replacement $\rho \to 1 - \rho$, which for the quantum well in quasi-equilibrium can be expressed as

$$
1 - \rho_{QW}(\varepsilon_{b,k}^{2D}) = \rho_{QW}(\varepsilon_{b,k}^{2D}) \exp \left(\frac{\varepsilon_{b,k}^{2D} - E_{F,b}^{eq}(w_b, T^{eq})}{k_B T^{eq}} \right).
\tag{23.33}
$$

For the QD scattering processes the out-scattering rates can thus be written as [72]:

$$
S_{b,m}^{\mathrm{cap,out}}(w_e, w_h, T^{eq}) = S_{b,m}^{\mathrm{cap,in}}(w_e, w_h, T^{eq}) \exp \left(\frac{\varepsilon_{b,m}^{QD} - E_{F,b}^{eq}}{k_B T^{eq}} \right)
\tag{23.34}
$$

$$S_b^{\text{rel,out}}(w_e, w_h, T^{\text{eq}}) = S_b^{\text{rel,in}}(w_e, w_h, T^{\text{eq}}) \exp\left(\frac{\varepsilon_{b,\text{GS}}^{\text{QD}} - \varepsilon_{b,\text{ES}}^{\text{QD}}}{k_{\text{B}} T^{\text{eq}}}\right), \qquad (23.35)$$

where $\varepsilon_{b,m}^{\text{QD}}$ denotes the energy of the localized QD state, with $m \in \{\text{GS}, \text{ES}\}$ distinguishing between GS and ES. The out-scattering of charge carriers thus becomes more probable at elevated charge-carrier temperatures [73,74]. Note that in the derivation of above expressions, only a quasi-equilibrium within the quantum well must be assumed without making assumptions about the QD occupations. Equations 23.34 and 23.35 are therefore also valid in nonequilibrium situations between QD and quantum well.

The resulting scattering contribution to the dynamics of the QD occupation probability is, therefore, written as

$$\left.\frac{\partial \rho_{b,\text{GS}}^{j}(z,t)}{\partial t}\right|_{\text{sc}} = S_{b,\text{GS}}^{\text{cap,in}}\left[1 - \rho_{b,\text{GS}}^{j}\right] - S_{b,\text{GS}}^{\text{cap,out}} \rho_{b,\text{GS}}^{j}$$
$$+ S_b^{\text{rel,in}} \rho_{b,\text{ES}}^{j}\left[1 - \rho_{b,\text{GS}}^{j}\right] - S_b^{\text{rel,out}} \rho_{b,\text{GS}}^{j}\left[1 - \rho_{b,\text{ES}}^{j}\right] \qquad (23.36)$$

$$\left.\frac{\partial \rho_{b,\text{ES}}^{j}(z,t)}{\partial t}\right|_{\text{sc}} = S_{b,\text{ES}}^{\text{cap,in}}\left[1 - \rho_{b,\text{ES}}^{j}\right] - S_{b,\text{ES}}^{\text{cap,out}} \rho_{b,\text{ES}}^{j}$$
$$- \frac{1}{2}\left\{S_b^{\text{rel,in}} \rho_{b,\text{ES}}^{j}\left[1 - \rho_{b,\text{GS}}^{j}\right] - S_b^{\text{rel,out}} \rho_{b,\text{GS}}^{j}\left[1 - \rho_{b,\text{ES}}^{j}\right]\right\}. \qquad (23.37)$$

The scattering contribution to the quantum-well equations can be calculated from charge-carrier number conservation. It is simply given by the total charge-carrier density captured in the QD states:

$$\left.\frac{\partial}{\partial t} w_b(z,t)\right|_{\text{sc}} = -2N^{\text{QD}} \sum_{j,m} v_m f(j) \left(S_{b,m}^{\text{cap,in}}(z,t)\left[1 - \rho_{b,m}^{j}(z,t)\right] - S_{b,m}^{\text{cap,out}}(z,t)\rho_{b,m}^{j}(z,t)\right). \qquad (23.38)$$

23.2.4 Fit Functions for Scattering Rates

The microscopic calculation of the QD charge-carrier scattering rates requires a summation over all possible electronic states in the DWELL system, and such a calculation is, therefore, quite expensive in terms of computation time. While the scattering rates can be implemented by means of lookup tables, having a simpler analytic expression would be advantageous.

A first look at the scattering rates in dependence of the reservoir densities w_b reveals for the capture rates a quadratic increase at low densities, and a transition to nearly linear increase at higher values of w_b. The relaxation rates, on the other hand, show a linear increase at first and subsequent saturation. This is depicted in Figure 23.6 (circles). Taking these characteristics into account, we fit the scattering rates, using the following functions

$$S_{b,m}^{\text{cap,in}}(w_b) = \frac{Aw_b^2}{B + w_b} \qquad (23.39)$$

$$S_{b,m}^{\text{rel,in}}(w_b) = \frac{Cw_b}{D + w_b}. \qquad (23.40)$$

The corresponding out-scattering rates are calculated via Equations 23.34 and 23.35. Table 23.1 gives the fitting parameters extracted from the microscopically calculated rates. The comparison shown in Figure 23.6 shows good agreement between the microscopically calculated rates and the fit functions. For high reservoir carrier densities the fits show a slight deviation, especially pronounced in the hole

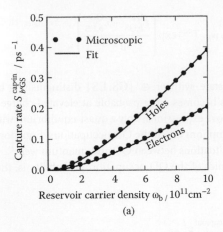
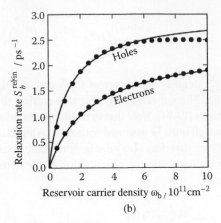

(a)   (b)

FIGURE 23.6 Fits of the quantum-dot scattering rates. Shown are (a) the ground-state direct-capture and (b) the intradot relaxation rate for electrons and holes, in dependence of the electron reservoir density w_b at $T = 300$ K. The microscopically calculated rates (circles) are shown together with the simple fit functions (lines). The fit parameters are given in Table 23.1.

TABLE 23.1   Fitting Parameters for the Scattering Rates

	Electrons		Holes	
	GS	ES	GS	ES
A (10^{-11}cm^2·ns^{-1})	32	54	60	79
B (10^{11}cm^{-2})	5.6	2.9	5.3	2.1
C (ns^{-1})		2400		3000
D (10^{11}cm^{-2})		2.7		1.2

Confinement energies: $\Delta E_e = 95$ meV, $\Delta_e = 60$ meV,
$\Delta E_h = 50$ meV, $\Delta_h = 25$ meV.

relaxation rate. These values, however, correspond to very strong electrical pumping. For currents typically used in these devices, the fit functions presented are in very good agreement with the microscopic calculations.

The relatively simple expressions extracted from the fits to the microscopically calculated scattering rates show that the common approach of capture rates which are directly proportional to the reservoir charge-carrier density [17,75] should even yield quantitatively acceptable results. The saturation of the intradot relaxation rates at elevated carrier densities suggests the use of constant relaxation rates when simple QD models are desired. Nevertheless, when the accurate description of dynamics over a large range of operating conditions is required, the microscopic description of the QD scattering should be preferred.

23.2.5 Modeling of Spontaneous Emission

The spontaneous emission created in optical amplifier devices will be subject to stimulated amplification when emitted along the propagation axis. This ASE can reach significant optical power levels and is important for the device characteristics and performance. Apart from adding an optical noise background to the device output, it can often become strong enough to influence the charge-carrier dynamics [76–78]. These effects will deteriorate the device performance and the optical signal quality.

A consistent description of the ASE is important. In general there exist two appropriate modeling approaches: the deterministic description of the ASE power spectral density in the frequency space [17,79] and the stochastic description in timedomain [43,78]. Here, we will employ the stochastic description, which simplifies the inclusion of time-varying input signals.

We, therefore, phenomenologically add an additional source term on the right-hand side of Equation 23.5, modeling the stochastic spontaneous emission added to the propagating electric field:

$$\frac{d}{dt} E_\pm(z, t') = S_\pm(z, t') + S_\pm^{sp}(z, t').$$ (23.41)

The electric field propagation along the one space-discretization section is again determined by integration of Equation 23.41 over the interval Δz:

$$E_\pm(z, t) \approx E_\pm(z \mp \Delta z, t - \Delta t) + \frac{\Delta t}{2} \left[S_\pm(z, t) + S_\pm(z \mp \Delta z, t - \Delta t) \right] + \int_{-\Delta t}^{0} S_\pm^{sp}\left(z \pm v_g \tau, t + \tau\right) d\tau.$$ (23.42)

The spontaneous emission source term must account for all optical transitions in the inhomogeneously broadened QD ensemble. Let $\eta_m^j(z, t)$ describe the spontaneous emission contribution of the jth subgroup of the mth localized state. We write for the electric field spontaneously added to the propagating field along Δz:

$$\int_{-\Delta t}^{0} S_\pm^{sp}\left(z \pm v_g \tau, t + \tau\right) d\tau \equiv \sum_{m,j} \eta_m^j(z, t).$$ (23.43)

The spontaneous emission of an optical transition has a finite linewidth given by its homogeneous broadening. The homogeneous linewidth of a given transition is directly linked to the dephasing time τ_2 of the microscopic polarization, with the linewidth simply given by $2\tau_2^{-1}$. In order to correctly implement the spectral properties of the ASE, the spontaneously emitted field $\eta_m^j(z, t)$ must, therefore, have the correct linewidth and center frequency. It is thus not possible to describe the spontaneous emission by white noise, which would produce a flat noise spectrum; it must instead be modeled using colored noise. We implement this colored noise by two-dimensional Ornstein–Uhlenbeck processes [80], which describe quantities which are driven by white noise but relax with a given rate γ toward zero. This relaxation rate leads to a finite "memory" of the process, which translates into a finite spectral width around a center frequency. The time evolution of each of the respective noise signals is modeled by the following stochastic differential equation:

$$\frac{d}{dt} \eta_m^j(z, t) = -(i\omega_m^j + \gamma)\eta_m^j(z, t) + \sqrt{D_{sp,m}^j(z, t)}\, \tilde{\xi}_m^j(z, t),$$ (23.44)

where $\tilde{\xi}(z, t)$ is a complex Gaussian white noise process, which is δ-correlated both in z and t. The relaxation rate of η_m^j is given by γ, and its center frequency by ω_m^j. The noise signal then fulfills the following properties [80]:

$$\langle \operatorname{Re} \eta_m^j(z, t) \rangle = \langle \operatorname{Im} \eta_m^j(z, t) \rangle = 0$$ (23.45)

$$\langle |\eta_m^j(z, t)|^2 \rangle = \frac{D_{sp,m}^j(z, t)}{\gamma}$$ (23.46)

$$\langle \eta_m^j(z, t)\eta_m^{j\,*}(z', t + \tau) \rangle \approx \frac{D_{sp,m}^j(z, t)}{\gamma} e^{-\gamma|\tau|} e^{i\omega_m^j \tau} \delta_{z,z'}.$$ (23.47)

Equation 23.47 is valid only under the assumption of a slowly varying noise amplitude $\partial_t D_{sp}(z,t) \ll \gamma$, such that within one correlation time γ^{-1} the spontaneous emission amplitude can be assumed as constant. We can use the Wiener–Khinchin-theorem [81] this relation can be used to calculate the power spectrum $S_{\eta_m^j}(z,\omega)$ of $\eta_m^j(z,t)$:

$$S_{\eta_m^j}(z,\omega) = \frac{1}{2\pi} \int_{-\infty}^{\infty} \langle \eta_m^j(z,t) \eta_m^{j\,*}(z,t+\tau) \rangle e^{-i\omega\tau} d\tau \tag{23.48}$$

$$= \frac{D_{sp,m}^j(z)}{\pi} \frac{1}{(\omega_m^j - \omega)^2 + \gamma^2}, \tag{23.49}$$

which yields a Lorentzian line shape with a FWHM of 2γ. We thus identify $\gamma = (\tau_2)^{-1}$, such that the noise linewidth equals the homogeneous linewidth of the QD transitions.

Using the noise correlation properties, the average power that is added to the electric field by the noise can be calculated. Combining Equations 23.42 and 23.43 and summarizing the deterministic source terms in a combined variable, $\widetilde{S}_{\pm}^{stim}$, yields for the electric field:

$$E_{\pm}(z,t) = E_{\pm}(z \mp \Delta z, t - \Delta t) + \widetilde{S}_{\pm}^{stim}(z,t) + \sum_{m,j} \eta_m^j(z,t) \tag{23.50}$$

$$\left\langle \left| E_{\pm}(z,t) \right|^2 \right\rangle = \left| E_{\pm}(z \mp \Delta z, t - \Delta t) + \widetilde{S}_{\pm}^{stim}(z,t) \right|^2 + \sum_{m,j} \left\langle \left| \eta_m^j(z,t) \right|^2 \right\rangle$$

$$= \left| E_{\pm}(z \mp \Delta z, t - \Delta t) + \widetilde{S}_{\pm}^{stim}(z,t) \right|^2 + \sum_{m,j} \tau_2 D_{sp,m}^j(z,t), \tag{23.51}$$

where we have used the zero mean property, $\langle \eta_m^j \rangle = 0$. On average, the spontaneous emission thus increases the squared modulus of the electric field along one space discretization step during the propagation time Δt by $\sum_{m,j} \tau_2 D_{sp,m}^j(z,t)$. Or, written in terms of a time derivative,

$$\frac{\partial}{\partial t} |E_{\pm}(z,t)|^2 \Big|_{sp} = \frac{\tau_2}{\Delta t} \sum_{m,j} D_{sp,m}^j(z,t). \tag{23.52}$$

In the photon picture, the average change of the electric field energy density due to the spontaneous emission can be calculated:

$$\frac{\partial}{\partial t} u(z,t) \Big|_{sp} = \frac{\varepsilon_{bg}\varepsilon_0}{2} \frac{\partial}{\partial t} |E_{\pm}(z,t)|^2 \Big|_{sp}$$

$$= \beta \frac{2N^{QD}\Gamma}{h^{QW}} \sum_{m,j} \nu_m f(j) \hbar\omega_m^j W_m \varrho_{e,m}^j(z,t) \varrho_{h,m}^j(z,t). \tag{23.53}$$

Here we model spontaneous emission by bimolecular recombination processes, given by the sum over all QD states. The spontaneous emission ratio β gives the fraction of photons spontaneously emitted into the waveguide mode. Comparing Equations 23.52 and 23.53 yields for the individual noise strengths

$$D_{sp,m}^j(z,t) = \frac{\Delta t}{\tau_2} \frac{2\beta\Gamma\hbar\omega_m^j 2N^{QD}}{\varepsilon_{bg}\varepsilon_0 h^{QW}} \nu_m f(j) \frac{1}{2} \left[R_{sp,m}^j(z,t) + R_{sp,m}^j(z \mp \Delta z, t - \Delta t) \right], \tag{23.54}$$

where the average of the spontaneous emission rate at the endpoints of the integration interval $[z, z \mp \Delta z]$ was taken, defined by

$$R^j_{sp,m}(z, t) := W_m \varrho^j_{e,m}(z, t) \varrho^j_{h,m}(z, t). \tag{23.55}$$

The spontaneous emission noise thus depends on the optical frequency and on the occupation of the individual QD subgroups. The resulting propagation equation for the electric field is thus given by

$$E_\pm(z, t) \approx E_\pm(z \mp \Delta z, t - \Delta t) + \frac{\Delta t}{2} \left[S_\pm(z, t) + S_\pm(z \mp \Delta z, t - \Delta t) \right] + \sum_{m,j} \eta^j_m(z, t). \tag{23.56}$$

23.3 Application Examples

The QDSOA model derived in the previous section will now be applied to an exemplary device. We implement Equations 23.11 through 23.17, which describe the active medium dynamics, along with Equation 23.56 to describe the electric field propagation including spontaneous emission noise. The charge-carrier scattering is described within the full microscopic framework as described in Equations 23.36 through 23.38, with the individual scattering rates depending on the reservoir carrier densities as well as the device temperature.

In this section we will at first investigate the static characteristics of the amplifier, focusing on the unique properties of the QD active medium. We thus calculate the pump-current dependent-gain spectra for the sample device, along with ASE spectra. Our modeling approach allows for a characterization over a large range of pump currents, accounting for the changing carrier dynamics by the nonlinear scattering rates. We analyze the performance of the QDSOA for optical signals centered on either the GS or ES energies, exploiting the broad gain spectrum due to the different localized states. Subsequently, we characterize the dynamic gain recovery after perturbation of the gain medium by a strong pulse. The gain recovery dynamics provide an important insight into the ultrafast amplification capabilities of an amplifier. Furthermore, we show how the internal charge-carrier dynamics imprint their signature onto the gain dynamics. The last part of this section addresses the phenomenon of coherent pulse-shaping by Rabi oscillations in QDSOAs, enabled by the comparably long dephasing time of the interband polarization. The interaction of ultra-short pulses with the QD gain medium, therefore, leads to strong modifications of the pulse shape. These modifications differ from classical descriptions and can be very complex due to the individual dynamics of QDs within the inhomogeneous ensemble.

23.3.1 Static Characterization of the QDSOA

In linear amplification applications, the device performance of SOAs is generally limited by two competing effects. On the one hand, the maximum achievable optical output power is limited by the charge carriers available for stimulated emission. The gain of the amplifier will, therefore, decrease when the optical power becomes too large. This effect is known as gain saturation. On the other hand, a too small optical signal will significantly reduce the signal-to-noise ratio, as the spontaneous emission background will dominate the output. Noise effects thus play an important role in the amplification of optical data signals [82,83]. A strong noise background will negatively impact the signal quality by distorting the corresponding optical output, and potentially corrupting the transmitted data stream. It is, therefore, important to investigate the ASE and gain of a given amplifier device to assess its suitability in a given application.

We implement the previously derived QD amplifier model to simulate the static characteristics of a specific device. We solve the delay-differential equation system using the simple forward-Euler method. The Gaussian white noise for implementing the spontaneous emission is generated by the Box–Muller algorithm [84]. For each spatial section of the amplifier device, we save a history array containing $\Delta t/dt$

entries (up to Δt before the current time t) of the electric field $E_\pm(z, t)$, the field source term $S_\pm(z, t)$, and the spontaneous emission rate $R_{\mathrm{sp},m}^j(z, t)$.

The modeled amplifier is a 3-mm-long DWELL structure, consisting of ten 5-nm-thick InGaAs quantum wells, each embedding a density of 3×10^{10} cm^{-2} InAs QDs with a shallow-etched, 4-μ-wide ridge waveguide. Refer to Figure 23.3 for an illustration of the modeled energy structure. The device parameters that will be used here and in the following sections are given in Table 23.2, unless otherwise noted.

Additional phenomenological dependencies of the dephasing time and the device temperature on the applied pump current are included to account for heating effects as well as carrier-induced dephasing. We assume a temperature that increases linearly with the applied current,

$$T_\ell(j) = 295\,\mathrm{K} + \frac{j}{14}\,\frac{\mathrm{K}}{\mathrm{mA}}\,. \tag{23.57}$$

The increase in temperature significantly alters the detailed balance relationship, Equations 23.34 and 23.35, which determines the out-scattering rates out of the localized QD states. This, in turn, determines the quasi-equilibrium distribution that is reached in the steady state. Furthermore, the dephasing time of the QD interband polarization is known to decrease under strong excitation and with temperature [27,28]. We thus introduce a current-dependent dephasing time,

$$\tau_2(j) = \frac{\tau_2^0}{1 + \frac{j}{300\,\mathrm{mA}}}\,, \tag{23.58}$$

TABLE 23.2   Model Parameters Used in the Simulations, unless Otherwise Stated.

Symbol	Value	Meaning
N^{QD}	3×10^{10} cm^{-2}	QD density per layer
a_L	10	Number of layers
h^{QW}	5 nm	QW layer height
w_{wg}	4μ	Waveguide width
n_{bg}	3.77	Background index
ΔE_{inh}	30 meV	QD inhomogeneous broadening FWHM
η	0.4	Current injection efficiency
A^S	0.7 ns^{-1}	QW linear recombination rate
B^S	50 nm^2 ns^{-1}	QW bimolecular recombination rate
W_m	1 ns^{-1}	QD spontaneous recombination rate
β	3.5×10^{-4}	Spontaneous emission ratio
$\langle \hbar\omega_{\mathrm{GS}} \rangle$	963 meV (1288 nm)	QD GS center emission
$\langle \hbar\omega_{\mathrm{ES}} \rangle$	1048 meV (1183 nm)	QD ES center emission
μ_m	0.6 nm e_0	QD transition dipole moment
τ_2	$200\,\mathrm{fs} \times \left(1 + \frac{j}{300\,\mathrm{mA}}\right)^{-1}$	QD polarization dephasing time
T_ℓ	$295\,\mathrm{K} + \frac{j}{14}\frac{\mathrm{K}}{\mathrm{mA}}$	Lattice temperature
Γ	0.045	Geometric confinement factor
$\Delta E_e(\Delta E_h)$	95 meV (50 meV)	Electron (hole) QD GS localization energy
$\Delta_e(\Delta_h)$	60 meV (25 meV)	Electron (hole) QD GS–ES energy spacing
Z	31	Number of space discretization steps
dt	1 fs	Numeric time step

QW, quantum well; QD, quantum dot; GS, ground state; ES, excited state; FWHM, full-width half maximum.

where τ_2^0 denotes the corresponding dephasing time at $j = 0$. The above effects lead to broadening and subsequent decrease of the optical gain at high pump currents, as frequently observed in experiments [85]. Note that the dephasing time can in principle be calculated from microscopic scattering processes similar to the formalism presented in Section 23.2.3. A rigorous treatment of these effects, however, requires microscopic approaches that go beyond the employed Born–Markov approximation [86–90], which are beyond the scope of this chapter.

With the above additions, we calculate the key characteristics of the amplifier device, starting with the small-signal optical gain spectrum. The small-signal gain $G(\omega)$ is calculated from the stationary QD distribution (i.e., without a disturbing optical signal) via

$$
G(\omega) = \exp\left[\frac{2}{v_g}\int_0^\ell g(z, t, \omega)\,dz\right], \tag{23.59}
$$

with the instantaneous amplitude gain

$$
g(z, t, \omega) = \frac{\hbar\omega\Gamma}{\varepsilon_0\varepsilon_{bg}h^{QW}}2N^{QD}\sum_{j,m}\nu_m f(j)\frac{\tau_2|\mu_m|^2}{2\hbar^2}\frac{\left[\rho_{e,m}^j(z, t) + \rho_{h,m}^j(z, t) - 1\right]}{1 + [\tau_2(\omega - \omega_m^j)]^2} \tag{23.60}
$$

determined from the steady-state solution to Equation 23.11. The resulting spectra are shown in Figure 23.7a, clearly showing the GS gain peak around 1300 nm, which saturates and decreases for higher pump currents, at which the ES gain becomes dominant. We plot the optical gain at the GS and ES transitions, at $\lambda = 1288$ nm and $\lambda = 1183$ nm, respectively, under the dependence of the pump current, shown in Figure 23.7b.

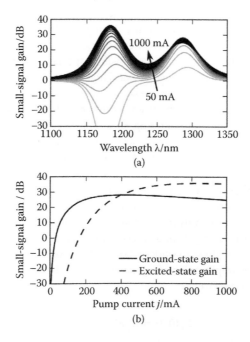

FIGURE 23.7 (a) Static gain spectra of the quantum dot semiconductor optical amplifier device for increasing pump currents up to $j = 1000$ mA in steps of 50 mA. (b) Small-signal gain at the ground (solid) and excited state (dashed) wavelengths ($\lambda_{GS} = 1288$ nm, $\lambda_{ES} = 1183$ nm) under the dependence of the pump current.

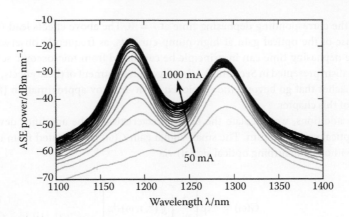

FIGURE 23.8 Simulated amplified spontaneous emission (ASE) spectra for increasing pump currents up to $j = 1000\,\text{mA}$ in steps of 50 mA.

Additionally, we can calculate the ASE spectra, shown in Figure 23.8. These are calculated by direct integration of the dynamic equations and subsequent Fourier transform of the electric field time series at the output facets. The resulting spectra can be used for a direct quantitative comparison with experimental data. The output power at the corresponding end facets is related to the electric field via

$$P_\pm^{\text{out}} = A_{\text{beam}} \frac{\varepsilon_0 n_{\text{bg}} c_0}{2} |E_\pm^{\text{out}}|^2 , \qquad (23.61)$$

where $A_{\text{beam}} = \frac{a_L h^{\text{QW}} w_{\text{wg}}}{\Gamma}$ is the beam area in the waveguide and $E_\pm^{\text{out}} = \{E_+(\ell), E_-(0)\}$ denotes the forward- and back-propagating electric fields at their respective end facets. The ASE spectra mimic the qualitative behavior of the gain spectra, with clearly distinguishable GS and ES peaks and a saturation for strong pump currents (see Figure 23.8, where increasing gray levels indicate higher pump current). In addition, an initial blueshift of the emission peaks is visible especially for the ES emission. This is due to state filling, i.e., the shift of the quasi-Fermi levels toward higher energies with increasing carrier densities. At low currents, only the lowest energy states are occupied by charge carriers, and only with increasing current can the higher lying energy states be filled. In the model, this is accounted for by the detailed balance relation between the in- and out-scattering rates (Equations 23.34 and 23.35), which ensures the relaxation toward a quasi-Fermi distribution in the steady state. The quasi-Fermi level shifts toward higher energies with increasing reservoir carrier density, thus leading to a higher occupation in higher energy states.

When an optical signal of significant power is amplified by the gain medium, its charge-carrier distribution can be substantially perturbed. This leads to a power-dependent response of the device and thus induces nonlinearities. In order to quantify this response, we simulate the amplifier under the injection of a constant optical input signal with varying power. As we have seen previously, the existence of localized QD states makes it possible for the amplifier to work on GS and ES wavelengths. We will, therefore, investigate the device performance under amplification of signals at either of the two corresponding wavelengths. We put in an electric field at the input facet as

$$E_+(0, t) = E_{\text{in}}^m e^{-i(\langle\omega^m\rangle - \omega)t} , \qquad (23.62)$$

with the frequency detuning $(\langle\omega^m\rangle - \omega)$ $(m \in \{\text{GS, ES}\})$ of the input signal relative to the carrier frequency chosen to yield signals centered on the energy of either QD state. After the propagation of the signal through the amplifier device and after a steady state is reached, we can simply evaluate the power-dependent device gain by the power ratio

$$G = \frac{|E_+(\ell, t)|^2}{|E_+(0, t)|^2}.$$

(23.63)

Figure 23.9a shows the GS gain in dependence on the achievable optical output power for different pump currents. The small-signal gain, i.e., at low input powers, shows the behavior that we have seen before in Figure 23.7, with a saturation and subsequent decrease of the gain for increasing current. In the linear amplification regime, for lower optical power, the gain is flat. Under these small-signal conditions, the optical power is not strong enough to perturb the gain medium appreciably. Once the optical power becomes large enough, the nonlinear or gain saturation regime is reached [10,11,79]. In Figure 23.9, the onset of the GS saturation regime increases from 18 dBm output power at $j = 200$ mA to 25 dBm at $j = 1000$ mA. Here, the number of charge carriers injected into the optically active QD states is not sufficient to replenish the QD states that are depleted by the signal, and the optical gain thus gradually decreases.

A general trend toward higher achievable output power with increasing current and corresponding shift of the nonlinear regime toward higher optical power can be observed both in the GS and ES (see Figure 23.9a and b). This is a direct consequence of the increased in-scattering rates and the higher reservoir charge-carrier density [10]. The faster and more efficient refilling of the QD states after depletion by the optical signal shifts the saturation regime toward higher power. The ES shows a transition to the nonlinear gain regime at lower optical power than the GS, as shown in Figure 23.9b. This can be understood by the weaker confinement of the QD ES relative to the reservoir, which increases the sensitivity of the ES occupation to changes in the reservoir charge-carrier density. Nevertheless, the ES allows a high gain and a sufficiently high saturation power to allow for its application in optical communication networks. Furthermore, the ultra-broad gain bandwidth of QDSOAs could be exploited for a simultaneous amplification of data signals with wide wavelength spacing. The low cross-talk between the GS and ES transitions makes an error-free amplification of two suitable data signals possible, as was recently shown [91].

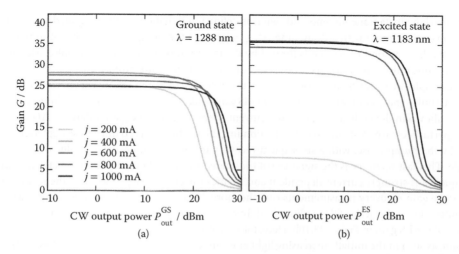

FIGURE 23.9 Optical gain G of the quantum-dot amplifier in dependence of the optical output power P_{out}^m for different pump currents j. Shown is the response to optical signals centered on (a) the ground state, and (b) the excited-state wavelength.

23.3.2 Gain Recovery Dynamics in QDSOAs

We now wish to characterize the dynamic performance of the modeled QD semiconductor amplifier. When optical amplifiers are used in optical data signal transmission, e.g., for propagation loss compensation, their dynamics must be fast enough to follow the signal symbol rate in order to minimize patterning effects and reduce transmission errors, as detailed in Chapter 25 on patterning effects. In experimental setups the gain recovery dynamics can be characterized by observing the time evolution of the optical gain after a perturbation of the active medium. This is commonly done in a pump-probe setup [19,92–94]. The principle is as follows: A strong and temporally narrow optical pump pulse is injected into the waveguide and depletes charge carriers during its amplification. A weaker probe pulse injected with a time delay Δt_{p-p} with respect to the pump pulse then experiences a different gain, depending on the perturbation of the carrier distribution after this delay time. By repeating the measurement for a range of Δt_{p-p} and measuring the probe pulse intensity, the time evolution of the gain after propagation of the pump pulse can be extracted experimentally.

The gain-recovery dynamics is obtained from the QDSOA model by simulating the amplifier in a pump-probe experiment. The perturbation of the gain medium is induced by a strong Gaussian pump pulse injected into the input facet at $z = 0$:

$$E_+(0, t) = E_{in} \exp\left[-4\ln 2\left(\frac{t - t_0}{\Delta_{FWHM}}\right)^2\right]e^{-i\delta\omega t},\tag{23.64}$$

with an input amplitude E_{in}, arrival time t_0, and pulse width Δ_{FWHM}. We additionally allow for a detuning $\delta\omega$ of the pump pulse with respect to the chosen optical reference frequency which we can use to tune the pump wavelength. In order to probe the gain after the pump pulse has passed the active medium, the probe pulse does not have to be modeled explicitly. Instead, we can directly calculate the gain that the probe would experience along the device:

$$G(\Delta t_{p-p}, \omega) = \exp\left[\frac{2}{v_g}\int_0^\ell g\left(z, t_0 + \Delta t_{p-p} + \frac{z}{v_g}, \omega\right)dz\right].\tag{23.65}$$

Here, we used the instantaneous amplitude gain, as defined in Equation 23.60, integrated along the device in a copropagating frame with a delay time Δt_{p-p} relative to the pump pulse. Note that in pump-probe experiments the temporal resolution is limited by the temporal width of the probe pulse. Measuring the integrated probe pulse power averages the extracted gain over the pulse profile.

As an example, we simulate the amplifier under the influence of a pump pulse with a width of $\Delta_{FWHM} = 300\,fs$ and a pulse energy of 1 pJ, centered on the GS gain peak. In Figure 23.10a the resulting gain recovery curves are shown in dependence of the pump current. Here, we show the gain difference ΔG with respect to the unperturbed case. A strong initial reduction in gain can be observed due to the depletion of charge carriers by the pump pulse, with a subsequent recovery toward the unperturbed value on a picosecond timescale. This fast recovery is the signature of the charge-carrier refilling by carrier scattering, which as a consequence is strongly current-dependent and mimics the current dependence of the scattering rates. Pump-probe gain recovery measurements can, therefore, be used to extract effective charge-carrier scattering timescales of the active QD medium [95]. In addition to the GS performance, we take a look at the dynamics of the ES gain in Figure 23.10b. Here, for $j = 100\,mA$, the pump-pulse induces an initial increase of the gain, as seen in the initially increasing light gray curve. At this current, the ES is below transparency, and the pump pulse is absorbed, which increases the inversion and thus the gain. For higher pump currents (increasing gray levels in Figure 23.10) the dynamics follows the same trends as the GS, albeit with a slower recovery. Here again the less efficient carrier refilling of the ES becomes apparent. The intricate scattering

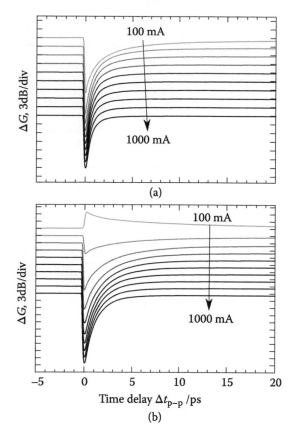

FIGURE 23.10 Gain recovery after perturbation with a strong pump pulse at $\Delta t_{p-p} = 0$. (a) Ground-state gain after excitation at the ground-state wavelength centered at $\lambda = 1288$ nm, and (b) excited-state gain after excitation at the excited-state wavelength centered at $\lambda = 1183$ nm. The current was increased from $j = 100$ mA to $j = 1000$ mA in steps of 100 mA. Shown is the gain difference with respect to the unperturbed case, with individual curves shifted vertically by 3 dB for improved readability.

dynamics between the different QD and reservoir states in the active medium, therefore, directly affects the performance of QD optical amplifiers, emphasizing the importance of its accurate modeling, as described in Section 23.2.3.

With the QDSOA model at hand, we can take a closer look at the dynamics of the QD gain medium. The gain recovery can be seen to consist of three main components, i.e., it can be fitted very accurately by a triexponential recovery, as shown in Figure 23.11a. Here, we do not show the fit for $t \lesssim 0.1$ ps as for small times the pulse is still interacting with the medium, which the fit cannot describe. We can attribute the three time constants to the intradot relaxation, being the fastest component, the charge-carrier capture from the reservoir, and a slow recovery of the whole system, governed mainly by the reservoir carrier lifetime. An experimental determination of the individual timescales, however, is usually difficult due to their narrow separation. This is already evident from the three effective timescales that determine the gain recovery. In principle, we would expect at least seven different timescales: GS and ES capture rates and intradot relaxation for both electrons and holes, along with the recovery of the reservoir. The three extracted timescales must, therefore, be interpreted as effective values.

Our model also allows us to analyze the charge-carrier dynamics within the gain medium directly, which is not possible within the experimental setup. We thus evaluate the variation of the carrier occupation in

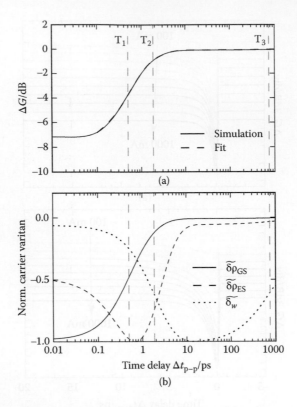

FIGURE 23.11 (a) Fit to the gain recovery after a perturbation of the ground state at $j = 500\,\text{mA}$. The three characteristic timescales, $\tau_1 = 0.49\,\text{ps}$, $\tau_2 = 1.8\,\text{ps}$, $\tau_3 = 750\,\text{ps}$, correspond to the intradot relaxation, charge-carrier capture, and reservoir refilling, respectively. (b) Time evolution of the normalized carrier variations with respect to their respective steady-state values in the GS, ES, and reservoir states.

the QD GS and ES as well as in the reservoir states at the output facet, by defining a normalized carrier variation,

$$\widetilde{\delta\rho}_m(t) = \frac{\sum_b \left[\rho_{b,m}(t) - \rho_{b,m}^0\right]}{\max\left|\sum_b \left[\rho_{b,m}(t) - \rho_{b,m}^0\right]\right|}, \tag{23.66}$$

where ρ_m^0 denotes the corresponding QD occupation in the unperturbed steady state. A similar expression for the reservoir density variation, $\widetilde{\delta w}$, can be written down. During the perturbation the normalized variations reach their minimum value of -1 and during recovery slowly grow back to a value of zero. The variations of the GS, ES, and reservoir carriers are plotted in Figure 23.11b, as solid, dashed, and dotted lines, respectively. We can clearly observe the three qualitative steps in the recovery of the gain medium we discussed earlier. The GS occupation is minimal right after the pulse has passed the device. On a timescale close to τ_1, the GS starts a fast recovery under a reduction of the ES occupation, illustrating the intradot relaxation from the ES states to the GS. These relaxation processes dominate the ultrafast gain recovery of QDSOAs. The ES then starts to recover with a timescale $\approx \tau_2$ due to charge-carrier capture from the reservoir, which is subsequently depleted. The timescale τ_3 then constitutes the recovery time of the reservoir, which equilibrates the whole system again.

For the modeled device, the two fast timescales of the GS gain recovery decrease from $\tau_1 = 1.4\,\text{ps}$, $\tau_2 = 7.5\,\text{ps}$ at a current of $100\,\text{mA}$ to $\tau_1 = 0.43\,\text{ps}$, $\tau_2 = 1.2\,\text{ps}$ at $j = 1000\,\text{mA}$ (not shown here). Together with the increasing saturation power at higher currents, a high operating current could be seen as optimal for high-power ultrafast applications. However, the saturation and decrease of the optical gain at elevated pump currents limits the range of possible operating conditions.

23.3.3 Rabi Oscillations in QD Semiconductor Amplifiers

When the quantum-mechanic phase is maintained over macroscopically accessible timescales, coherent effects can induce phenomena that significantly differ from classical predictions. The precondition for this is a sufficiently long dephasing time of the quantum-mechanic coherence. The comparably long lifetime of the interband polarization in semiconductor QDs makes them promising candidates for applications in quantum-optics [96]. Recently, coherent pulse propagation in macroscopic semiconductor devices has been observed in quantum-cascade lasers [97] and quantum-dash [30] and QD semiconductor amplifiers [32], relying on ultrashort, strong optical pulses. It is important to note that in the experiments using optical amplifiers, the measurements were performed at room temperature, opening up possible future quantum-coherent applications using uncooled devices.

In this section, we present a theoretical description of Rabi oscillations induced by ultrashort pulses in a QDSOA at room temperature [32]. Rabi oscillations denote a periodic exchange of energy between the optical field and the active medium, which leads to characteristic modifications of the optical pulse traveling through the amplifier device. The principle of Rabi oscillations can be derived from the dynamic equation for a single QD subgroup:

$$\frac{\partial}{\partial t}p_m^j(z,t) = -\left[i(\omega_m^j - \omega) + \frac{1}{\tau_2}\right]p_m^j - i\frac{\mu_m}{2\hbar}\left(\rho_{e,m}^j + \rho_{h,m}^j - 1\right)E(z,t), \quad (23.67)$$

$$\frac{\partial}{\partial t}\rho_{b,m}^j(z,t) = \frac{1}{\hbar}\text{Im}\left[p_m^j\mu_m^*E^*(z,t)\right] - W_m\rho_{e,m}^j\rho_{h,m}^j + \frac{\partial}{\partial t}\rho_{b,m}^j\bigg|_{\text{sc}}. \quad (23.68)$$

We reduce the above equations by introducing the inversion $d = (\rho_{e,m}^j + \rho_{h,m}^j - 1)$ and neglecting losses. We also limit ourselves to a resonant excitation, i.e., $\omega = \omega_m^j$:

$$\frac{\partial}{\partial t}p_m^j(z,t) = -i\frac{\mu_m}{2\hbar}d(z,t)E(z,t), \quad (23.69)$$

$$\frac{\partial}{\partial t}d(z,t) = \frac{2}{\hbar}\text{Im}\left[\mu_m^*E^*(z,t)p_m^j(z,t)\right]. \quad (23.70)$$

For an initial inversion of d_0 and zero polarization, the above equations have the solution

$$p_m^j(z,t) = -\frac{i}{2}\sin\Theta(z,t), \quad (23.71)$$

$$d(z,t) = d_0\cos\Theta(z,t), \quad (23.72)$$

with the pulse area defined as

$$\Theta(z,t) = \int_{-\infty}^{t} \frac{\mu_{\text{GS}}}{2\hbar}\left|E(z,t')\right|dt'. \quad (23.73)$$

Equation 23.72 shows that, with a proper choice of the pulse area, it is possible, e.g., to invert the charge-carrier distribution by choosing $\Theta = \pi$. When choosing $\Theta = n2\pi, n \in \mathbb{N}$, i.e., integer multiples of 2π, the system returns to its initial state after the exciting pulse has passed through. This important property shows that the coherent interaction between the optical field and the active material is a reversible process, as long as the polarization dephasing time is much larger than the pulse width. In most cases, however, this condition is not strictly fulfilled, leading to an additional damping of the Rabi oscillations[48].

We now apply our QD amplifier model to the amplification of strong optical pulses. We use the same pulse width of 300 fs as in the previous section, but with higher input amplitudes and we evaluate the output pulse shape as well as the integrated optical gain, as defined in Equation 23.65. For different input pulse areas Θ, the field amplitude at the output facet is shown in Figure 23.12a. At a small pulse area $\Theta = 0.1\pi$ (23 fJ pulse energy, light gray line) the pulse retains its Gaussian shape during propagation. For increasing pulse energies (darker lines in Figure 23.12), a pronounced shoulder at the leading edge of the pulse appears and the pulse shape noticeable diverts from its original Gaussian envelope. For $\Theta = 2\pi$ a pronounced dip in amplitude is visible, and a second dip appears for $\Theta = 5\pi$ (black line), where a complex pulse shape is obtained. A look into the dynamics of the optical gain, shown in Figure 23.12b, reveals a strong decrease in gain after injection of the optical pulse, as already seen in the previous section. For strong pulses, the gain exhibits oscillations and intermittently becomes negative, i.e., absorbing (see black line in Figure 23.12). This is a clear indication of the occurrence of Rabi oscillations, i.e., the coherent interaction between the microscopic polarization and the inversion. For an incoherent interaction, i.e., neglecting the coherent dynamics of the polarization, the optical gain cannot be reduced below transparency in optically pumped media. The coherent dynamics thus imprints a clear signature onto the QD amplifier behavior, which can differ strongly from conventional expectations.

We take a closer look at the dynamics of the amplifier under the amplification of strong pulses in Figure 23.13. The dynamics of the microscopic polarization and the carrier occupation across the inhomogeneously broadened QD GS at the output facet are visualized as gray scale values in Figure 23.13b and c, respectively. The optical pulse can be seen to induce a complex response of the gain medium. The QD

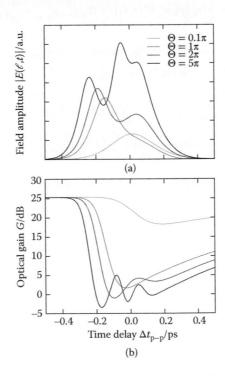

FIGURE 23.12 Rabi oscillations in the QD semiconductor optical amplifier. (a) Output electric field amplitudes for different input pulse areas $\Theta \in \{0.1\pi, 1\pi, 2\pi, 5\pi\}$, corresponding to pulse energies of 0.023, 2.3, 9.4, 58 pJ, respectively. (b) Integrated optical gain in dependence of the time delay $\Delta t_{\mathrm{p-p}}$. The pump current was set to $j = 200$ mA.

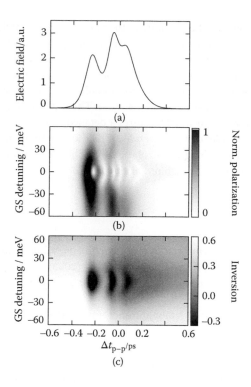

FIGURE 23.13 Quantum-dot (QD) ground-state dynamics for an input pulse area $\Theta = 5\pi$. (a) Output electric field amplitude. (b) Time evolution of the microscopic polarization amplitude p_m^j in dependence of the QD GS transition energy. The color code denotes the normalized polarization amplitude. (c) QD ground-state carrier dynamics in dependence of the QD GS transition energy. The color code denotes the QD subgroup inversion. The pump current was set to $j = 200\,\text{mA}$.

subgroups within the ensemble which are resonant to the pulse (detuning of 0 meV), exhibit pronounced Rabi oscillations visible in both the polarization and inversion. The outer subgroups show this behavior to a lesser extent but with a different temporal profile. The resulting macroscopic polarization that drives the pulse is a superposition of the individual microscopic polarization amplitudes across the whole QD ensemble (see Equation 23.13). The response of the gain medium to such strong optical pulses can, therefore, be very complex and go beyond what a simple two-level system would predict. This underlines the importance of taking both homogeneous as well as inhomogeneous effects into account when modeling light–matter interaction in QD devices in the high-power and ultrafast regime.

23.4 Conclusion

In this chapter, we have presented a model for QDSOAs that takes into account the intricate scattering dynamics between the localized QD states and the surrounding carrier reservoir states. The Coulomb scattering, which dominates the carrier dynamics, can be microscopically calculated to yield Boltzmann-like terms describing the charge-carrier scattering, which can be implemented numerically in terms of lookup tables, or approximated by simplified fit functions. The description of the electric field propagation along the device is modeled using a delay-differential equation approach, which allows a separation of numeric time step and spatial discretization length for an efficient integration algorithm.

The complex charge-carrier dynamics determine both the static and dynamic behavior of QDSOAs and is responsible for their ultrafast gain recovery. An accurate modeling of these dynamics is, therefore, necessary for a quantitatively accurate description of these optoelectronic devices. The simulations of an exemplary QDSOA device show that the signature of the internal charge-carrier dynamics can be directly seen in dynamic gain recovery measurements. The strong pump-current dependence of the scattering timescales lead to the existence of optimal operating conditions when either high gain or ultrafast gain recovery is desired. The long dephasing time of QDs compared to other semiconductor materials leads to a macroscopically accessible measurement of quantum-coherence manifesting itself in the break-up of ultrashort strong optical pulses, even at room temperature. The induced Rabi oscillations lead to a complex response of the QD gain medium which requires a detailed modeling of the polarization dynamics across the inhomogeneously broadened QD ensemble. The thoroughly nonclassical response of the QD medium to ultrashort pulses due to long-lived coherence might pave the way to novel quantum-optical devices using semiconductor QDs as the medium of choice.

References

1. P. Harrison, *Quantum Wells, Wires and Dots: Theoretical and Computational Physics of Semiconductor Nanostructures*. Chichester, UK: John Wiley & Sons (2005).
2. K. Lüdge, *Nonlinear Laser Dynamic—From Quantum Dots to Cryptography*. Weinheim: Wiley-VCH (2012).
3. D. Bimberg, Quantum dot based nanophotonics and nanoelectronics, *Electron. Lett.*, 44 (2008), p. 168.
4. J. J. Coleman, The development of the semiconductor laser diode after the first demonstration in 1962, *Semicond. Sci. Technol.*, 27 (2012), p. 090207.
5. D. Bimberg, D. Arsenijevic, G. Larisch, H. Li, J. A. Lott, P. Moser, H. Schmeckebier, and P. Wolf, Green nanophotonics for future datacom and Ethernet networks, *Proc. SPIE.*, 9134 (2014), pp. 913402–913913.
6. T. D. Steiner, *Semiconductor Nanostructures for Optoelectronic Applications, Semiconductor Materials and Devices Library*. Norwood, MA: Artech House (2004).
7. B. Lingnau, Nonlinear and Nonequilibrium Dynamics of Quantum-Dot Optoelectronic Devices, PhD thesis, Technical University of Berlin (2015).
8. B. Lingnau, W. W. Chow, E. Schöll, and K. Lüdge, Feedback and injection locking instabilities in quantum-dot lasers: A microscopically based bifurcation analysis, *New J. Phys.*, 15 (2013), p. 093031.
9. T. Akiyama, N. Hatori, Y. Nakata, H. Ebe, and M. Sugawara, Pattern-effect-free amplification and cross-gain modulation achieved by using ultrafast gain nonlinearity in quantum-dot semiconductor optical amplifiers, *Phys. Stat. Sol. B*, 238 (2003), pp. 301–304.
10. T. W. Berg, J. Mørk, and J. M. Hvam, Gain dynamics and saturation in semiconductor quantum dot amplifiers, *New J. Phys.*, 6 (2004), p. 178.
11. A. V. Uskov, E. P. O'Reilly, M. Lämmlin, N. N. Ledentsov, and D. Bimberg, On gain saturation in quantum dot semiconductor optical amplifiers, *Opt. Commun.*, 248 (2005), p. 211.
12. T. Akiyama, H. Kuwatsuka, N. Hatori, Y. Nakata, H. Ebe, and M. Sugawara, Symmetric highly efficient (~0dB) wavelength conversion based on four-wave mixing in quantum dot optical amplifiers, *IEEE Photon. Technol. Lett.*, 14 (2002), pp. 1139–1141.
13. N. Majer, K. Lüdge, and E. Schöll, Maxwell-Bloch approach to four-wave mixing in quantum dot semiconductor optical amplifiers, In *IEEE Proceedings of the 11th International Conference on Numerical Simulation of Optical Devices (NUSOD)*, edited by J. Piprek, pp. 153–154. Rome (2011).
14. O. Qasaimeh, Theory of four-wave mixing wavelength conversion in quantum dot semiconductor optical amplifiers, *IEEE Photon. Technol. Lett.*, 16 (2004), pp. 993–995.
15. H. Schmeckebier, C. Meuer, D. Arsenijevic, G. Fiol, C. Schmidt-Langhorst, C. Schubert, G. Eisenstein, and D. Bimberg, Wide-range wavelength conversion of 40-Gb/s NRZ-DPSK signals using a 1.3-μm quantum-dot semiconductor optical amplifier, *IEEE Photon. Technol. Lett.*, 24 (2012), pp. 1163–1165.

16. G. Contestabile, A. Maruta, S. Sekiguchi, K. Morito, M. Sugawara, and K. Kitayama, Cross-gain modulation in quantum-dot SOA at 1550 nm, *IEEE J. Quant. Electron.*, 46 (2010), pp. 1696–1703.

17. J. Kim, M. Laemmlin, C. Meuer, D. Bimberg, and G. Eisenstein, Theoretical and experimental study of high-speed small-signal cross-gain modulation of quantum-dot semiconductor optical amplifiers, *IEEE J. Quant. Electron.*, 45 (2009), pp. 240–248.

18. C. Meuer, H. Schmeckebier, G. Fiol, D. Arsenijevic, J. Kim, G. Eisenstein, and D. Bimberg, Cross-gain modulation and four-wave mixing for wavelength conversion in undoped and p-doped 1.3-μm quantum dot semiconductor optical amplifiers, *IEEE Photon. J.*, 2 (2010), pp. 141–151.

19. P. Borri, W. Langbein, J. M. Hvam, F. Heinrichsdorff, M. H. Mao, and D. Bimberg, Ultrafast gain dynamics in InAs–InGaAs quantum-dot amplifiers, *IEEE Photon. Technol. Lett.*, 12 (2000), 594–596.

20. S. Dommers, V. V. Temnov, U. Woggon, J. Gomis, J. Martinez-Pastor, M. Lämmlin, and D. Bimberg, Complete ground state gain recovery after ultrashort double pulses in quantum dot based semiconductor optical amplifier, *Appl. Phys. Lett.*, 90 (2007), p. 033508.

21. M. Lämmlin, GaAs-Based Semiconductor Optical Amplifiers with Quantum Dots as an Active Medium, PhD thesis, Technical University of Berlin (2006).

22. N. Majer, S. Dommers-Völkel, J. Gomis-Bresco, U. Woggon, K. Lüdge, and E. Schöll, Impact of carrier-carrier scattering and carrier heating on pulse train dynamics of quantum dot semiconductor optical amplifiers, *Appl. Phys. Lett.*, 99 (2011), p. 131102.

23. N. Majer, K. Lüdge, and E. Schöll, Cascading enables ultrafast gain recovery dynamics of quantum dot semiconductor optical amplifiers, *Phys. Rev. B.*, 82 (2010), p. 235301.

24. M. van der Poel, E. Gehrig, O. Hess, D. Birkedal, and J. M. Hvam, Ultrafast gain dynamics in quantum-dot amplifiers: Theoretical analysis and experimental investigations, *IEEE J. Quant. Electron.*, 41 (2005), pp. 1115–1123.

25. C. Meuer, C. Schmidt-Langhorst, R. Bonk, H. Schmeckebier, D. Arsenijevic, G. Fiol, A. Galperin, J. Leuthold, C. Schubert, and D. Bimberg, 80 Gb/s wavelength conversion using a quantum-dot semiconductor optical amplifier and optical filtering, *Opt. Express*, 19 (2011), pp. 5134–5142.

26. P. Borri, W. Langbein, S. Schneider, U. Woggon, R. L. Sellin, D. Ouyang, and D. Bimberg, Ultralong dephasing time in InGaAs quantum dots, *Phys. Rev. Lett.*, 87 (2001), p. 157401.

27. P. Borri, W. Langbein, S. Schneider, U. Woggon, R. L. Sellin, D. Ouyang, and D. Bimberg, Exciton relaxation and dephasing in quantum-dot amplifiers from room to cryogenic temperature, *IEEE J. Sel. Top. Quantum Electron.*, 8 (2002), pp. 984–991.

28. T. Koprucki, A. Wilms, A. Knorr, and U. Bandelow, Modeling of quantum dot lasers with microscopic treatment of Coulomb effects, *Opt. Quant. Electron.*, 42 (2011), pp. 777–783.

29. P. Borri, W. Langbein, S. Schneider, U. Woggon, R. L. Sellin, D. Ouyang, and D. Bimberg, Rabi oscillations in the excitonic ground-state transition of InGaAs quantum dots, *Phys. Rev. B*, 66 (2002), p. 081306(R).

30. A. Capua, O. Karni, G. Eisenstein, and J. P. Reithmaier, Rabi oscillations in a room-temperature quantum dash semiconductor optical amplifier, *Phys. Rev. B*, 90 (2014), p. 045305.

31. H. Kamada, H. Gotoh, J. Temmyo, H. Ando, and T. Takagahara, Exciton Rabi oscillation in single isolated quantum dots, *Phys. Stat. Sol. A*, 190 (2002), pp. 485–490.

32. M. Kolarczik, N. Owschimikow, J. Korn, B. Lingnau, Y. Kaptan, D. Bimberg, E. Schöll, K. Lüdge, and U. Woggon, Quantum coherence induces pulse shape modification in a semiconductor optical amplifier at room temperature, *Nature Commun.*, 4 (2013), p. 2953.

33. T. H. Stievater, X. Li, D. G. Steel, D. Gammon, D. S. Katzer, D. Park, C. Piermarocchi, and L. J. Sham, Rabi oscillations of excitons in single quantum dots, *Phys. Rev. Lett.*, 87 (2001), p. 133603.

34. A. Icsevgi and W. E. Lamb, Propagation of light pulses in a laser amplifier, *Phys. Rev.*, 185 (1969), pp. 517–545.

35. S. Schneider, P. Borri, W. Langbein, U. Woggon, J. Förstner, A. Knorr, R. L. Sellin, D. Ouyang, and D. Bimberg, Self-induced transparency in InGaAs quantum-dot waveguides, *Appl. Phys. Lett.*, 83 (2003), pp. 3668–3670.

36. T. W. Berg and J. Mørk, Quantum dot amplifiers with high output power and low noise, *Appl. Phys. Lett.*, 82 (2003), pp. 3083–3085.

37. T. Erneux, E. A. Viktorov, P. Mandel, T. Piwonski, G. Huyet, and J. Houlihan, The fast recovery dynamics of a quantum dot semiconductor optical amplifier, *Appl. Phys. Lett.*, 94 (2009), p. 113501.

38. S. B. Kuntze, A. J. Zilkie, L. Pavel, and J. S. Aitchison, Nonlinear state-space model of semiconductor optical amplifiers with gain compression for system design and analysis, *J. Lightwave Technol.*, 26 (2008), pp. 2274–2281.

39. M. Radziunas and H. J. Wünsche, Dynamics of multi-section DFB semiconductor laser: Traveling wave and mode approximation models, *Proc. SPIE*, 4646 (2002), p. 27.

40. J. Javaloyes and S. Balle, Multimode dynamics in bidirectional laser cavities by folding space into time delay, *Opt. Express*, 20 (2012), pp. 8496–8502.

41. W. H. Press, B. P. Flannery, S. A. Teukolsky, and W. T. Vettering, *Numerical Recipes*, 3rd ed. Cambridge: Cambridge University Press (2007).

42. M. Radziunas, Numerical bifurcation analysis of the traveling wave model of multisection semiconductor lasers, *Physica D*, 213 (2006), pp. 98–112.

43. M. Rossetti, P. Bardella, and I. Montrosset, Time-domain travelling-wave model for quantum dot passively mode-locked lasers, *IEEE J. Quant. Electron.*, 47 (2011), p. 139.

44. M. Rossetti, P. Bardella, and I. Montrosset, Modeling passive mode-locking in quantum dot lasers: A comparison between a finite- difference traveling-wave model and a delayed differential equation approach, *IEEE J. Quant. Electron.*, 47 (2011), p. 569.

45. H. Y. Liu, M. Hopkinson, C. N. Harrison, M. J. Steer, R. Frith, I. R. Sellers, D. J. Mowbray, and M. S. Skolnick, Optimizing the growth of 1.3 μm InAs/InGaAs dots-in-a-well structure, *J. Appl. Phys.*, 93 (2003).

46. Q. T. Vu, H. Haug, and S. W. Koch, Relaxation and dephasing quantum kinetics for a quantum dot in an optically excited quantum well, *Phys. Rev. B*, 73 (2006), p. 205317.

47. W. Lei, M. Offer, A. Lorke, C. Notthoff, C. Meier, O. Wibbelhoff, and A. D. Wieck, Probing the band structure of InAs/GaAs quantum dots by capacitance-voltage and photoluminescence spectroscopy, *Appl. Phys. Lett.*, 92 (2008), p. 193111.

48. M. O. Scully, *Quantum Optics*. Cambridge: Cambridge University Press (1997).

49. W. W. Chow and S. W. Koch, *Semiconductor-Laser Fundamentals*, Berlin: Springer (1999).

50. H. Haug and S. W. Koch, *Quantum Theory of the Optical and Electronic Properties of Semiconductors*, 2nd ed. Singapore: World Scientific (2004).

51. W. C. W. Fang, C. G. Bethea, Y. K. Chen, and S. L. Chuang, Longitudinal spatial in homogeneities in high-power semiconductor lasers, *IEEE J. Sel. Top. Quantum Electron.*, 1 (1995), pp. 117–128.

52. J. N. Fehr, M. A. Dupertuis, T. P. Hessler, L. Kappei, D. Marti, F. Salleras, M. S. Nomura, B. Deveaud, J.-Y. Emery, and B. Dagens, Hot phonons and Auger related carrier heating in semiconductor optical amplifiers, *IEEE J. Quant. Electron.*, 38 (2002), p. 674.

53. P. Sotirelis and K. Hess, Electron capture in GaAs quantum wells, *Phys. Rev. B*, 49 (1994), pp. 7543–7547.

54. M. Vallone, Quantum well electron scattering rates through longitudinal optic-phonon dynamical screened interaction: An analytic approach, *J. Appl. Phys.*, 114 (2013), p. 053704.

55. R. Heitz, H. Born, F. Guffarth, O. Stier, A. Schliwa, A. Hoffmann, and D. Bimberg, Existence of a phonon bottleneck for excitons in quantum dots, *Phys. Rev. B*, 64 (2001), p. 241305(R).

56. K. Schuh, P. Gartner, and F. Jahnke, Combined influence of carrier-phonon and coulomb scattering on the quantum-dot population dynamics, *Phys. Rev. B*, 87 (2013), p. 035301.

57. G. D. Mahan, *Many-Particle Physics*. New York, NY: Plenum (1990).

58. A. Wojs, P. Hawrylak, S. Fafard, and L. Jacak, Electronic structure and magneto-optics of self-assembled quantum dots, *Phys. Rev. B*, 54 (1996), p. 5604.

59. H. Haug and S. W. Koch, Semiconductor laser theory with many-body effects, *Phys. Rev. A*, 39 (1989), p. 1887.

60. H. Haug and S. Schmitt-Rink, Electron theory of the optical properties of laser-excited semiconductors, *Prog. Quant. Electron.*, 9 (1984), p. 3.
61. W. W. Chow and S. W. Koch, Theory of semiconductor quantum-dot laser dynamics, *IEEE J. Quant. Electron.*, 41 (2005), pp. 495–505.
62. H. Haug and A. P. Jauho, *Quantum Kinetics in Transport and Optics of Semiconductors*. Berlin: Springer (1996).
63. E. Malic, K. J. Ahn, M. J. P. Bormann, P. Hövel, E. Schöll, A. Knorr, M. Kuntz, and D. Bimberg, Theory of relaxation oscillations in semiconductor quantum dot lasers, *Appl. Phys. Lett.*, 89 (2006), p. 101107.
64. T. R. Nielsen, P. Gartner, and F. Jahnke, Many-body theory of carrier capture and relaxation in semiconductor quantum-dot lasers, *Phys. Rev. B*, 69 (2004), p. 235314.
65. L. Bányai, Q. T. Vu, B. Mieck, and H. Haug, Ultrafast quantum kinetics of time-dependent RPA-screened Coulomb scattering, *Phys. Rev. Lett.*, 81 (1998), pp. 882–885.
66. R. Binder, D. Scott, A. E. Paul, M. Lindberg, K. Henneberger, and S. W. Koch, Carrier-carrier scattering and optical dephasing in highly excited semiconductors, *Phys. Rev. B*, 45 (1992), p. 1107.
67. F. X. Camescasse, A. Alexandrou, D. Hulin, L. Bányai, D. B. Tran Thoai, and H. Haug, Ultrafast electron redistribution through Coulomb scattering in undoped GaAs: Experiment and theory, *Phys. Rev. Lett.*, 77 (1996), pp. 5429–5432.
68. M. G. Kane, Nonequilibrium carrier-carrier scattering in two-dimensional carrier systems, *Phys. Rev. B*, 54 (1996), pp. 16345–16348.
69. D. B. Tran Thoai and H. Haug, Coulomb quantum kinetics in pulse-excited semiconductors, *Z. Phys. B*, 91 (1992), pp. 199–207.
70. E. Schöll, *Nonequilibrium Phase Transitions in Semiconductors*. Berlin: Springer (1987).
71. K. Lüdge and E. Schöll, Quantum-dot lasers—Desynchronized nonlinear dynamics of electrons and holes, *IEEE J. Quant. Electron.*, 45 (2009), pp. 1396–1403.
72. K. Lüdge, Modeling quantum dot based laser devices. In *Nonlinear Laser Dynamics–From Quantum Dots to Cryptography*, Chapter 1, pp. 3–34. Weinheim: Wiley-VCH (2012).
73. M. Rossetti, A. Fiore, G. Sek, C. Zinoni, and L. Li, Modeling the temperature characteristics of InAs/GaAs quantum dot lasers, *J. Appl. Phys.*, 106 (2009), p. 023105.
74. J. Urayama, T. B. Norris, H. Jiang, J. Singh, and P. Bhattacharya, Temperature-dependent carrier dynamics in self-assembled InGaAs quantum dots, *Appl. Phys. Lett.*, 80 (2002), pp. 2162–2164.
75. T. Erneux, E. A. Viktorov, and P. Mandel, Time scales and relaxation dynamics in quantum-dot lasers, *Phys. Rev. A*, 76 (2007), p. 023819.
76. P. Baveja, D. Maywar, A. Kaplan, and G. P. Agrawal, Self-phase modulation in semiconductor optical amplifiers: Impact of amplified spontaneous emission, *IEEE J. Quant. Electron.*, 46 (2010), pp. 1396–1403.
77. T. W. Berg and J. Mørk, Saturation and noise properties of quantum-dot optical amplifiers, *IEEE J. Quant. Electron.*, 40 (2004), pp. 1527–1539.
78. A. M. de Melo and K. Petermann, On the amplified spontaneous emission noise modeling of semiconductor optical amplifiers, *Opt. Commun.*, 281 (2008), pp. 4598–4605.
79. C. Meuer, J. Kim, M. Lämmlin, S. Liebich, A. Capua, G. Eisenstein, A. R. Kovsh, S. S. Mikhrin, I. L. Krestnikov, and D. Bimberg, Static gain saturation in quantum dot semiconductor optical amplifiers, *Opt. Express*, 16 (2008), p. 8269.
80. C. W. Gardiner, *Handbook of Stochastic Methods*. New York, NY: Springer (1985).
81. N. Wiener, Generalized harmonic analysis, *Acta Math.*, 55 (1930), pp. 117–258.
82. R. Bonk, T. Vallaitis, J. Guetlein, C. Meuer, H. Schmeckebier, D. Bimberg, C. Koos, and J. Leuthold, The input power dynamic range of a semiconductor optical amplifier and its relevance for access network applications, *IEEE Photon. J.*, 3 (2011), pp. 1039–1053.

83. S. Wilkinson, B. Lingnau, J. Korn, E. Schöll, and K. Lüdge, Influence of noise on the signal properties of quantum-dot semiconductor optical amplifiers, *IEEE J. Sel. Top. Quantum Electron.*, 19 (2013), p. 1900106.

84. G. E. P. Box and M. E. Muller, A note on the generation of random normal deviates, *Ann. Math. Statist.*, 29 (1958), pp. 610–611.

85. H. Shahid, D. Childs, B. J. Stevens, and R. A. Hogg, Negative differential gain due to many body effects in self-assembled quantum dot lasers, *Appl. Phys. Lett.*, 99 (2011), p. 061104.

86. E. Goldmann, M. Lorke, T. Frauenheim, and F. Jahnke, Negative differential gain in quantum dot systems: Interplay of structural properties and many-body effects, *Appl. Phys. Lett.*, 104 (2014), p. 242108.

87. S. W. Koch, T. Meier, F. Jahnke, and P. Thomas, Microscopic theory of optical dephasing in semiconductors, *Appl. Phys. A*, 71 (2000), pp. 511–517.

88. M. Lorke, T. R. Nielsen, J. Seebeck, P. Gartner, and F. Jahnke, Influence of carrier-carrier and carrier-phonon correlations on optical absorption and gain in quantum-dot systems, *Phys. Rev. B*, 73 (2006), p. 085324.

89. H. H. Nilsson, J. Z. Zhang, and I. Galbraith, Homogeneous broadening in quantum dots due to Auger scattering with wetting layer carriers, *Phys. Rev. B*, 72 (2005), p. 205331.

90. H. Tahara, Y. Ogawa, F. Minami, K. Akahane, and M. Sasaki, Long-time correlation in non-Markovian dephasing of an exciton-phonon system in InAs quantum dots, *Phys. Rev. Lett.*, 112 (2014), p. 147404.

91. H. Schmeckebier, B. Lingnau, S. König, K. Lüdge, C. Meuer, A. Zeghuzi, D. Arsenijevic, M. Stubenrauch, R. Bonk, C. Koos, C. Schubert, T. Pfeiffer, and D. Bimberg, Ultra-broadband bidirectional dual-band quantum-dot semiconductor optical amplifier. In *Optical Fiber Communication Conference and Exposition* (2015), p. Tu3I.7.

92. A. Girndt, A. Knorr, M. Hofmann, and S. W. Koch, Theoretical analysis of ultrafast pump-probe experiments in semiconductor amplifiers, *Appl. Phys. Lett.*, 66 (1995), p. 550.

93. E. P. Ippen and C. V. Shank, Techniques for measurement, edited by S. L. Shapiro, In *Ultrashort Light Pulses: Picosecond Techniques and Applications*, pp. 83–122. Berlin: Springer (1977).

94. J. M. A. Mecozzi and J. Mørk, Theory of heterodyne pump-probe experiments with femtosecond pulses, *J. Opt. Soc. Am. B*, 13 (1996), pp. 2437–2452.

95. P. Borri, S. Schneider, W. Langbein, and D. Bimberg, Ultrafast carrier dynamics in InGaAs quantum dot materials and devices, *J. Opt. A: Pure Appl. Opt.*, 8 (2006), p. S33.

96. T. Brandes, Coherent and collective quantum optical effects in mesoscopic systems, *Phys. Rep.*, 408 (2005), pp. 315–474.

97. H. Choi, V. M. Gkortsas, L. Diehl, D. Bour, S. Corzine, J. Zhu, G. Höfler, F. Capasso, F. X. Kärtner, and T. B. Norris, Ultrafast Rabi flopping and coherent pulse propagation in a quantum cascade laser, *Nature Photon.*, 4 (2010), pp. 706–710.

24

Wave Mixing Effects in Semiconductor Optical Amplifiers

Simeon N.
Kaunga-Nyirenda

Michal Dlubek

Jun Jun Lim

Steve Bull

Andrew Phillips

Slawomir Sujecki

and

Eric Larkins

24.1 Introduction

The growing demand for telecommunications services continues to push optical transmission rates beyond 10 G-symbols/s and to use more efficient modulation formats (e.g., quadrature phase shift keying [QPSK], differential QPSK [DQPSK]) to cope with the ever-increasing demand for bandwidth. The growing demand for bandwidth is also driving a move to perform high bandwidth signal processing functions directly in the optical domain. All-optical signal processing generally requires nonlinear optical functionality. Semiconductor optical amplifiers (SOAs) are attractive for such functions because of their strong nonlinear optical response. Optical signal processing applications that employ SOAs include wavelength conversion, phase conjugation, optical switching, Boolean logic functions, all-optical recovery, and all-optical demultiplexing. SOAs offer large bandwidth, ease of integration, compactness, low power consumption, electrical pumping/control, and potentially low cost. Interest in SOAs continues to grow as photonic integrated circuit (PIC) integration densities continue to increase, with applications in signal leveling and signal regeneration or integration with electronics as optical interconnects for high-speed cross-chip communications. SOAs are also useful in electrically reconfigurable optical switching/routing networks (Williams et al., 2008). For these applications, however, SOAs may introduce nonlinear signal impairments that adversely affect system performance, and these impairments must be managed.

This chapter focuses on the numerical modeling and simulation of carrier plasma–induced nonlinear optical phenomena in SOAs. The chapter starts with a brief description of nonlinear processes in SOAs and their applications. We then discuss SOA modeling, paying particular attention to challenges encountered in the modeling and simulation of nonlinear effects. An efficient optical model based on a one-dimensional (1D) bidirectional traveling wave algorithm is then presented. The optical model is coupled to the semiconductor material through light–matter interactions. The electrical properties are simulated using a simplified 1D (unipolar) electrical model, which only takes into account the dynamics of the total carrier density in the active medium. The unipolar model assumes flat quasi-Fermi levels and charge neutrality in order to remove the additional computational complexity associated with the use of a self-consistent bipolar electrical model with band bending (due to charge imbalance in the well). Nevertheless, while a bipolar model will certainly improve the overall accuracy of the model, the focus of this work was the development and demonstration of an efficient model for the nonlinear optical response. For this purpose, the simplicity of the unipolar model is an advantage.

The material polarization is represented by an effective susceptibility that self-consistently includes the carrier-induced perturbations of the net optical gain and refractive index spectra, both of which are functions of the optical field. No particular functional dependence on the optical field is assumed. This allows a general representation of the light–matter interaction under quasi-equilibrium conditions, which is applicable for all optical intensity levels. The most widely used power series expansion of the polarization is usually implemented with a single nonlinear term, which is only applicable when the optical intensity is sufficiently weak. The gain and spontaneous emission spectra are calculated using the band structure details of the semiconductor. These spectra were appropriately broadened to account for the spectral broadening of the carrier energies in the bands due to carrier lifetime dependent dephasing and various scattering events. The refractive index spectrum is obtained from the broadened gain spectrum using a nonlinear Kramers–Kronig transformation.

The need to perform simulations in the time domain poses a particular challenge for the inclusion of the frequency-dependent complex material polarization (i.e., gain and refractive index dispersion) and spontaneous emission spectra. We use recursive digital filter functions to accurately represent the spectrally dependent material quantities in the time-domain simulation. This method captures the asymmetric nature of the spectral material response and can also be used to incorporate frequency responses obtained from experimental measurements as well as complicated responses that cannot easily be represented analytically. The use of spatiotemporal filter coefficients ensures that physical effects such as spatial-hole burning and spectral changes in the gain caused by band-filling and bandgap renormalization effects are correctly included. A further advantage of the recursive filter method is that the filter coefficients can be obtained offline from the main simulation engine and tabulated, improving the computational efficiency and minimizing the resources required by the simulations. The refractive index spectra are also represented more accurately than approaches using a constant linewidth enhancement factor (at best a small-signal representation of the Kramers–Kronig relation, which ignores the spectral dependence of the index changes).

The chapter concludes with a case study exploring carrier density modulation (CDM) caused by optical wave beating in SOAs. We use the time-domain model developed to obtain an understanding of wave mixing processes in SOAs and the dependence of CDM on different operating parameters. The simulation results are compared with experimental results from a commercial buried heterostructure (BH) 1550 nm InGaAs multiple quantum well (MQW)-SOA made by the Centre for Integrated Photonics (SOA-NL-OEC-1550).

24.2 Review of Nonlinear Phenomena in SOAs and Applications

Nonlinear optical processes occur when the complex dielectric response of the material depends on the strength of the optical field passing through the medium (Boyd, 2007). When an optical wave interacts with the semiconductor gain medium, there is a shift in the carriers of the gain medium, resulting in a field-induced dipole moment that becomes the source of an electromagnetic field at the microscopic level. This interaction between the optical field and the material is through the induced macroscopic material polarization (the polarization term in the wave equation; see Section 24.4.1). This macroscopic polarization is the sum of the distinct microscopic polarizations

$$P(\vec{r}, t) = \frac{2}{V_k} \sum_k d_{cv}(k)p(k, \vec{r}, t), \tag{24.1}$$

where V_k is the normalized volume, $d_{cv}(k)$ is the optical dipole matrix element, and p is the interband polarization. k, $\vec{r}$, and t are the electron wave vector, position vector, and time, respectively. The random high-speed temporal fluctuations of the microscopic dipole elements are eliminated by spatial averaging over the incoherent dipole ensemble. Only coherent (i.e., externally driven) fluctuations remain and their response can be described with a macroscopic polarization (Jackson, 1975). This induced macroscopic polarization, $\tilde{P}^{\pm}$, is a frequency-dependent quantity that is a function of the total optical field, $\tilde{E}$, interacting with the medium via the time-varying carrier distributions. In the frequency domain, the polarization is related to the total electric field by

$$\tilde{P}^{\pm}(\omega, N, \tilde{E}) = \varepsilon_0 \chi_{\text{eff}}(\omega, N, \tilde{E})\tilde{E}^{\pm}, \tag{24.2}$$

where $\tilde{P}(\omega)$ and $\tilde{E}(\omega)$ are the Fourier transforms of $P(t)$ and $E(t)$, respectively, and ω is the angular frequency. N is the total carrier density in the active region. The effective susceptibility, χ_{eff} takes into account all processes, both linear and nonlinear. No particular functional dependence on the optical field is assumed in Equation 24.2. When the optical field is small, the polarization can be expanded into a power series of the optical field and the effective susceptibility is

$$\chi_{\text{eff}} = \sum_n \chi^{(n+1)} \tilde{E}^n : n = 0, 1, 2 \ldots \tag{24.3}$$

At low optical intensities, the optical properties of the material tend to behave in a linear manner. However, as the optical intensity increases, the strong nonlinear response, which results from the interaction between strong optical signals and the gain medium, can manifest itself as self- and cross-gain modulation (SGM/XGM), self-, and cross-phase modulation (SPM/XPM), and wave mixing. SOAs are optoelectronic devices whose behavior is governed by the material properties of the amplifying medium and the light propagation through it. These are strongly coupled and influenced by the operating conditions, e.g., input power levels, pumping current and form of the input. The nonlinear processes arise because of the dynamics of the excited carrier populations (and their distributions in real and momentum space) as the material interacts with the optical radiation. These nonlinear phenomena are detrimental to applications that rely on device linearity but can be exploited for optical signal processing and other applications requiring all-optical functionality.

CDM, carrier heating (CH), and spectral hole burning (SHB) are the most important processes giving rise to nonlinear optical phenomena in semiconductor devices.

24.2.1 Carrier Density Modulation

CDM is a coherent interband process that results from the dependence of the stimulated carrier recombination rate on the optical intensity. Stimulated emission and absorption processes, and to some extent the (amplified) spontaneous emission process, modify the total carrier densities in the gain medium, leading to changes in the total gain spectrum and the refractive index of the semiconductor material. A strong optical input signal will increase the stimulated emission rate and decrease the carrier density in the active material. This reduces the gain and perturbs the refractive index, which affect the propagation of both the perturbing signal and other co-/counterpropagating signals. This is the basis for effects such as SGM, XGM, SPM and XGM. When at least two waves distinguishable in frequency, polarization, or wave vector propagate through the SOA, wave beating can excite the material by modulating the carrier density. This modulation in turn creates dynamic gain and index gratings, which are responsible for creating new frequency components.

24.2.2 Carrier Heating

Carrier heating (CH) results from modulation of the carrier energy distributions in the valence and/or conduction bands (sub-bands for quantum wells [QWs] and wires). CH is caused by processes that increase the temperatures of the carrier distributions relative to the lattice temperature, including carrier injection, stimulated emission, and free carrier absorption (FCA). Carrier injection adds high energy ("hot") carriers to the carrier plasma, while stimulated emission removes "cold" carriers close to the band edge. FCA transfers energy from photons to carriers, moving them to higher energies within the bands. Such carriers quickly thermalize and transfer energy to other carriers by carrier–carrier scattering, thereby increasing the plasma energy density or temperature (Uskov et al., 1994).

24.2.3 Spectral Hole Burning

SHB is also a form of modulation of the energy distributions of the carriers within the valence and/or conduction bands (or sub-bands for QWs and wires). While CH perturbs the entire carrier plasma, the plasma can usually still be described with a quasi-equilibrium distribution with an increased plasma temperature. SHB, on the other hand, is an energetically localized perturbation of the plasma, which cannot be described with a quasi-equilibrium distribution. SHB occurs when carriers in the states supporting stimulated recombination are consumed faster than they are replenished by intraband scattering. This creates a spectral hole in the intraband carrier distribution relative to the corresponding thermal carrier distribution.

In contrast to CDM, CH, and SHB are intraband processes. SHB is due to changes in the carrier distribution away from Fermi–Dirac distributions and is responsible for gain saturation. (The gain compression coefficient, frequently used to describe gain saturation at high power, is a manifestation of SHB.) CH results in a Fermi–Dirac distribution that has a higher temperature than that of the lattice. CDM, CH, and SBH all create perturbations that result in the modulation of the gain/absorption and refractive index spectra, whose dynamics also create spatiotemporal index and gain gratings that can diffract waves propagating in the medium to generate new frequency components. Although the gain and index gratings caused by CDM are larger, CH and SHB contribute to the nonlinear optical response. CDM dominates up to frequencies limited by the carrier replenishment rate. CH and SHB are important at higher frequencies, but CDM remains important.

24.2.4 Other Nonlinear Effects

Other processes, e.g., optical Kerr effect, two-photon absorption (TPA), and second harmonic generation (SHG), also contribute to the nonlinear optical response, but their contribution is smaller compared to that of CDM and CH, and are neglected in this work. The inclusion of the Kerr effect in simulations is complicated by the lack of causality. TPA is a multiphoton process that is resonantly enhanced for photon energies of the order of half the bandgap of the semiconductor material. Instantaneous coherent processes like the optical Kerr effect, TPA and SHG are not discussed further here, but are important for some applications (e.g., all-optical gates) at high optical power densities. The reader is referred to Boyd (2007) or other texts on nonlinear optics for more information.

24.2.5 Summary of Important Nonlinear Applications

Optical nonlinearities cause signal impairments in linear applications, but have been exploited to great benefit in the field of functional photonics. SOA nonlinearities are used for all-optical signal processing to eliminate the need for optical–electrical–optical (OEO) conversion in communication networks. Processing tasks such as demultiplexing, clock recovery, regeneration, and routing can be undertaken entirely in the optical domain at higher speeds than are possible with electronics (Manning et al., 1997). One of the key applications of nonlinear processes in SOAs is wavelength conversion.

Wavelength converters increase the flexibility and performance of all-optical networks based on wavelength division multiplexing (WDM) (Kovačević and Acampora, 1995). All-optical wavelength converters are essential for coping with increasing data rates. They increase the network performance by reducing the effects of lightpath blocking due to the wavelength continuity constraints in optical packet and optical circuit switching networks. Of the different approaches to all-optical wavelength conversion, SOA-based wavelength converters have attracted the greatest interest because of their compactness, easy integration, and strong nonlinear response. Different approaches have been used for SOA-based wavelength conversion, including XGM, XPM, and four-wave mixing (FWM). FWM is the most flexible approach because it has a large extinction ratio and retains the phase information of the carrier signal, making it bitrate independent and intrinsically transparent to modulation format (a useful attribute in optical networks). FWM-based SOA wavelength converters also have the potential for processing ultra-high speed analog and digital signals using ultrafast processes like SHB and CH. Spectral inversion also occurs during the FWM-conversion process. FWM-converted signals with inverted spectral distribution have been used for dispersion compensation by positioning a nonlinear SOA at the center of the link (Yanhua et al., 2003).

Another important application of nonlinear processes in SOA is optical performance monitoring in optical networks. The increasing complexity of optical networks and the emergence of new types of traffic and data protocols have rendered digital signal monitoring in the electronic domain impractical (Dlubek, 2008). As a result, optical signal monitoring methods that measure the analogue parameters in the optical domain are required for transparent optical networks. Optical sampling using FWM in SOAs has been used to generate signal histograms for bit error rate (BER) estimation (Dlubek, 2008).

24.3 Challenges in Modeling Nonlinear Effects

SOA modeling is a sophisticated subject due to the number and complexity of processes that take place simultaneously in the device. The dynamic nature of optical nonlinearities requires the use of time-domain models. Time-domain simulations are needed to determine the carrier densities (i.e., drift-diffusion, generation/recombination effects, etc.), whereas the calculation of the complex index needs to be done in the frequency domain. Optical propagation can be done in the frequency domain, but spatiotemporal carrier density (and complex index) variations make time-domain models more tractable.

The challenge for time-domain simulations is the inclusion of the frequency-dependent complex material polarization (gain and refractive index dispersion). Several attempts have been made to include dispersion in time-domain simulations: single-pole Lorentzian filter (Pratt and Carroll, 2000), finite impulse filter (FIR) theory (Jones et al., 1995; Toptchiyski et al., 1999), a first-order (Durhuus et al., 1992) and a second-order (Das et al., 2000) Taylor series expansion about the gain peak wavelength with constant coefficients. The second-order Taylor series expansion and the Lorentzian profile are both inadequate for the asymmetric material gain spectrum. Furthermore, the use of constant filter coefficients is inadequate for representing carrier-induced shifts in the gain spectrum. Other approaches include the spectrum slicing technique (SSM) (Park et al., 2005) and effective Bloch equations (EBE). The SSM method can handle the asymmetric gain spectrum, but is not suited to modeling nonlinear interactions (Park et al., 2005). The EBE fits the gain and refractive index spectra using a series of Lorentzians. A single Lorentzian is used in Ning et al. (1997), which is only useful for signals with a very narrow spectrum. Accurate implementation for high bandwidth signals requires the inclusion of more Lorentzian terms, making the model computationally intensive.

24.4 Improved Modeling of Nonlinear Phenomena in SOAs

The theoretical model described here is applied to an MQW SOA fabricated with a BH (Figure 24.1). The operation of the SOA depends on the interaction of the optical waves with the material. Processes such as optical gain, optical absorption, and changes in the refractive index must be appropriately coupled to the optical model in order to develop numerical models that accurately represent the operation of the actual device. This section describes the different components of a coupled optoelectronic simulation tool for nonlinear SOAs.

24.4.1 Optical Model

The optical field propagation in the SOA active region is described by Maxwell's equations

$$\nabla \times \nabla \times \vec{E} + \frac{n^2}{c^2}\frac{\partial^2 \vec{E}}{\partial t^2} = -\frac{1}{\varepsilon_0 c^2}\left(\frac{\partial^2 \vec{P}}{\partial t^2} + \frac{\partial \vec{j}}{\partial t}\right), \qquad (24.4)$$

where n is the refractive index of the unexcited material (background index), c is the speed of light in vacuum, ε_0 is the free-space permittivity, $\vec{P}(x, y, z, t)$ is the macroscopic polarization of the medium, and $\vec{j}(x, y, z, t)$ is the current density. It is assumed that the SOA is designed to support only the fundamental mode with a transverse field distribution $U(x, y)$, which is assumed to be wavelength independent and normalized such that $\int\int |U(x, y)|^2 \, dxdy = 1$. For simplicity, all effective and modal parameters (e.g., the transverse confinement factor) are assumed to have been obtained at this point, so that only the longitudinal spatial direction needs to be considered. The strong waveguiding of the BHs further confines the optical field in the transverse directions beyond the vertical confinement of the separate confinement heterostructure (SCH) (Agrawal and Dutta, 1986).

The variation of the injection current with respect to time is very small compared to that of the optical fields, so it is justifiable to assume

$$\frac{\partial j}{\partial t} = \frac{\partial (\sigma E)}{\partial t} \approx 0. \qquad (24.5)$$

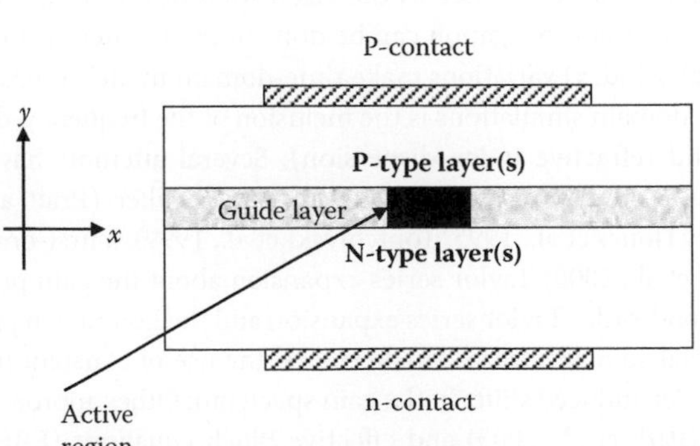

FIGURE 24.1 Schematic of a buried heterostructure (BH) semiconductor optical amplifier (SOA).

An alternative way of looking at this assumption is that optical fields are not usually generated directly by free currents and charges (Liu, 2005). Under these assumptions, the wave equations take the same form as those in a medium free of sources. The assumption in Equation 24.5 is plausible at optical frequencies, but this term may contribute in nonlinear optical beat-frequency devices, e.g., microwave and terahertz (THz) devices. This assumption is also not *a priori* valid for plasmonic devices (i.e., metals). For a straight and symmetric waveguide, the fields can be assumed to remain linearly polarized during propagation through the device. For a single polarization component, this implies that $\nabla \cdot \vec{E} = \nabla \cdot \vec{P} = 0$ and the left-hand side of Equation 24.1 is $\nabla \times \nabla \times \vec{E} = \nabla \left(\nabla \cdot \vec{E} \right) - \nabla^2 \vec{E} = -\nabla^2 \vec{E}$, so the scalar wave equation can be used.

Zero facet reflectivities are the target in traveling-wave SOAs. However, facets with antireflection (AR) coatings still exhibit residual reflectivities, which can result in the formation of an optical cavity with longitudinal resonances. Only real amplitude reflection coefficients are considered in this work. Although any wave reflected at the facet may initially be small, it will be amplified as it back-propagates along the waveguide such that it is not negligible compared to the forward field. Thus, reflected fields contribute to carrier depletion, gain compression, and refractive index change in the active medium. This affects the amplification, pulse shape, and phase of the output field. Reflected fields can also give rise to delayed replicas or echoes of the bits, which act as noise. Fields generated by ASE may also be nonnegligible compared to forward travelling fields. In the general case, the total optical field is comprised of both backward and forward propagating fields. Counterpropagating fields are also used deliberately in some applications relying on XGM and/or XPM.

Dropping the vector nature and using the slowly varying envelope approximation, the total optical field in the active region can be written as

$$E(x, y, z, t) = \frac{1}{2} U(x, y) \left\{ \psi^+(z, t) \exp \left[i \left(\omega t - kz \right) \right] + \psi^-(z, t) \exp \left[i \left(\omega t + kz \right) \right] + c.c. \right\}. \tag{24.6}$$

Taking into account that multiple signals may be input into the SOA, this can be rewritten as

$$E(x, y, z, t) = U(x, y) \sum_q \left\{ \psi_q^+(z, t) \exp \left[i \left(\omega_q t - k_q z \right) \right] + \psi_q^-(z, t) \exp \left[i \left(\omega_q t + k_q z \right) \right] + c.c. \right\}, \tag{24.7}$$

where $\lambda_q = \frac{c}{f_q}$ is the center wavelength of the qth signal and $\omega_q = 2\pi f_q$. From Equations 24.6 and 24.7, the total forward and backward slowly varying components are

$$\psi^+(z, t) = \sum_q \psi_q^+(z, t) \exp \left\{ i \left[\left(\omega_q - \omega_{\text{ref}} \right) t - \left(k_q - k_{\text{ref}} \right) z \right] \right\} + c.c. \tag{24.8}$$

$$\psi^-(z, t) = \sum_q \psi_q^-(z, t) \exp \left\{ i \left[\left(\omega_q - \omega_{\text{ref}} \right) t + \left(k_q - k_{\text{ref}} \right) z \right] \right\} + c.c., \tag{24.9}$$

where $k_{\text{ref}} = \frac{n \omega_{\text{ref}}}{c}$ is the optical propagation constant in the longitudinal direction and ω_{ref} is a reference frequency for the baseband transformation of all the signals propagating through the active medium. $\psi^+(z, t)$ and $\psi^-(z, t)$ are the forward and backward propagating complex envelopes of the optical field. The induced polarization can be written in a similar form as the field

$$P(x, y, z, t) = \frac{1}{2} U(x, y) \left\{ P^+(z, t) \exp \left[i (\omega t - kz) \right] + P^-(z, t) \exp \left[i (\omega t + kz) \right] + c.c \right\}, \tag{24.10}$$

where $P^+(z, t)$ and $P^-(z, t)$ are the slowly varying forward and backward propagating complex polarizations. Substitution of Equations 24.7 and 24.10 into the scalar version of the wave Equation 24.4 and applying the slowly varying envelope approximation (SVEA) and paraxial approximation leads to the following propagation equations

$$\frac{n}{c}\frac{\partial \psi^{\pm}}{\partial t} \pm \frac{\partial \psi^{\pm}}{\partial z} = i\frac{\Gamma_{xy}k_{ref}}{2\varepsilon_0 n^2}P^{\pm} + \psi_{sp}^{\pm}(z,t),$$ (24.11)

which are subject to the boundary conditions

$$\psi^{+}(0,t) = \sqrt{R_1}\psi^{-}(0,t) + \sqrt{1 - R_1}\psi_{in1}(0,t)$$ (24.12)

$$\psi^{-}(L,t) = \sqrt{R_2}\psi^{+}(L,t) + \sqrt{1 - R_2}\psi_{in2}(L,t),$$ (24.13)

where $\psi_{in1}(0,t)$ and $\psi_{in2}(L,t)$ are optical inputs injected through the end facets. These facets have power reflectivities R_1 and R_{21}, respectively. The last term in Equation 24.11 is the local spontaneous noise contribution to the total propagating field. For simplicity, k_{ref} is referred to as k from here onward.

The transverse confinement factor, Γ_{xy}, in 24.11 is given by

$$\Gamma_{xy} = \frac{\int_{-w/2}^{w/2}\int_{-d/2}^{d/2}|U(x,y)|^2\,dy\,dx}{\int_{-\infty}^{\infty}\int_{-\infty}^{\infty}|U(x,y)|^2\,dy\,dx},$$ (24.14)

where w and d are the active region width and thickness, respectively.

24.4.2 Electrical Model

The carrier dynamics are calculated using a simplified 1D carrier density model based on the evolution of the total carrier density in the active region (White et al., 1998)

$$\frac{\partial N(z,t)}{\partial t} = \frac{\eta J(z,t)}{ed} - \gamma_{nr}(N)N - \frac{i\Gamma_{xy}}{4\hbar}\left\{(\psi^{+})(P^{+})^{*} + (\psi^{-})(P^{-})^{*} - c.c.\right\},$$ (24.15)

where J is the total injected current density, e is the electron charge, γ_{nr} represents nonradiative recombination (generally carrier dependent), and η is the current injection efficiency. Carrier diffusion effects have been neglected in Equation 24.15 since the longitudinal electrical mesh spacing is larger than the carrier diffusion length. Thus, the total current density is only due to the external source from the electrodes. The current density from the external source is determined as in Dai et al. (1997)

$$J_{ext}(z,t) = \frac{V_{bias} - V_{jcn}(z,t)}{r_s},$$ (24.16)

where r_s is the series resistance (in units of Ωm^{-2}) of the p- and n-layers on either side of the active region and the p- and n-contact resistance. V_{jcn} is the voltage drop across the active layer (p–n junction) and relates to the difference between the two quasi-Fermi levels as

$$V_{jcn}(z,t) = \frac{E_{Fn}(z,t) - E_{Fp}(z,t)}{e},$$ (24.17)

where E_{Fn} and E_{Fp} are the electron and hole quasi-Fermi levels, respectively. The quasi-Fermi levels are calculated within the gain model under the commonly used assumption of charge neutrality. While charge neutrality is a valid assumption in bulk material, it may not be a valid assumption for QWs, especially at high current pumping levels. A significant charge imbalance in the QW may strongly affect the optical properties of the surrounding optical guiding layers (SCH layers) and the entire waveguide due to the carriers injected into the region (Tolstikhin, 2000). Nevertheless, charge neutrality is assumed in order to use the unipolar model to simplify the analysis, which should be acceptable for moderate injection currents. E_{Fn} and E_{Fp} are estimated using the Joyce–Dixon approximation (Joyce and Dixon, 1977)

$$E_{Fn,p} = k_B T \left[\ln\left(\frac{N}{N_{c,v}}\right) + \frac{1}{2}\left(\frac{N}{N_{c,v}}\right) + \frac{1}{24}\left(\frac{N}{N_{c,v}}\right)^2 - 0.0000347\left(\frac{N}{N_{c,v}}\right)^3 \right], \tag{24.18}$$

where N_c and N_v are the effective densities of states for electrons and holes, k_B is the Boltzmann constant, and T is the temperature. The nonradiative processes considered in Equation 24.15 are Shockley–Read–Hall (SRH) and Auger recombination. The recombination rates are calculated under the charge neutrality condition.

24.4.3 Light–Matter Interactions

The interaction between light and the semiconductor gain medium is represented by the material (interband) polarization through an effective susceptibility (Equation 24.2). A general description of the polarization that is applicable for all levels of optical intensity is necessary. The effective susceptibility consists of the carrier-dependent changes in net gain and refractive index, both of which are functions of the optical field

$$\chi_{\text{eff}}\left(\omega, N, \tilde{E}\right) = -\frac{1}{k}\left\{ i\left[g_{net}\left(\omega, N, \tilde{E}\right) - \left(\frac{1 - \Gamma_{xy}}{\Gamma_{xy}}\right)\alpha_{conf}\right] + k\Delta n\left(\omega, N, \tilde{E}\right) \right\}, \tag{24.19}$$

where α_{conf} represents the optical losses in the confinement regions. α_{conf} is needed because $1 - \Gamma_{xy}$ of the optical energy is located in the confinement layers outside the gain material. Equation 24.19 allows the induced nonlinear polarization to be described in a general fashion, without any assumptions based on its functional dependence on the optical field. This approach is more straightforward and physically robust than the common truncated series expansion of the polarization based on Equation 24.3. With this representation, the band structure is taken account of through the determination of the gain and refractive index changes.

To complete the analysis, a model is required for the material gain, spontaneous emission, and refractive index spectra. A parabolic model is used for the conduction band, while the valence band structure is calculated using a four-band **k·p**-model (Vurgaftman et al., 2001). The gain and spontaneous emission rates are obtained using Equations 24.20 and 24.21, respectively (Zory, 1993),

$$g\left(\hbar\omega\right) = \frac{1}{\hbar\omega}\frac{\bar{n}_g\pi e^2\hbar}{n^2 c\varepsilon_0 m_0^2}\sum_{ic=iv}|M_T|^2\rho_{red}^{mD}\left(f_n - f_p\right) \tag{24.20}$$

$$R_{sp}\left(\hbar\omega\right) = \frac{1}{\hbar\omega}\frac{\pi e^2\hbar}{n^2\varepsilon_0 m_0^2}\sum_{ic=iv}\left|M_{avg}\right|^2\rho_{red}^{mD}f_n\left(1 - f_p\right) \tag{24.21}$$

$$\left|M_{avg}\right|^2 = \frac{1}{3}\sum_{\substack{all\ 3 \\ polarizations}}|M_T|^2. \tag{24.22}$$

$f_{n,p}(E)$ is a Fermi–Dirac distribution function and gives the occupational probability of the electron of energy state E. ρ_{red}^{mD} is the reduced density of states, with $m = 2$ for QW material, and $m = 3$ for bulk.

$|M_{\mathrm{T}}|^2$ is the transition matrix element and determines the probability and hence strength of interaction between the states at energy E_{e} and E_{h} in the presence of an electromagnetic field.

Carrier lifetime dependent dephasing and intraband scattering processes (carrier–carrier, carrier–phonon, etc.) broaden the range of energy states that may participate in the optical transitions. This energy broadening is taken into account phenomenologically by directly broadening the optical spectra (Zory, 1993). Instead of broadening the gain spectrum directly, the spontaneous emission spectrum was broadened using a hyperbolic secant lineshape function (Lim et al., 2007)

$$L\left(E_{\mathrm{eh}} - \hbar\omega\right) = \frac{\tau_{\mathrm{in}}}{\hbar\pi} \sec h\left(\frac{E_{\mathrm{eh}} - \hbar\omega}{\hbar/\tau_{\mathrm{in}}}\right), \tag{24.23}$$

where τ_{in} is the intraband relaxation time mainly due to scattering processes, and E_{eh} is the transition energy. The gain spectrum is obtained from the broadened spontaneous emission spectrum, $R_{\mathrm{sp,broad}}\left(\hbar\omega\right)$, through

$$g\left(\hbar\omega\right) = \frac{1}{(\hbar\omega)^2} \frac{3\pi^2\hbar^3 c^2}{2\bar{n}^2} R_{\mathrm{sp,broad}}\left(\hbar\omega\right) \left[1 - \exp\left(\frac{\hbar\omega - \Delta E_{\mathrm{F}}}{k_{\mathrm{B}} T}\right)\right]. \tag{24.24}$$

The form of Equation 24.24 ensures that the gain passes through zero exactly at the quasi-Fermi level separation ΔE_{F}.

Gain compression due to SHB is included phenomenologically by introducing a constant gain compression factor, $\varepsilon_{\mathrm{shb}}$

$$g\left(\hbar\omega\right) = \frac{g\left(N, T\right)}{1 + \varepsilon_{\mathrm{shb}} S\left(\hbar\omega\right)}, \tag{24.25}$$

where $S\left(\hbar\omega\right)$ is the density of photons with energy $\hbar\omega$.

The carrier-induced changes in the refractive index spectrum are calculated from the changes in the gain/absorption spectrum through the nonlinear Kramers–Kronig relations (Hutchings et al., 1992) to ensure that the absorption and index changes are self-consistent. This is expressed as

$$\Delta n\left(\hbar\omega; \xi\right) = -\frac{c}{2\pi\omega} \int_{-\infty}^{\infty} \frac{\Delta g\left(\hbar\omega'; \xi\right)}{\omega' - \omega} d\omega', \tag{24.26}$$

where ξ is the external excitation, e.g., an electromagnetic field or temperature. This form is appropriate for different nonlinear processes, e.g., self-action effects (nonlinear refraction) and cascaded processes (Boyd, 2007). In this work, the carrier-induced changes in the gain/absorption spectrum only include band-filling effects and bandgap renormalization, but CH and SHB effects can also be included.

24.4.4 Inclusion of Material Dispersion in the Time-Domain Model

The challenge for time-domain simulations is the inclusion of the frequency-dependent complex material polarization (i.e., gain and refractive index dispersion). In this work, recursive digital filters are used to accurately represent the spectrally dependent material quantities in the time-domain model (Kaunga-Nyirenda et al., 2010). The most general linear digital filter takes a sequence of input points and produces a sequence of output points by the following difference equation (Smith, 2002):

$$y[n] = a_0 x[n] + a_1 x[n - 1] + a_2 x[n - 2] + \cdots + b_1 y[n - 1] + b_2 y[n - 2] + \dots, \tag{24.27}$$

where n is the time index, and $x[]$ and $y[]$ represent the input and output, respectively. The a's and b's that define the filter are called recursive coefficients, with $b_0 \equiv 1$ corresponding to the sample being calculated.

Digital filtering allows the computation of long convolution integrals of two functions in the time domain for discretely sampled signals. The optimization of the filter coefficients involves the fast Fourier transform. Hence, it can be argued that the overhead incurred in obtaining the filter coefficients at every spatial point and time step is much larger than Fourier transforming the field and multiplying by the frequency dependent quantities directly, followed by inverse Fourier transforming back into the time domain. This computational overhead is, however, avoided by obtaining the filter coefficients offline from the main simulation and storing them in tables. For a given set of parameters (e.g., carrier density and temperature), the gain and refractive index changes can be calculated and the filter coefficients obtained for each case and stored in tables. The filter coefficients can then be retrieved during simulation with negligible additional computational burden. This approach is not possible with the direct energy–time transformation approach, which must be done during the simulation.

Figure 24.2 shows the fitted gain and refractive index spectra for an eight-pole filter after applying the filter to an impulse input and taking the Fourier transform of the resulting impulse response. A good fit is observed between the filter response and the desired response. The number of poles is dictated by the index spectrum, which has a sharp turning point and is less smooth than the gain spectrum. The number of poles for the filters for both spectra has to be the same for ease of implementation in the field update equation. Therefore, the number of poles chosen was one that gave a satisfactory fit for the index spectral profile, i.e., 8 (Figure 24.2).

This method accurately captures the asymmetric spectral response of the material polarization (Figure 24.2) and can be used to incorporate frequency responses from experimental measurements as well as complicated responses, not easily represented analytically. The use of spatiotemporal filter coefficients allows correct inclusion of spatial-hole burning and shifts of the gain spectrum due to band-filling and bandgap renormalization.

24.4.5 Numerical Implementation

The optical propagation (Equation 24.11) and carrier density (Equation 24.15) are coupled through the material Equations 24.19 through 24.21. These two equations are solved self-consistently by dividing the active region into small enough (shorter than the diffusion length) sections, such that quantities can be considered constant within each section. The field propagation equations are marched in time using the Lax differencing scheme, which allows use of a larger time step than is allowed by the explicit differencing scheme. The stability condition of the Lax differencing scheme is given by the Courant condition $(c\Delta t/n\Delta z) \leq 1$ (Press et al., 2002). The carrier density equation is integrated using a fourth-order Runga–Kutta method. All carrier-dependent quantities are estimated at the midpoint of each section, while the optical fields are evaluated at the section boundaries (Kaunga-Nyirenda et al., 2010). This ensures that the photons traveling to the right and to the left in a particular section both interact with the same carrier-related quantities (Wong and Carroll, 1987).

24.5 Case Study: FWM

FWM studies in nonlinear media have focused largely on the optical characteristics of mixing processes and experimental observations in the optical domain (Agrawal, 1988; Mukai and Saitoh, 1990; Uskov et al., 1994; Darwish et al., 1996; Gong et al., 2004), with emphasis mainly on the issue of conversion efficiency (Mukai and Saitoh, 1990). Very little attention has been paid to the characteristics of wave mixing in the electrical domain. Electrical measurements have been reported before for semiconductor laser diodes (SLDs) and a simplified model based on frequency modulation was presented in Nietzke et al. (1989). Although the operational characteristics of SLDs and SOAs are different, the material physics is the same, so that CDM and resulting wave mixing have their origins in the same physical processes. The carrier density is usually clamped to its threshold value in SLDs, limiting the modulation depth. On the other hand,

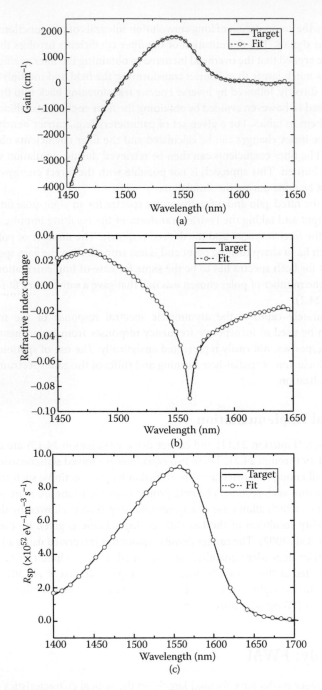

FIGURE 24.2 Fits for (a) gain spectrum, (b) refractive index change, and (c) spontaneous emission rate using recursive filters. The solid lines are the spectral quantity to be fitted, while the dotted lines are fits using the recursive filter. Number of poles for the filter is 8. Fits are obtained from the impulse responses of the filters.

large carrier density excursions are common in SOAs. In SLDs, the presence of strong optical feedback (i.e., large reflections at the facets) complicates the carrier density dynamics by introducing relaxation oscillations (Nietzke et al., 1989). A proper understanding of the characteristics of CDM is, therefore, likely to be stronger and clearer with SOAs (Dlubek et al., 2010). The model developed in Section 24.4 is used to study the characteristics of CDM over different operating conditions. No *a priori* assumptions are made about the existence or form of the CDM—it is obtained directly from the physics.

24.5.1 CDM Characteristics

The CDM characteristics in wave mixing in SOAs are studied using the model developed earlier. The input to the SOA is of the form

$$E_{in}(0, t) = \left\{ \psi_p(0, t) + \psi_s(0, t) \exp\left(-j\Omega t\right) \right\} \exp\left(-j\omega_{ref}t\right), \tag{24.28}$$

where $\Omega = 2\pi f_d$ and $f_d = \left| f_p - f_s \right|$ is the detuning frequency between the inputs (referred to as pump and probe). The pump and probe laser frequencies are $f_p = \omega_p/2\pi$ and $f_s = \omega_s/2\pi$, respectively. The time dependence of the pump and probe slowly varying fields is a general form and allows the use of modulated input signals. The input (Equation 24.28) enters through the boundary conditions (Equations 24.12 and 24.13).

The parameters used in the simulations are given in Table 24.1.

Indium gallium arsenide (InGaAs) was used as the active material for the quantum wells. The barrier material was strained InGaAs (−0.67% strain) (Kelly et al., 1996). This matches the material system for the device used in the experiments reported in Dlubek et al. (2010). The indium composition in the InGaAs active material was taken to be 0.62. This reproduced the device characteristics as reported in its data sheet and Kelly et al. (1996). This device has a nonlinear geometry, (Figure 24.3) so an effective length with a

TABLE 24.1   Simulation Parameters Used in the Study of Wave Mixing

Symbol	Description	Value (Units)
L	SOA length	1.1 mm (Lealman, 2009)
W	SOA width	1.3 μm (Lealman, 2009)
D	SOA thickness	0.1 μm (Lealman, 2009)
Δz	Longitudinal grid spacing	10 μm
Γ_{xy}	Optical confinement factor	0.18 (Lealman, 2009)
R_1	Input facet reflectivity	0.05% (Kelly et al., 1996)
R_2	Output facet reflectivity	0.05% (Kelly et al., 1996)
α_{int}	Internal loss	2.0 cm^{-1}
A	Linear gain coefficient	2.0×10^{-20} m^2
N_0	Carrier density at transparency	1.7521×10^{24} m^{-3}
τ_s	Effective carrier lifetime	0.7 ns
α_{LWEF}	Linewidth enhancement factor	−3
A	Linear radiative recombination coefficient	1.0×10^7 s^{-1}
B	Bimolecular recombination coefficient	5.6×10^{-16} m^{3}s^{-1}
C	Auger recombination coefficient	3.0×10^{-41} m^{6}s^{-1}
$\bar{n}_g$	Group index of refraction for the mode	3.6
n	Index of refraction of active region	3.3
H	Current injection efficiency	0.7
ε_{shb}	Gain compression factor	5.0×10^{-17} cm^3
τ_{in}	Intraband relaxation time	0.2 ps

Source: Agrawal, G.P., *Journal of the Optical Society of America B*, 5, 147–159, 1988.
Note: Some phenomenological parameters (carrier density at transparency, linear gain coefficient, effective carrier lifetime, linewidth enhancement factor, internal loss) are required as input to Agrawal's truncated series approximation (TSA) model.

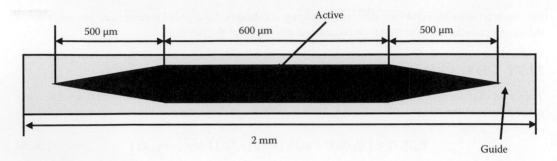

FIGURE 24.3 Sketch of the nonlinear geometry of a large spot semiconductor optical amplifier (SOA).

confinement factor equal to that of the untapered region was used in the simulations. All simulations were performed at 300 K. The carrier temperature was assumed to remain constant and equal to that of the lattice.

The CDM amplitude is obtained from the solution of the carrier density equation. A measurable quantity related to this is the variation in the injection current or voltage across the device, since the temporal fluctuations in the stimulated recombination rate result in temporal fluctuations in the current to resupply the carriers. The time series for the CDM amplitude is obtained at any spatial point along the longitudinal propagation direction as

$$\Delta N(z, t) = N(z, t) - N_{SS}(z), \tag{24.29}$$

where N_{SS} is the carrier density determined by the carrier injection, nonradiative recombination, and the amplified spontaneous emission (ASE) rates in the absence of external optical injection. The spectrum of the CDM amplitude is obtained by taking the discrete Fourier transform of Equation 24.29. Modulation of the carrier density results in modulation of the quasi-Fermi levels—and hence of the voltage drop across the active region (Equation 24.17). The spectrum of the injection current comes from DFT of Equation 24.16, while that of the voltage from Equation 24.17.

Simulated input and output optical spectra (from discrete Fourier transforms of the optical wave time series), and carrier density spectra (from discrete Fourier transform of Equation 24.29) are shown in Figure 24.4. The detuning of the two input signals was 2 GHz and the bias voltage was adjusted so that the current was 150 mA. The CW input optical fields had power levels of −0.7 and −0.8 dBm, respectively. The reference frequency was taken as the average of the pump and probe frequencies. An identical CDM spectra to Figure 24.4c was obtained when the reference frequency coincided with the pump frequency, confirming that the CDM spectra do not depend on the choice of the reference frequency used for the slowly varying envelope approximation. For the rest of the results presented below, the reference frequency coincided with the frequency of the pump. Figure 24.4 shows that the carrier density pulsates at the detuning frequency (beat frequency) of the pump and probe. The central peak in the carrier density spectrum is the static carrier density. The carrier population pulsations observed in Figure 24.4 are consistent with the experimental results (Nietzke et al., 1989).

The earlier simulations were repeated but with one of the optical input signals increased from −0.8 dBm to +4.7 dBm. Smaller secondary peaks are now also observed in Figure 24.4c at the multiples of the detuning frequency and these are enhanced as one of the input powers is increased (Figure 24.5).

The higher-order components (small secondary peaks at multiples of the detuning frequency) in Figure 24.5 are due to beating between the newly generated FWM products and the input beams or to higher-order harmonics in the CDM due to the nonlinearity of the carrier equation. Nonlinearities in the carrier equation are caused by the carrier density-dependent carrier lifetime and the saturation of the stimulated recombination rate. The peaks at the detuning frequency disappeared in the carrier density

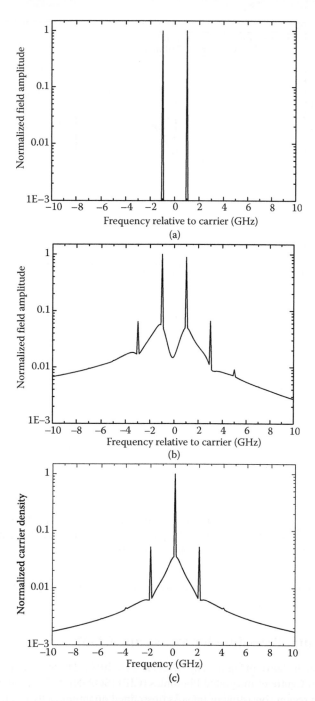

FIGURE 24.4 Simulated spectra (a), input optical field (b), output optical field, and (c) carrier density. Parameters: detuning 2 GHz, pump power −0.7 dBm, probe power −0.8 dBm, bias current 150 mA.

spectrum when one of the input signals was set to zero, confirming that the CDM is caused by the interaction of the input waves. This is consistent with the experimental results reported in (Dlubek et al., 2010). In addition to the primary peak, a weaker peak was observed in the measured CDM spectrum at twice the detuning frequency. This second harmonic had a bandwidth larger than the primary harmonic by a factor

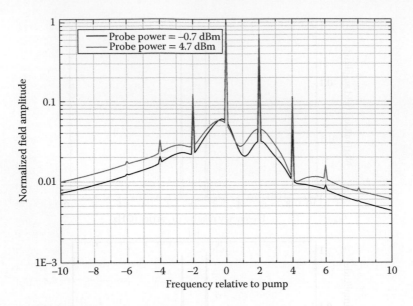

FIGURE 24.5 Optical spectra for the same parameters as in Figure 24.4 for two different probe powers. The pump signal is aligned to the reference signal, chosen around the gain peak.

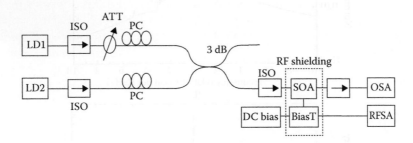

FIGURE 24.6 Carrier density modulation (CDM) measurement setup. ISO, isolator; ATT, variable optical attenuator; PC, polarization controller; 3 dB–50:50 coupler; OSA, optical spectrum analyzer; RFSA, radio frequency spectrum analyzer (From Dlubek, M.P. et al., *Optics Communications*, 283, 1481–1484, 2010).

of 1.5, indicating that it is the result of beating between the original waves and a new wave, which is itself broadened compared to the input waves.

24.5.2 Experimental Procedure for Measuring CDM

The experimental setup is shown in Figure 24.6 (Dlubek et al., 2010). The SOA used in the experiment was a nonlinear SOA from Centre of Integrated Photonics (CIP) (SOA-NL-OEC-1550 operating in 1550 nm region) with an active region consisting of InGaAs unstrained quantum wells and InGaAs barriers (Kelly et al., 1996). Two standard telecommunications diode lasers, LD1 and LD2, both a nominal wavelength of 1547.02 nm, were used as probe and pump, respectively. The frequency detuning between the two lasers was obtained by temperature tuning, so that it did not exceed 3 GHz (the bandwidth of the radio frequency (RF) spectrum analyzer). An optical spectrum analyzer was used to monitor the frequency detuning. A BiasT connector was used to uncouple the constant bias current to the SOA from the RF component from the device. Polarization controllers were used to adjust the polarization states of the interacting waves. Further experimental details are reported in Dlubek et al. (2010).

Before measuring the CDM spectrum, the reference beating spectrum of the two lasers (LD1 and LD2) was measured by heterodyning them on a standard 10 Gb/s photodetector. The 3 dB bandwidth of the heterodyne spectrum was ~40 MHz, suggesting that the laser linewidths were ~20 MHz. The spectral shape was approximately Lorentzian, as expected (Dlubek et al., 2010).

24.5.3 Dependence on Operating Conditions

The strength of interband effects depends on the rate of carrier replenishment into the active region. Thus, the CDM amplitude should decrease as the frequency detuning increases, since the carrier density cannot follow the fast oscillations of the stimulated emission rate. This was tested by simulating the FWM in the SOA for detuning frequencies up to 1 THz (by varying f_d in Equation 24.28), with all other parameters the same as in earlier simulations. Figure 24.7 shows the CDM spectra for different detunings. The peak at 2 GHz is a second harmonic for the 1 GHz detuning.

To obtain a clearer view of the frequency dependence of the CDM amplitude, the information in Figure 24.7 is extracted by obtaining the amplitude of the first harmonic (indicated against the detuning in Figure 24.7b) and plotting it against the detuning frequency (log scale), as shown in Figure 24.8. The CDM amplitude decreases sharply as the detuning frequency increases. The CDM amplitude at 1 THz is almost three orders of magnitude smaller than the value for a few GHz. The stimulated emission rate fluctuates at a much faster rate than the carriers injected into the active region. As a result, the carrier density cannot respond and the modulation efficiency decreases as the frequency detuning increases. Bream (2006) has also shown that the contribution of CDM to the refractive index perturbation is still significant even at frequencies well above the modulation bandwidths of the QW carrier distributions. As observed in Figure 24.8, although CDM continues to decrease with increasing detuning, it still contributes significantly to very fast nonlinear optical processes (even for detuning up to 1 THz) (Bream, 2006). Dynamic CH and SHB have not been taken into account in the model. It is interesting to note the similarity in the dependence of the CDM amplitude on detuning frequency and the dependence of amplitude of the conjugate signal on the detuning frequency observed in Figure 24.5a of Uskov et al. (1994). However, it is also worth noting that above 1 THz (~4 meV), acoustic phonon scattering should become relevant, while longitudinal optical (LO) phonon scattering should also be important above 9 THz. Thus, care is needed when extrapolating results to 1 THz and beyond, as other carrier scattering mechanisms can be resonantly excited.

Figure 24.9 shows the variation of the normalized CDM amplitude with bias current. The CDM in each case were normalized with respect to their maximum amplitude. As expected, the CDM amplitude increases with bias current. The detuning was adjusted to 1.23 GHz to match the experiments. In Agrawal's seminal paper on FWM (Agrawal, 1988), the carrier density rate equation is expressed as a power series involving the harmonics of the frequency detuning. The truncated series approximation (TSA1) curve in Figure 24.9 was calculated using Agrawal's model, where the power series is truncated at the first harmonic (Agrawal, 1988). The simulated and the experimental results both show that the CDM amplitude saturates as the bias current is increased further. Furthermore, a comparison of the experimental and simulated curves shows that the CDM approaches saturation at a faster rate in the experimental measurements than is observed in simulations. The CDM from measurements and truncated series assumption show the largest amplitude at 0 mA (experiment) and very close to 0 mA (TSA1).

The qualitative behavior is similar in both experiment and simulation. The model predicts the behavior of CDM better than the TSA1 in the gain region. At 0 mA, the rate of absorption is at its maximum and the net gain is equal to the absorption coefficient. The SOA behaves like an unbiased diode with an internal field and hence acts as a square-law detector. As the bias current increases from 0 mA toward transparency at around 35 mA, the rate of absorption decreases while the rate of emission increases. This results in an overall decrease in the net absorption and explains why the CDM amplitude decreases as the current increases from 0 mA toward transparency. Above transparency, the net gain increases. This leads to an increased rate of stimulated emission relative to absorption, which results in an increase in the CDM amplitude, saturating at about 150 mA. This is consistent with the measured gain-current dependence

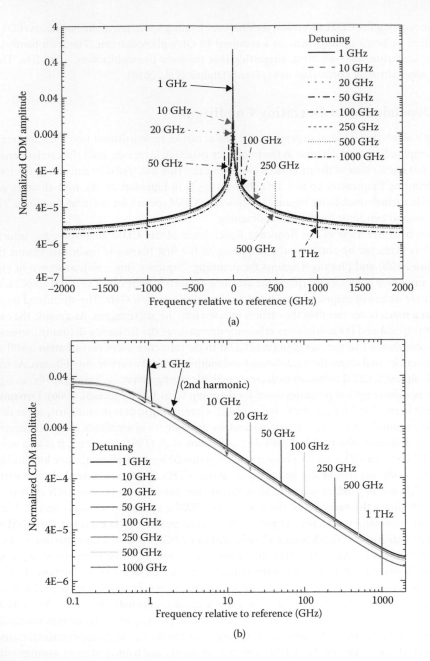

FIGURE 24.7 Normalized CDM amplitude for different detunings (a) full spectra and (b) positive spectra (log scale).

shown in Figure 24.10. The discrepancy between simulations and experiments in the absorption region and low bias current above transparency (0–50 mA) in Figure 24.9 can be attributed to the use of the unipolar model and the effects of ASE. The I–V characteristics of a diode appear most nonlinear near zero bias. The unipolar model is based on the linear approximation of the current injection (Equations 24.16 and 24.17), which does not include the nonlinear I–V characteristics, hence in the deviation for currents between zero and transparency. Spontaneous emission, and the subsequent ASE, have not been included in the propagating field. The effects of ASE on carrier dynamics are more pronounced at low (electrical)

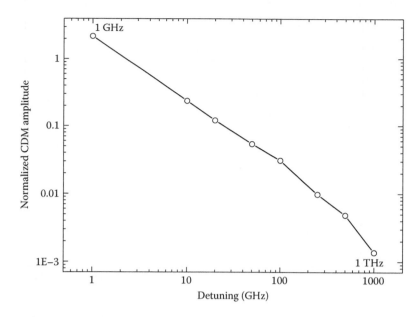

FIGURE 24.8 Normalized CDM amplitude at the first harmonic for different detunings (frequency in log scale).

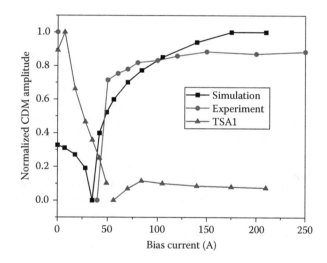

FIGURE 24.9 Normalized CDM amplitude at the first harmonic for different bias current, detuning 1.23 GHz, pump power −0.8 dBm, probe power −0.7 dBm. Truncated series approximation (TSA1) based on Agrawal's model (Agrawal, 1988), truncated at the first harmonic.

bias and optical input power level, since the rate of spontaneous emission can be comparable to the rate of stimulated emission.

The absence of experimental data points between 0 and approximately 50 mA in Figure 24.9 can be attributed to the low level of CDM, such that its amplitude was below the noise level. Although there is significant CDM in the absorption regime, it is not useful for many applications (apart from heterodyne detection, as discussed in Section 24.5.2) and both experiments and simulations confirm that there are no detectable/measurable optical mixing products.

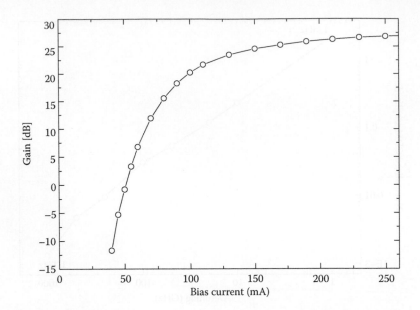

FIGURE 24.10   Measured small signal gain. Dependence on bias current for input power −28 dBm.

The effectiveness of CDM is expected to increase with input power, as the interaction between the different signals should be enhanced. As the optical power intensity increases in the active region, carrier depletion, and gain saturation effects should ultimately prevent any further increase in the CDM amplitude. (Signal distortion due to clipping as the carrier density approaches its transparency value should also create more signal components in the frequency spectrum.) Figure 24.11a and b show the simulated and measured CDM amplitude dependence on pump power for two probe powers (−0.8 dBm and +4.7 dBm). The detuning between the pump and probe was 1.23 GHz and the bias voltage was adjusted to give a bias current of 150 mA. The CDM amplitude increased with input power up to a point. The figures also reveal that the largest CDM is produced when the pump and probe signals have approximately equal power. This is a key result that cannot be determined from previous works, which rely on the assumption that the pump power is much larger than the probe power.

There is good qualitative agreement between simulation and experiment. However, there is some discrepancy in the amplitude of the CDM obtained using simulations and experiment, as shown in Figures 24.9 and 24.11. Apart from the limitations of the unipolar electrical and 1D traveling wave optical models, some of these differences are due to errors in the material parameters of the SOA, as these were not provided by the manufacturer. This discrepancy may also result from a lack of knowledge of the device's parasitic impedances and layer resistivities, which will also affect the RF signal measured at the device terminals (after the BiasT in Figure 24.6).

The normalized CDM amplitude is generally higher for the lower probe power (−0.8 dBm), as shown in Figure 24.12, where the dependence of the simulated CDM amplitude on the pump power is plotted for two probe powers (−0.8 dBm, +4.7 dBm). The other parameters are the same as those used in the simulations for Figure 24.11. When one beam is much weaker than the other, the CDM amplitude decreases as the beating is reduced due to carrier depletion and gain saturation by the strong beam.

24.6 Summary and Conclusions

We have developed a suitable time-domain model to investigate dynamic nonlinear optical effects in SOAs. The model takes into account the most important interband processes and minimizes the number of phenomenological parameters. This simulation tool was used to show how wave mixing processes manifest

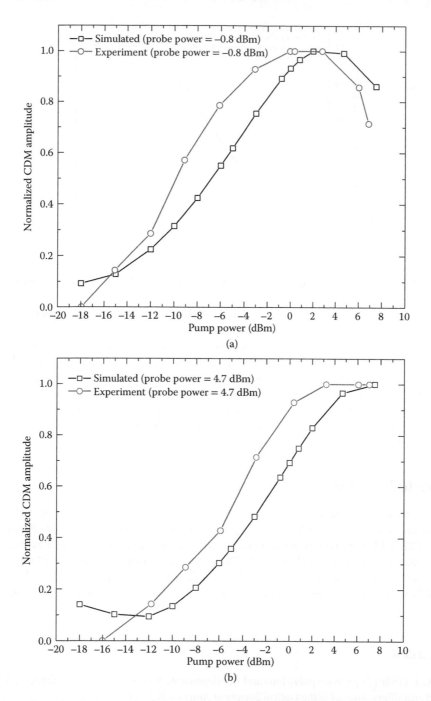

FIGURE 24.11 Normalized CDM amplitude dependence on pump power for different probe power levels: (a) probe power −0.8 dBm and (b) probe power +4.7 dBm. Detuning 1.23 GHz, bias current 150 mA.

themselves in SOAs, and discuss their impact on existing and potential applications. This is key to understanding and modeling nonlinear optical signal impairments. Furthermore, the knowledge obtained from these studies can be used in the design of new devices and/or for optimizing existing ones that rely on similar nonlinear optical effects, e.g., high-density PICs. The nonlinear phenomena are also accessible

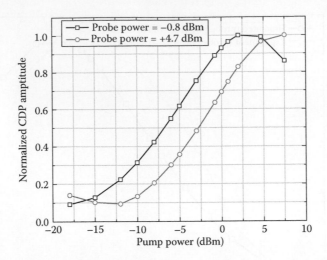

FIGURE 24.12　Simulated CDM amplitude (normalized) for two probe levels as in Figure 24.11a and b.

through electrical measurements, as well as through FWM and SHG measurements, providing additional opportunities for the validation of the simulation tool.

Other key outcomes of this work are the general description of the nonlinear polarization and the demonstration of the inclusion of nonlinear optical effects using the recursive filter functions. These can be used in more advanced tools, such as those with higher dimensionality (2D or 3D), those with a dynamic bipolar electrical model, and those with more accurate time domain-optical models (e.g., which properly include the spectral effects in the propagation and waveguiding). Finally, the general description of the nonlinear polarization means this approach can be readily adapted to include other contributions to the nonlinear polarization.

Acknowledgments

S.N. Kaunga-Nyirenda would like to thank the Commonwealth Scholarship Commission (United Kingdom) for financial support. This work was also supported in part by the Higher Education Funding Council of England (HEFCE) with a HERMES Fellowship and the UK Engineering and Physical Sciences Research Council (EPRSC) under grant GR/T09309/01. The authors are also grateful to the CIP for useful information and discussion about the SOA. The authors would also like to thank Prof. M. Wale (Oclaro) and Prof. K. Williams (TU Eindhoven) for useful discussions on the simulation requirements for active devices in PICs.

References

Agrawal, G.P. (1988) Population pulsations and nondegenerate four-wave mixing in semiconductor lasers and amplifiers. *Journal of the Optical Society of America B*, **5**, 147–159.

Agrawal, G.P. and Dutta, N.K. (1986) *Long-Wavelength Semiconductor Lasers*. New York, NY: Van Nostrand Reinhold.

Boyd, R.W. (2007) *Nonlinear Optics*. Cambridge, MA: Academic Press.

Bream, P.J. (2006) *Nonequilibrium Carrier Dynamics and Gain in Semiconductor Quantum Wells*. Nottingham: School of Electrical and Electronic Engineering.

Dai, Z., Michalzik, R., Unger, P., and Ebeling, K.J. (1997) Numerical simulation of broad-area high-power semiconductor laser amplifiers. *IEEE Journal of Quantum Electronics*, **33**, 2240–2254.

Darwish, A.M., Ippen, E.P., Le, H.Q., Donnelly, J.P., and Groves, S.H. (1996) Optimization of four-wave conversion efficiency in the presence of nonlinear loss. *Applied Physics Letters*, **69**, 737–739.

Das, N.K., Yamayoshi, Y., and Kawagushi, H. (2000) Analysis of basic four-wave mixing characteristics in a semiconductor optical amplifier by finite-difference beam propagation method. *IEEE Journal of Quantum Electronics*, **36**, 1184–1192.

Dlubek, M.P. (2008) *Optical Performance Monitoring for Amplified Spontaneous Emission Related Issues in Transparent Optical Networks*. Nottingham: School of Electrical and Electronic Engineering.

Dlubek, M.P., Kaunga-Nyirenda, S.N., Phillips, A.J., Sujecki, S., Harrison, I., and Larkins, E.C. (2010) Experimental verification of the existence of optically induced carrier pulsations in SOAs. *Optics Communications*, **283**, 1481–1484.

Durhuus, T., Mikkelsenm, B., and Stubkjaer, K.E. (1992) Detailed dynamic model for semiconductor optical amplifiers and their crosstalk and intermodulation distortion. *Journal of Lightwave Technology*, **10**, 1056–1064.

Gong, P.-M., Hsieh, J.-T., Lee, S.-L., and Wu, I. (2004) Theoretical analysis of wavelength conversion based on four-wave mixing in light-holding SOAs. *IEEE Journal of Quantum Electronics*, **40**, 31–40.

Hutchings, D.C., Sheik-bahae, M., Hagan, D.J., and Stryland, E.W. (1992) Kramers-Kronig relations in nonlinear optics. *Optical and Quantum Electronics*, **24**, 1–30.

Jackson, J.D. (1975) *Classical Electrodynamics*. New York, NY: John Wiley & Sons.

Jones, D.J., Zhang, L.M., Carroll, J.E., and Marcenac, D.D. (1995) Dynamics of monolithic passively mode-locked semiconductor lasers. *IEEE Journal of Quantum Electronics*, **31**, 1051–1058.

Joyce, W.B., and Dixon, R.W. (1977) Analytic approximations for the Fermi energy of an ideal Fermi gas. *Applied Physics Letters*, **31**, 354–356.

Kaunga-Nyirenda, S.N., Dlubek, M.P., Phillips, A.J., Lim, J.J., Larkins, E.C., and Sujecki, S. (2010) Theoretical investigation of the role of optically induced carrier pulsations in wave mixing in semiconductor optical amplifiers. *Journal of Optical Society of America B*, **27**, 168–178.

Kelly, A.E., Lealman, I.F., Rivers, L.J., Perrin, S.D., and Silver, M. (1996) Polarisation insensitive, 25 dB gain semiconductor laser amplifier without antireflection coatings. *Electronics Letters*, **32**, 1835–1836.

Kovačević, M., and Acampora, A. (1995) Benefits of wavelength translation in all-optical clear-channel networks. *IEEE Journal of Selected Areas in Communications*, **14**, 868–880.

Lealman, I.F. (2009) Centre for Integrated Photonics (CIP), B55, Adastral Park (Private Communication).

Lim, J.J., MacKenzie, R., Sujecki, S., Sadeghi, M., Wang, S.M., Wei, Y.Q., Gustavsson, J.S., Larsson, A., Melanen, P., Sipilä, P., Uusimaa, P., George, A.A., Smowton, P.M., and Larkins, E.C. (2007) Simulation of double quantum well GaInNAs laser diodes. *IET Optoelectronics*, **1**, 259–265.

Lim, J.J., Sujecki, S., Lang, L., Zhang, Z., Paboeuf, D., Pauliat, G., Lucas-Leclin, G., Georges, P., MacKenzie, R.C.I., Bream, P., Bull, S., Hasler, K.-H., Sumpf, B., Wenzel, H., Erbert, G., Thestrup, B., Petersen, P.M., Michel, N., Krakowski, M., and Larkins, E.C. (2009) Design and simulation of next-generation high-power, high-brightness laser diodes. *IEEE Journal of Selected Topics in Quantum Electronics*, **15**, 993–1008.

Liu, J. (2005) *Photonic Devices*. Cambridge: Cambridge University Press.

Manning, R.J., Ellis, A.D., Poustie, A.J., and Blow, K.J. (1997) Semiconductor laser amplifiers for ultrafast all-optical signal processing. *Journal of the Optical Society of America B*, **14**, 3204–3216.

Mukai, T., and Saitoh, T. (1990) Detuning characterisitcs and conversion efficiency of nearly degenerate four-wave mixing in a 1.5-μm travelling-wave semiconductor laser amplifier. *IEEE Journal of Quantum Electronics*, **26**, 865–875.

Nietzke, R., Panknin, P., Elsasser, W., and Gobel, E.O. (1989) Four-wave mixing in GaAs/AlGaAs semiconductor lasers. *IEEE Journal of Quantum Electronics*, **25**, 1399–1406.

Ning, C.Z., Indik, R.A., and Moloney, J.V. (1997) Effective Bloch equations for semiconductor lasers and amplifiers. *IEEE Journal of Quantum Electronics*, **33**, 1543–1550.

Park, J., Li, X., and Huang, W.P. (2005) Comparative study of mixed frequency-time-domain models of semiconductor optical amplifiers. *IEE Proceedings in Optoelectronics*, **152**, 151–159.

Pratt, E.M., and Carroll, J.E. (2000) Gain modelling and particle balance in semiconductor lasers. *IEE Proceedings in Optoelectronics*, **147**, 77–82.

Press, W.H., Teukolsky, S.A., Vetterling, W.T., and Flannery, B.P. (2002) *Numerical Recipes in C++: The Art of Scientific Computing*. Cambridge: Cambridge University Press.

Smith, S.W. (2002) *Digital Signal Processing: A Practical Guide for Engineers and Scientists*. Poway, CA: California Technical Publishing.

Tolstikhin, V.I. (2000) Carrier charge imbalance and optical properties of separate confinement heterostructure quantum well lasers. *Journal of Applied Physics*, **87**, 7342–7348.

Toptchiyski, G., Kindt, S., Petermann, K., Hillinger, E., Diez, S., and Weber, H.G. (1999) Time-domain modeling of semiconductor optical amplifiers for OTDM applications. *Journal of Lightwave Technology*, **17**, 2577–2583.

Uskov, A., Mork, J., and Mark, J. (1994) Wave mixing in semiconductor laser amplifiers due to carrier heating and spectral-hole burning. *IEEE Journal of Quantum Electronics*, **30**, 1769–1781.

Vurgaftman, I., Meyer, J.R., and Ram-Mohan, L.R. (2001) Band parameters for III–V compound semiconductors and their alloys. *Applied Physics Review*, **89**, 5815–5875.

White, J.K., Moloney, J.V., Gavrielides, A., Kovanis, V., Hohl, A., and Kalmus, R. (1998) Multilongitudinal-mode dynamics in a semiconductor laser subject to optical injection. *IEEE Journal of Quantum Electronics*, **34**, 1469–1473.

Williams, K.A., Aw, E.T., Wang, H., Penty, R.V., and White, I.H. (2008) Physical layer modelling of semiconductor optical amplifier based terabit/second switch fabrics. *International Conference of Numerical Simulation of Optoelectronic Devices*, Nottingham, United Kingdom.

Wong, Y.L., and Carroll, J.E. (1987) A Travelling-wave rate equation analysis for semiconductor lasers. *Solid-State Electronics*, **30**, 13–19.

Yanhua, H., Bandyopadhyay, S., Spencer, P.S., and Shore, K.A. (2003) Polarization-independent optical spectral inversion without frequency shift using a single semiconductor optical amplifier. *IEEE Journal of Quantum Electronics*, **39**, 1123–1128.

Zory, P.S., Jr. (ed.) (1993) *Quantum Well Lasers*. Cambridge, MA: Academic Press.

25

Semiconductor Optical Amplifier Dynamics and Pattern Effects

Zoe V. Rizou

and

Kyriakos E. Zoiros

25.1 Introduction

Since their advent in the second half of the 1980s (O'Mahony, 1988), semiconductor optical amplifiers (SOAs) have evolved technologically to the point that they have become key elements for the development of optical communications circuits, systems, and networks. Due to their attractive features of low power consumption, compactness, broad gain bandwidth, and ability for integration with affordable cost, the multifunctional potential of SOAs has been exploited in data amplification (Zimmerman and Spiekman, 2004) and processing (Mørk et al., 2003) in the optical domain and, more recently, for data encoding as well (Udvary and Berceli, 2010).

Motivated by the widespread employment of SOAs and the concomitant need to assist their design and support the implementation of the diverse applications they are destined to serve, the purpose of this chapter is to present in a concise and comprehensible manner the basic processes which govern the dynamic behavior of SOAs. Furthermore, it aims at describing how these processes can theoretically be taken into account in order to model their impact on the SOA response. It is concerned with the significant phenomenon of the pattern effect which manifests when the SOA gain is modulated, either optically by a single data pulse train or electrically by digital information superimposed on the SOA current. More specifically, the chapter topics which are addressed with regard to the SOA dynamics are outlined as follows:

Section 25.2: SOA Dynamics Background—Types, physical origin, conditions for manifestation, and qualitative impact on SOA response.

Section 25.3: SOA Model Formulation—SOA gain dynamics modeling theoretical formulation and numerical/closed form solutions.

Section 25.4: SOA Model Simplification—Ways of simplification and implications.

Section 25.5: SOA Optical Gain Modulation—Characterization and implications.

Section 25.6: Pattern Effect I—Direct Optical Amplification: Performance limitations and methods for mitigation.

Section 25.7: SOA Electrical Gain Modulation—Characterization and implications.
Section 25.8: Pattern Effect II—Direct Current Modulation: Performance limitations and methods for mitigation.

25.2 SOA Dynamics Background

Figure 25.1a depicts the standard configuration of an SOA (Mørk et al., 2003). Light is coupled via one of the facets into an intrinsic region of semiconductor material, which is sandwiched between a hole-dominant, that is, p-doped, and electron-dominant, that is, n-doped, cladding layer of higher bandgap energy and slightly lower refractive index. This structure allows confinement of a large number of free carriers in a small active volume, which hence exhibit a large density. By forward biasing the heterojunction through the injection of an electric current being below the lasing threshold point, the population of the carriers can be sufficiently inverted so that optical gain is established at a rate that prevails absorption. Then when signal photons of suitable energy travel through the active area, they stimulate radiative recombinations of electrons and holes and are coherently (i.e., with the same polarization, frequency, and phase) amplified as they propagate along the formed waveguide and exit from the other side of the device. In order to avoid unwanted back reflections and oscillations inside the cavity, the reflectivity of the SOA facets is reduced to very low values of less than 1×10^{-5} using antireflection coatings, tilted waveguides and tapered waveguides (Connelly, 2002). These technological means allow us to achieve a single-pass controllable amplification as high as 30 dB and with ripples-free parabolic-like spectrum profile, which can extend from 1300 to 1600 nm by changing the chemical composition of III–V group semiconductor materials. In particular, in this chapter we consider InGaAsP/InP-based SOAs, which operate in the 1550 nm telecommunications window where the attenuation of signals transmitted via optical fibers is lowest.

When a lightwave beam of optical frequency ν is injected into an SOA with a photon energy, $h\nu$, being larger than the bandgap energy, $E_g = E_c - E_v$, which, as shown in Figure 25.1b, separates the conduction and valence bands having energy at their bottom and top E_c and E_v, respectively, it triggers through band-to-band transitions stimulated emission and depletes the pumped carriers (Diez, 2000; Vacondio, 2011).

The reduction in the density of the excited carriers has then two consequences. First, the produced gain is reduced up to the point where it becomes saturated. Second, a set of dynamic processes tend to

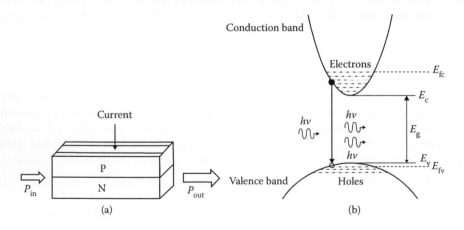

FIGURE 25.1　(a) Schematic representation of typical semiconductor optical amplifier (SOA) structure. (b) Simplified band diagram of semiconductor material. E_g, bandgap energy given by the difference between the energy at the bottom of the conduction band, E_c, and the energy at the top of the valence band, E_v; E_{fc}, quasi-Fermi level in the conduction band, and E_{fv}, quasi-Fermi level in the valence band.

bring the SOA gain back to its original level. These processes, which govern the SOA gain recovery, can be categorized in two cases of transitions (Occhi, 2002):

- Intraband transitions, which affect the distribution of carriers in the energy bands but leave the overall carrier concentrations unchanged. Varying the carrier density affects this distribution, which is further modified by spectral hole burning (SHB), free carrier absorption (FCA), carrier heating (CH), and carrier cooling.
- Interband transitions, which are related to the exchange of carriers between the energy bands and hence result in changes in the carrier density. These transitions are determined by electrical pumping, stimulated emission, absorption, spontaneous emission, nonradiative recombinations and two-photon absorption (TPA).

As electrons need not change band but are rearranged within the latter, the time constants of intraband transitions are much shorter than those of interband transitions. For this reason, the former are referred to as ultrafast effects. Given the time constants of these effects, they tend to be important compared with interband dynamics when the spectrum of the input signal is very wide, as in short pulse amplification, or when many channels are amplified in the context of wavelength division multiplexing (WDM) (Vacondio, 2011).

The SOA gain dynamics are directly coupled with those of free carriers (Occhi, 2002). Figure 25.2 qualitatively illustrates the temporal evolution of the free carrier distribution (energy versus density) in the conduction band (CB) of an SOA, which operates in the gain regime (Hall et al., 1994) subject to an ultrashort optical pulse that belongs to a time-varying signal (Occhi, 2002; Mørk et al., 2003; Wang, 2008). Prior to the pulse arrival, the carrier density within the CB is in quasi-equilibrium (Saleh and Teich, 1991), which occurs when the relaxation times for transitions within an energy band are much shorter than the relaxation time versus the other energy band. In this situation, the carrier distribution within the considered band is graphically shown in Figure 25.2a and is described by the product of the density of states, which is increased away from the band edge ($E = E_c$) at a rate that depends on the effective masses of electrons, with the probability that a given energy level, $E > E_c$, is occupied by an electron (Diez, 2000; Connelly, 2002). This probability is described by the Fermi function, $1/[\exp((E-E_{fc})/k_B T) + 1]$, where E_{fc} is the quasi-Fermi level, T is the temperature of the medium at thermal equilibrium and k_B is Boltzmann's constant (so that $k_B T$ defines the thermal energy, which equals 26 meV at $T = 300$ K). The position of the quasi-Fermi levels in the conduction and valence bands, E_{fc} and E_{fv}, respectively, is determined by the pumping rate (current injection), and if the latter is sufficiently large so that their separation exceeds the bandgap energy then the semiconductor medium provides gain and hence acts as an amplifier (Diez, 2000). Such a condition can be satisfied only if the quasi-Fermi levels lie, as shown in Figure 25.1, inside the conduction and valence bands, respectively, which for a semiconductor happens only under strong biasing (Vacondio, 2011). Note here that since the electrons and holes in the SOA are interrelated because of charge neutrality, it is sufficient to refer only to the conduction band (Schubert, 2004).

The natural quasi-equilibrium state is altered when pulses of duration in a subpicosecond timescale enter the SOA. This happens due to intraband effects which manifest according to the following phenomena: First, the induced stimulated emission responsible for signal amplification causes carriers to recombine around a narrow range of energies defined by pulse photon energy. This opens a hole in the carrier distribution, which is referred to as "SHB" and provokes a deviation from the normal Fermi distribution (Figure 25.2b). As a consequence, a localized reduction in the number of carriers at the transition energies occurs, while the total (electron/hole) carrier density within the band is reduced by the stimulated emission. For high-input pulse energies (>1 pJ) (Tang and Shore, 1998) TPA may also take place where, as the name implies, two photons are simultaneously absorbed and an electron is transferred from the valence band to a high energy level in the conduction band due to the high photon density in the active region (Hall et al., 1994; Mørk et al., 2003). In addition, a free carrier absorbs a photon and moves to a higher energy within the same band, according to the so-called FCA (Hall et al., 1994). After these effects have manifested on a timescale practically of few tens of femtoseconds (fs) (Diez, 2000) and the optical pulse has left

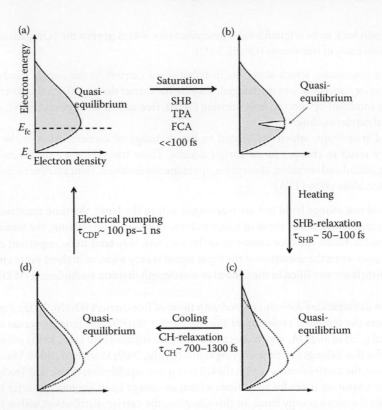

FIGURE 25.2 Evolution of carrier distribution induced in the conduction band (CB) of SOA by an ultrashort strong optical pulse. The shaded areas under the solid lines are proportional to the total carrier concentration in the CB (Wiesenfeld, 1996). The quasi-equilibrium is described by a Fermi distribution of quasi-Fermi level E_{fc}, which exceeds the minimum energy of the CB, E_c, in accordance with Figure 25.1b. SHB, spectral hole burning; CH, carrier heating; FCA, free carrier absorption; TPA, two-photon absorption, which occurs only for high pulse energies. The time constants values are typical for bulk SOAs, and may vary by more than a factor of two. (From Diez S, 2000, All-optical signal processing by gain-transparent semiconductor switches. PhD dissertation, Technical University of Berlin.)

the SOA, the Fermi distribution is restored via a process known as carrier–carrier scattering (Mørk et al., 2003), and the time required for this to happen is determined by the constant τ_{SHB}, which is known as SHB relaxation time, with values typically several tens of femtoseconds (50–100 fs). Since this parameter is finite, the governing SHB essentially sets the timescale during which the quasi-equilibrium Fermi distribution is established among the carriers in a band. Additionally, free carriers at energy levels lower than the average carrier density in the band are removed by stimulated emission or are transferred to higher levels due to the contribution of FCA and (possibly) TPA. As a result, the average carrier temperature is elevated and becomes higher than before the pulse arrival (Hall et al., 1994; Mørk et al., 2003). This transient increase of the electron and hole temperatures is called "CH" and impacts the carrier distribution (Figure 25.2c). The characteristic time, τ_{CH}, required for carriers to release their excess energy through phonon emission and cool down to the lattice temperature is of the order of several hundred femtoseconds (700–1300 fs). After this temperature relaxation time has elapsed, the starting Fermi distribution has been reestablished (Figure 25.2d), but the carrier density is still reduced compared to the stationary state (Figure 25.2a). Then the interband effect of carrier density pulsation (CDP) takes the lead and lasts between some hundreds of picoseconds and a few nanoseconds, depending on the dimensions and material of the active region as well as the SOA operating conditions. During this interval, which is defined by the respective time constant, τ_{CDP}, the carrier density, which had been decaying due to the stimulated emission, is increased again toward its original level owing to the resupply of electrons to the SOA via electrical pumping and the refilling of the respective bands. Finally, the carrier distribution is recovered to the quasi-equilibrium state (Figure 25.2a).

The distinctive evolution of the above processes can be conveniently viewed on the temporal profile of the SOA gain response, which is coupled to the carrier density. In practice, the SOA intraband and inter-band gain dynamics temporal characteristics and their dependence on operating parameters are directly disclosed by means of the pump-probe technique (Hall et al., 1994). This is a time-domain measurement technique where the SOA gain is perturbed by a strong optical pulse called "pump" and sampled at different delays, τ_{delay}, by a weak pulse called "probe," as illustrated in Figure 25.3a. The strong pump pulse drives the SOA under test into nonlinearity. The weak probe pulse cannot affect but instead experiences the altered SOA properties (Vallaitis, 2010). In order to capture the full extent of the SOA gain dynamics, the pulse repetition rate is of the order of several tens of MHz so that only one pump pulse at a time travels through the SOA and the gain of the latter is fully recovered before the arrival of the next pulse. Furthermore, ultrashort, that is, femtosecond-wide (Mørk et al., 2003) optical pulses are employed because their width determines the measurement's temporal resolution, which must be as high as possible to be able to extract the maximum information. A typical outcome of such measurement is graphically shown in Figure 25.3b, where the probe transmission has been plotted against the pump–probe delay, which is varied between a few ps to a few hundreds of ps (positive delay means that the pump precedes the probe pulse) [Occhi, 2002; Mørk et al., 2003; Wang, 2008; Vacondio, 2011]. The probe transmission is proportional to the SOA gain, so the curve monitors the SOA gain dynamic behavior.

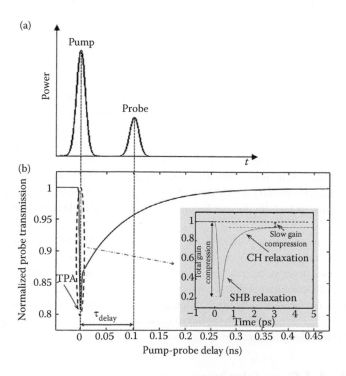

FIGURE 25.3 (a) Pump-probe technique. (From Vallaitis T., *Ultrafast Nonlinear Silicon Waveguides and Quantum Dot Semiconductor Optical Amplifiers*, Germany: Univ Karlsruhe, 2010.) (b) SOA typical evolution of (normalized) probe transmission as a function of pump–probe delay. The depicted curve does not represent exactly the defined SOA gain but rather its convolution with the pump and probe pulses. The timescale of the intraband gain dynamics is depicted in the zoomed-in inset. (From Occhi L, *Semiconductor optical amplifiers made of ridge waveguide bulk InGaAsP/InP: Experimental characterisation and numerical modelling of gain, phase, and noise*, PhD dissertation, Swiss Federal Institute of Technology in Zurich, 2002)

Initially, the pump pulse compresses the gain by an amount defined by the ratio of the unsaturated to the minimum level of the probe transmission, which occurs at the point where the carrier depletion is maximized by the pump peak power (Occhi, 2002). This gain compression is governed by SHB and (if enough photons are present in the SOA) TPA (Hall et al., 1994). Due to the finite resolution of the employed measurement technique (Hall et al., 1994), these ultrafast physical mechanisms are interpreted as instantaneous and thus overlap at zero time delay in the probe transmission (Schreieck, 2001). Afterward, the gain recovers within the interval required for the probe transmission to rise from 10% to 90% of the asymptotically approached maximum level. This recovery occurs according to the following separate processes. More specifically, in the first hundred femtoseconds timescale after the lowest compressed gain point SHB becomes the dominant nonlinear effect. Then CH takes over for a timescale of ~1 ps, which reestablishes the Fermi carrier distribution in the two energy bands. In the last time regime, the interband effects become pronounced and lead to slow gain compression defined by the unsaturated probe level, which is restored due to the applied electrical pumping, divided by the long-lasting probe transmission level (Occhi, 2002).

The impact of different SOA and signal operating conditions on the aforementioned gain dynamics has been investigated and assessed through experimental observations and numerical simulations. From the interpretation and analysis of the results obtained from both approaches, the following conclusions have been drawn (Occhi, 2002):

1. Both total and slow gain compression increase with the pump energy. For the total gain compression, this happens because the lattice temperature in the active region is elevated with higher pump energy. For the slow gain compression on the other hand, this variation is attributed to the reduction of the free carrier density as a by-product of the increased stimulated emission. Moreover, when the pulse energy is increased, it takes longer time for the gain compression associated with the intraband effects to recover.

2. The magnitudes of the two defined types of gain compression are affected differently by the width of the pump pulses. More specifically, the narrower the pump pulses, the larger the compression associated with the intraband effects. In contrast, the slow gain compression becomes smaller, even for input pulse energies higher than the SOA input saturation energy, for which the SOA gain is dropped by 3 dB against the small-signal gain. However, this is not possible for wider pulses, for example in the 100 ps range, since then the slow gain compression is much larger due to the much higher contribution of the interband effects to the SOA saturation.

3. Both gain compressions increase with current density because of the corresponding increase in gain. However, the fast gain recovery time behaves differently from the recovery time resulting from interband mechanisms. In fact, the former becomes longer for higher bias currents, while the latter is decreased.

4. The total gain compression is practically the same for all investigated device lengths in the case of an input energy equal to the SOA input saturation energy. The increase of the input energy beyond this point causes a larger gain reduction in longer SOAs, since their gain is higher compared to the shorter ones, and accordingly results in higher total gain compression. On the contrary, the fast gain recovery time is not influenced by the SOA longitudinal dimension. This finding does not hold too for the recovery time associated with the interband dynamics, since it decreases appreciably with the increase of the SOA length (Girardin et al., 1998).

25.3 SOA Model Formulation

Many theoretical treatments have been conducted aiming at explaining the operation, understanding the dynamics, predicting the behavior, evaluating the performance, and optimizing the working conditions of SOAs. For this purpose, both analytical and numerical models of different complexity level have been formulated. The more elaborate ones provide a thorough insight into what happens inside the SOA. This is done by microscopically describing the propagation of optical pulses and the change of the SOA carrier dynamics due to interaction with light. This approach provides extensive information at the expense of increased hardware and software resources. On the other hand, it is also practically desirable to maintain a good balance between mathematics and computer power. This can be achieved by viewing and treating the physical processes that manifest within an SOA from a phenomenological perspective (Toptchiyski, 2002).

The theoretical description of the operation of an SOA that is driven by a strong, ultrafast data pulse stream should be as realistic as possible while simultaneously being characterized by versatility and computational efficiency. In this effort, it is necessary to set the valid framework within which this task will be accomplished, which is based on the following key points. First, for SOA applications in which the employed pulse widths are below the critical value of 10 ps the influence of the intraband carrier processes on nonlinear gain compression is significant and must be included in the description of the SOA saturation under pulsed operation (Borri et al., 1999). Moreover, of the possible relevant phenomena, the effect of TPA can be disregarded because it becomes significant for pulses that are hundreds of femtoseconds wide and have energy larger than 1 pJ (Tang and Shore, 1998). In fact, in most cases of interest the width and energy of pulses launched into an SOA >0.1 ps and <1 pJ, respectively. For this reason, the main effects whose contribution is taken into account are CH and SHB. Second, the SOA small-signal gain, internal loss, and saturation energy are treated as wavelength independent parameters, since usually the spectral width of picosecond optical pulses is at least two orders of magnitude smaller than the gain bandwidth (Ning et al., 1997). The wavelength dependence of these SOA parameters must be taken into account when SOAs are employed for multichannel amplification (Jennen et al., 2001) or for nonlinear optical signal processing purposes, where more than one signal propagates and/or interacts within the active medium (Gutiérrez-Castrejón et al., 2000). The modeling approach that is usually adopted in this case is to approximate the spectral profile of the material gain coefficient per unit length with a parabolic function of wavelength, where the peak of the latter is a linear function of the carrier density (O'Mahony, 1988). The scope of this chapter, however, concerns single signal amplification and SOA linear applications, so high-order polynomial approximations of the gain spectrum would only increase the complexity of SOA modeling without offering essential information for the SOA performance. In addition to this reasoning, the central wavelength of the input signal launched into the SOA is assumed to be set at the peak of the SOA gain spectrum (in the vicinity of 1550 nm), which renders the modal gain (defined in Equation 25.2) independent of the wavelength. Third, the gain and group-velocity dispersion are neglected for pulse widths in the picosecond range and SOAs that are several hundreds of micrometers long (Agrawal and Olsson, 1989a; Tang and Shore, 1998). Fourth, the polarization of the input light is linear and preserved during propagation through the SOA, which is treated as a polarization independent element since the polarization sensitivity of practical devices is below 0.5 dB (Morito et al., 2003). Fifth, the residual facet reflectivities are practically negligible and cannot cause gain ripples, as enabled by the relevant technological evolution in the design and fabrication of SOAs devices (Zimmerman and Spiekman, 2004). Sixth, the SOA has the form of a traveling wave amplifier whose active region dimensions allow the support of only a single wave-guide, while it behaves as an isotropic device, which means that the susceptibility tensor is scalar (Gutiérrez-Castrejón et al., 2000).

The SOA response to an optical beam of electrical field $\vec{E}$ that propagates along the SOA largest dimension (that is, its length), which defines the z-axis in a Cartesian system of coordinates, is described by the carrier density rate equation (Tang and Shore, 1998)

$$\frac{\partial N}{\partial \tau} = \frac{I}{qV} - \frac{N}{\tau_{carrier}} - \frac{1}{\Gamma \sigma \hbar \omega} \frac{g \left| \vec{E} \right|^2}{1 + \varepsilon \left| \vec{E} \right|^2}, \tag{25.1}$$

where $N = N(z, \tau)$ is the carrier density, I is the injection current, q is the electron charge, $V = wdL$ is the volume of the active region whose characteristic geometries have a cross section (wd) of the order of 1 μm^2 and a length (L) of 0.5 to 2 mm, $\tau_{carrier}$ is the carrier lifetime that is linked to the total recombination rate of carriers as specified in Gutiérrez-Castrejón and Duelk (2007), Γ is the confinement factor that accounts for the transverse character of the optical waves and the spread of the optical mode outside the active region of the amplifier (Agrawal and Olsson, 1989a), σ is the mode cross section ($= wd/\Gamma$), $\hbar$ is the reduced Planck's constant, ω is the angular frequency of the electromagnetic radiation, and ε is the gain compression factor, described right below in Equation 25.13. The parameter $g = g(z, \tau)$ stands for the modal gain that is assumed to vary linearly with N as

$$g = \Gamma a_N (N - N_{tr}), \tag{25.2}$$

where a_N is the SOA differential gain and N_{tr} is the carrier density at transparency. Since, from Equation 25.2, $N = g/\Gamma a_N + N_{tr}$, Equation 25.1 can be transformed, after some algebraic manipulations, into the following gain equation

$$\frac{\partial g}{\partial \tau} = \frac{g_{ss} - g}{\tau_{carrier}} - \frac{g \left| \vec{E} \right|^2}{U_{sat} \left(1 + \varepsilon \left| \vec{E} \right|^2 \right)}, \tag{25.3}$$

where $g_{ss} = \Gamma a_N N_{tr}(I/I_{tr} - 1) = \frac{\ln(G_{ss})}{L}$ is the coefficient of the nominal SOA small-signal power gain, G_{ss}, per unit length, $I_{tr} = qVN_{tr}/\tau_{carrier}$ is the injected current required for transparency, and $U_{sat} = \hbar\omega\sigma/a_N$ is the SOA intrinsic saturation energy, which for typical values $d = 250$ nm, $w = 2$ μm, $\Gamma = 0.48$ and $a_N = 3.3 \times 10^{-20}$ m^2 at an operating wavelength $\lambda = 1550$ nm is approximately 1 pJ. Alternatively, $U_{sat} = P_{sat}\tau_{carrier}$, where P_{sat} is the saturation power of the SOA material, which in steady-state operation mode is approximately 0.7 times the SOA output saturation power (Connelly, 2002).

The dynamical evolution of an input light whose polarization is linear and maintained as it traverses the SOA obeys Maxwell's wave equation

$$\nabla^2\vec{E} - \frac{1}{c^2} \left(n_b^2 + \chi \right) \frac{\partial^2\vec{E}}{\partial\tau^2} = 0 \tag{25.4}$$

where c is the speed of light in vacuum, and $n_b^2 \approx 1 + \Re\{\chi_0\}$ is the background refractive index, with $\Re\{\chi_0\}$ denoting the real part of the medium susceptibility in the absence of external pumping by current injection so that it accounts for material absorption (Schubert, 2004), while

$$\chi = -\frac{\bar{n} c}{\omega\Gamma}g \left(\frac{1}{1 + \varepsilon \left| \vec{E} \right|^2} + \alpha_{CDP} + \alpha_{CH} \frac{\varepsilon \left| \vec{E} \right|^2}{1 + \varepsilon \left| \vec{E} \right|^2} \right) \tag{25.5}$$

is the susceptibility parameter (Gutiérrez-Castrejón et al., 2000), which represents the contribution of the charge carriers inside the active region, where $\bar{n}$ is the effective mode index, $\iota = \sqrt{-1}$ is the imaginary number, and the parameters α_{CDP} and α_{CH} are described below in Equation 25.22.

The electrical field of the linearly polarized light that is inserted in the SOA and propagates along its longitudinal axis, z, is $E(x, y, z, \tau) = \hat{x}\frac{1}{2}F(x, y)A(z, \tau) \exp[\iota(\beta z - \omega\tau)]$, where $\hat{x}$ is the polarization unit vector, $F(x, y)$ is the transverse mode distribution, $A(z, \tau)$ is the complex envelope of the optical pulse and $\beta = \bar{n}\omega/c$ represents the propagation constant. According to the procedure in Tang and Shore (1998) and Gutiérrez-Castrejón et al. (2000), which is based on the slowly varying envelope approximation and the integration over the transverse dimensions, an equation that describes the propagation of the optical pulse in the SOA can be obtained from Equations 25.4 and 25.5

$$\frac{\partial A(z, \tau)}{\partial z} + \frac{1}{\upsilon_{SOA}} \frac{\partial A(z, \tau)}{\partial \tau} = \frac{1}{2} \frac{gA(z, \tau)}{1 + \varepsilon|A|^2} - \frac{1}{2} \left[\alpha_{CDP}g - \alpha_{CH} \frac{\varepsilon g |A|^2}{1 + \varepsilon |A|^2} \right] A(z, \tau). \tag{25.6}$$

This equation has actually been derived from semiclassical density-matrix equations by adiabatical approximation of the interband polarization dynamics (Mørk and Mecozzi, 1996).

The time measured in a static reference (laboratory) coordinate system, τ, is transformed into the local time t measured in a reference frame moving with the pulse at the group velocity inside the amplifier, v_{SOA}:

$$t = \tau - \frac{z}{v_{\text{SOA}}}. \tag{25.7}$$

Thus, in this moving coordinate system, the pulse is centered on $t = 0$ at every plane along the amplifier. According to Equation 25.7, the transformation of a general function $f(z, \tau)$ to $f(z, t)$ leads to the following expressions for the time and spatial derivatives (Jennen et al., 2001):

$$\frac{\partial f(z, \tau)}{\partial \tau} = \frac{\partial f(z, t)}{\partial t} \cdot \frac{\partial t}{\partial \tau} = \frac{\partial f(z, t)}{\partial t} \tag{25.8}$$

$$\frac{\partial f(z, \tau)}{\partial z} = \frac{\partial f(z, t)}{\partial z} + \frac{\partial f(z, t)}{\partial t} \cdot \frac{\partial t}{\partial z} = \frac{\partial f(z, t)}{\partial z} - \frac{1}{v_{\text{SOA}}} \cdot \frac{\partial f(z, t)}{\partial t}. \tag{25.9}$$

Separating further "A" into amplitude and phase terms,

$$A = \sqrt{\text{Const} \times P} \exp(\iota \varphi) \Rightarrow P = |A|^2 \big/ \text{Const}. \tag{25.10}$$

"Const" is a normalization coefficient that relates the units between amplitude and power so that $A(z, t)$ is expressed in units of $W^{1/2}$, while $P = P(z, t)$ and $\varphi = \varphi(z, t)$ are the power and the phase of the traveling optical pulse, respectively. Applying Equations 25.8 and 25.9 together in Equation 25.6 and replacing in Equations 25.3 and 25.6 the squared modulus of the total electrical field, $|\vec{A}|^2 = \vec{E} \cdot \vec{E}^* = |A|^2$, where the symbol $*$ denotes the complex conjugate, the following set of partial coupled differential equations that govern the power, P, gain, g, and phase, φ, evolution of a strong optical signal that propagates and is amplified in an SOA is obtained (Tang and Shore, 1998)

$$\frac{\partial P(z, t)}{\partial z} = \frac{g(z, t)}{1 + \varepsilon P(z, t)} P(z, t) \tag{25.11}$$

$$\frac{\partial g(z, t)}{\partial t} = \frac{g_{\text{ss}} - g(z, t)}{\tau_{\text{carrier}}} - \frac{g(z, t)}{1 + \varepsilon P(z, t)} \frac{P(z, t)}{U_{\text{sat}}} \tag{25.12}$$

$$\frac{\partial \varphi(z, t)}{\partial z} = -\frac{1}{2} \left[\alpha_{\text{CDP}} g(z, t) - \alpha_{\text{CH}} \frac{\varepsilon g(z, t) P(z, t)}{1 + \varepsilon P(z, t)} \right] \tag{25.13}$$

The term $\varepsilon = \varepsilon_{\text{CH}} + \varepsilon_{\text{SHB}}$ combines the contribution to the SOA gain compression from CH and SHB when these two effects are considered instantaneous. This assumption holds when the width of the launched optical pulses exceeds 1 ps, and is therefore valid under most practical situations encountered in optical communications. In this case, the impact of CH and SHB on the gain compression is taken into account through the term $1 + \varepsilon P$, where the factor εP is connected to a stationary carrier density above transparency, at which the SOA gain is unity and hence leaves intact a signal that passes through (Occhi, 2002; Wang, 2008).

Equations 25.11 and 25.12 form a system with unknowns the power and gain of the signal that perturbs the SOA nonlinear optical properties. This system cannot be solved in closed form, but only numerically. For this purpose, and in order to account for the spatial and temporal dependence denoted by the variables (z, t), L is divided into discrete segments, each of length $\Delta z = \frac{L}{m}$, as shown in Figure 25.4, while the pulses are sliced over their period, τ_{period}, at uniform intervals $\Delta t = \frac{\tau_{\text{period}}}{k}$, where k is a non-zero integer so that their profile can be reconstructed at the output. Now, since the driving signal propagates through one amplifier segment during the sampling time, the space and time infinitesimal elements are linked to each

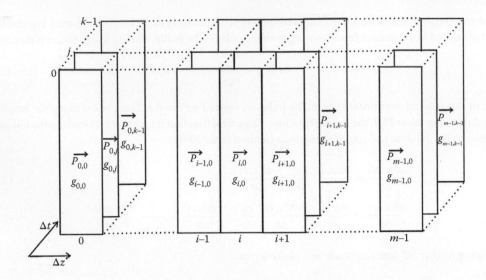

FIGURE 25.4 Dividing the SOA into discrete longitudinal (i) sections and calculating in each one of them the value of the signal power (P) and gain (g) for each sample of the pulse (j) based on the knowledge of the preceding value. This stepwise process is repeated in the formed spatio-temporal grid in an iterative manner until all required values are calculated.

other through $\Delta z = v_{SOA} \Delta t$ (Toptchiyski, 2002). The direct expansion gives

$$\Delta t = \frac{\Delta z}{v_{SOA}} = \frac{L}{m \cdot v_{SOA}} = \frac{\tau_{transit}}{m} = \frac{\tau_{period}}{k} \Rightarrow \frac{k}{m} = \frac{\tau_{period}}{\tau_{transit}}, \tag{25.14}$$

which means that the number k cannot be chosen independently from m. For example, if $\tau_{period} = 100$ ps and $\tau_{transit} = \frac{L}{v_{SOA}} = 12$ ps, the specific ratio equals 100:12. This means that the combination of k and m, or their selection, in order to satisfy this requirement, is not unique. Still, this task can be facilitated by taking into account that the number of the longitudinal sections must be larger than 10 (Wong and Blow, 2003). Otherwise, the SOA gain dynamics induced by the input lightwave signal are erroneously quantified. Thus, if $m = 120$, which is a more than sufficient value, then $k = 1000$. This pair allows obtaining a fairly accurate solution of the above equations with the use of simple first-order Euler differentiation. The latter is an efficient arithmetic method when studying in the time domain the nonlinear interactions between picosecond optical pulses whose envelope has a piecewise varying nature, as is the case for a Gaussian shape, and semiconductor active waveguide devices, provided that the sampling rate is high enough (Chi et al., 2001). More precisely, for each discrete space, i, and time section, j, the solution is expressed by

$$P_{i+1,j} = P_{i,j} + \Delta z \frac{dP}{dz}\Big|_{i,j}, \quad \text{for} \quad i = 0, 1, 2, \ldots, m-1 \quad \text{and} \quad j = 0, 1, 2, \ldots, k-1 \tag{25.15}$$

$$g_{i,j+1} = g_{i,j} + \Delta t \frac{dg}{dt}\Big|_{i,j}, \quad \text{for} \quad i = 0, 1, 2, \ldots, m-1 \quad \text{and} \quad j = 0, 1, 2, \ldots, k-1. \tag{25.16}$$

In other words, for an incoming signal traveling in the longitudinal direction measured from the left facet of the SOA, the knowledge of the values of $P(z, t)$ and $g(z, t)$ as well as of their derivatives (denoted by $|_{i,j}$) at a given step (i, j) of the algorithm, together with the fact that Equations 25.11 and 25.12 are essentially coupled in terms of the power and gain, allows us to calculate the next values $P(z + \Delta z, t)$ and $g(z, t + \Delta t)$ required at steps $(i + 1, j)$ and $(i, j + 1)$, respectively. This process is depicted in Figure 25.4 and is repeated in an iterative manner until all values are calculated in the spatio-temporal grid $i(\Delta z) \times j(\Delta t)$. For this purpose, it is necessary to define and apply the appropriate boundary and initial conditions. The setting of the boundary condition is straightforward since the profile of the launched optical power is expressed in analytical form depending on the particular pulse data format and shape. For example, for a return-to-zero on-off keying signal, the power of each temporal sample at the SOA input facet is given by (Pauer et al., 2001) $P_{0,j} = P(t) = \sum_n \{C_n\} p\left(j\Delta t - n\tau_{\text{period}}\right)$, where n is the number of bits contained in the data sequence length, the code $\{C_n\}$ is either "1" or "0" with equal probability ½, and $p(t)$ is the pulse shape. For Gaussian pulses (Agrawal and Olsson, 1989a), for example, $p(t) = P_{\text{peak}} \exp\left[-4 \ln 2(t - \tau_{\text{period}}/2)^2/\tau_{\text{FWHM}}^2\right]$, where P_{peak} is the peak power and τ_{FWHM} is the full-width at half-maximum (FWHM), which occupies some fraction of the operating period that defines the duty cycle (Pauer et al., 2001). The setting of the initial condition, on the other hand, is more elaborate, since it requires considering the flow of the SOA gain dynamics change by the continuously arriving optical pulses (Tang et al., 2000). More specifically, in each distance segment of the SOA, the first temporal segment in the leading edge of the first arriving pulse experiences an unsaturated, small-signal gain, so that $g_{i,0}|_1 = g_{\text{ss}}$. Then, by induction, in each distance segment, the first temporal segment in the leading edge of the following pulse, which reaches the SOA after τ_{period}, sees an initial, recovered gain given by $g_{i,0}|_n = g_{\text{ss}} + \left(g_{i,k-1}|_{n-1} - g_{\text{ss}}\right) \exp\left(-\frac{\tau_{\text{period}}}{\tau_{\text{carrier}}}\right)$, where $g_{i,k-1}|_{n-1}$ is the gain perturbed by the last temporal segment in the trailing edge of the preceding pulse at the same distance segment.

Apart from this arithmetic procedure, other more elaborate techniques can also be employed for numerically solving modified-type nonlinear Schrödinger equations as Equation 25.6, such as the well-known fourth-order Runge-Kutta, or the finite-difference beam propagation method (FDBPM), where the central-difference approximation is applied in the time domain and trapezoidal integration is executed over the spatial section in an iterative manner (Razaghi et al., 2009). This method has enabled us to accurately acquire information for the SOA characteristics, which include the pulse shape, spectrum, and chirp, in the presence of the main ultrafast nonlinear phenomena that critically affect the SOA gain dynamics (Hosseini et al., 2011).

25.4 SOA Model Simplification

The model deployed for simulating the operation of an SOA has been formulated on the basis of the partial coupled differential Equations 25.11 through 25.13. These constitute a system which, due to the existence of the temporal and longitudinal variables, t and z, respectively, is two-dimensional (2D) so that its solution is rather cumbersome from a computational standpoint. However, if the SOA is treated as a "black box" characterized by its impulse response to any arbitrary input optical signal, the dependence on the spatial variable can be dropped and the problem can be reduced to a one-dimensional (1D) one described by an ordinary differential equation in which time is the only independent variable, enabling us to greatly simplify the computational complexity (Cassioli et al., 2000). This reduction is realized starting from the local changes of the carrier density distribution, which occur due to the interaction between the propagating field and the semiconductor material. These changes are the result of the intraband (CH and SHB) and interband (CDP) effects, while nonlinear processes with characteristic times shorter than SHB, such as TPA, are ignored because their effect is comparatively not significant. The gain and the refractive index of the semiconductor depend on these locally perturbed densities, so each local density distribution is associated with a distinct term that contributes to the total gain. The gain then can be expanded to the sum of the respective contributions of the three different physical processes considered. The temporal evolution of the three quantities is described by equal number of rate equations, which in turn depend on the photon density profile and therefore are coupled to the equation that describes the propagation of the field along the waveguide. The key thus for reducing the order of the modeling system to 1D is the integration of both sides of the partial differential equations over the entire SOA length. This transformation leads to the following compact expression for the total device gain (Cassioli et al., 2000)

$$G(t) = \exp[h(t)], \tag{25.17}$$

where at each point of the pulse profile, $h(z, t) = \int_0^z \frac{g(z',t)}{1+\varepsilon P(z',t)} dz'$ represents the power gain integrated over the length of the SOA, which hence is treated as a spatially concentrated device. This dimensionless coefficient incorporates the contribution of each mechanism that affects the SOA gain:

$$h(t) = h_{CDP}(t) + h_{CH}(t) + h_{SHB}(t). \tag{25.18}$$

More specifically, h_{CDP} represents the gain associated with the interband effects and its temporal evolution is described by the following first-order ordinary differential equation

$$\frac{dh_{CDP}}{dt} = -\frac{h_{CDP}}{\tau_{carrier}} - \frac{1}{P_{sat}\tau_{carrier}} [G(t) - 1] P(t) + \frac{\ln G_{ss}}{\tau_{carrier}}. \tag{25.19}$$

h_{CH} is the gain associated with CH, which means that it describes the gain variation due to the mechanisms that heat the carrier distribution in the conduction band, such as FCA and TPA, as well as due to changes of the carrier density. The relevant rate equation has the form

$$\frac{dh_{CH}}{dt} = -\frac{h_{CH}}{\tau_{CH}} - \frac{\varepsilon_{CH}}{\tau_{CH}} [G(t) - 1] P(t) \tag{25.20}$$

where ε_{CH} is the CH nonlinear gain compression factor. The first term on the right-hand side characterizes the relaxation to the lattice temperature once the exciting pulse has left the SOA and before the arrival of the next pulse. The second term combines all the effects that heat-up the carrier distribution. The typical values of τ_{CH} for InGaAsP/InP-based SOAs with gain peak around 1.55 μm lie between 0.5 and 1 ps, while ε_{CH} is between 0.28×10^{-23} and 4.4×10^{-23} m^3 (Occhi, 2002).

Finally, the rate equation for the SHB contribution is

$$\frac{dh_{SHB}}{dt} = -\frac{h_{SHB}}{\tau_{SHB}} - \frac{\varepsilon_{SHB}}{\tau_{SHB}} [G(t) - 1] P(t) - \frac{dh_{CH}}{dt} - \frac{dh_{CDP}}{dt}, \tag{25.21}$$

where ε_{SHB} is the SHB nonlinear gain compression factor. Reported values of τ_{SHB} in the conduction band for InGaAsP/InP-based SOAs with maximum gain centered around 1.55 μm are between 30 and 250 fs, while ε_{SHB} has been found to be between 0.14×10^{-23} and 1.7×10^{-23} m^3 (Occhi, 2002).

The above dynamic processes affect not only the gain properties but also the refractive index of the SOA within the active region (Wiesenfeld, 1996). This happens because the free carriers contribute substantially to the SOA optical susceptibility, which is complex in nature. The imaginary part of this parameter is related to the gain of the SOA, and the real part is related to the refractive index (Eiselt et al., 1995). The gain and refractive index variations are not independent but coupled via the Kramers–Kronig relationship, according to which the refractive index modulation can be calculated from the gain variation (Hutchings et al., 1992). However, this calculation requires us to integrate over the whole wavelength spectrum the gain coefficient, which means that the latter must be known for a wide range of optical frequencies. Furthermore, the specific integral must be solved numerically, which increases the model computational time significantly. A more convenient manner to describe the refractive index behavior relies on the alpha factor, also known as the Henry factor (Henry, 1982), which links the changes of the active layer refractive index to those of the gain when the carrier density is varied. The important benefit offered by the employment of this factor is that the refractive index dynamics are given directly by the gain dynamics and hence are quantified much faster than with the Kramers–Kronig relations. Moreover, the adoption of the alpha-factor approach is practically realizable since it links the gain and the phase variations associated with the changes of the refractive index, where both are directly measurable. The phase of the optical field at the SOA output can be expressed as a linear function of the same variables which describe the optical gain:

$$\varphi(t) = -0.5 \left[\alpha_{CDP} h_{CDP}(t) + \alpha_{CH} h_{CH}(t) + \alpha_{SHB} h_{SHB}(t) \right], \qquad (25.22)$$

where α_{CDP}, α_{CH}, and α_{SHB} are the phase-amplitude coupling coefficients of CDP, CH, and SHB, respectively. Among these alpha factors, α_{CDP} is the traditional linewidth enhancement factor related to CDP, with typical values 3–12 (Agrawal and Olsson, 1989a). α_{CH} is the alpha factor associated with CH and varies especially near the band edge, where the depletion of cooler carriers takes place (Giller et al., 2006). However, usually signal photon energies influence carriers well inside the band and so, α_{CH} can be approximated by a constant with indicative values ranging between 1 and 4.5, as reported in the literature. Finally, α_{SHB} is the alpha factor linked to SHB and is nearly zero because SHB happens symmetrically around the central wavelength (Mecozzi and Mørk, 1997). Thus, the Kramers–Krönig integral over the whole spectrum is nearly zero and the change of the refractive index is very small near the peak wavelength.

Every phase variation at the SOA output is always accompanied by a certain amount of chirp (Agrawal and Olsson, 1989a). This is the instantaneous frequency deviation relative to the optical beam frequency, $\Delta\nu$, which is provoked because the phase, φ, of the amplified signal is not constant in time. Mathematically, it is given by

$$\Delta\nu(t) = -\frac{1}{2\pi}\frac{\partial\varphi}{\partial t}. \qquad (25.23)$$

The set of rate Equations 25.19 through 25.21 for the h_j (j = CDP, CH, SHB) functions is numerically solved for a given optical power to the SOA input, $P(t)$, which is the only excitation of the system. Then, Equations 25.17, 25.22 and 25.23 give the overall power gain, phase variation, and chirp that the input signal experiences through the SOA, respectively. In this manner, results have been obtained for the temporal, spectral, and chirp profile of the amplified pulses (Hussain et al., 2010).

Equations 25.11 through 25.13 can be further combined and reduced for optical pulses going into the SOA, whose width exceeds the critical limit of 10 ps, above which the influence of the CH and the SHB effects is not significant (Borri et al., 1999). This means that in this case it is not necessary to take into account these intraband processes. Then, by following the algebraic steps detailed in Mecozzi and Mørk (1997), but setting, ε to null and $h_{CDP} \equiv h(t)$, the differential equation for the SOA response is simplified to the well-known form that holds when the SOA gain saturation is caused only by the depletion of the carrier density due to stimulated emission (Agrawal and Olsson, 1989a)

$$\frac{dh(t)}{dt} = \frac{\ln(G_{ss}) - h(t)}{\tau_{carrier}} - \frac{P(t)}{U_{sat}}\left\{ \exp\left[h(t) \right] - 1 \right\}. \qquad (25.24)$$

This equation can be solved analytically, provided that the width of the launched optical pulses is much shorter than the SOA carrier lifetime. This condition holds in practice, since in real SOA devices, the values of the latter parameter are typically of the order of 100 ps and above. Thus, closed-form expressions can be obtained for the variation of the SOA gain in the rapid saturation and slow recovery regions. These expressions allow us to explicitly describe the pulse profile at the SOA output when this element is driven by multiple pulses of alternating binary content, as it happens in real optical networks that are based on optical time division multiplexing (OTDM) (Hamilton et al., 2002).

25.5 SOA Optical Gain Modulation

The influence of critical parameters, which include the SOA small-signal gain and carrier lifetime as well as the FWHM and energy of the input pulses, on the SOA output characteristics has been investigated and evaluated for the case where the change of the SOA gain dynamics is provoked by a 10-Gb/s data-modulated signal onto itself in the context of straightforward amplification (Zoiros et al., 2007). The results reveal that due to the continuous arrival of pulses, which cause the SOA saturation properties to be more strongly modified than for a single pulse, the requirements for the SOA small-signal gain and the input pulse energy are more stringent than those for the isolated pulse amplification studied elsewhere (Agrawal and Olsson, 1989a), while those for the FWHM and carrier lifetime are also tight but to a lesser extent. For example, Figure 25.5 shows that the SOA small-signal gain must be an order of magnitude lower than that for a single pulse or else the output pulse deviates from its normal symmetrical form (Zoiros et al., 2007). Moreover, Figure 25.6 shows that the pulse energy, U_{in}, must be a small fraction of the SOA saturation energy, U_{sat} (Zoiros et al., 2007). These trends together suggest that the SOA must be biased to operate in the low-saturation regime in order to preserve the shape of the input signal at the SOA exit. This condition essentially determines the dynamic range of the inserted optical signal, which is allowed for distortionless pulse amplification, and its validity is further supported by evidence given in Section 25.6. In contrast, when the focus is on the chirp and how it can be tailored in order to be exploited for pulse compression (Agrawal and Olsson, 1989b; Zoiros et al., 2005), the investigation that was conducted reveals that the

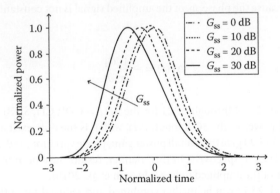

FIGURE 25.5   Shape of amplified pulses versus SOA small-signal gain, G_{ss}. The arrow indicates the G_{ss} direction of increase.

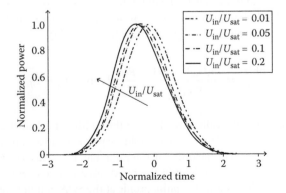

FIGURE 25.6   Shape of amplified pulses versus input pulse energy U_{in}, normalized to SOA saturation energy U_{sat}. The arrow indicates the U_{in}/U_{sat} direction of increase.

SOA small-signal gain and the energy of the incoming pulses must be increased to such extent so that together they heavily saturate the SOA. Then, the reduction of the instantaneous phase shift (Ueno et al., 2002) and of the accompanying chirp due to the alternating bit pulses and the concomitant inability of the SOA to timely recover between them can be efficiently mitigated. In this case, the chirp acquires the form shown in Figure 25.7, which is the required one for the intended purpose (Agrawal and Olsson, 1989a; Zoiros et al., 2007). In fact, the linearly increasing part of the chirp curve in the time domain is situated in the center of the pulse where most of its power is contained, while the chirp magnitude is sufficiently enhanced. Therefore, if the SOA is operated according to these conditions, then it can be ensured that the amplified optical pulses exhibit an appropriate power and chirp profile.

On the other hand, if the FWHM is such that its difference from the pulses repetition interval is smaller than the SOA gain recovery time, then the strain imposed on the SOA gain dynamics is heavier. This happens because the SOA responds more slowly to the rapidly varying logical content of the wider pulses, which worsens the pattern effect described in the next section (Kim et al., 2007). In this demanding case, Equation 25.24 must obligatorily be solved by a numerical approach. This can be done in a step-wise manner by sampling the optical pulse over its period at discrete temporal intervals, approximating the time derivative by a finite difference and applying the appropriate initial conditions to account for the different gain at the beginning of each new bit depending on how it has been perturbed during the previous one (Botsiaris et al., 2007). The knowledge then of h(t) allows us to calculate the electric field of the amplified data, $E_{SOA}(t) = E_{data}(t) \exp\left[\frac{1}{2}\left(1 - j\alpha_{CDP}\right) h(t)\right]$, where, according to Equation 25.10, $|E_{SOA}(t)|^2 = P_{SOA}(t)$ and $|E_{data}(t)|^2 = P(t)$ (Agrawal and Olsson, 1989a), as well as the chirp from $\Delta\nu(t) = -(1/4\pi)\left[\alpha_{CDP}dh(t)/dtdt\right]$, where the derivative of h(t) is directly taken from the right-hand of Equation 25.24. With this approach, the changes that are incurred on the amplified pulse trains and eye diagrams for different SOA saturation conditions can be correctly captured, as verified by comparison to experiments. For this purpose, the degree of pattern effect is modified by altering the maximum launched signal power and the SOA small-signal gain. Figure 25.8 depicts the amplified pulse patterns when the peak power of the marks successively drives the SOA into the low, medium, and deep saturation region, respectively. As the input peak power is progressively doubled, the peak amplitude fluctuations become more pronounced to an extent that depends on whether the marks are preceded either by one or more spaces or by other marks (Zoiros et al., 2008). In contrast, as the input peak power is decreased, these variations are smoothed and the associated pattern effect de-escalates. Similarly, when the SOA small-signal gain is halved, the peak amplitudes become less uneven, as illustrated by comparing Figure 25.9 to Figure 25.8c. The accurate reproduction of the form of the pattern effect which manifests on RZ pulses (Zoiros et al., 2009, 2010a,b), together with the good qualitative matching between the left- and right-hand sides of these figures, designates that the experimental trend for the SOA response is properly modeled. This also holds

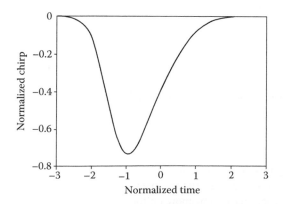

FIGURE 25.7 Profile of SOA output chirp for heavily saturated SOA.

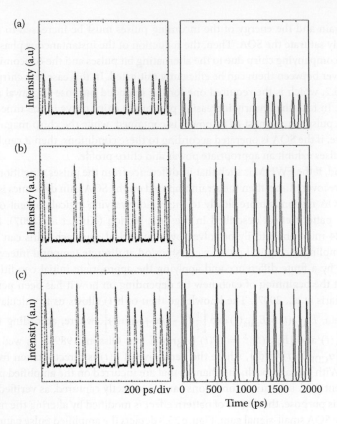

FIGURE 25.8 Experimental (left) and theoretical (right) amplified data pulse trains for (a) low, (b) medium, and (c) deep SOA saturation.

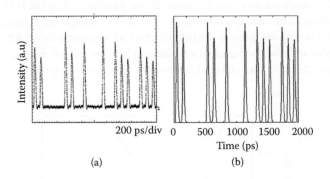

FIGURE 25.9 Experimental (a) and theoretical (b) amplified data pulse trains for halved SOA small-signal gain compared to Figure 25.8(c).

for the respective eye diagrams shown in Figure 25.10. In this case, the asymmetric subenvelopes that occur for a high-input peak power in Figure 25.10c gradually disappear as this parameter is decreased, in Figure 25.10a and b, and the eye diagrams become more uniform, which is an improvement that can be monitored too in the simulated pseudo-eye diagrams obtained as detailed in Gutiérrez-Castrejón et al. (2001).

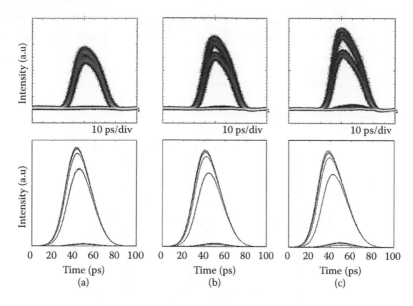

FIGURE 25.10 Experimental (top) and theoretical (bottom) eye diagrams at SOA output for (a) low, (b) medium, and (c) deep SOA saturation.

25.6 Pattern Effect I: Direct Optical Amplification

The distinctive properties of SOAs have revived interest for direct signal amplification purposes, which have traditionally been the primary target of SOAs (Connelly, 2002). As shown in Figure 25.11, the intention is to exploit SOAs as booster amplifiers to enhance the signal power before being launched into an optical link (Kim et al., 2003) as in-line amplifiers to compensate for transmission losses along the link (Boscolo et al., 2006) and as preamplifiers so that the received signal has sufficient power and can be correctly detected (Mynbaev and Scheiner, 2001; Singh, 2011).

In these applications, the profile of the output signal should ideally be an amplified replica of the input signal. In practice, the combination of the power and duration of the input data signal can be such that the SOA is heavily saturated. Concurrently, the SOA gain recovery time is finite with typical values of the order of few hundreds of picoseconds. Thus, in most practical cases, it exceeds the pulses repetition period, which scales inversely with the data rate. Under these driving conditions, the SOA gain is not perturbed in a regular fashion but according to the binary content of each bit slot. As can be schematically seen in Figure 25.12, if the SOA is excited by a train of distinguishable pulses, that is, "1" and "0," whose period, τ_{period}, is such that they arrive faster than the interval available for carrier replenishment, $\tau_{recovery}$, then the gain variation is not uniform since it is dropped for a "1" and partially recovered for a "0" (Manning et al., 1997). This mode of operation has a negative impact on the SOA response, as it can be highlighted by means of the following representative scenarios which concern two "1"s which are separated by a "0." First, in case of a row of marks, the "1" which follows immediately after the "1" that has initially been inserted in the SOA and hence modified its optical properties from a higher level, suffers a reduced gain and this continues to happen for every subsequent "1" with respect to the preceding "1" until an equilibrium is reached or the sequence is broken by a "0." Second, in case of multiple "0"s, which allow the gain to rise substantially, the leading "1" experiences a higher gain than the trailing "1," given that the latter is preceded by only one "0," while it encounters a comparatively lower gain if there are continuous "1"s that prevent the gain from completely recovering. In either case, the amplified output suffers from peak-to-peak amplitude fluctuations, as can characteristically be seen in Figure 25.13.

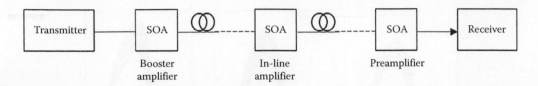

FIGURE 25.11   SOA direct amplification applications.

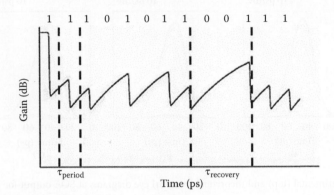

FIGURE 25.12   Evolution of SOA gain in response to data train input.

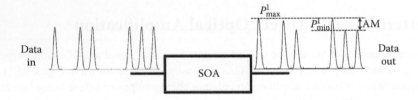

FIGURE 25.13   Data amplification in SOA and pattern-dependent output distortion, with amplitude modulation (AM) definition.

This undesirable phenomenon is called "pattern effect" (Manning et al., 1997) and constitutes a major obstacle in the effort to employ conventional SOAs in their classical role. This happens due to the significant deterioration of several metrics that characterize the quality of the amplified signal, which include the amplitude modulation (AM), the extinction ratio (ER), and the input power dynamic range (IPDR). The AM is defined as AM (dB) $= 10\log\left(P^1_{max}/P^1_{min}\right)$, where P^1_{max} and P^1_{min} are the maximum and minimum peak power of the marks in the amplified data stream, respectively (Figure 25.13). The AM quantifies the degree of uniformity of the marks and should be as low as possible but it is considered acceptable for lightwave telecommunications systems if it is lower than 1 dB (Zoiros et al., 2008). The ER is defined as ER (dB) $= 10\log\left(P^1_{avg}/P^0_{avg}\right)$, where P^1_{avg} and P^0_{avg} are the average power of the marks and spaces in the amplified data stream, respectively (Figure 25.16). The ER quantifies how distinct the marks are from the spaces and it should be as high as possible but it is considered acceptable if it is well over 10 dB (Hinton et al., 2008). Finally, the IPDR is defined as the range where the Q-factor, which in the thermal noise limit, where the amplitude fluctuations due to the pattern effect act as noise variance on the marks, is defined as Q $= P^1_{avg}/\sigma^1_{std}$, with σ^1_{std} being the standard deviation of the peak powers of the marks, is over six (Agrawal, 2002). The negative impact of the SOA pattern effect on these metrics has been investigated and evaluated and the main outcome is compiled and presented in the following in Figures 25.14 and 25.15 (Rizou et al., 2015a). In Figure 25.14, the SOA output suffers from intense peak amplitude fluctuations

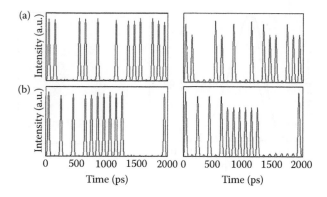

FIGURE 25.14 (a) SOA input data pattern with initial AM = 0.35 dB (left) and corresponding output (right). (b) SOA input data pattern which contains a string of consecutive "0"s (left) and corresponding output (right).

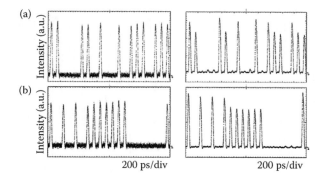

FIGURE 25.15 Corresponding experimental patterns of Figure 25.14.

both when the framed input data waveform has an initial AM = 0.35 dB (Figure 25.14a, left) or contains a string of consecutive "0"s (Figure 25.14b, left). In both cases, the AM level at the SOA exit is intolerable, as it amounts to AM = 1.3 dB (Figure 25.14, right). This behavior is comparable to the experimental one in Figure 25.15 with regard to pulse profile and amplitude wandering. Moreover, in Figure 25.16, the eye diagram at the SOA output is degenerated into secondary traces whose borders deviate from the principal one both in magnitude and shape, while the baseline defined by the spaces is less than 11 dB away from the average level of the marks (Rizou et al., 2013).

In Figure 25.17, on the other hand, the part of the chirp at the SOA output which is negative at the leading edge of each mark, that is, 'red chirp' (Agrawal and Olsson, 1989a), suffers from strong peak-to-peak excursions similar to those observed at the corresponding pulse positions, while the part of the chirp which is positive at the trailing edge of each mark, that is, 'blue chirp', exhibits a much weaker dependence than its red chirp counterpart. Physically, this happens because the red chirp is related to carrier depletion and gain compression, while the blue chirp is linked to carrier regeneration and gain recovery (Girault et al., 2008). This means that the magnitude of the induced chirp is proportional to the rate of change of the carrier concentration and subsequently of the gain. For this reason, it would be rationally expectable that the pattern effect due to the irregular gain variation would be imposed on both types of chirp. However, as can be seen in Figure 25.17, this holds almost exclusively for the red chirp, which exhibits strong peak fluctuations. For the blue chirp, in contrast, an increase in the peaks occurs only for successive "1"s, since then the carrier recombination rate is enhanced due to the weaker carrier depletion, but still this increase is so small that it is hardly noticed. This pronounced discrepancy in the pattern dependence is attributed to

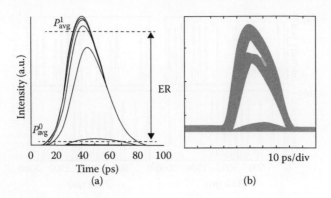

FIGURE 25.16 Theoretical (a) and experimental (b) eye diagrams at SOA output, with extinction ratio (ER) definition.

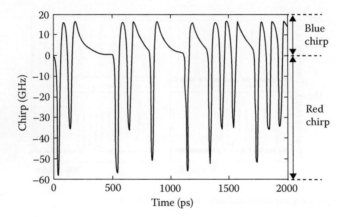

FIGURE 25.17 Chirp variation at SOA output.

the different timescale on which the two chirp phenomena take place. In fact, the carrier depletion and the accompanying phase increase in the leading edge of the pulse is somewhat faster compared to the much slower carrier replenishment and the concomitant phase decrease in its trailing edge, which gives rise to an approximately constant amount of blue chirp. Therefore, the explanation of the pattern effect can also be given in the frequency domain through the red chirp. For this purpose, its form in Figure 25.17 is correlated (Wang et al., 2007) with the corresponding gain change in Figure 25.18.

More specifically, the rising edge of the first "1" at the left-most edge of the pattern window acquires the largest red chirp among all the marks owing to the fact that, for the same energy, its gain is dropped from the maximum possible level, that is, that of the unsaturated state, and as a by-product the incurred gain difference is highest. Similarly, the next largest peak is that of the third "1," since it is preceded by a run of three "0"s. In contrast, the lower the carrier density, or equivalently the greater the extent of gain compression, the slower the additional carriers are depleted, thereby resulting in decreased gain variation and accordingly less red chirp, as it is the case for consecutive "1"s where the SOA cannot fully recover in between them. Finally, in Figure 25.19, the Q-factor is permissible only in saturation region A, where the SOA gain (inset) is reduced up to just 3 dB (Rizou et al., 2013). Thus, the IPDR cannot be extended to regions B and C, since the SOA pattern effect is aggravated as the gain is decreased by more than 3 and 6 dB, respectively. This limitation deteriorates the optical signal-to-noise ratio (OSNR), impedes the maximization of the output power and results in closer amplifier spacings. In order to combat all these pattern-induced

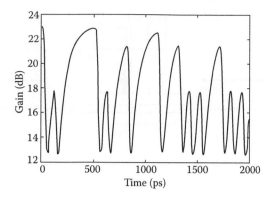

FIGURE 25.18 Instantaneous SOA gain variation.

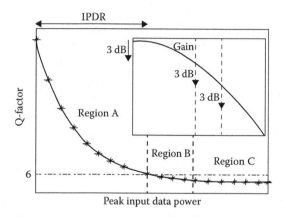

FIGURE 25.19 Variation of Q-factor and corresponding SOA gain (inset) versus peak input data power. The horizontal dotted line denotes the lower limit set for the Q-factor. Defined SOA saturation regions: A = low, B = medium, and C = deep.

performance degradations and restore the quality of the amplified signal, various methods have been proposed and employed. These methods involve intervening on the SOA design, material, and structure and applying sophisticated line coding and interferometric and filtering configurations (see relevant references in Rizou et al., 2013). However, this is a special subject that should be separately addressed elsewhere.

25.7 SOA Electrical Gain Modulation

In addition to straightforward amplification and all-optical signal processing, there has recently been intense research interest in SOAs for use as external modulators. The motivation behind this activity is to be able to provide both data amplification and modulation so as to overcome limitations imposed by other optical modulators (Udvary and Berceli, 2010) while enabling the implementation of various applications (see relevant references in Zoiros et al., 2015). The basic configuration that is exploited for this purpose is shown in Figure 25.20 (Rizou et al., 2015b). The SOA bias current, which is fixed at some point where the *P–I* curve is linear (Udvary and Berceli, 2010), is modulated by an electrical data signal. This modulation alters the SOA gain analogously, which is perceived by a lightwave beam of constant power over time (CW). In this manner, the SOA current modulation is mapped on the CW signal, which thus carries the original information that has now been converted into optical form. According to this physical process, an

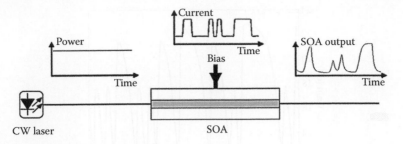

FIGURE 25.20   Basic configuration of directly modulated SOA.

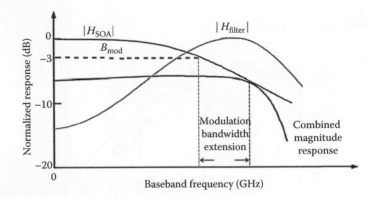

FIGURE 25.21   Schematic description of SOA electrical response characteristic $|H_{\mathrm{SOA}}|$, transfer function of a filter $|H_{\mathrm{filter}}|$ and the combined magnitude response.

exact copy of the applied electrical signal should be transferred at the SOA output. However, this does not happen when the SOA modulation bandwidth, B_{mod}, is smaller than the bandwidth that corresponds to the rate and format of the applied excitation, B_{exc}. More specifically, due to the SOA finite differential carrier lifetime, τ_d, B_{mod} is limited to the order of 1 GHz. In fact, the small-signal frequency response analysis procedure, where the SOA rate equations are linearized and solved analytically by assuming that there are only small perturbations around an operating point (König, 2014), reveals that B_{mod} scales inversely with τ_d according to $B_{\mathrm{mod}} = 1/(2\pi\tau_d)$ and the SOA electrical response exhibits the low-pass filter characteristic qualitatively shown in Figure 25.21 (Wang, 2008). On the other hand, the latter parameter depends on the data format (Agrawal, 2002), and for non-return-to-zero (NRZ) pulses is $B_{\mathrm{exc}} = B_{\mathrm{rep}}/2$, where B_{rep} is the repetition rate. Thus, if B_{rep} is increased so that B_{exc} exceeds B_{mod}, the performance of the directly modulated SOA becomes progressively pattern-dependent and eventually poor.

25.8 Pattern Effect II: Direct Current Modulation

The characteristics of the encoded pulse stream are degraded as the rate of SOA direct modulation is increased due to the concomitant aggravation of the associated pattern effect. In fact, Figure 25.22 shows that with a five-fold increase from 1 Gb/s to 5 Gb/s the amplitude deviations between marks, between spaces, and between marks and spaces are intensified (compare left- to right-hand side). Moreover, the pseudo-eye diagram becomes asymmetrical due to the long rise and fall times and tends to close (compare left to right-hand side of Figure 25.23). Finally, the chirp suffers from analogous peak amplitude fluctuations and transient distortions (compare left to right-hand side of Figure 25.24). Nevertheless, these impairments can be mitigated and the SOA be directly modulated at an extended data rate than that enabled

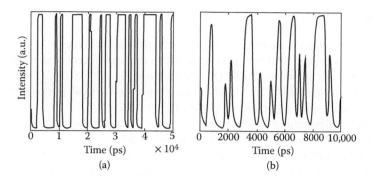

FIGURE 25.22 Encoded data pattern at directly modulated SOA output for (a) 1 Gb/s, (b) 5 Gb/s.

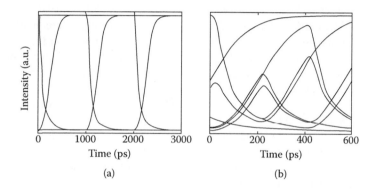

FIGURE 25.23 Pseudo-eye diagram of encoded signal at directly modulated SOA output for (a) 1 Gb/s, (b) 5 Gb/s.

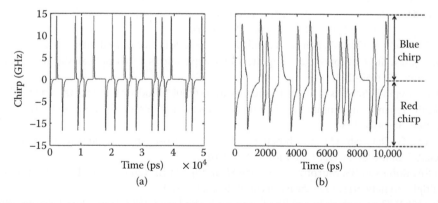

FIGURE 25.24 Chirp of encoded signal at directly modulated SOA output for (a) 1 Gb/s, (b) 5 Gb/s.

by its limited modulation bandwidth by means of post optical notch filtering. This technique can be implemented with different technologies (see relevant references in Zoiros and Morel, 2014), which have in common the fact that they produce a filter characteristic whose slope is in the opposite direction to that of the SOA, as seen in Figure 25.21. This counteracts the SOA response and so the combined transfer function becomes independent of the direct modulation frequency for a wider range than for the SOA only. In this manner, the quality characteristics of the encoded signal at the directly modulated SOA output can be significantly improved so that they are appropriate for supporting the target applications.

References

Agrawal GP (2002) *Fiber-Optic Communication Systems.* New York, NY: Wiley.

Agrawal GP and Olsson NA (1989a) Self-phase modulation and spectral broadening of optical pulses in semiconductor laser amplifiers. *IEEE J Quantum Electron.* 25:2297–2306.

Agrawal GP and Olsson NA (1989b) Amplification and compression of weak picosecond optical pulses by using semiconductor-laser amplifiers. *Opt Lett.* 14:500–502.

Borri P et al. (1999) Measurement and calculation of the critical pulsewidth for gain saturation in semiconductor optical amplifiers. *Opt Commun.* 164:51–55.

Boscolo S, Bhamber R, Turitsyn SK, Mezentsev VK, and Grigoryan VS (2006) RZ-DPSK transmission at 80 Gbit/s channel rate using in-line semiconductor optical amplifiers. *Opt Commun.* 266:656–659.

Botsiaris C, Zoiros KE, Chasioti R, and Koukourlis CS (2007) Q-factor assessment of SOA-based ultrafast nonlinear interferometer. *Opt Commun.* 278:291–302.

Cassioli D, Scotti S, and Mecozzi A (2000) A time-domain computer simulator of the nonlinear response of semiconductor optical amplifiers. *IEEE J Quantum Electron.* 36:1072–1080.

Chi JWD, Chao L, and Rao MK (2001) Time-domain large-signal investigation on nonlinear interactions between an optical pulse and semiconductor waveguides. *IEEE J Quantum Electron.* 37:1329–1336.

Connelly MJ (2002) *Semiconductor Optical Amplifiers.* Dordrecht: Kluwer Academic Publishers.

Diez S (2000) All-optical signal processing by gain-transparent semiconductor switches. PhD dissertation, Technical University of Berlin.

Eiselt M, Pieper W, and Weber HG (1995) SLALOM: Semiconductor laser amplifier in a loop mirror. *J Lightwave Technol.* 13:2099–2112.

Giller R, Manning RJ, and Cotter D (2006) Gain and phase recovery of optically excited semiconductor optical amplifiers. *IEEE Photon Technol Lett.* 18:1061–1063.

Girardin F, Guekos G, and Houbavlis A (1998) Gain recovery of bulk semiconductor optical amplifiers. *IEEE Photon Technol Lett.* 10:784–786.

Girault G et al. (2008) Analysis of bit rate dependence up to 80 Gbit/s of a simple wavelength converter based on XPM in a SOA and a shifted filtering. *Opt Commun.* 281:5731–5738.

Gutiérrez-Castrejón R and Duelk M (2007) Modeling and simulation of semiconductor optical amplifier dynamics for telecommunication applications. In: *Computer Physics Research Trends* (Bianco SJ, ed.), pp. 89–124. New York, NY: Nova Science Publishers.

Gutiérrez-Castrejón R, Occhi L, Schares L, and Guekos G (2001) Recovery dynamics of cross-modulated beam phase in semiconductor amplifiers and applications to all-optical signal processing. *Opt Commun.* 195:167–177.

Gutiérrez-Castrejón R, Schares L, Occhi L, and Guekos G (2000) Modeling and measurement of longitudinal gain dynamics in saturated semiconductor optical amplifiers of different length. *IEEE J Quantum Electron.* 36:1476–1484.

Hall KL, Lenz G, Darwish AM, and Ippen EP (1994) Subpicosecond gain and index nonlinearities in InGaAsP diode lasers. *Opt Commun.* 111:589–612.

Hamilton SA, Robinson BS, Murphy TE, Savage SJ, and Ippen EP (2002) 100 Gb/s optical time-division multiplexed networks. *J Lightwave Technol.* 20:2086–2100.

Henry CH (1982) Theory of the linewidth of semiconductor lasers. *IEEE J Quantum Electron.* 18:259–264.

Hinton K, Raskutti G, Farrell PM, and Tucker RS (2008) Switching energy and device size limits on digital photonic signal processing technologies. *IEEE J Sel Top Quantum Electron.* 14:938–945.

Hosseini SR, Razaghi M, and Das NK (2011) Analysis of ultrafast nonlinear phenomena's influences on output optical pulses and four-wave mixing characteristics in semiconductor optical amplifiers. *Opt Quant Electron.* 42:729–737.

Hussain K, Pradhan R, and Datta PK (2010) Patterning characteristics and its alleviation in high bit-rate amplification of bulk semiconductor optical amplifier. *Opt Quant Electron.* 42:29–43.

Hutchings DC, Sheik-Bahae M, Hagan DJ, and Van Stryland EW (1992) Kramers–Kronig relations in nonlinear optics. *Opt Quant Electron.* 24:1–30.

Jennen J, De Waardt H, and Acket G (2001) Modeling and performance analysis of WDM transmission links employing semiconductor optical amplifiers. *J Lightwave Technol.* 19:1116–1124.

Kim NY, Tang X, Cartledge JC, and Atieh AK (2007) Design and performance of an all-optical wavelength converter based on a semiconductor optical amplifier and delay interferometer. *J Lightwave Technol.* 25:3730–3738.

Kim Y et al. (2003) Transmission performance of 10-Gb/s 1550-nm transmitters using semiconductor optical amplifiers as booster amplifiers. *J Lightwave Technol.* 21:476–481.

König S (2014) *Semiconductor Optical Amplifiers and mm-Wave Wireless Links for Converged Access Networks.* Karlsruhe Series in Photon & Communication (Leuthold J, Freude W, Koos C, eds.), Vol. 14. Karlsruhe: University of Karlsruhe.

Manning RJ, Ellis AD, Poustie AJ, and Blow KJ (1997) Semiconductor laser amplifiers for ultrafast all-optical signal processing. *J Opt Soc Am B.* 14:3204–3216.

Mecozzi A and Mørk J (1997) Saturation effects in nondegenerate four-wave mixing between short optical pulses in semiconductor laser amplifiers. *IEEE J Sel Top Quantum Electron.* 3:1190–1207.

Morito K, Ekawa M, Watanabe T, and Kotaki Y (2003) High-output-power polarization-insensitive semiconductor optical amplifier. *J Lightwave Technol.* 21:176–181.

Mørk J and Mecozzi A (1996) Theory of the ultrafast response of active semiconductor waveguides. *J Opt Soc Am B.* 13:1803–1816.

Mørk J, Nielsen ML, and Berg TW (2003) The dynamics of semiconductor optical amplifiers: Modeling and applications. *Opt Photon News.* 14:42–48.

Mynbaev DK and Scheiner LL (2001) *Fiber-Optic Communications Technology.* New Jersey, NJ: Prentice-Hall.

Ning CZ, Indik RA, and Moloney JV (1997) Effective Bloch equations for semiconductor lasers and amplifiers. *IEEE J Quantum Electron.* 33:1543–1550.

Occhi L (2002) Semiconductor optical amplifiers made of ridge waveguide bulk InGaAsP/InP: Experimental characterisation and numerical modelling of gain, phase, and noise. PhD dissertation, Swiss Federal Institute of Technology, Zurich.

O'Mahony MJ (1988) Semiconductor laser optical amplifiers for use in future fiber systems. *J Lightwave Technol.* 6:531–544.

Pauer M, Winzer PJ, and Leeb WR (2001) Bit error probability reduction in direct detection optical receivers using RZ coding. *J Lightwave Technol.* 19:1255–1262.

Razaghi M, Ahmadi V, and Connelly MJ (2009) Comprehensive finite difference time-dependent beam propagation model of counter-propagation picosecond pulses in a semiconductor optical amplifier. *J Lightwave Technol.* 27:3162–3174.

Rizou ZV, Zoiros KE, Hatziefremidis A, and Connelly MJ (2013) Design analysis and performance optimization of a Lyot filter for semiconductor optical amplifier pattern effect suppression. *IEEE J Sel Top Quantum Electron.* 19:1–9.

Rizou ZV, Zoiros KE, Hatziefremidis A, and Connelly MJ (2015a) Performance tolerance analysis of birefringent fiber loop for semiconductor optical amplifier pattern effect suppression. *Appl Phys B.* 119:247–257.

Rizou ZV, Zoiros KE, and Houbavlis T (2015b) Operating speed extension of SOA external modulator using microring resonator. *Proceedings of 36th Progress in Electromagnetics Research Symposium,* Prague, Czech Republic, July 6–9.

Saleh BEA and Teich MC (1991) *Fundamentals of Photonics.* New York, NY: John Wiley & Sons.

Schreieck RP (2001) Ultrafast dynamics in InGaAsP/InP optical amplifiers and mode locked laser diodes. PhD dissertation, Swiss Federal Institute of Technology in Zurich.

Schubert C (2004) Interferometric gates for all-optical signal processing. PhD dissertation, Technical University of Berlin.

Singh S (2011) An approach to enhance the receiver sensitivity with SOA for optical communication systems. *Opt Commun.* 284:828–832.

Tang JM and Shore KA (1998) Strong picosecond optical pulse propagation in semiconductor optical amplifiers at transparency. *IEEE J Quant Electron.* 34:1263–1269.

Tang JM, Spencer PS, Rees P, and Shore KA (2000) Ultrafast optical packet switching using low optical pulse energies in a self-synchronization scheme. *J Lightwave Technol.* 18:1757–1764.

Toptchiyski GO (2002) Analysis of all-optical interferometric switches based on semiconductor optical amplifiers. PhD dissertation, Technical University of Berlin.

Udvary E and Berceli T (2010) Improvements in the linearity of semiconductor optical amplifiers as external modulators. *IEEE Trans Microw Theory Tech.* 58:3161–3166.

Ueno Y, Nakamura S, and Tajima K (2002) Nonlinear phase shifts induced by semiconductor optical amplifiers with control pulses at repetition frequencies in the 40–160-GHz range for use in ultrahigh-speed all-optical signal processing. *J Opt Soc Am B.* 19:2573–2589.

Vacondio F (2011) On the benefits of phase shift keying to optical telecommunication systems. PhD dissertation, Laval University of Canada.

Vallaitis T. (2010) *Ultrafast Nonlinear Silicon Waveguides and Quantum Dot Semiconductor Optical Amplifiers*. Karlsruhe Series in Photon & Communication (Leuthold J, Freude W, eds.), Vol. 7. Karlsruhe: University of Karlsruhe.

Wang J (2008) *Pattern Effect Mitigation Techniques for All-Optical Wavelength Converters Based on Semiconductor Optical Amplifiers. Karlsruhe Series in Photon & Communication* (Leuthold J and Freude W, eds.), Vol. 3. Germany: Univ Karlsruhe.

Wang J et al. (2007) Pattern effect removal technique for semiconductor optical amplifier based wavelength conversion. *IEEE Photon Technol Lett.* 19:1955–1957.

Wiesenfeld JM (1996) Gain dynamics and associated nonlinearities in semiconductor optical amplifiers. In: *Current Trends in Optical Amplifiers and Their Applications* (Lee TP, ed.), pp. 179–222. Singapore: World Scientific.

Wong WM and Blow KJ (2003) Travelling-wave model of semiconductor optical amplifier based non-linear loop mirror. *Opt Commun.* 215:169–184.

Zimmerman DR and Spiekman LH (2004) Amplifiers for the masses: EDFA, EDWA, and SOA amplest for metro and access applications. *J Lightwave Technol.* 22:63–70.

Zoiros KE, Chasioti R, Koukourlis CS, and Houbavlis T (2007) On the output characteristics of a semiconductor optical amplifier driven by an ultrafast optical time division multiplexing pulse train. *Optik.* 118:134–146.

Zoiros KE, Houbavlis T, and Moyssidis M (2005) Complete theoretical analysis of actively mode-locked fiber ring laser with external optical modulation of a semiconductor optical amplifier. *Opt Commun.* 254:310–329.

Zoiros KE, Janer CL, and Connelly MJ (2010a) Semiconductor optical amplifier pattern effect suppression for return-to-zero data using an optical delay interferometer. *Opt Eng.* 49:085005-085005-4. doi:10.1117/1.3481137.

Zoiros KE and Morel P (2014) Enhanced performance of semiconductor optical amplifier at high direct modulation speed with birefringent fiber loop. *AIP Adv.* 4:077107. doi:10.1063/1.4889869.

Zoiros KE, Morel P, Hamze M (2015) Performance improvement of directly modulated semiconductor optical amplifier with filter-assisted birefringent fiber loop. *Microw Opt Technol Lett.* 57:2247–2251.

Zoiros KE, O'Riordan C, and Connelly MJ (2009) Semiconductor optical amplifier pattern effect suppression using Lyot filter. *Electron Lett.* 45:1187–1189.

Zoiros KE, O'Riordan C, and Connelly MJ (2010b) Semiconductor optical amplifier pattern effect suppression using a birefringent fiber loop. *IEEE Photon Technol Lett.* 22:221–223.

Zoiros KE, Siarkos T, and Koukourlis CS (2008) Theoretical analysis of pattern effect suppression in semiconductor optical amplifier utilizing optical delay interferometer. *Opt Commun.* 281:3648–3657.

Index

Printed and bound by CPI Group (UK) Ltd, Croydon, CR0 4YY
24/10/2024
01778292-0016